ENVIRONMENTAL CATALYSIS

ENVIRONMENTAL CATALYSIS

VICKI H. GRASSIAN

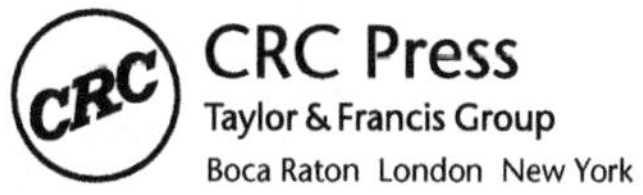

CRC Press is an imprint of the
Taylor & Francis Group, an **informa** business
A TAYLOR & FRANCIS BOOK

CRC Press
Taylor & Francis Group
6000 Broken Sound Parkway NW, Suite 300
Boca Raton, FL 33487-2742

First issued in paperback 2019

ISBN-13: 978-1-57444-462-9 (hbk)
ISBN-13: 978-0-367-39274-1 (pbk)
Library of Congress Card Number 2005044025

Library of Congress Cataloging-in-Publication Data

Grassian, Vicki H.
Environmental catalysis / Vicki H. Grassian.
p. cm.
Includes bibliographical references and index.
ISBN 1-57444-462-X (alk. paper)
1. Catalysis--Environmental aspects. 2. Environmental chemistry. I. Title.

QD505G73 2005
541'.395--dc22 2005044025

Visit the Taylor & Francis Web site at
http://www.taylorandfrancis.com

and the CRC Press Web site at
http://www.crcpress.com

Preface

There is currently no comprehensive book on the topic of environmental catalysis. With increasing interest in the environment and solutions to environmental problems, the time is right for the publication of a book on this topic. This book covers concepts, applications, techniques, and methods related to the area of environmental catalysis. From the contents of this book, it is anticipated that readers will gain an understanding of important issues and problems in Environmental catalysis, a knowledge of the state-of-the-art techniques and methods used to study environmental interfaces and environmental catalysis, and an appreciation that there are many opportunities for researchers and students in this area.

It is our opinion that a comprehensive book on environmental catalysis is needed for several reasons. First, there is currently no book that covers the breadth of topics presented here. Environmental catalysis will be explored from three different perspectives: (i) natural systems (air, water, and soils), (ii) environmental remediation, and (iii) green chemical processing (for the first time in any book). Current funding initiatives by the National Science Foundation, Environmental Protection Agency, and the Department of Energy highlight the importance of these areas from the perspective of federal funding priorities. The appreciation of catalysis in environmental remediation and in natural systems is a fairly recent phenomenon. Active clean-up programs at national laboratories (including superfund sites) emphasize the need for the development of this area. Truly interdisciplinary efforts are needed to solve the complex environmental problems that we face today and most likely in future years. This is happening slowly with the development of environmental centers supported by state and federal funds across the United States. This book reflects the interdisciplinary nature of environmental catalysis in that the contributing authors come from a variety of disciplinary departments including chemistry, atmospheric science, plant and soil science, civil and environmental engineering, chemical engineering, and geoscience.

Environmental catalysis science encompasses the study of catalysts and catalytic reactions that impact the environment. The general area of understanding environmental phenomena at a microscopic level is one of increasing scientific interest. Environmental catalysis and environmental molecular surface science are important and expanding areas of current research. Several of the chapters in this book focus on environmental molecular surface science because adsorption, surface reactions, and desorption play key roles in many environmental catalytic reactions.

In general, the subject can be divided into studies of natural systems and engineered systems and the book is divided accordingly. A Wien diagram of the three sections of the book is shown below. Section I covers environmental catalysis primarily in natural systems in chapters dealing with environmental catalysis in air, water, and soil. Sections II and III focus on engineered systems. Section II emphasizes the use of abiotic and biochemical catalysts for use in environmental remediation. Section III focuses on the use of these catalysts in green chemical processing.

The chapters in this book represent a cross-section of the activities in each of the three areas of environmental catalysis as defined by these sections. A combination of theory, computation, analysis, and synthesis is discussed in these chapters. Brief overviews of each of these sections are presented below.

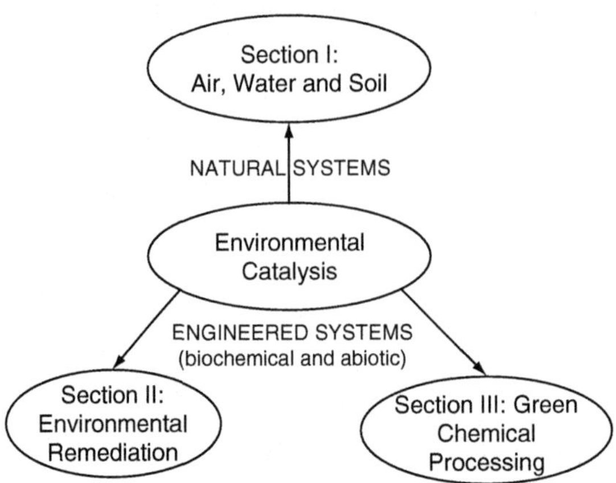

OVERVIEW OF SECTION I — ENVIRONMENTAL CATALYSIS IN AIR, WATER, AND SOIL

Perhaps the most widely known environmental catalytic process is the destruction of stratospheric ozone by atomic chlorine whose elucidation led to the Nobel Prize in chemistry in 1995. More recently the role of heterogeneous catalysis occurring on ice, dust, and mineral surfaces has been recognized. Along these lines, several chapters focus on surface catalysis and surface adsorption of gases on airborne particles present in the troposphere. Hudson and Tolbert in Chapter 6 (Uptake of Trace Species of Ice: Implications for Cirrus Clouds in the Upper Troposphere) examine the uptake of several trace atmospheric gases, including nitric acid, methanol, acetone, and acetaldehyde, on ice thin films which are used as laboratory surrogates of cirrus clouds. Johnson and Grassian discuss the role of mineral dust in the troposphere in Chapter 5 (Environmental Catalysis in the Earth's Atmosphere: Heterogeneous Reactions on Mineral Dust Aerosol) and the potential catalytic role that these particles play in ozone decomposition and the surface hydrolysis of dinitrogen oxides. In Chapter 7 (Surface Chemistry at Size-Selected, Aerosolized Nanoparticles), the oxidation of size-selected soot nanoparticles is examined. In this chapter by Roberts, several new techniques for studying the chemistry of size-selected individual particles are discussed.

Environmental catalysis in soils and in contact with groundwater is the focus of Chapters 1 through 4. In Chapter 1 (Metal and Oxyanion Sorption on Naturally Occurring Oxide and Clay Mineral Surfaces), Sparks describes the adsorption of metals and oxyanions on mineral surfaces. Schoonen and Strongin look at the environmental catalysis of electron transfer reactions at mineral surfaces in Chapter 2 (Catalysis of Electron Transfer Reactions at Mineral Surfaces). Martin focuses on the rates and mechanisms of iron and manganese oxide dissolution and precipitation. As discussed by Martin in Chapter 3 (Precipitation and Dissolution of Iron and Manganese Oxides), in a nontraditional sense, a catalyst can be defined as any agent that increases the rate of a desired process including oxidation, precipitation, and dissolution, regardless of whether the agent is recycled or consumed. Many of the reactions described in this book follow the more general definition put forth by Martin. In Chapter 4 (Applications of Nonlinear Optical Techniques for Studying Heterogeneous Systems Relevant in the Natural Environment), Voges, Al-Abadleh, and Geiger discuss several applications of the use of nonlinear optical techniques for studying environmental catalysis and heterogeneous systems on processes relevant to natural environments. Several important examples using techniques such as sum frequency generation and second harmonic generation are discussed.

OVERVIEW OF SECTION II — ENVIRONMENTAL CATALYSIS IN REMEDIATION

Catalysis is a key technology for the treatment of emissions in the United States. The automotive catalytic converter is well known to virtually everyone living in the industrialized world. The importance of catalysis for reducing human impact on the environment was emphasized in the reports from two workshops cosponsored by the National Science Foundation and the Department of Energy: "Basic Research Needs for Environmentally Responsive Technologies of the Future" and "Basic Research Needs for Vehicles of the Future." Catalysis can do this in two ways, by treating the emissions that result from human activities and by reducing the quantity of waste by-products through improvements in efficiency. Section II focuses to a large extent on the treatment of emissions and represents the largest section of the book.

An overview of the treatment of NO_x emissions in stationary sources is discussed by Curtin in Chapter 8 (Selective Catalytic Reduction of NO_x). Several other chapters are also devoted to the environmental catalysis of NO_x emissions. In Chapter 9 (Surface Science Studies of DeNO$_x$ Catalysts), Rodriguez takes a surface science approach to understanding NO_x reactions in heterogeneous catalysis. Chapters 10 and 12 focus on the use of state-of-the-art theory and computational modeling to better understand the factors controlling NO_x remediation. In Chapter 10 (Fundamental Concepts in Molecular Simulation of NO_x Catalysis) by Schneider, the fundamental molecular level details of NO_x reactions on oxide surfaces determined by ab initio quantum theory are described. McMillan, Broadbelt, and Snurr use computational analysis to model NO_x emissions abatement using zeolite catalysts in Chapter 12 (Theoretical Modeling of Zeolite Catalysis: Nitrogen Oxide Catalysis over Metal-Exchanged Zeolites).

Zeolites as environmental catalysts are also discussed by Larsen in Chapter 11 (Applications of Zeolites in Environmental Catalysis). Besides the use of zeolites in NO_x decomposition, Larsen discusses several other uses of zeolites in environmental catalysis including their use in photooxidation reactions. Photocatalysis in environmental remediation is also a focus of Chapters 13 through 15. In Chapter 13 (The Organic Chemistry of TiO_2 Photocatalysis of Aromatic Hydrocarbons), Jenks describes the photocatalytic degradation of aromatic and related compounds on titanium dioxide. His approach to this problem is from an organic chemistry perspective and is focused on detailed mechanisms of these complex reactions. In contrast, Pilkenton and Raftery take a more materials and physical approach to their studies on photocatalytic oxidation reactions on titanium dioxide in Chapter 14 (*In Situ* Solid-State NMR Studies of Photocatalytic Oxidation Reactions). In the last chapter on TiO_2 photocatalysis, Chapter 15 (Beyond Photocatalytic Environmental Remediation: Novel TiO_2 Materials and Applications), Agrios and Gray discuss a number of promising new environmental applications of TiO_2 photocatalysis including their use in solar cells.

Nanoscience and nanotechnology as well as biotechnology are the focus of the last four chapters of Section II. The use of nanoparticles in environmental remediation and even smaller molecular clusters are discussed in Chapters 16 and 17. Ranjit, Medine, Jeevanandam, Martyanov, and Klabunde in Chapter 16 (Nanoparticles in Environmental Remediation) show how oxide nanoparticles can be used in environmental remediation including the decontamination of chemical warfare agents and in photocatalysis. In Chapter 17 (Toward a Molecular Understanding of Environmental Catalysis: Studies of Metal Oxide Clusters and Their Reactions), Bernstein and Matsuda use small metal oxide clusters to better understand the interaction of pollutant molecules such as SO_2 and NO_2 on more extended oxide surfaces.

The last two chapters in Section II focus on using biocatalysis in environmental remediation. Biocatalysis refers to primarily the chemical reactions catalyzed by enzymes. An enormous effort has been devoted to the understanding of enzyme catalysis motivated by the foremost desire to improve human health. As a result of this effort, the detailed atomic structure and function has been worked out for the enzymes that control a number of the key reactions of biological systems. Perhaps the best studied and understood example is the cytochrome P450 enzyme involved in the metabolism of oxygen. Recently there has been an accelerating research effort to understand the biocatalytic systems of microbes in nature. These organisms appear to play a critical role in the cleansing of water in the natural environment. In Chapter 18 (Biocatalysis in Environemental Remediation–Bioremediation), Parkin discusses the general uses of biological processes that can be used for remediation purposes. He then focuses on biocatalysis in the environmental remediation of chlorinated organics from groundwater. Nyman, Middleton Williams, and Criddle focus on the *in situ* remediation of metals in the environment using biocatalysis in Chapter 19 (Bioengineering for the *In Situ* Remediation of Metals).

OVERVIEW OF SECTION III — ENVIRONMENTAL CATALYSIS IN GREEN CHEMICAL PROCESSING

In engineered systems, catalysis is the essential process technology for the production of transportation fuels and most chemicals. This section begins with Chapter 20 (Selective Oxidation) by Watson and Ozkan. As they state, selective oxidation reactions are used extensively in the chemical industry as a variety of useful chemical intermediates. The minimization of any by-products such as carbon monoxide and carbon dioxide is key in these reactions. In Chapter 21 (Environmental Catalysis in Organic Synthesis), Xiao goes on to discuss environmental catalysis in organic synthesis. Since organic synthesis is essential to the production of many chemical compounds including pharmaceuticals, it is important that organic chemists take the lead in the area of environmental catalysis. This lead role is seen in the chemistry described in Chapter 22 as well as Chapter 23. In Chapter 22 (Catalytic Reactions of Industrial Importance in Aqueous Media) Jiang and Li discuss the use of water as a favored solvent in green chemistry. They also discuss the twelve principles of green chemistry that have been laid out.

Other environmentally benign solvents and biphasic catalysts are also discussed in Chapters 23 and 24. In Chapter 23 (Zeolite-Based Catalysis in Supercritical CO_2 for Green Chemical Processing), Adewuyi describes the use of supercritical carbon dioxide in combination with zeolite catalysts as environmentally friendly in chemical processing. In Chapter 24 (Green Biphasic Homogeneous Catalysis), Jessop and Heldebrant expand upon the use of alternative solvents for use in homogeneous catalysis so as to better recover and recycle catalysts based on heavy metals.

In the last chapter of this section, Chapter 25 (Green Chemical Manufacturing with Biocatalysis), Stewart discusses the important role that biocatalysis can play in the chemical industry. Stewart presents case studies that highlight the potential for enzymes to be used in place of traditional synthetic methodology to address the goals of green chemistry.

FUTURE OUTLOOK

Environmental catalysis is clearly an intellectual frontier that will require the input of scientists from several far-ranging fields including chemistry, environmental science, geoscience, surface physics, atmospheric science, and engineering. With increasing groundwater pollution, increasing particulates in the atmosphere, and the increasing need to remove pollutants from industrial and automotive sources, it is clear that there will be plenty of issues and questions in environmental catalysis that will need to be addressed and answered in the future. We hope that this book inspires some readers to rise to the challenges that we are likely to face in the times ahead.

Vicki H. Grassian
Professor, Departments of Chemistry and Chemical and Biochemical Engineering
University of Iowa

Peter C. Stair
Professor, Department of Chemistry
Director, Environmental Molecular Science Institute
Northwestern University

Editor

Vicki H. Grassian received her B.S. in chemistry from the State University of New York at Albany in 1981. Professor Grassian completed her M.S. at Rensselaer Polytechnic Institute in 1982 and obtained her Ph.D. from the University of California-Berkeley in 1987. At Berkeley, she was advised by Professors Earl Muetterties and George Pimentel. Professor Grassian was a postdoctoral scientist at Colorado State University (1988) and a research associate at the University of California-Berkeley (1989). In 1990, she joined the University of Iowa as an assistant professor. Professor Grassian is currently a full professor in the Department of Chemistry and holds a joint appointment in the Department of Chemical and Biochemical Engineering. At the University of Iowa, Professor Grassian received a faculty-scholar award (1999–2001), a distinguished achievement award (2002), and a James Van Allen Natural Sciences Faculty Fellowship (2004). Her research interests are in the areas of heterogeneous atmospheric chemistry, environmental remediation and catalysis, environmental molecular surface science, nanoscience and nanotechnology. Professor Grassian has written over 100 peer-reviewed publications and book chapters. In 2003, she received a special 2-year creativity award from the National Science Foundation to support her research.

Contributors

Yusuf G. Adewuyi
Department of Chemical Engineering
North Carolina A&T State University
Greensboro, North Carolina

Alexander G. Agrios
Institute for Environmental Catalysis
Department of Civil and Environmental Engineering
Northwestern University
Evanston, Illinois

Hind A. Al-Abadleh
Department of Chemistry
Northwestern University
Evanston, Illinois

Elliot R. Bernstein
Department of Chemistry
Colorado State University
Fort Collins, Colorado

Linda J. Broadbelt
Institute for Environmental Catalysis
Department of Chemical and Biological Engineering
Northwestern University
Evanston, Illinois

Craig S. Criddle
Department of Civil and Environmental Engineering
Stanford University
Stanford, California

Teresa Curtin
Materials and Surface Science Institute
University of Limerick
Limerick, Ireland

Franz M. Geiger
Department of Chemistry
Northwestern University
Evanston, Illinois

Vicki H. Grassian
Departments of Chemistry and Chemical and Biochemical Engineering
and
Center for Global and Regional Environmental Research
University of Iowa
Iowa City, Iowa

Kimberly A. Gray
Institute for Environmental Catalysis
Department of Civil and Environmental Engineering
Northwestern University
Evanston, Illinois

David J. Heldebrant
Department of Chemistry
Queen's University
Kingston, Ontario, Canada

Paula K. Hudson
Department of Chemistry and Biochemistry
CIRES, University of Colorado
Boulder, Colorado

Pethaiyan Jeevanandam
Department of Chemistry
Kansas State University
Manhattan, Kansas

William S. Jenks
Department of Chemistry
Iowa State University
Ames, Iowa

Philip G. Jessop
Department of Chemistry
Queen's University
Kingston, Ontario, Canada

Nan Jiang
Department of Chemistry
Tulane University
New Orleans, Louisiana

Elizabeth R. Johnson
Department of Chemistry
University of Iowa
Iowa City, Iowa

Chao-Jun Li
Department of Chemistry
McGill University
Montreal, Quebec, Canada

Kenneth J. Klabunde
Department of Chemistry
Kansas State University
Manhattan, Kansas

Sarah C. Larsen
Department of Chemistry
University of Iowa
Iowa City, Iowa

Scot T. Martin
Division of Engineering and Applied Sciences
Harvard University
Cambridge, Massachusetts

Igor N. Martyanov
Department of Chemistry
Kansas State University
Manhattan, Kansas

Yoshiyuki Matsuda
Department of Chemistry
Colorado State University
Fort Collins, Colorado

Scott A. McMillan
Institute for Environmental Catalysis
Department of Chemical and Biological Engineering
Northwestern University
Evanston, Illinois

Gavin Medine
Department of Chemistry
Kansas State University
Manhattan, Kansas

Jennifer L. Nyman
Department of Civil and Environmental Engineering
Stanford University
Stanford, California

Umit S. Ozkan
Department of Chemical Engineering
The Ohio State University
Columbus, Ohio

Gene F. Parkin
Department of Civil and Environmental Engineering
University of Iowa
Iowa City, Iowa

Sarah Pilkenton
H.C. Brown Laboratory
Department of Chemistry
Purdue University
West Lafayette, Indiana

Daniel Raftery
H.C. Brown Laboratory
Department of Chemistry
Purdue University
West Lafayette, Indiana

Koodali T. Ranjit
Department of Chemistry
Kansas State University
Manhattan, Kansas

Jeffrey T. Roberts
Department of Chemistry
University of Minnesota–Twin Cities
Minneapolis, Minnesota

Jose A. Rodriguez
Department of Chemistry
Brookhaven National Laboratory
Upton, New York

William F. Schneider
Department of Chemical and Biomolecular Engineering
Department of Chemistry and Biochemistry
University of Notre Dame
Notre Dame, Indiana

Martin A. Schoonen
Department of Geosciences
The State University of New York
at Stony Brook
Stony Brook, New York

Randall Q. Snurr
Institute for Environmental Catalysis
Department of Chemical and Biological
Engineering
Northwestern University
Evanston, Illinois

Donald L. Sparks
Department of Plant
and Soil Sciences
University of Delaware
Newark, Delaware

Jon D. Stewart
Department of Chemistry
University of Florida
Gainesville, Florida

Daniel R. Strongin
Department of Chemistry
Temple University
Philadelphia, Pennsylvania

Margaret A. Tolbert
Department of Chemistry and Biochemistry
CIRES, University of Colorado
Boulder, Colorado

Andrea B. Voges
Department of Chemistry
Northwestern University
Evanston, Illinois

Rick B. Watson
Department of Chemical Engineering
The Ohio State University
Columbus, Ohio

Sarah M. Williams
Department of Civil and Environmental
Engineering
Stanford University
Stanford, California

Jianliang Xiao
Department of Chemistry
Liverpool Centre for Materials and
Catalysis
University of Liverpool
Liverpool, United Kingdom

Table of Contents

Section I

Environmental Catalysis in Air, Water, and Soil

1 Metal and Oxyanion Sorption on Naturally Occurring Oxide and Clay Mineral Surfaces

Donald L. Sparks
Department of Plant and Soil Sciences, University of Delaware

CONTENTS

1.1 INTRODUCTION

The sorption (retention) of metals including alkali (e.g., K), alkaline earth (e.g., Ca), transition metals (e.g., Cd and Ni), and oxyanions (e.g., phosphate, arsenate, and selenate) on soil mineral and organic constituents is one of the most important processes in controlling the fate, transport, and bioavailability of metals and oxyanions in soil and water environments. This review discusses various aspects of metal and oxyanion sorption on clay minerals and metal oxides and hydroxides, referred henceforth as metal–(oxyhydr)oxides, including the role of surface functional groups, types of surface complexes and products, macroscopic and molecular scale assessments of sorption, surface precipitation, and the kinetics of sorption processes.

Adsorption, surface precipitation, and polymerization are all examples of sorption, which is a general term that is used when the retention mechanism at a surface is unknown. There are various metal sorption mechanisms that could occur at soil mineral surfaces involving both physical and chemical processes (Figure 1.1).

Adsorption can be defined as the accumulation of a substance or material at an interface between the solid surface and the bathing solution. Adsorption can include the removal of solute (a substance dissolved in a solvent) molecules from the solution, solvent (continuous phase of a solution, in which the solute is dissolved) from the solid surface, and attachment of the solute molecule to the surface. Adsorption does not include surface precipitation (the formation of a three-dimensional phase product on a surface) or polymerization (formation of small multinuclear inorganic species such as dimers or trimers) processes. Before proceeding

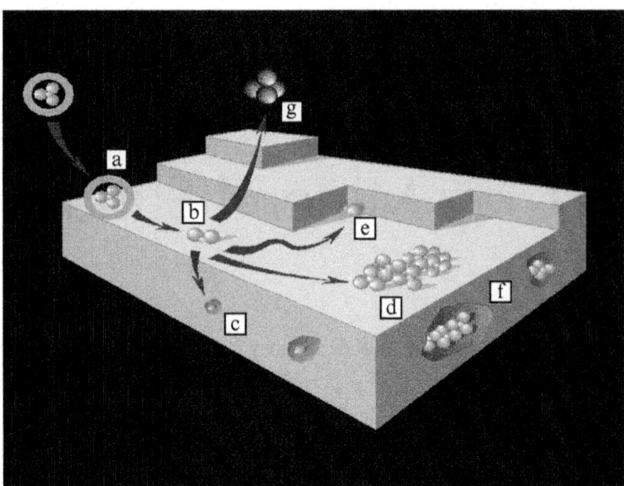

FIGURE 1.1 Various mechanisms of sorption of an ion at the mineral–water interface: (1) adsorption of an ion via formation of an outer-sphere complex (a); (2) loss of hydration water and formation of an inner-sphere complex (b); (3) lattice diffusion and isomorphic substitution within the mineral lattice (c); (4) and (5) rapid lateral diffusion and formation either of a surface polymer (d), or adsorption on a ledge (which maximizes the number of bonds to the adatom) (e). Upon particle growth, surface polymers end up embedded in the lattice structure (f); finally, the adsorbed ion can diffuse back in solution, either as a result of dynamic equilibrium or as a product of surface redox reactions (g). (From L Charlet and A Manceau. In: J Buffle and HP van Leeuwen (eds). *Environmental Particles.* Boca Raton: Lewis Publishers, 1993, pp. 117–164. With permission.)

any further, it would be useful to define a number of terms pertaining to retention (adsorption/sorption) of metal ions. The adsorbate is the material that accumulates at an interface, the solid surface on which the adsorbate accumulates is referred to as the adsorbent, and the ion in solution that has the potential of getting adsorbed is the adsorptive. If the general term sorption is used, the material that accumulates at the surface, the solid surface, and the ion in solution that can be sorbed are referred to as sorbate, sorbent, and sorptive, respectively [1, 2].

The most important adsorbents/sorbents for metals and oxyanions in natural systems such as soils and sediments are clay minerals and metal–(oxyhydr)oxides (Table 1.1). These sorbents exhibit significant surface area and surface charge that play a pivotal role in ion sorption. The surface charge can be negative and invariant with pH in the case of constant-charge clay minerals such as montmorillonite and vermiculite. The constant charge results from ionic substitution in the clay structure (in the past referred to as isomorphic substitution). The surface charge can also be variable becoming more negative with increased pH as surface functional groups (see discussion below) on clay minerals and metal–(oxyhydr)oxides deprotonate, and more positive as pH decreases as surface functional groups protonate. Clay minerals such as kaolinite and Al– and Fe–oxides are considered variable charge minerals.

Adsorption is one of the most important chemical processes in soils. It determines the quantity of plant nutrients, metals, oxyanions, pesticides, and other organic chemicals that are retained on soil surfaces and therefore is one of the primary processes that affects the transport of nutrients and contaminants in soils and sediments. Adsorption also affects the electrostatic properties, for example, coagulation and settling, of suspended particles and colloids. Both physical and chemical forces are involved in adsorption of solutes from solution. Physical forces include van der Waals forces (e.g., partitioning) and electrostatic outer-sphere complexes (e.g., ion exchange). Chemical forces result from short-range interactions

TABLE 1.1
Properties of Important Clay Mineral and Metal–(Oxyhydr)Oxide Sorbents in Natural Systems

Mineral	Chemical Formula	Specific Surface Area ($m^2\ g^{-1}$)
Clay Minerals		
Kaolinite	$Si_4Al_4O_{10}(OH)_8$	7–30
Halloysite	$Si_4Al_4O_{10}(OH)_8{\cdot}2H_2O$	10–45
Pyrophyllite	$Si_8Al_4O_{20}(OH)_4$	65–80
Talc	$Si_8Mg_6O_{20}(OH)_4$	65–80
Montmorillonite	$Si_8(Al, Fe^{2+}Mg)_4O_{20}(OH)_4$	600–800
Dioctahedral vermiculite	$(Si,Al)_8(AlMgFe^{3+})_4O_{20}(OH)_4$	50–800
Trioctahedral vermiculite	$(Si,Al)_8Mg_6O_{20}(OH)_4$	600–800
Muscovite	$(Si,Al)_8Al_4O_{20}(OH)_4$	60–100
Biotite	$(Si,Al)_8(Mg,Fe,Al)_6O_{20}(OH)_4$	40–100
Chlorite	$(Si,Al)_8(Al,Mg)_4O_{20}(OH)_4$ $[(Mg,Al)_2(OH)_6]$	25–150
Allophane	Variable (hydrous aluminosilicates with a molar Si/Al ratio of 1.2 to 1:1)	100–800
Metal–(Oxyhydr)Oxides		
Aluminum (oxyhydr)Oxides		100–220
Bayerite	α-$Al(OH)_3$	
Boehmite	γ-$AlOOH$	
Corundum	α-Al_2O_3	
Diaspore	α-$AlOOH$	
Gibbsite	γ-$Al(OH)_3$	
Iron (oxyhydr)Oxides		70–250
Akaganeite	β-$FeOOHCl$	
Ferrihydrite	$Fe_5HO_8{\cdot}4H_2O$	
Feroxyhyte	δ-$FeOOH$	
Goethite	α-$FeOOH$	
Hematite	α-Fe_2O_3	
Lepidocrocite	γ-$FeOOH$	
Maghemite	γ-Fe_2O_3	
Magnetite	$Fe(II)Fe_2O_4$	
Manganese (oxyhydr)Oxides		5–360
Birnessite	δ-MnO_2	
Manganite	γ-$MnOOH$	
Pyrolusite	β-MnO_2	

Source: DL Sparks. *Environmental Soil Chemistry*. 2nd ed. San Diego: Academic Press, 2002. With permission.

that include inner-sphere complexation that involves a ligand exchange mechanism, covalent bonding, and hydrogen bonding [1, 2].

1.2 SURFACE FUNCTIONAL GROUPS AND SURFACE COMPLEXATION

Surface functional groups on metal–(oxyhydr)oxide and clay mineral surfaces play a significant role in adsorption processes. A surface functional group is a "chemically reactive molecular unit attached at the boundary of a solid with the reactive groups of the unit exposed to the soil solution" [3]. Surface functional groups can be organic (e.g., carboxyl, carbonyl, phenolic) or inorganic molecular units. The major inorganic surface functional groups in soils are the siloxane surface associated with the plane of oxygen atoms bound to the silica tetrahedral layer of a phyllosilicate (clay mineral) and hydroxyl groups that are associated with the edges of inorganic minerals such as kaolinite, amorphous materials, and metal–(oxyhydr)oxides.

A cross section of the surface layer of a metal oxide is shown in Figure 1.2. In Figure 1.2a the surface is unhydrated and has metal ions that are Lewis acids and that have a reduced coordination number. The oxide anions are Lewis bases. In Figure 1.2b, the surface metal ions coordinate to H_2O molecules, forming a Lewis acid site, and then a dissociative chemisorption (chemical bonding to the surface) leads to a hydroxylated surface (Figure 1.2c) with surface OH groups [1, 2, 4].

The surface functional groups can be protonated or deprotonated by adsorption of H^+ and OH^-, respectively, as shown below:

$$S{-}OH + H^+ \longrightarrow S{-}OH_2^+ \tag{1.1}$$

$$S{-}OH \longrightarrow S{-}O^- + H^+ \tag{1.2}$$

Here the Lewis acids are denoted by S and the deprotonated surface hydroxyls are Lewis bases. The water molecule is unstable and can be exchanged for an inorganic or organic anion (Lewis base or ligand) in the solution which then bonds to the metal cation. This process is called ligand exchange [1, 2, 4].

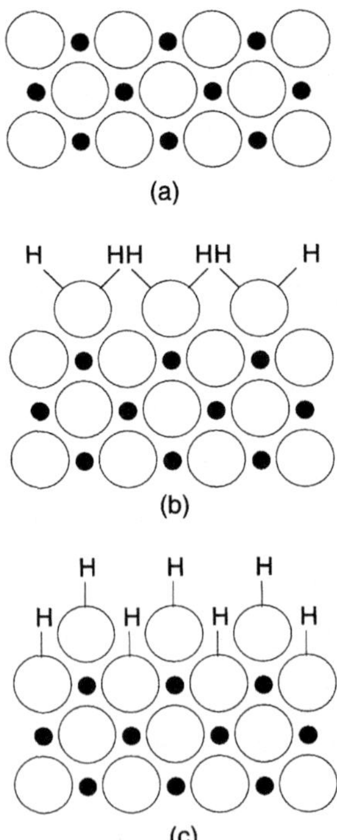

FIGURE 1.2 Cross section of the surface layer of a metal oxide. (●) Metal ions, (O) oxide ions. (a) The metal ions in the surface layer have a reduced coordination number and exhibit Lewis acidity. (b) In the presence of water, the surface metal ions may coordinate H_2O molecules. (c) Dissociative chemisorption leads to a hydroxylated surface. (From PW Schindler. In: MA Anderson and AJ Rubin (eds). *Adsorption of Inorganics at Solid–Liquid Interfaces*. Ann Arbor, MI: Ann Arbor Science, 1981, pp 1–49. With permission.)

The Lewis acid sites are present not only on metal–(oxyhydr)oxides such as on the edges of gibbsite or goethite, but also on the edges of clay minerals such as kaolinite. There are also singly coordinated OH groups on the edges of clay minerals. At the edge of the octahedral sheet, OH groups are singly coordinated to Al^{3+} and at the edge of the tetrahedral sheet they are singly coordinated to Si^{4+}. The OH groups coordinated to Si^{4+} dissociate only protons; however, the OH coordinated to Al^{3+} dissociate and bind protons. These edge OH groups are called silanol (SiOH) and aluminol (AlOH), respectively [1, 2, 5].

Spectroscopic analyses of the crystal structures of metal–(oxyhydr)oxides and clay minerals show that different types of hydroxyl groups have different reactivities. Goethite (α-FeOOH) has four types of surface hydroxyls whose reactivities are a function of the coordination environment of the O in the FeOH group (Figure 1.3). The FeOH groups are A-, B-, or C-type sites, depending on whether the O is coordinated with one, three, or two adjacent Fe(III) ions. The fourth type of site is a Lewis acid-type site, which results from chemisorption of a water molecule on a bare Fe(III) ion. Sposito [3] has noted that only A-type sites are basic, i.e., they can form a complex with H^+ and A-type and Lewis acid sites can release a proton. The B- and C-type sites are considered unreactive. Thus, A-type sites can be either a proton acceptor or a proton donor (i.e., they are amphoteric). The water coordinated with Lewis acid sites may be a proton donor site, i.e., an acidic site [2].

Clay minerals have both aluminol and silanol groups. Kaolinite has three types of surface hydroxyl groups: aluminol, silanol, and Lewis acid sites (Figure 1.4).

When the interaction of a surface functional group with an ion or molecule present in the soil solution creates a stable molecular entity, it is called a surface complex. The overall reaction is referred to as surface complexation. There are two types of surface complexes, outer-sphere and inner-sphere, that can be formed. Figure 1.5 shows surface complexes between metal cations and siloxane ditrigonal cavities on 2:1 clay minerals. Such complexes can also occur on the edges of clay minerals. If a water molecule is present between the surface

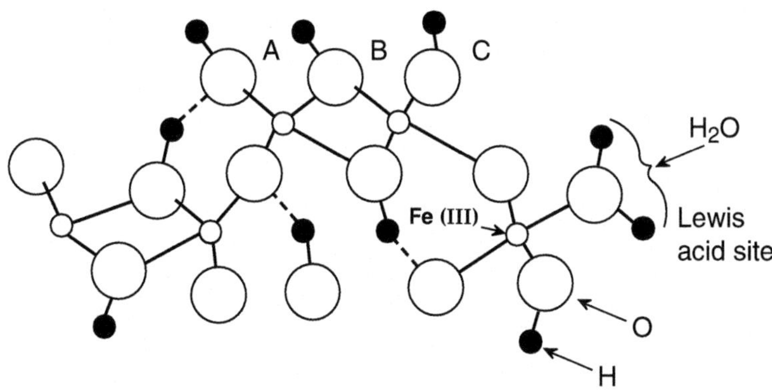

FIGURE 1.3 Types of surface hydroxyl groups on goethite: hydroxyls singly (A-type), triply (B-type), and doubly (C-type) coordinated to Fe (III) ions (one Fe–O bond not represented for type B and C groups); and a Lewis acid site (Fe(III) coordinated to an H_2O molecule). The dashed lines indicate hydrogen bonds. (From G Sposito. *The Surface Chemistry of Soils*. New York: Oxford University Press, 1984. With permission.)

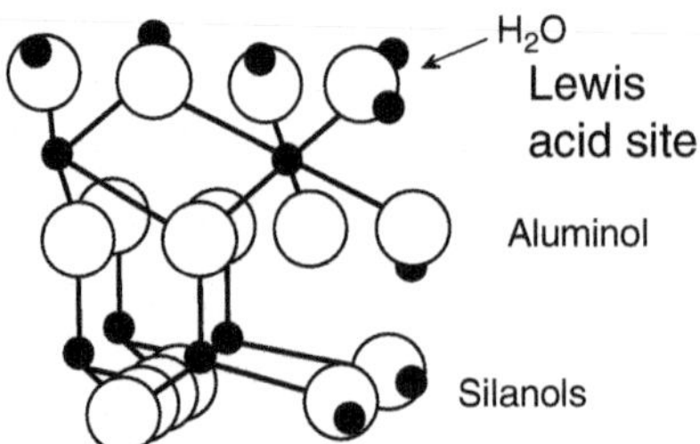

FIGURE 1.4 Surface hydroxyl groups on kaolinite. Besides the OH groups on the basal plane, there are aluminol groups, Lewis acid sites (at which H_2O is absorbed), and silanol groups, all associated with ruptured bonds along the edges of the kaolinite. (From W Stumm (ed.). *Aquatic Surface Chemistry*. New York: Wiley (Interscience), 1987. With permission.)

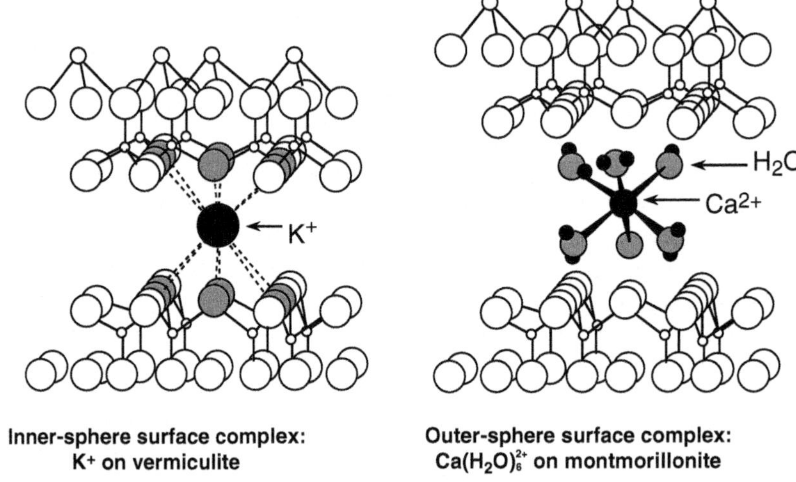

FIGURE 1.5 Examples of inner- and outer-sphere complexes formed between metal cations and siloxane ditrigonal cavities on 2:1 clay minerals. (From W Stumm (ed.). *Aquatic Surface Chemistry*. New York: Wiley (Interscience), 1987. With permission.)

functional group and the bound ion or molecule, the surface complex is termed outer-sphere [5]. If there is no water molecule present between the ion or molecule and the surface functional group to which it is bound it is an inner-sphere complex. Inner-sphere complexes can be monodentate (metal is bonded to only one oxygen) and bidentate (metal is bonded to two oxygens) and mononuclear and binuclear [2].

Outer-sphere complexes involve electrostatic coulombic interactions and are thus weak compared to inner-sphere complexes in which the binding is covalent or ionic. Outer-sphere complexation is usually a rapid process that is reversible, and adsorption occurs only on surfaces that are of opposite charge to the adsorbate [2].

Inner-sphere complexation is usually slower than outer-sphere complexation; it is often not reversible. Inner-sphere complexation can increase, reduce, neutralize, or reverse the charge on the sorptive regardless of the original charge. Adsorption of ions via inner-sphere complexation can occur on a surface regardless of the original charge. It is important to remember that outer- and inner-sphere complexation can, and often do, occur simultaneously [2].

The effects of ionic strength (*I*) on sorption are often used as indirect evidence for the formation of an outer-sphere or inner-sphere complex [6]. For example, strontium [Sr(II)] sorption on γ-Al_2O_3 is highly dependent on the ionic strength of the background electrolyte, $NaNO_3$, while Co(II) sorption is unaffected by changes in the ionic strength (Figure 1.6). The lack of *I* effect on Co(II) sorption would suggest formation of an inner-sphere complex, which is consistent with findings from molecular scale spectroscopic analyses [7–9]. The strong dependence of Sr(II) sorption on *I*, suggesting outer-sphere complexation, is also consistent with spectroscopic findings [10].

1.3 MACROSCOPIC ASSESSMENT OF METAL AND OXYANION SORPTION

Sorption of metal cations is pH-dependent and is characterized by a narrow pH range where sorption increases to nearly 100%, traditionally known as an adsorption edge (Figure 1.7). The pH position of the adsorption edge for a particular metal cation is related to its hydrolysis or acid–base characteristics. In addition to pH, the sorption of metals is dependent on sorptive concentration, surface coverage, the type of the sorbent, and reaction time [2].

One can measure the relative affinity of a metal cation for a sorbent or the selectivity. The properties of the metal cation, sorbent, and the solvent all affect the selectivity. Table 1.2 provides affinity sequences for alkali metal cations, alkaline earth cations, divalent first-row transition metal, and divalent heavy metal cations based on experimental sorption data from the literature. With monovalent alkali metal cations, electrostatic interactions predominate and the general order of selectivity is $Li^+ < Na^+ < K^+ < Rb^+ < Cs^+$ [11]. This order is related to the size of the hydrated radius. The ion in the above group with the smallest hydrated radius,

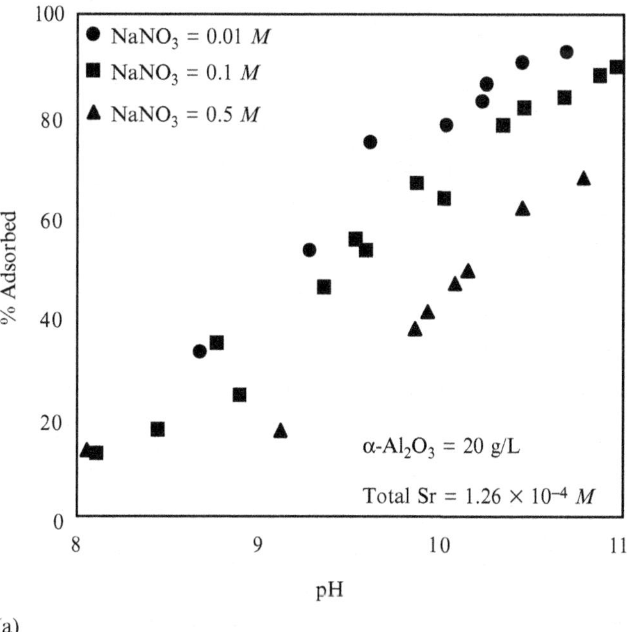

FIGURE 1.6 Effect of increasing ionic strength on pH adsorption edges for (a) a weakly sorbing divalent metal, Sr(II) and (b) a strongly sorbing divalent metal ion, Co(II). (From LE Katz and EJ Boyle-Wight. In: HM Selim and DL Sparks (eds). *Physical and Chemical Processes of Water and Solute Transport/Retention in Soils*. Madison, WI: Soil Science Society of America, 2001, pp. 213–256. With permission.)

continued

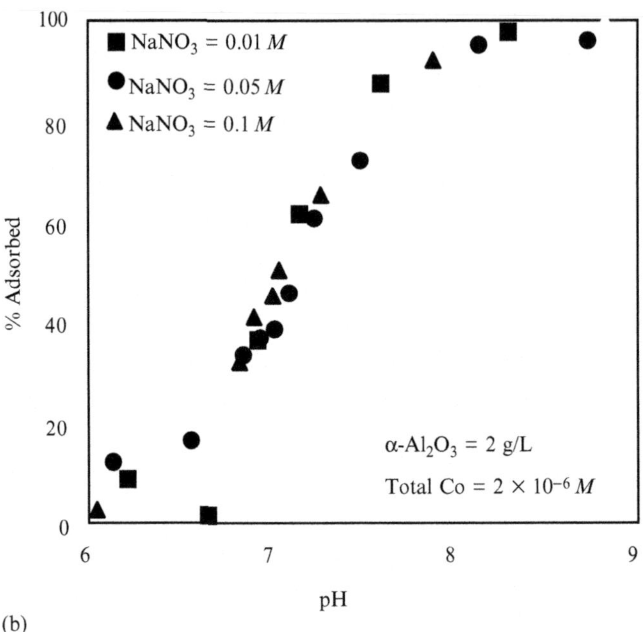

FIGURE 1.6 *continued*

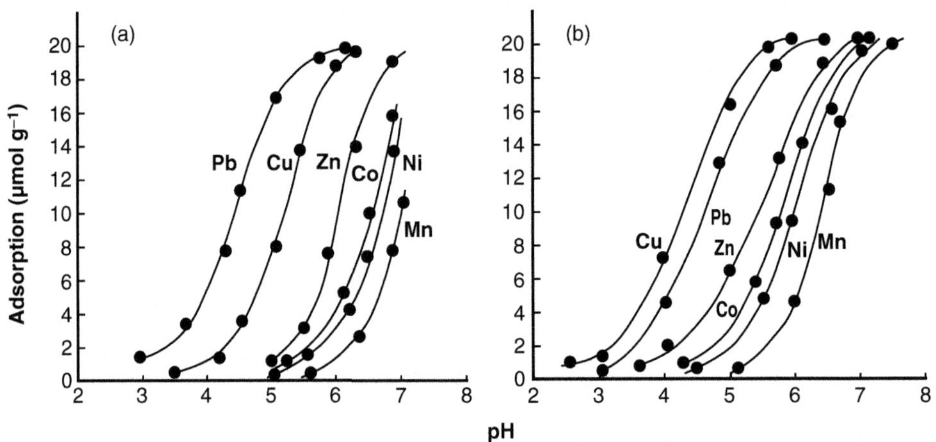

FIGURE 1.7 Sorption of a range of metals on (a) hematite and (b) goethite when they were added at a rate of 20 μmol g^{-1} of adsorbate. The values for the pK_1 for dissociation of the metals to give the monovalent MOH^+ ions are Pb, 7.71; Cu, 8; Zn, 8.96; Co, 9.65; Ni, 9.86; and Mn, 10.59. (From RM McKenzie. *Aust J Soil Res* 18:61–73, 1980. With permission.)

Cs^+, can approach the closest to the surface and be bound most tightly at the surface. However, in some hydrous oxides, the reverse order is often observed (Table 1.2). The reason for this selectivity is not well understood, but may be related to the effect of the solid on water that is present on the metal–(oxyhydr) oxide surface [11] or to variation in the solution matrix.

With divalent ions there is little consistency in the selectivity order. The correlation between the pK_{ads} and the first hydrolysis constant of the metal cations [12] has been used to predict the affinity of first-row transition metal cations for metal–(oxyhydr)oxide surfaces [7] as:

TABLE 1.2
Affinity Sequences for Alkali, Alkaline Earth, Transition Metal, and Heavy Metal Cations on Metal–(Oxyhydr)Oxide Surfaces

Affinitiy Sequence	Metal–(Oxyhydr)Oxide	Ref.
Alkali Cations		
Cs>Rb>K>Na>Li	Silica gel	[85]
K>Na>Li	Silica gel	[86]
Cs>K>Li	Silica glass	[87]
Cs>K>Na>Li	Pyrogenic silica	[88]
Li>Na>K	Silica	[89]
K>Na>Li	Hydrated alumina	[90]
Li>K~Cs	Hematite	[91]
Li>Na>K~Cs	Fe(III)–oxide	[92]
Cs~K>Na>Li	Magnetite	[93]
Li>Na>Cs	Rutile	[94]
Alkaline-Earth Cations		
Ba>Ca	Hydrous iron oxide	[95]
Ba>Ca>Sr>Mg	HFO gel	[15]
Mg>Ca>Sr>Ba	Hematite	[91]
Mg>Ca>Sr>Ba	Al–oxide gel	[15]
Mg>Ca>Sr>Ba	Hydrous γ-Al_2O_3	[17]
Ba>Sr>Ca	α-Al_2O_3	[96]
Ba>Sr>Ca>Mg	MnO_2	[97]
Ba>Sr>Ca>Mg	δ-MnO_2	[98]
Ba>Ca>Sr>Mg	Silica	[99]
Ba>Sr>Ca	α-SiO_2	[100]
First-Row Transition Metal and Heavy Metal Cations		
Pb>Zn>Cd	HFO	[101]
Zn>Cd>Hg	Ferric hydroxide	[102]
Pb>Cu>Zn>Ni>Cd>Co	HFO gel	[15]
Hg>Pb>Cu>Zn>Cd>Ni>Co>Mn	HFO	[103]
Cr(III)>Pb>Cu>Zn>Cd~Ni	HFO	[104]
Cu>Zn>Co>Mn	α-FeOOH	[105]
Cu>Pb>Zn>Co>Cd	α-FeOOH	[106]
Cu>Zn>Ni>Mn	Magnetite	[94]
Cu>Pb>Zn>Ni>Co>Cd	Al-oxide gel	[15]
Cu>Co>Zn>Ni	MnO_2	[107]
Co>Cu>Ni	Hydrous Mn-oxide	[108]
Pb>Zn>Cd	Hydrous Mn-oxide	[101]
Co~Mn>Zn>Ni	δ-MnO_2	[98]
Cu>Zn>Co>Ni	δ-MnOOH	[109]
Co>Cu>Zn>Ni	α-MnO_3	[109]
Co>Zn	δ-MnO_2	[110]
Zn>Cu>Co>Mn>Ni	Silica gel	[111]
Zn>Cu>Ni~Co>Mn	Silica gel	[112]

Source: Adapted from DG Kinniburgh and ML Jackson. In: MA Anderson and AJ Rubin (eds.). *Adsorption of Inorganics at Solid-Liquid Interfaces*. Ann Arbor, MI: Ann Arbor Science, 1981, pp. 91–160 and GE Brown, Jr. and GA Parks. *Int Geol Rev* 43:963–1073, 2001. With permission.

$$Cu^{2+} \sim Pb^{2+} > Zn^{2+} > Co^{2+} > Ni^{2+} > Cd^{2+}$$

This order is in agreement with the Irving–Williams series, which indicates the ability of metal cations to be coordinated by organic Lewis bases [13, 14]. It is also consistent with experimental data for first-row metal transition metal cation sorption on hydrous Al– and Fe(III)–oxide gels by Kinniburgh et al. [15] and on iron and manganese oxides by McKenzie [16]. Stumm [1] and Hayes and Katz [7] noted that the affinity of divalent alkaline earth cations to form outer-sphere complexes on metal–(oxyhydr)oxide surfaces can be correlated with their ionic radius values, which results in the following prediction,

$$Mg^{2+} > Ca^{2+} > Sr^{2+} > Ba^{2+}$$

This affinity has been noted for sorption of Mg^{2+}, Ca^{2+}, and Ba^{2+} on γ-Al_2O_3 (17) and for the four cations on hydrous Al-oxide gel [15]. It is also consistent with the prediction that harder Lewis acids (the metal cations) prefer hard Lewis bases (surface functional groups) [14]. Divalent transition and heavy metal cations, both of which are often sorbed as inner-sphere complexes (Table 1.2), are more strongly sorbed than alkaline earth cations.

Anion sorption varies with pH, usually increasing with pH and reaching a maximum close to the pK_a for anions of monoprotic conjugate acids (a compound that can donate one proton) and slope breaks have been observed at pK_a values for anions of polyprotic (a compound that can donate more than one proton) conjugate acids [18]. This relationship is traditionally referred to as an adsorption envelope (Figure 1.8). In Figure 1.8 one sees that for silicate and fluoride, sorption increased with pH and reached a maximum near the pK_a (pK_1 in Figure 1.8) of the acid, then decreased with pH. For fluoride the dominant species were HF and F^- and for silicate the species were H_4SiO_4 and $H_3SiO_4^-$. With selenite and phosphate, sorption decreased with increasing pH, with the pH decrease more pronounced above the pK_2. The ion species for selenite and phosphate were $HSeO_3^-$ and SeO_3^{2-}, and $H_2PO_4^-$ and

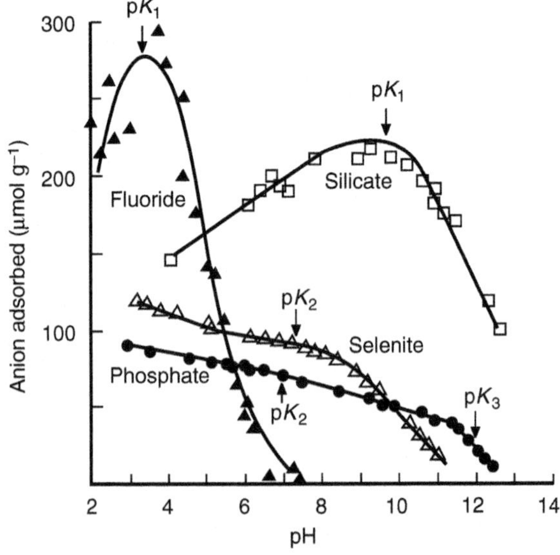

FIGURE 1.8 Sorption of a range of anions on goethite. Two samples of goethite were used and the level of addition of sorbing ion differed between the different ions. (From FJ Hingston, AM Posner, and JP Quirk. *J Soil Sci* 23:177–192, 1972. With permission.)

HPO_4^{2-}, respectively [19]. An example of the correlation between adsorption maxima (pH of inflection) and the pK_a values for conjugate acids is shown in Figure 1.9.

In summary, until recently, the majority of studies on metal and oxyanion sorption on natural sorbents have been macroscopic and have investigated the role that various environmental effects such as solution pH, metal/anion concentration, background electrolyte, and time have on sorption. Based on the large number of studies that have been reported, the following generalizations can be made [14]: (1) increasing sorption of cations or anions on sorbents with decreasing particle size in suspended solids or sediments can be inferred as sorption or surface precipitation; (2) cation sorption on metal–(oxyhydr)oxide and (alumino)-silicate sorbents increases as pH increases; (3) pH decreases with cation adsorption, which suggests that protons are released by one or more mechanisms, including ligand exchange reactions in which the sorptive cation replaces protons on the sorbent, or hydrolysis reactions in which protons are released from the hydration sphere of the aqueous cations as they adsorb or before adsorption when they are in the electrical double layer; (4) anion adsorption on metal–(oxyhydr)oxide and (alumino)silicate surfaces increases as pH is lowered; (5) pH increases with anion adsorption, indicating that hydroxide ions are released or protons are taken up during adsorption; (6) adion (anion or cation) sorption in excess of the available reactive surface sites as sorbate concentration increases or with steep increases in uptake with increasing pH (or decreasing pH with anions) can be interpreted as multinuclear complexation or precipitation of a three-dimensional solid that incorporates the sorbing cation or anion; (7) inhibition of cation or anion sorption or the release of sorbed cations or anions with increasing indifferent background electrolyte concentration can be inferred as nonspecific adsorption in the diffuse layer, or outer-sphere adsorption; (8) cation or anion sorption

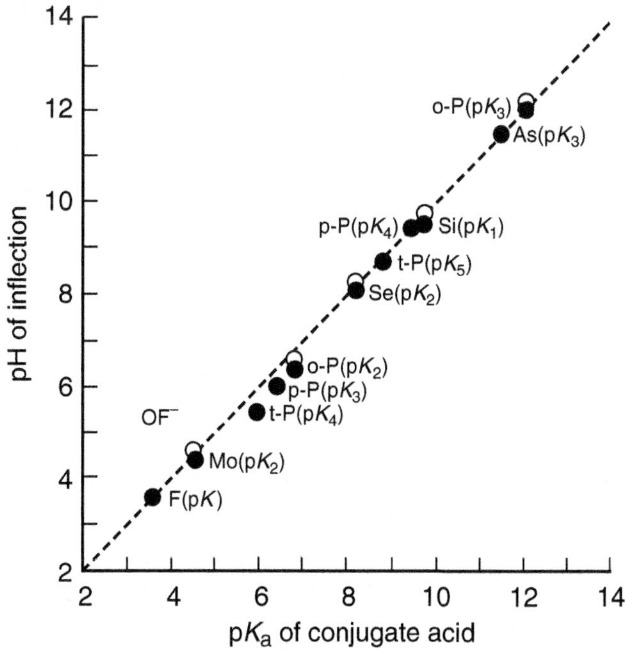

FIGURE 1.9 Relationship between pK_a and pH at the change in slope of sorption envelopes. Sorbents: ●, goethite; ○, gibbsite. Sorbates: F, fluoride; Mo, molybdate; t-P, tripolyphosphate; p-P, pyrophosphate; o-P, orthophosphate; Se, selenite; Si, silicate; As, arsenate. (From FJ Hingston, AM Posner, and JP Quirk. *J Soil Sci* 23:177–192, 1972. With permission.)

independent of the indifferent background electrolyte concentration can be interpreted as primarily inner-sphere complexation but may include some outer-sphere complexation; (9) the sorption of a cation (anion) well below (above) the point of zero charge (pzc) of the sorbent, in which case the sorbent is positively (negatively) charged, can be interpreted as the formation of strong covalent bonds between the sorbate and sorbent; and (10) sorption of a cation (anion) well above (below) the pzc of the sorbent, when the sorbent is negatively (positively) charged, can be interpreted to mean weak electrostatic bonding between the sorbate and sorbent.

1.4 MOLECULAR SCALE INVESTIGATIONS ON METAL AND OXYANION SORPTION

To definitively determine the type of surface complex or product that metals and oxyanions form on natural materials, one must employ molecular scale investigations. Over the past decade or more, *in-situ* (under natural conditions where water is present) molecular scale analytical techniques such as x-ray absorption fine structure spectroscopy (XAFS) and Fourier transform infrared spectroscopy (FTIR) have been employed to determine the type of metal and oxyanion surface complexes and products that form on minerals and soils (Table 1.3).

It is not the intent of this review to provide background on XAFS and FTIR. The reader can consult a number of definitive reviews that discuss the experimental aspects of XAFS [20–27] and FTIR [28–30]. However, since XAFS has been the most widely used method for determining metal and oxyanion complexation/precipitation products on clay minerals and metal–(oxyhydr)oxides, I do wish to provide a brief background on the technique and the unique information one can obtain.

Arguably, the application of synchrotron radiation to investigate contaminant binding mechanisms and speciation in natural materials has revolutionized the environmental and earth sciences. One of the most widely used synchrotron based techniques is XAFS. XAFS is a general term encompassing several energies around an absorption edge for a specific element, namely the pre-edge, near-edge (XANES), and extended portion (EXAFS). Each region provides specific information on an element depending on the selected energy range, making XAFS an element-specific technique. In the XANES region, electron transitions lead to an absorption edge from which chemical information of the target element, such as oxidation state, can be determined. EXAFS can provide the identity of the ligands surrounding the target element, specific bond distances, and coordination numbers of first and second shell ligands. This information is extremely useful in that it provides quantitative information on the geometry, composition, and mode of attachment of a metal/oxyanion ion at a sorbent interface. XAFS is an *in-situ* technique meaning the sample does need to be dried or placed under high vacuum and it can be applied to solid or liquid materials that are amorphous or crystalline [31]. One limitation of bulk XAFS techniques is that one is determining the average local chemical environment on a surface. Thus, where more than one type of surface product forms (outer-sphere and inner-sphere complexes, monodentate and multidentate complexes, or mononuclear and multinuclear complexes), XAFS detects only the primary type of surface product or the average type of product since it sums over all geometric configurations of the target atom [14]. Other disadvantages of XAFS spectroscopy are that it is relatively insensitive to weakly backscattering atoms such as carbon in the presence of oxygen, which makes it of limited utility to study metal–organic ternary surface complexes or metal ions bound to microbial cell surfaces or humic materials interacting with aqueous solutions. For some of these types of studies, FTIR spectroscopy can be useful [14].

Several hundred studies, using molecular scale spectroscopies, a number of which are cited in Table 1.3 and in reviews by Brown and Parks [14] and Brown and Sturchio [27], have provided a plethora of information on metal and oxyanion sorption on natural materials

TABLE 1.3
Sorption Mechanisms for Metals and Oxyanions on Clay Minerals and Metal–(Oxyhydr)Oxides

Metal	pH	Sorbent	Sorption Mechanism	Molecular Probe	Ref.
Cd(II)	7.4–9.8	Manganite	Inner-sphere	XAFS	[113]
Co(II)	8.1	Al_2O_3	Multinuclear complexes (low loading)	XAFS	[8]
			Co–Al hydroxide surface precipitates (high loading)		
	6.8–9	Silica	Co-hydroxide precipitates	XAFS	[114]
	5.3–7.9	Rutile	Small micronuclear complexes (low loading)	XAFS	[114]
			Large multinuclear complexes (high loading)		
	7.8	Kaolinite	Co–Al hydroxide surface precipitates	XAFS	[57]
	4.0	Humic substances	Inner-sphere	XAFS	[115]
Cr(III)	4	Goethite, hydrous ferric oxide	Inner-sphere and Cr-hydroxide surface precipitates	XAFS	[49]
	6	Silica	Inner-sphere monodentate (low loading)	XAFS	[45]
			Cr-hydroxide surface precipitates (high loading)		
	8	γ-Al_2O_3	Inner-sphere (<2 hours)	XAFS	[116]
			Hydroxo (2 hours–1 week)		
			Bridged Cr (III) dimers to high-order polymers		
Cu(II)	6.5	Bohemite	Inner-sphere (low loading)	EPR, XAFS	[117]
			Outer-sphere(high loading)		
	4.3–4.5	γ-Al_2O_3	Inner-sphere bidentate	XAFS	[118]
	5	Ferrihydrite	Inner-sphere bidentate	XAFS	[119]
	5.5	Silica	Cu-hydroxide clusters	XAFS, EPR	[120]
	4.4–4.6	Amorphous silica	Inner-sphere monodentate	XAFS	[118]
	4–6	Soil humic substance	Inner-sphere	XAFS	[121]
	4.2, 6.8	Montmorillonite	Outer-sphere (permanent charge sites)	XAFS	[122]
			Dimer surface complexes (edge sites)		
Ni	7.5	Pyrophyllite, kaolinite, gibbsite, and mont-morillonite	Mixed Ni–Al hydroxide (LDH)	XAFS	[55]
			Surface precipitates		
	7.5	Pyrophyllite	Mixed Ni–Al hydroxide (LDH) surface precipitates	XAFS	[53]
	7.5	Pyrophyllite-montmo-rillonite mixture (1:1)	Mixed Ni–Al hydroxide (LDH)	XAFS	[59]

(continued)

TABLE 1.3
Sorption Mechanisms for Metals and Oxyanions on Clay Minerals and Metal–(Oxyhydr)Oxides—*continued*

Metal	pH	Sorbent	Sorption Mechanism	Molecular Probe	Ref.
	6–7.5	Illite	Mixed Ni-Al hydroxide (LDH) surface precipitates at pH >6.25	XAFS	[123]
	7.5	Pyrophyllite (in presence of citrate and salicylate)	Ni–Al hydroxide (LDH) surface precipitates	DRS	[66]
	7.5	Gibbsite/amorphous silica mixture	γ-$Ni(OH)_2$surface precipitate transforming with time to Ni-phyllosilicate	XAFS-DRS	[124]
	7.5	Gibbsite (in presence of citrate and salicylate)	α-Ni hydroxide surface precipitate	DRS	[66]
	7.5	Soil clay fraction	Ni–Al hydroxide surface precipitate	XAFS	[56]
	7.5	Kaolinite/humic acid	Ni–Al hydroxide surface precipitate (1 wt % HA coating)		[123]
i			$Ni(OH)_2$ surface precipitate (5 wt % HA coating)		
Pb(II)	6	γ-Al_2O_3	Inner-sphere monodentate mononuclear	XAFS	[41]
	6.5	γ-Al_2O_3	Inner-sphere bidentate (low loading)	XAFS	[126]
			Surface polymers (high loading)		
	7	α-alumina (0001 single crystal)	Outer-sphere	Grazing incidence XAFS (GI-XAFS)	[127]
		α-alumina (1I02 single crystal)	Inner-sphere	Grazing incidence XAFS(GI-XAFS)	
	6 and 7	Al_2O_3 powders	Inner-sphere bidentate mononuclear (low loading)	XAFS	[9]
			Dimeric surface complexes (high loading)		
	6–8	Goethite and hematite powders	Inner-sphere bidentate, binuclear	XAFS	[128]
	Variable	Goethite	Inner-sphere (low loading)	XAFS	[48]
	3–7	Goethite (in presence of SO_4^{2-})	Inner-sphere bidentate due to ternary complex formation	XAFS, ATR-TIR	[129]
	5 and 6	Goethite (in absence and presence of SO_4^{2-})	Inner-sphere bidentate	XAFS, ATR-TIR	[130]
			Inner-sphere bidentate mononuclear (pH 6) [in absence of SO_4^{2-}]		
			Inner-sphere bidentate binuclear due to ternary complex formation [in presence of SO_4^{2-}]		

	5.7	Goethite (in presence of CO_3^{2-})	Inner-sphere bidentate	XAFS, ATR-FTIR	[131]
	4.5–6.3	Hydrous amorphous SiO_2	Inner-sphere mononuclear (pH <4.5)	XAFS	[132]
			Covalent polynuclear species (4.5–5.6)		
			Pb-Pb dimers (>6.3)		
	5	Ferrihydrite	Inner-sphere bidentate	XAFS	[119]
	4.5–6.5	Ferrihydrite	Inner-sphere nodentate/bidentate, pH 4.5	XAFS	[135]
			Inner-sphere bidentate edge sharing, pH >5		
	3.5	Birnessite	Inner-sphere mononuclear	XAFS	[133]
	6.7	Manganite	Inner-sphere mononuclear	XAFS	[133]
	6.77	Montmorillonite	Inner-sphere	XAFS	[134]
	6.31–6.76	Montmorillonite	Inner-sphere and outer-sphere	XAFS	[134]
	4.48–6.40	Montmorillonite	Outer-sphere	XAFS	[134]
Sr(II)	7	Ferrihydrite	Outer-sphere	XAFS	[136]
		Kaolinite, amorphous silica, goethite	Outer-sphere	XAFS	[137]
Zn(II)	7–8.2	Alumina powders	Inner-sphere bidentate (low loading)	XAFS	[138]
			Mixed metal–Al hydroxide surface precipitates (high loading)		
	7	γ-Al_2O_3 single crystals	Inner-sphere bidentate (*in-situ*); polynuclear surface complexes (ex-situ; in a humidified N_2 atmosphere)	Grazing incidence XAFS	[139]
	6.17–9.87	Manganite	Multinuclear hydroxo-complexes or Zn-hydroxide phases	XAFS	[113]
	7.5	Pyrophyllite	Mixed Zn-Al hydroxide surface precipitates	XAFS	[60]
	5.1–7.52	Amorphous SiO_2 gibbsite (high surface area)	Inner-sphere monodentate	XAFS	[31]
			Inner-sphere bidentate		
		Gibbsite (low surface area)	Ni–Al hydroxide (LDH) surface precipitate		
Oxyanion					
Arsenite	5.5, 8	γ-Al_2O_3	Inner-sphere bidentate binuclear and outer-sphere	XAFS	[140]
	5.8	$Fe(OH)_3$	Inner-sphere	ATR-FTIR, DRIFT	[141]
	5.5	Goethite	Inner-sphere bidentate binuclear	ATR-FTIR (dry)	[142]
	7.2–7.4	Goethite	Inner-sphere bidentate binuclear	XAFS	[143]
	5, 10.5	Amorphous Fe-oxides	Inner-sphere and outer-sphere	ATR-FTIR and Raman	[144]
		Amorphous Al-oxides	Inner and outer-sphere	ATR-FTIR and Raman	[144]
Arsenate [As(V)]	5, 9	Amorphous Al- and Fe-oxides	Inner-sphere	ATR-FTIR and Raman	[144]

(continued)

TABLE 1.3
Sorption Mechanisms for Metals and Oxyanions on Clay Minerals and Metal-(Oxyhydr)Oxides—*continued*

Metal	pH	Sorbent	Sorption Mechanism	Molecular Probe	Ref.
	5.5	Gibbsite	Inner-sphere bidentate binuclear	XAFS	[145]
	4, 8, 10	γ-Al_2O_3	Inner-sphere bidentate binuclear	XAFS	[146]
	4.5, 7.8	γ-Al_2O_3	Inner-sphere bidentate bi nuclear and possibly Al-arsenate-like surface precipitates at longer reaction times (up to 1 year)	XAFS	[147]
	5.5	Goethite	Inner-sphere bidentate binuclear	ATR-FTIR	[142]
	6	Goethite	Inner-sphere bidentate binuclear	XAFS	[147]
	5, 8	$Fe(OH)_3$	Inner-sphere	ATR-FTIR, DRIFT-FTIR	[30]
	8	Goethite	Inner-sphere bidentate, binuclear, inner-sphere monodentate	XAFS	[148]
	6, 8, 9	Goethite	Inner-sphere monodentate (low loading) Inner sphere bidentate binuclear (high-loading)	XAFS	[149]
	7	Green Rust Lepidocrocite	Inner-sphere bidentate binuclear (high loading)	XAFS	[150]
	5–8	Hydrous manganese oxide	Inner-sphere bidentate	XAFS	[151]
Boron (Trigonal) $[B(OH)_3]$ and tetrahedral $[B(OH)_4^-]$	7, 11	Amorphous $Fe(OH)_3$	Inner-sphere	ATR-FTIR, DRIFT-FTIR	[152]
	7, 10	Amorphous $Al(OH)_3$	Inner-sphere	ATR-FTIR, DRIFT-FTIR	[152]
	6.5, 9.4, 10.4	Hydrous ferric oxide	Outer-sphere and inner-sphere	ATR-FTIR	[34]
Carbonate	4.1–7.8	Amorphous Al- and Fe-oxides Gibbsite Goethite	Inner-sphere momnodentate	ATR-FTIR	[152]
	5.2–7.2	γ-Al_2O_3	Inner-sphere monodentate	ATR-FTIR and DRIFT-FTIR	[153]
	4–9.2	Goethite	Inner-sphere monodentate	ATR-FTIR	[154]
	4.8–7	Goethite	Inner-sphere monodentate	ATR-FTIR	[155]
Chromate [Cr(VI)]	5, 6	Goethite	Inner-sphere bidentate mononuclear [pH 5, 5 m*M* Cr(VI)] Inner-sphere bidentate binuclear [pH 6, 3 m*M* Cr(VI)] Inner-sphere monodentate [pH 6, 2 m*M* Cr(VI)]	XAFS	[149]

Phosphate	4–11	Boehmite	Inner-sphere	MAS-NMR	[156]
	3–12.8	Goethite	Inner-sphere monodentate	DRIFT-FTIR	[157]
	4–8	Goethite	Inner-sphere bidentate and monodentate	ATR-FTIR	[158]
	4–9	Ferrihydrite	Inner-sphere nonprotonated bidentate binuclear (pH >7.5)		[159]
			Inner-sphere protonated (pH 4–6)		
Selenate [Se(VI)]	4	Goethite	Outer-sphere	XAFS	[36]
	Variable	Goethite	Inner-sphere monodentate (pH <6)	ATR-FTIR and Raman	[160]
			Outer-sphere (pH >6)		
		Al-oxide	Outer-sphere		
	3.5–6.0	Hematite	Inner-sphere monodentate (pH = 3.0)	XAFS	[37]
		Goethite	Outer- and inner-sphere (pH = 6.0)		
	3.5–6.0	Hydrous/ferric oxide	Outer- (pH = 3.0) and inner-sphere (pH = 6.0)	XAFS	[37]
	3.5–6.7	Goethite $Fe(OH)_3$	Inner-sphere binuclear	XAFS	
Selenite [Se(IV)]	4	Goethite	Inner-sphere bidentate	XAFS	[36]
	3	Goethite $Fe(OH)_3$	Inner-sphere bidentate	XAFS	[37]
	5–8	Hydrous manganese oxide	Inner-sphere monodentate/bidentate mononuclear	XAFS	[151]
Sulfate	3.5–9	Goethite	Outer-sphere and inner-sphere modentate (pH <6)	ATR-FTIR	[33]
			Outer-sphere (pH >6)	ATR-FTIR and Raman	
	Variable	Goethite	Inner-sphere monodentate (pH <6), Outer-sphere (pH >6)	ATR-FTIR	[160]
		Al-Oxide	Outer-sphere ATR-FTIR	ATR-FTIR	
			Outer-sphere		
	3–6	Hematite	Inner-sphere monodentate		[32]

Source: Adapted from DL Sparks. *Environmental Soil Chemistry*. 2nd ed. San Diego: Academic Press, 2002. With permission.

including details on structure, stiochiometry, attachment geometry (innersphere vs. outer-sphere, monodentate vs. bidentate or tridentate), the presence of multinuclear complexes and precipitate phases, and the presence of ternary surface complexes when complexing ligands are present in solution [27]. The type of surface complexes on clay minerals and metal–(oxyhydr)oxides that occur with low atomic number metals such as Al, B, Ca, Mg, S and Si are difficult at present to determine with XAFS under *in-situ* conditions. However, major advances are now made in the area of soft x-ray XAFS spectroscopy that will enable one to determine *in situ* the types of surface complexes that form with these metals. Additionally, *in-situ* FTIR spectroscopic techniques such as attenuated total reflectance FTIR (ATR-FTIR) are quite amenable to study light metals such as S and B [32–34].

As one can observe from the studies reported in Table 1.3, environmental factors such as pH, surface loading, ionic strength, type of sorbent, and time all affect the type of sorption complex or product. An example of this is shown for Pb sorption on montmorillonite over an ionic strength (I) of 0.006 to 0.1 and a pH range of 4.48 to 6.77 (Table 1.4). Employing XAFS analysis, at a pH of 4.48 and an I value of 0.006, outer-sphere complexation on basal planes in the interlayer regions of the montmorillonite predominated. At a pH of 6.77 and I of 0.1, inner-sphere complexation on edge sites of montmorillonite was most prominent, and at pH of 6.76, I of 0.006 and pH of 6.31, I of 0.1, both inner- and outer-sphere complexations occurred. These data are consistent with other findings that inner-sphere complexation is favored at higher pH and ionic strength. Clearly, there is a continuum of adsorption complexes that can exist in soils [2].

Based on molecular scale studies such as those shown in Table 1.3, one can predict that alkaline earth cations, Mg^{2+}, Ca^{2+}, Sr^{2+}, and Ba^{2+}, primarily form outer-sphere complexes while the divalent first-row transition metal cations, Mn^{2+}, Fe^{2+}, Co^{2+}, Ni^{2+}, Cu^{2+}, and Zn^{2+}, and the divalent heavy metal cations such as Cd^{2+}, Hg^{2+}, and Pb^{2+} primarily form inner-sphere complexes. At higher metal loadings and higher pH values, sorption of metals such as Co, Cr, Ni, and Zn on phyllosilicates and metal–(oxyhydr)oxides can result in the formation of surface precipitates (Table 1.3). The formation of these multinuclear and precipitate phases will be discussed in more detail later.

Although there are currently experimental limitations to using *in-situ* molecular scale techniques to directly determine the type of surface complexes that the anions NO_3^-, Cl^-, and ClO_4^- form on mineral surfaces, one can propose that they are sorbed as outer-sphere complexes and sorbed on surfaces that exhibit a positive charge. Sorption is sensitive to ionic strength. Some researchers have also concluded that SO_4^{2-} [35] can be sorbed as an outer-sphere complex; however, there is other evidence that SO_4^{2-} can also be sorbed as an inner-

TABLE 1.4
Effect of I and pH on Type of Pb Adsorption Complexes on Montmorillonite

I (M)	pH	Removal from Solution (%)	Adsorbed Pb(II) (mmol kg^{-1})	Primary Adsorption Complex[a]
0.1	6.77	86.7	171	Inner-sphere
0.1	6.31	71.2	140	Mixed
0.006	6.76	99.0	201	Mixed
0.006	6.40	98.5	200	Outer-sphere
0.006	5.83	98.0	199	Outer-sphere
0.006	4.48	96.8	197	Outer-sphere

Source: DG Strawn and DL Sparks. *J Colloid Interface Sci* 216:257–269, 1999. With permission.
[a]Based on results from XAFS data analysis.

sphere complex (Table 1.3). There is direct spectroscopic evidence to show that selenate can be sorbed both as an outer-sphere and as an inner-sphere complex, depending on environmental factors [2, 36, 37].

Most other anions such as molybdate, arsenate, arsenite, selenite, phosphate, and silicate appear to be strongly sorbed as inner-sphere complexes, and sorption occurs through a ligand exchange mechanism (Table 1.3). The sorption maximum is often insensitive to ionic strength changes. Sorption of anions via ligand exchange results in a shift in the pzc of the sorbent to a more acid value [2].

1.5 SURFACE PRECIPITATION OF METALS

As the amount of metal cation or anion sorbed on a surface (surface coverage or loading, which is affected by the pH at which sorption occurs) increases, sorption can proceed from mononuclear adsorption to surface precipitation (a three dimensional phase). There are several thermodynamic reasons for surface precipitate formation: (1) the solid surface may lower the energy of nucleation by providing sterically similar sites [38]; (2) the activity of the surface precipitate is <1 [39]; and (3) the solubility of the surface precipitate is lowered because the dielectric constant of the solution near the surface is less than that of the bulk solution [40]. There are several types of surface precipitates. They can arise via polymeric metal complexes (dimers, trimers, etc.) that form on mineral surfaces and via the sorption of aqueous polymers [41]. Homogeneous precipitates can form on a surface when the solution becomes saturated and the surface acts as a nucleation site. When adsorption attains monolayer coverage sorption continues on the newly created sites causing a precipitate on the surface [38–42]. When the precipitate consists of chemical species derived from both the aqueous solution and dissolution of the mineral, it is referred to as a coprecipitate. The composition of the coprecipitate varies between the original solid and a pure precipitate of the sorbing metal. The ionic radius of the sorbing metal and sorbent ions must be similar for the formation of coprecipitates. Thus Co(II), Mn(II), Ni(II), and Zn(II) form coprecipitates on sorbents containing Al(III) and Si(IV) but not Pb(II) which is considerably larger (1.20 Å). Coprecipitate formation is most limited by the rate of mineral dissolution, rather than the lack of thermodynamic favorability [43, 44]. If the formation of a precipitate occurs under solution conditions that would, in the absence of a sorbent, be undersaturated with respect to any known solid phase, this is referred to as surface-induced precipitation [8].

Thus there is often a continuum between surface complexation (adsorption) and surface precipitation. This continuum depends on several factors: (1) ratio of the number of surface sites vs. the number of metal ions in solution; (2) the strength of the metal–oxide bond; and (3) the degree to which the bulk solution is undersaturated with respect to the metal hydroxide precipitate [38]. At low surface coverages surface complexation (e.g., outer- and inner-sphere adsorption) tends to dominate. As surface loadings increase, nucleation occurs and results in the formation of distinct entities or aggregates on the surface. As surface loadings increase further, surface precipitation becomes the dominant mechanism (Figure 1.10). For example, Fendorf et al. [45] and Fendorf and Sparks [46] used XAFS, FTIR, and high-resolution transmission electron microscopy (HRTEM) to study Cr(III) sorption on Si-oxide. At low Cr(III) surface coverage ($<20\%$), adsorption was the dominant process with an inner-sphere monodentate complex forming. As Cr(III) surface coverage increased ($>20\%$) surface precipitation occurred and was the dominant process. Table 1.3 shows that for a number of ions, as surface coverage increases, surface precipitates form.

Using *in-situ* XAFS, it has been shown by a number of scientists that multinuclear metal hydroxide complexes and surface precipitates Co(II), Cr(III), Cu(II), Ni(II), and Pb(II) can form on metal oxides, phyllosilicates, soil clays, and soils [8, 41, 44, 45, 47–61]. These metal hydroxide phases occur at metal loadings below a theoretical monolayer coverage and in a pH

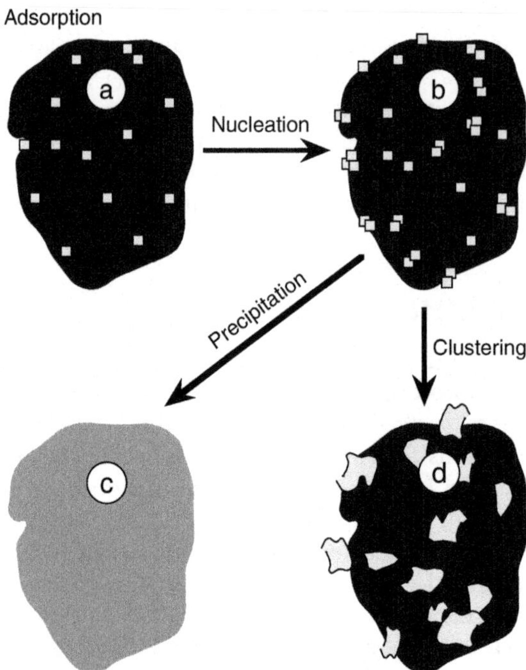

FIGURE 1.10 An illustration of metal ion sorption reactions on (hydr)oxide. (a) At low surface coverage, isolated site binding (adsorption) is the dominant sorption mechanism; (b) with increased metal loading. M–hydroxide nucleation begins; (c) further increases in metal loadings result in surface precipitation or (d) surface clusters. (From SE Fendorf. Ph.D. Dissertation, University of Delaware, Newark, 1992.)

range well below the pH where the formation of metal hydroxide precipitates would be expected according to the thermodynamic solubility product [2, 62].

Scheidegger et al. [55] were the first to show that sorption of metals, such as Ni, on an array of phyllosilicates and Al-oxide, could result in the formation of mixed metal–Al hydroxide surface precipitates which appear to be coprecipitates. The precipitate phase shares structural features common to the hydrotalcite group of minerals and the layered double hydroxides (LDH) observed in catalyst synthesis. The LDH structure is built of stacked sheets of edge-sharing metal octahedra containing divalent and trivalent metal ions separated by anions between the interlayer spaces (Figure 1.11). The general structural formula can be expressed as $[Me^{2+}_{1-x}\ Me^{3+}_{x}\ (OH)_2]^{x+}\cdot(x/n)A^{n}\text{–}mH_2O$, where for example, Me^{2+} could be Mg(II), Ni(II), Co(II), Zn(II), Mn(II), and Fe(II) and Me^{3+} is Al(III), Fe(III), and Cr(III) [8]. The LDH structure exhibits a net positive charge x per formula unit which is balanced by an equal negative charge from interlayer anions A^{n-} such as Cl^-, Br^-, I^-, NO_3^-, OH^-, ClO_4^-, and CO_3^{2-}; water molecules occupy the remaining interlayer space [63, 64]. The minerals takovite, $Ni_6Al_2(OH)_{16}CO_3\cdot H_2O$ and hydrotalcite, $Mg_6Al_2(OH)_{16}CO_3 \cdot H_2O$ are among the most common natural mixed-cation hydroxide compounds containing Al [64]. Figure 1.11 shows a Ni–Al LDH phase [2].

XAFS data, showing the formation of Ni–Al LDH phases on soil components, are shown in Figure 1.12 and Table 1.5 [55]. Radial structure functions (RSFs), collected from XAFS analyses, for Ni sorption on pyrophyllite, kaolinite, gibbsite, and montmorillonite were compared to the spectra of crystalline $Ni(OH)_2$ and takovite. All spectra showed a peak at $R \approx 0.18$ nm, which represents the first coordination shell of Ni. A second peak representing the

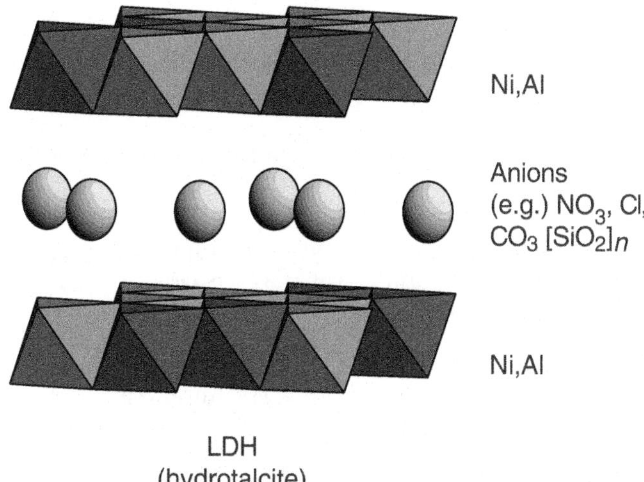

FIGURE 1.11 Structure of Ni–Al LDH showing brucite-like octahedral layers in which Al^{3+} substitutes for Ni^{2+} creating a net positive charge that is balanced by hydrated anions in the interlayer space. (From AC Scheinost, RG Ford, and DL Sparks. *Geochim Cosmochim Acta* 63:3193–3203, 1999. With permission.)

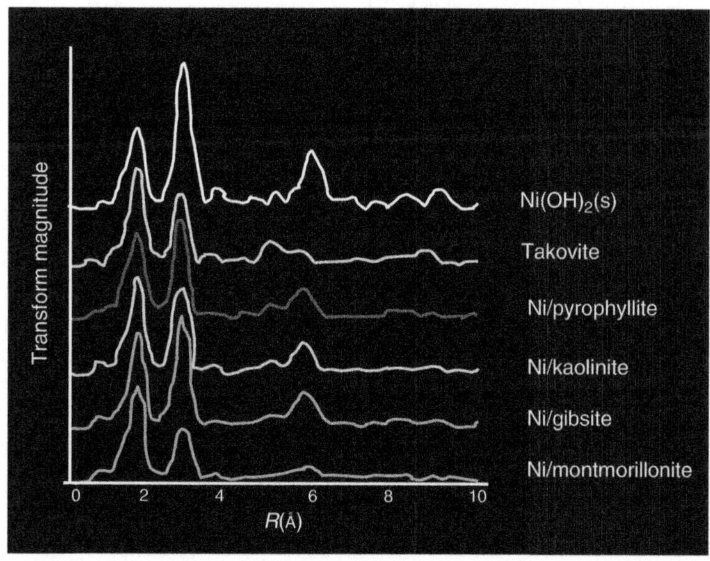

FIGURE 1.12 Radial structure functions (RSFs) produced by forward Fourier transforms of Ni sorbed on pyrophyllite, kaolinite, gibbsite, and montmorillonite compared to the spectrum of crystalline $Ni(OH)_2(s)$ and takovite. The spectra are uncorrected for phase shift. (From AM Scheidegger, GM Lamble, and DL Sparks. *J Colloid Interface Sci* 186:118–128, 1997. With permission.)

second Ni shell was observed in the spectra of the Ni sorption samples and takovite (Figure 1.12). The structural parameters, derived from XAFS analyses, for the various sorption samples and takovite and $Ni(OH)_2$ are shown in Table 1.6. In the first coordination shell Ni is surrounded by six O atoms, indicating that Ni(II) is in an octahedral environment. The Ni–O distances for the Ni sorption samples are 2.02 to 2.03 Å and similar to those in takovite (2.03 Å).

TABLE 1.5
Structural Information Derived From XAFS Analysis for Ni Sorption on Various Sorbents and for Known Ni Hydroxides[b]

	Γ (μmol/m^2)	Ni–O			Ni–Ni			Ni–Si/Al			N(Ni)/N(Si/Al)
		R (Å)	N	$2\sigma^2$	R (Å)	N	$2\sigma^2$	R (Å)	N	$2\sigma^2$	
$Ni(OH)_2$		2.06	6.0	0.011	3.09	6.0	0.010				
Takovite		2.03	6.0	0.01	3.01	3.1	0.009	3.03	1.1	0.009	2.8
Pyrophyllite	3.1	2.02	6.1	0.01	3.00	4.8	0.009	3.02	2.7	0.009	1.8
Kaolinite	19.9	2.03	6.1	0.01	3.01	3.8	0.009	3.02	1.8	0.009	2.2
Gibbsite	5.0	2.03	6.5	0.01	3.02	5.0	0.009	3.05	1.8	0.09	2.7
Montmorillonits	0.35	2.03	6.3	0.01	3.03	2.8	0.011	3.07	2.0	0.015	1.4

Source: AM Scheidegger, GM Lamble, and DL Sparks. *J Colloid Interface Sci* 186:118–128, 1997. With permission.
[a]Interatomic distances (*R*, nm), coordination numbers (*N*), and Debye–Waller factors ($2\sigma^2$, nm^2). The reported values are accurate to within $R \pm 0.002$ nm, $N_{(Ni-O)} \pm 20\%$, $N_{(Ni-Ni)} \pm 40\%$, and $N_{(Ni-Si/Al)} \pm 40\%$.

TABLE 1.6
Effect of Residence Time on Pb Desorption From a Matapeake Soil

Residence Time (days)	Sorbed Pb (mmol kg^{-1})	Desorbed Pb (mmol kg^{-1})	Percentage Pb Desorbed
1	54.9	27.9	50.8
10	60.0	28.7	47.
32	66.1	30.5	46.1

Source: DG Strawn and DL Sparks. *Soil Sci Soc Am J* 64:144–2000. With permission.

The Ni–O distances in crystalline $Ni(OH)_2(s)$ are distinctly longer (2.06 Å in this study). For the second shell, best fits were obtained by including both Ni and Si or Al as second-neighbor backscatter atoms. Since Si and Al differ in atomic number by 1 (atomic number = 14 and 13, respectively), backscattering is similar. They cannot be easily distinguished from each other as second-neighbor backscatters. There are 2.8 (for montmorillonite) to 5.0 (gibbsite), Ni second-neighbor (N) atoms, indicative of Ni surface precipitates. The Ni–Ni distances for the sorption samples were 3.00 to 3.03 Å, which are similar to those for takovite (3.01 Å), the mixed Ni–Al LDH phase, but much shorter than those in crystalline $Ni(OH)_2$ (3.09 Å). There are also 1.8 to 2.7 Si/Al second-neighbor atoms at 3.02 to 3.07 Å The bond distances are in good agreement with the Ni–Al distances observed in takovite (3.03 Å) [2].

Mixed Co–Al and Zn–Al hydroxide surface precipitates also can form on aluminum-bearing metal oxides and phyllosilicates [8, 57, 58, 60]. This is not surprising as Co^{2+}, Zn^{2+}, and Ni^{2+} all have similar radii to Al^{3+}, enhancing substitution in the mineral structure and formation of a coprecipitate. However, surface precipitates have not been observed with Pb^{2+}, as Pb^{2+} is too large to substitute for Al^{3+} in mineral structures [2].

Metal hydroxide precipitate phases can also form in the presence of non-Al-bearing minerals [65]. Using diffuse reflectance spectroscopy (DRS), which is quite sensitive for discriminating between Ni—O bond distances, it was shown that α-$Ni(OH)_2$ formed upon Ni^{2+} sorption to talc and silica (Figure 1.13).

The mechanism for the formation of metal hydroxide surface precipitates is not clearly understood. It is clear that the type of metal ion determines whether metal hydroxide surface

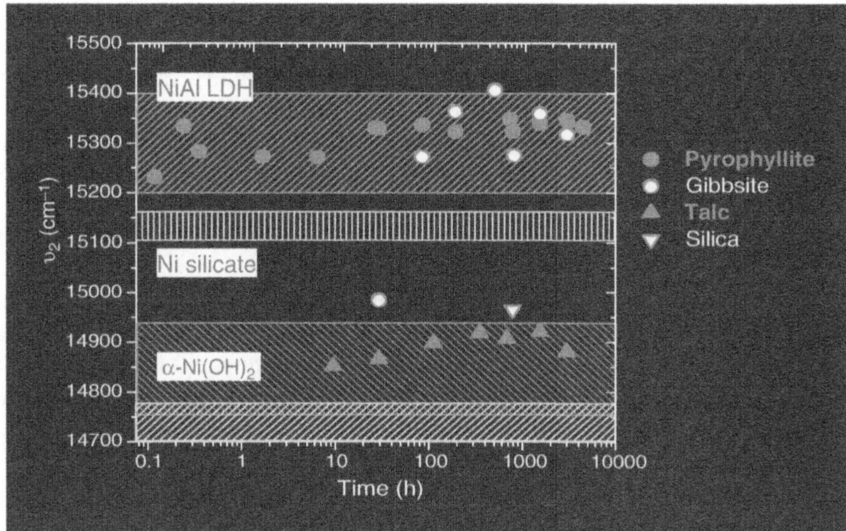

FIGURE 1.13 Fitted ν_2 band positions of the Ni reacted minerals (dots and triangles) over time using diffuse reflectance spectroscopy. The ν_2 band is attributed to crystal field splitting induced within the incompletely filled 3d electronic shell of Ni^{2+} through interaction with the negative charge of nearest-neighbor oxygen ions. For talc and silica, the ν_2 band was at ~ 14,900 cm^3, indicating the formation of a α-$Ni(OH)_2$ phase while for Al-containing pyrophyllite and gibbsite, it appeared at ~15,300 cm^{-1}, indicating a Ni–Al LDH phase. (From AC Scheinost, RG Ford, and DL Sparks. *Geochim Cosmochim Acta* 63:3193–3203, 1999. With permission.)

precipitates form, and the type of surface precipitate formed, i.e., metal hydroxide or mixed metal hydroxide, is dependent on the sorbent type. The precipitation could be explained by the combination of several processes [66]. First, the electric field of the mineral surface attracts metal ions, for example, Ni, through adsorption, leading to a local supersaturation at the mineral–water interface. Second, the solid phase may act as a nucleation center for polyhydroxy species and catalyze the precipitation process [43]. Third, the physical properties of water molecules adsorbed at the mineral surface are different from free water [5], causing a lower solubility of metal hydroxides at the mineral–water interface. With time Al, which is released by weathering of the mineral surface, slowly diffuses into the octahedral layer of the mineral and partially replaces the metal (e.g., Ni) in the octahedral sites. A Ni–Al LDH is formed, which is thermodynamically favored over α-Ni hydroxide [2].

The formation of metal hydroxide surface precipitates appears to be an important way to sequester metals. As the surface precipitates age, metal release is greatly reduced. Thus, the metals are less prone to leaching and being taken up by plants. This is due to silication of the interlayer of the LDH phases, creating precursor phyllosilicate surface precipitates. More details on the metal release rates from surface precipitates and the mechanisms for the metal sequestration are discussed later in this chapter [2].

1.6 KINETICS OF METAL AND OXYANION SORPTION

1.6.1 Rate-Limiting Steps and Time Scales

A number of transport and chemical reaction processes can affect the rate of sorption reactions. The slowest of these will limit the rate of a particular reaction. The actual chemical reaction (CR) at the surface, for example, adsorption, is usually very rapid and not rate-limiting. Transport processes (Figure 1.14) include: (1) transport in the solution phase, which is rapid, and in the laboratory, can be eliminated by rapid mixing; (2) transport across a liquid film at the particle–liquid interface (film diffusion [FD]); (3) transport in liquid filled macropores (>2 nm), all of which are nonactivated diffusion processes and occur in mobile regions; (4) diffusion of a sorbate along pore wall surfaces (surface diffusion); (5) diffusion of sorbate occluded in micropores (<2 nm) (pore diffusion); and (6) diffusion processes in the bulk of the solid, all of which are activated diffusion processes. Pore and surface diffusion can be referred to as interparticle diffusion, while diffusion in the solid is intraparticle diffusion [2].

Metal and oxyanion sorption reactions occur over a wide time scale ranging from milliseconds to days and months. The type of soil component can drastically affect the reaction rate. For example, sorption reactions are often more rapid on clay minerals such as kaolinite and smectites than on vermiculitic and micaceous minerals. This is in large part due to the availability of sites for sorption. For example, kaolinite has readily available planar external sites and smectites have primarily internal sites that are also quite available for retention of sorbates. Thus, sorption reactions on these soil constituents are often quite rapid, even occurring on time scales of seconds and milliseconds [2, 67].

On the other hand, vermiculite and micas have multiple sites for retention of metals including planar, edge, and interlayer sites, with some partially to totally collapsed mica sites. Consequently, sorption and desorption reactions on these sites can be slow, tortuous, and mass transfer controlled. Often, an apparent equilibrium may not be reached even after several days or weeks. Thus, with vermiculite and mica, sorption can involve two to three different reaction rates: high rates on external sites, intermediate rates on edge sites, and low rates on interlayer sites [68, 69].

1.6.2 Residence Time Effects on Metal and Oxyanion Sorption

The kinetics of heavy metal and oxyanion sorption on soil components such as clay minerals and metal hydr(oxides) and soils is typically characterized by a biphasic process in which sorption is initially rapid followed by slow reactions (Figure 1.15). The rapid step, which

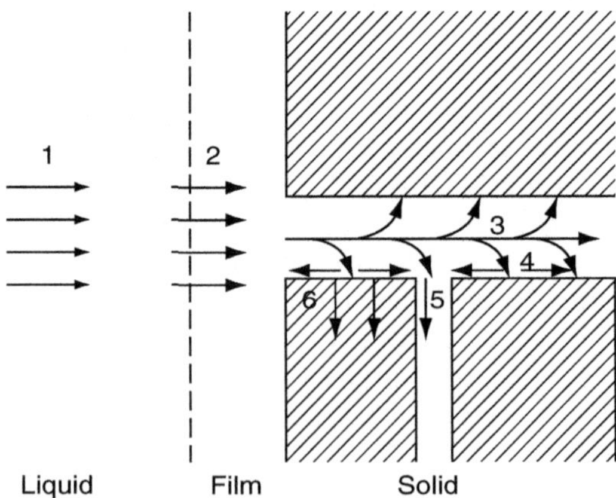

FIGURE 1.14 Transport processes in solid–liquid soil reactions-nonactivated processes: (1) transport in the soil solution, (2) transport across a liquid film at the solid–liquid interface, (3) transport in a liquid-filled macropore; activated processes, (4) diffusion of a sorbate at the surface of the solid, (5) diffusion of a sorbate occluded in a micropore, (6) diffusion in the bulk of the solid source. (From C Aharoni and DL Sparks. In: DL Sparks and DL Suarez (eds). *Rates of Soil Chemical Processes*. Madison, WI: Soil Science Society of America, 1991, pp. 1–18. With permission.)

occurs over milliseconds to hours, can be ascribed to CR and FD processes and in some cases surface precipitation [2, 44]. During this rapid reaction process, a large portion of the sorption may occur. For example, in Figure 1.15a one sees that ~90% of the total Ni sorbed on kaolinite and pyrophyllite occurred within the first 24 hr. For Pb sorption on a Matapeake soil, 78% of the total Pb sorption occurred in 8 min. Following the initial fast reaction, slow sorption continued, but only about 1% additional Pb was sorbed after 800 hr (Figure 1.15b). Figure 1.15c shows a biphasic reaction for As(V) sorption on ferrihydrite. Within 5 min, a majority of the total sorption had occurred. Slow sorption continued for at least 192 hr.

The mechanisms for the slow sorption are not well understood but have been ascribed to diffusion phenomena, sites of lower reactivity, and surface nucleation or precipitation [62, 70, 71]. Bruemmer et al. [72] studied Ni^{2+}, Zn^{2+}, and Cd^{2+} sorption on goethite, a porous Fe oxide that has defect structures in which metals can be incorporated to satisfy charge imbalances. It was found at pH 6 that as reaction time increased from 2 hr to 42 days (at 293 K), adsorbed Ni^{2+} increased from 12% to 70% of total sorption, and total Zn^{2+} and Cd^{2+} sorption over the same time increased by 33% and 21%, respectively. Metal uptake was hypothesized to occur by a three-step mechanism: (1) adsorption of metals on external surfaces, (2) solid-state diffusion of metals from external to internal sites, and (3) metal binding and fixation at positions inside the goethite particle.

The amount of contact time between metals and oxyanions and soil sorbents (residence time) can dramatically affect the degree of desorption, depending on the sorbate and sorbent.

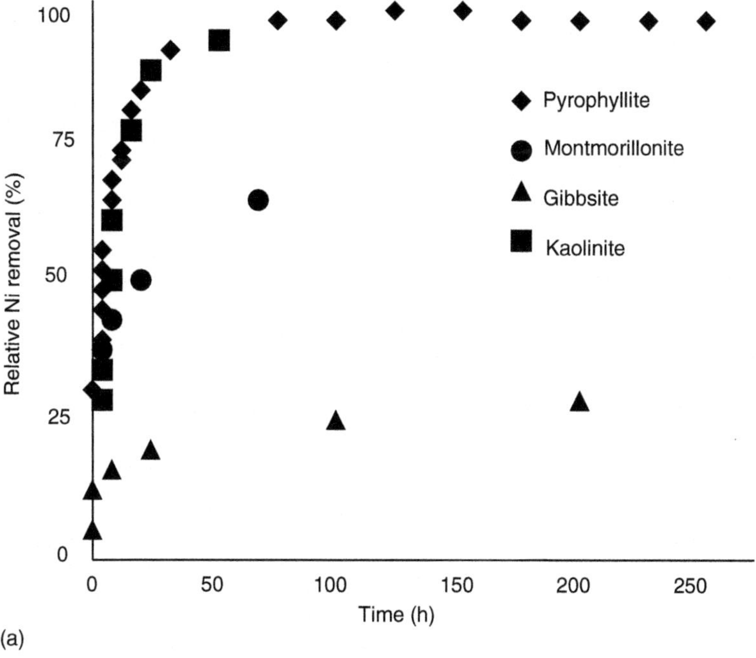

FIGURE 1.15 Kinetics of metal and oxyanion sorption on soil minerals and soil. (a) Kinetics of Ni sorption (%) on pyrophyllite (◆), kaolinite (■), gibbsite (▲), and montmorillonite (●) from a 3 m*M* Ni solution, an ionic strength $I = 0.1\ M$ $NaNO_3$, and a pH of 7.5. (From AM Scheidegger and DL Sparks. *Chem Geol* 132:157–164, 1996. With permission.). (b) Kinetics of Pb sorption on a Matapeake soil from a 12.25 m*M* Pb solution, an ionic strength $I = 0.05\ M$, and a pH of 5.5. (From DG Strawn and DL Sparks. *Soil Sci Soc Am J* 64:144–156, 2000. With permission). (c) Kinetics of As(V) sorption on ferrihydrite at pH 8.0 and 9.0. (From CC Fuller, JA Davis, and GA Waychunas. *Geochim Cosmochim Acta* 57:2271–2282, 1993. With permission.)

continued

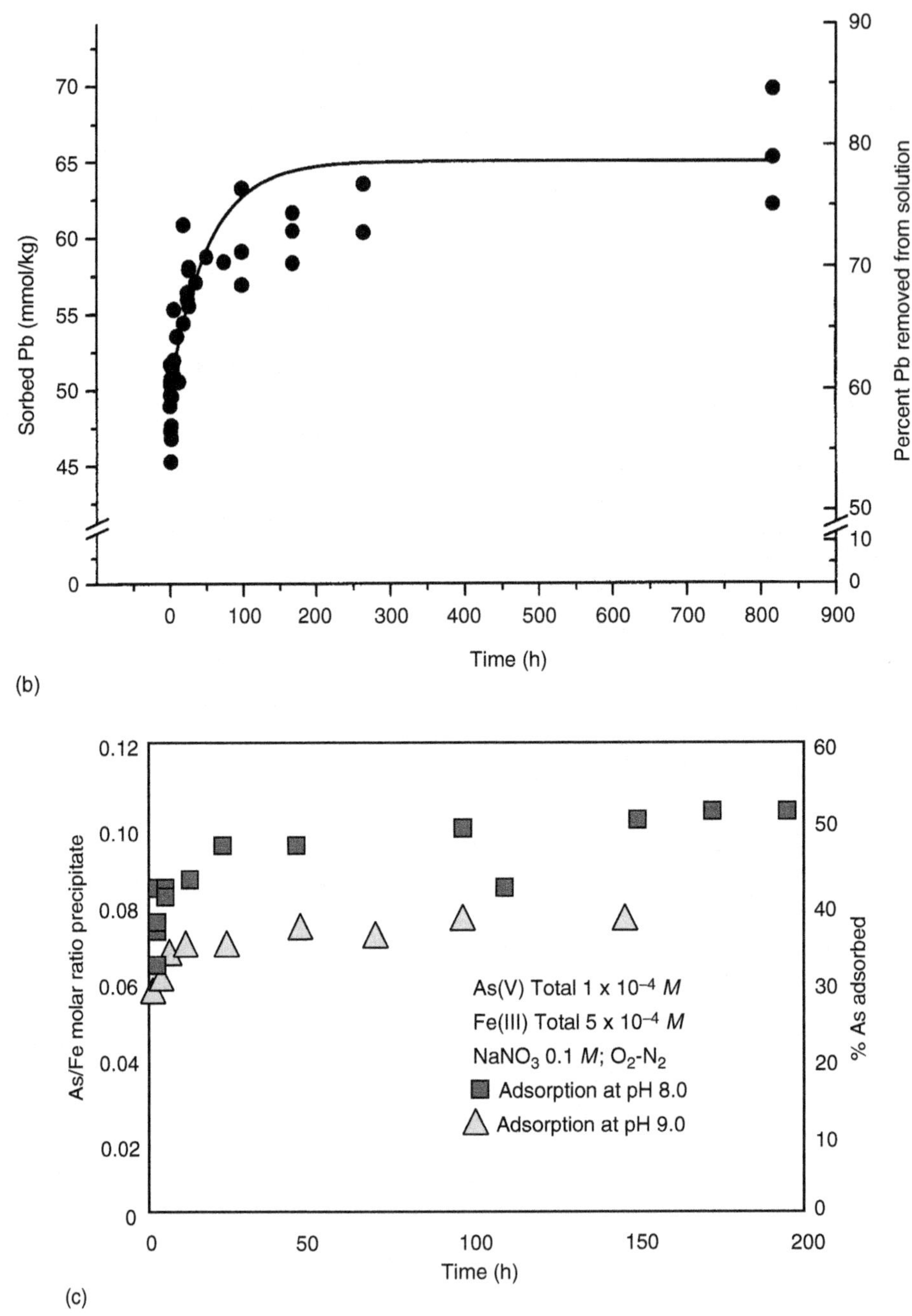

FIGURE 1.15 *continued*

Examples of this are shown in Figure 1.16 and Table 1.6. In Figure 1.16 the effect of residence time on Pb^{2+} and Co^{2+} desorption from hydrous Fe–oxide (HFO) was studied. With Pb, between pH 3 and 5.5, there was a minor effect of residence time (over 21 weeks) on Pb^{2+} desorption, with only minor hysteresis occurring (hysteresis varied from <2% difference between sorption and desorption to ~10%). At pH 2.5, Pb^{2+} desorption was complete within a 16-hr period and was not affected by residence time (Figure 1.16a). In a soil where

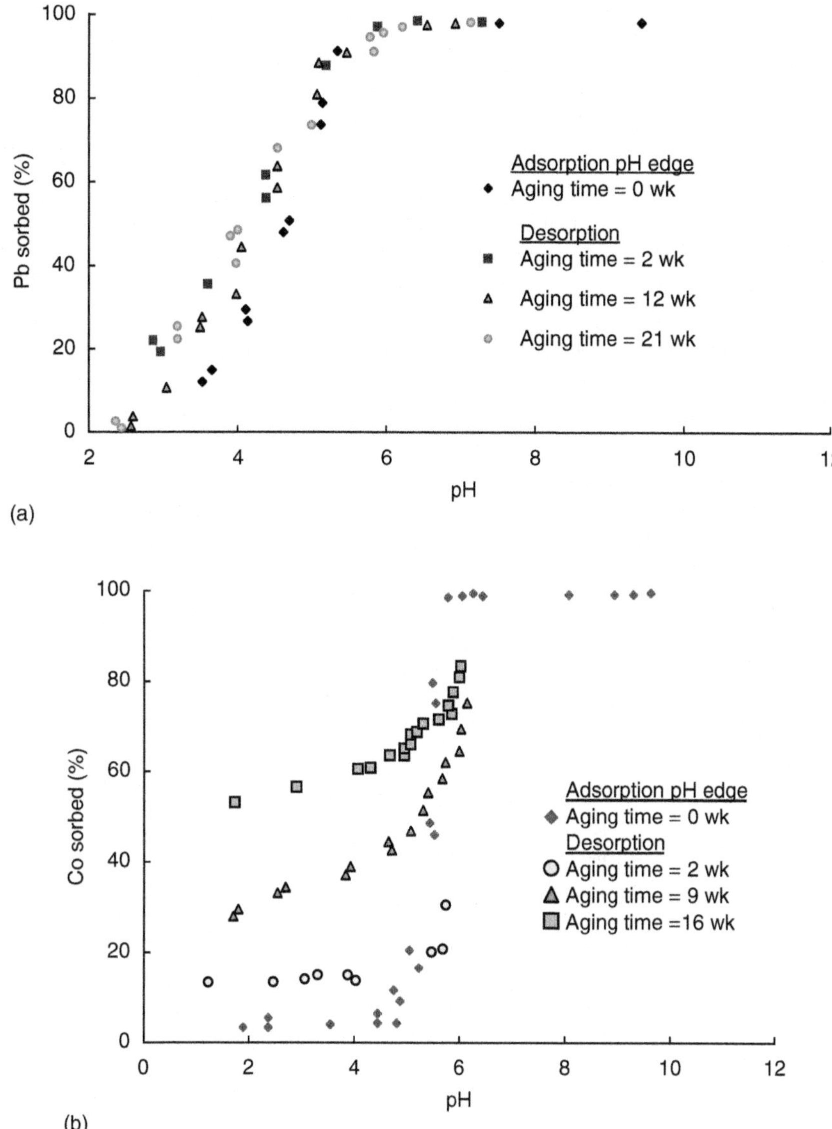

FIGURE 1.16 Fractional sorption–desorption of (a) Pb^{2+} and (b) Co^{2+} to hydrous Fe(III) as a function of pH and HFO–Pb^{2+} (a) and HFO–Co^{2+} (b) aging time. (From CC Ainsworth, JL Pilou, PL Gassman, and WG Van Der Sluys. *Soil Sci Soc Am J* 58:1615–1623, 1994. With permission.)

2.1% Soil organic matter (SOM) was present, residence time had little effect on the amount of Pb desorbed, but marked hysteresis was observed at all residence times (Table 1.6). This could be ascribed to the strong metal–soil organic matter complexes that occur and perhaps to diffusion processes [2].

With Co^{2+}, extensive hysteresis was observed over a 16-week residence time (Figure 1.16b) and the hysteresis increased with residence time. After a 16 week residence time, 53% of the Co^{2+} was not desorbed, and even at pH 2.5, hysteresis was observed. The extent of Co reversibility with residence time was attributed to Co incorporation into a recrystallizing solid by isomorphic substitution and not to micropore diffusion. Similar residence time effects on Co desorption have been observed with both Fe- and Mn-oxides [73] and with soil clays [74].

1.6.3 Kinetics of Metal Hydroxide Surface Precipitation/Dissolution

In addition to diffusion processes, the formation of metal hydroxide surface precipitates and subsequent residence time effects on natural sorbents can greatly affect metal release and hysteresis. It has generally been thought that the kinetics of formation of surface precipitates was slow. However, recent studies have shown that metal hydroxide precipitates can form on time scales of minutes. In Figure 1.17 one sees that mixed Ni–Al hydroxide precipitates formed on pyrophyllite within 15 min, and they grew in intensity as time increased. Similar results have been observed with other soil components and with soils [44, 56].

The formation and subsequent "aging" of the metal hydroxide surface precipitate can have a significant effect on metal release. In Figure 1.18 one sees that as residence time ("aging") increases from 1 hr to 2 years, Ni release from pyrophyllite, as a percentage of total Ni sorption, decreased from 23% to ~0%, when HNO_3 (pH 6.0) was employed as a dissolution agent for 14 days. This enhanced stability is due to the transformation of the metal–Al hydroxide precipitates to a metal–Al phyllosilicate precursor phase as residence time increases. This transformation occurs via a number of steps (Figure 1.19). There is diffusion of Si originating from weathering of the sorbent into the interlayer space of the LDH, replacing the anions such as NO_3^-. Polymerization and condensation of the interlayer Si slowly transforms the LDH into a precursor metal–Al phyllosilicate. The metal stabilization that occurs in surface precipitates on Al-free sorbents (e.g., talc) may be due to Ostwald ripening, resulting in increased crystallization [2, 61].

Thus, with time, one sees that metal sorption on soil minerals can often result in a continuum of processes from adsorption to precipitation to solid phase transformation (Figure 1.20), particularly in the case of certain metals such as Co, Ni, and Zn. The formation of metal surface precipitates could be an important mechanism for sequestering metals in soils

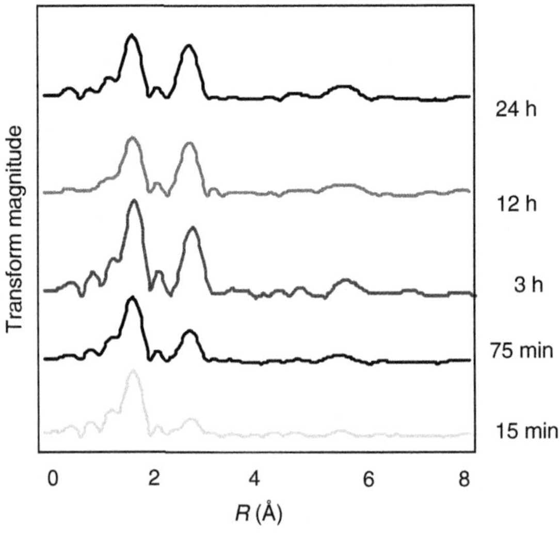

FIGURE 1.17 Radial structure functions (derived from XAFS analyses) for Ni sorption on pyrophyllite for reaction times up to 24 h, demonstrating the appearance and growth of the second shell (peak at ≈2.8 Å) contributions due to surface precipitation and growth of a mixed Ni–Al hydroxide phase. (From AM Scheidegger, DG Strawn, GM Lamble, and DL Sparks. *Geochim Cosmochim Acta* 62:2233–2245, 1998. With permission.)

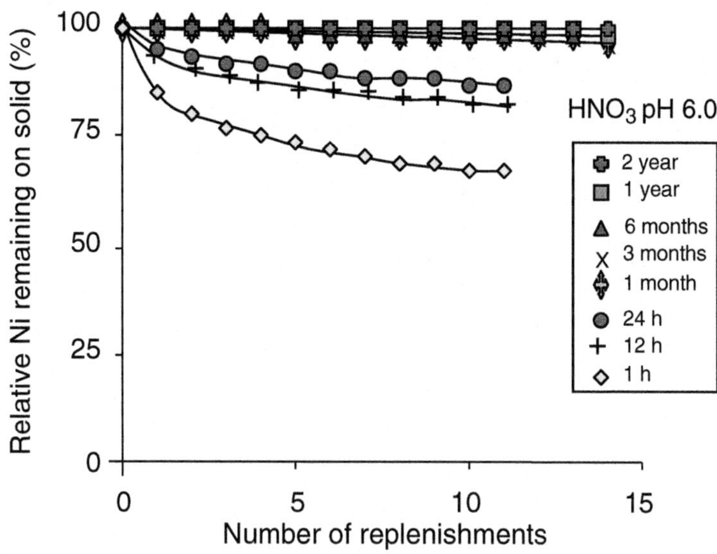

FIGURE 1.18 Dissolution of Ni from surface precipitates formed on pyrophyllite at residence times of 1 h to 2 years. The figure shows the relative amount of Ni^{2+} remaining on the pyrophyllite surface following extraction for 24-h periods (each replenishment represents a 24-h extraction) with HNO_3 at pH 6.0. (From KG Scheckel, AC Scheinost, RG Ford, and DL Sparks. *Geochim Cosmochim Acta* 4:2727–2735, 2000. With permission.)

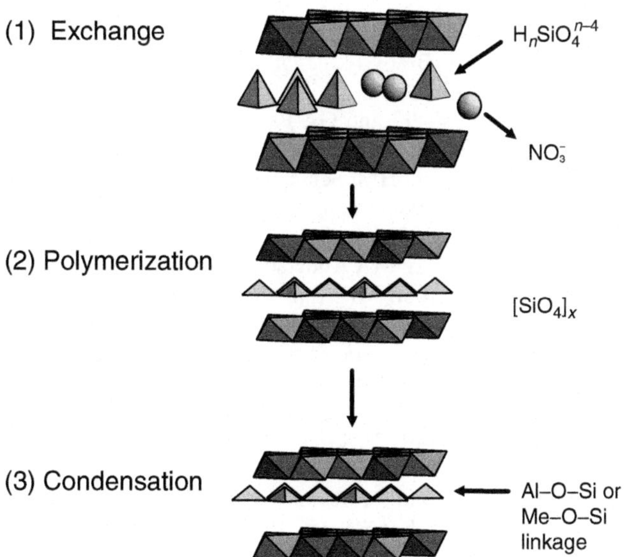

FIGURE 1.19 Hypothetical reaction process illustrating the transformation of an initially precipitated Ni–Al LDH to a phyllosilicate like phase during aging. The initial step involves the exchange of dissolved silica for nitrate within the LDH interlayer followed by polymerization and condensation of silica onto the octahedral Ni–Al layer. The resultant solid possesses structural features common to 1:1 and 2:1 phyllosilicates. (From RG Ford, AC Scheinost, and DL Sparks. In: DL Sparks (ed.). *Adv Agron* 2001, pp 41–62. With permission.)

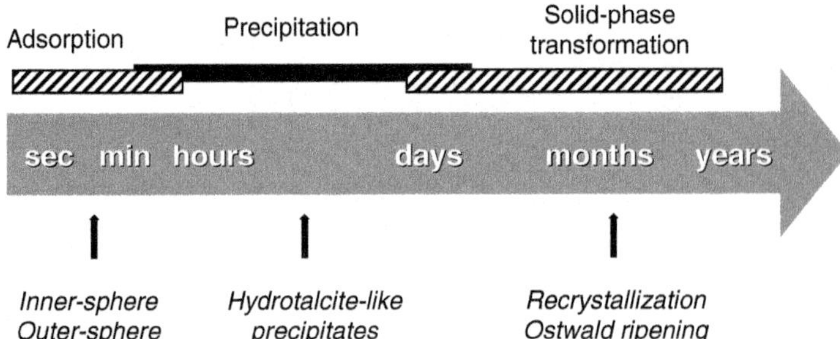

FIGURE 1.20 Changes in sorption processes with time showing a continuum from adsorption to precipitation to solid phase transformation. (From DL Sparks. *Environmental Soil Chemistry*. 2nd ed. San Diego: Academic Press, 2002. With permission.)

such that they are less mobile and bioavailable. Such products must be considered when modeling the fate and mobility of metals such as Co^{2+}, Mn^{2+}, Ni^{2+}, and Zn^{2+} in soil and water environments [2].

REFERENCES

1. W Stumm. *Chemistry of the Solid–Water Interface*. New York: Wiley, 1992.
2. DL Sparks. *Environmental Soil Chemistry*. 2nd ed. San Diego: Academic Press, 2002.
3. G Sposito. *The Surface Chemistry of Soils*. New York: Oxford University Press, 1984.
4. W Stumm (ed.). *Aquatic Surface Chemistry*. New York: Wiley (Interscience), 1987.
5. G Sposito. *The Chemistry of Soils*. New York: Oxford University Press, 1989.
6. KF Hayes, JO Leckie. *ACS Symp Ser* 323:114–141, 1986.
7. KF Hayes and LE Katz. In: PV Brady (ed.). *Physics and Chemistry of Mineral Surfaces*. Boca Raton: CRC Press, 1999, pp. 147–223.
8. SN Towle, JR Bargar, GE Brown, Jr., and GA Parks. *J Colloid Interface Sci* 187:62–82, 1997.
9. JR Bargar, GE Brown, Jr., and GA Parks. *Geochim Cosmochim Acta* 61:2617–2637, 1997.
10. LE Katz and EJ Boyle-Wight. In: HM Selim and DL Sparks (eds). *Physical and Chemical Processes of Water and Solute Transport/Retention in Soils*. Madison, WI: Soil Science Society of America, 2001, pp. 213–256.
11. DG Kinniburgh and ML Jackson. In: MA Anderson and AJ Rubin (eds). *Adsorption of Inorganics at Solid–Liquid Interfaces*. Ann Arbor, MI: Ann Arbor Science, 1981, pp. 91–160.
12. PW Schindler. *Oesterreichische Chemie-Zeitschrift* 86:141–147, 1985.
13. AF Williams. *A Theoretical Approach to Inorganic Chemistry*. Berlin: Springer-Verlag, 1979.
14. GE Brown, Jr. and GA Parks. *Int Geol Rev* 43:963–1073, 2001.
15. DG Kinniburgh, ML Jackson, and JK Syers. *Soil Sci Soc Am J* 40:796–799, 1976.
16. RM McKenzie. *Aust J Soil Res* 18:61–73, 1980.
17. CP Huang and W Stumm. *J Colloid Interface Sci* 43:409–420, 1973.
18. FJ Hingston. In: MA Anderson and AJ Rubin (eds). *Adsorption of Inorganics at Solid–Liquid Interfaces*. Ann Arbor, MI: Ann Arbor Science, 1981, pp. 51–90.
19. NJ Barrow. *Adv Agron* 38:183–230, 1985.
20. GE Brown, Jr. In: MF Hochella and AF White (eds). *Mineral–Water Interface Geochemistry*. Washington, DC: Mineralogical Society of America, 1990, pp. 309–353.
21. GE Brown, Jr., GA Parks, and PA O'Day. In: DJ Vaughan and RAD Pattrick (eds). *Mineral Surfaces*. London: Chapman and Hall, 1995, pp. 129–183.
22. SE Fendorf and DL Sparks. In: DL Sparks (ed.). *Methods of Soil Analysis: Part 3. Chemical Methods and Processes*. Madison, WI: Soil Science Society of America, 1996, pp. 377–416.
23. PM Bertsch and DB Hunter. In: PM Huang (ed.). *Future Prospects for Soil Chemistry*. Madison, WI: Soil Science Society of America, 1998, pp. 103–122.

24. SE Fendorf. In: DG Schulze, JW Stucki, and PM Bertsch (eds). *Synchrotron X-ray Methods in Clay Science*. Boulder, CO: The Clay Mineral Society, 1999, pp. 20–67.
25. PA O'Day (1999), *Rev Geophys*, *37*:249–274.
26. DG Schulze, JW Stucki, and PM Bertsch (eds). *Synchrotron X-ray Methods in Clay Science*. Boulder, CO: The Clay Minerals Society, 1999.
27. GE Brown, Jr, and NC Sturchio. In: PA Fenter, ML Rivers, NC Sturchio, and SR Sutton (eds). *Applications of Synchrotron Radiation in Low-Temperature Geochemistry and Environmental Science*. Washington, D.C.: The Mineralogical Society of America, 2002, pp. 1–115.
28. CT Johnston, G Sposito, and WL Earl. In: J Buffle and HP van Leeuwen (eds). *Environmental Particles*. Boca Raton: Lewis Publishers, 1993, pp. 1–36.
29. K Nakamoto (ed.). *Infrared and Raman Spectra of Inorganic and Coordination Compounds*. Part A: Theory and Applications in Inorganic Chemistry. 5th ed. New York: John Wiley & Sons, 1997.
30. DL Suarez, S Goldberg, and C Su. In: DL Sparks and TJ Grundl (eds). *Mineral–Water Interfacial Reactions: Kinetics and Mechanisms*. Washington, D.C.: American Chemical Society, 1998, pp. 136–178.
31. DR Roberts, RG Ford, and DL Sparks. *J Colloid Interface Sci* 263:364–376, 2003.
32. SJ Hug. *J Colloid Interface Sci* 188:415–422, 1997.
33. D Peak, RG Ford, and DL Sparks. *J Colloid Interface Sci* 218:289–299, 1999.
34. D Peak, GW Luther, and DL Sparks. *Geochim Cosmochim Acta* 67:2551–2560, 2003.
35. PC Zhang and DL Sparks. *Soil Sci Soc Am J* 54:1266–1273, 1990.
36. KF Hayes, AL Roe, GE Brown, Jr., KO Hodgson, JO Leckie, and GA Parks. *Science* 238:783–786, 1987.
37. A Manceau and L Charlet. *J Colloid Interface Sci* 164:87–93, 1994.
38. MB McBride. In: GH Bolt, MFD Boodt, MHB Hayes, and MB McBride, (eds). *Interactions at the Soil Colloid–Soil Solution Interface*. Dordrecht: Kluwer Academic Publishers, 1991, pp. 149–176.
39. G Sposito. In: JA Davis and KF Hayes (eds). *Geochemical Processes at Mineral Surfaces*. Washington, DC: American Chemical Society, 1986, pp. 217–229.
40. PA O'Day, GE Brown, and GA Parks. *J Colloid Interface Sci* 165:269–289, 1994.
41. CJ Chisholm-Brause, PA O'Day, GE Brown, Jr., and GA Parks. *Nature* 348:528–530, 1990.
42. KJ Farley, DA Dzombak, and FMM Morel. *J Colloid Interface Sci* 106:226–242, 1985.
43. MB McBride. *Environmental Chemistry of Soils*. New York: Oxford University Press, 1994.
44. AM Scheidegger, DG Strawn, GM Lamble, and DL Sparks. *Geochim Cosmochim Acta* 62:2233–2245, 1998.
45. SE Fendorf, GM Lamble, MG Stapleton, MJ Kelley, and DL Sparks. Environ Sci Technol 28:284–289, 1994.
46. SE Fendorf and DL Sparks. *Environ Sci Technol* 28:290–297, 1994.
47. CJ Chisholm-Brause, AL Roe, KF Hayes, GE Brown, Jr., GA Parks, and JO Leckie. *Geochim Cosmochim Acta* 54:1897–1909, 1990.
48. AL Roe, KF Hayes, CJ Chisholm-Brause, GE Brown, Jr., GA Parks, KO Hodgson, and JO Leckie. *Langmuir* 7:367–373, 1991.
49. L Charlet and A Manceau. *J Colloid Interface Sci* 148:443–458, 1992.
50. PA O'Day, GA Parks, and GE Brown, Jr. *Clays Clay Miner* 42:337–355, 1994.
51. JR Bargar, GE Brown, and GA Parks. *Physica B* 209:455–456, 1995.
52. C Papelis and KF Hayes. *Colloid Surfaces* 107:89, 1996.
53. AM Scheidegger, GM Lamble, and DL Sparks. *Environ Sci Technol* 30:548–554, 1996.
54. AM Scheidegger and DL Sparks. *Chem Geol* 132:157–164, 1996.
55. AM Scheidegger, GM Lamble, and DL Sparks. *J Colloid Interface Sci* 186:118–128, 1997.
56. DR Roberts, AM Scheidegger, and DL Sparks. *Environ Sci Technol* 33:3749–3754, 1999.
57. HA Thompson, GA Parks, and GE Brown Jr. *Clays Clay Min* 47:425–438, 1999.
58. HA Thompson, GA Parks, and GE Brown, Jr. *Geochim Cosmochim Acta* 63:1767–1779, 1999.
59. EJ Elzinga, and DL Sparks. *J Colloid Interface Sci* 213:506–512, 1999.
60. RG Ford and DL Sparks. *Environ Sci Technol* 34:2479–2483, 2000.
61. KG Scheckel and DL Sparks. *Soil Sci Soc Am J* 65:719–728, 2001.
62. AM Scheidegger and DL Sparks. *Soil Sci* 161:813–831, 1996.
63. R Allmann. *Chimia* 24:99–108, 1970.

64. RM Taylor. *Clay Miner* 19:591–603, 1984.
65. AC Scheinost, RG Ford, and DL Sparks. *Geochim Cosmochim Acta* 63:3193–3203, 1999.
66. NU Yamaguchi, AC Scheinost, and DL Sparks. *Soil Sci Soc Am J* 65:729–736, 2001.
67. DL Sparks. *Kinetics of Soil Chemical Processes*. San Diego, CA: Academic Press, 1989.
68. PM Jardine and DL Sparks. *Soil Sci Soc Am J* 48:39–45, 1984.
69. RNJ Comans and DE Hockley. *Geochim Cosmochim Acta* 56:1157–1164, 1992.
70. DL Sparks. In: PM Huang, N Senesi and J Buffle (eds). *Structure and Surface Reactions of Soil Particles*. New York, NY: John Wiley and Sons, 1998, pp. 413–448.
71. DL Sparks. In: PM Huang, DL Sparks, and SA Boyd (eds). *Future Prospects for Soil Chemistry*. Madison, WI: Soil Science Society of America, 1999, pp. 81–102.
72. GW Bruemmer, J Gerth, and KG Tiller. *J Soil Sci* 39:37–52, 1988.
73. CA Backes, RG McLaren, AW Rate, and RS Swift. *Soil Sci Soc Am J* 59:778–785, 1995.
74. RG McLaren, CA Backes, AW Rate, and RS Swift. *Soil Sci Soc Am J* 62:332–337, 1998.
75. L Charlet and A Manceau. In: J Buffle and HP van Leeuwen (eds). *Environmental Particles*. Boca Raton: Lewis Publishers, 1993, pp. 117–164.
76. PW Schindler. In: MA Anderson and AJ Rubin (eds). *Adsorption of Inorganics at Solid–Liquid Interfaces*. Ann Arbor, MI: Ann Arbor Science, 1981, pp. 1–49.
77. FJ Hingston, AM Posner, and JP Quirk. *J Soil Sci* 23:177–192, 1972.
78. SE Fendorf. Ph.D. Dissertation, University of Delaware, Newark, 1992.
79. C Aharoni and DL Sparks. In: DL Sparks and DL Suarez (eds). *Rates of Soil Chemical Processes*. Madison, WI: Soil Science Society of America, 1991, pp. 1–18.
80. DG Strawn and DL Sparks. *Soil Sci Soc Am* J 64:144–156, 2000.
81. CC Fuller, JA Davis, and GA Waychunas. *Geochim Cosmochim Acta* 57:2271–2282, 1993.
82. CC Ainsworth, JL Pilou, PL Gassman, and WG Van Der Sluys. *Soil Sci Soc Am J* 58:1615–1623, 1994.
83. KG Scheckel, AC Scheinost, RG Ford, and DL Sparks. *Geochim Cosmochim Acta* 64:2727–2735, 2000.
84. RG Ford, AC Scheinost, and DL Sparks. In: DL Sparks (ed.). *Adv Agron* 2001, pp. 41–62.
85. F Helfferich. *Ion Exchange*. Ann Arbor, MI: University Microfilms International, 1962.
86. HT Tien. *J Phys Chem* 69:350–352, 1965.
87. I Altug and ML Hair. *J Phys Chem* 71:4260–4263, 1967.
88. RP Abendroth. *J Colloid Interface Sci* 34:591–596, 1970.
89. FE Bartell and Y Fu. *J Phys Chem* 33:676–687, 1929.
90. SC Churms. *J South Afr Chem Inst* 19:98–107, 1966.
91. A Breeuwsma and J Lyklema. *Disc Faraday Soc* 522:3224–3233, 1971.
92. F Dumont and A Watillon. *Disc Faraday Soc* 52:352–360, 1971.
93. B Venkataramani, KS Venkateswarlu, and J Shankar. *J Colloid Interface Sci* 67:187–194, 1978.
94. YG Bérubé and PL de Bruyn. *J Colloid Interface Sci* 27:305–318, 1968.
95. JD Kurbatov, JL Kulp, and E Mack. *J Am Chem Soc* 67:1923–1929, 1945.
96. Y Belot, C Gailledreau, and R Rzekiecki. *Health Phys* 12:811–823, 1966.
97. JP Gabano, P Étienne, and JF Laurent. *Electrochim Acta* 10:947–963, 1965.
98. JW Murray. *Geochim Cosmochim Acta* 39:635–647, 1975.
99. TF Tadros and J Lyklema. *J Electroanal Chem* 22:1, 1969.
100. MA Malati and SF Estefan. *J Colloid Interface Sci* 24:306–307, 1966.
101. RR Gadde and HA Laitinen. *Anal Chem* 46:2022–2026, 1974.
102. E Brunnix. *Phillips Res Rep* 30:177–191, 1975.
103. DA Dzombak and FMM Morel. *Surface Complexation Modeling. Hydrous Ferric Oxide*. New York: Wiley, 1990.
104. JO Leckie. Personal communication, 2001.
105. H Grimme. *Z Pflanzenernähr Düng Bodenkunde* 121:58–65, 1968.
106. EA Forbs, AM Posner, and JP Quirk. *J Soil Sci* 27:154–166, 1976.
107. A Kozawa. *J Electrochem Soc* 106:552–556, 1959.
108. DJ Murray, TW Healy, and DW Fuerstenau. In: WJ Weber, Jr. and E. *Adsorption from Aqueous Solution*. Washington, DC: American Chemical Society, 1968, pp. 74–81.
109. RM McKenzie. *Geoderma* 8:29–35, 1972.
110. P Loganthan and RG Burau. *Geochim Cosmochim Acta* 37:1277–1293, 1973.

111. F Vydra and J Galba. *Colln Czech Chem Comm* 34:3471–3478, 1969.
112. K Taniguechi, M Nakajima, S Yoshida, and K Tarama. *Nippon Kagaku Zasshi* 91:525–529, 1970.
113. L Bochatay, P Persson, and S Sjoberg. *J Colloid Interface Sci* 229:584–592, 2000.
114. PA O'Day, CJ Chisholm-Brause, SN Towle, GA Parks, GE Brown, Jr. *Geochim Cosmochim Acta* 60:2515–2532, 1996.
115. K Xia, W Bleam, and PA Helmke. *Geochim Cosmochim Acta* 61:2223–2235, 1997.
116. JP Fitts, GE Brown, and GA Parks. *Environ Sci Technol* 34:5122–5128, 2000.
117. FJ Weesner and WF Bleam. *J Colloid Interface Sci* 196:79–86, 1997.
118. SF Cheah, GE Brown, and GA Parks. *J Colloid Interface Sci* 208:110–128, 1998.
119. AC Scheinost, S Abend, KI Pandya, and DL Sparks. *Environ Sci Technol* 35:1090–1096, 2001.
120. K Xia, A Mehadi, RW Taylor, and WF Bleam. *J Colloid Interface Sci* 185:252–257, 1997.
121. K Xia, W Bleam, and PA Helmke. *Geochim Cosmochim Acta* 61:2211–2221, 1997.
122. JD Morton, JD Semrau, and KF Hayes. *Geochim Cosmochim Acta* 65:2709–2722, 2001.
123. EJ Elzinga and DL Sparks. *Soil Sci Soc Am J* 65:94–101, 2001.
124. KG Scheckel and DL Sparks. *J Colloid Interface Sci* 229:222–229, 2000.
125. M Nachtegaal and DL Sparks. *Environ Sci Technol* 37:529–534, 2003.
126. DG Strawn, AM Scheidegger, and DL Sparks. *Environ Sci Technol* 32:2596–2601, 1998.
127. JR Bargar, SN Towle, GE Brown, and GA Parks. *Geochim Cosmochim Acta* 60:3541–3547, 1996.
128. JR Bargar, GE Brown, and GA Parks. *Geochim Cosmochim Acta* 61:2639–2652, 1997.
129. JD Ostergren, TP Trainor, JR Bargar, GE Brown, and GA Parks. *J Colloid Interface Sci* 225:466–482, 2000.
130. EJ Elzinga, D Peak, and DL Sparks. *Geochim Cosmochim Acta* 65:2219–2230, 2001.
131. JD Ostergren, GE Brown, Jr., GA Parks, and P Persson. *J Colloid Interface Sci* 225:483–493, 2000.
132. EJ Elzinga and DL Sparks. *Environ Sci Technol* 36:4352–4357, 2002.
133. CJ Matocha, DL Sparks, JE Amonette, and RK Kukkadapu. *Soil Sci Soc Am J* 65:58–66, 2001.
134. DG Strawn and DL Sparks. *J Colloid Interface Sci* 216:257–269, 1999.
135. P Trivedi, JA Dyer, and DL Sparks. *Environ Sci Technol* 37:908–914, 2003.
136. L Axe and PR Anderson. *J Colloid Interface Sci* 185:436–448, 1997.
137. N Sahai, SA Carroll, S Roberts, and PA O'Day. *J Colloid Interface Sci* 222:198–212, 2000.
138. TP Trainor, GE Brown, and GA Parks. *J Colloid Interface Sci* 231:359–372, 2000.
139. TP Trainor, JP Fitts, AS Templeton, D Grolimund, and GE Brown. *J Colloid Interface Sci* 244:239–244, 2001.
140. Y Arai, EJ Elzinga, and DL Sparks. *J Colloid Interface Sci* 235:80–88, 2001.
141. DL Suarez, S Goldberg, and C Su. In: DL Sparks and TJ Grundl (eds). *Mineral–Water Interfacial Reactions: Kinetics and Mechanisms*. Washington, D.C.: American Chemical Society, 1998, pp. 136–178.
142. XH Sun and HE Doner. *Soil Sci* 161:865–872, 1996.
143. BA Manning, SE Fendorf, and S Goldberg. *Environ Sci Technol* 32:2383–2388, 1998.
144. S Goldberg and CT Johnston. *J Colloid Interface Sci* 234:204–216, 2001.
145. ACQ Ladeira, VST Ciminelli, HA Duarte, MCM Alves, and AY Ramos. *Geochim Cosmochim Acta* 65:1211–1217, 2001.
146. Y Arai and DL Sparks. *Soil Sci* 167:303–314, 2002.
147. SE O'Reilly, DG Strawn, and DL Sparks. *Soil Sci Soc Am J* 65:67–77, 2001.
148. GA Waychunas, BA Rea, CC Fuller, and JA Davis. *Geochim Cosmochim Acta* 57:2251–2269, 1993.
149. SE Fendorf, MJ Eick, PR Grossl, and DL Sparks. *Environ Sci Technol* 31:315–320, 1997.
150. SR Randall, DM Sherman, and KV Ragnarsdottir. *Geochim Cosmochim Acta* 65:1015–1023, 2001.
151. AL Foster, GE Brown, and GA Parks. *Geochim Cosmochim Acta* 67:1937–1953, 2003.
152. CM Su and DL Suarez. *Environ Sci Technol* 29:302–311, 1995.
153. H Wijnja and CP Schulthess. *Spectrochim Acta Part A: Molec Biomolec Spectrosc* 55:861–872, 1999.
154. M Villalobos and JO Leckie. *Geochim Cosmochim Acta* 64:3787–3802, 2000.
155. H Wijnja and CP Schulthess. *Soil Sci Soc Am J* 65:324–330, 2001.
156. WF Bleam. *Adv Agron* 46:91–155, 1991.
157. P Persson, N Nilsson, and S Sjoberg. *J Colloid Interf Sci* 177:263–275, 1996.

158. ML Tejedor-Tejedor and MA Anderson. *Langmuir* 124:79–110, 1990.
159. Y Arai and DL Sparks. *J Colloid Interface Sci* 241:317–326, 2001.
160. H Wijnja and CP Schulthess. *J Colloid Interface Sci* 229:286–297, 2000.

2 Catalysis of Electron Transfer Reactions at Mineral Surfaces

Martin A. Schoonen[1,2] *and Daniel R. Strongin*[2,3]
[1]Department of Geosciences, Stony Brook University
[2]Center for Environmental Molecular Science, Stony Brook University
[3]Department of Chemistry, Temple University

CONTENTS

2.1 INTRODUCTION

Electron transfer (ET) reactions are important in environmental chemistry as toxicity, bioavailability, and fate are often dependent on the electronic state of pollutants. Two examples to illustrate this are: (a) the transformation of aqueous hexavalent chromium (Cr^{VI}) to trivalent chromium (Cr^{III}) and (b) the dehalogenation of carbon tetrachloride (CCl_4). Cr^{VI} is far more toxic than Cr^{III} and is mobile in aerobic environments [1]. The transformation of the hexavalent species to Cr^{III}, however, leads to the formation of a Cr^{III}–hydroxide under most conditions, which effectively immobilizes the chromium and limits its bioavailability. The reduction of chromium in soils is facilitated, and in some cases catalyzed, on mineral surfaces [2]. CCl_4 is a common pollutant in groundwater as a result of its use as an organic solvent [1]. Through successive ET steps, CCl_4 can be transformed to dissolved carbon dioxide and chloride. ET reactions are often slow in homogeneous solutions, but can be promoted by photochemical processes, enzymatic action, or the presence of mineral surfaces. In this chapter, the role of mineral surfaces in promoting ET reactions is examined. Only non-photochemical processes are considered in this chapter, and photochemical reactions are examined elsewhere in this book. Following a brief review of the key concepts in the mechanism of ET reactions, several types of ET reactions involving mineral surfaces will be presented. Wherever possible, examples involving various types of pollutants are used to illustrate the concepts presented in this chapter.

2.2 BACKGROUND

As the term *electron transfer* implies, ET reactions require the transfer of electrons between an electron donor (reductant) and an electron acceptor (oxidant). The change in Gibbs free energy for the reaction dictates whether the reaction can proceed spontaneously under a given set of conditions. However, whether a chemical reaction, which is thermodynamically favorable (exergonic, i.e., $\Delta_r G < 0$), proceeds at a significant rate depends on the kinetics of ET between the donor and the acceptor. Given that the energetics of the formation of a solvated electron is exceedingly high, ET proceeds by transferring electrons via a mechanism that does not involve the formation of a free solvated electron. The implication is that it is necessary for reactants in a homogeneous ET reaction to form a complex that allows for the transfer of electrons from donor to acceptor. This is a key aspect of ET reactions that is not always appreciated and differentiates ET processes from acid–base reactions, where the formation of free protons and their subsequent transfer during a chemical reaction is energetically feasible. As will be discussed below, for many ET reactions the formation of a suitable reaction complex or rearrangement coupled with electron transfer within this complex is the rate-limiting step. As a consequence, the kinetics of many ET reactions can be slow or inhibited. In contrast, acid–base reactions are universally fast because there is no requirement for the formation of reaction complex between the proton donor and acceptor.

Understanding at a molecular level what chemical or physical properties of the acceptor, donor, and complex control the flow of charge in ET in homogenous reactions is a crucial step in understanding how mineral surfaces can promote ET reactions. Toward this end, in the following section, we review the mechanism and kinetics of ET in homogeneous solution to set the stage for a discussion of the mechanisms by which mineral surfaces can facilitate ET reactions.

2.3 MECHANISM AND KINETICS OF ET IN HOMOGENEOUS SOLUTIONS

ET in homogeneous solutions requires the transfer of electrons from an electron donor to an acceptor. At a first glance it may appear to be an intractable problem to determine which orbitals are involved in an ET reaction. For example, molecular oxygen, a common electron acceptor, has 12 electrons occupying seven molecular orbitals (Figure 2.1). However, molecular orbital theory provides a useful model to analyze the ET process. As the distance between the two reactants decreases, their molecular orbitals start to interact. ET cannot occur from the interaction between two filled orbitals or two unfilled orbitals. ET can occur by the interaction of a filled orbital with an unfilled orbital. Of the possible interactions between filled and unfilled orbitals, the interaction between the highest occupied molecular orbital (HOMO) of the donor and the lowest unoccupied molecular orbital (LUMO) of the acceptor is energetically most favorable.

In principle, the transfer of up to two electrons can be accomplished through the formation of a single transition state complex. However, one-electron transfer reactions dominate. The transfer of one electron creates a very reactive intermediate (radical) that will readily extract a second electron to complete a two-electron transfer reaction. For example, oxidation of HS^- with O_2 is thought to create superoxide and a HS radical species in an one-electron transfer step (1) [3]. Each of these two species will react further. An example of such a subsequent reaction is given in reaction (2.2).

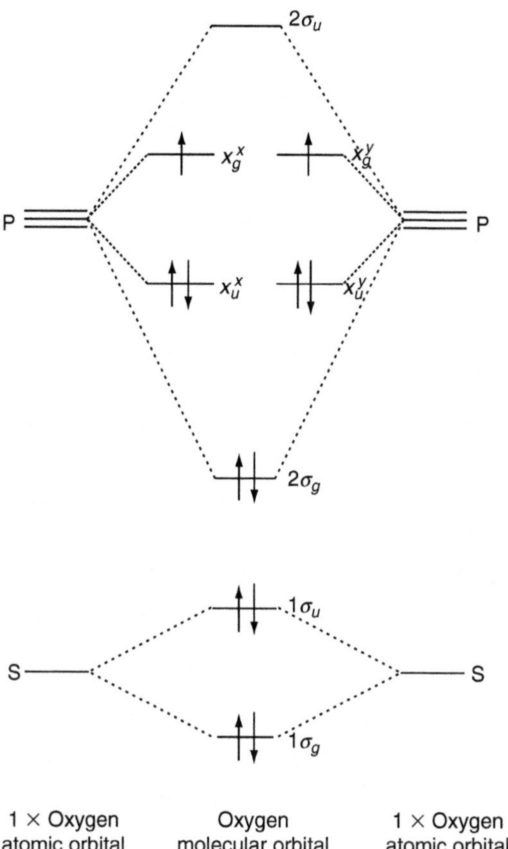

FIGURE 2.1 Molecular orbital diagram for molecular oxygen, O_2, in the ground state.

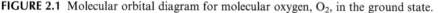

$$HS^{-}(aq) + O_2(aq) \longrightarrow O_2^{-\bullet} + HS^{\bullet} \tag{2.1}$$

$$O_2^{-\bullet} + HS^{\bullet} \longrightarrow S + HO_2^{-} \tag{2.2}$$

Processes that involve the overall transfer of more than two electrons require more than one transition state complex. One of these elementary steps will be rate limiting and dictate the kinetics of the process.

The arrangement and electronic structure of the intermediate complex in an elementary ET reaction dictates the rate. In general, it is illustrative to define two types of arrangements: (a) inner-sphere complexes and (b) outer-sphere complexes (see Figure 2.2). Inner-sphere complexes are an arrangement in which the donor and acceptor share a common ligand, while outer-sphere complexes are an arrangement where the coordination spheres of the donor and acceptor remain intact during the ET process. ET between metal complexes has received considerable attention and detailed reaction mechanisms have been established for many such reactions. In contrast, many ET reactions have not been studied in great detail or are not amenable to detailed studies. However, out of the vast body of work on the mechanisms of ET reactions — albeit mostly on ET reactions between metal complexes — some general trends have emerged that will be briefly introduced here.

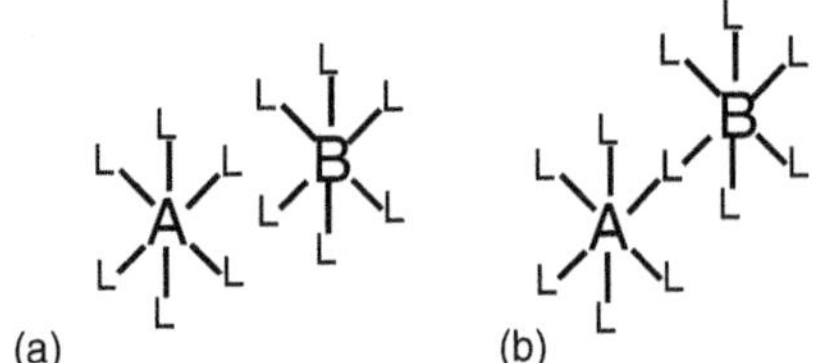

FIGURE 2.2 Schematic of (a) outer-sphere and (b) inner-sphere intermediate reaction complex in an elementary ET reaction.

2.3.1 Mechanism and Rate of Outer-Sphere Reactions

ET involving an outer-sphere mechanism is a four-step process. The first step is the formation of a complex between the two reactants (Figure 2.3 step I). Secondly, the complex needs to undergo a reorganization that leads to the formation of an activated complex in which the energy level of the donor's HOMO and that of the acceptor's LUMO are equal (Figure 2.3 step II). After ET (Figure 2.3 step III), the complex relaxes and dissociates to give the final products (Figure 2.3 step IV). For many reactions it is the second step or reorganization step that is rate limiting. For outer-sphere ET reactions involving metals this step is largely associated with a lengthening of the metal–ligand distances in the acceptor side of the complex and shortening of metal–ligand distances in the donor side of the complex. For example, consider the transfer of an electron from Fe^{2+} to a donor to form Fe^{3+}. In such a situation, the metal–water distance for Fe^{2+} is 0.13 Å larger than that with Fe^{3+} as the metal center [4]. The reorganization within the complex is an endergonic process and contributes to the activation energy of the ET reaction. Differences in the MO symmetry of the orbitals involved also play an important role. ET between orbitals with π^* symmetry (i.e., antibonding with electron density oriented away from the internuclear axis of the molecule) is energetically less endergonic than exchange between MOs with σ^* symmetry (i.e., antibonding with electron density concentrated along the inter-atomic axis of the molecule). For example, the rate constant for ET between $Co(H_2O)_6^{2+}$ and $Co(H_2O)_6^{3+}$ is much lower compared to the ET reactions involving Ru^{II}/Ru^{III} and Fe^{II}/Fe^{III} [5]. This is due to the fact that the Co^{II}/Co^{III} ET involves the transition from a high-spin electron configuration, $(\pi^*)^5\ (\sigma^*)^2$, to a low-spin configuration, $(\pi^*)^6$, via an intermediate $(\pi^*)^5\ (\sigma^*)^1$ configuration. Hence, in this scenario, the ET step involves a transfer between two orbitals with σ^* symmetry. Note that the nature of the ligand–metal bonding dictates whether the Co complex is in the high-spin versus low-spin configuration. With CN^- as a ligand, both oxidation states of Co will be in a low-spin configuration. The possibility of having both a high-spin and low-spin electronic configuration exists only for transition metals with four to seven d-shell electrons, which includes the environmentally important metals such as chromium, manganese, and iron.

Marcus developed a theoretical framework for predicting the rates of outer-sphere ET reactions. This theory, for which he received the Nobel Prize in 1992, addresses how the reorganization energy depends on the overall energetics of the reaction and relates the rate of the reaction to the rates of the respective self-exchange reactions. (Self-exchange reactions are ET reactions in which two complexes of the same metal, for example $Fe(H_2O)_6^{2+}$ and $Fe(H_2O)_6^{3+}$, exchange an electron.) Marcus theory predicts that the activation energy for outer-sphere ET reactions is smaller for more exergonic reactions. Specifically, Marcus theory predicts that for a first-order electron transfer process the Gibbs energy of activation, ΔG^*, can be expressed as

$$\Delta G^* = \frac{(\Delta_r G^o + \lambda)^2}{4\lambda} \tag{2.3}$$

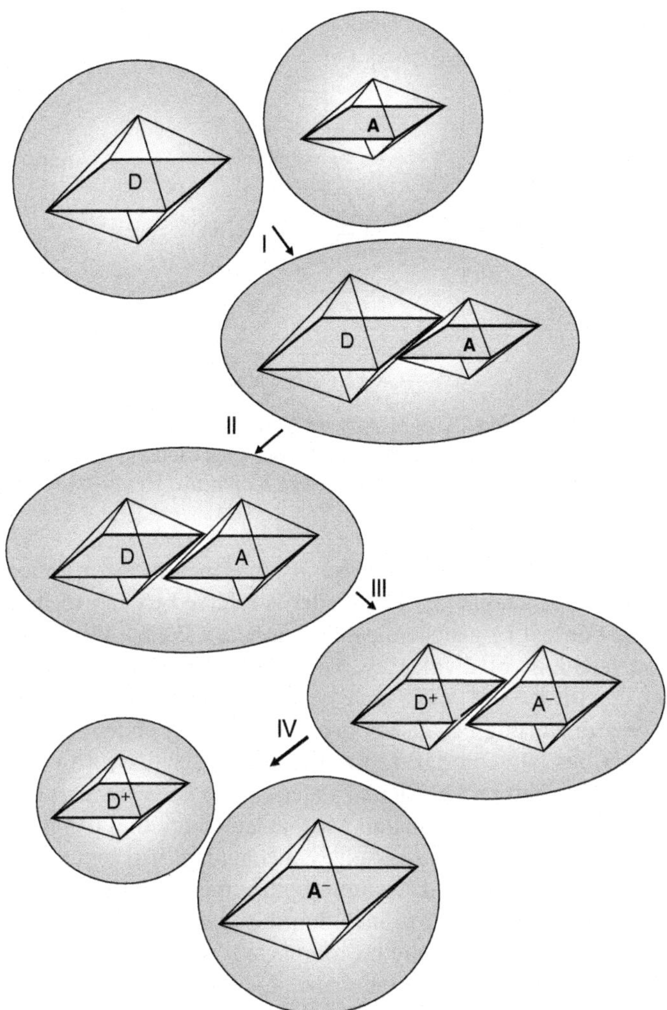

FIGURE 2.3 Diagram showing four-step process involved in outer-sphere ET reaction. Step I: formation of encounter complex; step II: reorganization within donor D and acceptor A coordination; step III: electron transfer; step IV: relaxation of coordination of A^- and D^+ coordination and breakup of successor complex into product. Note the changes in bond length as D loses an electron and A gains an electron.

where $\Delta_r G^o$ is the standard Gibbs energy in going from reactant to product and λ is a reorganization energy associated with Marcus theory (see Figure 2.4). The parameter λ is associated with the energy required to carry out the molecular rearrangements necessary to take a reactant complex DA (formed from donor D and acceptor A) to the equilibrium geometry of the product D^+A^-. Also included in this energy are the molecular rearrangements of the solvent around DA. The rate constant, k, for the process can be expressed as

$$k = Z \exp\{-\Delta G^* / \mathrm{RT}\} \tag{2.4}$$

where Z is the collision frequency of D and A in solution [6]. The implication is that ET reactions with a higher thermodynamic driving force (i.e., a more negative $\Delta_r G^o$) will have a higher reaction rate constant due to a smaller Gibbs energy of activation (see Equations (2.3)

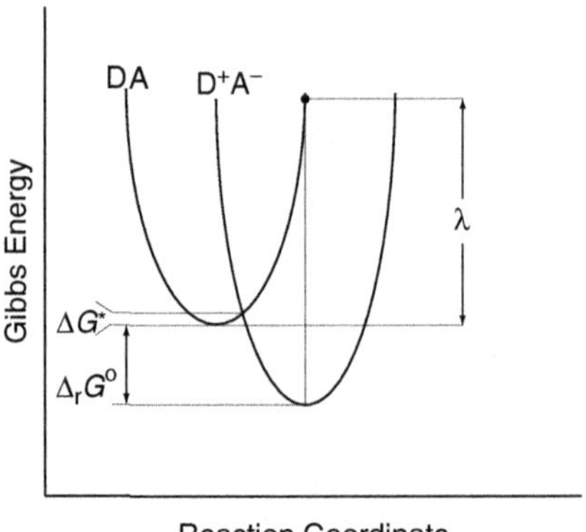

FIGURE 2.4 Schematic of the energetics involved with Marcus theory. The Gibbs energy of activation, ΔG^*, is the energy needed to form the transition state from the reactants in their equilibrium position. For a review, see RA marcus and N Sutin, *Biochimica et Biophysica Acta* 811(3):265–322, 1985.

and (2.4)). A high thermodynamic driving force also compensates for unfavorable symmetry matches between the two reactants [5]. The reader is referred to an excellent introduction to Marcus theory in the context of geochemical reactions by Wehrli [4]. The rate of the outer-sphere oxygenation reactions of divalent V, Cr, Mn, Fe, Co aquo and hydroxo ions (i.e., $M(H_2O)_{6-x}(OH)_x^{2-x}$, with $x = 0$, 1, or 2) have been calculated using ab initio calculations [7]. This approach is very promising because it provides an independent method for estimating the rate of these reactions. By comparing the theoretical results with experimentally determined rates one can infer whether a particular reaction proceeds via an inner-sphere or outer-sphere mechanism. In their study, Rosso and Morgan conclude that oxygenation of Fe, V, Co hexaquo complexes proceeds via an outer-sphere mechanism. An inner-sphere mechanism is inferred for Mn and Cr complexes and the aquo hydroxo complexes of Fe. In the future, this theoretical approach may be extended to include other ligands. This is of interest because in natural waters iron and other metals may be complexed by ligands other than just OH^-. For example, in a solution with chloride, carbonate, and sulfate concentrations equivalent to those in world average river water, $FeCO_3^0$ is an important species.

2.3.2 Mechanism and Rate of Inner-Sphere Reactions

ET involving an inner-sphere mechanism is also a four-step process, but it involves the formation of a reaction intermediate with a shared ligand. The first step is the formation of the reaction complex. This step is a substitution reaction in which one of the ligands is exchanged for a ligand bound to the other reactant (Figure 2.5). ET reactions take place through an inner-sphere mechanism only if the following two conditions are met [5]: (i) one of the reactants must have a ligand that can bind to two metal centers simultaneously and (ii) one of the metals must have a labile ligand that can be replaced by a bridging ligand. The formation of a new set of molecular orbitals involving the two metal centers and the bridging ligand is often the key step in the ET reaction mechanism. As pointed out by Haim [8], the new molecular orbitals bring the metals close together and mediate the ET. There are two ways in which the bridge can mediate the transfer. The electron transfer may be coupled with a transfer of the bridging ligand in the opposite direction (i.e., Donor + Acceptor-X$\rightarrow$ Donor-X$^{\oplus}$+Acceptor$^-$). Nonreducible bridging ligands such as Cl^- operate by this mechanism. Formation of a σ-bonding molecular orbital in the activated complex promotes the transfer of the electron via the bridging ligand. Large organic bridge molecules can be reduced to a radical and then transfer the electron to the acceptor. In the second mechanism,

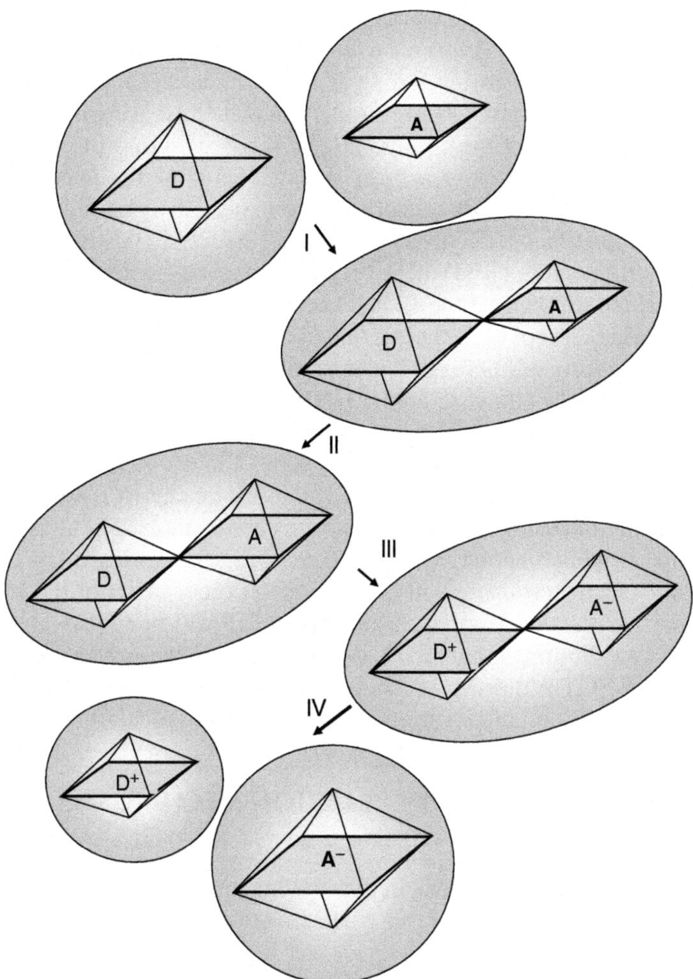

FIGURE 2.5 Diagram showing four-step process involved in inner-sphere ET reaction. Step I: formation of encounter complex; step II: reorganization within donor, D, and acceptor, A, coordination; step III: electron transfer; step IV: relaxation of coordination of A^- and D^+ coordination and breakup of successor complex into product. Note the changes in bond length as D loses an electron and A gains an electron.

the metals do not interact and there is no overlap of metal center orbitals. The rate of inner-sphere ET reactions between metal complexes is often higher than the comparable outer-sphere reaction [5]. Many reactions have not been studied in detail to know whether they proceed via an inner-sphere or an outer-sphere ET mechanism.

For bimolecular reactions between reactants that lack a strongly bound hydration shell the notion of inner-sphere ET versus outer-sphere ET is mute. Most anions and molecular species have a loose hydration shell that allows for rapid exchange. For example, neither reactants in reaction (2.1) (HS^- and O_2) will have tightly bound water molecules. Consequently, it is expected that the activated complex can be approximated considering only the interaction between these two species. One approach, introduced into the geochemistry community by Luther [3, 9, 10], is to use frontier orbital theory (FOT) to predict the molecular structure of the activated complex. FOT is widely used to predict the structure of activated complexes in bimolecular reactions involving organic molecules [11]. The structure of the activated complex is predicted by considering the interaction of the donor's HOMO and acceptor's LUMO. Other interactions are energetically less favorable as illustrated in Figure 2.6. Figure 2.6 is not only relevant to sort out the interacting orbitals for a given donor–acceptor pair, but it also illustrates that a given donor is expected to have strong interaction with those acceptors that have a LUMO at an energy level close to that of the donor's HOMO, provided that there is sufficient orbital overlap. The orbital overlap is a function of the symmetry of the interaction orbitals. As discussed above $\pi^*\rightarrow\pi^*$ or $\sigma^*\rightarrow\sigma^*$ transitions are expected to be much more facile than $\pi^*\rightarrow\sigma^*$ or $\sigma^*\rightarrow\pi^*$ transitions. If the energy difference between HOMO and LUMO is more than about 6 eV then the resulting MO will lie very close in energy to the orbital that lies lowest in energy [12]. In principle this can be the LUMO or HOMO (Figure 2.6b,c). However, with a HOMO–LUMO gap in excess of 6 eV, there is only ET if the LUMO lies lower (Figure 2.6c) [11].

One important limitation of FOT is that it can only be used to predict the activated complex on the basis of MO structure of the reactants [13]. This limits its usefulness to reactions in which the activated complex is expected to be more closely related to the reactants than the products. As postulated by Hammond [14], the transition states for exothermic reactions tend to be reactant-like, while the transition states for endothermic reactions tend to be product-like. FOT may also be of limited use in multi-electron transfer reactions. The electronic structure of the activated complex changes upon ET [11, 13]. Hence, the electronic structure of the activated complex after a one-electron transfer will be different. If the transfer of the first electron in a multi-electron transfer reaction is rate limiting then FOT will be useful. However, if a later elementary step is rate limiting, then FOT is expected to have limited predictive value. A practical limitation of FOT is that MO configurations for reactants can be readily calculated for the gas phase, but inclusion of solvation effects makes the calculations more complex.

2.4 INFLUENCE OF SURFACES ON REACTION MECHANISMS AND REACTION RATES

Surfaces can promote chemical reactions in two fundamentally different ways. The two ways are: (i) concentrating reactants and lowering the activation energy of the rate limiting step or

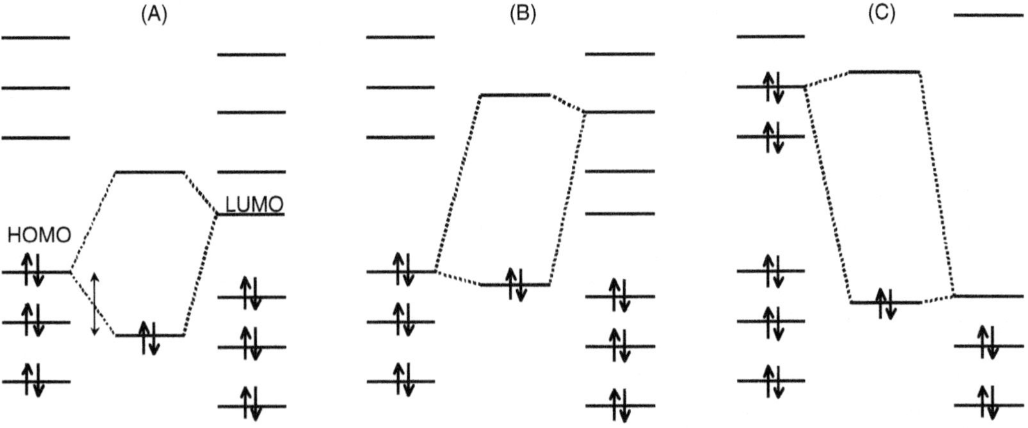

FIGURE 2.6 Schematic MO diagrams for the interaction between the frontier orbitals of two reactants. In A, a new MO is formed with contributions from the highest occupied molecular orbital (HOMO) and the lowest unoccupied molecular orbital (LUMO) of the two reactants. Note that the interaction of the HOMO and LUMO is more energetically favorable than any other combination (B). When there is a large energy difference between HOMO and LUMO, the resulting new MO will lie close in energy to either the LUMO or HOMO. In C, the new MO lies close to the LUMO.

(ii) inducing a fundamentally different reaction mechanism from the one operating in homogeneous solutions. The key concepts associated with these two ways will be discussed in this section. In the next section, several examples relevant to environmental chemistry will be discussed.

2.4.1 Concentration of the Reactants and Lowering the Activation Energy

Intuitively one may expect that simple concentration of reactants onto a surface would promote a reaction. However, theoretical studies show that this is typically not the case. For a bimolecular reaction there are two cases that one can consider: (a) reaction between a surface-bound species and a dissolved species and (b) reaction between two surface-bound species. Concentration of reactants is of course not specific to ET reactions.

Theoretical study of the first case shows that sorption of one of the reactants onto a surface is expected to decrease the normalized reaction rate by as much as a factor of 50 [15]. Adsorption of one of the reactants, but not the second one is quite common in mineral–solution systems. For example, interactions of charged species with mineral surfaces are often governed by the surface charge on the mineral. Hence, in a reaction involving an anion and a cation only one of the ionic species is expected to be concentrated on the surface. Similarly, oxygenation reactions of ionic species would most likely fall in this category. As estimated by Astumian and Schelly [15], the rate of the reaction can only be promoted if the sorption of one of the reactants leads to a lowering of the activation energy by at least 10 kJ/mol relative to the homogeneous reaction. In other words, without some specific interaction between the sorbate and sorbent, the concentration of one of the reactants on a surface will only decrease the reaction rate. The rates for reactions in which one of the reactant is sorbed are expected to follow more of a Langmuir–Rideal type of rate law [15, 16].

The sorption and reaction of two or more surface-bound chemical species and their transformation into a desorbed product (i.e., Langmuir–Hinshelwood kinetics) is more prevalent for gaseous reactants than aqueous reactant. A major reason for this circumstance is due to the charge development on mineral surfaces that inhibits the concurrent surface adsorption and reaction between cations and anions. In general, the rate of reaction between two sorbed species will be a function of their concentration, but this relationship can be complex because species can competitively or noncompetitively adsorb. An illustrative example of these concepts is the oxidation of CO by O_2 to CO_2 over a quartz surface [i.e., $CO(g) + 1/2O_2(g) \rightarrow CO_2(g)$] that can be represented with the following reaction step [17]:

$$> + O_2 = {>}O_2,\ K_1 = k_1/k_{-1} \tag{2.5}$$

$$> + CO = {>}CO,\ K_2 = k_2/k_{-2} \tag{2.6}$$

$${>}O_2 + {>}CO \longrightarrow {>}CO_2 + {>}O,\ k_3 \tag{2.7}$$

$${>}O + {>}CO \longrightarrow {>}CO_2,\ k_4 \tag{2.8}$$

$${>}CO_2 \longrightarrow CO_2 + {>},\ k_5 \tag{2.9}$$

where > denotes a surface site on quartz and k represents the rate constants for the particular reaction steps. At relatively low surface concentrations of adsorbed CO and O_2 where vacant surface sites dominate, the rate of CO_2 production will increase with increasing reactant concentration. However, at higher concentrations of CO, which has a strong adsorption energy on quartz, the rate law for the reaction can be given as:

$$\frac{d[CO_2(g)]}{dt} = \frac{2k_3K_1[O_2(g)]}{K_2[CO(g)]} \tag{2.10}$$

where the negative reaction order associated with CO reflects its strong adsorption and blocking of surface sites for O_2 adsorption. In this case the catalytic effect is largely due to the low activation energy for the reaction between adsorbed O_2 and CO (see reaction (2.7)) relative to the same reaction without a catalytic substrate (e.g., the homogeneous case) [17].

Noncomplementary reactions, which are reactions that involve more than one ET step, are likely to be faster if the process takes place on a surface. The notion is that the products of a first step may be reacting rapidly with another site after they migrate over the surface. For example, a three-electron transfer with adsorbed Fe^{II} or structural Fe^{II} would have to involve three different sites on the surface. For example, a recent study claims that the rate-limiting step in the reduction of CCl_4 with Fe^{II} sorbed on goethite is a termolecular reaction complex in which a CCl_4 molecule interacts with two adjacent Fe^{II}_{ads} sites [18].

2.4.2 Fundamentally Different Reaction Mechanism

While concentration by chemisorption can promote a reaction, in general, it occurs with a lowering of the activation energy (and change in mechanism) relative to its value in the homogeneous case. Surfaces of semiconducting and metallic minerals, for example, may promote ET reactions in a way that alleviates the necessity for reactants A and B to interact directly. The concept, discussed in more detail elsewhere [19], is illustrated in Figure 2.7. In this process, the donor transfers electrons to the surface and the acceptor species extracts electrons from the surface. This process alleviates the necessity for two electroactive species to form an activated complex. As explained earlier, a symmetry mismatch between the donor's HOMO and acceptor's LUMO may prevent the facile exchange of electrons in a homogeneous reaction. Through interaction with a suitable mineral this symmetry barrier can be lifted and electrons can pass through the solid from the donor to the acceptor. This ET mechanism may be quite common, but we are only aware of one set of studies involving minerals that have demonstrated this in detail. Xu and coworkers [20, 21] studied the reaction between dissolved thiosulfate and molecular oxygen. These two species do not react in solution, but react rapidly in the presence of a suitable mineral catalyst (pyrite or zinc sulfide doped with a transition metal). Xu and coworkers proposed that thiosulfate injects two electrons into the conduction band of the metal-sulfide mineral and that dioxygen accepts electrons at a different surface site. The kinetics for the reaction follows a Langmuir–Hinshelwood rate law as may be expected for a reaction that involves sorption on a surface [22].

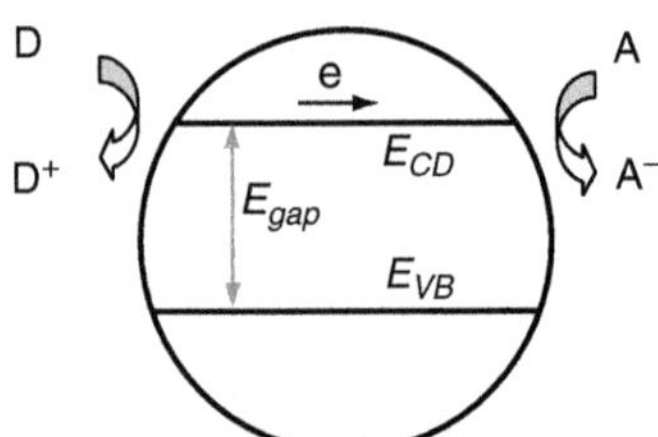

FIGURE 2.7 Schematic diagram of ET involving a semiconductor as an electron conduit. D is a donor species, A is an acceptor. E_{gap} is the band gap energy; E_{CD} refers to the absolute energy of the conduction band edge; E_{VB} refers to the absolute energy of the band edge of the conduction band. Diagram based on work by Xu and Schoonen [21].

The concept of net ET involving different donating and accepting surface sites is not new. The presence of cathodic and anodic sites on metals and semiconductors is a well-established concept in electrochemistry [16, 23]. The only difference here is that the solid is not corroded in the process as its oxidation state is not altered. Further support for the concept of interaction via multiple, nonequivalent sites on semiconductor minerals comes from recent theoretical calculations [24]. A schematic illustration of the surface picture invoked by these calculations is shown in Figure 2.8. It is important to point out that most of the reactivity of a surface is related to minority sites, such as defects, kinks, edges, and other perturbations of an ideal surface. Furthermore, different crystallographic orientations may have different reactivities. The importance of imperfections and crystallographic orientation catalytic activity and specificity is well documented for synthetic single crystal metals and alloys [25]. It is estimated that perhaps as little as 0.1% of the total surface is involved in the reaction. Mineral surfaces have not been studied to the same level of detail, but recent work on pyrite by our research group shows that much of its reactivity is due to the presence of sulfur-deficient defects [26–30]. In addition experimental results indicate that there is a difference in reactivity between the (1 1 1) and (1 0 0) orientation [31]. These experimental results on pyrite have been corroborated by recent theoretical calculations [32].

2.5 SPECIFIC EXAMPLES

In this section a limited number of specific examples of environmentally relevant reactions are discussed. The presentation of the examples is organized into two groups. The first group of examples is one in which species sorbed onto a mineral surface react. In these examples, the surface is not a reactant. In the second group of examples, the mineral surface is a reactant. The examples in this second group are mostly surface precipitation reactions. The emphasis in

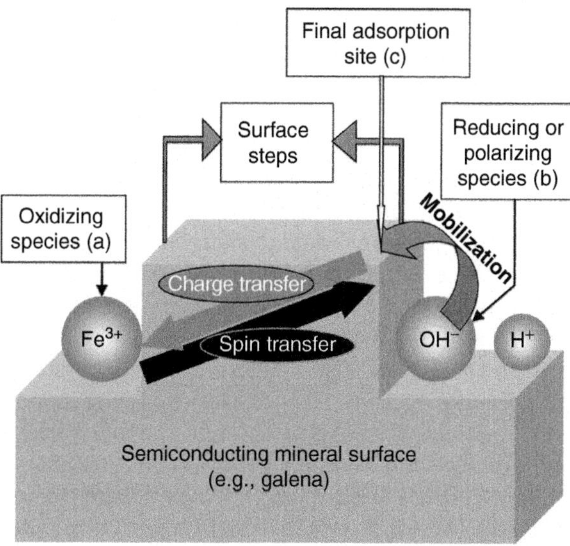

FIGURE 2.8 A schematic diagram of the proximity effect that has been proposed to be important for ET on semiconductor surfaces. Electron can be transported through the mineral so that oxidizing and reducing species do not need to be directly interacting. The proximity effect on semiconducting mineral surfaces: a new aspect of mineral surface reactivity and surface complexation theory. The figure is reprinted from Becker, U., Rosso, K.M. and Hochella, M.F. *Geochim. Cosmochim. Acta* 65: 2641–2649, 2001, with permission from Elsevier.

this section will be on a discussion of the mechanism of the reaction. For many reactions rate laws have been obtained, but the reaction mechanisms have not been conclusively established. Hence, it is not always clear whether the surface simply promotes a reaction that could, in principle, proceed in a homogeneous fashion or if it induces a fundamentally different reaction mechanism.

2.5.1 ET Reactions among Sorbed Species

Catalysis of reactions between sorbed electroactive species has been demonstrated in a number of studies over the last two decades. In these reactions, the surface serves as a sorbent of one or two of the species. In many cases the analogous homogeneous reactions do not proceed at a significant rate. Three types of ET reactions that hinge on a reaction among sorbed species are presented here: Oxygenation of sorbed metal species by molecular oxygen, bimolecular reactions with sorbed Fe^{II} as the electron donor, and oxidative coupling reactions involving aromatic compounds.

2.5.1.1 Oxygenation of Sorbed Metal Species

Oxygenation of metals is an important process that often leads to drastic changes in the mobility and bioavailability of metals. The rate of oxygenation of dissolved metals, such as Mn^{2+} and Fe^{2+}, depends on their speciation in solution and the presence of mineral surfaces can drastically accelerate the rate of oxygenation. The rate of oxygenation of ferrous iron serves as an example [4, 7]. The rate of oxygenation of $Fe(H_2O)_6^{2+}$ is very slow. By contrast the rate of oxygenation increases by six orders of magnitude if the dominant iron species is $Fe(H_2O)_5OH^+$, rather than $Fe(H_2O)_6^{2+}$. With $Fe(H_2O)_4(OH)^0$, the oxygenation is close to 12 orders of magnitude faster than with $Fe(H_2O)_6^{2+}$ as the dominant species. The acidity and composition of the solution will dictate the distribution of the Fe^{II} species in solution (see Figure 2.9). Fe^{II} sorbed onto goethite (FeOOH) as an inner-sphere complex has a rate of oxygenation that is nearly six orders of magnitude faster than for the $Fe(H_2O)_6^{2+}$ complex. As pointed out by Wehrli [4, 33], coordination of a Fe^{II} center with OH ligands derived either from solution or from the surface facilitates the electron transfer between O_2 and the metal center. The effect of other ligands on the rate of this process has not been studied. For example, in natural waters, a significant fraction of the total dissolved Fe^{II} may be present as $FeHCO_3^+$ (Figure 2.9b).

The nature of the substrate influences the rate of oxygenation. This effect is illustrated by a study of the rate of Mn^{II} oxidation on goethite (α-FeOOH), lepidocrosite (γ-FeOOH), silica (SiO_2), and alumina (δ-Al_2O_3) [34]. In homogeneous solutions, Mn^{II} oxidation by O_2 is extremely slow [34]. The presence of metal oxide surface promotes the reaction. However, the reaction rate constants drop-off in the following order: γ-FeOOH $\approx$ α-FeOOH $>$ SiO_2 $>$ δ-Al_2O_3 (see Table 2.1). The rates for the two iron oxy-hydroxides investigated by Davies and Morgan show little difference in rate. This suggests that the crystal structure of the bulk material does not exert a significant amount of control on the reaction. The decrease in rate from Fe^{III} to Si^{IV} and Al^{III} as metal centers is most likely due to the electronic configuration of the reaction intermediate formed on the surface [34]. However, there is some debate about the interpretation of the macroscopic rate data [35–37]. The debate centers on whether Mn(II) oxidation is accelerated by the formation of Mn^{III} surface precipitates as the process proceeds.

The rate laws for oxygenation of adsorbed metal ions are consistent with a bimolecular surface reaction between the metal sorbate and dissolved molecular oxygen. The rate laws have the following general form [4]:

$$-d[Me_{ads}]/dt = k[Me_{ads}][O_2(aq)] \tag{2.11}$$

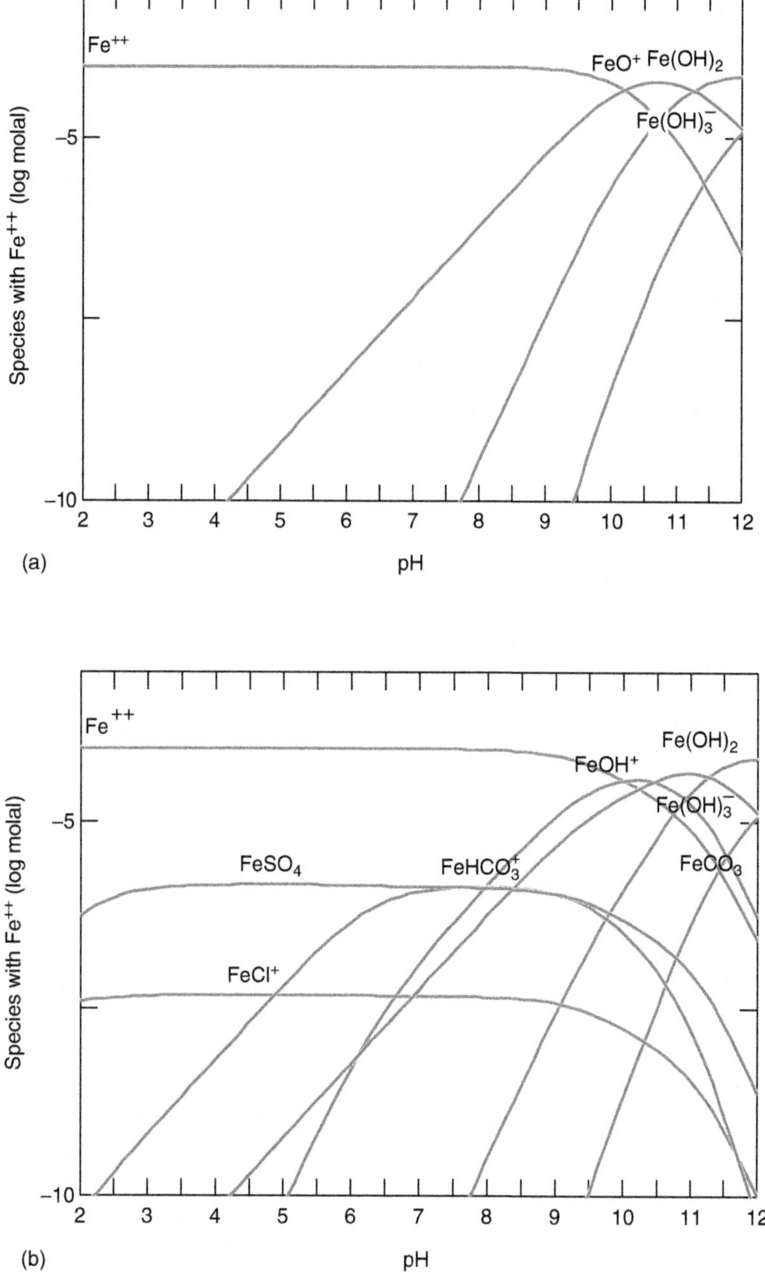

FIGURE 2.9 Fe^{II} speciation in water (a) and world average river water (b). World average river water concentrations taken from Berner and Berner [97].

The oxygenation reaction can either be an inner-sphere or an outer-sphere reaction. An inner-sphere reaction mechanism has been proposed for the oxygenation reaction of adsorbed Mn^{II} [34]. By contrast, oxygenation of V^{IV}, Fe^{II}, and Cu^{I} appears to proceed via an outer-sphere process [4, 33, 38]. While oxygenation of metals in solution has been studied using ab initio

TABLE 2.1
Rates of Oxygenation of Mn(II) Sorbed onto Selected Surfaces

Solid[a]	k^b (atm^{-1} min^{-1})
Lepidocrosite (γ-FeOOH)	0.052
Goethite (α-FeOOH)	0.042
Silica (SiO_2)	0.010
Alumina (δ-Al_2O_3)	<0.005

[a]Data from Davies and Morgan, their Table VIII [34].
[b]Rate law $d[Mn(II)]/dt = k\{(\alpha MO)_2\text{–}Mn\}\cdot A\cdot P_{O_2}$, where $\{(\alpha MO)_2\text{–}Mn\}$ is the concentration of the Mn(II) surface complex in mole/g; A is the specific surface area of the mineral.

calculations, this approach has not yet been extended to sorbed metals as far as we know. This would be a useful next step.

2.5.1.2 Reactions Involving Sorbed Fe(II) as Electron Donor

Ferrous iron is a common electron donor in anoxic environments [39, 40]. While dissolved ferrous iron is a relatively weak electron donor, ferrous iron adsorbed onto metal oxides is a much better electron donor. Adsorbed onto the surface, ferrous iron is postulated to form a Me–O–Fe^{II} species, where Me represents the metal of the underlying mineral (in natural environments Me is either Fe^{III}, Al^{III}, Si^{IV}, or Ti^{IV}). Sorption of ferrous iron onto a metal surface has the effect of changing the coordination around the Fe^{II} center. Bonding to the surface makes the ferrous iron a better donor because electron density from the underlying metal center is transferred through the bridging oxygen atom to the bound Fe^{II} [4, 33]. In essence, this increases its ability to reduce sorbed species. The reducing power of adsorbed Fe^{II} can be cast into a redox potential and compared to the redox potential of the Fe^{2+}(aq)/ Fe^{3+}(aq) couple. Table 2.2 is a compilation of estimates of the redox potential of Fe^{II} sorbed on several different mineral phases. Included in this compilation are the $Fe(OH)^+$(aq)/ Fe^{3+}(aq) couple, the $Fe(OH)_2^0$(aq)/Fe^{3+}(aq) couple, and an estimate of the green rust–goethite redox couple. Green rust (nominally $Fe_4{}^{II}Fe_2{}^{III}SO_4 \cdot 3H_2O$, but variable in composition) is a mixed valence iron hydroxide precipitate that can be seen as an endmember for a system where Fe^{II} is absorbed onto a Fe^{III} hydroxide phase. Green rust is a strong reductant and has been shown to reduce CCl_4 [41]. Note that Fe^{II} bound to other metal centers yields a reducing power that lies somewhere between the aqueous couple and that for the Fe^{II}–Fe^{III}(solid) systems. It should be pointed out that the redox potentials for sorbed Fe^{II} are only estimates. Comparison of two estimates for the same Fe^{II}–SiO_2 couple using exactly the same type of SiO_2 yields two values that differ by 0.2 V [42] (see Table 2.2).

The interaction of aqueous Fe^{II} with Fe^{III} hydroxide and oxide phases has been the subject of several recent studies. The interaction of Fe^{II} with hematite and goethite involves a fast initial uptake step followed by much slower processes that convert some of the adsorbate into irreversibly bound iron. In the first step an inner-sphere surface complex is formed. The fate of some of the complexed Fe^{II} in the second step is not clear. One possibility is that some of the complexes coalesce into surface clusters, possibly leading to the formation of a surface precipitate [43]. An interesting observation made by Pecher and coworkers [44] is that the reactivity of the surface-bound Fe^{II} increased after this second step was completed. This suggests that in this second step the electronic structure of the surface-bound Fe^{II} is different from the initial, isolated inner-sphere complex. Detailed surface science studies will

TABLE 2.2
Estimated Redox Potentials in Fe^{II}–Mineral Systems[a]

System	E_h^0 (pH 7)	Ref.
Fe^{II}-SiO_2	0.42±0.13	[55]
Fe^{II}-SiO_2	0.23	[42]
Fe^{II}-hematite (α-Fe_2O_3)	0.48±0.13	[55]
Fe^{II}-lepidocrosite	0.44±0.13	[55]
Fe^{II}-goethite (bidentate)	0.36	[4]
Fe^{II}-goethite (monodentate)	0.34	[4]
Fe-silicates (structural)	0.33–0.52	[95]
Green rust/magnetite	−1.113	[41]
Green rust/goethite	−0.645	[41]
Fe^{3+}/Fe^{2+}	0.77	[96]
$Fe(OH)^{2+}$(aq)/$Fe(OH)^{+}$(aq)	0.37	[42]
$Fe(OH)_2{}^{+}$(aq)/$Fe(OH)_2^0$(aq)	−0.145	[42]

[a] E_h^0 reported here is the redox potential versus the standard hydrogen electrode at pH 7 at standard state conditions (25°C, 1 atm).

be needed to better understand the nature of the irreversibly bound Fe^{II} on Fe^{III} oxide and hydroxide surfaces. While characterization of the state of these surfaces needs to be carried out, it is pointed out that Fe^{II} on Fe^{III} oxide and hydroxide clusters or nanostructures supported on substrates may play a role in the experimentally observed electron transfer. For example, it has been shown in the heterogeneous catalysis community that vanadia supported on oxide supports shows a striking dependence for alkane dehydrogenation reactions on particle size [45]. This phenomenon has been attributed to an electronic effect where the ease of reduction of the catalytic vanadia particle is a function of particle size [46]. The implications of this in geochemical reactions is that nanoprecipitates may exhibit redox properties that are not present in larger structures.

Surface-bound Fe^{II} has shown to be capable of reducing nitrate [47], arsenate [48], mercury [48], monochloramine [49], agricultural chemicals [42, 50], halogenated carbon compounds [18, 44, 48], uranyl [48, 51–54], chromate [55], pertechnetate [56], and N-substituted aromatics [39, 57–60]. Kinetic data for many of these studies are summarized in Table 2.3. The rate of reduction of these pollutants in these heterogeneous systems is in principle a combination of parallel reaction pathways. Assuming a bimolecular reaction between a pollutant P (dissolved and sorbed) and Fe(II) (dissolved and sorbed), the following generic rate law can be formulated [42]:

$$\begin{aligned} -\mathrm{d}[P]/\mathrm{d}t = {} & k_1[Fe^{2+}]_{aq}[P]_{aq} + k_2[Fe(OH)^{+}]_{aq}[P]_{aq} + k_3[Fe(OH)_2^0]_{aq}[P]_{aq} \\ & + k_4[Fe^{II}]_{surf}[P]_{aq} + k_5[Fe^{2+}]_{aq}[P]_{surf} + k_6[Fe(OH)^{+}]_{aq}[P]_{surf} \\ & + k_7[Fe(OH)_2^0]_{aq}[P]_{surf} + k_8[Fe^{II}]_{surf}[P]_{surf} \end{aligned} \tag{2.12}$$

If any of the Fe^{II} species or P species is negligible then this equation can be simplified. The concentration of sorbed Fe^{II} is also pH dependent and varies with the substrate. Even two different batches of the same mineral can lead to significant difference in the uptake of Fe^{II} as a function of pH [42]. If the pollutant is a molecular species its surface concentration may be insignificant and all terms with P_{surf} as reactant can be eliminated. For example, Strathmann and Stone [42] showed in an extensive study of the reduction of three pesticides

TABLE 2.3
Rate Constants for Reactions Involving Adsorbed Fe(II)

Substrate	Electron Acceptor	k	Ref.
Bimolecular Reaction Constants ($-k[Fe]_{surf}[P]_{aq}$) ($mol^{-1}\ s^{-1}$)			
Hematite (α-Fe_2O_3)	Uranyl (UO_2^{2-})	0.11 ± 0.0013	[54]
Magnetite (Fe_3O_4)	4-Chloro nitrobenzene	23	[53]
Hematite (1)(α-Fe_2O_3)	Oxamyl	2.12 ± 0.13	[42]
Hematite (2)(α-Fe_2O_3)	Oxamyl	2.9 ± 0.13	[42]
Goethite (α-FeOOH)	Oxamyl	0.74 ± 0.072	[42]
Anatase/rutile (TiO_2)	Oxamyl	2.24 ± 0.23	[42]
Rutile (TiO_2)	Oxamyl	1.36 ± 0.32	[42]
Anatase (TiO_2)	Oxamyl	0.017 ± 0.002	[42]
Kaolinite (clay)	Oxamyl	0.30 ± 0.027	[42]
Kaolinite (clay)	Oxamyl	0.20 ± 0.024	[42]
Alumina (γ-Al_2O_3)	Oxamyl	0.44 ± 0.044	[42]
Silica (SiO_2(am))	Oxamyl	20.19 ± 2.36	[42]
Silica (SiO_2(am))	Oxamyl	27.11 ± 2.08	[42]
Quartz (SiO_2)	Oxamyl	$8.52\ \pm 0.139$	[42]
Goethite (α-FeOOH)	Cr(VI)	2×10^3	[55]
Lepidocrocite (γ-FeOOH)	Cr(VI)	5×10^3	[55]
Silica (SiO_2(am))	Cr(VI)	8×10^3	[55]
Pseudo-First-Order Reaction Rate Constants ($-k[P]_{aq}$) (s^{-1})			
Goethite (α-FeOOH)	$CHBrCl_2$	0.36×10^{-6}	[44]
Goethite (α-FeOOH)	$CHBr_2Cl$	0.81×10^{-6}	[44]
Goethite (α-FeOOH)	$CHBr_3$	1.33×10^{-6}	[44]
Goethite (α-FeOOH)	CCl_4	4.44×10^{-6}	[44]
Goethite (α-FeOOH)	Hexachloroethane	13.9×10^{-6}	[44]
Goethite (α-FeOOH)	$CFBr_3$	1.41×10^{-4}	[44]
Goethite (α-FeOOH)	$CBrCl_3$	9.94×10^{-4}	[44]
Goethite (α-FeOOH)	CBr_2Cl_2	31.38×10^{-4}	[44]
Magnetite (Fe_3O_4)	Nitrobenzene	$0.31 \pm (0.067)$	[57]
Magnetite (Fe_3O_4)	2-Chloro nitrobenzene	$0.96 \pm (0.063)$	[57]
Magnetite (Fe_3O_4)	3-Chloro nitrobenzene	$0.81 \pm (0.053)$	[57]
Magnetite (Fe_3O_4)	4-Chloro nitrobenzene	$0.38 \pm (0.01)$	[57]
Goethite (α-FeOOH)	4-Chloro nitrobenzene	$2.2 \pm (0.73) \times 10^{-4}$	[59]
Goethite (α-FeOOH)	TNT	$4.6 \pm (1.2) \times 10^{-3}$	[59]
Phologopite (mica)	Mercury(II)	1.93×10^{-4}	[48]
Goethite (α-FeOOH)	4-Chloro nitrobenzene	$6.55 \times 10^{-6} (s^{-1}\ m^{-2})$	[60]
Silica (SiO_2(am))	4-Chloro nitrobenzene	$3.60 \times 10^{-9}\ (s^{-1}\ m^{-2})$	[60]
Alumina (γ-Al_2O_3)	4-Chloro nitrobenzene	$8.11 \times 10^{-9}\ (s^{-1}\ m^{-2})$	[60]
Montmorillonite (clay)	4-Chloro nitrobenzene	$2.03 \times 10^{-8}\ (s^{-1}\ m^{-2})$	[60]
Termolecular Reaction Rate Constants ($-k[Fe]_{surf}^2[P]_{aq}$~$mol^{-2}\ s^{-1}$)			
Goethite (α-FeOOH)	Carbon tetrachloride	42 ± 5	[18]

that the rate of the reaction can be adequately described by a rate law that only retained three terms:

$$-d[P]/dt = k_1[Fe^{2+}]_{aq}[P]_{aq} + k_3[Fe(OH)_2^0]_{aq}[P]_{aq} + k_4[Fe^{II}]_{surf}[P]_{aq} \tag{2.13}$$

In many studies a pseudo-first-order reaction rate constant is derived (see Table 2.3). In essence, the conditions in these studies are the ones where the concentration of the pollutant species, P, is limiting the reaction rate. Increasing the concentration of the initial pollutant concentration leads to a proportional increase in reaction rate (i.e., $k = k_4[Fe^{II}]$ and all other terms in Equation (2.13) are negligible). The other extreme condition is one where the reaction rate is limited by the availability of sorbed Fe^{II} species. Under these conditions the reaction rate will be independent of the pollutant concentration (i.e., zero order with respect to [P]). In between these extremes, the reaction order with respect to the pollutant will change from one to zero [22].

In the study of dehalogenation of CCl_4, a termolecular reaction surface mechanism was proposed. Amonette and coworkers [18] studied CCl_4 reduction by Fe^{II} sorbed onto goethite. They derived a rate law that is second order in the sorbed iron and first order with respect to the pollutant:

$$-d[CCl_4]/dt = -k[Fe(II)]^2_{surf}[CCl_4] \tag{2.14}$$

The rate constant for this reaction is given in Table 2.3. The investigators postulate that the rate-limiting step is a two-electron transfer step involving a reaction complex where CCl_4 is coordinated by two adjacent iron sites. Given that an elementary step with a two-electron transfer is far less likely than a one-electron transfer, it is perhaps possible that sorbed CCl_4 undergoes a sequence of two one-electron reactions without any surface diffusion. Perhaps it might be possible with vibrational spectroscopy to evaluate these proposed mechanisms. For example, advances in Fourier transform IR make now possible the study of surface reactions *in situ* and in real time [61].

Although there is a considerable body of kinetic data on reactions with sorbed Fe^{II} (Table 2.3) it remains difficult to compare rates reported by different laboratories. The formation of colloidal $Fe(OH)_3$ during the treatment of the substrate with $Fe^{II}(aq)$ and differences in surface area, surface site density, and surface structure make it difficult to compare among laboratories. Data reported within a single study show that the structure of the electron acceptor contributes to the difference in rate. For example, the reaction rate among different halogenated methane compounds varies by roughly a factor of 10^4 [44]. Surface-sensitive spectroscopic studies are needed to provide the molecular-scale insights necessary to develop an understanding of the elementary steps involved in these types of reactions. For example, it is not clear for any of the reactions listed in Table 2.3 whether the intermediate reaction complex is an inner-sphere or an outer-sphere reaction complex between the adsorbed Fe^{II} and the pollutant.

2.5.1.3 Oxidative Coupling of Aromatic Compounds

Oxidative coupling is an important natural and environmental process in which aromatic compounds couple to form polymers [62]. Xenobiotic compounds, such as phenol, chlorinated phenols, aniline, and other hydroxylated aromatic compounds, can undergo oxidative coupling reactions. The process, commercially used to synthesize certain resins [63], is initiated by the transformation of an aromatic compound into a free radical. The radical then reacts either with other radicals or reacts with pre-existing dimers or polymers, forming eventually insoluble organic matter. An example of an oxidative coupling process is given in Figure 2.10. In this example, oxidative coupling with 2,4-dichlorophenol as the sole substrate is considered. In the first step four possible radicals are formed. Dimerization among the four different radicals leads to ten different products. In Figure 2.10, only two out of the ten possible reactions are shown. The first example is a self-coupling reaction and the second example is a cross-coupling reaction. Dimers can undergo further coupling reactions to form multimers. With multiple substrates present in contaminated water, the radicals of different parent compounds can cross-couple leading to a large number of different products. Oxidative coupling reactions can lead to dehalogenation of chlorinated compounds [64] (see self-coupling reaction in Figure 2.10).

Oxidative coupling is also responsible for the formation of humic substances in nature. In the formation of humic substances, natural phenolic compounds are the reactants. In contaminated soils or aquifers, xenobiotic compounds can condensate with natural soil organic matter, immobilizing the compound and contributing to natural attenuation [65]. In essence, xenobiotic compounds can be incorporated into natural humic substances by this process.

Oxidative coupling reactions are promoted by the presence of minerals, such as Fe, Al, Mn oxides and hydroxides as well as clays [62]. The manganese oxide birnessite (δ-MnO_2) and Mn-hydroxides, both common components in soils, have received considerable attention [66–71]. The interaction of substituted phenols with Mn-oxides has been studied in detail [68, 69]. The rate-limiting step in this process is a one-electron transfer step that occurs between the phenolic adsorbate and an Mn^{III} or Mn^{IV} center. It is likely that the formation of an inner-sphere surface complex facilitates this reaction step. The reduced Mn center can dissolve and expose an underlying metal center or it can be reoxidized and regenerated. A comparison of oxidation rates for various substituted chlorophenols indi-

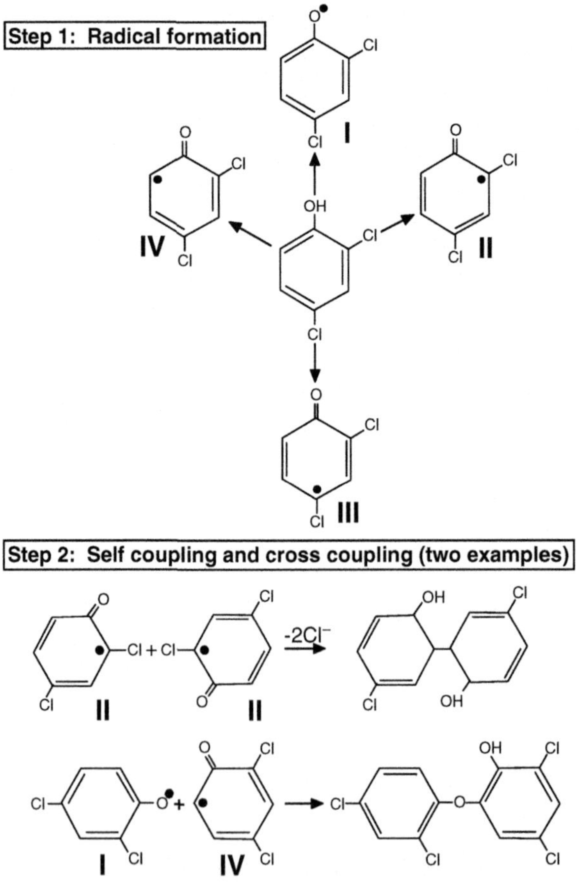

FIGURE 2.10 Example of an oxidative coupling reaction. The first step involves the formation of a radical. In the second step radicals can self couple (e.g., reaction between two II monomers) or cross couple (e.g., reaction between monomers I and IV). This figure is based on work by Dec and Bollag [64].

cates that the molecular structure of the adsorbate exerts a significant control on the rate. For example, oxidation of monochlorophenols is about six times slower with the chloro-substituent in the meta position than in the ortho position. This rate difference grows to an order of magnitude when compared to monochlorophenols with the chloro-substituent in the ortho position [69]. The reaction rate is first order in sorbate concentration and surface area of the Mn phase present.

2.5.2 Surface Precipitates

It is relatively straightforward to model the physical properties of a solution containing aqueous species to predict the conditions in which a specific precipitate with a known composition will form. The situation can be considerably more complex if the aqueous phase is in contact with a particular mineral surface. The number, charge, and type (i.e., step, kink, terrace etc.) can control the precipitation dynamics of initially solvated species. In this section we divide surface precipitation into two categories. The first category includes precipitation reactions where the oxidation states of the relevant species are unchanged, and in the second, which is perhaps more relevant to the theme of this chapter, includes surface precipitation involving redox reactions. The former category, however, is important to introduce because many of the chemical principles related to these precipitation reactions are relevant to the redox chemistry of the latter reactions. Both processes have tremendous implications for the environment, since they can determine the fate of benign and toxic materials in soils and groundwater.

It is generally well accepted that the presence of a chemically active substrate can induce precipitation reactions at solute concentrations that are undersaturated with regard to any solid phase. Not surprisingly, there has been a significant theoretical and experimental effort to understand how surfaces (often minerals) alter or even control the precipitation process. In one of the earliest efforts to provide a surface picture of how surfaces can influence precipitation chemistry, James and Healy [72], argued that the high electric field in the interfacial surface-solution region effectively could lead to a lowering of the solubility product of a particular compound relative to its solubility product in bulk solution. In essence they propose that the hydrolysis constant for a particular solid would be lowered in the interfacial region. Later models, such as one by Farley et al. [73], proposed that surface-induced precipitation was due in large part to the formation of solid solutions on the surface between substrate and sorbent species. These phases are unique to the interfacial region and would have a reduced solubility relative to pure sorbent phases. This particular mechanism would naturally be only operative where the substrate had a finite solubility under the conditions in question. The experimental verification of this model has been aided by the widespread use of synchrotron based x-ray absorption fine structure (XAFS) spectroscopy, which can be used to identify amorphous as well as crystalline precipitate phases [74–76]. A particularly instructive example of this phenomenon was presented by Towle et al. [77], who showed with a variety of advanced surface science techniques that Co^{II}-precipitates formed on a Al_2O_3 substrate at undersaturated conditions (relative to bulk Co(II) solids) were ternary phases due to the intermixing of Co and substrate Al^{3+} ions. A similar conclusion was made by Scheidegger et al. [78] who showed that mixed Ni/Al precipitates developed from the sorption of Ni on clays and aluminum oxide. The formation of solid solution precipitates is not a necessity and ultimately depends on such factors as the solubility of the substrate. A recent study by Waychunas et al. [79] investigated the sorption of Zn^{II} with ferrihydrite at conditions where the Zn^{II} aqueous concentrations was well below where any known Zn^{II} precipitate in a bulk solution could form. These researchers showed using extended XAFS that Zn hydroxide formed on the ferrihydrite surface under these undersaturated conditions, and there was no evidence of a mixed Zn–Fe precipitate.

So far theory and examples of precipitation reactions have been given that do not require changes in the oxidation states of the mineral substrate or sorbent. Many surface-induced precipitation reactions involve ET. In general, ET from the mineral substrate to the sorbent or from the sorbate to mineral can lead to a change in oxidation state of the sorbent that has a profound effect on the solubility of the sorbing material (or substrate). This change in oxidation state can be accompanied by the formation of a surface precipitate containing only the sorbed material or combined with material from the mineral substrate. There has been a significant amount of recent effort in understanding this phenomena, since it directly relates to the removal and mobilization of toxins in the environment. For example, Cr^{VI} and U^{VI} are aqueous toxins that can be removed from solution by ET between substrate and sorbent, leading to the formation of relatively insoluble Cr^{III} and U^{IV} precipitate [80–83]. Environmentally beneficial chemistry also can result from the ET from sorbent to mineral substrate, as in the case of the conversion of the toxic As^{III} to the less toxic As^{V} species on specific mineral substrates [84, 85]. It is noted that these types of reactions are in general stoichiometric and not catalytic. There is typically no cycling of oxidation states, which would be required to have the turnover necessary to carry out catalysis.

Two environmentally relevant mineral substrates that partake in ET reactions resulting in the formation of precipitate are ferrous-containing materials and Mn-oxides. Cr^{VI} reduction on minerals, such as magnetite and biotite, and As^{III} oxidation on Mn-oxides serve to illustrate some of the chemical principles fundamental to ET-induced precipitation.

Kendelewicz et al. [86] investigated the redox chemistry associated with the reaction of chromate solutions with Fe_3O_4 using NEXAFS and photoelectron spectroscopy. It was shown that the reduction of Cr^{VI} to Cr^{III} on the magnetite surface resulted in the formation of a completely oxidized iron (i.e., $Fe_2\,O_3$) interlayer. The eventual thickness of the Cr^{III} layer probably is determined by at least two factors. First, as the Cr(III) layer grows the transport of electrons to carry out the reduction of Cr^{VI} sorbed on the surface is decreased. Second, the growth of the precipitate depends on the transport of electrons from the Fe^{II} depleted interfacial region to the overlayer. As this depletion region grows, ET is presumably impeded. Understanding the controls on the kinetics of such ET processes through mineral phases (in this case an interfacial regions) is important. Fe^{II}/Fe^{III} electron hopping may be important in the transport of "reducing electrons" through the interfacial region in this example. This type of phenomenon is also presumably important in corrosive reactions on minerals. For example, Eggleston et al. [87] modeled pyrite oxidation by suggesting that oxide patches (Fe^{III}-bearing regions) acted as electron conduits for the oxidation of Fe^{II} on the perimeter of the oxide patch and the reduction of molecular oxygen. Perhaps, not surprisingly this theory is developing a framework by which to understand such phenomena [88].

ET reactions at mineral surfaces can also be sensitive to the structure of the reaction site. The reduction of Cr^{VI} on ferrous-containing phyllosilicate minerals, for example, shows a significant difference in redox chemistry depending on where the sorbed species interacts on the mineral. Ilton and Veblen [81] showed that the reduction of Cr^{VI} to Cr^{III} readily occurred on the edge–fluid interface of biotin (an Fe^{II} silicate), but no reduction occurred on the basal plane of the mineral. Recent theoretical research suggests that the hopping of electrons from Fe^{II} to Fe^{III} can occur along the octahedral sheets of annite, which would be consistent with reduction of sorbed species at edge sites [89].

An environmentally relevant mineral substrate that has a particularly rich redox chemistry that has been shown to induce surface precipitation is Mn-oxide in soils. In particular, there have been numerous studies of the birnessite phase (MnO_2), which is able to oxidize environmental toxins, such as As^{III} oxyanions, to As^{V}-containing species that are more easily sequestered by solid surfaces [90]. Such redox chemistry is important for the removal of mobile As^{III} from both natural and drinking water. Several studies have investigated the ET steps occurring in this chemistry [90–93]. The overall picture appears to consist of at least two steps represented in the following reaction steps [91]:

$$2Mn^{IV}O_2 + H_3AsO_3 + H_2O = 2Mn^{III}OOH + H_3AsO_4 \tag{2.15}$$

$$2MnOOH + H_3AsO_3 + 4H^+ = 2Mn^{2+} + H_3As^VO_4 + 3H_2O \tag{2.16}$$

The As^V resulting from the 2-ET shows significant release into solution, but also in part resides on the birnessite surface. Studies postulate, based on EXAFS and XANES, the existence of an inner-sphere bidentate binuclear complex (i.e., $>2Mn–As^{III}$) [92] and/or a krautite-like precipitate phase (i.e., $Mn^{II}HAs^VO_4XH_2O$) [93], respectively. Recent experiment and theory [24] suggests that ET occurring in a semiconducting phase (MnOOH) may result in the reduction of species relatively far away from the injection site. This phenomenon may result in a relatively heterogeneous redox chemistry on mineral surfaces. In the oxidation of As^{III}, the reduction of Mn^{III} to Mn^{II} and its desorption into solution may occur multiple lattice positions away from the initial As–Mn adsorption complex. Such a proximity effect (see Figure 2.8) [24] has been proposed to be occurring during the oxidation of Cr^{III} on a manganite surface [94].

2.6 CONCLUSIONS

ET processes on mineral surfaces play a dominant role in environmental chemistry. An understanding of redox chemistry on mineral surfaces is essential to understand the sequestration and decomposition of toxic species in aqueous solutions. Chemical examples have been presented that illustrate both catalytic and stoichiometric reactions that are driven by ET. Several research areas that deserve further attention become apparent after a review of this topic.

First, there are mechanistic issues that need to be addressed. Examples have been given in this chapter that involve the reactivity of sorbed Fe^{2+} for ET on mineral surfaces for the destruction of environmental toxins. Mechanistic information has largely been inferred from kinetic arguments. Molecular-level evidence, for example, detailing whether these reactions observe Langmuir–Rideal or Langmuir–Hinshelwood kinetics, or in many cases whether the reaction is catalytic, is not available. Molecular-level information is needed to understand the process in detail. This would provide the basis for new or improved environmental remediation strategies that may, for example, involve tailored surfaces. Experimental techniques sensitive to the chemical structure of the mineral–solution interface, such as nonlinear optical techniques (e.g., sum-frequency generation) and attenuated total reflection-Fourier transform infrared spectroscopy, should find increasing use in this area.

Another area that deserves attention is the morphology and reactivity of precipitate phases on mineral surfaces. Synchrotron-based techniques have been instrumental in characterizing the composition of these phases, but the chemical reactivity of the precipitate deserves attention. These precipitates, such as Mn-oxides overlayers in soils are expected to be nanodimensional in part. Hence, the electronic and resulting redox properties of these nanoparticles may be quite different from the bulk mineral substrates that are often used to model environmental surfaces. In short, the ability of these phases to oxidize or reduce environmental species may be quite different than their bulk counterparts. This area will be pushed forward by the ever-increasing number of synthetic routes for the production of well-defined nanostructures that are representative of those in the environment.

Finally, continued theoretical effort is essential for the understanding of environmental ET processes associated with bulk and nano-mineral surfaces. There are obviously a vast myriad of chemistry that deserves to be addressed by research. A theoretical framework to understand, for example, how ET occurs between mineral phases or between minerals (including nanophases) and sorbed species is needed to develop a predictive capability in evaluating processes that are inaccessible to inquiry by experimental methods.

ACKNOWLEDGMENTS

M.A.S. and D.R.S. gratefully appreciate the continued support from the Department of Energy-BES (grants DEFG029ER14633 and DEFG0296ER14644, respectively) and support from the National Science Foundation through the Center for Environmental Molecular Science (CEMS) that has allowed the writing of this chapter.

REFERENCES

1. CW Fetter. *Contaminant Hydrology*, 2nd ed. Englewood Cliffs, NJ: Prentice-Hall, 1998, 500pp.
2. B Deng and AT Stone. *Environ. Sci. Technol.* 30: 2484–2494, 1996.
3. GW Luther. In: *Aquatic Chemical Kinetics*, Stumm, W. (Ed.), New York: Wiley, 1990, pp. 173–198.
4. B Wehrli. In: *Aquatic Chemical Kinetics*, Stumm, W. (Ed.), New York: Wiley, 1990.
5. KF Purcell and JC Kotz. *Inorganic Chemistry*. Philadelphia: W.B. Saunders, 1977, 1116pp.
6. JI Steinfeld, JS Francisco, and WL Hase. *Chemical Kinetics and Dynamics*. Upper Saddle River: Prentice-Hall, 1999, 518pp.
7. KM Rosso and JJ Morgan. *Geochim. Cosmochim. Acta* 66: 4223–4233, 2002.
8. A Haim. *Acc. Chem. Res.* 8: 265, 1975.
9. GW Luther. *Geochim. Cosmochim. Acta* 51: 3193–3201, 1987.
10. G Luther, J Wu, and JB Cullen. In: *Aquatic Chemistry: Interfacial and Interspecies Processes*, CP Huang, CR O'Melia and JJ Morgan. (Eds.), Washington D.C.: American Chemical Society, 1995, pp. 135–156.
11. I Fleming. *Frontier Orbitals and Organic Chemical Reactions*. New York: Wiley Interscience, 1976, 249pp.
12. G Klopman. *J. Am. Chem. Soc.* 90: 223, 1968.
13. NT Anh and F Maurel. *New J. Chem.* 21: 861–871, 1997.
14. GS Hammond. *J. Am. Chem. Soc.* 77: 334, 1955.
15. RD Astumian and ZA Schelly. *J. Am. Chem. Soc.* 106: 304–308, 1984.
16. KJ Lardler. *Chemical Kinetics*, 3rd Ed. New York: HarperCollins Publishers, Inc., 1987.
17. AC Lasaga. *Kinetic Theory in the Earth Sciences*. Princeton: Princeton University Press, 1998, 811pp.
18. JE Amonette, DL Workman, DW Kennedy, JS Fruchter, and YA Gorby. *Environ. Sci. Technol.* 34: 4606–4613, 2000.
19. Y Xu and MAA Schoonen. *Am. Mineral.* 85: 543–556, 2000.
20. Y Xu, MAA Schoonen and DR Strongin. *Geochim. Cosmochim. Acta* 60: 4701–4710, 1996.
21. Y Xu and MAA Schoonen. *Geochim. Cosmochim. Acta* 59: 4605–4622, 1995.
22. MAA Schoonen, Y Xu, and DR Strongin (1998), *J. Geochem. Explor.* 62: 201–215.
23. JOM Bockris and SUM Khan. *Surface Electrochemistry*. New York, NY: Plenum Press, 1993, 1014pp.
24. U Becker, KM Rosso, and MF Hochella. *Geochim. Cosmochim. Acta* 65: 2641–2649, 2001.
25. GA Somorjai. *Introduction to Surface Chemistry*. New York: John Wiley, 1994, 667pp.
26. J Guevremont, DR Strongin, and MAA Schoonen. *Surf. Sci.* 391: 109–224, 1997.
27. JM Guevremont, DR Strongin, and MAA Schoonen. *Langmuir* 14: 1361–1366, 1998.
28. JM Guevremont, DR Strongin, and MAA Schoonen. *Am. Mineral.* 83: 1246–1255, 1998.
29. AR Elsetinow, JM Guevremont, DR Strongin, and MAA Schoonen. *Am. Mineral.* 85: 623–626, 2000.
30. MJ Borda, AR Elsetinow, DR Strongin, and MA Schoonen. *Geochim. Cosmochim. Acta* 67: 935–939, 2003.
31. JM Guevremont, AR Elsetinow, DR Strongin, J Bebié, and MAA Schoonen. *Am. Mineral.* 83: 1353–1356, 1998.
32. A Stirling, M Bernasconi and M Parrinello. *J. Chem. Phys.* 119: 4934–4939, 2003.

33. B Wehrli, B Sulzberger, and W Stumm. *Chem. Geol.* 78: 167–179, 1989.
34. SHR Davies and JJ Morgan. *Colloids Surfaces A: Physicochem. Eng. Aspects* 129: 63–77, 1989.
35. JJ Morgan and SHR Davies. *Geochim. Cosmochim. Acta* 62: 361–363, 1998.
36. JL Junta-Rosso, MF Hochella, and JD Rimstidt. *Geochim. Cosmochim. Acta* 61: 149–160, 1997.
37. JJ Rosso, MF Hochella, and JD Rimstidt. *Geochim. Cosmochim. Acta* 62: 365–368, 1998.
38. B Wehrli and W Stumm. *Geochim. Cosmochim. Acta* 53: 69–77, 1989.
39. K Rugge, TB Hofstetter, SB Haderlein, PL Bjerg, S Knudsen, C Zraunig, H Mosbaek, and TH Christensen. *Environ. Sci. Technol.* 32: 23–31, 1998.
40. W Stumm and B Sulzberger. *Geochim. Cosmochim. Acta* 56: 3233–3257, 1992.
41. M Erbs, HCB Hansen, and CE Olsen. *Environ. Sci. Technol.* 33: 307–311, 1999.
42. TJ Strathmann and AT Stone. *Geochim. Cosmochim. Acta* 67: 2775–2791, 2003.
43. BH Jeon, BA Dempsey, and WD Burgos. *Environ. Sci. Technol.* 37: 3309–3315, 2003.
44. K Pecher, SB Haderlein, and RP Schwarzenbach. *Environ. Sci. Technol.* 36: 1734–1741, 2002.
45. AT Bell. *Science* (Washington, DC) 299: 1688–1691, 2003.
46. K Chen, AT Bell and E Iglesia. *J. Catal.* 209: 35–42, 2002.
47. J Sorensen and L Thorling. *Geochim. Cosmochim. Acta* 55: 1289–1294, 1991.
48. L Charlet, D Bosbach, and T Peretyashko. *Chem. Geol.* 190: 303–319, 2002.
49. PJ Vikesland and RL Valentine. *Environ. Sci. Technol.* 36: 512–519, 2002.
50. J Cervini-Silva, J Wu, RA Larson and JW Stucki. *Environ. Sci. Technol.* 34: 915–917, 2000.
51. EJ O'loughtin, SD Kelly, RE Cook, R Csencsits, K.H. Kemmer. Environmental Science & Technology 37(4): 721–727, 2003.
52. L Charlet, E Liger, and P Gerasimo. *J. Environ. Eng.* 124: 25 (26pp), 1998.
53. L Charlet, E Silvester, and E Liger. *Chem. Geol.* 151: 85–93, 1998.
54. E Liger, L Charlet and P Van Cappellen. *Geochim. Cosmochim. Acta* 63: 2939–2955, 1999.
55. IJ Buerge and S Hug. *Environ. Sci. Technol.* 33: 4285–4291, 1999.
56. D Cui and T Eriksen. *Environ. Sci. Technol.* 30: 2259–2262, 1996.
57. J Klausen, S Trober, S Hadenlein, and R Schwarzenbach. *Environ. Sci. Technol.* 29: 2396–2404, 1995.
58. CG Heijman, E Grieder, C Holliger, and RP Schwarzenbach. *Environ. Sci. Technol.* 29: 775–783, 1995.
59. TB Hofstetter, CG Heijman, SB Haderlein, C Holliger, and RP Schwarzenbach. *Environ. Sci. Technol.* 33: 1479–1487, 1999.
60. CA Schultz and T Grundl. *J. Environ. Sci. Technol.* 34: 3641–3648, 2000.
61. MJ Borda, DR Strongin, and MA Schoonen. *Spectrochim. Acta, Part A* 59: 1103–1106, 2003.
62. J-M Bollag. *Environ. Sci. Technol.* 26: 1876–1881, 1992.
63. AS Hay. *J. Polym. Sci., Part A: Polym. Chem.* 36: 505–517, 1998.
64. J Dec and J-M Bollag. *Environ. Sci. Technol.* 28: 484–490, 1994.
65. H Selig, TM Keinath, and WJ Weber. *Environ. Sci. Technol.* 37: 4122–4127, 2003.
66. AT Stone and JJ Morgan. *Environ. Sci. Technol.* 18: 450–456, 1984.
67. AT Stone and JJ Morgan. *Environ. Sci. Technol.* 18: 617–624, 1984.
68. AT Stone. *Environ. Sci. Technol.* 21: 979–988, 1987.
69. H-J Ulrich and AT Stone. *Environ. Sci. Technol.* 23: 421–428, 1989.
70. J Dec and J-M Bollag. *Environ. Sci. Technol.* 28: 484–490, 1994.
71. J-W Park, J Dec, J-E Kim, and J-M Bollag. *Environ. Sci. Technol.* 33: 2028–2034, 1999.
72. RO James and TW Healy. *J. Colloid Interf. Sci.* 40: 53–64, 1972.
73. KJ Farley, DA Dzombak, and FMM Morel. *J. Colloid Interf. Sci.* 106: 226–242, 1985.
74. GA Waychunas, CC Fuller, JA Davis, and JJ Rehr. *Geochim. Cosmochim. Acta* 67: 1031–1043, 2003.
75. GE Brown, Jr., *Rev. Mineral.* 23: 309–363, 1990.
76. CJ Chisholm-Brause, PA O'Day, GE Brown., Jr, and GA Parks. *Nature* (London) 348: 528–531, 1990.
77. SN Towle, JR Bargar, GE Brown., Jr, and GA Parks. *J. Colloid Interf. Sci.* 187: 62–82, 1997.
78. AM Scheidegger, GM Lamble, and DL Sparks. *J. Colloid Interf. Sci.* 186: 118–128, 1997.
79. GA Waychunas, CC Fuller, and JA Davis. *Geochim. Cosmochim. Acta* 66: 1119–1137, 2002.
80. D Grolimund, TP Trainor, JP Fitts, T Kendelewicz, P Liu, SA Chambers, and GE Brown, Jr., *J. Synchrotr. Radiat.* 6: 612–614, 1999.

81. ES Ilton and DR Veblen. *Geochim. Cosmochim. Acta* 58: 2777–2788, 1994.
82. ES Ilton, CO Moses, and DR Veblen. *Geochim. Cosmochim. Acta* 64: 1437–1450, 2000.
83. K Idemitsu, K Obata, H Furuya, and Y Inagaki. *Mater. Res. Soc. Symp. Proc.* 353: 981–988, 1995.
84. M Bissen and FH Frimmel. *Acta Hydrochim. Hydrobiol.* 31: 9–18, 2003.
85. DW Oscarson, PM Huang, WK Liaw, and UT Hammer. *Soil Sci. Soc. Am. J.* 47: 644–648, 1983.
86. T Kendelewicz, P Liu, CS Doyle, and GE Brown. *Surf. Sci.* 469: 144–163, 2000.
87. CM Eggleston, J Ehrhardt, and WA Stumm. *Am. Mineral.* 81: 1036–1056, 1996.
88. KM Rosso, DMA Smith, and M Dupuis. *J. Chem. Phys.* 118: 6455–6466, 2003.
89. KM Rosso and ES Ilton. *J. Chem. Phys.* 119: 9207–9218, 2003.
90. MJ Scott and JJ Morgan. *Environ. Sci. Technol.* 29: 1898–1905, 1995.
91. HW Nesbitt, GW Canning, and GM Bancroft. *Geochim. Cosmochim.Acta* 62: 2097–2110, 1998.
92. BA Manning, SE Fendorf, B Bostick, and DL Suarez. *Environ. Sci. Technol.* 36: 976–981, 2002.
93. C Tournassat, L Charlet, D Bosbach, and A Manceau. *Environ. Sci. Technol.* 36: 493–500, 2002.
94. RM Weaver, MF Hochella, and ES Ilton. *Geochim. Cosmochim. Acta* 66: 4119–4132, 2002.
95. AF White and A Yee. *Geochim. Cosmochim. Acta* 49: 1263–1275, 1985.
96. W Stumm and JJ Morgan. *Aquatic Chemistry: Chemical Equilibria and Rates in Natural Waters*, 3rd ed., New York: Wiley Interscience, 1995, 1022pp.
97. EK Berner and RA Berner. *Global Environment*. Upper Saddle River, NJ: Prentice-Hall, 1996, 376pp.

3 Precipitation and Dissolution of Iron and Manganese Oxides

Scot T. Martin
Division of Engineering and Applied Sciences, Harvard University

CONTENTS

3.1 INTRODUCTION

Iron and manganese are the first and third most abundant transition metals on the Earth's crust (5.6×10^4 ppm and 9.5×10^2 ppm, respectively) [1]. The redox chemistries of iron II–III and manganese II–IV have important roles and impacts in the environment [2–5]. In contrast, the second most abundant transition metal, titanium (5.7×10^3 ppm), occurs only in the IV oxidation state. Abundant crustal aluminosilicates, such as clays and quartz, also lack a dynamic redox chemistry in natural waters [6].

Redox chemistry strongly influences the precipitation and dissolution of Fe and Mn solid phases (Figure 3.1) [7–10]. Aqueous Fe(II) and Mn(II) are significant in natural waters only in the absence of O_2. Insoluble Fe(III) and Mn(III/IV) oxides form under oxic conditions. Their solubilities limit the aqueous concentrations of Fe and Mn species.

The interconversions among redox states and physical states, while often thermodynamically favorable, are frequently slow in the absence of catalysis. For example, aqueous solutions of Mn(II) in the presence of O_2 at pH $= 8.4$ are exoergic toward oxidation, yet the uncatalyzed reaction proceeds slowly across years [11]. Surfaces, ligands, and other metals have varying

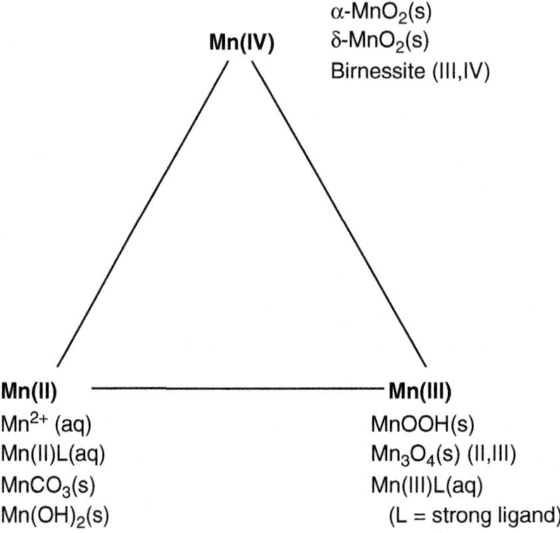

FIGURE 3.1 Common dissolved and precipitated iron and manganese species in several oxidation states.

levels of catalytic activity for Fe(II) and Mn(II) oxidation [4, 12]. As a second example, both the rate and the pathway of the crystallization of supersaturated solutions is influenced catalytically by a range of factors, including foreign surfaces, surface active ions, pH, and the magnitude of supersaturation [4, 9]. As a final example, iron and manganese oxide solids exposed to undersaturated solutions have a free energy driving force favorable to dissolution, but the uncatalyzed rates are often slow, especially in relation to the demand rate of biological organisms. Catalyzed pathways, which usually entail the formation of surface complexes between active ligands and oxide surface groups, are generally necessary to meet biological demand. In these examples, the distinction is blurred between a true catalyst, which is recycled during a reaction, and a stoichiomeric agent, which is consumed. As defined in this chapter, a catalyst is any agent that increases the rate of a desired process, including oxidation, precipitation, and dissolution, regardless of whether the agent is recycled or consumed.

Quantifying and predicting Fe and Mn cycling is separately motivated within at least four scientific communities [13]:

(1) Microbes such as *Geobacter* or *Shewanella*, reduce Fe(III) and Mn(III/IV) oxides in the absence of more favorable terminal electron acceptors such as O_2 or NO_3^-, [14–16]. The reduction reaction is coupled with hydrocarbon oxidation to complete the energy-producing metabolic pathways.

(2) In natural waters and soil zones having high dissolved oxygen concentrations, Fe and Mn acquisition at concentrations sufficient for enzymatic function [17–20] challenges microbes and plants [21–24]. The enzymes developed when aqueous Fe(II) and Mn(II) were abundant prior to oxygenation of the early Earth atmosphere [25]. Cellular Fe_T and Mn_T concentrations of 10^{-4} M [26] are 10^5 greater than typical ocean waters [27] and 10^2 greater than typical acidic soil solutions [6]. Organic ligands like oxalate, which are common biological exudates, and siderophores, which are tailored biological molecules, increase the dissolution rates of Fe(III) and Mn(III/IV) oxides and increase Fe and Mn bioavailability [28–30].

(3) The carbon, nitrogen, sulfur, oxygen, and phosphorous geochemical cycles require oxidation/reduction at many steps. Fe and Mn cycling is a key thermodynamic regulator and kinetic catalyst in natural waters [2, 4, 5].

(4) The transport and fate of heavy metal pollutants in natural waters is strongly affected by Fe and Mn oxide precipitation and dissolution [31]. Heavy metals, especially in sediments or groundwater, adsorb on Fe and Mn oxide surfaces [31–33]. The heavy metals are also incorporated in the Fe and Mn oxide matrix as impurities when precipitation occurs or when new mixed metal/Fe and metal/Mn coprecipitates are possible [31, 34–38]. As such, the cycling of redox conditions in natural waters and the associated precipitation and dissolution of Fe and Mn oxides lead to the cyclical uptake and release of heavy metal pollutants [39–44]. Movement of heavy metals through soils is also increased when these metals adsorb onto iron oxide colloids [45–47] and retarded when they adsorb onto iron and manganese oxide coatings on coarse grains, such as quartz sand.

This chapter focuses on the rates and mechanisms of Fe and Mn oxide precipitation and dissolution. This chapter is organized as follows. Thermodynamic driving forces, illustrated by pE–pH diagrams, are shown to constrain oxidation, precipitation, and dissolution (Section 3.2). Kinetic pathways for aqueous Fe(II) and Mn(II) oxidation by O_2 proceed by have both homogeneous and heterogeneous mechanisms (Sections 3.3 and 3.4). Dissolution occurs by proton-promoted, ligand-promoted, reductive, and synergistic pathways (Section 3.5). Modern molecular techniques provide an increasing basis for mechanistic descriptions and predictions of oxidation, precipitation, and dissolution (Section 3.6). Rather than reviewing all available experimental techniques [48], three informative examples are chosen for Section 3.6, including infrared spectroscopy, atomic force microscopy (AFM), and X-ray absorption spectroscopy (XAS). Beyond the scope of this chapter are photochemical reactions [49–51], biological reactions [16, 52–55], field studies [4, 56–58], and computational chemistry [59].

3.2 THERMODYNAMIC DRIVING FORCES

The free energy driving forces relating the various Fe and Mn aqueous and solid species (Figure 3.1) are represented in pE–pH diagrams (Figure 3.2) [9, 10, 60]. The pE axis represents the equilibrium partial pressures of O_2 (increasing at high pE) and H_2 (increasing at low pE) in the system and hence the oxidizing or reducing power of the environment. Specifically, pE is defined as:

$$
\begin{aligned}
\mathrm{p}E &= -\mathrm{pH} - \tfrac{1}{2}\log P_{\mathrm{H_2}} \\
\mathrm{p}E &= 20.77 - \mathrm{pH} + \tfrac{1}{4}\log P_{\mathrm{O_2}}
\end{aligned}
\tag{3.1}
$$

The dashed lines show the water stability region for 1 atm of gases. Above the top line, water should thermodynamically form O_2 if oxygen partial pressure is below 1 atm. Similarly, below

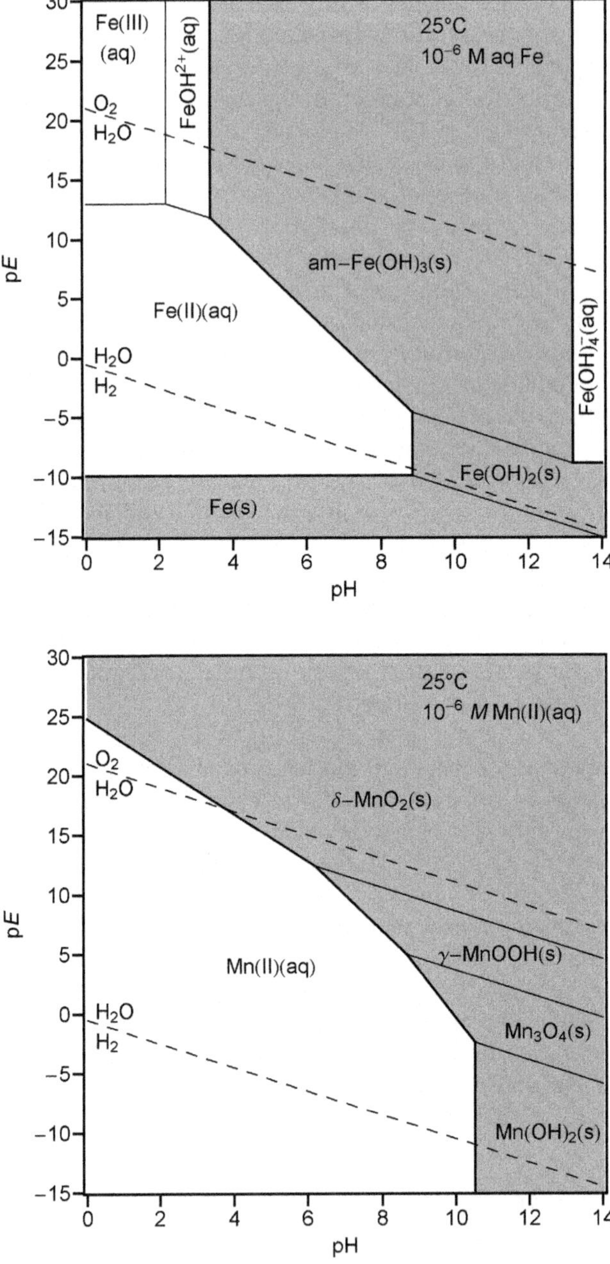

FIGURE 3.2 p*E*–pH diagrams of iron and manganese species [5, 10, 60]. The water stability field is shown for $P_{O_2} = 1$ atm and $P_{H_2} = 1$ atm.

the bottom line, water should thermodynamically form H_2 if hydrogen partial pressure is below 1 atm. Natural environments have a range of p*E*/pH values: oceans (p*E* = O_2 saturated, pH = 8), surface waters of lakes and rivers (p*E* = O_2 saturated, pH = 4 to 6), mine waters (p*E* = O_2 saturated, pH = 1 to 3), groundwater and sediments (p*E* = 0 to 3, pH = 7 to 9), and swamps (p*E* = 0 to −3, pH = 5 to 7) [9, 61, 62].

In the aqueous phase in the Fe system, there is an interchange among aqueous Fe(II) and Fe(III) redox species with decreasing pE. There is also a shift among the dominant hydrolysis species, such as from $[Fe(H_2O)_6]^{3+}$ to $[Fe(H_2O)_5OH]^{2+}$ with changing pH. In contrast, $[Mn(H_2O)_6]^{2+}$ is the unique dominant species at the common pE and pH values of natural waters: Hydrolysis of Mn^{2+}(aq) begins only for pH $>$ 10 and the stabilization of aqueous Mn(III) requires strong ligands [63].

Figure 3.2 is drawn for 10^{-6} M Fe_T (top) and 10^{-6} M Mn_T (bottom) at 25°C, where Fe_T is the sum of all species including for example Fe^{2+}(aq), Fe^{3+}(aq), $FeOH^{2+}$(aq), and the various solids. Mn_T is similarly defined. The lines in the diagram shift to form smaller gray regions and larger white regions for decreasing Fe_T and Mn_T. The white pE–pH regions show where Fe_T and Mn_T are thermodynamically speciated entirely in the aqueous phase. For instance, if the prescription of 10^{-6} M Fe_T or Mn_T initially includes solid species (e.g., as prepared in the laboratory or in an aquifer subject to seasonal pE–pH cycles) in the white regions, then it is predicted that these solids will dissolve. The dissolution rate is, however slow compared to the timescales of days, weeks, and months usually relevant to the environment (Section 3.5).

The gray pE–pH regions conversely show where some (viz. near the boundary lines) or most (viz. moving inward in the gray region) of 10^{-6} Fe_T or Mn_T should thermodynamically include solid phases, with a small amount of aqueous Fe and Mn species in equilibrium (e.g., 10^{-9} to 10^{-15} aq Fe and Mn). However, the precipitation rate may be slow (see Sections 3.3 and 3.4). For instance, in the absence of catalysis, aq Mn(II) at pE $=$ 10 and pH $=$ 8 persists for years in solution, even though oxidation and precipitation of Mn oxide solid phases is thermodynamically favored. By increasing pH, supersaturation increases and Mn(II) oxidizes and precipitates. However, the first solid to form is generally the least favored thermodynamically, a phenomenon described as Ostwald's rule of stages. For instance, Mn(III) solids such as MnOOH form initially even when the free energy of formation for $Mn^{IV}O_2$ is greater. The Mn^{IV} oxides form only after long aging times of $Mn^{III}OOH$ [64]. For a similar reason, in the diagram for Fe oxides, amorphous $Fe(OH)_3$(s) is employed in the analysis rather than thermodynamically favored but slowly forming hematite (α-Fe_2O_3) and goethite (α-FeOOH).

The pE–pH diagrams are useful for establishing the thermodynamically favorable pathways for transformations. However, predicting the transformation rates requires a kinetic analysis (see Sections 3.3–3.5).

3.3 RATES OF HOMOGENEOUS OXIDATION

Aqueous Fe(II) and Mn(II) are oxidized by reaction with dissolved O_2 [5, 12, 65, 66]. Thermodynamic analysis shows that the species Fe(III) is thermodynamically favorable for p$E >$ 13.2 at acidic pH (Figure 3.2). However, the reaction of Fe(II) with O_2 is not observed in the absence of catalysis at low pH. Only for pH $>$ 6 is the rate is appreciable, with a lifetime of approximately 1 day in water equilibrated with air at 25°C at pH $=$ 6 (Figure 3.3). For increasing pH, the lifetime rapidly decreases, with the value of 30 min at pH $=$ 7. For more alkaline pH, the reaction is even faster, but the dominant pathway shifts from homogeneous to heterogeneous (see Section 3.4.2) because the reaction product $Fe(OH)_3$(aq) rapidly polymerizes to form am-$Fe(OH)_3$(s), thus providing a reactive surface [67, 68].

The pH dependence of the reaction rate for pH $<$ 7 is explained by the mechanism shown in Table 3.1 [69]. With increasing pH, the dominant aqueous species shifts from $[Fe(H_2O)_6]^{2+}$ to $[Fe(H_2O)_5OH]^+$ to $[Fe(H_2O)_4(OH)_2]^0$, each of which having a progressively more rapid water-exchange rate and faster bimolecular reaction rate constant with O_2 [70]. The overall reaction rate is the sum of parallel pathways (Table 3.1 and Figure 3.3). The rate is also catalyzed by other dissolved species such as Cu^{2+}, Fe^{3+}, Al^{3+}, Co^{2+}, and Mn^{2+} [71–73].

Compared to Fe(II)(aq), the reaction of Mn(II)(aq) with O_2 is at least 10^6 times slower at circumneutral pH (Table 3.1 and Figure 3.3). Only for pH $>$ 8 does the reaction rate become

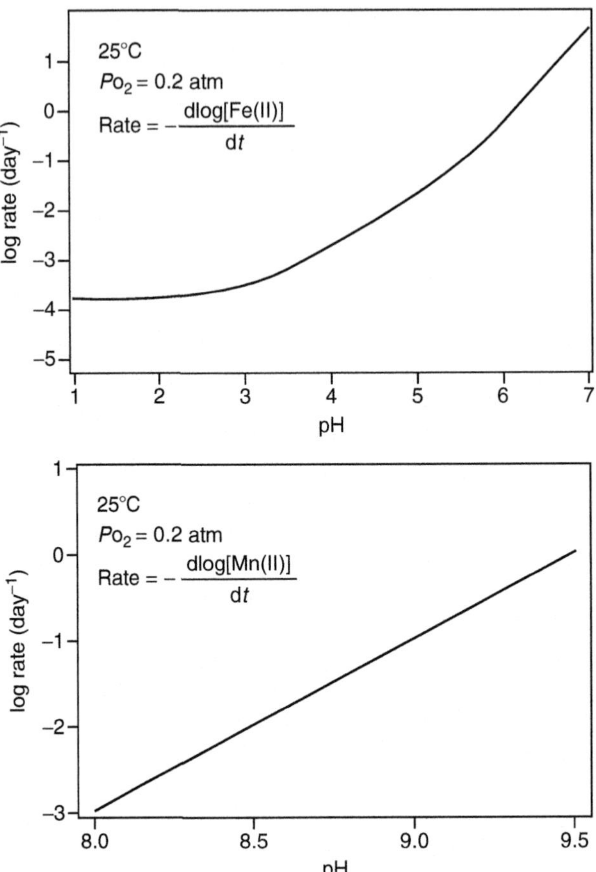

FIGURE 3.3 Homogeneous phase oxidation rates of aqueous Fe(II) and Mn(II) by O_2 [5, 12].

TABLE 3.1
Rate Equations for Homogeneous Phase Oxidation of Aqueous Fe(II) and Mn(II) by O_2 [5, 12]. The product $Mn(OH)_2^+$ rapidly polymerizes, forming a Mn oxide precipitate such as MnOOH(s).

Homogeneous Fe^{2+} oxidation

$Fe^{2+} + O_2 \rightarrow Fe^{3+} + O_2^-$	$\log k_0 = -5.1\ (M^{-1}\ s^{-1})$
$Fe(OH)^+ + O_2 \rightarrow Fe(OH)^{2+} + O_2^-$	$\log k_1 = +1.4\ (M^{-1}\ s^{-1})$
$Fe(OH)_2 + O_2 \rightarrow Fe(OH)_2^+ + O_2^-$	$\log k_2 = +6.9\ (M^{-1}\ s^{-1})$
$[Fe(OH)^+] = K_1[Fe^{2+}]/[H^+]$	$\log K_1 = -9.5\ (M)$
$[Fe(OH)_2] = \beta_2[Fe^{2+}]/[H^+]^2$	$\log \beta_2 = -20.6\ (M^2)$
$[O_2] = K_H P_{O_2}$	$\log K_H = -2.9\ (M\ atm^{-1})$

$$-(d[Fe^{2+}]/dt)_{homo} = k_0[O_2][Fe^{2+}] + k_1[O_2][Fe(OH)^+] + k_2[O_2][Fe(OH)_2]$$
$$= (k_0[Fe^{2+}] + k_1K_1\ [Fe^{2+}]/[H^+] + k_2\beta_2\ [Fe^{2+}]/[H^+]^2)K_H P_{O_2}$$

Homogeneous Mn^{2+} oxidation

$Mn(OH)_2 + O_2 \rightarrow Mn(OH)_2^+ + O_2^-$	$\log k_2 = +1.7\ (M^{-1}\ s^{-1})$
$[Mn(OH)_2] = \beta_2[Mn^{2+}]/[H^+]^2$	$\log \beta_2 = -22\ (M^2)$

$$-(d[Mn^{2+}]/dt)_{homo} = k_2[O_2][Mn(OH)_2]$$
$$= (k_2\beta_2[Mn^{2+}]/[H^+]^2)K_H P_{O_2}$$

appreciable. The reaction proceeds through the aqueous $Mn(OH)_2$ species, although the bimolecular rate constant of $Mn(OH)_2$ with O_2 is $10^{5.2}$ lower than that of $Fe(OH)_2$. The reaction product (Mn(III)), in the absence of strongly complexing ligands, rapidly polymerizes to form Mn oxide solids [67, 68], which catalyze further Mn(II) oxidation (see Section 3.4.2). Hence, separating homogeneous from heterogeneous pathways in Mn(II) oxidation is difficult because they occur simultaneously under most experimental conditions.

3.4 RATES OF HETEROGENEOUS OXIDATION

The rate of Mn(II) and Fe(II) oxidation by O_2 is catalyzed by metal oxide surfaces (>S) [12, 72, 74–76]. These surfaces are terminated by hydroxyl groups (>SOH), which bind Mn(II) and Fe(II) as $(>SO)_2Mn$ and $(>SO)_2Fe$. The inner-sphere surface complexes promote rapid oxidation, just as OH ligands do for the homogeneous complexes (see Section 3.3). The catalysis occurs both on foreign surfaces (e.g., Mn(II) on FeOOH) (see Section 3.4.1) and also for the special case of autocatalysis (e.g., Mn(II) on MnOOH producing additional MnOOH) (see Section 3.4.2).

3.4.1 Mineral Surfaces

Reaction rates at surfaces [77, 78] are given either as the conversion rate per unit surface area of the foreign surface (mol m^{-2} s^{-1}) or as the conversion rate per liter of a particulate suspension (M s^{-1}) [13]. The latter is the basic observable in experiments employing particulates, whereas the former is a more intrinsic measure, which can be estimated for a suspension of known loading (g L^{-1}) and specific surface area (m^2 g^{-1}) (Table 3.2). For single crystals, the conversion rate per unit surface is measured directly. For comparing the relative importance of homogeneous versus heterogeneous oxidation rates under a specific set of conditions, the heterogeneous rate expressed as (M s^{-1}) is more convenient because these units are the same as for homogeneous oxidation rates (Table 3.1). Heterogeneous oxidation rates in natural waters commonly exceed homogeneous oxidation rates.

Rate equations for heterogeneous oxidation are summarized in Table 3.2, where k is the oxidation rate coefficient (M^{-1} s^{-1}), $[>SOFe^{2+}]$ and $[>SOMn^{2+}]$ are the respective binuclear surface concentrations of adsorbed Fe(II) and Mn(II) (mol m^{-2}), K_H is the Henry's law partition coefficient of O_2 (M atm^{-1}), P_{O_2} is the partial pressure of O_2 (atm), A is the particulate surface area (m^2), and V is the container volume of the aqueous particulate suspension (L). $[>SOFe^{2+}]$ and $[>SOMn^{2+}]$, which depend on Fe_T/Mn_T and pH, are typically quantified by measurements of adsorption isotherms. Examples of heterogeneous rate coefficients are $\log k = 0.7$ (M^{-1} s^{-1}) for Fe^{2+}/O_2 reaction on FeOOH and $\log k = -1.55$ (M^{-1} s^{-1}) for Mn^{2+}/O_2 reaction on Al_2O_3.

In many real-world applications, aqueous concentrations of Fe(II) and Mn(II) are known rather than surface concentrations $[>SOFe^{2+}]$ and $[>SOMn^{2+}]$. It is convenient to recast the heterogeneous oxidation rate laws in terms of aqueous concentrations by the use of adsorption isotherm equations. For incomplete surface coverage not sufficient to significantly perturb the surface charge, surface concentration and aqueous concentration are related by a simple mass action equilibrium law quantified by an intrinsic binding constant β_s of the surface complex. This more detailed rate law is provided in Table 3.2. There, [>SOH] is the site density of OH groups on the metal oxide surface (mol m^{-2}). The first term gives the intrinsic chemical reactivity (i.e., the rate coefficient). The second term relates the surface concentration of adsorbed manganese to the aqueous concentration of manganese (i.e., a 3D to 2D transformation). The third term expresses the aqueous oxygen concentration. The fourth term scales the specific surface reactivity (mol m^{-2} s^{-1}) to the particulate suspension reactivity (M s^{-1}).

TABLE 3.2
Rate Equations for Catalytic Heterogenous Oxidation of Aqueous Fe(II) and Mn(II) by O_2 on Mineral Surfaces (5, 12, 78). The surface group $>SOM^{2+}$ is binuclear (i.e., $(>FeO)_2Fe^{2+}$).

Heterogeneous Fe^{2+} oxidation	
Rate per mineral surface area (mol m^{-2} s^{-1})	$-(d[>SOFe^{2+}]_{(mol/m^2)}/dt)_{hetero} = k[>SOFe^{2+}]_{(mol/m^2)}(K_H P_{O_2})$ Example: log k = 0.7 (M^{-1} s^{-1}) for Fe^{2+}/O_2 reaction on FeOOH
Rate of release in a particulate suspension (M s^{-1})	$-(d[Fe^{2+}]/dt)_{hetero} = k[>SOFe^{2+}]_{(mol/m^2)}(K_H P_{O_2})(A_{(m^2)}/V_{(L)})$ Example: as above
Detailed formulation	$-(d[Fe^{2+}]/dt)_{hetero} = k(\beta_s[Fe^{2+}][>SOH]_{(mol/m^2)}/[H^+]^2)(K_H P_{O_2})(A_{(m^2)}/V_{(L)})$
Heterogeneous Mn^{2+} oxidation	
Rate per mineral surface area (mol m^{-2} s^{-1})	$-(d[>SOMn^{2+}]_{(mol/m^2)}/dt)_{hetero} = k[>SOMn^{2+}]_{(mol/m^2)}(K_H P_{O_2})$ Examples: log k = −0.16 (M^{-1} s^{-1}) for Mn^{2+}/O_2 reaction on FeOOH log k = −1.55 (M^{-1} s^{-1}) for Mn^{2+}/O_2 reaction on Al_2O_3
Rate of release in a particulate suspension (M s^{-1})	$-(d[Mn^{2+}]/dt)_{hetero} = k[>SOMn^{2+}]_{(mol/m^2)}(K_H P_{O_2})(A_{(m^2)}/V_{(L)})$ Examples: as above
Detailed formulation	$-(d[Mn^{2+}]/dt)_{hetero} = k(\beta_s[Mn^{2+}][>SOH]_{(mol/m^2)}/[H^+]^2)(K_H P_{O_2})(A_{(m^2)}/V_{(L)})$ Example: for Mn^{2+}/O_2 reaction on FeOOH, log k = −0.16 (M^{-1} s^{-1}), β_s = $10^{-12.7}$ M, and [>SOH] = 1.6×10^{-5} mol m^{-2}

3.4.2 Autocatalysis

A special category of heterogeneous oxidation occurs when the product of the oxidation further accelerates the reaction rate [79, 80]. For example, the oxidation of Mn(II) produces MnOOH(s), as follows:

$$O_2 + 4Mn^{2+} + 6H_2O \xrightarrow[\text{heterogeneous}]{\text{homogeneous or}} 4MnOOH(s) + 8H^+ \tag{3.2}$$

As the reaction proceeds, the MnOOH(s) surface area and hence the heterogeneous reaction rate increase. The rate laws of autocatalysis (Table 3.3) are less precise than those of heterogeneous reactions on foreign mineral surfaces. Detailed descriptions for the autocatalysis pathways are hindered both by the complexities of separating homogeneous from heterogeneous pathways and by limitations in characterizing the increasing mineral surface area and the altering mineral phases during reaction.

3.5 DISSOLUTION RATES

Iron and manganese oxide solids dissolve at the rates shown in Figure 3.4 (center; right) when the contacting aqueous solution is strongly undersaturated (i.e., no back reaction from precipitation). The dissolution rates depend on many factors [2, 4, 13, 81–83]. For instance, the rates increase with acidic pH. There are also several parallel pathways having differing dissolution rates, including proton-promoted (see Section 3.5.1) (slowest), ligand-promoted (see Section 3.5.2), reductive (see Section 3.5.3), and synergistic (see Section 3.5.4) (fastest). The stoichiometries and the relative rates of these reaction pathways are given in Table 3.4. The rates further depend on crystalline phase: amorphous $Fe(OH)_3$ dissolves at least ten times faster than γ-FeOOH. The rates also depend on initial chemical or physical preparation and

TABLE 3.3
Rate Equations for the Autocatalytic Heterogenous Oxidation of Aqueous Fe(II) and Mn(II) by O_2 [5, 12, 99]

Heterogeneous autocatalytic Fe^{2+} oxidation

Autocatalytic rate ($M\ s^{-1}$)	$-\left(\frac{d[Fe^{2+}]}{dt}\right)_{autocatalytic} = k[FeO_x(s)][Fe^{2+}]$ FeO_x is poorly characterized and can contain several mineral phases (e.g., amorphous $Fe(OH)_3$ or FeOOH)

Heterogeneous autocatalytic Mn^{2+} oxidation

Autocatalytic rate ($M\ s^{-1}$)	$-\left(\frac{d[Mn^{2+}]}{dt}\right)_{autocatalytic} = k[MnO_x(s)][Mn^{2+}]$ Example: $k_s = 5 \times 10^{18}\ M^{-4}\ day^{-1}$ where MnO_x is poorly characterized and can contain several mineral phases (e.g., MnOOH, Mn_3O_4, MnO_2, or birnessite)

The oxidation products of Fe(II) and Mn(II) precipitate, thus increasing with time the particulate surface area and hence reaction rates. As compared to Table 3.2, the formulation in this table is less precise due to poor characterization of the precipitate product and its time evolution.

TABLE 3.4
Stoichiometry and Relative Dissolution Rates of the Proton-Promoted, Ligand-Promoted, Reductive, and Synergistic Pathways [2, 4, 83]. Photoreductive pathways are omitted.

Iron (III) oxide dissolution

	Proton-promoted
Slowest	$FeOOH + 3H^+ \rightarrow Fe(III)(aq) + 2H_2O$
↓	Ligand-promoted
	$(FeOOH){>}Fe^{III}{-}OH + L^- + H^+ \rightarrow (FeOOH){>}Fe^{III}{-}L + H_2O \rightarrow$ $FeOOH + Fe(III){-}L(aq)$
	Examples: L^- = oxalate, malonate, citrate
	Reductive
	$FeOOH + e^- + 3H^+ \rightarrow Fe(II)(aq) + 2H_2O$
	Examples: see below for manganese
	Synergistic
Fastest	$(FeOOH){>}Fe^{III} - L + e^- \rightarrow FeOOH + Fe(II)(aq) + L^-$
	Examples: L^- = oxalate, malonate, or citrate; e^- = ascorbate or Fe^{2+}

Manganese (III, IV) oxide dissolution

	Reductive
	$MnOOH + e^- + 3H^+ \rightarrow Mn(II)(aq) + 2H_2O$
	$MnO_2 + 2\,e^- + 4H^+ \rightarrow Mn(II)(aq) + 2H_2O$
	Examples: e^- = ascorbate, hydroquinone, dithionite ($S_2O_4^{2-}$), H_2S, pyrogallol

often pass through initial transients of rapid dissolution, which are at least ten times faster than the steady-state dissolution rates [84, 85].

Although the rates depend on sample crystallinity and preparation and are quite variable on a linear scale, on a log scale the differences among samples are less apparent. With this caveat, organizational statements are possible. For example, γ-FeOOH dissolves more slowly than amorphous $Fe(OH)_3$; iron oxides dissolve more slowly than manganese oxides; reductive

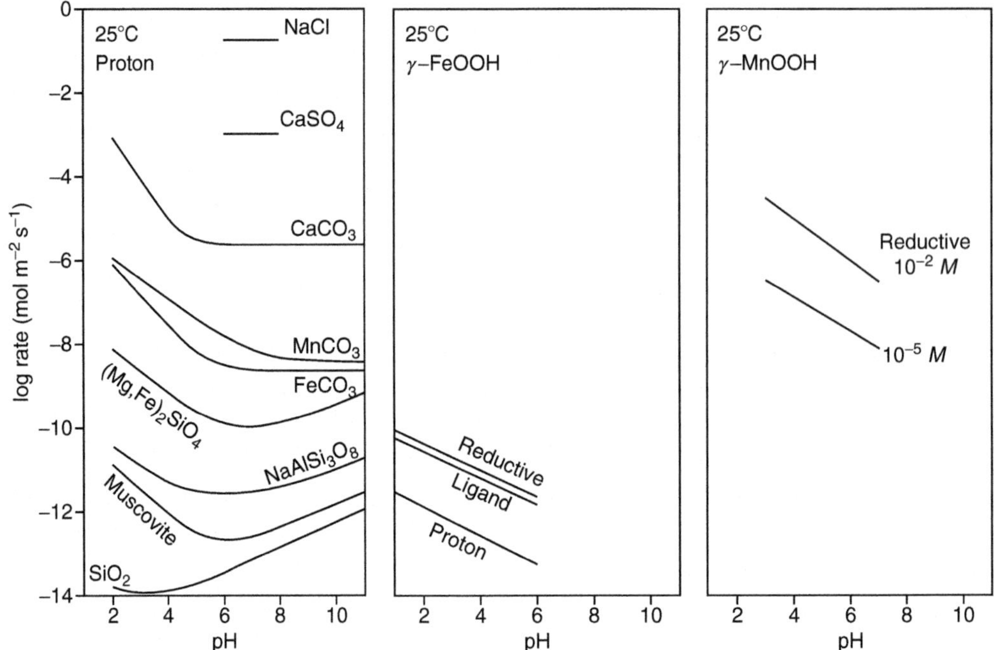

FIGURE 3.4 Comparison of pH-dependent dissolution rates of iron and manganese oxides with other common minerals [86, 99, 132–139]. Rates apply when the aqueous solution concentrations are far from equilibrium, i.e., strongly undersaturated. (Ascorbic acid rates are shown for reductive dissolution.)

dissolution is faster than ligand-promoted dissolution; and proton-promoted dissolution is the slowest of all. The dissolution rates of iron and manganese oxides can also be compared with the dissolution rates of other minerals (Figure 3.4, left versus center and right). Iron oxides dissolve at rate comparable to chain and sheet aluminosilicates. Manganese oxides dissolve at rates comparable to carbonates, although the precise rate depends strongly on reductant concentration.

3.5.1 Proton-Promoted

Protons increase dissolution rates [86], which is rationalized by a catalytic role in depolymerization. For example, dimers such as $[Fe_2(OH)_2]^{4+}$, $[Al_2(OH)_2]^{4+}$, and $[(VO)_2(OH)_2]^{2+}$ decompose at the compound rate: $k = k_1 + k_2[H^+]$, where k is the pseudo first-order rate coefficient, k_1 is the H_2O reaction pathway, and k_2 is the H^+ reaction pathway [87–90]. Protons accelerate the rate by attaching to the oxygen in the hydroxyl groups bridging the metals, thus removing electron density and weakening the bond strength of the metal–oxygen linkage:

H
OH
$(H_2O)_4$ — Fe Fe — $(H_2O)_4$
OH

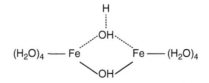

Minerals such as FeOOH are regarded as an infinite n-mer extension of $[Fe_2(OH)_2]^{4+}$. Mineral dissolution is then a stepwise depolymerization, and protons have the role of weakening bonds and thus increasing dissolution rates [83]. The log of the rate is proportional to the log of the surface proton concentration ($>SOH_2^+$).

3.5.2 Ligand-Promoted

Ligands, binding as inner-sphere complexes to the surface groups of iron and manganese oxides, increase dissolution rates [30, 83, 86, 91–93]. The increase is proportional to the ligand surface concentration and the ligand-binding strength. The rate law is: $R = k_L$ [>SL] where R is the dissolution rate (mol m^{-2} s^{-1}), k_L is the rate coefficient (s^{-1}), and [>SL] is the surface concentration of the ligand. For a homologous series of surface structures, the rate coefficient is proportional to the adsorption strength of the ligand, e.g., $k_L = 4.5 \times 10^{-10} K_L - 1.1 \times 10^{-6}$ where K_L is the Langmuir binding constant for the series oxalate, glutarate, and malonate binding to hematite [93]. The relationship occurs because strong electron donation by the oxygen-ligand to a surficial metal atom removes electron density in the bonds between the metal atom and the oxygen atoms of the mineral lattice, thus weakening the bond and lowering the energy barrier for the dissolution of the metal atom [82]. In this regard, Mn and Fe cations bind much more strongly to oxygen than nitrogen-containing ligands [9]. Furthermore, ligands effective for promoting dissolution have two or more functional groups capable of chelation to form inner-sphere bidentate mononuclear complexes. An example is oxalate (see Figure 3.6, $n = 0$). In contrast, ligands forming bidentate binuclear complexes such as phosphate or borate stabilize the surface against attack by H^+ and H_2O and thus reduce (inhibit) the dissolution rate. Monodentate ligands such as acetate do not perceptibly affect the dissolution rate.

3.5.3 Reductive

Reductants rapidly accelerate iron and manganese oxide dissolution [94–99]. Examples of reductants are ascorbic acid, hydrogen sulfide, and phenols. A reductant typically forms an inner-sphere complex at the surface, though not always so. When an electron is transferred to a Mn(III/IV) oxide, a surficial Mn(II) ion locked inside an oxide lattice is formed. Because Mn(II) oxides are much more soluble than the corresponding higher oxides (Figure 3.2), rapid Mn(II) depolymerization occurs, which is followed by release to the aqueous phase of Mn(II). Fe(III) oxides are similarly reduced, followed by the release of aqueous Fe(II).

The Mn oxide solids are strong oxidants capable of oxidizing common organic matter:

$$MnO_2(s) + 2e^- + 4H^+ \longrightarrow Mn^{2+} + 2H_2O, \; pE^0 = 20.8, \; E^0_{1/2} = +1.23\,V \qquad (3.3)$$

$$MnOOH(s) + e^- + 3H^+ \longrightarrow Mn^{2+} + 2H_2O, \; pE^0 = 25.4, \; E^0_{1/2} = +1.50\,V \qquad (3.4)$$

$$H_2CO(aq) + 2MnO_2(s) + 4H^+ \longrightarrow 2Mn^{2+} + CO_2(aq) + 3H_2O, \; \Delta pE^0 = 18.4, \; E^0 = +1.09\,V \qquad (3.5)$$

The thermodynamic driving force of the reaction is often a good predictor of rate for a homologous series of reactants (i.e., a linear free energy relationship). For example, for a series of substituted phenols like *p*-methylphenol at 10^{-4} M and pH $= 4.4$, the Hammett constant σ is a predictor of reductive dissolution rate R (mol m^{-2} s^{-1}) of manganese oxide [95]:

$$\log_{10} R = -7.79 - 3.63\sigma \qquad (3.6)$$

The correlation holds because the Hammett constant is also a predictor of reduction potential, which is the driving force for manganese oxide dissolution. Equation (3.6) is derived over a $\Delta\sigma$ range of 0.66, so the effect on log R is substantial.

3.5.4 Synergistic

Some species interact cooperatively to increase the dissolution rate above the sum of the individual dissolution rates. For example, Fe(III) oxides dissolve more rapidly in the presence of aqueous Fe(II) and oxalate than in the presence of either separately [100]. The Fe(II)–oxalato aqueous complex is a strong reductant, which rapidly reduces the iron oxide through reductive dissolution. Another example is the rapid dissolution of iron oxides in the presence of a ligand–reductant pair, such as oxalate and ascorbate [101]. Ascorbate reduces Fe(III) to Fe(II) at the surface of the Fe(III) oxide, while oxalate forms an inner-sphere complex at the surface and rapidly dissolves Fe(II). A final example is a surface and aqueous ligand pair, such as oxalate and a trihydroxamate siderophore, desferrioxamine B (DFO-B) [102]. At low concentrations, oxalate adsorbs to the iron oxide surface. However, for sufficiently low oxalate concentrations, the free energy driving force toward dissolution is small because the total aqueous solubility of an iron oxide is low. The DFO-B, however, binds aqueous iron very strongly. Addition of DFO-B provides a sink for aqueous iron. Under these conditions, in one step oxalate dissolves Fe(III) as a Fe(III)–oxalato complex, which rapidly hands off the Fe(III) to DFO-B. Oxalate is then free to recycle to the iron oxide surface, leading to further dissolution. The rates in all three examples are proportional to the surface concentrations of the reactive species.

3.5.5 Master Equation

The overall dissolution rate R (mol m^{-2} s^{-1}) is the sum of several process, as represented conceptually in the equation below [83]:

$$R = k_{\mathrm{H}}[> \mathrm{SOH}_2^+]^n + k_{\mathrm{OH}}[> \mathrm{SO}^-]^m + k_{\mathrm{aq}} + \sum_a K_{\mathrm{L,a}}[> \mathrm{SL}_a] - \sum_b k_{\mathrm{I},b}[> \mathrm{SI}_b] + \sum_c k_{\mathrm{R},c}[> \mathrm{SR}_c] + \sum_d k_{\mathrm{M},d}[> \mathrm{SM}_d] \quad (3.7)$$

for pathways mediated by proton (H), hydroxide (OH), aqueous (aq), a ligands (L), b inhibitors (I), c reductants (R), and d metal ions (M). Under tested applications, not more than two or three terms dominate the overall dissolution rate. Hence, Equation (3.7) is not to be interpreted literally in its complexity.

3.6 MOLECULAR ENVIRONMENTAL CHEMISTRY

The growth of molecular environmental chemistry in recent years provides a renewed basis for testing and constraining models of mineral precipitation and dissolution [48, 103–106]. The left-hand side of Equations (3.6) or (3.7) and the quantitative formulations given in Tables 3.1 to 3.3 provide examples of macroscopic quantities, which in these cases are the measurable dissolution rates. From the dependencies of these direct macroscopic observables on factors such as pH or reductant concentration, molecular descriptions of surface processes, such as given in the right hand side of Equation (3.7), are constructed by inference. These molecular inferences from macroscopic observables are the common approach for the studies described in Sections 3.2 to 3.5. In the latest research (see Sections 3.6.1 to 3.6.3), direct measurement of the molecular quantities, which is always preferable to inference, is an accelerating trend and provides a manner for further improving mechanistic descriptions and thus our predictive capability of macroscopic quantities. Of the numerous available techniques [48, 105], three illustrative examples include infrared (IR) spectroscopy (Section 3.6.1), AFM (Section 3.6.2), and XAS (Section 3.6.3).

3.6.1 Infrared (IR) Spectroscopy

Inner-sphere surface complexes, whose concentrations appear as [>SX] in the right hand side of Equation (3.7), are directly observable by infrared spectroscopy [93, 107–111]. Infrared absorption occurs at the resonance frequencies of atomic vibrations. The number of vibrations and their wave number positions depend upon the symmetry of the surface complex and its local chemical environment.

Duckworth and Martin [93] employ a homologous series of dicarboxylic acids, including oxalate, malonate, succinate, glutarate, and adipate, to study the effect of chain length on the nature of the surface complexation structures formed on hematite. The infrared spectra (Figure 3.5) lead to the deduction of the structure of the inner-sphere surface complexes

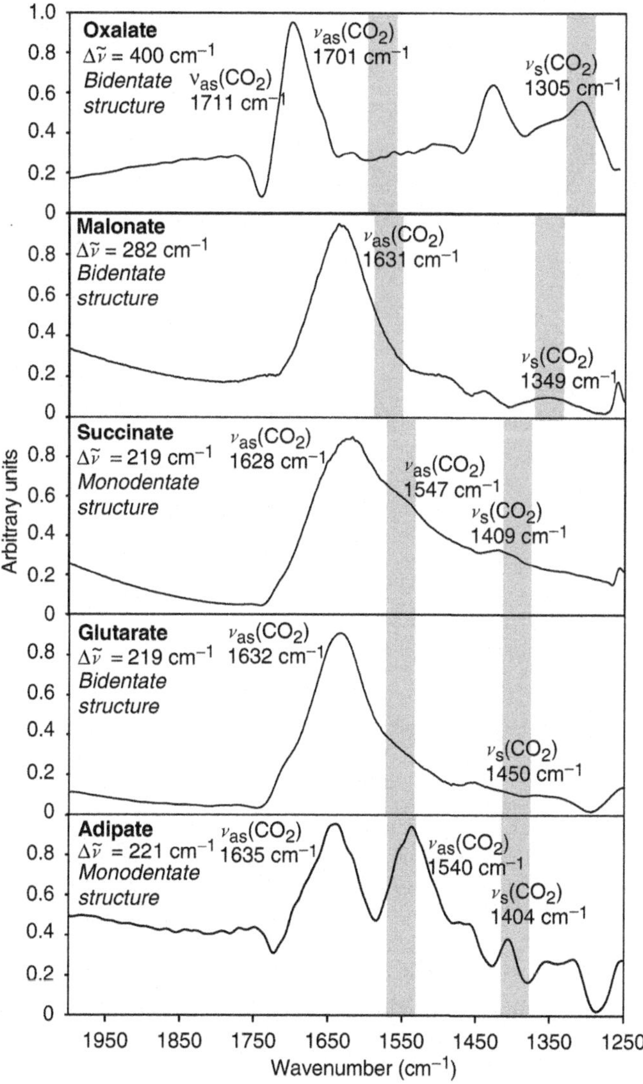

FIGURE 3.5 Infrared spectra of the surface-adsorbed complexes of dicarboxylic acids on hematite at pH = 5.0 [93]. $\Delta\tilde{\nu}$ is the difference between $\nu_{as}(CO_2)$ and $\nu_s(CO_2)$. For comparison, gray bars indicate the absorption regions of $\nu_{as}(CO_2)$ and $\nu_s(CO_2)$ vibrations of the aqueous species. The bidentate structures cause faster ligand-promoted dissolution than the monodentate structures (Table 3.5).

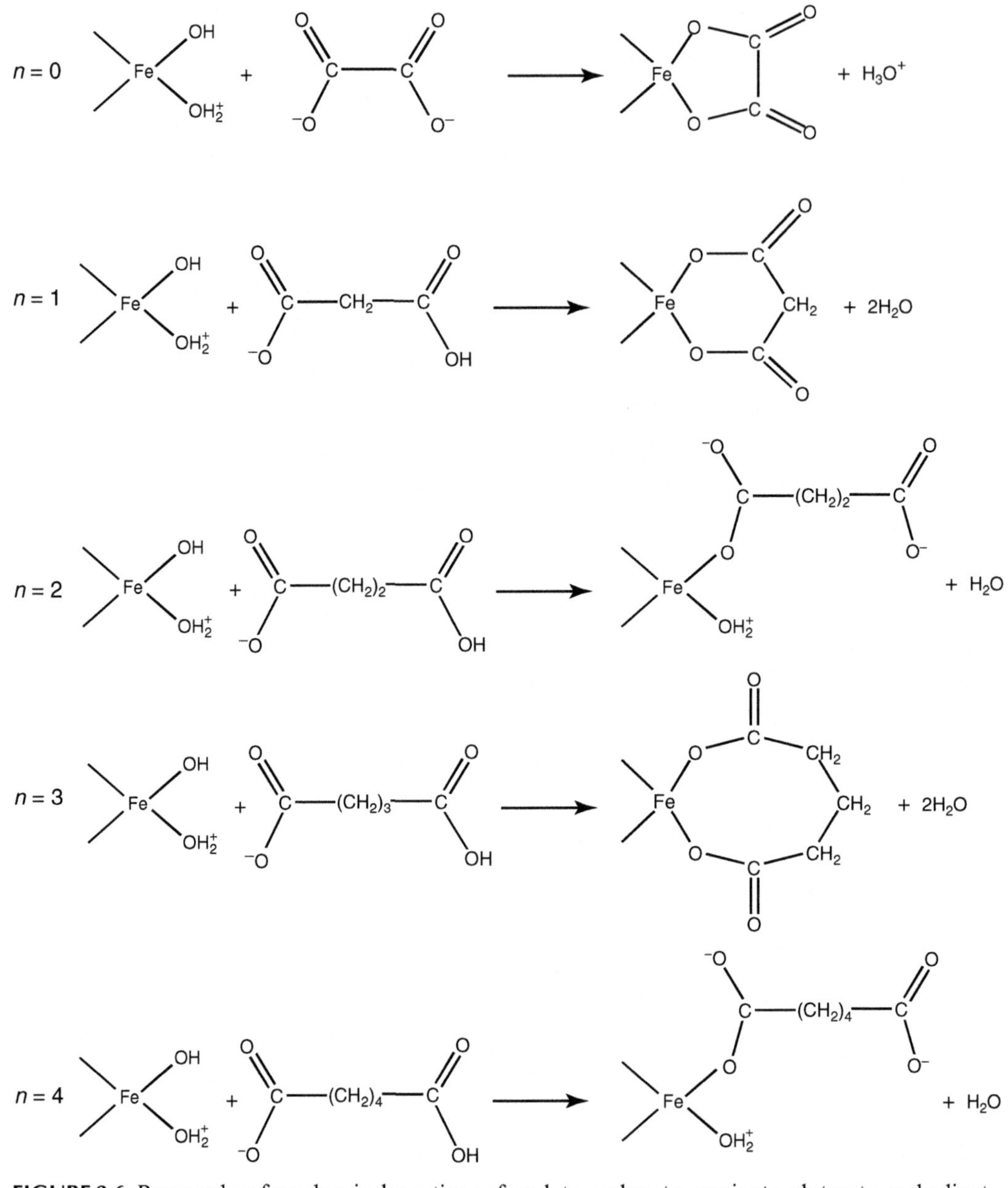

FIGURE 3.6 Proposed surface chemical reactions of oxalate, malonate, succinate, glutarate, and adipate to a hematite surface hydroxyl group at pH = 5.0 [93]. The bidentate structures cause faster ligand-promoted dissolution than the monodentate structures (Table 3.5).

(Figure 3.6). Oxalate, malonate, and glutarate form bidentate surface complexation structures, while succinate and adipate form monodentate structures. The ligand-promoted dissolution rates are also measured, and the three bidentate ligands have dissolution rates at least 10 to 100 times faster than the monodentate structures (Table 3.5). These results illustrate the gains possible from molecular level insights. For instance, rationalizing the difference in the dissolution rate of glutarate versus that of succinate is guesswork when only macroscopic measurements are available.

TABLE 3.5
Physical Data and Experimental Results for the Ligand-Promoted Dissolution of Iron Oxide (Hematite) by Dicarboxylic Acids [93]

Ligand	Formula	pK_{a1}	pK_{a2}	pH	Rate ($mol\ m^{-2}\ s^{-1}$)	Rate constant, k (s^{-1})	Langmuir binding constant, K (M^{-1})
Oxalate (0)	$^{-}OOCCOO^{-}$	1.25	4.27	5.0	1.0×10^{-10}	1.5×10^{-5}	$30{,}000 \pm 3000$
Malonate (1)	$^{-}OOC(CH_2)COO^{-}$	2.85	5.70	5.0	1.4×10^{-11}	2.0×10^{-6}	3000 ± 300
Succinate (2)	$^{-}OOC(CH_2)_2COO^{-}$	4.42	5.42	5.0	$<2.8 \times 10^{-12}$	$<4.1 \times 10^{-7}$	2700 ± 300
Glutarate (3)	$^{-}OOC(CH_2)_3COO^{-}$	4.34	5.43	5.0	3.3×10^{-11}	4.8×10^{-6}	7200 ± 700
Adipate (4)	$^{-}OOC(CH_2)_4COO^{-}$	4.21	5.64	5.0	$<2.8 \times 10^{-12}$	$<4.1 \times 10^{-7}$	N/A

The numbers in parentheses give the carbon chain length, n, in $^{-}OOC(CH_2)_nCOO^{-}$. Also given are the Langmuir binding constants, K (M^{-1}), of the inner-sphere surface complexes, which are determined by analysis of infrared spectra. Conditions: 5 mM acetate buffer, pH = 5.0, 2 g L^{-1} hematite, 25°C.

3.6.2 Atomic Force Microscopy (AFM)

The master equation for dissolution (Equation (3.7)), which is derived from macroscopic observations, assumes each surface site is equivalent. Under this treatment, the surface is a terrace of infinite extent. In reality, manganese and iron oxide surfaces have microtopography, including for example terraces, steps, kinks, and pits [99, 112, 113]. Terrace ions locked into place by five bonds to nearest neighbors are released more slowly than ions at kink positions having only three nearest neighbors [114]. Moreover, point defects, line defects, and other dislocations dissolve rapidly. A master equation derived solely from macroscopic observations is thus inherently limited in its accuracy and range of application.

The atomic force microscope (AFM) allows measurement of mineral surface topography under aqueous conditions (Figure 3.7) [115, 116]. A sharp tip with a curvature radius of approximately 10 nm is rastered laterally across the surface, and vertical deflections of the tip provide the topography of the underlying surface [117]. The microscopic data collected in time series allow new models of mineral dissolution to be formulated and constrained [118]. The dynamic range of the AFM, which is limited by the scanning rate on the upper side and the tip lifetime on the lower side, is convenient for studying minerals having dissolution rates of 10^{-6} to 10^{-10} mol m^{-2} s^{-1} [119]. Hence, manganese oxides are easily studied (Figure 3.4), whereas iron oxides dissolve too slowly for AFM work, except under extreme conditions. Manganite (MnOOH) reductive dissolution shows overall step retreat (e.g., circled region in Figure 3.7), while certain areas of the surface are simultaneously reconstructing to minimize surface energy by filling in pits (e.g., central arrow in Figure 3.7). Scaling of the complex time-dependent changes shown in Figure 3.7 to the surface-averaged (i.e., macroscopic) dissolution rate is an outstanding research challenge.

AFM is also useful for studying precipitation. Junta and Hochella [120, 121] report on the heterogeneous oxidation of Mn(II) at hematite, goethite, and albite surfaces by O_2 for $7.8 < pH < 8.7$. For the specified reaction conditions, manganese oxide precipitates form preferentially at the steps and kinks of the substrates, which could suggest a higher rate of heterogeneous oxidation at those microtopographic features. Alternatively, the oxidation may occur widely on the terrace, and the oxidized monomeric products may diffuse rapidly to more stable step and kink positions.

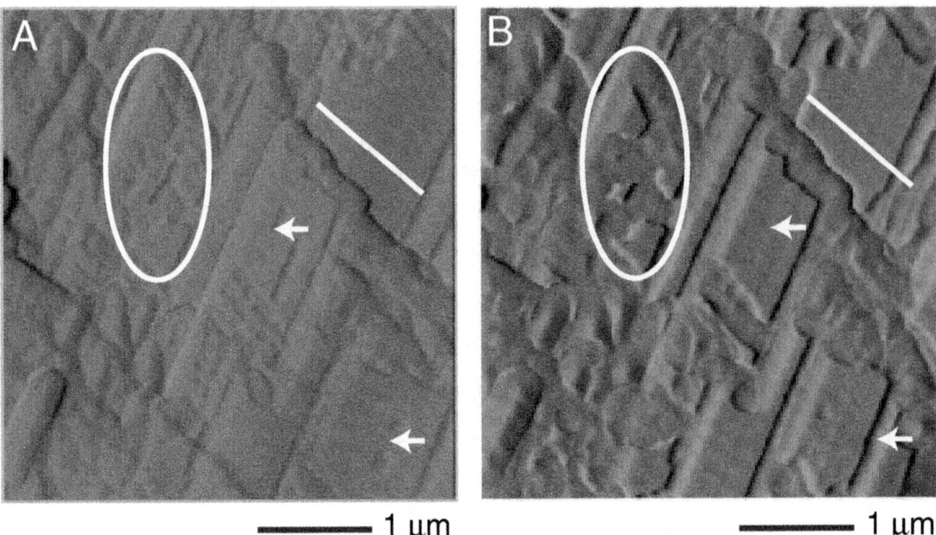

FIGURE 3.7 AFM deflection-mode micrographs of surfaces changes observed *in situ* at 298 K during exposure of manganite (γ-MnOOH) to ascorbic acid at pH = 3.2 (10 m*M* $NaNO_3$) [99]. (A) Prior to exposure to reductant. (B) Same as *A* after exposure to 1 m*M* reductant for 95 min. In height images (not shown), *z*-scale is 125 nm.

3.6.3 X-Ray Absorption Spectroscopy (XAS)

XAS responds to the local coordination environment and chemical oxidation state of the inner-sphere surface complexes of heavy metals on iron and manganese oxides [122–126]. X-ray absorption is modulated at energies near and above an absorption edge because the final quantum mechanical state of the system, which is a hole in the sample and a photoelectron in vacuum, is affected by scattering and constructive/destructive interference of the ejected electron at its de Broglie wavelength by coherent shells of neighboring atoms. Experiments must be conducted on synchrotron beam lines to obtain the high intensity monochromatic x-rays necessary for XAS studies.

The interaction of arsenic with manganese oxide surfaces provides an example of the important molecular information possible with XAS studies [127, 128]. The macroscopic observation is that As(III) reductively dissolves Mn(IV) oxides to form As(V). Understanding the mechanisms and pathways in greater molecular detail and hence quantifying the reductive dissolution rates in greater accuracy is important because As(III) is more mobile in aquifers than As(V). X-ray absorption near-edge spectroscopy (XANES) shows that when aqueous As(III) adsorbs to the surface of MnO_2, the surface complex is As(V). The extended x-ray absorption fine-structure (EXAFS) results show that the inner-sphere complex is bidentate binuclear, $(>Mn^{IV}O)_2As^{V}OOH$ (Figure 3.8). High-precision, quantitative structural information on local ordering is also derived from EXAFS measurements: the As–Mn distance is 3.22 Å. The precise distances coupled with computational modeling of optimized structures allow an investigator to rule out a host of other possibilities, e.g., bidentate mononuclear binding, mixed Mn/As co-precipitates, Mn(II)/Mn(III) structures, and As(III) structures. The only optimized structure providing good agreement with the measurements (Figure 3.8, left) is shown in Figure 3.8 (right). XAS studies show that bidentate binuclear structures are also formed by arsenic on iron oxides [129–131].

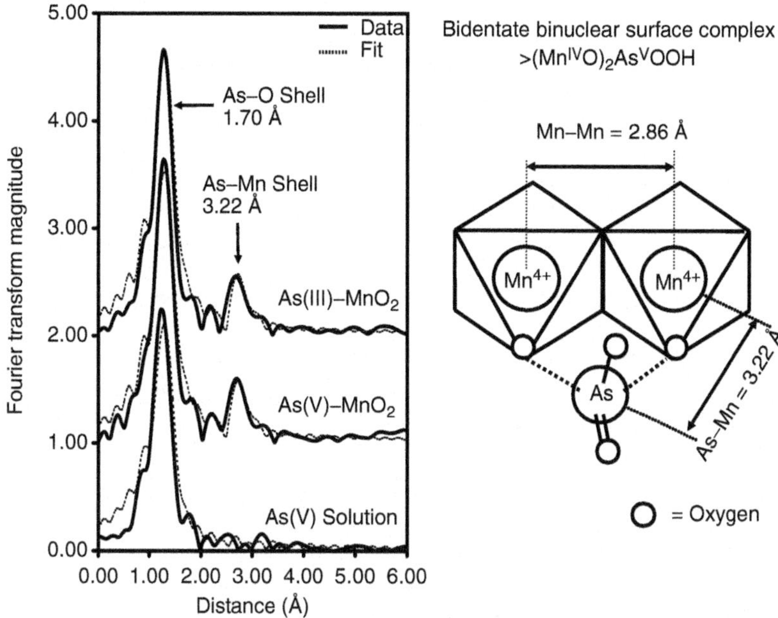

FIGURE 3.8 Structural information for surface complexes from synchrotron-based extended x-ray absorption fine structure (EXAFS) measurements [127] (left). Radial structure functions (not phase corrected) for As(III)- and As(V)-treated synthetic birnessite (MnO_2) and 1.0 m*M* As(V) solution. Dashed lines are the fits to the experimental RSF data and peaks correspond to As–O and As–Mn atomic shells around the As atom (right). Structural diagram of MnO_2 crystallite showing possible linkages between an arsenate ion (As(V) tetrahedron) and a pair of edge-linked MnO_6 octahedra. As–Mn interatomic distance is 3.22 Å, suggesting a bidentate binuclear complex. (Adapted from BA Manning, SE Fendorf, B Bostick, and DL Suarez. *Environ. Sci. Technol.* 36: 976–981, 2002.)

3.7 CONCLUDING REMARKS

Because the uncatalyzed rates of Fe and Mn oxide precipitation and dissolution are often slow, an understanding of the catalyzed pathways and mechanisms is necessary for quantitative predictions of Fe and Mn transformations and the associated impacts in the environment. Tables 3.1 to 3.5 and Figure 3.3 and Figure 3.4 summarize the known rate information. Aqueous Fe(II) and Mn(II) oxidation by O_2 is catalyzed heterogeneously by other mineral surfaces. The oxidized products polymerize and precipitate, first as higher energy oxides having little long-range order and after aging as lower energy crystalline oxides. When exposed to undersaturated aqueous conditions, the oxides dissolve at rates dependent upon catalytic additives, including proton-promoted, ligand-promoted, reductive, and synergistic dissolution pathways. Molecular techniques increasingly provide a detailed mechanistic description of these processes. Under favorable circumstances, these new descriptions are detailed enough that they can be modeled by ab initio and semiempirical methods, allowing for direct comparison of experimental observations and computational results and thus for the further improvement and refinement of the latter.

ACKNOWLEDGMENTS

S.T.M. is grateful for support received from the Chemical Sciences, Geosciences, and Biosciences Division of the Office of Basic Energy Sciences in the U.S. Department of Energy and the American Chemical Society Petroleum Research Fund.

REFERENCES

1. PA Cox. *The Elements on Earth: Inorganic Chemistry in the Environment.* New York: Oxford University Press, 1995, 287pp.
2. AT Stone and JJ Morgan. In: *Aquatic Surface Chemistry*, W Stumm, ed. New York: Wiley, 1987, pp. 221–254.
3. P Huang. In: *Rates of Soil Chemical Processes*, D Sparks and D Suarez, eds. Madison, WI: Soil Science Society of America, 1991, pp. 191–230.
4. RM Cornell and U Schwertmann. *The Iron Oxides: Structure, Properties, Reactions, Occurrence, and Uses.* New York: VCH, 1996, 573pp.
5. JJ Morgan. In: *Metal Ions in Biological Systems*, A Sigel and H Sigel, eds. New York: Marcel Dekker, 2000, pp. 1–34.
6. G Sposito. *The Chemistry of Soils.* New York: Oxford University Press, 1989, 277pp.
7. U Schwertmann and RW Fitzpatrick. In: *Biomineralization Processes of Iron and Manganese*, HCW Skinner and RW Fitzpatrick, eds. Cremlingen-Destedt: Catena Verlag, 1992, pp. 7–30.
8. JB Dixon and HCW Skinner. In: *Biomineralization Processes of Iron and Manganese*, HCW Skinner and RW Fitzpatrick, eds. Cremlingen-Destedt: Catena Verlag, 1992, pp. 7–30.
9. W Stumm and JJ Morgan. *Aquatic Chemistry.* New York: Wiley, 1996, 1022pp.
10. JI Drever. *The Geochemistry of Natural Waters.* Upper Saddle River, NJ: Prentice-Hall, 1997, 436pp.
11. D Diem and W Stumm. *Geochim. Cosmochim. Acta* 48: 1571–1573, 1984.
12. B Wehrli. In: *Aquatic Chemical Kinetics: Reaction Rate Processes in Natural Waters*, W Stumm, ed. New York: Wiley, 1990, pp. 311–336.
13. MA Blesa, PJ Morando, and AE Regazzoni. *Chemical Dissolution of Metal Oxides.* Boca Raton, FL: CRC Press, 1994, 401pp.
14. KH Nealson and D Saffarini. *Annu. Rev. Microbiol.* 48: 311–343, 1994.
15. HL Ehrlich. *Earth Sci. Rev.* 45: 45–60, 1998.
16. DR Lovley. In: *Environmental Microbe-Metal Interactions*, DR Lovley, ed. Washington, DC: ASM Press, 2000, pp. 3–30.
17. H Sigel, ed. *Metal Ions in Biological Systems: Iron in Model and Natural Compounds.* New York: Marcel Dekker, 1978, 417pp.
18. A Sigel and H Sigel, eds. *Metal Ions in Biological Systems: Iron Transport and Storage in Microorganisms, Plants, and Animals.* New York: Marcel Dekker, 1998, 775pp.
19. A Sigel and H Sigel, eds. *Metal Ions in Biological Systems: Manganese and its Role in Biological Processes.* New York: Marcel Dekker, 2000, 761pp.
20. I Bertini, A Sigel, and H Sigel, eds. *Handbook on Metalloproteins.* New York: Marcel Dekker, 2001, 1108pp.
21. DJ Horvath. *Geol. Soc. Am. Bull.* 83: 451–462, 1972.
22. EA Curl and B Truelove. *The Rhizosphere.* New York: Springer-Verlag, 1986, 174pp.
23. RK Vempati and RH Loeppert. *J. Plant Nutr.* 11: 1557–1576, 1988.
24. H Marschner. *Mineral Nutrition of Higher Plants.* San Diego: Academic Press, 1995, 889pp.
25. JF Banfield and KH Nealson, eds. *Geomicrobiology: Interactions between Microbes and Minerals.* Washington, DC: Mineralogical Society of America, 1997, 448pp.
26. LA Finney and TV O'Halloran. *Science* 300: 931–936, 2003.
27. FMM Morel and NM Price. *Science* 300: 944–947, 2003.
28. H Marschner, V Romheld, and I Cakmak. *J. Plant Nutr.* 10: 1175–1184, 1987.
29. H Marschner, M Treeby, and V Romheld. *Z. Pflanzenernahr. Bodenk.* 152: 197–204, 1989.
30. AT Stone. In: *Geomicrobiology: Interactions between Microbes and Minerals*, JF Banfield and KH Nealson, eds. Washington, DC: Mineralogical Society of America, 1997, pp. 309–344.
31. EA Jenne. In: *Trace Inorganics in Water*, RA Baker, ed. Washington, DC: American Chemical Society, 1968, pp. 337–387.
32. DA Dzomback and FMM Morel. *Surface Complexation Modeling.* New York: Wiley, 1990, 393pp.
33. P Trivedi and L Axe. *Environ. Sci. Technol.* 35: 1779–1784, 2001.
34. JD Hem. In: *Particulates in Water*, MC Kavanaugh and JO Leckie, eds. Washington, DC: American Chemical Society, 1980, pp. 45–72.

35. JD Hem, CJ Lind, and CE Roberson. *Geochim. Cosmochim. Acta* 53: 2811–2822, 1989.
36. JD Hem and CJ Lind. *Geochim. Cosmochim. Acta* 55: 2435–2451, 1991.
37. NL Dollar, CJ Souch, GM Filippelli, and M Mastalerz. *Environ. Sci. Technol.* 35: 3608–3615, 2001.
38. PE Kneebone, PA O'Day, N Jones, and JG Hering. *Environ. Sci. Technol.* 36: 381–386, 2002.
39. L Sigg, M Sturm, and D Kistler. *Limnol. Oceanogr.* 32: 112–130, 1987.
40. TA Jackson. *J. Geochem. Explor.* 52: 97–125, 1995.
41. KA Hudson-Edwards. In: *Environmental Mineralogy: Microbial Interactions, Anthropogenic Influences, Contaminated Land and Waste Management*, JD Cotter-Howells, LS Campbell, E Valsami-Jones, and M Batchelder, eds. The Mineralogical Society of Great Britain and Ireland, 2000, pp. 207–226.
42. B Muller, L Granina, T Schaller, A Ulrich, and B Wehrli. *Environ. Sci. Technol.* 36: 411–420, 2002.
43. MJ La Force, CM Hansel, and S Fendorf. *Soil Sci. Soc. Am. J.* 66: 1377–1389, 2002.
44. M Taillefert, BJ MacGregor, JF Gaillard, CP Lienemann, D Perret, and DA Stahl. *Environ. Sci. Technol.* 36: 468–476, 2002.
45. Y Ouyang, D Shinde, RS Mansell, and W Harris. *Crit. Rev. Environ. Sci. Technol.* 26: 189–204, 1996.
46. AB Kersting, DW Efurd, DL Finnegan, DJ Rokop, DK Smith, and JL Thompson. *Nature* 397: 56–59, 1999.
47. BD Honeyman. *Nature* 397: 23–24, 1999.
48. GE Brown, VE Henrich, WH Casey, DL Clark, CM Eggleston, AR Felmy, DW Goodman, M Gratzel, G Maciel, MI McCarthy, KH Nealson, DA Sverjensky, MF Toney, and JM Zachara. *Chem. Rev.* 99: 77–174, 1999.
49. WG Sunda, SA Huntsman, and GR Harvey. Photoreduction of manganese oxides in seawater and its geochemical and biological implications. *Nature* 301: 234–236, 1983.
50. B Sulzberger. In: *Aquatic Chemical Kinetics*, W Stumm, ed. New York: Wiley, 1990, pp. 401–429.
51. B Sulzbergerand H Laubscher. In: *Aquatic Chemistry: Interfacial and Interspecies Processes*, CP Huang, CR O'melia, and JJ Morgan, eds. Washington, DC: American Chemical Society, 1995, pp. 1–32.
52. M Silver, HL Ehrlich, and KC Ivarson. In: *Interactions of Soil Minerals with Natural Organics and Microbes*, PM Huang and M Schnitzer, eds. Wisconsin: SSSA, 1986, pp. 497–519.
53. KW Mandernack, J Post, and BM Tebo. *Geochim. Cosmochim. Acta* 59: 4393–4408, 1995.
54. BM Tebo, WC Ghiorse, LG van Waasbergen, PL Siering, and R Caspi. In: *Geomicrobiology: Interactions between Microbes and Minerals*, JF Banfield and KH Nealson, eds. Washington, DC: Mineralogical Society of America, 1997, pp. 225–266.
55. FH Chapelle. *Ground-Water Microbiology and Geochemistry*. New York: John Wiley, 2001, 477pp.
56. JL Schnoor. In: *Aquatic Chemical Kinetics*, W Stumm, ed. New York: Wiley, 1990, pp. 475–504.
57. WH Casey, JF Banfield, HR Westrich, and L McLaughlin. *Chem. Geol.* 105: 1–15, 1993.
58. AE Kehew. *Applied Chemical Hydrogeology*. Upper Saddle River, NJ: Prentice-Hall, 2001, 368pp.
59. RT Cygan and JD Kubicki, eds. *Molecular Modeling Theory: Applications in the Geosciences*. Washington, DC: Mineralogical Society of America, 2001, 531pp.
60. DG Brookins. *Eh-pH Diagrams for Geochemistry*. Berlin: Springer-Verlag, 1988, 176pp.
61. RM Garrels and CL Christ. *Solutions, Minerals, and Equilibria*. New York: Harper & Row, 1965, 450pp.
62. D Langmuir. *Aqueous Environmental Geochemistry*. Upper Saddle River, NJ: Prentice-Hall, 1997, 600pp.
63. CF Baes and RE Mesmer. *The Hydrolysis of Cations*. New York: Wiley, 1976, 490pp.
64. JD Hem and CJ Lind. *Geochim. Cosmochim. Acta* 47: 2037–2046, 1983.
65. PC Singer and W Stumm. *Science* 167: 1121, 1970.
66. FJ Millero. *Physical Chemistry of Natural Waters*. New York: Wiley, 2001, 654pp.
67. J Livage, M Henry, and C Sanchez. *Prog. Solid. St. Chem.* 18: 259–341, 1988.
68. M Henry, JP Jolivet, and J Livage. *Struct. Bond.* 77: 153–206, 1992.
69. FJ Millero. *Geochim. Cosmochim. Acta* 49: 547–553, 1985.
70. KM Rosso and JJ Morgan. *Geochim. Cosmochim. Acta* 66: 4223–4233, 2002.
71. W Stumm and GF Lee. *Industrial Eng. Chem.* 53: 143–146, 1961.
72. I Matsui. Lehigh University, thesis, 1973.

73. RW Coughlin and I Matsui. *J. Catal.* 41: 108–123, 1976.
74. SHR Davies. In: *Geochemical Processes at Mineral Surfaces*, JA Davies and KF Hayes, eds. Washington, DC: American Chemical Society, 1986, pp. 487–502.
75. B Wehrli, B Sulzberger, and W Stumm. *Chem. Geol.* 78: 167–179, 1989.
76. B Wehrli, G Friedl, and M Alain. In: *Aquatic Chemistry: Interfacial and Interspecies Processes*, CP Huang, CR O'melia, and JJ Morgan, eds. Washington, DC: American Chemical Society, 1995, pp. 111–134.
77. W Sung and JJ Morgan. *Geochim. Cosmochim. Acta* 45: 2377–2383, 1981.
78. SHR Davies and JJ Morgan. *J. Colloid Inter. Sci.* 129: 63–77, 1989.
79. H Tamura, K Goto, and M Nagayama. *Corrosion Sci.* 16: 197–207, 1976.
80. W Sung and J Morgan. *Environ. Sci. Technol.* 14: 561–568, 1980.
81. JG Hering and W Stumm. *Rev. Miner.* 23: 427–465, 1990.
82. W StummIn: CP Huang, CR O'melia, and JJ Morgan, eds. *Aquatic Chemistry: Interfacial and Interspecies Processes*. Washington, DC: American Chemical Society, 1995, pp. 1–32.
83. W Stumm. *Colloid Surface A: Phys. Eng.* 120: 143–166, 1997.
84. SD Samson and CM Eggleston. *Geochim. Cosmochim. Acta* 64: 3675–3683, 2000.
85. SD Samson and CM Eggleston. *Environ. Sci. Technol.* 32: 2871–2875, 1998.
86. B Zinder, G Furrer, and W Stumm. *Geochim. Cosmochim. Acta* 50: 1861–1869, 1986.
87. B Lutz and H Wendt. *Berich Bunsen Gesell* 74: 372, 1970.
88. B Wehrli, E Wiel, and G Furrer. *Aquat. Sci.* 52: 3–31, 1990.
89. M Birus, N Kujundzic, and M Pribanic. *Prog. React. Kinet.* 18: 171–271, 1993.
90. G Lente and I Fabian. *Inorg. Chem.* 38: 603, 1999.
91. G Furrer and W Stumm. *Geochim. Cosmochim. Acta* 50: 1847–1860, 1986.
92. AT Stone, A Torrents, J Smolen, D Vasudevan, and J Hadley. *Environ. Sci. Technol.* 27: 895, 1993.
93. OW Duckworth and ST Martin. *Geochim. Cosmochim. Acta* 65: 4289–4301, 2001.
94. AT Stone and JJ Morgan. *Environ. Sci. Technol.* 18: 617–624, 1984.
95. AT Stone. *Environ. Sci. Technol.* 21: 979–988, 1987.
96. AT Stone. *Geochim. Cosmochim. Acta* 51: 919–925, 1987.
97. AT Stone, KL Godteredsen, and B Deng. In: *Chemistry of Aquatic Systems: Local and Global Perspectives*, G Bidoglio and W Stumm, eds. New York, Springer Publishing Co., 1994, pp. 337–374.
98. T Schmidt. *New Phytol.* 141: 1–26, 1999.
99. YS Jun and ST Martin. *Environ. Sci. Technol.* 37: 2363–2370, 2003.
100. D Suter, C Siffert, B Sulzberger, and W Stumm. *Naturwissenschaften* 75: 571–573, 1988.
101. S Banwart, S Davies, and W Stumm. *Colloid Surf.* 39: 303–309, 1989.
102. SF Cheah, SM Kraemer, J Cervini-Silva, and G Sposito. *Chem. Geol.* 198: 63–75, 2003.
103. MF Hochella. *Rev. Miner.* 23: 87–132, 1990.
104. DL Sparks and TJ Grundl, eds. *Mineral-Water Interfacial Reactions: Kinetics and Mechanisms*. Washington, DC: American Chemical Society, 1998, 438pp.
105. PA O'Day. *Rev. Geophys.* 37: 249–274, 1999.
106. GE Brown. *Science* 294: 67, 2001.
107. MI Tejedor-Tejedor and MA Anderson. *Langmuir* 2: 203–210, 1986.
108. MI Tejedor-Tejedor, E Yost, and MA Anderson. *Langmuir* 6: 979, 1990.
109. SJ Hug and B Sulzberger. *Langmuir* 10: 3587–3597, 1994.
110. ST Martin, JM Kesselman, DS Park, NS Lewis, and MR Hoffmann. *Environ. Sci. Technol.* 30: 2535–2542, 1996.
111. SJ Hug. *J. Colloid Interface Sci.* 188: 415–422, 1997.
112. WK Burton, N Cabrera, and FC Frank. *Philos. Trans. R. Soc. London. Ser. A* 243: 299–358, 1951.
113. C Eggleston, S Higgins, and P Maurice. *Environ. Sci. Technol.* 32: 456A–459A, 1998.
114. IV Markov. *Crystal Growth for Beginners*. Singapore: World Scientific, 1995, 422pp.
115. AJ Gratz, S Manne, and PK Hansma. *Science* 251: 1343–1346, 1991.
116. PE Hillner, AJ Gratz, S Manne, and PK Hansma. *Geology* 20: 359–362, 1992.
117. VJ Morris, AR Kirby, and AP Gunning. *Atomic Force Microscopy for Biologists*. London: Imperial College Press, 1999, 332pp.
118. OW Duckworth and ST Martin. *Geochim. Cosmochim. Acta* 67: 1787–1801, 2003.

119. PM Dove and FM Platt. *Chem. Geol.* 127: 331–338, 1996.
120. JL Junta and MF Hochella Jr. *Geochim. Cosmochim. Acta* 58: 4985–4999, 1994.
121. J Junta, MF Hochella, Jr., and D Rimstidt. *Geochim. Cosmochim. Acta* 61: 149–159, 1997.
122. EE Koch, ed. *Handbook of Synchrotron Radiation.* Amsterdam: North-Holland, Elsevier 1983, pp. 560
123. DC Koningsberger and R Prins, eds. *X-ray Absorption: Principles, Applications, Techniques of EXAFS, SEXAFS, and XANES.* New York: Wiley, 1987, 673pp.
124. KF Hayes, AL Roe, GE Brown, KO Hodgson, JO Leckie, and GA Parks. *Science* 238: 783–786, 1987.
125. GE Brown, Jr. and GA Parks. *Int. Geol. Rev.* 43: 963–1073, 2001.
126. PA Fenter, ML Rivers, NC Sturchio, and SR Sutton, eds. *Applications of Synchrotron Radiation in Low-Temperature Geochemistry and Environmental Science.* Washington, DC: Mineralogical Society of America, 2002, 579pp.
127. BA Manning, SE Fendorf, B Bostick, and DL Suarez. *Environ. Sci. Technol.* 36: 976–981, 2002.
128. AL Foster, GE Brown, and GA Parks. *Geochim. Cosmochim. Acta* 67: 1937–1953, 2003.
129. GA Waychunas, BA Rea, CC Fuller, and JA Davis. *Geochim. Cosmochim. Acta* 57: 2251–2269, 1993.
130. GA Waychunas, JA Davis, and CC Fuller. *Geochim. Cosmochim. Acta* 59: 3655–3661, 1995.
131. S Fendorf, MJ Eick, P Grossl, and DL Sparks. *Environ. Sci. Technol.* 31: 315–320, 1997.
132. B Simon. *J. Cryst. Growth* 52: 789–794, 1981.
133. E Busenberg and LN Plummer. *U.S. Geol. Surv. Bull.* 1578, 1986.
134. HU Sverdrup. *The Kinetics of Base Cation Release due to Chemical Weathering.* Lund: Lund University Press, 1990, 245pp.
135. RA Wogelius and JV Walther. *Geochim. Cosmochim. Acta* 55: 943–954, 1991.
136. JI Drever. *Geochim. Cosmochim. Acta* 58: 2325–2332, 1994.
137. JV Walther. *Am. J. Sci.* 296: 693–728, 1996.
138. AA Jeschke, K Vosbeck, and W Dreybrodt. *Geochim. Cosmochim. Acta* 65: 27–34, 2001.
139. OW Duckworth and ST Martin. *Geochim. Cosmochim. Acta*, vol. 68, 2004, pp. 607–621.

4 Applications of Nonlinear Optical Techniques for Studying Heterogeneous Systems Relevant in the Natural Environment

Andrea B. Voges, Hind A. Al-Abadleh, and Franz M. Geiger
Department of Chemistry, Northwestern University

CONTENTS

4.1 INTRODUCTION

Surfaces and interfaces are ubiquitous in the environment. They range from liquid or solid atmospheric particulate matter commonly found in both the lower and upper atmosphere to mineral oxides buried under an aqueous phase in soil environments. Recently, naturally occurring systems involving such interfaces have received much attention [1–8] as they can: (1) alter the chemical composition of the atmosphere, soils, and oceans via chemical transformation reactions; (2) change the chemical behavior of condensed phase surfaces; (3) change the physical behavior of the condensed phase surfaces on which the heterogeneous processes occur; and (4) change the optical properties of aerosols and thereby influence the radiative balance of the atmosphere. Through these four roles, surfaces and interfaces can have profound implications for chemical transport, reactivity, and energy budgets in soil and atmospheric environments.

Traditionally, laboratory investigations of surfaces and interfaces have been restricted to the low total pressure environments that are necessary for operating the instrumentation commonly used for surface studies. Naturally occurring heterogeneous processes, however, take place either in the presence of a gas phase that is characterized by pressures between approximately 760 Torr (atmospheric boundary layer) and 1–10 Torr (stratosphere), or under a condensed phase such as water. A number of surface science techniques that can operate under these conditions have been applied to studying interfacial phenomena found in nature, and the reader is referred to a recent review by Brown et al. [8] and additional reviews in the November 2003 issue of *Chemical Reviews* [9]. In this chapter, the contributions that surface nonlinear optical (NLO) techniques can make in the field of environmental surface science are highlighted.

NLO techniques are based on the frequency conversion of visible or infrared (IR) light in a noncentrosymmetric medium, such as a surface or interface [10–12]. The vast majority of environmentally relevant liquid and solid bulk materials above and below any interface are centrosymmetric insulators and do not give rise to NLO signals. The only important requirement for generating NLO signals from heterogeneous systems is optical transparency of the bulk phases with respect to the optical frequencies that are applied and generated. Thus, NLO approaches have provided researchers with the ability to investigate a wide range of heterogeneous processes, including processes occurring at buried interfaces such as liquid–liquid [13–15] or liquid–solid systems [16–22] as well as other surfaces and interfaces that are specifically important in environmental chemistry and physics [23–30].

The applicability of NLO techniques for studying various interfacial phenomena has been reviewed by several authors [17, 31–33]. This chapter will specifically discuss the applicability of NLO techniques for studying interfacial phenomena found in nature. In Section 4.1, the operational principle behind NLO techniques is described, including some key experimental considerations. Section 4.2 reviews the literature of NLO measurements that focus on gas–liquid interfaces and is followed in Section 4.3 by a discussion on how buried interfaces can be studied by NLO techniques. Gas–solid interfaces are reviewed in Section 4.4, and Section 4.5 will focus on a select group of special topics. Section 4.6 will present an outlook into the future of NLO techniques in the field of environmental sciences.

4.1.1 Surface Studies in the UV–Vis Region: Second Harmonic Generation

In second harmonic generation (SHG), two photons of the same fundamental frequency are concertedly coupled into one photon at the doubled frequency [10–12]. This NLO effect becomes important when high-intensity light fields, produced for example by lasers, are present in noncentrosymmetric media, including surfaces and interfaces [10, 12, 17, 34–36]. For instance, if an 800 nm pulsed laser beam is directed with the appropriate polarization onto an aqueous–solid interface at the appropriate angle, a 400 nm SHG field will propagate collinearly with the reflected 800 nm probe light from the water–solid interface (see Figure 4.1a). In general, the SHG field is generated within the region that is characterized by large changes in dielectric constant as one crosses the interface along the surface normal. The width of the interfacial region that is probed by NLO techniques depends on the nature of the bulk media above and below the interface, as well as the size, orientation, and concentration profile of the interfacial species, and can be as thin as a few solvent molecules, as seen in the case of the water–cyclohexane interface [19, 37, 38].

4.1.2 SHG in the Absence of Adsorbates

For incident probe light of a given polarization, wavelength, and pulse energy, the second harmonic intensity, I_{SHG}, is determined by the second-order nonlinear susceptibility of the interface, $\chi^{(2)}_{int}$. Under the electric dipole approximation, [11] the nonlinear bulk susceptibility of insulating centrosymmetric media such as bulk water, most environmentally relevant mineral oxides and air, is zero. Thus, in this approximation, SHG from centrosymmetric media is not

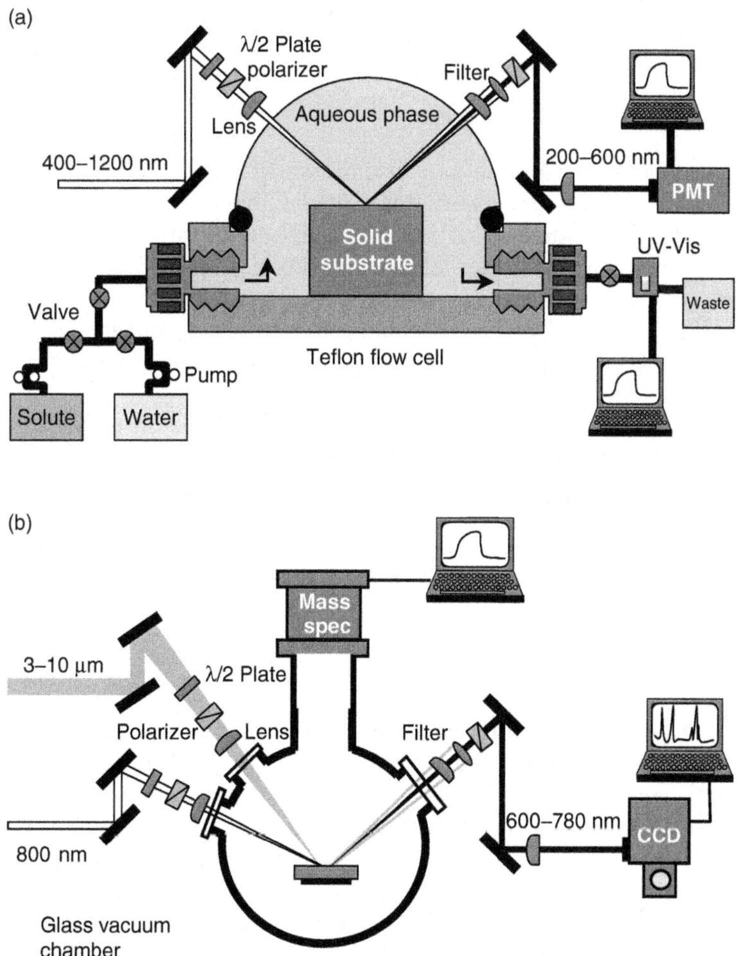

FIGURE 4.1 Two possible experimental geometries for studying environmentally important surfaces using NLO techniques: (a) SHG setup for probing a liquid–solid interface, and (b) SFG setup for probing a gas–solid interface.

allowed. It should be noted, however, that the electric dipole approximation can break down in metals, where quadrupolar contributions can become important [39].

Since the interface between any two bulk media is inherently noncentrosymmetric, $\chi^{(2)}_{int}$ is nonzero in this region and SHG is allowed. The nonlinear susceptibility $\chi^{(2)}$ is a third-rank tensor that can be viewed similarly to the NMR chemical shift tensor, in that it can be used to extract detailed structural and geometric information about the interface under investigation. The tensor elements within $\chi^{(2)}$ can be probed individually by carrying out polarization-resolved SHG measurements. With such measurements, the structure and symmetry of the interface can be studied. For instance, SHG polarization studies can test whether the arrangement of surface species is isotropic or anisotropic within the interfacial plane. When probing the water–air interface, which is fully isotropic within the surface plane, with light whose polarization is parallel to the plane of incidence (p-in polarization), the SHG response will be polarized mainly along the same direction. This p-in/p-out polarization combination results in a nonzero SHG response. With probe light whose polarization is perpendicular to the plane of incidence (s-in polarization), the SHG response along the same direction is zero. Thus, the s-in/s-out polarization combination yields zero SHG photons. Polarization-resolved NLO experiments make it possible to investigate structural changes at an interface that might occur during, for instance, dissolution or adsorption of foreign species.

4.1.3 Probing Adsorbates with SHG

When probing with the appropriate polarization, small but nonzero surface SHG signals are produced at any interface between bulk media that are transparent at the optical frequencies present at the interface. If the frequency of the probe light is tuned to be in resonance with an electronic transition of a surface species, the SHG signal intensity can be enhanced by electronic resonances of the interfacial species with either the probe or the second harmonic light fields. Resonantly enhanced SHG measurements allow for the investigation of heterogeneous processes with surface selectivity and molecular specificity (vide infra).

In the case of resonantly enhanced SHG, $\chi^{(2)}_{int}$ consists of a nonresonant and a resonant contribution, $\chi^{(2)}_{NR}$ and $\chi^{(2)}_{R}$, respectively. As is the case for the overall nonlinear interfacial susceptibility, $\chi^{(2)}_{NR}$ and $\chi^{(2)}_{R}$ are nonzero due to the fact that a noncentrosymmetric environment is probed. $\chi^{(2)}_{R}$ can be modeled as the product of the number density of molecules adsorbed on a surface, N_{ads}, and the molecular hyperpolarizability, $\alpha^{(2)}$, averaged over all molecular orientations, according to

$$\sqrt{I_{SHG}} \sim \chi^{(2)}_{int} = \chi^{(2)}_{NR} + \chi^{(2)}_{R} = \chi^{(2)}_{NR} + N_{ads}\left\langle \alpha^{(2)} \right\rangle \tag{4.1}$$

Equation (4.1) thus provides the link between the macroscopic interfacial susceptibility and the molecular-level hyperpolarizability. This Equation also illustrates how SHG can be used to perform kinetic studies of heterogeneous processes by monitoring the square root of the SHG intensity as a function of time. After subtracting the nonresonant background contribution from the bare interface (no adsorbates present), the square root of the SHG signal intensity is proportional to the number of adsorbed species at the interface. This procedure is well established for transparent substrates in water or air, as $\chi^{(2)}_{NR}$ is usually much smaller than $\chi^{(2)}_{R}$ [17].

The perturbation expansion for the second-order nonlinear molecular polarizability, $\alpha^{(2)}$, contains terms related to electronic transitions in the molecules and can be described by electric dipole transition terms such as:

$$\overset{\leftrightarrow(2)}{\alpha_{ijk}} = -\frac{4\pi^2 e^3}{h^2}\sum_{b,c}\frac{\langle a|\vec{\mu}_i|b\rangle\cdot\langle b|\vec{\mu}_j|c\rangle\cdot\langle c|\vec{\mu}_k|a\rangle}{(\omega-\omega_{\mathrm{ba}}+\mathrm{i}\Gamma_{\mathrm{ba}})(2\omega-\omega_{\mathrm{ca}}+\mathrm{i}\Gamma_{\mathrm{ca}})} \quad (4.2)$$

Here, Γ_{ba} and Γ_{ca} are damping coefficients for avoiding a singularity in Equation (4.2) and describe the line widths of the electronic transitions in the surface species under investigation, $\vec{\mu}_i$ is the electric dipole transition moment of state i, a, b, and c represent the ground, intermediate, and final states respectively, and the summation is over all excited states [10, 11]. As ω, the frequency of the input light, or 2ω, the second harmonic frequency, approach a natural resonance frequency of the adsorbate, ω_{ba} or ω_{ca}, the nonlinear molecular hyperpolarizability, and thus the SHG efficiency, increases. Experimentally, the electronic resonance is generally restricted to two-photon resonances as photobleaching and breakdown of the interfacial species will eventually occur if adsorbates possess resonances at the fundamental probe light frequency. Therefore, if one wishes to probe the presence of an adsorbate that has a resonance at 400 nm, one would carry out a resonantly enhanced SHG experiment that uses 800 nm probe light. If one wishes to spectroscopically identify the adsorbate, one would tune the probe light to different wavelengths, which would map out the two-photon resonance of the adsorbate with the fundamental light field. Thus, SHG allows for interface-specific spectroscopic studies.

The absolute orientation of the adsorbates and the underlying interface can be determined with SHG phase measurements, in which the constructive and destructive interference of the interfacial SHG field with an SHG field generated using a sample with well-known $\chi^{(2)}$, usually single crystalline quartz, is measured [12, 40]. Similarly to the polarization-resolved nonresonant SHG measurements (vide supra), tilt and azimuthal angles for describing the orientation of surface-bound species can be measured using resonantly enhanced polarization-resolved SHG experiments. As has recently been discussed by Simpson and Rowlen [41, 42], orientation effects on the SHG signals are important to consider for many heterogeneous systems when carrying out adsorption isotherms or kinetic studies.

4.1.4 Surface Studies in the IR Region: Sum Frequency Generation

In contrast to SHG, which is commonly applied to probe NLO responses in the UV–Vis wavelength regime, sum frequency generation (SFG) [43, 44] is applied in the IR frequency region, probing adsorbate vibrational modes that are both IR and Raman active. One usually combines an IR probe pulse with a visible pump pulse at the surface of interest and thus generates the SFG signal, which appears in the visible frequency regime (see Figure 4.1b). The spectroscopic capabilities of SFG are apparent when considering the following expression for the SFG intensity [10]:

$$I_{\mathrm{SFG}} \propto \left|\chi^{(2)}_{\mathrm{NR}} + \sum_{\nu}\chi^{(2)}_{R\nu}\cdot \mathrm{e}^{\mathrm{i}\gamma_\nu}\right|^2 \quad (4.3)$$

In this expression, the resonantly enhanced nonlinear susceptibility consists of contributions from each vibrational mode ν, and the contributions are coupled to one another by their relative phases γ_ν. For each mode ν, the nonlinear susceptibility, $\chi^{(2)}_{R\nu}$, can be expressed by the following equation:

$$\chi^{(2)}_{R\nu} \propto \frac{A_K M_{IJ}}{\omega_\nu - \omega_{\mathrm{IR}} + \mathrm{i}\Gamma_\nu} \quad (4.4)$$

where A_K is the Raman transition probability, M_{IJ} is the IR transition moment, I, J, K refer to the surface coordinate system, ω_ν is the IR frequency of mode ν, ω_{IR} is the frequency of the IR probe light, and Γ_ν is the damping coefficient for mode ν that, like the one employed in Equation (4.2), avoids singularities in Equation (4.3) and can be used to describe the natural linewidth of mode ν. Equation (4.4) contains the SFG selection rules, which result in the fact that only modes that are both IR and Raman active can be observed.

Similarly to SHG, SFG allows for structural, spectroscopic, and kinetic measurements. For structural studies, the individual tensor elements of the nonlinear susceptibility are probed using polarization combinations of the three light fields that are involved in SFG. In general, the polarizations of the three light fields are provided in the order: (1) SFG signal, (2) visible pump, and (3) IR probe light. For example, the SFG polarization combination *ssp* denotes that the s-polarized SFG signal intensity is obtained from an interface that is probed using an s-polarized pump field and a p-polarized IR probe field.

Equation (4.4) is used to spectroscopically identify adsorbates in the IR region. For example, one can monitor the symmetric C-H stretch of a methyl group in a surface-active organic species at an interface by overlapping, in both time and space, a 3.478 μm IR (2875 cm^{-1}) probe pulse with an 800 nm (12,500 cm^{-1}) pump pulse at the interface. If the organic species is present at the surface, and if the IR resonance occurs at 3000 cm^{-1}, a resonantly enhanced SFG signal will be produced whose frequency is the sum of the IR probe and pump light frequencies. The SFG signal thus appears at 15,500 cm^{-1} or 645 nm. At the insulator surfaces that are important in environmental systems, the image dipole formation that is commonly observed in metals [45] does not occur and vibrational modes (including phonon modes) that are oriented parallel or perpendicular to the substrate surface can be detected. Nanosecond and picosecond IR laser pulses have the bandwidths necessary for high-resolution surface IR studies, however, each laser has to be tuned to each individual wave number. While automated and computer-controlled IR laser systems are available, scanning SFG setups are restricted to monitoring no more than a few simultaneously recorded IR features.

The commercial development of femtosecond solid-state laser technology during the 1990s allowed for facile generation of broadband tunable femtosecond laser pulses in the IR region of the electromagnetic spectrum. From Heisenberg's uncertainly principle, the uncertainty (or breadth) in the energy (or IR wavelength) of an ultrashort Fourier transform-limited IR laser pulse increases as the pulse duration is shortened. In order to observe a methyl stretch of a surface-bound organic species, one could use a 100-fsec-long IR pulse centered at 3000 cm^{-1}. This IR pulse has a bandwidth of over a hundred wave numbers and therefore ranges from about 2800 to 3200 cm^{-1}. Such broadband IR pulses can be used to generate broadband sum frequency signals, which for the methyl group range from 637 to 654 nm and are still centered around 645 nm. This approach, called broadband sum frequency generation (BBSFG) [18], allows for the generation of broadband surface IR spectra of up to 1000 cm^{-1} bandwidth depending on IR pulse duration [46, 47] within a single laser shot.

If the SFG intensity of a particular mode has been found to be proportional to surface coverage, kinetic studies can be carried out. These types of measurements are similar to the ones possible in SHG (vide supra). However, kinetic studies based on SFG are expected to be even more sensitive to coverage-dependent structural changes in the adsorbates than kinetic experiments based on SHG. Recording adsorption isotherms with various polarization combinations can circumvent this problem as such measurements can assess how each mode probed in the SFG experiment depends on surface coverage.

4.1.5 Experimental Considerations

The high energy light fields that are necessary for NLO signal generation from interfaces are readily generated using pulsed laser systems. The use of optical parametric conversion of laser frequencies produced by commercially available lasers allows for the generation of tunable laser light in a range extending from wavelengths shorter than 400 nm to those longer than

10 μm while maintaining short pulse durations. This means that one can readily study any surface-active species that has electronic resonances in the UV–Vis as short as 200 nm or vibrational transitions between 4000 and 1000 cm^{-1}. In general, low energy pulses in low picosecond time regime are short enough to avoid interfacial heating [49].

In a typical NLO set-up, laser pulses with a known polarization, frequency, and energy are focused onto the surface of interest and the scattered NLO signal, be it SHG or SFG, is collected using a recollimating lens (see Figure 4.1). While reflection geometries like the ones shown are most common, transmission geometries can also be used [48]. After rejection of the reflected probe or pump light fields with appropriate optical and spatial filters, the NLO signal is sent through a set of optics for polarization analysis, spectrally filtered using a monochromator and sent into a photomultiplier tube (PMT). For low signal intensities, the PMT signal is amplified and detected using a single photon counter. If high pulse energies are available, and if thermal damage to the interface is negligible, the intensities of the nonlinear signals can be quite large and direct signal detection can be carried out using oscilloscopes. A liquid nitrogen cooled charge-coupled device (CCD) detector attached to a spectrograph is commonly used for the detection of BBSFG spectra with a time resolution that is generally limited by the repetition rate of the laser system. The spectral resolution is generally determined by the pulse duration of the pump light field, however, SFG in combination with Fourier transformation of the signal can remove this limitation [50].

Laser damage [49] to the sample is usually assessed using a variety of methods, the most important being the determination of the damage threshold level [39]. This level is determined by measuring the NLO signal intensity from the interface as a function of incident laser power (here, laser power refers to the energy per pulse multiplied by the pulse duration). The SHG signal intensity follows a square dependence on incident energy, so doubling the incident laser power will result in four times more SHG photons in the same time frame. The SFG signal intensity is linearly dependent on both the IR probe and the visible pump power. The correct power dependencies break down at high pulse energies due to the higher-order NLO effects that usually precede laser damage, in which all odd-term contributions, such as third harmonic generation, are allowed in centrosymmetric media. A second indication that one indeed follows a second-order NLO effect is the proper spectral response. For instance, a 120 fsec 800 nm pulse has a bandwidth of 8 nm and the bandwidth of the 400 nm SHG signal is reduced by the square root of two (to about 6 nm) assuming the addition of two Gaussian pulses. Before laser damage occurs, one usually detects a broadening of the NLO bandwidth for high input powers and eventually begins to generate white light, which corresponds to high signal intensities at arbitrary wavelengths. These concerns apply to all NLO studies, and the proper power and spectral dependence of the surface signals should be verified when planning NLO experiments.

4.2 GAS–LIQUID INTERFACES

Characterizing and studying gas–liquid interfaces in the presence and absence of adsorbate species is crucial for understanding the interaction of atmospheric trace gases with environmentally important condensed liquid phases. As the most abundant liquid in our environment is water, the discussion will begin by reviewing how NLO techniques have been applied for studying water molecules at neat air–water interfaces. Due to their polar nature, interfacial water molecules can assume various orientations in the presence of polar or nonpolar species, and they can facilitate heterogeneous reactions that would not occur in their absence. An understanding of heterogeneous reactions that are of environmental importance thus begins with an understanding of interfacial water molecules. Therefore, this section starts with a review on SFG spectroscopy applied to neat water surfaces. Then, the interaction of atmospherically relevant acids, bases, and anhydrides with water, and the adsorption of organic compounds to the aqueous–air interface will be reviewed.

4.2.1 Neat Water Surfaces

Water is ubiquitous and therefore it is not surprising that many environmentally relevant processes can occur at air–water interfaces. Examples include the sequestering of gas-phase species in water, production of volatile species during heterogeneous reactions, phase transitions such as the melting of glaciers, and the chemistry and physics of atmospheric aerosols. In addition, the presence of interfacial water influences a wide variety of heterogeneous processes at aqueous–solid interfaces, including the dissolution of mineral oxides in soils, adsorption, reaction, and desorption processes of surface-active species in soils and the atmosphere, the formation of cloud condensation nuclei, and aerosol particle growth.

Several NLO studies focusing on the structure of interfacial water molecules have been carried out [51–56]. The interfacial structure of neat water molecules in contact with air, organics, and various solid phases such as quartz has been studied using SFG and results are summarized in several reviews [31–33]. By probing the vibrational frequency of the OH stretching modes in surface water molecules, SFG provides information about the structure of the hydrogen bonding network at the interface. Most SFG studies are performed with *ssp* polarization in order to probe surface vibrational modes with dipole components perpendicular to the surface.

The first SFG spectrum of the neat air–water interface was reported by Shen and coworkers as a function of temperature [51]. The SFG spectra show stretching frequencies at 3680, 3400, and 3200 cm^{-1}, suggesting the presence of different types of OH groups with different degrees of hydrogen bonding. The 3680 cm^{-1} frequency was assigned to free or dangling OH groups that are oriented away from the liquid surface. SFG has also helped to illuminate the percentage of these free OH groups above the surface plane. Calculations from SFG spectra of both the air–water and aqueous methanol solutions–air interfaces indicate that about 20% of the water molecules have an OH bond above the surface plane of the bulk water [51, 57]. Polarization-resolved experiments and calculations that are based upon a delta function for characterizing orientation distributions indicate that the free OH bond tilts away from the surface of the bulk liquid at approximately 38° from the surface normal [51]. This is consistent with the fact that geometry dictates that a water molecule, which has its molecular dipole aligned with, and its HOH plane perpendicular to, the interface, would have an OH bond tilted 38° from the surface normal. Experimentally, a 180° phase difference between the free and bonded OH stretches confirms that, on average, the free and bonded OH bonds are directionally opposed [51].

The SFG spectrum of the air–water interface obtained by Raymond et al. [58], is shown in Figure 4.2. The assignment of the free OH peak from the air–water interface is now widely accepted, and there is, in general, remarkably good qualitative agreement among the published SFG spectra of air–water interfaces (see Table 4.1). Based on recent experimental and theoretical advances in this area, it is now possible to analyze the SFG response from air–water interfaces in great detail. The various methods used for interpreting the SFG spectral features that correspond to fully hydrogen-bonded water molecules include the comparison of SFG with Raman spectra, ab initio molecular dynamics calculations on ice and water clusters, and SFG isotopic studies using D_2O and HOD [58–61].

One interpretation of the broad series of bands between 3100 and 3500 cm^{-1} that correspond to the hydrogen-bonded region found in SFG spectra from air–water interfaces states that the lower frequency peaks arise from strongly hydrogen-bonded water molecules, including the ones that possess free OH groups. The higher frequency peak (3400 cm^{-1}) is then due to more weakly bonded water molecules [56]. Another interpretation assigns the lower frequency spectral features around 3150 cm^{-1} to an "ice-like" structural and hydrogen bond

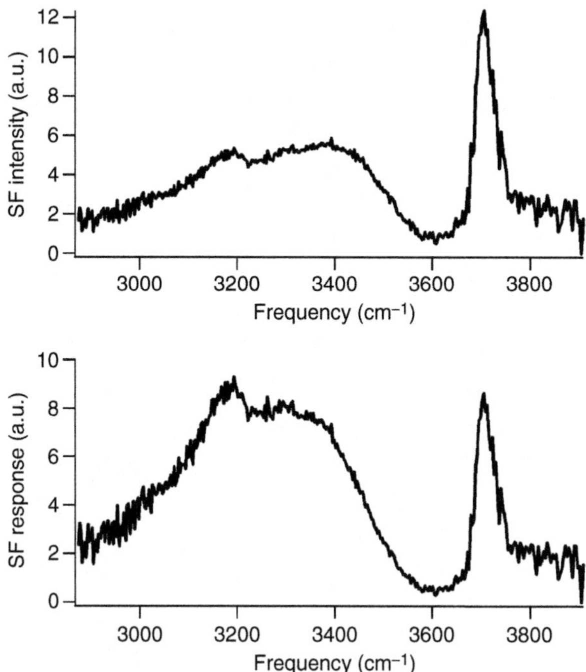

FIGURE 4.2 Scanning SFG spectra of water molecules at the air–H_2O interface (a) normalized to the SFG spectrum of GaAs(1 0 0), (b) after correction. Spectra are recorded in the OH stretching region of water and show features due to stretches assigned to free and hydrogen-bonded OH groups. (From EA Raymond, TL Tarbuck, MG Brown, and GL Richmond. *J. Phys. Chem. B* 107(2): 546–556, 2003, with permission. Copyright 2003 American Chemical Society.)

environment for the interfacial water molecules [51]. This interpretation of the lower frequency spectral features would be consistent with the symmetric stretch of symmetrically bonded water molecules in the lower frequency regime. The higher frequency spectral feature is then attributed to a more disordered, or "liquid-like," structural and hydrogen bond environment for the interfacial water molecules [44]. Table 4.1 presents a summary of the vibrational modes that are found in interfacial water molecules and highlights the sensitivity of these vibrational modes to the interfacial environment and the presence of co-adsorbates.

4.2.2 Surfaces of Aqueous Electrolyte Solutions

Following the previous discussion on molecular structure at the air–water interface, this section will focus on how NLO techniques can be applied to study the molecular structure of water molecules at surfaces of aqueous electrolyte solutions. Aqueous solutions in the environment can contain a large variety of inorganic and organic species at various concentration levels. In general, the amount and chemical identity of the solutes depend on the origin and history of the mobile phase (usually water or air) and on the geographic location and history of the stationary phase (usually soil or frozen water). Physical and chemical processes that occur at aqueous surfaces can be influenced by surface-active solutes, which until recently have been considered to consist mainly of organic species. Recent NLO measurements have shown that anions and cations can have profound impacts on the structure of interfacial water, and so the question arises whether they should also be considered to be surface-active. Electrolytes can have an impact on natural phenomena as

TABLE 4.1
SFG Studies of Water Molecules at Different Interfaces[a]

Interface	Polarization (SFG, VIS, IR)	Solute (m*M*; Mole Fraction, $\bar{x}$)	Ionic Strength (Electrolyte, m*M*)	pH	ν(OH) (cm^{-1})	Assignments[b]	Refs.
H_2O–CaF_2	*ssp*	n.a.	n.r.	2.9, 5.1	3160, 3450	OH in sym. environment, highly ordered water-like, low H bonding order	[267]
	ssp	n.a.	n.r.	6.4	Zero-signal	n.a.	[267]
	ssp	n.a.	n.r.	9.3–13.7	3160, 3657	OH in sym. environment, highly ordered Ca-OH	[267]
	ssp	[NaF] = 0.96–10 m*M*	n.r.	13.2	3160, 3657	OH in sym. environment, highly ordered Ca-OH	[267]
H_2O–CCl_4	*ssp*	n.a.	n.r.	2.45–7	3160	OH-SS	[112]
					3500	OH-AS	
					3669	Free-OH	
	ssp	n.a.	n.r.	9.87	3160	OH-SS	[112]
					3420	OH-AS	
					3680[c]	Free OH	
	ssp	[SDS] = 1.0 m*M*	n.r.	n.r.	3200	OH in sym. environment, highly ordered	[53]
	ssp	[SDS] = 4×10^{-2}–8.3×10^{-3} m*M*	n.r., 10	n.r.	3200–3250[d]	OH-SS in sym. environment, ice-like	[55, 73]
					3400–3450	OH-SS in asym. environment, water-like	
H_2O–hexane	*ssp*	n.a.	n.r.	n.r.	3460	H-bonded OH	[112]
					3669 (W)	Free OH	
H_2O–quartz	*ssp*	n.a.	NaCl, 10–50	1.5–12.3[c]	3200[e]	Coupled OH-SS in sym. environment	[112]
					3450[e]	OH-SS in asym. environment	
					3600 (VW)	OH-AS in asym. environment	
H_2O–quartz	*ppp*, *pss*, *sps*	n.a.	n.r.	5.6	3450	OH-SS in asym. environment	[112]
					3600	OH-AS in asym. environment	
H_2O–air	*ssp*	n.a.	n.a.	n.r.	3694	Free OH	[58, 59]
					3420	Uncoupled donor OH	
					3310, 3200	OH in sym. environment	
	ssp	n.a.	n.a.	n.r.	3150	OH in sym. environment	[70]
					3450	OH-AS in asym. environment; OH-AS	
					3700	Free-OH	

ssp	n.a.	n.a.	n.r.	3250	OH-SS, ice-like	[72]
				3450	Weakly coupled OH stretches	
				3550 (VW)	Coupled OH-AS; OH stretch of H_2O coordinated to free OH	
				3700	Free OH	
				3760	Unassigned	
ssp	$\bar{x}_{HCl} = 0.01–1$	n.a.	n.r.	3150[f]	S H-bonded H_2O	[69, 70]
				3450[f]	OH-AS in asym. environment; OH-AS	
				3700[f]	Free OH	
ssp	$\bar{x}_{HNO_3} = 0.005–0.4$	n.r.	n.r.	3150[f]	OH-SS in sym. environment	[68]
				3400[f]	OH-AS in asym. environment; OH-AS	
				3700	Free-OH	
ssp	$\bar{x}_{H_2}SO_4 = 0.01–0.9$	n.r.	n.r.	3200[f,g]	OH-SS in sym. environment	[65, 66]
				3450[f,g]	OH in sym. and asym. environment	
				3710[f,g]	Free-OH	
ssp	$\bar{x}_N = 0.01–0.2$; N = HCl, HNO_3, H_2SO_4	n.r.	n.r.	3150[f,g]	OH-SS in sym. environment	[71]
				3400[f,g]	OH in asym. environment; OH-AS in sym. environment	
				3700[f,g]	Free OH	
ppp, pss	$\bar{x}_{NH_3} = 0.3$	n.a.	n.r.	Not distinct	n.a.	[62, 64]
ssp	$\bar{x}_{NH_3} = 0.06–0.3$	n.a.	n.r.	3000–3500	H-bonded H_2O	[62]
				3700[f] (W)	Free OH	
sps	$\bar{x}_{NH_3} = 0.06–0.3$	n.a.	n.r.	3400	H-bonded H_2O	[62, 64]
ssp	$\bar{x}_N = 0.01–0.5$; N = H_2SO_4, Cs_2SO_4, Li_2SO_4, K_2SO_4	n.r.	n.r.	3000–3300[f,g]	OH-SS in sym. environment	[27, 67]
				3300–3600[f,g]	OH in asym. environment; OH-AS in asym. environment	
				3700[f,g]	Free OH	
ssp	$\bar{x}_N = 0.01–0.2$; N = NaCl, $NaNO_3$, $KHSO_4$	n.r.	n.r.	3150[f,g]	OH-SS in sym. environment	[71]
				3400[f,g]	OH in asym. environment; OH-AS in sym. environment	

(*continued*)

TABLE 4.1
SFG Studies of Water Molecules at Different Interfaces[a]—*continued*

Interface	Polarization (SFG, VIS, IR)	Solute (m*M*; Mole Fraction, $\bar{x}$)	Ionic Strength (Electrolyte, m*M*)	pH	ν(OH) (cm^{-1})	Assignments[b]	Refs.
H_2O–CaF_2	*ssp*	n.a.	n.r.	2.9, 5.1	3160, 3450	OH in sym. environment, highly ordered water-like, low H bonding order	[265]
					3700[f,g]	Free OH	
	ssp	$\bar{x}_N = 0.015$; N = NaF, NaCl, NaBr, NaI	n.r.	n.r.	3250[f]	OH-SS, ice-like	[72]
					3450[f]	Weakly coupled OH stretches	
					3700	Free OH	
					3760	Unassigned	
	ssp	$\bar{x}_{glycerol} = 0.05$–$1.0$	n.a.	n.r.	3710[f]	Free OH	[266]
	ssp	[SDS] = 0.050–1.000 m*M*	n.r., 4–4000	n.r.	3200–3250[e,f]	Coupled OH-SS in sym. environment, ice-like	[73]
					3400–3450[e,f]	OH-SS in asym. environment, water-like	
Ice–quartz	*ssp*	n.a.	n.a.	n.a.	3200	Coupled OH-SS in sym. environment	[73]
					3469	OH-SS in asym. environment	
Ice–air	*ssp*	n.a.	n.a.	n.a.	3695	Free OH	[28, 267]
					3150	H-bonded OH	
	ppp	n.a.	n.a.	n.a.	3695	Free OH	[28, 267]
					3150	H-bonded OH	
	sps, sss, pps, spp, psp	n.a.	n.a.	n.a.	VW	n.a.	[199]

[a] All experiments were performed with scanning SFG systems except for those from Ref. [72].
[b] As reported by authors.
[c] Orientation of water molecules flips about 180°.
[d] Intensity varies with temperature.
[e] Intensity varies with pH and ionic strength.
[f] Intensity varies with concentration of solute.
[g] Intensity varies with solute.
n.a., not applicable; n.r., not reported; SDS, sodium dodecyl sulfate; W, weak; VW, very weak; OH-SS, symmetric stretch of H-bonded H_2O; OH-AS, asymmetric stretch of H-bonded H_2O.

diverse as transport across cell membranes and reaction of gas-phase atmospheric molecules [56] as is reflected in the wide variety of aqueous solutions studied using NLO techniques.

Ammonia is the most abundant alkaline compound found in the atmosphere and is involved heavily in neutralizing strong inorganic acids [6]. These reactions play a key role in numerous atmospheric processes, actively regulating the concentration of important gas-phase species such as HNO_3 and HCl [6]. SFG was utilized to acquire the first direct experimental surface vibrational spectrum of aqueous ammonia solutions [62] and provides evidence for the existence of an ammonia–water surface complex first hypothesized in 1928 [63]. The SFG spectrum of an aqueous ammonia solution taken with *ssp* orientation showed two distinct features [62]; a large sharp peak at $3312\,cm^{-1}$ that was assigned to the NH (ν_1) symmetric stretch of ammonia and a weaker peak just above $3200\,cm^{-1}$ that was assigned to an overtone of the ν_4 stretch, the asymmetric angle deformation mode. At low concentrations of NH_3 ($x = 0.14$), the free OH peak of water was completely suppressed. In addition, a decrease in the signal intensity in the broad hydrogen bonding region (3200 to $3400\,cm^{-1}$) indicates interference among the hydrogen bonding interactions. The position and shape of the ν_1 peak indicated the presence of an ammonia–water complex [62]. The nonsymmetric nature of the ν_1 SFG spectral feature was suggested to be due to the rotational structure of the Q-band, which is presented as further evidence of the ammonia–water complex. Furthermore, the transmission IR spectrum and the SFG spectrum are redshifted by the same amount from the gas-phase spectrum of NH_3. SFG studies aimed at determining the molecular orientation of NH_3 on the aqueous surface were also performed [64]. Results indicated an average molecular tilt angle of $25° \leq \theta \leq 38°$ from the surface normal with a twist angle of less than 10°. These SFG studies have been the cornerstone for the recent advances in a more complete understanding of the aqueous ammonia surface.

As is the case with bases, acids are extremely important in the lower and upper atmosphere. For instance, the majority of the condensed phase in the stratosphere consists of sulfuric acid solutions with mole fractions of sulfuric acid between 0.1 and 0.4 [27]. The surfaces of these acid particles, known as stratospheric aerosols, are thought to act as cloud condensation nuclei for polar stratospheric clouds and play a significant role in heterogeneous stratospheric chemistry at midlatitudes [5]. Many aspects of the surfaces of aqueous H_2SO_4 solutions have been studied using SFG.

Shultz and coworkers investigated structural changes at the air–water interface upon addition of salts and inorganic acids [33, 56]. Aqueous solutions of H_2SO_4 [27, 65–67], HNO_3 [68], HCl [69, 70], alkali sulfates [27, 67, 71], NaCl [71], and $NaNO_3$ [71] were investigated using SFG. While briefly discussed below, the reader is also referred to the recent review by Shultz et al. [33]. In general, the results indicate that acids have a stronger effect on the orientation of interfacial water than the corresponding salts. Depending on the acid or salt concentration, associated ion complexes or molecular species can penetrate to the surface monolayer and suppress the free OH peaks. For instance, Baldelli et al. [65] showed that the free OH peak of water at approximately $3710\,cm^{-1}$ can be completely suppressed when probing a solution that contains a 0.4 mole fraction of sulfuric acid. Lower concentrations of H_2SO_4 were shown to result in an enhancement of the spectral features in the hydrogen-bonding region (3100 and $3600\,cm^{-1}$) [65]. These effects were attributed to the uptake of surface water molecules by H_2SO_4 molecules and the formation of H_2SO_4–H_2O complexes with no preferred orientation. In solutions with sulfuric acid mole fractions between 0.01 and 0.9, SFG spectra revealed that interfacial sulfuric acid does not have free OH groups directed into the gas phase [65]. In order to determine the relevance of laboratory studies conducted at 273 K to stratospheric conditions, Schnitzer et al. [66] supercooled H_2SO_4 to 216 K in order to attain the temperature range relevant for stratospheric sulfate aerosols (205 to 240 K) [66]. When the liquid surfaces were compared using SFG, no difference was seen between the spectra at the two temperatures, confirming the applicability of laboratory studies of this interface performed at warmer temperatures.

The effect of alkali metal SO_4^{2-}/HSO_4^- salt solutions on the degree of hydrogen bonding within the interfacial water molecules was also investigated [27, 67]. Although results varied slightly depending upon the specific alkali metals used, trends were consistent with concentration studies of H_2SO_4, where an increase in the cation concentration resulted in increased intensities of the spectral features in the hydrogen bonding region while reducing the free OH peak [27]. From this work, it is evident that the presence of the electric double layer can impact the orientation and structure of surface water molecules; the preferential partitioning of anions over cations establishes an electric double layer at the surface of the solution. This double layer creates a negative potential at the interface that reorients submerged water molecules. The resulting water structure is more ordered and gives rise to the increase in the SFG spectral features near $3200\,cm^{-1}$ [67].

HCl, another environmentally important acid, is known to be a key reactant in several atmospheric chlorine-activation reactions [5]. However, the current understanding of the reaction mechanisms that involve HCl lacks molecular detail regarding the state of adsorbed HCl, particularly whether it is dissociatively or molecularly adsorbed [69]. SFG was used to obtain the surface IR spectrum of neat HCl, i.e., condensed HCl, for which a dominant peak was seen at $2800\,cm^{-1}$ [69]. This peak was assigned to the stretch of surface HCl by comparison with gas-phase Raman and IR spectra and indicates that HCl must be present on the surface in clusters between three and four molecules in size. Molecular HCl was not observed on the surface of HCl aqueous solutions, solutions of HCl in H_2SO_4, or HCl on Pyrex [69, 70]. This contradicts the implications of surface tension measurements that HCl adsorbs on aqueous surfaces in a molecular fashion [70]. SFG spectra obtained from surfaces of aqueous HCl solutions show that the intensity of the spectral features in the hydrogen bonding region is affected by the HCl concentration in solution. At low HCl concentrations, the two lower frequency features assigned to hydrogen-bonded water increase in intensity but do not shift in the frequency domain. This is consistent with a net orienting effect of the electric double layer on the surface water molecules. At higher concentrations, all of the SFG features, including the peak corresponding to free OH groups, are broadened and reduced in intensity. This is noted to resemble the FTIR spectrum of multiple hydrate species, in which the lack of distinct spectral features is attributed to the presence of hydrated protons.

In a recent complimentary study of other electrolyte species, Liu et al. [72] examined the effect of low concentrations (0.015 mole fraction) of sodium halides on the SFG spectra of interfacial water molecules. It was observed that NaF and NaCl did not greatly distort the SFG spectrum of the neat air–water interface. In contrast, NaBr and NaI were reported to alter the spectral features in the OH stretching region, which resulted in increased signal intensities in the hydrogen bonding region of the SFG spectra. These studies suggest that certain anions could preferentially populate liquid water surfaces, which would have profound consequences on the heterogeneous chemistry of deliquesced sea salt particles.

4.2.3 Surface Potential and Surface pK_a

As mentioned in the previous section, charged species at interfaces can set up an interfacial potential whose magnitude is highly relevant to the propensity of solutes or solvent molecules to interact with the interface. As a simple example, one might consider the adsorption of a polar molecule from the gas phase onto an aqueous surface that contains a positively charged surface species. Following Coulomb's law, the polar adsorbate will align itself along the direction of the interfacial potential. In a situation where the adsorbate contains a functional group with a partial negative charge, the interfacial potential will align the adsorbate such that the functional group with the negative partial charge points towards the interface.

Adsorption of such a species is hindered by electrostatics if the interfacial potential is reversed, made possible for instance by changing the bulk pH in the case of surface-active acids, bases, and salts.

Using a charged surfactant, namely sodium dodecyl sulfate (SDS), Gragson and Richmond [73] studied how water molecules at the air–water interfaces can undergo such potential-dependent alignment. The SFG response in the OH stretching region at the air–water interface was monitored while varying the interfacial potential by changing the surface charge density, ionic strength, and temperature. The authors determined that as the potential is increased, a large electrostatic field is produced in the interfacial region by the (charged) surfactant-induced alignment of interfacial water molecules from "water-like" to "ice-like" structures. Examination of the SFG spectra as a function of SDS concentration, and thus as a function of the electrostatic field produced, reveals the formation of more strongly hydrogen-bonded interfacial water molecules. SFG signal intensity enhancements are a result of the alignment of water molecules at the interface, which reaches a maximum at a surface potential of about 260 mV. The decrease in surface potential observed by increasing the ionic strength is manifested by a dramatic intensity decrease in the spectral peak that corresponds to the "ice-like" OH groups. This indicates an overall weakening of the hydrogen bonding network. The same trend has been observed by increasing the temperature at constant ionic strength.

The interfacial potential of heterogeneous systems that involve acid–base equilibria have been studied by Shen and coworkers [74], Eisenthal and coworkers [75–77], and Eggleston and coworkers [78]. For an acid–base reaction such as

$$HA^+ + H_2O \rightleftharpoons A + H_3O^+ \tag{4.5}$$

the acid dissociation constant, K_a, can be extracted from SHG signal recorded as a function of bulk pH. The relative concentration of the acid–base species determines the interfacial charge density, which is then combined with a Guoy–Chapman electric double layer model describing the charged interface [77]. The potential-dependent SHG signal intensity can be described by a $\chi^{(3)}$ process, in which the square root of the SHG signal intensity represents the total SHG electric field, $E_{2\omega}$, that is proportional to the second-order polarization $P^{(2)}$ from the surface, and is expressed as

$$E_{2\omega} \propto P^{(2)} = \chi^{(2)} E_\omega E_\omega + \chi^{(3)} E_\omega E_\omega \Phi(0) \tag{4.6}$$

where E_ω is the applied electric field, $\chi^{(2)}$ is the second-order susceptibility contribution from the interfacial water and the acid–base species, $\chi^{(3)}$ is the third-order susceptibility contribution due to the polarized and aligned water molecules in the double layer, and $\Phi(0)$ is the interfacial potential. The example that demostrates the $\chi^{(3)}$ technique most elegantly is based on the pioneering work of Eisenthal and co-workers on solid-liquid interfaces [79]. SHG intensities from the silica–water interface have been measured as a function of pH (pH 2 to 12) in the presence of NaCl as a supporting electrolyte (see Figure 4.3) [30, 79, 80]. The acid–base equilibria of the surface silanol sites can be represented by

$$> SiOH_2^+ + OH^- \rightleftharpoons > SiOH + H_2O \rightleftharpoons > SiO^- + H_3O^+ \tag{4.7}$$

where >Si indicates silicon atoms bound to bulk oxygen atoms in the quartz lattice. Using a dual-pK_a model, two pK_a values of 4.5 and 8.5 were determined for the above equilibrium. The same results were obtained by Higgins et al. [78] who extended the pH range towards a pH of 1. These results are consistent with a point of zero charge (PZC) for silica–water interfaces between 2 and 3.5 [2]. SHG measurements of PZCs for other oxide interfaces under water, such as corundum–water interfaces, have been reported by Stack et al. [81] in the pH

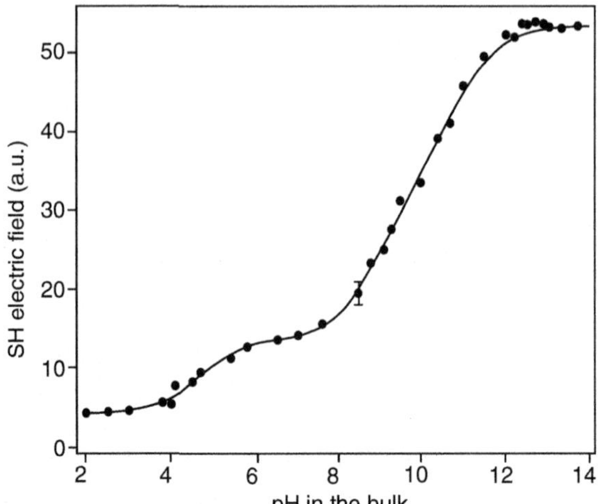

FIGURE 4.3 Variation of the SH field with bulk pH at the silica–water interface. The line through the experimental data is a theoretical fit from which two pK_a values at 4.5 and 8.5 are obtained. See text for details. (From S Ong, X Zhao, and KB Eisenthal. *Chem. Phys. Lett.* 191(3–4): 327–335, 1992. Copyright 1992 with permission from Elsevier.)

range of 5 to 6. The PZC for corundum seems to be lower than those measured using other techniques [82, 83].

In a related study, Shen and coworkers recorded the SFG spectra of water at the interface with quartz as a function of bulk pH [84]. Over the pH range covered, free OH groups are suppressed. This is accredited to the hydrophilic interaction of water with the quartz surface. The peaks at 3400 and 3200 cm^{-1} have been shown to vary in strength with pH, suggesting that the structure of the hydrogen-bonded interfacial water molecules is dependent on pH.

Kim and Cremer [85] investigated structural changes of interfacial water molecules at the water–quartz interface upon layer-by-layer adsorption of charged polymers. Polymers investigated were polydiallyldimethylammonium chloride (PDDA), which is positively charged, and the sodium salt of polystyrene sulfonic (PSS) acid, which is negatively charged. SFG studies focused on measuring the intensity of the OH stretching region of hydrogen-bonded water molecules as a function of pH between 3.8 and 8.0 at a constant ionic strength of 32 m*M*. The SFG intensities in this region are found to decrease with decreasing pH. Upon adsorption of the positively charged PDDA to the water–quartz interface, this trend is fully reversed and the change is attributed to an intensity increase of the oscillators that correspond to strongly hydrogen-bonded interfacial water molecules. Adsorption of the negatively charged PSS results in reversal of the observed pH trend yet again. These observations are consistent with a reversal of the interfacial potential in the presence of positively or negatively charged adsorbed polymers.

Many aqueous species in the natural environment exist in a charged state. Other solutes can exist in a neutral state at one pH and become positively or negatively charged as the pH is changed. Similarly to the experiments that focused on determining the surface pK_a values for the silica–water or the corundum–water interface, NLO techniques are ideal for determining pK_a values of ionizable interfacial species other than >SiOH or >AlOH groups. For instance, pH-dependent SHG measurements of *p*-hexadecyl aniline at the air–water interface show that the pK_a value lies at 3.6 ± 0.2 versus 5.3 ± 0.1 in the bulk [75], and indicate that the acid–base equilibrium at the interface is shifted towards the energetically favorable neutral form [75]. In a related study, the pK_a value of the long chain alkyl amine $CH_3(CH_2)_{21}NH_3^+$ has been measured at the air–water interface as a function of surface monolayer density and bulk electrolyte concentration [77]. The measured surface potential was again modeled using the Gouy–Chapman model. The model yields pK_a values ranging from 8.2 to 10.2 as the surface area varies from 22 to 100 Å^2/molecule. After accounting for surface inhomogeneities, a single pK_a value of 10.1 $\pm$ 0.2 was obtained for all alkyl amine surface coverages. Furthermore, the pK_a values were not found to depend on the bulk electrolyte concentration, which was varied from 1 to 100 mM.

4.2.4 Organic Species at Aqueous Surfaces

Discussion of aqueous interfaces should not be considered complete without the inclusion of recent NLO studies that involve organic species at the air–water interface. Surface-bound organic species are important components of heterogeneous processes in the environment. Organic molecules that possess double bonds on one end and a hydrophilic group on the other are commonly found in the lower atmosphere as partially oxidized organic species originating from biogenic emissions [86–88]. Partially oxidized organic molecules can adsorb onto the surfaces of liquid or solid atmospheric particulate matter via their polar end and can thereby render the particle surface hydrophobic. Surface processing of the nonpolar double bond by atmospheric oxidants such as O_3 or OH can then result in the formation of polar functional groups where the nonpolar double bonds were previously located. The freshly formed polar aerosol particle surface can take up water and form a larger liquid aerosol particle [89, 90].

Organic adlayers on the surfaces of soil particles exposed to air are equally important, as many organic molecules contain polar functional groups that allow the organic species to be surface-active [91]. Trace gases could interact with surface-bound organic species in quite a different manner than that expected from the neat metal oxide–air interface. Furthermore, the adsorption of atmospheric water vapor on organically coated mineral oxides can be different from the adsorption of atmospheric water vapor on uncoated mineral oxides [89]. Thus, understanding organic adlayers and the impact they have on the kinetics and thermodynamics of surface processes that involve pollutants and trace species can allow for an improved description of real-world chemistry.

Organic adlayers on liquid water have also been studied from many different perspectives using NLO techniques. Many studies have sought to assess the orientation of various molecules adsorbed to the air–water interface [91–97]. Other studies carried out on organic species at air–water interfaces have focused on measurements of adsorption kinetics and thermodynamics [98–100], static and dynamic solvation (vide infra) [101, 102], and the time-dependence of adlayer structure [103].

Although adsorption is normally considered to be the transfer of a gas phase molecule to a state in which it is bound to either a solid or a liquid, the term can be used to describe the transfer of molecules from a bulk solution to the solution–air interface. Adsorption from solution was used to study the adsorption of dimethyl sulfoxide (DMSO) to liquid–air interfaces [104]. DMSO is known to be an intermediate in the atmospheric oxidation of dimethyl sulfide (DMS) to methanesulfonic acid (MSA) and ultimately H_2SO_4, however, the understanding of the overall reaction pathway is only partially complete [5]. A more thorough description of the air–aqueous DMSO interface would provide an important step towards a comprehensive picture of these reactions. Resonantly enhanced SHG measurements on DMSO at the air–water interface by Karpovich and Ray [105] showed that surface concentrations of DMSO increase with increasing mole fraction of DMSO, x_{DMSO}, in solution and saturate when $x_{DMSO} = 0.065$ as with a classic monolayer-limited adsorption process. Applying the Langmuir equation to the recorded isotherm, the Gibbs free energy of adsorption, ΔG_{ads}, was calculated to be around -12 kJ/mol at 293 K. As these experiments were performed at equilibrium, the partitioning of DMSO between the gas phase, the liquid phase, and the liquid–vapor interface was assessed by using the solvation and adsorption energies exclusively. The DMSO surface residence time was derived from the evaporation rate constant, which was obtained as follows [104]: the solvation free energy, ΔG_{solv}, is understood as the energy required to transfer one mole of solute into water with no change in volume and was calculated according to $\Delta G_{solv} = RT \ln(K_H RT)$, where K_H is the Henry's law solubility constant and has a value of 1.052×10^5 mol/(kg atm). This yielded a ΔG_{solv} of −36 kJ/mol for DMSO in water. The free energy barrier for solvation, $\Delta G^{\ddagger}$, was calculated according to $\Delta G^{\ddagger} = RT \ln(\alpha/1 - \alpha)$, where α is the mass accommodation coefficient of DMSO by water ($\alpha = 0.05$) as determined by De Bruyn et al. [105], and was found to be 7.4 kJ/mol. Addition of the solvation and adsorption energies lead to a value for the free energy of evaporation, ΔG_{evap}, of −48 kJ/mol. Transition state theory was then used to estimate the evaporation rate constant, k_{evap}, given by the relation $k_{evap} = (k_b T/h) \exp(\Delta G_{evap}/RT)$. The residence time of DMSO at the liquid–vapor interface was estimated to be tens of microseconds by taking the inverse of k_{evap}.

In addition to the resonantly enhanced SHG measurements on DMSO by Karpovich and Ray [104], SFG experiments by Allen et al. [26] on DMSO adsorption from solution to the air–water interface identified DMSO spectroscopically by the symmetric methyl stretch mode of the surface-bound species, which was reported to have a 20 cm^{-1} bandwidth. Using chemical activities as opposed to the mole fractions used in the SHG work by Karpovich and Ray [104], Allen et al. [26] obtained a value of −15 kJ/mol for ΔG_{ads} from adsorption isotherm measurements. This value was compared to a ΔG_{ads} of -19.8 ± 0.4 kJ/mol obtained by applying the Langmuir adsorption model to surface tension measurements of DMSO at low activities (<0.006). The SFG experiments also established a molecular orientation for surface-bound DMSO of 55° from surface normal [26]. Additionally, SFG spectra showed a blueshift of the methyl symmetric stretch as a function of decreasing DMSO concentration. This blueshift was reported to be consistent with a decrease in the electronic interactions between neighboring sulfur and methyl groups as the surface population of DMSO molecules decreases.

In related experiments by Goh and coworkers with inorganic compounds, nonequilibrium SO_2 adsorption to the air–aqueous phase interface from solution was studied using SHG [100]. By comparing the resonantly enhanced SHG data to gas phase absorption peaks in the UV–Vis region, the surface species was identified to be SO_2. Adsorption isotherms recorded using resonantly enhanced SHG resulted in a ΔG°_{ads} from solution of approximately −14 kJ/mol at 298 K and a ΔG°_{ads} from the gas of approximately −23 kJ/mol; both free energies are acknowledged to be upper limits for the liquid-to-surface and the gas-to-surface adsorption free energies, as the experiments were not performed under equilibrium conditions with respect to SO_2.

Adsorption from solution has been studied in a variety of other ways as well. Rasing et al. [98] monitored the adsorption from a saturated solution of sodium-dodecylnapthalene-sulfonate (SDNS) in 2% aqueous NaCl. The surface concentration of SDNS was measured as a function of time using SHG. It was determined that SDNS adsorption occurred in two different time regimes, the first resulting from bulk diffusion control and the second from statistical control of surfactant molecule adsorption, i.e., the sticking probability. In a different set of experiments, Messmer and coworkers used SFG to elucidate the details of the competitive adsorption of 2,4,7,9-tetramethyl-5-decyn-4,7-diol and linear alkane sodium salts [99]. The experimental data were compared to various models for mixed surfactant behavior that included kinetic, steric–kinetic, and electrostatic effects. It was determined that such models could provide rough estimates of the actual competitive adsorption processes observed.

A study performed of palmitic acid ($C_{15}H_{31}COOH$) spread on the air–water interface [103] demonstrates the ability of NLO techniques to access diffusion processes occurring at interfaces. Palmitic acid was dissolved in hexane and spread onto a 10 m*M* *p*-nitrophenol (PNP) aqueous solution. PNP was used as a chromophore because previous studies had revealed that its surface population, and hence the resulting signal intensity, is dependent on the surface coverage of the fatty acid [106]. The SHG signals showed fluctuations in the signal intensities that could be statistically differentiated from noise. These signal fluctuations were suggested to originate from islands of high-density, liquid-like structures of the palmitic acid adlayer at the interface. Periods of low signal intensity were comparable to the ones obtained from the bare air–aqueous PNP interface and were proposed to be due to a low-density, gas-like structure of the palmitic acid adlayer at the interface. As the concentration of the palmitic acid increased, a higher SHG signal intensity was obtained, indicating an increase of liquid-like islands drifting along the interface. Assuming that only one liquid-like cluster existed in the sampled region at any particular time, a diffusion constant of approximately 10^{-8} cm^2/s for palmitic acid on the air–water interface was calculated [103]. This experiment highlights the ability of NLO techniques to assess precise information about the behavior of molecular or ionic species at interfaces.

The applicability of SHG for measuring adsorption isotherms, for extracting kinetic data and for obtaining orientational information from polarization-resolved SHG measurements was discussed in the introduction. Yet, the adsorption of molecules on surfaces can be complicated by the possibility that molecular orientation of adsorbates is coverage dependent [42]. Simpson and Rowlen [41, 42] carried out a comprehensive theoretical treatment and experimental investigation for extracting the extent to which molecular orientation may change as a function of surface coverage. This method is also known as orientation-insensitive SHG. The key element to this technique is the use of apparent average tilt angle of the molecular orientation axis with respect to the surface normal (i.e., orientation angle) as a guide for determining the polarization rotation angle of the fundamental beam at which the SHG signal is minimally dependent on orientation. This also has been tested against the molecular second-order nonlinear tensor element dominant for a particular system. For example, theoretical simulations have shown that for an apparent orientation angle between 0° and 50°, and a molecular hyperpolarizability that is oriented along the dipole direction of the molecule, a polarization rotation angle of 63° in a total internal reflection geometry results in minimal dependence of the SHG signal on molecular orientation.

Molecular orientation at liquid–air interfaces can also be studied using SFG. For instance, Ma and Allen [107] studied the effect of water molecules on the orientation of methanol molecules at the methanol-air interface. Surface spectra were acquired with BBSFG and the bulk methanol solution was monitored using Raman and FTIR spectroscopies. The symmetric methyl stretch of methanol was observed to blueshift between the bulk spectrum and the surface spectrum. The authors correlated this shift with an increased tendency of surface methanol groups to accept hydrogen from the surface and subsurface water molecules. Studies focusing on the symmetric methyl stretch were combined with partially deuterated methanol (CH_3OD) experiments to reveal that the surface methanol molecules become less ordered at bulk methanol mole fractions above 0.57.

While the above discussion emphasizes the variety of aqueous interfacial systems, it should be stated that NLO techniques are also readily applied to liquids other than water. For instance, Baldelli [107] applied SFG to study the ionic liquid 1-butyl-3-methylimidazolium *bis*-trifluoromethylsulfonamide at room temperature. Even in the case of benzene and mesitylene, organic liquids with zero and near-zero hyperpolarizabilites, surface-induced dipoles allow for resonant NLO detection of interfacial molecules [109]. The surface studies discussed here only highlight the broad range of applications to which NLO techniques may be applied.

4.3 BURIED AQUEOUS INTERFACES

The past decade has witnessed a rapid growth of NLO investigations on liquid–liquid and solid–liquid interfaces. These interfaces are also commonly called "buried interfaces." Buried aqueous interfaces play major roles in the physics and chemistry of heterogeneous nucleation, which impacts the growth and phase transitions of atmospheric aerosols [7], groundwater chemistry and physics [110], and corrosion and catalysis [22], to name a few examples. In this section, specific examples focusing on interfacial processes such as solvation, reactions, and structure of adsorbates are presented.

While interfaces in the natural environment can be highly complex, the studies reviewed in this work focus on well-defined heterogeneous systems and provide a good starting point for studying surfaces and interfaces that are environmentally relevant. A general strategy for studying natural interfaces is to begin with well-defined systems that can be expanded in chemical and structural complexity. If one wishes to understand how humic substances interact with mineral oxide–water interfaces, one could argue that studies such as the ones reviewed in this section would provide a reasonable, well-controlled first-order approximation for the impact of charged functional groups commonly found in humic substances. Likewise, the interface of seawater with crude oil, which is important in marine oil spills, could be modeled by studying the octane–water interface as discussed here and then expanding the complexity of this interface by gradually introducing ionic species and other organic molecules.

4.3.1 Aqueous–Liquid Interfaces

The structure of water at immiscible aqueous–liquid interfaces is interesting from a fundamental perspective as well as in the context of environmental chemistry. Pollutants commonly found in fuel can partition between the aqueous phase and the fuel phase and thereby enter an aquifer or the marine environment. The molecular-level processes that drive pollutant partitioning at the aqueous-liquid interface are largely unknown and NLO techniques are well suited for studying these problems.

Richmond and coworkers used SFG to study the structure of water at the water–CCl_4 interface and reported the existence of free OH groups at 3669 cm^{-1} as well as hydrogen-bonded water molecules with a broad OH spectral feature centered around 3450 cm^{-1} [31, 111]. This finding is in contrast to the air–water interface, which shows two maxima in the hydrogen bonding region between 3200 and 3400 cm^{-1} (vide supra), and was attributed to weak H_2O–H_2O interactions at the interface with CCl_4. SFG studies based on isotopic dilution, polarization-dependent SFG measurements, and the SFG signal dependence on pH showed that hydrogen bonding at the interface is weak between adjacent water molecules, yet these interactions result in substantial orientation of weakly hydrogen-bonded interfacial water molecules [31, 111].

The same authors reported SFG spectra obtained from CCl_4 in contact with an aqueous phase containing different millimolar concentrations of SDS at a constant ionic strength [31, 53–55]. The SFG intensity in the OH stretching region that corresponds to hydrogen-bonded water molecules was found to increase in the presence of the surfactant. Spectral features observed in the hydrogen-bonded "ice-like" region centered around 3200 cm^{-1} indicate that the presence of SDS strengthens the hydrogen bonding interactions among interfacial water molecules. Isotopic dilution experiments from the CCl_4-water interface with 1.0 mM SDS were also carried out with various mixtures of H_2O and D_2O at room temperature [53]. Upon decreasing the mole fraction of H_2O in samples mixed with D_2O, a blueshift of about 120 cm^{-1} in the "ice-like" spectral feature is observed (from 3200 to 3320 cm^{-1}). This shift was attributed to the intermolecular decoupling of the OH oscillators upon dilution with D_2O. In the presence of the anionic surfactant at room temperature, the structure of interfacial water is found to be similar to that of ice and follows a tetrahedral arrangement.

NLO measurements focusing on the potential-dependence of the alignment of interfacial water molecules at the CCl_4-water interface were carried out as well [73]. These measurements were performed in the presence of SDS. The SFG field was found to depend nonlinearly on the surface potential, which was determined from surface tension measurements. This deviation from linearity was also observed for air–water interfaces in the presence of SDS (vide supra). In both cases, the deviation from linearity results from the alignment of the water transition dipole moments at the interface with the polarization vector of the IR probe light [73]. A maximum alignment of interfacial water molecules was observed to occur at an interfacial potential of about 160 mV. In addition, the temperature dependence of the SFG intensity is about three times greater than that observed for the air–water interface, where it was observed to diminish as temperature increases.

In addition to having an impact on interfacial potential, organic molecules commonly found in soil environments can contain functional groups with complexing or chelating capabilities that efficiently sequester inorganic anions and cations. Furthermore, such organic functional groups can facilitate the transport of anions and cations from a hydrophobic soil environment into a hydrophilic soil environment. Nochi et al. [112] carried out resonantly enhanced SHG measurements between 480 and 650 nm to study the spectra and orientation of alkali metal–crown ether complexes at the heptane–water interface at room temperature. The aqueous solution contained micromolar concentrations of [2-hydroxy-5-(4nitrophenylazo) phenyl]-methyl-15-crown-5 (azoprobe 1) and 0.10 M alkali metal chloride at a pH of 11.8. Azoprobe 1-alkali metal complexes in bulk water show a UV–Vis absorption maximum around 494 nm due to the $\pi \rightarrow \pi^*$ transitions, whereas a clear redshift is observed in the SHG spectra at the heptane-water interface, where the λ_{max} was found to be around 541 nm. This redshift was attributed to the negative solvatochromism (vide infra) at the interface between two liquids of different polarities. Compared to the SHG intensity from Li^+ complexes or the free ligand, the formation of the Na^+ and K^+ complexes resulted in a threefold enhancement of the SHG signal intensity. Polarization-resolved SHG spectra were used to gain information on the orientation of these complexes at the interface, and it was found that Na^+ and K^+ complexes lie flatter at the interface, while the Li^+ complexes and the free ligand assume a "lift-up" orientation at the interface. The results of this study show alkali-metal recognition by complexing agents at an immiscible interface and demonstrate the potential of NLO techniques to study complexing and chelating processes involving organic and inorganic species.

4.3.2 Aqueous–Solid Interfaces

Aqueous–solid interfaces are exceedingly common in the environment and can range from mineral oxide–water interfaces in soil environments to ice–water interfaces in lakes and the atmosphere. These interfaces can have profound effects on chemical transport phenomena as well as phase transitions in the environment. For instance, mineral dissolution, the melting of ice, and the binding of nutrients or pollutants to solid phase materials are interfacial processes that involve aqueous–solid interfaces. Of particular interest are not only the fundamental principles underlying the physical and chemical phenomena that can occur at such interfaces, but also the impact that environmentally important co-adsorbed organic and inorganic molecules can have on such interfacial processes. This section begins with a discussion of inorganic solids under aqueous solutions, which will be followed by a review of organic solids under aqueous solutions.

4.3.2.1 Inorganic Solids under Aqueous Solution

The interface of water with inorganic solids is particularly important for soil and atmospheric chemistry [89, 113]. In the atmosphere, liquid aerosol formation and freezing rates are commonly accelerated by mineral oxide particles suspended in the aerosol, which has profound implications for the atmospheric radiative budget [114]. Mineral oxides in soils are commonly in contact with groundwater, which may carry nutrients and pollutants that can undergo heterogeneous chemical transformation reactions at, or simply bind to, mineral oxide–water interfaces [2]. As discussed in Section 4.2.3, silica–water and corundum–water interfaces have been studied in order to obtain surface pK_a and PZC values, as this information provides insight into how a mineral oxide–water interface in soil will interact with adsorbates as a function of pH.

Besides investigating mineral oxide–water interfaces in the absence of inorganic or organic solutes, SHG and SFG can also be used for studying the adsorption of environmentally relevant species, be they ions or molecules, from the aqueous phase. Few studies have been reported where adsorption isotherms of environmentally important species have been measured on water–oxide interfaces using NLO techniques [29, 30, 80]. Recently, however, the adsorption isotherm and kinetics of adsorption/desorption of chromate, an important toxic metal contaminant in groundwater, on fused quartz–water interfaces at room temperature have been investigated using SHG [29, 30]. The SHG spectrum of Cr(vi) at the quartz–water interface has been acquired for concentration levels similar to those found in contaminated soils (see Figure 4.4). Adsorption isotherms (see Figure 4.5) have been measured at various pH values between 4 and 9 in a chromate concentration range from 10^{-6} to $10^{-4}M$. By applying the Langmuir adsorption model, chromate-binding constants near 10^5 M^{-1} were obtained in a pH range between 4 and 9. The chromate adsorption–desorption behavior was found to be reversible with adsorption and desorption rate constants of 3×10^3 s^{-1} M^{-1} and 0.9×10^{-3} s^{-1}, respectively. Furthermore, studies are underway with other metal contaminants on fused quartz surfaces decorated with organic adlayers in order to mimic geosorbent surfaces that contain surface-bound naturally occurring polar organic species.

In a related SHG study, Higgins et al. [80] investigated the adsorption of a dicarboxylic aromatic acid, 5,6-carboxy-X-rhodamine, on crystalline (α-Al_2O_3(0 0 0 1)) and polycrystalline corundum samples at a pH of 5 and 3.5 and at concentrations up to 120 μM. The adsorption process was monitored as a function of solution pH, with the maximum adsorption observed at a pH of 3.5. Although the authors did not report the PZC of the alumina–water interfaces used in their study, the adsorption results are in quantitative agreement with experimental batch studies. Furthermore, the authors used SHG to record the adsorption isotherm of an herbicide, glyphosate, on amorphous silica at a pH of 3.8 and an ionic strength of 0.01 M $NaNO_3$. In these studies, the adsorption isotherms of glyphosate and the dicarboxylic acid appear Langmuirian, but free adsorption energies were not reported.

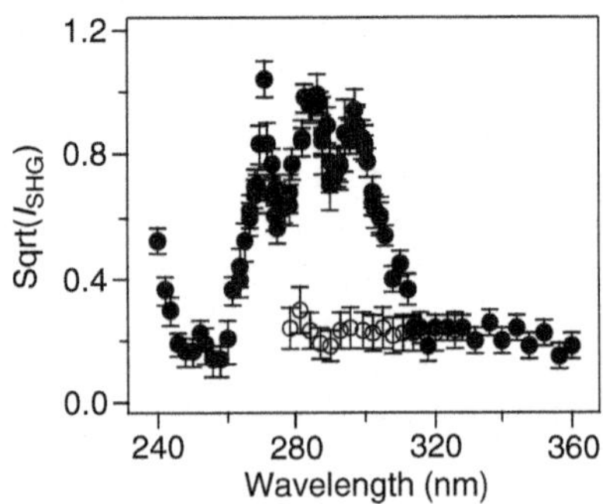

FIGURE 4.4 Normalized surface SHG spectrum of a quartz–water interface at pH 7 (empty circles) and a quartz–water interface exposed to a 5×10^{-5} M CrO_4^{2-} solution at pH 7 (filled circles). The polarization combination is p-in/p-out. (From AL Mifflin, KA Gerth, BM Weiss, and FM Geiger. *J. Phys. Chem. A* 107(32): 6212–6217, 2003, with permission. Copyright 2003 American Chemical Society.)

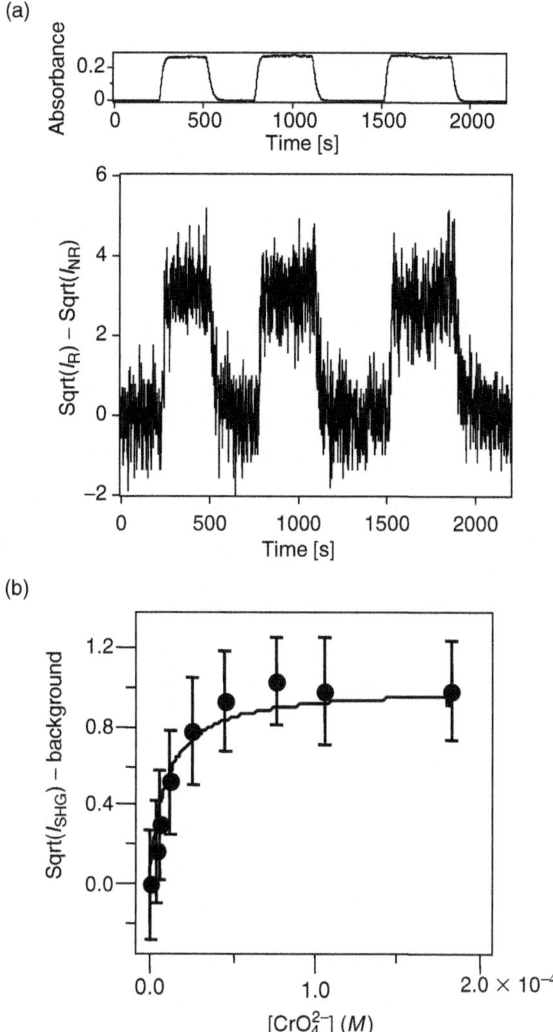

FIGURE 4.5 Cr(vi) adsorption on fused silica–water interface. (a) Top: CrO_4^{2-} bulk absorbance versus time trace at 372 nm (left) for three consecutive chromate adsorption and desorption experiments at pH 7. Bottom: Time dependence of the resonant SHG signal from Cr(vi) binding at the quartz-water interface recorded simultaneously with the absorbance measurements. λ_{SHG} = 290 nm; [CrO_4^{2-}] = 7×10^{-5} *M*. The polarization combination was p-in/p-out. (From AL Mifflin, KA Gerth, BM Weiss, and FM Geiger. *J. Phys. Chem. A* 107(32): 6212–6217, 2003, with permission. Copyright 2003 American Chemical Society.) (b) Adsorption isotherm of Cr(vi) at the quartz–water interface at pH 7 bulk solution. (From AL Mifflin, KA Gerth, and FM Geiger. *J. Phys. Chem. A* 107(45): 9620–9627, 2003, with permission. Copyright 2003 American Chemical Society.)

Surfactant molecules at aqueous–solid interfaces are also of interest in soil chemistry, as they can be used as a proxies for studying charged organic species at mineral oxide–water interfaces. In general, the adsorption of surfactants to aqueous–solid interfaces can alter the surfactant properties as these molecules contain a long hydrophobic hydrocarbon chain on one end and a polar hydrophilic group on the other end. Adsorption of surfactants from an aqueous phase onto the surface of salts or minerals is often driven by the electrostatic interactions between the charged headgroups and the oppositely charged adsorption sites.

Adsorption is initiated by monomer adsorption followed by aggregation of the surfactant molecules through dispersion and van der Waals forces. As a result, surface charge in both the interfacial region and the solid surface changes, which in turn disturbs the intermolecular hydrogen bonding of water molecules [115, 116]. Using SFG spectroscopy at the quartz-liquid interface, Miranda et al. [115] studied the chain conformation of self-assembled monolayers (SAMs) composed of a cationic surfactant, namely dioctadecyl dimethyl ammonium chloride (DOAC). In contrast to what was observed when the liquid phase consisted of nonpolar solvents, the SFG intensity could not be distinguished from the noise level when the liquid phase was water or methanol, another polar species. This observation has been attributed to interactions between water and DOAC, which tend to force the solute chains into a contracted form and result in randomly oriented CH_2 and CH_3 groups in a DOAC monolayer with a large number of defects.

Using SFG, Becraft et al. [116] studied the adsorption of an anionic surfactant, SDS, from aqueous solution onto the surface of fluorite (CaF_2). They recorded both the OH and the CH vibrational modes arising from interfacial water molecules and the hydrophobic tail of SDS, respectively. The SDS bulk concentrations ranged from 6.1×10^{-5} to 3×10^{-3} M. The work was carried out at a pH of 5.1, which is below the PZC of CaF_2–water interface (pH 6) [117], making the interfacial charge positive at this pH. The adsorption of SDS on the CaF_2–water interface was found to change the charge at that interface and, depending on the SDS bulk concentrations, resulted in a reorientation of the interfacial water molecules. The bulk SDS concentration at which the positively charged CaF_2 surface sites were neutralized by the negatively charged SDS headgroups at the interface was defined as the effective PZC for the CaF_2–H_2O–SDS interface. In general, the orientation of interfacial water molecules at the effective PZC should be disordered. The effective PZC was thus measured by determining the SDS bulk concentration that resulted in a decrease of the SFG intensity in the OH region to nearly zero. This concentration was found to be just above 2×10^{-4} M. Below this concentration, SDS adsorbs as monomers and above it, a bilayer of SDS is formed. As discussed by Moore et al. [118], the interpretation of SFG spectra is challenging, and this has been demonstrated in the case of SDS adsorption on the CaF_2–water interface. It was suggested that SFG be used in conjunction with complementary studies. Additionally, attention should be paid to the symmetry of, and the phase relations between, the various stretching and bending modes since they are coupled through Equation (4.3).

As many mineral oxide surfaces in soils are covered, at least partially, with organic adlayers, this section will continue with an SFG study that focuses on the self-assembly of organic adlayers at interfaces. SAMs are often used as model systems for monolayer structures and interfacial reactions as they form spontaneously during the adsorption of a molecule with a headgroup that has a particular affinity to the solid substrate. As reviewed by Ulman [119] and Schreiber [120], thiol SAMs (R-SH) and disulfide SAMs (R-S-S-R) on gold substrates have received great attention. *In situ* SFG spectroscopy has been utilized for studying the self-assembly of *n*-alkanethiol [121] and dioctadecyl disulfide on gold [122]. With the proper choice of molecules that contain headgroups for selectively forming a chemical bond with oxide surfaces, one could model environmentally relevant polar organic molecules adsorbed to aqueous–solid interfaces that are important in nature. Several studies on SAM formation from the liquid phase can serve as a guideline for designing experiments that focus on the heterogeneous chemistry and physics of environmentally important interfaces in the presence of organic and inorganic pollutants or nutrients.

In a recent SFG study by Messmer and coworkers [123], the adsorption from anhydrous hexadecane and CCl_4 as well as conformational changes of *n*-octadecyltrichlorosilane (OTS) SAMs on fused silica surfaces were followed *in situ* (Figure 4.6). The effect of water in the deposition solution on the growth of the OTS film was also investigated. Three peaks in the CH stretching region are of interest and have been used as indicators for the degree of order in

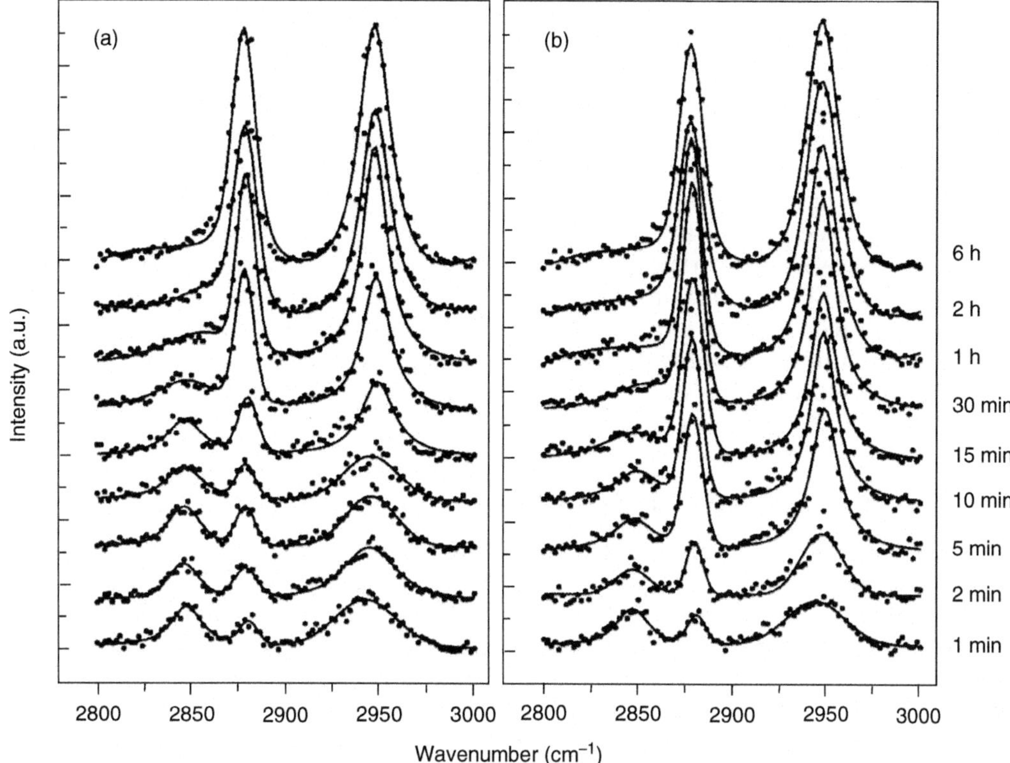

FIGURE 4.6 *ssp*-SFG spectra of OTS adsorption on fused silica surfaces from a solvent mixture of (a) low water content and (b) high water content recorded as a function of time. The spectra show changes in the alkyl chain conformations upon adsorption and formation of the monolayer. See text for details. (From Y Liu, LK Wolf, and MC Messmer. *Langmuir* 17(14): 4329–4335, 2001, with permission. Copyright 2001 American Chemical Society.)

the monolayer. These spectral signatures are around 2850 cm^{-1} (CH_2 symmetric stretch), 2878 cm^{-1} (CH_3 symmetric stretch), and 2946 cm^{-1} (a Fermi resonance between the CH_3 symmetric stretch and the CH_2 asymmetric stretch). The study showed that the conformation of the alkyl chains of OTS monolayers adsorbed to fused silica changes in three stages: initially, an increase in the surface coverage results in disordered islands, a change from a fully gauche disordered to an all-*trans* ordered conformation follows, and finally a much slower adsorption process is observed. Water in the solvent was found to facilitate the deposition of OTS and resulted in more ordered films than those prepared from solvents with low water content. While the scope of this work was to understand a heterogeneous silanization reaction and not an environmentally relevant surface reaction, studies such as the one carried out by Messmer and coworkers could be applied in principle to a variety of geochemically and atmospherically important surface processes involving organic species and can be used to yield detailed information on the adsorption kinetics of polar organic molecules at mineral oxide surfaces.

One of the most important interfaces between water and inorganic solids is the ice–water interface. The ice–water interface possesses a low surface energy, which opens the question of why foreign molecules adsorb to this interface [124]. Physical and chemical processes occurring at the ice–water interface play crucial roles in melting and freezing and can have direct implications for the survival of marine animals at temperatures where seawater is not yet frozen but at which the animal's blood would freeze. For instance, the Arctic winter flounder fish has developed antifreeze proteins (AFPs) that block ice crystal growth in its blood. Motivated by this intriguing capability, Bouchez and Hicks [125] studied the adsorption of an AFP to the water–single crystalline hexagonal ice I interface at 0°C. Using SHG in a total internal reflection geometry, the adsorption of AFP was found to follow its antifreeze activity defined by the temperature difference between the melting and freezing point of an ice crystal in the presence of AFP. For comparison, the authors also studied 2,2′-dihydroxy-1,1′ binaphthol (BN) adsorption to the ice–water interface, with results which are consistent with the replacement of interfacial water molecules upon BN adsorption. The antifreeze activity of the AFP was then argued to involve not only adsorption, but also restricted water access to the ice–water interface in the presence of irreversibly adsorbed AFP, which then suppresses further ice growth. This study demonstrates that NLO techniques allow for sophisticated investigations of complex molecules interacting with highly dynamic interfaces.

4.3.2.2 Organic Solids under Aqueous Solution

Like inorganic solids, solid organic compounds are important in the natural environment. A major component of soils is natural organic matter (NOM) [2, 126]. NOM extracted from soils and sediments can best be described as a complex mixture of macromolecules [127] with fulvic and humic acids as two key constituent. These acids can contain a significant number of COOH, OH, and C=O groups, which make the acids hydrophilic in nature. Laboratory studies focusing on the interaction of trace metal contaminants with mineral surfaces and NOM have been reviewed recently [113, 128, 129]. Modeling studies of NOM at oxide–water interfaces have been performed as well [130–134]. Gustafsson et al. [135] highlighted the importance of including NOM in models simulating metal binding to soils. Trace metals that are of interest include Cr, Co, Ni, As, Zn, and Cd, which tend to form different types of complexes with soil components depending on their oxidation state. Furthermore, dissolved organic matter (DOM) is known to enhance the weathering of minerals (i.e., dissolution and replacement of minerals) in the environment [126–136]. DOM species can also adsorb on soil minerals [130, 137–149] and, via their functional groups, be involved in adsorption and complexation reactions with aqueous metal contaminants or nutrients [128–129, 135]. In general, the adsorption process readily occurs on these interfaces in the pH range of soils (pH 3 to 8) and is usually expressed in terms of electrostatic interactions, complexation, charge neutralization, and ligand-exchange reactions between functional groups of DOM and adsorption sites on minerals. These types of interactions are often affected by temperature, pH, and ionic strength. Main complications addressed in these studies are the complexity of the structure of DOM and NOM in general and the heterogeneity of the molecular environment of the functional groups.

Using NLO techniques, it is possible to study similarly complex molecules at interfaces. One such example is the characterization of polymer surfaces, especially surfaces of charged polymers, under aqueous solutions. Such studies can provide a great deal of information regarding surface-bound NOM in soils. SFG and SHG have been shown to be well suited for studying buried interfaces of biological [15] and industrial [22] importance. Similar successes can be expected for the case of environmentally important organic molecules at the aqueous–mineral oxide interface. In an SFG study by Wang et al. [150], the adsorption of proteins from solution onto a polystyrene film supported on a CaF_2 prism was studied *in situ*. SFG spectra were collected in the carbonyl stretching range (1500–1950 cm^{-1}) for different surface-bound proteins in air, water, and buffer solution. Differences in interfacial protein coverage, orientation, and secondary protein structure were observed for air–polymer, water–polymer, and buffer solution–polymer interfaces.

Other complex interfacial systems can be studied as well. For instance, Salafsky and Eisenthal [151] showed that polarization-resolved SHG studies can be applied to study the orientation of a dye molecule (4-[5-methoxyphenyl-2-oxazolyl]pyridinium methanesulfonate (4PyMPO-MeMs) adsorbed from aqueous solution to a phospholipid bilayer supported on a glass prism. The concentration of the SHG dye used in these experiments was 10 μM at a pH of 7. The dye molecule was found to be oriented at $\theta = 19^\circ$ and 33° with respect to the surface normal in the presence and absence of the supported bilayer, respectively. In another study, Watry and Richmond [152] employed polarization-resolved SFG for monitoring the CH stretches of various hydrophilic and hydrophobic amino acids adsorbed at the CCl_4–D_2O interface. Spectral fits combined with surface tension measurements were used to show that amino acid monolayers with hydrophobic side chains pack more tightly than those with hydrophilic side chains at the liquid interface. Both of the aforementioned studies demonstrate the feasibility of NLO techniques for investigating molecules of a complex nature at various buried interfaces.

4.4 GAS–SOLID INTERFACES

Naturally occurring solid atmospheric particulate matter is found throughout the lower atmosphere [5, 6, 153]. Motivated by the fact that ice surfaces can catalyze chlorine-activation reactions in the stratosphere [154], several groups are now investigating the impact that ice and other solid particulate matter can play in the upper and lower troposphere [155]. Intense research is focused on mineral dust particles [89], sea salt particles [156], ice particles in cirrus clouds [157], and soot [158]. Many tropospheric and stratospheric reactions that are slow in the gas phase can take place, and may in fact be catalyzed, by atmospheric particulate matter. Furthermore, co-adsorbed water appears to play a crucial role in heterogeneous reactions that involve tropospheric aerosols, as is best demonstrated by the fact that halogen-activation reactions involving HNO_3 and sea salt particles do not occur unless some interfacial water is present on the salt surface to catalyze the reaction [159]. As discussed in Section 4.2.4, partially oxidized organic molecules can play a key role in the physical chemistry of solid atmospheric particulate matter [89].

4.4.1 Mineral Oxides and Salts

Satellite measurements often show very large clouds of dust over Northern Africa, Southern Europe, and Central Asia [160–166]. These clouds can have up to weeklong lifetimes and routinely reach both the east and west coasts of the continental United States [167–168]. On the local scale, dust clouds can be linked to agricultural activities [172] and adverse effects on health [174–176]. Dust clouds contain micron to submicron size mineral dust particles that originate from soil erosion during the dry seasons [5, 6]. These mineral dust particles are mainly composed of silica and silicate minerals but also contain oxides such as Al_2O_3, Fe_2O_3, CaO, and MgO as well as carbonates such as $CaCO_3$ and $MgCO_3$ [153, 167]. Due to their optical properties and their potential for aerosol formation via heterogeneous nucleation in the upper and lower troposphere, mineral dust particles are believed to impact global climate by influencing the atmospheric radiative budget on the global scale [5, 6]. Furthermore, such particles are believed to participate in key steps involving the microphysics of aerosol and cloud formation [7, 114, 177, 178].

NLO techniques have been used to characterize mineral oxide surfaces directly as well as their interactions with water and environmentally prevalent organic molecules. As dicussed in Section 4.2.3, the charge density and interfacial potential for mineral oxide-water interfaces can be determined using $\chi^{(13)}$ technique. What follows here is a discussion of selected metal oxides (SiO_2, TiO_2, and mica) and other contributions that NLO techniques have made to the current understanding of their surface structure and reactivity. The current NLO approaches for studying SiO_2, TiO_2, and mica are applicable to other metal oxides such as Fe_2O_3 and Al_2O_3, and would yield additional data that could be applied directly to improve the current understanding of heterogeneous systems in the environment.

A large component of soils and dust particles, SiO_2 is the most abundant naturally occurring metal oxide and the metal oxide most studied by NLO techniques. While the characteristics of SiO_2 have been investigated with SHG [179], scientific queries have focused mainly on its role as a support for organic molecules. Polarization-resolved SHG measurements have allowed for the determination of tilt angles of dye molecules adsorbed to the surface. Orientation data for a few selected molecules are summarized in Table 4.2. These molecules are important as they contain several functional group motifs that are commonly

TABLE 4.2
Orientation of Molecules on a Silica–Air Interface

Adsorbate	SHG System	Tilt[a]	Molecular Structure	Ref.
Rhodamine 6G	347 nm, (energy–pulse n.r.), 10 ns	34°	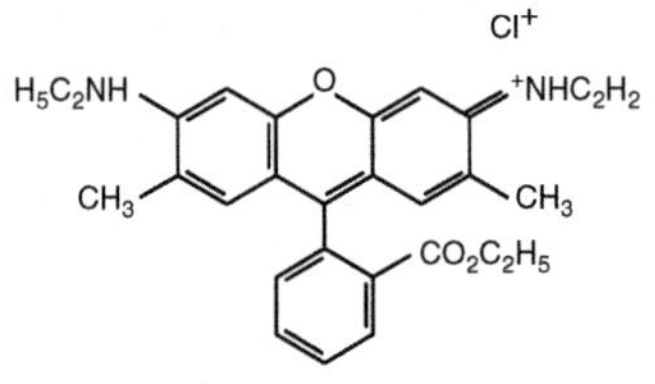 H_5C_2NH, O, Cl^+, $^+NHC_2H_2$, CH_3, CH_3, $CO_2C_2H_5$	[270]
	780 nm, <30 J–pulse, 80 fs	51(2)°		[271]
Rhodamine 110	347 nm, 1 mJ, 10 ns	34°	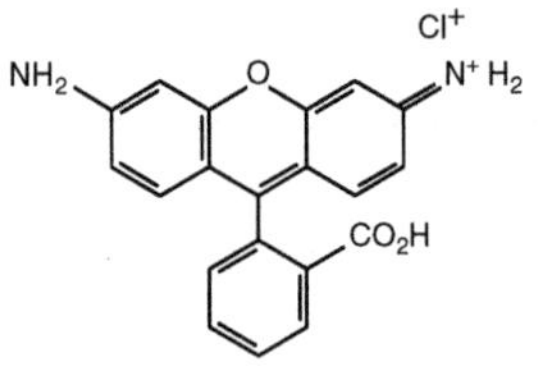 NH_2, O, Cl^+, $N^+ H_2$, CO_2H	[270]
Malachite green chloride	800 nm, <10 J–pulse, 100 fs	8.2(5)°	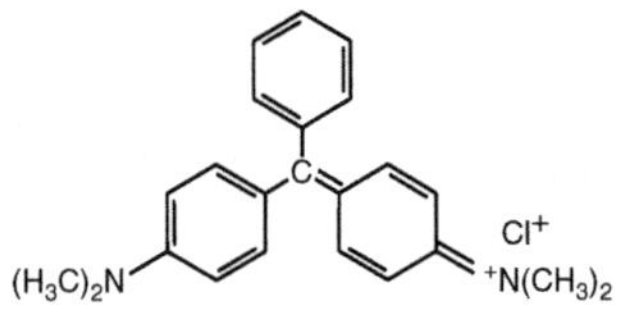 C, $(H_3C)_2N$, Cl^+, $^+N(CH_3)_2$	[272]
Methylene blue	532 nm, 585 nm, 615 nm, 5 mJ–cm^2, 10 ns	58°	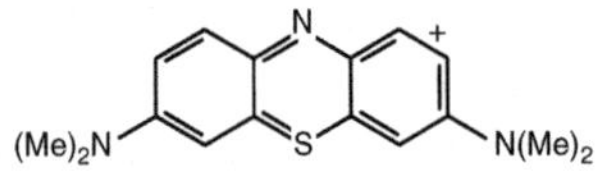 N, +, S, $(Me)_2N$, $N(Me)_2$	[273]
p-Nitrobenzoic acid	532 nm, (energy–pulse n.r.), 10 ps	70(3)°	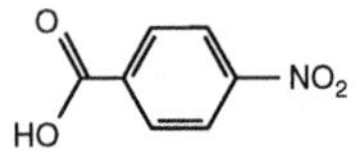 O, HO, NO_2	[40]

[a]Angles given from the surface normal.
n.r., not reported.

found in NOM, specifically fulvic acids. The data obtained from dyes adsorbed on glass can thus provide a starting point for developing NLO studies focused on the role that polar aromatic organic molecules have on heterogeneous soil chemistry. In the context of organic molecules at mineral oxide surfaces, two SHG studies should be mentioned that focus on the formation of aromatic imines via silicon–oxygen coupling directly at the surface [180, 181]. In these studies, imine-coupled multilayers of aromatic molecules were characterized by measuring the SHG signal intensity dependence on incident angle. Similar measurements were carried out by Bakiamoh and Blanchard [182, 183], who used the incident angle-dependent SHG measurements to show increased disorder in multilayer structures of long-chain aliphatic molecules with ionic terminal groups used for growing the multilayers. Thus far, the interactions of atmospheric species with SiO_2 have not yet been studied, and this area of research would represent a significant step towards understanding heterogeneous atmospheric chemistry in the context of mineral dust particles.

In the past two decades, TiO_2 [184] has become one of the most extensively studied metal oxides due to its many applications in both heterogeneous photocatalysis for environmental remediation and solar cell technologies. While many surface science techniques have been used to study TiO_2, there is still considerable need for surface selective and molecularly specific insights. At the most fundamental level, SHG spectroscopy has been used to show that TiO_2 has a 3.4 eV electronic excitation between surface states [185]. Furthermore, the surface of TiO_2 is known to change between a hydrophobic and a hydrophilic state when irradiated with UV light. A recent SFG study suggested that this switch is due to photooxidation of hydrocarbon films on the surface of the TiO_2 (see Figure 4.7) [185]. SFG spectra recorded by Wang et al. [186] showed two bands in the CH stretching region (2800 to $3025\,cm^{-1}$) that were not detected using FTIR due to the fact that the FTIR spectrum is generally dominated by the broad band of hydrogen-bonded OH in that frequency region. The CH peaks found in the SFG study did not disappear upon evacuation of the sample chamber, indicating strong adsorption, but disappeared upon irradiation with UV light. The CH stretching signature returned when the UV-cleaned surface was exposed to ambient conditions for approximately 1 day. The system was then tuned to the free OH stretching region that ranges from 3500 to $3800\,cm^{-1}$. Examining the UV-irradiated TiO_2 film, multiple

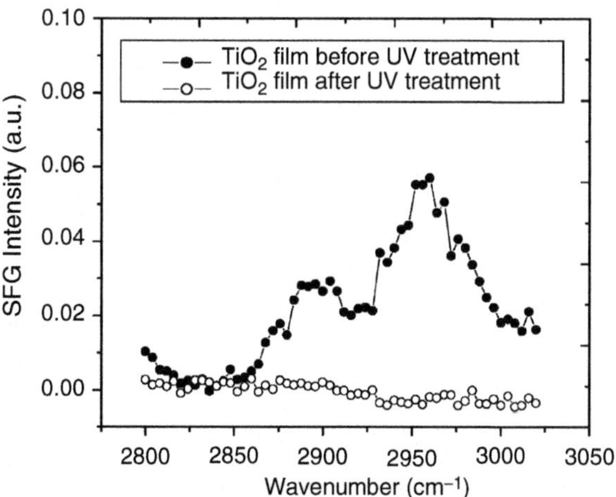

FIGURE 4.7 *ssp*-SFG spectra of a TiO_2 film before and after irradiation with UV light. Spectra are recorded in the CH stretching region and show surface cleanup upon UV irradiation. (From C-Y Wang, H Groenzin, and MJ Shultz. *Langmuir* 19(18): 7330–7334, 2003, with permission. Copyright 2003 American Chemical Society.)

broad spectroscopic features were found in this region [186]. These features were not found to be as strong on freshly prepared, nonirradiated films. Control experiments indicate that samples with strong free OH stretching bands after UV irradiation showed no SFG response in that spectral region after the addition of methanol, as would be expected from the suppression of free OH stretch in the presence of an organic layer. These studies solved the long-standing problem of why photoactivation of TiO_2 is required for many photochemical reactions on that substrate. Additional SHG work on TiO_2 by Shultz et al. [187] shows that surface oxygen vacancies can be repaired with low exposures (10^{-5} to 10^{-7} Torr) of N_2O. Corresponding x-ray photoelectron spectroscopy (XPS) studies showed no increase in the N-1s signal, suggesting that the N_2O, unlike O_2, which is proposed to heal defect sites molecularly, dissociates to heal oxygen vacancies. While the focus of this study was on the mechanism of defect repair, it revealed both structural and heterogeneous reactivity information useful to the environmental community and laid important groundwork for further research.

Another important mineral oxide is mica, which is a silica-based rock-forming mineral in the environment. Miranda et al. [188] applied SFG to probe the structure of surface water molecules adsorbed to mica substrates. It was found that at room temperature, the structure of a water monolayer is consistent with an ordered "ice-like" film. This result has direct implications for the interaction of gas phase species with mica-rich minerals in high relative humidity (RH) environments and suggests that the extent of adsorption and reaction involving such water-coated mica surfaces could be comparable to the adsorption onto ice.

In a related study by Mugele et al. [189], an 8-carbon and an 11-carbon alcohol were sandwiched between two atomically flat mica surfaces and SFG was used to characterize the structure of the alcohol monolayers covering the mica surfaces. The SFG experiments were complimented by surface force measurements of the confined liquids. The SFG spectrum revealed the formation of CH_3-terminated monolayers with the OH groups binding to the substrate at the mica–air interface. The alcohol adlayers were shown to have a polar tilt angle between 20° and 40°. The SFG spectra also indicate that the methylene groups in the alcohols are not arranged in an all-*trans* configuration, as evident from the intensity of the symmetric CH_2 stretch at $2835\,cm^{-1}$.

As mentioned above, water films on mineral oxide surfaces can be crucial for driving surface chemistry in environmentally important heterogeneous systems. With thicknesses varying between one and several monolayers, water films are likely to be strongly influenced by the surface structure of the underlying substrates to which they are bound. SFG experiments on the highly hydrophilic surface of mica have shown the existence of a water layer with a tight hydrogen bonding network at room temperature [188]. At 90% RH the SFG spectrum did not show resonances in the free OH or OD stretching frequency regions, indicating that on average, the dipoles of the adsorbed water molecules point towards the hydrophilic mica surface. At higher RH values (97%) the water film appeared to become a multilayer, as free OH peaks were detected.

In the introduction to this section, it was mentioned that ionic compounds are important in heterogeneous tropospheric chemistry, as they are mainly found in the form of sea salt particles. With stoichiometric ratios of either one or two halide anions per cation in bulk salts, surface-bound acids, acid anhydrides, and mixed halogens can release large amounts of chlorine and bromine into the gas phase, where they can impact the tropospheric ozone and OH cycles [5]. While atmospherically relevant surface reactions involving sea salt have not yet been carried out using NLO techniques, such experiments can be planned and executed easily. The SHG work by Zink et al. [190] on the interaction of water with alkaline earth halides can be used as a guideline. In these studies, polished single crystals of CaF_2 and BaF_2 maintained under UHV were exposed to H_2O doses ranging from 0.5 to 10^7 L (one Langmuir, 1L, is defined as the exposure of a unit area to 10^{-6} Torr of a gas for one second). In these nonresonant SHG experiments, H_2O adsorption was observed as a decrease in the

SHG signal and a change in the SHG polarization. The observed coverage dependence of the azimuthal SHG anisotropy was used to suggest that initial water adsorption onto both substrates was epitaxial, i.e., occurred in registry with the symmetry of the underlying substrate. For BaF_2, the epitaxial growth continued at higher coverages, in contrast to CaF_2, where additional water uptake resulted in an isotropic in-plane surface symmetry. These patterns of ordered and disordered adsorption were supported by molecular dynamics simulations [191]. This molecular-level detail is important for understanding phase transitions in salts, which are often hygroscopic, or the propensity of a mineral to interact with a foreign species in the presence of water vapor.

4.4.2 Ice

In addition to the ionic solids described in the earlier sections, molecular solids are highly important in the environment, especially ice and carbon soot. This section focuses on the surface chemistry and physics of ice. While ice surfaces are now known to play a key role in a wide variety of natural phenomena [192–194], with an important example being halogen-activation reactions that are catalyzed by ice surfaces [195, 196], a full characterization of the properties of ice surfaces and accurate predictions of their impact on natural systems remains difficult. NLO techniques can serve as powerful tools for probing the surface structure of ice, and allow for real time monitoring of catalytic reactions on the surface of the ice.

One of the ongoing debates in ice surface physics and chemistry is the dependence of ice surface structure on temperature, which is difficult to determine due to the fact that ice is a high vapor pressure substrate that is incompatible with surface techniques that require UHV conditions. In 1859, Faraday [197] first proposed the existence of a liquid-like layer on an ice surface below the bulk melting temperature, with direct implications for atmospheric chemistry, glacier flows, and thundercloud electrification. One way to track the surface structure of ice during the premelting [198] stage is to monitor the presence of the free OH bonds, which can be detected using SFG, as a function of temperature.

Monitoring the free OH stretch at $3695\,cm^{-1}$ with SFG, Wei et al. [28, 199] showed that disordering of the basal (0 0 0 1) planes of ice can be detected at 200 K. The disorder was reported to increase with increasing temperature. Polarization-resolved SFG measurements with *ssp* and *ppp* beam polarizations were used to determine an approximate orientation distribution for the free OH bonds, characterized by the OH tilt angle θ_M. θ_M was then related to an orientational order parameter, S, such that when $\theta_M \rightarrow 0$, $S \rightarrow 1$, corresponding to perfect orientational order. Total disorder would correspond to an S value of 0 resulting from $\theta_M = 90°$. Disorder was reported to set in above 200 K. Notably, the surface structure remained distinct from that of liquid water, suggesting that it would be incorrect to view the disordered surface layer of the ice as simply a thin film of liquid water. These findings agree with molecular dynamics simulations that, using both classical and first principles approaches, show the onset of disorder in the top layers of the interfacial region at 200 K [200]. It should be noted that the 200 K onset temperature for ice premelting falls within the low end of the wide range of onset temperatures, which have been reported to be as high as a few fractions of a degree below the ice melting point [201].

In addition to helping to elucidate the structure of the ice–vapor interface at temperatures near the melting point, NLO techniques can also be used for monitoring surface adsorption and desorption with molecular specificity. For example, resonantly enhanced SHG was used to study the uptake of HOCl, a species involved in stratospheric ozone depletion, on ice surfaces under stratospherically relevant temperature conditions [24]. HOCl surface lifetimes were obtained by measuring the first-order desorption rate constants, k_{des}, followed by Arrhenius and weighted transition state theory treatments. This resulted in an activation energy, $E^{\ddagger}_{des}$, of HOCl desorption from ice of $36 \pm 2\,kJ/mol$, an activation entropy of desorption, $\Delta S^{\ddagger}_{des}$, of -90 ± 20 J/(mol K) and an activation enthalpy of desorption, $\Delta H^{\ddagger}_{des}$, of 31 ± 4 kJ/mol. These thermodynamic parameters resulted in a free activation energy of desorption, $\Delta G^{\ddagger}_{des}$, of 48 ± 8 kJ/mol for desorption of HOCl from ice at 185 K. Using the Arrhenius prefactor to obtain the lifetime of a surface vibration, τ_{ν} [202], a 4 s HOCl surface lifetime τ at 185 K was then calculated according to $\tau = \tau_{\nu} e^{(E^{\ddagger}_{des} - RT)}$. Under the experimental conditions, this surface lifetime corresponds to a surface coverage for HOCl between 0.01% and 0.1% of a monolayer [203, 204]. Additional experiments showed that HNO_3, a common component of stratospheric ice particles, acts as a molecular glue to lengthen HOCl surface lifetimes by at least one order of magnitude and facilitate heterogeneous stratospheric ozone depletion chemistry [24].

In addition to providing information on adsorption and desorption properties, SHG can be used to further characterize chemical reactions occurring on the ice surface. SHG was used to study the hydrolysis of chlorine nitrate, $ClONO_2$, another important species in ice-catalyzed stratospheric ozone depletion, on ice [23, 24]. The formation of HOCl was found to occur on a 1 to 10 min timescale. The reaction was shown to be autocatalytic in HOCl, while the HNO_3 coproduct was found to poison the ice surface.

4.4.3 Polymers

As mentioned in the introduction to Section 4.3.2.2, NOM often consists of biopolymers that contain functional groups for binding nutrients, toxic metals, and organic pollutants. NOM can be adsorbed to mineral oxides, but can also be present in the form of insoluble NOM, which may include hydrophobic organic matter (HOM). While NLO techniques have been used for studying the surfaces of functionalized polymers in the context of their industrial applications, they provide highly useful information for beginning to understand the environmental significance of heterogeneous processes occurring at the HOM–air interface.

Somorjai and coworkers prepared samples of polymer films by either solvent casting or spin coating, followed by annealing for several hours to remove the solvent [205, 206]. Different interfaces were studied using SFG, and scanning force microscopy (SFM) was applied for monitoring the friction and the elastic modulus of the polymer surface in contact with different environments. This combination of techniques allowed for a correlation of the surface structure with the mechanical properties of the polymer surfaces. Air–polymer interfaces with varying degrees of polymer crystallinity were examined. The random packing of polymer chains and the disorder of the polymer surface were evidenced by larger bandwidths of the SFG spectral features. The average orientation of the methylene groups in the hydrocarbon backbone was found to be around 42° in the case of the polymer with the highest degree of crystallinity (70–75%), and around 59° in the case of the polymer with the lowest degree of crystallinity (2%). These angles indicate that the hydrocarbon backbone tends to lie parallel to the surface for maximum interaction with underlying chains.

In another set of experiments, SFG studies on polymers mixed with antioxidation additives showed that the additives preferentially segregate to the polymer surface. SFG spectra of the different examined polymers were also recorded as a function of temperature above and below the glass transition temperature, T_g, of the polymer. Spectra reflect changes in the surface structure at T_g. For polypropylene samples below T_g (glassy state), it was found that the hydrocarbon backbone at the surface was more ordered and more polar than above T_g (rubbery state), where CH_2 groups were randomly oriented, reflecting the disorder of the chains. The reader is referred to a recent review by Somorjai and coworkers for further applications [207].

4.4.4 High-Pressure CO Adsorption and Oxidation

The oxidation of CO is highly relevant in the context of treating industrial emissions and will serve as one example where NLO techniques have revolutionized laboratory approaches that focus on mechanistic studies of industrial processes that are environmentally important. UHV studies on single crystal and thin film model catalyst surfaces [208, 209] are usually performed to obtain fundamental information such as the binding energies of the various adsorption sites, the abundance and chemical identity of active sites, the relative rates of the elementary reaction steps, and reaction probabilities. Yet, reaction conditions for studying the heterogeneous function of CO oxidation catalysts in UHV systems are much different from those found in industrial catalytic reactors. An elevated pressure reactor coupled to a UHV surface analysis system can bridge the pressure and materials gap between model catalytic surfaces and complex real catalysts [210–216]. When combined with SFG, one can perform *in situ* studies of catalysts operating under industrial reaction conditions. Several studies have employed low and high pressure SFG as a means for identifying surface species, including intermediates, that are involved in the catalytic oxidation of CO. Substrates include metals [217–221], oxide thin films grown on metal single crystal surfaces [222], and polycrystalline metal foils [223]. In addition, high pressure SFG has been shown to be a promising technique for studying catalytic reactions on oxide-supported metallic nanoparticles and clusters [224, 225]. Other processes utilizing SFG in combination with a high-pressure reaction chamber include the hydrogenation of ethylene [226] and propylene [227] on Pt(1 1 1) surfaces and the hydrogenation and dehydrogenation of cyclohexene on Pt(1 0 0) and Pt(1 1 1) [228–231].

In the studies focusing on CO, the stretching mode of surface-bound CO was monitored as a function of coverage and temperature. The results of these studies are summarized in Table 4.3. It is important to note that the CO stretch downshifts when going from macro-

TABLE 4.3
Summary of Studies on CO Adsorption and Oxidation Using SFG

Surface	CO Pressure (Torr)	CO Exposure (L)	Surface Coverage	Temperature (K)	ν(C–O) (cm^{-1})	Refs.
Pt(1 1 1)	n.a.	0.5–12	Up to 0.5	300	2083(2)–2093(2)	[220]
	n.a.	n.a.	0.5	150	2103(2), 1854(2)	[220]
	7.6×10^{-8}–1.52×10^{2}	n.a.	n.a.	230	2097	[222]
	1×10^{-7}–7×10^{2}	n.a.	n.r.	295	1845, 2095–2045[a]	[221, 230]
	1×10^{2}, mixture[b,d]	n.a.	n.r.	590–1100	2090–2045[a]	[221, 230]
	4×10^{1}, mixture[b,d]	n.a.	n.r.	540–600	2090–2240[a]	[221, 230]
Sputtered-Pt(1 1 1)	n.a.	0.5–7.5	n.r.	300	2072–2093	[220]
Pt(1 1 0)-(1×2)	n.a.	0.5–15	Up to 1.0	300	2065–2094	[220]
Ni(1 0 0)	n.a.	0.5–10	0.1–0.7	100	2017–2063[c]	[222]
Rh(1 1 1)	10^{-7}–1×10^{2}	n.a.	n.r.	125	1875, 2085	[221, 230]
	7×10^{2}	n.a.	n.r.	125	1875, 2020–2025	[221, 230]
Pt polycrys. foil	7.6×10^{-1}	n.a.	n.r.	300–660	2096(4)–2057(5)	[225]
Pt–Al_2O_3 with: 3 nm Pd particles	7.6×10^{-8}–1.52×10^{2}	n.a.	n.r.	190–300	1976, 2106	[226]
6 nm Pd particles	7.6×10^{-8}–1.52×10^{2}	n.a.	n.r.	190–300	1983–1971, 2103–2112[b,c]	[226]
NiO(1 1 1)–Ni(1 1 1)	1×10^{-7}	n.a.	n.r.	140	2144	[224]
Ni(1 1 1)	n.a.	8	n.r.	140	2074	[224]

[a]Coverage dependent.
[b]Pressure dependent.
[c]Temperature dependent.
[d]Denotes mixture of gases equal to O_2 = 40 Torr and total = 740 Torr in He.
n.a., not applicable; n.r., not reported.

scopically flat metal surfaces to metal nanoparticles. In the context of heterogeneous processes that are important in the natural environment, this is a crucial finding as the fraction of nano-sized particles in colloids and aerosols can be high. The fact that the CO stretch frequency depends on whether CO is adsorbed on a flat surface or a nanoparticle's surface suggests that particle size may affect chemical behavior as well, and this topic will be discussed below.

4.5 SPECIAL TOPICS

In this section, four special applications involving NLO techniques and how they can be applied for studying environmentally important surfaces and interfaces are discussed. Beginning with an overview of how solvation at interfaces can be quantified using SHG, a brief review on how dynamic processes involving interfacial solvation can be studied using NLO techniques will follow. After a discussion on microparticles, including colloids and how their surfaces can be studied in the context of pollutant binding, this section finishes with a brief review on how interfacial chirality can be probed using NLO techniques.

4.5.1 Solvation at Interfaces

When interacting with adsorbates, interfaces can be viewed as two-dimensional solvents. For instance, an atmospheric trace gas species can bind to the surface of a liquid or solid aerosol particle and, once adsorbed, be considered solvated by surface sites. Likewise, in soil environments, surface sites on a mineral oxide can either strip water molecules that form the solute hydration sphere and solvate the adsorbed species from the solid condensed phase side, or bind some of the hydrating water molecules and lead to net adsorption of the solute. In both cases, solvation of the foreign species by surface sites is an important step for adsorption and the subsequent fate of the surface-bound species. Several NLO studies have been published that examine interfacial solvation to various degrees [19, 37, 231, 232].

Using resonantly enhanced SHG, Walker and coworkers determined the solvent polarity of aqueous–alkane and silica–organic solvent interfaces using chromophores such as 4-aminobenzophenone (4ABP), *para*-nitrophenol (PNP), and dimethyl-*p*-nitrophenol (dmPNP) [37, 233, 234]. They found that subtle alterations of solute structure could alter significantly the solute's local solvation environment at the same interface. In the case of a hydrophilic interface such as silica, substrate–solvent interactions were found to be the dominant influence on the interfacial solvent polarity. For example, weak interaction of nonpolar solvents with silica surfaces results in a stronger interaction of surface dipoles with adsorbed solutes creating more polar environments at the interface than in the bulk. Additionally, solvent structure was found to affect the interfacial polarity through its size and packing ability.

Walker and coworkers [19, 38, 235] also applied resonantly enhanced SHG to specially designed solvatochromic surfactants that can act as "molecular rulers" at different liquid–liquid interfaces in order to measure the interfacial width between two immiscible liquids. Solvatochromism describes the shift in an electronic UV-Vis absorbance band of a solute due to the polarity of a solvent in which the solute is dissolved. The surfactants are based on *p*-nitroanisole and consist of anionic sulfate groups attached to hydrophobic solvatochromic probes by alkyl spacers of various lengths [235]. The bulk solution excitation wavelength monotonically shifts towards the red as solvent polarity increases (see Figure 4.8) [38]. With the assumption that the anionic headgroup remains solvated in the aqueous phase, varying the alkyl spacer length should allow further penetration into the organic phase. Hence, monitoring the probe excitation wavelength as a function of the alkyl spacer length measures the distance required for solvent polarity to change from an aqueous to an organic limit [235]. The SHG spectra of *p*-nitroanisole, C_n-rulers ($n \geq 2$) adsorbed to water–cyclohexane and

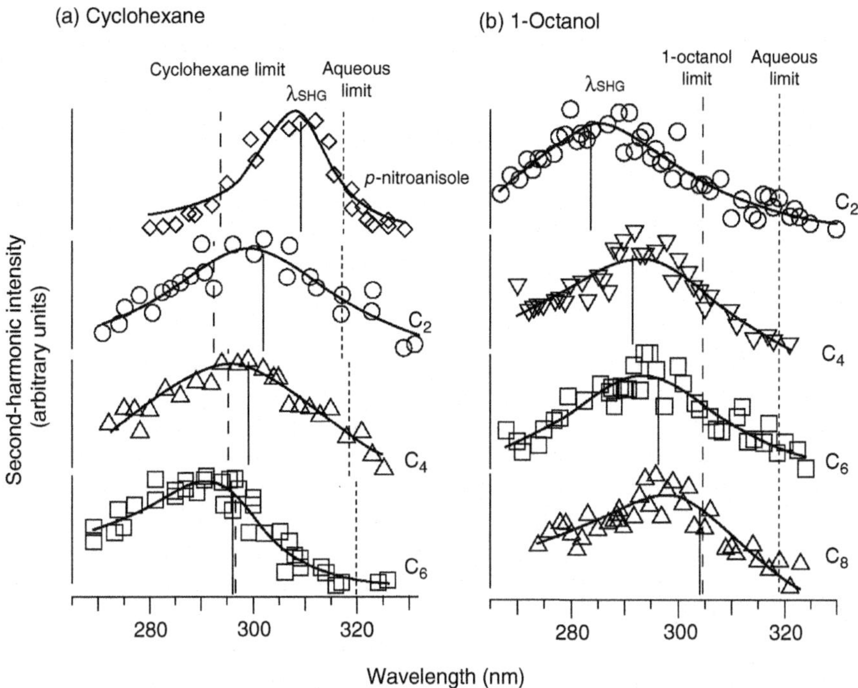

FIGURE 4.8 SHG spectra of molecular rulers of various lengths adsorbed at the (a) cyclohexane–water and (b) water–1-octanol interfaces. The solid vertical line corresponds to SHG maxima and dashed lines correspond to the excitation maxima in bulk solution. The figure shows the solvatochromic behavior of adsorbed molecular rulers as interfacial polarity changes. (From WH Steel and RA Walker. *Nature* 424: 296–299, 2003, with permission.)

water–1-octanol interfaces at less than 10% surface coverage are consistent with the expected solvatochromic behavior. SHG polarization measurements show that the orientation of the chromophores varies between 45° and 51° and between 36° and 39° relative to the surface normal for the water–cyclohexane and the water–1-octanol interfaces, respectively.

In another solvation study, Zhang et al. [231] applied SFG to study how the surface structure of polymers that contain both hydrophilic and hydrophobic groups respond to varying degrees of hydration by taking the polymer out of water and into air. One of the polymers studied was a 100 μm thick BioSpan-S film, which possesses both hydrophobic (poly(dimethylsiloxane), PDMS) and hydrophilic (ether and urethane segments) components. The SFG peak intensities of each hydrophilic and hydrophobic component depend on the environment with which the polymer is in contact. SFG spectra were recorded at 300 K in air and in D_2O and complimented the contact angle measurements as a function of hydration time. By comparison to reference polymers representative of the hydrophobic and hydrophilic component of the polymer under investigation, a band at 2919 cm^{-1} was assigned to the methyl stretch of PDMS and bands at 2851 and 2785 cm^{-1} were assigned to the symmetric stretch of CH_2 groups of biospan. As the hydration time increased, weakening of the former band and strengthening of the latter ones suggested surface reconstruction as well as structural changes within the polymer surface. Water contact angle measurements were carried out and reflect the degree of hydrophilicity of the surface, which was found to be reversible.

Hommel and Allen [235] applied SFG to investigate the rearrangement of 1-methyl naphthalene (1-MN) at the surface of upon addition of low concentrations of water. The surface and bulk spectra of 1-MN were obtained using BBSFG and Raman spectroscopy, respectively. Using the *ssp*-polarization combination, the following peaks were identified: an overtone of the methyl bending mode at 2865 cm^{-1}, a methyl symmetric stretching mode at 2908 cm^{-1}, an additional bending overtone at 2932 cm^{-1}, a methyl asymmetric stretching mode at 2967 cm^{-1}, an overtone of the aromatic C=C stretching mode at 3010 cm^{-1}, and the aromatic CH stretching mode at 3061 cm^{-1}. Upon addition of water at low concentrations (1:336, water:1-MN), the CH_3 symmetric stretch peak intensity increases while the CH stretch peak intensity decreases. This is attributed to the rearrangement of surface-bound 1-MN such that more of the methyl groups point upwards. The reorientation was found to be reversible with dehydration.

Molecular rearrangement in the presence and absence of water was also studied by Chen and coworkers, who used *in situ* SFG to follow the orientation of the ester side chain methyl groups of a series of poly(methacrylate) molecules at air and water interfaces [237, 238]. The polymers include poly(methyl methacrylate) (PMMA), poly(*n*-butyl methacrylate) (PBMA), and poly(*n*-octyl methacrylate) (POMA). In these polymers, the length of the alkyl side chain and hence the distance of the methyl group under investigation from the polymer backbone is varied [238]. All polymers studied showed surface restructuring behavior upon contact with water. For example, the methyl groups were found to be oriented normal to the PBMA surface in air, and to lie towards the surface when in contact with water with an orientation angle larger than 58° to the surface normal. Chen and coworkers also studied polymer–polymer interfaces [239]. In one example, the PBMA methyl groups at the interface with polystyrene (PS) were found to have a range of possible orientations in between the two extremes of the PBMA–air and PBMA–water interfaces. These studies demonstrate that solvent molecules and interfacial polarity can have profound impacts on the structure of interfacial species which could, in turn, impact the adsorption behavior of foreign molecules or ions.

4.5.2 Dynamics

Besides these static solvation experiments, SHG in a pump–probe setup has proven to be a promising technique for monitoring the dynamics of interfacial solvation, which usually occurs on the sub-picosecond timescale [240]. The ultrafast solvation dynamics of a molecular probe (IR144: tricarbocyanine) and the dynamic relaxation times have been measured at the silica–acetonitrile and silica–butanol interfaces [240]. At both interfaces, interfacial polarity was found to be smaller than the polarity of the bulk solvents and the time constants extracted from the time-resolved measurements were determined to be 1.05 ± 0.14 and 1.20 ± 0.15 ps for silica–acetonitrile and silica–butanol interfaces, respectively. Both of these observations have been attributed to the interactions of the solvent molecules with the silanol groups of the silica surface.

Time-resolved SHG has also been used to chart the static and dynamic solvatochromic effects of the probe molecule coumarin 314 (C314) adsorbed at the air–water interface [102]. Using a tunable optical parametric generator in a pump–probe setup, the spectrum of C314 was collected and showed a maximum SHG response at 419 ± 3 nm. Solvation was shown to proceed with two time constants, $\tau_1 \sim 200$ fs and $\tau_2 = 1.2$ ps, that were independent of C314 interfacial orientation. The effect of the surfactant stearic acid, $CH_3(CH_2)_{16}COO^-$, on C314 adsorbed at the air–water interface was also investigated [102]. It was noted that the molecular orientation of the C314 dye depends on the charge of the carboxylate–carboxyl group; the C314 tilt angle from the surface normal was found to be $30 \pm 5°$ in the presence of the anionic surfactant monolayer, and $41 \pm 3°$ in the presence of the neutral surfactant monolayer. It was also shown that both the neutral and anionic forms resulted in a blueshift of about 12 nm in the C314 spectral peak.

In natural environments, interfaces are often necessarily charged, and in biology specifically, essential reactions depend on transport through charged cell membranes or protein folding based on hydrophilic functional groups. In order to further understand the unique properties of this class of interfaces, solvation dynamics experiments similar to those mentioned above were carried out at a negatively charged interface, namely SDS monolayers at the air–water interface [241]. Solvation dynamic times τ_1 and τ_2 were studied individually as a function of surface charge, and found to show unique behaviors with increasing surface charge. The faster τ_1 component was found to be unaffected until the surface charge reached 100 Å^2/elemental charge (close to the physiological range for cellular membranes) when it was found to slow to 600 ± 70 fs in comparison with the bare air–water interface. The slower τ_2 component was affected even at low surface charges, slowing to 4.4 ± 0.9 ps at 500 Å^2/elemental charge and 5.4 ± 1.1 ps at 100 Å^2/elemental charge. The different behavior of the two time constants as a function of surface charge suggests that these two solvation modes reflect two separate motions of the hydrogen-bonded water network. Solvation dynamics were compared to SFG spectra and a connection was drawn between the spectral feature centered at 3200 cm^{-1} and the slower solvation timescale.

4.5.3 Colloids

Colloids are prevalent in soils. Their surfaces can act as heterogeneous chemical reactors or simply bind solutes. Colloid surfaces are charged and the interaction of foreign species with the colloid surface will depend on the charged state of the interface. Recent pioneering work shows that it is possible to generate and collect NLO signals from microscopic particles suspended in solution if the particle size is comparable to the coherence length of the NLO process (see Figure 4.9) H Wang, ECY Yan, E Borguet, and KB Eisenthal. *Chem. Phys. Lett.* 259(1–2): 15–20, 1996 [151, 242, 245]. Examples include emulsions [242], colloids [186, 243, 246], and spherical liposomes and phospholipid bilayers [151, 244, 247]. Surface potential measurements [242] and experiments focusing on the transfer of interfacial molecules from one type of colloid to another have also been performed [248]. One commonly uses either a straight-through or a right-angle optical geometry and collects the scattered SHG photons after filtering out the fundamental light and correcting the signal intensity for the hyper-Rayleigh signal contribution. Polarization-resolved studies are possible in principle by recovering the polarization information through angle-resolved signal detection.

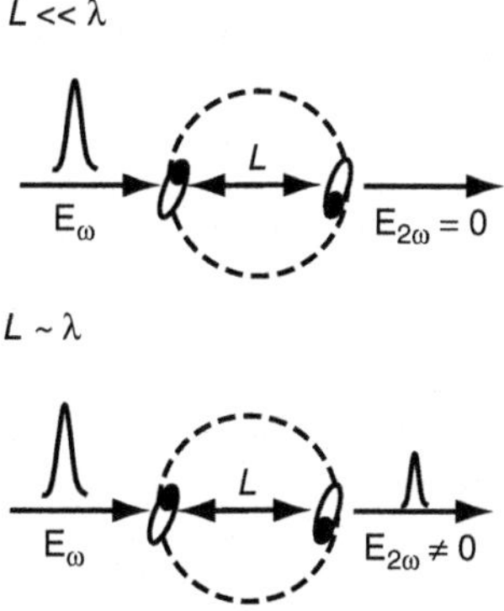

FIGURE 4.9 Schematic of NLO signal generation from a centrosymmetric system consisting of two identical molecules separated by a distance *L*. (From H Wang, ECY Yan, E Borguet, and KB Eisenthal. *Chem. Phys. Lett.* 259(1–2): 15–20, 1996. Copyright 1996 with permission from Elsevier.)

In a related set of experiments, Ma and Allen [249] reported diffuse reflection SFG spectra obtained from a flat bed of 1–5 μm SDS particles in air (see Figure 4.10). The average nonlinear susceptibility was used for the spectral fitting procedures and, for the leveled particle sample, was considered to be the sum of the nonlinear susceptibilities of the individual microscopic particle surfaces. Similarly to the NLO studies of colloids in solution, it should be pointed out that the extraction of the polarization information for this disordered particle ensemble is possible in principle.

The recent development of NLO signal generation from microscopic particles in solution represents a significant step towards studying environmentally important heterogeneous systems, as one can now, for instance, follow the adsorption of a solute onto individual colloid particles suspended in solution. This is especially important in the context of chemical transport in soil environments, where pollutants and nutrients can bind to colloid surfaces and be transported as far as the colloid sedimentation rate, the adsorbate lifetime, and the underground water flow rate allow. This work could be expanded to study the surface chemistry of aerosols in the gas phase as they are exposed to trace gas species.

The ability to monitor heterogeneous processes on submicron size particles would allow for experiments that take into account the fact that the surfaces of particles, if they are small enough, possess curvature and thus a surface energy that can be quite different from that of a macroscopically flat surface. This effect is known as the Kelvin effect [250], which can have a direct impact on the physics and chemistry of the atmospheric and soil environments [7, 8]. Experiments carried out on macroscopically flat surfaces thus only apply to relatively large particles with diameters of several hundred nanometers or larger. In contrast, when following heterogeneous processes on particles with diameters below 100 nm, the Kelvin effect should be included, which is possible through NLO signal generation from microscopic particles.

4.5.4 Chiral Surfaces

Many humic and fulvic acids common in NOM contain stereocenters that could preferentially bind one adsorbate enantiomer over another. This is especially important given the fact that concentrations of endocrine disrupting chemicals [251, 252], which often contain chiral functional groups, can be high enough to be of environmental concern [253, 254]. NLO techniques

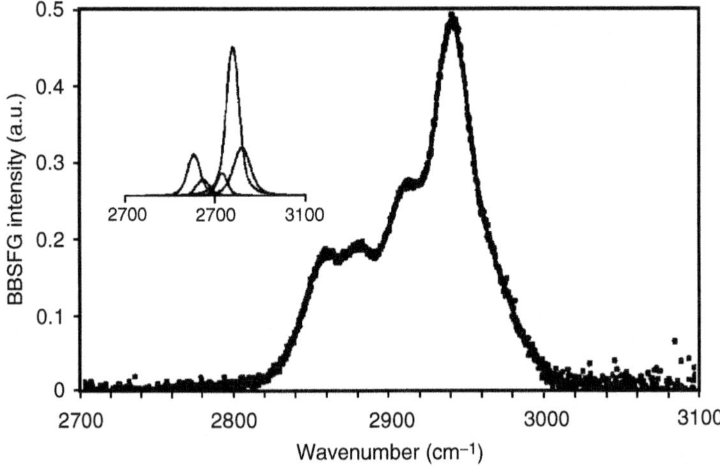

FIGURE 4.10 Diffuse reflectance BBSFG spectrum of SDS particles. Calculated fit is shown in the inset. (From G Ma and HC Allen. *J. Am. Chem. Soc.* 124(32): 9374–9375, 2002, with permission. Copyright 2002 American Chemical Society.)

have the capability to selectively probe chiral species at surfaces and interfaces, demonstrating the potential for understanding binding mechanisms for chiral adsorbates of environmental relevance, including endocrine disruptors.

Hicks and coworkers [255–257] described the theory behind the applicability of SHG to studying chiral monolayers. Briefly, when circularly polarized (cp) light is directed onto an interface that contains a chiral adsorbate, the NLO response is different if the interface is probed with left versus right circularly polarized light. These differences can be multiple orders of magnitude larger than the ones observed in linear circular dichroism, where the effect is usually below 0.1%. For example, left circularly polarized light yielded greater signals than the right circularly polarized light for the R-enantiomer of *S*-(2,2′-dihydroxy-1,1′–binaphthyl) (BN) [258]. For resonantly enhanced NLO measurements, the difference in the SHG intensities of left versus right circularly polarized light also depends on the wavelength of the SHG signal. This method is related to circular dichroism (CD) spectroscopy and thus is referred to as SHG-CD [259].

In a slightly different experimental approach based upon optical rotatory dispersion (ORD), one can obtain the SHG-ORD angle, ϕ, which yields additional information regarding the chirality of surface species. SHG-ORD was used for determining the handedness of R- and S-BN at air–water, air–quartz, and water–CCl_4 interfaces kept at various temperatures. SHG-ORD rotation angles were extracted from the SHG spectra for the R- and S-species at the different interfaces studied and found to be 17.0°, 18.4°, and −13° for the R-enantiomer at air–water, air–quartz, and water–CCl_4 interfaces, respectively. Within 0.2°, the ORD rotation angles for the S-enantiomer were found to be identical to the ones obtained from the R-enantiomer but, most importantly, opposite in sign. The SHG-ORD for BN was also found to be independent of the density of surface species.

A variety of developing studies take advantage of the sensitivity of SHG to chiral environments. In recent work by Burke et al. [260] the chiral sensitivity observed in SHG was attributed to orientational effects alone. This was experimentally supported by strong chiral dichroic ratios in SHG signals obtained from achiral chromophores adsorbed to chiral surfaces. In other work, the chiral SHG response has been optically isolated and utilized to monitor the adsorption of melittin, a peptide found in bee venom, with a planar-supported lipid bilayer of 1-palmitoyl-2-oleoyl-*sn*-glyerco-3-phosphotidylcholine [261]. This particular work highlights the potential of chiral SHG techniques to detect adsorption and interfacial phenomena in extremely complex systems such as those found in biological environments. In summary, the extraordinary sensitivity of NLO techniques to chirality could be applied to a variety of environmentally important heterogeneous processes involving chiral surfaces and also chiral adsorbates. As optically active species are scarce in the atmosphere, surface studies of chiral systems would be applicable mainly to problems focused on soil chemistry.

4.6 OUTLOOK

Current projections indicate that there will be 7.5 billion people on the planet by the year 2020 [264]. In the context of the increasing energy needs and the associated substantial increases in environmental pollution, the grand challenges in science and engineering include pollution prevention and the development of effective environmental remediation strategies. These issues require molecular-level knowledge about the interfaces at which pollutant-binding mechanisms operate, as well as detailed information about the kinetics and thermodynamics of pollutant binding. NLO techniques provide a straightforward means for obtaining interface specific information in real time, *in situ*, and under the concentration and partial pressure conditions commonly found in soils and the atmosphere. While relatively costly to set up, NLO approaches have unique capabilities and are highly effective when used in conjunction with other surface science techniques.

SHG and SFG are expected to become more common due to the relative ease with which one can now operate a laser system. Furthermore, the importance of environmentally relevant heterogeneous systems will gain more and more recognition as our ability to study them with surface selective probes increases. Examples already discussed in this chapter provide a good starting point for devising innovative strategies that can be used for studying catalytic heterogeneous processes at environmentally relevant surfaces and interfaces. These can include the NLO approaches discussed in this chapter, and may be expanded to include two-dimensional spectroscopies based on SFG. Two-dimensional SFG can be viewed as the optical analog to two-dimensional NMR spectroscopy and opens the possibility for investigating complex molecules in complex heterogeneous environments. It is expected that the application of NLO techniques to environmental problems will provide the surface- and interface-specific kinetic, thermodynamic, and spectroscopic data necessary for developing improved pollutant remediation strategies and for advancing chemical transport models.

Atmospheric and geochemical transport models are becoming increasingly complex, and heterogeneous processes are now incorporated explicitly in such models [160, 263, 264]. However, surface-specific experiments that are carried out under environmentally representative temperature, concentration, and pressure conditions are scarce. In this context, NLO techniques offer several unique advantages that allow for real-time measurements as well as *in situ* spectroscopic and structural studies directly on surfaces. The detailed laboratory data derived from surface-specific real-time NLO measurements will become increasingly necessary for describing heterogeneous processes on environmentally important surfaces and interfaces in chemical transport models. Such data can include heterogeneous reaction rate constants as well as surface lifetimes of reactants, intermediates, and products. Likewise, equilibrium heterogeneous binding constants and surface spectra of reactants, intermediates, and products are necessary in next generation chemical transport models. Combined with such models, laboratory data derived from NLO measurements have the capacity to advance the understanding of the impact that human activities have on the environment, and can enable us to address environmental pollution in this century on a detailed molecular level.

ACKNOWLEDGMENTS

Support through Northwestern University Institute for Environmental Catalysis (NSF grant # NSF-9810378) and the ACS-PRF (38960-G5S) is gratefully acknowledged. A.B.V. wishes to acknowledge support by NASA Headquarters under the Earth System Science Fellowship Grant NGT-530456. The authors wish to thank Catherine Schmidt for her editorial work.

REFERENCES

1. FMM Morel and JG Hering. *Principles and Applications of Aquatic Chemistry*, New York: Wiley-Interscience, 1993.
2. D Langmuir. *Aqueous Environmental Geochemistry*, New Jersey: Prentice-Hall, 1997.
3. RP Schwarzenbach, PM Gschwend, and DM Imboden. *Environmental Organic Chemistry*, 2nd ed. Hoboken, NJ: John Wiley & Sons, 2003.
4. W Stumm and JJ Morgan. 3rd ed. *Aquatic Chemistry*, New York: Wiley-Interscience, 1996.
5. BJ Finlayson-Pitts and JN Pitts Jr. *Chemistry of the Upper and Lower Atmosphere*, New York: Academic Press, 2000.
6. JH Seinfeld and SN Pandis. *Atmospheric Chemistry and Physics*, New York: John Wiley & Sons, 1998.
7. ST Martin. *Chem. Rev.* 100(9): 3403–3453, 2000.
8. GE Brown, VE Henrich, WH Casey, DL Clark, C Eggleston, A Felmy, DW Goodman, M Gratzel, G Maciel, MI McCarthy, KH Nealson, DA Sverjensky, MF Toney, and JM Zachara. *Chem. Rev.* 99(1): 77–174, 1999.

9. AR Ravishankara, ed. Atmospheric Chemistry: Long-Term Issues. *Chem. Rev.* 103(12), 2003.
10. YR Shen. *The Principles of Nonlinear Optics*, New York: John Wiley & Sons, 1984.
11. RW Boyd. *Nonlinear Optics*, New York: Academic Press, 1992.
12. TF Heinz. *Nonlinear Surface Electromagnetic Phenomena in Nonlinear Surface Electromagnetic Phenomena*, H.E. Ponath and G.I. Stegeman, eds. Amsterdam: Elsevier, 1991, pp. 353–410.
13. RA Walker, JC Conboy, and GL Richmond. *Langmuir* 13(12): 3070–3073, 1997.
14. RA Walker, JA Gruetzmacher, and GL Richmond. *J. Am. Chem. Soc.* 120(28): 6991–7003, 1998.
15. RA Walker, BL Smiley, and GL Richmond. *Spectroscopy* 14(1): 18–29, 1999.
16. TP Petralli-Mallow, AL Plant, ML Lewis, and JM Hicks. *Langmuir* 16(14): 5960–5966, 2000.
17. KB Eisenthal. *Chem. Rev.* 96(4): 1343–1360, 1996.
18. LJ Richter, TP Petralli-Mallow, and JC Stephenson. *Opt. Lett.* 23(20): 1594–1596, 1998.
19. WH Steel, F Damkaci, R Nolan, and RA Walker. *J. Am. Chem. Soc.* 124(17): 4824–4831, 2002.
20. X Zhang, O Esenturk, and RA Walker. *J. Am. Chem. Soc.* 123(43): 10768–10769, 2001.
21. XC Zhang and RA Walker. *Langmuir* 17(15): 4486–4489, 2001.
22. CT Williams and DA Beattie. *Surf. Sci.* 500(1–3): 545–576, 2002.
23. FM Geiger, CD Pibel, and JM Hicks. *J. Phys. Chem. A* 105(20): 4940–4945, 2001.
24. FM Geiger, AC Tridico, and JM Hicks. *J. Phys. Chem. B* 103(39): 8205–8215, 1999.
25. FM Geiger and JM Hicks. *Laser Techniques in Surface Science* San Diego, CA: SPIE: Bellingham, WA, 1998.
26. HC Allen, DE Gragson, and GL Richmond. *J. Phys. Chem. B* 103(4): 660–666, 1999.
27. S Baldelli, C Schnitzer, DJ Campbell, and MJ Shultz. *J. Phys. Chem. B* 103(14): 2789–2795, 1999.
28. X Wei, PB Miranda, and YR Shen. *Phys. Rev. Lett.* 86(8): 1554–1557, 2001.
29. AL Mifflin, KA Gerth, and FM Geiger. *J. Phys. Chem. A* 107(45): 9620–9627, 2003.
30. AL Mifflin, KA Gerth, BM Weiss, and FM Geiger. *J. Phys. Chem. A* 107(32): 6212–6217, 2003.
31. MR Watry, MG Brown, and GL Richmond. *Appl. Spectrosc.* 55(10): 321A–340A, 2001.
32. PB Miranda and YR Shen. *J. Phys. Chem. B* 103(17): 3292–3307, 1999.
33. MJ Shultz, C Schnitzer, D Simonelli, and S Baldelli. *Int. Rev. Phys. Chem.* 19(1): 123–153, 2000.
34. YR Shen. *Nature* 337(6207): 519–525, 1989.
35. YR Shen. *Surf. Sci.* 299(1–3): 551–562, 1994.
36. KB Eisenthal. Equilibrium and dynamic processes at interfaces by second harmonic and sum frequency generation. In: *Annual Review of Physical Chemistry*, HL Strauss, GT Babcock, and SR Leone, eds. Annual Reviews: Palo Alto, CA, 1992, pp. 627–661.
37. WH Steel and RA Walker. *J. Am. Chem. Soc.* 125(5): 1132–1133, 2003.
38. WH Steel and RA Walker. *Nature* 424: 296–299, 2003.
39. ML Sandrock, CD Pibel, FM Geiger, and CA Foss. *J. Phys. Chem. B* 103(14): 2668–2673, 1999.
40. TF Heinz, HWK Tom, and YR Shen. *Phys. Rev. A* 28(3): 1883–1885, 1983.
41. GJ Simpson and KL Rowlen. *Anal. Chem.* 72(15): 3399–3406, 2000.
42. GJ Simpson and KL Rowlen. *Anal. Chem.* 72(15): 3407–3411, 2000.
43. YR Shen. *Pure Appl. Chem.* 73(10): 1589–1598, 2001.
44. GL Richmond. *Chem. Rev.* 102(8): 2693–2724, 2002.
45. S Baldelli, N Markovic, P Ross, YR Shen, and G Somorjai. *J. Phys. Chem. B* 103(42): 8920–8925, 1999.
46. EL Hommel and HC Allen. *Anal. Sci.* 17(1): 137–139, 2001.
47. EL Hommel, G Ma, and HC Allen. *Anal. Sci.* 17(11): 1325–1329, 2001.
48. X Wei, SC Hong, AI Lvovsky, H Held, and YR Shen. *J. Phys. Chem. B* 104(14): 3349–3354, 2000.
49. JM Hicks, LE Urbach, EW Plummer, and HL Dai. *Phys. Rev. Lett.* 61(22): 2588–2591, 1988.
50. JA McGuire, W Beck, X Wei, and YR Shen. *Opt. Lett.* 24(24): 1877–1879, 1999.
51. Q Du, R Superfine, E Freysz, and YR Shen. *Phys. Rev. Lett.* 70(15): 2313–2316, 1993.
52. DE Gragson and GL Richmond. *Langmuir* 13(18): 4804–4806, 1997.
53. DE Gragson and GL Richmond. *J. Chem. Phys.* 107(22): 9687–9690, 1997.
54. DE Gragson and GL Richmond. *J. Phys. Chem. B* 102(20): 3847–3861, 1998.
55. DE Gragson and GL Richmond. *J. Phys. Chem. B* 102(3): 569–576, 1998.
56. MJ Shultz, S Baldelli, C Schnitzer, and D Simonelli. *J. Phys. Chem. B* 106(21): 5313–5324, 2002.
57. Q Du, E Freysz, and YE Shen. *Science* 264(5160): 826–828, 1994.
58. EA Raymond, TL Tarbuck, MG Brown, and GL Richmond. *J. Phys. Chem. B* 107(2): 546–556, 2003.

59. EA Raymond, TL Tarbuck, and GL Richmond. *J. Phys. Chem. B* 106(11): 2817–2820, 2002.
60. A Morita and JT Hynes. *Chem. Phys.* 258(2–3): 371–390, 2000.
61. CI Ratcliffe and DE Irish. *J. Phys. Chem.* 86(25): 4897–4905, 1982.
62. D Simonelli, S Baldelli, and MJ Shultz. *Chem. Phys. Lett.* 298(4–6): 400–404, 1998.
63. OK Rice. *J. Phys. Chem.* 32(4): 583–592, 1928.
64. D Simonelli and MJ Shultz. *J. Chem. Phys.* 112(15): 6804–6816, 2000.
65. S Baldelli, C Schnitzer, MJ Shultz, and DJ Campbell. *J. Phys. Chem. B* 101(49): 10435–10441, 1997.
66. C Schnitzer, S Baldelli, and MJ Shultz. *Chem. Phys. Lett.* 313(3–4): 416–420, 1999.
67. S Baldelli, C Schnitzer, MJ Shultz, and DJ Campbell. *Chem. Phys. Lett.* 287(1–2): 143–147, 1998.
68. C Schnitzer, S Baldelli, DJ Campbell, and MJ Shultz. *J. Phys. Chem. A* 103(32): 6383–6386, 1999.
69. S Baldelli, C Schnitzer, and MJ Shultz. *J. Chem. Phys.* 108(23): 9817–9820, 1998.
70. S Baldelli, C Schnitzer, and MJ Shultz. *Chem. Phys. Lett.* 302(1–2): 157–163, 1999.
71. C Schnitzer, S Baldelli, and MJ Shultz. *J. Phys. Chem. B* 104(3): 585–590, 2000.
72. D Liu, G Ma, LM Levering, and HC Allen, *J. Phys. Chem. B* 108(7): 2252–2260, 2004.
73. DE Gragson and GL Richmond. *J. Am. Chem. Soc.* 120(2): 366–375, 1998.
74. XD Xiao, V Vogel, and YR Shen. *Chem. Phys. Lett.* 163(6): 555–559, 1989.
75. X Zhao, S Subrahmanyan, and KB Eisenthal. *Chem. Phys. Lett.* 171(5–6): 558–562, 1990.
76. X Zhao, S Ong, H Wang, and KB Eisenthal. *Chem. Phys. Lett.* 214(2): 203–207, 1993.
77. H Wang, X Zhao, and KB Eisenthal. *J. Phys. Chem. B* 104(37): 8855–8861, 2000.
78. SR Higgins, AG Stack, KG Knauss, C Eggleston, and G Jordan. Probing molecular-scale adsorption and dissolution-growth processes using nonlinear optical and scanning probe methods suitable for hydrothermal applications. In: *Water-Rock Interactions, Ore Deposits, and Environmental Geochemistry: A Tribute to David A*, RE Hellmann, ed. The Geochemical Society, 2002.
79. S Ong, X Zhao, and KB Eisenthal. *Chem. Phys. Lett.* 191(3–4): 327–335, 1992.
80. SR Higgins, A Stack, and CM Eggleston. *Mineral. Mag.* 62A: 616–617, 1998.
81. AG Stack, SR Higgins, and CM Eggleston. *Geochim. Cosmochim. Acta* 65(18): 3055–3063, 2001.
82. M Kosmulski. *Geochim. Cosmochim. Acta* 67(2): 319–320, 2003.
83. A Stack, SR Higgins, and CM Eggleston. *Geochim. Cosmochim. Acta* 67(2): 321–322, 2003.
84. Q Du, E Freysz, and YR Shen. *Phys. Rev. Lett.* 72(2): 238–241, 1994.
85. J Kim and PS Cremer. *J. Am. Chem. Soc.* 122(49): 12371–12372, 2000.
86. R Atkinson. *Atmos. Environ.* 34(12–14): 2063–2101, 2000.
87. HJ Tobias and PJ Ziemann. *J. Phys. Chem. A* 105(25): 6129–6135, 2001.
88. JJ Orlando, B Noziere, GS Tyndall, GE Orzechowska, SE Paulson, and Y Rudich (2000), *J. Geophys. Res.-Atmos.*, *105*(D9): 11561–11572.
89. CR Usher, AE Michel, and VH Grassian. *Chem. Rev.* 103(12): 4883–4940, 2003.
90. V Vaida and JE Headrick. *J. Phys. Chem. A* 104(23): 5402–5412, 2000.
91. JM Hicks, K Kemnitz, KB Eisenthal, and TF Heinz. *J. Phys. Chem.* 90(4): 560–562, 1986.
92. SR Goates, DA Schofield, and CD Bain. *Langmuir* 15(4): 1400–1409, 1999.
93. X Zhuang, PB Miranda, D Kim, and YR Shen. *Phys. Rev. B* 59(19): 12632–12640, 1999.
94. DE Gragson, BM McCarty, and GL Richmond. *J. Phys. Chem.* 100(34): 14272–14275, 1996.
95. R Edgar, JY Huang, R Popovitz-Biro, K Kjaer, WG Bouwman, PB Howes, J Als-Nielsen, YR Shen, M Lahav, and L Leiserowitz. *J. Phys. Chem. B* 104(29): 6843–6850, 2000.
96. V Vogel, CS Mullin, and YR Shen. *Langmuir* 7(6): 1222–1224, 1991.
97. GR Bell, ZX Li, CD Bain, P Fischer, and DC Duffy. *J. Phys. Chem. B* 102(47): 9461–9472, 1998.
98. T Rasing, T Stehlin, YR Shen, MW Kim, and P Valint. *J. Chem. Phys.* 89(5): 3386–3387, 1988.
99. MC Henry, YJ Yang, RL Pizzolatto, and MC Messmer. *Langmuir* 19(7): 2592–2598, 2003.
100. DJ Donaldson, JA Guest, and MC Goh. *J. Phys. Chem.* 99(23): 9313–9315, 1995.
101. AV Benderskii and KB Eisenthal. *J. Phys. Chem. B* 105(28): 6698–6703, 2001.
102. D Zimdars and KB Eisenthal. *J. Phys. Chem. B* 105(18): 3993–4002, 2001.
103. X Zhao, MC Goh, S Subrahmanyan, and KB Eisenthal. *J. Phys. Chem.* 94(9): 3370–3373, 1990.
104. DS Karpovich and D Ray. *J. Phys. Chem. B* 102(4): 649–652, 1998.
105. WJ De Bruyn, JA Shorter, P Davidovits, DR Worsnop, MS Zahniser, and CE Kolb (1994), *J. Geophys. Res.-Atmos.*, *99*(D8): 16927–16932.
106. X Zhao, MC Goh, and KB Eisenthal. *J. Phys. Chem.* 94(6): 2222–2224, 1990.
107. G Ma and HC Allen. *J. Phys. Chem. B* 107(26): 6343–6349, 2003.

108. S Baldelli. *J. Phys. Chem. B* 107(25): 6148–6152, 2003.
109. EL Hommel and HC Allen. *Analyst* 128(6): 750–755, 2003.
110. DL Sparks, ed. D.L. Sparks. 2nd ed. 1999: *Soil Physical Chemistry* Boca Raton: CRC Press.
111. LF Scatena, MG Brown, and GL Richmond. *Science* 292(5518): 908–912, 2001.
112. K Nochi, A Yamaguchi, T Hayaskita, T Uchida, and N Teramae. *J. Phys. Chem. B* 106(38): 9906–9911, 2002.
113. HA Al-Abadleh and VH Grassian. *Surf. Sci. Rep.* 52(3–4): 63–162, 2003.
114. B Zuberi, AK Bertram, T Koop, LT Molina, and MJ Molina. *J. Phys. Chem. A* 105(26): 6458–6464, 2001.
115. PB Miranda, V Pflumio, H Saijo, and YR Shen. *Chem. Phys. Lett.* 264(3–4): 387–392, 1997.
116. KA Becraft, FG Moore, and GL Richmond. *J. Phys. Chem. B* 107(16): 3675–3678, 2003.
117. HS Choi. *Can. Metall. Q.* 2: 410–414, 1963.
118. FG Moore, KA Becraft, and GL Richmond. *Appl. Spec.* 56(12): 1575–1578, 2002.
119. A Ulman. *Chem. Rev.* 96: 1533, 1996. Add issue ; 4 and final page; 1554
120. F Schreiber. *Prog. Surf. Sci.* 65(5–8): 151–256, 2000.
121. N Himmelhaus, F Eisert, M Buck, and M Grunze. *J. Phys. Chem. B* 104(3): 576–584, 2000.
122. CS-C Yang, LJ Richter, JC Stephenson, and KA Briggman. *Langmuir* 18(20): 7549–7556, 2002.
123. Y Liu, LK Wolf, and MC Messmer. *Langmuir* 17(14): 4329–4335, 2001.
124. NH Fletcher. *The Chemical Physics of Ice* London: Cambridge University Press, 1970.
125. CM Bouchez and JM Hicks. *Proc. SPIE Int. Soc. Opt. Eng.* 2547: 153–163, 1995.
126. VP Evangelou. *Environmental Soil and Water Chemistry* New York: John Wiley & Sons, 1998.
127. WJ Weber, EJ Leboeuf, TM Young, and W Huang (2001), *Wat. Res.*, *35*(4): 853–868.
128. GE Brown and GA Parks. *Int. Geo. Rev.* 43(11): 963–1073, 2001.
129. LA Warren and EA Haack. *Earth-Sci. Rev.* 54(4): 261–320, 2001.
130. H Van de Weerd, WH Van Riemsdijk, and A Leijnse. *Environ. Sci. Technol.* 33(10): 1675–1681, 1999.
131. KK Au, AC Penisson, SL Yang, and CR O'Melia. *Geochim. Cosmochim. Acta* 63(19–20): 2903–2917, 1999.
132. LK Koopal, WH van Riemsdijk, and DG Kinniburgh. *Pure Appl. Chem.* 73(12): 2005–2016, 2001.
133. JD Kubicki, GA Blake, and SE Apitz. *Geochim. Cosmochim. Acta* 61(5): 1031–1046, 1997.
134. H van de Weerd, WH van Riemsdijk, and A Leijnse (2002), *Water Resour. Res.*, *38*(8), 33–1 – 33–19.
135. JP Gustafsson, P Pechova, and D Berggren. *Environ. Sci. Technol.* 37(12): 2767–2774, 2003.
136. SA Welch and WJ Ullman. *Geochim. Cosmochim. Acta* 57(12): 2725–2736, 1993.
137. JA Davis. *Geochim. Cosmochim. Acta* 46(11): 2381–2393, 1982.
138. D Vasudevan, EM Cooper, and OL Van Exem. *Environ. Sci. Technol.* 36(3): 501–511, 2002.
139. K Namjesnik-Dejanovic and PA Maurice. *Geochim. Cosmochim. Acta* 65(7): 1047–1057, 2000.
140. P Reiller, V Moulin, F Casanova, and C Dautel (2002), *Appl. Geochem.*, *17*(12): 1551–1562.
141. AK Aufdenkampe, JI Hedges, JE Richey, AV Krusche, and CA Llerena. *Limn. Oceanogr.* 46(8): 1921–1935, 2001.
142. S Jayasundera and A Torrents. *J. Environ. Eng.-Asce* 123(11): 1162–1165, 1997.
143. R Kretzschmar, H Holthoff, and H Sticher. *J. Colloid Inter. Sci.* 202(1): 95–103, 1998.
144. JD Kubicki, LM Schroeter, MJ Itoh, BN Nguyen, and SE Apitz. *Geochim. Cosmochim. Acta* 63(18): 2709–2725, 1999.
145. K Namjesnik-Dejanovic, PA Maurice, GR Aiken, S Cabaniss, YP Chin, and MJ Pullin. *Soil Sci.* 165(7): 545–559, 2000.
146. CH Specht and FH Frimmel. *Phys. Chem. Chem. Phys.* 3(24): 5444–5449, 2001.
147. CH Specht, MU Kumke, and FH Frimmel (2000), *Water Res.*, *34*(16): 4063–4069.
148. J Hur and MA Schlautman. *J. Colloid Inter. Sci.* 264(2): 313–321, 2003.
149. M Meier, K Namjesnik-Dejanovic, PA Maurice, YP Chin, and GR Aiken. *Chem. Geol.* 157(3–4): 275–284, 1999.
150. J Wang, MA Even, X Chen, AH Schmaier, JH Waite, and Z Chen. *J. Am. Chem. Soc.* 125(33): 9914–9915, 2003.
151. JS Salafsky and KB Eisenthal. *Chem. Phys. Lett.* 319(5–6): 435–439, 2000.
152. MR Watry and GL Richmond. *J. Phys. Chem. B* 106(48): 12517–12523, 2002.
153. VH Grassian. *J. Phys. Chem. A* 106(6): 860–877, 2002.
154. S Solomon (1988), *Rev. Geophys.*, *26*(1): 131–148.

155. AR Ravishankara. *Science* 276(5315): 1058–1065, 1997.
156. BJ Finlayson-Pitts and JC Hemminger. *J. Phys. Chem. A* 104(49): 11463–11477, 2000.
157. MA Zondlo, PK Hudson, AJ Prenni, and MA Tolbert. *Annu. Rev. Phys. Chem.* 51: 473–499, 2000.
158. VH Grassian. *Int. Rev. Phys. Chem.* 20(3): 467–548, 2001.
159. HC Allen, JM Laux, R Vogt, BJ Finlayson-Pitts, and JC Hemminger. *J. Phys. Chem.* 100(16): 6371–6375, 1996.
160. Y Zhang and GR Carmichael. *J. Appl. Meteorol.* 38(3): 353–366, 1999.
161. F Raes, T Bates, F McGovern, and M Van Liedekerke. *Tellus Ser. B-Chem. Phys. Meteorol.* 52(2): 111–125, 2000.
162. M de Reus, J Strom, J Curtius, L Pirjola, E Vignati, F Arnold, HC Hansson, M Kulmala, J Lelieveld, and F Raes (2000), *J. Geophys. Res.-Atmos.*, *105*(D20): 24751–24762.
163. M de Reus, F Dentener, A Thomas, S Borrmann, J Strom, and J Lelieveld (2000), *J. Geophys. Res.-Atmos.*, *105*(D12): 15263–15275.
164. JR Parrington. *Science* 220(4598): 666, 1983.
165. JR Parrington, WH Zoller, and NK Aras. *Science* 220(4593): 195–197, 1983.
166. Visible Earth: http:–/visibleearth.nasa.gov/Atmosphere/Aerosols/Dust_Ash.html. 2002, Goddard Space Flight Center.
167. JM Prospero, K Barrett, T Church, F Dentener, RA Duce, JN Galloway, H Levy, J Moody, and P Quinn. *Biogeochemistry* 35(1): 27–73, 1996.
168. ACaDB NASA Code 916. Chinese Dust Strom, April 1998: http://toms.gsfc.nasa.gov/aerosols/china_1998.html, 1998.
169. D Pimentel, C Harvey, P Resosudarmo, K Sinclair, D Kurz, M McNair, S Crist, L Shpritz, L Fitton, R Saffouri, and R Blair. *Science* 269(5223): 464–465, 1995.
170. D Pimentel, C Harvey, P Resosudarmo, K Sinclair, D Kurz, M McNair, S Crist, L Shpritz, L Fitton, R Saffouri, and R Blair. *Science* 267(5201): 1117–1123, 1995.
171. SW Trimble and P Crosson. *Science* 289(5477): 248–250, 2000.
172. BJ Wienhold, A Luchiari, and R Zhang. *Ann. Arid Zone* 39(3): 333–346, 2000.
173. BJ Wienhold, JF Power, and JW Doran. *Soil Sci.* 165(1): 13–30, 2000.
174. MJ Nieuwenhuijsen and MB Schenker. *Am. Ind. Hyg. Assoc. J.* 59(1): 9–13, 1998.
175. M Schenker. *Environ. Health Perspect.* 108: 661–664, 2000.
176. MB Schenker, MR Orenstein, CL Saiki, and SJ Samuels. *Am. J. Respir. Crit. Care Med.* 159(3): A297–A297, 1999.
177. M Baker. *Nature* 413(6856): 586–587, 2001.
178. HR Prupacher and JD Klett. *Microphysics of Clouds and Precipation* Boston: Kluwer Academic Press, 1997.
179. P Godefroy, W de Jong, CW van Hasselt, MAC Devillers, and T Rasing. *Appl. Phys. Lett.* 68(14): 1981–1983, 1996.
180. XQ Zhang, XZ You, SH Ma, and Y Wei. *J. Mater. Chem.* 5(4): 643–647, 1995.
181. JH Moon, JH Kim, KJ Kim, TH Kang, B Kim, CH Kim, JH Hahn, and JW Park. *Langmuir* 13(16): 4305–4310, 1997.
182. SB Bakiamoh and GJ Blanchard. *Langmuir* 17(11): 3438–3446, 2001.
183. SB Bakiamoh and GJ Blanchard. *Langmuir* 18(16): 6246–6253, 2002.
184. U Diebold. *Surf. Sci. Rep.* 48(5–8): 53–229, 2003.
185. E Kobayashi, T Wakasugi, G Mizutani, and S Ushioda. *Surf. Sci.* 402–404: 537–541, 1998.
186. C-Y Wang, H Groenzin, and MJ Shultz. *Langmuir* 19(18): 7330–7334, 2003.
187. AN Shultz, WM Hetherington, III, DR Baer, L-Q Wang, and MH Engelhard. *Surf. Sci.* 392(1–3): 1–7, 1997.
188. PB Miranda, L Xu, YR Shen, and M Salmeron. *Phys. Rev. Lett.* 81(26): 5876–5879, 1998.
189. F Mugele, S Baldelli, GA Somorjai, and M Salmeron. *J. Phys. Chem. B* 104(14): 3140–3144, 2000.
190. JC Zink, J Reif, and E Matthias. *Phys. Rev. Lett.* 68(24): 3595–3598, 1992.
191. B Wassermann, J Reif, and E Matthias. *Phys. Rev. B* 50(4): 2593–2597, 1994.
192. MJ Molina, T-L Tso, LT Molina, and FC-Y Wang. *Science* 238(4831): 1253–1257, 1987.
193. MA Tolbert, MJ Rossi, R Malhotra, and DM Golden. *Science* 238(4831): 1258–1260, 1987.
194. PJ Crutzen and F Arnold. *Nature* 324(6098): 651–655, 1986.
195. S Solomon (1999), *Rev. Geophys.*, *37*(3): 275–316.

196. JPD Abbatt. *Chem. Rev.* 103(12): 4783–4800, 2003.
197. M Faraday. *Philos. Mag.* 17: 162, 1859.
198. NH Fletcher. *Philos. Mag.* 18(156): 1287–1300, 1968.
199. X Wei, PB Miranda, C Zhang, and YR Shen. *Phys. Rev. B* 66(8): 085401/085401–085401/085413, 2002.
200. C Girardet and C Toubin. *Surf. Sci. Rep.* 44(7–8): 159–238, 2001.
201. VF Petrenko and RW Whitworth. *Physics of Ice* New York: Oxford University Press, 1999.
202. DE Brown and SM George. *J. Phys. Chem.* 100(38): 15460–15469, 1996.
203. R Oppliger, A Allanic, and MJ Rossi. *J. Phys. Chem. A* 101(10): 1903–1911, 1997.
204. DR Hanson and AR Ravishankara. *J. Phys. Chem.* 96(6): 2682–2691, 1992.
205. DH Gracias, Z Chen, YR Shen, and GA Somorjai. *Accounts Chem. Res.* 32(11): 930–940, 1999.
206. Z Chen, R Ward, Y Tian, F Malizia, D Gracias, YR Shen, and GA Somorjai. *J. Biomed. Mater. Res.* 62(2): 254–264, 2002.
207. Z Chen, YR Shen, and GA Somorjai. *Ann. Rev. Phys. Chem.* 53: 437–465, 2002.
208. D Brinkley and T Engel. *J. Phys. Chem. B* 104(42): 9836–9841, 2000.
209. D Brinkley and T Engel. *Surf. Sci.* 415(3): L1001–L1006, 1998.
210. DW Goodman. *Surf. Rev. Lett.* 2(1): 9–24, 1995.
211. J Szanyi and DW Goodman. *Rev. Sci. Instrum.* 64(8): 2350–2352, 1993.
212. DR Kahn, EE Petersen, and GA Somorjai. *J. Catal.* 34(2): 294, 1974.
213. DW Blakely, EI Kozak, BA Sexton, and GA Somorjai. *J. Vac. Sci. Technol.* 13(5): 1091–1096, 1976.
214. DW Goodman, RD Kelley, TE Madey, and JT Yates Jr. *J. Catal.* 63(1): 226–234, 1980.
215. CT Campbell and MT Paffett. *Surf. Sci.* 139(2–3): 396–416, 1984.
216. JJ Vajo, W Tsai, and WH Weinberg. *Rev. Sci. Instrum.* 56(7): 1439–1442, 1985.
217. J Miragliotta, RS Polizotti, P Rabinowitz, SD Cameron, and RB Hall. *Appl. Phys. A* 51(3): 221–225, 1990.
218. C Klünker, M Balden, S Lehwald, and W Daum. *Surf. Sci.* 360(1–3): 104–111, 1996.
219. X Su, PS Cremer, YR Shen, and GA Somorjai. *J. Am. Chem. Soc.* 119(17): 3994–4000, 1997.
220. G Rupprechter, T Dellwig, H Unterhalt, and H-J Freund. *Top. Catal.* 15(1): 19–26, 2001.
221. GA Somorjai and G Rupprechter. *J. Phys. Chem. B* 103(15): 1623–1638, 1999.
222. A Bandara, S Dobashi, J Kubota, K Onda, A Wada, K Domen, H Hirose, and SS Kano. *Surf. Sci.* 387(1–3): 312–319, 1997.
223. H Härle, A Lehnert, U Metka, H-R Volpp, L Willms, and J Wolfrum. *Chem. Phys. Lett.* 293(1–2): 26–32, 1998.
224. T Dellwig, G Rupprechter, H Unterhalt, and H-J Freund. *Phys. Rev. Lett.* 85(4): 776–779, 2000.
225. X Su, J Jensen, MX Yang, MB Salmeron, YR Shen, and GA Somorjai. *Faraday Discus.* 105: 263–274, 1997.
226. PS Cremer, X Su, YR Shen, and GA Somorjai. *J. Am. Chem. Soc.* 118(12): 2942–2949, 1996.
227. PS Cremer, BJ McIntyre, M Salmeron, YR Shen, and GA Somorjai. *Catal. Lett.* 34(1–2): 11, 1995.
228. GA Somorjai, X Su, KR McCrea, and KB Rider. *Top. Catal.* 8(1–2): 23–34, 1999.
229. X Su, KY Kung, J Lahtinen, YR Shen, and GA Somorjai. *J. Mol. Catal. A.: Chem.* 141(1–3): 9–19, 1999.
230. X Su, K Kung, J Lahtinen, RY Shen, and GA Somorjai. *Catal. Lett.* 54(1–2): 9–15, 1998.
231. D Zhang, RS Ward, YR Shen, and GA Somorjai. *J. Phys. Chem. B* 101(44): 9060–9064, 1997.
232. H Wang, E Borguet, and KB Eisenthal. *J. Phys. Chem. B* 102(25): 4927–4932, 1998.
233. X Zhang and RA Walker. *Langmuir* 17(15): 4486–4489, 2001.
234. X Zhang, MM Cunningham, and RA Walker. *J. Phys. Chem. B* 107(14): 3183–3195, 2003.
235. CL Beildeck, WH Steel, and RA Walker. *Langmuir* 19(12): 4933–4939, 2003.
236. EL Hommel and HC Allen. *J. Phys. Chem. B* 107(39): 10823–10828, 2003.
237. J Wang, SE Woodcock, SM Buck, C Chen, and Z Chen. *J. Am. Chem. Soc.* 123: 9470–9471, 2001.
238. C Chen, J Wang, MA Even, and Z Chen. *Macromolecules* 35(21): 8093–8097, 2002.
239. J Wang, Z Paszti, MA Even, and Z Chen. *J. Am. Chem. Soc.* 124(24): 7016–7023, 2002.
240. XM Shang, AV Benderskii, and KB Eisenthal. *J. Phys. Chem. B* 105(47): 11578–11585, 2001.
241. AV Benderskii and KB Eisenthal. *J. Phys. Chem. A* 106(33): 7482–7490, 2002.
242. ECY Yan, Y Liu, and KB Eisenthal. *J. Phys. Chem. B* 102(33): 6331–6336, 1998.

243. H Wang, T Troxler, A-G Yeh, and H-L Dai. *Langmuir* 16(6): 2475–2481, 2000.
244. JS Salafsky and KB Eisenthal. *J. Phys. Chem. B* 104(32): 7752–7755, 2000.
245. JI Dadap, J Shan, KB Eisenthal, and TF Heinz. *Phys. Rev. Lett.* 83(20): 4045–4048, 1999.
246. ECY Yan and KB Eisenthal. *J. Phys. Chem. B* 103(29): 6056–6060, 1999.
247. ECY Yan and KB Eisenthal. *Biophys. J.* 79(2): 898–903, 2000.
248. ECY Yan, Y Liu, and KB Eisenthal. *J. Phys. Chem. B* 105(36): 8531–8537, 2001.
249. G Ma and HC Allen. *J. Am. Chem. Soc.* 124(32): 9374–9375, 2002.
250. PW Atkins. 6th ed. *Physical Chemistry* Oxford: Oxford University Press, 1998.
251. T Colborn. *Environ. Health Perspect.* 103: 135–136, 1995.
252. H Fang, W Tong, LM Shi, R Blair, R Perkins, W Branham, BS Hass, Q Xie, SL Dial, CL Moland, and DM Sheehan. *Chem. Res. Toxicol.* 14(3): 280–294, 2001.
253. TA Hanselman, DA Graetz, and AC Wilkie. *Environ. Sci. Technol.* 37(24): 5471–5478, 2003.
254. S Jobling, M Nolan, CR Tyler, G Brighty, and JP Sumpter. *Environ. Sci. Technol.* 32(17): 2498–2506, 1998.
255. JM Hicks. *ACS Symposium Series 810*. Oxford: Oxford University Press, 2002.
256. JD Byers, HI Yee, and JM Hicks. *J. Chem. Phys.* 101(7): 6233–6241, 1994.
257. HI Yee, JD Byers, and JM Hicks. *Proc. SPIE—Int. Soc. Opt. Eng.* 2125: 119–131, 1994.
258. T Petralli-Mallow, T Maeda-Wong, JD Byers, HI Yee, and JM Hicks. *J. Phys. Chem.* 97(7): 1383–1388, 1993.
259. JD Byers, HI Yee, T Petralli-Mallow, and JM Hicks. *Phys. Rev. B.* 49(20): 14643–14647, 1994.
260. BJ Burke, AJ Moad, MA Polizzi, and GJ Simpson. *J. Am. Chem. Soc.* 125(30): 9111–9115, 2003.
261. MA Kriech and JC Conboy. *J. Am. Chem. Soc.* 125(5): 1148–1149, 2003.
262. SG Benka. *Phys. Today* April: 38–39, 2002.
263. A Tabazadeh and RP Turco (1993), *J. Geophys. Res.-Atmos.*, *98*(D7): 12727–12740.
264. D Rai, JM Zachara, LE Eary, CC Ainsworth, JE Amonette, CE Cowan, RW Szelmeczka, CT Resch, RL Schmidt, DC Girvin, and SC Smith. *Chromium Reactions in Geological Materials*, Electric Power Research Institute, 1988.
265. KA Becraft and GL Richmond. *Langmuir* 17(25): 7721–7724, 2001.
266. S Baldelli, C Schnitzer, MJ Shultz, and DJ Campbell. *J. Phys. Chem. B* 101(23): 4607–4612, 1997.
267. X Wei and YR Shen. *Appl. Phys. B: Lasers Opt.* 74(7–8): 617–620, 2002.
268. TF Heinz, CK Chen, D Ricard, and YR Shen. *Phys. Rev. Lett.* 48(7): 478–481, 1982.
269. T Kikteva, D Star, Z Zhao, TL Baisley, and GW Leach. *J. Phys. Chem. B* 103(7): 1124–1133, 1999.
270. T Kikteva, D Star, and GW Leach. *J. Phys. Chem. B* 104(13): 2860–2867, 2000.
271. DA Higgins, SK Byerly, MB Abrams, and RM Corn. *J. Phys. Chem.* 95(18): 6984–6990, 1991.
272. H Wang, ECY Yan, E Borguet, and KB Eisenthal. *Chem. Phys. Lett.* 259(1–2): 15–20, 1996.

5 Environmental Catalysis in the Earth's Atmosphere: Heterogeneous Reactions on Mineral Dust Aerosol

Elizabeth R. Johnson[1] *and Vicki H. Grassian*[1,2,3]
Departments of [1]Chemistry and [2]Chemical and Biochemical Engineering, and the [3]Center for Global and Regional Environmental Research, University of Iowa

CONTENTS

5.1 INTRODUCTION — MINERAL DUST AEROSOL: A SOURCE OF POTENTIALLY CATALYTIC REACTIVE SURFACES IN THE ATMOSPHERE

There is a great deal of interest in particulate matter in the atmosphere as reactions that take place on the surface of these particles can alter the chemical balance of the atmosphere [1–7]. Because mineral dust makes up a significant fraction of the tropospheric aerosol mass, there is the potential that surface reactions involving mineral dust aerosol may be important in tropospheric chemistry [7, 8]. Surface reactions of trace atmospheric gases on mineral dust aerosol can provide reaction pathways that differ from those in the gas phase. Additionally, surface adsorption on mineral dust can be a sink or loss mechanism for molecules from the gas phase.

Mineral dust aerosol originates from soil particles that have been mobilized by strong wind currents and entrained into the atmosphere. Since these particles are eroded soils, their chemical composition is similar, if not identical, to crustal rock [9]. The Earth's crust is dominated by silicon and aluminum oxides. Several studies (see Ref. [9] and references therein) on the elemental content of windblown dust originating in various locations around the world report that mineral dust is approximately 60% SiO_2 and 10–15% Al_2O_3. The percentage of other oxides, such as Fe_2O_3, MgO, and CaO, are slightly more varied and depend on source location.

Although the average chemical composition may be similar in locations around the world, the mineralogy of dust particles can be quite varied. Common minerals found in dust aerosol include quartz, feldspars, micas, chlorite, kaolinite, illite, smectite, palygorskite, calcite, dolomite, gypsum, halite, opal, and mixed layer clays [10]. The chemical composition of these minerals is given in Table 5.1 [11, 12]. Within a given size distribution of transported aerosol, the larger, coarser particles are typically composed of quartz, feldspars, and carbonates, and the smaller, finer particles are often clays or micas. As the dust is transported farther away from a source region, the overall composition tends to become enriched with clays as the larger quartz particles fall out of the atmosphere through gravitational settling.

A majority of the mineral dust aerosol comes from desert regions in Asia and Africa. Mineral dust aerosol originating from North Africa has a slightly different elemental content compared to Asian dust, and therefore different mineral content and properties. For instance, Asian dusts tend to be grayish in color whereas dusts from African regions have more brown, yellow, and red tones, which is due in large part to the higher iron oxide (hematite) content in

TABLE 5.1
Chemical Formulae for Common Clays and Minerals in Dust[a]

Mineral	Formula[a]
Calcite	$CaCO_3$
Chlorite	$A_{5-6}Z_4O_{10}(OH)_8^b$
Corundum	α-Al_2O_3
Dolomite	$CaMg(CO_3)_2$
Feldspars	$WZ_4O_8^c$
Gypsum	$CaSO_4{\cdot}2H_2O$
Halite	NaCl
Hematite	α-Fe_2O_3
Illite	$(K,H_3O)(Al,Mg,Fe)_2(Si,Al)_4O_{10}[(OH)_2,H_2O]$
Kaolinite	$Al_4Si_4O_{10}(OH)_8$
Magnesite	$MgCO_3$
Montmorillonite (Smectite)	$(Na,Ca)_{0.33}(Al,Mg)_2Si_4O_{10}(OH)_2{\cdot}nH_2O$
Mica	$W(X,Y)_{2-3}Z_4O_{10}\ (OH,F)_2^d$
Opal	$SiO_2{\cdot}nH_2O$
Palygorskite	$(Mg,Al)_2Si_4O_{10}(OH){\cdot}4H_2O^e$
Quartz	SiO_2

[a]*Source*: Refs. [11] and [12].
[b]Typically A = Al, Fe, Li, Mg, Mn and/or Ni; Z = Al, B, Si, and/or Fe.
[c]Typically W = Na, K, Ca, and/or Ba; Z = Si and/or Al.
[d]Typically W = K or Na; X and Y = Al, Mg, Fe^{2+}, Fe^{3+}, Li; Z = Si and Al.
[e]From Ref. [9].

African soils. Iron oxide has very different optical properties compared to most other minerals [13]. Furthermore, even the mineral content of dust originating within the Saharan desert varies regionally and differences are observed whether a plume came from the northern, western, central, or southern region of the desert [14]. Investigations report that dust from the northern Sahara is abundant in illite [15, 16], as well as carbonates, chlorite, palygorskite, and montmorillonite [17] whereas the southern Sahara and Sahel regions contain more kaolinite and hematite [15–17].

Claquin et al. have attempted to model the mineralogy of atmospheric dust sources [18]. Prospero et al. have identified a "dust belt" in northern latitudes [19]. The "dust belt" extends from the west coast of North Africa through the Middle East, Central and South Asia all the way to China. The "dust belt" consists of the largest and most persistent sources of dust in the world. Knowledge of these dust sources and understanding the variations in regional mineralogy will help provide insight into the chemical reactivity of mineral dust particles from different source regions.

5.2 POSSIBLE TYPES OF SURFACE REACTIONS ON MINERAL DUST

Several potentially important reaction mechanisms involving mineral dust are discussed here. These different reaction mechanisms are presented in such a way that they can be incorporated into atmospheric chemistry models. The importance of atmospheric chemistry models in understanding heterogeneous reactions on mineral dust aerosol is discussed in Section 5.3.

Generalized schemes for the different reaction mechanisms involving trace atmospheric gases and mineral dust aerosol are shown in Figure 5.1. Scheme 5.1 represents the nonreactive reversible adsorption of a gas, G, taken up by a surface according to

$$\mathrm{G} + \mathrm{site} \leftrightarrows \mathrm{G(a)} \tag{5.1}$$

Assuming a simple Langmuir-type adsorption mechanism, the coverage of G on the surface, θ, will depend on the partial pressure of G in the atmosphere, the temperature and the Langmuir equilibrium constant, K. The equilibrium constant is defined as the ratio of the rate of adsorption onto surface sites divided by the rate of desorption. This type of interaction for the adsorption of gases onto stratospheric particles has been previously discussed by Tabazadeh and Turco [20]. Under steady-state conditions, the equilibrium coverage is given as

$$\theta = \frac{\mathrm{KP}}{1 + \mathrm{KP}} \tag{5.2}$$

and the uptake or sticking coefficient, γ, is related to the coverage according to the equation

$$\gamma = \gamma_0(1 - \theta) \tag{5.3}$$

where γ_0 is the initial uptake coefficient. Thus, the uptake coefficient will be coverage and time-dependent as the surfacc approaches an equilibrium coverage. It is important to note that Scheme 1 in Figure 5.1 also implies that as mineral dust particles are transported in the atmosphere from region to region the adsorbed gas will be released from the particle, i.e., there will be desorption from the surface, when there is a decrease in the partial pressure of that gas. Thus, mineral dust potentially represents a viable way to transport molecules through the atmosphere and can be a source of gas-phase pollutants. This may be especially important in the case of semivolatile organic compounds.

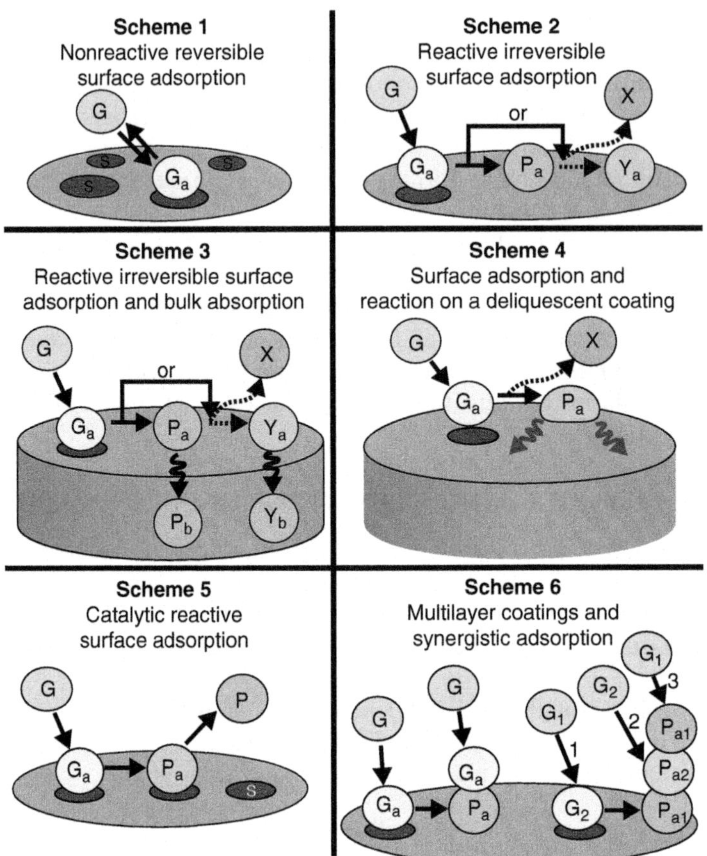

FIGURE 5.1 Schematic diagram of different reaction schemes for uptake of gases on mineral dust. Various chemical components include surface sites (S), gas-phase reactants (G), and product species (P, X, and Y), which can be released into the gas phase, adsorbed to the surface (subscripted a), or incorporated into the bulk (subscripted b). Numbers located along arrows represent sequential reactions while subscripted numbers signify different species of the same phase (i.e., G_1 is a different reactant gas than G_2). These schemes can be incorporated into atmospheric chemistry models. See text for further details.

Scheme 2 in Figure 5.1 represents the adsorption of a gas molecule followed by surface reaction, with and without the formation of gas-phase products, according to reactions (5.4a) and (5.4b), respectively,

$$G(g) + \text{site} \longrightarrow G(a) \longrightarrow P(a) \tag{5.4a}$$

or

$$G(g) + \text{site} \longrightarrow G(a) \longrightarrow Y(a) + X(g) \tag{5.4b}$$

In this case, a constraint on the adsorption needs to be considered because as surface sites become blocked in this mechanism by surface-bound products (P or Y, with and without the formation of gas-phase products), the reaction should cease. This case was considered earlier by Tabazadeh and Turco [20] and more recently by Ammann et al. [21]. In the study by

Ammann et al., observed uptake coefficients on solid surfaces were parameterized in terms of reversible adsorption followed by surface reaction with a surface site that could be, for example, a preadsorbed molecule. Ammann et al. also showed that the coverage, and therefore time-dependent uptake could be used to model laboratory data. This model was specifically applied to reactions of ozone with organic-coated soot and organic aerosol, but should be applicable to mineral dust chemistry as well.

Scheme 3 in Figure 5.1 represents the reversible adsorption of a gas molecule followed by surface reaction, with and without the formation of gas-phase products and absorption of the product into the bulk of the surface. In this case, surface sites are regenerated and the reactivity of the mineral dust particle will be constrained by the total volume of the particle and not just the number of available surface sites.

Scheme 4 in Figure 5.1 represents surface adsorption of a gas molecule on a deliquescent film coating the dust particle, followed by solvation into the aqueous phase, diffusion and reaction within the film, with and without the formation of gas-phase products. In this case, the uptake may potentially be described by gas uptake onto an aqueous droplet of similar composition. The modeling of gas uptake into aqueous droplets has been previously reviewed [22, 23].

Scheme 5 represents a catalytic reaction, i.e., the rate of the reaction can be described to continuously occur in the atmosphere without saturation. However, it should be noted that even for catalytic reactions, the reactive uptake coefficient can decrease over time until a steady-state uptake coefficient is reached, i.e., some of the more reactive sites may become less active over time.

Scheme 6 represents an even greater level of complexity whereby multilayer coatings can form on the surface and synergistic adsorption can also occur, i.e., an adsorbed molecule may assist in the adsorption of another molecule by providing a reactive site (e.g., adsorbed C=C for ozone reaction) or a medium (e.g., adsorbed H_2O for HNO_3 uptake). Multilayer coatings and synergistic adsorption was considered by Mozurkewich in his work on modeling adsorption and chemistry important in polar stratospheric clouds [24]. The reader is referred to that paper for additional details.

The above discussion examines reaction mechanisms that may need to be included in atmospheric chemistry models. As discussed below, it is important that the models increase the level of sophistication of reaction mechanisms involving heterogeneous reactions on mineral dust if they are to accurately depict the chemistry of the troposphere.

5.3 THE ROLE OF MODELING ANALYSIS, LABORATORY STUDIES, AND FIELD MEASUREMENTS IN UNDERSTANDING SURFACE REACTIONS IN THE ATMOSPHERE

An understanding of the Earth's atmosphere involves field measurements, modeling analyses, and laboratory studies. These three endeavors provide the necessary information needed to sort out the complexity of chemical processes in the Earth's atmosphere. Atmospheric chemistry models are a very powerful way to evaluate the effects that heterogeneous reactions on aerosols may have on the atmosphere. These models can simultaneously take into account aerosol properties (e.g., particle composition and size distributions), gas-phase and heterogeneous chemistry, meteorology and transport processes. Therefore, computer modeling is becoming increasingly important as it can be used to explain the impact that a particular event or emission of a specific species may have on the balance of atmospheric processes and climate. Atmospheric chemistry models utilize data obtained from laboratory and field measurements. Models can be useful tools in aiding the future direction of laboratory or field investigations. For example, sensitivity analyses can be used to determine which reactions are most sensitive to the value of the rate and will have the greatest impact on other coupled processes. In addition, models have the potential to provide predictive capabilities.

The role of mineral dust aerosol as a reactive surface in the troposphere was first discussed in detail in Dentener et al. [25]. In this modeling study, the impact of heterogeneous chemistry on the nitrogen, sulfur, and photochemical oxidant cycles was investigated. The mineral dust reactions of interest included those with SO_2, HO_2, N_2O_5, HNO_3, and O_3. Because the gas-phase chemistry in the troposphere is very complex, a three-dimensional model was utilized to integrate heterogeneous reactions and to simulate sources, transport, and removal processes of mineral dust. Heterogeneous reactions were treated as pseudo-first-order processes in which the rate coefficient (k_j) was calculated according to the following equation:

$$k_j = \int_{r_1}^{r_2} k_{d,j}(r)n(r)\mathrm{d}(r). \tag{5.5}$$

where $n(r)\mathrm{d}(r)$ is the number density of particles between r (aerosol radius) and $r + \mathrm{d}r$, in units of cm^{-4}, and $k_{d,j}$ is a size-dependent mass transfer coefficient in $\mathrm{cm}^3\ \mathrm{s}^{-1}$. The term $k_{d,j}$ is a function of the mass accommodation coefficient (α), where α is defined as the probability that a gas–surface collision will result in the adsorption of the gas molecule by the surface. Often α is used to describe gas collisions with aqueous surfaces because it accounts for the solvation of a gas molecule into an aqueous phase. However, when considering the uptake coefficient on a solid surface without consideration of solvation, the reaction probability or sticking coefficient (γ) is typically defined as the ratio of the number of gas molecules that are lost from the gas phase per second to the total number of gas–surface collisions per second. Therefore, $k_{d,j}$ was determined from

$$k_{d,j} = \frac{4\pi r D_j V}{1 + K_n(\lambda + 4(1-\gamma)/3\gamma)} \tag{5.6}$$

where D_j is the gas-phase molecular diffusion coefficient in $\mathrm{cm}^2\ \mathrm{s}^{-1}$, K_n is the Knudsen number ($=\lambda/r$), λ is the effective mean free path of a gas molecule in air, and V is a ventilation coefficient. The variable γ is the important parameter under consideration in the current review because its value is determined from laboratory experiments.

Another way to express the pseudo-first-order rate constant for the loss of a gas onto an aerosol surface through adsorption processes, as described by Preszler Prince et al. [26], is

$$k = \frac{\gamma(\bar{c}/4)[C_{\mathrm{mass}}]S}{1 + (\gamma/f(K_n))} \tag{5.7}$$

where $\bar{c}$ is the mean speed of the gas-phase species in $\mathrm{m\ s}^{-1}$, C_{mass} is the concentration of the sample in $\mathrm{g\ m}^{-3}$, and S is the specific surface area of the aerosol in $\mathrm{m}^2\ \mathrm{g}^{-1}$ [26]. The λ and K_n terms are defined in the same way as in Equation (7.6). Equation (7.7) takes into account the gas diffusion to the particle through the Fuchs–Sutugin correction term, $f(K_n)$, where

$$f(K_n) = \frac{K_n(K_n + 1)}{0.75 + 0.283K_n} \tag{5.8}$$

In the flow regime where the mean free path of the gas is much larger than the particle radius and in the limit where $\gamma \ll K_n$, Equation (5.7) simplifies to

$$k = \gamma(\bar{c}/4)[C_{\mathrm{mass}}]S \tag{5.9}$$

and the rate constant is only dependent on the reactive uptake and the gas kinetic collision rate. Conversely, when K_n is relatively small compared to γ, then by following a method similar to the one by Lovejoy and Hanson [27],

$$k = \frac{D[C_{\text{mass}}]S}{r} \tag{5.10}$$

where D is the diffusion coefficient in $m^2\ s^{-1}$, arriving from the relationship that $\lambda = 2D/\bar{c}$. Therefore in this case, the rate constant is limited by the diffusion of the gas to the aerosol surface.

To gain an understanding of the overall effect that heterogeneous chemistry can have on the atmosphere, Dentener et al. included reaction probabilities for multiple species within the nitrogen, sulfur, and photochemical oxidant cycles. The γ values that were used in the model included those for HO_2 (0.1), N_2O_5 (0.1), HNO_3 (0.1), O_3 (5×10^{-5}), and SO_2 (3×10^{-4} at RH<50% and 0.1 at RH>50%). Typically, the authors had to estimate the magnitude of γ based on the uptake of the particular gas-phase species on other surfaces that may not be representative of mineral dust, which added a high level of uncertainty into the model.

Laboratory studies soon followed after the work of Dentener et al. [25] to better quantify the uptake of atmospheric gases on mineral dust. These studies have recently been reviewed by Usher et al. [8]. Newer data from recent laboratory studies are now being incorporated into modeling studies and modeling analyses of chemical processes in the Earth's atmosphere [28–35]. However, the laboratory studies have shown that the chemistry of trace atmospheric gases can be quite complex and in many cases cannot be described by a constant pseudo-first-order uptake that is invariant to transport time, changing surface composition, or changing reaction mechanism as the aerosol ages. This complexity is yet to be incorporated into atmospheric chemistry models. For example, laboratory studies have shown that the mineralogy of the dust is an important factor in dust reactivity [36, 37]. That is to say, not all mineral dust particles will react similarly, e.g., calcite reacts much differently than hematite, quartz, and other oxides. Dust mineralogy must be included in the models if the influence of dust on the chemical balance of the troposphere can truly be discerned.

It is often the results of field studies that spur atmospheric chemistry models to interpret the data by invoking new pathways and reaction mechanisms, many times those that involve heterogeneous reactions. One of the largest field campaigns was completed in Spring 2001 to measure trace gases and particulate matter in Asia, Aerosol Characterization Experiment (ACE-Asia). (See Ref. [38] for a description of the ACE-Asia field campaign.) Particularly noteworthy about this field campaign is the focus on tropospheric aerosol and the suite of instruments used to measure gases and aerosol. Although a wealth of information regarding tropospheric aerosol has in the past been obtained from filter-based measurements, recent developments in single particle mass spectrometry [39–44] and scanning electron microscopy of single particles [45–50] are important and provide much needed detailed information about tropospheric aerosol on a particle-by-particle basis.

In the next two sections, several potentially important catalytic reactions on mineral dust are discussed. The catalytic decomposition of ozone on mineral dust aerosol is the focus of Section 5.4 and the hydrolysis of several dinitrogen oxides on mineral dust aerosol, including N_2O_4 and N_2O_5, is the focus of Section 5.5. A discussion of field measurements, modeling analysis, and laboratory studies of these reactions are presented below.

5.4. CATALYTIC DESTRUCTION OF OZONE ON MINERAL DUST AEROSOL

Ozone reactions in the atmosphere are of great interest because of the important and beneficial role that ozone plays in the stratosphere. In fact, it is the stratospheric ozone hole that made it clear that heterogeneous reactions were important in the Earth's atmosphere [1, 2]. Although ozone plays a beneficial role in the stratosphere, the presence of ozone in the troposphere is considered to be quite harmful to human health and is an important pollutant molecule in the troposphere whose concentration is monitored by the U.S. Environmental Protection Agency. The focus of this chapter is on heterogeneous reactions in the troposphere and, as discussed below, one potential role of mineral dust aerosol is to decompose ozone in the troposphere.

5.4.1 Field Measurements

Several field studies have reported low ozone mixing ratios in regions containing high mass concentrations of mineral dust particulates [51, 52]. Probably the most notable example of this anticorrelation was observed by de Reus et al. in 1997 during the ACE-2 campaign in the vicinity of the Canary Islands [52]. Through the use of a box model simulation, it was determined that the dust surface provided a sink for the direct removal of ozone, causing a maximum of 50% ozone depletion, agreeing with profiles from the field data, but it was noted that there was large uncertainty attributed to the uncertainty in the value of the reactive uptake coefficient.

5.4.2 Modeling Analysis

As discussed earlier, in 1996, Dentener et al. used a global three-dimensional model for evaluating the direct loss of ozone due to heterogeneous reactions with mineral dust, based on measurements of deposition rates to soils, sand, and other terrestrial surfaces [25]. By estimating a range of reactive uptake coefficients, the importance of the heterogeneous loss of ozone in the troposphere was simulated. Using a value of the reactive uptake coefficient of 2×10^{-4} as an upper limit, O_3 loss would approach up to 20% in regions of high dust concentrations, whereas the use of a "best guess" reactive uptake coefficient of 5×10^{-5}, 2% to 6% ozone loss would be due to direct O_3–dust interactions. However, there was high uncertainty in this prediction due to the lack of information on the reactive uptake of ozone onto mineral surfaces, a point discussed by Jacob [22].

More recent modeling studies have evaluated the impact of heterogeneous reactions on mineral dust tropospheric ozone levels [29–34]. Some of these, in agreement with Dentener et al. [25], have indicated that the direct decomposition of ozone on mineral dust is an important process for ozone loss [29, 32, 33] whereas other modeling studies have indicated that the direct loss of ozone on mineral dust aerosol is not an important loss mechanism for tropospheric ozone [34]. Instead it is suggested that due to coupled processes, heterogeneous reactions of nitrogen oxides can influence tropospheric ozone levels [30, 34]. Bian and Zender recently published a modeling study on the relative impact of direct ozone destruction on mineral dust and "HNO_3-induced" destruction of ozone on mineral dust, i.e., indirect ozone loss as a result of HNO_3 uptake on dust and coupled gas-phase processes, and conclude that these rates may in fact be comparable under certain conditions [32]. Thus, it is evident that there are still some questions that remain from the different modeling analysis related to the direct loss of ozone to mineral dust. The laboratory studies discussed below suggest that the heterogeneous reaction of ozone on mineral dust is complex. The detailed mechanism of ozone decomposition depends on the mineralogy of the dust as well as the surface composition of the mineral dust particles.

5.4.3 Laboratory Studies

The catalytic destruction of ozone on oxide and other surfaces has been of great interest for many years. In an early laboratory study by Suzuki et al., some of the salient details of O_3–dust interactions were investigated [53]. Although this investigation did not offer much quantitative rate information or suggest a mechanism, the relative results showed fundamentally the variations in reactivity for various mineral oxides. Using a UV absorption monitor, it was reported that the relative O_3 reactivity of SiO_2, α-Fe_2O_3, Fe_3O_4, α-Al_2O_3, and natural sea sand collected in Japan, which was further separated into an "iron sand" component, to compare with the selected iron oxide compounds, and a "remainder sand" component. For the reaction on natural sand, the decomposition rate decreased with decreasing ozone concentrations and increased proportionally with increasing volume of sand. At the highest experimental concentration of ozone, the decomposition rate was initially fast before slowing to a much lower steady-state value; however, the decomposition rate was nearly constant throughout the exposure at the lowest concentration of ozone. Additionally, the "iron sand" had a similar reactivity to Fe_3O_4, a major component in the "iron sand", and decomposed O_3 at a faster rate than the natural sand and the "remainder sand", this suggests that iron oxide as Fe_3O_4 more effectively destroys ozone than the other phases present in the sand. It was concluded that iron-containing particles were more effective in catalyzing ozone decomposition relative to alumina and silica particles.

Klimovskii et al. reported on the decomposition of O_3 on γ-Al_2O_3 [54]. A reactive uptake coefficient of 10^{-4} was calculated using the geometric area as the surface area of the sample. It was also observed that as the ozone concentration decreased, there was a simultaneous increase in the pressure of O_2. A mechanism was proposed based on one suggested earlier by Golodets [55]:

$$O_3(g) + () \longrightarrow O(a) + O_2 \tag{5.11}$$

$$O_3(g) + O(a) \longrightarrow 2O_2(g) + () \tag{5.12}$$

where () represents an active surface site, such as anion vacancies. Reaction (5.12) was suggested as the rate-determining step. The resulting net reaction for ozone destruction would be

$$2O_3(g) \longrightarrow 3O_2(g) \tag{5.13}$$

Moreover, the authors note that adsorbed oxygen, O(a), can also combine to form O_2, according to the following reaction:

$$O(a) + O(a) \longrightarrow O_2(g) + 2() \tag{5.14}$$

In a series of published papers, Alebic-Juretic et al. studied ozone reactions on environmentally and atmospherically relevant compounds including silica gel, alumina, calcite, TiO_2, Saharan sand, wood ash, coal fly ash, pollen, and sodium halides [56–58]. Using a fluidized bed reactor, ozone uptake on these various samples was studied and the data were evaluated. The relative efficiency for ozone removal by the various compounds at steady state was found to be

$$Al_2O_3 > \text{wood ash} > \text{silica gel} > \text{Saharan sand} = \text{calcite} \tag{5.15}$$

At steady state, the concentration detected at the exit of the fluidized bed reactor loaded with Al_2O_3 was 58% of the initial ozone concentration, while after a certain time period Saharan sand and calcite were found not to decrease ozone concentrations very much.

The kinetics and mechanism of ozone destruction on various types of solid particulates were considered later by Alebic-Juretic et al. [58]. It was indicated that the elimination of ozone from an air stream is not a straightforward process. Rather, the process was described as involving the adsorption to a surface and successive chemical or catalytic transformations at the active sites until they are fully occupied or inactivated, which was observed experimentally through the loss of ozone and the gradual decrease in the effectiveness of a particular substance to ozone removal. In most cases, the process could not be described by first-order kinetics, rather Alebic-Juretic et al. [57, 58] compared these reactions on powders to investigations of ozone on soot, which presented similar rapid initial loss of O_3 followed by a slower rate of ozone reaction. The process, including steps where a swift initial decomposition is followed by slower reactions with surface species and inactivation of reactive surface sites, is consistent with that observed by Klimovskii et al. and described by Equations (5.11) through (5.14) [54].

Michel et al. measured the reactive uptake of ozone onto powders chosen as models for atmospheric dust and a sample of authentic dust using a Knudsen cell reactor [59, 60]. Samples of Saharan sand and China loess showed a rapid initial drop in ozone loss, followed by a decrease in reactivity lower steady-state value, similar to what was previously observed by others [53, 56]. The difference between the authentic dusts and the mineral oxide compounds was thought to be caused by possible organic contaminants on the authentic samples. The presence of organic contaminants has been recently proposed by Grøntoft to explain a very fast initial reaction of ozone on surfaces present in the environment [61].

Values of the initial uptake coefficients for the oxide samples, based on the specific surface areas of the powders, were reported as $5 \pm 3 \times 10^{-5}$ for SiO_2, $8 \pm 5 \times 10^{-5}$ for α-Al_2O_3 and $1.8 \pm 0.7 \times 10^{-4}$ for α-Fe_2O_3. The greater reactivity of the iron oxide is in agreement with what was observed by Suzuki et al. [53]. The value obtained for alumina is considerably greater than what was determined by Hanning-Lee et al., who calculated a much lower uptake coefficient of 2×10^{-10} for another alumina phase, γ-Al_2O_3, at 22°C [62]. The authentic dust samples resulted in uptake values of $4 \pm 2 \times 10^{-6}$ for sieved Saharan sand, 2.7 $\pm$ 0.9×10^{-5} for China loess, and $6 \pm 3 \times 10^{-5}$ for ground Saharan sand. Differences in the reactive uptake coefficient for Saharan sand were attributed to differences in sample treatment, as the sieving may have changed the elemental composition of the particles in the sample by concentrating certain mineral phases and excluding others. Therefore, the reactive uptake coefficient obtained for the ground sand was considered as an upper limit. A calculation of ozone molecules lost per unit BET surface area of the samples confirmed that many more O_3 molecules were destroyed than can reasonably be accommodated on the surface for a stoichiometric reaction. Thus, Michel et al. concluded that ozone destruction on these surfaces was in fact catalytic in nature.

In a later study by Michel et al., the reactivity of a clay compound (kaolinite) as a model for mineral dust was measured [60]. In addition, the pressure, time, and temperature dependencies of the reaction of ozone on several different minerals were examined in more detail. The initial uptake of ozone onto the clay compound was $3 \pm 1 \times 10^{-5}$, which is on the order of the values that were determined for the authentic mineral dusts from Asia and Africa. The initial uptake of ozone onto α-Fe_2O_3 and α-Al_2O_3 was studied as a function of ozone partial pressure between 3 and 30 μTorr (corresponding to 2 to 40 ppb); however, neither compound displayed a significant dependence of initial γ within the experimental ozone pressure range that was measured. This result suggested that the initial adsorption of ozone to the surface can be described as a first-order process in this pressure range. The dependence of initial reactive uptake coefficient on temperature was examined in order to probe the energy of activation for the reaction and showed a weak temperature dependence of $\gamma_{o,BET}$ for α-Al_2O_3 corresponding to an activation energy of $7 \pm 4\,kJ\,mol^{-1}$. Similar activation energies were reported in earlier catalysis studies [63, 64].

In order to confirm the catalytic nature of the ozone reaction with mineral oxides, the dependence of the reactive uptake on ozone exposure time was measured in experiments with α-Al_2O_3, α-Fe_2O_3, and Saharan sand, whereby each sample was exposed to ozone for extended periods of time on the order of hours. For the mineral oxides, the capacity to decompose ozone decreased over the initial 2–3 h, and after this time, the decomposition settled into steady-state uptake with no indication that the rate of ozone uptake would approach zero. Under steady-state conditions, uptake values for α-Fe_2O_3 and α-Al_2O_3 were 2.2×10^{-5} and 7.6×10^{-6}, respectively. This decrease of approximately 90% of the initial uptake is significant, but the samples continued to take up ozone, on the order of 3×10^{14} molecules per minute with a total O_3 consumption between 1×10^{17} and 2.5×10^{17} molecules. This decrease in the efficiency of mineral oxides to decompose ozone agrees with previous reports of ozone destruction by metal oxides for use as environmental catalysts to destroy ozone [63–66] and by Alebic-Juretic et al. with similar mineral oxides [56–58]. The sample of Saharan sand also exhibited a considerable decrease in reactivity over time and with expectations that the sand would continue reacting in the same manner as the mineral oxides, the extrapolated steady-state value was estimated to be 6×10^{-6}. Michel et al. concluded that although the rate decreased over time, the catalytic nature of the reaction of ozone on mineral particles may make this heterogeneous reaction important to tropospheric chemistry in comparison with other ozone loss mechanisms.

Uptake of O_3 onto mineral dust using a Knudsen cell reactor was also investigated by Hanisch and Crowley [67]. In this report, the reaction of ozone with Saharan sand from the Cape Verde Islands was investigated as a function of sample treatment and ozone concentration using a Knudsen reactor. A quick initial drop was observed in the ozone signal followed by a recovery somewhat toward the baseline to a nonzero steady-state value, which is similar to the authentic dust samples explored in the Knudsen cell study of Michel et al. [59, 60] and with other methods utilized by Suzuki [54] and Alebic-Juretic [56–58]. Also in agreement with Michel et al., Hanisch and Crowley report that the ozone uptake was dependent on the BET surface area of the sample. Therefore, to take mass-dependence into account, the reported reactive uptake coefficients were calculated with a pore diffusion model. Additionally, the reactive uptake coefficient was dependent on the ozone concentration, and at lower ozone concentrations, the reactive uptake coefficients were an order of magnitude greater than those at higher concentrations and ranged from 2.2×10^{-6} to 4.8×10^{-5}. These values are in relatively good agreement with the reactivity determined for Saharan sand with a similar ozone concentration by Michel et al. [59]. Based on the reported results, possible mechanisms of ozone destruction on the sand surface were discussed. An ozone molecule attaches to an active surface site through a terminal oxygen atom and dissociates into gas-phase O_2 and a surface-bound O atom. Either this O atom can react with another incoming ozone molecule to produce two gas-phase O_2 molecules or the O atoms can migrate at appropriate temperatures and also react to form O_2. In the experiments done by Hanisch and Crowley discussed above, O_2 production was monitored. The values for γ as determined here would be in the lower limit of values that might make O_3 loss on dust surfaces important (as proposed by Dentener et al. [25], with $\gamma > 10^{-5}$), but since photochemical processes would be faster during the day, the authors suggested this mechanism may have greater importance at night.

In order to further evaluate the potential importance of the heterogeneous reactions versus other loss mechanisms due to chemical and photochemical processes in the atmosphere, a simple analysis described by Li et al. for organic compounds [68] is applied here to ozone. This analysis is not a substitute for the more elegant three-dimensional atmospheric chemistry models described previously [25, 29–34]; however, it does allow for a rapid evaluation of the relative rates of heterogeneous reactions compared to some of the other chemical and photochemical loss mechanisms for ozone in the troposphere.

Table 5.2 lists various loss mechanisms for ozone through gas-phase reactions and photolysis. These gas-phase processes include ozone reaction with hydrogen, nitrogen, and chlorine oxides, i.e., HO_x, NO_x, and ClO_x reactions. These reactions are of course bimolecular in

TABLE 5.2
Ozone Loss through Gas-Phase Reactions and Photolysis

Reaction	Second-Order Rate Constant (cm^3 $molecules^{-1}s^{-1}$)[a]
$O_3 + h\nu \rightarrow O_2 + O(^3P)$	10^{-5}–10^{-3} s^{-1} [b]
$O_3 + h\nu \rightarrow O_2 + O(^1D)$	10^{-6}–10^{-3} s^{-1} [b]
$O + O_3 \rightarrow 2O_2$	8.0×10^{-15}
$H + O_3 \rightarrow OH + O_2$	2.9×10^{-11}
$OH + O_3 \rightarrow HO_2 + O_2$	7.3×10^{-14}
$HO_2 + O_3 \rightarrow OH + 2O_2$	1.9×10^{-15}
$NO + O_3 \rightarrow NO_2 + O_2$	1.9×10^{-14}
$NO_2 + O_3 \rightarrow NO_3 + O_2$	3.2×10^{-17}
$HNO_2 + O_3 \rightarrow HNO_3 + O_2$	$<5.0 \times 10^{-19}$
$Cl + O_3 \rightarrow ClO + O_2$	1.2×10^{-11}
$OClO + O_3 \rightarrow$ products	3.0×10^{-19}
$ClO + O_3 \rightarrow ClOO + O_2$	$<1.4 \times 10^{-17}$
$ClO + O_3 \rightarrow OClO + O_2$	$<1.0 \times 10^{-18}$

[a]Ref. [69].
[b]Range of first-order photolysis rate constant from Ref. [70].

nature and the second-order rate constants for each of the reactions are given in Table 5.2. In addition, first-order photolysis rate constants are given for the photodissociation of ozone. As described in Li et al., these ozone loss mechanisms for gas-phase processes can be compared to heterogeneous loss of ozone on mineral dust aerosol by first calculating a pseudo-first-order reaction kinetics for the bimolecular reactions. Table 5.3 gives the concentrations of the ozone reactant partner assumed for each reaction and the calculated pseudo-first-order rate constants. Pseudo-first-order rate constants were only calculated for the most important tropospheric reactions. From the pseudo-first-order rate constants, a first-order lifetime can be determined. These lifetimes are listed in the last column of Table 5.3.

To compare the gas-phase loss mechanisms to the heterogeneous loss of ozone, the rates of heterogeneous reactions are expressed in terms of a pseudo-first-order mass transfer rate constant. The calculation includes an aerosol size distribution (vide infra). From this pseudo-first-order mass transfer rate constant, the lifetime of ozone molecules due to heterogeneous uptake can be determined. The data for these calculations are shown in Figure 5.2 as a double-y plot. The pseudo-first-order mass transfer rate constant and lifetimes are plotted as a function of γ, the heterogeneous uptake coefficient. Since the total surface area available for reaction plays a role in heterogeneous chemistry, the pseudo-first-order mass transfer rate constant must be calculated using an assumed mineral dust aerosol size distribution and particle density. This is shown as the dotted line in Figure 5.2. The specific mathematical details, which describe this distribution, are given in Ref. [68].

The importance of heterogeneous reactions is determined by whether or not they are competitive with the other loss processes. Using the range of values reported for γ given in Table 5.4 (with the exception of the outlier value of 2×10^{-8}) and the range of concentrations of the ozone reactant partners, lifetimes of these different processes can be compared. It is seen from the plot in Figure 5.2 that the heterogeneous loss of ozone on mineral dust is competitive with several reactions including the HO_x reactions with ozone and some of the NO_x reactions. At the lower end of values for γ, the heterogeneous loss of ozone becomes less competitive.

TABLE 5.3
Pseudo-First-Order Rate Constants and Lifetimes for Selected Ozone Loss Mechanisms

Reaction	Concentration of Reactant Partner (molecules cm^{-3})[a]	Pseudo-First-Order Order Rate Constant (s^{-1})	Lifetime (s)
$O_3 + h\nu \rightarrow O_2 + O(^3P)$		10^{-5}–10^{-3}	1000–1.0×10^5
$O_3 + h\nu \rightarrow O_2 + O(^1D)$		10^{-6}–10^{-3}	1000–1.0×10^6
$OH + O_3 \rightarrow HO_2 + O_2$	5.0×10^6 (day)	3.65×10^{-7}	2.74×10^6
	$\leq 2 \times 10^5$ (night)	1.46×10^{-8}	6.85×10^7
$HO_2 + O_3 \rightarrow OH + 2\,O_2$	3×10^8	5.7×10^{-7}	1.75×10^6
$NO + O_3 \rightarrow NO_2 + O_2$	2.5–250 $\times 10^{11}$ [b]	4.75–475 $\times 10^{-3}$	2.1–210
	5–250 $\times 10^9$	9.5–475 $\times 10^{-5}$	0.21–1.1 $\times 10^4$
	0.5–2 $\times 10^9$	0.95–3.8 $\times 10^{-5}$	2.6–10.5 $\times 10^4$
$NO_2 + O_3 \rightarrow NO_3 + O_2$	2.5–250 $\times 10^{11}$ [b]	0.08–8 $\times 10^{-4}$	1.25–12.5 $\times 10^4$
	5–250 $\times 10^9$	0.16–8 $\times 10^{-6}$	1.25–62.5 $\times 10^5$
	0.5–2 $\times 10^9$	1.6–6.4 $\times 10^{-8}$	1.56–6.25 $\times 10^7$

[a] *Source*: Ref. [71].
[b] Concentration ranges given are for urban/suburban regions, rural regions, and remote tropical forest/remote marine regions, respectively. *Source*: Ref. [72].

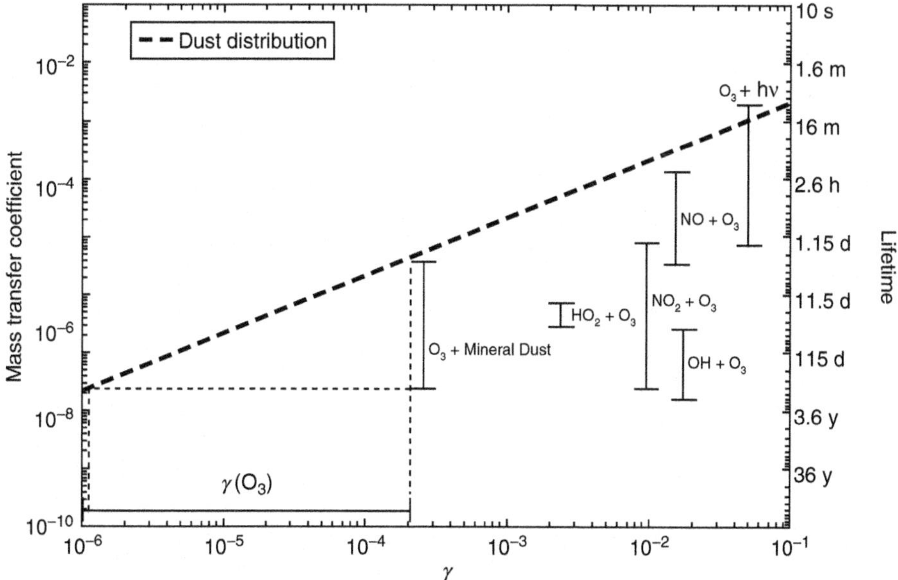

FIGURE 5.2 A comparison of the heterogeneous uptake of ozone on mineral dust aerosol with other ozone loss mechanisms is shown above. In this comparison, the lifetimes associated with these various processes, i.e., heterogeneous uptake, gas-phase chemistry, and photochemistry, are calculated using the data given in Table 5.2 to Table 5.4. The ranges of the lifetimes determined for these various processes are given for the range of values given in Table 5.2 to Table 5.4. As shown in the plot, heterogeneous uptake (O_3 + Mineral Dust) is competitive with several ozone loss mechanisms as the calculated lifetimes are comparable.

Although the above analysis can be used to begin the understanding of the direct decomposition reaction of ozone on mineral dust, some recent results suggest that ozone dissociation may be more complicated than the laboratory studies discussed above suggest.

TABLE 5.4
A Summary of Reported Literature Values of the Heterogeneous Uptake Coefficients for Ozone on Oxides, Clays, and Authentic Dusts

Mineral Dust	Uptake Coefficient	Initial or Steady-State Uptake	Ref.
α-Al_2O_3	$(1.4 \pm 0.3) \times 10^{-4}$	Initial	Michel et al. [60]
	7.6×10^{-6}	Steady state	Michel et al. [60]
γ-Al_2O_3	10^{-4}	—[a]	Klimovskii et al. [54]
	2×10^{-10}	—[a]	Hanning-Lee et al. [62]
α-Fe_2O_3	$(1.8 \pm 0.7) \times 10^{-4}$	Initial	Michel et al. [59]
	2.2×10^{-5}	Steady state	Michel et al. [60]
SiO_2	$(5 \pm 3) \times 10^{-5}$	Initial	Michel et al. [59]
China Loess	$(2.7 \pm 0.9) \times 10^{-5}$	Initial	Michel et al. [59]
Saharan Dust (Cape Verde)	3.0×10^{-5}	Initial	Hanisch and Crowley [57]
	7.0×10^{-6}	Steady state	Hanisch and Crowley [57]
Saharan Sand	$(6 \pm 3) \times 10^{-5}$ (ground)	Initial	Michel et al. [59]
	0.6–1.1×10^{-5} (ground)	Steady state	Michel et al. [59]
	$(4 \pm 2) \times 10^{-5}$ (sieved)	Initial	Michel et al. [59]
Clay (Kaolinite)	$(3 \pm 1) \times 10^{-5}$	Initial	Michel et al. [60]

[a]Unclear if the reported values are for intial or steady-state uptake

This complexity arises from the fact that there is little known about the molecular level details of the surface composition of these particles in the atmosphere. It is clear from studies involving single-particle analysis that mineral dust particles are often coated with organic, aqueous, or inorganic coatings [73–75]. Particles that contain coatings of nitrate, sulfate, or organic matter will presumably exhibit different reactivities toward trace gases, and the chemistry that takes place between the coating of the particle and gas-phase species can change the fate of the particle by altering its optical properties through differences in hygroscopicity or chemical composition, or by affecting its ability to act as a cloud condensation nucleus.

Grøntoft observed the decomposition of ozone on various indoor surfaces including the reaction of ozone on concrete floor tiles, a surface containing a variety of mineral oxide components [61]. A concrete floor tile was exposed to 40 ppb ozone for more than 2 days, followed by repeat exposures 6 and 7 months after the initial experiment. Observations showed that the deposition rate to the surface decreased between the first, second, and third experiments. The deposition to the surface never reached steady state by the end of the first experiment, though approached it by the end of the second, and relatively quickly reached a steady state in the third. The data correlated well with a first-order model, which supports the supposition that the heterogeneous reaction of ozone with solid surfaces is first order under these conditions of constant humidity and pH. As noted previously, Grøntoft proposed that oxidation of an organic contaminant could explain the fast initial deposition rates compared to later rates. These data suggest that atmospheric aging may be an important process in ozone reactions on environmentally relevant surfaces.

Usher et al. undertook a more detailed investigation of the effects of mineral dust surface coatings on the uptake of ozone [76]. In these studies, the uptake of ozone on laboratory-processed particles was examined. Mineral oxide particles representative of mineral dusts were pretreated *ex situ* either with a reactive inorganic or organic species and then reacted with ozone in a Knudsen cell reactor. Changes in the reactivity of the oxide particles toward ozone were observed and these changes were found to depend on the detailed chemical nature of the reactant coating.

Particles containing a nitrate coating exhibited an approximately 70% decrease in reactivity as compared to the pure Al_2O_3 surface. In contrast, particles pretreated with SO_2 to yield surface SO_3^{2-} exhibited enhanced reactivity toward ozone as compared with pure Al_2O_3; the initial reactive uptake coefficient increased approximately 30%. The SO_3^{2-} groups on the surface were oxidized to form SO_4^{2-} by ozone, and this reaction represents a mechanism for the oxidation of SO_2 on surfaces to form sulfate.

In addition to aging of aerosol by the presence of inorganic species, Usher et al. probed the processing of mineral dust with organic coatings through monitoring the reaction of ozone with particles of SiO_2 that were either functionalized with octenyltrichlorosilane or octyltrichlorosilane. When the particles were coated with the alkyl compound, the mass spectral signal within the Knudsen apparatus showed the anticipated result of decreased reactivity toward ozone. With a coverage of $2 \pm 1 \times 10^{14}$ per cm^2 for the alkyl compound on the surface of the SiO_2 particles, the reactivity decreased approximately 40% from that of unreacted SiO_2. Since there were still active sites available for reaction with ozone, these sites accounted for the observed uptake in the signal, given that alkane species are unreactive toward ozone. On the other hand, the reactivity of alkene-functionalized silica was increased by 40%. The initial uptake value onto this surface was a combination of uptake by reactive double bonds and reactive sites on the bare silica surface.

From the surface coverages and the change in reactivity, a site-specific uptake was calculated for these different coatings. The values of heterogeneous uptake coefficient calculated for these different sites are given in Table 5.5 for reactions of ozone on these coated particles. In the case of the organic-coated particles, similar values of the heterogeneous uptake coefficient have been measured by Rudich and coworkers for reaction of ozone with organic compounds bound to glass surfaces [77, 78].

It is important to note that since nearly all of the ozone decomposition studies have been done under dry conditions, near 0% RH, there is some question to whether mineral oxide surfaces in the presence of water vapor, for example from 20% to 90% RH, would be active toward ozone decomposition. Iron oxide for example has been shown to be one of the most reactive components of mineral dust [53, 59, 60]. As discussed by Al-Abadleh and Grassian in a recent review article [79], the nature of oxide surfaces depends on the relative humidity of the ambient air. Under relative humidity conditions appropriate for the troposphere, the oxide surface is covered with surface hydroxyl groups and adsorbed water. Since it is often proposed that metal cation sites, also referred to as Lewis acid sites, and oxygen vacancies are the active

TABLE 5.5
Summary of Reaction Mechanisms and Calculated Site-Specific Uptake Coefficients[a]

Reaction	Uptake coefficient
I. Ozone Reaction with HNO_3-Pretreated α-Al_2O_3	
$NO_3^-(a) + O_3(g) \rightarrow$ No Reaction	$\gamma(NO_3^-(a)) < 6 \times 10^{-7}$
II. Ozone Reaction with SO_2-Pretreated α-Al_2O_3	
$SO_3^{2-}(a) + O_3(g) \rightarrow SO_4^{2-}(a) + O_2(g)$	
$HSO_3^-(a) + O_3(g) \rightarrow HSO_4^-(a) + O_2(g)$	$\gamma(SO_3^{2-}/HSO_3^-(a)) = 2.4 \pm 0.8 \times 10^{-4}$
III. Ozone Reaction with C_8-Alkene-Functionalized SiO_2	
$SiO—Si—(CH_2)_5—CH{=}CH_2(a) + O_3(g) \rightarrow$	ozonide ring (O–O–O) on $H_2C—CH$-R (a)
$\gamma C{=}C(a)) = 1.0 \pm 0.3 \times 10^{-4}$	
IV. Ozone Reaction with C_8-Alkane-Functionalized SiO_2	
$SiO—Si—(CH_2)_6—CH_{3(a)} + O_{3(g)} \rightarrow$ No Reaction	$\gamma C—C(a)) < 1 \times 10^{-7}$

[a] *Source*: Ref. [76].

sites in oxide catalysis, as these sites are no longer present in the presence of water vapor there may be a decrease in reactivity of ozone on the surface of mineral oxides at higher relative humidity.

Mogili et al. have recently examined the heterogeneous reaction kinetics of ozone on α-Fe_2O_3 at different relative humidities in an environmental reaction chamber [80]. The results of this study show that the reactivity decreases with increasing RH. Figure 5.3 shows the decay kinetics for gas-phase ozone after the introduction of α-Fe_2O_3 at <1% RH and at 20% RH. As can be seen from the kinetic data, there is a decrease in the rate of ozone loss at higher relative humidity. The calculated uptake coefficient decreased by a factor of six at the higher RH. These data clearly show that RH is an important factor in ozone uptake on mineral dust aerosol.

5.5 TROPOSPHERIC FORMATION OF HONO AND HNO_3: CATALYTIC HYDROLYSIS OF N_2O_3, N_2O_4, AND N_2O_5 ON MINERAL DUST AEROSOL

As noted above, mineral dust aerosol particles can be coated with a layer or thin film of water. Thus hydrolysis reactions may be important on these surfaces. Hydroylsis reactions of nitrogen oxides are the focus of this section. Under tropospheric conditions of temperature and relative humidity, there is evidence that N_2O_3, N_2O_4, and N_2O_5, undergo hydrolysis on aerosol surfaces according to reactions (5.16)–(5.18) to yield nitric and nitrous acid, HNO_3, and HONO:

$$N_2O_3 + H_2O \longrightarrow HONO + HONO \qquad (5.16)$$

$$N_2O_4 + H_2O \longrightarrow HONO + HNO_3 \qquad (5.17)$$

$$N_2O_5 + H_2O \longrightarrow HNO_3 + HNO_3 \qquad (5.18)$$

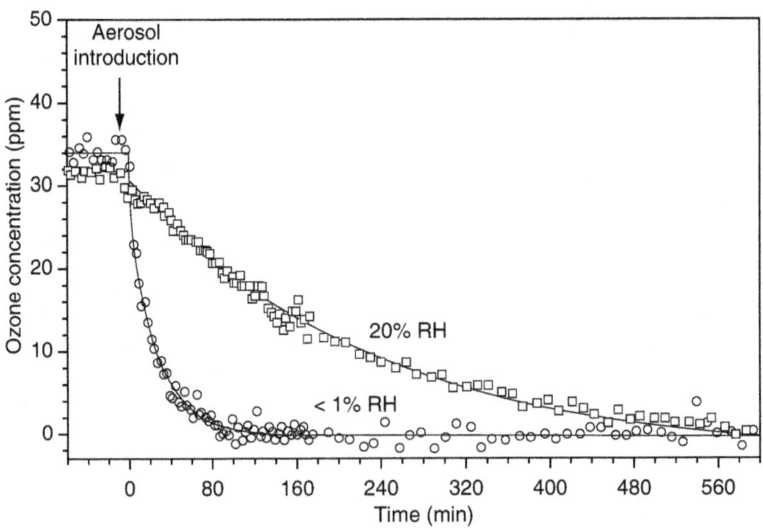

FIGURE 5.3 Gas-phase decay of ozone monitored in an environmental reaction chamber after the introduction of α-Fe_2O_3 oxide at <1% and 20% RH. Adsorbed water has an impact on the rate of this reaction, the first-order heterogeneous uptake coefficient decreases by a factor of six at the higher relative humidity.

Because NO, NO_2, and NO_3 are all radicals, they can associate to form higher molecular weight dinitrogen compounds according to reactions (5.19)–(5.21):

$$NO + NO_2 \longrightarrow N_2O_3 \tag{5.19}$$

$$NO_2 + NO_2 \longrightarrow N_2O_4 \tag{5.20}$$

$$NO_2 + NO_3 \longrightarrow N_2O_5 \tag{5.21}$$

These dinitrogen oxides are formed in the troposphere with different propensities. The measured equilibrium constants at 298 K are 2.1×10^{-20}, 2.8×10^{-19}, and 3.1×10^{-11} for reactions (5.19), (5.20), and (5.21), respectively [69].

One of the reasons why these reactions are of great interest is because gaseous nitrous acid is a product. HONO serves as a major source of OH radicals in the troposphere, especially in polluted urban environments [81, 82]. OH radicals are formed through photolysis of HONO [83]. Hydroxyl radical is the most important trace species in atmospheric oxidation mechanisms [84, 85]. While the importance of nitrous acid as a source of OH radicals is recognized, its sources are not well understood. It has been suggested that surface-catalyzed reactions of nitrogen oxides may be responsible for as much as 95% of all the HONO found in the troposphere [86–88]. Two review articles discuss the evidence presented for heterogeneous reaction of NO_2 to produce gas-phase HONO in humid environments [89, 90]. The reaction proceeds over a wide variety of surfaces including those of glass, sodium halides, metals, and metal oxides.

Studies to investigate the importance of heterogeneous hydrolysis reactions of dinitrogen compounds on mineral dust aerosol surfaces in field, modeling, and laboratory investigations are discussed below.

5.5.1 Field Studies

The results of several field studies have suggested that aerosol surfaces can mediate HONO formation via reaction (5.17). A study by Andrés-Hernández et al. using differential optical absorption spectroscopy (DOAS) was done to investigate the origin of nighttime concentrations of tropospheric HONO at several European locations; in Italy (Ispra and Milan) and in Switzerland (Claro) [91]. The measured increase in the HONO to NO_x ratio during the night indicated that HONO formation was occurring.

Analysis of data collected in Ispra, a nonurban environment, showed an improvement in all correlations when aerosol surface chemistry involving reaction (5.17) was included, suggesting that the role of aerosol surfaces is quite significant. For data collected in Milan, a polluted, urban environment, correlations did not improve with the addition of aerosols to the analysis. This suggests that aerosol surfaces have a weak influence on HONO formation in heavily polluted areas, but the results in each of these areas could be strongly impacted by emissions of trace gases. Because of this possibility, final results were obtained from data collected in Claro, where the boundary layer has the lowest influence and HONO formation via reaction (5.17) would be easiest to identify. In fact in Claro, a correlation of the aerosol surface chemistry involving reaction (5.17) and the HONO to NO_x ratio was found. This was presented as strong evidence for the heterogeneous formation of HONO on atmospheric aerosols.

A field study in New Zealand over the city of Christchurch reported close correlations between the ratio of HONO concentrations to NO_2 concentrations and aerosol densities [92]. These results suggest significant HONO formation potentially occurred on aerosol surfaces. The highest concentrations of HONO existed during nights when general pollution levels, especially those of suspended particulate matter, were also high. The correlation between the HONO to NO_2 ratio and aerosol density existed for the complete data set and also for individual pollution episodes.

Another field study, this time conducted at the continental boundary layer near Berlin during the Berliner Ozone Experiment (BERLIOZ), was done to monitor nitrate radical concentrations [93]. DOAS was used to quantify the significance of NO_3 to NO_x removal in the troposphere. It was found that a major sink for the nitrate radical occurs indirectly through N_2O_5 hydrolysis on aerosol surfaces. By assuming that this is a valid loss pathway for NO_x species, it was found that conversion of NO_x to HNO_3 via N_2O_5 is comparable in importance to daytime conversion by OH and NO_2. Through their measurements, the researchers were able to estimate an uptake value of 0.07 for N_2O_5 on aerosol surfaces near the city of Falkenberg. Again, as in the other studies, the type of aerosol was not defined in this study.

A more definitive link between heterogeneous reactions of NO_2 (N_2O_4) and HONO formation with *mineral dust* aerosol was found in a recent field study in Phoenix, Arizona. In the summer of 2001, Wang and coworkers used DOAS to measure HONO and NO_2 concentrations in Phoenix [94]. The HONO:NO_2 ratio rarely exceeded 3%, except during two nocturnal dust storms, when it increased to around 18%. To show the influence of the dust storms on the HONO to NO_2 ratio, the data collected from the nights of the storms was correlated to "normal" nights, when different weather patterns made elevated dust concentrations unlikely and the relative humidity remained below 50%. The dramatic increase in the secondary HONO to NO_2 ratio suggests that the conversion of NO_2 to HONO is extremely efficient on mineral dust particles via reaction (5.17). The particle composition of the Phoenix dust storms was determined to be similar to other mineral dusts found in different parts of the world.

5.5.2 Modeling Studies

Some of the field studies noted above used modeling analysis to better understand measured HONO levels. It was noted that inclusion of the heterogeneous hydrolysis of N_2O_4 in the model helped explain the observations, namely it could simulate the measured HONO:NO_2 ratio. In other modeling studies, Dentener and Crutzen predicted concentrations of NO_x and O_3 to be 45% and 10% lower than those predicted by models using only gas-phase reactions when reaction (5.18), the hydrolysis of N_2O_5 on ammonium sulfate aerosol surfaces, was included in a three-dimensional global tropospheric model. Dentener and Crutzen assumed an uptake coefficient (γ) of 0.1 [95] in the model.

In a comparison study performed by Tie et al., distributions of NO_x and O_3 analyzed during tropospheric ozone production about the spring equinox (TOPSE) were compared to the calculations of a global chemical/transport model (Model for Ozone And Related chemical Tracers [MOZART]) [96]. More specifically, researchers analyzed the effect of sulfate aerosols on N_2O_5 hydrolysis. Results of this comparison showed that when heterogeneous reaction (5.18) is excluded from calculations, the model greatly overestimates NO_x concentrations, especially at the high latitudes of the Northern Hemisphere in the winter and spring. When reaction (5.18) was included in the model, its calculations predicted NO_x concentrations similar to those observed during TOPSE, indicating that the heterogeneous reaction plays a significant role in tropospheric NO_x chemistry.

A value of γ equal to 0.1 was then used in another study by Dentener et al. that explored the role of mineral dust as a reactive surface in the troposphere [25]. In their work, Dentener and coworkers assumed that the same uptake coefficient for N_2O_5 on sulfate aerosols could also be used for mineral dust aerosols. Although the uptake coefficient of 0.1 is a high value and may overestimate the amount of N_2O_5 removed by mineral dust aerosols, it was suggested that the use of this value may be justified because the production of N_2O_5 is most effective at night, when relative humidity is at a maximum. It was also suggested that during this time, mineral aerosols would have some water content that would help enhance the hydrolysis reaction. Since the 1996 modeling studies of Dentener et al., there have been other modeling studies that have used an uptake coefficient of $\gamma = 0.1$ for the N_2O_5 hydrolysis reaction on mineral dust [30] whereas lower values between 1×10^{-3} to 0.05 have been used by others [29, 34].

Because there is no consensus or basis for these values of the uptake coefficients several questions arise. Where do these values for γ used in modeling studies come from? What is known about hydrolysis reactions of nitrogen oxides on mineral dust from laboratory studies? As discussed in the next section, very little is known about the hydrolysis of N_2O_5 on surfaces representative of mineral dust whereas there have been some studies of N_2O_3 and many more of N_2O_4 hydroylsis.

5.5.3 Laboratory Studies

A survey of the literature shows that as of 2004 laboratory investigations of the heterogeneous hydrolysis of N_2O_5 are limited to sulfate, bisulfate, nitrate, NaCl, sea salt, and organic aerosol surfaces. Although laboratory studies on these other aerosol surfaces provide valuable insight into how parameters such as relative humidity, phase, temperature, organic coatings, and ionic character affect the reactivity of N_2O_5 with the particle surface, to date there are currently no published reports on the heterogeneous chemistry of N_2O_5 on mineral dust. Thus, a summary of the N_2O_5 hydrolysis on some of these other aerosol surfaces are discussed and the potential implication of the significance of these results on the reaction of N_2O_5 on mineral dust is surmised.

Kane et al. examined how changes in relative humidity can determine the amount of N_2O_5 taken up by sulfate aerosol surfaces [97]. For ammonium sulfate and bisulfate aerosols, reactivity increases with relative humidity and water content above the efflorescence point, where the particles form aqueous aerosols. Below the efflorescence point the particles are dry and in the solid phase, reactivity is significantly reduced.

Thornton et al. examined the effect of relative humidity, phase, and particle size on the uptake of N_2O_5 on submicron malonic acid aerosols [98]. Similar to that of the sulfate aerosols, phase plays an important role for N_2O_5 hydrolysis on organic aerosol. At higher relative humidity, above approximately 50%, the organic particle becomes aqueous, and uptake is determined to be the same order of magnitude as that reported for inorganic aerosols that have also undergone deliquescence. At low relative humidity, the particle is "dry" and in the solid phase, uptake of N_2O_5 is found to be extremely small. From their work, Thornton and coworkers recommend that a single uptake coefficient should be used for N_2O_5 hydrolysis on neutral aqueous aerosols, with a value ranging from 0.03 to 0.05, depending on the relative humidity.

While organic aerosols show reactivity towards N_2O_5, inorganic particles with an organic coating show a greatly reduced activity. Folkers and associates studied how coating aqueous NH_4HSO_4 particles with films composed of the ozonolysis products of α-pinene affected the uptake of N_2O_5 [99]. Their results showed that the reaction probability decreased by more than an order of magnitude when compared to uncoated aerosols. Calculations determined the main cause for a reduction in the amount of hydrolysis that occurred could be due to lower solubility of gaseous N_2O_5 in the organic layer.

In some aerosols the ionic character of the water layer on the particle surface can determine the efficiency of N_2O_5 hydrolysis on the surface. Nitrate ions have been shown to inhibit the hydrolysis reaction, due to a favorable recombination reaction as shown in the following reaction:

$$NO_2^+ + NO_3^- \longrightarrow N_2O_5 \tag{5.22}$$

which can reduce the uptake coefficient by as much as an order of magnitude compared to other aqueous aerosols at lower relative humidities. Work by Hallquist et al. showed that as relative humidity increases, the uptake coefficient for N_2O_5 also increased for sodium nitrate aerosols [100]. This result correlates with a decrease in the activity of the nitrate ions present in the aerosol. The results of by Hallquist et al. are in agreement with the proposed mechanism for N_2O_5 hydrolysis discussed by Wahner [101, 102].

The nitrate effect seen in studies of sodium nitrate aerosols may also be an important factor in the uptake of N_2O_5 on mineral dust. Calcium carbonate, an important component of mineral dust, effectively reacts with HNO_3 to form calcium nitrate, as shown in reaction (5.23) [103].

$$CaCO_3(s) + 2HNO_3(g) \longrightarrow Ca(NO_3)_2(s) + CO_2(g) + H_2O(g) \tag{5.23}$$

Adsorbed water on the carbonate surface enhances its reactivity with nitric acid [104]. The increased formation of nitrate in the adsorbed water layer could prevent possible uptake of N_2O_5. However, this is yet to be explored in laboratory studies. Clearly these studies are needed to better understand the heterogeneous hydrolysis of N_2O_5 on mineral dust aerosol.

Whereas there are no laboratory studies of N_2O_5 hydrolysis on mineral dust aerosol, there have been a number of laboratory studies of N_2O_4 hydrolysis [17]. As discussed above, an important feature of the N_2O_4 hydrolysis reaction is that one of the products is gas-phase HONO. Goodman et al. used FTIR and UV–Vis to investigate this reaction on hydrated silica particles, providing an *in-situ* characterization of the reaction products, which included gas-phase HONO and surface-bound HNO_3 [105]. While this study employed silica particles as a reactive surface, reaction (5.17) could potentially occur on any hydrated mineral dust surface. Mineral dust particles, especially those containing metal oxides, should have an adsorbed water layer present on the particle surface under atmospheric conditions, thus providing a potentially reactive surface on which HONO formation can occur [86].

A majority of the studies of HONO formation on surfaces has been done by Finlayson-Pitts and co-workers. A recent review by Finlayson-Pitts et al. does an excellent job of summarizing much of that work [90]. Primarily based on studies using glass surfaces (which contain silicates), a new mechanism for the hydrolysis of NO_2 (N_2O_4) on surfaces was proposed. This new mechanism is summarized in the schematic diagram in Figure 5.4.

The mechanism has several key steps. In the first step, gas-phase NO_2 forms N_2O_4 on the aerosol surface. N_2O_4 is the important precursor species in the reaction. Once adsorbed, N_2O_4 may then interact with water molecules, undissociated HNO_3 molecules (formed in the reaction of NO_2 with hydrated silica surfaces), or with HNO_3–H_2O complexes or hydrates. N_2O_4 isomerizes to form surface asymmetric $ONONO_2$, which subsequently autoionizes to form $NO^+NO_3^-$. In a final step, $NO^+NO_3^-$ reacts with water to generate HONO and HNO_3. Secondary reactions of HONO are the likely source of the gas-phase NO that is also sometimes observed. While the majority of the work done by Finlayson-Pitts and coworkers explored these reactions on silica surfaces, these results may potentially be extended to mineral dust aerosols, which also contain silicates as well as other components.

N_2O_3 has also been suggested as a precursor to HONO in the troposphere through reaction (5.16), but until recently it had not been reported to adsorb on solids at room temperature. Using FTIR spectroscopy, Mochida and Finlayson-Pitts were able to observe N_2O_3 adsorbed on a porous glass surface at room temperature, using NO and NO_2 as starting material [106]. Once the appearance of adsorbed N_2O_3 had been verified, it was tested to see if it reacted with water vapor to form HONO, as expected in reaction (5.16). While gas-phase HONO was observed, the main N_2O_3 peak at $1870\,cm^{-1}$ showed no significant change with the addition of water vapor to the glass surface. This result may be due to several factors. If the amount of N_2O_3 lost from the surface is small, it may not be observed. Gas-phase NO and NO_2 may also rapidly replenish any N_2O_3 lost from the surface. Another conclusion that can be drawn from this result is that N_2O_3 is not an intermediate in HONO formation, but rather

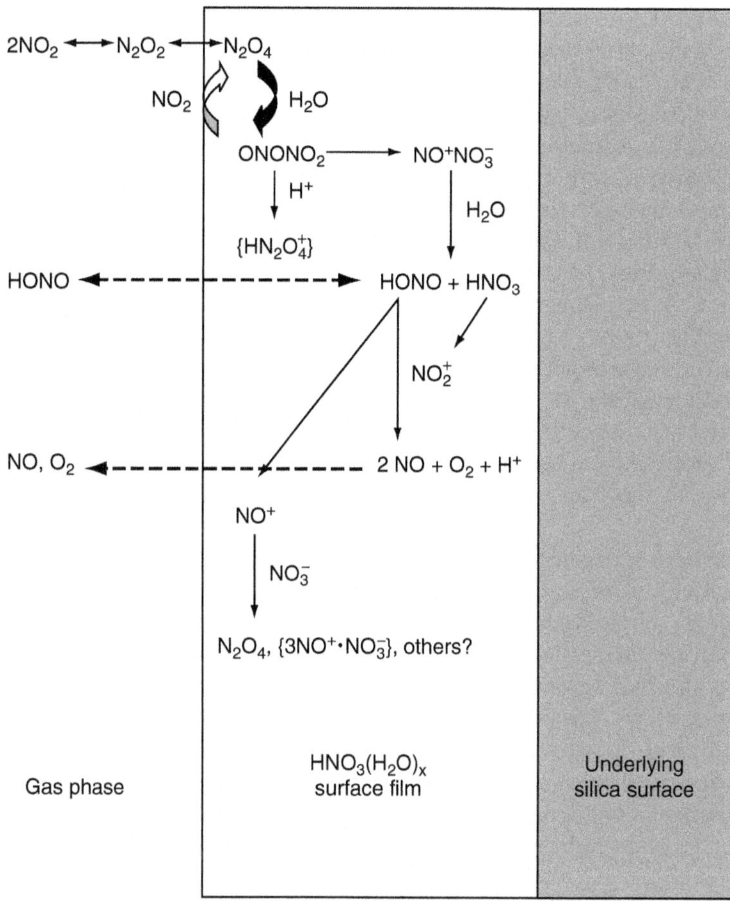

FIGURE 5.4 Schematic diagram of proposed mechanism of heterogeneous hydrolysis of NO_2. (From B. J. Finlayson-Pitts, L.M. Wingen, A. L. Sumner, D. Syomin, and K. A. Ramazan, *Phys. Chem. Chem. Phys.*, 2003, 5, 223–242. Reproduced by permission of the PCCP Owner Societies.)

HONO is formed directly from reaction between NO and NO_2. While a dependence on NO for HONO formation suggests that N_2O_3 hydrolysis may be an important step, the study by Mochida and Finlayson-Pitts indicates that its atmospheric importance to this reaction is minimal.

5.6 CONCLUSIONS CONCERNING HETEROGENEOUS REACTIONS ON MINERAL DUST AEROSOL IN THE TROPOSPHERE: FUTURE STUDIES AND FURTHER IMPLICATIONS

Although only two potentially important classes of reactions involving mineral dust were discussed here — ozone decomposition and dinitrogen oxide hydrolysis — many of the complexities associated with mineral dust reactions were revealed. In Section 5.1, it was noted that particle bulk mineralogy was very important in the chemical reactivity of mineral dust aerosol. However it is also quite clear from Sections 5.4 and 5.5 of this chapter that another important issue related to the reactivity of mineral dust, perhaps even more important than particle bulk mineralogy, is the chemical composition of the particle surface. If one takes a metal oxide particle as a model for mineral dust, it well known that the surface, after exposure to water vapor, is truncated with hydroxyl groups and the hydroxylated surface readily adsorbs water at relative humidities found in the troposphere [79]. There is also the possibility that there can be deliquescent layer around the mineral dust particle surface [104, 107–110] due to a coating or a deliquescent salt. For example if one wants to fully understand the hydrolysis of N_2O_5 on mineral dust aerosol, a laboratory study might include the uptake of N_2O_5 under three conditions as defined in Figure 5.5. Under dry conditions <5% RH, where the surface may be truncated with hydroxyl groups and little adsorbed water, the reaction should be minimal and γ_{DRY} is expected to be negligible. Under conditions more appropriate of the troposphere, there may be anywhere from one to four monolayers (ML) of water and γ_{ML} may be a low value, perhaps close to that of an crystalline salt or organic aerosol, but smaller than γ_{AQ} the case where there is a deliquescent coating. As discussed in Section 5.5, the nature of the deliquescent coating, e.g., nitrate salt coating, might also affect the heterogeneous hydrolysis of N_2O_5. All of these scenarios may be possible and should be looked at in a focused laboratory study. Incorporating the complexities found in the laboratory study into atmospheric chemistry models will be a challenge. However, this needs to be done at some level if these models can accurately describe the chemistry of the troposphere.

As discussed here, inorganic, organic, and aqueous coatings on mineral oxide surfaces will impact the reactivity of these particles. Atmospheric aerosols may have coatings of varying thickness possibly containing multiple components. From these laboratory studies, it is clear that the gas–particle interaction will depend on the nature of the molecular coating of the particle surface and that there are many issues that need to be resolved before atmospheric reactions of ozone and particulate matter containing oxide particles can be completely understood and properly assessed by atmospheric chemistry models.

For the most part, the heterogeneous reactivity of mineral dust aerosol was discussed here from the perspective of changing the chemical balance of the atmosphere. However, there are potentially even greater implications of the chemistry of mineral dust aerosol. The diagram in

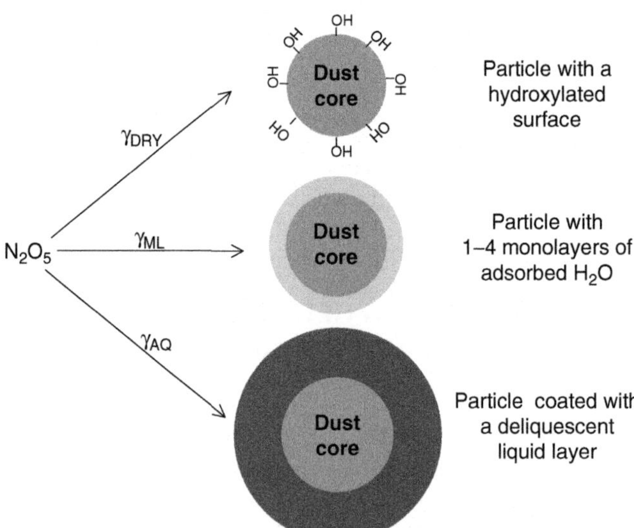

FIGURE 5.5 N_2O_5 uptake and hydrolysis on mineral dust particles under dry conditions, particle coated with one to four monolayers of water and the particle coated with a deliquescent layer. Even if the same dust core is present, it is expected that the uptake coefficient, γ, will be different and will depend on the composition of the surface layer.

Figure 5.6 shows some of the effects that mineral dust aerosol has on global processes. These processes include climate, biogeochemical cycles, and health. Like atmospheric aerosols in general, mineral aerosol may affect local and global climate through the absorption and scattering of solar radiation [111–117]. When aerosol particles absorb and scatter radiation themselves, the resulting radiative forcing is deemed "direct," whereas if the particles influence the optical properties of clouds, then the radiative forcing is "indirect." Positive radiative forcing values, measured in $W\,m^{-2}$, result in a warming effect on Earth's surface, and conversely, negative forcing has a cooling effect. Heterogeneous chemistry on surfaces will alter the optical properties of mineral dust particles. However little is known about the link between mineral dust chemistry and how it impacts the climate effects of mineral dust aerosol through direct and indirect radiative forcing. This is an important area worthy of further study.

Airborne mineral dust can have numerous repercussions on human health, the most notably and most dangerous being the effects of inhaled particles on the human respiratory system [118]. The collapse of the World Trade Center towers in New York City on September 11, 2001 released aerosols containing a wide range of mineral components, and respiratory complaints were reported by not only the workers and volunteers at or near the WTC site, but later by residents downwind of the site [119]. Additionally, long exposures to quartz particles could lead to silicosis. However, despite the respiratory conditions that can be induced, typical inhalation of mineral dust experienced by humans rarely leads directly to death [120]. Mineral dust aerosol particles may provide surfaces whereby other hazardous chemical species can be adsorbed or be produced through reaction. The extent to which the surface chemistry of mineral dust aerosol can impact health effects is currently unknown and should be investigated in more detail.

In addition to the climate and health effects discussed above, airborne soil particles may have significant effects on biogeochemical cycles through the global transport, deposition, and accretion of these particles. Soils which entrain the deposited airborne particles may become enriched in nutrients that otherwise are not present in native soils. The effects of mineral dust deposition into ocean waters have become a topic of interest in recent years (see reviews in Refs. [112, 121, 122]). It has been estimated that annually 360 to 500 Tg of mineral dust is deposited into the oceans [112, 123] with approximately 50% of the total deposition

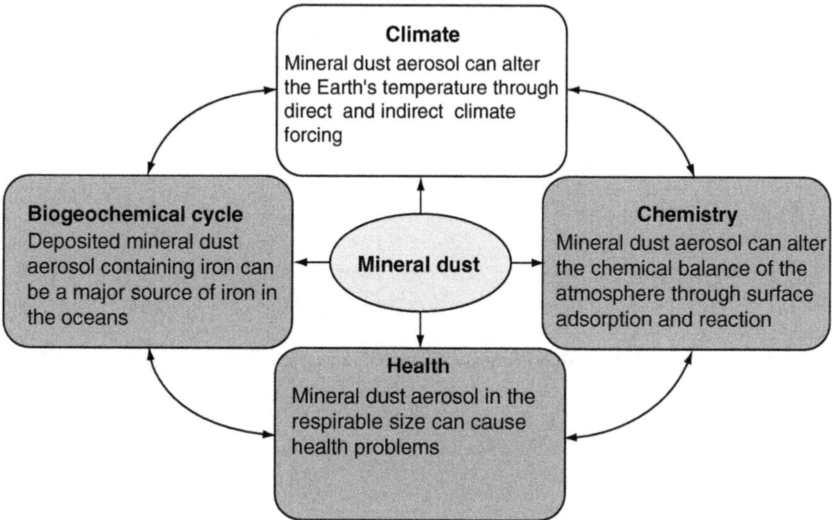

FIGURE 5.6 The connection between mineral dust in the atmosphere and its role in several global processes including climate, biogeochemical cycles, and health is shown in this schematic diagram.

occurring in the North Atlantic Ocean. Wind-transported mineral dust plays a role in supplying soluble iron to the oceans [124–126], providing micronutrients to biological species [127], such as phytoplankton [128], and ultimately influencing the iron budget of the upper ocean [129]. How the chemical and photochemical processing of the mineral aerosols may reduce Fe(III) to a more soluble Fe(II) species an important question. The solubility of iron in dust has been investigated in multiple studies [124–126, 130–133]. In one investigation by Zhu et al., samples of airborne Saharan sand were collected on filters in Barbados followed by the quantification of the soluble Fe(II) content of the aerosol through a spectrophotometric method in which the Fe(II) was complexed with an absorbing dye molecule [126]. They concluded that the dust had a total iron content of 3.4%, with only 6.2% of the total iron in soluble form and rapidly oxidized to the Fe(III) form. However, they also noted "the dispersed state of the oxidized product may make it more available for chemical and biological processing" suggesting that this needs further consideration.

The impact of air pollution and the heterogeneous chemistry of pollutant molecules on mineral dust and how this may impact other global processes have very recently been addressed in two modeling studies. Meskhidze et al. propose that air pollution, in particular SO_2 emissions, can be linked to iron bioavailability in mineral dust through heterogeneous chemistry [134]. The link between SO_2 emissions, iron mineral chemistry in the atmosphere, and iron as a nutrient for ocean life is important from several perspectives including the alteration of ocean photosynthesis. In another study, Fan et al. propose that air pollution and heterogeneous chemistry on the mineral dust can impact the lifetime of mineral dust in the atmosphere [35]. They hypothesize that surface coatings on the dust are highly soluble and can make the dust aerosol effect cloud condensation nuclei. This implies that there will be an increase in dust deposition. Thus, these modeling studies suggest that the heterogeneous chemistry of mineral dust with anthropogenic gases has far-reaching implications that will impact the Earth's hydrosphere, biosphere, and atmosphere. It is important that future laboratory and field studies be designed to better understand these linkages.

REFERENCES

1. CE Kolb, DR Worsnop, MS Zahniser, P Davidovits, LF Keyser, M-T Leu, MJ Molina, DR Hanson, and AR Ravishankara. *Adv Ser Phys Chem* 3:771–875, 1995.
2. MJ Molina, LT Molina, and CE Kolb. *Annu Rev Phys Chem* 47:327–367, 1996.
3. MO Andreae and PJ Crutzen. *Science* 276:1052–1058, 1997.
4. AR Ravishankara. *Science* 276:1058–1065, 1997.
5. SM Kreidenweis. In: DL Macalady, ed. *Perspectives in Environmental Chemistry*. New York: University Press, 1998, 257–270.
6. BJ Finlayson-Pitts and JC Hemminger. *J Phys Chem A* 104:11463–11477, 2000.
7. VH Grassian. *Int Rev Phys Chem* 20:467–548, 2001.
8. CR Usher, AE Michel and VH Grassian. *Chem Rev* 103:4883–4940, 2003.
9. AS Goudie and NJ Middleton. *Earth-Sci Rev* 56:179–204, 2001.
10. K Pye. *Aeolian Dust and Dust Deposits*. London: Academic Press, 1987.
11. LG Berry, B Mason, and RV Dietrich. *Mineralogy*, 2nd ed. New York: W.H. Freeman and Company, 1983.
12. JW Anthony, RA Bideaux, KW Bladh, and MC Nichols. *Handbook of Mineralogy*. Tucson, AZ: Mineral Data Publishing, 1995.
13. IN Sokolik and OB Toon (1999), *J Geophys Res*, *104*:9423–9444, doi: 10.1029/1998JD200048.
14. E Molinaroli. In S Guezoni and R. Chester, eds. *The Impact of Desert Dust Across the Mediterranean*. Dordrecht: Kluwer, 1996.
15. R Chester, H Elderfield, JJ Griffin, LR Johnson, and RC Padgham, *Mar Geol* 13:91–105, 1972.
16. S Caquineau, A Gaudichet, L Gomes, M-C Magonthier, and B Chatenet (1998), *Geophys Res Lett*, *25*:983–986, doi: 10.1029/98GL00569.

17. D Sarnthein, J Thiede, U Pflaumann, H Erlenkeuser, D Futterer, B Koopman, H Lange, and E Seibold. In: U von Rad, K Hinz, M Sarnthein, and E Seibold, eds. *Geology of the Northwest African Continental Margin*. Berlin: Springer, 1982.
18. T Claquin, M Schulz, and YJ Balkanski (1999), *J Geophys Res*, *104*:22243–22256, doi: 10.1029/1999JD900416.
19. JM Prospero, P Ginoux, O Torres, SE Nicholson, and TE Gill (2002), *Rev Geophys*, *40*(1):1002, doi: 10.1029/2000GR000095.
20. A Tabazadeh and RP Turco (1993), *J Geophys Res*, *98*:12727–12740, doi: 10.1029/93JD00947.
21. M Ammann, U Pöschl, and Y Rudich. *Phys Chem Chem Phys* 5:351–356, 2003.
22. D Jacob. *Atmos Environ* 34:2131–2159, 2000.
23. CE Kolb, P Davidovits, JT Jayne, Q Shi, and DR Worsnop. *Prog React Kinetics Mechanism* 27:1–46, 2002.
24. M Mozurkewich (1993), *Geophys Res Lett*, *20*:355–358, doi: 10.1029/93GL00745.
25. FJ Dentener, GR Carmichael, Y Zhang, J Lelieveld, and PJ Crutzen (1996), *J Geophys Res*, *101*:22869–22890, doi: 10.1029/96JD01818.
26. A Preszler Prince, JL Wade, VH Grassian, PD Kleiber, and MA Young. *Atmos Environ* 36:5729–5740, 2002.
27. ER Lovejoy and DR Hanson. *J Phys Chem* 99:2080–2087, 1995.
28. GM Underwood, CH Song, M Phadnis, GR Carmichael, and VH Grassian (2001), *J Geophys Res*, *106*:18055–18066, doi: 10.1029/2000JD900552.
29. H Bian and CS Zender (2003), *J Geophys Res*, *108*(D21), 4672, doi: 10.1029/2002JD003143.
30. RV Martin, DJ Jacob, RM Yantosca, M Chin, and P. Ginoux (2003), *J Geophys Res*, *108*(D3), 4097, doi: 10.1029/2002JD002622.
31. H Liao, PJ Adams, SH Chung, JH Seinfeld, LJ Mickley, and DJ Jacob (2003), *J Geophys Res*, 108, doi: 10.1029/2001JD001260.
32. H Bian and CS Zender (2004), *J Geophys Res* (submitted).
33. Y Tang, GR Carmichael, G Kurata, I Uno, RJ Weber, C-H Song, SK Guttikunda, J-H Woo, DG Streets, C Wei, AD Clarke, B Huebert, and BL Anderson (2004), *J Geophys Res*, *109*:D19521, doi: 10.1029/2003JD003806.
34. SE Bauer, Y Balkanski, M Schulz, DA Hauglustaine, and F Dentener (2004), *J Geophys Res*, *109*: doi: 10.1029/2003GL003868.
35. S-M Fan, LW Horowitz, H Levy II, and WJ Moxim (2004), *Geophys Res Lett*, *31*:L02104, doi: 10.1029/2003GL018501.
36. BJ Krueger, VH Grassian, MJ Iedema, JP Cowin, and A Laskin. *Anal Chem* 75:5170–5179, 2003.
37. BJ Krueger, VH Grassian, JP Cowin, and A Laskin. *Atmos Environ* 38:6251–6261, 2004.
38. BJ Huebert, T Bates, PB Russell, G Shi, YJ Kim, K Kawamura, G Carmichael, and T Nakajima (2003), *J Geophys Res*, *108*(D23):8633, doi: 10.1029/2003JD003550.
39. DT Suess and K Prather. *Chem Rev* 99:3007–3036, 1999.
40. DB Kane, B Oktem, and MV Johnston. *Aerosol Sci Tech* 34:520–527, 2001.
41. MV Johnston and AS Wexler. *Anal Chem* 67:A721–A726, 1995.
42. JW Morris, P Davidovits, JT Jayne, JL Jimenez, Q Shi, CE Kolb, DR Worsnop, WS Barney, and G Cass (2002), *Geophys Res Lett*, *29*:1357, doi: 10.1029/2002GL014692.
43. DM Murphy and DS Thomson. *Aerosol Sci Tech* 22:237–249, 1995.
44. E Woods, GD Smith, Y Dessiaterik T Baer, and RE Miller. *Anal Chem* 73:2317–2322, 2003.
45. PR Buseck and M Posfai. *Proc Natl Acad Sci* 96:3372–3379, 1999.
46. PR Buseck, DJ Jacob, M Posfai, J Li, and JR Anderson. *Int Geol Rev* 42:577–593, 2000.
47. LA DeBock and RE Van Grieken. In KR Spurny, ed., *Analytical Chemistry of Aerosols*, Boca Raton, FL: Lewis Publishers, 1999, pp. 243–276.
48. CU Ro, J Osan, I Szaloki, KY Oh, H Kim, and RE Van Grieken. *Environ Sci Technol* 34:3023–3030, 2000.
49. A Laskin, MJ Iedema, and JP Cowin. *Aerosol Sci Tech* 37:246–260, 2003.
50. EA Reid, J S Reid, M. M Meier, M R Dunlap, S S Cliff, A Broumas, K Perry, and H. Maring (2003), *J Geophys Res*, *108*(D19), 8591, doi: 10.1029/2002JD002935.

51. JM Prospero, R Schmitt, E Cuevas, DL Savoie, WC Graustein, KK Turekian, A Volz-Thomas, A Diaz, SJ Oltmans, and HH Levy II (1995), *Geophys Res Lett*, *22*:2925–2928, doi: 10.1029/95GL02791.
52. M de Reus, F Dentener, A Thomas, S Borrmann, J Ström, and J Lelieveld (2000), *J Geophys Res*, *105*:15263–15276, doi: 10.1029/2000JD900164.
53. S Suzuki, Y Hori, and K Osamu. *Bull Chem Soc Jpn* 52:3103–3104, 1979.
54. AO Klimovskii, AV Bavin, VS Tkalich, and AA Lisachenko. *React Kinet Catal Lett* 23:95–98, 1983.
55. GI Golodets, *Heterogeneous Catalytic Reactions Involving Molecular Oxygen*. Elsevier: New York, 1983.
56. A Alebic-Juretic, T Cvitas, and L Klasinc. *Ber Bunsenges Phys Chem* 96:493–495, 1992.
57. A Alebic-Juretic, T Cvitas, and L Klasinc. *Environ Monit Assess* 44:241–247, 1997.
58. A Alebic-Juretic, T Cvitas, and L Klasinc. *Chemosphere* 41:667–670, 2000.
59. AE Michel, CR Usher, and VH Grassian (2002), *Geophys Res Lett*, *29*:1665, doi: 10.1029/2002GL014896.
60. AE Michel, CR Usher, and VH Grassian. *Atmos Environ* 37:3201–3211, 2003.
61. T Grøntoft. *Atmos Environ* 36:5661–5670, 2002.
62. MA Hanning-Lee, BB Brady, LR Martin, and JA Syage (1996), *Geophys Res Lett*, *23*:1961–1964, doi: 10.1029/96GL01808.
63. R Radhakrishnan and ST Oyama. *J Catal* 199:282–290, 2001.
64. W Li and ST Oyama. *J Am Chem Soc* 120:9047–9052, 1998.
65. B Dhandapani and ST Oyama. *Appl Catal B: Environ* 11:129–166, 1997.
66. C Heisig, W Zhang, and ST Oyama. *Appl Catal B: Environ* 14:117–129, 1997.
67. F Hanisch and JN Crowley. *Atmos Chem Phys* 3:119–130, 2003.
68. P Li, KA Perreau, E Covingtion, GC Carmichael, and VH Grassian (2001), *J Geophys Res*, *106*:5517–5529, doi: 10.1029/2000JD900573.
69. SP Sander, RR Friedl, AR Ravishankara, DM Golden, CE Kolb, MJ Kurylo, RE Huie, VL Orkin, MJ Molina, GK Moortgat, and BJ Finlayson-Pitts. Chemical Kinetics and Photochemical Data for Use in Stratospheric Modeling, Evaluation number 14, JPL Pub 02-25, 2003.
70. S Madronich and S Flocke. The role of solar radiation in atmospheric chemistry. In: *Handbook of Environmental Chemistry*, P. Boule, ed., Heidelberg: Springer-Verlag, 1998, pp. 1–26.
71. JH Seinfeld and SN Pandis. *Atmospheric Chemistry and Physics*: *From Air Pollution to Climate Change*. New York, NY: Wiley-Interscience, 1998, pp. 71–74 and 250–253.
72. Committee on Tropospheric Ozone, National Research Council. Rethinking the Ozone Problem in Urban and Regional Air Pollution. National Academy Press: Washington, DC, 1991.
73. S-H Lee, DM Murphy, DS Thomson, and AM Middlebrook (2002), *J Geophys Res*, *107*(D1):4003, doi: 10.1029/2000JD000011).
74. CA Noble and KA Prather. *Environ Sci Technol* 30:2667–2680, 1996.
75. DM Murphy, DS Thomson, and MJ Mahoney. *Science* 282:1664–1669, 1998.
76. CR Usher, AE Michel, D Stec, and VH Grassian. *Atmos Environ* 37:5337–5347, 2003.
77. ER Thomas, GJ Frost, and Y Rudich (2001), *J Geophys Res*, *106*:3045–3056, doi: 10.1029/2000JD900595.
78. T Moise and Y Rudich. *J Phys Chem A* 106:6469–6476, 2002.
79. HA Al-Abadleh and VH Grassian. *Surface Sci Rep* 52:63–161, 2003.
80. P Mogili, MA Young, and VH Grassian. in preparation, 2005.
81. B Alicke, U Platt, and J Stutz (2002), *J Geophys Res*, *107*(D22):8196, doi: 10.1029/2000JD000075.
82. J Stutz, B Alicke, and A Neftel (2001), *J Geophys Res*, *107*(D22):8192, doi:10.1029/2001JD000390.
83. DL Baulch, RA Cox, PJ Crutzen, RF Hampson, JA Kerr, J Troe, and RR Watson. *Phys Chem Ref Data* 11:327–496, 1982.
84. WH Chan, RJ Nordstrom, JG Calvert, and JH Shaw. *Environ Sci Technol* 10:674–682, 1976.
85. RA Cox. *J Photochem* 25:1984, 43–48.
86. JN Pitts Jr., E Sanhueza, R Atkinson, WPL Carter, AM Winer, GW Harris, and CN Plum. *Int J Chem Kinet* 16:919–939, 1984.
87. R Svensson, E Ljungstrom, and O Lindqvist. *Atmos Environ* 21:1529–1539, 1987.
88. ME Jenkin, RA Cox, and DJ Williams. *Atmos Environ* 22:487–498, 1988.

89. G Lammel and JN Cape. *Chem Soc Rev* 25:361–370, 1996.
90. BJ Finlayson-Pitts, LM Wingen, AL Sumner, D Syomin, and KA Ramazan. *Phys Chem Chem Phys* 5:223–242, 2003.
91. MD Andrés-Hernández, J Notholt, J Hjorth, and O Schrems. *Atmos Environ* 30:175–180, 1996.
92. AR Reisinger. *Atmos Environ* 34:3865–3874, 2000.
93. A Geyer, B Alicke, S Konrad, T Schmitz, J Stutz, and U Platt (2001), *J Geophys Res*, *106*(8):8013–8025, doi: 10.1029/2000JD900681.
94. S Wang, R Ackermann, CW Spicer, JD Fast, M Schmeling, and J Stutz (2003), *Geophys Res Lett*, *30*(11):1595, doi: 10.1029/2003GL017014.
95. FJ Dentener and PJ Crutzen (1993), *J Geophys Res*, *98*(D4):7149–7163, doi: 10.1029/92JD02979.
96. X Tie, L Emmons, L Horowitz, G Brasseur, B Ridley, E Atlas, E Stround, P Hess, A Klonecki, S Madronich, R Talbot, and J Dibb (2003), *J Geophys Res*, *108*(D4):8364, doi: 10.1029/2001 JD001508.
97. SM Kane, F Caloz, and M Leu. *J Phys Chem A* 105:6465–6470, 2001.
98. JA Thornton, CF Braban, and JPD Abbatt. *Phys Chem Chem Phys* 5:4593–4603, 2003.
99. M Folkers, TF Mentel, and A Wahner (2003), *Geophys Res Lett*, *30*(12):1664, doi: 10.1029/2003GL017168.
100. M Hallquist, DJ Stewart, SK Stephenson, and RA Cox. *Phys Chem Chem Phys* 5:3453–3463, 2003.
101. A Wahner, TF Mentel, M Sohn, and J Stier (1998), *J Geophys Res*, *103*(23):31103–31112.
102. TF Mentel, M Sohn, and A Wahner. *Phys Chem Chem Phys* 1:5451–5457, 1999.
103. AL Goodman, GM Underwood, and VH Grassian (2000), *J Geophys Res*, *105*(23):29053–29064, doi: 10.1029/2000JD900396.
104. BJ Krueger, VH Grassian, A Laskin, and JP Cowin (2003), *Geophys Res Lett*, *30*:1148, doi: 10.1029/2002GL016563.
105. AL Goodman, GM Underwood, and VH Grassian. *J Phys Chem A* 103:7217–7223, 1999.
106. M Mochida and BJ Finlayson-Pitts. *J Phys Chem A* 104:8038–8044, 2000.
107. JH Han and ST Martin (1999), *J Geophys Res*, *104*(D3):3543–3553, doi: 10.1029/1998JD100072.
108. HM Hung, A Malinowski, and ST Martin. *J Phys Chem A* 107(9):1296–1306, 2003.
109. JH Han, HM Hung, and ST Martin (2002), *J Geophys Res*, *107*(D10):4086, doi: 10.1029/2001 JD001054.
110. A Laskin, TW Wietsma, BJ Krueger, VH Grassian (2005), *J Geophys Res* (submitted).
111. IN Sokolik, DM Winker, G Bergametti, DA Gillette, G Carmichael, YJ Kaufman, L Gomes, L Schütz, and JE Penner (2001), *J Geophys Res*, *106*:18015–18028, doi: 10.1029/2000JD900498.
112. JM Prospero. In: S Guerzoni, and R Chester, eds. *The Impact of Desert Dust Across the Mediterranean*, Kluwer: Dordrecht, 1996.
113. X Li, H Maring, D Savoie, K Voss, and JM Prospero. *Nature* 380:416–419, 1996.
114. I Tegen, AA Lacis, and I Fung. *Nature* 380:419–422, 1996.
115. C Moulin, CE Lambert, F Dulac, and U Dayan. *Nature* 387:691–694, 1997.
116. P Alpert, YJ Kaufman, Y Shay-El, D Tanre, A da Silva, S Schubert, and JH Joseph. *Nature* 395:367–370, 1998.
117. RL Miller and IJ Tegen. *J Climate* 11:3247–3267, 1998.
118. WL Eschenbacher, GJ Kullman, and CC Gomberg. In: RL Harris, ed. *Patty's Industrial Hygiene.* 5th ed. New York: John Wiley & Sons, 2000, Vol. 1.
119. S Levin, R Herbert, G Skloot, J Szeinuk, A Teirstein, D Fischler, D Milek, G Piligian, E Wilk-Rivard, and J Moline. *Am J Ind Med* 42:545–547, 2002.
120. JC Wagner, K McConnochie, AR Gibbs, and FD Pooley. In: A Parker and JE Rae, eds. *Environmental Interactions of Clays.* Berlin: Springer, 1998.
121. JM Prospero. *Proc Natl Acad Sci USA* 96:3396–3403, 1999.
122. JM Prospero. In: V Ittekkot, P Schafer, S Honjo, and PJ Depetri, eds. *Particle Flux in the Ocean.* Chichester: Wiley & Sons, 1996; Vol. SCOPE Report 57.
123. RA Duce. In: RJ Charlson and J Heintzenberg, ed. *Aerosol Forcing of Climate.* Chichester: Wiley & Sons, 1995.
124. G Zhuang, RA Duce, and DR Kester (1990), *J Geophys Res*, *95*(C9):16207–16216.
125. X Zhu, JM Prospero, DL Savoie, FJ Millero, RG Zika, and ES Saltzman (1993), *J Geophys Res*, *98*(D5):9039–9046, doi: 10.1029/93JD00202.

126. XR Zhu, JM Prospero, and FJ Millero (1997), *J Geophys Res*, *102*:21297–21305, doi: 10.1029/97JD01313.
127. JH Martin and SE Fitzwater. *Nature* 331:341–343, 1988.
128. JJ Walsh and KA Steidinger (2001), *J Geophys Res*, *106*(C6):11597–11612, doi: 10.1029/1999JC000123.
129. IY Fung, SK Meyn, I Tegen, SC Doney, JG John, and JKB Bishop (2000), *Global Biogeochem Cycles*, *14*:281–295.
130. G Zhuang, Z Yi, RA Duce, and PR Brown (1992), *Global Biogeochem Cycles*, *6*:161–173.
131. LJ Spokes, TD Jickells, and B Lim. *Geochim Cosmochim Acta* 58:3281–3287, 1994.
132. KV Desboeufs, R Losno, F Vimeux, and SJ Cholbi (1999), *Geophys Res*, *104*:21287–21299.
133. KV Desboeufs, R Losno, and JL Colin. *Atmos Environ* 35:3529–3537, 2001.
134. N Meskhidze, WL Chameides, A Nenes, and G Chen (2003), *Geophys Res Lett*, *30*(21):2085, doi: 10.1029/2003GL018036.

6 Uptake of Trace Species by Ice: Implications for Cirrus Clouds in the Upper Troposphere

Paula K. Hudson and Margaret A. Tolbert
Department of Chemistry and Biochemistry, CIRES, University of Colorado, Boulder

CONTENTS

6.1 INTRODUCTION

Ice particles in the upper troposphere can be in the form of cirrus clouds, subvisible cirrus, or aircraft condensation trails (contrails). The cirrus and subvisible cirrus are highly variable with particle sizes ranging from 2 to 3000 μm diameter [1] and surface area densities ranging from 20 to 20,000 $\mu m^2/cm^3$ [2]. Further, studies indicate that 30% of the Earth is covered in cirrus at any time and that subvisible cirrus may be present in the tropics 75% of the time [2]. The aircraft contrail particles are typically smaller in size and have higher number densities than natural cirrus [3]. One satellite study over Europe showed that contrails may be present in the sky 60% of the time, with coverages ranging from <0.1% to 8% [4]. Although the exact conditions necessary for cirrus and contrail nucleation are not yet known, these particles will provide ample sites for heterogeneous chemical reactions in the upper troposphere.

Nitric acid is a key constituent in the troposphere. As a reservoir for NO_x species ($NO_x = NO + NO_2$) it can affect the reactivity of nitrogen oxides in the troposphere. Heterogeneous loss of nitric acid to ice has been previously studied with laboratory predictions of monolayer coverages under atmospheric conditions. However, with conflicting field measurements and laboratory results, the interaction of nitric acid with ice must be examined again. The presence of nitric acid on cirrus could also affect the heterogeneous chemistry of other species.

In the upper troposphere, gas-phase organic species such as methanol, acetone, and acetaldehyde, can form odd hydrogen species ($HO_x = OH + HO_2$) through reaction with the hydroxyl radical, OH, or by photolysis. Because OH is responsible for most of the oxidizing chemistry that takes place in the troposphere, understanding its sources and sinks is crucial. However, for methanol, acetone, and acetaldehyde, the balance of their sources and sinks is incomplete. Heterogeneous processes on cirrus clouds, either as sinks for the gas-phase species or as reactive sites, could help resolve this issue.

This chapter focuses on studies of the interaction of the upper tropospheric species (nitric acid, methanol, acetone, and acetaldehyde) with ice representative of cirrus clouds. Experiments were performed using a Knudsen cell flow reactor coupled to a Fourier transform infrared-reflection absorption spectroscopy (FTIR-RAS) setup. In this chapter, we describe the use of this setup to determine equilibrium coverages of condensed species and condensed-phase products formed as a result of the uptake. Implications for cirrus clouds in the upper troposphere are discussed.

6.2 EXPERIMENTAL

6.2.1 Reaction Chamber

The experimental setup, shown schematically in Figure 6.1, combines a Knudsen cell flow reactor [5] and FTIR-RAS [6] to simultaneously monitor gas and condensed phases, respectively. The apparatus consists of a single stainless steel chamber that houses a vertically mounted gold substrate upon which thin films are deposited. A potassium chloride (KCl) window is located on each side of the chamber for passage of the IR beam. The chamber is pumped by a 240 L/sec turbomolecular pump. Chamber pressures are monitored using an ionization gauge and an absolute capacitance manometer (baratron), while gas-phase constituents are analyzed with an electron-impact quadrupole mass spectrometer (UTI-100 C).

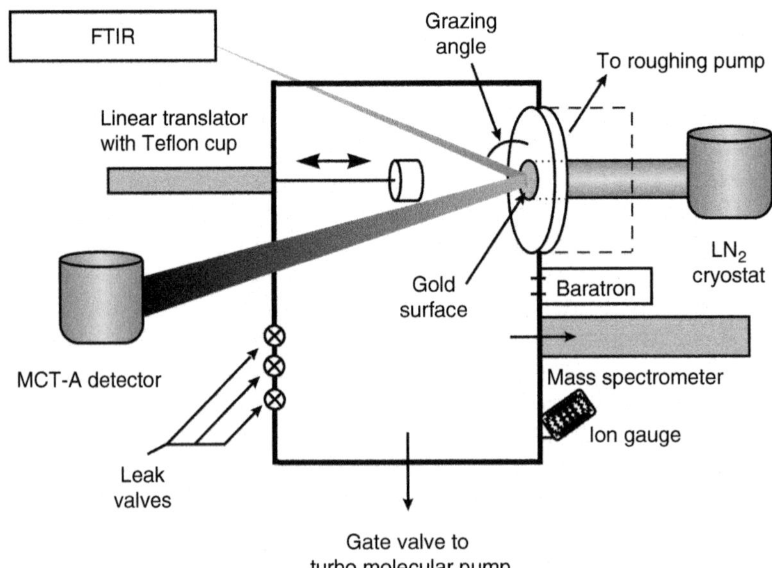

FIGURE 6.1 Experimental setup showing the Knudsen cell flow reactor and FTIR-RAS.

Partial pressures of the gas-phase species are determined by calibrating the mass spectrometer signals to the ion gauge and absolute capacitance manometer. Water partial pressures are read directly from the absolute capacitance manometer. The partial pressures of gas-phase constituents monitored are much smaller than the water vapor pressure, and therefore, contribute little to the absolute capacitance manometer value. Three leak valves are used for the introduction of water, nitric acid, and the other organic species: methanol, acetone, and acetaldehyde into the chamber. A gate valve separates the chamber from a turbomolecular pump. The gate valve contains a 0.17-cm^2 orifice (A_h) for the determination of accurate molecular flows during an experiment when the gate valve is closed [7]. A Teflon cup, having a volume of 7 cm^3 and an o-ring seal, is located on a pneumatic linear translator directly across from the gold substrate. When the translator is extended completely, the Teflon cup covers the gold substrate isolating it from the rest of the chamber. Tests have shown that when the cup is closed and the surface temperature is decreased to $T = 160$ K, the pressure of water vapor outside the cup can be increased to over 1000 times the vapor pressure of ice at $T = 160$ K without depositing an ice film on the substrate. This is to say that when the Teflon cup is closed, the composition of the film deposited on the gold substrate will remain constant regardless of the chamber activity.

The optically flat gold substrate ($d = 2.54$ cm) is cooled by thermal contact with a liquid nitrogen-filled cryostat, and resistively heated to a desired temperature ($T = 120$ to 220 K) using a Minco heater controlled by a Eurotherm temperature programmer. The substrate temperature is measured using three Type T (copper–constantan) thermocouples mounted to the back of the gold substrate using a thermally conductive epoxy. Temperature gradients across the surface are <0.5 K. The heating–cooling assembly is housed in a differentially pumped stainless steel sleeve with a Teflon seal between the sleeve and gold substrate. In this setup, the gold substrate is the only cold surface in the chamber.

6.2.2 Determination of Surface Coverage

Because the chamber was designed to function as a Knudsen cell flow reactor, the number of molecules lost to the film surface upon exposure of the gas-phase species can be readily calculated. Surface coverages are calculated by first calibrating the mass spectrometer signal of each gas-phase species to the ionization gauge and absolute capacitance manometer. To perform a calibration, the gas-phase species is introduced into the room temperature chamber and the pressure is incrementally increased from $P = 1 \times 10^{-7}$ to 1×10^{-4} Torr. The linear relationship between the mass spectrometer signal and the ion gauge and absolute capacitance readings is then determined. In this way, the mass spectrometer signal can be converted to an absolute pressure. Calculations of steady-state flow rates are determined by using the Knudsen effusion relation:

$$F = \frac{PA_h}{(2\pi mkT)^{0.5}} \tag{6.1}$$

where P is the partial pressure of the species of interest, A_h is the effective area of the gate valve orifice (0.17 cm^2), m is the molecular weight of the species, k is the Boltzmann constant, and T is the temperature of the gas-phase species ($T = 298$ K). When the ice film is exposed to the gas phase and uptake occurs, the partial pressure is lower and thus the flow out of the chamber is lower. If an equilibrium coverage of the species on the film is established, the steady-state flow of molecules returns to its original value because the additional loss to the ice film is no longer present.

The number of molecules lost from the gas phase is simply the integrated area of the signal as a function of exposure time. The surface coverage of the species (molecules/cm^2) is calculated by dividing the number of molecules lost from the gas phase by the geometric surface area of the ice which is assumed to be that of the gold substrate, $A_s = 5.07\,cm^2$. For ice films deposited at $T < 200\,K$, this is a good approximation. At $T > 200\,K$, vapor-deposited ice films are rough and have a larger surface area. For this reason, the true surface area of the ice films was experimentally determined through BET measurements using butane as an adsorbate [8]. We found that the surface area for ice at $T = 211$ to $219\,K$ is approximately 2.27 times larger than the smooth film. Therefore, the surface area of the rough film must be 2.27 times larger than the smooth film, yielding an ice film surface area of $11.51\,cm^2$. In experiments where condensation of the adsorbate occurs (i.e., no recovery of the signal), coverages are not calculated. Uptake is unlimited and therefore the surface multilayer coverage is dependent on exposure.

6.2.3 FTIR-RAS

We use a Nicolet Magna-IR 550 Spectrometer in a reflectance geometry to monitor condensed-phase species. Infrared radiation reflected at grazing angles off a metal surface allows for enhanced sensitivity at the point of reflection for polarization parallel to the plane of incidence [6]. The incoming IR beam is reflected off the gold substrate at a grazing angle of 83° from surface normal and detected by a mercury cadmium telluride-A (MCT A) detector. The IR spectrum from 4000 to $400\,cm^{-1}$ is monitored throughout each experiment.

The purpose of using FTIR-RAS is twofold. First, the IR is used to monitor film growth and stability. For ice films, the integrated area under the OH stretching region from 3500 to $3000\,cm^{-1}$ is monitored with time. Film thickness is determined as a function of this integrated area. Once the film has reached a given thickness, the water vapor pressure is adjusted until the integrated area under this region is constant as a function of time. Knowledge of the vapor pressure required to keep a film stable also allows for an internal temperature calibration using calculated water vapor pressure relationships to temperature [9]. For HNO_3-doped ice films, the entire spectrum from 4000 to $400\,cm^{-1}$ is monitored for film stability.

Second, the IR is used to detect phase and composition of ice and HNO_3-doped ice films as well as adsorbates on these films and potential reaction products. The composition of HNO_3-doped ice films is determined by comparing spectra to previously published reference spectra. Because FTIR-RAS has monolayer sensitivity to strong absorbers, we are able to spectroscopically detect small coverages of adsorbates and gain information on the type of adsorption present. By comparing the spectra taken before and after exposure to a gas-phase species, we are able to look for possible condensed-phase reaction products. The ability to simultaneously monitor gas- and condensed-phase species is a very powerful tool.

Reference spectra for the films used in the present study are shown in Figure 6.2. An ice film at $T = 200\,K$ is shown in Figure 6.2a and illustrates the characteristic OH stretch from ~3500 to $3000\,cm^{-1}$ and the ice libration near $850\,cm^{-1}$. At temperatures above $T = 165\,K$, a constant flow of water vapor must be established to maintain a constant film thickness during the experiment. Below 165 K, the vapor pressure of the ice film is comparable to, or less than, that of the background water in the chamber, so film desorption is negligible. By monitoring the integrated area of the OH stretching region, the H_2O partial pressure is adjusted until a stable area is achieved with no growth or evaporation of ice. A typical film, approximately 50 nm thick, is considered stable when the integrated area under the OH stretching region changes by <0.5% over 100 sec.

Figure 6.2b shows a typical spectrum of an ice film with monolayer (ML) HNO_3 coverage. To achieve this film, an ice film at $T = 200\,K$ is exposed to pressures of $HNO_3 \leq 1 \times 10^{-6}$

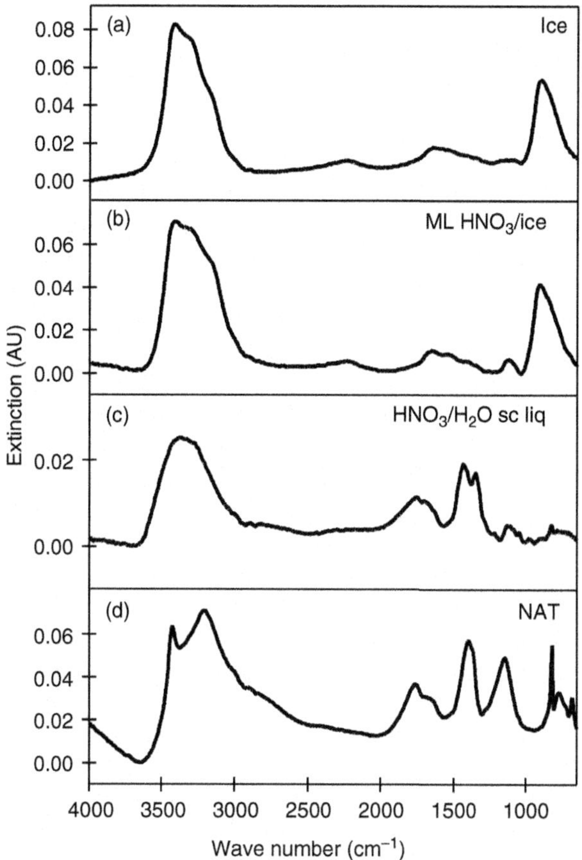

FIGURE 6.2 Reference FTIR-RAS spectra of (a) ice, (b) ice with monolayer HNO_3 coverage, (c) supercooled HNO_3/H_2O solution, and (d) NAT.

Torr until the growth of the NO_3^- asymmetric stretches are just barely discernible in the IR around 1435 and 1350 cm^{-1}. Prolonged exposure of the film to low pressures of HNO_3 does not result in any further changes. Figure 6.2c shows a supercooled HNO_3/H_2O solution. To form a supercooled HNO_3/H_2O solution, extended exposure of HNO_3 to the ice film at T = 200 K at higher HNO_3 pressures is required. The conversion to the supercooled solution is denoted by an increase in the two NO_3^- peaks at 1435 and 1350 cm^{-1} and a rounding or smoothing out of the OH stretching region from 3500 to 3000 cm^{-1} as well as the loss of the ice libration at 850 cm^{-1}. Figure 6.2d shows a nitric acid trihydrate (NAT) film. NAT growth is accomplished in one of the two ways. In the first, an ice film is deposited at T = 190 K. Once a constant film thickness is established, the ice is exposed to HNO_3 forming a supercooled solution. With small decreases in H_2O pressure, NAT crystallizes out of solution as evidenced by the two NO_3^- peaks combining to one peak around 1395 cm^{-1}, a sharp NO_3^- peak at 820 cm^{-1}, and a sharp peak around 3425 cm^{-1} in the H_2O region indicating a crystalline solid. The second way of forming NAT involves depositing an ice film at 175 K. Exposure of ice to HNO_3 at this low temperature results in immediate NAT crystallization. The NAT film temperature is then increased to 190 K. No changes in reactivity were noted upon varying the NAT film preparation technique.

6.3 CASE STUDIES

6.3.1 The Uptake of HNO_3 by Ice

6.3.1.1 Background

Nitrogen oxides play an important role in many atmospheric processes, some of which are summarized in Figure 6.3. Nitrogen oxides are divided into two classes, active ($NO_x = NO + NO_2$) and reservoir species ($NO_y = NO_x + N_2O_5 + HNO_3 + ClONO_2 + PAN + \dots$). The NO_x species play key roles in photochemical ozone production. Nitric acid is thought to be the most important reservoir of nitrogen oxides. It has the ability to tie up NO_x into a less-photochemically active form that, in turn, shifts the chemical reactivity of nitrogen oxides in the troposphere. Understanding the balance between NO_x and HNO_3 species is critical to understanding the oxidative capacity of the troposphere. When compared to atmospheric measurements, current models overestimate the ratio of HNO_3 to NO_x by factors of 2 to 10 [10]. This implies either a missing sink of HNO_3, a missing source of NO_x species, or a missing conversion from HNO_3 to NO_x. Many laboratory studies have investigated heterogeneous pathways to identify a process that would correct this imbalance in models.

Heterogeneous reactions of nitrogen oxides on soot [11, 12], sulfate aerosols [10, 13], dust, and biomass burning particles [14, 15] have been proposed to be important in the troposphere. Because cirrus clouds are prevalent and have large surface areas, ice is a natural surface to consider for heterogeneous processes. For example, recent work has shown that chlorine activation on cirrus in the upper troposphere may be important for global ozone loss [16].

A number of studies have previously examined the heterogeneous removal of HNO_3 on ice using a variety of techniques. Laboratory results and models thereof predict coverages of 1×10^{14} to 1×10^{15} molecules/cm^2 over a wide range of temperature and pressure [17–20]. There is also a limited amount of field data for HNO_3 on cirrus. Weinheimer et al. [21] find submonolayer coverages of particulate NO_y on cirrus ice. Because gas-phase HNO_3 concentrations were not measured in this study [21], it is difficult to quantitatively compare the data with laboratory observations. However, good agreement is found with the laboratory data assuming that only 10% to 20% of the atmospheric NO_y was in the form of HNO_3. Meilinger et al. [22] report <0.01 ML coverage of adsorbed NO_y and measured gas-phase HNO_3 concentrations of 25 to 110 pptv in a cirrus cloud. They state that according to current laboratory results this coverage is too small. They are able to resolve the conflict by the partitioning of HNO_3 into ternary solutions at the low temperatures of their observations ($T = 196$ K). However, they further state that pressure-dependent uptake may also explain the results [22]. One difficulty in comparing the laboratory data to field data is the vastly different temperature and pressure conditions under which the data were obtained. Here we attempt to determine a relationship between gas-phase HNO_3 pressure and the resulting coverage on ice films for laboratory measurements to be comparable to atmospheric conditions.

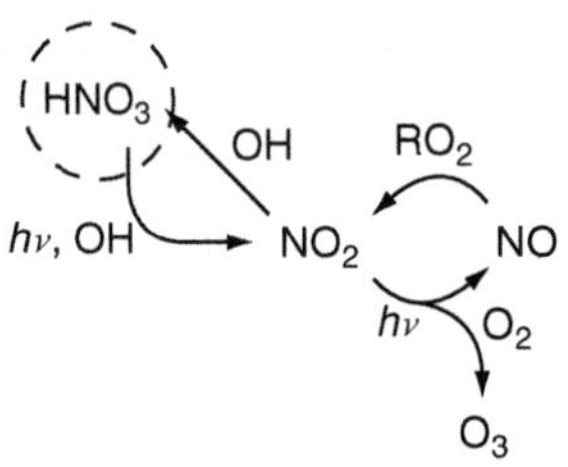

FIGURE 6.3 Schematic of key nitrogen oxide reactions in the troposphere. NO_x ($= NO + NO_2$) is involved in the production of tropospheric ozone. HNO_3 is an important reservoir for nitrogen oxides because it ties up NO_x into a less photochemically active form.

Our study has examined the uptake of nitric acid on ice over a range of upper tropospheric temperatures ($T = 210$ to 220 K) using two different nitric acid pressures ($P_{HNO_3} = 8.5 \times 10^{-7}$ and 1.7×10^{-6} Torr). In this study, we find sharp increases in nitric acid coverage to amounts greater than a monolayer at the lowest temperatures studied. We also find an increase in HNO_3 coverage on ice with increased HNO_3 pressure. The strong temperature and pressure dependencies have not been observed in previous laboratory studies.

The observation of coverages greater than a monolayer forces a revision of the monolayer model that had been applied to the previous laboratory data [19]. Here, a multilayer model, the Frenkel–Halsey–Hill (FHH) model, has been fit to the data from this study. The model has also been applied to a previous laboratory measurement of nitric acid uptake on ice particles and agreement is found to within a factor of 2. Extrapolations of the model to determine coverage using conditions representative of field observations by Weinheimer et al. [21] and Meilinger et al. [22] have been successful. Comparison with more recent field observations [23] is also excellent.

6.3.1.2 Results and Discussion

Figure 6.4 shows three different types of uptake observed upon exposure of ice films to nitric acid using different nitric acid pressure and ice film temperature conditions. The left-hand panels of Figure 6.4a–c show HNO_3 flow as a function of exposure time to the ice film. At $t = 0$ sec, indicated by the arrow, the Teflon cup is retracted and a drop in the nitric acid flow is observed indicating the uptake of nitric acid onto the ice film. The magnitude of the drop in

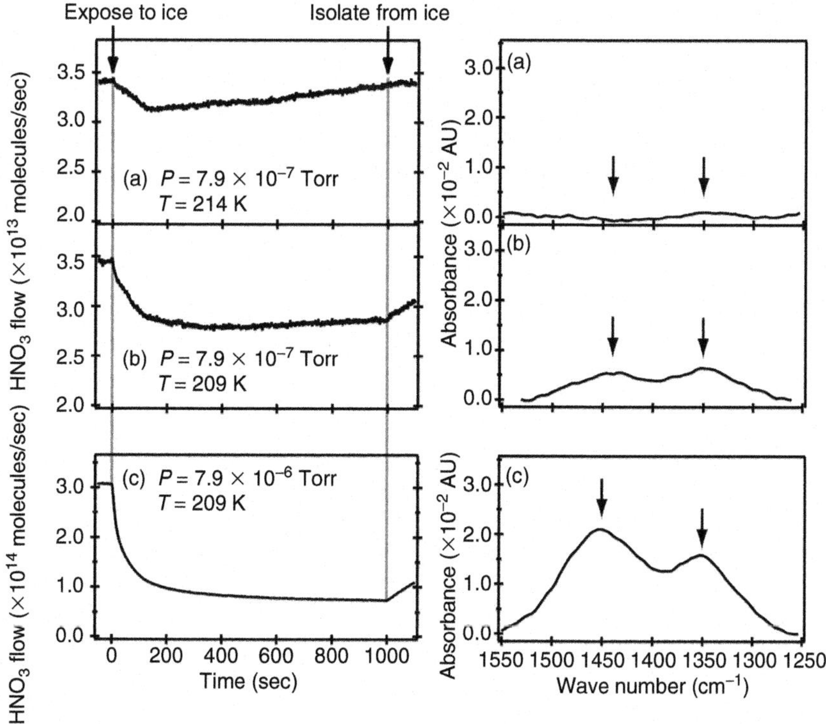

FIGURE 6.4 The left panel shows the HNO_3 mass spectrometer signal calibrated to flow upon exposure to ice films at different HNO_3 pressures and ice film temperatures. Three uptake regimes are shown (a) small limited uptake, (b) large limited uptake, and (c) unlimited uptake. The right panels show FTIR-RAS spectra of the asymmetric nitrate stretch region for the same uptake regimes.

flow varies as a function of ice film temperature as well as nitric acid pressure. Note that the nitric acid flow in Figure 6.4a and b is on the same scale and that in Figure 4c is an order of magnitude greater.

In Figure 6.4a ($P_{HNO_3} = 7.9 \times 10^{-7}$ Torr, $T = 214$ K) a small decrease in the nitric acid flow is observed after exposure to the ice film followed by a gradual recovery to its original level. This indicates that saturation coverage of HNO_3 on ice has been obtained. A slightly larger drop is observed in the HNO_3 flow in Figure 6.4b ($P_{HNO_3} = 7.9 \times 10^{-7}$ Torr, $T = 209$ K) followed by a gradual increase in the flow rate. Although not observed on this timescale, with longer exposures the HNO_3 flow continues to return to its original value. For future reference, Figure 6.4a and b will be referred to as small and large limited uptake, respectively. When the ice film temperature is kept constant at $T = 209$ K, and the initial HNO_3 pressure is increased by an order of magnitude, as shown in Figure 6.4c, $P_{HNO_3} = 7.9 \times 10^{-6}$ Torr, $T = 209$ K, an even larger drop in signal is observed followed by a continually decreasing signal. This type of uptake is also observed when $P_{HNO_3} = 7.9 \times 10^{-7}$ Torr and $T = 200$ K although data are not shown. This type of uptake will be referred to as unlimited uptake and will later be identified as originating from the formation of a supercooled HNO_3/H_2O solution. After $t = 1000$ sec, the Teflon cup is again extended over the film, isolating it from the nitric acid. At this point, no change in HNO_3 flow is observed in Figure 6.4a because a steady-state coverage has been reached. The HNO_3 flow begins to return to its original level in Figures 6.4b and c when the cup is closed because, prior to isolation, the ice was still removing nitric acid from the gas phase. With time, the HNO_3 flow returns to its original value after the ice surface has been isolated.

Previous studies of HNO_3 on ice have observed different uptake behavior depending on the phase of HNO_3/H_2O formed as a result of the HNO_3 exposure pressure and ice film temperature. Zondlo et al. [20] observed enhanced nitric acid uptake on thin ice films with the formation of a supercooled HNO_3/H_2O solution. An HNO_3/H_2O phase diagram supported this result. To help explain the three types of uptake observed in Figure 6.4, we also have examined these regimes with respect to the HNO_3/H_2O phase diagram.

Figure 6.5 shows the HNO_3/H_2O phase diagram. The open squares are supercooled HNO_3/H_2O solution vapor pressures as determined by Hanson [24] with solid lines fit to

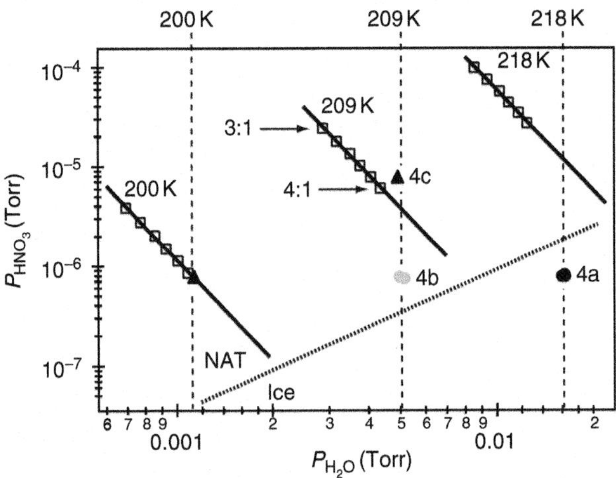

FIGURE 6.5 HNO_3 and H_2O phase diagram with vapor pressures of supercooled HNO_3/H_2O solution data (open squares) and the NAT/ice equilibrium line. The solid circles are the small limited uptake (black 4a) and large limited uptake (gray 4b) regimes. The solid triangles are the unlimited uptake (4c) regime. Above the line through the open squares a supercooled solution can thermodynamically form.

the data. For HNO_3 and water pressures above the supercooled HNO_3/H_2O solution line at a given temperature, it is thermodynamically possible to form a supercooled solution, even though conditions favor the most thermodynamically stable form, NAT. For HNO_3 and H_2O pressures below the supercooled HNO_3/H_2O solution line for a given temperature, it is not possible to form supercooled solutions. It is apparent that both the small and large limited uptake experiments (black and gray solid circles) are well below their corresponding supercooled HNO_3/H_2O solution line at $T = 214$ and 209 K, respectively. Thermodynamically, the uptake of HNO_3 on ice in the small and large limited uptake (Figure 6.4a and b) cannot be due to formation of a supercooled solution. The enhanced uptake from small to large limited uptake is not due to a phase change from formation of a supercooled solution.

The HNO_3 pressure and ice film temperature conditions of the unlimited uptake experiment (solid triangles, Figure 6.4c) are above the supercooled solution line in Figure 6.5 at the corresponding temperature of $T = 209$ K. Because it is above the supercooled solution line, the unlimited uptake of HNO_3 observed in Figure 6.4c is likely due to formation of a supercooled HNO_3/H_2O solution. Similarly, at $P_{HNO_3} = 7.9 \times 10^{-7}$ Torr, when the ice film temperature is decreased to $T = 200$ K, as discussed previously, the experimental point falls on the supercooled HNO_3/H_2O solution line corresponding to $T = 200$ K.

Because we can simultaneously monitor the infrared spectra, we can further understand the differences in HNO_3 uptake between the three uptake regimes. The right-hand panels of Figure 6.4 show the IR spectra in the region of the asymmetric NO_3^- peaks (1500 to 1250 cm^{-1}) for the three uptake regimes. For all three spectra in Figure 6.4, a baseline has been subtracted for easier interpretation. No NO_3^- peaks are present in Figure 6.4a, which is to be expected due to the small submonolayer coverage ($\Theta = 4 \times 10^{14}$ molecules/cm^2). The coverage of HNO_3 is equivalent in Figure 6.4b and c, $\Theta = 1.3 \times 10^{15}$ molecules/cm^2. It can be seen that for the large limited uptake in Figure 6.4b, the lower wave number peak at 1340 cm^{-1} is larger than the higher wave number peak at 1440 cm^{-1}. The reverse is true for the supercooled HNO_3/H_2O solution in Figure 6.4c.

We have compared our supercooled HNO_3/H_2O solution spectra with that of Zondlo et al. [25]. Dividing the lower wave number peak amplitude (1.5) by the higher wave number peak amplitude (2) yields a value of 0.75. This is in close agreement with the ratio of 0.76 from Zondlo et al. [25] for a supercooled 4:1 H_2O:HNO_3 solution. This composition also agrees well with the phase diagram in Figure 6.5. Zondlo et al. [25] verify their spectra with spectra derived from optical constants from Toon et al. [26]. Although the large limited uptake and unlimited uptake experiments are performed in the NAT stability regime, it can be seen in Figure 6.4b and c that neither of these spectra are representative of the NAT spectra shown in Figure 6.2d.

Figure 6.6 summarizes the nitric acid experiments at $P_{HNO_3} = 8.5 \times 10^{-7}$ Torr integrated as coverage, in both the small and large limited uptake regimes. The area of the drops and recoveries in Figure 6.4a and b are integrated as a function of time yielding the number of molecules lost to the surface in the small limited uptake regime. As was discussed before, coverage is calculated by dividing the number of molecules lost to the surface by the surface area of the ice film. For the rough films deposited from $T = 211$ to 219 K, the surface area is 11.51 cm^2. The uptake in the small limited regime can be calculated directly because a steady-state coverage has been reached. These experiments are shown in Figure 6.6 as solid circles. However, in the cases of the large limited uptake, the signal does not return to its original flow within the timescales of our experiments (Figure 6.4b). For these cases, integrated coverages depend on the length of exposure (i.e., the longer exposure yields a greater coverage.) As mentioned previously, due to experimental complications, the large limited uptake coverages are only used as lower limits. The time-dependent coverages are shown in Figure 6.6 as open symbols as a function of time. The arrows over the four experiments in Figure 6.6 denote that the coverage is still increasing after $t = 2000$ sec of nitric acid exposure to the ice film. It can

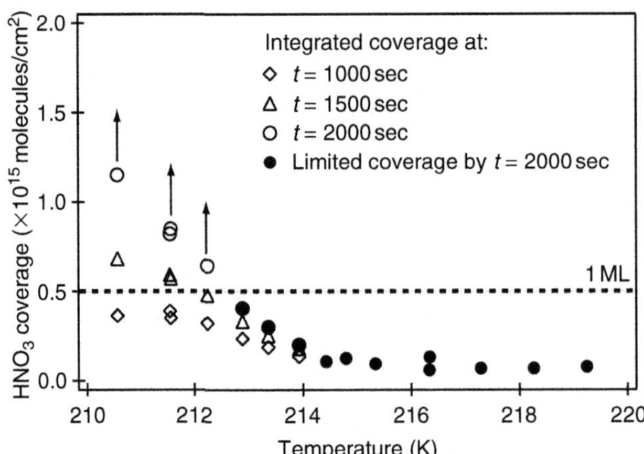

FIGURE 6.6 Nitric acid coverage as a function of ice film temperature and exposure time. The open symbols represent coverages that have not yet saturated. The arrows indicate that the coverages are still increasing. The solid symbols represent coverages that are saturated by $t = 2000$ sec.

be seen that approximate monolayer coverages form on ice around $T = 213$ K and that the coverage on ice decreases as temperature increases. At temperatures, $T \leq 213$ K, and $P_{HNO_3} = 8.5 \times 10^{-7}$ Torr, coverages greater than one monolayer form.

Figure 6.7 shows HNO_3 coverage data at $t = 2000$ sec as a function of ice film temperature at two pressures $P_{HNO_3} = 8.5 \times 10^{-7}$ Torr (open and solid circles) and 1.7×10^{-6} Torr (open and solid squares) from this study. The solid lines in Figure 6.7 represent a multilayer FHH model that has been applied to the data. In this model it is assumed that adsorbed molecules have a distribution of energies for the first layer of adsorbed gas and intermolecular attractions exist. Additionally, in the multilayer region, a more gradual decrease in the

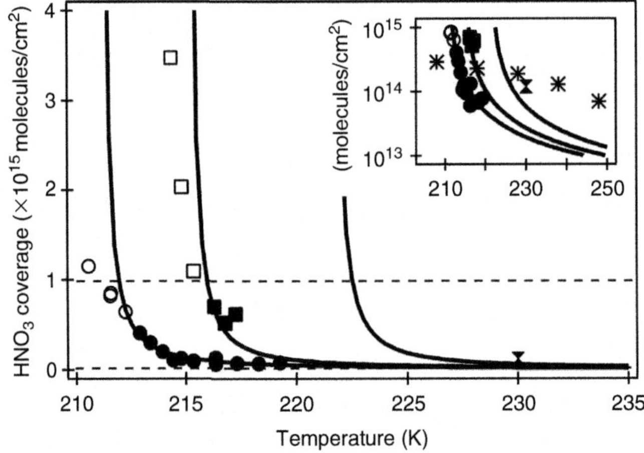

FIGURE 6.7 Nitric acid coverage as a function of ice film temperature at three different pressures, $P = 8.5 \times 10^{-7}$ Torr (solid circles), $P = 1.7 \times 10^{-6}$ Torr (solid squares) from the study, and $P = 5.0 \times 10^{-6}$ Torr (solid hourglass) from Arora et al. [18]. The curve through all three data sets is from the FHH model that was fit to the lowest pressure data. The pressure was adjusted for the two higher pressure curves. The boxed region is duplicated in the inset on a semi-log plot with added Abbatt data.

adsorption energy occurs as the film approaches its vapor pressure. The form of the FHH model is

$$\left(\frac{\Theta_{\text{measured}}}{\Theta_{\text{max}}}\right) = \left\{\frac{A}{\ln\left(\frac{P^{\circ}}{P}\right)}\right\}^{B} \tag{6.2}$$

where P is the pressure of nitric acid used and P° corresponds to the vapor pressure of a 35 wt% HNO_3/H_2O solution (6.5:1 H_2O:HNO_3 solution) determined by

$$\log_{10} P^{\circ}(\text{Torr}) = \left(\frac{-3431.4}{T(\text{K})}\right) + 10.184 \tag{6.3}$$

The maximum monolayer coverage (Θ_{max}) is calculated using the liquid density (ρ) at 293 K, and the molecular weight (MW), to approximate the radius for HNO_3 as if it were a sphere:

$$\frac{4}{3}\pi r^3 = \frac{\text{MW}}{\rho N_{\text{A}}} \tag{6.4}$$

The calculated radius is then used to approximate the maximum monolayer surface coverage:

$$\Theta_{\text{max}} = \frac{1}{\pi r^2} \tag{6.5}$$

Using this method, we find $\Theta_{\text{max}} = 5 \times 10^{14}$ molecules/cm^2 for HNO_3. The coefficients A and B in Equation (6.2) have been determined by a best fit to the $P_{HNO_3} = 8.5 \times 10^{-7}$ Torr data yield values of $A = 0.25$ and $B = 1.3$. The model was only fit to the $P_{HNO_3} = 8.5 \times 10^{-7}$ Torr data. The curves through the $P_{HNO_3} = 1.7 \times 10^{-7}$ Torr data and the Arora et al. [18] data $P_{HNO_3} = 5 \times 10^{-6}$ Torr, were created by changing only the pressure term, P, in Equation (6.2). The curve is not expected to fit the open symbols for either pressure as these coverages are still increasing as a function of exposure time as shown in Figure 6.6. The inset in Figure 6.7 shows the data on a log scale with extended temperatures. For comparison, data from Abbatt [17] (asterisks) from $T = 208$ to 248 K has been added for $P_{HNO_3} = (6 - 9) \times 10^{-7}$ Torr. Recent data obtained using a system similar to Abbatt gave similar results for HNO_3 coverage [27]. The agreement between our FHH model and the Arora et al. [18] data on ice particles using the FHH model is within a factor of 2. This agrees well with the overall uncertainty they state in their results [18]. The agreement with the data of Abbatt [17] and Hynes et al. [27] is less favorable.

6.3.1.3 Atmospheric Implications

The data described here, and modeled by the FHH multilayer model, can be used to estimate coverages of HNO_3 on cirrus clouds. This model has been applied to NO_y measurements in a lee-wave cloud from Weinheimer et al. [21] and NO_y measurements in cirrus from Meilinger et al. [22]. The coverages for Weinheimer et al. [21] and Meilinger et al. [22] have been determined using values in Table 6.1. The condensed-phase NO_y is converted to a number of molecules NO_y/cm^3 air by using the ideal gas law at the pressure and temperature provided in Table 6.1. This value is then divided by the surface area of the cloud stated in the references, also provided in the table, to yield a condensed phase range of coverages. The range of coverages for the two data sets is shown as vertical solid lines in

TABLE 6.1
Data Compiled from Weinheimer et al. [21] and Meilinger et al. [22]

Ref.	Temperature (K)	Pressure (mbar)	Gas-Phase HNO_3 (pptv)	Calculated Gas-Phase HNO_3 ($\times 10^{-8}$ Torr)	Condensed NO_y (pptv)	Surface Area ($\mu m^2/cm^3$)	Calculated NO_y Coverage ($\times 10^{13}$ molec/cm^2)
Weinheimer et al. [21]	209[a]	193[b]	25–75[c]	0.36–1.09	25–75[c]	2000[c]	0.84–2.51
Meilinger et al. [22]	196[d]	200[e]	25–110[d,f]	0.38–1.65	2–8[d,f]	419[d]	0.35–1.40

[a]Jensen et al. [40].
[b]Baumgardner and Gandrud [41].
[c]Weinheimer et al. [21].
[d]Meilinger et al. [22].
[e]Schiller et al. [42].
[f]Feigl et al. [43].

Figure 6.8. Also shown is the FHH model calculated at three HNO_3 pressures using the coefficient determined from Equations (6.2) and (6.3). The pressures used for the FHH model, shown as dashed curved lines outlining a shaded area in Figure 6.8, cover the pressure range to include both Weinheimer et al. [21] and Meilinger et al. [22], $P_{HNO_3} = 3.5 \times 10^{-9}$ to 1.7×10^{-8} Torr.

The Weinheimer et al. [21] measurements fall above and through the outlined region of possible gas-phase HNO_3 pressures while the Meilinger et al. [22] data fall below the outlined region. The Weinheimer et al. [21] data are difficult to directly compare because they have no direct measurements of gas-phase HNO_3, but only of gas-phase NO_y. Using previous laboratory results of Zondlo et al. [20] and Abbatt [17], Weinheimer et al. [21] infer complete loss of gas-phase HNO_3 to the wave cloud thus determining a gas-phase HNO_3/NO_y ratio of 10% to 20% (25 to 75 pptv). We believe the field observations should always fall at or below

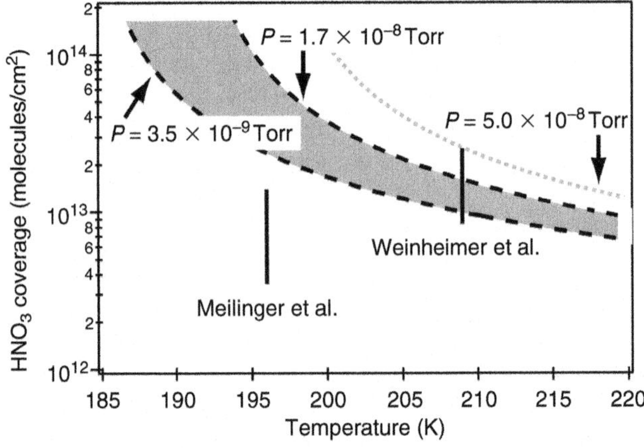

FIGURE 6.8 Semi-log plot of HNO_3 coverage as a function of ice film temperature with modeled data from the FHH model at $P = 3.5 \times 10^{-9}$ and 1.7×10^{-8} Torr. An additional pressure curve at $P = 5.0 \times 10^{-8}$ Torr is the small dashed line. The two vertical lines indicate the range in coverage outlined by Weinheimer et al. [21] ($T = 209$ K) and Meilinger et al. [22] ($T = 196$ K).

the FHH model. In order for this to be true, the HNO_3 pressure would need to be $P_{HNO_3} = 5.0 \times 10^{-8}$ Torr, the small dashed line in Figure 6.8, which is consistent with a HNO_3/NO_y ratio of 50% (375 pptv). This is quite reasonable as explained in Weinheimer et al. [21]. Another possible explanation for the measurements to be above the model could be from the enhanced uptake due to co-condensation of HNO_3 and H_2O in the rapid formation of the lee-wave cloud.

The Meilinger et al. [22] data fall below the range of pressures outlined in that measurement. This is a more expected result. The model comes from experiments that have equilibrium coverages of nitric acid on ice. Coverages in the atmosphere are not likely to be at equilibrium due to a number of reasons including slow gas-phase diffusion of HNO_3 to the cloud, or exposure of the cloud to the gas-phase HNO_3 may be short relative to the time needed to reach equilibrium at the time it was measured. The model can act as an upper limit of HNO_3 adsorption onto a cirrus cloud. The FHH model has very recently been applied to evaluate NO_y coverages on cirrus ice particles from several airborne field campaigns [23]. Specifically, Kondo et al. [23] report NO_y coverages on ice as a function of temperature and find dramatically increasing coverage with decreasing temperature. They find that the data are well represented by the FHH model presented here. In the future, as more complete field data become available, the model presented herein will be able to be tested even more rigorously.

6.3.2 The Interaction of Methanol, Acetone, and Acetaldehyde with Ice

6.3.2.1 Background

Our understanding of the chemistry of the upper troposphere has undergone significant revisions in recent years. One of the most dramatic examples of this is the recognition of the important role that oxygenated hydrocarbons play in this part of the atmosphere. Species such as methanol (CH_3OH), acetone ($(CH_3)_2C(O)$), and acetaldehyde (CH_3CHO) are abundant in the upper troposphere, with measured mixing ratios of 400 to 900 pptv, 300 to 3000 pptv, and 30 to 100 pptv, respectively [28–31].

A schematic of some of the known tropospheric chemistry of oxygenated organics is illustrated in Figure 6.9. Their destruction is dominated by reaction with OH and photolysis, while their sources include direct biogenic emissions and biomass burning. The destruction of oxygenated hydrocarbons results in the formation of HO_x ($HO_x = OH + HO_2$). The OH radical is responsible for most of the chemistry in the troposphere. Thus, understanding the sources and sinks for OH is vital to assessing the future oxidative capacity of our atmosphere.

For most of the troposphere, ozone photolysis in the presence of water vapor is the major source of HO_x:

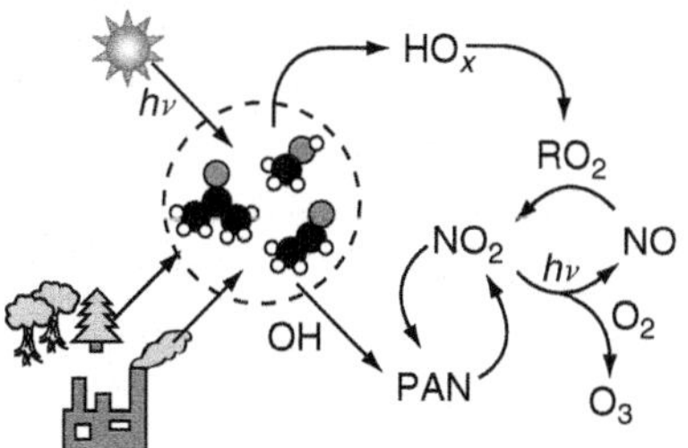

FIGURE 6.9 A schematic of some of the known tropospheric chemistry of oxygenated organics. Oxygenated organics are emitted from both natural and anthropogenic sources. Photochemistry in the atmosphere degrades these compounds to produce HO_x and PAN.

$$O_3 + h\nu \longrightarrow O(^1D) + O_2 \qquad (\lambda < 310\,\text{nm}) \qquad \text{(R6.1)}$$

$$O(^1D) + H_2O \longrightarrow 2OH \qquad \text{(R6.2)}$$

However, in the upper troposphere, the mixing ratio of water is sufficiently low to make this mechanism inefficient. It has recently been recognized that HO_x production from oxygenated hydrocarbons is important, providing as much HO_x as reactions (R6.1) and (R6.2) in some regions of the atmosphere [29]. Calculations have shown that the oxidation of acetaldehyde and methanol in biomass burning plumes leads to a net production of 0.8 HO_x and the oxidation of acetone creates a net source of 3.6 HO_x [32]. Furthermore, recent modeling studies suggest that acetone photooxidation provides a global source of HO_x to the upper troposphere while the presence of aldehydes and other oxygenated hydrocarbons enhances HO_x locally in the tropics [33].

In addition to having a major impact on the HO_x cycle, oxygenated hydrocarbons may influence NO_y in the troposphere by reactions of their decomposition products or by direct reactions with NO_y species. For example, one decomposition product of acetone, the peroxyacetyl radical $CH_3C(O)O_2$, can react with NO_2 to form peroxyacetyl nitrate (PAN):

$$CH_3C(O)O_2 + NO_2 + M \longrightarrow PAN + M \qquad \text{(R6.3)}$$

PAN is of great atmospheric interest due to its role in transporting NO_x to rural areas where it can contribute to smog formation.

Direct heterogeneous reactions involving oxygenated hydrocarbons and NO_y species could be important in converting HNO_3 into NO_x. As mentioned above, atmospheric models consistently overpredict the HNO_3/NO_x ratio by factors of 2 to 10 [11, 34, 35]. One previous attempt to reconcile this imbalance relied on reaction of formaldehyde with HNO_3 on tropospheric aerosols or clouds according to [10]:

$$HCHO + HNO_3 \longrightarrow HONO + HCOOH \qquad \text{(R6.4)}$$

While laboratory studies have shown that this reaction does occur on sulfuric acid surfaces [36, 37], the rate appears to be too slow to completely solve the HNO_3/NO_x dilemma [38]. It is possible that heterogeneous reactions of other oxygenated organic species like methanol, acetone, and acetaldehyde may help alleviate this problem.

Although the abundance and importance of oxygenated hydrocarbons in the upper troposphere are now recognized, a complete understanding of their sources and sinks is still lacking. For example, a recent study reports that methanol sources are more than twice as large as known sinks [30]. It further states that methanol was found to be less abundant in the upper troposphere relative to acetone implying unknown, possibly heterogeneous sinks for methanol [30]. A recent study has proposed that the heterogeneous conversion of methanol to formaldehyde on cirrus could be an additional sink for methanol [39]. Further measurements in the Pacific show values of acetone and acetaldehyde two to ten times larger than models can predict [31]. To help develop a more complete understanding of the budgets of methanol, acetone, and acetaldehyde in the upper troposphere, we have examined possible heterogeneous loss of these species on various thin films representative of upper tropospheric cirrus clouds.

6.3.2.2 Results and Discussion

The uptake of methanol, acetone, and acetaldehyde on ice at atmospherically relevant temperatures, $T > 200\,\text{K}$, was below our limit of detection of 10^{12} molecules/cm^2. Therefore, experiments were performed at lower temperatures where significant uptake was observed. Figure 6.10 shows typical uptake curves for methanol on ice at $T = 180$, 167.5, and 160 K. A constant flow of methanol, 1.6×10^{13} molecules/sec ($P_{\text{methanol}} = 2.7 \times 10^{-7}$ Torr), was established below $t = 450$ sec. At $t = 450$ sec, the Teflon cup was retracted exposing the ice film to the methanol. All three experiments show a significant drop in methanol signal upon exposure to the ice film.

After the initial drop, all signals "saturate," i.e., they recover to their original level. While each of these films was isolated from the methanol at different times, there was no change in signal upon isolation with any of them. The recovery time until saturation varies with the temperature of the ice film with the coldest experiment having the longest recovery time. The area of the loss and recovery of the methanol signal is used to calculate the number of molecules lost from the gas phase and, in turn, the coverage of methanol on the ice film. Therefore, the largest coverage occurs at the coldest temperatures and the smallest coverage occurs at the warmest temperatures. At temperatures $T \leq 149$ K, condensation of methanol on the ice film occurs. This appears as a drop in the signal with no recovery. While submonolayer coverages of methanol on ice are below the detection limit of the FTIR-RAS, the condensation of multilayers of methanol on the ice is observed in the IR.

Uptake curves similar to methanol were obtained for acetone ($P_{\text{acetone}} = 3.1 \times 10^{-7}$ Torr) and acetaldehyde ($P_{\text{acetaldehyde}} = 3.0 \times 10^{-7}$ Torr). The saturated coverage results for methanol (solid squares), acetone (open triangles), and acetaldehyde (solid circles) experiments as a function of ice film temperature are shown in Figure 6.11. For all three species, coverage increases with decreasing temperature. At the lowest temperatures, multilayers for all species grow before a complete monolayer is deposited. Therefore, the maximum coverage reported is that of the largest submonolayer coverage obtained, before multilayer growth begins. Measured methanol coverages range from $\Theta = 3 \times 10^{14}$ molecules/cm^2 at $T = 150$ K to our lower limit of detection of $\Theta = 10^{12}$ molecules/cm^2 at $T = 180$ K. Acetone coverage is similar but its maximum coverage shifts to a lower temperature of $T = 140$ K.

Because the coverages for these organics on ice at atmospherically relevant temperatures are below our detection limit, we have attempted to model the observed uptake and extrapolate to higher temperatures. We use a Langmuir model to fit our experimental data. Briefly, the Langmuir model assumes an equilibrium between gas and adsorbed phase species:

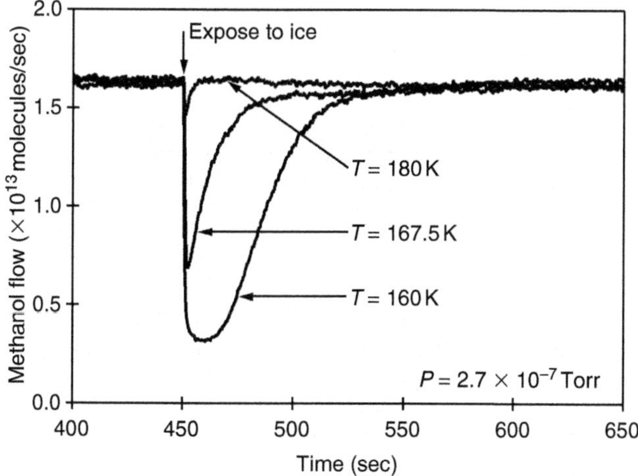

FIGURE 6.10 Mass spectrometer signal, calibrated to methanol flow, upon exposure of the ice films to methanol at the temperatures noted.

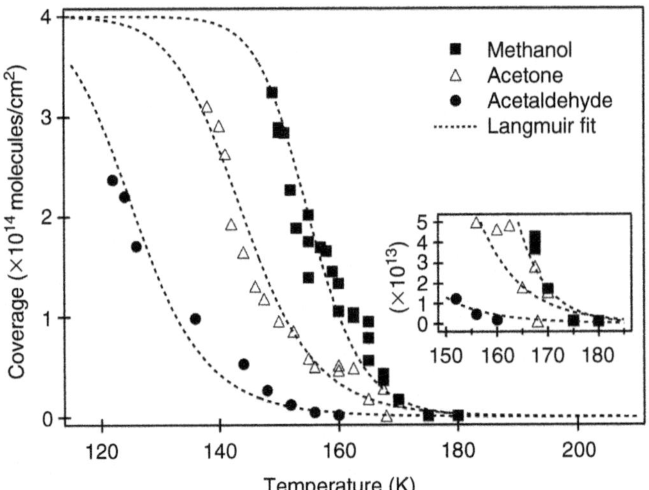

FIGURE 6.11 Coverage uptake curves for methanol (solid squares), acetone (open triangles), and acetaldehyde (solid circles) as a function of ice film temperature. The dashed lines are Langmuir best fits.

$$\text{Organic(g)} \xleftrightarrow{\text{K}} \text{Organic(ad)} \tag{6.6}$$

Under these conditions, fractional surface coverage $\Theta = \Theta_{\text{measured}}/\Theta_{\text{max}}$ is given by

$$\Theta = \frac{KP_{\text{A}}}{1 + (KP_{\text{A}})} \tag{6.7}$$

where P_{A} is the pressure of the organic species referenced to STP (1 atm). The maximum monolayer coverage, Θ_{max}, was calculated for the three organic species using Equations (6.4) and (6.5). We find an average $\Theta_{\text{max}} = 4 \times 10^{14}$ molecules/cm^2 for all three species. The Langmuir fits to the coverage data on a linear scale are shown as the dashed curves in Figure 6.11. While the Langmuir fits model the data quite well, they tend to overpredict the coverages at the higher temperatures for all three species, as shown in the inset in Figure 6.11.

In addition to studies on pure ice, we also measured the uptake of acetone, acetaldehyde, and methanol on an ice film with monolayer HNO_3 coverage, a supercooled 4:1 H_2O:HNO_3 solution, and a NAT film. All films were at $T = 200$ K with the exception of the NAT film that was at $T = 190$ K. For acetone and acetaldehyde, no uptake was observed at $T = 200$ K on any of these surfaces. For methanol, however, the supercooled HNO_3/H_2O solution showed a small uptake on the order of $\Theta = 4 \times 10^{12}$ molecules/cm^2.

The enhanced uptake for methanol on the supercooled solution at $T = 200$ K most likely results because adsorption onto the surface of a liquid is more efficient than onto a crystalline solid. Additionally, the uptake may not be limited to only the surface. Diffusion into the liquid, as opposed diffusion into the crystalline solid, may also be an enhanced process.

No gas-phase reaction products were observed with the reactions of any of the species on ice or nitric acid/ice films. Condensed-phase products may have formed; however, coverages were below our limits of detection.

6.3.2.3 Atmospheric Implications

The data described here can be used to estimate surface coverages for methanol, acetone, and acetaldehyde on ice clouds in the upper troposphere and our results are summarized in

TABLE 6.2
Predicted Coverages from Langmuir Fit for Measured Atmospheric Mixing Ratios and Calculation of Percent Gas-Phase Loss to a 20,000 μm²/cm³ Cloud from Calculated Coverages

	Mixing Ratio (pptv)	Θ at 210 K and 300 mbar (molec/cm²)	Loss from Gas Phase for a 20,000 μm²/cm³ Cloud (%)
Methanol	900	3×10^{10}	0.1
Acetone	350	6×10^{10}	0.2
Acetaldehyde	80	5×10^{9}	0.1

Table 6.2. We obtain upper limits to the experimental coverages on cirrus clouds by extrapolating the fits to $T = 210$ K and assuming atmospherically relevant pressures of $P_{methanol} = 2.0 \times 10^{-7}$ Torr, $P_{acetone} = 7.9 \times 10^{-8}$ Torr, and $P_{acetaldehyde} = 1.8 \times 10^{-8}$ Torr for methanol, acetone, and acetaldehyde, respectively. These extrapolations predict coverages of $\Theta = 3.0 \times 10^{10}$ molecules/cm² (7.5×10^{-5} ML) for methanol, $\Theta = 5.6 \times 10^{10}$ molecules/cm² (1.4×10^{-4} ML) for acetone, and $\Theta = 5.2 \times 10^{9}$ molecules/cm² (1.3×10^{-5} ML) for acetaldehyde. For all three species the presence of HNO_3 on the ice surface, when it remains in a crystalline form, does not enhance the uptake at atmospherically relevant temperatures. Zondlo et al. [20] showed that under conditions of high HNO_3 partial pressures, addition of HNO_3 to an ice film results in a supercooled solution. Similar results were shown in Section 6.3.1.2. With the formation of a supercooled HNO_3/H_2O solution, the uptake of methanol increases at $T = 200$ K. However, even with this enhanced uptake the expected coverage remains on the order of $\Theta = 10^{12}$ molecules/cm² (3.8×10^{-2} ML) at $T = 200$ K, and does not represent a large surface coverage. Therefore for the compositions of cirrus clouds explored, we do not expect significant heterogeneous chemistry of methanol, acetone, or acetaldehyde with the possible exception of methanol on supercooled HNO_3/H_2O solution.

Number densities of cirrus clouds are highly variable ranging anywhere from ~20 to 20,000 μm²/cm³. However, because the expected coverages are so small even the highest number density clouds will not act as significant sinks for these organic molecules. The predicted losses for the gas-phase species onto a 20,000 μm²/cm³ cloud are included in Table 6.2. It can be seen that in all cases, <0.2% of the organics are lost from the gas phase, even for a very large surface area cirrus cloud. Therefore, cirrus clouds are not likely to act as a significant sink for methanol, acetone, or acetaldehyde. Future studies will explore the possible loss of these species to other atmospheric surfaces in an attempt to resolve their uncertain atmospheric budgets.

REFERENCES

1. DR Dowling and LF Radke. *J Appl Meteorol* 29:970–978, 1990.
2. OB Toon. Heterogeneous Chemistry in the Upper Troposphere. Workshop on Modeling Heterogeneous Chemistry of the Lower Stratosphere/Upper Troposphere, Le Bischenberg, Bischoffsheim, France, October 21–23, 1996.
3. JF Gayet, G Febvre, G Brogniez, H Chepfer, W Renger, and P Wendling. *J Atmos Sci* 53:126–138, 1996.
4. U Schumann and P Wendling. *Air Traffic and the Environment*. Berlin: Springer-Verlag, 1990.
5. DM Golden, GN Spokes, and SW Benson. *Angew Chem Internat Edit* 12:534–546, 1973.
6. RG Greenler. *J Chem Phys* 44:310–315, 1966.

7. SB Barone, MA Zondlo, and MA Tolbert. *J Phys Chem A* 101:8643–8652, 1997.
8. PK Hudson, JE Shilling, MA Tolbert, and OB Toon (2002), *J Phys Chem A*, *106*:9874–9882.
9. J Marti and K Mauersberger (1993), *Geophys Res Lett*, *20*:363–366.
10. RB Chatfield (1994), *Geophys Res Lett*, *21*:2705–2708.
11. DA Hauglustaine, BA Ridley, S Solomon, PG Hess, and S Madronich (1996), *Geophys Res Lett*, *23*:2609–2612.
12. DJ Lary, AM Lee, R Toumi, MJ Newchurch, M Pirre, and JB Renard (1997), *J Geophys Res*, *102*:3671–3682.
13. A Fried, BE Henry, JG Calvert, and M Mozurkewich (1994), *J Geophys Res*, *99*:3517–3532.
14. TM Miller and VH Grassian (1998), *Geophys Res Lett*, *25*:3835–3838.
15. A Tabazadeh, MZ Jacobson, HB Singh, OB Toon, JS Lin, RB Chatfield, AN Thakur, RW Talbot, and JE Dibb (1998), *Geophys Res Lett*, *25*:4185–4188.
16. S Borrmann, S Solomon, JE Dye, B Juo, and JPD Abbatt (1996), *Geophys Res Lett*, *23*:2133–2136.
17. JPD Abbatt (1997), *Geophys Res Lett*, *24*:1479–1482.
18. OP Arora, DJ Cziczo, AM Morgan, JPD Abbatt, and RF Niedziela (1999), *Geophys Res Lett*, *26*:3621–3624.
19. A Tabazadeh, OB Toon, and EJ Jensen (1999), *Geophys Res Lett*, *26*:2211–2214.
20. MA Zondlo, SB Barone, and MA Tolbert (1997), *Geophys Res Lett*, *24*:1391–1394.
21. AJ Weinheimer, TL Campos, JG Walega, FE Grahek, BA Ridley, D Baumgardner, CH Twohy, B Gandrud, and EJ Jensen (1998), *Geophys Res Lett*, *25*:1725–1728.
22. SK Meilinger, A Tsias, V Dreiling, M Kuhn, C Feigl, H Ziereis, H Schlager, J Curtius, B Sierau, F Arnold, M Zoger, C Schiller, and T Peter (1999), *Geophys Res Lett*, *26*:2207–2210.
23. Y Kondo, OB Toon, H Irie, B Gamblin, M Koike, N Takegawa, MA Tolbert, PK Hudson, AA Viggiano, LM Avallone, AG Hallar, BE Anderson, GW Sachse, SA Vay, DE Hunton, JO Ballenthin, and TM Miller (2003), *Geophys Res Lett*, *30*(4), 1154, doi: 10.1029/2002GLO16539.
24. DR Hanson (1990), *Geophys Res Lett*, *17*:421–423.
25. MA Zondlo, SB Barone, and MA Tolbert. *J Phys Chem A* 102:5735–5748, 1998.
26. OB Toon, MA Tolbert, BG Koehler, AM Middlebrook, and J Jordan (1994), *J Geophys Res*, *99*:25631–25654.
27. RG Hynes, MA Fernandez, and RA Cox (2002), *J Geophys Res*, *107*(D24), 4797, doi: 10.1029/2001JD001557.
28. F Arnold, V Burger, B DrosteFanke, F Grimm, A Krieger, J Schneider, and T Stilp (1997), *Geophys Res Lett*, *24*:3017–3020.
29. HB Singh, M Kanakidou, PJ Crutzen, and DJ Jacob. *Nature* 378:50–54, 1995.
30. H Singh, Y Chen, A Tabazadeh, Y Fukui, I Bey, R Yantosca, D Jacob, F Arnold, K Wohlfrom, E Atlas, F Flocke, D Blake, N Blake, B Heikes, J Snow, R Talbot, G Gregory, G Sachse, S Vay, and Y Kondo (2000), *J Geophys Res*, *105*:3795–3805.
31. H Singh, Y Chen, A Staudt, D Jacob, D Blake, B Heikes, and J Snow. *Nature* 410:1078–1081, 2001.
32. R Holzinger, C Warneke, A Hansel, A Jordan, W Lindinger, DH Scharffe, G Schade, and PJ Crutzen (1999), *Geophys Res Lett*, *26*:1161–1164.
33. JF Muller and G Brasseur (1999), *J Geophys Res*, *104*:1705–1715.
34. L Jaegle, DJ Jacob, Y Wang, AJ Weinheimer, BA Ridley, TL Campos, GW Sachse, and DE Hagen (1998), *Geophys Res Lett*, *25*:1705–1708.
35. HB Singh, D Herlth, R Kolyer, L Salas, JD Bradshaw, ST Sandholm, DD Davis, J Crawford, Y Kondo, M Koike, R Talbot, GL Gregory, GW Sachse, E Browell, DR Blake, FS Rowland, R Newell, J Merrill, B Heikes, SC Liu, PJ Crutzen, and M Kanakidou (1996), *J Geophys Res*, *101*:1793–1808.
36. LT Iraci and MA Tolbert (1997), *J Geophys Res*, *102*:16099–16107.
37. JT Jayne, DR Worsnop, CE Kolb, E Swartz, and P Davidovits. *J Phys Chem* 100:8015–8022, 1996.
38. J Crawford, D Davis, J Olson, G Chen, S Liu, H Fueberg, J Hannan, Y Kondo, B Anderson, G Gregory, G Sachse, R Talbot, A Viggiano, B Heikes, J Snow, H Singh, and D Blake (2000), *J Geophys Res*, *105*:19795–19809.
39. L Jaegle, DJ Jacob, WH Brune, I Faloona, D Tan, BG Heikes, Y Kondo, GW Sachse, B Anderson, GL Gregory, HB Singh, R Pueschel, G Ferry, DR Blake, and RE Shetter (2000), *J Geophys Res*, *105*:3877–3892.

40. EJ Jensen, OB Toon, A Tabazadeh, GW Sachse, BE Anderson, KR Chan, CW Twohy, B Gandrud, SM Aulenbach, A Heymsfield, J Hallett, and B Gary (1998), *Geophys Res Lett*, *25*:1363–1366.
41. D Baumgardner and BE Gandrud (1998), *Geophys Res Lett* 25:1129–1132.
42. C Schiller, A Afchine, H Eicke, C Feigl, H Fischer, A Giez, P Konopka, H Schlager, F Tuitjer, FG Wienhold, and M Zoger (1999), *Geophys Res Lett*, *26*:2219–2222.
43. C Feigl, H Schlager, H Ziereis, J Curtius, F Arnold, and C Schiller (1999), *Geophys Res Lett*, *26*:2215–2218.

7 Surface Chemistry at Size-Selected, Aerosolized Nanoparticles

Jeffrey T. Roberts
Department of Chemistry, University of Minnesota–Twin Cities

CONTENTS

7.1 INTRODUCTION

The atmosphere abounds in all types of aerosol particles, including combustion particles (e.g., soot), mineral dust, sea salt, organic particles, and aqueous inorganic acids [1, 2]. It would be difficult to overstate the significance of atmospheric aerosol particles. They are the source of heterogeneous processing in the atmosphere. Aerosols influence the planet's radiative balance, and as such they play an important role in the global climate. Air visibility is determined by aerosol loading. Finally, the inhalation of aerosol particles almost certainly impacts human health.

Chemistry is central to many, if not all, aerosol-related phenomena, from particle nucleation and growth to heterogeneous processing to biological effects. In some cases, especially for systems involving liquid particles, the relevant chemistry may be largely bulk-phase, but in others the most chemically important properties are probably associated with the particle–gas interface. In thinking about the particle–gas interface, it is important to consider that most of the surface area of a typical aerosol is carried by the nanosize fraction, i.e., by those particles having diameters <100 nm [3, 4]. Given the manifold importance of aerosols, as well as the possibility of size-dependent chemical properties in the nanometer-size regime, it is surprising how little understanding is there of the surface chemistry of nanometer-sized aerosol particles.

For example, almost nothing is known about the chemical states of surface atoms: the bonding to carbon in an organic particle, the oxidation states of the metals in an inorganic particle, etc. Measurements have been reported for particles that are impacted onto grids or other immobilizing substrates, but impaction may cause chemical state changes [5]. Clearly, a good set of tools for studying nanoaerosol surface chemistry awaits development.

Over the past several years, I and my coworkers have been working to develop a set of tools for studying surface chemistry at aerosolized nanoparticles [6, 7]. The primary focus has been on probing authentic aerosol particles, rather than deposited particles or laboratory surrogates for particles. Measurements have emphasized particles that are <50 nm in diameter, for which size-dependent effects are anticipated to be most pronounced, and they have been carried out in a size-selected fashion [6, 7]. The work described in the chapter concerns soot and soot-like nanoparticles, including particles generated in a diesel engine [7], but the methods we are beginning to develop could be applied to virtually any nanometer-sized aerosol particle. Our hope is that our work will advance the overall understanding of the surface chemical properties of atmospheric particles. More generally, we hope that it will lead to the development of new methods for studying surface reactions at any nanometer-sized aerosol particle.

This rest of this chapter is divided into three parts. Section 7.2 provides a description the aerosol flow reactor that we use to study aerosol surface chemistry, including a discussion of the differential mobility analyzer (DMA), which is an essential tool for creating streams of size-selected particles. In Section 7.3, we review our work on the kinetics and mechanisms of molecular oxygen reacting with soot, and we put that work into the context of the broader literature on combustion particle oxidation. Finally, in Section 7.4, we consider possible future applications of the aerosol flow reactor.

7.2 EXPERIMENTAL METHODS

7.2.1 Overview

Figure 7.1 shows, in block diagram form, essential elements of our "nanoaerosol toolkit." Particles are extracted from the aerosol sample (the reactor is compatible with virtually any atmospheric pressure source) and passed through a device called a DMA. The use and operation of DMAs is described in Section 7.2.2; briefly, a DMA is where size classification occurs [8,9]. After leaving the DMA, particles move through an atmospheric pressure flow reactor that is designed for maximal flexibility in terms of temperature and gas-phase composition. Finally, the particle stream is sent to one of three analysis modules: (i) a second DMA, in which chemistry-induced particle size changes are measured; (ii) a UV spectrometer, in which photoionization probabilities are measured as a function of wavelength, and (iii) a particle impactor, in which particles can be deposited for postreaction transmission electron microscopy (TEM) analysis.

A schematic diagram of one configuration of the aerosol reactor is shown in Figure 7.2. The aerosol source in this case is an ethene diffusion flame; the analysis module is a second DMA (DMA-2). Particles are extracted from the tip of the flame through a 1 mm diameter orifice. Once inside the aerosol system, the particles are diluted and cooled by a 20 L min^{-1} flow of carrier gas and swept into a ^{210}Po bipolar diffusion neutralizer. The neutralizer is necessary because, as explained below, charged particles are classified on the basis of electric mobility. When particles pass through the diffusion neutralizer, they undergo low-energy collisions with ions produced by the radioactive Kr source. The result is an ensemble of aerosol particles with an equilibrium charge distribution [10]. The equilibrium distribution depends on many variables, including particle size. For the systems described in this work, experimental conditions were such that ~90% of all particles acquire one elementary unit of charge in the neutralizer, either positive or negative. The rest are either neutral or carry multiple charges.

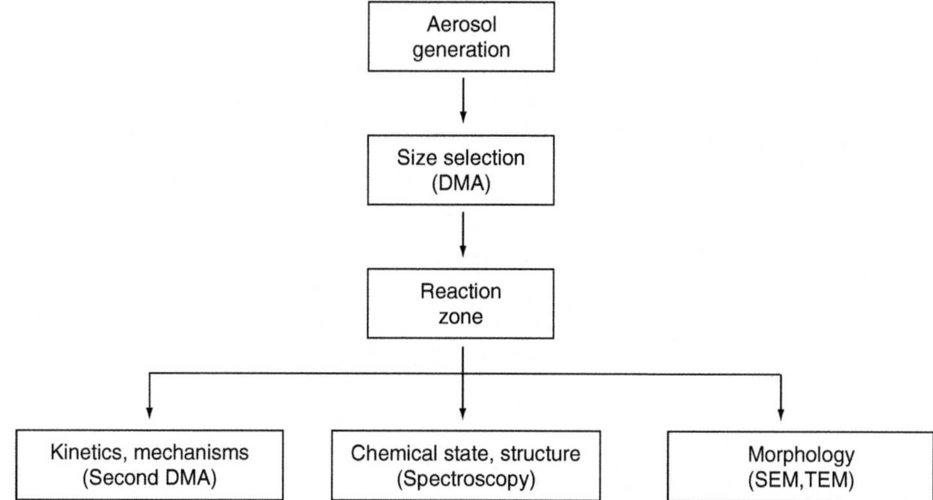

FIGURE 7.1 A block diagram of the "nanoaerosol toolkit."

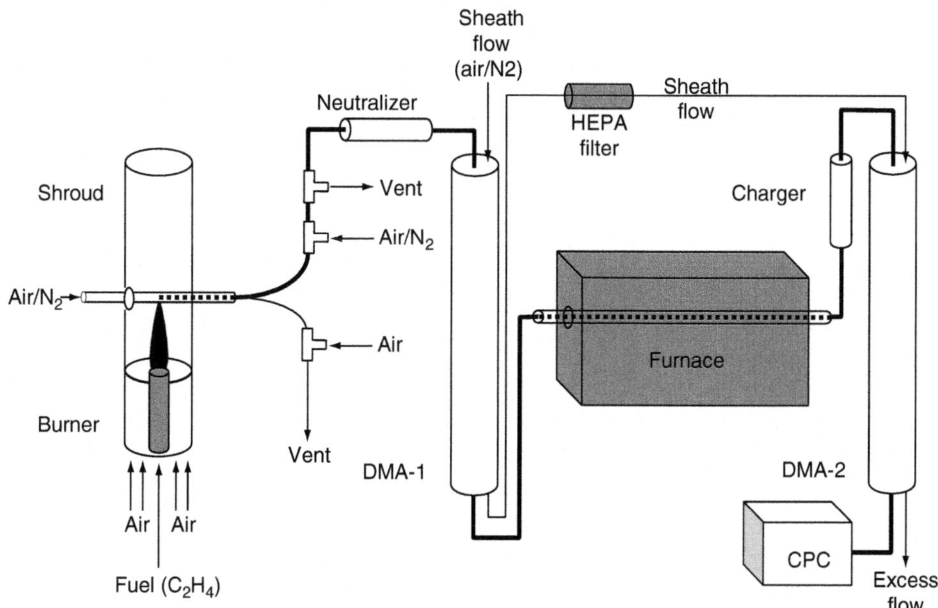

FIGURE 7.2 A schematic diagram of the nanoaerosol reactor operating in one of many possible configurations. (From KJ Higgins, H Jung, DB Kittelson, JT Roberts, and MR Zachariah. *J Phys Chem A*, 106: 96–103, 2002. With permission.)

7.2.2 Creating a Stream of Size-Selected Particles

The creation of a stream of size-selected aerosol particles is accomplished using a DMA, a cross-sectional, schematic diagram of which is shown in Figure 7.3. Key elements are a center rod upon which a variable voltage is applied and an outer cylinder that is held at ground potential. Charged particles are swept into the DMA via a carrier gas at near-atmospheric pressure. Particles have a tendency to follow the stream lines of the carrier gas, but because

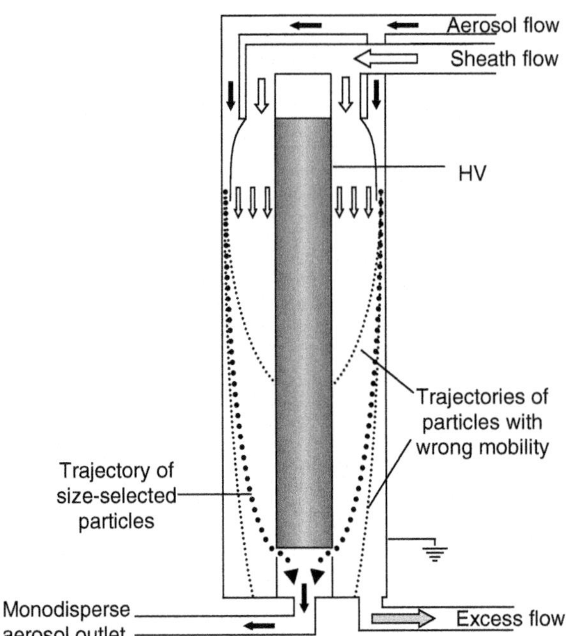

FIGURE 7.3 A schematic diagram of a differential mobility analyzer (DMA). Particles with the correct electric mobility pass through the DMA. Those with mobilities outside the desired range either hit and stick to the center rod electrode or go past the exit orifice and are carried away in the excess flow. (From KJ Higgins, H Jung, DB Kittelson, JT Roberts, and MR Zachariah. *J Phys Chem A*, 106: 96–103, 2002. With permission.)

they are electrically charged they also migrate against the electric field setup between the inner and outer cylinders [8, 9]. In the free-molecular and transition regimes, where the particle diameter is smaller than or comparable to the mean free path of the gas molecules (~65 nm at room temperature and atmospheric pressure), electric mobility is inversely proportional to the mass transfer rate to a particle [11]. This in turn is proportional to the gas-accessible surface area. As a result, the trajectory of a particle through a DMA is determined by its electric mobility, which for a nanometer-sized particle is directly related to the surface area-to-charge ratio. Particles of the desired electric mobility pass out of the DMA through an orifice at the end of the center rod; others are deposited inside the DMA or exit the DMA through an excess flow valve.

The most convenient way of detecting nanoparticles that exit a DMA is with a condensation particle counter (CPC) [12]. CPCs work by allowing supersaturated *n*-butanol vapor to condense on the particles in an aerosol stream them until they grow to a size (ca. 1 μm) that scatters visible radiation efficiently. Particles are detected by a single-particle counting optical counter, which operates on the principle of light scattering. Under optimum conditions, CPCs are zero-background detectors capable of detecting single particles as small as 2 nm in diameter.

A DMA can be operated in two modes. The center rod voltage can be stepped up, in which case the DMA is a tool for determining the particle size distribution of an aerosol sample [9]. Alternatively, the center rod voltage can be fixed. The DMA then passes a stream of particles that all have the same electric mobility [8]. Figure 7.4 illustrates both of these capabilities. In one measurement, soot particles were sampled from an ethene flame, passed through a bipolar particle neutralizer, and sent through a DMA operated in scanning mode.

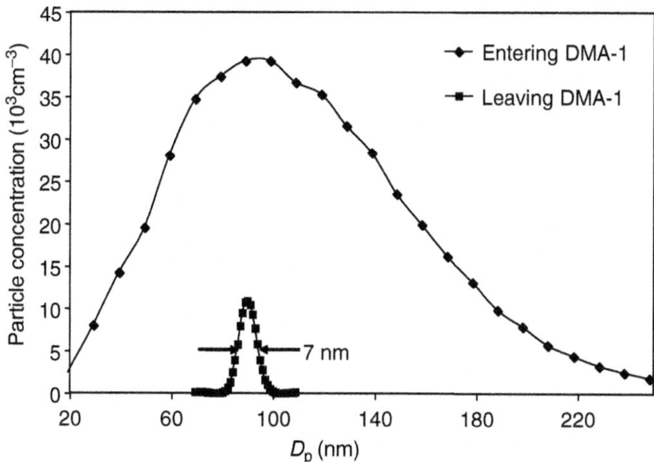

FIGURE 7.4 Two scans illustrating the twin capabilities of the SMA: as a tool for measuring the size distribution function of an aerosol and as a method for preparing a monodisperse stream of aerosol particles. The sample in this case was a flame soot aerosol, sampled from the tip of an ethane diffusion flame. (From KJ Higgins, H Jung, DB Kittelson, JT Roberts, and MR Zachariah. *J Phys Chem A*, 106: 96–103, 2002. With permission.)

The broad distribution peaks at a mobility diameter, D_p, equal to roughly 100 nm. In a second measurement, the particles were passed through the same DMA, but with the center rod voltage set to transmit particles of $D_p = 90$ nm. The aerosol stream was passed through a second DMA, and the size distribution was measured. The distribution, which peaks at 90 nm as expected, is very narrow, with a full-width-at-half-maximum of 7 nm.

DMAs and CPCs are commercially available. Current DMA and CPC models allow for the classification and detection of particles down to 3 nm mobility diameter at concentrations as low as 1 cm^{-3} and as high as 10^6 cm^{-3}.

7.2.3 Three Methods for Studying Aerosol Surface Chemistry

7.2.3.1 Tandem-DMA: Surface Kinetics and Mechanisms

Reactions that result in changes in particle mobility diameter can be studied using tandem-DMA (T-DMA). In T-DMA, size-selected aerosol streams are passed through a reaction zone and then directed into a second DMA, where their new size distributions are measured. The technique has been extensively used in the past to study hygroscopic and deliquescent properties of aerosol particles [13, 14], and to study aerosol condensation and evaporation [15, 16]. As will be shown below, T-DMA can deliver powerful *chemical* insights, and can be harnessed as a tool for investigating kinetics and mechanisms of nanoaerosol reactions. The only requirement is that the change in D_p (ΔD_p) be measurable by a DMA. The DMAs used in the work described in this chapter are capable of measuring diameter changes as small as 1% of the initial particle diameter. For small enough particles ($D_p \leq 10$ nm), this translates into potential monolayer sensitivity.

Figure 7.5 shows two examples of T-DMA scans, in this case for the in-air oxidation of soot particles extracted from an ethene flame.[6] In both cases, DMA-1 was set to transmit particles of $D_p = 130$ nm. The size-selected particle streams were sent though a variable temperature furnace of length 30 cm with dry air as the carrier gas. The gas flow rates were such that the residence time of a particle in the hot zone was ~2 s. For one of the T-DMA

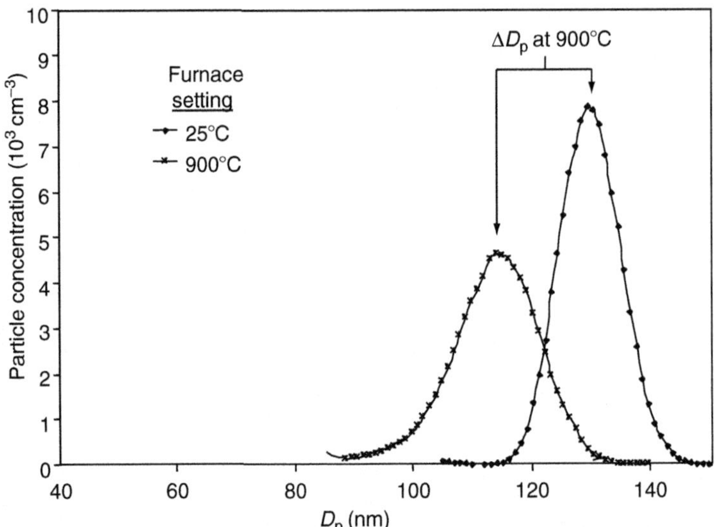

FIGURE 7.5 Tandem differential mobility scans of flame soot in dry air. Particles were sampled from the tip of an ethane diffusion flame and mobility classified ($D_p = 130$ nm) with DMA-1. The particles then passed through a furnace, which was set either to room temperature or to 900°C. The change in mobility diameter, ΔD_p, may be used to determine the rate of soot oxidation in air.

scans, the furnace was off, and the gas temperature through the entire system was constant at 25°C. The peak particle mobility diameter measured by DMA-2 was 130 nm, as expected. In the second case, the furnace was set to a peak temperature of 900°C. Because the carrier gas consisted of ~20% O_2, the soot particles were partially oxidized, presumably to CO and H_2O. The particles shrank to a peak mobility diameter of 114 nm, for $\Delta D_p = 16$ nm. In Section 7.3, we will show how ΔD_p values can be analyzed to obtain quantitative kinetic information and, in some cases, mechanistic information.

7.2.3.2 Photoelectron Spectroscopy

One of the most critical needs in aerosol science is the development of new spectroscopic probes, particularly of the aerosol–gas interface. There are at present no molecularly incisive methods for spectroscopically probing the surfaces of nanoaerosol particles. Optical methods have been developed to investigate surface and bulk structure in micron-sized particles [17–20]. The problem is that these methods, which are generally based on infrared absorption or Raman scattering, are not easily extended to nanoparticles, mostly because of signal-to-noise limitations.

We are developing a new photoemission-based method to study the surfaces of nanoaerosol particles at atmospheric pressure. The method draws heavily on ideas first published by Siegmann and coworkers that has also been used by others [11d, 21–27]. The method exploits the fact that the inelastic mean free path of a low-energy electron at 1 atm is ~1 μm. Photoelectrons that originate from a micro- or macroscopic object scatter back toward the object, eventually reneutralizing it, as shown schematically in Figure 7.6. However, for a nanoparticle suspended in a gas, the reneutralization probability is essentially zero. As a result, photoionization probabilities of nanoaerosol particles can in principle be measured at atmospheric pressure. Photoemission is surface-sensitive because electrons that originate from the interior of a particle have a lower probability of escaping into the gas phase than

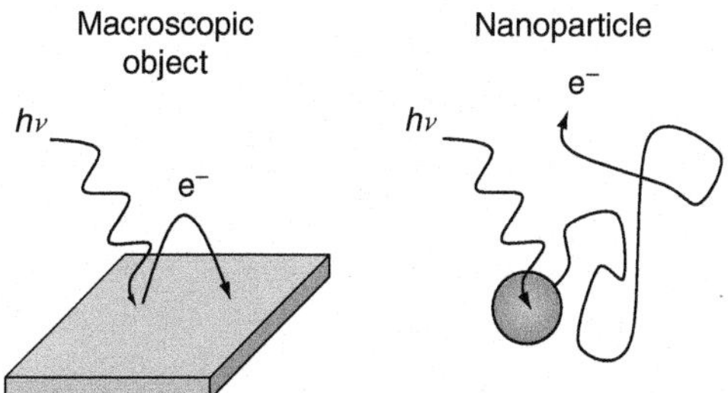

FIGURE 7.6 Photoionization from a micro- or macroscopic object versus photoionization from an aerosolized nanoparticle.

do those that originate from the surface. The electron escape depth depends on its energy and the medium that it penetrates, but for 10 eV electrons in a solid, a typical escape depth is ~5 Å.

The aerosol spectrometer works as follows. A size-selected particle beam is created with a DMA. The resulting particles are then sent through first through an aerosol neutralizer and then through an electrostatic precipitator. The result is a neutral, monodisperse particle beam, which is swept by its carrier gas through a reaction zone and then through an ultraviolet-transparent irradiation cell. Particles are irradiated, while in the tube, by a deuterium lamp/grating monochromator system, with the photon energy scanned between 4.5 eV (near the soot work function) and 8 eV. The particle stream is passed through a second DMA set to transmit only particles that were charged due to photoionization. Particles are to be detected by a CPC. One of the most attractive aspects of this experiment is that it requires no electrometer or elaborate electron detection scheme. Rather, photoionization will be monitored by detecting the number of charged particles passed by a DMA. This makes detection simple: CPCs are extraordinarily sensitive instruments, capable of measuring effective particle currents as low as 10^{-18} A. Moreover, photoionization of the carrier gas, should it occur, will produce no false counts at the detector.

Our initial work is focusing on measurements of the valence-level ionization threshold, Φ_{particle}. This quantity is analogous to the bulk work function, Φ_{bulk}, which is rigorously defined only for an infinitely sized object. Measurements will be made as functions of particle size and processing history. Chemical processing is expected to be important because it dictates, in part, the composition and functionality of a particle surface, which in turn determine reactivity. For instance, the oxidation of a soot particle could result in the conversion of a hydrocarbon-like surface into one that is largely terminated by oxygen-containing functional groups. Such changes are believed to result in the transformation of a hydrophobic into a hydrophilic surface, and they probably influence the ability of soot nanoparticles to act as condensation nuclei for cloud particles. These changes should also lead to shifts in Φ_{particle} [28].

Future work will concern changes in Φ_{particle} associated with the adsorption of hydrocarbons to soot, and on the potential application of aerosol photoelectron spectroscopy (APES) as a semiquantitative tool for measuring adsorbate coverages. This, in turn, would allow for the measurement of adsorption and desorption kinetics at atmospheric pressure, something that cannot currently be accomplished. Spectral structure above the threshold will be investigated as well. This has the potential of providing information about valence electronic structure, for instance the average hybridization state of carbon.

7.2.3.3 Transmission Electron Microscopy

The particles in an aerosol are generally characterized by broad size and shape distributions [1–3]. Figure 7.7 shows several TEM images of soot particles that were collected by passing a carbon-film TEM grid through the tip of an ethene flame. The particles display many morphologies,

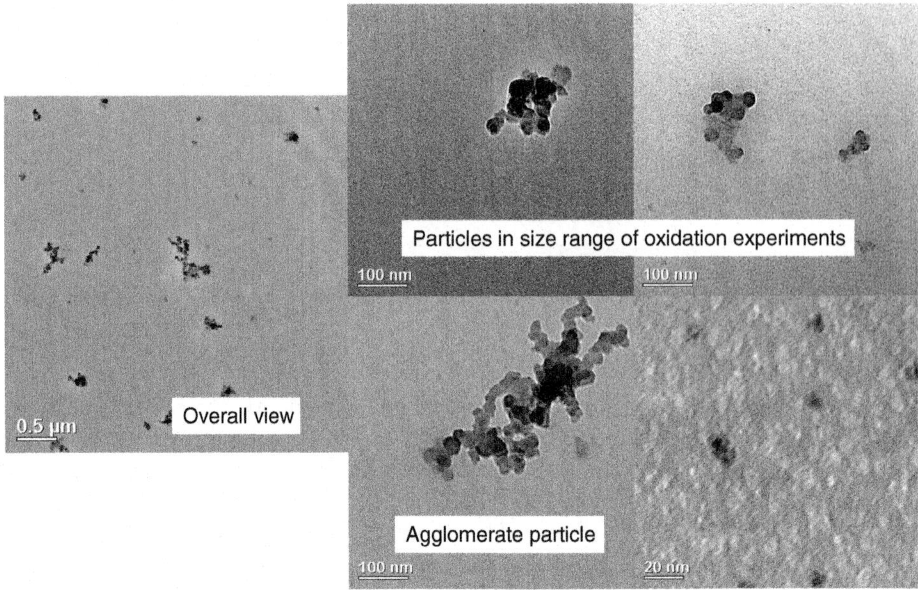

FIGURE 7.7 TEM images of ethene soot particles on a carbon grid. The images are at different magnifications so as to give an overview of different particle morphologies, and to show typical morphologies of particles in the size range investigated in this work (D_p = 40–130 nm). In this size range, most soot particles are fairly dense aggregates of smaller (~20 nm diameter) primary soot particles. (From KJ Higgins, H Jung, DB Kittelson, JT Roberts, and MR Zachariah. *J Phys Chem A*, 106: 96–103, 2002. With permission.)

from spherical "primary" particles having diameters up to ~50 nm, to complex, micron-sized aggregates of fused primary particles. An obvious question is, how does the morphology of a soot nanoparticle change when it undergoes reaction? There is some discussion in the literature on this point. For instance, it has been suggested that oxidation by O_2 or OH result in the formation of weakly cohesive soot aggregates [29, 30].

We have used TEM to study how the structures of soot — both primary particles and aggregates — change as they oxidize. Experiments were conducted using the apparatus shown in Figure 7.2 with dry air as the carrier gas, except that the second DMA was replaced with a collection plate on which particles were deposited for TEM analysis. Figure 7.8 shows several representative images of particles of original mobility diameter 130 nm [6]. The particle samples were collected immediately after mobility classification (Figure 7.8a), and after passing through the furnace set at 800°C and 900°C (Figures 7.8b and c, respectively). As expected, oxidation in the furnace causes the particles to shrink, with the amount of size reduction at 900°C exceeding that at 800°C. What is more interesting is that the morphology of a particle appears to be largely preserved during initial oxidation, an observation which is confirmed by analyzing the images of statistically significant numbers of particles. In particular, the average number of primary particles in an aggregate is unchanged, as is the mean number of connection points between primary particles, at least for in-air oxidation up to 900°C of particles 50–130 nm in mobility diameter. It is as though oxidation causes the particles shrink uniformly, rather than through preferential gasification of the most exposed primary particles in an aggregate. The aggregates appear to undergo some densification in the furnace, as revealed by an analysis of the mean cross-sectional area, but at this point it is not known whether densification occurs via a purely thermal process, or whether it is assisted by oxidation. More experiments are underway to determine which explanation is correct.

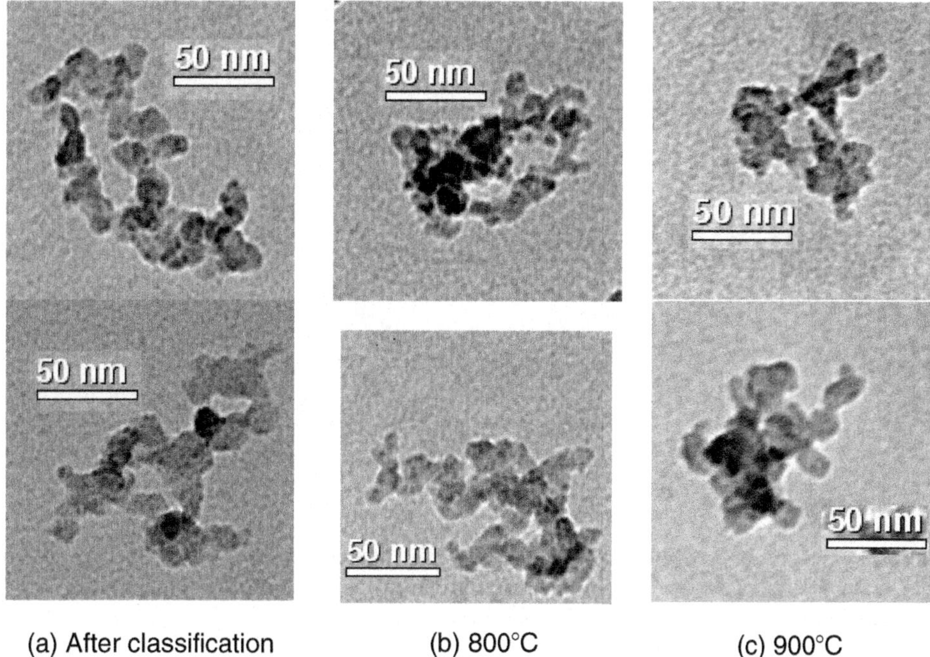

FIGURE 7.8 TEM images of ethene soot particles sampled in the aerosol flow reactor. The particles were of original mobility diameter 130 nm. The images refer to particles collected (a) immediately after mobility classification by DMA-1, (b) after passing through the furnace at a setting of 800°C, and (c) after passing through the furnace at a setting of 900°C.

7.3 KINETICS AND MECHANISMS OF SOOT OXIDATION

7.3.1 Ethene Soot

There is a rich literature on the high-temperature oxidation of soot and soot surrogates [31], beginning with early work on carbon filaments [32, 33], carbon or graphite rods [34], coal char [35], and flame soot [36–39]. More recent work has used various synthetic carbon blacks as models for soot [40] and employed a variety of approaches, including shock tubes [41, 42], thermogravimetric analysis, and immobilized beds, to study their oxidation. Studies using authentic soot have generally been confined to techniques that rely on collecting and immobilizing the soot, or on light scattering or absorption within a sooting flame.

Using T-DMA, we have studied the oxidation of freshly generated soot particles *in the aerosol state* [6]. In these experiments, particles were extracted from the tip of an ethene diffusion flame [43] and sampled through a small orifice [23, 36], using the aerosol flow reactor shown in Figure 7.2. The carrier gas was either dry, filtered air, pure N_2, or an O_2/N_2 mixture of continuously variable composition. Oxidation took place within a quartz tube (1.0 cm i.d., 120 cm long) that was housed inside a tube furnace. Representative T-DMA scans for the oxidation of 130 nm initial mobility diameter particles by dry air at furnace settings between 25°C and 975°C are shown in Figure 7.9. The scans change with increasing furnace temperature in three obvious ways: (1) the peak heights decrease, (2) the positions of the peak maxima decrease, and (3) the peaks broaden and in some cases adopt what appear to be bimodal shapes. As will be explained below, only the second of these changes, i.e., the decrease in most probable mobility diameter, is largely associated with particle oxidation. Peak broadening is mostly a consequence of the fact that gas flow in the reactor is laminar. Particles have a

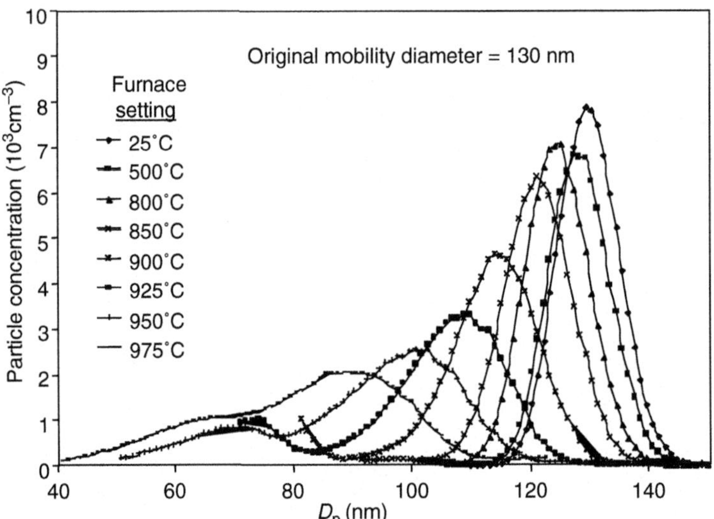

FIGURE 7.9 T-DMA scans of ethene soot in air for furnace settings of 25–975°C. The original particle mobility diameter was 130 nm. (From KJ Higgins, H Jung, DB Kittelson, JT Roberts, and MR Zachariah. *J Phys Chem A*, 106: 96–103, 2002. With permission.)

tendency to follow the stream lines of the carrier gas, and so the particles in an aerosol sample experience a distribution of residence times in the furnace region, leading to differing amounts of oxidative loss. Also, the apparent bimodal distributions in some of the T-DMA scans is an artifact associated with the blending of peaks derived from doubly charged particles. The decreasing peak heights are mostly a result of peak broadening, but also a consequence of themophoretic loss to the reactor walls as the particles experience a steep negative temperature gradient on leaving the furnace.

When N_2 rather than dry air is used as the carrier gas, results are very different. Figure 7.10 shows three sets of T-DMA scans, with each set corresponding to a different initial D_p, either 40, 93, or 130 nm. For each set, three scans are shown: one with N_2 as the carrier gas and the furnace off, one with N_2 as the carrier gas and the furnace set to 1100°C, and one with dry air as the carrier gas and the furnace set to 1100°C. Some size reduction occurs at 1100°C in pure nitrogen gas, when no oxidation is possible, but the amount of size reduction is quite modest, ~5 nm in diameter. Nonoxidative shrinkage is attributed either to the evaporation of semivolatile material or to thermally induced restructuring of aggregates to more dense structures. Switching the carrier gas from nitrogen to dry air results in a dramatic change at this furnace setting: the particles disappear. We conclude that this behavior is associated with oxidation, in which the solid particles are converted to gas-phase products. At 1100°C, the oxidation rate is apparently so rapid that the particles are either quantitatively oxidized or shrink to sizes that are too small to detect with a CPC.

By comparing T-DMA scans obtained in dry to those obtained in nitrogen, changes in mobility diameter that result from oxidation can be separated from those that result from nonoxidative processing. If it is assumed the nonoxidative and oxidative size reduction components are additive, then $\Delta D_p^{O_2}$, the change in mobility diameter caused by oxidation, can be obtained from the equation:

$$\Delta D_p^{O_2} = \Delta D_p - \Delta D_p^{N_2} \tag{7.1}$$

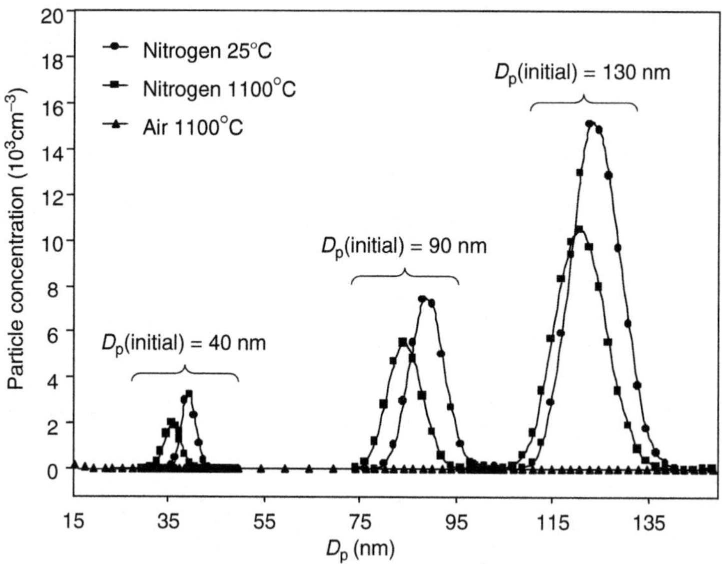

FIGURE 7.10 The effect of carrier gas (dry air versus nitrogen) on T-DMA scans for particles of initial mobility diameter 40, 90, and 130 nm. Two furnace settings were used: 25°C (for which only the nitrogen scans are shown) and 1100°C. The disappearance of particles in the 1100°C scans when air is the carrier gas is the result of soot oxidation by molecular oxygen. (From KJ Higgins, H Jung, DB Kittelson, JT Roberts, and MR Zachariah. *J Phys Chem A*, 106: 96–103, 2002. With permission.)

where ΔD_p is the change in mobility diameter in air measured at some furnace setting and $\Delta D_p^{N_2}$ is the change in mobility diameter measured in nitrogen at the same setting. For a surface-reaction-limited process that is first-order in oxygen pressure, the rate of particle oxidation, expressed as dm/dt, the change in mass with respect to time, may be written:

$$-\frac{dm}{dt} = -\sigma Z_{O_2} A T^{1/2} e^{-E_a/RT} \tag{7.2}$$

where σ is the particle surface area, Z_{O_2} is the oxygen collision rate with the surface, A is the preexponential factor, and E_a the activation energy. If it is assumed that the soot particles are well represented as spheres of physical diameter D_p, then Equation (7.2) reduces to:

$$-\frac{dD_p^{O_2}}{dt} = -\frac{2}{\rho} Z_{O_2} A T^{1/2} e^{-E_a/RT} \tag{7.3}$$

where ρ is the density of soot. Equation (7.3) makes it possible to relate an experimental observable, $\Delta D_p^{O_2}$, to the Arrhenius parameters A_{nm} and E_a.

We studied the in-air oxidation of soot particles of original mobility diameter 40, 90, and 130 nm over a broad range of temperatures, from 25°C to 1100°C. We also carefully measured the nonuniform temperature fields in the furnace, and we took into account the fact that gas expansion effects cause the gas flow velocities to vary with furnace position. Experimental measurements were then fit to Equation (7.3). Figure 7.11 plots both measured values and model predictions of $\Delta D_p^{O_2}$ versus furnace setting. Agreement between the data and the model is remarkable, especially considering the broad temperature range (nearly 300°C) over which data were obtained. The size reduction measurements data could all be fit to a single value of E_a, 164 kJ mol^{-1}. Adequate fits required that the preexponential be allowed to vary with

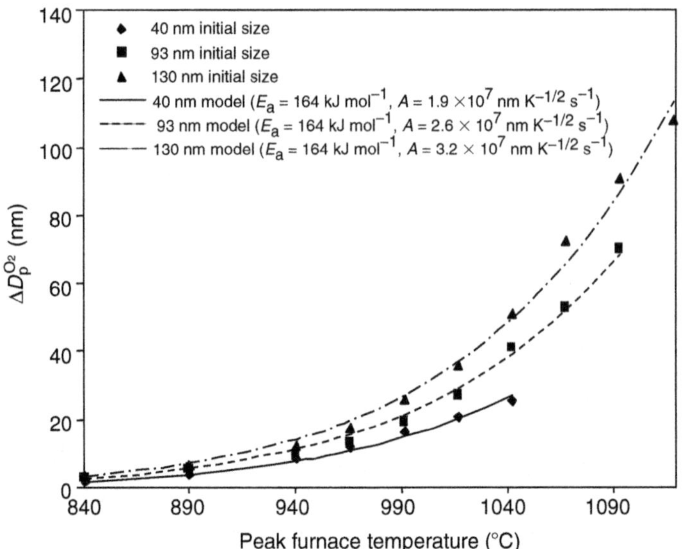

FIGURE 7.11 A comparison of measured $\Delta D_p^{O_2}$ values versus predictions of the kinetic model described in the text. (From KJ Higgins, H Jung, DB Kittelson, JT Roberts, and MR Zachariah. *J Phys Chem A*, 106: 96–103, 2002. With permission.)

initial mobility diameter, but the difference between the lowest and highest fitted values was less than a factor of two, with values of A varying from 1.9×10^7 nm $K^{1/2}$ s^{-1} (40 nm) to 3.2×10^7 nm $K^{1/2}$ s^{-1} (130 nm).

The work described above represents what we believe is the first work on the in-air oxidation of a freshly generated, aerosolized soot nanoparticles. Nevertheless, there is an abundant literature on the oxidation of immobilized soot particles or soot surrogates. The activation energy we obtained using T-DMA is in good agreement with that measured by others, as are the total oxidation rates that we measure [31, 32, 34, 41, 44]. Because of the different ways in which reaction rate expressions are defined, it is difficult to compare the measured preexponential factors of different studies. One question that arises in the present study is the origin of the size-dependent preexponential factor that we measure. One possibility is that particles of different initial mobility diameter have different effective densities. If this were the case, then the assumption of a constant ρ value in Equation (7.3) would lead to an *apparent* size-dependence in the preexponential factor. Interestingly, TEM measurements, including those shown above, show that the soot particles are transitioning in the 40–130 nm range from nearly spherical primary particles to more complex (and less dense) aggregates [30]. This is consistent with the fact that our measured A factors increase with mobility diameter in this range.

7.3.2 Diesel Soot

Diesel particles differ from soot particles generated using pure hydrocarbon fuels. For instance, elements such as sulfur, which is present in diesel fuel, and metals from lubricating oil and cylinder abrasion can be incorporated into diesel particles and alter the chemical properties of the particles [5, 45, 46]. For this reason, we decided to study diesel soot oxidation using T-DMA. Details of the experiments are presented elsewhere [7]. Briefly, we employed an apparatus that was essentially identical to that shown in Figure 7.2, except that diffusion burner was replaced with a particle sampler that was 25 cm downstream of the turbocharger exit of a four-cylinder, four-cycle diesel engine. The output of the engine was coupled to a dynamometer for load control.

T-DMA measurements on diesel soot show that there are at least two different particle classes in the diesel soot samples. The particles of one class oxidize in a way that resembles flame soot. Figure 7.12 shows a series of scans corresponding to in-air oxidation of particles of initial mobility diameter 40 nm generated under 50% load conditions. The particles shrink as the furnace temperature increases, with ΔD_p equal to 1 nm at a furnace setting of 500°C (the lowest setting studied other than room temperature) and ~15 nm at 1000°C. Measurements made using nitrogen as the carrier gas establish that virtually all of the diameter decrease at 500°C is associated with nonoxidative processing. At higher furnace settings oxidative processing is the dominant contributor to ΔD_p. For instance, at a furnace setting of 1000°C $\Delta D_p^{N_2}$ is ~3 nm, larger than that at 500°C, but only 15% of the diameter change measured when air is used as the carrier gas. Results for other initial mobility diameter particle sizes sampled under 50% and 75% engine load conditions were similar.

Although most of the particles in a diesel aerosol sample appear to oxidize normally in the furnace, some disappear. This can be seen by comparing how the T-DMA peak heights decrease with increasing furnace setting in the diesel scans (Figure 7.12) to the way they decrease in the ethene soot scans (Figure 7.9). No comparable decrease in peak heights is observed when pure nitrogen is the carrier gas, and so we conclude that particle disappearance is oxygen-assisted. Smaller particles are more likely to disappear in this fashion, as shown by Figure 7.13, which plots the fractional penetration of particles leaving DMA-2 versus furnace setting for different initial mobility diameters and engine loads. Fractional penetration is defined as the ratio of the integrated area of a peak at a specified furnace setting to that when the furnace is set to room temperature. Above room temperature, the fractional penetration must be less than unity, because some particles will be lost to diffusion or thermophoresis. However, losses should be identical, or nearly so, at constant furnace setting and initial mobility diameter. From Figure 7.13, it can be seen that this is not the case. For example, for the 50% and 75% load particles at a furnace setting of 500°C, the fractional

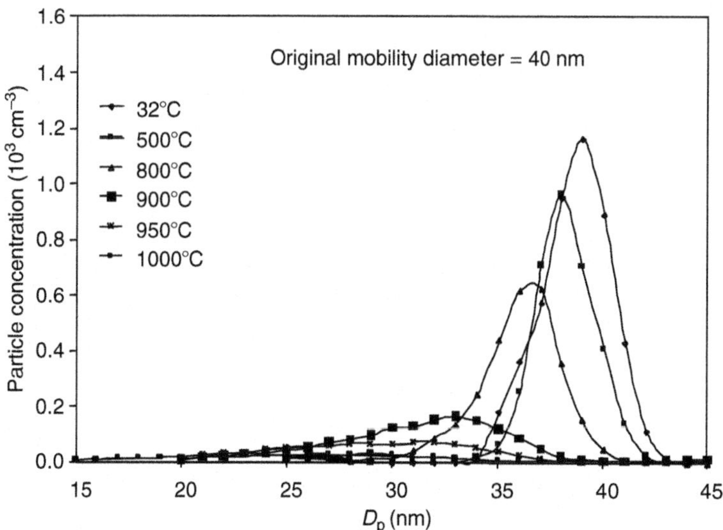

FIGURE 7.12 T-DMA scans of diesel soot in air for furnace settings of 32°C–1000°C. The original particle mobility diameter was 40 nm. (From KJ Higgins, H Jung, DB Kittelson, JT Roberts, and MR Zachariah. *Environ Sci Technol*, 37: 1949–1954, 2003. With permission.)

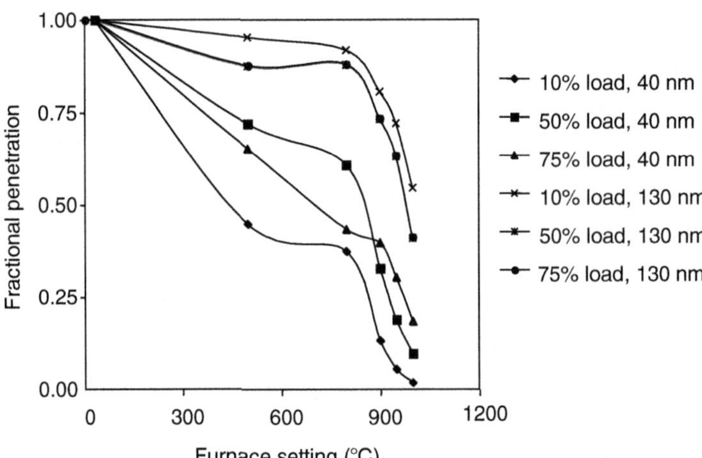

FIGURE 7.13. Particle losses (as measured by fractional penetration, which is defined in the text) for several initial mobility diameter and engine load combinations. The 65% loss between furnace settings of 32°C and 500°C is attributed to the complete evaporation of particles composed of volatile hydrocarbons and sulfates. The sharp loss observed above 800°C for all diameter–engine load combinations is not well understood. (From KJ Higgins, H Jung, DB Kittelson, JT Roberts, and MR Zachariah. *Environ Sci Technol* 37: 1949–1954, 2003. With permission.)

penetration is ~0.7 at an initial mobility diameter of 40 nm and ~0.9 at initial mobility diameters of 90 and 130 nm. Above 800°C, the total particle counts and therefore the fractional penetration values drop sharply for all sizes. What these results indicate is that a substantial portion of the 40 nm particles (i.e., those that disappear in the furnace) are nucleation mode particles composed of volatile or readily oxidized material.

The situation is even more complex for particles generated under 10% engine load conditions. Partly, this may be seen by examining ΔD_p values for particles that do not disappear. At a furnace setting of 500°C, ΔD_p is ~4 nm for 10% load particles, irrespective of the carrier gas, compared to ~1 nm for the higher load particles. At 1000°C, ΔD_p for the low load particles is ~11 nm with nitrogen as the carrier gas, compared to 3 nm for the 50% and 75% load particles. The much larger nonoxidative size decrease at 500°C is attributed to the large amount of volatile material that, in addition to forming the large peak of nucleation mode particles described above, condenses onto the particles in the dilution and cooling regions of the sampling system.

Data from all furnace settings and initial particle sizes for each engine load were fit together to determine a set of Arrhenius oxidation parameters for each of the three engine load data sets. We assumed the same rate expression as was used for ethene soot (Equation (7.2)). Table 7.1 lists the parameters of the best fits to the data. Under 50% and 75% engine load conditions, larger particles oxidize slightly more rapidly than smaller ones. The reason for this is unrelated to the activation energies, which are essentially size independent at 109 and 108 kJ mol^{-1} for 50% and 75% load particles, respectively. Rather, the preexponential factor is size-dependent, as it is for flame soot. Rate parameters for the 50% and 75% load particles are so close as to be equal within experimental errors. The parameters for the 10% load particles are somewhat different, but it is unclear whether the difference is meaningful, because the data set for 10% load particles was considerably smaller and the resulting analysis correspondingly more uncertain.

It is informative to compare the oxidation rates and corresponding Arrhenius parameters for flame and diesel soot. At equivalent temperatures, diesel particles oxidize more slowly, by

TABLE 7.1
Fitted Kinetic Parameters for Diesel Soot Oxidation (Parameters are Defined in the Text)

Engine Load	10%	50%	75%
E_a (kJ mol^{-1})	114	109	108
$A_{40\,nm}$ ($\times 10^4$ nm K$^{-1/2}$ s^{-1})	8.8	3.1	2.6
$A_{90\,nm}$ ($\times 10^4$ nm K$^{-1/2}$ s^{-1})	7.7	4.5	3.8
$A_{130\,nm}$ ($\times 10^4$ nm K$^{-1/2}$ s^{-1})	11.2	5.4	4.8

up to a factor of four depending on the furnace setting. The may seem surprising given that the activation energy for diesel soot oxidation (~110 kJ mol^{-1}, depending on engine load) is significantly lower than that for flame soot (164 kJ mol^{-1}). However, this difference is more than offset by the preexponential factors, which are three orders of magnitude lower than those for flame soot. For a surface reaction, the value of the preexponential factor is determined in part by the detailed reaction transition state and in part by the coverage of surface reaction sites. It is difficult to rationalize the huge differences in A factors for flame and diesel soot oxidation in terms of different transition states. We therefore suggest that the active site coverage is much lower on diesel soot, but that these sites are much more reactive (as reflected by the activation energy) than they are on flame soot. An explanation which would be consistent with such behavior is catalysis by some trace compound or element.

The study that is most relevant to this work is that of Miyamoto et al. [44], in which the oxidation rates of "usual" diesel soots were compared to those of diesel soots containing an oxidation catalyst that had been incorporated by means of a fuel additive. The temperature range studied in this work was ~500–700°C. An enhanced oxidation rate was observed for the catalyst-laden soots in the initial phases of oxidation. In the case of a calcium-based catalyst, the activation energy was lowered from 191 kJ mol^{-1} (in usual soot) to 48 kJ mol^{-1}. Moreover, the preexponential factor was also lowered, by roughly seven orders of magnitude. Although the Miyamoto results are more dramatic than the differences reported herein, they can help explain our observations. It is well known that metals, including calcium, are found in varying quantities in diesel soot as a result of additives to the lubricating oil or to the diesel fuel itself [47]. In fact, previous studies of diesel particle oxidation have suggested the possibility of oxidation rate enhancement due to the presence of inorganic impurities such as metals or metal oxides [48]. The effect of such catalysts would be to lower the activation energy for oxidation, while the effect on the observed oxidation rate would depend on a number of variables, including catalyst concentration (i.e., the number of surface reaction sites).

Assuming that our explanation for the soot data is correct, it appears that, in the absence of a catalyst, diesel particles are inherently slower to oxidize than flame soot particles. If diesel soot behaved similarly to flame soot, we would expect the oxidation rate for diesel soot to be as fast or faster than the flame soot, depending on the presence of a catalyst. In fact, the oxidation rate is slower. This reason for this behavior remains a mystery, and its resolution awaits further study.

7.4 FUTURE PROSPECTS

The prospects, not to mention the need, for future research in nanoaerosol surface chemistry are very bright. Most of the surface area of a typical atmospheric aerosol is found in the nanometer-sized particle fraction. For laboratory studies of surface-catalyzed or surface-promoted processing in the atmosphere, the best approaches are therefore likely to be those that focus on nanoparticles. The subject will be important in other fields as well. For instance, many potential applications of nanostructured materials will probably require the development of methods to synthesize nanoparticles via gas-to-particle conversion, and to chemically manipulate the surfaces of these particles while they are in an aerosolized form.

The work described in this chapter represents what is, to the best of our knowledge, the first application of T-DMA to study the *kinetics* and *mechanisms* of aerosol reactions. The DMA may seem a blunt tool for gaining molecular-level insights into reactivity, but current model mobility analyzers are capable of measuring diameter changes as small as 2 Å, which for virtually all adsorbates is less than one molecular layer thick. Given recent progress in instrument design and optimization, future DMA models are likely to be capable of measuring ΔD_p values as small as 1 Å or even less. Thus, one can imagine studying adsorption–desorption kinetics, the activation barriers to dissociative adsorption, or even kinetic isotope effects of aerosol reactions involving the scission or formation of a bond to hydrogen.

As an illustration of T-DMA, we considered the in-air oxidation of soot nanoparticles. There is huge potential for T-DMA as a tool for studying other broad classes of aerosol reactions, but its application will not be without difficulty. T-DMA requires that the gas flow and temperature fields in the reactor be extremely well characterized. This will be especially true for reactions that are confined to a single monolayer, because changes in particle diameter will be very small. For reactions that are not highly thermally activated, the flow reactor will need to be designed so that the contact time between particle and gas-phase reactants is well defined and controlled. Finally, for studies of very small particles, which are subject to diffusional wall losses, strategies may need to be developed to concentrate the particles in the reactor or to synthesize or create aerosol samples of adequate particle number densities. None of these problems is insurmountable, but they do point out some of the thinking that must go into the design of any size-selected aerosol surface chemistry experiment.

One limitation of T-DMA is that it provides no specific chemical information. T-DMA delivers no insights, for instance, into the hybridization or oxidation states of atoms at the surface of an aerosol particle. In fact, there are currently no online aerosol techniques that provide this kind of information. Until this problem is remedied, there will be huge limits on our depth of understanding of aerosol surface chemistry. Some of this kind of information is obtainable using electron microscopy or an electron-based spectroscopic method. However, these require that the particles be deposited onto substrates, and the very act of depositing a particle can change its chemical state. Moreover, many atmospheric aerosol particles are composed, at least in part, of volatile or semivolatile material. Vacuum-based methods cannot be used.

Our initial studies of APES are quite promising, and they suggest that APES has the potential of becoming a widely applicable diagnostic tool for probing chemical states in aerosolized particles. However, measurements of the appearance potential in the valence region, however helpful, will be intrinsically difficult to interpret. More thinking is needed about how best to probe the surface states of atoms and molecules at an aerosolized nanoparticle. The development of a suite of widely applicable, molecularly incisive tools for studying aerosol surfaces remains one of the most important problems in modern aerosol science.

ACKNOWLEDGMENTS

This material is based on work supported by the National Science Foundation under Grant No. 0094911. The author also acknowledges the contributions of his longtime collaborator, Michael Zachariah. None of these experiments would have been possible without the hard work of a superb group of students and postdoctoral associates, most notably Henry Ajo, Brian Ford, Kelly Higgins, and Amanda Nienow.

REFERENCES

1. WC Hinds. *Aerosol Technology: Properties, Behavior, and Measurement of Airborne Particles*. 2nd ed. New York: John Wiley, 1999.
2. JH Seinfeld and SN Pandis. *Atmospheric Chemistry and Physics: From Air Pollution to Climate Change*. New York: John Wiley, 1998.
3. R Jaenicke. In: RM Harrison and RV Grieken, eds. *Atmospheric Particles*. New York: John Wiley, 1998, pp. 1–28.
4. KT Whitby and GM Sverdup. *Adv Environ Sci Technol* 9: 477–517, 1980.
5. See, for example: ADH Clague, JB Donnet, TK Wang, and JCM Peng. *Carbon* 37: 1553–1556, 1999.
6. KJ Higgins, H Jung, DB Kittelson, JT Roberts, and MR Zachariah. *J Phys Chem A* 106: 96–103, 2002.
7. KJ Higgins, H Jung, DB Kittelson, JT Roberts, and MR Zachariah. *Environ Sci Technol* 37: 1949–1954, 2003.
8. (a) BYH Liu, DYH Pui. *J Colloid Interface Sci* 47: 155–171, 1974. (b) EO Knutson and KT Whitby. *J Aerosol Sci* 6: 443–451, 1975.
9. (a) EO Knutson and KT Whitby. *J Aerosol Sci* 6: 453–460, 1975. (b) EO Knutson. In: BYH Liu, ed. *Fine Particles: Aerosol Generation, Measurement, and Sampling*. New York: Academic Press, 1976, pp 740–762. (c) WA Hoppel. *J Aerosol Sci* 9: 41–54, 1978. (d) Y Kousaka, K Okuyama, and M Adachi. *Aerosol Sci Technol* 4: 209–225, 1985.
10. NA Fuchs. *Geophis Pura Appl* 56: 185–193, 1963.
11. (a) P Meakin, B Donn, and GW Mulholland, *Langmuir* 5: 510–518, 1989. (b) A Schmidt-Ott, U Baltensperger, HW Gäggeler, and DT Jost. *J Aerosol Sci* 21: 711–717, 1990. (c) SN Rogak, U Baltensperger, and RC Flagan. *Aerosol Sci Technol* 14: 447–458, 1991. (d) A Keller, A Fierz, M Siegmann, HC Siegmann, and A Filippov. *J Vac Sci Technol A* 19: 1–8, 2001.
12. PH McMurry. *Aerosol Sci Technol* 33: 230–240, 1990.
13. BYH Liu, DYH Pui, KT Whitby, DB Kittelson, Y Kousaka, and RL McKenzie. *Atmos Environ* 12: 99–104, 1978.
14. PH McMurry, H Takano, and GR Anderson. *Environ Sci Technol* 17: 347–352, 1983.
15. DJ Rader and PH McMurry. *J Aerosol Sci* 17: 771–787, 1986.
16. B Franz, T Eckhardt, and T Kauffeldt. *J Aerosol Sci* 31: 415–426, 2000.
17. GG Hoffmann, B Oelichmann, and B Schrader. *J Aerosol Sci* 22: S427–S430, 1991.
18. KH Fung and IN Tang. *J Aerosol Sci*, 23: 301–307, 1992.
19. JX Zhang and PM Aker. *J Chem Phys* 99: 9366–9375, 1993.
20. R Vehring, CL Aardal, G Schweiger, and EJ Davis. *J Aerosol Sci* 29: 1045–1061, 1998.
21. H Burtscher, L Scherrer, HC Siegmann, A Schmidt-Ott, and B Federer. *J Appl Phys* 55: 3787–3791, 1982.
22. R Niessner. *J Aerosol Sci* 17: 705–714, 1986.
23. M Kasper, K Siegmann, and K Sattler. *J Aerosol Sci* 28: 1569–1578, 1997.
24. T Jung, H Burtscher, and A Schmidt-Ott. *J Aerosol Sci* 19: 485–490, 1988.
25. R Niessner, H Schroeder, H Robers, and KL Kompa. *J Aerosol Sci* 19: 491–500, 1988.
26. M Kasper, K Sattler, K Siegmann, U Matter, and HC Siegmann. *J Aerosol Sci* 30: 217–225, 1999.
27. M Kasper, A Keller, J Paul, K Siegmann, and HC Siegmann. *J Electron Spectrosc* 98–99: 83–93, 1999.
28. U Müller, H Burtscher, and A Schmidt-Ott. *Phys Rev B* 38: 7814–7816, 1998.
29. JB Howard and JP Longwell. Formation mechanisms of PAH and soot in flames. Proceedings of 7th International Symposium on Polynuclear Aromatic Hydrocarbons: Formation, Metabolism, and Measurement, 1983, pp 27–62.
30. H Jung, DB Kittelson, and MR Zachariah. *Combust Flame* 136: 445–456, 2004.
31. BR Stanmore, JF Brilhac, and P Gilot. *Carbon* 39: 2247–2268, 2001.
32. RF Strickland-Constable. *Trans Faraday Soc* 1944: 40–46.
33. G Blyholder, JS Binford, and H Eyring. *J Phys Chem* 62: 263–267, 1958.
34. JR Walls and RF Strickland-Constable. *Carbon* 1: 333–338, 1964.
35. M Starsinic, RL Taylor, PL Walker, and PC Painter. *Carbon* 21: 69–74, 1983.
36. KB Lee, MW Thring, and JM Beer. *Combust Flame* 6: 137–145, 1962.

37. A Feugier. *Combust Flame* 19: 249–256, 1972.
38. A Garo, G Prado, and J LaHaye. *Combust Flame* 79: 226–233, 1990.
39. T Schafer, F Maus, H Bockhorn, and F Fetting. *Z Naturforsch, A: Phys Sci* 50: 1009–1022, 1995.
40. P Ciambelli, M D'Amore, V Palma, and S Vaccaro. *Combust Flame* 99: 413–421, 1994.
41. C Park and JP Appleton. *Combust Flame* 20: 369–379, 1973.
42. P Cadman and RJ Denning. *J Chem Soc Faraday Trans* 92: 4159–4165, 1996.
43. RJ Santoro, HG Semerjian, and RA Dobbins. *Combust Flame* 51: 203–218, 1983.
44. N Miyamoto, H Zhixin, and O Hideyuki. SAE Technol Pap Ser, 1988: No. 881224.
45. JW Frey and M Corn. *Nature* 216: 615–616, 1967.
46. KA Berube, TP Jones, BJ Williamson, C Winters, AJ Morgan, and RJ Richards. *Atmos Environ* 33: 1599–1614, 1999.
47. AF Ahlström and CUI Odenbrand. *Carbon* 3: 475–483, 1989.
48. K Otto, MH Sieg, M Zinbo, and L Bartosiewicz. SAE Technol Pap Ser, 1980: No. 800336.

Section II

Environmental Catalysis in Remediation

8 Selective Catalytic Reduction of NO_X

Teresa Curtin
Materials and Surface Science Institute, University of Limerick

CONTENTS

8.1 INTRODUCTION

Nitrogen oxides (NO_x) are very simple molecules that are naturally present in the atmosphere. The natural sources of these oxides include nitrogen fixation by lightening, volcanic activity, and microbial activity. For a long time, there was little concern about these nitrogen oxides. However, the levels of NO_x have now increased to an extent that they have become an extremely important family of air polluting compounds. The increase in NO_x emissions is due to the significant rise in anthropogenic (human) activities. This chapter will discuss the problems associated with these pollutants, their sources and currently available NO_x abatement technologies. In particular, selective catalytic reduction (SCR) of NO_x using ammonia will be discussed.

8.2 SOURCES AND EFFECTS OF NITROGEN OXIDE EMISSIONS

8.2.1 Nitrogen Oxides (NO_x)

What is NO_x? NO_x is a collective term used when referring to nitrogen oxides. Several nitrogen oxides exist and Table 8.1 lists these compounds with some of their general properties. The two oxides that are of primary concern regarding pollution (general air quality) are NO and NO_2 as these are emitted to the atmosphere in very large quantities. These two oxides will be the focus of this chapter.

Nitrogen oxides play an important role in the photochemistry of the troposphere and the stratosphere. Anthropogenic nitric oxides are composed mainly of NO and this species is easily oxidized to NO_2 by ozone or hydroperoxide radicals ($HOO^{\bullet}$). These reactions are represented by Equations (8.1) and (8.2), respectively.

$$NO + O_3 \longrightarrow NO_2 + O_2 \tag{8.1}$$

$$NO + HOO^{\bullet} \longrightarrow NO_2 + OH^{\bullet} \tag{8.2}$$

NO_2 can then react with hydroxy radicals to form nitric acid (Equation (8.3)). This water-soluble acid is washed out of the atmosphere and thus contribute to acidification [1].

$$2NO_2 + OH^{\bullet} \longrightarrow HNO_3 \tag{8.3}$$

NO_x and volatile organic compounds react photochemically in the lower atmosphere to produce peroxyacetyl nitrate (PAN), peroxybenzoyl nitrate (PBN), and other trace oxidizing agents eventually leading to the formation of smog. NO_x is also considered to be a serious health hazard for humans. For example, NO_2 can severely irritate the mucous membrane. When in contact with body moisture, nitrous and nitric acids are formed and attack the walls of the alveoli in the lungs leading to respiratory problems. Therefore, regulation of the amount of NO_x emitted into the atmosphere is of great importance both in terms of reducing environmental damage and minimizing hazards to health.

8.2.2 Sources of NO_x

The largest source of NO_x emissions is from combustion processes. Nitrogen oxides are emitted when fossil fuels are burned to produce steam for electric power or when gasoline is used to power automobiles. In fact, automobiles and other mobile sources are responsible for over half of the NO_x emitted from anthropogenic sources in Europe in 2000. This can be seen from Figure 8.1. Of a total of 9497 thousand tonnes of NO_x emitted, over 40% comes from stationary sources such as power plants, industrial boilers, petroleum refineries, etc. A small

TABLE 8.1
Nitrogen Oxides (NO_x)

Name	Formula	Properties
Nitric oxide	NO	Colorless, odorless gas
Nitrogen dioxide	NO_2	Pungent reddish-brown gas
Nitrous oxide	N_2O	Colorless gas, slightly sweet odor
Nitrogen trioxide	N_2O_3	Unstable at room temperature
Nitrogen pentoxide	N_2O_5	Colorless solid

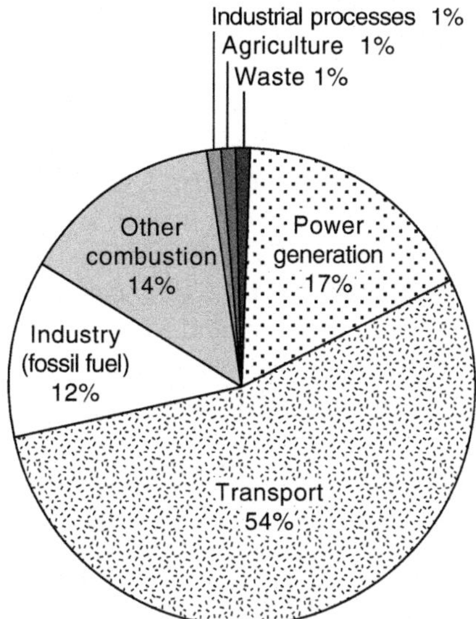

FIGURE 8.1 Major sources of NO_x in Europe in 2000. Adapted from B Gugele and M Ritter. *Annual European Community CLRTAP Emission Inventory 2000, Technical Report 91*, European Environment Agency, 2000, p. 28.

percentage of emissions result from industrial processes (e.g., chemical production and metal production), agriculture (soil and agricultural residues), and waste incineration [2].

8.2.2.1 NO_x from Fuel Combustion

There are three mechanisms of NO_x formation during combustion — thermal, prompt, and fuel NO_x. An understanding of these mechanisms is vital when designing technologies for NO_x emission reduction.

8.2.2.1.1 Thermal NO_x

This method of NO_x formation was first proposed by Zeldovich [3] and involves the reaction of gaseous N_2 and gaseous O_2 at combustion conditions. The overall reactions are presented as follows:

$$O_2 \longleftrightarrow 2O \tag{8.4}$$

$$O + N_2 \longleftrightarrow NO + N \tag{8.5}$$

$$N + O_2 \longleftrightarrow NO + O \tag{8.6}$$

Both thermodynamics and kinetics are important to the formation of thermal NO_x, so concentration, temperature, and residence time influence the amount of NO_x produced. Therefore, the formation of NO_x can be limited by reducing the flame temperature ($<1300°C$) and reducing the residence time.

8.2.2.1.2 Prompt NO_x

Prompt NO_x is formed when molecular nitrogen (N_2) in the air reacts with hydrocarbon free radicals. Prompt NO_x occurs in fuel-rich regions where the hydrocarbon free radicals are more likely to be found. In general, considerably more NO is formed by thermal fixation than by the generation of prompt NO_x. Prompt NO_x can be reduced by operating at lower temperatures and highly oxidizing combustion conditions.

8.2.2.1.3 Fuel NO_x

Some fuels may contain compounds possessing N—H or N—C bonds. Examples include pyridine, quinoline, and amine type compounds. During the combustion process these nitrogen-containing organics decompose into compounds such as HCN, NH_3, or free radicals such as $NH^\bullet$ and $CN^\bullet$. All of these compounds ultimately form NO_x [4]. Fuel NO_x is independent of the flame temperature at normal combustion temperatures and is insensitive to the nature of the organic nitrogen compound. The amount of fuel NO_x formed will therefore depend on the amount of nitrogen-containing compounds in the original fuel. Fuel oil typically contains 0.1% to 0.5% nitrogen while coal could contain up to 1.6% nitrogen. Fuel NO_x typically falls in the range of 20% to 50% of the total NO_x emissions depending on the combustion conditions [5]. Fuel NO_x can be limited by decreasing the concentration of bound nitrogen in the fuel or by operating the burner in fuel-rich conditions.

8.3 NO_x CONTROL TECHNOLOGIES

There are two primary categories of control techniques for NO_x emissions. These are

- Combustion control
- Flue gas treatment

Combustion control is a primary treatment method for controlling NO_x emissions since efforts are made to minimize the level of NO_x formation during the combustion process. Flue gas treatment is a secondary treatment method whereby the flue gas, containing the produced NO_x, is treated using an add-on technology.

8.3.1 Combustion Control

Combustion control measures are now considered to be an integral part of newly built power plants and they can also be retrofitted to existing plants whenever NO_x emissions need to be reduced. Combustion control techniques can vary from simple changes in operating conditions to the design of sophisticated staged burners. All take advantage of the thermodynamic and kinetic processes involved in forming NO_x. Some reduce the peak flame temperature and some reduce the oxygen concentration in the primary flame zone, both of which ultimately result in lower NO_x formation. Another approach is to reconvert already formed NO_x back to nitrogen and oxygen. Very often more than one control technique is used in combination to achieve the desired reduction in NO_x emission levels. The following is a brief description of some of the combustion control techniques currently available for reducing NO_x formation (see Figure 8.2). More detailed information on these technologies can be obtained from other sources [6–9].

1. *Injection of water or steam:* Injection of steam or water results in the lowering of the peak temperature therefore leading to lower levels of thermal NO_x formation.
2. *Flue gas recirculation:* A portion of the flue gas is recirculated back into the combustion air. This flue gas is relatively depleted in oxygen therefore the oxygen concentration in the combustion zone is decreased and in addition the overall temperature is decreased.
3. *Reduced air preheat:* Combustion air is sometimes heated by the exiting flue gas. This tends to raise the combustion temperature leading to thermal NO_x formation. Reducing the preheat of the combustion air leads to lower efficiency, but it can limit NO_x generation.

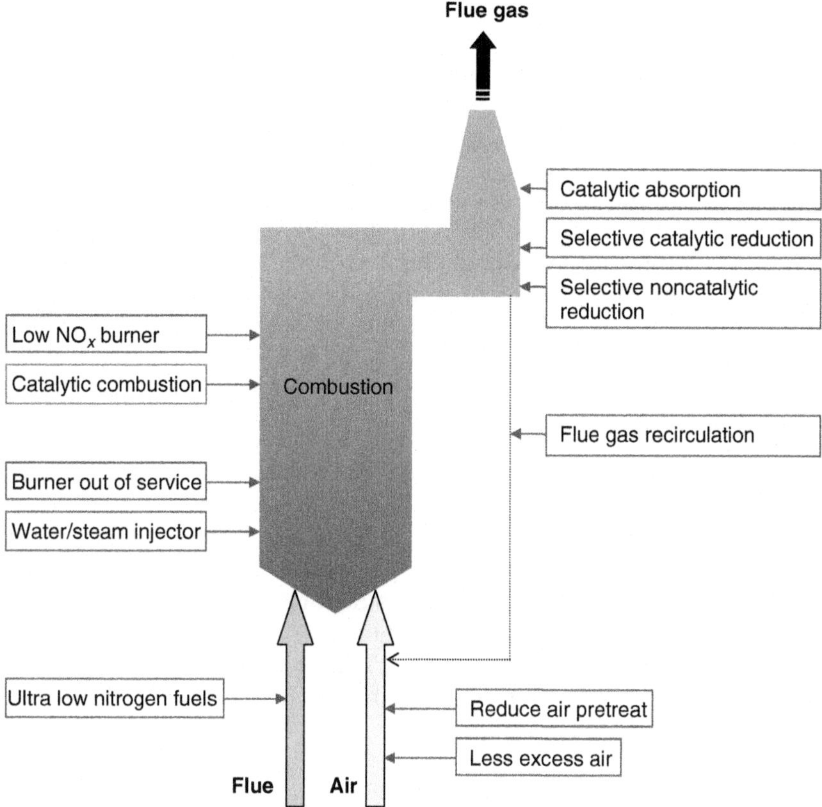

FIGURE 8.2 Possibilities for NO_x reduction from a combustion chamber.

4. *Less excess air:* This simply involves cutting back on the excess oxygen sent to the combustion zone.
5. *Burner out of service:* For multiburner equipment, selected burners can be taken out of service. This simply involves cutting fuel but continuing to supply combustion air. This is an inexpensive way to stage the combustion air thus leading to lower combustion temperatures.
6. *Low NO_x burners:* These are also commonly referred to as staged burners. These can be staged air or staged fuel burners.
   (a) *Staged air:* The primary flame is burned with excess fuel and low levels of air. This is achieved by, for example, dividing the combustion air into two streams. The first stream is mixed with fuel generating fuel-rich conditions. The second air stream is introduced downstream of the primary flame where the temperature is lower and here combustion is completed.
   (b) *Staged fuel:* Fuel is introduced in two locations. Some of the fuel is mixed with excess air in the first zone forming a hot primary flame. These conditions promote the formation of thermal NO_x. Additional fuel is added outside the primary flame zone forming a low oxygen zone. Here some of the NO_x is converted back to nitrogen and oxygen.
7. *Catalytic combustion:* This approach uses a catalyst that allows combustion to occur below NO_x formation temperatures. It is not often used because the long-term performance of these systems is still under investigation. However, it can be a very effective way of reducing NO_x formation [10].

8. *Low nitrogen fuels:* These fuels can avoid NO_x that results from the oxidation of nitrogen contained in conventional fuel. Natural gas and light distillate oils are considered low nitrogen fuels. Coke (quenched char from coal) is also considered a low nitrogen fuel since the nitrogen volatile fraction of the coal is removed during the production of coke.

Some characteristics and efficiencies of these combustion control NO_x reduction technologies are summarized in Table 8.2. The NO_x reduction efficiencies vary depending on the technique. However, with strict laws demanding ever-lower levels of emitted NO_x, frequently combustion modification is not sufficient to meet the demands of these stringent emission laws. Flue gas treatment, in many cases, is the only technology that meets legislation requirements.

8.3.2 Flue Gas Treatments

Some well-known flue gas treatment technologies for NO_x control are as follows:

- *Selective catalytic reduction (SCR):* NO_x is selectively reduced (using a reductant such as ammonia) to nitrogen and water using a catalyst. Nonselective catalytic reduction can also be used but this is less desirable than SCR because oxygen in the flue gas stream as well as NO_x are consumed by the reductant.
- *Selective noncatalytic reduction:* NO_x emissions in the flue gas are converted into nitrogen and water by injecting urea (NH_2CONH_2) or ammonia. The reactions, in a simplified form, are as follows [12]:

$$2NO + NH_2CONH_2 + \frac{1}{2}O_2 \longrightarrow 2N_2 + CO_2 + 2H_2O \tag{8.7}$$

$$4NO + 4NH_3 + O_2 \longrightarrow 4N_2 + 6H_2O \tag{8.8}$$

Because the optimum temperature for selective noncatalytic reduction is achieved at 870–1200°C, the reagent (urea or ammonia) is introduced in the upper part of the boiler or in the ducts. Potential problems with this process include incomplete mixing of ammonia with the hot flue gas and improper temperature control. If the temperature is too low, unreacted ammonia will be released; if the temperature is too high ammonia will oxidize to form NO_x.

TABLE 8.2
Characteristics of NO_x Abatement Techniques

NO_x Abatement Technique	Mode of Operation	Efficiency
Water injection	Reduced peak temperature	40–70%
Flue gas recirculation	Reduced peak temperature	40–80%
Reduced air pretreat	Reduced peak temperature	25–65%
Less excess air	Reduced peak temperature	1–15%
Burner out of service	Reduced peak temperature	30–60%
Low NO_x burners	Reduced residence time at peak temperature	30–50%
Selective catalytic reduction	Chemical reduction of NO_x	70–90%
Selective noncatalytic reduction	Chemical reduction of NO_x	25–50%
Catalytic absorption	Chemical reduction of NO_x and absorption of SO_2	60–90%
Catalytic combustion	Low temperature combustion (gas and liquid fuels only)	>90%

- *Sorption:* Some dry and wet sorption processes have been reported for the simultaneous removal of NO_x and SO_x ($SO_2 + SO_3$) from flue gas streams. In wet scrubbing, additives have to be added to the scrubbing system to convert insoluble NO into a form that can be removed by the scrubbing liquid. These additives include strong oxidizing agents such as $KMnO_4$ and $NaClO_2$ [13]. An example of dry sorption uses a copper oxide catalyst (e.g., copper oxide on alumina). Copper oxide reacts with SO_x to form copper sulfate. Both copper oxide and copper sulfate catalyze the selective reduction of NO_x in the presence of ammonia. The catalyst is periodically regenerated to reconvert the copper sulfate back to copper oxide yielding a SO_2-rich stream that can be further processed to form elemental sulfur or H_2SO_4 [14].

One of the most successful technologies used for the abatement of NO_x in flue gas streams is the SCR process. The remainder of this chapter will focus on this technology. Issues such as important operating parameters and difficulties associated with the system will also be discussed.

8.4 SELECTIVE CATALYTIC REDUCTION

8.4.1 SCR Reactions

The process of SCR is based on a very simple chemical reaction, which involves the reaction of NO_x and ammonia in the presence of oxygen to form nitrogen and water. The technology involves the injection of ammonia into the NO_x containing flue gas. This flue gas then passes through a catalyst, which facilitates the reaction of ammonia with NO_x forming harmless nitrogen and water. In practice, several reactions are known to occur but the SCR of NO_x is best represented by the following two equations, Equations (8.9) and (8.10). Since the concentration of NO_2 in combustion flue gasses is relatively low, the dominant reaction is represented by Equation (8.9).

$$4NO + 4NH_3 + O_2 \longrightarrow 4N_2 + 6H_2O \tag{8.9}$$

$$2NO_2 + 4NH_3 + O_2 \longrightarrow 3N_2 + 6H_2O \tag{8.10}$$

In general, the reaction requires temperatures between 250°C and 450°C in the presence of oxygen. The term "selective" describes the ability of ammonia to react selectively with NO_x, instead of being oxidized by oxygen. A number of other reactions, some undesirable, can also take place during the SCR of NO with ammonia. Some of these are listed as follows.

$$6NO + 4NH_3 \longrightarrow 5N_2 + 6H_2O \tag{8.11}$$

$$8NO + 2NH_3 \longrightarrow 5N_2O + 3H_2O \tag{8.12}$$

$$4NO + 4NH_3 + 3O_2 \longrightarrow 4N_2O + 6H_2O \tag{8.13}$$

Under certain conditions the "nonselective" oxidation of ammonia may take place.

$$4NH_3 + 3O_2 \longrightarrow 2N_2 + 6H_2O \tag{8.14}$$

$$2NH_3 + 2O_2 \longrightarrow N_2O + 3H_2O \tag{8.15}$$

$$4NH_3 + 5O_2 \longrightarrow 4NO + 6H_2O \tag{8.16}$$

A simplified diagram of the SCR system is presented in Figure 8.3. The composition of a flue gas is very much dependant on the fuel. For example, a typical flue gas from a coal-fired

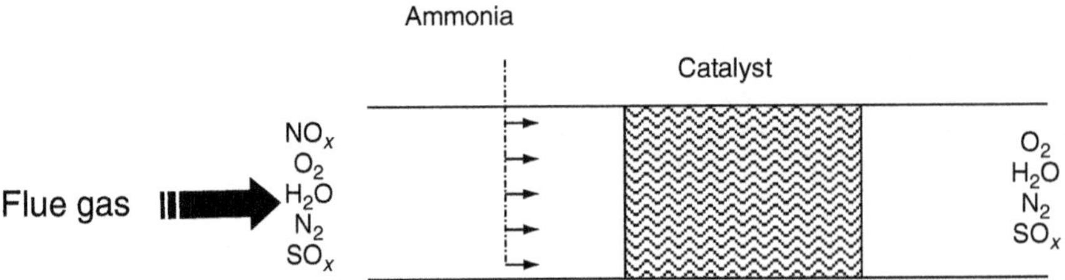

FIGURE 8.3 Simple schematic of SCR reactor.

utility plant would be composed of 150–1000 ppm NO_x, 5% O_2, 8% H_2O, 13% CO_2, 200–2000 ppm SO_2, dust, and trace amounts of other components such as alkali metals. The flue gas from a natural gas fired plant would contain more oxygen, less NO_x, and significantly lower amounts of SO_2 and dust. Ammonia is injected into this flue gas before it passes through the catalyst. While in contact with the catalyst, NO_x is converted into harmless nitrogen and water. It is important that the catalyst does not oxidize the SO_2 in the flue gas to SO_3 as this would lead to the subsequent formation of ammonium sulfate salts (Equations (8.17)–(8.19)). These sulfate salts may deposit onto the catalyst, if the temperature is not high enough, or deposit on the equipment downstream of the catalytic reactor where the temperature is cooler [15]. In addition, SO_3 may result in the formation of sulfuric acid (Equation (8.20)), which would cause corrosion problems downstream of the reactor.

$$2SO_2 + O_2 \longrightarrow 2SO_3 \tag{8.17}$$

$$SO_3 + 2NH_3 + H_2O \longrightarrow (NH_4)_2SO_4 \tag{8.18}$$

$$SO_3 + NH_3 + H_2O \longrightarrow NH_4HSO_4 \tag{8.19}$$

$$SO_3 + H_2O \longrightarrow H_2SO_4 \tag{8.20}$$

8.4.2 Type of Catalyst

Several catalysts are reported to be active for the SCR of NO_x with ammonia. The most well-known and most commonly used catalyst is vanadia well dispersed on a titania support (V_2O_5/TiO_2) with tungsten oxide or molybdenum oxide. Vanadia (V_2O_5) is the active ingredient and is responsible for converting NO_x into nitrogen. However, it is also responsible for the undesired oxidation of SO_2 (Equation (8.17)). Therefore, the vanadia content of the catalyst is generally kept low (0.3–1.5%) especially in the presence of high SO_2 concentrations. The titania (anatase) support is very stable and weakly and reversibly sulfated in the presence of SO_2. WO_3 (10 wt.%) is added to increase acidity, increase activity, increase the thermal stability of the catalyst, and limit the oxidation of SO_2. Catalysts containing MoO_3 (6 wt.%) instead of WO_3 are also available. These catalysts, although less active than the analogous V_2O_5–WO_3/TiO_2 samples, are more tolerant to arsenic, a common catalyst poison [16]. V_2O_5/TiO_2 operates best at temperatures between 300°C and 400°C.

Another catalyst used for the SCR of NO_x is supported noble metal, in particular, supported platinum. Noble metal catalysts were the first catalysts to be developed for the SCR process but are now limited for use in low-temperature applications (230–270°C). These catalysts are selective for the SCR of NO but also promote the unwanted oxidation of sulfur dioxide. In addition, when operating above their temperature range they oxidize ammonia to NO. Some low-temperature SCR catalysts are available commercially. One such example includes that developed by Shell International Chemie, which is based on vanadia and operates in the temperature range 140–250°C [17]. Zeolites, such as iron-exchanged mordenite, have also been reported as successful SCR catalysts. They operate at higher temperatures (up to 600°C) that the vanadia and noble metal based catalysts [18].

Commercially, the catalyst can be obtained in a variety of different geometric shapes. These include monolith and plate type structures. Typical examples of these structures are presented in Figure 8.4 [19]. Monoliths are inorganic oxides or metals made into a honeycomb type structure containing equally sized parallel channels. The channels can exist in a variety of geometric shapes but in general the cross-section is usually square. The size of the channel openings are chosen based on the characteristics of the flue gas and typically can vary from 10 to 200 cpsi (cells per square inch). The use of monolithic structures as a support for the active catalyst offers several advantages. These include

- Lower pressure drop due to large open frontal area with thin parallel channels
- Superior attrition resistance
- Lower tendency to fly ash plugging
- High external surface per unit volume of reactor
- Good mechanical properties
- Good mass transfer properties.

SCR honeycomb monoliths are prepared by extruding a paste of the catalyst components with water and binders [20]. The plate catalyst is formed by depositing the catalytic paste onto stainless steel parallel plates. In general, the surface of the metal has to be pretreated so that the subsequent addition of the catalyst will be facilitated. The distance between plates varies

FIGURE 8.4 Typical plate and honeycomb type monoliths. Reprinted from H Gutberlet and B Schallert. *Catal Today* 16: 207–236, 1993, with permission from Elsevier.

between 4 and 10 mm. The SCR catalyst can also be coated onto a ceramic monolith (for example cordierite — $2MgO{\cdot}2Al_2O_3{\cdot}5SiO_2$) or a metal monolith. These coated ceramic structures are usually employed in dust free flue streams. Figure 8.5 presents a diagrammatic representation of these SCR catalytic structures.

8.4.3 The SCR Reactor

A typical SCR reactor is presented in Figure 8.6. The plate or honeycomb elements containing the catalytic material are placed side by side and built into modules. A module can consist of 8 to 81 elements depending on the catalyst type. These modules are then placed into the reactor to form catalysts layers [16]. The reactor usually has more than one layer of catalyst modules therefore the flue gas passes through the layers in stages. Ammonia, either as an aqueous solution or anhydrous liquid is injected into the flue gas stream. A dummy layer is installed upstream of the catalyst layers for flow straightening and uniform distribution.

8.4.4 The Positioning of the SCR Reactor

The flue gas leaving a fossil fuel powered boiler is subjected to a series of treatments other than SCR to purify the gas stream before its release to the environment. Depending on the fuel used, treatments such as dust and SO_2 removal may be required. Therefore, the SCR reactor can be installed in several different positions in the flue gas stream. Some of the most commonly used configurations are:

- *High dust*: The SCR unit is placed immediately after the boiler.
- *Low dust*: The SCR unit is situated after the dust removal system.
- *Tail end*: The SCR is located downstream of dust and SO_2 removal units.

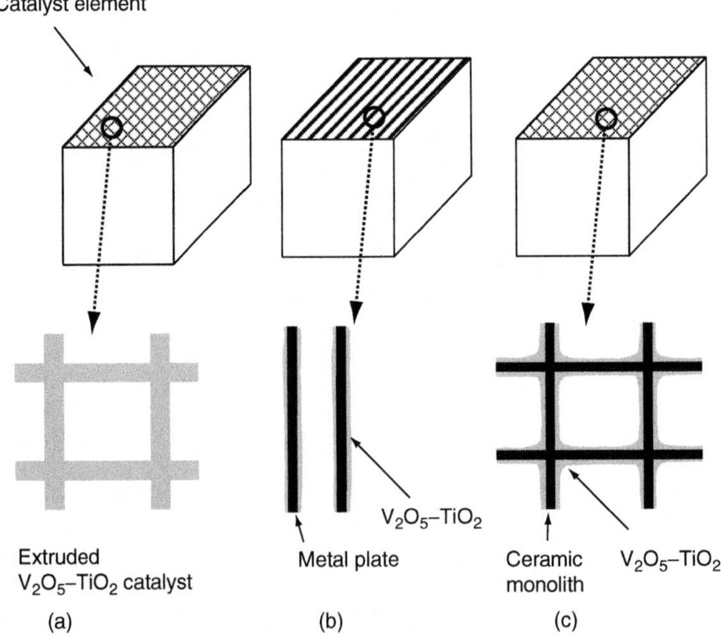

FIGURE 8.5 Various types of V_2O_5/TiO_2 catalytic structures. (a) Extruded V_2O_5/TiO_2 catalyst, (b) plate-type catalyst, and (c) V_2O_5/TiO_2 coated onto a ceramic monolith.

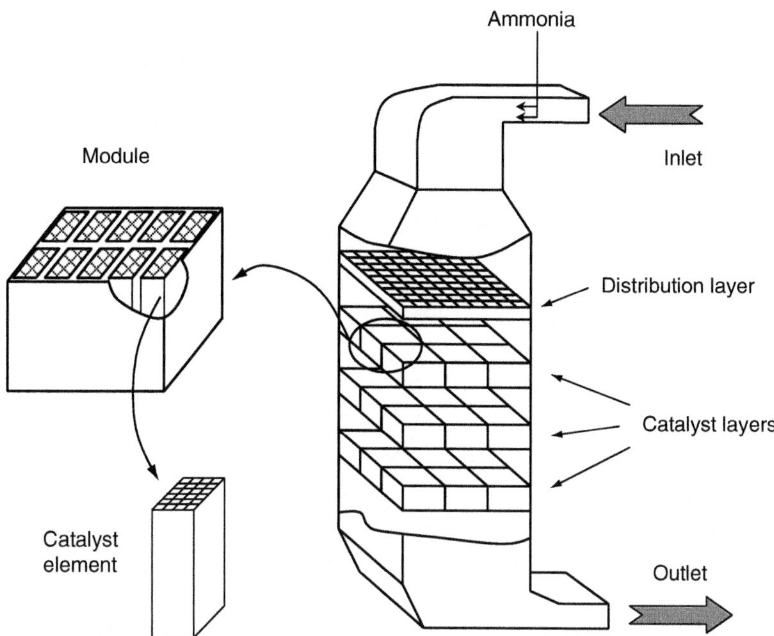

FIGURE 8.6 The SCR reactor. (Reprinted from P Forzatti. *Appl Catal A: General* 222: 221–236, 2001, with permission from Elsevier).

The choice of arrangement will depend on a number of factors including the characteristics of the flue gas, the space available, and economics [15, 21].

8.4.4.1 High-Dust Configuration

In the high-dust configuration, the SCR unit is located between the economizer (exit of the boiler) and the air preheater where the temperature is ideal for catalyst activity (300–400°C). This is presented schematically in Figure 8.7. The advantage of this arrangement is that the flue gasses are sufficiently hot and do not require any further heating. However, the temperature can vary, depending on the operating conditions of the boiler, which may influence NO_x removal efficiency. Dust and SO_2 are removed downstream of the SCR unit using, typically, electrostatic precipitation (ESP) and flue gas desulfurization (FGD), respectively. The SCR unit has to be designed to handle the dust so it must have large channel openings or holes (9–11 cpsi) to prevent plugging. This results in a larger overall reactor volume.

8.4.4.2 Low-Dust Configuration

In low-dust configuration the SCR unit is situated downstream of the electrostatic precipitator and upstream of the air preheater and FGD. Therefore, the gas going through the SCR unit is relatively dust free but still contains SO_2. The channel openings in the SCR catalysts are usually between 11 and 50 cpsi.

8.4.4.3 Tail-End Configuration

In the tail-end configuration, the SCR reactor is located downstream of the air preheater, particulate collector, and SO_2 removal unit. Accordingly, catalysts with higher geometric surface area (200 cpsi) and higher vanadia content are used since SO_2 oxidation is not an

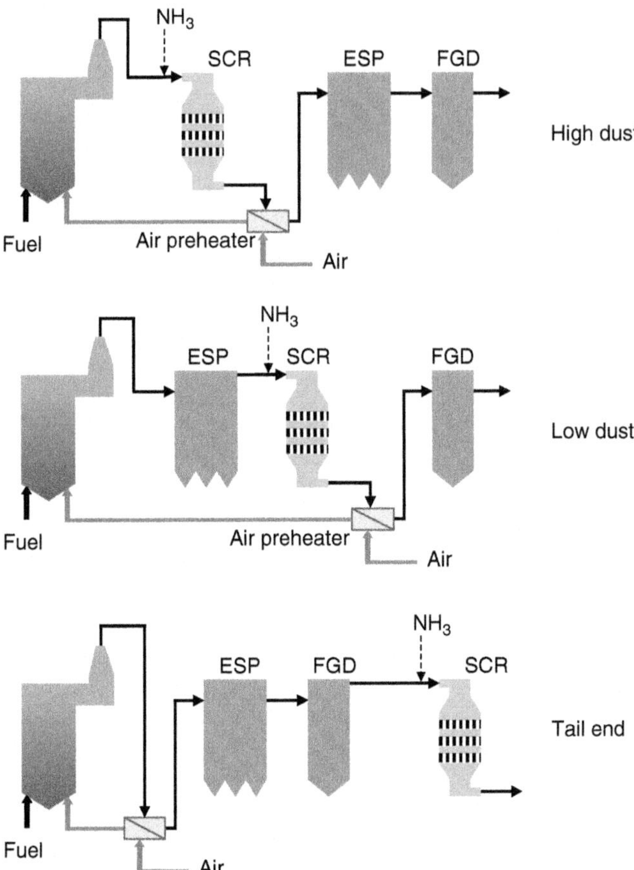

FIGURE 8.7 Possible configurations of SCR unit (SCR, selective catalytic reduction; ESP, electrostatic precipitation; FGD, flue gas desulfurization).

important issue. This results in lower catalyst volume and a reduced deterioration rate of the catalyst. However, in the tail-end system the temperature of the flue gas is too low for the catalytic reaction and has to be raised to a higher temperature (300–350°C) depending on the flue gas concentrations. The gas heating requirement will add to the operational cost of the tail-end arrangement but the fact that the temperature can be kept constant is an advantage.

8.4.5 Difficulties Associated with the System

The level of NO_x removed will depend on a number of factors including the quantity of ammonia introduced into the system. From Equation (8.9), ammonia reacts with NO in a 1:1 ratio, therefore, NO_x removal increases with increasing levels of ammonia. However, if too much ammonia is added or if the activity of the catalyst declines, some of the ammonia will pass through the catalyst without reacting. This is commonly known as "ammonia slip." Because ammonia itself is a harmful pollutant, there are limits to the amount of ammonia slip allowed for SCR reactors. This is usually less than 5 to 10 ppm. In addition, unreacted ammonia may react with any sulfur trioxide (SO_3) present in the flue gas stream to form ammonium bisulfate (NH_4HSO_3), which may plug and corrode downstream equipment. Ammonium sulfate may also deposit onto the surface of the catalyst but this can be avoided by keeping the temperature above 300°C. If the reactor temperature drops below 200°C explosive ammonium nitrate may form.

Sophisticated equipment is required for ammonia distribution into the flue stream. Ammonia is introduced through a series of nozzles so that it is entrained over the entire area of the flue gas duct. Ammonia can be in the anhydrous (pure) or aqueous form. Anhydrous ammonia is least costly and requires less storage space but is more difficult to handle. It is stored in pressure tanks as a liquefied gas and is vaporized and diluted with air before injection into the flue gas stream. Aqueous ammonia (diluted with 70% to 80% water) can be stored in atmospheric pressure tanks. It can be injected directly into the flue gas or it can be vaporized, mixed with air and then injected.

Catalyst erosion is a problem when there is a significant amount of particulates in the gas stream. This is particularly true for honeycomb catalysts. However, development of honeycomb-hardened catalysts has helped to prevent this facial erosion. In the case of plate catalysts, erosion is not a problem since the metal support is usually made of steel. The catalysts are configured into structured grids to minimize dust accumulation. Some catalyst beds are fit with stream-operated soot blowers to remove any dust build-up.

Deactivation of the catalyst can occur over extended periods of use. The main causes of catalyst deactivation are

- *Thermal damage*: With V_2O_5/TiO_2 catalyst, excessively high temperatures may convert the high surface area form of TiO_2, anatase, into a low surface area form, called rutile. This transformation results in irreversible deactivation.
- *Catalyst poisoning*: Various constituents of the fuel or lubricating oil may lead to catalyst deactivation. In coal-fired burners, the major catalyst poisons are alkali metals (Na, K) and arsenic. These metals bind permanently to the catalyst eventually inhibiting reaction. In diesel engines, phosphorous from lubricating oil accumulates on the catalyst surface leading to deactivation.
- *Pore blockage*: Dust build up and deposition of ammonium sulfate can result in the blocking of catalysts pores. This results in reduced access to active sites within the catalyst, which results in a fall in overall activity.

Despite the potential problems associated with the SCR process it is a proven technology for the reduction of NO_x emissions from stationary sources with a catalyst lifetime of 5 to 9 years. It can be used for the efficient control of NO_x emissions from gas-, oil-, and coal-fired power plants. In fact, SCR is one of the few technologies capable of efficiently removing NO_x from high sulfur coal. The by-products formed are nonpolluting and the technology can be retrofitted to existing plants.

8.4.6 Current Developments

SCR is a relatively mature technology that has been successfully employed to reduce NO_x emissions from stationary sources. Despite the success of SCR, continuous efforts are made to improve the process. The following are some issues that are currently under investigation.

- Develop the SCR catalyst further so that it will exhibit high activity for the removal of NO_x but present low activity for the oxidation of SO_2.
- Develop a technology that would address the release of unconverted NH_3. This would involve developing an ammonia destruction catalyst for installation after the SCR unit [22, 23]. Such a technology would allow increased NO_x removal efficiencies since tight control of ammonia injection would not be an issue.
- Understanding deactivation, particularly by flue gas constituents.

- Further develop reliable catalysts that can be used for low temperature (150–300°C) and high temperature (up to 600°C) use.
- To further improve the understanding of the reaction mechanism and interaction of species with various components of the catalysts.
- Extend the application of the technology to mobile sources such as diesel trucks.

REFERENCES

1. SE Manahan. *Fundamentals of Environmental Chemistry*, 2nd ed. Boca Raton, FL: Lewis Publishers, 2001, pp. 554–562.
2. B Gugele and M Ritter. *Annual European Community CLRTAP Emission Inventory 2000, Technical Report 91*, European Environment Agency, 2000, p. 28.
3. YA Zeldovich. *Acta Physicochimica USSR* 21 (4): 557–628, 1947.
4. RJ Heinsohn and RL Kabel. *Sources and Control of Air Pollution*. New York: Prentice-Hall, 1990, p. 308.
5. N deNevers. *Air Pollution Control Engineering*. New York: McGraw-Hill, 1995, pp. 378–394.
6. KB Schnelle and CA Brown. *Air Pollution Control Technology Handbook*. Boca Raton, FL: CRC Press, 2002, pp. 241–255.
7. H Bosch and F Janssen. *Catal Today* 2 (4): 369–532, 1987.
8. Clean Air Technology Center, North Carolina. US EPA Technical Bulletin, Nitrogen oxides (NO_x), why and how they are controlled. EPA 456/F-99-006R: 1–57,1999.
9. SC Wood. *Chem Eng Process* 90: 32–38, 1994.
10. P Forzatti. *Catal Today* 62: 51–65, 2000.
11. R Carroni, V Schmidt, and T Griffin. *Catal Today* 75: 287–295, 2002.
12. LJ Muzio, GC Quartucy, and JE Cichanowicz. *Int J Env Poll* 17 (1–2): 4–30, 2002.
13. H Chu, TW Chein, and SY Li. *Sci Tot Env* 275: 127–135, 2001.
14. G Centi, BK Hodnett, P Jaeger, C Macken, M Marella, M Tomaselli, G Paparatto, and S Perathoner. *J Mater Res* 10 (3): 553–561, 1995.
15. RM Heck and RJ Farrauto. *Catalytic Air Pollution Control: Commercial Technology*, 2nd ed. New York: Wiley Interscience, 2002, pp. 306–333.
16. P Forzatti. *Appl Catal A: General* 222: 221–236, 2001.
17. CJG van der Grift, AF Woldhuis, and OL Maaskant. *Catal Today* 27: 23–27, 1996.
18. TQ Long and RT Yang. *J Catal* 188 (2): 332–339, 1999.
19. H Gutberlet and B Schallert. *Catal Today* 16: 207–236, 1993.
20. IM Lachman and JL Williams. *Catal Today* 14: 317–329, 1992.
21. RM Heck. *Catal Today* 53: 519–523, 1999.
22. T Curtin and S Lenihan. *Chem Commun* 11: 1280–1281, 2003.
23. T Curtin, F O'Regan, C Deconinck, N Knuttle, and BK Hodnett. *Catal Today* 55 (1–2): 189–195, 2000.

9 Surface Science Studies of $DeNO_x$ Catalysts

Jose A. Rodriguez
Department of Chemistry, Brookhaven National Laboratory

CONTENTS

9.1 INTRODUCTION

The destruction or removal of nitrogen oxides ($DeNO_x$ process) is a very important issue in environmental catalysis [1, 2]. Nitrogen oxides (NO, NO_2, and N_2O) are formed during combustion reactions in automotive engines and industrial plants when the nitrogen present in air–fuel mixtures reacts with oxygen [1]. In addition, the burning of oil-derived fuels with N-containing impurities [3] also leads to the formation of nitrogen oxides [1, 2]. Subsequent release of the NO_x combustion products into the atmosphere contributes to the generation of smog and constitutes a serious health hazard for the respiratory system [1]. Furthermore, in the air, the NO_x species undergo oxidation and react with water, producing acid rain that corrodes monuments and kills vegetation [1]. Thus, to reduce environmental pollution, it is necessary to use catalysts or sorbents to remove or trap the nitrogen oxides before they reach the atmosphere. Although $DeNO_x$ operations have been a major concern for the last three decades, recently, research in this area has intensified [2]. In part, this is motivated by the fact that there is evidence that emission of the noxious NO_x vapors into the atmosphere has increased worldwide [4]. New and more stringent environmental regulations emphasize the need for more efficient catalysts or sorbents for preventing or controlling NO_x emissions [2, 4]. There is still no universally acceptable solution to this major problem in environmental chemistry [2–7].

Depending on the nature of the combustion process, various approaches for NO_x removal have been developed [2, 4, 5, 7]. Some of the catalytic approaches proposed involve the reaction of NO_x with CO, selective catalytic reduction with NH_3 or hydrocarbons, or the direct decomposition of nitric oxides. Currently, the removal of nitric oxides from automobile exhaust is accomplished by the so-called three-way catalysts [2, 6], which contain noble metals like Pt, Pd, or Rh as the active phase, and ceria as an additive due to its ability to store or release oxygen depending on the composition of the feed [8]. Future exhaust catalysts could consist of a mixture of BaO/MgO and Pt [9]. Pure oxides are also useful for $DeNO_x$ operations in many applications [9, 10–12].

In general, the $DeNO_x$ catalysts and sorbents used in the industry are complex systems, and there is a need to obtain a fundamental understanding of their behavior to optimize their performance [2, 4, 6–8]. Well-defined single-crystal surfaces are ideal models to study the chemistry associated with $DeNO_x$ processes [10, 13–20]. During the last 25 years surface scientists have developed a variety of techniques that make the study of the structure, composition, and chemical bonding in a surface monolayer at a molecular level possible [13–24]. Low-energy electron diffraction (LEED), field ion microscopy (FIM), x-ray photoelectron diffraction (XPD), and scanning tunneling microscopy (STM) provide complimentary information about adsorption sites and the morphology of a layer of adsorbed species [20, 22–24]. The strength of the interactions between a chemisorbed atom or molecule and a solid surface can be probed using thermal desorption mass spectroscopy (TDS) [20, 22]. Ultraviolet and x-ray photoelectron spectroscopies (UPS and XPS), inverse photoemission (IPS), work function measurements ($\Delta\varphi$), and photoemission of adsorbed xenon (PAX) give information about the oxidation states and electronic interactions in gas–solid interfaces [10, 14, 18–20, 23, 24]. Molecular-beam based techniques can be used to study the dynamics of surface reactions [13, 21, 24], while reaction kinetics under high-pressure conditions can be investigated in reactors attached to ultrahigh vacuum chambers [15, 16, 22].

In this chapter, we review the existing literature that deals with the adsorption and chemistry of NO_x molecules on well-defined surfaces of metals and oxides. We focus our attention on systems that have a direct link to those used in industrial or commercial applications. Examples that are provided demonstrate the relevance of single-crystal studies for modeling the behavior of high-surface-area supported catalysts or sorbents. Following this approach, one can obtain a fundamental knowledge that can be used for improving existing technologies or developing new ones.

9.2 ADSORPTION AND REACTION OF NO_X MOLECULES ON METAL SURFACES

Many metals are able to catalyze the decomposition of NO_x species and their reduction with hydrogen, hydrocarbons, or ammonia [5, 22]. Thus, a very large fraction of the $DeNO_x$ catalysts used in the industry contains a metal as the active phase [4–7]. Furthermore, for the production of nitric acid, ammonia is frequently passed over a noble metal catalyst and is oxidized to nitric oxide. The NO is further oxidized to NO_2, which is adsorbed in water to form HNO_3 [5]. In this section, we will examine the chemistry of NO, NO_2, and N_2O over metal surfaces. Of these molecules, the most studied is NO thus far.

9.2.1 NO ADSORPTION AND REACTIONS

The NO molecule has an unpaired electron in its $2\pi^*$ orbital [25, 26]. When present on a metal, NO can either donate its $2\pi^*$ electron to the surface or it can accept electron density from the surface into the partially filled $2\pi^*$ orbital [10, 11, 25, 26]. This leads to different types of adsorption geometries (bonding via N or O, with the N—O molecular axis vertical, tilted, or parallel to the substrate) and a very rich surface chemistry [26]. Previous studies found that the susceptibility of NO to dissociate on a metal surface varied with the position of the substrate in the periodic table [27]. It was pointed out that surfaces of Pd, Pt, Cu, Ag, and Au should not be able to dissociate NO, while dissociation of the molecule should occur on the rest of the transition metals. More recent studies of NO adsorption show a complex situation [26]. For several metal elements (Ru, Co, Rh, Ir, Ni, Pd), whether a surface dissociates NO often depends on the temperature of the system, surface coverage, crystal plane, and the concentration of surface defects [26]. The three-way catalysts used for the treatment of automotive exhaust contain Pt, Pd, or Rh, or alloys of these metals [4–6]. The $DeNO_x$ activity of Rh is much higher than those of Pd and Pt, but due to the high cost of Rh, future $DeNO_x$ systems could contain only Pt or Pd [4, 7]. Interesting differences are observed in the chemistry of NO on surfaces of Rh, Pt, and Pd.

9.2.1.1 NO Chemistry on Rh Surfaces

Studies examining the interaction of NO with single-crystal Rh(1 0 0), Rh(1 1 0), and Rh(1 1 1) show molecular adsorption at low temperatures, and dissociation upon heating [13, 14, 28–30]. At high coverage, a fraction of the adsorbed NO desorbs molecularly, and the remaining NO decomposes to adsorbed N and O atoms. Figure 9.1 shows the TDS spectra taken after exposing a Rh(1 0 0) surface to NO at 100 K [28]. The desorption products were NO (m/e = 30), N_2 (m/e = 28), O_2 (m/e = 32), and traces of a CO impurity (m/e = 28). For low NO coverages (<0.2 ML), N_2 and O_2 are the only desorption products, indicating that NO dissociates completely [28]. For $\theta_{NO} > 0.2$ ML, part of the NO desorbs in a peak near 440 K, implying that dissociation becomes hindered by the increased occupation of the surface. Simultaneous with NO desorption at ~440 K there is a peak for desorption of N_2. Similar trends are observed in thermal desorption spectra for NO on a Rh(1 1 1) surface [31].

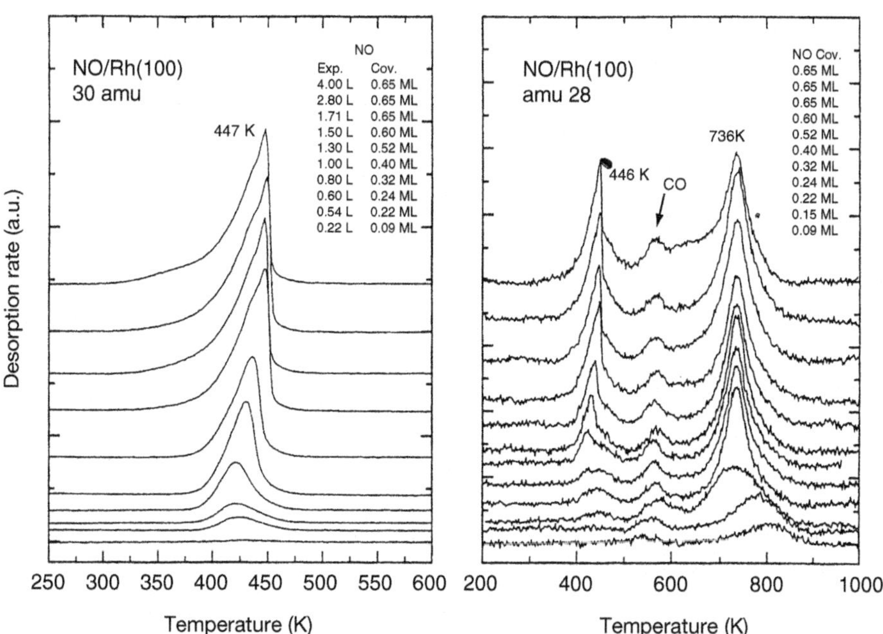

FIGURE 9.1 Thermal desorption spectra of NO (m/e = 30, left panel) and N_2 (m/e = 28, right panel) obtained after exposing the Rh(1 0 0) crystal to various doses of NO (given in langmuir, L) at 100 K. (From A Siokou, RM van Hardeveld, and JW Niemantsverdriet. *Surf. Sci.* 402–404: 110, 1998. With permission.)

The first peak for N_2 desorption is attributed to a disproportionation reaction with a N—NO intermediate [13, 28]:

$$NO_{ads} + N_{ads} \longrightarrow N\text{—}NO_{ads} \longrightarrow O_{ads} + N_{2,\ gas} \tag{9.1}$$

The line shape of the TDS spectra strongly suggests that the rate of catalytic reduction of NO is determined by the mechanism of N_2 formation on the surface [13].

Several studies have investigated the reaction between NO and CO on rhodium surfaces using different techniques and conditions [13–15, 32, 33]. The Arrhenius plots for NO reduction by CO on Rh(1 0 0) and Rh(1 1 1) are shown in Figure 9.2 for equal partial pressures of the reactants [15]. Significantly different activation energies (Rh(1 1 1): 29 kcal/mol; Rh(1 0 0): 24 kcal/mol) and specific activities are observed on the two surfaces, indicating that this reaction is sensitive to the structure of the catalyst surface [15]. A similar conclusion can be reached after comparing the behavior of the {NO + CO} reaction on Rh(1 1 0) and Rh(1 1 1), with the reaction occurring 1.3 to 6.3 times faster on the Rh(1 1 0) surface [33]. This has been attributed to a slightly more facile NO dissociation process on the more open (1 1 0) surface [33]. Furthermore, the {NO + CO} reaction exhibits substantially different kinetic behavior on a high surface area Rh/Al_2O_3 catalyst and on Rh(1 1 1) [34].

The dependence of the {NO + CO} reaction rate on the partial pressures of the reactants is shown in Figure 9.3 [15]. The data indicate little, if any, variation of the rate with changing

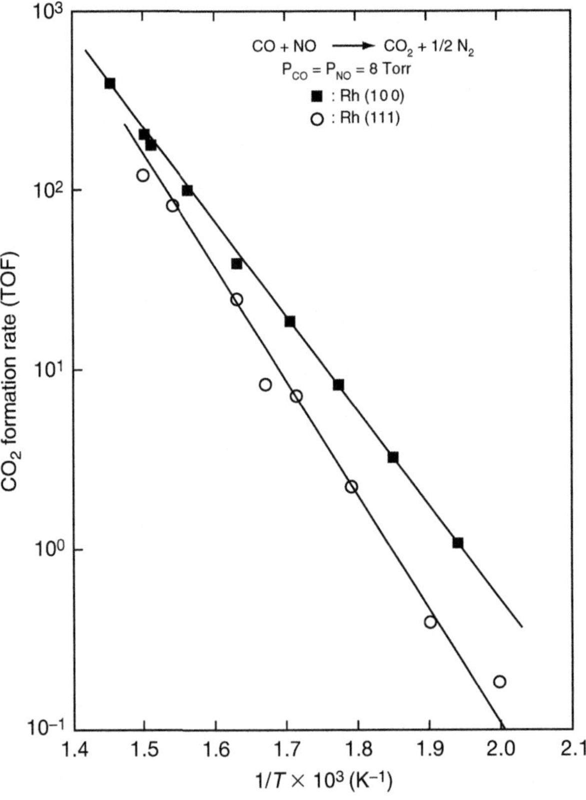

FIGURE 9.2 Arrhenius plot of the CO_2 formation rate from the {NO + CO} reaction on Rh(1 1 1) and Rh(1 0 0). $P_{CO} = P_{NO} = 8$ Torr. (From CHF Peden, DW Goodman, DS Blair, PJ Berlowitz, GB Fischer, and SH Oh. *J. Phys. Chem.* 92: 1563, 1988. With permission.)

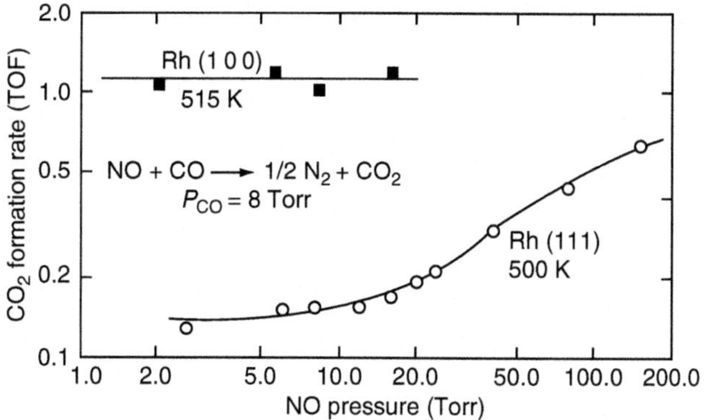

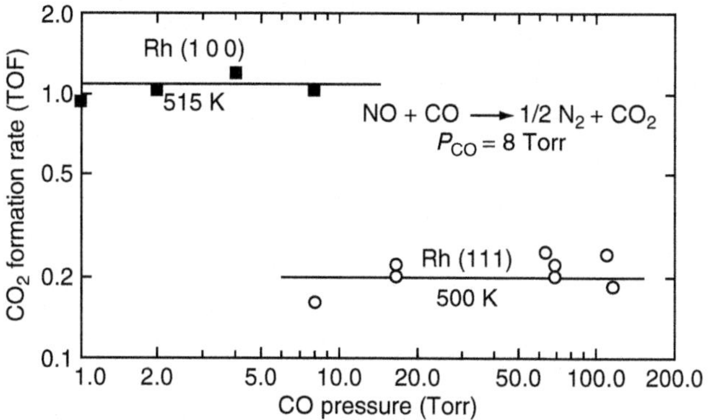

FIGURE 9.3 NO and CO partial pressure dependencies of the {NO + CO} reaction on Rh(1 1 1) and Rh(1 0 0) catalysts. (From CHF Peden, DW Goodman, DS Blair, PJ Berlowitz, GB Fischer, and SH Oh. *J. Phys. Chem.* 92: 1563, 1988. With permission.)

CO pressures on either surface. Similar NO pressure independence is observed over the two surfaces for NO pressures below 20 Torr. On Rh(1 1 1), the reaction rate becomes positive order at higher NO partial pressures. The kinetic data are usually interpreted in terms of the following set of elementary processes [13, 33, 34]:

$$CO_{gas} \longrightarrow CO_{ads} \tag{9.2}$$

$$NO_{gas} \longrightarrow NO_{ads} \tag{9.3}$$

$$NO_{ads} \longrightarrow N_{ads} + O_{ads} \tag{9.4}$$

$$NO_{gas} + N_{ads} \longrightarrow ON{-}N_{ads} \longrightarrow N_{2,\,gas} + O_{ads} \tag{9.5}$$

$$2N_{ads} \longrightarrow N_{2,\,gas} \tag{9.6}$$

$$NO_{ads} + N_{ads} \longrightarrow N_2O_{gas} \tag{9.7}$$

$$CO_{ads} + O_{ads} \longrightarrow CO_{2,\,gas} \tag{9.8}$$

The relative importance of the steps (9.4) to (9.6) seems to change depending on the experimental conditions and the surface of rhodium used [13, 33]. Figure 9.4 shows specific rates of CO_2, N_2O, and N_2 formation for the {NO + CO} reaction over Rh(1 1 1) as a function of the

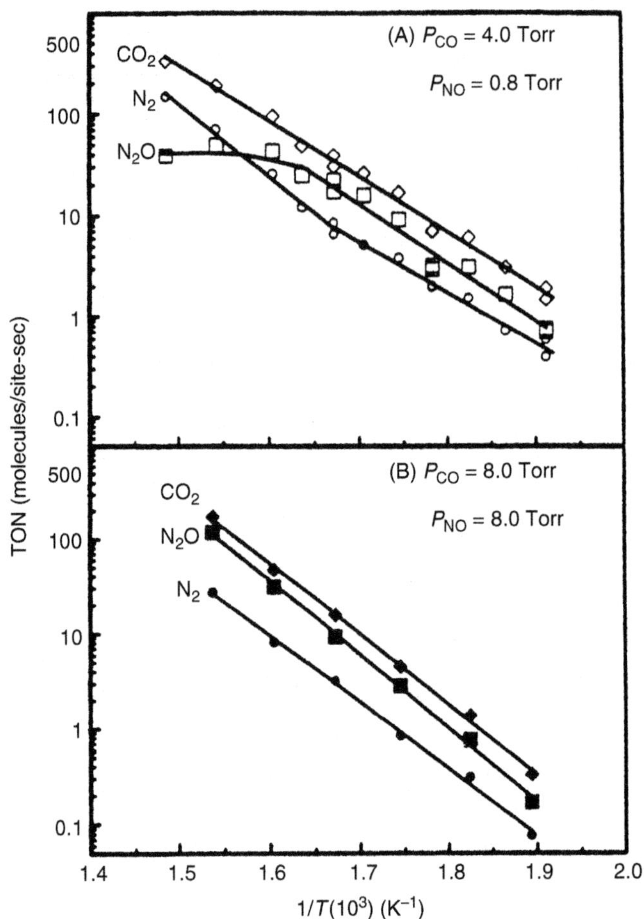

FIGURE 9.4 Specific rates of CO_2, N_2O, and N_2 formation for the {NO + CO} reaction over Rh(1 1 1) as a function of $(1/T)$ using: (A) $P_{NO} = 0.8$ Torr and $P_{CO} = 4$ Torr and (B) $P_{NO} = P_{CO} = 8$ Torr. (From H Permana, KYS Ng, CHF Peden, SJ Schmieg, DK Lambert, and DN Belton. *J. Catal.* 164: 194, 1996. With permission.)

inverse temperature [32]. *In situ* infrared measurements indicate that changes in the surface coverages of NO and CO correlate well with changes in N_2O selectivity [32]. Below 635 K, where N_2O formation is favored, NO dominates on the surface. Above 635 K, where N_2 formation is preferred, CO is the majority surface species [32].

Based on the recent kinetic experiments with {NO + CO} molecular beams on Rh(1 1 1), a model has been proposed for the mechanism of nitrogen monoxide reduction that involves the growth of atomic nitrogen surface islands [13, 35–37]. It is suggested that there is a preferential reaction between the N atoms on the edges of those islands and new incoming NO molecules to form N—NO intermediates, which subsequently dissociate to produce molecular nitrogen. In Monte Carlo simulations it was found that an Eley–Rideal mechanism that includes a NO(gas) + N(ads) $\longrightarrow$ N_2(gas) + O(ads) step is absolutely necessary to be able to sustain a steady-state catalytic regime [38].

Under some conditions of temperature and pressure, oscillatory behavior has been observed for the reduction of NO by carbon monoxide, hydrogen, or ammonia on rhodium surfaces [26, 39, 40]. This phenomenon has been studied in more detail on platinum surfaces and will be examined in the next section.

9.2.1.2 NO Chemistry on Pt Surfaces

In general, the dissociation of NO on platinum surfaces is more difficult than on rhodium surfaces [26]. For the {NO + CO} reaction, platinum catalysts are less active than rhodium catalysts [39, 40]. Previous studies for the {NO + CO} reaction on Pt(1 1 1) and (1 1 0) surfaces concluded that the reaction proceeds by a Langmuir–Hinshelwood mechanism [41]. A similar mechanism involving reactions (9.2) to (9.4), (9.6), and (9.8) has been proposed in more recent studies for the Pt(1 0 0) surface [42], with the dissociation of NO according to reaction [4] as the rate-limiting step. At low pressures, N_2O production is negligible compared to CO_2 or N_2 production [42].

Kinetic oscillations for the {NO + CO} reaction have been seen on supported Pt catalysts [43] and on Pt(1 0 0) surfaces [26, 40, 42, 44, 45]. On Pt(1 0 0), the reaction of NO and CO exhibits two branches of steady-state production with the occurrence of kinetic oscillations [42]. The structure of the surface during the oscillations was monitored with LEED (see Figure 9.5). The clean Pt(1 0 0) surface is reconstructed under ultrahigh vacuum conditions into a hexagonal (hex) surface [17, 47]. CO adsorption on the Pt(1 0 0)–(hex) surface occurs without inducing a (hex) $\longrightarrow$ (1 $\times$ 1) back reconstruction [17, 48]. On the other hand, NO adsorption immediately lifts the (hex) reconstruction [17, 49]. For the {NO + CO} reaction, the (1 $\times$ 1) $\longrightarrow$ (hex) transition of the surface structure was observed to take place *only* in one

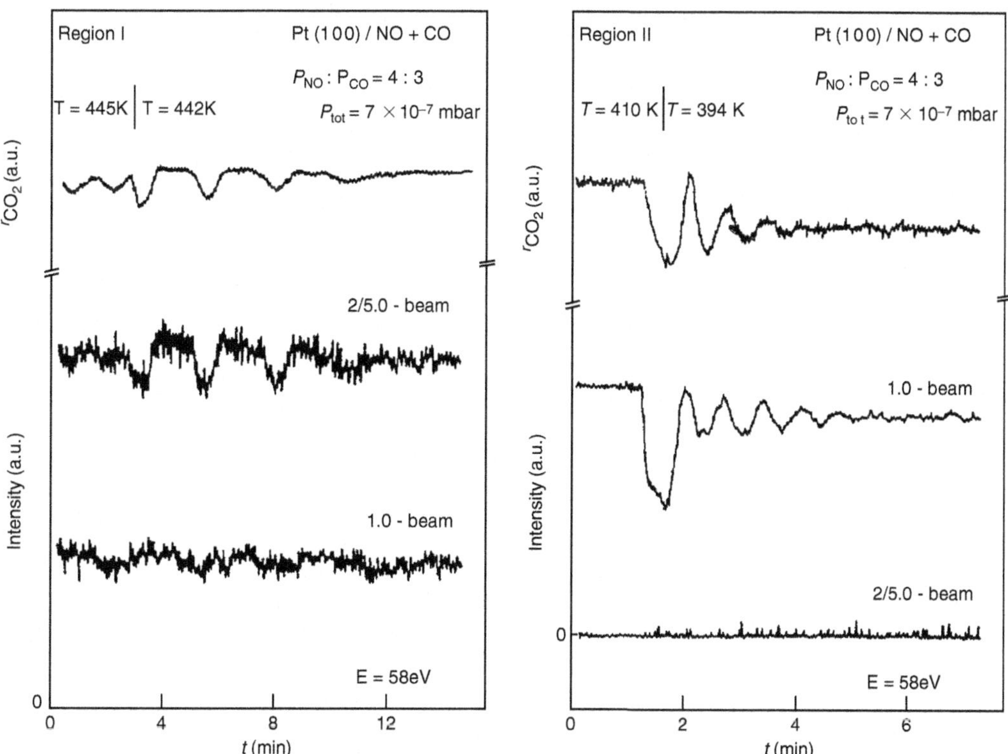

FIGURE 9.5 Changes in surface structure during kinetic oscillations as monitored by video LEED. Region I (left-side panel): Changes in hex structure intensity occur in parallel to the reaction-rate oscillations indicating the participation of the (hex) $\rightleftarrows$ (1 $\times$ 1) phase transition in the oscillation mechanism. Region II (right-side panel): The LEED intensity of the (hex) structure is always zero; hence the phase transition (hex) $\rightleftarrows$ (1 $\times$ 1) is not involved in the mechanism of the oscillations. (From T Fink, JP Dath, R Imbihl, and G Ertl. *J. Chem. Phys.* 95: 2109, 1991. With permission.)

of the two regions of reaction rate oscillations (see Figure 9.5) [42]. This implies that the $(1 \times 1) \longrightarrow$ (hex) transition is of minor relevance to the mechanism responsible for the oscillations. In the region where the (1×1) structure is maintained, studies of photoemission electron microscopy (PEEM) show that the oscillations proceed unsynchronized without exhibiting macroscopic rate variations [45]. Instead, there is spatiotemporal pattern formation. The patterns seem to be dominated by periodic wave trains, which become unstable at lower temperatures, giving rise to spiral waves and irregularly shaped reaction fronts [45]. The experiments were modeled by a set of coupled differential equations [42]. The occurrence of oscillations can be rationalized in terms of a periodic sequence of autocatalytic "surface explosions" and the restoration of an adsorbate-covered surface [42].

Kinetic oscillations have been observed for the $\{NO + H_2\}$ reaction on Pt(1 0 0) and polycrystalline Pt surfaces [40, 50]. Typical results for Pt(1 0 0) are shown in Figure 9.6 [50]. All three N-containing reaction products, N_2, NH_3, and N_2O, are formed via dissociation of NO according to reaction (9.4), followed by reactions (9.6) and (9.7), and the hydrogenation of N adatoms into ammonia [50]. The $\{NO + H_2\}$ reaction has also been studied on Pt(1 1 1) and Pt(1 1 0), but oscillations were not found. In fact, the reaction rates over these surfaces were much lower than those over Pt(1 0 0) as should be expected on the basis of the corresponding activities for NO dissociation [40]. Kinetic oscillations have also been reported for the $\{NO + NH_3\}$ reaction on a Pt(1 0 0) substrate [40]. Theoretical studies have provided possible explanations for the oscillatory behavior of the $\{NO + H_2\}$ and $\{NO + NH_3\}$ reactions on the Pt(1 0 0) surface [50–52]. The availability of a vacant site near the adsorbed NO and N may play a central role in the selectivity toward N_2O and N_2 formation.

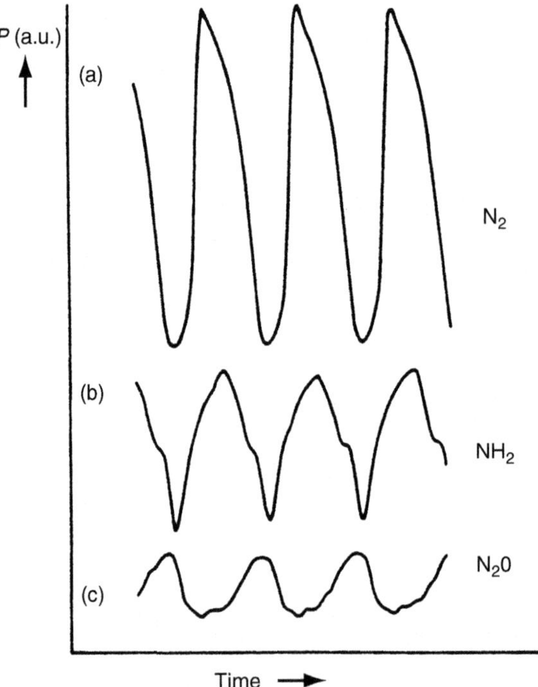

FIGURE 9.6 Variations of the N_2, NH_3, and N_2O partial pressures during the $\{NO + H_2\}$ reaction over Pt(1 0 0) as a function of time. $T = 460\,K$, $p(NO)$ and $p(H_2) = 3 \times 10^{-6}$ mbar. (From PD Cobden, J Siera, and BE Nieuwenhuys. *J. Vac. Sci. Technol. A* 10: 2487, 1992. With permission.)

9.2.1.3 NO Chemistry on Pd Surfaces

Like platinum, palladium is a less active catalyst for the {NO + CO} reaction than rhodium [39]. However, among the metals used in the three-way catalysts, palladium is the best for accelerating the rate of the {NO + H_2} reaction [39]. Most of the surface science studies dealing with these two reactions have been done on single-crystal faces of Rh and Pt [13, 16, 40, 46]. The ability of a Pd surface to dissociate NO is highly structure-sensitive [21]. On a perfect Pd(1 1 1) surface, there is no dissociation of NO [53]. Molecular adsorption and dissociation strongly compete on Pd(1 0 0) and Pd(1 1 0) surfaces [54, 55]. The degree of dissociation is much higher on stepped Pd(1 1 1), Pd(1 1 2), Pd(3 3 1), and Pd(3 2 0) surfaces [21, 53, 56, 57]. The NO decomposition products are N and O adatoms, which form N_2 or N_2O through reactions (9.1), (9.6), and (9.7).

Recent studies have examined the interaction of NO with Pd nanoparticles (sizes ranging from 2.8 to 45 nm) supported on a well-defined MgO(1 0 0) substrate [21]. Above room temperature NO interacts weakly with the oxide support. The results of molecular beam experiments indicate that the physisorbed molecules can diffuse toward the Pd nanoparticles and become chemisorbed [21]. Once chemisorbed, the NO molecules either dissociate into N and O adatoms, or desorb. The activation energy for desorption is close to those found on Pd(1 1 1) and Pd(1 0 0). This agreement is not surprising since Pd particles epitaxially grown on MgO(1 0 0) mainly exhibit (1 1 1) and (1 0 0) facets [21]. The dissociation of NO on the supported palladium leads to the formation of N_2 (by association of two nitrogen adatoms), without production of N_2O. From the transient kinetics of N_2 desorption, it was concluded that strongly bound nitrogen adatoms coexist on the surface of the Pd nanoparticles with loosely bound nitrogen species [21].

9.2.2 Adsorption and Reactions of N_2O and NO_2 on Metals

Nitrous oxide is an AB_2 triatomic molecule with an asymmetric shape (NNO). It is a closed shell system (i.e., not a radical molecule like NO), but the energy necessary for the cleavage of the N—O bond is relatively small, ~40 kcal/mol [10]. Little work has been done concerning the adsorption and decomposition of nitrous oxide on transition metal surfaces [58–65]. The molecule dissociates upon adsorption on Ni(1 0 0) and Ni(1 1 0) at 300 K [59, 60], yielding gaseous N_2 and adsorbed atomic oxygen. Molecular adsorption on nickel was found at temperatures below 200 K [59]. On Ru(0 0 0 1) at 100 K, N_2O dissociates at small exposures and adsorbs molecularly at high exposures [61]. On the more open Ru(1 0 $\bar{1}$ 0) surface, nitrous oxide adsorbs dissociatively with both atomic nitrogen and oxygen chemisorbed [62]. No evidence for N_2O decomposition on Pt(1 1 1) was found at temperatures below 1000 K [63]. In this system, to induce N_2O dissociation, the surface temperature must be elevated above 1000 K [64], or the adsorbed layer must be irradiated with photons [65]. A clear structural effect has been observed for the adsorption of N_2O on rhodium surfaces [58]. On Rh(1 1 0), nitrous oxide adsorbs at room temperature and decomposes into gaseous N_2 and O adatoms. On the other hand, Rh(1 1 1) is rather inert to the incoming N_2O, and little adsorption and decomposition was found on this surface even when the dosed gas was heated up to 900 K [58].

Nitrogen dioxide is a radical and a strong oxidizing agent [66]. Compared to NO, relatively few studies have been focused on the interaction of NO_2 with surfaces of transition metals [66–77]. In inorganic complexes, NO_2 can be present as a ligand that is bonded through the nitrogen end (η^1-N), the oxygen atoms (η^2-O,O), or a combination of nitrogen and oxygen (η^2-N,O) (see Figure 9.7). All of these bonding configurations have been observed when NO_2 is adsorbed on metals: η^1-N bonding on Ru(0 0 0 1) (67), η^2-O,O bonding on Ag(1 1 0) [69] and Au(1 1 1) [67], and η^2-N,O bonding on Pt(1 1 1) (71) and Pd(1 1 1) [76]. NO_2 displays a rich chemistry on metal surfaces. In principle, the molecule can undergo partial ($NO_2 \longrightarrow NO + O$) or full decomposition ($NO_2 \longrightarrow N + 2O$), or it can react with

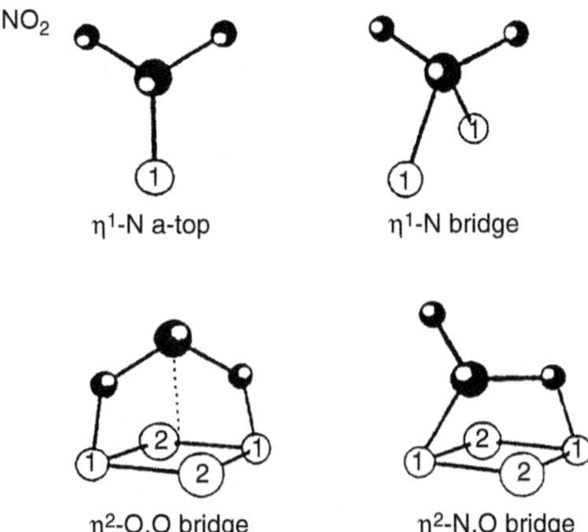

FIGURE 9.7 Bonding configurations of NO_2 on metal surfaces. (From T Jirsak, M Kuhn, and JA Rodriguez. *Surf. Sci.* 457: 254, 2000. With permission.)

O present on the surface to form nitrate species ($NO_2 + O \longrightarrow NO_3$). At 300 K, nitrogen dioxide dissociates to adsorbed NO and O on Ru(0 0 0 1) [67], Rh(1 1 1) [70], Pd(1 1 1) [76], Pt(1 1 1) [71], Ag(1 1 1) [75], and Ag(1 1 0) [69]. Mo(1 1 0) [66] and W(1 1 0) [73] completely dissociate NO_2 into N and O atoms at room temperature. Molecular adsorption of NO_2 has been observed on Pd(1 1 1) [76] and Pt(1 1 1) [71] at 100 to 150 K. At these low temperatures, Ru(0 0 0 1), Rh(1 1 1), Ag(1 1 1), Ag(1 1 0), and Mo(1 1 0) partially dissociate the molecule (i.e., adsorbed NO as an intermediate), whereas W(1 1 0) is still able to induce full decomposition (i.e., N and O only products). In general, NO_3 formation by reaction of NO_2 with O adatoms is difficult on transition metal surfaces [67, 70, 71, 76]. Substantial amounts of NO_3 formed by this reaction pathway have only detected on Ag(1 1 0) [69] and polycrystalline Zn [77]. The behavior observed for the NO_2/Au(1 1 1) system is unique in the sense that nitrogen dioxide does not dissociate on this substrate [74].

The dosing of NO_2 at 500 to 700 K yields very large amounts of adsorbed oxygen plus gaseous NO or N_2, or both, on Ru(0 0 0 1), Rh(1 1 1), Pd(1 1 1), Pt(1 1 1), Ag(1 1 1), and Mo(1 1 0) surfaces [66, 68, 70, 76]. In this respect, NO_2 is a much better oxidizing agent than molecular oxygen. This correlates well with the difference in the strength of the N—O ($NO_{2,gas} \longrightarrow NO_{gas} + O_{gas}$, $\Delta H = 73$ kcal/mol) and O—O bonds ($O_{2,gas} \longrightarrow 2O_{gas}$, $\Delta H = 119$ kcal/mol) in these species [66]. For the cases of Mo(1 1 0), Ru(0 0 0 1), and Pd(1 1 1), the oxidation goes well beyond the surface and films of MoO_2, RuO_x, and PdO_x are formed [66, 68, 76].

9.3 ADSORPTION AND REACTION OF NO_x MOLECULES ON OXIDE SURFACES

Oxides are frequently utilized in $DeNO_x$ operations [1, 2, 4–9]. Traditionally, oxides were used as sorbents for trapping NO_x molecules [1, 4, 9]. But they can also act as active catalysts, or as "supports" for metal catalysts [5, 12]. In this respect, two classical examples are metal-doped MgO, a catalyst for the decomposition of N_2O [10], and the important role of CeO_2 or ZrO_2–CeO_2 mixtures for the efficient performance of the three-way catalysts [2, 6, 8]. In principle, the behavior of a NO_x molecule on an oxide surface is usually complex, since the adsorbate can interact with the metal cations, where N—O bond cleavage may occur, or the interaction could be with the O centers producing stable nitrite or nitrate species. Furthermore, the metal elements are able to form a large diversity of oxide compounds [12, 78]. These can adopt a vast number of structural geometries, and at an electronic level they can exhibit metallic character or behave as semiconductors or insulators [78]. It is well documented that the chemical reactivity of a metal oxide is affected by the size of its bandgap and the degree of ionicity in its metal–oxygen bonds [79]. These electronic properties can be modified after doping an oxide with a second metal [10, 79], and mixed-metal oxide catalysts are frequently used in industrial applications [2, 5, 12]. Obtaining a fundamental knowledge of the factors that control the chemical properties of oxide surfaces is a challenge for modern science and a prerequisite for the rational design of $DeNO_x$ catalysts. In the last 5 to 10 years, there have been significant advances in methods for the preparation and control of well-defined oxide surfaces [20, 79, 80, 81]. These advances have made possible a better understanding of the chemistry of NO_x molecules on oxide surfaces.

9.3.1 NO Chemistry on Oxide Surfaces

The adsorption of NO has been investigated on MgO(1 0 0) [20, 82], metal-doped MgO(1 0 0) [10, 83], NiO(1 0 0) [20], TiO_2(1 1 0) [84], Cr_2O_3(0 0 1) [10], Cr_2O_3(1 1 1) [85], CeO_2(1 1 1) [19], YSZ(1 0 0) [86], and $SrTiO_3$(1 0 0) [87]. NO bonds weakly to most of these surfaces. A highly ionic system like MgO(1 0 0) displays the smallest NO adsorption energy, ~5 kcal/mol [20]. On the other hand, systems that contain metal–oxygen bonds that are not fully ionic, Cr_2O_3(1 1 1) or $SrTiO_3$(1 1 0), interact better with NO [85, 87].

In both TiO_2 and $SrTiO_3$, the titanium atoms have a *formal* oxidation state of +4. However, theoretical calculations indicate that the positive charge for Ti in $SrTiO_3$ is almost half of that found in TiO_2 [87, 88]. The strong metal↔oxygen↔metal interactions that drastically modify the electron density of the Ti atoms in $SrTiO_3$ have a substantial impact in the chemical reactivity of these metal cations. Experiments of TDS show that at small coverages, NO desorbs at much higher temperatures from $SrTiO_3$(1 0 0), ~260 K, than from TiO_2(1 1 0), ~130 K [87]. From these experiments, one can estimate NO adsorption energies of 16 kcal/mol on $SrTiO_3$(1 0 0) and 8 kcal/mol on TiO_2(1 1 0). Recent theoretical studies for a series of ABO_3 perovskites (A = Ca, Sr, Li, K, and Na; B = Ti, Zr, Nb) show very strong metal↔oxygen↔metal interactions that drastically modify the electron density of the transition-metal cations (B) with respect to TiO_2, ZrO_2, and NbO_2 [88]. Thus, when dealing with the design or performance of ABO_3 perovskites as $DeNO_x$ catalysts, a simple extrapolation of the catalytic properties of the individual AO and BO_2 oxides may not be a reliable approach [89].

Figure 9.8 displays the valence photoemission spectra for pure MgO(1 0 0), and a series of $TM_xMg_{1-x}O$(1 0 0) systems containing a small amount (x = 0.06–0.07) of Zn, Ni, Fe, and Cr [90]. Doping induces the appearance of occupied electronic states *above* the valence band of MgO. The position of these states depends on the nature of the metal element. In the case of $Zn_{0.06}Mg_{0.94}O$(1 0 0), the new states almost overlap with the valence band of the host oxide. On the other hand, for $Cr_{0.07}Mg_{0.93}O$(1 0 0), the Cr-induced states are located well above the MgO valence band [90]. As we will see below, the differences in the electronic properties of MgO(1 0 0) and $TM_xMg_{1-x}O$(1 0 0) have an impact on the chemical properties of these oxides.

The bottom panel in Figure 9.9 shows NO-TDS data acquired after saturating the surface of MgO(1 0 0) with NO at 80 K [83]. A small amount of NO (~0.2 ML) is present on imperfections of the MgO(1 0 0) surface, has an adsorption energy of ~8 kcal/mol, and desorbs around 140 K. Figure 9.9 also displays the TDS data for NO/$Ni_{0.06}Mg_{0.94}O$(1 0 0) and NO/$Cr_{0.07}Mg_{0.93}O$(1 0 0) systems. Photoemission measurements indicate that NO does not dissociate on these surfaces [10]. The NO-TDS spectra for NO/$Ni_{0.06}Mg_{0.94}O$(1 0 0) have a broad peak near 240 K that can be assigned to Ni-bonded NO [83]. An adsorption energy of ~15 kcal/mol can be estimated from the TDS data for NO on Ni sites of $Ni_{0.06}Mg_{0.94}O$(1 0 0).

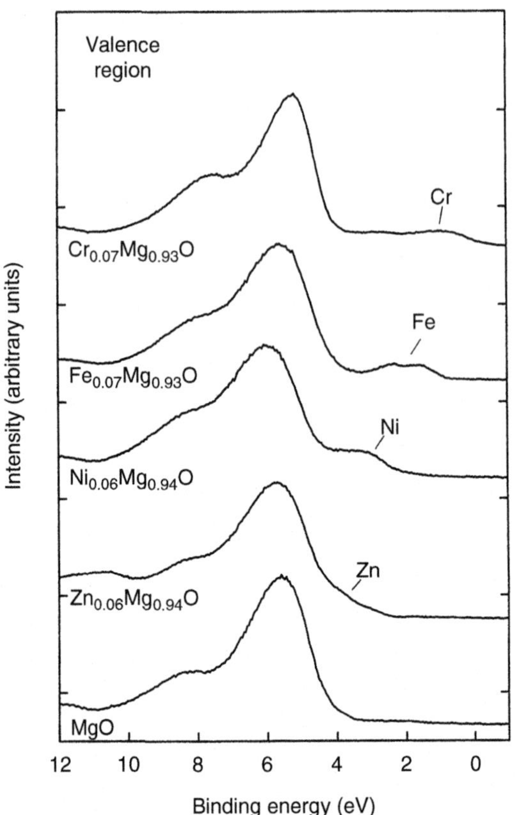

FIGURE 9.8 Valence photoemission spectra for MgO(1 0 0) and a series of $TM_xMg_{1-x}O(1\ 0\ 0)$ systems. (From JA Rodriguez, T Jirsak, L González, J Evans, M Perez, and A. Maiti. *J. Chem. Phys.* 115: 10914, 2001. With permission.)

An even larger NO adsorption energy is found for Cr sites of $Cr_{0.07}Mg_{0.93}O(1\ 0\ 0)$. The NO-TDS spectra for the NO/$Cr_{0.07}Mg_{0.93}O(1\ 0\ 0)$ system (top panel in Figure 9.9) exhibit features around room temperature that correspond to Cr-bonded NO and are associated with an adsorption energy of ~20 kcal/mol [83].

A comparison of the results in Figures 9.8 and 9.9 points to a strong link between the electronic properties of the $TM_xMg_{1-x}O(1\ 0\ 0)$ systems and the NO adsorption energy. An identical result is found in DF slab calculations [10]. The extent of band(oxide)–orbital(adsorbate) mixing varies substantially from one dopant to another [10]. This can be explained using simple models for adsorbate bonding [25, 79]. For a radical molecule like NO, one can get an approximate expression for the bonding energy (Q) derived from the interaction between the lowest unoccupied molecular orbital (LUMO) of the adsorbate and the occupied states (OS) of an oxide

$$Q\alpha(E_{\rm OS} - E_{\rm LUMO}) \tag{9.9}$$

where $E_{\rm OS}$ and $E_{\rm LUMO}$ are the energies of the oxide occupied states and the adsorbate LUMO, respectively. Clearly, there is a relationship between the energy position of the dopant electronic levels and the chemical reactivity of a mixed-metal oxide. This relationship is not linear, because one must also consider the number of states with low stability provided by a dopant agent [10]. In addition, several factors can affect the strength of a metal–NO

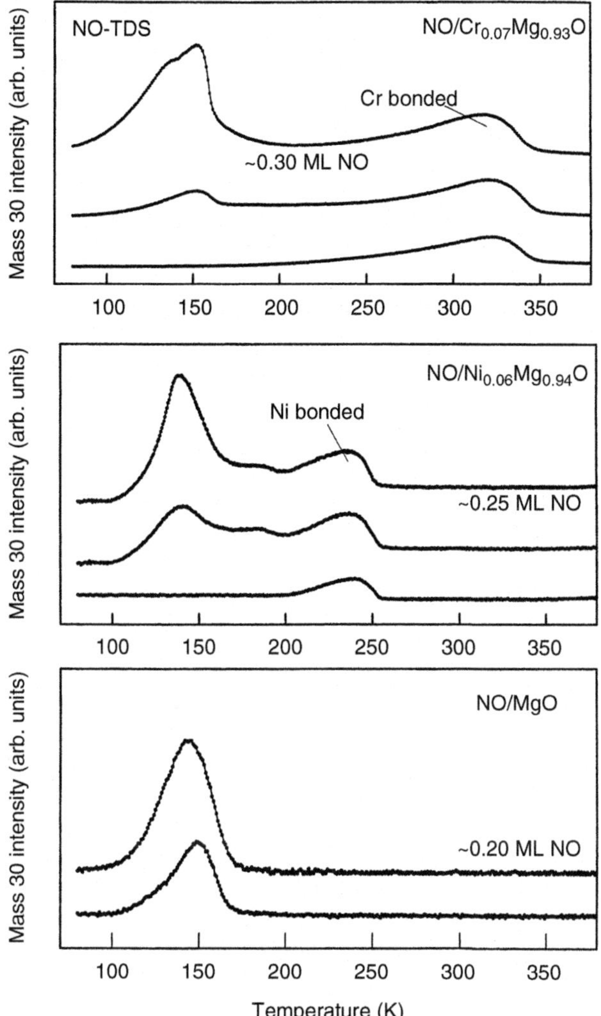

FIGURE 9.9 NO thermal desorption spectra for MgO(1 0 0), $Ni_{0.06}Mg_{0.94}O$(1 0 0), and $Cr_{0.07}Mg_{0.93}O$(1 0 0). The NO molecule was dosed at 80 K. In the final step of each experiment the surface was saturated with NO. (From JA Rodriguez, T Jirsak, M Pérez, L González, and A Maiti. *J. Chem. Phys.* 114: 4186, 2001. With permission.)

bond [25, 91, 92]. As we will see below, Equation (9.9) also applies when dealing with the adsorption of NO_2. In general, an important parameter to consider when designing a mixed-metal oxide catalyst is the relative position of the electronic states introduced by the dopant agent [79].

The chemistry of NO on a metal oxide surface is very sensitive to the presence of defects and O vacancies [19, 20, 82, 84, 87, 93]. Figure 9.10 shows N 1s photoemission spectra for the adsorption of NO at 80 K on stoichiometric MgO(1 0 0) and on a surface with O vacancies, MgO_{1-x} [93]. On MgO(1 0 0), a single peak is observed that corresponds to adsorbed NO. In contrast, the spectrum for the MgO_{1-x} surface shows the coexistence of NO and N_2O. The nitrous oxide is formed by interaction of NO with defect sites of MgO [82, 93]. O vacancies also favor the transformation of NO into N_2O on CeO_2(1 1 1) [19], TiO_2(1 1 0) [84], and $SrTiO_3$(1 0 0) [87]. The adsorption of NO on defective MgO_{1-x}(1 0 0) and $SrTiO_{3-x}$(1 0 0)

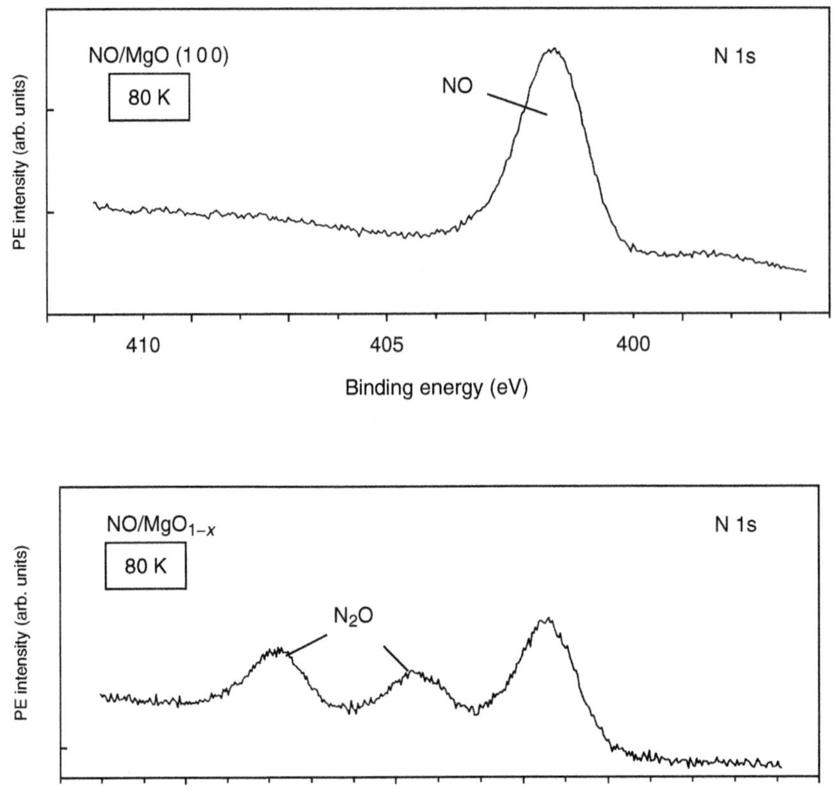

FIGURE 9.10 N 1s photoemission spectra for the adsorption of NO at 80 K on stoichiometric MgO(1 0 0) and on a surface with O vacancies, MgO_{1-x}. (From JA Rodriguez, T Jirsak, JY Kim, JZ Larese, and A Maiti. *Chem. Phys. Lett.* 330: 475, 2000. With permission.)

have been examined with density functional calculations [87, 94]. These studies show a similar mechanism for the formation of N_2O. Initially an NO molecule adsorbs on a vacancy site through its O end (see top of Figure 9.11). A substantial charge is transferred from the oxide to the NO molecule, and there is a large (~0.2 Å) elongation in the N—O bond. The dissociation of this bond is facilitated by the addition of a second NO molecule and formation of an ON—NO dimer (bottom of Figure 11). In essence, O vacancies are removed through a $2NO_{ads} \longrightarrow O_{ads} + N_2O_{ads}$ reaction [87, 94]. The removal of the O vacancies in the oxide surface can be followed with core and valence level photoemission [82, 87, 93].

9.3.2 N_2O Chemistry on Oxide Surfaces

Figure 9.12 shows the N 1s spectra for the adsorption of the same amount of N_2O on MgO (1 0 0), $Zn_{0.06}Mg_{0.94}O$(1 0 0), $Cr_{0.08}Mg_{0.92}O$(1 0 0), and Cr_2O_3(0 0 0 1) at 80 K [10]. The top spectrum displays the two peaks characteristic of adsorbed N_2O. The MgO(1 0 0) surface was not able to dissociate N_2O and the molecule desorbed intact upon heating to 200 K [10]. No dissociation was observed upon N_2O adsorption on $Zn_{0.06}Mg_{0.94}O$(1 0 0) at 80 K, but during heating from 80 to 250 K a signal for desorption of N_2 was detected. In the case of N_2O/$Cr_{0.08}Mg_{0.92}O$(1 0 0) and N_2O/Cr_2O_3(0 0 0 1), a third N 1s peak appears near 399 eV that can be assigned to N_x species produced by decomposition of N_2O on Cr centers of these oxide

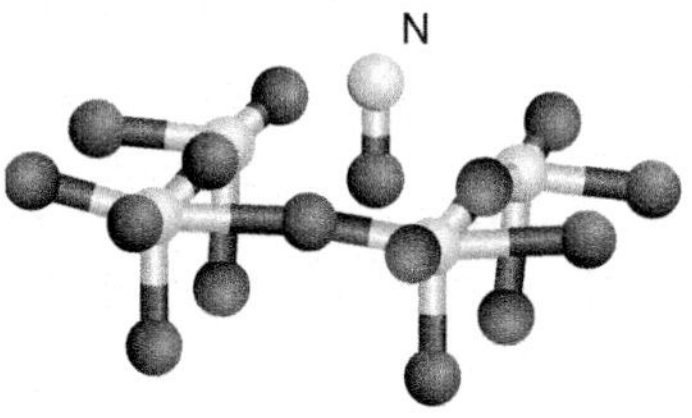

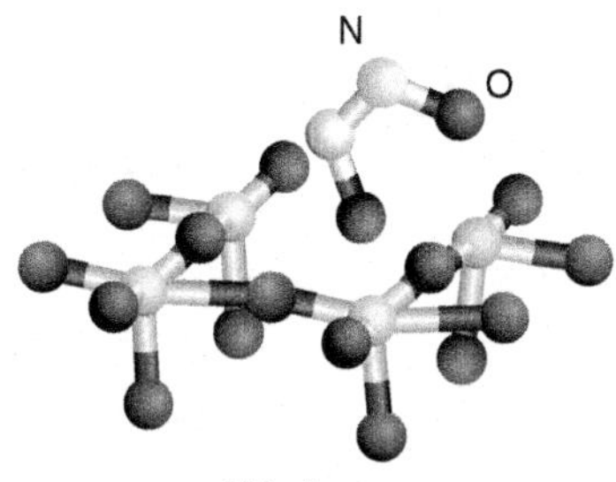

FIGURE 9.11 Adsorption geometries for NO (top) and ON—NO (bottom) on a reduced $SrTiO_{3-x}$(1 0 0) surface. (From JA Rodriguez, S Azad, LQ Wang, J García, A Etxeberria, and L González. *J. Chem. Phys.* 118: 6562, 2003. With permission.)

surfaces at 80 K [10]. Adsorption and reaction studies indicate that the activity of the mixed-metal oxides for breaking N—O bonds in N_2O increases in the following sequence [10]: MgO(1 0 0) $<$ $Zn_{0.06}Mg_{0.94}O$(1 0 0) $<$ $Cr_{0.07}Mg_{0.93}O$(1 0 0). This trend agrees qualitatively with studies for the catalytic decomposition of N_2O at 870 to 1170 K (for example, see Figure 9.13), which show $Cr_xMg_{1-x}O$ as a very good catalyst and pure MgO as a relatively poor catalyst [10]. The trends in Figure 9.12 and Figure 9.13 reflect very well variations in the electronic properties of the oxides (see Figure 9.8). Since the $Cr_xMg_{1-x}O$ has occupied electronic states located well above the valence band of MgO, it interacts better with the unoccupied orbitals of N_2O which are N—O antibonding [10, 79]. This point will be discussed in more detail below when comparing the behavior of N_2O, NO, and NO_2.

9.3.3 NO_2 Chemistry on Oxide Surfaces

NO_2 is more reactive than NO or N_2O. On metal surfaces, adsorbed NO_2 usually undergoes partial ($NO_2 \longrightarrow NO + O$) or complete dissociation ($NO_2 \longrightarrow N + 2O$). These decomposition processes are not so easy on oxides due to the ionicity of these compounds. The positive charge on metal cations in general makes an electron transfer into the unoccupied orbitals of NO_2 and the subsequent breaking of N—O bonds difficult [10, 11, 95, 96]. On oxides, adsorbed NO_2 may be a consequence of the adsorption of NO_2 on metal cations or the reaction of NO with O sites [9–11]. The reaction of NO_2 with O sites can produce NO_3, a species that can also be a consequence of the disproportionation of two NO_2 molecules ($2NO_2 \longrightarrow NO_3 + NO$) on metal cations [96]. The nitrogen oxides (NO, NO_2, and NO_3) are neither strong Lewis acids nor bases, and a combination of electron transfer (redox) and acid–base interactions can occur during the coadsorption of NO_2 with NO or NO_3 on oxides [11, 95]. The net effect of these adsorbate–adsorbate interactions is an increase in the adsorption energy of the NO_x species [11, 95].

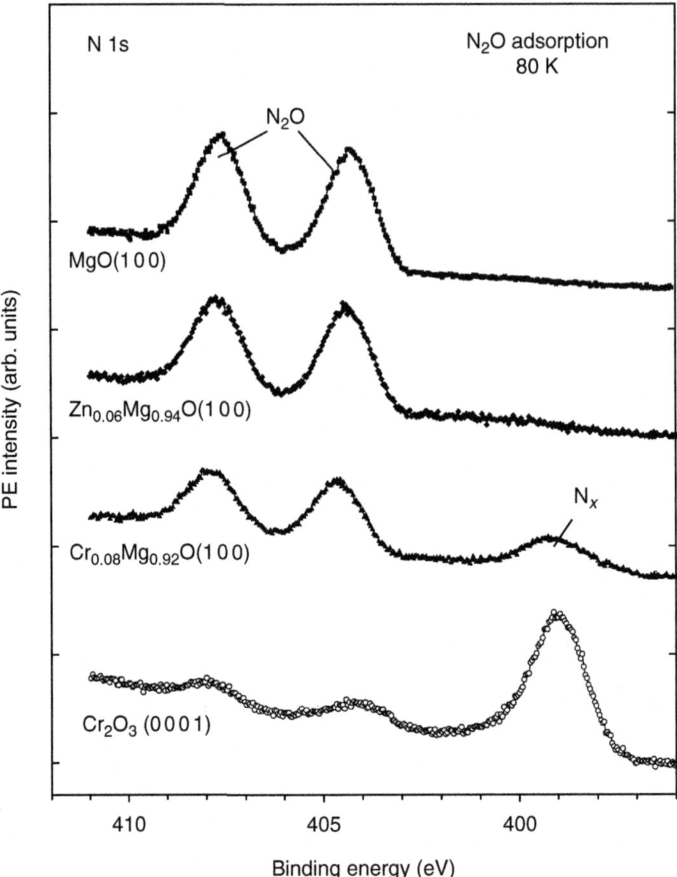

FIGURE 9.12 N 1s photoemission spectra recorded after adsorbing ~0.15 ML of N_2O on MgO(1 0 0), $Zn_{0.06}Mg_{0.94}O$(1 0 0), $Cr_{0.07}Mg_{0.93}O$(1 0 0), and Cr_2O_3(0 0 0 1) surfaces at 80 K. (From JA Rodriguez, M. Pérez, T. Jirsak, L. González, A. Maiti, and JZ Larese. *J. Phys. Chem. B* 105: 5497, 2001. With permission.)

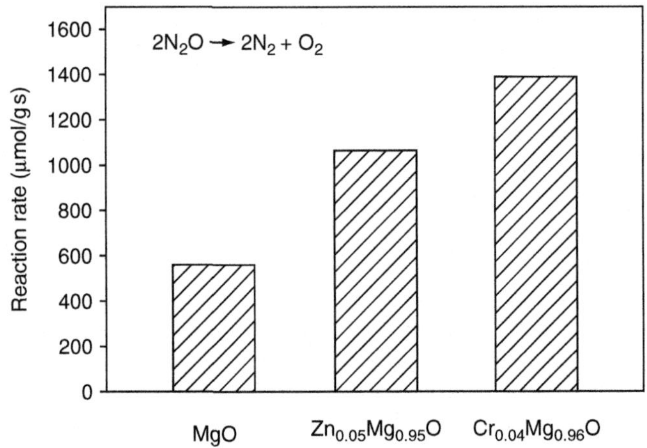

FIGURE 9.13 Steady-state rate for the $2N_2O \longrightarrow 2N_2 + O_2$ reaction on MgO, $Zn_xMg_{1-x}O$, and $Cr_xMg_{1-x}O$ catalysts ($x \leq 0.05$). $T = 900$ K, $P_{N_2O} = 0.2$ atm, $P_{He} = 0.8$ atm. (From JA Rodriguez, M. Pérez, T. Jirsak, L. González, A. Maiti, and JZ Larese. *J. Phys. Chem. B* 105: 5497, 2001. With permission.)

Figure 9.14 displays the N K-edge XANES spectra recorded after dosing NO_2 to powders of titania (anatase and rutile phases), MgO, CeO_2, Cr_2O_3, Fe_2O_3, CuO, and ZnO at 300 K [96]. A comparison with the corresponding spectra for the KNO_3 and KNO_2 standards indicates that NO_3 is the dominant species on the oxides, with a small amount of adsorbed NO_2. Following a common line of thought [97], one could assume that the nitrate is formed by direct interaction of NO_2 with O atoms in the oxide surface. This hypothesis was tested by adsorbing NO_2 on the (1 1 0) face of rutile [96]. Figure 9.15 displays the geometry of the TiO_2(1 1 0) surface. It exposes pentacoordinated Ti atoms, with O atoms in in-plane and bridging positions. The O atoms in the bridging positions are ideal for the adsorption of NO_2 via N and formation of a nitrate species. However, experiments of photoemission indicate that at small doses of NO_2 or very low temperatures (~100 K), NO_2 interacting with Ti sites is the only NO_x species present on the TiO_2(1 1 0) surface [96]. To induce the formation of substantial amounts of NO_3, one needs large coverages of NO_2 and high temperatures [96]. This behavior indicates that the nitrate is probably formed by a disproportionation reaction ($2NO_2 \longrightarrow NO_3 + NO$) on the metal cations and not by direct reaction of NO_2 with the O centers. The results of DF calculations support this mechanism for the formation of nitrate on TiO_2(1 1 0) and MgO(1 0 0) [10, 96]. Furthermore, after adsorbing high coverages of nitrogen dioxide on the Zn-terminated face of ZnO, NO_3 is detected on the surface and can be formed

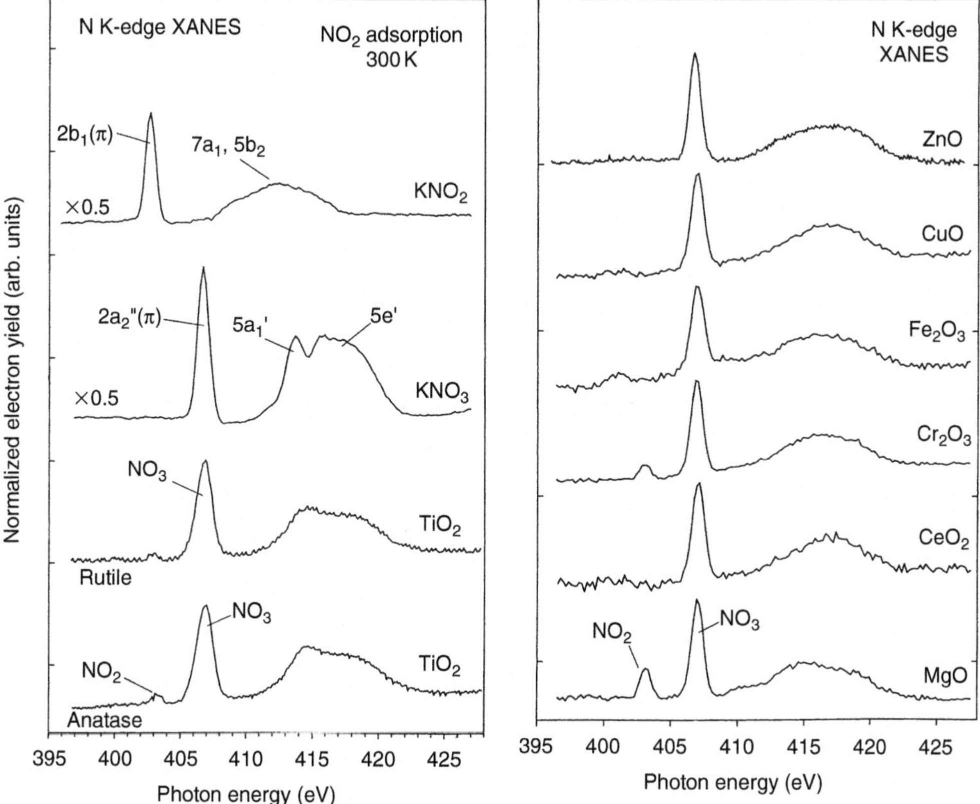

FIGURE 9.14 N K-edge XANES results for the interaction of NO_2 powders of TiO_2 (in rutile and anatase phases), MgO, CeO_2, Cr_2O_3, Fe_2O_3, CuO, and ZnO. For comparison, the corresponding spectra for KNO_2 and KNO_3 are also included. (From JA Rodriguez, T Jirsak, G Liu, J Hrbek, J Dvorak, and A Maiti. *J. Am. Chem. Soc.* 123: 9597, 2001. With permission.)

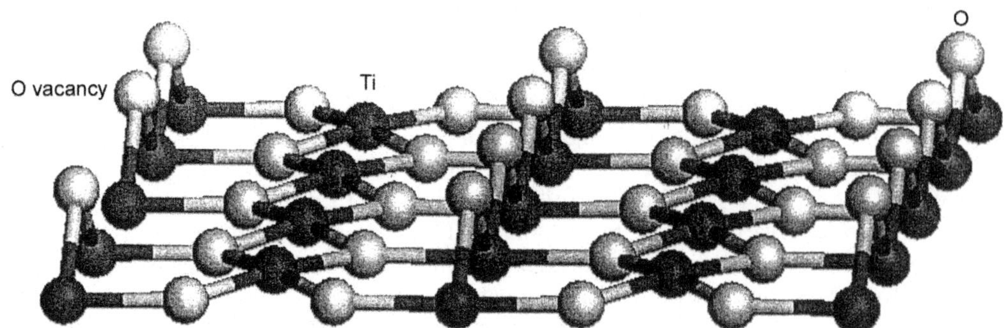

FIGURE 9.15 Schematic model for the $TiO_2(1\ 1\ 0)$ surface.

only via a disproportionation of NO_2 since O sites of the oxide are not exposed to the adsorbate [98].

NO_2 is extremely reactive toward O vacancies present in oxides [18, 96]. On oxide surfaces that are partially reduced, the NO_2 molecule dissociates into O adatoms and NO that usually desorbs into gas phase if the temperature is >300 K. For some oxide surfaces that are heavily reduced, NO_2 undergoes full decomposition. The interaction of NO_2 with O vacancies is so strong that this adsorbate can induce the migration of such defects from the bulk to the surface of an oxide [96]. At the end, any metal center that is present in the surface and has a relatively small coordination number (or low oxidation state) eventually will get oxidized or nitrated [18, 96].

In $DeNO_x$ processes, NO_2 frequently interacts with mixed-metal oxides [2, 4–6]. An interesting correlation has been found between the electronic properties of $TM_xMg_{1-x}O$ (1 0 0) and the reactivity of these oxides toward NO_2 [10]. Figure 9.16 shows the N 1s photoemission spectra obtained after depositing a small amount, ~0.2 ML, of NO_2 on MgO(1 0 0), $Zn_{0.06}Mg_{0.94}O(1\ 0\ 0)$, $Cr_{0.07}Mg_{0.93}O(1\ 0\ 0)$, and $Cr_2O_3(0\ 0\ 0\ 1)$ at 80 K [10]. For NO_2 on MgO(1 0 0), a single peak is found near 404 eV that corresponds to chemisorbed NO_2. A similar peak is seen for NO_2 on $Zn_{0.06}Mg_{0.94}O(1\ 0\ 0)$. In contrast, the $Cr_{0.07}Mg_{0.93}O$ (1 0 0) surface is much more reactive and nearly half of the NO_2 dissociates into NO and O adatoms. Finally, for NO_2 on $Cr_2O_3(0\ 0\ 0\ 1)$, there is a complete transformation of the adsorbate into NO, as has also been reported for the $NO_2/Cr_2O_3(1\ 1\ 1)$ system (1 0 0). A key issue for the design of mixed-metal oxide catalysts is the fact that Cr is able to induce occupied electronic states with a low stability after orbital hybridization within a matrix of MgO (Figure 9.8). Figure 9.17 displays the calculated energies for the valence and conduction bands of MgO, the electronic states induced by Zn and Cr, and the molecular orbitals of NO, NO_2, and N_2O [10]. NO and NO_2 are radical molecules with low-lying empty orbitals (2π, $6a_1$). The higher in energy the electronic states induced by the dopant agent (Zn, Cr, Ni, etc), the more favorable an electron transfer and interaction with the $NO(2\pi)$ and $NO_2(6a_1)$ orbitals (see Equation (9.9)). In the case of N_2O, one has a closed-shell system with unoccupied orbitals that are very high in energy. For this situation [25, 79], the bonding energy (Q) derived from the interaction between the LUMO of the adsorbate and the OS of an oxide varies according to

$$Q\alpha(\beta_{\text{LUMO-OS}})^2/(E_{\text{LUMO}} - E_{\text{OS}}) \tag{9.10}$$

where $\beta_{\text{LUMO-OS}}$ is the resonance integral for the interacting levels, and E_{OS} and E_{LUMO} are the energies of the oxide occupied states and the adsorbate LUMO, respectively. Here, the closer the energy of the dopant-induced OS to the LUMO of N_2O, the stronger the bonding

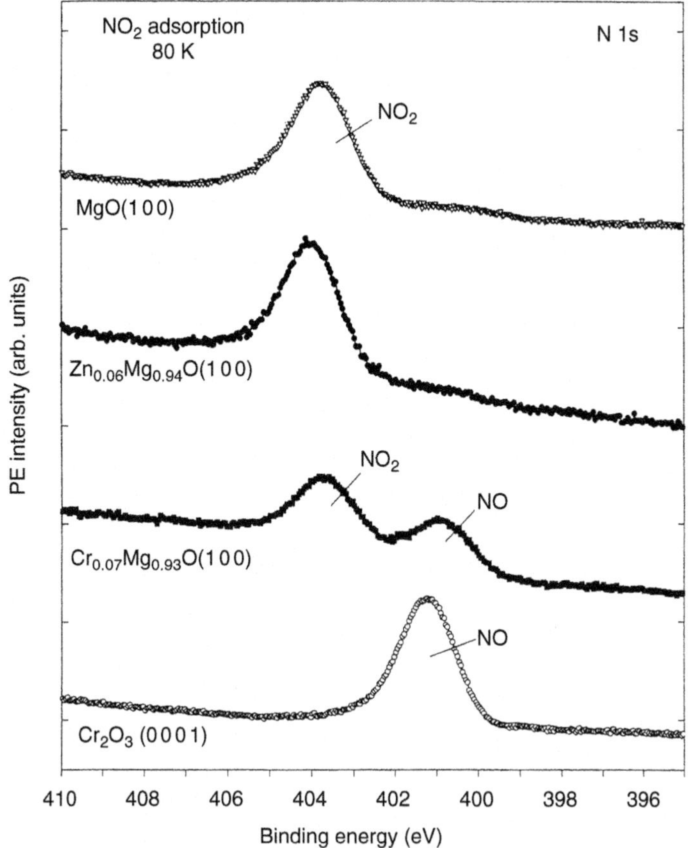

FIGURE 9.16 N 1s photoemission spectra for the adsorption of ~0.2 ML of NO_2 on MgO(1 0 0), $Zn_{0.06}Mg_{0.94}O(1\ 0\ 0)$, $Cr_{0.07}Mg_{0.93}O(1\ 0\ 0)$, and $Cr_2O_3(0\ 0\ 0\ 1)$ at 80 K. (From JA Rodriguez, M Pérez, T Jirsak, L González, A Maiti, and JZ Larese. *J. Phys. Chem. B* 105: 5497, 2001. With permission.)

interactions between adsorbate and oxide and the easier the cleavage of the N—O bond. Thus, when developing mixed-metal oxides for $DeNO_x$ operations, the general idea is to find metal dopants that upon hybridization within a host oxide produce occupied electronic states located well above the valence band of the host oxide.

9.4 CONCLUSION

The experiments described above for the interaction of NO_x molecules with well-defined metal and oxide surfaces show how complex can be the chemistry associated with $DeNO_x$ operations. Using the modern techniques of surface science many new interesting phenomena have been discovered, and as a result models traditionally used to describe or explain $DeNO_x$ processes have been modified. The progress in this area has been very impressive, but clearly more research is necessary to produce the type of knowledge that is necessary for a rational design of $DeNO_x$ catalysts.

ACKNOWLEDGMENTS

Some of the studies described above here were done in collaboration with S. Azad, Z. Chang, J. Dvorak, D. Fischer, L. González, D.W. Goodman, J. Hrbek, T. Jirsak, J.Z. Larese, G. Liu,

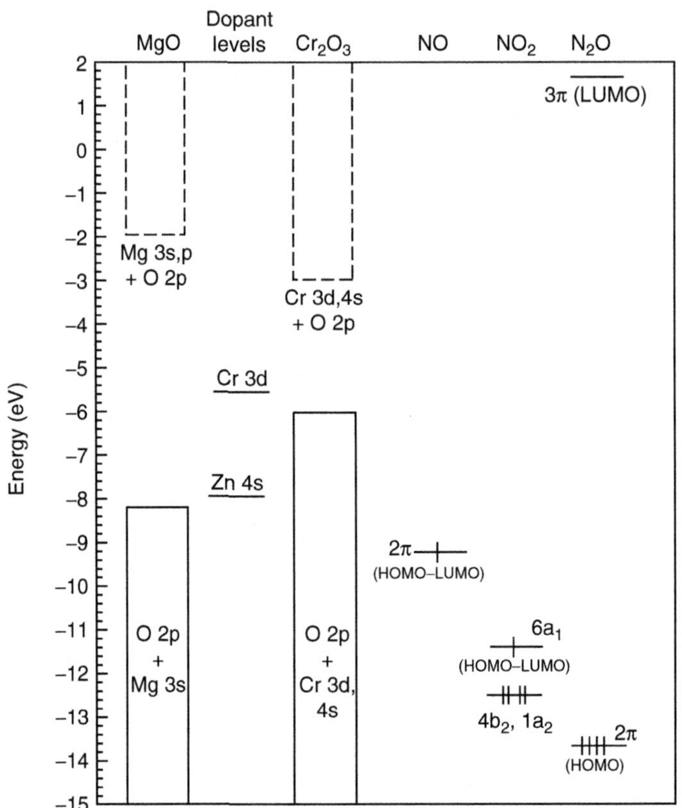

FIGURE 9.17 Energy range covered by the bands of pure MgO and Cr_2O_3. The empty and occupied states are indicated by dashed and solid lines, respectively. For comparison, the positions for the dopant levels in $Cr_{0.07}Mg_{0.93}O$(1 0 0) and $Zn_{0.06}Mg_{0.94}O$(1 0 0), plus molecular orbital energies for NO, NO_2, and N_2O are also included. All the energies are reported with respect to the vacuum level. (From JA Rodriguez, M Pérez, T Jirsak, L González, A Maiti, and JZ Larese. *J. Phys. Chem. B* 105: 5497, 2001. With permission.)

A. Maiti, C.H.F. Peden, M. Pérez, and L.Q. Wang. Many thanks to all of them. The research done at Brookhaven National Laboratory was supported by the US Department of Energy, Division of Chemical Sciences.

REFERENCES

1. AC Stern, RW Boubel, and DB Turner. *Fundamentals of Air Pollution*, 2nd ed. New York: Academic Press, 1984.
2. M Shelef and RW McCabe. *Catal. Today* 62: 35, 2000.
3. JG Speight. *The Chemistry and Technology of Petroleum*, 2nd ed. New York: Marcel Dekker, 3. 1991.
4. L Kirby. *The Control and Prevention of Acid Rain*. New York: Prentice-Hall, 2001.
5. JM Thomas and WJ Thomas. *Principles and Practice of Heterogeneous Catalysis*. New York: VCH Publishers, 1997.
6. KC Taylor. *Catal. Rev. Sci. Eng*. 35: 457, 1993.
7. K Almusaiteer, R. Krishnamurthy, and SS Chuang. *Catal. Today* 55: 291, 2000.
8. A Trovarelli. *Catal. Rev. Sci. Eng*. 38: 439, 1996.

9. PJ Schmitz and RJ Baird. *J. Phys. Chem. B* 106: 4172, 2002.
10. JA Rodriguez, M. Pérez, T. Jirsak, L. González, A. Maiti, and JZ Larese. *J. Phys. Chem. B* 105: 5497, 2001.
11. WF Schneider. *J. Phys. Chem. B* 108: 273, 2004.
12. HH Kung. *Transition Metal Oxides: Surface Chemistry and Catalysis*. New York:12. Elsevier, 1989.
13. F Zaera and CS Gopinath. *Phys. Chem. Chem. Phys.* 5: 646, 2003.
14. SB Schwartz, GB Fischer, and LD Schmidt. *J. Phys. Chem.* 92: 389, 1988.
15. CHF Peden, DW Goodman, DS Blair, PJ Berlowitz, GB Fischer, and SH Oh. *J. Phys. Chem.* 92: 1563, 1988.
16. JA Rodriguez and DW Goodman. *Surf. Sci. Rep.* 14: 1, 1991.
17. MY Smirnov, D Zemlyanov, VV Gorodetskii, and EI Vovk. *Surf. Sci.* 414: 409, 1998.
18. G Liu, JA Rodriguez, J Hrbek, J Dvorak, and CHF Peden. *J. Phys. Chem. B* 105: 7762, 2001.
19. SH Overbury, DR Mullins, DR Huntley, and L. Kundakovic. *J. Catal.* 186: 296, 1999.
20. HJ Freund. *Faraday Discuss.* 114: 1, 1999.
21. L Piccolo and CR Henry. *Surf. Sci.* 452: 198, 2000.
22. GA Somorjai. *Introduction to Surface Chemistry and Catalysis*. New York: Wiley, 1994.
23. G Ertl and J Kuppers. *Low Energy Electrons and Surface Chemistry*. Weinheim, Germany: VCH Publishers, 1985.
24. DP Woodruff and TA Delchar. *Modern Techniques of Surface Science*. New York: Cambridge University Press, 1986.
25. E Shustorovich and RC Baetzold. *Science* 227: 876, 1985.
26. WA Brown and DA King. *J. Phys. Chem. B* 104: 2578, 2000.
27. G Broden, TN Rhodin, C Bruncker, R Benbow, and Z Hurych. *Surf. Sci.* 59: 593, 1976.
28. A Siokou, RM van Hardeveld, and JW Niemantsverdriet. *Surf. Sci.* 402–404: 110, 1998.
29. M Bowker, Q Guo, and RW Joyner. *Surf. Sci.* 257: 33, 1991.
30. L Bugyi and F Solymosi. *Surf. Sci.* 258: 5, 1991.
31. TW Root, GB Fisher, and LD Schmidt. *J. Chem. Phys.* 85: 4687, 1986.
32. H Permana, KYS Ng, CHF Peden, SJ Schmieg, DK Lambert, and DN Belton. *J. Catal.* 164: 194, 1996.
33. CHF Peden, DN Belton, and SJ Schmieg. *J. Catal.* 155: 204, 1995.
34. SH Oh, GB Fischer, JE Carpenter, and DW Goodman. *J. Catal.* 100: 360, 1986.
35. CS Gopinath and F. Zaera. *J. Catal.* 200: 270, 2001.
36. F Zaera, S Wehner, CS Gopinath, JL Sales, V Gargiulo, and G Zgrablich. *J. Phys. Chem. B* 105: 7771, 2001.
37. F. Zaera and CS Gopinath. *J. Chem. Phys.* 116: 1129, 2002.
38. V Bustos, CS Gopinath, R Uñac, F Zaera, and G Zgrablich. *J Chem. Phys.* 114: 10927, 2001.
39. TB Kobylinski and BW Taylor. *J. Catal.* 33: 376, 1974.
40. NMH Janssen, PD Cobden, and BE Nieuwenhuys. *J. Phys.: Condens. Mat.* 9: 1889, 1997.
41. RM Lambert and CM Comrie. *Surf. Sci.* 46: 61, 1974.
42. T Fink, JP Dath, R Imbihl, and G Ertl. *J. Chem. Phys.* 95: 2109, 1991.
43. F Schuth and E Wicke. *Ber. Bunsen. Phys. Chem.* 93: 491, 1989.
44. SB Schwartz and LD Schmidt. *Surf. Sci.* 206: 169, 1988.
45. G Veser and R Imbihl. *J. Chem. Phys.* 100: 8483, 1994.
46. R Imbihl. *Prog. Surf. Sci.* 44: 185, 1993.
47. JJ McCarroll. *Surf. Sci.* 53: 297, 1975.
48. RJ Behm, PA Thiel, PR Norton, and G Ertl. *J. Chem. Phys.* 78: 7437, 1983.
49. P Gardner, M Tushaus, R Martin, and AM Bradshaw. *Surf. Sci.* 240: 112, 1990.
50. PD Cobden, J Siera, and BE Nieuwenhuys. *J. Vac. Sci. Technol. A* 10: 2487, 1992.
51. SJ Lombardo, T Fink, and R Imbihl. *J. Chem. Phys.* 48: 5526, 1993.
52. M Gruyters, AT Pasteur, and DA King. *J. Chem. Soc. Faraday Trans.* 92: 2941, 1996.
53. RD Ramsier, Q Gao, H Neergaard-Waltenburg, KW Lee, OW Nooij, L Lefferts, and JT Yates. *Surf. Sci.* 320: 209, 1994.
54. S Sugai, H Watanabe, T Kioka, H Miki, and K Kawasaki. *Surf. Sci.* 259: 109, 1991.
55. RG Sharpe and M Bowker. *Surf. Sci.* 360: 21, 1996.
56. HD Schmick and HW Wassmuth. *Surf. Sci.* 123: 471, 1982.
57. PW Davies and RM Lambert. *Surf. Sci.* 110: 227, 1981.

58. Y Li and M Bowker. *Surf. Sci.* 348: 67, 1996.
59. DA Hoffman and JB Hudson. *Surf. Sci.* 180: 77, 1987.
60. R Sau and JB Hudson. *J. Vac. Sci. Technol. A* 18: 607, 1981.
61. Y Kim, JA Schreffels, and JM White. *Surf. Sci.* 114: 349, 1982.
62. R Klein and R Siegel. *Surf. Sci.* 92: 337, 1980.
63. NR Avery. *Surf. Sci.* 131: 501, 1983.
64. LA West and GA Somorjai. *J. Vac. Sci. Technol.* 9: 668, 1972.
65. J Kiss, D Lennon, SK Jo and JM White. *J. Phys. Chem.* 95: 8054, 1991.
66. T Jirsak, M Kuhn, and JA Rodriguez. *Surf. Sci.* 457: 254, 2000.
67. U Schwalke, JE Parmeter, and WH Weinberg. *J. Chem. Phys.* 84: 4036, 1986.
68. J Hrbek, DG van Campen, and IJ Malik. *J. Vac. Sci. Technol. A* 13: 1409, 1995.
69. DA Outka, RJ Madix, GB Fischer, and C DiMaggio. *Surf. Sci.* 179: 1, 1987.
70. T Jirsak, J Dvorak, and JA Rodriguez. *Surf. Sci.* 436: L683, 1999.
71. ME Bartram, RG Windham, and BE Koel. *Surf. Sci.* 184: 57, 1987.
72. KA Peterlinz and SJ Siberner. *J. Phys. Chem.* 99: 2817, 1995.
73. JC Fuggle and D Menzel. *Surf. Sci.* 79: 1, 1979.
74. ME Bratram and BE Koel. *Surf. Sci.* 213: 137, 1989.
75. G Polzonetti, P Alnot, and CR Brundle. *Surf. Sci.* 238: 226, 1990.
76. DT Wickham, BA Banse, and BE Koel. *Surf. Sci.* 243: 83, 1991.
77. JA Rodriguez, T Jirsak, J Dvorak, S Sambasivan, and D Fischer. *J. Phys. Chem. B* 102: 319, 2000.
78. VE Henrich and PA Cox. *The Surface Science of Metal Oxides*. Cambridge: Cambridge University Press, 1994.
79. JA Rodriguez. *Theor. Chem. Acc.* 102: 117, 2002.
80. DW Goodman. *Chem. Rev.* 95: 523, 1995.
81. M Barteau. *Chem. Rev.* 96: 1413, 1996.
82. CM Kim, CW Yi, BK Min, AK Santra, and DW Goodman. *Langmuir* 18: 5651, 2002.
83. JA Rodriguez, T Jirsak, M Pérez, L González, and A Maiti. *J. Chem. Phys.* 114: 4186, 2001.
84. DC Sorescu, CN Rusu, and JT Yates. *J. Phys. Chem. B* 104: 4408, 2000.
85. C Xu, M Hassel, H Kuhlenbeck, and HJ Freund. *Surf. Sci.* 258: 23, 1991.
86. RM Ferrizz, T Egami, GS Wong, and JM Vohs. *Surf. Sci.* 476: 9, 2001.
87. JA Rodriguez, S Azad, LQ Wang, J García, A Etxeberria, and L González. *J. Chem. Phys.* 118: 6562, 2003.
88. JA Rodriguez, A Etxeberria, L González, and A Maiti. *J. Chem. Phys.* 117: 2699, 2002.
89. JA Rodriguez. *Catal. Today* 85: 177, 2003.
90. JA Rodriguez, T Jirsak, L González, J Evans, M Perez, and A. Maiti. *J. Chem. Phys.* 115: 10914, 2001.
91. A Rochefort and R Fournier. *J. Phys. Chem.* 100: 13506, 1996.
92. F Illas, N Lopez, JM Ricart, A Clotet, JC Conesa, and M Fernández-García. *J. Phys. Chem. B* 102: 8017, 1998.
93. JA Rodriguez, T Jirsak, JY Kim, JZ Larese, and A Maiti. *Chem. Phys. Lett.* 330: 475, 2000.
94. C DiValentin, G Pacchioni, S Abbet, and U Heiz. *J. Phys. Chem. B* 106: 7666, 2002.
95. M Miletic, JL Gland, WF Schneider, and KC Hass. *J. Phys. Chem. B* 107: 157, 2003.
96. JA Rodriguez, T Jirsak, G Liu, J Hrbek, J Dvorak, and A Maiti. *J. Am. Chem. Soc.* 123: 9597, 2001.
97. TJ Dines, CH Rochester, and AM Ward. *J. Chem. Soc. Faraday Trans.* 87: 643, 1991.
98. JA Rodriguez, T Jirsak, S Chaturvedi, and J Dvorak. *J. Mol. Catal. A* 47: 167, 2001.

10 Fundamental Concepts in Molecular Simulation of NO_x Catalysis

William F. Schneider
Department of Chemical and Biomolecular Engineering, Department of Chemistry and Biochemistry
University of Notre Dame

CONTENTS

10.1 WHY NO_X CATALYSIS?

The control of nitrogen oxide emissions from combustion sources is a long-standing and important challenge for environmental protection. Nitrogen forms a uniquely large number of oxides [1], including amongst others nitrous oxide (N_2O), a potent greenhouse gas, nitric oxide (NO), and nitrogen dioxide (NO_2). The last two (collectively, NO_x) are noxious, contribute directly to the formation of acid rain, and are key ingredients in the production of photochemical smog (ozone) and particulates [2, 3]. For these reasons, NO_x is one of the six "criteria" pollutants used to establish national air quality standards in the United States [4], and NO_x emissions from mobile and stationary sources are closely monitored and regulated [5].

NO_x is an unavoidable by-product of combustion in air. NO is the primary combustion-generated component of NO_x, and its formation can be viewed in the context of its equilibrium with N_2 and O_2:

$$N_2(g) + O_2(g) \Longleftrightarrow 2NO(g) \tag{10.1}$$

This equilibrium is overwhelmingly reactant dominated at ambient conditions, but equilibrium NO concentrations rise rapidly with increasing temperature — as shown below, these concentrations exceed 100 ppm in air at combustion temperatures of 2000 K or more. Further, at high temperatures a number of NO-forming reaction pathways become kinetically accessible, including the direct oxidation of N_2 by O and OH radicals generated during combustion (“thermal” NO), reactions of fuel hydrocarbon radicals with N_2 (“prompt” NO), and direct oxidation of nitrogen-containing fuel constituents [6]. As exhaust gases quickly cool, these reactions are quenched and the NO generated at high temperature is kinetically “frozen in.” The fundamental goal of NO_x catalysis, then, is to reverse the NO-generating reactions at lower temperatures than they occur spontaneously.

NO_x produced during combustion of nearly stoichiometric air–fuel mixtures, as is typical of modern gasoline internal combustion engines, can be converted with high efficiency back to N_2 with so-called “three-way catalysts” (TWCs) [7] comprising precious metals, in particular Rh [8, 9], dispersed on an oxide support. The effectiveness of the TWC is directly linked to the ability to balance the amounts of oxidizing (NO_x and O_2) and reducing (CO, H_2, and hydrocarbon) gases presented to the catalyst [10]. As the TWC has little ability to discriminate between reactions of NO_x and O_2 with reductant, it has limited utility for NO_x reduction to N_2 in the presence of excess O_2, and the search for NO_x removal catalysts effective under these “lean” conditions continues to attract much attention [11]. Lean-burn gasoline and diesel engines are inherently more fuel-efficient than their stoichiometric counterparts [12], but the absence of a satisfactory NO_x control technology is a serious impediment to their broader adoption.

Practical NO_x removal under excess O_2 conditions can be accomplished either by selective catalytic reduction (SCR) to N_2 or by cyclic storage and reductive regeneration with a lean NO_x trap (LNT). Various transition metal oxide and metal-exchanged zeolite catalysts are active for NO_x SCR with NH_3-based or hydrocarbon-based reductants [11]. The plasma-assisted [13, 14] and thermal [6, 15, 16] lean NO_x reduction processes are interesting variations on this theme in which electrical or thermal energy is used to convert reductant into more reactive and NO_x-selective forms. In the LNT [17], basic metal oxides are added as NO_x adsorbents to what is essentially a three-way catalyst. Under lean conditions, NO_x is stored as nitrates on these oxides; during periodic rich regeneration, the nitrates decompose and release NO_x for reduction to N_2 on the TWC components of the trap.

Viewed from a technology perspective, these various NO_x removal approaches are distinctly different in design and application. When viewed from a molecular science perspective, however, these distinctions dissolve into common questions about NO_x adsorption, desorption, and reaction at the interfaces of gas and solid metal or metal oxide phases — in fact, the same questions that dominate molecular environmental NO_x chemistry. Further, because NO_x reduction is invariably performed in a background dominated by N_2 and often O_2, this catalytic chemistry is closely linked to, and often limited by, gas-phase thermodynamics and kinetics. Molecular simulation has and will continue to play a major role in elucidating these processes. In this chapter, standard approaches to understanding the gas-phase thermodynamics and kinetics of NO_x chemistry will first be reviewed. Some of the technical details of applying modern first principles simulation tools to NO_x chemistry will then be discussed, followed by three examples to illustrate the application of these tools to NO_x adsorption, oxidation, and decomposition in the gas-phase and on model metal oxide, metal, and zeolite catalysts.

10.2 GAS-PHASE NO_x THERMODYNAMICS AND KINETICS

NO, NO_2, and in fact all nitrogen oxides have positive free energies of formation at standard temperature and pressure (Table 10.1), and thus decomposing any of these to their elemental constituents (N_2 and O_2) would appear to be a straightforward problem in catalysis. In fundamental ways, however, NO_x catalysis distinctly differs from, for instance, catalytic CO or hydrocarbon oxidation for environmental protection. Because the formation energies of the nitrogen oxides are similar and small, the thermodynamic driving force for many relevant reactions are also small or highly sensitive to temperature, total pressure, and background

TABLE 10.1
Standard Thermodynamic Properties of Gaseous Nitrogen Oxides and Related Molecules at 1 bar and 298.15 K [19]

	$\Delta_f H^\circ_{298}$ (kcal mol^{-1})	S°_{298} (cal mol^{-1} K^{-1})	$\Delta_f G^\circ_{298}$ (kcal mol^{-1})
N_2	0	45.7957	0
O_2	0	49.033	0
N	112.97	36.640	108.87
O	59.555	38.490	55.389
NO	21.58	50.373	20.70
NO_2	7.911	57.371	12.252
N_2O	19.61	52.572	24.90
NO_3	17.00	60.375	27.75
N_2O_3	19.80	73.743	33.40
N_2O_4	2.17	72.749	23.37
N_2O_5	2.70	82.827	28.21

N_2 and O_2 concentrations. Compounding these limitations are the very high conversion efficiencies necessary to meet real-world emission targets, as well as the inherently slow kinetics of several important gas-phase NO_x reactions.

In thinking about molecular simulation of environmental NO_x catalysis it is thus helpful to first consider the thermodynamics and kinetics of the underlying gas-phase chemistry [18]. The NO decomposition reaction, which is the conceptually simplest approach to NO removal, provides a useful case study:

$$2\mathrm{NO(g)} \longleftrightarrow \mathrm{N_2(g)} + \mathrm{O_2(g)} \tag{10.2}$$

From commonly available thermodynamic tables [19], the forward reaction is exothermic ($\Delta H^\circ_{298} = -43.2\,\mathrm{kcal\ mol^{-1}}$) and, as expected for a reaction that conserves total mole number, has a small standard entropy change ($\Delta S^\circ_{298} = -6\,\mathrm{cal\ mol^{-1}\ K^{-1}}$). The standard Gibbs free energy at room temperature is thus

$$\Delta G^\circ_{298} = \Delta H^\circ_{298} - T\Delta S^\circ_{298} = -41.4\,\mathrm{kcal\ mol^{-1}} \tag{10.3}$$

ΔH° and ΔS° are nearly temperature independent, so the temperature at which the standard free energy change goes to zero, or equivalently at which the equilibrium shifts from product to reactant dominated, can be reasonably well approximated as:

$$T(\Delta G^\circ = 0) \approx \Delta H^\circ_{298}/\Delta S^\circ_{298} = 7200\,\mathrm{K} \tag{10.4}$$

Up to very high temperatures, then, NO is thermodynamically unstable with respect to the elements. Despite these favorable thermodynamics, the uncatalyzed reaction (10.2) is exceedingly slow until very high temperatures (>2000 K, see Ref. [20]) for mechanistic reasons that will be further discussed below. The reaction is notoriously unresponsive to typical catalyst materials [11, 21]; Cu-exchanged zeolites are among the few catalysts found to have appreciable NO decomposition activity [22].

Poorly performing catalysts are not the only obstacle to NO decomposition as a NO_x remediation strategy. To be environmentally effective and practically useful, any catalytic NO_x removal system must be able to decrease NO concentrations to quite low levels (less than

10 ppm by volume* in a background dominated by N_2 and, in particular for lean combustion applications, O_2. Under these distinctly nonstandard thermodynamic conditions, the equilibrium concentration of NO can become appreciable [18]. As an example, at 773 K in air ($P_{N_2} \approx$ 0.79 atm and $P_{O_2} \approx$ 0.21 atm), the equilibrium NO partial pressure according to Reaction (10.2) can be calculated as follows:

$$K^{eq} = \exp[-(\Delta H^\circ - T\Delta S^\circ)] \approx 8 \times 10^{10} \tag{10.5}$$

$$P_{NO} = \sqrt{\frac{P_{N_2}P_{O_2}}{K^{eq}}} \approx 1 \times 10^{-6}\,\text{atm} \tag{10.6}$$

The equilibrium NO amount thus approaches within 10% of the target emissions value. Put another way, a perfectly effective NO decomposition catalyst could at best decrease the NO concentration in a predominantly N_2/O_2 gas stream at 773 K to 1 ppm; if operating on a gas containing 100 ppm NO, the maximum possible conversion to N_2 is 99%, or on 10 ppm NO, 90% conversion. For this reason, an "approach to equilibrium" measure is a more appropriate gauge of catalytic NO decomposition activity than is simple NO conversion [18]. Further, by the principle of microscopic reversibility [23], a catalyst ideally effective for Reaction (10.2) acting on an N_2/O_2 gas stream containing less than 1 ppm NO would tend to *increase* the NO concentration in the gas stream, which is clearly an undesirable attribute.

The details of this thermodynamic analysis depend on specific conditions of temperature, pressure, and gas composition. Free energy minimization methods [24], as available in standard thermochemical software packages such as HSC Chemistry [25], provide a more rigorous and general way to treat individual or coupled NO_x equilibria [18]. Using tabulations of individual species free energies, these methods determine equilibrium system compositions as a function of initial composition and target temperature, pressure, or other constraints. For example, consider a catalyst that is active not only for Reaction (10.2) but also for NO oxidation and disproportionation:

$$2NO(g) + O_2(g) \longleftrightarrow 2NO_2(g) \tag{10.7}$$

$$3NO(g) \longleftrightarrow N_2O(g) + NO_2(g) \tag{10.8}$$

Figure 10.1(a) shows the calculated equilibrium concentrations of NO, NO_2, and N_2O (plotted on a log scale) as a function of temperature at 1 atm total pressure in nominally 79% N_2 and either 21% O_2 or 0.01% O_2 and 20.99% Ar. The equilibrium concentrations of all three nitrogen oxides rise with temperature and are uniformly greater at the higher O_2 concentration. While their equilibrium concentrations are always far less than that of the background N_2 and O_2, they do reach environmentally significant (>1 ppm) levels as low as 500°C.

These equilibrium curves place lower bounds on achievable NO_x and N_2O concentrations through catalyzed Reactions (10.2), (10.7), and (10.8) (or any other set of linearly independent reactions spanning the same composition space). To put these results into context, they can be contrasted with the limitations imposed on CO oxidation by the CO/CO_2 equilibrium in O_2:

---

*An order of magnitude estimate in an automotive context can be made as follows. Allowable NO_x emissions are approaching 0.02 g/mile averaged over a standard drive cycle; this much NO_x emitted as NO occupies a volume of 0.016 L/mile at 1 atm and 298 K. A vehicle with a 2 L displacement, four-stroke engine turning at 1500 rpm and traveling 60 mile/h pumps (1500 rpm/2) × (2 L) × (1 min/mile) = 1500 L/mile of air and fuel. Of this, the allowable NO content is thus 0.016 L/1500 L = 10 ppm. In practice, to compensate for transient operation and short residence times, much lower concentrations must be attainable at steady state.

$$2CO(g) + O_2(g) \longleftrightarrow 2CO_2(g) \tag{10.9}$$

Figure 10.1(b) compares the equilibrium NO_x concentrations (sum of NO and NO_2) in the O_2-excess and O_2-deficient cases above with the equilibrium CO concentrations in nominally 15% CO_2 and either 21% or 0.01% O_2 and balance inert Ar. The minimum achievable CO concentrations are seen to be much lower than the NO_x concentrations at temperatures below 1000°C regardless of O_2 conditions — the greater driving force for Reaction (10.9) ($\Delta H^{\circ}_{298} = 135.3\,\text{kcal mol}^{-1}$) compared to Reaction (10.2) implies a much smaller constraint on CO oxidation than NO_x decomposition catalysis at practically interesting conditions.

Finally, NO_x thermodynamics place constraints on the net rate of NO_x-destroying reactions. Taking the NO decomposition reaction as an example again, as the NO concentration

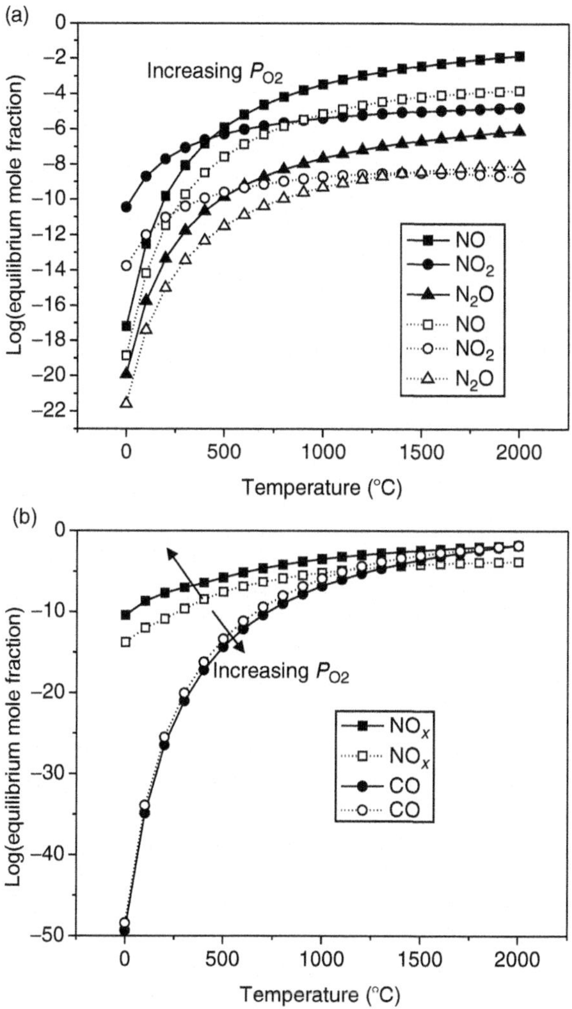

FIGURE 10.1 (a) Thermodynamic equilibrium concentrations of NO, NO_2, and N_2O in nominally 79% N_2 and 21% O_2 (solid lines) and 79% N_2, 0.01% O_2, and balance Ar (dashed lines). The equilibrium NO_x concentrations are nonnegligible, especially in high O_2 backgrounds. (b) Comparison of equilibrium NO_x concentrations from (a) to equilibrium CO concentration in 21% O_2, 15% CO_2, and balance Ar (solid line) and in 0.01% O_2, 15% CO_2, and balance Ar (dashed line). Equilibrium CO concentrations are lower than NO_x except under O_2-starved and high-temperature conditions.

in a gas mixture approaches its equilibrium value, the forward and reverse rates of Reaction (10.2) necessarily approach one another, and the net decomposition rate vanishes [26]:

$$-\frac{dP_{NO}}{dt} = \text{rate}_f - \text{rate}_r \longrightarrow 0 \text{ as } P_{NO} \longrightarrow P_{NO}^{eq} \tag{10.10}$$

The form of this approach towards equilibrium depends on the underlying rate laws and details of the reaction mechanism. For example, if the forward and reverse reactions were bimolecular, then the net NO destruction rate would be:

$$-\frac{dP_{NO}}{dt} = k_f P_{NO}^2 - k_r P_{N_2} P_{O_2} = k_f \left(P_{NO}^2 - \frac{P_{N_2} P_{O_2}}{K^{eq}} \right) \tag{10.11}$$

The forward rate of NO consumption decreases quadratically with decreasing P_{NO} (the forward rate would be 100 times greater at 100 ppm NO than 10 ppm), while, because near equilibrium both P_{N_2} and P_{O_2} are much greater than P_{NO}, the reverse rate of NO production is essentially constant. At equilibrium these rates balance; near equilibrium the net rate is reduced relative to points distant from equilibrium.

These examples are intended to illustrate the insight to be gained from consideration of gas-phase thermodynamics and kinetics. More elaborate and detailed analyses are possible; the goal of first principles simulation is to extend the ideas to composition and reactivity at catalytic surfaces.

10.3 ELECTRONIC STRUCTURE SIMULATIONS FOR NO_x CATALYSIS

While the gas-phase thermodynamics and kinetics of NO_x chemistry are reasonably well established, the same cannot be said for catalytic processes involving NO_x. Density functional theory (DFT) [27–29] has emerged in the past 15 years as a powerful technique for providing reliable molecular-level descriptions of heterogeneous chemistry. DFT is not a complete or fully satisfactory theory of molecular electronic structure, but it is more generally applicable and computationally tractable than more elaborate methods and, appropriately applied, can provide quantitative insights into adsorption and catalytic reactivity. General-purpose DFT software packages such as Vasp [30], ADF [31], and DMol [32] are widely available and extensively applied. These and other DFT packages vary widely in computational implementation. Here we review only the most salient principles of DFT simulation necessary to understand its application to NO_x catalysis.

Figure 10.2 illustrates the key inputs and outputs of a DFT simulation. These calculations begin with a molecular-level definition of the system of interest, including the identity and positions of the atoms $\mathbf{R}_N$ and, in the case of a periodic simulation [33], the size and shape of the simulation supercell, given by the three lattice vectors **a**, **b**, and **c**. The Hohenberg–Kohn theorem [34] guarantees that the total energy E of this system is a universal functional of the electron distribution, $\rho(\mathbf{r})$:

$$E = E\,[\rho; \mathbf{R}_N, \ \mathbf{a}, \ \mathbf{b}, \ \mathbf{c}] \tag{10.12}$$

The energy and electron distribution can be found by expressing ρ in terms of one-electron molecular orbitals ψ_i:

$$\rho = \sum_i |\psi_i|^2 \tag{10.13}$$

and solving for these ψ_i as eigenfunctions of the Kohn–Sham [35] equation:

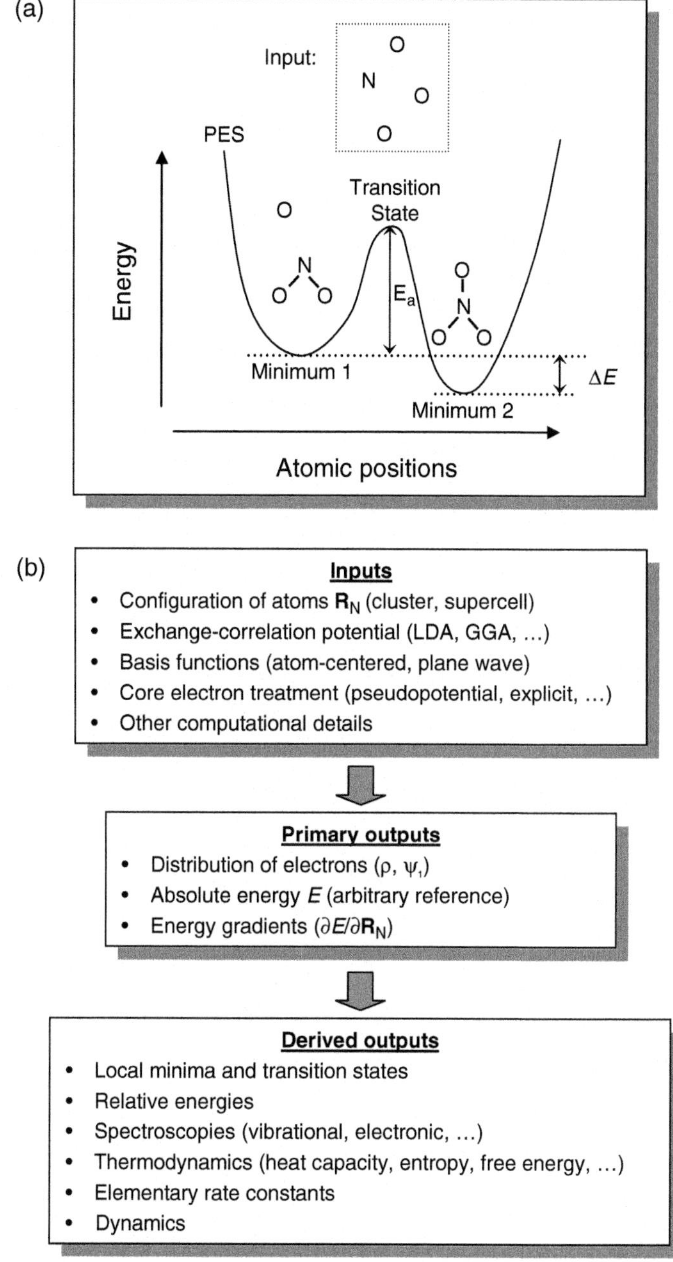

FIGURE 10.2 (a) Schematic potential energy surface as might be derived from a DFT simulation. (b) Flow diagram of inputs and outputs from a DFT simulation.

$$\epsilon_i \psi_i = (t_i + \upsilon_{\text{Coulomb}}[\rho; \mathbf{R}_\text{N}, \mathbf{a},\ \mathbf{b},\ \mathbf{c}] + \upsilon_{\text{XC}}[\rho])\psi_i \tag{10.14}$$

The forms of the kinetic (t_i) and Coulomb ($\upsilon_{\text{Coulomb}}$) operators are known; all the unknown or difficult-to-describe many-body quantum mechanical parts of the electron interactions are collected in the exchange–correlation functional, υ_{XC}. Common approaches to treating υ_{XC} include the local density approximation (LDA, [35, 36]), various generalized gradient approximations (GGAs, [37–41]), and "hybrid" functionals incorporating exact exchange [42, 43]. Equations (10.12)–(10.14) are typically solved by expanding the ψ_i in a basis of atom-centered or plane-wave functions φ_α:

$$\psi_i(\mathbf{r}) = \sum_\alpha C_{\alpha i}\varphi_\alpha(\mathbf{r}) \tag{10.15}$$

and iterating the expansion coefficients $C_{\alpha i}$ to self-consistency. The variation of the calculated E with $\mathbf{R}_\mathrm{N}$ or $\mathbf{a}$, $\mathbf{b}$, and $\mathbf{c}$ defines a potential energy surface (PES), and from this minimum energy configurations, transition states, vibrational spectra, thermodynamic, and other properties of interest can be determined (Figure 10.2). State-of-the-art software and computers allow DFT calculations of this sort to be performed routinely on systems containing on the order of hundreds atoms.

A reasonable place to start the discussion of NO_x catalysis simulation is with the molecules themselves. Figure 10.3 plots the energies ε_i and isosurfaces of the valence molecular orbitals ψ_i of NO and NO_2 [44] calculated within the spin-polarized GGA [40, 41] using atom-centered basis functions as implemented within the Dmol code [32]. (Electron spin is included explicitly in these calculations to properly treat the paramagnetic ground states of the odd-electron molecules.) These energy diagrams are helpful in understanding the structure and

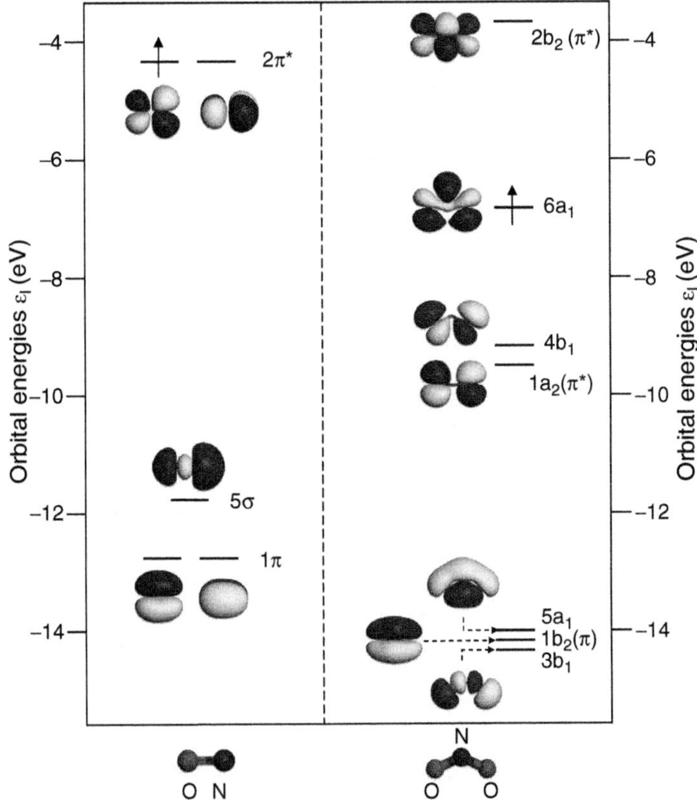

FIGURE 10.3 Molecular orbital energy diagrams for gas-phase NO and NO_2. Plotted are the spin-up orbital energies from spin-polarized PW91-GGA calculations along with molecular orbital isosurfaces to show orbital shapes and bonding characters. The orbitals are labeled by their symmetry representations within the $C_{\infty v}$ (NO) and C_{2v} (NO_2) point groups.

bonding of molecules and in anticipating their gas-phase and catalytic reactivity [45]. For instance, NO has three fully occupied bonding orbitals (5σ and both components of the 1π) and a single electron in the antibonding 2π* orbital, implying a $^2\Pi$ ground state and a net bond order of 2.5. The electronic structure of bent NO_2 is more complicated, involving a mixture of bonding, nonbonding, and antibonding σ and π orbitals that yield an N—O bond order of approximately 1.5. The lone electron in the highest lying 6a_1 orbital implies a $^2A_1$ ground state. The presence of high-lying partially occupied levels suggests NO and NO_2 to be both relatively easily oxidized and reduced, and as seen below, electron sharing and electron transfer both contribute significantly to NO_x catalytic chemistry.

Table 10.2 compares the GGA-calculated geometries (energy minimizing structures) and vibrational spectra of NO and NO_2 and several other small molecules [44, 46] with experimental values [46, 47]. As seen here, the GGA tends to systematically overestimate bond lengths by 1% to 2% but nicely reproduces qualitative structure trends. Computed vibrational frequencies are also systematically overestimated by a small amount. DFT frequencies are typically computed based on a harmonic approximation to the true PES and thus are not strictly comparable to the experimentally observed anharmonic frequencies. The harmonic approximation and exchange–correlation errors in DFT often cancel such that calculated and observed frequencies are similar. The LDA performs comparably to the GGA for structures and vibrational frequencies [48].

Computed reaction energies provide a more stringent test of DFT. Table 10.3 compares the observed reaction enthalpies with calculated energies of Reactions (10.2) and (10.7) within the spin-polarized GGA [44]. Reaction energies are computed from the difference between product and reactant electronic energies from Equation (10.12):

TABLE 10.2
Spin-Polarized GGA-PW91 Equilibrium Structures, Harmonic Vibrational Frequencies, Zero-Point Vibrational Energies [44] and Zero-Point Corrected Bond Dissociation Energies Compared with Observed Structures, Fundamental Frequencies [47], and Bond Enthalpies [140] in Parenthesis [46][a]

	Ground Electronic State	Equilibrium Structures (Å)		Vibrational Frequencies (cm^{-1})		ZPVE (kcal mol^{-1})	Bond Dissociation Energy (kcal mol^{-1})
NO	$^2\Pi$	N—O	1.163	$\nu_{N—O}$	1888	2.70	169.9
			(1.151)		(1876)	(2.71)	(150.0)
NO_2	$^2A_1$	N—O	1.208	ν_{asym}	1646	5.34	97.7[b]
			(1.194)		(1617)		(72.8)
		∠O—N—O	133.3°	ν_{sym}	1346		
			(133.8°)		(1320)		
				ν_{bend}	741		
					(750)		
N_2	$^1\Sigma_g^+$	N—N	1.107	$\nu_{N—N}$	2368	3.39	241.8
			(1.094)		(2331)	(3.36)	(224.9)
O_2	$^3\Sigma_g^-$	O—O	1.225	$\nu_{O—O}$	1540	2.20	142.4
			(1.207)		(1556)	(2.25)	(117.9)

[a]Structures and vibrational frequencies as computed in the local orbital approximation with DMol [32]. Bond energies as calculated with Vasp [30] using PAW potentials and a 400 eV plane-wave cutoff.
[b]ON—O bond energy.

TABLE 10.3
Comparison of Observed [25] and GGA-Calculated NO_x Reaction Energies (kcal mol^{-1})

	$2NO \longrightarrow N_2 + O_2$	$2NO + O_2 \longrightarrow 2NO_2$
ΔH°_{298} (experiment)	−43.2	−27.3
Calculated		
ΔE_{rxn}	−46.0	−53.7
ΔZPVE	0.2	3.1
ΔH (0 ⟶ 298 K)	0.0	−1.4
Total (ΔH°_{298} calculated)	−45.8	−52.0

$$\Delta E_{rxn} = \sum_{products} E - \sum_{reactants} E \tag{10.16}$$

Reaction energies calculated this way neglect quantum mechanical zero-point vibrational energies (ZPVE) as well as any thermal (non-0 K) effects. ZPVE can be estimated reasonably well from calculated harmonic vibrational frequencies (Table 10.1), and similarly the translational, rotational, and vibrational contributions to enthalpy and entropy can be approximated from the calculated structures and vibrational frequencies using standard statistical mechanical formulas [49]. As seen in Table 10.3 these corrections tend to be small, and since vibrational spectra are expensive to calculate for most systems of interest, the ZPVE and enthalpy corrections are frequently neglected unless higher accuracy is demanded by the particular application.

The DFT electronic energy ΔE_{rxn} is the largest potential source of error in the calculation of reaction energies. Unlike the properties discussed above, ΔE_{rxn} is quite sensitive to the performance of v_{XC}. Reactions that involve large differences in exchange and correlation, for instance reactions that do not conserve electron spin or that involve substantial electronic rearrangements can exhibit large energy error. The NO oxidation reaction (Reaction 10.7) is one such example in which spin is not conserved from reactants (four unpaired spins) to products (two unpaired spins), and the PW91-GGA-calculated reaction energy is in error by 25 kcal mol^{-1}. Bond dissociation energies calculated within this approximation are similarly large (Table 10.2 and Ref. [46]). In contrast, the NO decomposition reaction (Reaction 10.2) conserves spin and bond order from reactants to products and, due to better error cancellation, has a much smaller net reaction energy error. The magnitude of these errors does depend on the choice of density functional (see Ref. [50] for one extensive comparison for small molecules), but the best performing functionals are typically parameterized against experimental data and are the most expensive to apply. Fortunately, many of the most interesting questions to be addressed in NO_x heterogeneous catalysis can be cast in a form in which the DFT errors cancel (e.g., comparisons of NO_x adsorption energies on similar oxides [51]), and it is often possible to apply corrections based on known small-molecule reaction energies. In general, absolute reaction energies should be tested and verified against experiment whenever possible.

10.4 REACTIONS ON METAL OXIDES: NO_x ADSORPTION

The adsorption of NO_x on a metal oxide surface is the heterogeneous reaction perhaps most broadly relevant to NO_x environmental catalysis. Adsorption and reaction on mineral dusts are critically important in environmental NO_x chemistry [52], and metal oxides are present as supports or active adsorbing materials in essentially all heterogeneous NO_x removal catalysts. Numerous adsorbates are known or proposed to arise from NO_x adsorption on metal oxide surfaces [53], and in general this chemistry is not well understood at the molecular level. Metal oxide surface reactions have traditionally been classified as either Lewis acid–base or oxidation–reduction [54], but as shown below, NO_x adsorbates can exhibit surprising combinations of these two even with the simplest metal oxides [51, 55]. Here we will use NO_2 adsorption on magnesium oxide (MgO) to illustrate some issues in simulating NO_x reactivity on metal oxide surfaces.

The bulk properties of a crystalline material like MgO are most directly probed within a periodic DFT framework. As shown on the left side of Figure 10.4, MgO has the simple cubic, or rock salt, structure, which can be represented (among other ways) as a supercell containing four MgO formula units. Using a projector-augmented wave (PAW) [56] treatment for the core (O 1s and Mg 1s–2s) electrons, a plane-wave basis with 400 eV cutoff, and the PW91-GGA, all as implemented in Vasp [30, 57], the MgO lattice constant and bulk modulus are calculated to be 4.248 Å and 149 GPa [51]. The calculated lattice constant slightly overestimates the experimental value of 4.211 Å [58], as is typical for the GGA, while the bulk modulus is approximately 10% less than experiment [59]. Similar lattice constant performance is found for the other alkaline earth oxides [51].

A heterogeneous catalyst can be represented in a molecular simulation either as a finite cluster of atoms (possibly embedded in some field meant to simulate the environment of an extended system) or as a semi-infinite slab. Illustrations of both types of models for the MgO(0 0 1) surface are shown on the right side in Figure 10.4: a cluster of ions arranged in two layers with overall stoichiometry $Mg_{25}O_{25}$, and a tetragonal supercell with stoichiometry $Mg_{16}O_{16}$, arrayed in three ionic layers. Each model has merits and demerits: the supercell approach more directly captures the long-range interactions present in extended compounds and avoids potentially unphysical edge effects inherent to finite clusters; the cluster approach avoids possibly unphysical interactions between periodic images of the supercell and provides

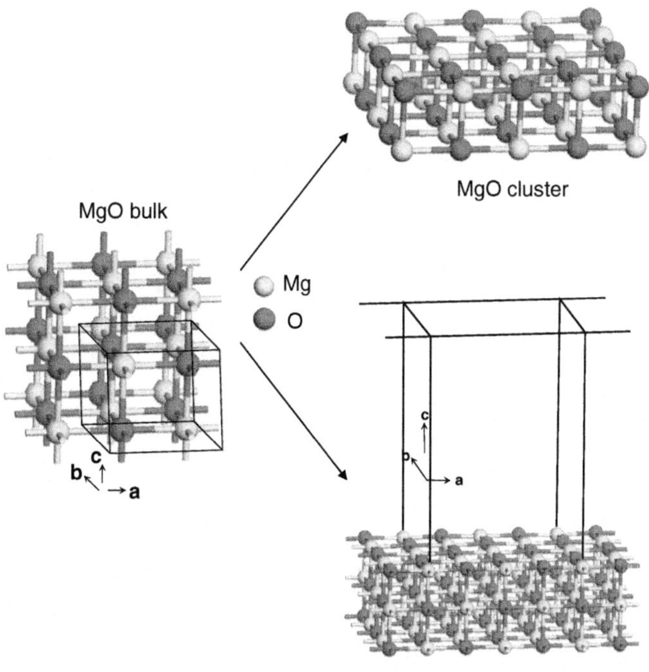

FIGURE 10.4 The bulk MgO rock salt crystal structure and cluster and supercell representations of the low-energy MgO(0 0 1) surface.

more flexibility in treating charged or excited systems. Both methods have been used productively to study NO_2 adsorption on MgO [44, 51, 55, 60–62].

The key parameters entering the construction of a slab model are the slab thickness, the lateral repeat dimensions (which determines the minimum separation, or coverage, of adsorbates), and the vacuum spacing. Optimal choices depend on the problem of interest; for MgO and the other alkaline earth oxides, three-layer slabs with a 2×2 lateral repeat and vacuum spacing approximately twice the slab thickness (as shown in Figure 10.4) are found to represent NO_x adsorption structures and energies well [55]. The surface energy provides an additional check on the reliability of slab models:

$$E_{\text{surface}} = (E_{\text{slab}} - nE_{\text{bulk}})/a \tag{10.17}$$

where E_{slab} is the energy of the slab, E_{bulk} is the bulk energy per formula unit, n is the number of formula units within the slab supercell, and a is the surface area of the slab within the supercell. The surface energy for the unrelaxed and relaxed MgO(0 0 1) surfaces calculated using the three-layer slab and the same computational parameters as described above are 0.92 and 0.90 J m^{-2} [51], in reasonable agreement with the experimental value of 1–1.2 J m^{-2} [63].

As the small difference in surface energies suggests, the relaxed MgO(0 0 1) surface is quite close to the ideal bulk termination. The (0 0 1) surface exposes five-coordinate Lewis acidic Mg^{2+} ions and Lewis basic O^{2-} ions, and MgO chemistry tends to be dominated by reactions of the latter. Even with this high-symmetry surface termination, the number of possible locations and orientations of an NO_2 (or similarly sized) adsorbate is very large, and chemical intuition is helpful in guiding the search for stable adsorption states. NO_2 is not a particularly good Lewis acid or base, but a reasonable guess at its reactivity towards MgO, confirmed by simulations, is of two types of adsorption modes, one in which NO_2 acts as a donor (base) towards Lewis acid surface sites:

basic NO_2:

$$\text{—Mg}^{2+}\text{—O}^{2-}\text{—Mg}^{2+}\text{—} + NO_2 \longrightarrow \text{—Mg}^{2+}\text{—O}^{2-}\text{—Mg}^{2+}\text{—} \text{ (NO}_2\text{ bridging via O atoms to both Mg}^{2+}\text{)} \tag{10.18}$$

and another in which NO_2 acts as an acceptor (acid) towards Lewis base surface sites:

acidic NO_2:

$$\text{—Mg}^{2+}\text{—O}^{2-}\text{—Mg}^{2+}\text{—} + NO_2 \longrightarrow \text{—Mg}^{2+}\text{—}\overset{NO_2\uparrow}{O^{2-}}\text{—Mg}^{2+}\text{—} \tag{10.19}$$

Figure 10.5 shows the lowest energy configurations obtained from DFT simulations for these two adsorption modes.

In the basic mode (Figure 10.5a), NO_2 preferentially adsorbs O-down and bridging two Mg^{2+} ions, with a GGA-calculated binding energy of -11 kcal mol^{-1} [51, 55]** again within the DFT model described above. This is a relatively weak association, and consistent with that the NO_2 structure is only slightly perturbed from the isolated molecule (Table 10.2). Molecular orbital interaction diagrams provide a convenient way to rationalize and predict adsorption phenomena on metal oxides [64], and the right side of Figure 10.5(a) shows a schematic

**Differing definitions of adsorption energies are used in the literature. In this work we define adsorption energies according to $\Delta E_{\text{adsorption}} = E_{\text{adsorbate+surface}} - (E_{\text{surface}} + E_{\text{adsorbate}})$, so that exothermic adsorption is a negative quantity.

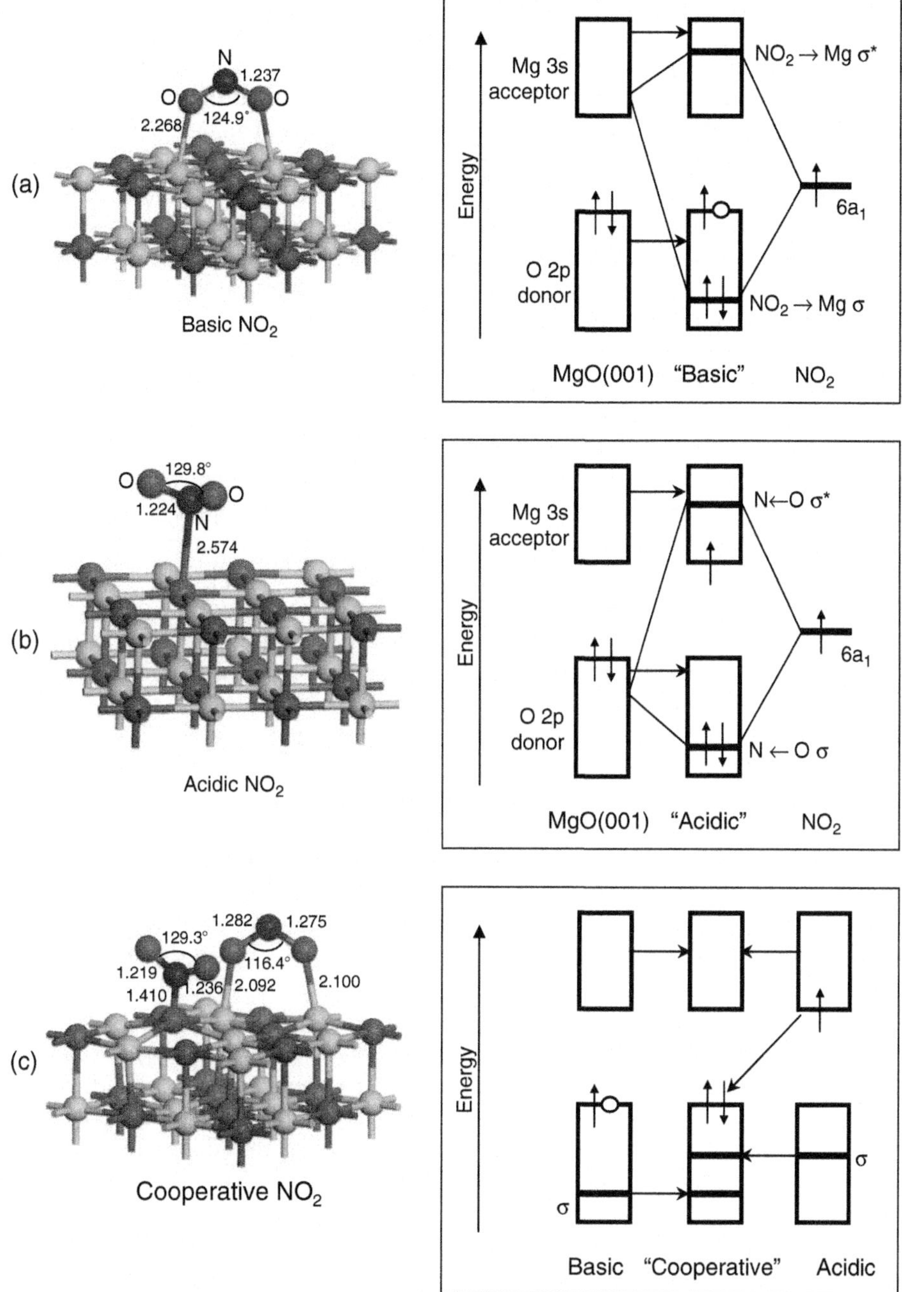

FIGURE 10.5 PW91-GGA-calculated geometries and qualitative molecular orbital interaction diagrams for (a) O-down, (b) N-down, and (c) cooperative NO_2 adsorption on an MgO(0 0 1) slab model. Bond distances reported in Ångstroms.

diagram for the NO_2 basic binding to MgO(0 0 1). MgO is an ionic and insulating main group oxide, and as such, its valence electronic structure includes a series of filled orbitals of mostly O 2p character (the valence band) separated by an energy gap from a series of vacant orbitals of mostly Mg 3s character (the conduction band). The filled valence set is available to donate charge and gives MgO its Lewis basic character, while the vacant conduction set is potentially available to accept charge. While in principle all the NO_2 orbitals shown in Figure 10.3 can mix with these MgO levels, the important features of the interaction can be obtained from the $6a_1$ level alone. In the basic adsorption configuration, this orbital interacts most directly with Mg^{2+}-centered acceptor levels to form filled bonding and vacant antibonding states. Because the $6a_1$ is only half-filled, the other bonding electron must come from MgO, and as a result an electron "hole" is created at the top of the valence band. As the bonding level is polarized more towards NO_2 than the MgO surface, NO_2 is partially reduced by the interaction, and the strength of the basic mode binding is thus in part a measure of the ability of NO_2 to oxidize or extract an electron from MgO. Mulliken population analysis, which provides a rough measure of charge sharing and transfer, confirms a transfer of approximately 0.26 e of charge to NO_2 [55, 61]. This charge transfer also accounts for the slight compression of the O—N—O angle and lengthening of the N—O bonds as the NO_2 becomes more NO_2^- -like [44]. As one moves down the alkaline earth oxide family, this charge transfer occurs more readily and the basic bonding mode commensurately increases in strength [51].

The association between MgO(0 0 1) and NO_2 in the acidic configuration is an even weaker — 3 kcal mol^{-1} within the GGA supercell DFT model. As shown in Figure 10.5(b), the minimum energy separation between NO_2 and surface O^{2-} is quite large, and the NO_2 molecule is only slightly distorted from its equilibrium gas-phase structure [51, 55, 60, 61, 65]. Surprisingly, then, an individual NO_2 molecule interacts more strongly with the acidic (Mg^{2+}) sites of MgO(0 0 1) than the basic (O^{2-}) ones, opposite to the case for CO_2 or SO_x [51, 60, 66–69]. An orbital analysis is rather tenuous for this weak MgO–acidic NO_2 interaction, but it is useful to consider in qualitative terms. Again, this interaction potentially involves a number of NO_2 orbitals, but for our purposes we can focus just on the $6a_1$ and its primary interaction with the filled O-centered levels of MgO. This interaction again produces both bonding and antibonding levels, the latter of which may or may not encroach on the metal oxide conduction band depending on the strength of adsorption. In the MgO(0 0 1) case the interaction is so weak that the antibonding level is primarily NO_2-based; as the interaction becomes stronger down the increasingly basic alkaline earth oxide family, the antibonding level will have progressively greater contribution from the metal cation-based conduction band. This binding mode, then, in part probes the ability of NO_2 to reduce or donate an electron to the metal oxide surface. The wide energy gap between filled and vacant MgO(0 0 1) levels makes NO_2 ineffective in donating electrons to this surface.

The real power of the orbital analysis becomes apparent when used to recognize that neither the basic nor the acidic binding mode is optimal for NO_2 — each involves some compromise between formation of strong bonding interactions and either oxidation or reduction of the MgO surface. An evidently more natural way to accommodate NO_2 on undefected MgO(0 0 1) is as pairs in acidic and basic configurations, one acting as an electron donor and the other as an acceptor. As shown on the right side of Figure 10.5(c), combining these two structures allows the energetically undesirable electron hole and excess electron from the basic and acidic configurations, respectively, to satisfy one another and increase the stability of the overall system. DFT calculations confirm this expectation. The left side of Figure 10.5(c) shows the GGA-calculated relaxed geometry of one such NO_2 "cooperative" pair comprised of two NO_2 within one MgO(0 0 1) 2×2 supercell. The structure of the cooperative pair is obviously much different from that of the isolated adsorbates. In particular, the acidic NO_2 relaxes dramatically towards and binds with the underlying surface O ion, forming a structure suggestive of a nitrate (NO_3^-) ion, while the basic NO_2 relaxes towards a nitrite (NO_2^-) structure. The total adsorption energy of this pair is calculated to be -31 kcal mol^{-1}, or more than twice that of the individual NO_2 in isolation [51, 55]. This enhanced adsorption is consistent with the observed chemisorption of NO_2 as mixed nitrite and nitrate on MgO(0 0 1) [61, 65]. Further, the calculated charge distribution about each adsorbed NO_2 is consistent with the transfer of nearly a full electron from the acidic to the basic site [44, 55]:

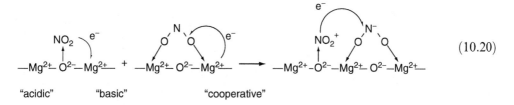

$$-Mg^{2+}-O^{2-}-Mg^{2+}-(NO_2,\ \text{"acidic"}) + -Mg^{2+}-O^{2-}-Mg^{2+}-(NO_2,\ \text{"basic"}) \longrightarrow -Mg^{2+}-O^{2-}-Mg^{2+}-O^{2-}-Mg^{2+}-(NO_2^{+}\ \cdots\ N^{-}O_2,\ \text{"cooperative"}) \tag{10.20}$$

The pairing up of electrons shown in Figure 10.5(c) underpins this enhancement in adsorption: if this pairing is destroyed by imposing a triplet electron configuration on the cooperative pair, one recovers the same structures and energetics as that of the isolated adsorbates.

Reaction (10.20) suggests an alternative analysis of the cooperative adsorption effect in terms of ionic, rather than neutral, adsorbates [44, 55]. NO_2 has both a relatively low ionization energy and high electron affinity, so that the energy to transfer an electron between two gas-phase molecules costs a relatively modest 169 kcal mol^{-1} [70, 71]:

$$2NO_2 \longrightarrow NO_2^+ + NO_2^- \tag{10.21}$$

NO_2^+ is isoelectronic with CO_2 and thus is expected to behave as a Lewis acid towards MgO:

$$-Mg^{2+}-O^{2-}-Mg^{2+}- + \overset{+}{O{=}N{=}O} \longrightarrow -Mg^{2+}-O^{2-}(-NO_2)^{+}-Mg^{2+}- \tag{10.22}$$

while NO_2^- acts as a Lewis base:

$$-Mg^{2+}-O^{2-}-Mg^{2+}- + NO_2^- \longrightarrow -Mg^{2+}(O)-O^{2-}-Mg^{2+}(O)-\ [\ddot{N}O_2] \tag{10.23}$$

A DFT cluster simulation is suitable for charged adsorbates, and Figure 10.6 shows the calculated NO_2^+ and NO_2^- adsorbate structures on MgO clusters evaluated at the GGA level with the local orbital DMol code [44]. Consistent with the charge transfer model, the structures of the ionic adsorbates are closer to those found in the NO_2 cooperative pair (Figure 10.5c) than to those of isolated NO_2 adsorbates (Figure 10.5a and b). The adsorption energies of the cation and anion (−106 and −38 kcal mol^{-1}, respectively) are greater than the isolated acidic and basic NO_2 as well (−10 and −15 kcal mol^{-1}, respectively, within the same cluster model).

These adsorption energy enhancements are large, but not large enough to overcome the energy cost to transfer an electron between two NO_2 — the net energy of Reactions (10.21)–(10.23) is approximately +25 kcal mol^{-1}. The missing piece of the analysis is the electrostatic stabilization provided by combining two oppositely charged adsorbates on the MgO surface:

$$[-Mg^{2+}-O^{2-}(NO_2)-Mg^{2+}-]^{+} + -Mg^{2+}-O^{2-}-Mg^{2+}-(\ddot{N}O_2) \longrightarrow -Mg^{2+}-O^{2-}(NO_2)^{+}-Mg^{2+}-O^{2-}-Mg^{2+}-(NO_2) \tag{10.24}$$

The energy of Reaction (10.24) is difficult to calculate directly, but by constructing an appropriate thermodynamic cycle and using the results described above, a value of −55 kcal mol^{-1} can be inferred for the cooperative structure shown in Figure 10.5(c) [55]. This large

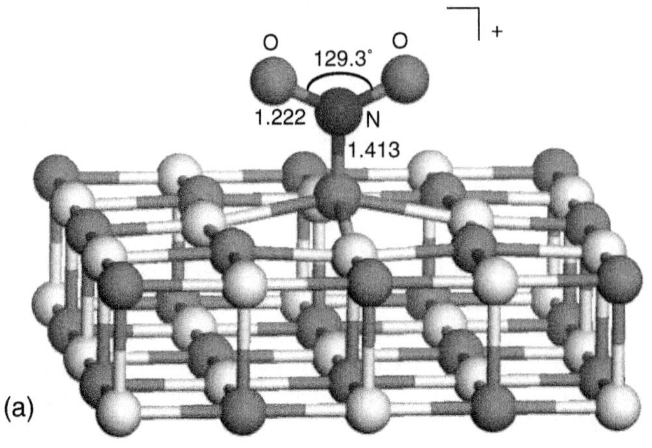

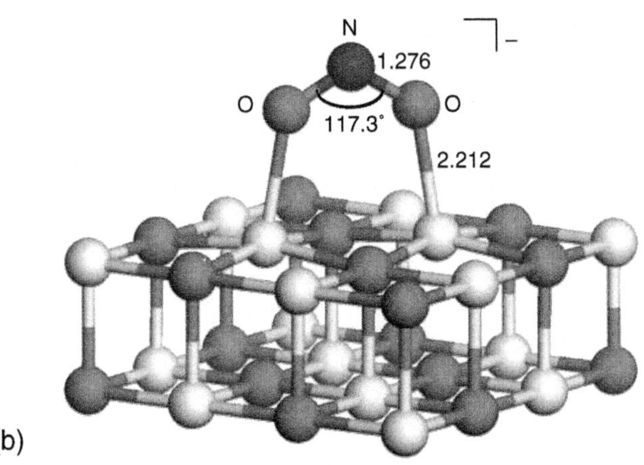

FIGURE 10.6 PW91-GGA-calculated geometries of (a) NO_2^+ and (b) NO_2^- adsorbed on $(MgO)_n$ clusters. The outermost Mg and O ions in each model are fixed at their bulk locations to simulate the constraints imposed by an extended crystal. Bond distances reported in Ångstroms.

lateral interaction is essential to the overall stability of the NO_2 cooperative pair. As expected for an electrostatic interaction, its strength diminishes with increasing adsorbate separation [55], but up to a separation of a couple lattice constants is great enough to preserve cooperative pairing.

NO_2 chemisorption on MgO can thus be understood in terms of the electronic structure of NO_2 and its ions and their interactions with the MgO surface. The discussion is not limited to NO_2 amongst NO_x or to MgO amongst metal oxides. Like NO_2, odd electron NO and NO_3 are readily oxidized or reduced, interact strongly with the MgO(0 0 1) surface as ions, and can form a number of cooperative pairs [55]. In particular, NO and NO_2 adsorbed as a cooperative pair provides a model for a nitrited MgO surface (Figure 10.7a), and similarly NO_2 and NO_3 as a cooperative pair represents a nitrated MgO surface (Figure 10.7c). In each case the cooperative pair is bound to the surface approximately twice as strongly as the individual adsorbates in isolation. The same analysis can be extended to the other alkaline earth oxides [51]; as shown in Figure 10.7 the absolute adsorption energies increase down the family while the cooperative enhancement remains approximately constant. The results for BaO are in

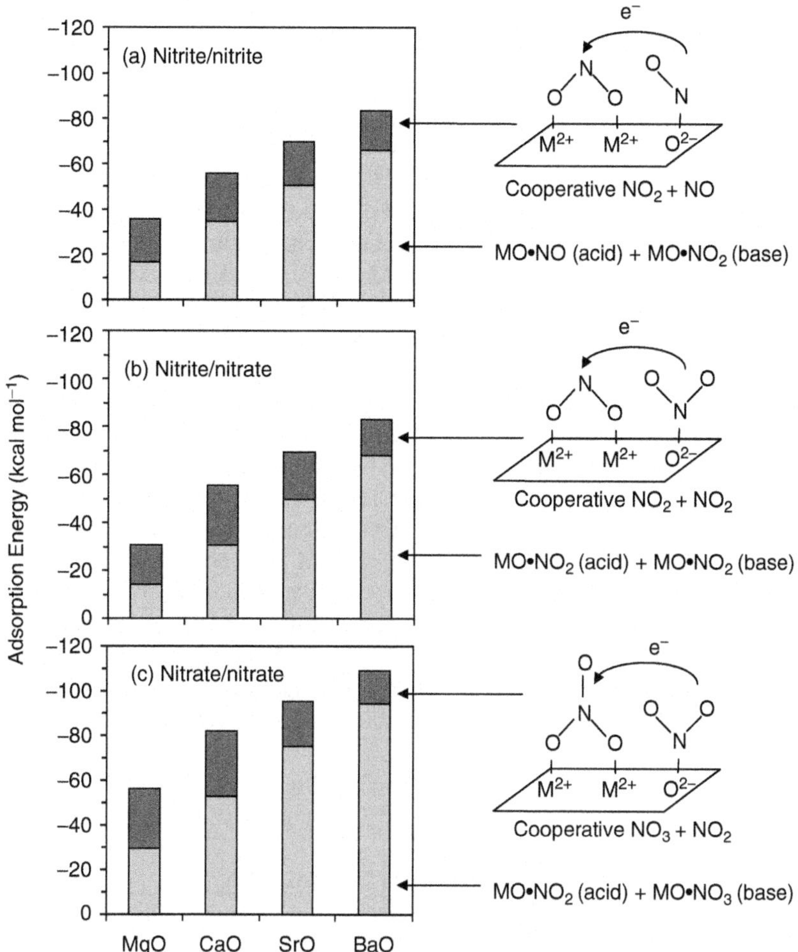

FIGURE 10.7 Sum of isolated NO_x adsorption energies (light bars) along the series of (0 0 1) alkaline earth oxide surfaces, and the additional stabilization imparted by cooperative adsorption (dark bars). Plots (a), (b), and (c) correspond to coadsorption of NO and NO_2, of two NO_2, and of NO_2 and NO_3, respectively.

particular in good agreement with adsorption energies inferred from modeling of NO_x adsorption kinetics [72].

The unusual electronic structure of NO_x results in a rich and nonintuitive adsorption chemistry on the alkaline earth oxides. We have covered only some of the basic concepts in simulating this chemistry here. Many more questions remain to be addressed, such as the effects of coverage, surface defects, other coadsorbates (e.g., water) in modifying NO_x adsorption chemistry, as well as modifications in the chemistry from the basic to transition metal [62] and mixed metal oxides. Further, a key practical question to consider is the connection between the surface models discussed here and the bulk-like products often observed in environmental catalytic applications. Preliminary results suggest surprisingly simple linear relation between the surface and bulk stability of adsorbed NO_x on the alkaline earth oxides, but the robustness and generality of this relationship requires further exploration [51].

10.5 REACTIONS ON METAL SURFACES: NO OXIDATION

NO oxidation:

$$2NO(g) + O_2(g) \longleftrightarrow 2NO_2(g) \quad (10.7)$$

This reaction is perhaps most fundamental to NO_x remediation, as the NO $\leftrightarrow$ NO_2 equilibrium plays a role both in the ability of NO_x to be stored and to be catalytically reduced. NO oxidation is an unusual reaction in that, as is clear from Figure 10.1, the equilibrium shifts from products to reactants at only 400°C in the presence of excess O_2 and at even lower temperatures at diminishing background O_2 levels. This low-temperature equilibrium reflects both the unfavorable standard entropy change of the reaction ($\Delta S^{\circ}_{298} = -35.0\,\text{cal mol}^{-1}\,\text{K}^{-1}$, Table 10.1) and the fact that two ON—O bonds (72.8 kcal mol^{-1} each) are only slightly stronger than one O=O double bond (117.9 kcal mol^{-1}), making Reaction (10.7) exothermic by only -27.3 kcal mol^{-1}. In contrast, the analogous CO oxidation reaction (Reaction 10.9) is exothermic by -135.3 kcal mol^{-1}. As a result, NO oxidation catalysis is more sensitive to the nature and strength of the interactions of the reactants and products with the catalyst surface than is the more thermodynamically preferred CO oxidation reaction. Further, gas-phase NO oxidation has an appreciable rate constant at modest temperatures, but the second-order dependence on NO makes the reaction slow at typically interesting concentrations [73].

Supported Pt is the prototypical NO oxidation catalyst, and several studies have explored the kinetics of NO oxidation over Pt particles [72, 74–76]. The reaction mechanism is not well established, but available evidence points towards a Langmuir–Hinshelwood type mechanism in which adsorbed NO and dissociated O_2 combine to form NO_2 [76]:

$$2Pt + O_2(g) \longleftrightarrow 2Pt\text{—}O \quad (10.25)$$

$$Pt + NO(g) \longleftrightarrow Pt\text{—}NO \quad (10.26)$$

$$Pt\text{—}O + Pt\text{—}NO \longleftrightarrow Pt\text{—}NO_2 \quad (10.27)$$

$$Pt\text{—}NO_2 \longleftrightarrow Pt + NO_2(g) \quad (10.28)$$

DFT simulations have been used to study this chemistry [77, 78] on the close-packed Pt(1 1 1) surface and, in particular, point out the sensitivity of the reaction parameters to the lateral interactions between neighboring adsorbates on the surface. In what follows we will use N, O, and NO_x adsorption on the Pt(1 1 1) surface to illustrate the simulation of NO_x reactivity on metal surfaces.

Pt (and all metals) differs from the alkaline earth oxides described above in having no bandgap and highly delocalized (metallic) bonding. For these reasons, metal surface reactions are most often simulated within a periodic supercell/slab model. By Bloch's theorem, the periodic supercell wave functions can be written as a product of supercell-invariant part and a periodic, plane-wave part:

$$\psi_i^k(\mathbf{r}) = \psi_i^{\text{o}}(\mathbf{r})e^{i\mathbf{k}\cdot\mathbf{r}} \quad (10.29)$$

where **k** is a wavevector within the first Brillouin zone of the periodic lattice [79]. Equation (10.29) means that the infinitely repeated supercell has associated with it an infinitely many **k**s and thus infinitely many wave functions. In practical supercell DFT calculations only a discrete number of **k** points can be sampled: fewer (or perhaps only one) **k**-points are necessary for large supercells and systems with large bandgaps, as with the alkaline earth oxide systems considered above, while more are required for smaller supercells and vanishing bandgaps. The **k**-points are generally chosen using symmetry-based conventions [80]. Bulk Pt has the face-centered cubic (fcc) structure, and Figure 10.8 shows the variation of the PW91-GGA lattice constant and total energy with number of symmetry-unique **k**-points, computed with Vasp [30] using PAW core potentials [57] and a 400 eV plane-wave cutoff. Convergence is slow and not smooth, requiring approximately 60 **k**-points for the lattice constant and more than 100 for the total energy. (By way of comparison, the analogous bulk MgO calculations in the primitive fcc cell converge with an order of magnitude fewer **k**-points, and the MgO surface calculations discussed above required only a single **k**-point.) The converged Pt lattice constant (2.818 Å) again slightly overestimates the experimental value of 2.774 Å [81].

Pt also differs from the alkaline earth oxides in its preferred surface termination. Electrostatics cause the highly ionic oxides to prefer a surface plane with a zero net charge, as provided by the (0 0 1) plane; in contrast, metallic Pt prefers the more densely packed (1 1 1) termination because it retains the maximum number of nearest neighbors to each surface atom. Figure 10.9(a) shows a representative hexagonal supercell for the Pt(1 1 1) surface. In this model a four-layer-thick slab of metal is separated from its periodic images by approximately 12 Å of vacuum, large enough to limit NO_x adsorbate interactions to one slab face. A 2×2 array of symmetry equivalent Pt are included per layer, and with these lateral dimensions the slab energy is well converged with a 7×7 **k**-point mesh (eight symmetry unique **k**-points) within the slab plane. The (1 1 1) surface relaxes little from the ideal bulk termination, and as with the oxides, it is common practice to fix the locations of the bottom-most layers of metal atoms at their bulk locations when simulating surface chemistry.

As shown in Figure 10.9(b), the Pt(1 1 1) surface presents several candidate binding sites for adsorbed NO_x species, including atop Pt, bridging two surface Pt, and two different types of threefold hollows (designated hcp and fcc to distinguish those with a Pt atom directly below the hollow from those with a Pt atom two layers below). Even more so than for the metal oxides, it is difficult to anticipate a *priori* preferred adsorbate sites and orientations on a

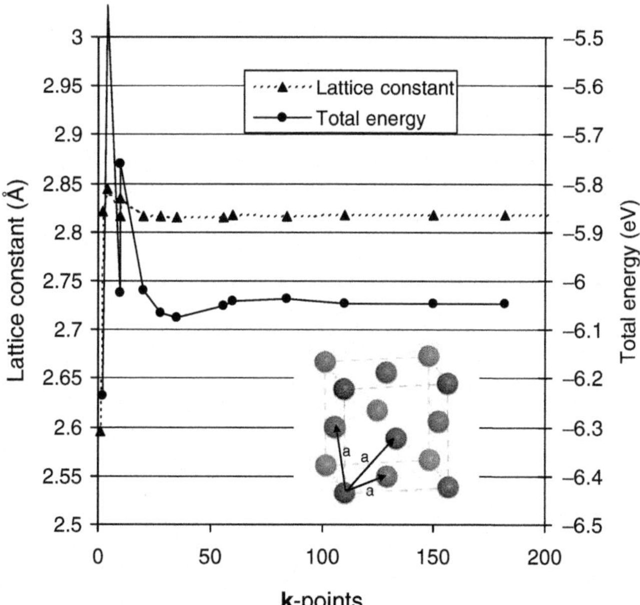

FIGURE 10.8 Convergence of DFT-calculated Pt bulk lattice constant and total energy with number of symmetry-unique **k**-points. PAW/PW91-GGA calculations performed using a primitive face-centered cubic (*fcc*) cell and Vasp. The inset shows the conventional fcc cell and the primitive fcc lattice vectors.

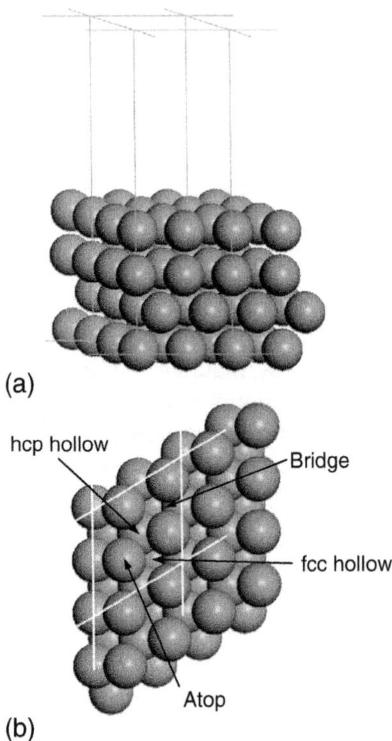

FIGURE 10.9 Perspective (a) and top (b) views of a four-layer, 2 × 2 supercell model of the Pt(1 1 1) surface, along with the locations of several high-symmetry adsorption sites.

metal surface. The chemical intuition that guides prediction of oxide adsorption is often less applicable to metals, and concepts like the d-band center that prove so useful for understanding trends in adsorption energy among metals [82] may not help in understanding adsorbate siting. Further, the energy differences between different adsorption sites can be small and difficult to predict accurately using DFT, as has been demonstrated in detail for CO adsorption on Pt(1 1 1) [83]. The available experimental and computational evidence indicates that NO prefers the fcc hollow site [84–87] while NO_2 preferentially bridges two Pt [78, 88, 89], and we will limit ourselves to those configurations here.

Figure 10.10 shows the PAW/PW91-calculated structures of adsorbed atomic N and O and molecular NO and NO_2 on the four-layer Pt(1 1 1) slab. All four interact strongly with the surface, and NO and NO_2 are significantly perturbed from their gas-phase geometries (Table 10.2). The interactions of these adsorbates with the metal surface are qualitatively different from those discussed above with the metal oxides, as can be understood from the schematic orbital interaction diagram shown in Figure 10.11. The discrete atomic Pt 5d orbitals smear into a broad band of states in Pt metal, with the Fermi level, i.e., the highest occupied energy level, in the middle of this band. The low-lying NO 5σ and 1π orbitals (Figure 10.3) interact rather weakly with Pt; rather, the Pt—NO bonding is dominated by mixing of the singly occupied NO 2π orbitals with the Pt $5d$ manifold. To a first approximation this interaction forms discrete bonding and antibonding orbitals, the former below the Fermi level and occupied, the latter above the Fermi level and vacant [82, 90–92]. Unlike the insulating oxides, the unpaired electron from NO is readily incorporated into the sea of Pt 5d energy states — the surface–adsorbate system is nonmagnetic and charge-transfer-

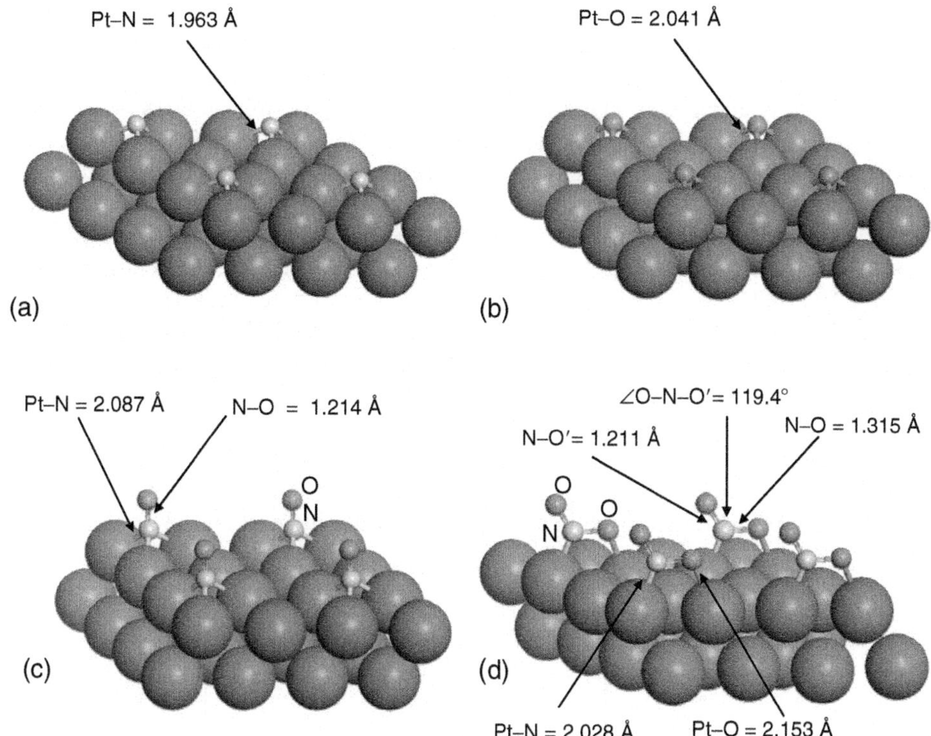

FIGURE 10.10 PAW/PW91-GGA calculated structures of (a) N, (b) O, (c) NO, and (d) NO_2 at 1/4 monolayer coverage on Pt(1 1 1). The bottom two Pt layers of the four-layer slab are fixed at the bulk locations and hidden for clarity.

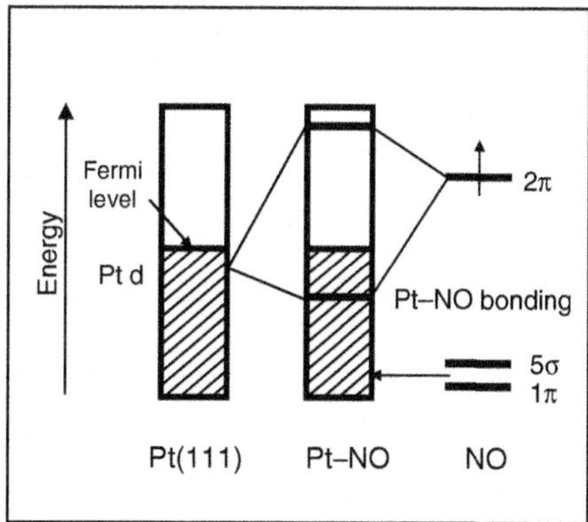

FIGURE 10.11 Qualitative molecular orbital interaction diagram for NO adsorption on Pt(1 1 1).

based cooperative effects are not expected to appreciably modify adsorption. NO binds normal to the (1 1 1) plane to maximize overlap of both NO 2π components with the metal surface; net donation of electron density from the Pt surface into these antibonding orbitals increases the N—O separation by 0.05 Å. NO_2 binds to the Pt(1 1 1) surface similarly; again mixing of its antibonding valence orbitals with the Pt 5d band produces a nonmagnetic and bond-expanded adsorbate.

Calculated adsorption energies can be sensitive to slab thickness as well as to surface coverage. Figure 10.12(a) shows the dependence of atomic N(^{4}S), O(^{3}P), and NO_x adsorption energies on Pt slab thickness at 1/4 monolayer coverage (one adsorbate for every four surface atoms of Pt). These adsorption energies are referenced to the spin-polarized, ground state atoms and molecules, simulated using large supercells. The absolute adsorption energies are uniformly larger on the three-layer slab but converge to within 1 kcal mol^{-1} at four Pt layers for all adsorbates except NO_2. The general energy trends are the same on all the slabs: the atoms bind most strongly to Pt(1 1 1), while NO binds somewhat more strongly than does NO_2 [78]. The large binding strength of the atoms accounts for the greater tendency of metals to dissociate NO_x [92] than for metal oxides to do so. The absolute binding energies tend to be overestimated by the PW91-GGA [93]: the calculated 1/4-ML NO adsorption energy of −45 kcal mol^{-1} can be compared with −25 kcal mol^{-1} inferred from the temperature programmed desorption (TPD) of NO from Pt(1 1 1) [94]; the calculated O(^{3}P) adsorption energy (−102 kcal mol^{-1}) overestimates the TPD-derived value of −75 to −80 kcal mol^{-1} [95, 96] by a similar amount.

The lateral interactions between adsorbed N, O, and NO_x are uniformly repulsive, and as shown in Figure 10.12(b), the absolute adsorption energies increase by several kcal mol^{-1} as

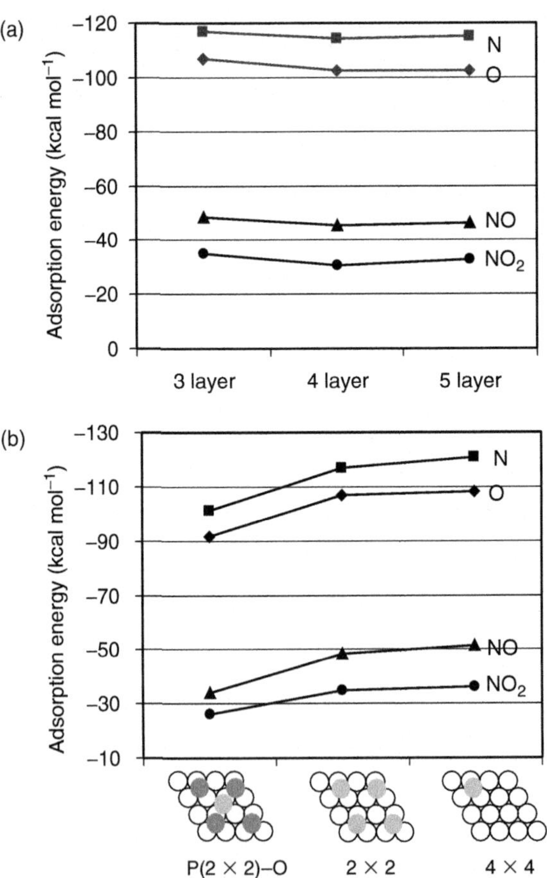

FIGURE 10.12 (a) PAW/PW91-GGA-calculated adsorption energies on Pt(1 1 1) as a function of slab thickness at 1/4-monolayer coverage. (b) PAW/PW91-GGA-calculated adsorption energies on a three-layer-thick Pt(1 1 1) slab as a function of surface coverage: 1/16-monolayer (4 × 4 supercell) and 1/4-ML (2 × 2 supercell) on an otherwise bare Pt(1 1 1) surface, and 1/16-ML on a Pt(1 1 1)-P(2 × 2)-O surface.

the surface coverage is reduced from 1/4 ML (evaluated using a three-layer-thick, 2 × 2 supercell) to 1/16 ML (from a three-layer-thick, 4 × 4 supercell). The long-range lateral interactions between atomic O can be modeled as a collection of surface-normal dipoles and the energy extrapolated to zero coverage [97, 98]. The difference between 1/16 ML and extrapolated zero coverage adsorption energies is estimated to be less than 1 kcal mol^{-1} [97]; likewise the N and NO_x adsorption energies are likely close to their low-coverage limits at 1/16 ML.

As noted above, the NO oxidation equilibrium is particularly sensitive to O_2 because of the close balance between ON—O and O=O bond strengths. One useful way to examine the relationship between the Pt(1 1 1) surface and NO_x chemistry, then, is to consider the thermodynamics of adsorbed species as a function of O_2 exposure. In principle, detailed adsorbate surface phase diagrams can be constructed from calculated, coverage-dependent adsorption energies and other thermodynamic parameters [99, 100], but for illustrative purposes we adopt a simpler model of relative adsorbate stability in two limits: low total surface coverage and high O coverage. For the low-coverage limit we use the 1/16 ML adsorption energies discussed above, but add 20 kcal mol^{-1} to all adsorption energies to account for the systematic overbinding of the PW91-GGA. For the high O coverage limit we start with the P(2 × 2)—O ordered phase that forms on the fcc lattice at modest O_2 exposures and ambient temperatures [101], simulated with four O adsorbates within a three-layer-thick, 4 × 4 supercell, and calculate the adsorption energies for N, O, and NO_x in one of the unoccupied fcc sites. Figure 10.12(b) shows that the absolute adsorption energies of N and NO_x are decreased on the order of 25 kcal mol^{-1} by destabilizing lateral interactions with surface O, while NO_2 is destabilized to a somewhat lesser extent. Sites other than the fcc may be preferred at this high O coverage (see Ref. [90]), and in fact the greater stabilization of NO_2 in part reflects the greater relaxation allowed to this lower symmetry adsorbate [78]. As with the low-coverage energies, we arbitrarily decrease the Pt(2 × 2)—O absolute binding energies by 20 kcal mol^{-1} to approximately correct for PW91-GGA overbinding errors.

These adsorption energies can be put on a single energy scale by combining them with the gas-phase O—O, N—O, and ON—O bond strengths (Table 10.2). Like the surface adsorption energies, the bond energies are overestimated by about 20 ± 5 kcal mol^{-1} by the PW91-GGA, and we use the experimental values in constructing the energy scale. Taking the zero of energy to be the bare Pt(1 1 1) (or P(2 ×2)—O) surface plus gas-phase NO + 1/2O_2, the relative adsorbate stabilities calculated are as shown in Figure 10.13. The surface coverage is clearly seen to have a large effect on the thermodynamically preferred surface species. In the low-coverage limit, the large Pt(1 1 1)—O and Pt(1 1 1)—N bond strengths favor dissociation of O_2 and NO_x, and the lowest energy surface state is adsorbed atomic N and O. Adsorbed molecular NO is slightly higher in energy and adsorbed NO_2 nearly 30 kcal mol^{-1} higher yet. Thus, the gas-phase exothermicity of Reaction (10.7):

$$2NO(g) + O_2(g) \longleftrightarrow 2NO_2(g) \tag{10.7}$$

is reversed on bare Pt(1 1 1), and in this low-coverage limit the surface prefers to dissociate NO and NO_2 rather than to oxidize NO to NO_2 [78]. The Pt(1 1 1) surface is known to dissociate NO_2 to NO and O at low coverage [89, 96, 102] but not to dissociate NO [92, 100, 103]. This discrepancy may in part arise from uncertainties in the thermodynamic picture in Figure 10.13, but it also likely reflects the dissociation kinetics playing a role in tipping the balance between dissociative and molecular adsorption [86].

The presence of surface O has a large effect on the relative surface stabilities, as shown on the right side of Figure 10.13. The decrease in the number of destabilizing lateral interactions clearly increases the stability of the more highly oxidized adsorbates in this limit. Dissociation of NO to the atoms becomes approximately thermoneutral. Dissociation of NO_2 to NO and O

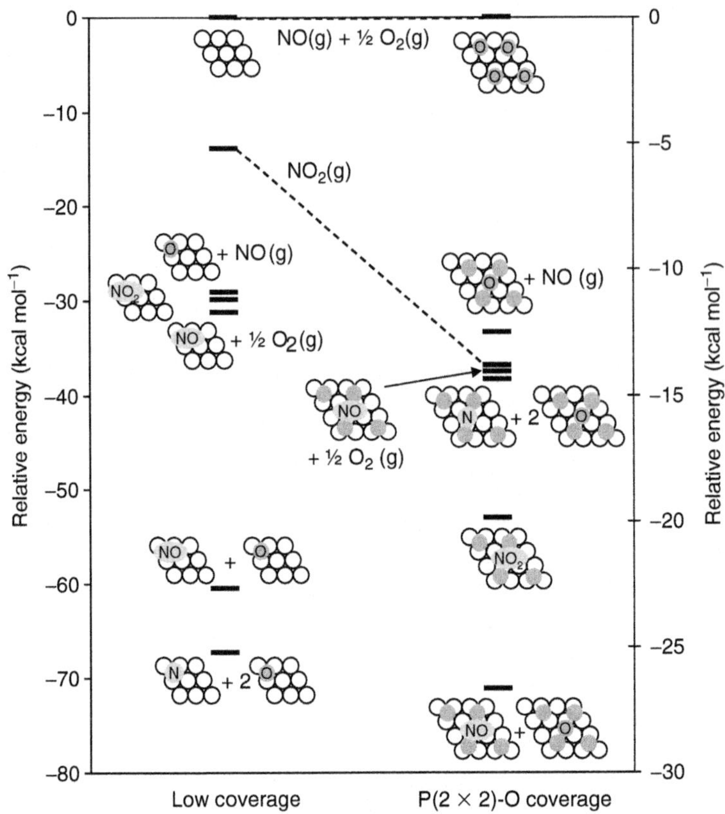

FIGURE 10.13 Stabilities of NO_x surface adsorbates on Pt(1 1 1) relative to gas-phase NO + $1/2O_2$ in the low-coverage and high-O-coverage limits. The single energy scale is constructed by adding 20 kcal mol^{-1} to the PAW/PW91-GGA absolute adsorption energies to compensate for systematic overbinding and combining with experimental O═O, N═O, and O—NO bond energies.

remains thermodynamically favored — reflecting the weakness of the ON—O bond relative to Pt(1 1 1)—O even in this limit — but the driving force is decreased from 30 to only several kcal mol^{-1}. The results are consistent with the observation that NO_2 exposure can induce O coverages >1/4 ML on Pt(1 1 1) [89]. NO oxidation to NO_2, to the extent it occurs on Pt(1 1 1), would appear to require even higher O coverage and concomitantly lower Pt—O bond energies.

Figure 10.13 is based on a number of crude approximations and is only intended to illustrate the range of possible behaviors of NO_x on Pt and other metal surfaces, and again these results say nothing about the kinetics or dynamics of any surface processes. For instance, the analogous CO oxidation on Pt(1 1 1) has been shown to occur at the boundaries of O-covered and CO-covered domains [104], and similar domain effects may operate in NO oxidation. Simulation of oxidation kinetics involves the calculation of (coverage-dependent) energy barriers and prefactors. The barriers to NO + O combination on Pt(1 1 1) have been investigated computationally by constraint [77] and more reliable nudged elastic band [78] methods, and the results coupled with kinetic Monte Carlo simulations to simulate coverage-dependent NO oxidation kinetics [78].

As with the metal oxide surface chemistry, NO_x chemistry on metal surfaces is both rich and not entirely intuitive. The discussion here illustrates some of the computational and physical factors to be considered in simulating the metal surface reactivity.

10.6 REACTIONS ON METAL-EXCHANGED ZEOLITES: NO DECOMPOSITION

As a final example of molecular simulation applied to NO_x catalysis, we return to the NO decomposition reaction:

$$\begin{matrix} O & & O & & O{=}O \\ \| & + & \| & \longrightarrow & + \\ N & & N & & N{\equiv}N \end{matrix} \tag{10.30}$$

As noted above, the homogeneous decomposition of gas-phase NO is immeasurably slow at ambient conditions, despite having a negative free energy of reaction and an apparently simple reaction pathway through a four-membered-ring transition state. This kinetic inertness derives from the fact that, along an adiabatic (electronic state preserving) potential energy surface in which two NO collide side-on to produce N_2 and O_2, the ground state NO orbitals evolve into excited state orbitals of N_2 and O_2 [105, 106]. The reaction is said to be "orbital symmetry forbidden:" the probability of hopping from the reactant to the product ground states is low and the reaction thus intrinsically slow. Rather, in the gas-phase and at high temperatures, NO decomposition is initiated by a disproportionation:

$$2NO(g) \longleftrightarrow N_2O(g) + O(^3P,\ g) \tag{10.31}$$

Reaction (10.31) is symmetry allowed [107] but has a large activation barrier ($E_a = 65.2\,\text{kcal mol}^{-1}$) [108], in part reflecting the high reaction endothermicity ($\Delta H^{\circ}_{298} = +36.0\,\text{kcal mol}^{-1}$, Table 10.1).

Traditional metal and metal oxide catalysts are generally ineffective towards NO decomposition [11, 109], but Cu-exchanged zeolites, in particular Cu-ZSM-5, have notably high catalytic activity [22, 110, 111]. Metal-exchanged zeolites straddle the boundary between homogeneous and heterogeneous catalysis: nanoporous, crystalline zeolites $M^{n+}_{x/n}[(AlO_2^-)_x(SiO_2)_{1-x}]$ are in one sense very high surface area silicon–aluminum oxide catalysts, but their catalytic reactivity tends to be associated with the charge-compensating M^{n+} cations that behave essentially as discrete reaction centers (Figure 10.14) [112–114]. Metal-exchanged zeolites thus have elements that suggest both supercell and cluster representations. In practice, the zeolite unit cells are large enough that suitably constructed clusters, possibly supplemented with an empirical embedding potential [115], are more generally useful for simulating catalytic reactivity [116].

The basic building blocks of a zeolite framework are corner-sharing SiO_4 and AlO_4 tetrahedra (or T-sites). The Al-substituted sites are locally anionic and Lewis basic; the catalytic activity of Cu-ZSM-5 is associated with Cu^+ ions bound to these basic T-sites. For brevity, we refer to a Cu^+ ion coordinated to the zeolite framework as ZCu. For the purposes of exploring NO decomposition mechanisms, we adopt an abbreviated ZCu model that lends itself to ready simulation and analysis. The model includes a single Al^{3+} T-site, its surrounding complement of O^{2-} anions, and terminating hydrogens to satisfy the oxygen valences:

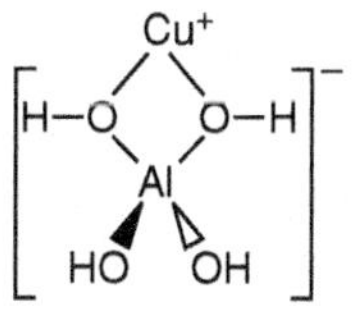

Cu^+
H–O O–H
Al
HO OH
$[\ \ ]^-$

The twofold Cu^+ coordination is well established from experiment [117, 118] and simulation [118–120], and this model can be thought of as one point along a continuum from models that include only the local Cu^+ coordination environment [48] to much more extensive

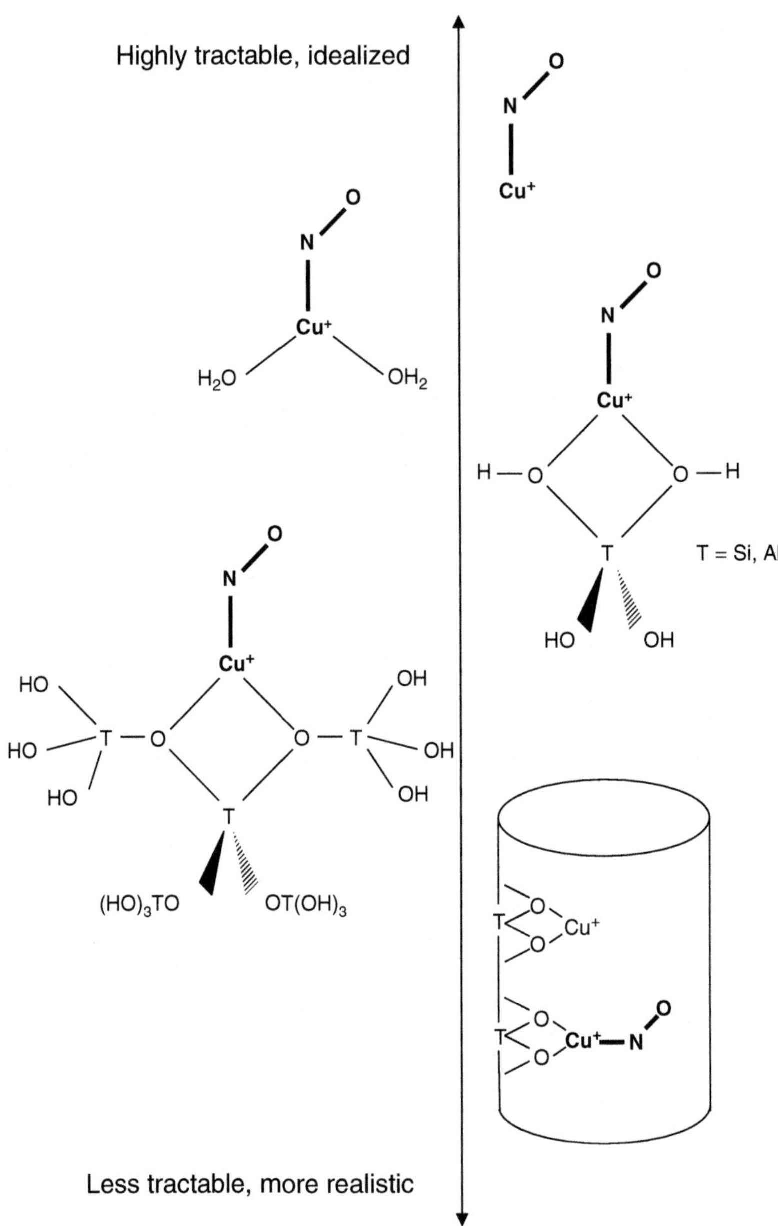

FIGURE 10.14 Schematic of highly idealized to highly detailed models possible for modeling an isolated Cu^+ ion exchanged into ZSM-5. The optimal choice depends on the nature of the questions to be addressed. Results reported here are based on the single T-site (center) model.

zeolite representations (Figure 10.14). The general trends in structure and bonding at the Cu site are largely robust to these approximations [121, 122].

The adsorption properties of the primary reactants and products of NO decomposition are readily evaluated within this "single T-site" model. The spin-polarized LDA structures, calculated using the local orbital code ADF [31], are shown in Figure 10.15 [107, 123]. Consistent with experiment, N_2 [124] and NO [125, 126] are found to form simple, N-bound coordination complexes with ZCu, including in particular both a mono- and dinitrosyl. The nature of this bonding can be understood by reference to the left side of Figure 10.16.

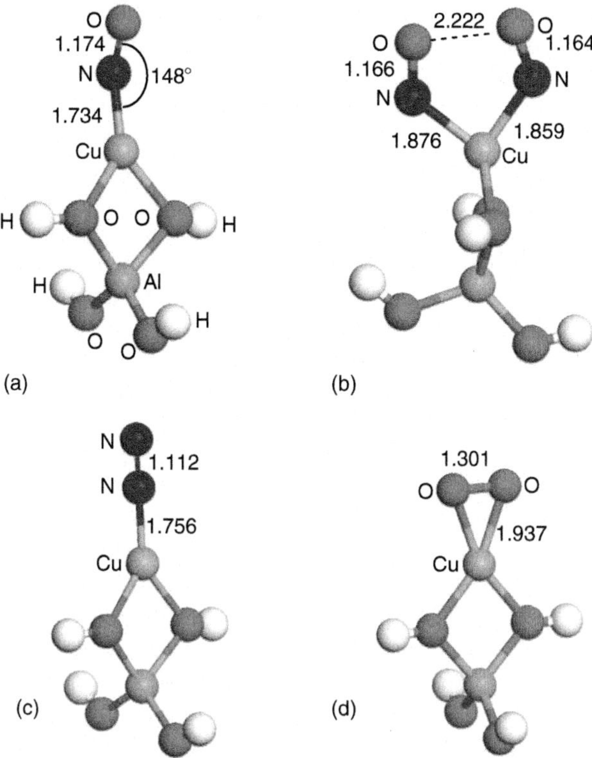

FIGURE 10.15 Spin-polarized LDA structures of ZCu-bound NO, N_2, and O_2, calculated using ADF. Bond distances reported in Ångstroms.

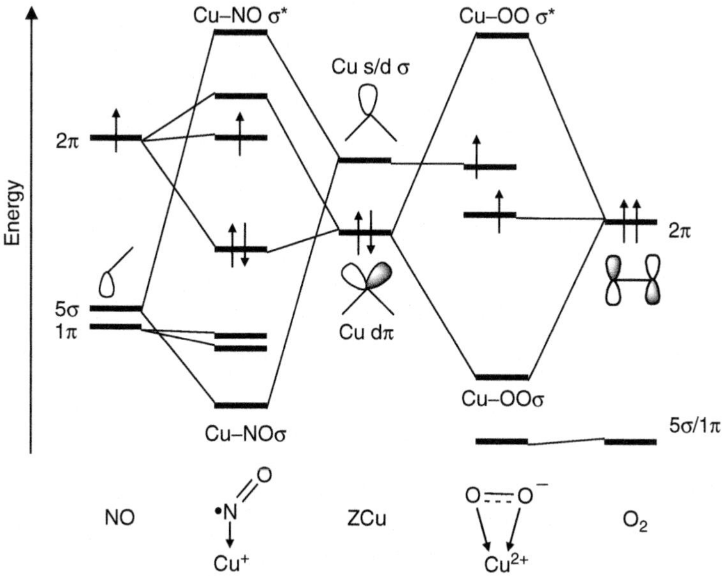

FIGURE 10.16 Schematic molecular orbital diagrams for the interaction of ZCu (center) with an NO adsorbate (left) and an O_2 adsorbate (right).

An isolated Cu^+ ion has a $3d^{10}4s^0$ electron configuration; twofold coordination to the zeolite framework mixes and splits the s and d levels, most importantly producing a 3d, π-symmetry highest occupied orbital and a 3d/4s, σ-symmetry lowest unoccupied orbital. The ZCu—NO bond is formed by charge donation from a hybrid of the filled NO 1π and 5σ orbitals (Figure 10.3) into the Cu σ level; this dative interaction is supplemented by back donation of charge from the filled Cu π into the antibonding NO 2π level, slightly elongating the N—O bond. Similarly N_2 or a second NO binds as a Lewis base to the ZCu site, coordinating with but not altering the nominal Cu^+ oxidation state. Adsorption induces a redshift in the N—O and N—N stretch modes. The magnitude of the calculated frequency shifts is sensitive to the zeolite model employed [127, 128], but the spin-polarized LDA frequencies combined with the single T-site model reasonably reproduce the observed trends [123].

The molecular orbitals of the more electronegative O_2 are all shifted downward in energy relative to NO (or N_2), so that the most important orbital interactions with ZCu are through the O_2 2π set. Charge donation here occurs primarily from ZCu to the adsorbate: O_2 binds side-on to ZCu, and the Cu—O_2 bond is formed through interaction of the filled Cu π level with the in-plane O_2 2π component. The out-of-plane 2π and Cu σ levels are less strongly perturbed by O_2 binding, and each becomes singly occupied in the final, paramagnetic product. The net result is an approximately one-electron oxidation of Cu^+ by O_2, i.e., a formally Cu^{2+}—O_2^- adduct. The square planar coordination about the Cu center is consistent with that expected for Cu^{2+}. The O—O bond is predicted to be lengthened approximately 0.1 Å and its stretch mode redshifted 300 cm^{-1} relative to gaseous O_2 [123]; this places the mode in the range of zeolite lattice vibrations, complicating its spectroscopic detection.

Two types of adsorption on ZCu can thus be distinguished, one coordinative and the other redox. The binding energies of the four adsorption complexes, calculated at the LDA geometries using the Becke exchange [37] and Perdew correlation [38] functionals, are $ZCu(NO)_2$ (-57 kcal mol^{-1}) > ZCuNO (-38) > $ZCuO_2$ (-28) > $ZCuN_2$ (-25). This particular combination of exchange and correlation functionals likely overestimates the absolute binding energies: more recent B3LYP calculations place the NO adsorption energy closer to -20 kcal mol^{-1}, for instance [122]. The $ZCu(NO)_2$ dinitrosyl is found to be particularly stable, consistent with experimental observation, and it is an obvious candidate intermediate along the pathway to catalytic NO decomposition [111]. However, as shown in Figure 10.15, the dinitrosyl shows no evidence of an incipient N—N bond. In fact, careful analysis of the dinitrosyl electronic structure shows that the orbital symmetry constraints that inhibit concerted gas-phase NO decomposition apply to adsorbed $(NO)_2$ as well [106], and thus the dinitrosyl is unlikely to contribute to NO decomposition chemistry, at least as an intermediate leading directly to N_2 and O_2.

The redox bonding of O_2 to ZCu suggests another possible intermediate along the NO decomposition pathway. NO forms a very weakly bound dimer in the gas-phase, but one-electron reduction to the $N_2O_2^-$ anion greatly enhances the N—N bonding [106]:

$$\text{O=N—N=O} \xrightarrow{e^-} [\text{O—N=N—O}]^- \quad (10.32)$$

An $N_2O_2^-$ intermediate coordinated to ZCu may be too unstable or transient to observe experimentally, but it is readily amenable to DFT simulation. Placing two NO O-down on ZCu and relaxing on the triplet (ZCu^{2+}— $N_2O_2^-$) potential energy surface yields the "hyponitrite" structure shown on the left side of Figure 10.17. As with $ZCuO_2$, the bonding here results from charge transfer from Cu to the adsorbate, and this adsorbate reduction drives the formation of a short N—N bond — encouraging evidence of its potential role in NO decomposition [107]. The metastable hyponitrite is approximately 17 kcal mol^{-1} higher in

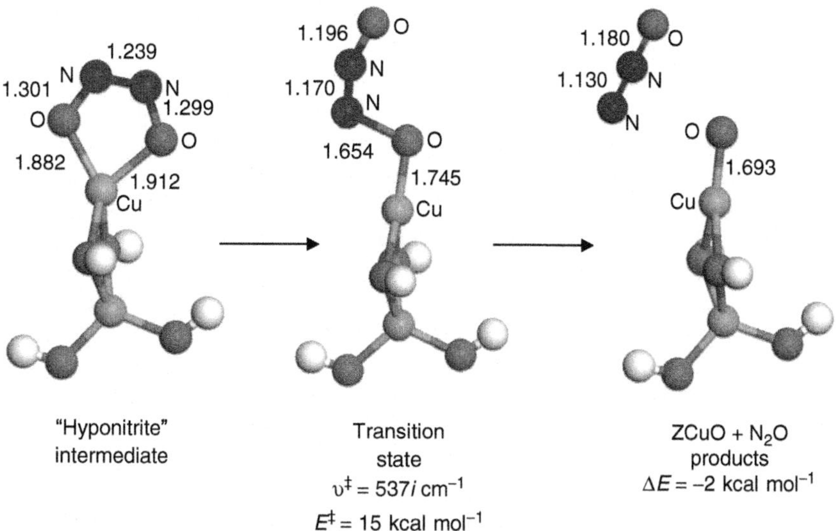

FIGURE 10.17 Spin-polarized LDA structures along a reaction pathway from the metastable, ZCu-bound hyponitrite intermediate through a transition state to ZCuO and N_2O. Bond distances reported in Ångstroms.

energy than the dinitrosyl within the model used here, implying that it would be a minor but not negligible fraction of the total concentration of "$ZCu(NO)_2$."

To determine whether the hyponitrite contributes directly or indirectly to NO decomposition catalysis requires a search for low barrier reaction pathways from the intermediate to relevant products. While the hyponitrite does not appear likely to decompose directly to N_2 and O_2 [106], in analogy with Reaction (10.31) a plausible reaction coordinate is one in which N_2O is liberated through cleavage of one O—N and one Cu—O bond.

$$ZCu(O_2N_2) \longrightarrow [ZCuO{-}NNO]^{\ddagger} \longrightarrow ZCuO + N_2O \tag{10.33}$$

A search for possible transition states along this coordinate on the LDA energy surface yields the structure at the center of Figure 10.17 [107]. LDA vibrational analysis confirms that this is a saddle point on the dissociative potential energy surface, i.e., that all vibrational frequencies are real except for one imaginary mode that points along the reaction coordinate. The Becke–Perdew GGA energy barrier with respect to the hyponitrite is calculated to be 15 kcal mol^{-1}. The overbinding that characterizes this exchange-correlation functional also leads to a general underestimation of barrier heights; a B3LYP evaluation yields a barrier approximately 13 kcal mol^{-1} higher [129]. In either case, the barrier is low relative to ZCu and two NO, and this is a thermodynamically and kinetically credible candidate step in the NO decomposition process.

Reaction (10.33) does not by itself form a complete catalytic cycle for NO decomposition. An obvious candidate reaction leading to N_2 and O_2 products combines N_2O with adsorbed O:

$$ZCuO + N_2O \longrightarrow [ZCuOO{-}NN]^{\ddagger} \longrightarrow ZCuO_2 + N_2 \tag{10.34}$$

Subsequent desorption of O_2 from ZCu would complete the cycle. This second O-atom transfer reaction is predicted to have an energy barrier of approximately 36 kcal mol^{-1} relative to reactants within the single T-site model [107, 123, 129].

Reaction barriers are expected to be particularly sensitive to the details of the zeolite model, and even in the absence of model uncertainties, the accurate calculation of barrier heights for arbitrary reactions remains a challenge for DFT (in fact for electronic structure methods in general). In the present context, the identification of reaction channels that contribute to NO decomposition is more useful than any precise reaction barrier, especially if the knowledge can be used to understand the factors that facilitate or inhibit catalytic activity. For instance, the simulation results help to explain the observation of low-temperature N_2O generation from Cu-ZSM-5-catalyzed NO decomposition [130–132]. Further, the ability of Cu to cycle between Cu^+ and Cu^{2+} oxidation states, which allows it to both drive N—N bond formation and to act as host to the product O atom and O_2 molecule, is clearly an essential ingredient to catalytic activity. Finally, the conceptual understanding that comes from the simulations form a basis for inferring behavior on other metal ions or metal ion aggregates.

As an example of this last point, recent experimental evidence supports the concept of successive O atom transfers to Cu as the basis of catalytic NO decomposition in Cu-ZSM-5 but indicates that catalytic activity is more likely associated with two exchanged Cu^+ in close proximity rather than the single exchange Cu captured in the single T-site model [130, 133–135]. Statistical analysis of the distribution of Al-substituted T-sites within ZSM-5 as a function of Si:Al ratio show that, at practically relevant ratios of around 20:1, the probability of two Al-substituted T-sites, and thus of two exchanged Cu^+, in close-enough proximity to act in concert for catalysis is actually quite high [136, 137]. Using a pair of single T-sites to describe the Cu pair coordination environment, both single-O-bridged and a number of di-O-bridged Cu pairs can be identified (Figure 10.18) and are quite strongly bound [138]:

$$\text{ZCuO} + \text{ZCu} \longrightarrow \text{ZCu—O—CuZ}, \ \Delta E = -60\,\text{kcal mol}^{-1} \tag{10.35}$$

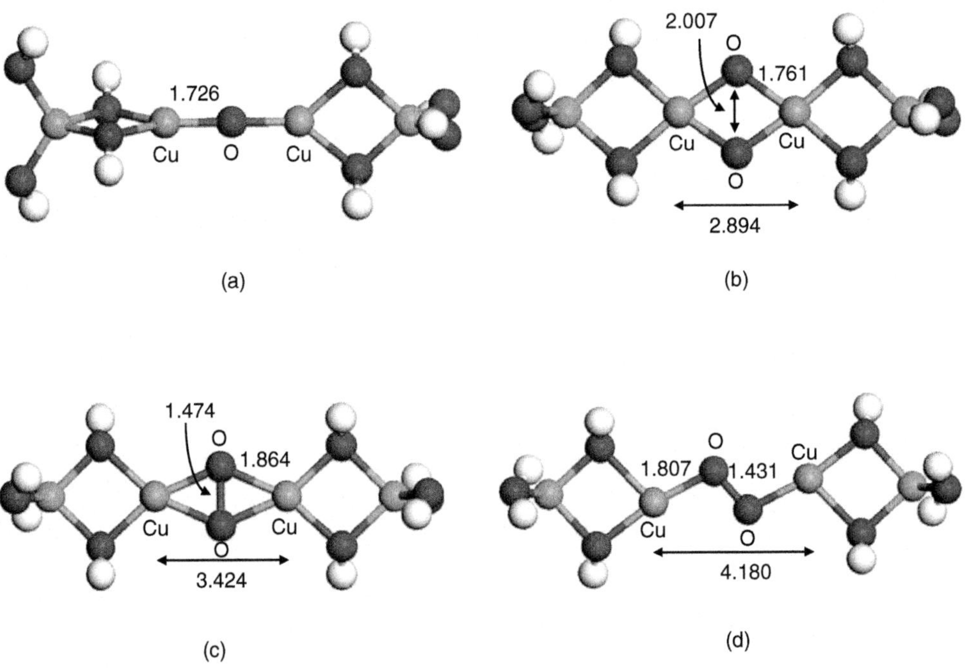

FIGURE 10.18 Spin-polarized LDA structures of O-bridged Cu pairs, calculated within the single T-site model. (a) Single O-bridged Cu-pair. (b)–(d) Three types of O_2-bridged Cu pairs observed at increasing Cu–Cu separation. Bond distances reported in Ångstroms.

$$ZCuO_2 + ZCu \longrightarrow ZCu{-}O_2{-}CuZ,\ \Delta E \sim -40\,\text{kcal mol}^{-1} \tag{10.36}$$

The di-O-bridged pair can exist in several isomeric forms depending on Cu separation, and this conformational flexibility allows the O_2-bridged pairs to retain their integrity even within the highly constrained environment of the zeolite lattice [139]. Importantly, the electronic structure of the bare and O-bridged Cu pairs are essentially the same as that of the single-Cu ZCu, ZCuO, and $ZCuO_2$ species, and thus the NO decomposition chemistry mapped out above on the single ZCu site is expected to find an exact parallel on the Cu pairs:

$$ZCuL\,CuZ + 2NO \longrightarrow ZCu{-}[O{-}N{=}N{-}O^{-}]{-}CuZ \longrightarrow ZCu{-}O{-}CuZ + N_2O \tag{10.37}$$

$$ZCu{-}O{-}CuZ + N_2O \longrightarrow ZCu{-}O_2{-}CuZ + N_2 \tag{10.38}$$

$$ZCu{-}O_2{-}CuZ \longrightarrow ZCuLCuZ + O_2 \tag{10.39}$$

The Cu pair sites are expected to be better electron donors than a single Cu, thus promoting Reactions (10.36) and (10.37) and inhibiting Reaction (10.38). In fact, evidence points to O_2 desorption as the rate-limiting step in NO decomposition on Cu-ZSM-5 [133–135].

10.7 FINAL OBSERVATIONS

Environmental NO_x catalysis is a particularly timely and fruitful area of research. Successes in the catalytic removal of NO_x have had a profoundly positive impact on environmental quality, and furthering these successes is key to making present NO_x emission reduction technologies increasingly robust and affordable and to realizing the energy efficiency benefits promised by newer lean combustion engines. On top of these societal drivers, the nitrogen oxides manifest a combination of curious properties that makes their chemistry both scientifically interesting and challenging:

- *Thermodynamic properties*: All nitrogen oxides are thermochemically unstable with respect to N_2 and O_2, but the balance between the oxides and their elements is delicate. NO_x equilibria are sensitive to environmental conditions, in particular to N_2 and O_2 concentrations and to temperature. The $NO \leftrightarrow NO_2$ equilibrium in particular is dynamic under practically relevant conditions.
- *Kinetic properties*: NO and NO_2 are kinetically inert with respect to elemental decomposition but readily interconvert and react with other gas-phase species.
- *Electronic structure*: NO, NO_2, and even NO_3 are rare example of stable, odd-electron free radicals. The molecules themselves are difficult to treat accurately within an electronic structure framework, and thus simulation of their homogeneous and heterogeneous chemistry is particularly challenging.
- *Heterogeneous reactivity*: As a consequence of their unusual electronic structures, NO_x molecules defy a single chemical classification. Molecular NO and NO_2 have both large electron affinities and low ionization potentials. In different environments, they can behave as one-electron acceptors or donors, as Lewis acids or bases.

In this chapter, we have illustrated how these characteristics come together to influence NO_x chemistry in different ways on different catalytic materials, and in particular how molecular electronic structure simulations based on DFT can be used to rationalize and anticipate this chemistry. The metal oxide, metal, and zeolite systems considered here exhibit distinctly different reactivates towards NO_x, but DFT simulations allow all of this chemistry to be described from a consistent theoretical foundation. The art of NO_x catalyst simulation — in fact of all heterogeneous chemical simulation — lies in framing appropriate questions and in constructing tractable yet sufficiently complete models to answer those questions. Many scientific and technical questions remain to be answered along the path to fully predictable and controllable environmental NO_x catalysis. The contributions of molecular simulation in realizing that vision will only increase with time.

ACKNOWLEDGMENTS

The author would like to sincerely thank the many collaborators and colleagues who have inspired and educated him in the theoretical and practical aspects of NO_x catalysis. In particular, Ken Hass, Chris Goralski, Alex Bogicevic, Staffan Ovesson, Benqt Lundqvist, John Gland, Marina Miletic, Bernhardt Trout, Xi Lin, Jim Adams, Rampi Ramprasad, Bryan Goodman, Debasis Sengupta, and Donghai Sun have all contributed in various ways to the conception, execution, analysis, and interpretation of the computational models, and their contributions are warmly acknowledged and greatly appreciated.

REFERENCE

1. N.N. Greenwood and A. Earnshaw. *Chemistry of the Elements*, 2nd ed. Oxford: Butterworth-Heinemann, 1997.
2. J.H. Seinfeld, R. Atkinson, R.L. Berglund, W.L. Chameides, W.R. Cotton, K.L. Demerjian, J.L. Elston, F. Fehsenfeld, B.J. Finlayson-Pitts, R.C. Harriss, C.E. Kolb, Jr., P.J. Lioy, J.A. Logan, M.J. Prather, A. Russell, and B. Steigerwald. *Rethinking the Ozone Problem in Urban and Regional Air Pollution*. Washington D.C.: National Academies Press, 1992.
3. B.J. Finlayson-Pitts and J.N. Pitts. *Chemistry of the Upper and Lower Atmosphere*. New York: Academic Press, 2000.
4. The Plain English Guide to the Clean Air Act Report No. EPA-400-K-93-001.U.S. Environmental Protection Agency, 1993.
5. Latest Findings on National Air Quality: 2002 Status and Trends Report No. EPA 454/K-03-001. U.S. Environmental Protection Agency, 2003.
6. J.A. Miller and C.T. Bowman. *Prog Energy Comb Sci* 15:287–338, 1989.
7. J.T. Kummer. *Prog Energy Comb Sci* 6:177–199, 1980.
8. J.T. Kummer. *J Phys Chem* 90:4747–4752, 1986.
9. M. Shelef and G.W. Graham. *Catal Rev-Sci Eng* 36:433–457, 1994.
10. M. Shelef and R.W. McCabe. *Catal Today* 62:35–50, 2000.
11. V.I. Pârvulescu, P. Grange, and B. Delmon. *Catal Today* 46:233–316, 1998.
12. M.A. Weiss, J.B. Heywood, E.M. Drake, A. Schafer, and F.F. AuYeung. On the Road in 2020: A Life-Cycle Assessment of New Automobile Technologies Report No. MIT EL 00-003. MIT, Energy Laboratory Report, 2000.
13. J.W. Hoard. *Soc Automot Eng* 2001:0185, 2001.
14. R.G. Tonkyn, S.E. Barlow, and J.W. Hoard. *Appl Catal B: Environ* 40:207–217, 2003.
15. R.K. Lyon. *Environ Sci Technol* 21:231–236, 1987.
16. D. Sun, W.F. Schneider, J.B. Adams, and D. Sengupta. *J Phys Chem A* 108:9365–9374, 2004.
17. N. Miyoshi, S. Matsumoto, K. Katoh, T. Tanaka, J.Harada, N. Takahashi, K. Yokota, M. Sugiura, and K. Kasahara. *Soc Automot Eng* 1995:0809, 1995.
18. C.T. Goralski and W.F. Schneider. *Appl Catal B: Environ* 37:263–277, 2002.
19. H.Y. Afeefy, J.F. Liebman, and S.E. Stein. Neutral Thermochemical Data, in: NIST Chemistry WebBook, NIST Standard Reference Database Number 69, P.J. Linstrom and W.G. Mallard, Eds., March 2003 (National Institute of Standards and Technology, Gaithersburg MD, 20899). http://webbook.nist.gov.
20. Q.L. Trung, D. Mackay, A. Hirata, and O. Trass. *Combust Sci Technol* 10:155, 1975.
21. M. Shelef, K. Otto, and H. Gandhi. *Atmos Environ* 3:107–122, 1969.
22. M. Iwamoto and H. Hamada. *Catal Today* 10:57–71, 1991.

23. J.H. Espenson. *Chemical Kinetics and Reaction Mechanisms*. New York: McGraw-Hill, 1981.
24. S.I. Sandler. *Chemical and Engineering Thermodynamics*, 2nd ed. New York: Wiley, 1989.
25. HSC Chemistry 5.0, Outokumpu Research Oy: Pori, Finland, 2002. www.outokumpu.com/hsc.
26. R. Aris. *Elementary Chemical Reactor Analysis*. Mineola: Dover, 1999.
27. W. Kohn, A.D. Becke, and R.G. Parr. *J Phys Chem* 100:12974–12980, 1996.
28. J.M. Seminario and P. Politzer, Eds., *Modern Density Functional Theory: A Tool for Chemistry*. Amsterdam, Elsevier, 1995.
29. R.G. Parr and W. Yang. *Density-Functional Theory of Atoms and Molecules*. New York: Oxford University Press, 1999.
30. G. Kresse and J.Furthmüller. *Comp Mat Sci* 6:15–50, 1996.
31. G. te Velde, F.M. Bickelhaupt, E.J.Baerends, C. Fonseca Guerra, S.J.A. van Gisbergen, J.G. Snijders, and T. Ziegler. *J Comput Chem* 22:931–967, 2001.
32. B. Delley. *J Chem Phys* 92:508–517, 1990.
33. M.C. Payne, M.P. Teter, D.C. Allan, T.A. Arias, and J.D. Joannopoulos. *Rev Mod Phys* 64:1045–1097, 1992.
34. P. Hohenberg and W. Kohn. *Phys Rev B* 136:864–871, 1964.
35. W. Kohn and L.J.Sham. *Phys Rev A* 140:1133–1138, 1965.
36. S.H. Vosko, L. Wilk, and M. Nusair. *Can J Phys* 58:1200–1211, 1980.
37. A.D. Becke. *Phys Rev A* 38:3098–3100, 1988.
38. J.P. Perdew and W. Yue. *Phys Rev B* 33:8800–8802, 1986.
39. J.P. Perdew. *Phys Rev B* 33:8822–8824, 1986.
40. J.P. Perdew, J.A. Chevary, S.H. Vosko, K.A. Jackson, M.R. Pederson, D.J.Singh, and C. Fiolhais. *Phys Rev B* 46:6671–6687, 1992.
41. J.P. Perdew and Y. Wang. *Phys Rev B* 45:13244–13249, 1992.
42. A.D. Becke. *J Chem Phys* 98:1372–1377, 1993.
43. A.D. Becke. *J Chem Phys* 98:5648–5652, 1993.
44. M. Miletic, J.L. Gland, W.F. Schneider, and K.C. Hass. *J Phys Chem B* 107:157–163, 2003.
45. T.A. Albright, J.K. Burdett, and M.H. Whangbo. *Orbital Interactions in Chemistry*. New York: Wiley-Interscience, 1985.
46. A. Stirling, I. Pápai, J.Mink, and D.R. Salahub. *J Chem Phys* 100:2910–2923, 1994.
47. K.P. Huber and G. Herzberg. Constants of Diatomic Molecules (data prepared by J.W. Gallagher and R.D. Johnson, III) in NIST Chemistry WebBook, NIST Standard Reference Database Number 69, P.J. Linstrom and W.G. Mallard, Eds., March 2003. (National Institute of Standards and Technology, Gaithersburg MD, 20899). http://webbook.nist.gov.
48. W.F. Schneider, K.C. Hass, R. Ramprasad, and J.B. Adams. *J Phys Chem* 100:6032–6046, 1996.
49. G.N. Lewis and M. Randall. *Thermodynamics*, 2nd ed. New York: McGraw-Hill, 1961 (Revised by K.S. Pitzer and L. Brewer).
50. L.A. Curtiss, K. Raghavachari, P.C. Redfern, and J.A. Pople. *J Chem Phys* 106:1063–1079, 1997.
51. W.F. Schneider. *J Phys Chem B* 108:273–282, 2004.
52. V.H. Grassian. *J Phys Chem A* 106:860–877, 2002.
53. K.I. Hadjivanov. *Catal Rev-Sci Eng* 42:71–144, 2000.
54. V.E. Henrich and P.A. Cox. *The Surface Science of Metal Oxides*. Cambridge: Cambridge University Press, 1994.
55. W.F. Schneider, K.C. Hass, M. Miletic, and J.L. Gland. *J Phys Chem B* 106:7405–7413, 2002.
56. P.E. Blöchl. *Phys Rev B* 50:17953–17979, 1994.
57. G. Kresse and J.Joubert. *Phys Rev B* 59:1758–1775, 1999.
58. R.W.G. Wyckoff. *Crystal Structures*. New York: Wiley-Interscience, 1963.
59. A.N. Baranov, V.S. Stepanyuk, W. Hergert, A.A. Katsnelson, A. Settels, R. Zeller, and P.H. Dederichs. *Phys Rev B* 66:155117–155117-4, 2002.
60. E.J.Karlsen, M.A. Nygren, and L.G.M. Pettersson. *J Phys Chem B* 107:7795–7802, 2003.
61. J.A. Rodriguez, T. Jirsak, J.-Y. Kim, J.Z. Larese, and A. Maiti. *Chem Phys Lett* 330:475–483, 2000.
62. J.A. Rodriguez, M. Pérez, T. Jirsak, L. González, A. Maiti, and J.Z. Larese. *J Phys Chem B* 105:5497–5505, 2001.

63. P.W. Tasker, in: *Advances in Ceramics*, Vol. 10, *Structure and Properties of MgO and Al_2O_3 Ceramics*, W.B. Kingery, Ed. Columbus: American Ceramic Society, 1988.
64. J.A. Rodriguez. *Theor Chem Accts* 107:117, 2002.
65. J.A. Rodriguez, T. Jirsak, S. Sambasivan, D. Fischer, and A. Maiti. *J Chem Phys* 112:9929–9939, 2000.
66. W.F. Schneider, J.Li, and K.C. Hass. *J Phys Chem B* 105:6972–6979, 2001.
67. G. Pacchioni, A. Clotet, and J.M. Ricart. *Surf Sci* 315:337–350, 1994.
68. G. Pacchioni, J.M. Ricart, and F. Illas. *J Am Chem Soc* 116:10152–10158, 1994.
69. G. Pacchioni. *Surf Sci* 281:207–219, 1993.
70. J.E. Bartmess, in: NIST Chemistry Webbook, NIST Standard Reference Database Number 69, P.J.Linstrom and W.G. Mallard, Eds. Gaithersburg MD: National Institutes of Standards and Technology, 2001.
71. S.G. Lias, in: NIST Chemistry Webbook, NIST Standard Reference Database Number 69, P.J.Linstrom and W.G. Mallard, Eds. Gaithersburg MD: National Institutes of Standards and Technology, 2001.
72. L. Olsson, H. Persson, E. Fridell, M. Skoglundh, and B. Andersson. *J Phys Chem B* 105:6895–6906, 2001.
73. J. Olbregts. *Int J Chem Kinet* 17:835–848, 1985.
74. E. Xue, K. Seshan, and J.R.H. Ross. *Appl Catal B: Environ* 11:65–79, 1996.
75. L. Olsson, B. Westerberg, H. Persson, E. Fridell, M. Skoglundh, and B. Andersson. *J Phys Chem B* 103:10433–10439, 1999.
76. L. Olsson and E. Fridell. *J Catal* 210:340–353, 2002.
77. R. Burch, S.T. Daniells, and P. Hu. *J Chem Phys* 107:2902–2908, 2002.
78. S. Ovesson, B.I. Lundqvist, W.F. Schneider, and A. Bogicevic. *Phys Rev B* 2005, in press.
79. N.W. Ashcroft and N.D. Mermin. *Solid State Physics*. New York: W.B. Saunders, 1976.
80. H.J.Monkhorst and J.D. Pack. *Phys Rev B* 13:5188–5192, 1976.
81. H.E. Swanson and E. Tatge. *National Bureau of Standards (US), Circular* 359:1–95, 1953.
82. B. Hammer and J.K. Nørskov. *Adv Catal* 45:71–129, 2000.
83. P.J.Feibelman, B. Hammer, J.K. Norskov, F. Wagner, M. Scheffler, R. Stumpf, R. Watwe, and J.Dumesic. *J Phys Chem B* 105:4018–4025, 2001.
84. H. Aizawa, Y. Morikawa, S. Tsuneyuki, K. Fukutani, and T. Ohno. *Surf Sci* 514:394–403, 2002.
85. J.F. Zhu, M. Kinne, T. Fuhrmann, R. Denecke, and H.-P. Steinruck. *Surf Sci* 529:384–396, 2003.
86. A. Bogicevic and K.C. Hass. *Surf Sci* 506:L237–L242, 2002.
87. Q. Ge and D.A. King. *Chem Phys Lett* 285:15–20, 1998.
88. M.E. Bartram, R.G. Windham, and B.E. Koel. *Langmuir* 4:240–246, 1988.
89. M.E. Bartram, R.G. Windham, and B.E. Koel. *Surf Sci* 184:57–74, 1987.
90. M.E. Bartram, B.E. Koel, and E.A. Carter. *Surf Sci* 1989:467–489, 1989.
91. R.A. van Santen and M. Neurock. *Catal Rev-Sci Eng* 37:557–698, 1995.
92. W.A. Brown and D.A. King. *J Phys Chem B* 104:2578–2595, 2000.
93. B. Hammer, L.B. Hansen, and J.K. Nørskov. *Phys Rev B* 59:7413–7421, 1999.
94. R.J.Gorte, L.D. Schmidt, and J.L. Gland. *Surf Sci* 109:367–380, 1981.
95. C.T. Campbell, G. Ertl, H. Kuipers, and J.Segner. *Surf Sci* 107:220–236, 1981.
96. D.H. Parker, M.E. Bartram, and B.E. Koel. *Surf Sci* 217:489–510, 1989.
97. X. Lin, W.F. Schneider, and B.L. Trout. *J Phys Chem B* 108:250–264, 2004.
98. X. Lin, K.C. Hass, W.F. Schneider, and B.L. Trout. *J Phys Chem B* 106:12575–12583, 2002.
99. K. Reuter and M. Scheffler. *Phys Rev Lett* 90:046103, 2003.
100. C. Stampfl, H.J.Kreuzer, S.H. Payne, H. Pfnür, and M. Scheffler. *Phys Rev Lett* 83:2993–2996, 1999.
101. N. Materer, U. Starke, A. Barbieri, R. Doll, K. Heinz, M.A. Van Hove, and G.A. Somorjai. *Surf Sci* 325:207–222, 1995.
102. D. Dahlgren and J.C. Hemminger. *Surf Sci* 123:L739–L742, 1982.
103. J.L. Gland and B.A. Sexton. *Surf Sci* 94:335–368, 1980.
104. G. Ertl. *J Mol Catal A: Chem* 182–183:5–16, 2002.
105. R.G. Pearson. *Symmetry Rules for Chemical Reactions: Orbital Topology and Elementary Processes*. New York: Wiley, 1976.

106. R. Ramprasad, K.C. Hass, W.F. Schneider, and J.B. Adams. *J Phys Chem B* 101:6903–6913, 1997.
107. W.F. Schneider, K.C. Hass, R. Ramprasad, and J.B. Adams. *J Phys Chem B* 101:4353–4357, 1997.
108. W. Tsang and J.T. Herron. *J Phys Chem Ref Data* 20:609–663, 1991.
109. Y. Li and W.K. Hall. *J Catal* 129:202–215, 1991.
110. G. Centi and S. Perathoner. *Appl Catal A* 132:179–259, 1995.
111. M. Shelef. *Chem Rev* 95:209–225, 1995.
112. R.A. Schoonheydt. *Catal Rev-Sci Eng* 35:129–168, 1993.
113. *Handbook of Zeolite Science and Technology*. New York: Marcel Dekker, 2003.
114. I.E. Maxwell. *Adv Catal* 31:1–76, 1982.
115. P. Treesukol, J.P. Lewis, J.Limtrakul, and T.N. Truong. *Chem Phys Lett* 350:128–134, 2001.
116. J. Sauer. *Chem Rev* 89:199–255, 1989.
117. B. Wichterlova, J.Dedecek, Z. Sobalik, A. Vondrova, and K. Klier. *J Catal* 169:194–202, 1997.
118. C. Lamberti, S. Bordiga, M. Salvalaggio, G. Spoto, A. Zecchina, F. Geobaldo, G. Vlaic, and M. Bellatreccia. *J Phys Chem B* 101: 344–360, 1997.
119. D. Nachtigallova, P. Nachtigall, M. Sierka, and J.Sauer. *Phys Chem Chem Phys* 1:2019–2026, 1999.
120. B. Wichterlova, J.Dedecek, and A. Vondrova. *J Phys Chem* 99:1065–1067, 1995.
121. K.C. Hass and W.F. Schneider. *Phys Chem Chem Phys* 1:639–648, 1999.
122. P. Treesukol, J.Limtrakul, and T.N. Truong. *J Phys Chem B* 105:2421–2428, 2001.
123. W.F. Schneider, K.C. Hass, R. Ramprasad, and J.B. Adams. *J Phys Chem B* 102:3692–3705, 1998.
124. S. Recchia, C. Dossi, R. Psaro, A. Fusi, and R. Ugo. *J Phys Chem B* 106:13326–13332, 2002.
125. A.W. Aylor, S.C. Larsen, J.A. Reimer, and A.T. Bell. *J Catal* 157:592–602, 1995.
126. M. Iwamoto, H. Yahiro, N. Mizuno, W.-X. Zhang, Y. Mine, H. Furukawa, and S. Kagawa. *J Phys Chem* 95:9360–9366, 1992.
127. R. Ramprasad, W.F. Schneider, K.C. Hass, and J.B. Adams. *J Phys Chem B* 101:1940–1949, 1997.
128. H.V. Brand, A. Redondo, and P.J.Hay. *J Phys Chem B* 101:7691–7701, 1997.
129. N. Tajima, M. Hashimoto, F. Toyama, A.M. El-Nahas, and K. Hirao. *Phys Chem Chem Phys* 1:3823–3830, 1999.
130. B. Modén, P. Da Costa, D.K. Lee, and E. Iglesia. *J Phys Chem B* 106:9633–9641, 2002.
131. M.V. Konduru and S.S.C. Chuang. *J Phys Chem B* 103:5802–5813, 1999.
132. Y. Li and J.N. Armor. *Appl Catal* 76:L1–L8, 1991.
133. M.H. Groothaert, K. Lievens, H. Leeman, B.M. Weckhuysen, and R.A. Schoonheydt. *J Catal* 220:500–512, 2003.
134. M.H. Groothaert, K. Lievens, J.A. van Bokhoven, A.A. Battiston, B.M. Weckhuysen, K. Pierloot, and R.A. Schoonheydt. *Chem Phys Chem* 4:626–630, 2003.
135. M.H. Groothaert, J.A. van Bokhoven, A.A. Battiston, B.M. Weckhuysen, and R.A. Schoonheydt. *J Am Chem Soc* 125:7629–7640, 2003.
136. B.R. Goodman, K.C. Hass, W.F. Schneider, and J.B. Adams. *Catal Lett* 68:85–93, 2000.
137. M.J.Rice, A.K. Chakraborty, and A.T. Bell. *J Catal* 186:222–227, 1999.
138. B.R. Goodman, W.F. Schneider, K.C. Hass, and J.B. Adams. *Catal Lett* 56:183–188, 1998.
139. B.R. Goodman, K.C. Hass, W.F. Schneider, and J.B. Adams. *J Phys Chem B* 103:10452–10460, 1999.
140. *CRC Handbook of Chemistry and Physics*, 84th ed. Boca Raton: CRC Press, 2003.

11 Applications of Zeolites in Environmental Catalysis

Sarah C. Larsen
Department of Chemistry, University of Iowa

CONTENTS

11.1 INTRODUCTION

Zeolites are crystalline, aluminosilicate molecular sieves that have pores of molecular dimensions [1, 2]. Zeolites are nontoxic and environmentally benign materials that can be synthesized with a wide range of pore sizes and topologies. The framework structures and properties of some common zeolites are shown in Figure 11.1 and Table 11.1, respectively. Zeolites are used in applications such as catalysis, chemical separations, and as ion exchangers [3]. Some types of zeolites are naturally occurring; there are 40 known naturally occurring zeolites and over 150 synthetic zeolites [4]. Zeolites are classified as microporous materials that have pore dimensions in the subnanometer to nanometer size range. The zeolite framework typically consists of SiO_4^{-4} and AlO_4^{-5} moieties in tetrahedral coordination with shared oxygen atoms. A zeolite composed entirely of SiO_4^{-4} tetrahedral units is neutral, whereas the introduction of aluminum into the framework introduces a negative charge that is charge-compensated by a cation, such as sodium, to maintain charge neutrality. The charge-compensating cation can be easily exchanged with other cations to introduce different selectivities or reactivities.

The faujasite zeolites, X and Y, are widely used as catalysts in the industry. Zeolites X and Y have the same structure but differ in silicon to aluminum ratio as shown in Table 11.1. The

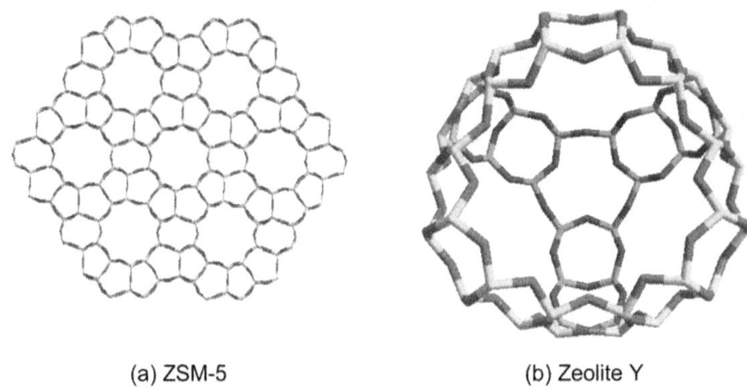

FIGURE 11.1 Structures of common zeolites: (a) ZSM-5 and (b) zeolite Y.

TABLE 11.1
Zeolites and their Si/Al Ratios and Pore Dimensions [5]

Zeolite Name	Si/Al	Pore Size (Å)
Zeolite A	1–1.5	4.1
Zeolite X	1–1.5	7.4
Zeolite Y	2–5	7.4
ZSM-5	~10–∞	5.1 × 5.5 [1 0 0]; 5.3 × 5.6 [0 1 0]
Ferrierite	~5	4.2 × 5.4 [0 0 1]; 3.5 × 4.8 [0 1 0]
Mordenite	~5	6.5 × 7.0 [0 0 1]; 2.6 × 5.7 [0 1 0]
Beta	>8	7.6 × 6.4 [1 0 0]; 5.5 × 5.5 [0 0 1]

faujasite structure (Figure 11.1a) consists of sodalite cages and 12-membered oxygen rings that form a three-dimensional channel system with 7.4 Å apertures. The zeolite ZSM-5 (or MFI) has a silicon to aluminum ratio >10. The framework of ZSM-5 is composed of straight 10-ring, elliptical channels (pore dimension: 5.3 × 5.6 Å) running along the [0 1 0] direction and sinusoidal 10-ring, elliptical channels (pore dimension: 5.1 × 5.5 Å) along the [1 0 0] direction [5]. The framework of zeolite Beta is similar in topology to ZSM-5, but the pore size is larger. The framework of Beta is composed of straight 12-ring channels (pore dimension: 5.5 × 5.5 Å) along the [0 0 1] direction and sinusoidal 12-ring, elliptical channels (pore dimension: 7.6 × 6.4 Å) along [1 0 0] direction [5]. Zeolites ferrierite and mordenite have silicon to aluminum ratios around 5 and have two-dimensional channel systems.

The zeolite chemical composition, framework topology, and pore size control the selectivity and reactivity of these materials. The characteristic pore sizes are the basis for the shape selectivity associated with zeolites. For example, molecules that are larger than the pore size will be excluded from the interior of the zeolite and molecules that are smaller than the pore size will be able to diffuse into the zeolite channels. The amazing shape selectivity of zeolites has led to a multitude of applications for zeolites in catalysis and separations. One example is gas purification where small molecules are able to enter the zeolite pores while the larger molecules flow freely through the zeolite bed without entering the zeolite pores [3]. A second example of shape selectivity involves the catalytic cracking of straight versus branched hydrocarbons such as *n*-hexane and 3-methylpentane [3]. The expected or intrinsic reactivity based on solution studies is that the branched hydrocarbon, 3-methylpentane, will be more reactive than the straight-chain hydrocarbon. The reactivity observed in zeolite 5A is actually reversed in that *n*-hexane is almost ten times more reactive than 3-methylpentane as shown in

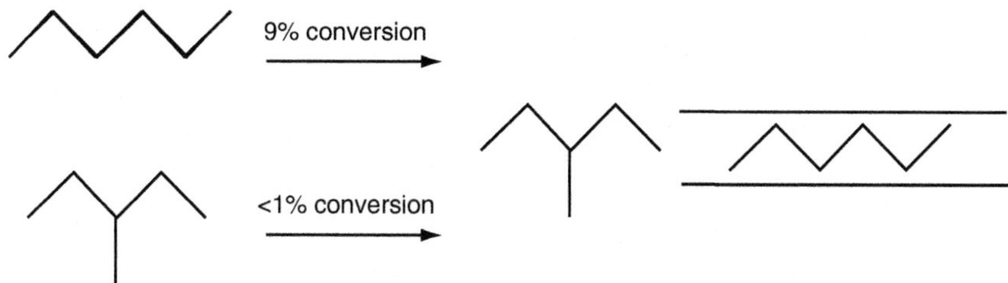

FIGURE 11.2 Illustration of the shape selectivity in HZSM-5 for *n*-hexane over 3-methylpentane [3].

Figure 11.2. The difference in reactivity can be attributed to the ease with which *n*-hexane is transported through the zeolite pores relative to 3-methylpentane.

Zeolites are perhaps best known in catalysis for their use in petroleum refining or catalytic cracking reactions in which petroleum is converted into smaller gasoline range paraffins and olefins. In these applications, the acid form of the zeolite is the active site for the conversion of hydrocarbons. In recent years, applications for catalysts related to environmental issues have been driven by legislation that has placed regulations on air and water pollutants. This has led to a growth in the development of catalytic processes designed to alleviate environmental problems. Zeolite catalysts have been used in some environmental applications [6], such as the production of lead-free octane enhancers and as phosphate-free ion exchangers for laundry detergents to increase the water softening ability [4] and have the potential to be used in many other applications in environmental catalysis. Some examples of zeolites use or potential use as environmental catalysts are listed in Table 11.2.

In general, applications for zeolites as environmental catalysts can be grouped into two major categories:

1. Reduction in the emissions of nitrogen oxides and volatile organic compounds (VOCs)
2. Environmentally benign synthesis and manufacturing.

These two general areas of zeolites as environmental catalysts will be reviewed in this chapter. In the first area related to the reduction of nitrogen oxides and VOCs, the applications of zeolites in the decomposition and selective catalytic reduction (SCR) of nitrogen oxides will be discussed with a specific focus on transition metal-exchanged zeolites. Discussion of the second area related to environmentally benign synthesis and manufacturing will focus on the selective oxidation and partial oxidation of hydrocarbons as an example of the environmentally benign synthesis of chemicals.

TABLE 11.2
Examples of Zeolites in Environmental Catalysis [4]

Zeolite Catalyst	Application
Copper, cobalt ZSM-5, and Beta	Nitrogen oxide selective catalytic reduction with hydrocarbons
Iron ZSM-5	Nitrogen oxide selective catalytic reduction with ammonia
Copper, cobalt, iron, ZSM-5, and mordenite	N_2O decomposition
High silica, hydrophobic ZSM-5, and zeolite Y	VOC removal
TS-1 (ZSM-5 with titanium substituted into the framework)	Caprolactam in an oxidation process

11.2 REDUCTION IN THE EMISSIONS OF NITROGEN OXIDES AND VOLATILE ORGANIC COMPOUNDS

The emission of nitrogen oxides (NO_x and N_2O) from stationary and automotive sources, such as power plants and lean-burn engines, is a major environmental pollution issue. NO_x leads to the production of ground-level ozone, photochemical smog, and acid rain; N_2O is a greenhouse gas, which contributes to global warming and participates in the depletion of stratospheric ozone. The catalytic reduction of nitrogen oxides to molecular nitrogen is an important environmental challenge for scientists and engineers.

VOCs are defined as stable organic compounds with vapor pressures of 0.1 mm Hg or greater at normal temperature and pressure [7]. VOCs include alcohols, ketones, aldehydes, aromatics, and halogenated hydrocarbons. VOCs are environmentally unacceptable because they contribute to ground-level ozone, photochemical smog, and the greenhouse effect and they can be toxic to the environment. The target of most abatement strategies is solvent use and industrial processes that produce large quantities of VOCs. Zeolites, such as high silica, hydrophobic zeolites have proven to be effective for adsorption from high-humidity dilute streams containing VOCs and subsequent combustion [6].

Transition metal-exchanged zeolite catalysts are effective catalysts for several potential applications in the reduction of nitrogen oxide and VOC emissions, such as:

- The direct decomposition of NO_x and N_2O to N_2 and O_2 [8–12]
- SCR of nitrogen oxides with hydrocarbons [13] or ammonia
- Catalytic combustion of VOCs [6].

For these applications, transition metal-exchanged zeolites often exhibit greater activities than traditional metal oxide catalysts. The catalytic activity of the transition metal-exchanged zeolite depends on the identity of the *zeolite host* and of the *exchanged transition metal.* For example, copper-exchanged ZSM-5 is active for the direct decomposition of NO_x, but copper-exchanged Y zeolite is not active [9]. Similarly, cobalt-exchanged ZSM-5 is active for the SCR of NO_x with CH_4 and for the ammoxidation of ethane to acetonitrile but copper-exchanged ZSM-5 is not active for these reactions [14, 15].

The zeolite provides a unique electronic environment for exchanged transition metal cations that depends on the framework composition and topology of the parent zeolite. The cations are located in coordinatively unsaturated sites that are accessible to reactant molecules and the cations are stabilized in specific coordination environments and oxidation states that may be essential for catalytic activity. Therefore, it is important to characterize the local electronic environment of the exchanged transition metals and to correlate changes in the local environment with catalytic activity.

Many different spectroscopic techniques have been used to characterize transition metal-exchanged zeolites. Magnetic resonance techniques, such as electron paramagnetic resonance (EPR) and solid-state nuclear magnetic resonance (NMR), have been used to investigate the catalytic activity of transition metal-exchanged zeolites. EPR spectroscopy has been used to probe the electronic environment of paramagnetic transition metals, such as copper (Cu^{2+}, d^9), vanadium (V^{4+}, d^1), and iron (Fe^{2+}, d^6; Fe^{3+}, d^5) exchanged into zeolites. The information obtained from the EPR experiments provides a molecular level understanding of the local electronic environment of transition metals exchanged into zeolites with different framework topologies and compositions. In addition, *in situ* EPR techniques have been developed for studying the role of transition metals as active sites in catalytic reactions. In related experiments, solid-state NMR has been used to investigate the reactions of reactant molecules in transition metal-exchanged zeolites. Surface species formed in the zeolites can be monitored with solid-state NMR in order to understand the details of fundamental reactivity.

Many other spectroscopic techniques [16] such as Fourier transform infrared (FTIR) spectroscopy [17–20], x-ray absorption spectroscopy [21], and Mössbauer spectroscopy [22] have also been used to investigate transition metal-exchanged zeolites.

In the following sections, the applications of zeolite catalysts for nitrogen oxide decomposition and SCR and for reduction of VOC emissions will be discussed. Magnetic resonance studies of these materials will be emphasized as well as the feasibility for practical application of the zeolite catalysts.

11.2.1 Direct Decomposition of Nitrogen Oxides

To date, copper-exchanged zeolites are the most promising zeolites for the direct decomposition of nitrogen oxides, such as NO_x and nitrous oxide (N_2O). Copper-exchanged ZSM-5 is unique in its demonstrated ability to catalyze the *direct* decomposition of NO_x into nitrogen and oxygen [8, 9]. In particular, Cu-ZSM-5 has the unique ability to desorb the product oxygen that is formed from the direct decomposition of NO_x. The product oxygen typically deactivates other traditional transition metal catalysts. Cu-ZSM-5 and Cu-Beta are also active for the direct catalytic decomposition of nitrous oxide into nitrogen and oxygen. Numerous studies have focused on evaluating the catalytic activity of copper-exchanged zeolites, and the local environment of copper in the zeolites has been probed using various spectroscopic techniques such as FTIR [17–20], x-ray adsorption near edge (XANES) [21], NMR [23–26], and EPR [27–30].

The catalytic activity of copper-exchanged zeolites is influenced by the zeolite structure. Determining the location and coordination of copper ions in the zeolite is crucial in understanding the role of copper in the decomposition of nitrogen oxides. EPR spectroscopy has been extensively used to probe the structural environment of paramagnetic copper sites in zeolites [31–35]. Several groups have studied Cu-ZSM-5 and Cu-Beta using EPR and pulsed EPR techniques [31–35]. In previous studies, EPR signals from copper-exchanged zeolite samples exposed to various treatments have been analyzed to provide information about the local environment of copper in the zeolite [31–35]. As shown in Figure 11.3a, the EPR spectrum for hydrated copper-exchanged Beta is broad and structureless at room temperature indicating a mobile $[Cu(H_2O)_5]^{2+}$ species in the zeolite. When the Cu-Beta sample is cooled and the motion is slowed down, the EPR spectrum changes as shown in Figure 11.3b and exhibits the characteristic Cu^{2+} hyperfine features due to the hyperfine coupling between the unpaired electron spin and the copper ($I = 3/2$) nuclear spin. After the Cu-Beta sample is dehydrated at elevated temperature, the EPR spectrum (Figure 11.3c) changes again and no motional effects are observed at room temperature or low temperature. The interpretation is that initially, the $Cu(H_2O)_5^{2+}$ species is located in the channel intersection and is mobile. After dehydration the copper complex is bound to the zeolite in a square planar geometry and is immobile. Similar EPR spectra are observed for Cu-ZSM-5.

The autoreduction of Cu^{2+} to Cu^{1+} in Cu-ZSM-5 during dehydration has also been monitored by EPR spectroscopy through the loss of Cu^{2+} signal intensity [34]. Approximately 40% to 60% of the copper is reduced through autoreduction and the signal intensity can be reversibly recovered by exposing the sample to water or nitric oxide. Based on the EPR results, the mechanism shown below for autoreduction was proposed [34].

$$[Cu^{2+}OH^-]^+ \overset{\Delta}{\longleftrightarrow} Cu^{1+} + OH \frac{[Cu^{2+}OH^-]^+ + OH \overset{\Delta}{\longleftrightarrow} Cu^{2+}O^- + H_2O}{2[Cu^{2+}OH^-]^+ \overset{\Delta}{\longleftrightarrow} Cu^{1+} + Cu^{2+}O^- + H_2O} \tag{11.1}$$

In the proposed scheme, the Cu^{2+} and $Cu^{2+}OH^-$ species are observable by EPR, but the $Cu^{2+}O^-$ species is not. The ability of the zeolite to stabilize these different copper species with

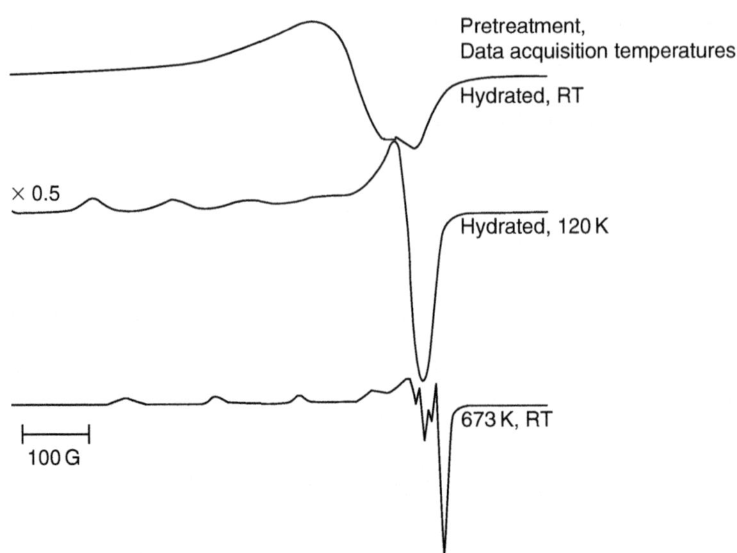

FIGURE 11.3 EPR spectra of copper-exchanged Beta zeolite: (a) hydrated obtained at room temperature, (b) hydrated obtained at 120 K, and (c) dehydrated at 673 K obtained at room temperature ($\nu_{EPR} = 9.44$ GHz).

different oxidation states and to facilitate the reduction of copper ions is critical to the catalytic activity of these materials for the decomposition of nitrogen oxides.

Despite the promising properties of copper-exchanged zeolites for the direct decomposition of nitrogen oxides, several obstacles such as inhibition by water, poisoning by sulfur dioxide, and relatively low activity and selectivity have prevented the practical application of these materials [6, 36]. However, if these challenges can be overcome, the widespread application of these transition metal-exchanged zeolite catalysts for direct nitrogen oxide decomposition may occur in the future.

11.2.2 Selective Catalytic Reduction of Nitrogen Oxides

Recently, the SCR of nitrogen oxides by hydrocarbons (SCR-HC) and NH_3 (SCR-NH_3) over transition metal-exchanged zeolites, particularly in the presence of oxygen, has attracted much interest for emission abatement applications in stationary sources, such as natural gas fueled power plants [4, 10, 37–39].

$$\text{SCR-HC: } 4NO + 4C_xH_y + (2x + y - 2)O_2 \longrightarrow 2N_2 + 4xCO_2 + 2yH_2O \quad (11.2)$$

$$\text{SCR-NH}_3\text{: } 4NO + 4NH_3 + O_2 \longrightarrow 4N_2 + 6H_2O \quad (11.3)$$

The advantage of SCR-HC for natural gas fueled power plants is that the hydrocarbon reductant is already present in the exhaust stream. SCR-HC of NO_x also shows promise for applications to lean-burn gasoline and diesel engines where noble-metal three-way catalysts are not effective at reducing NO_x in the presence of excess oxygen [40]. It has recently been reported that FeZSM-5 is active for the SCR of NO with ammonia (SCR-NH_3) [40–43]. NO_x SCR reactions on zeolite catalysts have been extensively studied using a broad range of spectroscopic and analytical methods. An overview of the application of copper- and cobalt-exchanged zeolites to SCR-HC and the application of iron-exchanged zeolites for SCR-NH_3 is provided in the next two sections.

11.2.2.1 Copper- and Cobalt-Exchanged Zeolites for SCR-HC

Different hydrocarbons and various metal ion-exchanged zeolite catalysts have been evaluated for SCR-HC. Several comprehensive reviews of SCR-HC on zeolites [6, 13, 36, 39, 44] are available in the literature so only an overview of this topic will be presented here. Although other metal-exchanged zeolite catalysts are also active for the SCR of NO_x, CuZSM-5 has been most intensively studied. Numerous studies have been conducted with much of the research focusing on the evaluation of catalyst performance, measurement of reaction kinetics, structural characterization, and spectroscopic identification of surface species. Several different reaction intermediates have been proposed for the conversion from NO_x to N_2, such as an adsorbed nitrogen oxide complex NO_y ($y \geq 2$) [45–49], a carbonaceous deposit [50], oxygenated hydrocarbons [51, 52], isocyanates [53], nitrite and nitrate complexes [54], and nitro- or nitroso compounds [47, 55–60].

Recently, the role of the surface nitroso complex was examined by Beutel et al. [46, 55]. It was reported that NO reacts with alkyl radicals, which have been formed by H-abstraction and the H-abstraction is mediated by the chemisorbed NO_y complex, to form nitro- or nitroso-alkanes [46, 55, 61]. The nitroso-alkanes are unstable and should immediately isomerize to form an oxime [62]. Beutel et al. have studied the interaction of the isomerization product of nitrosopropane, ^{14}N-labeled acetone oxime, with ^{15}NO on CuZSM-5 [47]. They observed the production of $^{14}N^{15}N$ and $^{14}N^{15}NO$ (or $^{15}N^{14}NO$) by the combination of FT-IR surface studies and mass spectrometry for gas-phase analysis. The isotopic labeling studies show that N—N bonds form via the interaction of gaseous NO with an adsorbed oxime complex.

Solid-state NMR is a powerful noninvasive probe that can be used to obtain valuable structural and mechanistic information about catalysts. Previous NMR studies have demonstrated numerous applications of NMR in which catalytic reactions on zeolites have been monitored and surface phenomena on supported metal catalysts have been studied [63–65]. Solid-state NMR was used to investigate the interaction of adsorbed acetone oxime and its hydrolysis products with NO on CuZSM-5 and HZSM-5, which are both active for SCR-HC [24]. Through the combination of ^{13}C and ^{15}N NMR and isotopic labeling, the carbon- and nitrogen-containing adsorbed surface species and gaseous products formed under conditions of thermodynamic equilibrium were monitored and identified.

The decomposition of isotopically labeled acetone oxime on HZSM-5 and on CuZSM-5 was monitored using ^{15}N and ^{13}C NMR [24]. The predominant reaction path for acetone oxime decomposition was different on HZSM-5 and on CuZSM-5. The differences are attributed to an enhancement of acid site chemistry on HZSM-5, which leads to the formation of acetic acid and methylamine by acid-catalyzed reactions. On CuZSM-5, acetone oxime primarily decomposes to acetone and hydroxylamine. Acetone oxime and its hydrolysis products further react to form N_2 and N_2O on both zeolites. Acetone oxime reacts with gas-phase NO on HZSM-5 and on CuZSM-5 to form a new N—N bond. If both nitrogen atoms came from the same source (i.e., ^{15}NO), the two N_2O peaks in the ^{15}N NMR spectrum would be expected to have the same intensity. As shown in Figure 11.4, the N_2O peaks have different intensities indicating that the nitrogen comes from two different sources: one from the adsorbed oxime complex and the other from gas-phase NO. As illustrated by the results in Figure 11.4, the ^{15}N NMR results show that the ^{15}NO bond of gas-phase ^{15}NO remains intact when it reacts with adsorbed acetone oxime to form $N^{15}NO$. The reaction of acetone oxime and NO to form N_2 and N_2O proceeds at room temperature on CuZSM-5, but not until 150°C on HZSM-5. ^{15}N NMR has provided important information, such as the identification and quantification of $^{15}N^{14}NO$ and $^{14}N^{15}NO$, which are not distinguished in mass spectrometry. In addition, NMR was utilized to determine the partition between $^{14}N^{15}NO$ and $^{15}N^{14}NO$ of N_2O. The results demonstrate the potential of solid-state NMR for studying surface species that may be important catalytic intermediates.

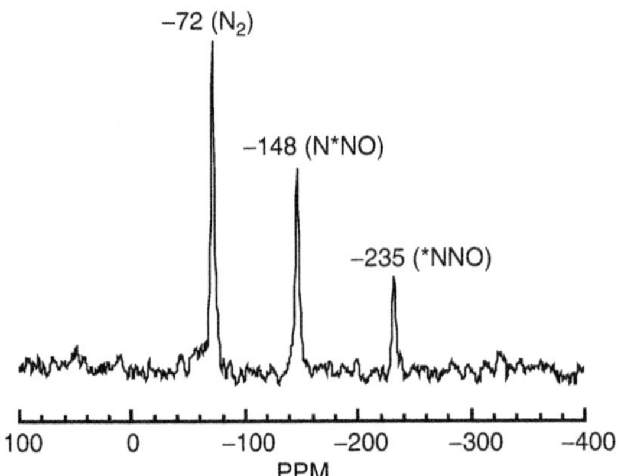

FIGURE 11.4 ^{15}N MAS NMR spectrum obtained at room temperature after the reaction of acetone oxime and ^{15}NO on CuZSM-5 at 150°C. The NMR spectrum was obtained at 7 T (^{15}N Larmor frequency = 30.425 MHz) with a spinning speed of 5 kHz and 25,000 scans were acquired. (Adapted from J Wu and SC Larsen. *J. Catal.* 182: 244–256, 1999. Copyright 1999, Elsevier Science. With permission.)

Complete conversion of NO to N_2 has been reported on CoZSM-5 at ~400°C with methane as a reductant. Methane is very appealing as a reductant because it is readily available in the exhaust stream for many applications, such as stationary sources powered with natural gas. The critical reactions to consider are the reduction of NO by methane and the combustion of methane:

$$2NO + CH_4 + O_2 \longrightarrow N_2 + 2H_2O + CO_2 \tag{11.4}$$

$$CH_4 + 2O_2 \longrightarrow 2H_2O + CO_2 \tag{11.5}$$

The detailed reaction mechanism is not well understood, although many ideas have been presented in the literature.

Hall and coworkers investigated several exchanged zeolites using various hydrocarbon reductants [49]. The hydrocarbons used as reductants were: CH_4, C_3H_8, i-C_4H_{10}, n-C_5H_{12}, neopentane (2,2-dimethylpropane), 3,3-dimethylpentane, 2,2,4-trimethylpentane, and 3,3-diethylpentane (neononane). Based on the average kinetic diameters of these molecules, only CH_4, C_3H_8, i-C_4H_{10}, and n-C_5H_{12} should be able to enter the pores of ZSM-5 (pore diameter ~ 5.1 to 5.6 Å). For example, neopentane has a kinetic diameter of 6.2 Å so it should not be able to appreciably diffuse into the pores of ZSM-5. Based on these factors, one would expect the SCR activity for the larger, branched reductants to be much lower than the activity for methane, propane, and n-pentane. However, Hall and coworkers reported only small changes in reactivity for different hydrocarbons suggesting that diffusion of the intact hydrocarbon reductant into the interior of the zeolite may not be essential for SCR of NO. For example, at 723 K the conversion of NO into N_2 over CoZSM-5 was 73% for n-pentane and 68% for the branched neopentane isomer. It is possible that combustion of the hydrocarbon could be taking place on the outer surface of the zeolite rather than on the internal surface of the zeolite. It is not clear whether the next step involves diffusion of smaller hydrocarbon fragments into the zeolite or diffusion of small molecule reactants such as NO, NO_2, O_2 out of the zeolite. Neononane was the exception to these trends: it showed a marked decrease in activity compared to the other hydrocarbons suggesting that transport factors are important for neononane.

Varying the zeolite from CoZSM-5 to Co-ferrierite, changed the reactivity pattern [49]. Methane was the more active reductant over Co-ferrierite while isobutane was the more active reductant on CoZSM-5. This has been attributed to a molecular sieving effect since ferrierite has a unidimensional ten-membered ring pore system as opposed to the three-dimensional ten-membered ring pore system of ZSM-5. Further data analysis by Hall and coworkers showed that the selectivity plots for CoZSM-5 for the different hydrocarbon reductants could all be correlated on a single curve, except for neononane [49]. The correlation in selectivity for all the hydrocarbon reductants implies that the same mechanism is responsible for SCR of NO in all the cases.

Despite the promising characteristics of copper- and cobalt-exchanged ZSM-5 for SCR-HC applications, these materials are currently too sensitive to water and sulfur dioxide and their overall activity is too low for practical application [39]. Some strategies for addressing these shortcomings include designing a composite catalyst system that contains a mixture of two or more catalytic materials. For example, in a recent patent [66], Toyota combined Co-Ba/HZSM-5 with Cu/HZSM-5 and achieved a high NO_x conversion over a wide temperature range. Other examples are available in the patent literature [39].

11.2.2.2 Iron-Exchanged Zeolites for SCR-NH_3

Iron-exchanged zeolites (e.g., FeZSM-5) have been shown to be promising catalysts for the SCR of NO_x with ammonia as the reductant [42, 43 ,67]. FeZSM-5 catalysts are active at high temperatures even at high levels of water vapor and SO_2 compared to other transition metal-exchanged zeolites. The broader range of temperatures for SCR-NH_3 activity and the ease of disposal are clear advantages for the use of iron-exchanged zeolites relative to V_2O_5 supported on TiO_2, which is used commercially for SCR-NH_3. Koebel et al. proposed that urea could be a potential reductant for the SCR of NO_x [68, 69]. Unlike ammonia, the handling, storage, and transport of urea are efficient and safe. In addition, urea is nontoxic even at high concentrations in aqueous solution. Due to these reasons and its extensive study in NO_x removal in diesel engines, urea could possibly replace ammonia as a reductant for SCR.

However, the catalytic activity of FeZSM-5 is strongly dependent on exchange procedure (solid state, sublimation, etc.), parent zeolite, and the pretreatment of the zeolite [70, 71]. Catalytic measurements and temperature programmed desorption (TPD) have been used to obtain valuable information about the kinetics and type of species resulting from NO_x SCR-NH_3 reactions on FeZSM-5 surfaces, as well as conversion of NO to N_2. Moreover, spectroscopic methods such as EPR and Mössbauer have been used to study the environment of iron cations (Fe^{2+}/Fe^{3+}) in the zeolite and to identify iron oxide aggregates. FT-IR spectroscopy has been used to identify the species formed on the FeZSM-5 surfaces during the NO_x SCR-NH_3 reaction.

The EPR spectrum of hydrated FeZSM-5 (Fe/Al $= 0.67$) prepared by sublimation is shown in Figure 11.5. EPR signals with g-values of 4.3, 5.6, and 6.3 were observed in the spectrum. Similar EPR signals have been observed for iron-exchanged zeolite samples prepared by other groups [72, 73]. Based on the results reported in the literature, the $g = 4.3$ signal is attributed to Fe^{3+} in tetrahedral coordination; the peaks at 5.6 and 6.4 are assigned to Fe^{3+} cations in distorted tetrahedral coordinations. Based on a comparison of the catalytic activity of samples prepared by various exchange methods, Yang and coworkers concluded that the Fe^{3+} in tetrahedral coordination ($g = 4.3$ signal) is responsible for the catalytic activity of FeZSM-5 for the SCR of NO_x with NH_3[74].

In related experiments, the reactions of adsorbed urea and NO on FeZSM-5 and HZSM-5 were investigated using ^{13}C and ^{15}N solid-state NMR [75]. The carbon- and nitrogen-containing products formed after thermal pretreatments at different temperatures on FeZSM-5 and HZSM-5 were identified. Urea hydrolyzed to form CO_2 and NH_3 on H- and FeZSM-5.

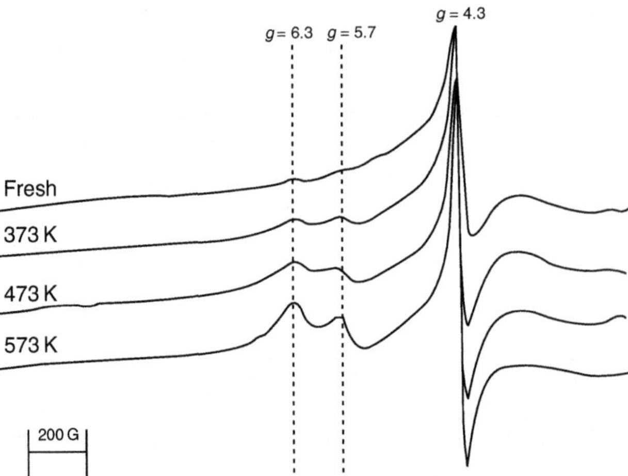

FIGURE 11.5 EPR spectrum of untreated FeZSM-5 (Fe/Al = 0.69) obtained at 120 K (ν_{EPR} = 9.44 GHz): (a) fresh, (b) after pretreatment at 373 K, (c) after pretreatment at 473 K, and (d) after pretreatment at 573 K.

The disappearance of urea and the formation of CO_2 and NH_3 were monitored by ^{13}C and ^{15}N MAS NMR. A broadened ammonia signal in the ^{15}N NMR spectra for both H- and FeZSM-5 was attributed to the formation of a strongly bound ammonia (NH_4^+) species adsorbed on Bronsted acid sites [74]. Reaction of urea, NO, and O_2 on FeZSM-5 resulted in the formation of CO_2, NH_3, and N_2. As shown in Figure 11.6, when oxygen was added to the urea/NO sample, CO_2 and NH_3 were observed as well as the formation of N_2 indicating the potential of these materials for SCR-NH_3 [75].

Iron-exchanged zeolites have great potential for SCR-NH_3 applications. In particular, FeZSM-5 is active at high space velocities and in the presence of water and sulfur dioxide [74].

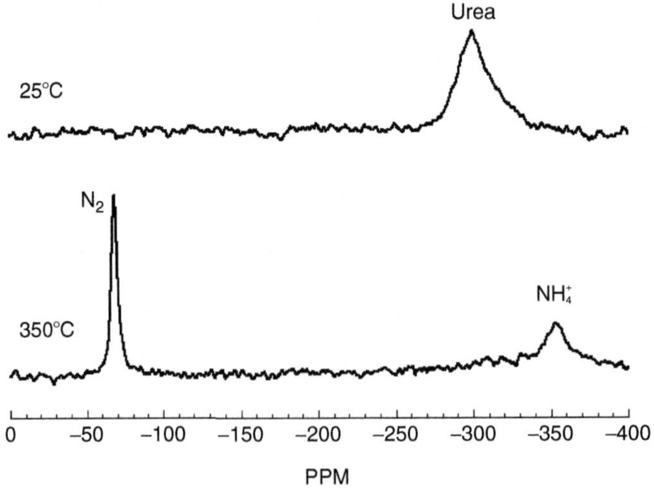

FIGURE 11.6 ^{15}N MAS NMR spectra of urea, ^{15}NO and O_2 on FeZSM-5 (Fe/Al = 0.11) after room temperature adsorption (top) and after heating to 350°C (bottom). NMR spectra were obtained at 7 T (^{15}N Larmor frequency = 30.425 MHz) with a spinning speed of 4 kHz. 25,000 scans were acquired.

However, from an environmental perspective, the best solution would be to develop a catalytic process that does not rely on ammonia as the reductant since there are problems associated with excess ammonia in the stream (also called ammonia slip) as well as with the transport and storage of ammonia. It is advantageous to use a reductant that is already in the exhaust stream such as hydrocarbons or to eliminate the reductant all together.

11.2.2.3 Other Zeolites Evaluated for the Reduction of Nitrogen Oxides

Many other zeolite systems have been evaluated for the reduction of nitrogen oxides [6, 39]. For example, in a recent review article on SCR-HC technologies with zeolite catalysts, platinum, palladium, gallium, indium, cerium, silver, nickel, manganese, rhodium, and H-form zeolites were discussed [39]. The strategy of combining metals has also been reviewed [39]. Most of these zeolite-based catalysts are plagued with similar difficulties that limit practical application such as a limited temperature range for catalytic activity and inhibition by water and sulfur compounds.

11.2.3 Catalytic Combustion of VOCs

Gas or liquid streams containing VOCs can be treated with zeolites that function as adsorbates or as oxidation catalysts [6]. For applications in which the stream contains high concentrations of VOCs, the zeolite is usually used to adsorb and separate the VOCs, which can then be incinerated or recycled. For dilute VOC streams, adsorption by zeolites can be coupled with catalytic oxidation. The advantage of zeolites relative to carbon-based materials is for applications involving dilute, humid streams. In these applications, hydrophobic high silica zeolites are most effective for adsorption [4, 6]. For catalytic oxidation of VOCs, palladium- and chromium-exchanged faujasite zeolites exhibit good performance for the oxidation of nonchlorinated and chlorinated VOCs, respectively [6].

11.3 ENVIRONMENTALLY BENIGN SYNTHESIS AND MANUFACTURING USING ZEOLITES

Zeolites are increasingly utilized as catalysts in various chemical processes and are improving the environmental acceptability of these processes. Zeolite catalysts may be used to replace strong acids such as HF, HCl, or H_2SO_4 in chemical processes and to decrease waste and energy use by providing more efficient catalytic routes. One example is the synthesis of caprolactam, an intermediate in the production of nylon and other synthetic fabrics using titanium-substituted ZSM-5, hydrogen peroxide, and ammonia [5]. In this process, the number of steps and the waste produced was decreased substantially by using the TS-1 zeolite as a catalyst. Another example is the synthesis of cumene in which several zeolite catalysts are used to replace hazardous reagents such as solid phosphoric acid [5].

In this section, the focus will be on the partial oxidation of hydrocarbons in cation-exchanged zeolites. The partial oxidation of hydrocarbons is significant to chemical industry because the products are used to convert petroleum hydrocarbon feedstocks into chemicals important in the polymer and petrochemical industries. Liquid-phase air oxidations are generally preferred by chemical industry because of the mild reaction conditions. Typically, the conversions of the oxidation processes are very low in order to maintain high selectivity. This is necessary because the desired partial oxidation products can easily be further oxidized under typical reaction conditions. A major motivation for the development of new oxidation routes is the desire to achieve high selectivities at high conversions. From an environmental perspective, it would also be desirable to eliminate the use of organic solvents, to use a clean, inexpensive oxidant, such as molecular oxygen and to use sunlight to catalyze the oxidation reaction.

The thermal and photooxidation of hydrocarbons in cation-exchanged zeolites proceed with remarkable selectivity using visible light or mild thermal conditions [76, 77]. Representative reactions shown in Figure 11.7 include the photooxidation of propylene to acrolein, toluene to benzaldehyde, and cyclohexane to cyclohexanone. Frei and coworkers [78] suggested that the oxidation of hydrocarbons in cation-exchanged Y zeolites proceeds through a hydrocarbon·O_2 charge transfer state as illustrated in Figure 11.8 for propylene. The stabilization of this charge transfer state by the zeolite leads to the remarkable product selectivity. The second step in the proposed mechanism involves transfer of a proton to O_2^- to form an allyl/$HO_2^{\bullet}$ radical pair that reacts further to yield the selective formation of a hydroperoxide at low temperatures near −100°C. In the case of alkenes, the hydroperoxide fragments upon warming to room temperature to form the corresponding aldehyde or ketone products and water. In the following sections, the partial oxidations of 1-alkenes, aromatics, and cyclohexane in zeolite Y are discussed.

11.3.1 Thermal and Photooxidation of Alkenes and Aromatics in Cation-Exchanged Zeolites

A variety of techniques have been used to investigate the factors that influence product formation and selectivity in the room-temperature photooxidation of 1-alkenes and aromatics

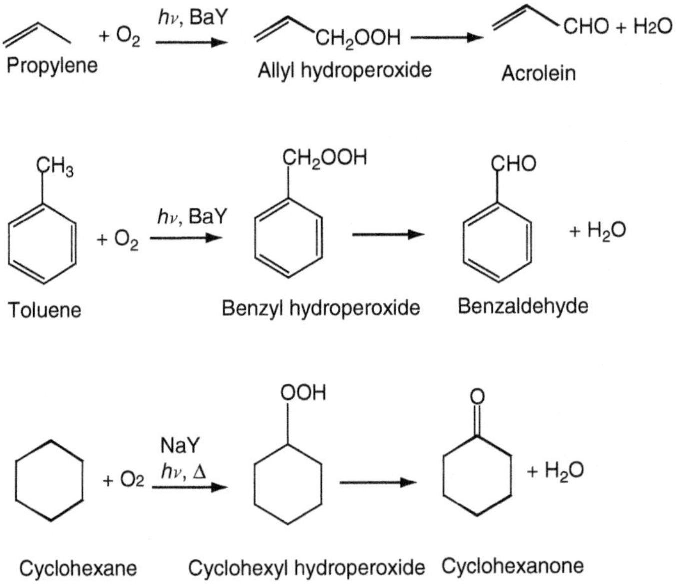

FIGURE 11.7 Examples of selective thermal and photooxidation reactions in cation-exchanged zeolites.

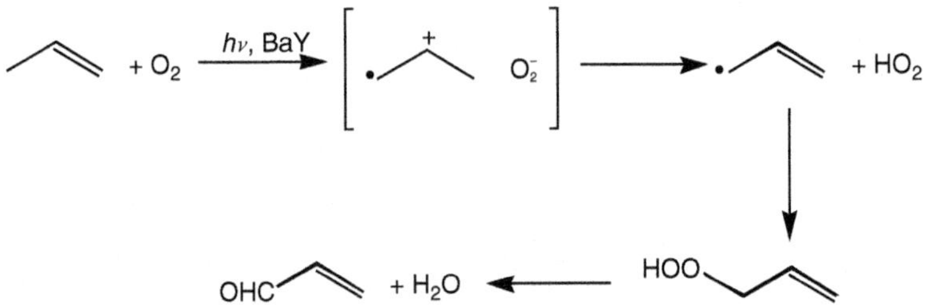

FIGURE 11.8 Proposed mechanism for the selective photooxidation of propylene in BaY zeolite.

in zeolites [79]. Several complementary *in situ* and *ex situ* methods for product analysis for determining the detailed product distribution have been used. Several factors were identified that influenced product selectivity and formation including excitation wavelength, temperature, hydrocarbon chain length, and the parent zeolite.

For propylene, 1-butene and 1-pentene loaded in BaY, irradiation with broadband visible light in the presence of molecular oxygen resulted in an initial excitation of an O_2·alkene complex that could go on to selectively form unsaturated aldehydes and ketones, through a hydroperoxide intermediate [79–81]. Other products such as epoxides and alcohols were formed when the hydroperoxide intermediate reacted with an unreacted parent alkene molecule as shown in Figure 11.9 for the propylene reactions. Nonselective saturated aldehydes were formed as well via a thermal ring-opening reaction of the epoxide in the zeolite [82]. The loss of selectivity was found to increase with increasing temperature and, at a given temperature, increasing chain length.

The main conclusions drawn from this study were that product formation and selectivity in the photooxidation of 1-alkenes in zeolites depended on more factors than previously recognized [79]. These factors include thermal reactions of the reactant and photoproduct molecules in the zeolite at ambient temperatures. Several reactions of 1-alkenes in cation-exchanged zeolites have not previously been identified. These include: epoxide ring opening, double-bond migration, and alkene polymerization. The reactions are proposed to occur at residual Bronsted acid sites that are present in various amounts in cation-exchanged zeolites. The colorimetric detection of acid sites in zeolites was used to qualitatively assess the number and relative strength of acid sites in our zeolite samples [83, 84].

For toluene and *p*-xylene photoxidation with visible light and molecular oxygen in different zeolite hosts, the effect of zeolite chemical composition and framework topology on the selectivity of toluene photooxidation reactions was investigated [85, 86]. Toluene photooxidation in BaY yielded benzyl-hydroperoxide as the primary photoproduct, with subsequent thermal reaction to form benzaldehyde in high yield (87%) [85, 86]. Benzyl alcohol is another product of the reaction; it is formed via a bimolecular reaction and its yield

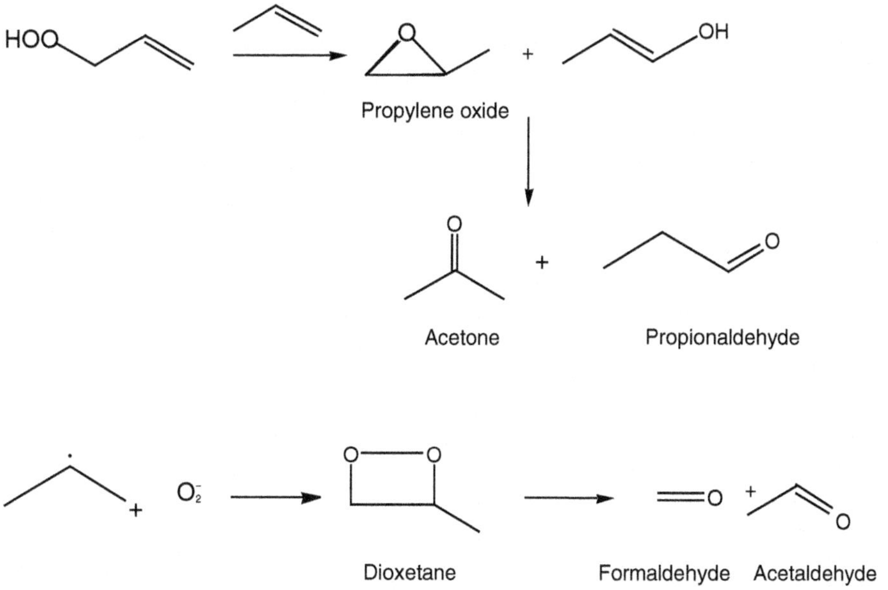

FIGURE 11.9 Nonselective reactions that occur at acid sites during the photooxidation of propylene on BaY.

increased with toluene loading. The toluene photooxidation reaction in zeolites, BaZSM-5 and BaHY, exhibited a significant loss of selectivity to benzaldehyde of 31% and 56% relative to 87% for BaY. The detailed product distributions are listed in Table 11.3. An explanation for the observed product distribution is that acid-catalyzed chemistry is responsible for the loss of selectivity and the formation of phenol, cresols, and condensation products in BaHY and BaZSM-5 relative to BaY. Similar selectivity is seen for paraxylene photooxidation in BaX, NaY, and BaY to yield paratolualdehyde with ~90% selectivity and a similar loss of selectivity is seen in the other host zeolites. It was found that the zeolites with similar quantities of Bronsted acid sites exhibited similar reactivity and selectivity patterns.

This work demonstrates that the presence of small amounts of acid sites can alter the product selectivity in the photooxidation of aromatic and alkenic hydrocarbons. Therefore, in reactions of this type, zeolite materials that have minimal acid sites for retaining high product selectivity will be an important requirement for environmentally benign synthesis using photooxidation reactions.

11.3.2 Kinetics of the Photo and Thermal Cyclohexane Oxidation Reaction in BaY and NaY

Current liquid-phase methods for the autoxidation of cyclohexane using soluble cobalt or manganese catalysts only exhibit acceptable product selectivity (>80%) when cyclohexane conversion is very low (<5%). Although more efficient processes are known, economic and safety considerations strongly favor oxidants, such as air or oxygen for use in large-scale production, and have led to the large-scale adoption of a cost-effective, yet seemingly inefficient production process.

Frei and coworkers demonstrated that the oxidation of cyclohexane to cyclohexanone with molecular oxygen occurs with very high selectivity under mild thermal and photochemical conditions in cation-exchanged zeolites, such as NaY [87]. An additional benefit of this approach is that the oxidation of cyclohexane by molecular oxygen in NaY is desirable from an environmental perspective; the high selectivity minimizes waste and the process utilizes the clean and inexpensive oxidant, molecular oxygen.

Frei and coworkers proposed a reaction mechanism for the oxidation of cyclohexane in NaY that involved a charge transfer complex, [(cyclohexane)$^+$ O_2^-]. The hypothesis is that the charge transfer complex is stabilized by the exchangeable cation in the zeolite and that this stabilization allows access to the charge transfer state by visible light irradiation or by thermal activation. In the next step of the reaction, a proton from the cyclohexane cation radical is abstracted by O_2^- to form HO_2 radical and a cyclohexyl radical. The HO_2 radical attacks the

TABLE 11.3
Photoproduct Distributions from *Ex Situ* GC Analysis after Toluene Photooxidation in BaY, BaZSM-5, and BaHY

Zeolite	BZ[a] (%)	BA[a] (%)	Phenol, Cresols (%)	Condensation Products (%)	Others (%)
BaY	87	4	3	2	4
BaZSM-5	31	1	31	15	22
BaHY	56	4	10	20	10

Adapted from AG Panov, et al., *J. Phys. Chem. B* 104: 5706–5714, 2000. Copyright 2000, American Chemical Society. With permission.
[a]BZ = benzaldehyde and BA = benzyl alcohol.

cyclohexyl radical to form cyclohexyl hydroperoxide under both photochemical and thermal conditions. Using FT-IR spectroscopy, the formation of cyclohexyl hydroperoxide that reacts thermally to form cyclohexanone and water was observed. Complete selectivity in the photo-oxidation of cyclohexane to cyclohexanone was reported at conversions as high as 40%, based on *in situ* FT-IR measurements [87].

It was determined that the oxidation of cyclohexane in zeolite Y could be accomplished both photochemically and thermally under extremely mild conditions with high yield and high selectivity [88, 89]. Conversions of ~50% were obtained at 65°C with complete selectivity to products originating with the hydroperoxide (cyclohexanone [68%], cyclohexanol [11%], and cyclohexyl hydroperoxide [21%]). Higher conversions could be obtained at higher temperatures, but most of our kinetic studies were conducted at 65°C or below. The kinetics of the thermal and photochemical oxidation of cyclohexane in zeolite Y were examined in detail. These investigations were accomplished using *ex situ* GC product analysis and *in situ* FT-IR and solid-state NMR spectroscopies. The results indicated that cyclohexyl hydroperoxide, cyclohexanone, and cyclohexanol (and water) are the main products of the thermal and photochemical oxidation of cyclohexane in BaY. The increase in the (cyclohexanol/cyclohexanone) branching ratio as a function of cyclohexane loading suggests that the cyclohexanol is formed by a bimolecular process. The overall percent conversion of cyclohexane decreases dramatically at cyclohexane loadings of greater than three cyclohexane molecules per supercage. Pronounced deuterium kinetic isotope effects (~5.5) as shown in Figure 11.10 were observed for both the thermal and photochemical cyclohexane oxidation reactions, indicating that a proton transfer step is the rate-determining step in the reaction mechanism. For the thermal oxidation of cyclohexane in BaY and NaY, activation energies of 62 ($\pm$10) and 85 ($\pm$3) kJ/mol, respectively, were obtained [88]. This lower activation energy for BaY relative to NaY is consistent with the proposed mechanism for the oxidation reaction in which the charge transfer complex is stabilized by the electric field in the zeolite at the cation site.

These studies show the potential of zeolite-based catalysts for applications in environmentally benign synthesis. Future challenges for this technology include increasing the

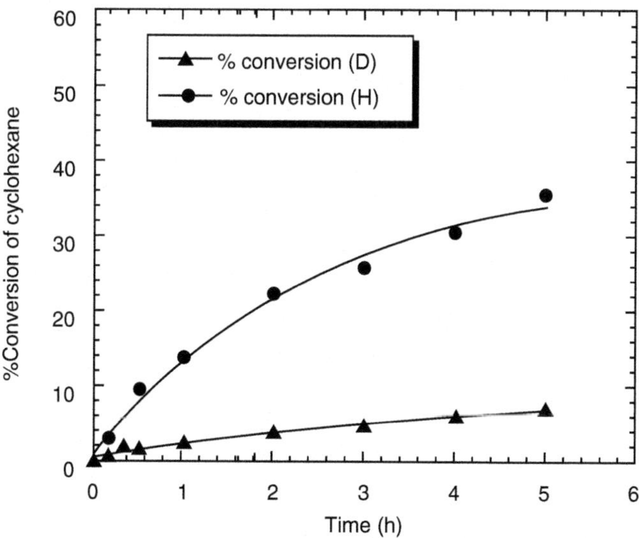

FIGURE 11.10 Kinetic data obtained for thermal oxidation of cyclohexane (mixture of h6 and d6) in BaY. (Reprinted from RG Larsen, AC Saladino, TA Hunt, JE Mann, M Xu, VH Grassian, and SC Larsen. *J. Catal.* 204: 440–449, 2001. Copyright 2001, Elsevier Science. With permission.)

conversion, removing the oxygenated product from the zeolite, and improving the quantum efficiency for the photochemical reactions.

11.4 FUTURE DIRECTIONS

A broad range of zeolite materials have been studied for applications in environmental catalysis ranging from nitrogen oxide reduction with transition metal-exchanged zeolites to partial oxidation reactions with cation-exchanged zeolites. The pore structures and topologies of zeolites provide a unique environment for catalysis. Shape selectivity and confinement both contribute to the unique reactivity of zeolite catalysts. For transition metal-exchanged zeolites, the dispersion of the transition metal in the zeolite is critical to the catalytic activity. The goal for the future is to design better environmental catalysts using the information that has been obtained from detailed spectroscopic studies of these promising zeolite catalysts.

ACKNOWLEDGMENTS

Financial support from the Environmental Protection Agency (R825304 and R829600), National Science Foundation (CHE-02048047), Petroleum Research Fund, and the University of Iowa for various aspects of this work is gratefully acknowledged. Dr. Patrick Carl, Dr. Alexander Saladino, Dr. Jianjun Wu, Dr. Yan Xiang, Dr. Conrad Jones, Gonghu Li, Jennifer Mann, Dr. Alexander Panov, Dr. Kari Myli, Dr. Russell Larsen, and Prof. Vicki Grassian are acknowledged for their contributions to this work.

REFERENCES

1. ME Davis. *Acc. Chem. Res.* 26: 111–115, 1993.
2. DW Breck. *Zeolite Molecular Sieves*. Wiley: New York, 1974.
3. BC Gates. *Catalytic Chemistry*. Wiley: New York, 1992.
4. BK Marcus and WE Cormier. *Chem. Eng. Prog.* 95: 47–53, 1999.
5. WM Meier, DH Olson and Ch. Baerlocher, *Atlas of Zeolite Structure Types*, 4th Edition. Elsevier: London, 1996.
6. G Delahay and B Coq in *Zeolites for Cleaner Technologies*, M Guisnet and JP Gilson, Eds. Imperial College Press: London, 2002.
7. JJ Spivey. *Ind. Eng. Chem. Res.* 26: 2165–2180, 1987.
8. M Iwamoto, H Furukawa, Y Mine, F Uemura, S-i Mikuriya, and S Kagawa. *J. Chem. Soc. Chem. Commun.* 1272–1273, 1986.
9. M Iwamoto, H Furukawa, and S Kagawa. In *New Developments in Zeolite Science and Technology*, Y Murakam, Ed. Elsevier Publishing Company: New York, 1986, p. 943.
10. M Iwamoto and H Hamada. *Catal. Today* 57–71, 1991.
11. Y Li and JN Armor. *Appl. Catal.* 76: L1–L8, 1991.
12. Y Li and JN Armor. *Appl. Catal. B: Environ.* 1: L21–L29, 1992.
13. M Shelef. *Chem. Rev.* 95: 209, 1995.
14. Y Li and JN Armor. *Chem. Commun.* 2013–2014, 1997.
15. Y Li and JN Armor. *J. Catal.* 173: 511–518, 1998.
16. B Wichtelova, Z Sobalik, and J Dedecek. *Appl. Catal. B: Environ.* 41: 97–114, 2003.
17. H-J Jang, WK Hall, and JL d'Itri. *J. Phys. Chem.* 100: 9416–9420, 1996.
18. J Szanyi and MT Paffett. *J. Catal.* 164: 232–245, 1996.
19. B Wichterlova, J Dedecek, Z Sobalik, A Vondrova, and K Klier. *J. Catal.* 169: 194–202, 1997.
20. T Cheung, SK Bhargava, M Hobday, and K Foger. *J. Catal.* 158: 301–310, 1996.
21. D-J Liu and HJ Robota. *Catal. Lett.* 21: 291–301, 1993.
22. WN Delgass, RL Garten, and M Boudart. *J. Catal.* 18: 90–98, 1970.
23. A Gedeon, JL Bonardet, and J Fraissard. *J. Phys. Chem.* 97: 4254–4255, 1993.

24. J Wu and SC Larsen. *J. Catal.* 182: 244–256, 1999.
25. J Wu and SC Larsen. *Catal. Lett.* 70: 43–50, 2000.
26. P Marturano, L Drozdova, A Kogelbauer, and R Prins. *J. Catal.* 192: 236–247, 2000.
27. M Anpo, M Matsuoka, Y Shioya, H Yamashita, E Giamello, C Morterra, M Che, HH Patterson, S Webber, S Ouellette, and MA Fox. *J. Phys. Chem.* 98: 5744–5750, 1994.
28. PJ Carl and SC Larsen. *J. Catal.* 182: 208–218, 1999.
29. E Giamello, D Murphy, G Magnacca, C Morterra, Y Shioya, T Nomura, and M Anpo. *J. Catal.* 136: 510–520, 1992.
30. C Oliva, E Selli, A Ponti, L Correale, V Solinas, E Rombi, R Monaci, and L Forni. *J. Chem. Soc. Faraday Trans.* 93: 2603–2608, 1997.
31. PJ Carl and SC Larsen. *J. Catal.* 182: 208–218, 1999.
32. PJ Carl and SC Larsen. *J. Phys. Chem. B* 104: 6568–6575, 2000.
33. PJ Carl, SL Baccam, and SC Larsen. *J. Phys. Chem. B* 104: 8848–8854, 2000.
34. SC Larsen, A Aylor, AT Bell, and JA Reimer. *J. Phys. Chem.* 98: 11533–11539, 1994.
35. AV Kucherov, AA Slinkin, DA Kondrat'ev, TN Bondarenko, AM Rubinstein, and KM Minachev. *Zeolites* 5: 320–324, 1985.
36. VI Parvulescu, P Grange, and B Delmon. *Catal. Today* 46: 233–316, 1998.
37. T Inui, S Kojo, T Yoshida, M Shibata, and S Iwamoto. *Stud. Surf. Sci. Catal.* 69: 355–364, 1991.
38. M Shelef. *Chem. Rev.* 95: 209–225, 1995.
39. Y Traa, B Burger, and J Weitkamp. *Micropor. Mesopor. Mater.* 30: 3–41, 1999.
40. BK Cho. *J. Catal.* 142: 418, 1993.
41. RQ Long and RT Yang. *J. Catal.* 188: 332–339, 1999.
42. RQ Long and RT Yang. *J. Am. Chem. Soc.* 121: 5595–5596, 1999.
43. AZ Ma and W Grunert. *Chem. Commun.* 71–72, 1999.
44. A Fritz and V Pitchon. *Appl. Catal. B: Environ.* 13: 1–25, 1997.
45. Y Li and JN Armor. *Appl. Catal. B* 3: 239, 1994.
46. T Beutel, BJ Adelman, G-D Lei, and WMH Sachtler. *Catal. Lett.* 32: 83, 1995.
47. T Beutel, B Adelman, and WMH Sachtler. *Catal. Lett.* 37: 125–130, 1996.
48. T Beutel, J Sarkany, G-D Lei, JY Yan, and WMH Sachtler. *J. Phys. Chem.* 100: 845–851, 1996.
49. F Witzel, GA Sill, and WK Hall. *J. Catal.* 149: 229, 1994.
50. JL d'Itri and WMH Sachtler. *Appl. Catal. B* 2: L7, 1993.
51. CN Montreuil and M Shelef. *Appl. Catal. B* 1: L1, 1992.
52. I Halasz, A Brenner, KYS Ng, and Y Hou. *J. Catal.* 161: 359–372, 1996.
53. Y Ukiso, S Sato, A Abe, and K Yoshida. *Appl. Catal. B* 2: 147, 1993.
54. VA Sadykov, SL Baron, VA Matyshak, GM Alikina, RV Bunina, A Ya Rozovskii, VV Lunin, EV Lunina, AN Kharlanov, AS Ivanova, and SA Veniaminov. *Catal. Lett.* 37: 157–162, 1996.
55. BJ Adelman, T Beutel, G-D Lei, and WMH Sachtler. *J. Catal.* 158: 327, 1996.
56. BJ Adelman, T Beutel, G-D Lei, and WMH Sachtler. *Appl. Catal. B* 11: L1, 1996.
57. C Yokoyama and M Misono. *J. Catal.* 150: 9, 1994.
58. RHH Smits and Y Iwasawa. *Appl. Catal. B* 6: L201, 1995.
59. EV Rebrov, AV Simakov, NN Sazonova, VA Rogov, and GB Barannik. *Catal. Lett.* 51: 27–40, 1998.
60. KCC Kharas. *Appl. Catal. B: Environ.* 2: 207–224, 1993.
61. AD Cowan, R Dumpelmann, and NW Cant. *J. Catal.* 151: 356, 1995.
62. D Barton and WD Ollis. *Comprehensive Organic Chemistry: The Synthesis and Reactions Organic Compounds*, Perganon Press: Oxford, 1979, vol 2, p. 391.
63. Marcel Dekker: New York.
64. CP Slichter. *Ann. Rev. Phys. Chem.* 37: 25–51, 1986.
65. TM Duncan. *Colloid Surf.* 45: 11–31, 1990.
66. T Mizuno, S Takeshima, K Sekizawa, and S Kasahara. Toyota Jidosha Kabuskiki Kaishi: European Patent Application, 1993.
67. RQ Long and RT Yang. *J. Catal.* 188: 332–339, 1999.
68. M Koebel, M Elsener, and G Madia. *Ind. Eng. Chem. Res.* 40: 52, 2001.
69. M Koebel, M Elsener, and M Kleeman. *Catal. Today* 59: 335, 2000.
70. X Feng and WK Hall. *Catal. Lett.* 41: 45–46, 1996.
71. WK Hall, X Feng, J Dumesic, and R Watwe. *Catal. Lett.* 1998: 13–19, 1998.

72. D Goldfarb, M Bernardo, KG Strohmaier, DEW Vaughan, and H Thomann. *J. Am. Chem. Soc.* 116: 6344–6353, 1994.
73. AV Kucherov, TN Montreuil, TN Kucherova, and M Shelef. *Catal. Lett.* 56: 173–181, 1998.
74. RQ Long and RT Yang. *J. Catal.* 194: 80–90, 2000.
75. CA Jones, D Stec, and SC Larsen. *J. Mol. Catal. A: Chem.* 212: 329–336, 2004.
76. VH Grassian and SC Larsen. Photooxidation of hydrocarbons in cation-exchanged zeolites, In *Handbook of Photochemistry and Photobiology*, HS Nalwa, Ed. American Scientific Publishers: Stevenson Ranch, CA, 2003.
77. H Frei, F Blatter, and H Sun. *Chemtech* 24–30, 1996.
78. F Blatter and H Frei. *J. Am. Chem. Soc.* 115: 7501–7502, 1993.
79. Y Xiang, SC Larsen, and VH Grassian. *J. Am. Chem. Soc.* 121: 5063–5072, 1999.
80. F Blatter, H Sun, and H Frei. *Catal. Lett.* 35: 1–12, 1995.
81. KB Myli, SC Larsen, and VH Grassian. *Catal. Lett.* 48: 199–202, 1997.
82. WF Hoelderich and N Goetz. In *9th International Zeolite Conference*, R von Ballmoos, JB Higgins, MMJ Treacy, Eds. Butterworth-Heinemann: Boston, MA, 1993, pp. 309–317.
83. VJ Rao, DL Perlstein, RJ Robbins, PH Lakshminarasimhan, H-M Kao, CP Grey, and V Ramamurthy. *Chem. Commun.* 269–270, 1998.
84. KJ Thomas and V Ramamurthy. *Langmuir* 14: 6687–6692, 1998.
85. AG Panov, KB Myli, Y Xiang, VH Grassian, and SC Larsen, Photooxidation of toluene in cation-exchanged zeolites. In *Green Chemical Syntheses and Processes*, P Anastas, LG Heine, and TC Williamson, Eds. American Chemical Society: Washington, DC, 2000, pp. 206–216.
86. AG Panov, RG Larsen, NI Totah, SC Larsen, and VH Grassian. *J. Phys. Chem. B* 104: 5706–5714, 2000.
87. H Sun, F Blatter, and H Frei. *J. Am. Chem. Soc.* 118: 6873–6879, 1996.
88. RG Larsen, AC Saladino, TA Hunt, JE Mann, M Xu, VH Grassian, and SC Larsen. *J. Catal.* 204: 440–449, 2001.
89. G Li, M Xu, SC Larsen, and VH Grassian. *J. Mol. Catal. A: Chem.* 194: 169–180, 2003.

12 Theoretical Modeling of Zeolite Catalysis: Nitrogen Oxide Catalysis over Metal-Exchanged Zeolites

Scott A. McMillan, Linda J. Broadbelt, and Randall Q. Snurr
Institute for Environmental Catalysis, Northwestern University
Department of Chemical and Biological Engineering, Northwestern University

CONTENTS

12.1 INTRODUCTION

Environmental catalysis is projected to grow faster than any other segment of the catalyst market in the next 20 years [1]. This growth will come both from traditional areas, such as emissions treatment, and from new areas that range from the automotive industry to pharmaceuticals. To sustain this growth and to meet more stringent environmental demands, researchers will require ever-increasing control and understanding of catalytic materials and processes at the molecular level. Computational modeling will play a major role in meeting these goals. In conjunction with improved methods for catalyst synthesis and characterization, modeling holds the potential one day to permit the *design* of catalysts with enhanced activity and selectivity while promoting environmentally sound chemical processes.

The combination of experimental and computational methods has been increasingly recognized as the most promising path to understanding catalysis, particularly for well-characterized systems such as zeolites [2]. Computational methods provide molecular-scale insight that may be difficult to attain via experiment and can be used as a screening tool to study materials that have not yet been synthesized. Models are invaluable for interpreting experimental observations, extrapolating outside of experimental conditions, and guiding and directing experimental efforts. However, models must be continuously reviewed and calibrated, sometimes revised, and occasionally discarded based on experimental data. Experimentation, thus, provides a check on the accuracy of computational methods and also identifies potential reaction mechanisms for computational study.

Microkinetic models [3], which relate molecular-level transformations to macroscopic measures such as reaction rates and selectivities, are an excellent example of the possible synthesis of experimental and theoretical studies. Microkinetic models use the entire postulated reaction mechanism without assuming a rate-limiting step and hence are applicable over a wide range of conditions where the rate-limiting step may change. These models track both gas-phase and surface species and can therefore become quite large. As a result, they require a large amount of information to obtain the rate constants for all elementary steps in the mechanism. This information can come from experimental sources, when available, complemented by various levels of theory. Although the level of detail that microkinetic models require is formidable, the method allows for bridging the gap between the disparate molecular and macroscopic time scales, and this allows computational results to be directly compared to experiment and to be used as a *predictive* tool.

Catalysis is controlled by a number of physical and chemical processes. In microporous catalysts, such as zeolites, reactants must first overcome any external mass transfer limitations in order to adsorb into a zeolite particle. Within the particle, reactants must diffuse to adsorption sites. At some sites, catalytic reactions will occur, converting reactants into products. The products will also diffuse to and adsorb on chemically active sites, perhaps undergoing further reactions, on their way out of the zeolite particle. The time and length scales of these processes vary over many orders of magnitude, presenting a formidable challenge to modeling. However, each process may be understood on its own time and length scale using appropriate computational techniques. An outline for coupling these individual methods in a hierarchical methodology is illustrated in Figure 12.1 [4], although the actual linkage among the different techniques is still a difficult challenge. The focus in this chapter will primarily be on modeling at the atomic scale, i.e., quantum chemical calculations. The use of quantum mechanics is naturally required if one wishes to examine the making and breaking of bonds that is at the heart of catalysis. The application of these methods in this chapter will focus on zeolite-based catalysis, particularly nitrogen oxide catalysis over transition metal-exchanged zeolites.

With increasing computational power, improvements in theory and computational algorithms, as well as user-friendly software, quantum chemical calculations are really becoming a full partner with experiment. Today it is fully accepted and expected that major research efforts in catalysis will include a modeling component, and quantum chemical calculations are fast becoming almost as routine as many spectroscopic methods. Thus, while there may be researchers who specialize in modeling (just as one finds specialists in Raman or Auger spectroscopies), one also finds increasing numbers of experimental groups that employ computational quantum chemistry, just as they might employ IR or NMR, as the need arises in their research. This chapter is aimed primarily at researchers who wish to obtain a short introduction to how computational quantum chemistry works and what kinds of results one can obtain. The examples also provide a brief review of modeling efforts for de-NO_x catalysis in zeolites.

12.2 COMPUTATIONAL QUANTUM CHEMICAL METHODS

Modern experimental techniques provide a wealth of information that enhances understanding of how catalysts function. For example, x-ray techniques and microscopy can characterize the bulk and surface structure of catalysts, and stable surface species can be observed with

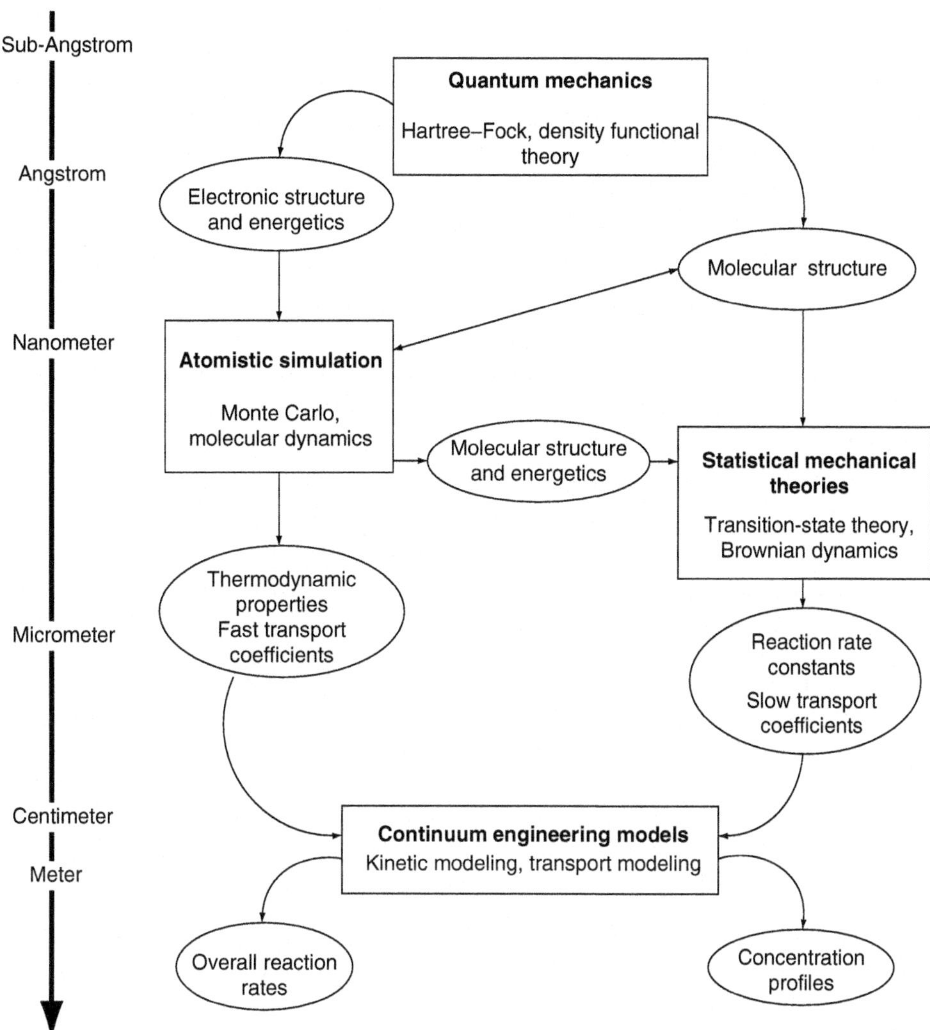

FIGURE 12.1 Illustration of catalysis modeling techniques used from the molecular scale to the macroscopic scale. The modeling techniques are labeled with rectangles and the calculated properties used to link the techniques into a hierarchical approach are denoted by ovals. (After LJ Broadbelt and RQ Snurr. *Appl Catal A* 200:23–46, 2000.)

infrared spectroscopy. More recently, computational chemistry and molecular modeling techniques have also been applied to study catalysis. These techniques can elucidate relevant catalytic processes ranging from the atomic scale to the macroscopic scale. Especially powerful are quantum chemical calculations, which solve the Schrödinger equation to yield the energy and electronic structure of a collection of atoms given only the identity of the atoms and their positions. These first principles or ab initio calculations can also find minimum-energy configurations for the atomic positions and predict quantities such as heats of formation, heats of reaction, bond strengths, infrared frequencies and modes, reaction transition states, and reaction rate constants. In addition, the calculations yield the full three-dimensional electron density, which can reveal interesting insights into bonding, for example.

While ab initio methods are extremely powerful, they are also computationally very expensive. The calculation time formally scales as somewhere from the fourth to the eighth power with respect to the number of electrons. Hence, the choice of system size is restricted to perhaps a dozen atoms for the most accurate ab initio methods to hundreds or thousands of atoms with considerably less accurate techniques. Progressing from these relatively small systems to predictions of a macroscopic property such as the conversion in a fixed-bed reactor requires a multiscale modeling approach as described above, where the output from smaller length scale models is used as input to more coarse-grained models allowing one to progress from the molecular scale to the reactor scale.

The properties of atoms and molecules depend strongly on their electronic structure. Quantum mechanics provides a mathematical formulation for the electronic behavior of systems of elementary particles. Quantum mechanics is discussed in many texts; two recent texts with an emphasis on applied computational chemistry are by Leach [5] and Jensen [6]. The fundamental equation of quantum mechanics is the Schrödinger equation. The time-independent Schrödinger equation is

$$\mathcal{H}\Psi = E\Psi \tag{12.1}$$

where $\mathcal{H}$ is the Hamiltonian, E is the electronic energy, and Ψ is the many-electron wave function. The Hamiltonian is an operator containing terms for the kinetic and potential energy of the electrons. One interpretation of the wave function is that the wave function at a point $\mathbf{r}$ when multiplied by its complex conjugate is the probability of finding the particle at $\mathbf{r}$ [5].

The Schrödinger equation is a partial differential eigenvalue equation. We wish to solve it for the energy E and the wave function Ψ, but it can be solved exactly only for some very simple problems. Even the helium atom cannot be solved analytically. Thus for larger systems, several simplifying assumptions must be made. The motion of electrons is correlated, which makes even numerical solutions of the Schrödinger equation difficult to obtain. Numerical solutions of the Schrödinger equation require a trade-off between accuracy and speed. The most accurate methods generally utilize the fewest approximations, while the reverse is true for the fastest methods. Semiempirical methods are the fastest methods that account for at least some of the electronic structure, but they make several severe approximations; these approximations are partially mitigated by parameterizing the method against a library of reference molecules [7–9]. Semiempirical methods are no longer widely used in catalysis and will not be discussed here.

Methods that use approximate forms of the wave function are usually termed wave function methods. The Hartree–Fock (HF) method is the simplest of these methods. The many-electron wave function is taken as a single Slater determinant of one-electron orbitals. Electron correlation is neglected as a result of this approximation, or equivalently, electrons "feel" only the average field generated by all of the other electrons. Computational time for HF formally scales as N^4, where N is the number of electrons, although linear scaling has been obtained recently. HF calculations are frequently used as the reference state for comparing other methods. Other wave function methods, such as configuration interaction (CI) and Møller–Plesset perturbation theory (MP2, MP3, etc.), improve upon the HF method by explicitly including electron correlation. CI utilizes a linear combination of Slater determinants to form the many-electron wave function. While these post-HF methods recover a significant fraction of the electron correlation energy, the computational scaling can be as poor as N^8.

Density functional theory (DFT) is an alternative approach that has computational scaling similar to HF but also recovers some of the electron correlation energy [10]. The conceptual basis for DFT was laid in the 1920s, but it was not put on a firm theoretical basis until the work of Hohenberg, Kohn, and Sham in the 1960s [11, 12]. They showed that the ground-state properties of a molecule were determined completely by the electron density. DFT includes electron correlation via the exchange-correlation functional, which is unknown, although many approximate functionals have been proposed and tested [6, 10]. Despite the lack of the exact exchange-correlation functional, DFT energies, geometries, and vibrational frequencies compare favorably to correlated wave function methods while having a computational cost comparable to HF. Development of improved functionals is an active area of research.

In addition to choosing a method for solving the Schrödinger equation, the wave function is usually approximated by means of a series expansion. The one-electron molecular orbitals are typically expanded as a linear combination of hydrogen-like atomic orbitals, which are in turn expanded as a series of mathematical basis functions. The functional form of the basis functions and the set of coefficients are termed a basis set. The basis set coefficients are fit using very high-level calculations on isolated atoms, ions, or small molecular fragments and remain fixed. A number of standard basis sets, most typically based on Gaussians, that yield good calculated properties have been tabulated for most of the periodic table [6]. Effective core potential (ECP) basis sets replace the core electrons of an atom with a potential field. The core potential prevents the valence electrons from collapsing into core orbitals. The advantage of ECP basis sets is that the relatively unreactive core electrons (i.e., the 1s through 2p electrons of a transition metal) are not treated explicitly, reducing the computational cost of the calculation with little or no impact on the calculated properties.

Hence, a quantum chemical calculation requires the choice of a method and basis set. These are typically denoted by an "alphabet soup" of abbreviations, which we will not go into detail here. The text by Leach provides a nice introduction [5]. Once these choices have been made, the Schrödinger equation is solved iteratively; when the average electron field does not change between iterations, the calculation has converged to a self-consistent field (SCF). The electronic energy corresponding to this state should be the minimum energy for this particular set of atomic positions. Occasionally, the SCF procedure may converge to a self-consistent wave function that does not correspond to the energy minimum. Most commonly this occurs for molecules that contain metal atoms; these systems may converge to a low-lying excited state rather than the true ground state due to the close energetic spacing of the metal d orbitals. The stability of the SCF wave function should be checked, and if found to be unstable, re-optimized to the true ground state prior to extracting any chemical properties [13].

The molecular structure may be determined by optimizing the atomic coordinates such that the electronic energy is minimized. Typically, this procedure follows the energy gradient downhill on the electronic energy surface to the nearest local minimum. At this set of atomic coordinates, the derivatives of the energy with respect to the coordinates of each atom are zero. Additionally, all of the calculated molecular frequencies, or the eigenvalues of the Hessian matrix, are positive at a geometrically converged calculation. Complex molecules may have multiple energy minima that complicate efforts to locate the true molecular structure. Stochastic optimization techniques may be used to overcome some of these obstacles (e.g., Refs. [14, 15]). Once the set of atomic coordinates is optimized and the wave function is stable, chemical properties may be extracted from the wave function. For example, the relative energy of two states may be used to compute their equilibrium populations, and the population of specific orbitals can be examined to determine the mechanism of interaction between two atoms.

A variety of commercial and academic software is available for performing quantum chemical calculations. The user must select the basis set and the method (sometimes called the "level of theory") from lists of choices. One of the key variables in this selection is the trade-off between chemical accuracy and the time required to perform the calculation. A graphical user interface may make it possible to draw an initial molecular structure, and the software can be asked to find the minimum-energy geometry or calculate the energy of a given geometry. After the main calculation, the user can visualize the structure, calculate IR frequencies, analyze bonding, etc. The ease of use and impressive graphics of the software can be seductive, but the user is still responsible for interpreting the results and for important choices that determine whether the results are meaningful or not.

12.3 SELECTIVE CATALYTIC REDUCTION OF NITROGEN OXIDES

The examples in this chapter concern catalysis of nitrogen oxides on transition metal-exchanged zeolites. As described in other chapters, nitrogen oxides are atmospheric pollutants that cause acid rain. Nitrogen forms several stable oxides, generally referred to as NO_x:NO (nitric oxide), NO_2 (nitrogen dioxide), N_2O (nitrous oxide), N_2O_3 (nitrogen trioxide), and N_2O_5 (nitrogen pentoxide). NO and NO_2 are the principal NO_x atmospheric pollutants [16, 17]. The vast majority of NO_x is formed during the combustion of fossil fuels for power generation (~60%) and transportation (~40%) [17].

The current technology for treating NO_x exhaust streams from the majority of automobiles is the three-way catalytic converter. Since its introduction in 1975, it has been included as standard on all new nondiesel automobiles [18]; emission control for diesel automobile engines, diesel electric turbines, and other lean-burn engines will be addressed in the next paragraph. The catalytic converter usually contains a combination of platinum, palladium, and rhodium. The three-way catalyst performs three functions: the oxidation of unburned hydrocarbons (HC) to CO_2 and H_2O, the oxidation of carbon monoxide (CO) to CO_2, and the reduction of NO_x to N_2. The conversion of all three components is plotted versus the air–fuel ratio in Figure 12.2. The air–fuel ratio is calculated in mass units; a stoichiometric air–fuel mixture has a ratio of ~14.7. Engines operating with excess fuel in the air–fuel mixture (i.e., the air–fuel ratio is <14.7) are said to run rich, while operation with excess air (i.e., the air–fuel ratio is >14.7) are lean-burning. Figure 12.2 shows that the

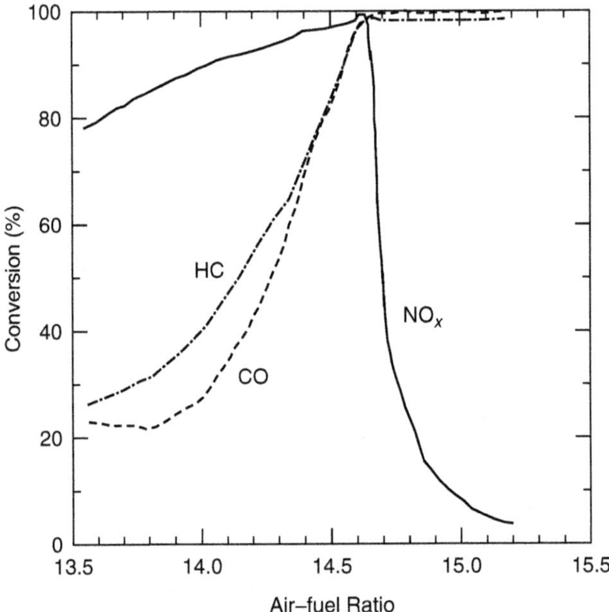

FIGURE 12.2 Conversion of hydrocarbons (HC), carbon monoxide (CO), and nitrogen oxides (NO_x) versus the air–fuel ratio over the three-way catalyst. The air–fuel ratio is expressed in mass units. (Adapted from M Shelef and RW McCabe. *Catal Today* 62:35–50, 2000.)

oxidation of CO and any unburned hydrocarbons is promoted by excess air, while NO_x reduction requires excess fuel to serve as the reductant. The conditions for simultaneous oxidation and reduction with the three-way catalyst are only satisfied near the stoichiometric ratio; this requirement for the effective operation of the automobile pollution control device constrains automobile engines to operate near the stoichiometric ratio despite the significant performance gains that can be obtained by operating in the lean-burn region.

The search for an effective catalyst for the selective reduction of nitrogen oxides (NO_x) in lean-burn exhaust streams is ongoing [18]. In excess air, the current catalytic converter reduces the excess oxygen instead of reducing NO_x, so a significant fraction of NO_x passes through the pollution control device untreated. A new catalyst is needed that can reduce NO_x under these conditions; the emission control catalyst must *selectively* reduce NO_x rather than oxidize the reductant with O_2 or further oxidize NO_x. The ideal catalyst would directly decompose NO_x to N_2 and O_2; this reaction is thermodynamically favored below 900°C, but, to date, no suitable catalyst has been found. Further, catalytic NO_x decomposition is most likely practically impossible due to the constraints of thermodynamic equilibrium and forthcoming NO_x emission standards [19]. The alternative is NO_x reduction with either ammonia or hydrocarbon reductants. The ammonia-based process is successful and is currently used to treat emissions from stationary sources [20–22]. However, ammonia is a toxic and hazardous chemical that is unsuitable for mobile sources such as passenger automobiles. More significantly, the infrastructure is simply not available to distribute ammonia in a manner similar to gasoline. For these reasons, selective catalytic reduction (SCR) using hydrocarbon reductants (HC-SCR) is the most promising emission control technology for mobile sources.

Almost 20 years ago, Iwamoto et al. [23] reported that copper exchanged into the zeolite MFI was an active NO decomposition catalyst. The reaction taking place was later shown to be reduction of NO by hydrocarbons in the emission stream, not the direct decomposition of NO. Nevertheless, this report demonstrated that metal-exchanged zeolites were promising materials for HC-SCR. Since the initial work with Cu-MFI, countless combinations of metals and supports have been examined while searching for an effective HC-SCR catalyst [21, 22, 24–28, and references therein]. Zeolites exchanged with transition metals have been very effective for treating model exhaust streams, i.e., without water and sulfur oxides [24]. The success of these efforts has widely varied, but catalysts possessing long-term stability and activity under realistic lean-burn exhaust conditions have been elusive. Preliminary data suggest that catalysts based on the SUZ-4 zeolite may have the requisite stability for practical applications [29].

Although the research efforts to find an HC-SCR catalyst have yet to produce a commercially viable catalyst, they have provided valuable fundamental insights. The recognized key to develop a viable lean-burn NO_x catalyst is an understanding of the molecular-scale processes occurring over the catalyst. Specific intermediates have been identified by spectroscopy, catalyst structures have been determined by x-ray techniques, and conversion and selectivity data have been collected under a variety of reaction conditions. More recently, computational chemistry and molecular modeling techniques have been used to study NO_x reduction catalysts. The immediate goals of all these efforts are to identify why some catalysts are so active for treating model lean-burn exhaust streams and to use that knowledge to aid the design and development of new and better catalysts.

12.4 THEORETICAL MODELING OF METAL-EXCHANGED ZEOLITES

12.4.1 The Nature of the Active Site

Zeolites are microporous, crystalline materials composed of TO_4 (T = Si or Al) tetrahedra. The tetrahedra can be combined in many ways with shared oxygen atoms, yielding a large number of zeolite structures, or topologies. A few of the known zeolite structures that have been studied extensively are shown in Figure 12.3. Zeolite pore sizes are on the order of Angstroms and are therefore very similar to the size of potential reactants, such as hydrocarbons, which makes zeolites effective molecular sieves [30]. The sieving property of zeolites is related primarily to physical adsorption and steric interactions [31].

The ability of zeolites to catalyze chemical reactions is closely related to the zeolite aluminum content. Aluminum-centered tetrahedra have an uncompensated electron and thus are not charge balanced. The charge is distributed over the framework oxygen atoms and is not localized on the aluminum atom. A cation is required to compensate the charge, which gives zeolites ion-exchange and solid acid properties. As a result, the number of aluminum-centered tetrahedra is a very important characteristic and is usually reported as the ratio of silicon to aluminum atoms. The only known limit on the aluminum distribution in zeolites is Löwenstein's rule [32], which states that two aluminum-centered tetrahedra cannot share an oxygen atom. Hence, a Si/Al ratio of 1 is a lower bound for all zeolite topologies.

From a general point of view, zeolites are ideal systems for combining experiment and computational quantum chemistry because the structures are crystalline and there is hope to synthesize well-defined active sites. Two big challenges in molecular modeling of transition metal-exchanged zeolites are closely related to the location of the zeolite aluminum atoms:

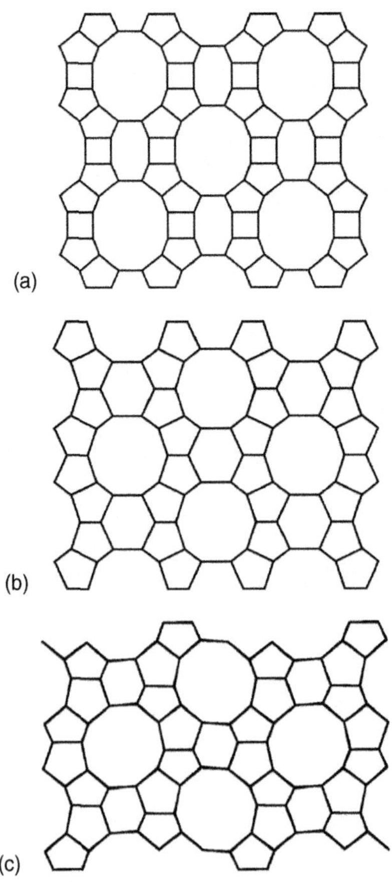

FIGURE 12.3 Illustration of the (a) mordenite (MOR), (b) ferrierite (FER), and (c) MFI zeolite structures. Vertices correspond to tetrahedral atoms (silicon or aluminum) and bonds correspond to oxygen atoms.

What is the state of the exchanged metal and where are metal cations located in the zeolite? Experiment and theory approach these questions from different perspectives. Experimental studies characterize the metal-exchanged zeolite *in situ* and attempt to characterize the type and location of the metal cations. Theory begins with the type and location of the metal cation and attempts to predict the properties that correspond to that particular environment.

Experimental studies have had some success answering both of these questions, especially the first one. For example, Sachtler and coworkers have used infrared spectroscopy (IR), temperature programmed desorption (TPD), and x-ray diffraction (XRD) to propose three types of exchanged cobalt species in MFI depending on the exchange conditions [33, 34]: isolated divalent cobalt cations that are only coordinated to the framework (Co^{2+}), multinuclear oxo ions (e.g., $[Co—O—Co]^{2+}$), and cobalt oxides (Co_3O_4). Isolated cobalt cations are the predominant form of exchanged cobalt when using aqueous ion exchange; the other types of cobalt are formed during solid-state and sublimation exchange procedures. The formation of $[Co—O—Co]^{2+}$ and Co_3O_4 species has also been observed by Raman spectroscopy [35].

Multiple extraframework sites exist for metal cations, depending on the type of zeolite [36]. The zeolite ligand field differs for each extraframework site, so techniques that probe the metal–ligand environment, such as diffuse reflectance ultraviolet–visible (UV–Vis) spectroscopy [37], can be used to distinguish cobalt in different sites. Based on differences in UV–Vis diffuse reflectance spectra, Wichterlová and coworkers have proposed extraframework sites for cobalt in the zeolites mordenite, ferrierite, and MFI [38–40]. Other detailed analyses of metal-exchanged zeolites prepared by aqueous ion exchange indicate that divalent metal cations are primarily isolated and without extraframework ligands after calcination [41–46]. Magnetic susceptibility and electron paramagnetic resonance results indicate that cobalt cations are exchanged in the divalent, high-spin electronic state [47, 48]. Seff and coworkers [49–51] have investigated the siting of metals in zeolites A, X, and Y with XRD; they observed isolated metal cations that are only coordinated to the zeolite framework and specific adsorbed molecules such as NO and H_2S. XRD investigations have also been performed to determine the extraframework locations of divalent copper and nickel cations [52–54].

Computational studies have also been used to probe the siting of metal cations in zeolites, although some have avoided this issue by using minimal representations of the zeolite that contain only the metal cation, one or two T atoms, and a few oxygen atoms (e.g., Refs. [55–59]). The location of isolated single aluminum substitutions and acid site characterization have been studied extensively by computation [60–63, and references therein]. For each unit of charge on the metal cation, a framework aluminum atom must be located nearby. Hence, for divalent cations, two aluminum tetrahedra must be in the vicinity of each other. While recent results indicate the importance of nearby aluminum centers [46], experimental techniques have not been able to determine the arrangement of aluminum atoms in ion-exchange sites. Meanwhile, computational studies containing different aluminum arrangements have only recently been reported [64–67]. The probability of finding such pairs may be small, especially in high Si–Al zeolites. The maximum divalent cation exchange ratio has been computed using reasonable metal bonding radii and a random distribution of aluminum [68–71]; the maximum calculated metal to aluminum ratio is significantly <0.5. However, catalysts that contain primarily divalent cations have been prepared with metal/Al ratios near 0.5, indicating that the aluminum distribution may not be completely random.

12.4.2 Cluster Models of Zeolite Active Sites

Many zeolites have unit cells that are too large for an entire unit cell to be considered in routine quantum chemical calculations. Computational studies of metal-exchanged zeolites have therefore generally utilized zeolite cluster models; a region of the zeolite framework is extracted and usually terminated by hydrogen atoms (e.g., Figure 12.4). The use of zeolite clusters introduces three potential errors: (1) the termination is artificial, (2) long-range contributions are neglected, and (3) the cluster size may be insufficient to describe the local environment. Previous cluster studies constrained the entire cluster to the crystallographic zeolite geometry, despite the substitution of an aluminum tetrahedron in place of a silicon tetrahedron [72, 73]. Lonsinger et al. [74] demonstrated that structural relaxation around the substituted aluminum atom was significant. They also reported that the range of the artificial termination effects was approximately two bonds from the terminating atoms. Sometimes clusters are considered as completely flexible, or only the terminating groups are constrained to their crystallographic positions in current computational studies. Fixing the terminating groups accounts for geometric restrictions on cluster relaxation that would come from the surrounding zeolite structure omitted in the cluster approach. The impact of the choice of terminating group, either —OH or —SiH_3, was studied by Curtiss and coworkers [75, 76]. Enlarging a SiH_3-terminated cluster by including additional terminating shells resulted in an oscillation in the proton affinity; the addition of a shell of OH increased the proton affinity, while additional shells of SiH_3 lowered the proton affinity [75]. The oscillation in the proton affinity is due to long-range electrostatic effects; the other calculated properties rapidly converged with respect to the cluster size.

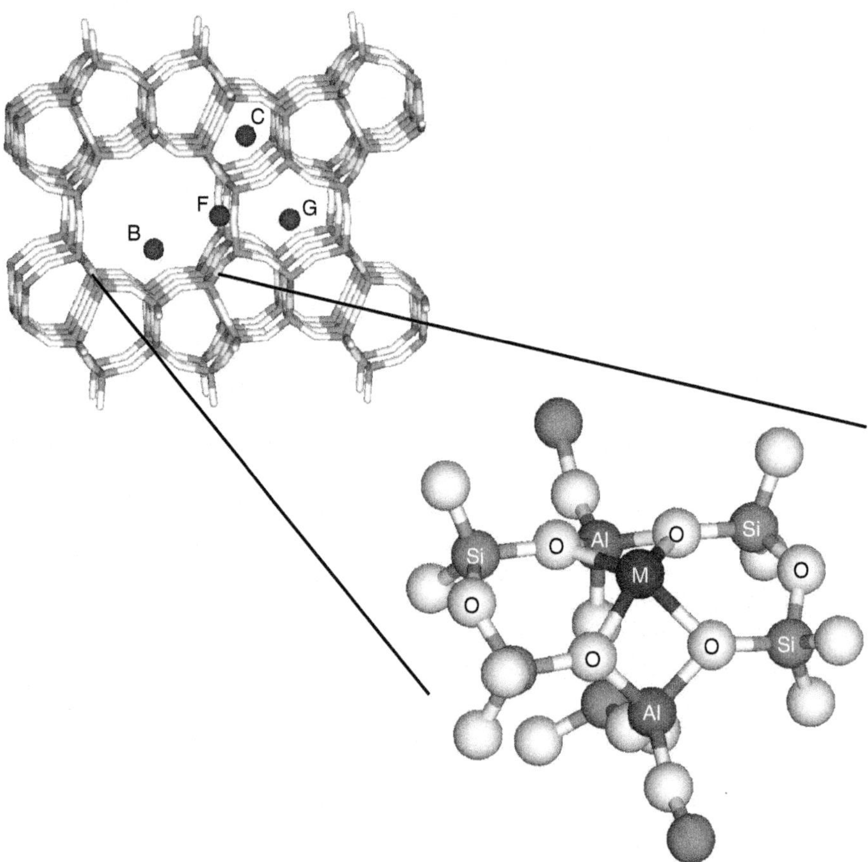

FIGURE 12.4 Extraction of a zeolite cluster representing the **B** extraframework site of ferrierite. Unless otherwise indicated, all dark gray atoms are silicon atoms and all light gray atoms are oxygen atoms. Terminating hydrogen atoms are not shown.

Long-range energetic contributions were incorporated by Cook et al. [77] with a Madelung potential field; the missing zeolite atoms immediately surrounding the cluster were replaced with fixed-point charges that were included in the electronic Hamiltonian. While the geometry of the cluster did not significantly change with the inclusion of long-range effects, the energy and charge distribution did. However, the inclusion of Madelung effects has fallen into disfavor since cluster termination effects begin to dominate and can lead to physically unrealistic results [78]. Truong and coworkers have developed an alternative methodology for embedding clusters in Madelung fields that does not suffer from cluster termination boundary effects [79–81]; the electric field of the missing atoms immediately adjacent to the cluster is replaced with a charged surface surrounding the cluster.

Other embedding methods have been proposed that allow the portion of the zeolite outside of the quantum mechanical cluster also to relax [82, 83]. Quantum mechanics is used to model a cluster embedded in a larger cluster or periodic representation of the zeolite calculated with molecular mechanics. Using this embedding scheme (QM/Pot), Brändle and Sauer [82] determined that about 35% of the heat of ammonia adsorption in H-faujasite is due to long-range interactions. The QM/Pot scheme was also found to be more reliable for incorporating long-range effects than a Madelung potential field [84]. Specific extraframework sites for monovalent copper in MFI have been investigated with QM/Pot [85].

Embedding minimizes, but does not eliminate, the errors associated with zeolite cluster models. These errors can be completely eliminated by performing periodic DFT calculations on the zeolite unit cell. Generally, these calculations are very computationally expensive and do not include geometry optimization [86, 87]. Geometries have been optimized for zeolites with relatively small unit cells; in these cases, embedded results have been found to agree well with the periodic calculations [88].

Despite these advances in embedded cluster techniques, discrete clusters are still commonly used (e.g., Ref. [89]). The cluster approximation is based on a simple premise, i.e., the chemical interactions and transformations that occur are local to the zeolite catalytic site. Advances in parallel computation and computer power permit the use of larger clusters than ever before. Theoretical studies on clusters that include 30 to 40 heavy atoms are now feasible; clusters of this size mitigate most of the adverse cluster termination effects, are sufficiently large to accurately describe the local zeolite environment, and recover some of the intermediate-range effects, especially for cases where the long-range effects should be minor. When long-range effects are expected to be significant, such as the adsorption of molecules with large dipoles (e.g., ammonia), embedding techniques such as QM/Pot will be more accurate. However, until recently, theoretical studies of zeolite catalysis almost universally used cluster models to investigate adsorption and reaction over metal-exchanged zeolites.

12.4.3 Influence of Metal–Zeolite Coordination Environment

The interaction of metal-exchanged zeolites with a wide range of probe molecules has been calculated with first principles (e.g., Refs. [56, 90–94]). Until recently, most computational studies focused on a single metal–zeolite environment. However, experimental evidence strongly suggests that the catalytic activity of a metal cation depends on the local environment [42, 95]. Li and Armor [95] observed that the turnover of NO to N_2 per cobalt cation increased with increasing cobalt loading (Figure 12.5), while Kaucký et al. [42] correlated the occupation of specific extraframework sites with the NO_x reduction activity at very low cobalt loadings.

Recently, several computational investigations have begun to address the role of metal–zeolite coordination environment on catalysis. The key feature of these studies is the application of a uniform computational methodology to several distinct metal–zeolite environments. Faster computers have made it possible to use cluster models big enough to capture differences among possible sites for the metal ions. For example, calculations on

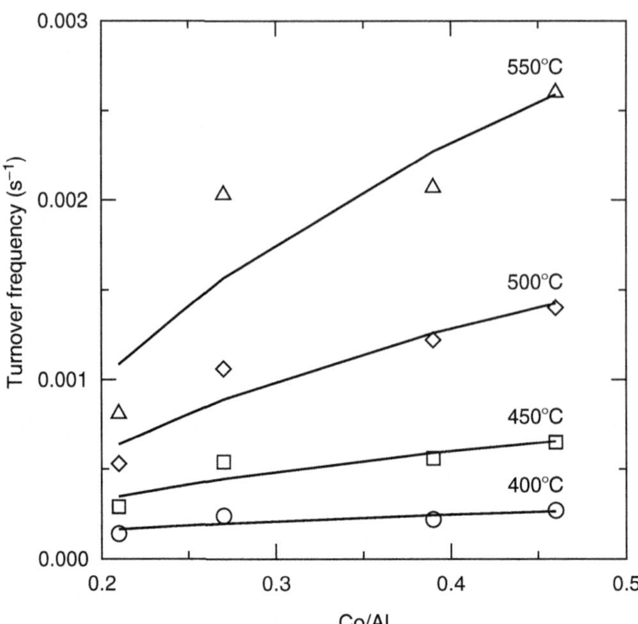

FIGURE 12.5 Turnover frequency of NO to N_2 per cobalt cation as a function of cobalt loading over cobalt-exchanged ferrierite. (Data from Y Li and JN Armor. *J Catal* 150:376–387, 1994.)

copper-exchanged zeolites revealed that the EPR spectrum depends on the number and relative position of the aluminum atoms in a six-membered ring [65]. Specific copper sites were proposed in zeolites A, Y, and ZK4 based upon the agreement between calculation and experiment.

A similar approach was taken to identify the extraframework position of cobalt in ferrierite. The coordination of cobalt to the ferrierite framework leads to new infrared bands in the 800 to 1000 cm^{-1} transmission window [43, 44]. The framework infrared modes perturbed by cobalt were calculated for eight possible cobalt–ferrierite environments and compared to experimental infrared signatures [67]. Coupled with a comparison of the calculated cobalt–ferrierite geometries to experimental EXAFS data and calculated cobalt-binding energies with experimental cobalt cation migration data, two particular sites were found to agree most closely with the experimental data. The fundamental conclusion of both of these studies is that the properties of metal cations exchanged into zeolites depend on the local zeolite environment. These differences can be exploited by combining calculation and experiment to help determine the location of metal cations in zeolites.

A central issue is how the local zeolite environment influences the catalytic activity of a metal cation. This question has been approached experimentally [42] but has not been addressed directly with theoretical calculations. Theory has the potential to make a large impact here since different zeolite environments may be studied individually, while experiments will always measure the average activity over all the occupied sites. While theory has not yet been applied to study NO_x catalytic activity as a function of zeolite environment, several groups have begun by exploring reactions such as NO_x adsorption and hydrocarbon chemistry [96–102].

One such study has examined the adsorption of methane in zinc-exchanged zeolites. The adsorption of methane to divalent zinc cations depends upon the zeolite ring size [101]. Adsorption is favored on zinc located in four-membered rings compared to five-membered rings. For both ring sizes, methane preferred a threefold adsorbed configuration. Zinc located in the four-membered ring is able to catalyze the dissociation of hydrogen with activation energy of 75 kJ/mol [98]. Zinc coordinated in the five-membered ring does not promote hydrogen dissociation. The energy required for the simultaneous adsorption of ethane and abstraction of a hydrogen atom was also a function of the zinc environment [102]. The energies varied by 37 kJ/mol and generally followed the same trend as the stability for hydrogen reduction of the zinc–zeolite environments. The interaction of zinc with probe molecules was observed to result in the partial extraction of the metal from the extraframework zeolite sites; the differences in the zinc stability and the degree of corresponding framework relaxation for each site were proposed to account for the observed differences in the interaction of probe molecules with zinc [98].

The structure of the metal–zeolite environment changes when interacting with a probe molecule compared to its initial state, as shown below using cobalt as an example:

$$(\text{Z—Co})_\text{i} + \text{X} \xrightarrow{\Delta E_\text{ads}} (\text{Z—Co})_\text{f}\text{—X} \tag{12.2}$$

$(\text{Z—Co})_\text{i}$ is the initial bare cobalt environment, and $(\text{Z—Co})_\text{f}\text{—NO}$ is the cobalt environment with an adsorbed molecule, X. While one particular structural change may be predominant (e.g., partially extracting the metal from the extraframework site as described earlier for zinc), many less obvious changes also occur and contribute to ΔE_ads [96]. The degree of structural change of the metal–zeolite environment induced by the adsorbate may be evaluated by computing the "strain" energy [96, 99]:

$$(\text{Z—Co})_\text{i} \xrightarrow{\Delta E_\text{strain}} (\text{Z—Co})_\text{f} \tag{12.3}$$

The strain reaction does not include any contributions due to the formation of metal–adsorbate bonds, but it does consider the geometry of the zeolite in the adsorbed configuration. A second reaction measuring the strength of the metal–adsorbate bond can be constructed,

$$(\text{Z—CO})_\text{f} + \text{X} \xrightarrow{\Delta E_\text{bond}} (\text{Z—Co})_\text{f}\text{—X} \tag{12.4}$$

such that

$$\Delta E_\text{ads} = \Delta E_\text{bond} + \Delta E_\text{strain} \tag{12.5}$$

The calculated adsorption energy of CO on copper exchanged in MFI and ferrierite varied by 54 and 67 kJ/mol, respectively, depending upon the copper extraframework environment [99]. The strain energy of the less favorable environments was 25 to 50 kJ/mol, while the strain energy of the preferred sites was always <8 kJ/mol. Similar characteristics were observed in the calculations of NO adsorption on cobalt in several ferrierite environments [96]. The so-called B and G sites are shown in Figure 12.6. NO preferred adsorption on cobalt in the B extraframework site over the G site by about 50 kJ/mol. The average strain energy required to alter the G site structurally was 33 kJ/mol compared to only 7 kJ/mol for the B site. Like the CO adsorption case, the difference in the strain energies only partially accounts for the preferential adsorption characteristics of NO. The formation of the cobalt–nitrogen bond (ΔE_bond) was also favored for cobalt in the B site. Calculating the coordination number of cobalt in each site revealed that cobalt was more coordinatively unsaturated in the B site, thus accounting for the stronger cobalt–nitrogen bond.

We further examined the characteristics of preferential adsorption of several other molecules on these two cobalt–ferrierite environments [100]. The overall preference for the B site

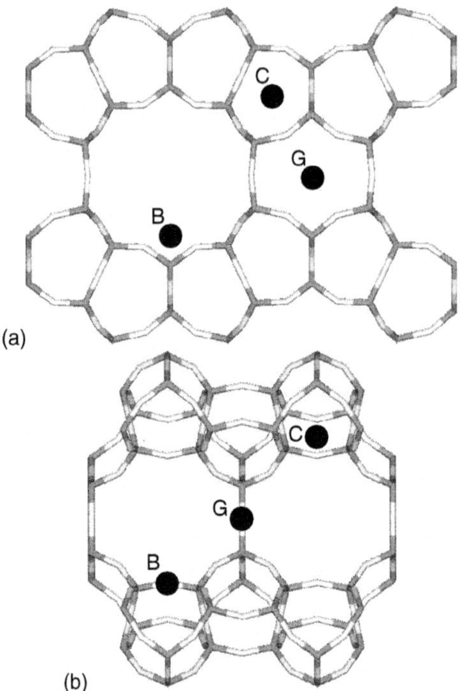

FIGURE 12.6 Location of extraframework sites B, C, and G in ferrierite (a) looking down the ten-ring channels and (b) looking down the eight-ring channels. The C site does not participate in NO_x reduction and was not considered.

compared to the G site based on adsorption energy differences varied from nearly negligible to about 40 kJ/mol depending upon the adsorbate. Figure 12.7 shows that the difference in the strain energy between the two sites also varied with the adsorbate. However, the difference in the bond energy was relatively constant regardless of the adsorbate. This suggests that adsorption is intrinsically favored by ~13 kJ/mol on cobalt in the B environment due to cobalt being more coordinatively unsaturated in this site. However, the adsorbates extract cobalt from the zeolite to different degrees, resulting in a variable strain energy. The two characteristics usually reinforce one another, making the B environment even more preferred for adsorption. However, in some cases, the strain effect reverses and negates the intrinsic preference of molecules for cobalt in the B environment. Additional effects were observed for molecules capable of forming secondary hydrogen bonds with the zeolite or highly polar molecules such as water.

Given these differences in adsorption between different sites, it seems likely that differences in catalytic activity will also be seen. This should be a fruitful area of future research. As pointed out above, these differences may be difficult to detect experimentally, and computational quantum chemistry may play an important role in their elucidation.

12.4.4 Reaction Pathway Analysis

Some authors have examined the energetics of all adsorbed reactants, products, and stable intermediates for proposed reaction pathways in zeolites. In some cases, they have also found the transition states for the elementary steps of the reaction. Most of the reaction pathway analysis of NO_x catalysis in zeolites has focused on decomposition pathways over small clusters of copper-exchanged zeolites. Trout et al. [103] calculated the structure and energetics

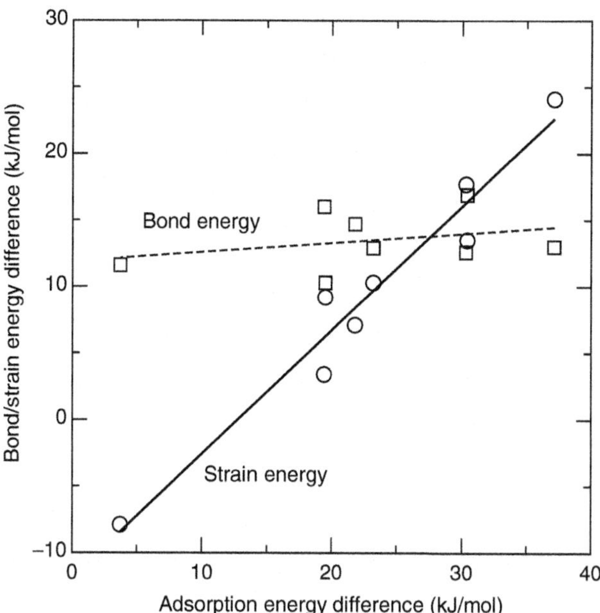

FIGURE 12.7 Trends in the difference of the bond (□) and strain (○) energies for the B and G environments in ferrierite compared to the total difference in the adsorption energies [100]. The molecules considered were NO, NO_2, N_2O, N_2, and CO.

of many stable NO_x intermediates on a copper-exchanged MFI cluster. The cluster contained a single aluminum atom in the T12 tetrahedral position of MFI and was a "three-shell" cluster: counting outward from the aluminum atom, a first shell of oxygen atoms, a second shell of silicon atoms, and a third shell of oxygen atoms terminated by hydrogen atoms comprise the zeolite cluster. The monovalent copper cation was coordinated to two of the zeolite oxygen atoms neighboring the aluminum atom. Based on the calculated thermodynamics of 28 elementary reactions, they proposed the overall NO_x decomposition pathway depicted in Figure 12.8.

Schneider et al. [57] explored the core NO decomposition reaction pathway (highlighted reactions in Figure 12.8) over a copper-exchanged zeolite cluster, consisting of only a single AlO_4 tetrahedron. The minimal cluster does not correspond to any particular zeolite so their calculated properties only provide a rough estimate of the corresponding activation energies in any particular zeolite. However, Schneider et al. located the transition states for the elementary reaction steps, and this is a computationally demanding task, much more so than calculating the stable geometries of the products and reactants. Hence, the use of minimal zeolite clusters is offset by their value for calculating barriers along the reaction coordinate. Two activation barriers were found for the core NO_x decomposition pathway, both involving the adsorbed atomic oxygen intermediate:

$$\mathrm{ZCu{-}ONNO} \longrightarrow \mathrm{ZCu{-}O} + \mathrm{N_2O}, \quad E_{\mathrm{act}} = 29\,\mathrm{kJ/mol} \tag{12.6}$$

$$\mathrm{ZCu{-}O} + \mathrm{N_2O} \longrightarrow \mathrm{ZCu{-}O_2} + \mathrm{N_2}, \quad E_{\mathrm{act}} = 151\,\mathrm{kJ/mol} \tag{12.7}$$

The relative stability of this intermediate was suggested to have a significant effect on the catalyst system, i.e., zeolites that form stable ZCu–O intermediates will promote N_2O formation but inhibit N_2 formation. Subsequent calculations established several pathways

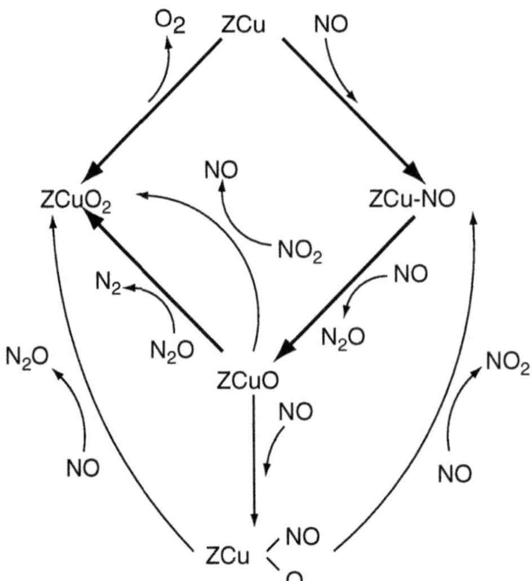

FIGURE 12.8 Nitrogen oxide decomposition pathway over copper-exchanged zeolites [103].

connecting ZCu and ZCu–O [104, 105]. Others have also investigated elementary NO_x decomposition reactions and overall reaction pathways over copper-exchanged zeolites (i.e., Refs. [59, 106]). More complex intermediates, such as $ZCu(N_2O_3)$, have also been proposed [90, 107]. While these species form stable intermediates, the decomposition barriers are large, which suggests they are not actively involved in the decomposition pathway.

While most of the computational contributions have focused on NO_x decomposition over copper-exchanged zeolites, other metals are also active for NO_x decomposition and reduction [24]. Recently, Ryder et al. [89] calculated the barrier for a metal oxo cation-(i.e., $[CoO]^+$) mediated N_2O decomposition reaction over cobalt- and iron-exchanged MFI (145 and 165 kJ/mol, respectively). This is consistent with the experimental observation that N_2O decomposition is more rapid over Co-MFI than Fe-MFI. The structure and orientation of stable adsorbed species, as well as reactive intermediates, can be resolved with theoretical calculations. The calculated infrared shifts of adsorbed N_2O were found to depend upon its orientation; adsorption through the nitrogen end produced a blueshift for both N–N and N–O stretching modes, in agreement with experiment [108]. Adsorption through the oxygen end produced a redshift in the N–O stretching mode that is not observed experimentally for these materials when exposed to N_2O.

12.4.5 Dealing with Electron Spin

In systems involving metal atoms, an important consideration when performing quantum chemical calculations is the electronic spin state. Along the N_2O reaction coordinate, Ryder et al. [89] determined that one particular electronic spin configuration was always the lowest energy. For iron, the potential energy surface corresponding to a total of five unpaired electrons (sextet) was lower in energy than either surface corresponding to one (doublet) or three (quartet) unpaired electrons. The quintet spin configuration was consistently preferred for cobalt. In this case, the alternative spin configurations do not directly contribute to the reaction energy surface; however, one cannot usually predict a priori which spin state will be lowest in energy so all reasonable spin states must be calculated.

In many cases, the overall spin state of the reactant and product configurations will not be the same. For example, the lowest energy spin configuration shifts from the singlet state to the triplet state during the direct dissociation of N_2O to form an oxidized copper atom [104, 105, 109]:

$$ZCu + N_2O \longrightarrow ZCuO + N_2 \tag{12.8}$$

The unoxidized copper–zeolite prefers the singlet spin state, while the triplet spin state is lower in energy for the oxidized species. The overall reaction energy surface is thus a combination of the singlet and triplet surfaces; this type of reaction is termed spin-forbidden. The activation barrier for such reactions may not correspond to a transition state on a single spin energy surface, but rather may be located based on an intersection point of the two energy sufaces. Accurately determining the crossing of two potential energy surfaces corresponding to different spin states, known as a conical intersection or avoided crossing, is a formidable task. At the geometric configuration corresponding to the crossing point, the electronic energies of the two spin surfaces will be identical. Extremely high-level quantum chemical methods are required to locate the conical intersection for two spin states and to compute the spin coupling efficiency, which determines the kinetic rate of the reaction [6, 110]. In practice, approximate spin correlation diagrams from DFT are useful for estimating activation barriers of spin-forbidden reactions [105].

Issues concerning the spin state should be expected for NO_x reactions over metal-exchanged zeolites. The metal cations usually possess unpaired electrons, as do many NO_x species such as NO and NO_2. Only the total spin is specified in a quantum chemical calculation; the software does not give the user the lowest energy spin state automatically. The expected localization of the unpaired electrons on particular atoms should be confirmed by checking the atomic spin densities. Cobalt cations exchanged in zeolites are known to be high spin (quartet) [48]; calculations confirm that about 2.6 unpaired electrons are localized on cobalt [67]. Similarly, NO, which carries a single unpaired electron, adsorbed on cobalt produces a triplet complex with the unpaired electrons localized on cobalt [96, 111]. The alternative spin states, singlet and quintet, were also calculated, but both were higher in energy than the triplet. In general, all reasonable spin states must always be considered.

12.5 CONCLUSIONS AND FUTURE DIRECTIONS

Nearly all the theoretical investigations into NO_x catalysis have focused on NO_x decomposition rather than reduction with hydrocarbons. The mechanism of activation of the hydrocarbon is not fully understood. While NO_x reduction is more difficult to study with computational methods than NO_x decomposition, SCR with hydrocarbons is the most promising solution for the treatment of lean-burn exhaust gases and needs to be addressed with theoretical models.

Sachtler and coworkers [33] have shown that binuclear oxo metal species ($[M—O—M]^{2+}$) are generally more active than isolated divalent metal cations. While some theoretical studies have begun to address this topic [97, 98, 112, 113], much more work is needed to fully understand catalysis over these species. The size of the binuclear species requires even larger zeolite clusters; the quantum mechanical–molecular mechanics embedding schemes will most likely be necessary for these calculations.

The theoretical modeling of zeolite catalysis has revealed a wealth of information about the siting and activity of exchanged metal cations. Advances in embedded modeling techniques that combine quantum chemical and molecular mechanics methods will provide a more thorough understanding of the interdependence of the type of metal cation, zeolite topology, and metal–zeolite environment on catalytic activity. These advances, when coupled with continually increasing computational resources, ensure that computational studies of zeolite catalysis will become even more useful in the future.

ACKNOWLEDGMENT

The authors acknowledge support from the Northwestern University Institute for Environmental Catalysis.

REFERENCES

1. J Oleson, V Haynes, B Ramaker, R Slough, H Whalen Jr., T Sciance, M Saft, D Jost, W Tamarelli, S Turner, D Brooks, D Artemis, and D Kellogg, Eds. *Technology Vision 2020: The US Chemicals Industry*. Washington, DC: American Chemical Society, 1996.
2. AT Bell. *Catal Today* 38:151–156, 1997.
3. JA Dumesic, DF Rudd, LM Aparicio, JE Rekoske, and A Treviño. *The Microkinetics of Heterogenous Catalysis*. Washington, DC: American Chemical Society, 1993.
4. LJ Broadbelt and RQ Snurr. *Appl Catal A* 200:23–46, 2000.
5. AR Leach. *Molecular Modelling: Principles and Applications*. Harlow, England: Longman, 1996.
6. F Jensen. *Introduction to Computational Chemistry*. Chincester, UK: John Wiley & Sons, 1999.
7. JJP Stewart. *J Comput Chem* 10:209–220, 1989.
8. JJP Stewart. *J Comput Chem* 10:221–264, 1989.
9. AD Bacon and MC Zerner. *Theo Chim Acta* 53:21–54, 1979.
10. RG Parr and W Yang. *Density-Functional Theory of Atoms and Molecules*. New York: Oxford, 1989.
11. P Hohenberg and W Kohn. *Phys Rev* 136:B864–B871, 1964.
12. W Kohn and LJ Sham. *Phys Rev* 140:A1133–1138, 1965.
13. R Bauernschmitt and R Ahlrichs. *J Chem Phys* 104:9047–9052, 1996.
14. NC Haubein, SA McMillan, and LJ Broadbelt. *J Chem Inf Comput Sci* 43:68–74, 2003.
15. SA McMillan, NC Haubein, RQ Snurr, and LJ Broadbelt. *J Chem Inf Comput Sci*, 43:1820–1828, 2003.
16. K Wark and CF Warner. *Air Pollution*. New York: Harper Collins, 1981.
17. N de Nevers. *Air Pollution Control Engineering*. New York: McGraw-Hill, 1995.
18. M Shelef and RW McCabe. *Catal Today* 62:35–50, 2000.
19. CT Goralski Jr and WF Schneider. *Appl Catal B* 37:263–277, 2001.
20. RM Heck. *Catal Today* 53:519–523, 1999.
21. G Busca, L Lietti, G Ramis, and F Berti. *Appl Catal B* 18:1–36, 1998.
22. P Forzatti and L Lietti. *Heterogen Chem Rev* 3:33–51, 1996.
23. M Iwamoto, H Furukawa, Y Mine, F Uemura, S Mikuriya, and S Kagawa. *J Chem Soc Chem Commun* 16:1272–1273, 1986.
24. Y Traa, B Burger, and J Weitkamp. *Micropor Mesopor Mater* 30:3–41, 1999.
25. M Misono. *Cat Tech* 2:53–69, 1998.
26. JN Armor. *Catal Today* 26:147–158, 1995.
27. R Burch, JP Breen, and FC Meunier. *Appl Catal B* 39:283–303, 2002.
28. MD Amiridis, TJ Zhang, and RJ Farrauto. *Appl Catal B* 10:203–227, 1996.
29. A Subbiah, K Cho Byong, RJ Blint, A Gujar, GL Price, and JE Yie. *Appl Catal B* 42:155–178, 2003.
30. BC Gates. *Catalytic Chemistry*. New York: John Wiley & Sons, 1992.
31. J Ǩarger and DM Ruthven. *Diffusion in Zeolites and Other Microporous Solids*. New York: John Wiley & Sons, 1992.
32. W Löwenstein. *Am Mineral* 39:92–96, 1954.
33. X Wang, HY Chen, and WMH Sachtler. *Appl Catal B* 26:L227–L239, 2000.
34. X Wang, H Chen, and WMH Sachtler. *Appl Catal B* 29:47–60, 2001.
35. H Ohtsuka, T Tabata, O Okada, LMF Sabatino, and G Bellussi. *Catal Today* 42:45–50, 1998.
36. WJ Mortier. *Compilation of Extra Framework Sites in Zeolites*. London: Butterworth Scientific, 1982.
37. G Kortüm, W Braun, and G Herzog. *Angew Chem Internat Edit* 2:333–341, 1963.
38. J Dědeček and B Wichterlová. *J Phys Chem B* 103:1462–1476, 1999.
39. D Kaucký, J Dědeček, and B Wichterlová. *Micropor Mesopor Mater* 31:75–87, 1999.
40. J Dědeček, D Kaucký, and B Wichterlová. *Micropor Mesopor Mater* 35–36:483–494, 2000.

41. L Drozdová, R Prins, J Dědeček, Z Sobalìk, and B Wichterlová. *J Phys Chem B* 106:2240–2248, 2002.
42. D Kaucký, A Vondrová, J Dědeček, and B Wichterlová. *J Catal* 194:318–329, 2000.
43. Z Sobalík, J Dědeček, B Wichterlová, L Drozdová, and R Prins. *J Catal* 194:330–342, 2000.
44. Z Sobalík, Z Tvarůžková, and B Wichterlová. *J Phys Chem B* 102:1077–1085, 1998.
45. Z Sobalík, Z Tvarůžková, and B Wichterlová. *Micropor Mesopor Mater* 25:225–228, 1998.
46. O Bortnovsky, Z Sobalík, and B Wichterlová. *Micropor Mesopor Mater* 46:265–275, 2001.
47. Y Li, TL Slager, and JN Armor. *J Catal* 150:388–399, 1994.
48. EM El-Malki, D Werst, PE Doan, and WMH Sachtler. *J Phys Chem B* 104:5924–5931, 2000.
49. YH Yeom, Y Kim, and K Seff. *J Phys Chem* 100:8373–8377, 1996.
50. S Zhen and K Seff. *Micropor Mesopor Mater* 39:1–18, 2000.
51. WV Cruz, PCW Leung, and K Seff. *Inorg Chem* 18:1692–1696, 1979.
52. MC Dalconi, G Cruciani, A Alberti, P Ciambelli, and MT Rapacciuolo. *Micropor Mesopor Mater* 39:423–430, 2000.
53. MP Attfield, SJ Weigel, and AK Cheetham. *J Catal* 170:227–235, 1997.
54. MP Attfield, SJ Weigel, and AK Cheetham. *J Catal* 172:274–280, 1997.
55. KC Hass and WF Schneider. *J Phys Chem* 100:9292–9301, 1996.
56. WF Schneider, KC Hass, R Ramprasad, and JB Adams. *J Phys Chem B* 100:6032–6046, 1996.
57. WF Schneider, KC Hass, R Ramprasad, and JB Adams. *J Phys Chem B* 101:4353–4357, 1997.
58. R Ramprasad, WF Schneider, KC Hass, and JB Adams. *J Phys Chem B* 101:1940–1949, 1997.
59. Y Yokomichi, T Yamaba, H Ohtsuka, and T Kakumoto. *J Phys Chem* 100:14424–14429, 1996.
60. MV Frash and RA van Santen. *Top Catal* 9:191–205, 1999.
61. RA van Santen. *Catal Today* 50:511–515, 1999.
62. B van de Graaf and SL Njo, KS Smirnov. In *Reviews in Computational Chemistry*, Vol. 14, chap. 3, KB Lipkowitz and DB Boyd, Eds. New York: Wiley, pp. 137–223. 2000.
63. A Goursot, B Coq, and F Fajula. *J Catal* 216:324–332, 2003.
64. D Nachtigallov'a, P Nachtigall, and J Sauer. *Phys Chem Chem Phys* 3:1552–1559, 2001.
65. K Pierloot, A Delabie, MH Groothaert, and RA Schoonheydt. *Phys Chem Chem Phys* 3:2174–2183, 2001.
66. A Delabie, K Pierloot, MH Groothaert, BM Weckhuysen, and RA Schoonheydt. *Phys Chem Chem Phys* 4:134–145, 2002.
67. SA McMillan, LJ Broadbelt, and RQ Snurr. *J Phys Chem B* 106:10864–10872, 2002.
68. X Feng and WK Hall. *Catal Lett* 46:11–16, 1997.
69. MJ Rice, AK Chakraborty, and AT Bell. *J Catal* 186:222–227, 1999.
70. MJ Rice, AK Chakraborty, and AT Bell. *J Catal* 194:278–285, 2000.
71. BR Goodman, KC Hass, WF Schneider, and JB Adams. *Catal Lett* 68:85–93, 2000.
72. JG Fripiat, P Galet, J Delhalle, JM André, JB Nagy, and EG Derouane. *J Phys Chem* 89:1932–1937, 1985.
73. EG Derouane and JG Fripiat. *Zeolites* 5:165–172, 1985.
74. SR Lonsinger, AK Chakraborty, DN Theodorou, and AT Bell. *Catal Lett* 11:209–218, 1991.
75. HV Brand, LA Curtiss, and LE Iton. *J Phys Chem* 96:7725–7732, 1992.
76. HV Brand, LA Curtiss, and LE Iton. *J Phys Chem* 97:12773–12782, 1993.
77. SJ Cook, AK Chakraborty, AT Bell, and DN Theodorou. *J Phys Chem* 97:6679–6685, 1993.
78. RA van Santen. *Catal Today* 38:377–390, 1997.
79. EV Stefanovich and TN Truong. *J Phys Chem B* 102:3018–3022, 1998.
80. JM Vollmer and TN Truong. *J Phys Chem B* 104:6308–6312, 2000.
81. JM Vollmer, EV Stefanovich, and TN Truong. *J Phys Chem B* 103:9415–9422, 1999.
82. M Brändle and J Sauer. *J Mol Catal A: Chem* 119:19–33, 1997.
83. J Sauer and M Sierka. *J Comput Chem* 21:1470–1493, 2000.
84. U Eichler, M Brändle, and J Sauer. *J Phys Chem B* 101:10035–10050, 1997.
85. D Nachtigallová, P Nachtigall, M Sierka, and J Sauer. *Phys Chem Chem Phys* 1:2019–2026, 1999.
86. M Brändle, J Sauer, R Dovesi, and N Harrison. *J Chem Phys* 109:10379–10389, 1998.
87. A Kessi and B Delley. *Int J Quantum Chem* 68:135–144, 1998.
88. JR Hill, CM Freeman, and B Delley. *J Phys Chem A* 103:3772–3777, 1999.
89. JA Ryder, AK Chakraborty, and AT Bell. *J Phys Chem B* 106:7059–7064, 2002.

90. X Solans-Monfort, V Branchadell, and M Sodupe. *J Phys Chem A* 104:3225–3230, 2000.
91. P Treesukol, J Limtrakul, and TN Truong. *J Phys Chem B* 105:2421–2428, 2001.
92. MJ Rice, AK Chakraborty, and AT Bell. *J Phys Chem A* 102:7498–7504, 1998.
93. L Rodriguez-Santiago, M Sierka, V Branchadell, M Sodupe, and J Sauer. *J Am Chem Soc* 120:1545–1551, 1998.
94. A Sierraalta, R Añez, and MR Brussin. *J Catal* 205:107–114, 2002.
95. Y Li and JN Armor. *J Catal* 150:376–387, 1994.
96. SA McMillan, LJ Broadbelt, and RQ Snurr. *J Catal* 219:117–125, 2003.
97. LAMM Barbosa, GM Zhidomirov, and RA van Santen. *Phys Chem Chem Phys* 2:3909–3918, 2000.
98. LAMM Barbosa, GM Zhidomirov, and RA van Santen. *Catal Lett* 77:55–62, 2001.
99. M Davidová, D Nachtigallová, R Bulánek, and P Nachtigall. *J Phys Chem B* 107:2327–2332, 2003.
100. SA McMillan, RQ Snurr, and LJ Broadbelt. *J Phys Chem B*, 107:13329–13335, 2003.
101. LAMM Barbosa, RA van Santen, and J Hafner. *J Am Chem Soc* 123:4530–4540, 2001.
102. AA Shubin, GM Zhidomirov, AL Yakovlev, and RA van Santen. *J Phys Chem B* 105:4928–4935, 2001.
103. BL Trout, AK Chakraborty, and AT Bell. *J Phys Chem* 100:17582–17592, 1996.
104. WF Schneider, KC Hass, R Ramprasad, and JB Adams. *J Phys Chem B* 102:3692–3705, 1998.
105. D Sengupta, JB Adams, WF Schneider, and KC Hass. *Catal Lett* 74:193–199, 2001.
106. N Tajima, M Hashimoto, F Toyama, AM El-Nahas, and K Hirao. *Phys Chem Chem Phys* 1:3823–3830, 1999.
107. X Solans-Monfort, V Branchadell, and M Sodupe. *J Phys Chem B* 106:1372–1379, 2002.
108. BR Wood, JA Reimer, and AT Bell. *J Catal* 209:151–158, 2002.
109. X Solans-Monfort, M Sodupe, and V Branchadell. *Chem Phys Lett* 368:242–246, 2003.
110. AHH Chang and DR Yarkony. *J Chem Phys* 99:6824–6831, 1993.
111. SK Park, V Kurshev, CW Lee, and L Kevan. *Appl Magn Reson* 19:21–33, 2000.
112. AL Yakovlev, GM Zhidomirov, and RA van Santen. *J Phys Chem B* 105:12297–12302, 2001.
113. BR Goodman, KC Hass, WF Schneider, and JB Adams. *J Phys Chem B* 103:10452–10460, 1999.

13 The Organic Chemistry of TiO_2 Photocatalysis of Aromatic Hydrocarbons

William S. Jenks
Department of Chemistry, Iowa State University

CONTENTS

13.1 INTRODUCTION

The use of semiconductor photocatalysts, particularly TiO_2, to mineralize aromatic compounds found in water is an attractive method to remove such pollutants. Exhaustive oxidation similar to combustion is achieved despite the aqueous environment, converting even the most otherwise persistent compounds such as polychlorobiphenyls (PCBs) into relatively harmless inorganic ions, CO_2, and water. Molecular oxygen is the oxidant, and energy is supplied when the catalyst absorbs light at wavelengths sufficient to promote electrons across its bandgap. Though this phenomenon is already in use in niche applications, certain problems will need to be worked out for wider application. The most important among them is to extend the useful range of light farther into the visible range to make better use of solar radiation. Nonetheless, this field has drawn intense research for decades. Mills' excellent review outlines much of this history and many of the applications of photocatalytic methods [1].

An important area of research that began early in the history of this chemistry, starting to gain prominence in the late 1990s, is the exploration of the pathways and mechanisms by which the degradations occur. These processes were of fundamental interest, and have become practical issues, because it has recently been shown that partial degradation of certain compounds leads to solutions that are more toxic than the original solution [2–7]. (Of course, as degradation nears completion, the toxicity returns to near-zero levels.) If solar irradiation is to be taken as the ideal light source, partial degradation may be a major issue to contend with in realistic applications. Thus, having a predictive model to define the chemistry of these processes and the relative rates of reactivity of different portions of representative molecules is extremely important.

In this chapter, the pathways of photocatalytic degradation of aromatic and closely related compounds are reviewed, along with a sufficient discussion of the fundamental aspects of TiO_2 photophysics to present this chemical process completely. Our approach is rooted in the organic chemistry of understanding reactive intermediates. However, despite the burgeoning research in the field over the 5 years or so before the writing of this review, there remain an enormous number of uncertainties. More information is available regarding pathways than mechanisms. One of the difficulties in determining detailed mechanisms for these degradations is their very complexity, from the sheer number of reactive intermediates to the uncertainty in some of their identities.

A further difficulty in mechanistic argument is the problem of sample bias. In virtually every case of partial degradations of particular pollutants, many compounds are produced. For chemical analysis, these compounds are usually separated by high-performance liquid chromatography (HPLC), gas chromatography (GC), or other chromatographic methods, but only a certain fraction of them are identified. Thus, we make arguments about the pathways of degradation based only on the compounds we know about, and have a tendency to ignore the peaks that remain unidentified. Furthermore, we have a tendency to concentrate on the large peaks. While unavoidable to an extent, this natural inclination ignores the possibility that certain compounds are destroyed faster than they are made, a phenomenon that has been documented in a few instances. This circumstance leads to very little buildup of the intermediate in question throughout the course of the degradation. Nonetheless, if we recognize these inherent difficulties, there are still many salient conclusions that can be drawn.

We will first review the standard methods by which many of these studies have been performed, and make some recommendations of procedure. We then discuss some mechanistic issues and the characteristic reactions of aromatic compounds under photocatalytic conditions. Within this context, we also discuss the nature of the primary oxidizing agents. Finally, we move to a survey of degradation chemistry, mostly of agricultural herbicides and pesticides that are characterized by certain functional groups not previously discussed in detail.

It should be noted that there are innumerable papers in the literature describing the destruction or removal of a particular compound or class of compounds from aqueous solution using photocatalytic methods or discussing their kinetics. However, only a limited set make a serious attempt to elucidate the stepwise pathway by which this occurs or determine the structures of isolable intermediates, which is the main focus of this chapter. Thus, there are many compounds whose destruction is known and we will not discuss them here. An excellent example of this problem of lack of molecular detail despite much analytical work is found for PCBs, whose destruction is documented with little investigation of the reaction intermediates.

Many heterogeneous photocatalyst systems besides TiO_2, which is our primary focus, are known. They range from platinized or transition metal doped TiO_2 to other semiconductors, to the use of additives such as bromate or persulfate. We do not discuss them here, despite their obvious importance in the development of this technique for practical applications. This is partly to keep the conditions as comparable as possible throughout the review, and to limit what might otherwise be an unreasonably large data set with only subtle differences. One class of catalyst systems whose degradation chemistry has been explicitly compared to TiO_2 elsewhere is the polyoxometallate group [8–11]. For a recent, more wide-ranging review of photocatalytic degradation by TiO_2, readers are referred to Bahnemann [12] or Hoffman [13]. A recent review of the degradation chemistry of aromatic compounds at the gas–solid interface is also available [14].

13.2 EXPERIMENTAL TECHNIQUES

13.2.1 Sample Composition

Experimental techniques for determining the pathways of degradations of aromatic compounds are straightforward and do not require expensive equipment that is not commonly available in university chemistry departments. Aside from choosing a target pollutant, the main choices to make are catalyst type and conditions. Most authors operate under slurry conditions, where TiO_2 particles are suspended in aqueous solution with concentrations in the range of 0.5–1.0 g L^{-1}. In order to disperse larger aggregates of catalyst, many authors immerse the flasks in ultrasonic cleaners for several minutes prior to irradiation. Since the rate of degradation is usually related directly to $[O_2]$, and oxygen is consumed by the reactions, it is good practice to keep the oxygen concentration constant throughout irradiations. The simplest way to do this is by continuously bubbling the solutions, either with air or with pure O_2. The latter is often preferred, simply to maximize the reaction efficiency.

The most commonly used catalyst has been Degussa P25. Other commonly used commercial preparations include Hombikat UV 100, and more recently, various materials from Millenium Chemicals, such as its PC50. Many authors report studies that take advantage of modified or home-made catalysts as well. Various schemes can also be used to coat TiO_2 films onto glass or other surfaces; this will not be discussed in any detail, since the devices themselves are often the focal point of these studies.

The initial concentrations of organic materials used in experiments of this type tend to be higher than the concentrations that would be found in realistic waste streams. A survey of studies indicates a usual range of 0.1–5 m*M*. Relatively high concentrations are often desired if identification of the early degradation products is the goal, in that the degradations should only be carried out to 10–15% conversion to ensure that primary degradation products are the main new components of the mixtures. Degradations carried out to higher conversions suffer from the introduction and buildup of downstream intermediates or products. It is often the case that the maximum concentration (as a function of degradation time) of a particular early degradation intermediate is only several percent of the initial concentration of the starting material, again explaining the relatively high starting concentrations. It is a standard procedure also to check the time-course of degradation product mixtures.

The choice of pH often strongly affects reaction mixtures and the regulation of pH can be critical to the reproducibility of the experiments. This can be due to changes in adsorption properties, dark reactions of the organics, and also effects of pH on the TiO_2 photophysics. (See, for example, Ref. [15].) Unbuffered suspensions of TiO_2 in water with organic materials are usually in the pH range of 4 to 5, but as the degradations proceed, the production of acid lowers the pH.

If either high or low pH is desired, it is often unnecessary to buffer the solutions, assuming that low conversions are going to be obtained. At pH 12, the hydroxide concentration is about 10 m*M*, so oxidations that produce 1 m*M* or less acid do not significantly affect the pH. Similarly, in the pH range of 2 to 3, there is sufficient acid already present that the pH does not ordinarily change much for reactions to modest conversion. Some authors advocate the use of perchloric acid for acidification because the perchlorate is thought to be a noninterfering counterion, but HCl, HNO_3, and other acids have been used as well.

For intermediate pH values, the optimum solution appears to be that used by Bahnemann et al., which involves an automated pH-stat that adds base as required [16]. This allows the regulation of the pH without buffer. Buffers, such as phosphate, may be successfully used as long as the ions do not interfere with the chemistry or adsorb significantly to the TiO_2.

13.2.2 Irradiation

Broadly speaking, TiO_2 begins to absorb light at wavelengths of about 400 nm and shorter, though the precise onset depends on the degree and phase of crystallinity of the material. Doped TiO_2 samples often are designed to absorb in the visible region, and even with unmodified TiO_2 there are some reported instances of charge-transfer absorption bands in the visible region with some electron-rich arenes. Cases of reactivity, based on electron transfer, of such absorption bands have been documented [17]. In general, mechanistic investigations of photocatalytic arene degradations are most concerned with events that are initiated by light absorption by TiO_2, rather than the arene itself, so some control of wavelengths is often necessary. In certain cases, this is straightforward because the arene does not absorb in the long wavelength region of TiO_2 absorption, but this is not always true. Often, the simplest solution is to check whether any observable photochemistry occurs on irradiation of solutions that contain all elements, excepting the photocatalyst, on timescales relevant to the rest of the experiments.

Several different kinds of light sources are reasonable and convenient. The two most common are based on either high-pressure Xe or Hg arc lamps or low-pressure Hg fluorescent tubes. High-pressure Xe or Hg lamps produce significant output from near 200 nm, through the UV and visible regions, and well into the IR region [18]. The commonly used lamps of $\geq$500 W power thus produce huge amounts of heat that must be filtered out, generally by passing the light through a water filter before it hits the sample. Filters [18] are commonly used to remove the deepest UV radiation as well. Pyrex, depending on its thickness, transmits less than 10% of wavelengths of about 275 nm and shorter, which is at least effective in removing the large 254-nm spike in Hg lamps.

For laboratory-scale investigations, low-pressure fluorescent tubes can be very convenient. Lamps that have coatings which emit broadly ($\pm$ ca. 40 nm) with centers at 300, 350, 365, and 420 nm are commercially available and are widely used with the Rayonet photoreactors from Southern New England Ultraviolet [19]. The last two are very useful for TiO_2 photolysis without additional filters or need for special glassware. They are considerably lower in wattage than the high-pressure lamps and emit very little heat. The lower wattage is compensated by its emission remaining entirely within a narrow spectral range. The Rayonet reactors are designed to have several lamps aligned in parallel surrounding the reaction vessel.

13.2.3 Analysis

Photocatalytic degradation of aromatic compounds, while ultimately providing only a few mineralization products, yields many compounds at the intermediate stages. Any given starting material may have several competing reactive pathways. Each of the primary products, in turn, may yield a number of different secondary products. The compounds in a reaction can quickly become quite numorous. Only as the compounds become oxidized to a greater extent — and the number of remaining carbons in each product becomes small — does the number of possible compounds shrink again.

As a result, in order to trace pathways, it is necessary to develop analytical methods that can separate large numbers of related materials efficiently, provide sufficient information to allow unambiguous identification, and allow for quantification. Furthermore, it is good practice, once these methods are established, to carry out degradations that trace the time-course of concentrations of the intermediate degradation products, whether or not they have been identified.

In general, it is more straightforward to show direct parent–daughter relationships by carrying out degradations of the target compound to only a few percent conversion. Under most circumstances, the majority of the new products will be the primary products of degradation of the starting materials. Each of these primary daughter products, once identified, can then be used as "starting material" for subsequent partial degradations, leading to a new set of primary products; these are of course secondary degradation products with respect to the original starting material. This can be repeated as often as necessary to trace downstream pathways. This iterative method is far superior to the speculation that one can make based on products seen from more extensive degradations of the original target material.

The two most commonly used methods for product analysis are GC and HPLC, coupled to one of several detection systems. The utility of these chromatographic methods, of course, derives from their ability to separate the inevitable mixtures of compounds before analysis, and from the reproducible retention times that can be catalogued against authentic samples. Each method has its advantages relative to the other.

HPLC is amenable to direct analysis of the aqueous mixtures after removal of the TiO_2, with or without a concentration protocol. The reactions of aromatic substrates necessarily produce highly functionalized compounds at the intermediate stages of degradation. Unfortunately, it is the experience of the authors that it is difficult to sufficiently resolve such compounds — especially highly oxygenated ones — by HPLC without engendering excessively long run times for other compounds in the mixture, though progress towards this has been made by other authors. HPLC can be counted on to transmit the compounds from the reaction mixture to the detector directly, without derivatization, an important advantage over GC.

Of the commonly used methods for detection with HPLC, the least attractive for this application is single- or dual-wavelength UV absorption. The UV absorption at particular wavelengths can be useful for quantitation, but only when the user can be quite confident that the retention time serves to uniquely identify the compounds. A step up for detection is to use HPLC coupled with full UV or UV–Vis absorption (i.e., a diode array detector), but even this has shortcomings. First, and perhaps most obviously, UV absorption data is a useful fingerprint, but does not supply much of structural information beyond certain functional group information. Second, and more insidiously, some of the compounds that are difficult to resolve by their retention times because of their structural similarity are also ones that will have nearly identical absorption spectra.

Coupling of HPLC to mass spectrometry (MS) provides what will probably be the best information to come from HPLC. Electrospray (ESI) and atmospheric pressure chemical ionization (APCI) are particularly soft ionization techniques and can generally be counted on to provide protonated or deprotonated molecular ions when it is effective in obtaining a signal. A disadvantage is that the minimal fragmentation means that less information is available for structural determination. Also, less extensive libraries are available for comparison to such data, compared to standard electron impact (EI) methods.

GC generally provides better temporal resolution of compounds than does HPLC, and it is commonly coupled to MS. A recent review discusses the use of solid-phase extraction (SPE) methods to concentrate samples before analysis [6].

Most commonly, EI and chemical ionization (CI), and quadrupoles are used as mass detectors, but ion traps and time-of-flight instruments are both available. Despite the obvious advantages of the very high resolution achievable with GC separation, there are drawbacks in this method as well. The principle one is that many compounds do not tolerate the relatively high temperatures required to elute the compounds through a gas chromatograph. As a result, such compounds are simply unobserved. They may stick to the injector port or the GC-column, or perhaps even worse, they may provide "phantom" thermal degradation products if degradation can be completed in the injector port.

We are unaware of any method that convincingly proves that such thermal degradations do not occur for any particular sample, though systematic variation of the injector port temperature can give some clues. However, fairly straightforward functionalization protocols (see, for example, Ref. [20]) can be used to minimize the possibility of missing important compounds. In the Jenks group's experience, compounds containing aldehyde and carboxylic acid groups are the most likely not to survive GC conditions without functionalization, and functionalization became universal practice for runs primarily concerned with product identification.

Many reasonable protocols for functionalization of reaction mixtures may be envisioned. The protocol adopted by Jenks group involves exhaustive silylation of reaction mixtures from which the catalyst and water have been removed. The chemistry consists of using $[(CH_3)_3Si]_2NH$ in the presence of catalytic $(CH_3)_3SiCl$ and pyridine [20]. This places $Si(CH_3)_3$ on all alcohols, phenols, and carboxylic acids. Typically, three runs are performed. The first involves functionalization and direct analysis of the mixture. The second and third runs are performed as usual, but with a reduction step before silylation. In the second run $NaBH_4$ is used, and in the third, $NaBD_4$. The borohydride (or borodeuteride) reduces ketone and aldehyde functionalities to alcohols, but does not affect carboxylic acids and their derivatives (e.g., esters).

GC–MS runs carried out with both EI and CI methods then lead to fragmentation patterns and molecular masses. The reduced mixtures often reveal compounds that do not survive the conditions of GC analysis without reduction before silylation. The comparison of CI masses obtained with $NaBD_4$ and $NaBH_4$ provides the number of reduced functional groups. Comparison of the corresponding CI and EI runs sometimes provides information about the position of the reduced ketone or aldehyde if the EI fragmentation pattern can be rationally interpreted. This reduces the number of aldehyde- or ketone-containing compounds that might have yielded a particular peak after reduction. Furthermore, whether a compound observed in the unreduced runs is observed in the reduced runs provides information about the nature of any oxygenated functional group it may have.

While these techniques, along with searching MS spectra against electronic libraries for matches, provide excellent foundations for making educated guesses about the identities of compounds, there is no substitute for obtaining authentic samples of candidate compounds. Authentic samples provide both data with regard to chromatographic behavior and the fingerprint information that can be obtained from MS data. This sometimes means that tedious organic synthesis must be done, but we believe that without it, the majority of product assignments done in the absence of authentic samples must be regarded as tentative.

13.3 MECHANISTIC ISSUES

13.3.1 Early Events on the Semiconductor Particle

The degradation conditions we are concerned with here are those in which the aromatic compounds are present as dilute "pollutants" in oxygenated (or at least aerated) aqueous solutions. We shall limit our discussion of mechanism to the use of TiO_2 particles as photocatalysts, though many other semiconductors can be used and the TiO_2 can be modified in different ways. However, our emphasis is on the reactive chemistry of the organic compound, so many of the ideas expressed here may be applicable, after sensible adjustment, to different catalyst systems. More detailed discussions of the fundamental processes of the semiconductor itself appear elsewhere in the primary and review literature [1, 12, 13, 21–34].

Irradiation of the TiO_2 particles with light of at least the energy of the semiconductor bandgap results in charge separation within the particle, with an electron getting promoted from the valence band to the conduction band. The resulting electron (e^-_{cb}) and hole (h^+_{vb}) rapidly migrate to trap sites within the particle, but can also simply recombine with release of heat on the nanosecond timescale. Traps of varying "depths" are almost certainly available within the particles, resulting in variable mobility of the holes and electrons. (See, for example, Ref. [26].) Recent studies by Emeline and Serpone provide a more sophisticated elaboration of this model [35–40].

$$TiO_2 + h\nu \longrightarrow e^-_{cb} + h^+_{vb}$$

$$e^-_{cb} + h^+_{vb} \longrightarrow \text{heat}$$

$$e^-_{cb} \longrightarrow e^-_{trap}$$

$$h^+_{vb} \longrightarrow h^+_{trap}$$

$$e^-_{trap} + O_{2,ads} \longrightarrow O^{\bullet -}_{2\,ads}$$

Electron–hole annihilation is the principle reason for the low quantum yields usually reported for semiconductor-mediated photocatalytic degradation. It is also the reason that most degradations are brought to a halt in the absence of O_2, because adsorbed molecular oxygen acts as an electron sink, isolating the hole for further reactivity. Other additives may be used as electron sinks [41–45], but in order to keep this chapter to a reasonable length, we have again decided not to cover this literature. It should also be pointed out that the use of additives presents a significant disadvantage for some practical applications.

In some instances, the organic compound may act as an electron acceptor rather than an electron donor (e.g., benzoquinone [46, 47], haloalkanes [48–50], and certain other electron-poor organics [5, 51, 52]). Emeline suggests that even chlorophenol may act as an electron acceptor when certain semiconductor transitions are excited [35]. However, being an electron acceptor is apparently more common with more easily reduced compounds, such as nitroarenes [53–65]. However, in most instances, O_2 is the primary electron sink.

Molecular oxygen plays a second critical role. In a hypothetical system containing only TiO_2, H_2O, and the organic component, the only reactions that could be catalyzed would be net hydrations. Thus, although the original organic pollutant could be destroyed, the total organic carbon would not be affected. In order to achieve mineralization, O_2 is required because, from an overall thermodynamic perspective, it is O_2, not the other reactive intermediates, that is the oxidant.

The transient-trapped hole and the nature of its reactions have been of greater controversy over the years. In many early references, the degradation chemistry is explained in terms of the reactivity of a hydroxyl radical, presumably formed by the oxidation of surface water groups. Matthews' work (for example, Ref. [66]) is among the earliest that suggested this, by determining products and comparing to authentic hydroxyl radical sources.

Also, much early work made the assumption that the hydroxyl radical was released into bulk solution. Although still not universally held (see, for example, Ref. [67]), the more widely taken position at this time is that the hydroxyl-like chemistry takes place at the surface or within a few monolayers, due mainly to the conclusion that the oxidative species are surface bound [25, 68–72]. Some product studies can be rationalized by invoking steric interactions in photocatalytic conditions that would not be particularly important in homogeneous solution [73].

Even taking the surface-bound nature of the hydroxyl chemistry for granted, a more complex picture regarding the reactivity has arisen. While many reactions seem hydroxyl-like, it became clear that certain reactions are initiated by oxidative single electron transfer (SET). (This is the subject of Section 13.4.) SET reactions between a substrate and hole obviously imply reactivity at or near the catalyst surface.

This surface-bound model for reactivity led to the widespread use of the Langmuir–Hinshelwood (LH) treatment for reaction rate as a function of concentration. However, distinct shortcomings have been discovered in the application of the LH model to photocatalytic degradation. Dark adsorption parameters, for example, are known not to correspond to those obtained from LH plots [74]. Rates can also go up and then down as a function of substrate concentration, rather than showing standard saturation kinetics. Alternate kinetic treatments have been advanced. Interested readers are referred to the analysis by Minero [75].

Returning to the primary photophysical and photochemical events, some authors have postulated two distinct types of reactive intermediates, i.e., a trapped hole and $HO^{\bullet}_{ads}$. On the basis of spectroscopic studies, others have argued that only a single important oxidizing species is available, which has been formulated as an adsorbed hydroxyl radical with alternate resonance forms [25, 69].

$$[\text{Ti-O}^{\bullet +}\text{-Ti}]\ \text{OH} \longleftrightarrow [\text{Ti-O-Ti}]\ \text{O}^{\bullet +}\text{H}$$

We have no reason to dispute this formulation, but do side with those that contend that there are at least two distinct types of reactivities, which we have referred to as due to HO^{g}_{ads} and h^+ as if they were different species. Bahnemann discusses this phenomenon as distinct trap sites, which are "deep" (reacting as hydroxyls) and "shallow" (reacting as h^+) [76]. A related, physically reasonable explanation for the distinct reactivity (vide infra) may be that there are different types of sites available on the TiO_2 particle surface. Some of them, whether because of the energetic depth of the trap, or the physical arrangement of the surface atomic structure, are amenable specifically to electron transfer reactions, whereas others result in hydroxyl-like reactivity.

Microscopic TiO_2 particles are generally neither fully crystalline nor free from surface defects that differ from the usual anatase or rutile configurations. They are calcined under different conditions; the whole purpose of this treatment is to alter the crystalline phase of the catalyst. It is thus reasonable to expect a distribution of different traps, surface sites, and boundary regions. It should be noted, for example, that Degussa P25, one of the most active TiO_2 preparations, contains substantial portions of both the rutile and anatase phases. It undoubtedly must also contain amorphous and boundary regions; it necessarily follows that multiple types of surface sites abound. A recent electron paramagnetic resonance (EPR)-based investigation indeed concludes that the phase-interfacial regions contribute to the relatively high efficiency of P25 and may contribute to the formation of surface "hot spots" of reactivity [77]. Nonetheless, we are unaware of any *direct* evidence for different types of sites correlating with different types of reactivities.

As will be discussed in more detail below, we believe that substrates are generally more susceptible to the SET chemistry if they possess functional groups such as phenols, alcohols, or carboxylic acids that afford strong adsorption. We postulate that these allow a certain fraction of the material to be chemisorbed with direct O–Ti bonding, as schematically illustrated below for 4-chlorocatechol [78–80], which facilitates the rapid electron transfer chemistry. In the absence of a substrate with direct O–Ti bonding, we further postulate that the simply has a greater chance of finding an appropriate surface hydroxyl trap that leads to the hydroxyl-like reagent.

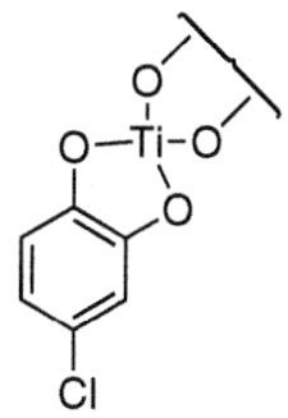

Finally, a word should be said about trapped electrons and superoxide ($O_2^{\bullet -}$) as reactive intermediates and their roles in initiating degradative chemistry. There are now several instances in which reductive electron transfer is viewed as the initial step in the degradation of particular compounds. There is not much evidence or argument about the nature of the electron in the same way as there has been controversy regarding the corresponding holes. However, in much the same way, it should be assumed that there are multiple types of sites at which reductive electron transfer may occur, and it may come to be that distinct types of reactivities are eventually distinguished.

There are several instances in which superoxide or its conjugate acid, the hydroperoxyl radical ($HOO^{\bullet}$, pK_a 4.5), are invoked as reactive intermediates involved with degradative chemistry. However, we are not aware of any cases in which direct reactivity of superoxide or $HOO^{\bullet}$ is thought to be the major reagent that initiates reactions with stable adsorbed substrates. It is possible, of course, that certain easily reduced compounds may use superoxide as an electron shuttle from which to receive an electron from the photoactivated TiO_2 particle.

13.3.2 Prototypical TiO_2-Photocatalyzed Reactions of Arenes

13.3.2.1 Common Oxidative Reactions

The most common reactions of arenes with photoactivated TiO_2 in oxygenated aqueous media are ring hydroxylation, hydroxyl substitution, and oxidation of aliphatic substituents. Radical-based dimerization and oligomerization also occurs. For example, see Ref. [81].

The most plausible mechanism for hydroxylation of the aromatic moiety begins with the hydroxyl adduct of the aromatic ring (however formed), and relies on subsequent addition of ubiquitous O_2 and elimination of $HOO^{\bullet}$ to restore aromaticity. This overall mechanism has been documented extensively by von Sonntag and others under radiolysis conditions where homogeneously dispersed $HO^{\bullet}$ is formed [82–85] and even in gas phase [86]. The E_i elimination of $HO_2^{\bullet}$ from smaller systems in reactions related to combustion has also been studied in great detail and is well understood [87–89]. This elimination is exothermic in cases where either a carbonyl or an aromatic nucleus is formed. The hydroxylation of benzene is illustrated as a prototype, without regard to the exact mechanism of formation of the initial cyclohexadienyl intermediate. However, it should be emphasized that this reaction is general for any number of systems of different substitution patterns and different aromatic nuclei [53, 56, 90–100].

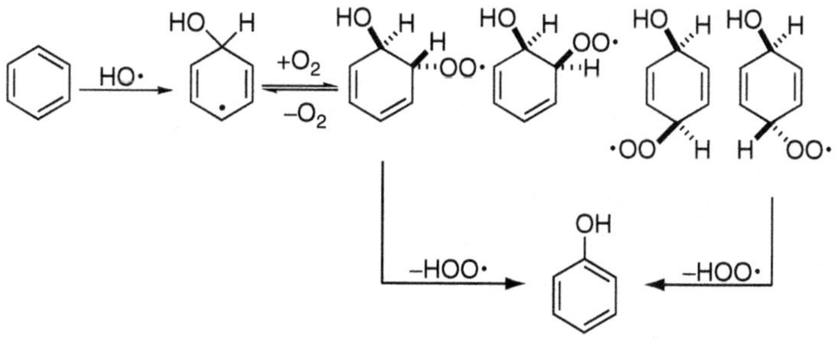

When the benzene is substituted, regioselectivity is an issue. There are many examples in the literature showing that the regioselectivity, which is determined in the step in which the cyclohexadienyl radical is formed, is analogous to that of electrophilic aromatic substitution. Thus, for example, the first hydroxylation of anisole [47] (methoxybenzene) occurs predominantly at the *ortho* and *para* positions, whereas the first hydroxylation of dimethyl phenylphosphonate occurs predominantly at the *meta* position [101]. This has sometimes been used to suggest that reaction is through a hydroxyl-like species, with the idea that $HO^{\bullet}_{ads}$ should be electrophilic [102]. Quite interestingly, the partial degradation of nitrobenzene provides *meta* and *para* hydroxylated products in very similar yields, though the redox chemistry that can occur at the nitro group makes the precise determination of mechanism more difficult to determine [56]. Hydroxylation of fluorophenols takes place at positions *ortho* and *para* to the hydroxyl group [103].

OCH_3 — TiO_2, O_2, hν →

% of hydroxylated product, low conversion, pH = 7: 41 (ortho), 1 (meta), 57 (para)

$P(O)(OCH_3)_2$ — TiO_2, O_2, hν →

% of hydroxylated product, low conversion, pH unregulated: 6 (ortho), 72 (meta), 21 (para)

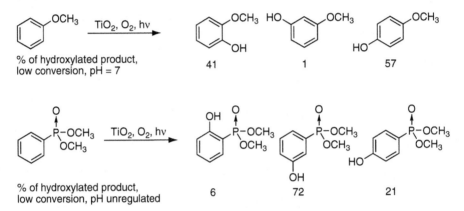

Unlike electrophilic aromatic substitution, however, there is also strong evidence for *ipso* attack and substitution, based on an addition–elimination mechanism whose details are also still investigated. For example, chloroarenes such as chlorophenols and PCBs are known to be subject to substitution reactions where hydroxyl groups replace chlorine atoms [20, 81, 104–125].

Perhaps the ultimate example of the competition between hydroxylation and HO substitution for Cl on an arene is found in Pramauro's work on 2,3,6-trichlorobenzoic acid [126]. Almost every plausible isomer was detected. Fluorophenols suffer the same reaction, also directing hydroxylation in *ortho/para* fashion [103]. By use of ^{18}O labeling, it has been shown that the transformation of anisole (methoxybenzene) to phenol can take place via *ipso* substitution under photocatalytic conditions [47]. This substitution reaction is also ubiquitous among many aromatic substrates and can also occur with other substituents as leaving groups, such as sulfonate [64, 65, 127–134].

One of the intriguing issues (see below) regarding these reactions is that apparently similar reaction conditions can give quite different results, as seen for example, in the ratio of 4-chlorocatechol vs. hydroquinone from 4-chlorophenol. A recent attempt to understand and predict this kind of chemoselectivity has been reported by Choi and coworkers, using B3LYP density functional methods to study the reactivity of various chlorinated dioxins with hydroxyl radicals [135]. In that study, they find that substitution for chloride gives more stable products, but the transition state for such attack is higher energy than that for attack at the oxygenated positions.

Density functional theory (DFT) computational methods have also been used to study the second half of the substitution reaction, that is, whether the second step is homolytic or heterolytic [136]. The calculations included continuum models for the water solvent, but still showed that the homolytic pathway was considerably lower in energy than the heterolytic dissociation. However, it would also be wise to compare these results to a third possibility. One may write a solvent-assisted heterolytic mechanism by which proton transfers to or from water molecules provide one or both of the ionic compounds as their conjugate acid or base in a single step concerted with scission.

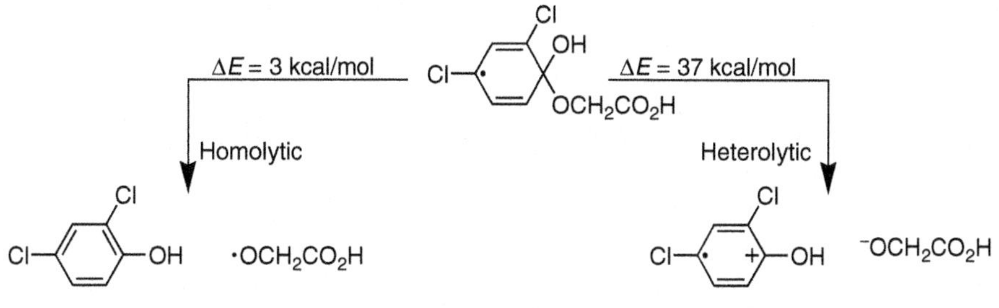

ΔE = 3 kcal/mol ΔE = 37 kcal/mol

Homolytic Heterolytic

•OCH_2CO_2H $^{-}OCH_2CO_2H$

The oxidation of aliphatic side chains will only be discussed in general terms. Usually, these reactions can be rationalized by hydroxyl radical-like reactivity. Alkyl radicals are formed by hydrogen abstraction (or possibly stepwise loss of electron and proton). Quite frequently, the observed products imply that the alkyl radicals have been trapped by O_2 prior to subsequent reactivity, though other fragmentation routes have been suggested. The peroxyl radicals themselves may suffer several fates, but the usual assumption is that Russell-type chemistry [137] occurs, leading to alcohols and carbonyls in most cases.

Kinetic isotope effects (KIEs) are expected for most mechanisms involving hydrogen abstraction, because removal of the H atom is usually the rate-determining step in such sequences. Related, though not equivalent, to KIEs are product selectivity studies. Here, isotopic substitution affects product mixtures, and the product ratios are related back to relative rates of key steps. Hydrogen–deuterium oxidation selectivities close to 3 for the oxidation of the methyl groups of anisole and trimethyl cyanurate have been reported [47, 138]. These are consistent with hydrogen abstraction but do not absolutely rule out a stepwise electron–proton abstraction if degenerate electron transfer is assumed to be very rapid, even on the particle surface. However, they are very similar to results obtained with other reagents, such as direct photolysis of H_2O_2, which should produce homogeneously dispersed hydroxyl radicals. HO• is not expected to undergo SET reactions in most cases because of the activation barrier induced by the high solvent reorganization energy, so these selectivities were taken as evidence of hydrogen abstraction as the primary process.

A counterexample, in which H/D selectivity is very close to 1.0 has also been found in the competition between demethylation of dimethyl phenylphosphonate and its hexadeuterated isotopolog [139]. This surprising result led to the speculation that there might be a stepwise mechanism for loss of H•. If electron abstractions were isotope-independent (expected) and effectively irreversible (less likely), then any KIE for the deprotonation of the radical cation would not be transmitted to relative product yields, since the potential KIE takes place after the product-determining step. This could be tested by using the D_3 isotopolog, because the loss of H^+ or D^+ would then not be after the product-determining step. Nonetheless, H/D selectivity remained in the range of 1.1, distinctly low for a primary KIE. Furthermore, the low product selectivity was also observed under other conditions that should generate homogeneous hydroxyl radicals or close equivalents. The possibility that the small product selectivity was reflective of an equilibrium isotope effect achieved by degenerate hydrogen abstraction can be eliminated because the mass spectral analysis routinely showed no isotope scrambling. Because approximately equal but low apparent KIEs are observed under several different sets of hydroxyl and hydroxyl-like conditions, there is as of yet no compelling evidence for a mechanism other than hydrogen abstraction, however unattractive it may be to explain the low observed product isotope selectivity.

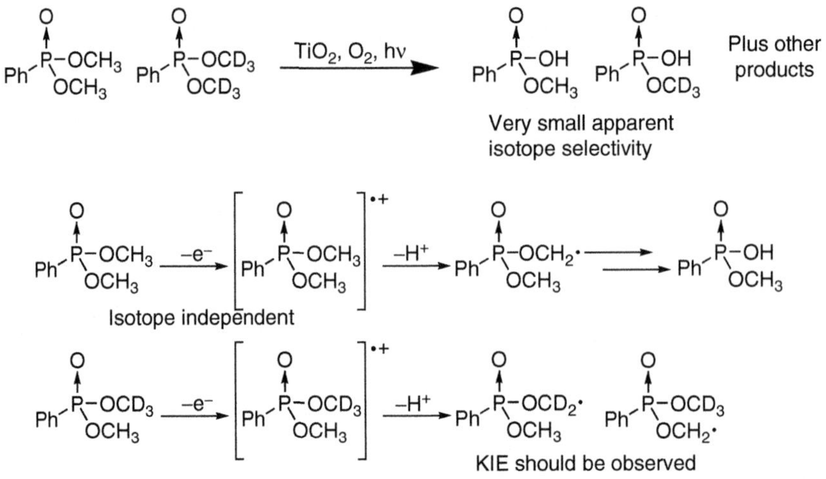

TiO_2, O_2, hν

Plus other products

Very small apparent isotope selectivity

−e⁻ −H⁺

Isotope independent

KIE should be observed

13.3.2.2 Less Common Reactions: Reductive Chemistry

Ordinarily, photocatalytic degradative conditions with unmodified TiO_2 in water are thought of as oxidative, but there are exceptions to this generalization [140]. Most of them are in the reduction of haloalkanes, but several have been observed or proposed in aromatic and related systems. Unfortunately, in nearly all reported cases, the observed chemistry is limited to the stable degradation intermediates or products, which universally have been reduced by two (or four) electrons, and the source of the second electron is a matter of speculation.

Perhaps the prototypical organic substrates that accept electrons, rather than react oxidatively are quinones, as reported by Richard [46]. The corresponding hydroquinones are produced and O_2 is not required for their conversion to hydroquinones. Benzoquinone can be used as an organic sacrificial electron acceptor in order to probe the mechanism of other photocatalytic reactions in the absence of molecular O_2 [47]. The Jenks group found that 1,2, 4-benzenetriol–hydroxybenzoquinone redox couple, despite oxidizing spontaneously and uncontrollably in (dark) aqueous solution except with the rigorous exclusion of O_2, appears to react as the triol under photocatalytic conditions because it is reduced so easily [20].

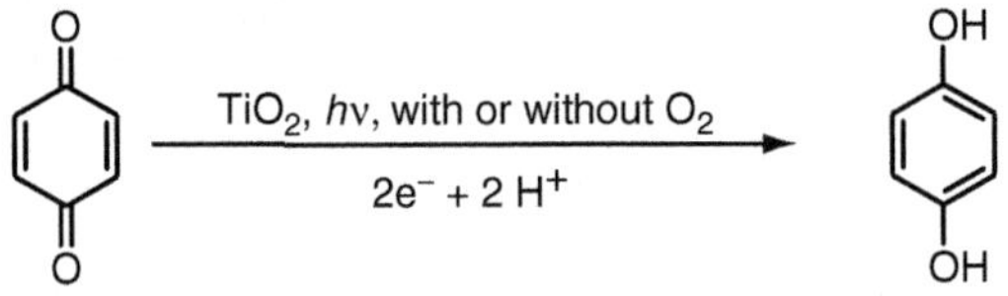

TiO_2, *h*ν, with or without O_2

$2e^- + 2H^+$

Among the most conspicuous examples of substrates being reduced in aromatic (and closely related) compounds outside the quinones are nitrogenous compounds. It is widely understood in organic chemistry that nitroarenes are easily reduced with chemical reagents and can often be good electron acceptors under many different conditions. It is thus not terribly surprising that reductive chemistry is observed here. Minero and coworkers reported the first study of nitrobenzene degradation [141], and showed that trace quantities of nitrosobenzene and aniline could be observed in the reaction mixture after partial degradation. Subsequent study showed that nitrosobenzene is degraded much more quickly than nitrobenzene [56]. This was assumed to be the reason that only a small steady-state concentration of nitrosobenzene could be built up. Furthermore, increasing the O_2 concentration by changing from air-saturated to O_2-saturated water decreased the rate of decomposition of nitrosobenzene. This is interpreted as evidence that O_2 and nitrosobenzene compete for a single reactive intermediate, which is assumed to be e^-_{cb}, since the usual primary purpose of the O_2 is to accept electrons from TiO_2. Ferry supported this conclusion, based on experiments using alcohols as potential sacrificial electron donors, concluding that nitrobenzene reacts directly with the photoactivated TiO_2 [142].

After full mineralization, the nitrogen from nitrobenzene was detected as nitrate and ammonium in a ratio of about 6:1. This was in line with a related report by Maurino [53]. Both ammonium and nitrate are, in fact, found on mineralization of most N-containing compounds, though often the nitrate grows in at the expense of ammonia towards the end of the mineralization process, indicating that the nitrogen is often released in a low oxidation state. For example, Prevot proposed [143] a hydrolytic scheme for the loss of nitrogen from chloramben [144].

Chloramben — TiO_2, O_2, hν → Mechanism to here unclear → [O] → H_2O →

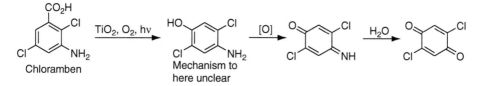

Other reactions can still compete with this reduction chemistry. The same sort of hydroxylation and substitution reactions noted above have been documented for various nitroarenes [145–147].

TiO_2, O_2, *hν* → Plus hydroxylated compounds

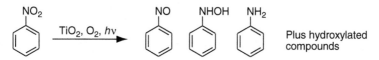

Another N-containing functional group subject to reduction is the azo functionality. For one of the simplest possible azo compounds, azobenzene (Ph—N=N—Ph), photolysis with TiO_2 in the absence of O_2 provided the two-electron reduction product Ph—NH—NH—Ph [148]. Possibly of wider practical interest is the degradation of common azo dyes under more typical aerated conditions. Several photoreactions have been studied, though few papers address the products. Among those, reduction products have been observed, such as with Acid Orange 7 (AO7), in which the corresponding aniline was positively identified [149]. It should be noted that another detailed study of AO7 degradation does not report any nitrogenous aromatic intermediates, suggesting that most of the nitrogen is lost as N_2, as also reported for some other azo dyes [60, 61, 150]. Galindo reported the degradation of a related azo dye, AO52, comparing H_2O_2 photolysis to TiO_2 [151]. With HO• generated from peroxide, substitution at the *ipso* positions leads to the major isolated intermediates, identified by comparison to authentic samples. In neutral to basic pH conditions, they hypothesize that similar hydroxyl attack is the mode of action with TiO_2, but at pH 4 and below, the azo dye is an electron acceptor.

Acid orange 7 — TiO_2, O_2, *hν* →

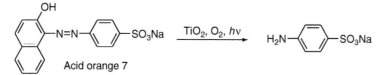

Muneer and coworkers have recently described a series of compounds in which reductive electron transfer is proposed to be the initial step after photoexcitation of the catalyst. They suggest that the imide hydrolysis of terbacil is initiated by formation of the radical anion, followed by trapping by the net elements of hydroxyl radical, yielding a net hydroxide attack overall [51]. Additionally, reduction of the endocyclic olefin position is observed. Similarly, diethyl phthalate is converted to the anhydride, presumably via the monoacid, under TiO_2-mediated photocatalytic degradation conditions [52]. A similar photoinduced hydrolysis mechanism was proposed. An alternative conventional mechanism also deserves examination. One of the methylenes might be oxidatively attacked (e.g., by hydrogen abstraction) and the alkyl group ultimately removed, also leading to the monoacid, and thus the anhydride after condensation. Finally, Muneer et al. propose a reductive mechanism for the formation of benzoquinone and hydroquinone from benzidine [152], though the latter is generally thought of as much more easily oxidized than reduced.

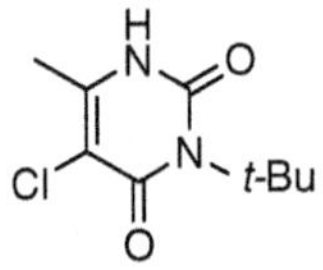

Terbacil

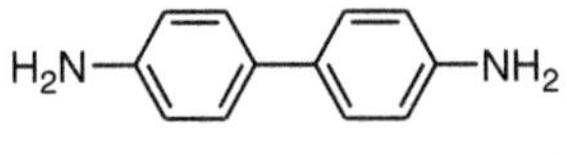

Benzidine

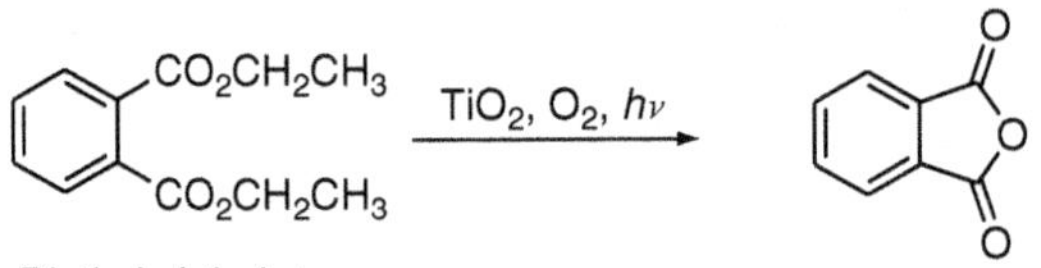

Diethyl phthalate

TiO_2, O_2, $h\nu$

Though not aromatic themselves, maleic and fumaric acids (and closely related derivatives) are ubiquitously observed on the photocatalytic degradation of aromatic compounds. A fairly firm case can be made for reductive SET in the case of these compounds in acidic media. At low pH, virtually all of the observed chemistry is *cis–trans* isomerization [153, 154]. Oh and coworkers report that this chemistry, like that of benzoquinone, does not require O_2, and is, in fact, more efficient in its absence. Furthermore, the *cis–trans* isomerization is suppressed by either addition of fluoride (which displaces the acid from the surface, vide infra) or by esterification, which prevents chemisorption. A mechanism based on reversible reductive SET is thus proposed.

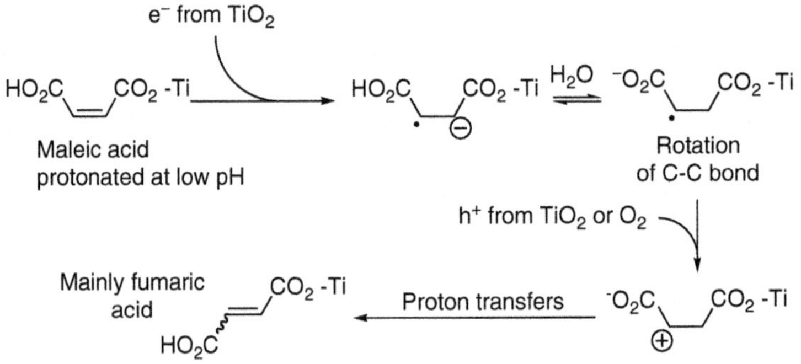

e^- from TiO_2

Maleic acid protonated at low pH

Rotation of C-C bond

h^+ from TiO_2 or O_2

Mainly fumaric acid

Proton transfers

13.3.3 Ring Opening of Aromatic Substrates

The opening of the ring of aromatic compounds (or their intermediate degradation products) under photocatalytic degradative conditions is one of the most important processes to understand. Getting down to acyclic compounds changes the type of subsequent reactivity and completely changes the nature of the compound itself. Several studies have addressed this point, and the most credible of them now agree on or at least imply the basic result that the key transformation is the net oxidation of a double bond such that each terminus is transformed to a carbonyl. This is illustrated for benzene for clarity, though benzene itself is not appreciably ring-opened.

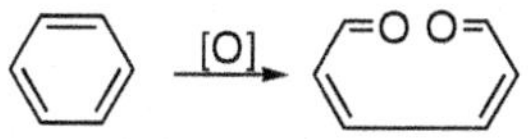

Among the first workers who addressed the ring-opening chemistry was Pichat [90]. Some earlier work identified acyclic compounds, but they were clearly not the primary ring-opened products. A seminal paper addressed the partial degradations of the three possible dimethoxybenzene isomers [155]. These were chosen for ease of handling, in that the methyl esters protected the phenolic groups, simplifying the GC–MS analysis and other technical issues. From the *meta* and *para* isomers, only compounds with five or fewer carbons were detected. However, from *ortho*-dimethoxybenzene, dimethyl (*E,E*)-muconate was detected and directly confirmed with authentic samples. Two isomers of the corresponding monomethyl ester were detected by analysis of the mass spectral data. While the Pichat group did not directly propose a mechanism, they pointed out two important facts: (1) that dimethoxybenzenes are easily oxidized to the radical cation, and (2) that they showed in a previous study [155] that elimination of superoxide from the reaction mixture with superoxide dismutase (SOD) significantly slows down the degradation, while elimination of H_2O_2 with a catalase does not. The implied pathway is illustrated:

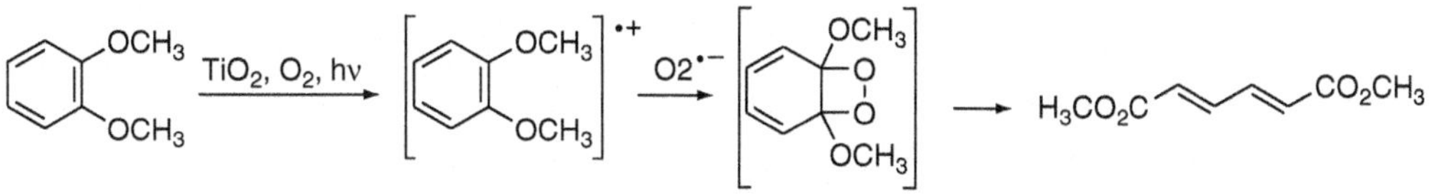

A slight puzzle here is the stereochemistry of the muconic ester, but it may be supposed that *cis–trans* isomerization occurs under the reaction or analysis conditions to give the most stable stereoisomer. From pyridine, analysis of the fragmentation pattern of a GC–MS peak yielded a tentative identification of a five-carbon, nitrogen-containing acyclic dial that might have been formed via analogous chemistry [133]. Another heteroaromatic compound, quinoline, however, provided an important result [93, 156].

Subjection of quinoline to photo-Fenton conditions, which produces hydroxyl radical-like conditions, produced several derivatives, of which the greater proportion was the result of functionalization of the more electron-rich benzene moiety [93, 156]. Compared to this, the amount of 5-hydroxyquinoline was halved when TiO_2 degradation was carried out at the same pH, and none of the quinone was observed. Raising the pH to 6 dramatically increased the quantity of *o*-aminobenzaldehyde and its formyl derivative. Addition of a SOD, which should lower the concentration of available superoxide, dropped the concentration of *o*-aminobenzaldehyde to trace levels. The conclusions drawn from this were that (a) $HO^•$ chemistry is not sufficient to explain the degradation of quinoline, and (b) superoxide is involved in the formation of *o*-aminobenzaldehyde.

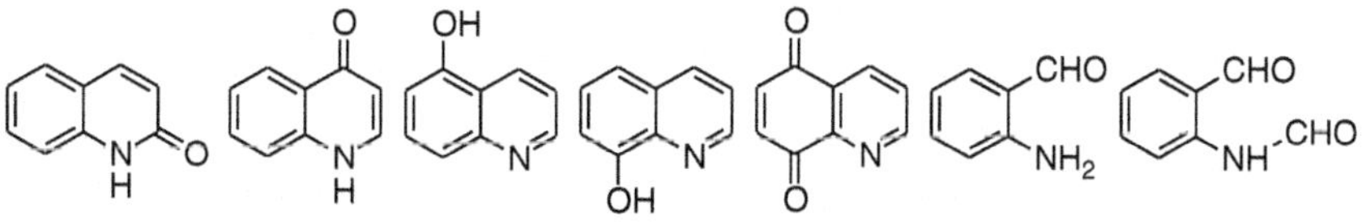

	2-quinolinone	4-quinolinone	4-hydroxyquinoline	8-hydroxyquinoline	quinoline-5,8-dione	o-aminobenzaldehyde	N-formyl-o-aminobenzaldehyde
Photo-Fenton pH 3	1.5	7	33	tr	20	6	tr
TiO_2 pH 3	2	7	16	tr	-	8	2.5
TiO_2 pH 6	3	10	2	tr	-	50	10
TiO_2 pH 6 + SOD	2.5	5	tr	tr	-	tr	10

Relative yields at low conversion of quinoline under various degradative conditions

It was argued that oxidative electron transfer from unprotonated quinoline adsorbed to TiO_2 provides a radical cation that preferentially reacts on the pyridine nucleus due to partial localization of the reactive odd electron center. A mechanism to get to *o*-aminobenzaldehyde that does not include *N*-formyl-*o*-aminobenzaldehyde was required, and the superoxide trapping interpretation was offered. Fundamentally, then, the authors argued that there were two types of pathways, those that depended on hydroxyl radical chemistry, and those that depend on oxidative electron transfer, with latter leading to ring opening in this case.

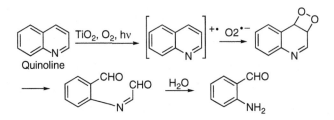

Recently, Wiest and Kamat have presented a related work on quinoline based on radiolysis experiments that generated HO•, SET oxidizing agents, and DFT calculations [157]. They argue that the energetics of the various hydroxyl radical adducts to quinoline are such that little selectivity should be observed, at least in free solution, which is not inconsistent with the fundamentals of the Pichat report. However, they argue that the quinoline radical cation should react with water on a timescale that makes its direct reaction with superoxide seem unreasonable. Thus, they argue that production of *o*-aminobenzaldehyde should depend on the decay pathway of the (net) hydroxyl adduct. No product study was available in this report. It should also be remembered that the radiolysis conditions are in homogeneous solution, while the photocatalytic reactions occur on or near the surface of the TiO_2 particle, which may introduce unusual effects due to local concentration effects or steric effects. Adsorbed substrates may have a different selectivity than those in homogeneous solution because of the electronic effects due to chemi- or physisorption. Nonetheless, the radiolysis work is important and may reveal details that have been incorrectly inferred from product work alone.

It must certainly be granted that there is no direct evidence for the formation of a dioxetane in the quinoline or dimethoxybenzene degradations by TiO_2. Indeed, even the imine precursor to *o*-aminobenzaldehyde is but a proposal. Nonetheless, the SOD experiments cannot be dismissed [158]. An alternative general ring-opening mechanism that allows for the hydration of the radical cation but also involved HOO• or $O_2^{\bullet -}$ might be formulated such that the carbon–carbon cleavage occurred by way of the dehydration of a hydroperoxide.

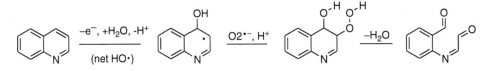

Direct evidence for the products indicated by the dioxetane (or hydroperoxide) mechanism comes from the work of the Jenks group [20, 159]. They showed that TiO_2-mediated degradation of 1,2,4-benzenetriol provides two major cleavage products, with the diacid (or, rather, tautomers of it) predominating. Independent synthesis of authentic samples with the correct stereochemistry provides definitive proof of structure. Many other, smaller compounds were also identified.

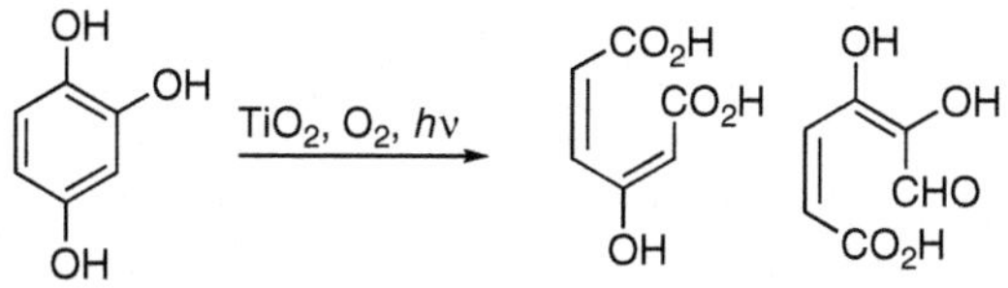

Other proposals for the ring-opening chemistry of phenol derivatives had appeared previously (see, for example, Ref. [12]), with varying degrees of evidence backing the suggestions. In most instances, detection and identification of the five- and six-carbon compounds had been limited to GC–MS analysis, rather than authentic samples. An earlier report of the degradation of 1,2,4-benzenetriol notes the observation of muconic acid (hexadienoic acid) and butadiene, in addition to fumaric acid, as identified by GC–MS analysis of the mixtures [160]. It must be noted that formation of both muconic acid and butadiene would require very unusual reduction reactions in addition to oxidative cleavage.

4-Chlorocatechol leads to degradation products in good analogy to those from benzenetriol, though ring opening is not predominant until a third hydroxyl is added [159]. The authors used a formulation of the dioxetane mechanism to explain the product formation, but the strength of this work is the exacting product identification, rather than mechanistic experiments. In a later publication [161], the degradations of several *O*-methylated derivatives of 1,2,4-benzenetriol and 4-chlorocatechol were reported, using both TiO_2 and other hydroxyl radical-type conditions. When observed, the six-carbon ring-opened compounds were again analogous. Here, Fenton conditions or H_2O_2 photolysis did not produce the ring-opened compounds, which was used as an argument in favor of oxidative electron transfer being the first step in the pathway, leading to ring opening under TiO_2 conditions.

In the instance of hydroxylated arenes, especially those like benzenetriol that have *ortho* hydroxyl groups, it is likely that the phenols are chemisorbed, which may influence the course of the reaction. A plausible mechanism under these conditions could include rapid deprotonation of the radical cation to yield a bound semiquinone. There is excellent evidence from the work of Foote [162] using KO_2 in DMSO that superoxide will react with semiquinones in homogeneous solution to produce cleavage products analogous to those reported by the Jenks group.

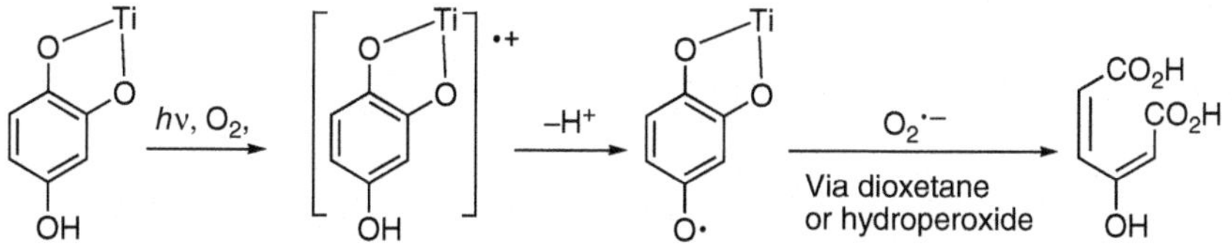

A few other papers report ring-opened products for related aromatic systems [92, 163]. The degradation of naphthalene leading to ring-opened products was first published by Das et al. in combination with some work on anthracene [164]. The primary ring-cleavage products from naphthalene were *ortho*-formylcinnamaldehydes, and the authors suggested that these were formed by electron transfer-initiated formation of the dioxetane. It was shown that naphthols did not lead to the formylcinnamaldehydes. Shortly thereafter, Theurich et al. reported several more degradation products, including the naphthols and 1,4-naphthoquinone [165]. These authors suggest that the *ortho*-formylcinnamaldehydes are derived from the corresponding *ortho*-naphthoquinone, though the *ortho*-quinone was not observed and such a reaction requires addition of the elements of H_2 across the carbon–carbon bond.

Soana et al. subsequently reported a similar set of products [166]. Identification by nuclear magnetic resonance (NMR) and comparison to authentic samples is secure. These researchers also reported that addition of SOD to the reaction mixture reduced the rate of disappearance of naphthalene, and especially suppressed the formation of the dialdehyde products [166], again implying the involvement of superoxide. Similarly, degradation by Fenton's reagent gave considerably more quinone and suppressed dialdehyde formation. These authors favor a hydroperoxide mechanism over a dioxetane, but allow that both are reasonable.

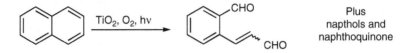

Additionally reported by Theurich and coworkers was the presence of phthalaldehyde. A mechanism for its formation was proposed that begins with electron transfer and involves formation of the six-centered endoperoxide (rather than the four-membered ring dioxetane), either as the radical cation or neutral. Homolysis of the O–O bond is followed by loss of acetylene.

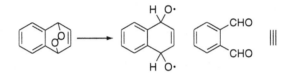

Theurich et al. and Das et al. agree that degradation of anthracene leads quickly to anthraquinone, and that subsequent oxidation is slower [163, 164]. Without drawing any mechanistic conclusions, Theurich et al. reported that phthalic acid is the main observed product. Hydroxylated products were also observed.

Degradation of benzofuran begins by chemical reaction occurring on the furan ring [96]. Hydroxylation of the furan double bond occurs at both positions. Salicylaldehyde and its hydroxylated analogs were all observed, rather than furo-annelated benzene degradation products. It was shown specifically that benzo-2-furanone leads to salicylaldehyde, but few other mechanistic conclusions could be drawn.

13.4 THE NATURE OF THE PRIMARY OXIDIZING AGENT

Two issues have been the topic of discussion in the mechanism of photocatalytic degradations since at least the mid-1980s. First, there is the question as to whether most if not all reactivity occurs at the surface of the TiO_2 particle, or whether solvated hydroxyl radicals are formed. It is now fairly widely accepted that at least the great majority of chemistry is not the result of solvated hydroxyl radicals, and this will not be discussed at length [69, 70, 167, 168]. The second question addresses the nature of the surface-bound oxidizing agents.

It was once the dominant interpretation of degradation patterns that their chemistry derived from hydroxyl radicals, whether adsorbed or solvated [169]. This largely derived from the similarity in product distributions between the TiO_2 reactions and Fenton chemistry or other hydroxyl-like conditions. However, in recent years, even the identity of hydroxyl radical as the predominant oxidant in Fenton chemistry has been strongly questioned. (See, for example, Refs. [170–183].) Moreover, several TiO_2-mediated reactions that are initiated by SET have been reported, among them the photo-Kolbe reaction that leads to decarboxylation of organic acids [184]. It has also been recognized that that most of the organic reactive intermediates that are attributed to hydroxyl radical reactivity can be formed by stepwise reactions in water that begin with electron abstraction.

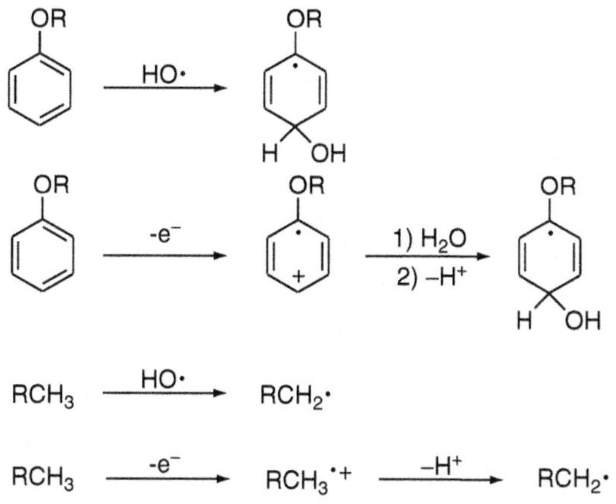

$$\mathrm{RCH_3} \xrightarrow{\mathrm{HO\cdot}} \mathrm{RCH_2\cdot}$$

$$\mathrm{RCH_3} \xrightarrow{-e^-} \mathrm{RCH_3^{\cdot +}} \xrightarrow{-\mathrm{H}^+} \mathrm{RCH_2\cdot}$$

Evidence cited in favor of the formation of hydroxyl radicals includes the formation of HO-adduct spin-traps detected by EPR [185–191]. However, it is not clear why this result does not have the same weakness as other conventional product study, in that a net HO• adduct might be formed by stepwise oxidation and hydration.

A later school of thought arose that advocated that perhaps all oxidative chemistry was initiated by SET. This derived mainly from successful microsecond timescale flash photolysis experiments showing intermediates derived from SET. (See for, example, Refs. [28] and [192].) In the absence of water, electron transfer is the clearly dominant pathway, as demonstrated by Fox and coworkers [167, 193–195].

Certain product studies in oxygenated aqueous solutions also appear to support oxidative SET as the exclusive mechanism of initiation. For example, reactions attributed to SET mechanisms were observed in the TiO_2-sensitized degradations of aromatic molecules designed to fragment on SET, but not from hydrogen abstraction [196]. Ranchella and coworkers examined the degradation of a *t*-butyl substituted benzyl alcohol, for which it was asserted that $HO^{\bullet}$ chemistry would lead to the corresponding ketone.

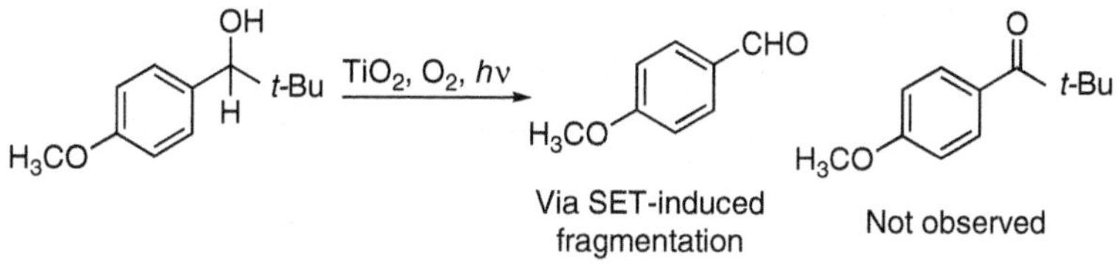

Via SET-induced fragmentation

Not observed

A unifying view that the question of h^+ vs. $HO^{\bullet}_{ads}$ oxidation is a moot question was advanced by Serpone on the basis of several physical studies [69, 197]. Among them, it was found that $HO^{\bullet}$, formed by radiolysis, reacted with TiO_2 surfaces irreversibly and with a rate constant approaching $10^{12}\ M^{-1}\ s^{-1}$ [198]. This was used to rule out the diffusion of hydroxyl radicals away from the TiO_2 surface. His formulation of the oxidizing species was Ti(IV)—O—Ti(IV)—$OH^{\bullet +}$. This was viewed as a single indistinguishable entity that fulfilled the requirement for being both the "surface-trapped hole" and the "adsorbed hydroxyl radical." An interesting corollary of these kinetic data is that hydroxyl radicals formed by other methods would probably also be trapped by TiO_2 to form $HO^{\bullet}_{ads}$ under normal conditions. This is a salient point because superoxide can be converted to homogeneous hydroxyl radicals by a series of secondary reactions.

$$2\,O_2^{\bullet-} + 2\,H^+ \longrightarrow H_2O_2 + O_2$$

$$H_2O_2 + e^- \longrightarrow HO^- + HO^\bullet$$

$$H_2O_2 + h\nu \longrightarrow 2HO^\bullet$$

Several studies have reported results, however, that appeal to the view that there are two different types of reactivities from photoactivated TiO_2. Though there are isolated reports of photophysical or kinetic experiments that may be interpreted this way, such as the ambiguous kinetics of aromatic amine radical cations reported by Fox [199], the great majority of support comes from product study. This dual-reactivity position is distinct from the one which suggests that all reactions, even those that appear to be hydrogen abstraction, are actually initiated by SET. In this context, most authors refer to hole oxidation and hydroxyl radical chemistry in one way or another. However, even if this view of dual reactivity is accepted at face value, the origin of this effect is not resolved. As alluded to previously, both hole trap energy depth and detailed surface structure or adsorption mode might be reasonable explanations. Furthermore, a two-mode model might also be a simplification of a distribution of reactivities, represented in the extremes of electron transfer and hydroxyl chemistry. Nonetheless, the two-mode model has proved useful in interpreting considerable data for degradations of aromatic compounds and may explain some other puzzling data, such as the ambiguous results of a Hammett study on the rates of degradations of substituted phenols [200].

Cunningham studied the degradation kinetics of a series of substituted benzoic acids, reporting the inadequacy of the usual Langmuir–Hinshelwood (LH) equations to adequately describe the kinetics of degradation [74]. In explaining the discrepancies, a hypothesis was advanced in which oxidation of a water cluster at the interface of a particle preceded electron tunneling as a substrate approached, but if the latter did not happen on an appropriate timescale, a hydroxyl radical would be formed. Either higher effective binding constants in the light than measured in the dark or release of $HO^\bullet$ from the surface was required to complete the model. While this interpretation is not currently prevailing, the concept of development of hydroxyl radicals with the advancement of time is an important one in the discussions of some other authors.

On of the earliest studies giving conclusions supporting dual reactivity based on product study did not use aromatic substrates, but rather acetate (CH_3CO_2H) [12]. It had been established that $HO^\bullet$ produced by radiolysis reacted with acetate at least mainly by hydrogen abstraction, yielding various oxidized two-carbon products in the presence of O_2. In contrast, electrochemical oxidation on metal electrodes provided one-carbon products, based on the Kolbe reaction. Photocatalytic treatment with TiO_2 provides both. It was thus concluded that both processes were active. A similar conclusion was later reached for acetates and chloroacetates using electrochemically activated TiO_2 electrodes [201].

A closely following report referring to differential product formation between holes and $HO^\bullet$ using aromatic substrates was that of Richard, who studied the oxidation of 4-hydroxybenzyl alcohol [202]. Oxidation to the aldehyde, conversion to hydroquinone, and hydroxylation were all observed in comparable yield under ordinary TiO_2 conditions. Addition of an alcohol suppressed the hydroxylation reaction completely and partly suppressed formation of hydroquinone. This was interpreted as the result of scavenging hydroxyl radicals and that there was a second "reagent", i.e., h_{vb}^+, that caused the formation of the aldehyde and hydroquinone. Although both the hole and the hydroxyl radical could in principle lead to the same chemistry, in practice they had different regiochemical selectivities, it was argued.

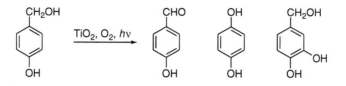

Consistent with this idea was the result that Fenton chemistry yielded the aldehyde predominantly, with some hydroxylation product. Conversion of a benzyl alcohol to hydroquinone probably requires more than a single step, but it is reasonable to think that the path may go through a photo-Kolbe reaction of the corresponding benzoic acid. Similar regioselectivity of reaction arguments are central to the two-mechanism hypothesis advanced by Pichat concerning quinoline, as discussed in Section 13.3 [93, 156].

Richard followed up her work with experiments on the oxidation of phenol [190]. The major primary products detected were *ortho*- and *para*-hydroquinone. Again, *i*-PrOH was added, with the result that *ortho*-hydroquinone was not detected. Hydrogen peroxide photolysis in the presence of O_2 provided mainly *ortho*-hydroquinone, and these results were interpreted to mean that phenol oxidation by h^+_{vb} led exclusively to *para*-hydroxylation, whereas $HO^\bullet$ led to both products. (Richard also did product study chemistry that is often cited as evidence that singlet oxygen, $^1O_2(^1\Delta_g)$, is not an oxidant in TiO_2 or ZnO chemistry [203–205].)

A nearly simultaneous example was the work of Stafford and coworkers on the degradation of 4-chlorophenol [109]. They modeled the chemistry of TiO_2 by various radiolysis techniques and characterized transient hydroxyl adducts by flash photolysis. Under reducing conditions, phenol was the major product. Near neutral pH, when hydroxyl radical was the major reactive intermediate, hydroquinone was formed in a ratio of about 5:1, or about 4:1 when O_2 was present. Several conditions of TiO_2 slurries were cited or measured in which various ratios of hydroquinone to 4-chlorocatechol were produced, generally favoring the latter, but with some variation. Azide radicals oxidize 4-chlorophenol by a SET pathway, but no aromatic compounds were observed in the product mixtures, except for some oligomers.

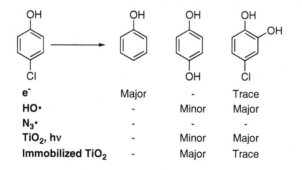

e^-	Major	-	Trace
HO•	-	Minor	Major
N_3•	-	-	-
TiO_2, hv	-	Minor	Major
Immobilized TiO_2	-	Major	Trace

Studies were cited in which immobilized films of TiO_2, rather than slurries, were irradiated to degrade chlorophenol, in which hydroquinone, rather than 4-chlorocatechol, was the predominant detected intermediate. It was emphasized that these discrepancies might be due to initial product formation, but might also be due to artifacts of differential adsorption or secondary degradation due to concentrations, surface area, etc., and this was later examined in more detail [106, 108]. Ultimately, the authors concluded that hydroxyl-mediated oxidation was not the exclusive mechanism of 4-chlorophenol oxidation under TiO_2 conditions.

A related approach was taken by Goldstein, who reported the oxidation of phenol by the strong one-electron oxidizing agent, $SO_4^{\bullet-}$, and radiolytically generated $HO^\bullet$, in comparison to colloidal TiO_2 [206]. Instead of O_2, Cu^{2+} was used as a secondary oxidizing agent. The two products of interest were *ortho*-hydroquinone (catechol) and *para*-hydroquinone, with a limiting ratio of approximately 2 achieved under certain conditions for each reagent system. The effect of pH on the yields at low pH is not the same for TiO_2 and radiolytically generated HO• it was thus concluded that free HO• is not the primary oxidant in the former system.

In the pH range of 2.0 to 3.3, the ratio of *ortho* to *para* hydroxylation was compared [206]. At ca. 10 m*M* phenol, there is a strong dependence on pH that is similar to that of the radiolytic system. At much higher concentration (>0.1 *M*), the pH profile for colloidal TiO_2 is more similar to that of the sulfate radical anion. It was thus concluded that, in the TiO_2 oxidations at low phenol concentration, $HO^{\bullet}_{acs}$ chemistry predominates, while SET is the predominant mechanism at high concentration, where adsorption to the TiO_2 particles is saturated.

These results, however, point out the very difficulties of the method of comparison between TiO_2 and related conditions. These problems are exacerbated when the comparisons are for regiochemistries of what is essentially the same reaction at different locations, such as hydroxylation of an aromatic ring. Rather large differences in product ratios can reflect very small energetic differences caused by the unanticipated or unaccounted minor flaws in reaction analogy. Differential adsorption of regioisomers, for example, of starting materials and products is obviously not relevant to methods without heterogeneous catalysts, such as Fenton chemistry or radiolysis. Secondly, at least under current thought, "hydroxyl radicals" that would be involved in TiO_2 chemistry are adsorbed to a macroscopic particle that also adsorbs the organic substrate. Thus, steric constraints and subtle electronic effects may affect the regiochemistry of hydroxylation. Similarly, there is some ambiguity with the role of the alcohol additives. Surely they are good $HO^{\bullet}$ scavengers, but they can also displace weakly bound substrates from the surface, and might also react with h^{-}_{vb}.

A study that used a probe based on completely different reactions at different parts of the molecule was the degradation of the herbicide 2,4-dichlorophenoxyacetic acid (2,4-D). Substrates labeled with $^{14}CO_2H$ and H/D isotope effects were used in reaching its conclusions [207]. At low pH, decarboxylation is the major primary process, with dichlorophenyl formate, dichlorophenol, and formaldehyde formed in about the same yield. Neither methanol nor *t*-BuOH has much effect, though there is a significant solvent KIE. At high pH, the yield of identifiable products is much lower, but dichlorophenol, formaldehyde, and glyoxylate are all observed in comparable yield. Additional products were postulated, but not identified. Similar product mixtures and pH rate profiles were obtained by other authors for 2,4-D and closely related compounds [208–210]. At high pH, alcohol additives decrease the rate of degradation significantly, and the isotope effect essentially disappeared [207]. These results were taken to indicate a change in major mechanisms with the increased pH. With Fenton's reagent, decarboxylation is not a major reaction and alcohols inhibit the reactions.

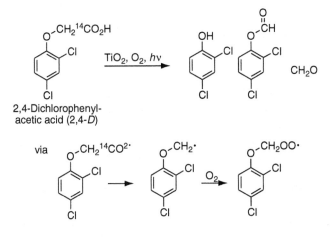

Carboxylic acids are generally well adsorbed to TiO_2 at low pH. (See, for example, Ref. [153].) The above evidence, taken in sum, was used to conclude that the decarboxylation of 2,4-D was initiated by SET to the TiO_2 particle and that another $HO^{\bullet}_{abs}$-based mechanism was responsible for other reactions: "Holes carry out electron transfer oxidation, preferring carboxyl over other functional groups," and, "The trapped hole . . . behaves like free HO• in that it abstracts hydrogen atoms and/or adds to the aromatic ring" [207]. Later researchers also supported the dual mechanism hypothesis based on extensions of this work, though they suggested that dichlorophenol derived from fragmentation of the radical cation centered in the aromatic ring [211].

An investigation by Li and coworkers examined the chemistry of a series of hydroxy- and methoxybenzenes, ranging from 1,2,4-trimethoxybenzene to 1,2,4-benzenetriol and an analogous series based on 4-chlorocatechol [161]. The essential hypothesis to be checked was whether the difference in adsorption mode caused by capping the hydroxyl groups with methyls would affect the chemistry. Previous results [20, 159, 212] suggested that the reactivity of the *ortho*-dihydroxyarenes towards ring opening might be facilitated by chemisorption (i.e., C—O—Ti linkages) and initiated by SET. With OCH_3 groups in lieu of OH groups, this sort of chemisorption would be blocked, but the compounds would be otherwise very similar. Treatment of these compounds with $HO^{\bullet}$-type conditions (e.g., H_2O_2 photolysis) resulted in demethylation and hydroxylation of rings, but very little if any ring opening [161]. However, the major primary products from TiO_2 chemistry for the series varied considerably. When all or most of the substituents are methoxy, then methyl oxidation and ring hydroxylation dominate. Conversely, ring opening predominates as the methyls are removed. Furthermore, addition of *i*-PrOH to the solutions selectively inhibits the reactions attributed to $HO^{\bullet}_{ads}$ chemistry.

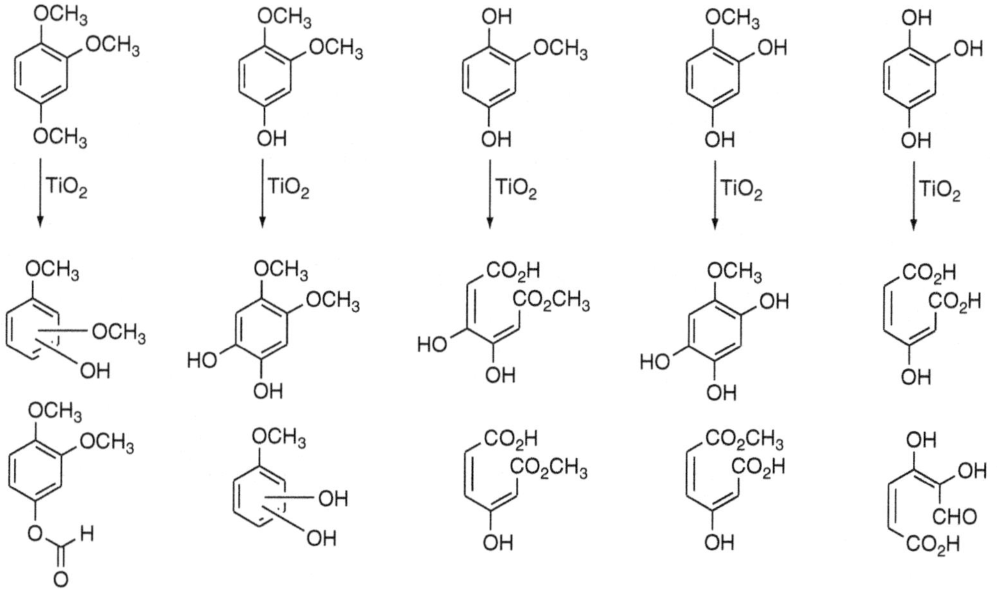

The important advance in this paper is the idea that the chemisorbed substrates are more effective at SET reactions than the nonspecifically adsorbed compounds. As a corollary, it was suggested that part of the reason for this is that the SET reaction must happen rapidly, as the SET reaction and finding a trap leading to $HO^{\bullet}_{ads}$ reactivity are in competition. This is consistent with but a refinement of the concepts suggested previously. It remains to be seen, however, whether it is correct that a hole that might lead to SET will actually evolve into one that leads to $HO^{\bullet}$ chemistry. In the Bahnemann model, this corresponds to a hole moving from a shallow to a deep trap. Alternatively, it may be that the trap site is more or less fixed, once localized. Current experiments do not definitively address this.

One final, but very important system that bears on this discussion, at least indirectly is that of fluorinated TiO_2, developed by Minero and Pelizzetti [71, 72]. Through alcohol inhibition experiments, they estimated that phenol reacts with TiO_2 to form hydroquinone and catechol approximately 90% by way of $HO^{\bullet}_{ads}$, and 10% by SET. The most interesting portion of this research, however, is the study of the TiO_2 system at low pH (approximately 2 to 4) in the presence of moderate concentrations of NaF. Under these circumstances, the fluoride displaces the surface hydroxyl groups, replacing them with Ti—F termini. This obviously changes the surface of the catalyst and its chemistry dramatically. Substrates that interact strongly with the OH groups or bind to Ti via their own O atoms will clearly have much lower affinity for this modified catalyst. Most importantly, the authors demonstrate via elegant steady-state kinetic experiments that this is a photocatalytic system that produces solvated $HO^{\bullet}$ radicals. Put simply, the TiO_2/F catalyst has similar oxidizing power to naked TiO_2, but nothing on the surface to trap holes. Thus, when nearby water molecules are oxidized, and they are not trapped by the TiO_2/F surface, they can diffuse away, in marked contrast to the naked TiO_2 catalyst. This concept has obvious applicability in mechanistic studies of photocatalysis reactions, even if it has limited applicability in real-world water reclamation projects.

13.5 SELECTED EXAMPLES OF PARTIAL DEGRADATION PATHWAYS FOR AROMATIC SYSTEMS

13.5.1 ATRAZINE AND SIMILAR TRIAZINE-CONTAINING COMPOUNDS

Atrazine is a broadleaf herbicide that is a common pollutant in agricultural groundwater. It was thus an attractive target for degradation, and several studies have appeared. The early work by Pelizzetti et al. [64] was groundbreaking.

The important structural feature of atrazine is the *sym*-triazine ring. Compared to benzenes, azabenzenes typically react as electron-poor substrates, with the triazine ring as an extreme example. Thus, a trend that will be seen below is that when given the choice between a comparatively electron-rich "normal" benzene ring and a triazine ring — even a heavily substituted one — hydroxylation generally occurs at the benzene ring, rather than substitution or other reactions at the triazine.

Atrazine itself, however, contains only the single aromatic ring, so the reactivity comparison is strictly between the alkyl groups and reactivity at the aromatic nucleus. Pelizzetti's group, mainly by comparison to authentic samples, identified several degradation intermediates. The first step of degradation is dominated by oxidation of the ethyl or isopropyl groups, as evidenced by time profiles and quantification, though some substitution of the Cl by OH does occur. No "early" substitution of the exocyclic nitrogens is observed. This was confirmed by later authors [213]. Given the regioselectivity of the oxidations of the alkyl chains, i.e., immediately next to the aniline-type nitrogen, it is tempting to speculate that this reaction occurs by stepwise oxidative SET, followed by loss of the appropriate proton, in that anilines are such good electron donors. However, there is no direct evidence on this point.

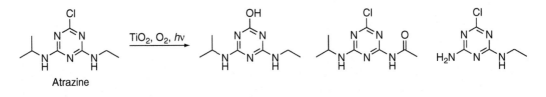

Atrazine $\xrightarrow{TiO_2,\ O_2,\ h\nu}$ (products)

An important and very unusual conclusion of Pelizzetti's original paper was that the endpoint of the degradation of atrazine by TiO_2 is not the same as for most other compounds; mineralization is not achieved. Instead, degradation terminates with cyanuric acid, a result duplicated by other laboratories for atrazine and other related triazine herbicides [64, 65, 129, 138, 213–216]. Pelizzetti reported that photolysis of pristine solutions of cyanuric acid in the presence of TiO_2 simply did not degrade. Similar results were later obtained with other *sym*-triazine-containing compounds, such as azo dyes [60] or sulfonylurea herbicides (Section 13.5.2).

TiO_2, O_2, $h\nu$

Minor

Major

Cyanuric acid

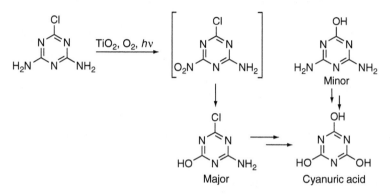

A major "later" intermediate in the degradation of atrazine and its related derivatives [64, 65] is diaminochlorotriazine. Its partial degradation produced mainly the product of HO substitution for NH_2, rather than substitution for Cl. About 90% of the recovered nitrogen was in the form of NO_3, and it was concluded (though no related intermediates were isolated) that the oxidation of the nitrogen to the nitro oxidation state occurred while still attached to the aromatic ring.

In a related context, the degradation of another small cyanuric derivative containing a NH_2 substituent has also been reported [215]. The isolated products showed a clear preference for the oxidation of the alkyl groups before removal of the nitrogen. One may speculate that the methyl group is removed by stepwise oxidation up to the acid, followed by a Kolbe-type reaction. Oxidation of the alkyl groups, both here and in atrazine, is favored over reactions of the ring-bound nitrogens.

TiO_2, O_2, $h\nu$

Ammelide

Cyanuric acid

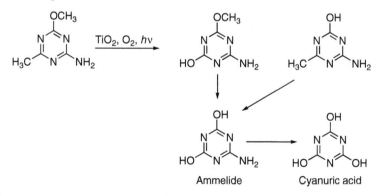

Naturally, the stability of cyanuric acid drew some attention, because of the rarity of such resistant compounds. Cyanuric acid is known to be unusually resistant to decomposition by other methods as well, which raised the possibility that there was a special chemical explanation [217]. For example, the carbon atoms are fully oxidized, so it must ultimately react either by hydrolysis, reduction, or by reactions of the nitrogens. The known biological degradations occur by way of hydrolysis [218–221].

The initial report by Pelizzetti suggested that cyanuric acid was stable to oxidation by Fenton conditions [64]. Since $HO^{\bullet}$ is a relatively unselective and very powerful oxidant of organic systems, this suggested a unique chemical (un)reactivity of this substrate. Tetzlaff tested the hypothesis that $HO^{\bullet}_{ads}$ might add to the aromatic ring, and be reversibly lost by using ^{18}O-labeled cyanuric acid [138]. If $HO^{\bullet}$ were to add and then be lost, or if $HO^{\bullet}$ were to add, and then the elements of water be lost to form the cyanuryl radical (which might then pick up an electron or H-atom from other sources), the ^{18}O would be exchanged out with unlabeled solvent–TiO_2 OH groups. Instead, the ^{18}O was fully retained over several days of irradiation. It was thus concluded that this was not the source of the unusual stability of cyanuric acid to TiO_2 conditions.

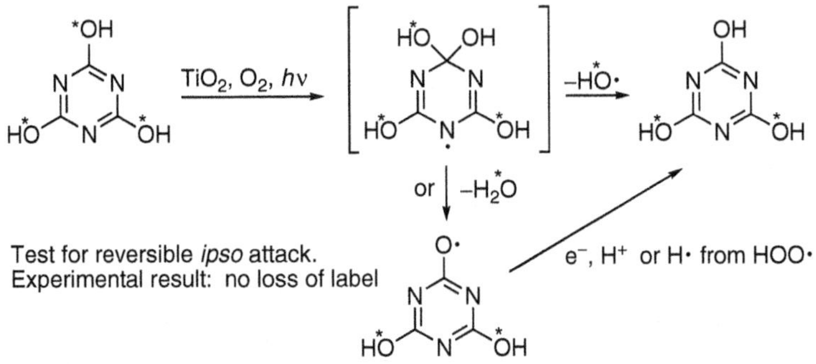

TiO_2, O_2, $h\nu$

$-H\overset{*}{O}\cdot$

or $-H_2\overset{*}{O}$

e^-, H^+ or H· from HOO·

Test for reversible *ipso* attack.
Experimental result: no loss of label

The same research group later reexamined this issue with a new hypothesis that interaction between cyanuric acid simply does not interact sufficiently with the TiO_2 surface to be degraded [101]. This was brought up as a reasonable possibility by a similar report that 4-*tert*-butylpyridine, a compound that should have no special inherent stability, is also stable to TiO_2-mediated photocatalytic degradation in some pH regimes [222].

It was reported that cyanuric acid is degraded under three sets of conditions: hydrogen peroxide photolysis, Fenton-type conditions, and TiO_2-mediated photocatalytic degradations at low pH in the presence of fluoride ion [223]. The latter conditions are expected to produce free hydroxyl radicals [71, 72] (see Section 13.4). Similar results were obtained for 4-*tert*-butylpyridine. The authors thus concluded that cyanuric acid can be degraded under conditions where it comes into contact with the appropriate oxidizing agents. However, because the cyanuric acid is naturally degraded more slowly than its intermediate products, none of the intermediates could be identified under any of the successful degradative conditions. Thus, although, it now appears that cyanuric acid can be degraded, the many interesting details, such as the role of its structure and that of its amide-like tautomer, isocyanuric acid, are yet to be revealed.

One other report suggests that cyanuric acid can be degraded. Gawlik and coworkers reported that a commercial catalyst known as PHOTOPERM, which contains 30% TiO_2 also degrades cyanuric acid [224]. It is tempting to speculate that this unique success among TiO_2 experiments (in the absence of fluoride) implies that the support is critically involved, perhaps as providing the binding site. Also interestingly, these authors report that addition of H_2O_2 to the mixture quenches the ability of the catalyst to degrade cyanuric acid. Kinetic traces imply that degradation occurs as soon as the H_2O_2 is consumed. No structural information about intermediates on the path to mineralization, or the fate of the nitrogen was available. The authors conclude that cyanuric acid is cleaved open by a SET-based mechanism analogous to that discussed in Section 13.3.3. They suggest that the inhibition is due to preferential consumption of the required superoxide by hydrogen peroxide over reaction with the cyanuric radical.

Cyromazine is another example of a triazine-containing pesticide. Its mode of action is as a growth regulator and larvicide. The remarkable feature of this compound, relative to atrazine, is the presence of the cyclopropyl ring. Most of the compounds along the path of its degradation have been identified by interpretation of liquid chromatography–MS (LCMS) data by Goutailler and coworkers [129]. The cyclopropyl group is oxidized in stepwise fashion, beginning with a ring opening. Several of these steps occur before any compounds are observed where OH has been substituted for either of the NH_2 groups. As with the other atrazine compounds, no nitro groups or other partially oxidized nitrogen functionalities were identified.

Cyromazine — TiO_2, O_2, $h\nu$ → Mainly stepwise oxidation to melamine

Melamine — Stepwise substitution of OH for NH → Cyanuric acid

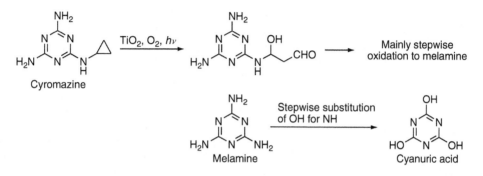

Though the discussion of alkyl oxidation has not been a major portion of this review, it is worth noting in this case because of the cyclopropyl group, which is opened in the initial degradation step. The methylene–cyclopropane rearrangement, in which a cyclopropylmethyl radical rapidly forms a homoallyl radical, is extremely general, and is ubiquitously used as evidence for radical intermediates in physical organic studies. It is tempting to speculate that loss of either an electron or hydrogen atom from the nitrogen would lead to analogous chemistry, forming an imine linkage. Further reaction of the resulting primary radical with O_2 or superoxide might then lead to the reported aldehyde functionality. The reported CHOH group alpha to the nitrogen would constitute a hydrolytically unstable hemiaminal, were it not for the triazine group that ought to make the nitrogen more amide-like and less amine-like, which is another unique feature of this set of compounds. Formation of melamine is prominent in the reaction mixture, even from the beginning, but the very observation [129] of stepwise side chain oxidation intermediates suggests that this particular hemiaminal is unusually stable.

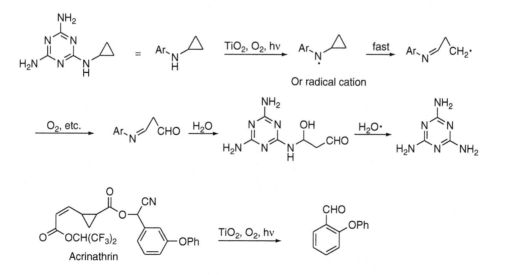

TiO_2, O_2, hv — Or radical cation — fast — O_2, etc. — H_2O — H_2O•

Acrinathrin — TiO_2, O_2, hv →

Goutailler also studied the degradation of a diazine insecticide, dicyclanil [225]. The cyclopropene and diazine draw obvious comparisons to cyromazine. Four parallel degradation pathways were proposed. Ring opening in the same pattern as cyromazine was detected, along with loss with the cyclopropyl group altogether.

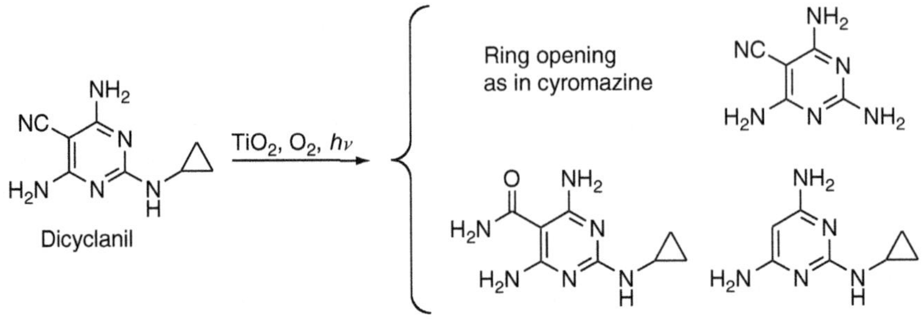

Dicyclanil

TiO_2, O_2, $h\nu$

Ring opening as in cyromazine

Two important reactions were also observed at the nitrile functionality. Catalytically induced hydrolysis led to the amide, and, apparently by reduction, the cyano group was removed. (In both cases, subsequent cyclopropane ring-opening products were reported.) In the other pathways, the same two reactions are observed. It is not entirely clear whether loss of the nitrile comes directly, or by way of oxidative (quasi-photo-Kolbe?) chemistry of the amide. The only other class of nitrile-containing compound with documented degradation products of which we are aware is acrinathrin, a pesticide, in which phenoxybenzaldehyde was identified [226]. However, this conversion represents, at least formally, only a hydrolysis and may involve little or no oxidative chemistry at the nitrile itself. No information about the opening of the cyclopropane ring was available, either.

In contrast to the triazine, however, the diazine ring in dicyclanil does undergo degradation. Proposed partial pathways from advanced intermediates are illustrated. Characteristic of the products is that they simply represent hydrolysis of the 2-carbon, which is already at the CO_2 oxidation state. It is not clear at this point whether the photocatalyst is required or whether this is a dark reaction, but the rest of the degradation clearly requires oxidative steps.

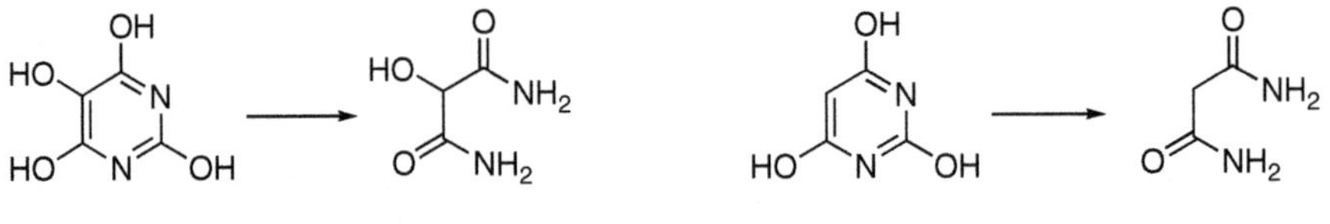

A few intermediates have been proposed, based on GC–MS interpretation, in the degradation of irgarol, another cyclopropene-substituted triazine herbicide [216]. Degradation of the two alkyl groups represents about 90% of the observed early intermediates, with the remainder resulting from either oxidation or reduction at sulfur. Interestingly, these authors propose that the cyclopropyl group is transformed to an ethyl, rather than the oxidative path proposed above. The mechanism of this reaction is less obvious, but clearly reductive if correct. It is clear that these cyclopropyl triazines represent an area where further investigation is well justified by the need to understand these reactions on a fundamental basis. They may also show utility in related ambiguous cases, providing a probe for radical localization by the ring-opening reaction.

Irgarol

TiO_2, O_2, $h\nu$

Major paths

Minor paths

Metamitron

TiO_2, O_2, $h\nu$

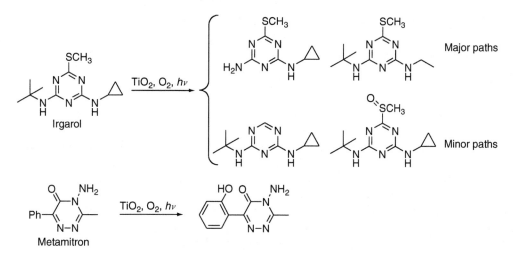

A brief report concerning the opening steps of the herbicide metamitron degradation using either colloidal TiO_2 or Degussa P25 has appeared [227]. Though the triazine is not of the *sym* variety, the electron-poor nature of the ring is still evident in that hydroxylation occurs on the phenyl ring. NMR characterization placed the hydroxyl group in the 2′ position; unfortunately no further downstream intermediates could be identified.

13.5.2 Sulfonylurea and Urea Herbicides

Like the carbons of cyanuric acid, the central carbon of the urea functional group is already fully oxidized. In fact, in water, the major tautomer of cyanuric acid ("isocyanuric acid") resembles a cyclic ureic trimer. Cleavage of this functional group formally requires only hydrolysis, though this might be accomplished by activation of the nitrogens through oxidization or reduction. Little is known about the details of this chemistry at the time of writing this chapter. However, the chemistry of several agriculturally relevant compounds has been investigated.

Sulfonylurea herbicides are characterized by three structural elements: a triazine group derived from cyanuric acid, a sulfonylurea bridge, and a second aromatic group. The sulfonyl group is another potential site of oxidation, and also serves to make the adjacent nitrogen proton rather acidic. Cinosulfuron is representative.

TiO_2, O_2, $h\nu$

Cinosulfuron

Two major regioisomers

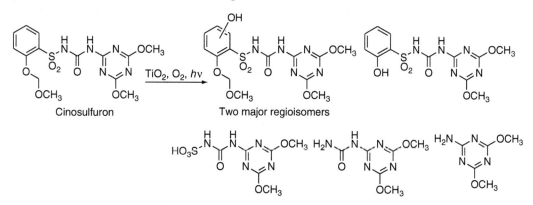

Several primary reactions are observed on treatment of cinosulfuron with TiO_2 and light in water [215, 228]. Consistent with smaller benzene-based aromatic systems, hydroxylation and oxidation of the alkyl portion of the methoxymethyl ether are competitive. Also apparently competitive are reactions in which the sulfonylurea bridge is broken. Though no intermediates were isolated that would imply that hydroxylation had not occurred in advance (i.e., the left-hand portion of the molecule as illustrated), the time-course of observation of the triazine-containing ureas suggest that they are initial products. No information is available on the mechanisms of urea scission; oxidation, reduction, and hydrolysis are all formal possibilities.

As before, oxidation of the nitrogen immediately next to the triazine ring is reserved for later stages. Ultimately, cyanuric acid remains, as predictable from abovediscussed results. The remaining nitrogen is quantitatively detected as nitrate. The sulfur is only partly accounted for as sulfate, but it is suggested that strong sulfate adsorption to the catalyst accounts for this [215].

Degradations of chlorsulfuron and thifensulfuron methyl apparently follow similar courses [128]. One difference is the report, however, of evolution of both ammonia and nitrate. Chloride is the most rapidly formed ion, followed by sulfate from chlorsulfuron, and hydroxylated compounds are observed. Degradation of the nontriazinic aromatic ring is rapid compared to processes at the triazine. A similar pattern of oxidation of the nontriazinic ring and intermediate functional groups has been reported for triazine-containing azo dyes as well [60].

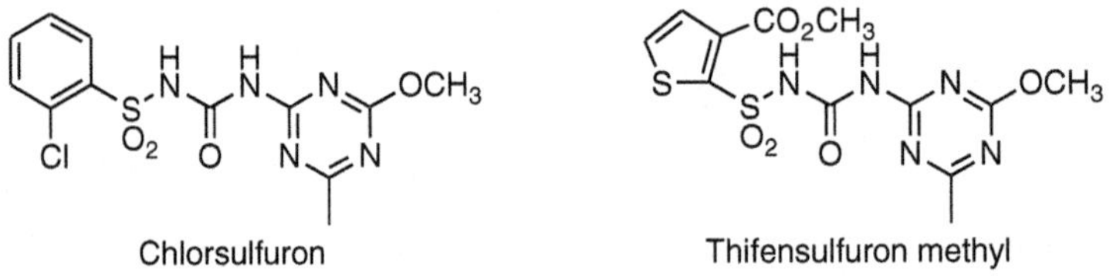

Chlorsulfuron

Thifensulfuron methyl

Monuron and diuron are older aromatic, urea-containing herbicides whose degradations have been studied [229]. Monuron produces 4-chlorophenol and 4-chlorophenyl isocyanate. The former is subsequently hydroxylated, and presumably degrades as discussed in Section 13.3. Hydroxylation of the isocyanate produces the corresponding benzoxazolone after cyclization. Interestingly, despite its water sensitivity, it is reported that the isocyanate, along with its cyclic derivative, are the two most abundantly detected intermediates. The authors propose that hydrolysis of the oxazolone, followed by decarboxylation will lead to the corresponding quinone imine, which would be subject to rapid degradation. The authors could not rule out the formation of 4-chloroaniline as another primary product, but it was also an impurity in the starting material.

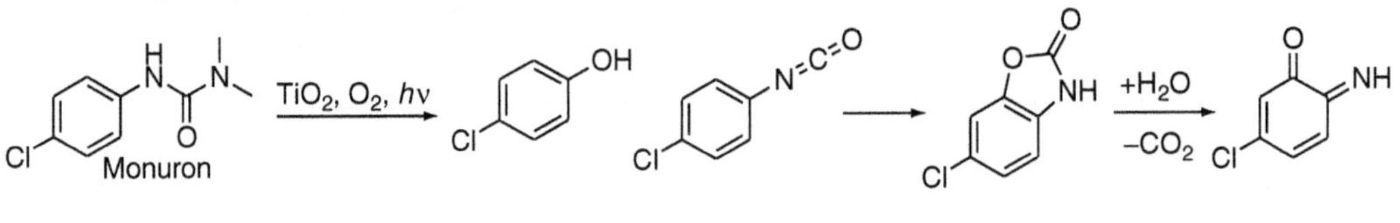

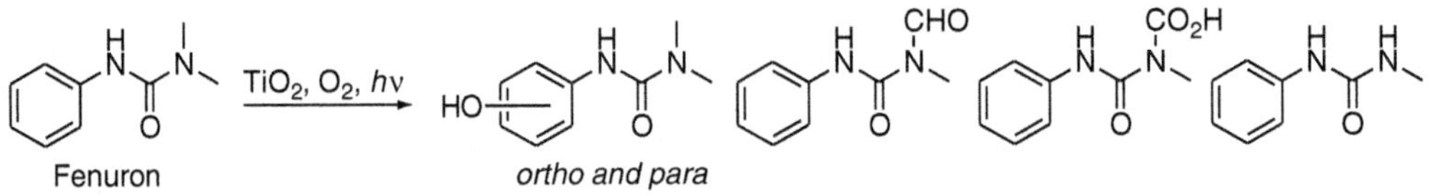

Monuron

Fenuron

ortho and para

The parent compound in this series is fenuron, which was studied by Richard [230]. Here, only conventional products are reported. At low pH, the ring hydroxylation is suppressed and only *N*-methyl oxidation products are observed. The loss of CO_2 to form *N*-methyl-*N*′-phenyl urea is a reaction that continues without catalysis in the dark.

Early stages in the degradation of diuron are proposed to involve the two possible substitutions of OH for Cl and oxidation of one of the *N*-methyl groups to a formate [2]. With colloidal TiO_2 suspensions, *N*-demethylation is an early reaction product [231].

A reasonable mechanism for this reaction is proposed. Hydrogen abstraction, followed by addition of O_2 and subsequent Russell chemistry [137] leads to oxidized methyl groups that are hydrolytically unstable but detectable. The second methyl group can be removed in the same way. Dechlorination and other aromatic functionalization were not addressed.

Diuron $\xrightarrow{TiO_2, O_2, h\nu}$ Ar-HN-C(O)-N(CH_2•)- $\xrightarrow{O_2}$ Ar-HN-C(O)-N(CH_2OO•)- $\xrightarrow{\text{Russell}}$ Ar-HN-C(O)-N(CH_2OH)- + Ar-HN-C(O)-N(CHO)- $\xrightarrow{H_2O}$ 3,4-dichlorophenyl-NH-C(O)-NH-

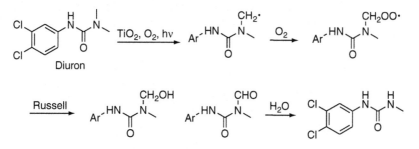

For a related series of phenylurea herbicides with different groups substituted on the aryl ring, no intermediates were identified in the course of a set of kinetic studies [232]. However, it was argued that the relative rates of their initial degradations was tied to the electron-donating ability of the aromatic substituents, due to the electrophilic nature of the presumed $HO^{\bullet}$ oxidizing agent. In support of this hypothesis, it was noted that the rate of disappearance correlated inversely with the dark adsorption equilibria.

13.5.3 Carbamate and Amide Herbicides and Pesticides

Carbamates are of course analogs of ureas, in which one of the amide-like nitrogens is replaced by an ester-like oxygen. As in the simpler carboxylic series, the oxygen "side" of this functional group is expected to be the more reactive with respect to substitution, hydrolysis, etc. Little is currently known about the initial degradative steps of this functionality, except that the phenols on which such carbamates can be based are observed as the major intermediate. Both ammonia and nitrate account for the nitrogen, but larger pieces are not known.

Carbaryl is an example of a carbamate-containing aromatic insecticide whose photocatalytic chemistry has been investigated [91]. The greatest proportion of nitrogen is found as ammonia in this degradation until the later stages, suggesting conversion to nitrate is a downstream process, and further, that the nitrogen is removed by pathways that do not involve its oxidation. However, at relatively early stages of the degradation, the total inorganic nitrogen recovered as either ammonia or nitrate is somewhat less than half of the nitrogen "missing" from the identified larger organic compounds. Thus, either small organic nitrogen-containing fragments are not accounted for or other inorganic nitrogen compounds (e.g., NH_2OH, hydroxylamine) are also present in large proportion.

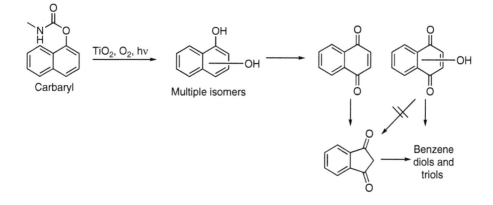

Carbaryl $\xrightarrow{TiO_2, O_2, h\nu}$ naphthalenediol (Multiple isomers) → 1,4-naphthoquinone + hydroxy-1,4-naphthoquinone → indane-1,3-dione → Benzene diols and triols

The study of the aromatic portion of carbaryl degradation was done with degradations also starting from some of the intermediates, so the proposed pathways can be seen as more sure than some, though a few identifications are based only on HPLC retention time data [91]. 1-Napthol itself was not observed, but degradations using it as starting material indicated that its degradation was faster than that of carbaryl itself, so it may be an intermediate between the pesticide and the diols. Independent degradation of the naphthoquinones confirmed that only the parent is the precursor to the indanedione, whose structure is confirmed by published mass spectra and coincidence with authentic samples. In particular, the authors report that 2-hydroxynapthoquinone, which might be expected to be an intermediate between naphthoquinone and indanedione, did not produce the latter compound. This remarkable ring contraction reaction is not observed in the chemistry of hydroquinone or benzoquinone. Neither the mechanism nor the oxidation state of the missing carbon is known. No other intermediates were observed until the compounds reached the level of benzene diols and triols.

A series of substituted *O*-aryl-*N*-methyl carbamate herbicides, all based on phenyl groups, was examined by Tanaka and his group [233]. The rate of disappearance qualitatively followed the electron richness of the aromatic ring, as gauged by the Hammett parameters for the substituents. As with carbaryl, the first intermediates observed were those corresponding to hydrolysis of the carbamate, i.e., the phenols. For the case of Ar = 3,4-dimethylphenyl, methylhydroquinone, and 1,2,4-benzenetriol were both observed.

13.5.4 Amide-Based Agricultural Chemicals

Of the classic carboxylic acid derivatives, the amide is the hardest to hydrolyze because it has a highly delocalized N—C—O structure that makes nucleophilic attack comparatively less favorable. Perhaps as a result of this, and in contrast to the carbamates, chemistry in the aliphatic portions of compounds derived from aniline (as opposed to phenol in the former case) has been observed to compete with release of the aromatic moiety.

Despite identification solely on mass spectral analysis, an extensive list of the degradation products from the herbicide propachlor has been developed by Konstantinou and coworkers [97]. Several pathways are competitive in the initial stages, including aromatic *para*-hydroxylation, the major initial product. Formation of a lactam is formally Friedel–Crafts chemistry [234]. Given the excellent reactivity of α-haloketones in S_N^2 reactions, it is not beyond imagination that this reaction could either be catalyzed by TiO_2 in the dark, or by reversible oxidative SET. The product in which hydroxylation accompanies cyclization is a major intermediate in the GC trace.

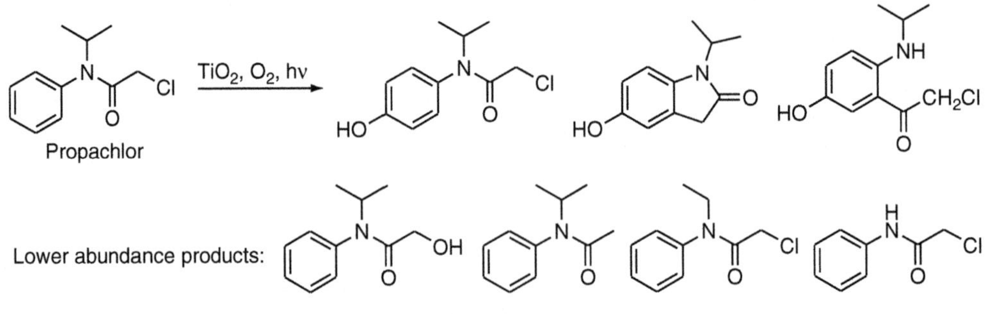

Propachlor —TiO_2, O_2, hν→ (HO-substituted products, CH₂Cl)

Lower abundance products:

The third major product is also intriguing. Shift of the carbonyl portion of the amide to the aromatic ring is known in the direct photolysis of amides as the photo-Fries reaction and goes by a radical mechanism beginning with homolytic scission of the N—(CO) bond. Here, given the hydroxylation, it presumably is a secondary product, but the mechanism by which it is catalyzed by TiO_2 is still not clear. Though the reaction by direct photolysis is documented to be much slower than in the presence of TiO_2, there is no other documentation of catalysis of photo-Fries-like chemistry using either oxidative or reductive SET sensitizers in the literature. Products corresponding to the Friedel–Crafts or photo-Fries reactions in the absence of hydroxylation are also observed in much smaller quantities (not shown).

Other low-abundance early products deriving from propachlor include those due to loss of fragments from the *N*-alkyl group and either reduction or substitution of the Cl atom. The loss of a relatively distal methyl group without oxidation at first seems unusual, but the isopropyl is positioned such that the carbon–carbon bond could be cleaved and maintain conjugation through the amide. The rest of the compounds identified, none of which involve ring opening, involve application of two or more of the abovementioned processes.

The same group had also reported on the degradation of the related herbicide propanil in the previous year [235, 236]. The intermediates proposed in this degradation are more conventional than those of propachlor. They include standard HO for Cl substitution and hydroxylation products, alkyl hydroxylation, and various substituted anilines. The reason or reasons for the stark contrast between the behavior of propanil and propachlor are not addressed, and do not appear to derive from artifacts driven by grossly different analytical methods. It is possible, perhaps even likely, that since a fairly large fraction of the published GC peaks are not identified, the photo-Fries-type products are actually in the mixture, but not identified.

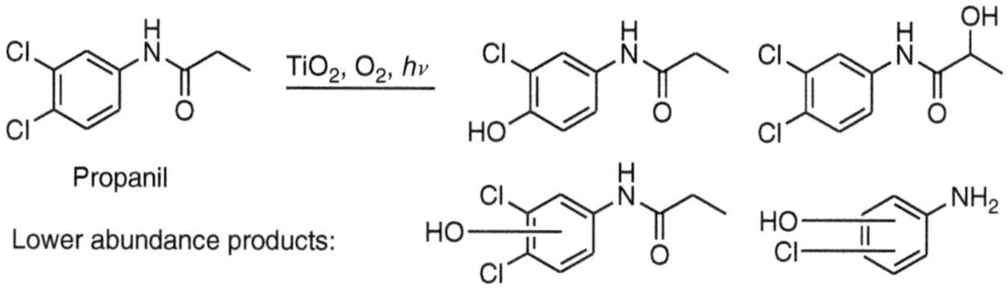

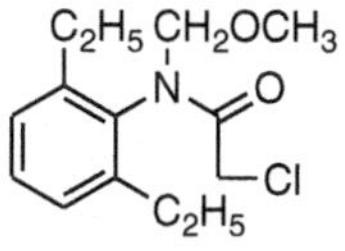

Propanil

TiO_2, O_2, $h\nu$

Lower abundance products:

Alachlor

Alachlor is an amide-containing herbicide with alkyl groups pendant on the aromatic ring such that the photo-Fries reaction is not likely to occur. An early report suggests that the initial reactions, as would be expected, include oxidation of the side chains and hydroxylation of the aromatic ring [237]. A second report notes four abundant but unidentified intermediates with rather anonymous EI–GC–MS fragmentation patterns after a moderate level of conversion [238].

13.5.5 Sulfur-Containing Analogs

The degradative chemistry of organophosphates and organophosphonates is subject to an extensive recent review, which compares results from several different degradative conditions and is discussed briefly in Section 13.3.2.1 [239]. From a mechanistic standpoint, the degradations centering on the phosphorus functionality occur by attack at the aryl or alkyl substituents, rather than at the phosphorus atom itself. Several sulfur-containing analogs of these compounds exist as agricultural chemicals as well, opening new reaction channels, in that sulfur is traditionally thought to be more oxidizable and more nucleophilic than oxygen, in analogous functional groups.

The thionophosphate family includes parathion, fenthion, and bromophos derivatives. Dichlorofenthion was studied by Konstantinou and coworkers [216]. An important electronic structural point needs to be made with regard to these compounds. Although commonly written as P=S double bonds, this is not an especially useful description. Unlike conventional double bonds, these compounds are dative bonds, i.e., there is a single bond, with an additional dipolar attraction due to the ylide-like manner of bonding, as is conventionally written for amine oxides. There is not any significant d-orbital involvement in the bonding. This is true for phosphonates, phosphates, phosphine oxides and their sulfur analogs, in addition to the common oxides of sulfur: sulfoxides, sulfones, sulfonates, sulfuric acid, etc. (The electronic structure of SO_2 is quite analogous to ozone, O_3, but thiocarbonyls $R_2C{=}S$ also have a true C=S double bond.)

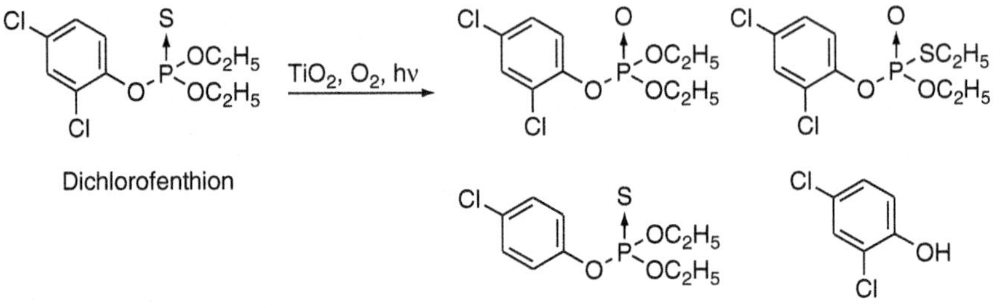

TiO_2, O_2, hν

Dichlorofenthion

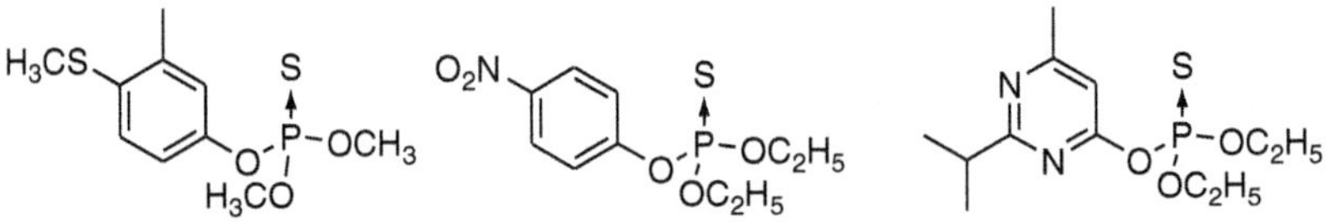

Fenthion Ethyl parathion Diazinon

Because of the formal $R_3P^+{-}S^-$ structure, the sulfur atom appears to be an extremely attractive focal point for oxidative reactions. However, the study of Konstantinou and coworkers shows that reaction at sulfur has competition with oxidation at other sites [216]. Ordinary reactions of the benzene nucleus are accompanied by two sulfur-related transformations. The thionophosphate is converted to the phosphate. Unfortunately, it is not known exactly how this happens. In general, such reactions can be hydrolytic. However, here, an oxidation of the sulfur might be the mechanism by which such hydrolysis is activated.

Isomerization of the thionophosphate to the thiophosphate in the absence of TiO_2 would probably be attributed to homolytic scission of the O—C bond, followed by recombination, though heterolytic mechanisms are formally possible. (The opposite reaction is observed in some alkyl thiocarbamates [6].) Again, no mechanistic information is available here, but the results are important and invite exploration.

The Albanis group also reports the partial degradation of fenthion and ethyl parathion [240]. Reactions analogous to those for dichlorofenthion are observed. Additionally, oxidation of the SCH_3 to the sulfoxide and sulfone occurs in the case of fenthion. Again, it is not straightforward to distinguish between hydroxyl attack and SET for these transformations from the products only. Diazinon degradation intermediates also show the competitiveness of other reasonable pathways with the oxygen for sulfur substitution reaction [241].

13.6 SUMMARY AND OUTLOOK

The most obvious conclusion one reaches after perusing this literature is clichéd: that the progress made is excellent, but that much more remains to be done. However, in this case, the cliché is apt. Considerable progress has been made in the identification of intermediates and pathways, particularly for small model systems. Where glaring holes exist is immediately obvious from Section 13.5. These authors have made pioneering contributions, but in most cases, little is really known about mechanism. In particular, there are several notable instances in which it is not clear whether photocatalytically assisted hydrolysis occurs (e.g., cleavage of ureas, exchange of O for S in thionophosphates) or whether an oxidative or reductive mechanism in order. Moreover, even the release of nitrogen from anilines is not understood on a fundamental level. Clearly the application of our efforts in these directions is necessary, not only for the fundamental reasons of understanding, but towards understanding necessary for the development of new, modified catalysts. Rational design based on mechanism, rather than empirical evaluation, ought to be the goal.

A related point can certainly be made that substrates whose chemistry in a standard set of conditions, e.g., slurries of Degussa P25, is well characterized can be used to evaluate other, modified catalysts. Certainly, this approach has been used with phenol and chlorophenol, particularly in evaluating overall degradation efficiency. However, in designing catalysts that might moderate the transient appearance of more toxic intermediates or be especially well suited for the degradation of certain types of compounds, a mechanistic understanding of the catalyst–substrate interaction is far superior.

ACKNOWLEDGMENTS

The untiring assistance of Youn-Chul Oh, particularly in the searching and gathering from the literature is most gratefully acknowledged. Dr. Oh has also been a particularly valuable contributor to the research output of this laboratory. Financial support for our efforts in photocatalysis has come from the National Science Foundation and the Center for Catalysis of IPRT, to whom we are also most grateful.

REFERENCES AND NOTES

1. A Mills and S Le Hunte, *J. Photochem. Photobiol. A* 108:1–35, 1997.
2. M Muneer, J Theurich, and D Bahnemann, *Res. Chem. Intermed.* 25:667–683, 1999.
3. WF Jardim, SG Moraes, and MMK Takiyama (1997), *Water Res*, *31*:1728–1732.
4. M Muneer, J Theurich, and D Bahnemann, *Proc. Electrochem. Soc.* 98–5:174–187, 1998.
5. M Muneer and D Bahnemann, *Water Sci. Technol.* 44:331–337, 2001.
6. IK Konstantinou and TA Albanis, *Appl. Catal.* B 42:319–335, 2003.
7. J Peller, O Wiest, and PV Kamat, *Environ. Sci. Technol.* 37:1926–1932, 2003.
8. A Mylonas, A Hiskia, E Androulaki, D Dimotikali, and E Papaconstantinou, *Phys. Chem. Chem. Phys.* 1:437–440, 1999.
9. A Hiskia, A Mylonas, and E Papaconstantinou, *Chem. Soc. Rev.* 30:62–69, 2001.
10. A Hiskia, E Androulaki, A Mylonas, S Boyatzis, D Dimoticali, C Minero, E Pelizzetti, and E Papaconstantinou, *Res. Chem. Int.* 26:235–251, 2000.
11. CL Hill, *Chem. Rev.* 98:1–2, 1998.
12. D Bahnemann, In: P Boule, ed. *Handbook of Environmental Chemistry*. Berlin, Germany: Springer, 1999, pp 285–351.
13. MR Hoffmann, ST Martin, W Choi, and DW Bahnemann, *Chem. Rev.* 95:69–96, 1995.
14. M Lewandowski and DF Ollis, *Molec. Supramolec. Photochem.* 10:249–282, 2003.
15. C Wang, DW Bahnemann, and JK Dohrmann, *Water Sci. Technol.* 44:279–286, 2001.
16. D Bahnemann, D Bockelmann, and R Goslich, *Solar Energy Mater.* 24:564–583, 1991.
17. AG Agrios, KA Gray, and E Weitz, *Langmuir* 19:1402–1409, 2003.
18. *Handbook of Organic Photochemistry*, Vol. 1. JC Scaiano, ed. Boca Raton, FL: CRC Press, 1989.
19. www.rayonet.org.
20. X Li, JW Cubbage, TA Tetzlaff, and WS Jenks, *J. Org. Chem.* 64:8509–8524, 1999.
21. AL Linsebigler, G Lu, and JT Yates, Jr., *Chem. Rev.* 95:735–758, 1995.
22. J Augustynski, *Struc. Bonding* 69:1–61, 1988.
23. PV Kamat, *Chem. Rev.* 93:267–300, 1993.

24. JM Kesselman, A Kumar, and NS Lewis, In: DF Ollis and Al-Ekabi, H, eds. *Photocatalytic Purification and Treatment of Water and Air*. Amsterdam: Elsevier Science, 1993, pp 19–37.
25. N Serpone, Pelizzetti, and H Hidaka, In: DF Ollis and Al-Ekabi, H, eds. *Photocatalytic Purification and Treatment of Water and Air*. Amsterdam: Elsevier Science, 1993, pp 225–250.
26. D Bahnemann, Photocatalytic detoxification of polluted waters. In *Handbook of Environmental Chemistry*, P Boule, ed. Springer: Berlin, Germany, 1999; Vol. 2L, pp 285–351.
27. E Pelizzetti and C Minero, *Electrochim. Acta* 38:47–55, 1993.
28. RB Draper and MA Fox, *Langmuir* 6:1396–1402, 1990.
29. MW Peterson, JA Turner, and AJ Nozik, *J. Phys. Chem.* 95:221–225, 1991.
30. JZ Zhang, *J. Phys. Chem. B* 104:7239–7253, 2000.
31. RF Howe and M Grätzel, *J. Phys. Chem.* 91:3906–3909, 1987.
32. OI Micic, Y Zhang, KR Cromack, AD Trifunac, and MS Thurnauer, *J. Phys. Chem.* 97:7277–7283, 1993.
33. C Minero, *Catal. Today* 54:205–216, 1999.
34. DW Bahnemann, M Hilgendorff, and R Memming, *J. Phys. Chem. B* 101:4265–4275, 1997.
35. A Emeline, A Salinaro, and N Serpone, *J. Phys. Chem. B* 104:11202–11210, 2000.
36. AV Emeline, SV Petrova, VK Ryabchuk, and N Serpone, *Chem. Mater.* 10:3484–3491, 1998.
37. AV Emeline, GV Kataeva, VK Ryabchuk, and N Serpone, *J. Phys. Chem. B* 103:9190–9199, 1999.
38. AV Emeline, V Ryabchuk, and N Serpone, *J. Photochem. Photobiol. A* 133:89–97, 2000.
39. AV Emeline, GN Kuzmin, D Purevdorj, VK Ryabchuk, and N Serpone, *J. Phys. Chem. B* 104:2989–2999, 2000.
40. A Emeline, A Salinaro, VK Ryabchuk, and N Serpone, *Int. J. Photoenergy* 3:1–16, 2001.
41. ST Martin, AT Lee, and MR Hoffmann, *Environ. Sci. Technol.* 29:2567–2573, 1995.
42. D Bahnemann, J Cunningham, MA Fox, E Pelizzetti, P Pichat, and N Serpone, Photocatalytic treatment of waters. In: *Aquatic and Surface Photochemistry*, GR Helz, RG Zepp, and DG Crosby, eds. Lewis Publishers: Boca Raton, 1994, pp 261–316.
43. EJ Wolfrum and DF Ollis, *Aquat. Surf. Photochem.* 451–465, 1994.
44. JL Muzyka and MA Fox, *J. Photochem. Photobiol. A* 57:27–39, 1991.
45. H Al-Ekabi, B Butters, D Delany, J Ireland, N Lewis, T Powell, and J Story, In: DF Ollis and Al-Ekabi, H, eds. *Photocatalytic Purification and Treatment of Water and Air*. Amsterdam: Elsevier, 1993, pp 321–335.
46. C Richard, *New J. Chem.* 18:443–445, 1994.
47. X Li and WS Jenks, *J. Am. Chem. Soc.* 122:11864–11870, 2000.
48. W Choi and MR Hoffmann, *J. Phys. Chem.* 100:2161–2169, 1996.
49. W Choi and MR Hoffmann, *Environ. Sci. Technol.* 29:1646–1654, 1995.
50. M Hilgendorff, M Hilgendorff, and DW Bahnemann, *J. Adv. Oxid. Technol.* 1:35–43, 1996.
51. M Muneer and D Bahnemann, *Appl. Catal. B* 36:95–111, 2002.
52. M Muneer, J Theurich, and D Bahnemann, *J. Photochem. Photobiol. A* 143:213–219, 2001.
53. V Maurino, C Minero, E Pelizzetti, P Piccinini, N Serpone, and H Hidaka, *J. Photochem. Photobiol. A* 109:171–176, 1997.
54. K Waki, J Zhao, S Horikoshi, N Watanabe, and H Hidaka, *Chemosphere* 41:337–343, 2000.
55. R Dillert, M Brandt, I Fornefett, U Siebers, and D Bahnemann, *Chemosphere* 30:2333–2341, 1995.
56. P Piccinini, C Minero, M Vincenti, and E Pelizzetti, *Catal. Today* 39:187–195, 1997.
57. JC D'Oliveira, C Guillard, C Maillard, and P Pichat, *J. Environ. Sci. Health, Part A* A28:941–962, 1993.
58. K Nohara, H Hidaka, E Pelizzetti, and N Serpone, *Catal. Lett.* 36:115–118, 1995.
59. K Nohara, H Hidaka, E Pelizzetti, and N Serpone, *J. Photochem. Photobiol. A* 102:265–272, 1997.
60. C Hu, JC Yu, Z Hao, and PK Wong, *Appl. Catal. B* 42:47–55, 2003.
61. M Stylidi, DI Kondarides, and XE Verykios, *Appl. Catal. B* 40:271–286, 2003.
62. G Liu and J Zhao, *New J. Chem.* 24:411–417, 2000.
63. N Serpone, P Calza, A Salinaro, L Cai, A Emeline, H Hidaka, S Horikoshi, and E Pelizzetti, *Proc. Electrochem. Soc.* 97-20:301–320, 1997.
64. E Pelizzetti, V Maurino, C Minero, V Carlin, ML Tosato, E Pramauro, and O Zerbinati, *Environ. Sci. Technol.* 24:1559–1565, 1990.
65. C Minero, V Maurino, and E Pelizzetti, *Res. Chem. Int.* 23:291–310, 1997.

66. RW Matthews, *J. Chem. Soc. Faraday Trans. 1* 80:457–571, 1984.
67. J-P Aycard, E Volanschi, M Hanch, H Zineddine, and TY N'Guessan, *J. Chem. Res., Synop.* 346, 1995.
68. N Serpone, E Pelizzetti, and H Hidaka, *Trace Met. Environ.* 3:225–250, 1993.
69. N Serpone, *Res. Chem. Intermed.* 20:953–992, 1994.
70. C Minero, F Catozzo, and E Pelizzetti, *Langmuir* 8:481–486, 1992.
71. C Minero, G Mariella, V Maurino, and E Pelizzetti, *Langmuir* 16:2632–2641, 2000.
72. C Minero, G Mariella, V Maurino, D Vione, and E Pelizzetti, *Langmuir* 16:8964–8972, 2000.
73. J Clarke, RR Hill, and DR Roberts, *J. Adv. Oxid. Technol.* 4:103–108, 1999.
74. J Cunningham and G Al-Sayyed, *J. Chem. Soc. Faraday Trans.* 86:3935–3941, 1990.
75. C Minero, V Maurino, and E Pelizzetti, *Molec. Supramolec. Photochem.* 10:211–229, 2003.
76. D Bahnemann, *Handb. Environ. Chem.* 2:285–351, 1999.
77. DC Hurum, AG Agrios, KA Gray, T Rajh, and MC Thurnauer, *J. Phys. Chem. B* 107:4545–4549, 2003.
78. ST Martin, JM Kesselman, DS Park, NS Lewis, and MR Hoffmann, *Environ. Sci. Technol.* 30:2535–2545, 1996.
79. D Vasudevan and AT Stone, *Environ. Sci. Technol.* 30:1604–1613, 1996.
80. R Rodríguez, MA Blesa, and AE Regazzoni, *J. Coll. Interface Sci.* 177:122–131, 1996.
81. J Theurich, M Lindner, and DW Bahnemann, *Langmuir* 12:6368–6376, 1996.
82. X-M Pan and C von Sonntag, *A. Naturforsch* 45b:1337–1340, 1990.
83. X-M Pan, MN Schuchmann, and C von Sonntag, *J. Chem. Soc. Perkin Trans. 2* 289–297, 1993.
84. C von Sonntag and H-P Schuchmann, *Angew. Chem. Int. Ed. Engl.* 30:1229–1253, 1991.
85. X Fang, X Pan, A Rahmann, H-P Schuchmann, and C von Sonntag, *Chem. Euro. J.* 1:423–429, 1995.
86. TH Lay, JW Bozzelli, and JH Seinfeld, *J. Phys. Chem.* 100:6543–6554, 1996.
87. JC Rienstra-Kracofe, WD Allen, and HFI Schaefer, *J. Phys. Chem. A* 104:9823–9840, 2000.
88. MS Stark, *J. Am. Chem. Soc.* 122:4162–4170, 2000.
89. IS Ignatyev, Y Xie, WD Allen, and HF Schaefer III, *J. Chem. Phys.* 107:141–155, 1997.
90. P Pichat, *Water Sci. Technol.* 35:73–78, 1997.
91. E Pramauro, AB Prevot, M Vincenti, and G Brizzolesi, *Environ. Sci. Technol.* 31:3126–3131, 1997.
92. P Pichat, L Cermenati, A Albini, D Mas, H Delprat, and C Guillard, *Res. Chem. Intermed.* 25:161–170, 2000.
93. L Cermenati, A Albini, P Pichat, and C Guillard, *Res. Chem. Intermed.* 26:221–234, 2000.
94. B Pal and M Sharon, *J. Mol. Catal. A: Chem.* 160:453–460, 2000.
95. M Sokmen, DW Allen, AT Hewson, and MR Clench, *J. Photochem. Photobiol. A* 141:63–67, 2001.
96. L Amalric, C Guillard, and P Pichat, *J. Photochem. Photobiol. A* 85:257–262, 1995.
97. IK Konstantinou, VA Sakkas, and TA Albanis (2002), *Water Res*, *36*:2733–2742.
98. K Okamoto, Y Yamamoto, H Tanaka, M Tanaka, and A Itaya, *Bull. Chem. Soc. Jpn.* 58:2023–2028, 1985.
99. K Okamoto, Y Yamamoto, H Tanaka, M Tanaka, and A Itaya, *Bull. Chem. Soc. Jpn.* 58:2015–2022, 1985.
100. C Minero, C Aliberti, E Pelizzetti, R Terzian, and N Serpone, *Langmuir* 7:928–936, 1991.
101. Y-C Oh and WS Jenks, *J. Photochem. Photobiol. A* 162:323–328, 2004.
102. L Amalric, C Guillard, E Blanc-Brude, and P Pichat (1996), *Water Res*, *30*:1137–1142.
103. C Minero, C Aliberti, E Pelizzetti, R Terzian, and N Serpone, *Langmuir* 7:928–936, 1991.
104. A Mills and P Sawunyama, *J. Photochem. Photobiol. A* 84:305–309, 1994.
105. U Stafford, KA Gray, and PV Kamat, *Chem. Oxid.* 4:193–204, 1997.
106. U Stafford, KA Gray, and PV Kamat, *Res. Chem. Intermed.* 23:355–388, 1997.
107. U Stafford, KA Gray, and PV Kamat, *Heterog. Chem. Rev.* 3:77–104, 1996.
108. U Stafford, KA Gray, and PV Kamat, *J. Catal.* 167:25–32, 1997.
109. U Stafford, KA Gray, and PV Kamat, *J. Phys. Chem.* 98:6343–6351, 1994.
110. KA Gray, P Kamat, U Stafford, and M Dieckmann. In *Mechanistic studies of chloro- and nitrophenolic degradation on semiconductor surfaces*, Aquat. Surf. Photochem, 1994, GR Helz, ed. Lewis, Boca Raton, Fla: 1994, pp 399–408.
111. U Stafford. Photocatalytic Oxidation of a Model Halogenated Aromatic Compound: a Mechanistic Study (Chlorophenol, Titanium Dioxide), University of Notre Dame, 1994.

112. G Al-Sayyed, J-C D'Oliveira, and P Pichat, *J. Photochem. Photobiol.* A 58:99–114, 1991.
113. H Al-Ekabi and N Serpone, *J. Phys. Chem.* 92:5726–5731, 1988.
114. H Al-Ekabi, N Serpone, E Pelizzetti, C Minero, MA Fox, and RB Draper, *Langmuir* 5:250–255, 1989.
115. A Mills, S Morris, and R Davies, *J. Photochem. Photobiol. A* 70:183–191, 1993.
116. KA Gray and U Stafford, *Res. Chem. Intermed.* 20:835–853, 1994.
117. M Barbeni, E Pramauro, E Pelizzetti, E Borgarello, M Graetzel, and N Serpone, *Nouv. J. Chim.* 8:547–550, 1984.
118. C Dong, C-W Chen, and S-S Wu, *Hazard. Ind. Wastes* 27:361–370, 1995.
119. C Dong and C-P Huang, *Adv. Chem. Ser.* 244:291–313, 1995.
120. CD Dong and CP Huang, In: DF Ollis and Al-Ekabi, H, eds. *Photocatalytic Purification and Treatment of Water and Air.* Amsterdam: Elsevier, 1993, pp 701–706.
121. J Cunningham and P Sedlak, *J. Photochem. Photobiol. A* 77:255–263, 1994.
122. J-C D'Oliveira, C Minero, E Pelizzetti, and P Pichat, *J. Photochem. Photobiol. A* 72:261–267, 1993.
123. Q Dai, N He, Y Guo, and C Yuan, *Chem. Lett.* 1113–1114, 1998.
124. FS Euetit, M Trillas, JG Euetit, and X Domènech, *J. Environ. Sci. Health* A29:1409–1421, 1994.
125. C Minero, E Pelizzetti, P Pichat, M Sega, and M Vincenti, *Environ. Sci. Technol.* 29:2226–2234, 1995.
126. AB Prevot and E Pramauro, *Talanta* 48:847–857, 1999.
127. V Maurino, C Minero, E Pelizzetti, and N Serpone, *NATO ASI Ser., Ser. 3* 12:707–718, 1996.
128. V Maurino, C Minero, E Pelizzetti, and M Vincenti, *Coll. Surf. A* 151:329–338, 1999.
129. G Goutailler, JC Valette, C Guillard, O Paisse, and R Faure, *J. Photochem. Photobiol. A* 141:79–84, 2001.
130. PL Yue and D Allen, In: DF Ollis and Al-Ekabi, H, eds. *Photocatalytic Purification and Treatment of Water and Air.* Amsterdam: Elsevier, 1993, pp 607–611.
131. KC Pugh, DJ Kiserow, JM Sullivan, and JH Grinstead, Jr., *ACS Symp. Ser.* 607:174–194, 1995.
132. FH Frimmel and DP Hessler, Photochemical Degradation of Triazine and Anilide Pesticides in Natural Waters. In *Aquatic and Surface Photochemistry*, GR Helz, RG Zepp, and Crosby, DG, eds. Lewis Publishers: Boca Raton, 1994.
133. C Maillard-Dupuy, C Guillard, H Courbon, and P Pichat, *Environ. Sci. Technol.* 28:2176–2183, 1994.
134. B Sangchakr, T Hisanaga, and K Tanaka, *J. Photochem. Photobiol. A* 85:187–190, 1995.
135. JE Lee, W Choi, BJ Mhin, and K Balasubramanian, *J. Phys. Chem. A* 108:607–614, 2004.
136. J Peller, O Wiest, and PV Kamat, *Chem. Eur. J.* 9:5379–5387, 2003.
137. GA Russell, *J. Am. Chem. Soc.* 79:3871–3877, 1957.
138. T Tetzlaff and WS Jenks, *Org. Lett.* 1:463–465, 1999.
139. Y-C Oh, Y Bao, and WS Jenks, *J. Photochem. Photobiol. A* 160:69–77, 2003.
140. E Pelizzetti and C Minero, *Colloids Surf. A* 151:321–327, 1999.
141. C Minero, E Pelizzetti, P Piccinini, and M Vincenti, *Chemosphere* 28:1229–1244, 1994.
142. JL Ferry and WH Glaze, *Langmuir* 14:3551–3555, 1998.
143. Neither the ketenimine nor the dichlorobenzoquinone were observed, but this portion of the mechanistic proposal appears quite reasonable.
144. B Prevot Alessandra, M Vincenti, A Bianciotto, and E Pramauro, *Appl. Catal. B* 22:149–158, 1999.
145. MS Vohra and K Tanaka (2002), *Water Res*, *36*:59–64.
146. K Tanaka, W Luesaiwaong, and R Hisanaga, *J. Mol. Catal. A: Chem.* 122:67–74, 1997.
147. W Spacek, R Bauer, and G Heisler, *Chemosphere* 30:477–484, 1995.
148. H Tada, M Kubo, Y-i Inubushi, and S Ito, *Chem. Commun.* 977–978, 2000.
149. K Tanaka, K Padermpole, and T Hisanaga (1999), *Water Res*, *34*:327–333.
150. C Hu, Y Tang, JC Yu, and PK Wong, *Appl. Catal. B* 40:131–140, 2003.
151. C Galindo, P Jacques, and A Kalt, *J. Photochem. Photobiol. A* 130:35–47, 2000.
152. M Muneer, HK Singh, and D Bahnemann, *Chemosphere* 49:193–203, 2002.
153. MI Franch, JA Ayllon, J Peral, and X Domenech, *Catal. Today* 76:221–233, 2002.
154. Y-C Oh, X Li, JW Cubbage, and S Jenks William, "Mechanisms of Catalyst Action in the TiO_2-mediated Photocatalytic Degradation of Maelic and Fumaric Acid." *Appl. Catal. B: Environmental* 2004, 54:105–114.

155. L Amalric, C Guillard, and P Pichat, *Res. Chem. Int.* 20:579–594, 1994.
156. L Cermenati, P Pichat, C Guillard, and A Albini, *J. Phys. Chem. B* 101:2650–2658, 1997.
157. AR Nicolaescu, O Wiest, and PV Kamat, *J. Phys. Chem. A* 107:427–433, 2003.
158. The SOD experiments have been successfully duplicated in the Jenks laboratory.
159. X Li, JW Cubbage, and WS Jenks, *J. Org. Chem.* 64:8525–8536, 1999.
160. C Bouquet-Somrani, A Finiels, P Graffin, and J-L Olivé, *Appl. Catal. B* 8:101–106, 1996.
161. X Li, JW Cubbage, and WS Jenks, *J. Photochem. Photobiol. A* 143:69–85, 2001.
162. Y Morooka and CS Foote, *J. Am. Chem. Soc.* 98:1510–1514, 1976.
163. J Theurich, DW Bahnemann, R Vogel, FE Ehamed, G Alhakimi, and I Rajab, *Res. Chem. Intermed.* 23:247–274, 1997.
164. S Das, M Muneer, and KR Gopidas, *J. Photochem. Photobiol. A* 77:83–88, 1994.
165. This is referred to in the text as an oxidation, which is presumably an oversight, since it is a reduction.
166. F Soana, M Sturini, L Cermenati, and A Albini, *J. Chem. Soc., Perkin Trans. 2* 699–704, 2000.
167. MA Fox and MT Dulay, *Chem. Rev.* 93:341–357, 1993.
168. PV Kamat and K Vinodgopal, In: V Ramamurthy and Schanze, K, eds. *Molecular and Supramolecular Photochemistry*. New York: Marcel Dekker, 1998, pp 307–350.
169. CS Turchi and DF Ollis, *J. Catal.* 122:178–192, 1990.
170. SH Bossmann, E Oliveros, S Goeb, S Siegwart, EP Dahlen, L Payawan, Jr., M Straub, M Woerner, and AM Braun, *J. Phys. Chem. A* 102:5542–5550, 1998.
171. T Kurata, Y Watanabe, M Katoh, and Y Sawaki, *J. Am. Chem. Soc.* 110:7472–7478, 1988.
172. I Yamazaki and LH Piette, *J. Am. Chem. Soc.* 113:7588–7593, 1991.
173. RG Zepp, BC Faust, and J Hoigné, *Environ. Sci. Technol.* 26:313–319, 1992.
174. ML Kremer, *Phys. Chem. Chem. Phys.* 1:3595–3605, 1999.
175. S Goldstein and D Meyerstein, *Acc. Chem. Res.* 32:547–550, 1999.
176. F Buda, B Ensing, MCM Gribnau, and EJ Baerends, *Chem. Eur. J.* 7:2775–2783, 2001.
177. F Gozzo, *J. Molec. Catal. A* 171:1–22, 2001.
178. JJ Pignatello, D Liu, and P Huston, *Environ. Sci. Technol.* 33:1832–1839, 1999.
179. PA MacFaul, DDM Wayner, and KU Ingold, *Acc. Chem. Res.* 31:159–162, 1998.
180. DA Wink, CB Wink, RW Nims, and PC Ford, *Environ. Health Pers. Suppl.* 102:11–15, 1994.
181. DA Wink, RW Nims, JE Saavedra, WE Utermahlen, Jr., and PC Ford, *Proc. Natl Acad. Sci. USA* 91:6604–6608, 1994.
182. C Walling, *Acc. Chem. Res.* 31:155–157, 1998.
183. DT Sawyer, A Sobkowiak, and T Matsushita, *Acc. Chem. Res.* 29:409–416, 1996.
184. I Izumi, FF Fan, and AJ Bard, *J. Phys. Chem.* 85:218–223, 1981.
185. CD Jaeger and AJ Bard, *J. Phys. Chem.* 83:3146–4152, 1979.
186. V Brezova and A Stasko, *J. Catal.* 147:156–162, 1994.
187. V Brezova, A Stasko, S Biskupic, A Blazkova, and B Havlinova, *J. Phys. Chem.* 98:8977–8984, 1994.
188. V Brezova, A Stasko, and L Lapcik, Jr., *J. Photochem. Photobiol. A* 59:115–121, 1991.
189. R Morelli, IR Bellobono, CM Chiodaroli, S Alborghetti, *J. Photochem. Photobiol. A* 112:271–276, 1998.
190. C Richard and P Boule, *New J. Chem.* 18:547–552, 1994.
191. A Blazkova, B Mezeiova, V Brezova, M Ceppan, and V Jancovicova, *J. Mol. Catal. A: Chem.* 153:129–137, 2000.
192. RB Draper and MA Fox, *J. Phys. Chem.* 94:4628–4634, 1990.
193. MA Fox, In: N Serpone and Pellizzetti, E, eds. *Photocatalysis: Fundamentals and Applications*. New York: John Wiley & Sons, 1989, pp 421–455.
194. MA Fox, *Adv. Electron Transfer Chem.* 1:1–53, 1991.
195. MA Fox, BL Lindig, and C-C Chen, *J. Am. Chem. Soc.* 104:5828–5829, 1982.
196. M Ranchella, C Rol, and GV Sebastiani, *J. Chem. Soc. Perkin Trans. 2* 311–315, 2000.
197. R Terzian, N Serpone, and MA Fox, *J. Photochem. Photobiol. A* 90:125–135, 1995.
198. D Lawless, N Serpone, and D Meisel, *J. Phys. Chem.* 95:5166–5170, 1991.
199. MA Fox and MT Dulay, *J. Photochem. Photobiol. A* 98:91–101, 1996.
200. KE O'Shea and C Cardona, *J. Org. Chem.* 59:5005–5009, 1994.
201. JM Kesselman, O Weres, NS Lewis, and MR Hoffmann, *J. Phys. Chem. B* 101:2637–2643, 1997.

202. C Richard, *J. Photochem. Photobiol. A* 72:179–182, 1993.
203. C Richard and P Boule, *J. Photochem. Photobiol. A* 84:151–152, 1994.
204. C Richard and P Boule, *Sol. Energy Mater. Sol. Cells* 38:431–440, 1995.
205. C Richard, P Boule, and JM Aubry, *J. Photochem. Photobiol. A* 60:235–243, 1991.
206. S Goldstein, G Czapski, and J Rabani, *J. Phys. Chem.* 98:6586–6591, 1994.
207. Y Sun and JJ Pignatello, *Environ. Sci. Technol.* 29:2065–2072, 1995.
208. M Trillas, J Peral, and X Domenech, *Appl. Catal. B* 5:377–387, 1995.
209. K Tanaka and KSN Reddy, *Appl. Catal. B* 39:305–310, 2002.
210. A Topalov, D Molnar-Gabor, MM Kosanic, and B Abramovic (2000), *Water Res. 34*:1473–1478.
211. K Djebbar and T Sehili, *Pestic. Sci.* 54:269–276, 1998.
212. JM Kesselman, NS Lewis, and MR Hoffmann, *Environ. Sci. Technol.* 31:2298–2302, 1997.
213. V Héquet, P Le Cloirec, C Gonzalez, and B Meunier, *Chemosphere* 41:379–386, 2000.
214. V Hequet, C Gonzalez, and P Le Cloirec (2001), *Water Res, 35*:4253–4260.
215. E Vulliet, C Emmelin, J-M Chovelon, C Guillard, and J-M Herrmann, *Appl. Catal. B* 38:127–137, 2002.
216. IK Konstantinou, TM Sakellarides, VA Sakkas, and TA Albanis, *Environ. Sci. Technol.* 35:398–405, 2001.
217. JA Wojtowicz, In: JI Kroschwitz, ed. *Kirk–Othmer Encyclopedia of Chemical Technology*. New York: Wiley, 1991, pp 835–851.
218. AM Cook, *FEMS Microbiol. Rev.* 46:93–116, 1987.
219. M-S Lai, AS Weber, and JN Jensen (1995), *Water Environ. Res, 67*:347–354.
220. C Ernst and H-J Rehm, *Appl. Microbiol. Biotechnol.* 43:150–155, 1995.
221. AM Cook, P Beilstein, H Grossenbacher, and R Hütter, *Biochem. J.* 231:25–30, 1985.
222. A Nedoloujko and J Kiwi (2000), *Water Res, 34*:3277–3284.
223. No obvious explanation is found for the different result with respect to Fenton conditions reported by the two groups.
224. BM Gawlik, A Moroni, IR Bellobono, and HW Muntau, *Global Nest Int. J.* 1:23–32, 1999.
225. G Goutailler, C Guillard, R Faure, and O Paisse, *J. Agric. Food Chem.* 50:5115–5120, 2002.
226. S Malto, J Blanco, AR Fernandez-Alba, and A Aguera, *Chemosphere* 40:403–409, 2000.
227. K Macounova, J Urban, H Krysova, J Krysa, J Jirkovsky, and J Ludvik, *J. Photochem. Photobiol. A* 140:93–98, 2001.
228. E Vulliet, J-M Chovelon, C Guillard, and J-M Herrmann, *J. Photochem. Photobiol. A* 159:71–79, 2003.
229. E Pramauro, M Vincenti, V Augugliaro, and L Palmisano, *Environ. Sci. Technol.* 27:1790–1795, 1993.
230. C Richard and S Bengana, *Chemosphere* 33:635–641, 1996.
231. K Macounová, H Krysová, J Ludvík, and J Jirkovsky, *J. Photochem. Photobiol. A* 156:273–282, 2002.
232. S Parra, J Olivero, and C Pulgarin, *Appl. Catal. B* 36:75–85, 2002.
233. K Tanaka, SM Robleto, T Hisnaga, R Ail, Z Ramli, and WA Bakar, *J. Molec. Catal.* 144:425–430, 1999.
234. The structure given in the figure does not correspond to the name given in the Table of MS library hits found in the corresponding Table. The Table lists the *N*-acetyl compound, rather than *N*-isopropyl. The former, however, does not make sense in this context, and we presume it to be a typographical error.
235. IK Konstantinou, VA Sakkas, and TA Albanis, *Appl. Catal. B* 34:227–239, 2001.
236. TA Albanis, IK Konstantinou, and VA Sakkas, *Water Sci.Technol.Water Supply* 2:225–232, 2002.
237. PN Moza, K Hustert, S Pal, and P Sukul, *Chemosphere* 25:1675–1682, 1992.
238. GA Penuela and D Barcelo, *J. Chromatogr. A* 754:187–195, 1996.
239. KE O'Shea, *Molec. Supramolec. Photochem.* 10:231–248, 2003.
240. VA Sakkas, DA Lambropoulou, TM Sakellarides, and TA Albanis, *Anal. Chim. Acta* 467:233–243, 2002.
241. VN Kouloumbos, DF Tsipi, AE Hiskia, D Nikolic, and RB van Breeman, *J. Am. Soc. Mass Spectrom.* 14:803–817, 2003.

14 *In Situ* Solid-State NMR Studies of Photocatalytic Oxidation Reactions

Sarah Pilkenton and Daniel Raftery
H. C. Brown Laboratory, Department of Chemistry, Purdue University

CONTENTS

14.1 INTRODUCTION

The widespread use of volatile organic compounds (VOCs) in domestic and industrial applications has led to a number of significant environmental problems including degraded air quality, groundwater contamination, global warming, and ozone depletion in the stratosphere. In recent decades, considerable attention has been focused on the detoxification of VOC pollutants. Currently available methods, such as incineration, carbon trapping, and thermal catalysis, all have serious technical or economic drawbacks that limit their usefulness. One promising alternative method for detoxification is semiconductor photocatalysis [1] due to a number of favorable factors: (1) photocatalytic oxidation reactions can proceed at ambient temperature and pressure, (2) semiconductor photocatalysts, such as TiO_2 and ZnO, are nontoxic, inexpensive, and chemically and physically stable, and (3) the excitation light sources can be sunlight or low-cost fluorescent lights. In addition to the application of photocatalytic processes to the detoxification of environmental pollutants [1, 2], photocatalytic processes also have found applications in solar energy conversion and in the environmentally benign synthesis of organic molecules [3, 4].

Photocatalytic processes exhibit a range of surface chemistry that requires the use of numerous analytical methods for full characterization. Precise identification and characterization of the reaction intermediates formed during photocatalytic oxidation reactions and careful analysis of reactants, products, and intermediates in the adsorbed state are critical for the determination of reaction mechanisms. Many studies have utilized analytical methods, such as temperature-programmed desorption and oxidation (TPD–TPO), GC, GC–MS, FT-IR, MS, or trapping agents for the detection of chemical species resulting from the photocatalytic oxidation of both chlorinated and nonchlorinated VOCs. Inhibition of further reactions during sampling was unavoidable in many cases. In some cases, *in situ* detection of intermediates without interruption of the reaction was employed using FT-IR spectroscopy [5–7]. While sensitive, FT-IR methods exhibit difficulties in direct quantitation. Some of these difficulties can be avoided by the complimentary use of GC–MS [8, 9], but careful and separate calibration runs are required for each intermediate.

To elucidate the complex surface chemistry involved in photocatalytic processes and provide complementary information to the techniques mentioned above, we introduced a new approach to the study of photocatalytic reactions, namely *in situ* solid-state nuclear magnetic resonance (SSNMR) spectroscopy [10]. SSNMR techniques, particularly *in situ* methods, have played an invaluable role in the study of a broad range of issues involved in heterogeneous catalysis due to the wealth of structural and dynamical information available via NMR [11, 12]. *In situ* SSNMR measurements can be made under batch-type conditions, which requires the catalysts and reactants to be sealed inside an NMR tube or specially designed rotor, or alternatively, the measurements can be made under flowing conditions using specially designed NMR probes [13–16]. SSNMR is well suited for *in situ* studies of the complex reaction chemistry in photocatalytic oxidation reactions due to its atomic specificity, high resolution, and quantitative capabilities; as well as its ability to study and differentiate surface and gas-phase species.

A number of researchers have used both solid-state and liquid-state NMR to study semiconductor photocatalysis. Liquid-state NMR can be used to identify photooxidation products in aqueous dispersions of the reactants and the catalyst [17–20]. In one approach, the aqueous phase is evaporated then dissolved in a deuterated solvent and standard NMR is used to identify the photooxidation products [21]. Also, liquid-state NMR has been used to analyze surface-bound reactants and intermediates produced in photooxidation reactions where the surface-bound molecules have been removed from the catalyst via extraction into a liquid phase [22]. Haw and coworkers used ^{13}C MAS SSNMR to identify the photoproducts of benzaldehyde or acetophenone with toluene in the zeolite catalysts, USY, and NaY [23].

This chapter describes the recent SSNMR experiments made in our laboratory and those of others to follow the *in situ* photocatalytic oxidation of both chlorinated and nonchlorinated VOCs. After a brief introduction to SSNMR concepts and methods, a description of the types of important information on photocatalysis that can be obtained using SSNMR studies is provided. As shown below, SSNMR experiments are used to identify new reactive intermediates and surface species, and to study the surface bonding of adsorbates, which has revealed much about the photocatalytic active sites. SSNMR has also been used to probe the surface morphology of a variety of catalysts and study the effect of surface morphology on photocatalytic activity. Additionally, *in situ* studies of photocatalytic reactions have been used to evaluate new photocatalysts [24–27]. Examples and studies of a number of different types of photocatalysts, including powdered semiconductor photocatalysts and supported metal oxide catalysts [10, 28–32], mixed metal oxide photocatalysts [27], and zeolite photocatalysts [23, 26] are provided for illustration.

14.2 A BRIEF INTRODUCTION TO SSNMR CONCEPTS

NMR spectroscopy provides highly detailed information about local molecular structure. The chemical shift is the primary means used to identify the structure of a molecule. As electrons circulate around a nucleus, small, local magnetic fields are generated, and these magnetic fields give rise to a resonance frequency, or chemical shift that depends on the chemical environment of the nucleus. These fields usually oppose the applied magnetic field, which results in the nucleus exposed to an applied magnetic field that is slightly different than the external magnetic field. The magnitude of the resultant field, B_0, can be written as

$$B_0 = B_{\mathrm{appl}}(1 - \sigma) \tag{14.1}$$

where B_{appl} is the applied magnetic field and σ is the screening constant, which is dependent on the density and spatial distribution of electron density around the nucleus. As the electronegativity of groups adjacent to the nucleus of interest increases the electron density or "shielding" is reduced, and the chemical shift of the nucleus will increase. For ^{13}C, the chemical shift region is roughly 0 to 200 ppm (parts per million). Some characteristic chemical shifts are the following: methyl carbons (0 to 30 ppm), methylene carbons (20 to 40 ppm), aromatic carbons (100 to 160 ppm), and carbonyl carbons (160 to 210 ppm).

In the solid state, the chemical shift of a molecule is dependent on the orientation of the molecule with respect to the external magnetic field. For example, the chemical shift anisotropy (CSA) for a nuclear spin in an axially symmetric molecular environment is described by [33]

$$\omega_{\mathrm{cs}} = \sigma_{\mathrm{iso}}\gamma B_0 + \tfrac{1}{2}\,\delta_{\mathrm{cs}}(3\cos^2\theta - 1) \tag{14.2}$$

The first term in Equation (14.2) gives the isotropic chemical shift value, and the second term in Equation (14.2) describes the effect of molecular orientation on the chemical shift, where δ_{cs} is the CSA width parameter and θ is the angle between the molecule or molecular segment and the external magnetic field. In liquid samples, rapid molecular tumbling averages the second term in Equation (14.2) to zero, leaving only a single sharp peak at the isotropic chemical shift. When the motion of the molecule is restricted, for example in a solid, the CSA is not averaged and results in a broad distribution of chemical shift values. The interaction strength, or broadening effect of the CSA, is often many tens of kHz for ^{13}C NMR.

A technique known as magic angle spinning (MAS) is applied to reintroduce the spatial averaging inherent in liquids, and thereby narrows the broad lineshapes in solids caused by the CSA. MAS introduces a modulation that has a $(3\cos^2\varphi - 1)$ dependence. This term is equal to zero when the angle φ is 54.74°, which is known as the "magic angle." In MAS, the sample is spun at an angle of 54.74° with respect to the z-component of the external magnetic field. If the sample is spun at a speed exceeding the magnitude of the CSA of the sample, only an isotropic resonance appears in the NMR spectrum. At spinning speeds smaller than the magnitude of the CSA, the spectrum is broken up into the isotropic peak and a series of modulation peaks called spinning sidebands that appear at multiples of the spinning speed. Figure 14.1 shows the proton-decoupled ^{13}C SSNMR spectra of glycine under static and MAS conditions at several different spinning speeds [34]. Despite an increase in spectral complexity, there is an advantage to slow MAS in that the spinning sideband pattern can be analyzed to yield the full CSA information. The sideband information yields a three-dimensional picture of the electron density surrounding the nuclear spin of interest that is not available in spectra composed of only isotropic chemical shifts.

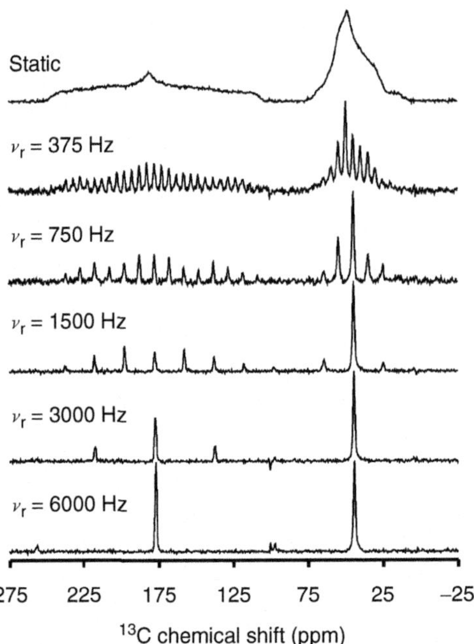

FIGURE 14.1 Proton-decoupled ^{13}C NMR spectra of glycine under static and MAS NMR conditions. (Reprinted from JR Smith. Ph.D. thesis, Purdue University, 2001. With permission.)

The second type of broadening observed in SSNMR spectra is due to static dipolar coupling between two magnetic nuclei, for example ^1H and ^{13}C. Equation (14.3) is the heteronuclear dipolar frequency [33]:

$$\omega_D^{IS} = -\left(\frac{\mu_0 h}{4\pi}\right)\left(\frac{\gamma^I \gamma^S}{r_{IS}^3}\right)(3\cos^2\theta - 1) \tag{14.3}$$

The dipolar coupling constant is given by the term $\gamma^I\gamma^S/r^3$, where r is the internuclear distance between the two nuclei, and θ is the angle between the two nuclei and the z-axis of the external magnetic field. The broadening effects of ^1H—^{13}C heteronuclear dipolar coupling strengths vary from 10 to 30 kHz. ^1H—^{13}C heteronuclear dipolar coupling can be eliminated through the use of proton decoupling. In heteronuclear decoupling, the protons are exposed to continuous high-power radio frequency (rf) irradiation, which eliminates (averages to zero) the ^1H heteronuclear couplings to the ^{13}C species.

Homonuclear dipolar coupling, such as ^1H—^1H, is described by an equation similar to that given in Equation (14.3). As in the case of heteronuclear dipolar coupling, homonuclear dipolar coupling is also dependent on the internuclear distance between the two nuclei, and the angle between the two nuclei and the z-axis of the external field. Homonuclear dipolar coupling strengths vary from as large as 100 kHz for ^1H—^1H dipolar couplings to 3 kHz for ^{13}C—^{13}C dipolar couplings. ^1H—^1H dipolar couplings typically are not averaged by MAS due to the magnitude of the coupling, although their effect on the ^{13}C MAS NMR spectrum is minimal.

An SSNMR experiment known as cross polarization (CP) is used to enhance the intensity of NMR signals from low-frequency nuclei, such as ^{13}C and ^{15}N, by exploiting the dipolar coupling of these nuclei to higher frequency nuclei, such as ^1H. CP is often explained using a thermodynamic picture, in which the abundant spins with a larger Larmor frequency, ^1H, are considered to have a cold spin temperature due to their large Boltzmann population factor and the ^{13}C (with a Larmor frequency four times less than ^{1}H) have a high-spin temperature [33]. Through an exchange of energy from the cold spins to the hot spins, the temperature of the ^{13}C spin system is decreased until the two systems approach a common spin temperature leading to an increase in the spin polarization of the ^{13}C spins. For the two systems to exchange energy, the ^{13}C spins must be coupled to the protons, and in solid samples the two are coupled via through space dipolar interactions. The difference in the Zeeman energies of the two systems does not allow energy exchange to occur easily in the laboratory frame. To solve this problem, and allow energy exchange to occur, a pair of rf fields is applied simultaneously at the resonance frequencies of the proton and ^{13}C spins such that their rf field-induced nutation rates are equal. The field strengths required to provide the energy matching condition satisfies the Hartmann–Hahn condition, given by

$$\omega_{1HH} = B_{1H}\gamma_{1H} = B_{1C}\gamma_{1C} \tag{14.4}$$

where B_{1H} and B_{1C} are the magnitudes of the rf magnetic fields applied to ^{1}H and ^{13}C, respectively. The polarization of the ^{13}C spin system can be increased to a maximum enhancement based on the ratio of their thermal Boltzmann polarizations, or γ_H/γ_C, which in the case of proton–carbon CP is a factor of 4. Faster ^{1}H relaxation compared to ^{13}C allows one to repeat the experiment more rapidly and provides additional time savings in the experiment. It should be noted that CP only works well when the molecular system is relatively stationary, so that CP also provides a tool to differentiate static and mobile species.

14.3 NMR METHODS

SSNMR studies of catalytic reactions are challenging for two main reasons. First, due to sensitivity concerns NMR is inherently a bulk technique and therefore requires a large sample. As a result, NMR surface studies are normally restricted to high surface area samples. In addition, to study catalytic surfaces, it is often necessary to isolate the experimental system from the atmosphere. In most cases, this typically requires the use of sealed samples.

It is of interest to study photocatalytic oxidation reactions *in situ* because the detection of intermediates and final products yield information about the active sites on the catalyst and information about the kinetics and mechanism of the photooxidation reaction may also be obtained. In order to study photocatalytic oxidation reactions using *in situ* SSNMR a number of criteria must be met. First, the sample must be irradiated inside the bore of the magnet, which is challenging. While it is possible to halt irradiation for each acquisition, this process is tedious when a reaction progression is desired. Second, the reaction conditions must be chosen. The photocatalytic oxidation reaction can be studied under either batch type or flowing conditions, however, ^{13}C-labeled reagents are expensive, therefore, batch-type conditions are preferable for *in situ* SSNMR experiments. Under batch conditions, the sample containing the catalyst and reactants is sealed. And finally, light must be transmitted through the walls of the rotor containing the sample. The *in situ* SSNMR experiments described below were performed using a home-built double resonance MAS NMR probe based on a design by Gay [35]. This design allows sealed glass NMR tubes to be spun at speeds up to 2.7 kHz, and a quartz rod light pipe was incorporated into the probe to bring the light to the sample through a 10-mm gap in the rf coil. There are also some commercial MAS probes that are capable of spinning sealed samples, however, light introduction is often inefficient in these probes.

Figure 14.2 shows apparatus for *in situ* SSNMR along with the optical–MAS probe head based on the design by Gay. Light from a 300 W Xe arc lamp (ILC Technology) is delivered to the sample region inside the magnet via a liquid-filled optical light guide (Oriel Corporation, Stratford, CT). The wavelength of light delivered to the sample is selected by the use of dichroic mirrors (Oriel Corporation) attached to the Xe arc lamp. A 70-mm-long polished

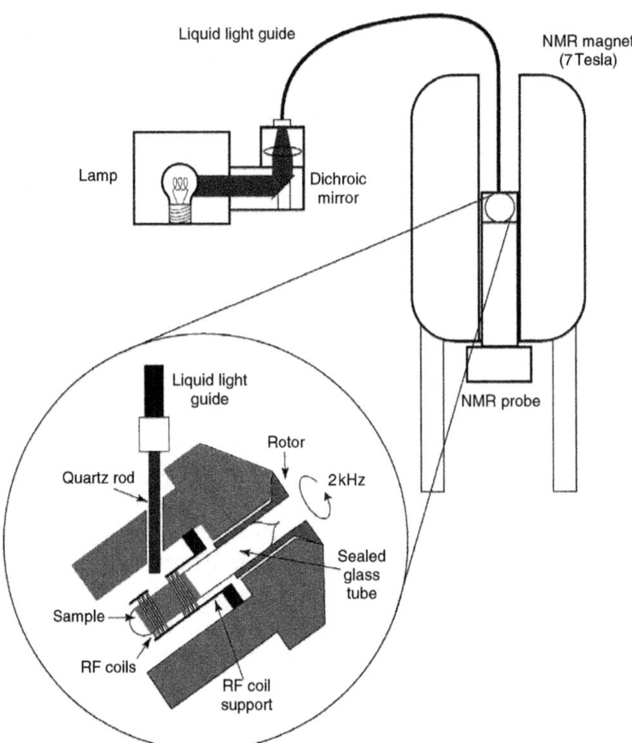

FIGURE 14.2 Schematic diagram of the *in situ* SSNMR apparatus and optical MAS NMR probehead.

suprasil quartz rod is attached to the end of the light guide to reduce rf pickup by the probe coil from the aluminum-encased light guide. The light from the quartz rod quickly diverges to cover the entire sample region. The near UV light power reaching the sample was measured to be 5 mW by standard ferrioxalate actinometry [36].

14.4 SAMPLE PREPARATION

Several considerations are important in the preparation of samples for *in situ* SSNMR studies. The catalyst must be small enough to fit inside an NMR tube, and the sample must be well balanced because it must spin stably at the magic angle. The sample should also be uniformly irradiated. Both semiconductor and zeolite photocatalysts are typically available as powders. These powders can easily be packed into an NMR tube, but both are effective light scatters, which leads to the development of dark regions in the interior of the sample where strongly bound reactants or intermediates will remain unaffected by irradiation [10].

Several different methods can be used to eliminate dark regions in the samples. The simplest solution is to coat a thin layer of the photocatalyst inside the NMR tube. However, this method greatly limits the amount of catalyst present in the sample, and makes the observation of surface species by NMR very difficult. Another option is to use supported photocatalysts. These supported catalysts must allow light to penetrate into the center of the sample, which limits the type of support that can be used. In particular, we have used two types of supported photocatalysts, monolayers of metal oxides deposited on transparent porous Vycor glass (PVG) and metal oxide or zeolite particles coated on optical microfibers. Detailed discussions of the synthesis of these supported photocatalysts are reported elsewhere [25–30]. Briefly, supported catalysts are prepared on a gas rack starting with PVG that is evaculated and calcined at high temperature. The supported catalyst is formed by chemical vapor deposition of a metal chloride or metal propoxide species and then gas-phase hydrolysis to form the supported metal oxide. Figure 14.3 shows an example of a monolayer metal oxide (V_2O_5) photocatalyst supported on PVG and sealed in a 5-mm NMR tube.

All samples are prepared using high-vacuum methods to reduce sample contamination. The catalyst is placed in an NMR tube and attached to a vacuum manifold. The catalyst is then pretreated by heating the catalysts under a vacuum of approximately 5×10^{-5} Torr, and then calcined under 1 atm of dry O_2. Once the heating cycle is complete, the catalyst is once again evacuated to a pressure of 5×10^{-5} Torr. ^{13}C-labeled organic compounds and O_2 are then loaded onto the catalyst using a liquid nitrogen trap, and the sample is flame sealed 10 to 12 mm above the catalyst. Additionally, the PVG catalysts are fitted with plastic endcaps that have an outer diameter that fits snugly against the inner wall of the NMR tube. The plastic endcaps allow samples prepared with the PVG catalysts to be spun at speeds as high as 2.7 kHz.

Optical fiber-supported catalysts are prepared by solution phase bonding using a suspension of the photocatalyst particles in a surfactant such as Triton X-100 (Aldrich). After several dipping and calcining steps, a bundle of particle-coated optical fibers were loaded into a 5-mm NMR tube and sample preparation preceded as for the other catalysts [24–26].

14.5 SSNMR STUDIES OF SURFACE SPECIES AND PHOTOOXIDATION REACTIONS ON TiO_2

In heterogeneous photocatalysis, reactions occur at the surface of a catalyst, therefore, it is important to understand the interaction of the reactants with the surface. SSNMR provides detailed information on a number of processes important to the understanding of photocatalytic oxidation. As discussed below, the use of probe molecules is useful to understand the adsorption of molecules on the photocatalyst and to identify the active sites on the catalyst. The effect of different catalyst morphologies on the formation of surface species and

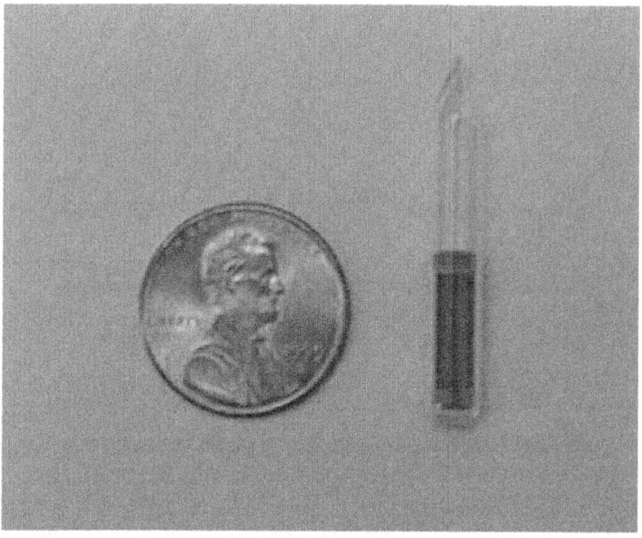

FIGURE 14.3 Photograph of V_2O_5 photocatalyst supported on PVG. (Reproduced from S Pilkenton, W Xu, and D Raftery. *Anal Sci* 17:125–130, 2001. Copyright 2001, The Japan Society for Analytical Chemistry. With permission.)

reactivity is also addressed. Finally, the formation and reactivity of surface-bound intermediates is also discussed.

14.5.1 Adsorption and Reactivity of Ethanol on TiO_2

The local coordination environments of the binding sites on both polycrystalline and single-crystal TiO_2 surfaces have been widely studied using the adsorption and thermal desorption or decomposition of ethanol, which is an often used as model reactant [37–44]. The presence of two distinct adsorption patterns of ethanol on TiO_2, hydrogen bonding of ethanol to surface hydroxyl groups and the dissociation of ethanol at coordinatively unsaturated Ti atoms to form surface-bound Ti-ethoxide species ($CH_3CH_2O^-$), has been observed using both infrared and TPD experiments [38–44]. SSNMR experiments, in particular the CP experiment discussed above, have also been used to study surface ethanol species [25, 28]. The CP–MAS NMR experiment is significantly influenced by molecular mobility while both mobile and surface-bound species are observed in the standard pulse acquire (Bloch decay) ^{13}C MAS NMR spectra, therefore, chemisorbed molecules can be easily distinguished from physisorbed or gaseous molecules using this approach. Figure 14.4 shows ethanol adsorbed on several different TiO_2 catalysts and pure PVG. There is a single resonance at approximately 17 ppm corresponding to the methyl carbon of ethanol, however, there are two resonances corresponding to the ethanol methylene carbon on the TiO_2 catalysts. The upfield resonance has been identified as the hydrogen-bonded ethanol species while the downfield resonance corresponds to the Ti-bound ethoxide species [25]. The chemical shifts of the Ti-ethoxide species vary from 67 to 78 ppm on the different catalysts, indicating that the electronic environments on each of the catalysts are different. ^{13}C chemical shift values of these species and others are included in Table 14.1.

Reactivity studies of the surface ethanol species using *in situ* SSNMR identified the active sites on the TiO_2 surface [25]. It was found that the intensity of the Ti-ethoxide resonance (~78 ppm) decreased with time while the resonance corresponding to the hydrogen-bonded ethanol species (59.4 ppm) remained relatively unchanged. It was concluded that the ethoxide is the photochemically active surface ethanol species. The coordinatively unsaturated Ti sites, which form ethoxide species upon addition of ethanol, are the photochemically active sites on the TiO_2–PVG catalyst, and their presence is critical for the effective photooxidation of ethanol.

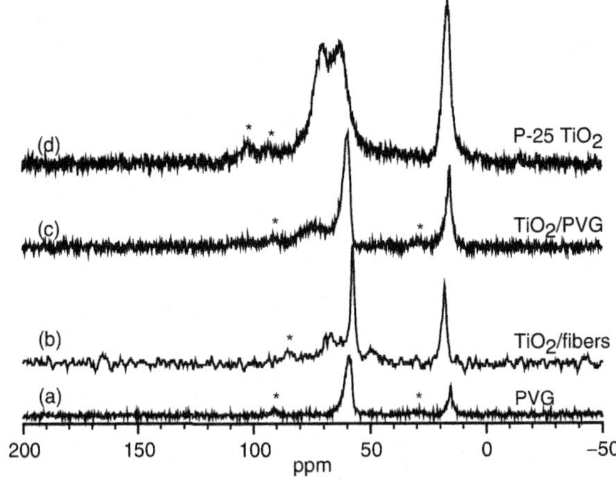

FIGURE 14.4 Proton-decoupled ^{13}C CP–MAS NMR spectra of ethanol adsorbed on different surfaces: (a) PVG, (b) TiO_2–optical microfibers, (c) TiO_2–PVG, and (d) Degussa P-25 TiO_2 powder. Note: asterisks represent spinning sidebands. (Reproduced from S Pilkenton, S-J Hwang, and D Raftery. *J Phys Chem B* 103:11152–11560, 1999. Copyright 1999, American Chemical Society. With permission.)

TABLE 14.1
^{13}C Chemical Shifts for Species Observed in Solid-State NMR Photocatalytic Studies

Chemical	Chemical Formula	^{13}C Chemical Shift (ppm)
Ethanol	CH_3CH_2OH	17.0, 56.9
Ti-ethoxide	$CH_3CH_2O—Ti$	16.1, 69.9 (Degussa P-25 TiO_2)
		17, 72 (Hombikant UV-100 TiO_2)
		17, 78 (Single monolayer TiO_2/PVG)
		17, 73 (Two monolayer TiO_2/PVG)
		17, 73 (Four monolayer TiO_2/PVG)
		18, 67 (TiO_2/optical microfibers)
Hydrogen-bonded ethanol	CH_3CH_2OH	16.1, 62.2 (Degussa P-25 TiO_2)
		17, 62.7 (Hombikant UV-100 TiO_2)
		17, 59.4 (Single monolayer TiO_2/PVG)
		17, 60.6 (Two monolayer TiO_2/PVG)
		17, 62.1 (Four monolayer TiO_2/PVG)
		18, 57.4 (TiO_2/optical microfibers)
Acetaldehyde	CH_3CHO	28.8, 198.7
Ti-acetate		21.9, 179.0
1,1-Diethoxyethane	$CH_3CH(OC_2H_5)_2$	14.5, 22.5, 61.6, 93.8
Acetic acid	CH_3COOH	19.2, 171.2
Formic acid	$HCOOH$	161.2
Carbon dioxide	CO_2	124.0
Ti-2-propoxide	$(CH_3)_2CO—Ti$	76.2 (Degussa P-25 TiO_2)
		78 (Hombikant UV-100 TiO_2)
Hydrogen-bonded 2-propanol	C_3H_8O	64.4 (Degussa P-25 TiO_2)
		70 (Hombikant UV-100 TiO_2)
		64.4 (TiO_2/PVG catalysts)
Mesityl oxide	$C_6H_{10}O$	19.3, 26.6, 123.9, 153.5, 197.7
Isobutylene	C_4H_8	140
Acetone	C_3H_6O	29.3, 205.5 (Degussa P-25 TiO_2)
		28.0, 215.0 (TiO_2/PVG catalysts)
Diacetone alcohol	$C_6H_{11}O_2$	54.7, 69.6
Propylene oxide	C_3H_6O	50
Ti-formate	$HCOO—Ti$	167 (Degussa P-25 TiO_2)
Carbon monoxide	CO	183.0
Trichloroethylene	C_2HCl_3	116.7, 124.0
Dichloroacetyl chloride	C_2HCl_3O	70.0, 167.5
Phosgene	CCl_2O	143.0, 150 (Degussa P-25 TiO_2)
Pentachloroethane	C_2HCl_5	79.5, 100.0
Dichloroacetic acid	$C_2H_2Cl_2O_2$	63.0, 167.5
Dichloroacetate	$Cl_2HCCOO—Ti$	64.0, 177.3 (Degussa P-25 TiO_2)
	$Cl_2HCCOO—Ti$	63.7, 175.0 (TiO_2/PVG)
	$Cl_2HCCOO—Si$	63.7, 162.1 (TiO_2/PVG)
Oxalyl chloride	$C_2Cl_2O_2$	159.0
Trichloroacetaldehyde	C_2HCl_3O	93.0, 177.0
Trichloroacetyl chloride	C_2Cl_4O	94.0, 164.0
Trichloroacetic acid	$C_2HCl_3O_2$	88.0, 164.0
Trichloroethan-2-ol		72.7, 84.7
Trichloroacetate	$Cl_3CCOO—Ti$	92.1, 171.0 (TiO_2/PVG)
	$Cl_3CCOO—Si$	89.2, 159.1 (TiO_2/PVG)
Dichloromethane	CH_2Cl_2	58.5
Trichloromethane	$CHCl_3$	78.5

It is also important to consider the effect of the presence of other molecules on the formation and reactivity of surface species. One molecule to consider in particular is water. Figure 14.5 shows ethanol adsorbed on dehydrated and partially hydrated Degussa P-25 TiO_2 powder. The presence of water inhibits the formation of the Ti-bound ethoxide species. *In situ* irradiation of ethanol on hydrated TiO_2 powder and hydrated TiO_2–PVG catalysts show that the presence of water retards the rate of oxidation of ethanol because ethanol and water compete for the active site on the catalyst [25].

14.5.2 The Effect of Surface Morphology

The difference in the surface morphology between bulk and supported catalysts often affects the activity of the catalyst toward the oxidation of a particular reactant. This effect can be seen in the study of the absorption and photooxidation of acetone and 2-propanol on TiO_2 powder and supported TiO_2 catalysts. The surface species and catalytic activities of five different TiO_2 catalysts were used in this study: Hombikat UV-100 TiO_2 powder, Degussa P-25 TiO_2 powder, a single monolayer of TiO_2 deposited on PVG (TiO_2–PVG), as well as, two monolayer and four monolayer TiO_2–PVG catalysts. The Hombikat UV-100 TiO_2 powder is 100% anatase while the Degussa P-25 TiO_2 powder is 70% anatase and 30% rutile. The one- and two-monolayer TiO_2–PVG catalysts are amorphous, while the four-monolayer TiO_2–PVG catalyst has an anatase structure as determined by x-ray diffraction [30].

Figure 14.6 shows the ^{13}C CP–MAS NMR spectrum of 2-propanol absorbed on the one-, two-, and four-monolayer TiO_2–PVG catalysts and pure anatase TiO_2 powder. Two surface species, a hydrogen-bonded 2-propanol species and a Ti-bound 2-propoxide species formed at coordinatively unsaturated Ti sites, are present on the anatase powder paralleling what occurs for ethanol adsorption. However, only the hydrogen-bonded 2-propanol species is present on the monolayer TiO_2–PVG catalysts, which leads to a much lower reaction rate. The 2-propoxide species does not form on the TiO_2–PVG catalysts for a number of reasons including steric effects, lower defect site availability on the TiO_2–PVG catalysts, and differences in local coordination environments [30]. In contrast to the 2-propanol studies, studies of acetone adsorption on the same catalysts [31] reveal a surface species with a chemical shift of 50 ppm, propylene oxide, that is only present on the TiO_2–PVG catalysts (see Figure 14.7).

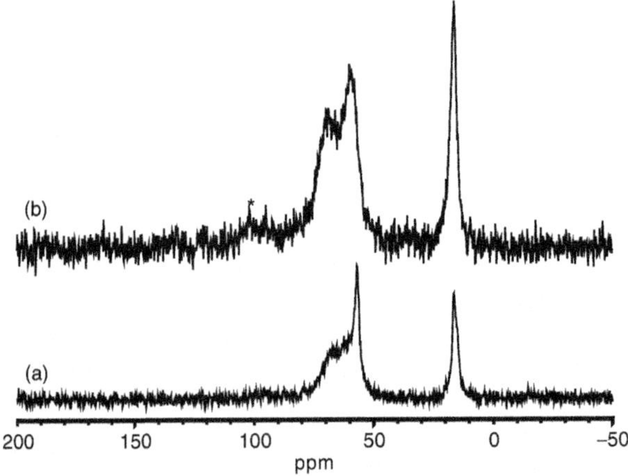

FIGURE 14.5 Proton-decoupled ^{13}C CP–MAS NMR spectra of 48 μmol of ethanol and 96 μmol of O_2 loaded onto (a) hydrated and (b) dehydrated Degussa P-25 TiO_2 powder. (Reproduced from S Pilkenton, S-J Hwang, and D Raftery. *J Phys Chem B* 103:11152–11560, 1999. Copyright 1999, American Chemical Society. With permission.)

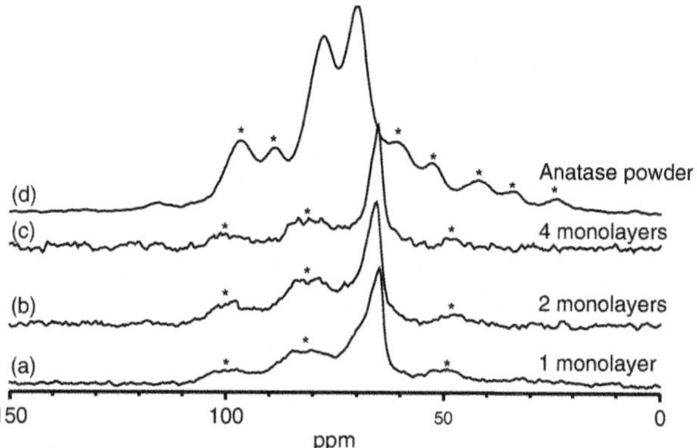

FIGURE 14.6 Proton-decoupled ^{13}C CP–MAS NMR spectra of 35 μmol of 2-propanol and 96 μmol of O_2 loaded onto (a) one-, (b) two-, and (c) four-monolayer TiO_2—PVG catalysts, and (d) pure anatase TiO_2 powder. (Reproduced from S Pilkenton, W Xu, and D Raftery. *Anal Sci* 17:125–130, 2001. Copyright 2001, The Japan Society for Analytical Chemistry. With permission.)

The different surface morphologies and subsequent formation of different surface species upon adsorption of a reactant have a dramatic effect on the rate of photocatalytic oxidation. The photocatalytic oxidation of 2-propanol on Degussa P-25 TiO_2 powder proceeds much more rapidly than on any of the monolayer TiO_2–PVG catalysts due to the presence of the Ti-bound 2-propoxide species. The 2-propoxide species is quickly oxidized to CO_2 while the photooxidation of the hydrogen-bonded 2-propanol proceeds through an acetone intermediate that is then slowly oxidized to CO_2 [32]. Scheme 14.1 and Scheme 14.2 show the mechanisms for the photocatalytic oxidation of 2-propanol on TiO_2 powder and TiO_2–PVG catalysts, respectively.

In contrast to the results observed for 2-propanol, the photooxidation of acetone proceeds faster on the TiO_2–PVG catalysts than the TiO_2 powder due to the formation of the propylene oxide surface species [31]. On TiO_2–PVG catalysts, propylene oxide is quickly oxidized to acetic acid and then to CO_2 and CO, while acetone that is hydrogen-bonded to the catalyst is

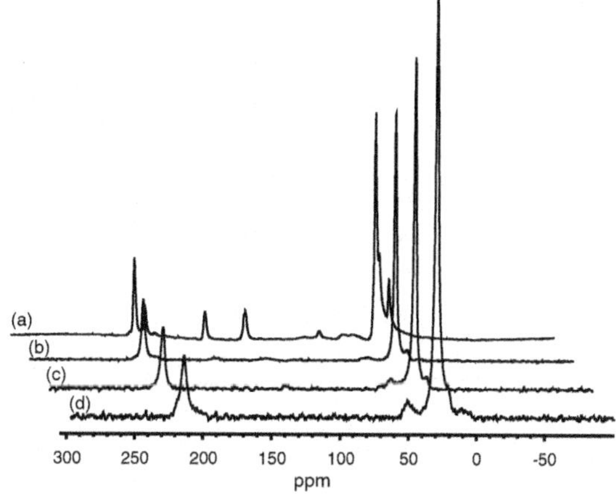

FIGURE 14.7 Proton-decoupled ^{13}C MAS NMR spectra of 32 μmol of acetone and 96 μmol O_2 loaded onto (a) Degussa P-25 TiO_2 powder, and the (b) one-, (c) two-, and (d) four-monolayer TiO_2—PVG catalysts. (Reproduced from W Xu and D Raftery. *J Catal* 204:110–117, 2001. Copyright 2001, Elsevier.)

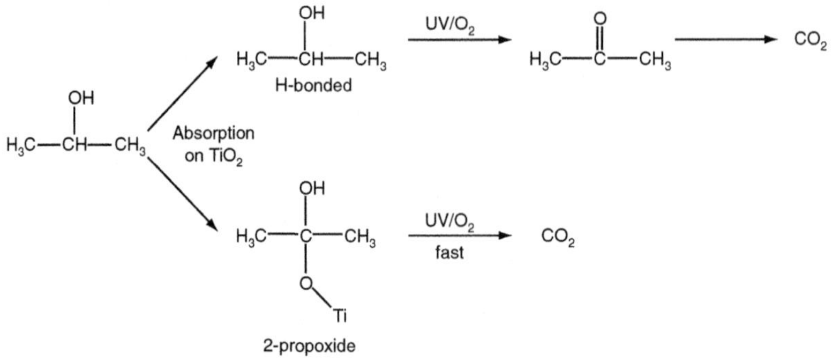

SCHEME 14.1

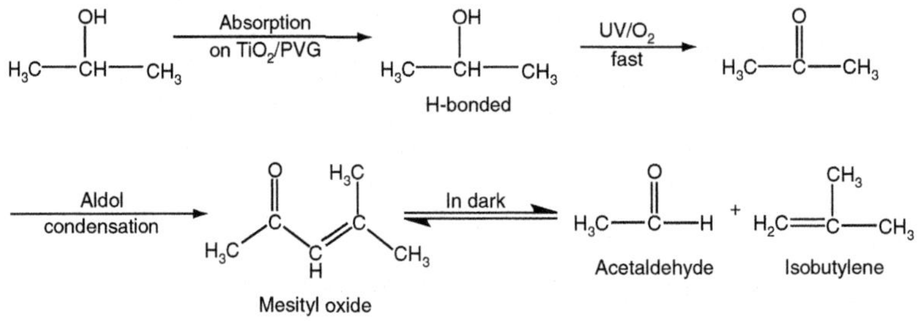

SCHEME 14.2

slowly oxidized to CO_2. Since the only type of acetone present on the TiO_2 powder is the hydrogen-bonded acetone species, the photooxidation reaction proceeds very slowly on the powdered catalyst. Scheme 14.3 and Scheme 14.4 show the mechanisms for photocatalytic oxidation of acetone using the TiO_2 powder and TiO_2–PVG catalysts, respectively. ^{13}C chemical shifts are shown in Table 14.1.

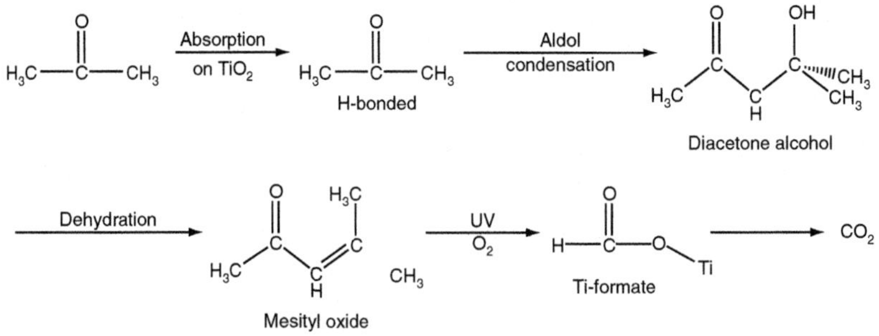

SCHEME 14.3

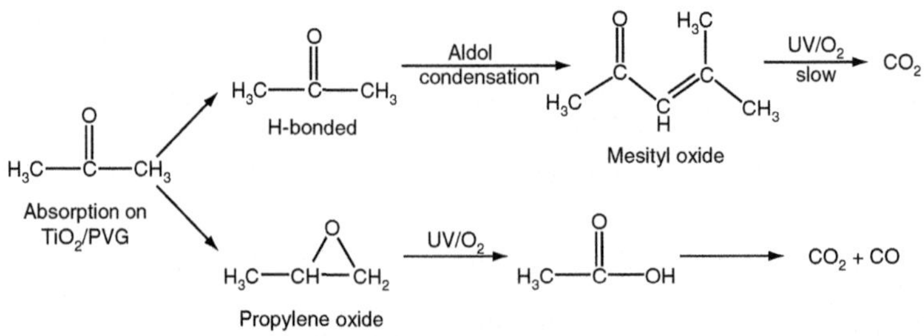

SCHEME 14.4

14.5.3 Formation and Characterization of Surface-Bound Intermediates During PCO

In the previous sections, we have focused on the formation of surface species upon the adsorption of reactants on TiO_2 catalysts. SSNMR is also useful in following: the formation of surface-bound intermediates, in particular, those formed during the photooxidation. For example, *in situ* SSNMR methods were used to follow the PCO of trichloroethylene (TCE) on Degussa P-25 TiO_2 powder and a single monolayer TiO_2–PVG catalyst [10]. *In situ* SSNMR studies of the photocatalytic oxidation of TCE on TiO_2 powder showed the formation of several long-lived volatile intermediates, including carbon monoxide (CO), dichloroacetyl chloride ($Cl_2CHCOCl$, DCAC), phosgene (CCl_2O), and pentachloroethane (C_2HCl_5), and their conversion to the final product CO_2 (Figure 14.8a). However, the carbon balance obtained by integrating the peak areas of the spectra shown in Figure 14.8a indicated a significant loss of signal (up to 50%). This loss in signal is due to the formation of surface-bound intermediates, which have a very low intensity due to their broad peaks in the ^{13}C Bloch Decay NMR spectra. Figure 14.8b shows the results of ^{13}C CP-MAS NMR experiments that reveal the presence of a surface-bound intermediate, dichloroacetate (Cl_2CHCOO^-), which forms from the reaction of DCAC with surface hydroxyl groups.

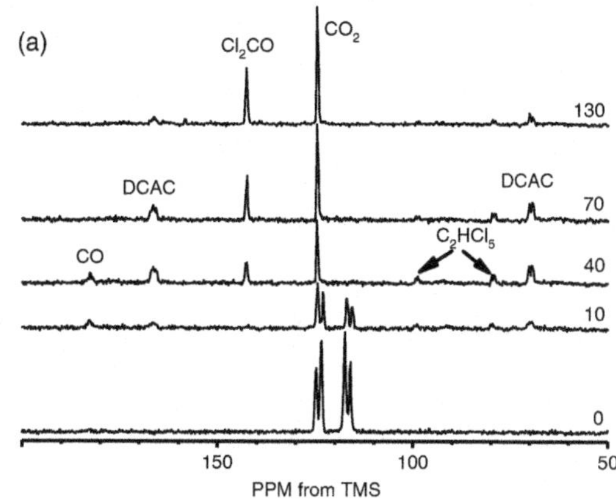

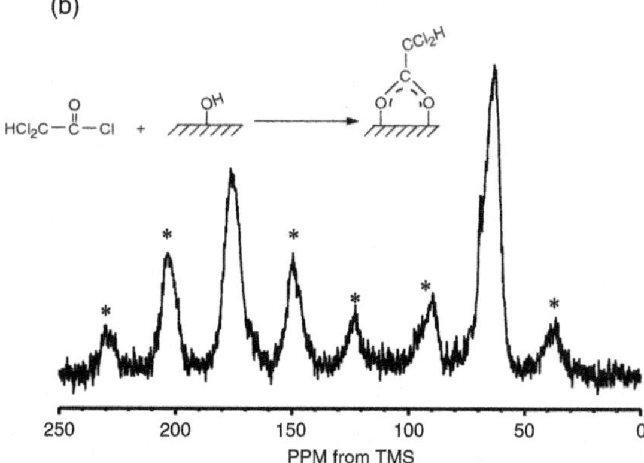

FIGURE 14.8 (a) Proton-decoupled ^{13}C MAS NMR spectra obtained during the *in situ* irradiation of 48 μmol of TCE and 60 μmol of O_2 on Degussa P-25 TiO_2 powder. (b) ^{13}C CP–MAS spectrum recorded after the UV light was turned off. (Reproduced from reference S-J Hwang, C Petucci, and D Raftery. *J Am Chem Soc* 119:7877–7878, 1997. Copyright 1997, American Chemical Society. With permission.)

This surface-bound intermediate did not oxidize further upon subsequent UV irradiation because DCAC could migrate to dark regions of the packed powder catalyst and react where it was inaccessible to UV irradiation.

In order to determine whether the surface-bound dichloroacetate could be photooxidized, a catalyst without dark regions was employed that consisted of a monolayer of TiO_2 dispersed on transparent PVG [29]. The photooxidation of TCE on the TiO_2–PVG catalysts produces the intermediates DCAC, oxalyl chloride ($Cl_2C_2O_2$), pentachloroethane, trichloroacetaldehyde (Cl_3CCHO), and trichloroacetalchloride (Cl_3CCOCl), and the final products CO_2 and phosgene. ^{13}C CP-MAS NMR spectra revealed the presence of two surface-bound intermediates, dichloroacetate and trichloroacetate (Cl_3CCOO^-), which are bound to the surface of the TiO_2–PVG catalyst through chemical bonds to both Si and Ti sites. Upon further UV irradiation, dichloroacetate was slowly photooxidized to form phosgene and CO_2. Although dichloroacetate could be completely oxidized, the other surface species found on the TiO_2–PVG catalyst, trichloroacetate, was very resistive to further destruction.

14.6 EVALUATION OF NEW SEMICONDUCTOR PHOTOCATALYSTS WITH SSNMR

In situ SSNMR can also be used to investigate the activity of other semiconductor photocatalysts. Not only can one use surface probe molecules to investigate reactivity, but multinuclear studies can be used to examine reactive sites on the catalyst surface directly. In this section, we will describe the study of new supported semiconductor photocatalysts including a supported TiO_2 photocatalyst and two mixed metal oxide photocatalysts.

14.6.1 TiO_2-Coated Optical Microfibers

As stated previously, the light scattering properties of TiO_2 make it difficult to irradiate the interior of the packed bed reactor using our *in situ* SSNMR experiments. To eliminate the dark regions, partially transparent supported catalysts are used. An alternative to the TiO_2–PVG catalysts described in the previous section is TiO_2 supported on optical microfibers. These catalysts were prepared by removing the cladding from 9 μm quartz optical fibers, then coating the fibers with a suspension of TiO_2 powder [24, 25]. Figure 14.9 shows SEM images of the uncoated and TiO_2-coated optical microfibers. The specific surface area of the catalyst was measured to be $7.4 \pm 0.1\,m^2\,g^{-1}$.

The photocatalytic activity of the TiO_2–optical microfiber catalysts was evaluated by observing the *in situ* photooxidation of TCE and ethanol using SSNMR. As in the photooxidation reactions observed on the other TiO_2 photocatalysts, the photooxidation of TCE using the optical microfiber catalyst results in the formation of DCAC, dichloroacetic acid, oxalyl chloride, trichloroacetaldehyde, and the final products — CO_2 and phosgene. ^{13}C chemical shifts of these species are provided in Table 14.1. The surface-bound intermediate, dichloroacetate, was completely photooxidized using the optical fiber catalysts demonstrating that no dark regions were present in the catalyst.

The photooxidation of ethanol on the TiO_2–optical microfiber catalysts was also studied using *in situ* SSNMR experiments [25]. These studies show that ethanol is photooxidized faster on the optical microfiber catalysts than the TiO_2-packed powder and TiO_2–PVG catalysts due to the ability to illuminate the center of the sample and the increased activity of the P-25 active sites compared to the monolayer TiO_2 sites. Also, the rate of photocatalytic oxidation is also enhanced by the reoccupation of defect sites left vacant after the photooxidation of ethoxide species with new ethanol molecules from the gas phase due to the multiple monolayers of ethanol present in the sample. A new intermediate, 1,1-diethoxyethane ($CH_3CH(OC_2H_5)_2$) was also observed in the photooxidation of ethanol on the TiO_2 optical

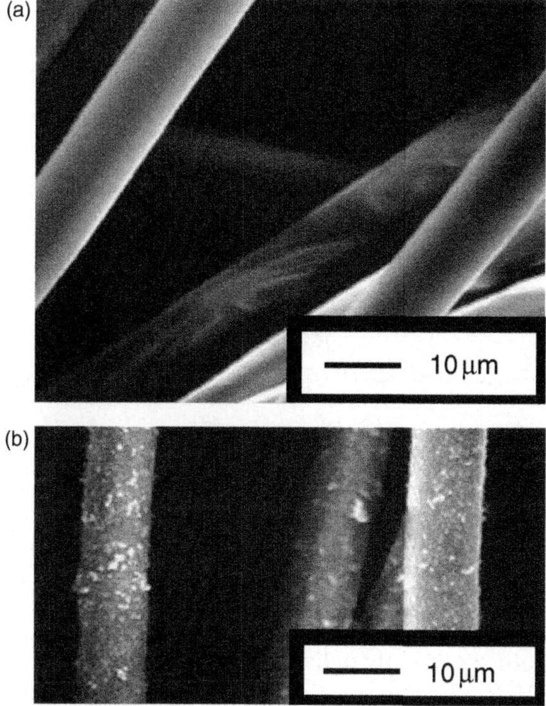

FIGURE 14.9 SEM micrographs of (a) uncoated and (b) TiO_2-coated optical microfibers. (Reproduced from S Pilkenton, S-J Hwang, and D Raftery. *J Phys Chem B* 103:11152–11560, 1999. Copyright 1999, American Chemical Society. With permission.)

microfiber photocatalysts that results from the multiple monolayer coverages of ethanol used in the experiment (see Figure 14.10).

14.6.2 V-Doped TiO_2 Photocatalyst

Due to the size of the TiO_2 bandgap (3.2 eV), ultraviolet light must be used to use TiO_2 as a photocatalyst. Only 2% of sunlight is composed of ultraviolet light, and so there has been an interest in extending the wavelength range of TiO_2 into the visible region. A number of groups have approached this problem by doping TiO_2 with metals such as chromium, platinum, vanadium, iron, molybdenum, osmium, rhenium, or rhodium [45–50]. A more recent approach to the preparation of visible photocatalysts uses the ion bombardment of TiO_2 with nitrogen [51]. We approached this problem by preparing two simple V-doped TiO_2 catalysts supported on PVG: The first was one monolayer of TiO_2 supported on a 0.5 wt% V–PVG catalyst (TiO_2–V–PVG), and a the second was 1.7 wt% V loading supported on one monolayer of TiO_2 on PVG (V–TiO_2–PVG). Both of these catalysts were prepared using simple wet chemical or CVD methods [27]. Their photocatalytic activities under both UV and visible irradiation were studied with SSNMR.

UV–visible absorption spectra of the V-doped TiO_2 photocatalysts exhibited a redshift into the visible spectral region compared to the undoped catalyst. ESR and ^{51}V SSNMR studies indicated that the active vanadium species present in the V-doped TiO_2 catalysts was a V^{4+} species. SSNMR studies of ethanol photooxidation on the TiO_2–V–PVG and V–TiO_2–PVG after 1 h of ex situ irradiation under visible light ($\lambda > 396$ nm) showed that

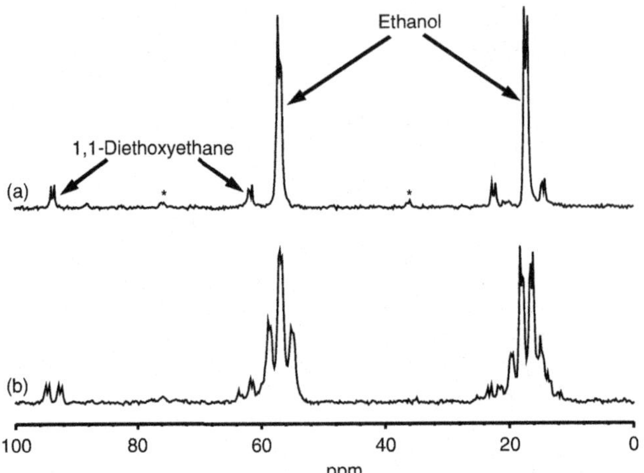

FIGURE 14.10 ^{13}C MAS NMR spectra showing the formation of 1,1-diethoxyethane during the photocatalytic oxidation of ethanol on the TiO_2—optical microfiber catalyst: (a) proton decoupled and (b) proton coupled. (Adapted from S Pilkenton, S-J Hwang, and D Raftery. *J Phys Chem B* 103:11152–11560, 1999. Copyright 1999, American Chemical Society. With permission.)

both catalysts were photocatalytically active with the V–TiO_2–PVG catalyst being less efficient than the TiO_2–V–PVG catalyst due to the presence of more V^{5+} sites on the V–TiO_2–PVG catalyst as observed with ^{51}V SSNMR (Figure 14.11b and e). Both catalysts are also photoactive under UV irradiation with photocatalytic activities comparable to the single monolayer TiO_2–PVG photocatalyst with the V–TiO_2–PVG photocatalyst, slightly less efficient than the TiO_2–V–PVG catalyst (Figure 14.11c and f).

14.6.3 Mixed SnO_2–TiO_2 Catalysts

There has also been interest in increasing the photocatalytic activity of TiO_2 under UV irradiation by doping TiO_2 with other metal oxides. TiO_2–SnO_2-coupled thin film coated

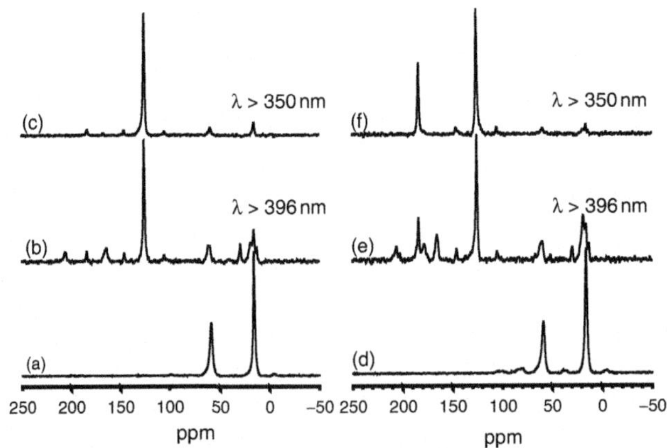

FIGURE 14.11 Proton-decoupled ^{13}C MAS NMR spectra of ethanol on the TiO_2–V–PVG catalyst (a) before, and after 1 h irradiation with (b) visible and (c) UV light. The corresponding spectra acquired using the V–TiO_2–PVG catalyst (d) before, and after 1 h irradiation with (e) visible and (f) UV light. (Adapted from S Klosek and D Raftery. *J Phys Chem B* 105:2815–2819, 2001. Copyright 2001, American Chemical Society. With permission.)

electrodes prepared using dispersions of TiO_2 and SnO_2 particles have been shown to be more efficient photocatalysts in the oxidation of azo dyes [52]. SnO_2-doped TiO_2 catalysts prepared using sol–gel methods [53, 54] and homogeneous precipitation methods [55] have also been shown to exhibit a higher photocatalytic activity in the photooxidation of VOCs and dyes. To study the reactivity of these mixed metal oxide photocatalysts using SSNMR, three SnO_2-doped TiO_2 monolayer photocatalysts were prepared using chemical vapor deposition of the metal chlorides on PVG: The first was a monolayer of SnO_2 deposited on a monolayer TiO_2–PVG catalyst (SnO_2–TiO_2–PVG), the second was TiO_2 deposited on a monolayer SnO_2–PVG catalyst (TiO_2–SnO_2–PVG), and the third consisted of SnO_2 and TiO_2 codeposited on a PVG substrate (SnO_2/TiO_2–PVG).

The photocatalytic activity of this family of photocatalysts was studied using *in situ* SSNMR (Figure 14.12). The photooxidation of ethanol on the SnO_2-doped TiO_2 photocatalysts shows that the SnO_2–TiO_2–PVG catalyst is much less reactive than the other photocatalysts. The TiO_2–SnO_2–PVG catalyst is slightly less reactive than the TiO_2–PVG catalyst while the SnO_2/TiO_2–PVG has a comparable activity to the TiO_2–PVG photocatalyst. ^{119}Sn SSNMR was used to confirm the presence of tin on each of the mixed metal oxide photocatalysts, and the ^{119}Sn NMR studies revealed that the oxidation state of the tin species on the photocatalysts was +4 [56].

14.7 SSNMR STUDIES OF ZEOLITE PHOTOCATALYSTS

Zeolite photochemistry has attracted a significant amount of interest because the transformations of organic material within the cavities and channels of the zeolite result in product distributions considerably different from those in solution [22, 57, 58]. When irradiated with visible light, large-pore alkali or alkaline-earth zeolites have shown to be very selective photocatalysts for the partial oxidation of small alkanes, olefins, and alkylbenzenes [59, 60].

Haw and coworkers studied the photoreaction of benzaldehyde or acetophenone with toluene on two FAU-type zeolites, ultrastable-Y (USY), and NaY under *ex situ* UV irradiation to understand the effect of zeolite acidity on the photoproducts produced [23]. The photoreaction between benzaldehyde and toluene on NaY results in the formation of 1,2-diphenylethanol. However, 1,2-diphenylethanol is unstable in the acidic environment on zeolite USY and

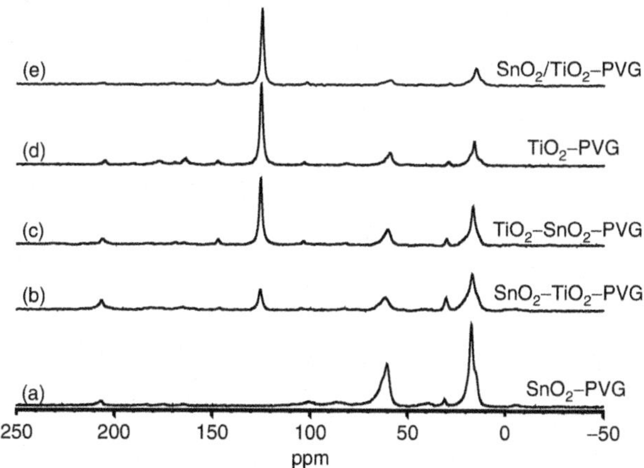

FIGURE 14.12 Proton-decoupled ^{13}C MAS NMR acquired after 5.5 h of UV irradiation of 48 μmol ethanol and 96 μmol O_2 on (a) SnO_2–PVG, (b) SnO_2–TiO_2–PVG, (c) TiO_2–SnO_2–PVG, (d) TiO_2–PVG, and (e) SnO_2/TiO_2–PVG. (Reproduced from S Pilkenton and D Raftery. *Solid State Nucl Magn Reson* 24:236–253, 2003. With permission.)

undergoes a Friedel–Crafts reaction with another toluene molecule to produce isomeric triphenylethane products (Figure 14.13). The products formed in the photoreaction between acetophenone and toluene were also affected by the acidity of the zeolite.

As in the case of semiconductor powders, zeolites also scatter light, therefore, it is advantageous to prepare supported zeolite photocatalysts. We have prepared BaY zeolite coated 9 μm optical microfibers using sol–gel methods, which allow *in situ* investigation [26]. Figure 14.14 shows SEM images of the zeolite-coated optical fiber catalyst. The photocatalytic activity of the supported zeolite photocatalyst under irradiation with visible light was compared to that of the powdered zeolite catalyst using *in situ* SSNMR. The photocatalytic activities of the supported and powdered photocatalysts were compared by observing the photooxidation of TCE and dichloromethane (CH_2Cl_2) [26]. Quantitative capabilities of SSNMR allow a direct means to determine the product yields in these reactions. In the case of TCE oxidation, 78% of the TCE present was oxidized with the zeolite-coated optical fiber catalyst after 100 min of irradiation resulting in the formation of dichloroacetyl chloride ($Cl_2CHCOCl$, DCAC), oxalyl chloride (ClCOCOCl, OC), trichloroethan-2-ol (Cl_2CHCH-ClOH, TCEOH), phosgene (CCl_2O), trichloroactaldehyde (CCl_3CHO, TCAA), and carbon monoxide (CO), while only 34% of the TCE present in the sample was oxidized after 533 min of irradiation using the powdered zeolite. ^{13}C chemical shifts of these species are provided in Table 14.1. Similarly, the photooxidation of methylene chloride was 20 times slower when the powdered zeolite catalyst was used compared to the supported zeolite photocatalyst. Figure 14.15 shows the ^{13}C MAS NMR spectra obtained before and after the photooxidation of

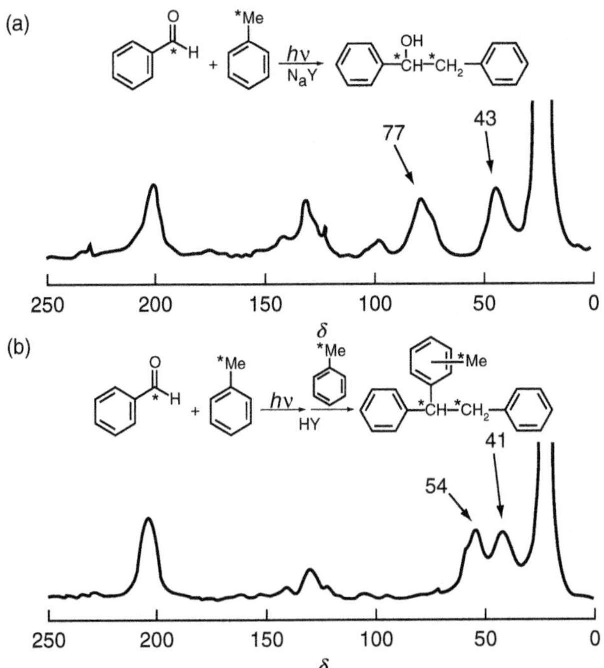

FIGURE 14.13 Proton-decoupled ^{13}C CP–MAS NMR spectra of the photoproducts produced after 4 h of UV irradiation of benzaldehyde and toluene on (a) NaY and (b) USY zeolites. (Reproduced from J Zhang, TR Krawietz, TW Skloss, and JF Haw. *Chem Commun* 685–686, 1997. Copyright 1997, Royal Society of Chemistry. With permission.)

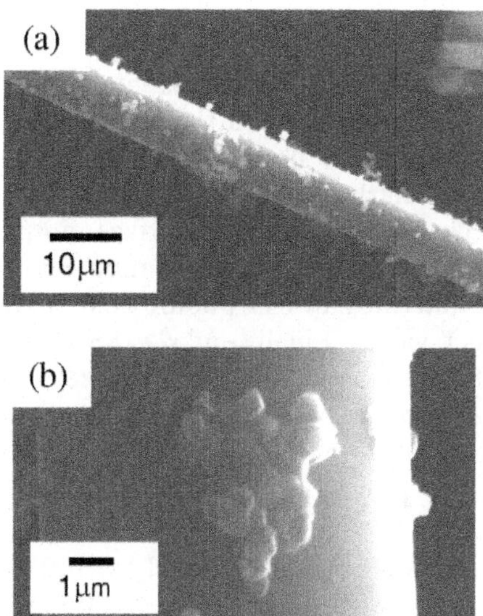

FIGURE 14.14 (a) SEM images of BaY zeolite-coated optical microfibers and (b) a magnified image of (a). (Reproduced from AR Pradhan, MA Macnaughtan, and D Raftery. *J Am Chem Soc* 122:404–405, 2000. Copyright 2000, American Chemical Society. With permission.)

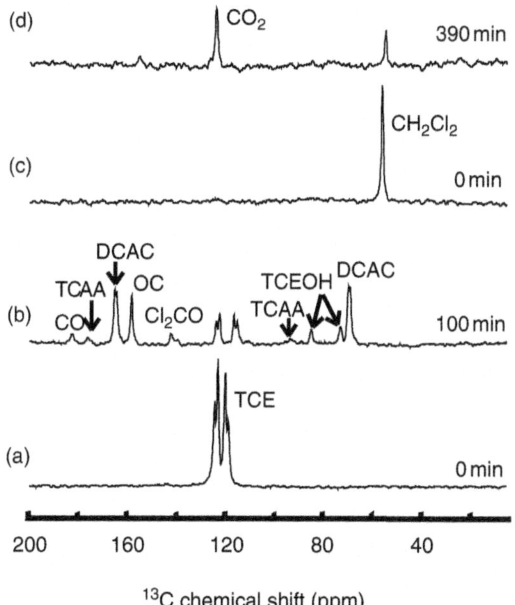

FIGURE 14.15 Proton-decoupled ^{13}C MAS NMR spectra of TCE and O_2 on the BaY-coated optical microfibers (a) before and (b) after 100 min of irradiation, and CH_2Cl_2 and O_2 on the BaY-coated optical microfibers (c) before and (d) after 390 min of irradiation. (Reproduced from AR Pradhan, MA Macnaughtan, and D Raftery. *J Am Chem Soc* 122:404–405, 2000. Copyright 2000, American Chemical Society. With permission.)

TCE and methylene chloride using the zeolite-coated optical microfibers. Optical microfibers were shown to be efficient and robust supports for carrying out photocatalytic reactions using both TiO_2 and zeolite-based photocatalysts [24, 25, 61].

14.8 CONCLUSIONS

SSNMR experiments provide both new and complementary information to that provided by other commonly used analytical methods, and therefore allows a better understanding of photocatalytic oxidation reactions on both semiconductor and zeolite photocatalysts. SSNMR experiments are particularly useful in acquiring information about the identity and reactivity of surface species and the effect of surface morphology on the formation of these species. *In situ* SSNMR studies of photooxidation reactions are also useful in the discovery of new long-lived intermediates in photooxidation reactions, and to determine reaction kinetics and product yields. Finally, *in situ* SSNMR studies of photocatalyzed reactions on new photocatalysts are useful in providing mechanistic information to enhance our ability to understand and improve photocatalytic oxidation reactions.

ACKNOWLEDGMENTS

Support for this work was provided by the National Science Foundation (CHE 97-33188, CAREER Grant), the donors of the Petroleum Research Fund, administered by the American Chemical Society, and the A.P. Sloan Foundation. The research contributions of current and former graduate students and postdoctoral researchers are also gratefully acknowledged.

REFERENCES

1. MR Hoffman, ST Martin, W Choi, and DW Bahnemann. *Chem Rev* 95:69–96, 1995.
2. AL Linsebigler, G Lu, and JT Yates Jr. *Chem Rev* 95:735–758, 1995.
3. MA Fox and MT Dulay. *Chem Rev* 93:341–357, 1993.
4. A Hagfeldt and M Grätzel. *Chem Rev* 95:49–68, 1995.
5. J Fan and JT Yates Jr. *J Am Chem Soc* 118:4686–4692, 1996.
6. LA Phillips and GB Raupp. *J Mol Catal* 77:297–311, 1992.
7. MD Driessen, AL Goodman, TM Miller, GA Zaharias, and VH Grassian. *J Phys Chem B* 102:549–556, 1998.
8. MR Nimlos, WA Jacoby, DM Blake, and TA Milne. *Evniron Sci Technol* 27:732–740, 1993.
9. WA Jacoby, MR Nimlos, DM Blake, RD Noble, and CA Koval. *Environ Sci Technol* 28:1661–1668, 1994.
10. S-J Hwang, C Petucci, and D Raftery. *J Am Chem Soc* 119:7877–7878, 1997.
11. EG Derouane, H He, SBD-A Hamid, D Lambert, and I Ivanova. *J Mol Catal A: Chem* 158:5–17, 2000.
12. JF Haw, JB Nicholas, T Xu, LW Beck, and DB Ferguson. *Acc Chem Res* 29:259–267, 1996.
13. M Hunger and T Horvath. *J Chem Soc Chem Commun* 1423–1424, 1995.
14. E MacNamara and D Raftery. *J Catal* 175:135–137, 1998.
15. C Keeler, JC Xiong, H Lock, S Dec, T Tao, and GE Maciel. *Catal Today* 49:377–383, 1999.
16. LK Carlson, PK Isbester, and EJ Munson. *Solid State Nucl Magn Reson* 16:93–102, 2000.
17. J Kiwi, C Pulgarin, P Peringer, and M Grätzel. *New J Chem* 17:487–494, 1993.
18. G Liu, T Wu, J Zhao, H Hidaka, and N Serpone. *Environ Sci Technol* 33:2081–2087, 1999.
19. A Topalov, D Molnár-Gárbor, M Kosanic, and B Abramovic. *Water Res 34*:1473–1478, 2000.
20. S Horikoshi, H Hidaka, and N Serpone. *J Photochem Photobiol A — Chem* 138:69–77, 2001.
21. KB Sherrard, PJ Marriott, RG Amiet, MJ McCormick, R Colton, and K Millington. *Chemosphere* 33:1921–1940, 1996.
22. Y Xiang, SC Larsen, and VH Grassian. *J Am Chem Soc* 121:5063–5072, 1999.
23. J Zhang, TR Krawietz, TW Skloss, and JF Haw. *Chem Commun* 685–686, 1997.
24. CV Rice and D Raftery. *Chem Commun* 895–896, 1999.
25. S Pilkenton, S-J Hwang, and D Raftery. *J Phys Chem B* 103:11152–11560, 1999.
26. AR Pradhan, MA Macnaughtan, and D Raftery. *J Am Chem Soc* 122:404–405, 2000.
27. S Klosek and D Raftery. *J Phys Chem B* 105:2815–2819, 2001.

28. S-J Hwang and D Raftery. *Catal Today* 49:353–361, 1999.
29. S-J Hwang, C Petucci, and D Raftery. *J Am Chem Soc* 120:4388–4397, 1998.
30. S Pilkenton, W Xu, and D Raftery. *Anal Sci* 17:125–130, 2001.
31. W Xu and D Raftery. *J Catal* 204:110–117, 2001.
32. W Xu and D Raftery. *J Phys Chem B* 105:4343–4349, 2001.
33. K Schmidt-Rohr and HW Spiess. *Multidimensional Solid-State NMR and Polymers.* New York: Academic Press, 1994.
34. JR Smith. Studying Conformational Polymorphism in Pharmaceutical Solids Using Multidimensional Solid-State NMR and Electronic Structure Calculations. Ph.D. thesis, Purdue University, 2001.
35. ID Gay. *J Magn Reson* 58:413–420, 1984.
36. CG Hatchard and CA Parker. *Proc Royal Soc London A* 235:518–536, 1956.
37. AV Kiselev and AV Uvarov. *Surf Sci* 6:399–421, 1967.
38. I Carrizosa and G Munuera. *J Catal* 49:174–188, 1977.
39. I Carrizosa and G Munuera. *J Catal* 49:189–200, 1977.
40. KS Kim, MA Barteau, and WE Farneth. *Langumir* 4:533–543, 1988.
41. VS Lusvardi, MA Barteau, and WE Farneth. *J Catal* 153:41–53, 1995.
42. VS Lusvardi, MA Barteau, WR Dolinger, and WE Farneth. *J Phys Chem* 100:18183–18191, 1996.
43. Y Suda, T Morimoto, and M Nagao. *Langmuir* 3:99–104, 1987.
44. L Gamble, LS Jung, and CT Campbell. *Surf Sci* 348:1–16, 1996.
45. E Borgarello, J Kiwi, M Grätzel, E Pelizzetti, and M Visca. *J Am Chem Soc* 104:2996–3002, 1982.
46. H Kisch, L Zang, C Lange, WF Maier, C Antonius, and D Meissner. *Angew Chem Int Ed* 37:3034–3036, 1998.
47. K Wilke and HD Breuer. *J Photochem Photobiol A* 121:49–53, 1999.
48. ST Martin, CL Morrison, and MR Hoffmann. *J Phys Chem* 98:13695–13704, 1994.
49. W Choi, A Termin, and MR Hoffman. *J Phys Chem* 98:13669–13679, 1994.
50. JC Yu, J Lin, and RWM Kwok. *J Photochem Photobiol A* 111:199–203, 1997.
51. R Asahi, T Morikawa, T Ohwaki, K Aoki, and Y Taga. *Science* 293:269–271, 2001.
52. (a) K Vinodgopal and PV Kamat. *Environ Sci Technol* 29:841–845, 1995. (b) K Vinodgopal, I Bedja, and PV Kamat. *Chem Mater* 8:2180–2187, 1996. (c) K Vinodgopal and PV Kamat. *Chemtech* 26:18–22, 1996.
53. J Lin, JC Yu, D Lo, and SK Lam. *J Catal* 183:368–372, 1999.
54. H Tada, A Hattori, Y Tokihisa, K Imai, N Tohge, and S Ito. *J Phys Chem B* 104:4585–4587, 2000.
55. L Shi, C Li, H Gu, and D Fang. *Mater Chem Phys* 62:62–67, 2000.
56. S Pilkenton and D Raftery. *Solid State Nucl Magn Reson* 24:236–253, 2003.
57. NJ Turro. *Pure Appl Chem* 58:1219–1229, 1986.
58. KB Yoon. *Chem Rev* 93:321–339, 1993.
59. F Blatter, H Sun, S Vasenkov, and H Frei. *Catal Today* 41:297–309, 1998.
60. H Frei, F Blatter, and H Sun. *Chemtech* 26:24–30, 1996.
61. AR Pradhan, MA Macnaughtan, and D Raftery. *Chem Mater* 14:3022–3027, 2002.

15 Beyond Photocatalytic Environmental Remediation: Novel TiO_2 Materials and Applications

Alexander G. Agrios and Kimberly A. Gray
Institute for Environmental Catalysis, Department of Civil and Environmental Engineering, Northwestern University

CONTENTS

15.1 INTRODUCTION

Environmental applications of photocatalysis using TiO_2 have attracted an enormous amount of research interest over the last three decades [1–4]. It is well established that slurries of TiO_2 illuminated with UV light can degrade to the point of mineralization almost any dissolved organic pollutant. Nevertheless, photocatalysis, particularly in aqueous media, has still not found widespread commercial implementation for environmental remediation. The main hurdle appears to be cost, which is high enough to prevent the displacement of existing and competing technologies by photocatalysis.

Still, TiO_2 photocatalysis is used industrially for certain niche applications. Work by Anderson and coworkers [5–8] has led to commercial systems for the destruction of gas-phase ethylene to prevent the premature ripening of fruit (commercialized by KES Science & Technology as "Bio-KES"). A number of companies have produced home systems for photocatalytic purification of air (e.g., Pionair) and water (e.g., ewater). A few companies produce photocatalytic systems for industrial water or air treatment (Purifics, Matrix Photocatalytic, Zentox/PHOTOX, Trojan Technologies). The scale of implementation of photocatalysis, however, remains much smaller than those of other advanced oxidation technologies (AOTs) such as UV–H_2O_2 or UV–O_3. Efforts to increase the practical utility of TiO_2 have proceeded on two fronts.

First, materials research is continuously underway to improve the fundamental performance of the catalyst. Much of the cost associated with photocatalysis derives from the energy costs of powering the UV illumination source. Intense illumination is required because the quantum efficiency, Φ, of TiO_2 is low (less than 1% under typical reaction conditions). The low Φ results from a high rate of recombination, whereby the photoexcited electron returns to the valence band, releasing the absorbed photon energy as heat. If a superior catalyst were designed with a markedly improved quantum efficiency, the overall cost of a photocatalytic process would be dramatically reduced. Improved catalyst design is derived from mechanistic investigations, which probe the pathways that drive photocatalysis and explain why certain preparations of TiO_2 are more photocatalytically effective than others.

Second, TiO_2 has been incorporated into novel processes and applications that capitalize on its strengths while de-emphasizing its weaknesses. For example, TiO_2 has been coated on various surfaces (including windows, windshields, and bathroom tiles) to render them self-cleaning under ambient light. Here, the photons are free, and sufficient reaction occurs to support the advertised function. The self-cleaning is assisted by the phenomenon of photo-induced hydrophilicity, which results from the illumination of TiO_2 films and improves the sheeting action of rainwater. Thus, the TiO_2 coating modifies surface properties favorably in addition to degrading oils. Another promising use of TiO_2 does not rely directly on its light-harvesting ability but rather on its performance as an electron shuttle. Despite a perception that it is somewhat inefficient in its use of absorbed light, TiO_2 is at the heart of highly efficient solar cells, where an adsorbed dye is responsible for light absorption and the TiO_2 merely conveys an excited electron from the dye to a conducting support.

Environmental research on TiO_2 began in earnest with reports of the photocatalytic splitting of water to yield H_2 and O_2. Because of the low efficiencies of water splitting, research interest gradually shifted to TiO_2-based treatment of polluted air and water. Today, TiO_2 research continues to evolve to include many environmental applications beyond the remediation of contaminated air or water. The new areas of TiO_2 research pertain to energy efficiency, ambient air quality improvement, and self-cleaning materials. In this chapter, we present some of the advances in the fabrication of TiO_2 materials, recent improvements in our understanding of the fundamental mechanisms that occur on illuminated TiO_2 surfaces, and discuss a number of emerging, novel uses for TiO_2 made possible by these discoveries. While our discussion of these topics is limited by space, our goal is to highlight a few examples that illustrate the new directions that TiO_2 catalysis is taking in improving environmental quality.

15.2 ADVANCES IN MATERIALS

While mixed-phase formulations of TiO_2 such as Degussa P25 have received much attention for their high photocatalytic activities, the pursuit of new preparations has been an area of vigorous research. Recent work has addressed both the preparation of TiO_2 nanoparticles with improved fundamental properties and the postsynthetic treatment of titania to produce catalytic systems with improved practical features. A wide variety of synthesis and treatment techniques have been attempted to yield improved photocatalytic or physical properties of TiO_2 particles. These include the preparation of high surface area particles, highly active mixed-phase catalysts, and visible light active doped titania and nanocomposites; the immobilization TiO_2 films (usually on glass substrates); the development of supported TiO_2 to allow efficient catalyst separation and recovery; and the modification of TiO_2 to alter the surface chemistry.

15.2.1 Sol–Gel Techniques

Much of the materials science research surrounding TiO_2 has been based on sol–gel preparations, which generally involve the hydrolysis of a Ti^{4+} salt (e.g., $TiCl_4$ or titanium tetraalkoxides) to form a nanoparticulate sol of TiO_2, often followed by evaporation or dialysis steps that lead to a gel. Selection of the anion in the Ti^{4+} salt combined with choices of numerous reaction variables including pH, water concentration, Ti concentration, addition of surfactants, etc., afford endless possibilities for controlling properties of the resulting sol or gel, especially particle size [9, 10]. Doped titania can be prepared by mixing the salts of Ti and the dopant before hydrolysis. In comparison to methods such as flame hydrolysis, sol–gel techniques have the advantages of producing relatively uniform particle size and exerting good control over particle properties, as well as carefully controlling ambient temperature and pressure conditions except for a final sintering step. Sol–gel syntheses are typically used to produce pure-phase anatase, although mixed-phase TiO_2 has also been prepared [11]. In addition, small particles (50 nm or less) of rutile have been synthesized [12, 13].

The preparation of a sol–gel often serves as a starting point for casting a film, through dip-coating [14], spin-coating [15], spray-painting [16], or screen-printing [16]. Generally, proper stabilization and crystallization of the film requires firing at high temperatures (ca. 500°C). This requires that the substrate be glass, metal, ceramic, or other such materials. However, a low-temperature mechanical sintering method has been developed where rollers subject the TiO_2 film to very high pressures [17–20]. This permits films to be formed on flexible plastic substrates, which is a prerequisite for a continuous manufacturing process.

In an effort to improve the surface area and morphology of TiO_2, Colón et al. [21] performed a standard sol–gel preparation but added activated carbon to the mix. TiO_2 nanoparticles form over the very high surface area (1400 m^2/g) carbon. The carbon is almost completely removed by combustion during calcination at 450°C, leaving pure anatase of high surface area. Changing the amount of carbon used significantly altered the resulting TiO_2 nanoparticles. A formulation where the mass of carbon used was five times that of TiO_2 resulted in a TiO_2 surface area of 117 m^2/g, compared to 13 m^2/g without any carbon. For comparison, Degussa P25 has a surface area of 50 m^2/g.

Photocatalytic reactors may be designed with the TiO_2 either in slurry or immobilized as a film. A slurry system requires a method for separating the TiO_2 from the treated water in order to reuse the catalyst; this is generally accomplished by centrifugation or filtration. Immobilization of the catalyst renders this step unnecessary, but can depress overall reaction rates due to mass transfer limitations. This has led to modifications of the slurry design including the coating of a TiO_2 film on beads of glass or silica [22, 23].

Amal and coworkers [24–27] have attempted to coat TiO_2 around a magnetic core for ease of separation of the catalyst from a suspension. Numerous complications arise. First, UV illumination of TiO_2-coated 40–70 nm magnetite particles (Fe_3O_4) causes dissolution of the magnetite. This was solved by encapsulating the magnetite in an insulating layer of SiO_2 before adding the titania coating [24, 26]. Second, when the magnetite particles are coated with TiO_2 using a conventional sol–gel method, the resulting TiO_2 film is amorphous. Photocatalytic activity requires crystalline TiO_2, which can be achieved by sintering the material at 450°C, but this also results in a substantial loss of magnetism of the core. An alternative method, expected to produce a crystalline TiO_2 layer but require heating only to 90°C in water, resulted in poor crystallization and disappointing photocatalytic performance [27].

15.2.2 Physical Vapor Deposition

Physical vapor deposition (PVD) describes a family of techniques that rely on physical rather than chemical processes to deposit a film. A starting material is vaporized by evaporation, sputtering, or ion bombardment, and condenses on a substrate. One widely studied technique is magnetron sputtering, where ions produced within a plasma are accelerated by an electric field toward a "target" material, which also serves as one electrode for the electric field. Atoms of the target are ejected and deposited on a collector within the sputtering chamber. TiO_2 films have been prepared by reactive sputtering, where Ti atoms are ejected from a Ti metal target into an atmosphere consisting of O_2 and Ar, where they react to form Ti oxides. The oxide particles are deposited on the collector, which is commonly glass or quartz. The technique gives reproducible, uniform, thin films. The characteristics of the film can be altered by adjusting parameters including O_2/Ar ratio [28], pressure of the gas [29], and temperature of the substrate [30], and by postsputtering treatments such as annealing [31]. The film itself can be bombarded with ions such as Xe^+ [32] or Ar^+ [33] as it grows, which can alter its stoichiometry. The ions generated in the plasma impinge on the film surface, with energies on the order of a few to tens of electron volts [34], which is small compared to the tens of keV used for ion bombardment. Pulsing the magnetron sputter process increases the amount of sputterable material and results in different film properties [35]. Other PVD techniques that have been used to make TiO_2 include filtered arc deposition (FAD) [36], ion-assisted deposition (IAD) [37], and plasma-assisted pulsed laser deposition (PA-PLD) [38].

Variable phase compositions can be obtained by PVD with different operational parameters. By sputtering Ti under a range of O_2 and Ar flow conditions, Guerin and Shah [39] produced a variety of titanium oxides including TiO_2, Ti_2O_3, TiO, and Ti. Generally, processes that produce low-energy particles (such as evaporation) result in amorphous or anatase phases, while high-energy particles (produced by sputtering or ion bombardment) result in anatase or rutile phases [35]. Low total pressures within the sputtering chamber favor rutile structures, while high pressures favor anatase [40]. These two findings are consistent with one another, since high gas pressures tend to attenuate particle energies through multiple collisions. The transition from the anatase phase to rutile appears to be quite abrupt [40], but sometimes mixed-phase titania are produced. Wiggins et al. [31] produced a mixed-phase TiO_2 by sputtering; subsequent annealing generally improved crystallinity and converted anatase to rutile at temperatures above 850°C. PVD allows the conditions of synthesis to be tuned in order to produce the desired size, phase, and stoichiometry, and to create nanocomposites.

15.2.3 Nitrogen Doping

Many dopants have been incorporated into TiO_2 in the hope of extending its photoactivity into the visible range and thereby promoting solar photocatalysis. Based on theoretical considerations, Asahi et al. [41] determined that nitrogen would be an ideal element with which to dope TiO_2 to produce a visible light-active photocatalyst. They prepared such a material by sputtering a TiO_2 film in a N_2/Ar atmosphere. The absorbance of this $TiO_{2-x}N_x$ was significant through wavelengths longer than 500 nm. In photocatalytic performance, $TiO_{2-x}N_x$ equaled an undoped TiO_2 sample under UV illumination but far outperformed TiO_2 in visible light as measured by the degradation of methylene blue.

A number of reports have followed up on the finding by Asahi et al. Sol–gel methods of N doping have been developed to replace the more complicated sputtering techniques. N-doped TiO_2 was produced by hydrolysis of $Ti(SO_4)_2$ in 28% NH_3 [42], or by direct amination of TiO_2 by reaction with triethylamine [43]. Both methods result in yellow, visible-light-active photocatalysts similar to that of Asahi et al. An advantage of the direct amination is that it can be applied to any TiO_2, including Degussa P25, which likewise takes on a yellow color and visible light activity. N-doping has also been achieved through a gas-phase reaction, where TiO_2 is heated to 550–600°C under a stream of NH_3 [44].

15.2.4 Other Methods

After TiO_2 particles are prepared, various postsynthesis modifications of the catalyst can be performed. Some examples are discussed below.

15.2.4.1 Mechanical Alteration of TiO_2

Samples of Degussa P25 and Hombikat UV-100 were altered by mechanical milling, with curious results [45]. The surface area of P25 increases somewhat with milling, from 50 to 70 m^2/g after 12 h, attributed to the breaking up of larger crystallites into small ones, but the surface area nearly returns to its nominal value after 25 h of milling. For Homikat, the surface area declined dramatically from 290 to 83 m^2/g after 180 h milling. This was rationalized as the plugging of micropores and densification of the secondary particles. X-ray diffraction (XRD) shows a phase change in P25 from the 80:20 anatase:rutile to 100% rutile after 25 h. Oddly, the Hombikat was transformed from 100% anatase into 100% brookite, a phase rarely encountered in TiO_2 syntheses. For both catalysts, photocatalytic activity decreased greatly with milling time, which is perhaps due to the phase changes.

15.2.4.2 Acid Pretreatment

Lewandowski and Ollis [46, 47] examined the effect of acid pretreatments of TiO_2 on its ability to degrade branched aromatics. Based on reports that cofeeding chlorinated aromatics (TCE or PCE) can improve degradation rates of aromatics, presumably because of chlorine radical formation, the authors hypothesized that adding chloride to the catalyst would have the same effect. HCl pretreatment of TiO_2 did increase transformation rates for toluene, but NaCl pretreatment did not. Therefore acidity is a required component of the enhancement mechanism, but acidity by itself does not enhance degradation rates: pretreatments with HF, HI, or HBr actually decrease toluene and *m*-xylene degradation rates. This is explained by thermodynamic calculations showing that TiO_2 is incapable of oxidizing F^- to $F^\bullet$. I^- and Br^- can be converted to radicals by TiO_2, but $I^\bullet$ and $Br^\bullet$ are incapable of abstracting hydrogen from the methyl groups of toluene or *m*-xylene. The role of H^+ remains unclear.

15.2.4.3 Loading with Metal Nanoclusters

TiO_2 nanoparticles can be capped with metal nanoclusters to improve photocatalytic performance. TiO_2–gold composites have received particular attention [48] since it was demonstrated that gold-coated TiO_2 electrodes have triple the photocurrent of an electrode without gold according to photocurrent action spectroscopy [49]. The photocurrent difference is even more pronounced (fourfold) under the imposition of a positive bias voltage. In this photoelectrochemical setting, where the electron is swept by the circuit, the improved incident photon-to-current efficiency (IPCE) of the TiO_2–Au was attributed to the action by gold to facilitate electron transfer between the photogenerated TiO_2 hole and redox-active species in the electrolyte. That is, gold accelerates the scavenging of TiO_2 valence-band (VB) holes, which reduces rates of recombination. However, the improvement in photocatalytic activity for the TiO_2–Au film is modest (10–15%) and the Au particles themselves are oxidized over prolonged irradiation [50, 51]. When TiO_2 particles are in suspension, rather than integrated into an electrochemical system, adsorbed gold nanoparticles accumulate electrons and thus shift the Fermi level of the TiO_2 by −22 mV [52].

Various methods exist for attaching gold clusters to TiO_2 particles. In the simplest, a film of TiO_2 on glass is immersed in a solution of gold colloid [49]. Typically, the TiO_2 nanoparticles used are small (20 nm) and the gold metal clusters are even smaller (5 nm) [52]. Gold particles derivatized with tetraoctylammonium bromide (TOAB) can be electrophoretically bonded to TiO_2 by the application of an electric field between the TiO_2 electrode and a counterelectrode [53]. Gold in the form of $HAuCl_4$ associates with TiO_2 in acidic solution due to electrostatic attraction between the negatively charged $AuCl_4^-$ ion and the positively charged TiO_2 surface. Subsequent reduction of the gold by $NaBH_4$ forms Au clusters directly on the semiconductor surface [54]. Alternatively, PVD (e.g., electron beam evaporation) of a gold metal sample can be used to coat a TiO_2 film with gold particles [55]. If the loading of Au is high (TiO_2:Au ratio < 1:1), laser light of 532 nm can cause morphological changes, e.g., the fusion of Au particles into larger clusters [56–58].

Silver particles have been formed on a TiO_2 surface by UV illumination of a TiO_2 film in contact with aqueous $AgNO_3$, causing photoreduction of Ag^+ to silver metal in the presence of a hole scavenger (formic acid) [59]. The electrochemical effects of the silver particles were minor compared to those of gold. However, the presence of silver metal improved photocatalytic efficiency up to a silver loading of about 1%, above which the obstruction of light by Ag outweighed any electrochemical benefits.

15.2.4.4 TiO_2–WO_3 Composites

Composites of TiO_2 and WO_3 have attracted interest due to two interesting properties. First, the conduction band (CB) of WO_3 can accept an electron from the CB of photoexcited TiO_2, and this electron is stable in WO_3 for a considerable period of time. Second, when an electron resides in the CB of WO_3, that material takes on a blue color. The first property, the storage of reductive energy, has been exploited for certain photocatalytic or corrosion-protection devices; the combination of both properties has led to photochromic devices.

A photoexcited electron on TiO_2 can be transferred to WO_3 through a conducting connection. However, the resulting negative charging of WO_3 and positive charging of TiO_2 must be counterbalanced, generally by reaction at the TiO_2 surface and intercalation of cations into WO_3. This can place important limitations on the charge-transfer process. When separate but electronically coupled films of TiO_2 and WO_3 are illuminated in dry air, minimal charge transfer occurs.

When a TiO_2 film and a WO_3 film are coated side-by-side on a single ITO electrode, immersed in 3 wt.% NaCl, and illuminated with UV, the WO_3 film becomes blue, indicating charge transfer from TiO_2 to WO_3 [60]. The color thus produced is stable indefinitely. But in pure water, colorization of WO_3 occurs only for that part of the film within about 1 mm of the TiO_2 film. The TiO_2 decharging process is believed to be the oxidation of water,

$$2H_2O + 4h^+ \longrightarrow O_2 + 4H^+$$

where the protons thus released then intercalate into WO_3 to balance the accumulated negative charge. The dependence of charge transfer upon spatial proximity of the two films is seen as evidence that the ionic conductivities of pure water and of the hydroxyl groups on the semiconductor surfaces are insufficient to permit long-range travel of the ionic species. A TiO_2–WO_3 composite film is readily colorized by UV even in pure water [60]. Novel applications of this composite are discussed below.

15.3 MECHANISTIC INVESTIGATIONS

A large body of research has focused on the fundamental processes that give rise to the surface chemistry, photocatalytic activity, and other properties of TiO_2. The results serve as the foundation for the fabrication of new materials, the engineering of photocatalytic systems, and the design of new applications. A few intriguing topics of study are reviewed below.

15.3.1 Second Harmonic Generation

Nonlinear laser spectroscopies including second harmonic generation (SHG) and sum frequency generation (SFG) are relatively recent optical probe techniques that have tremendous potential for examining interfaces [61, 62]. Commonly used spectroscopic techniques such as UV–Vis, Raman, FTIR, etc. are not ideally suited to probing semiconductor–liquid interfaces when the interfacial species of interest are also present in the bulk, since the bulk signal typically overwhelms the signal from the interface. Under sufficiently strong electric fields, as can be generated by a laser, second-order polarization terms become high enough to generate an observable second-order light wave. In SHG, illumination at a frequency ω gives rise to a wave of frequency 2ω. At the intersection of two laser beams of frequencies ω_1 and ω_2, SFG involves the generation of a light wave of frequency $\omega_1 + \omega_2$.

The key property of SHG and SFG is that the second-order processes are electric dipole forbidden wherever centrosymmetry, i.e., a center of inversion, exists, as is the case in the bulk of any gas, liquid, amorphous solid, or centrosymmetric crystal [61]. If only these types of phases are present, asymmetry exists only at the boundaries of phases, and thus SHG–SFG report exclusively on the interface, as no signal is generated in the bulk. In the case of TiO_2, rutile is centrosymmetric but anatase is not. Still, the SH signal for surface species in anatase systems can be maximized by operating at wavelengths far from the bulk resonance but close to the resonances of surface species of interest [63]. Calculating differences in phase between the bulk anatase and surface signals can allow the further separation of the two contributions to the observed SH signal.

SHG was used to observe the charge-transfer complex formed between catechol and anatase, and to measure an adsorption isotherm for catechol by direct *in situ* measurement of the catechol surface concentration [63]. Corn and coworkers took advantage of the electric field-induced second harmonic (EFISH) resulting from a space–charge layer at the surface of a TiO_2 electrode to measure the flat-band potential of the electrode, which coincides with the least EFISH and hence a minimum SHG [64]. In addition, the slow [65] and fast [66] kinetics of migration of photoproduced holes were followed, since the holes reduce EFISH by combining with surface electrons and reducing the space-charge potential. SHG–SFG is a powerful new tool for *in situ* studies of the TiO_2 surface that is only beginning to realize its potential.

15.3.2 Anatase–Rutile Interactions

It has long been empirically observed that mixed-phase preparations of TiO_2 — that is, containing both anatase and rutile — tend to exhibit higher photocatalytic activities than pure-phase TiO_2. The best-known example of this phenomenon is Degussa P25, which consists of about 70–80% anatase and the remainder rutile, with traces of brookite and amorphous phases, and which has set the standard for photocatalytic activity [1]. Furthermore, when Bacsa and Kiwi [67] prepared TiO_2 samples with a range of anatase:rutile ratios, the highest photoactivity was obtained with a 70:30 ratio, which is similar to that of P25. Ohno et al. [68] found a very different optimal ratio, of 10:90, but still noted that mixed-phase TiO_2 preparations significantly outperformed pure-phase anatase or rutile. While mixed-phase catalysts appear to be higher-activity materials, the conflicting results regarding the optimal anatase–rutile ratio point to the fact that the relative proportions of phases may be less critical in controlling the catalyst activity than factors such as the size, morphology, and conditions of synthesis.

The reasons for the improved performance of such preparations has been unclear. It was hypothesized that rutile acts as a sink for the electrons generated in anatase, serving to physically separate the electron and hole and thereby depress rates of recombination [1, 69] (see Figure 15.1a). This model is consistent with the fact that the band edges of rutile lie within those of anatase; i.e., the potential of the CB edge of anatase is more negative than that of rutile. However, recent EPR evidence shows that just the opposite occurs: rutile undergoes bandgap activation, and electrons are shuttled from rutile to anatase sites, which must be of lower energy [70] (see Figure 15.1b). This implies that one or more trap sites exsist on anatase at potentials more positive than the CB edge of either anatase or rutile. This was recently confirmed by a photoacoustic spectroscopy study of anatase, which found trap sites on anatase at an average of 0.8 eV below the CB edge [71].

Recent research has suggested that the size and morphology of rutile and anatase nanocrystals are critical to the separation and enhanced activity of mixed-phase catalysts like Degussa P25. An emerging model of P25 particles describes an atypically small rutile core surrounded by anatase crystallites [70]. Catalytic "hot spots" are believed to exsist at the intersection of the two phases, where distorted geometry gives rise to unique surface chemistry [70]. Further EPR evidence shows that most recombination in the mixed-phase P25 occurs not within the lattice but at surface sites on both anatase and rutile phases [72], as illustrated in Figure 15.1(b).

15.3.3 Quantum Size Effects

It is possible to synthesize semiconductor particles that are sufficiently small that quantum effects become apparent in the photoelectrochemical properties. Such particles called "quantum dots," are generally smaller than 10 nm [1]. One important quantum effect is that the small number of orbitals in each particle results in narrower bands than in large particles, with the result that the distance between the VB and CB — i.e., the bandgap — increases. Thus, quantum-sized TiO_2 has a shorter-wavelength absorption threshold than bulk TiO_2, and the

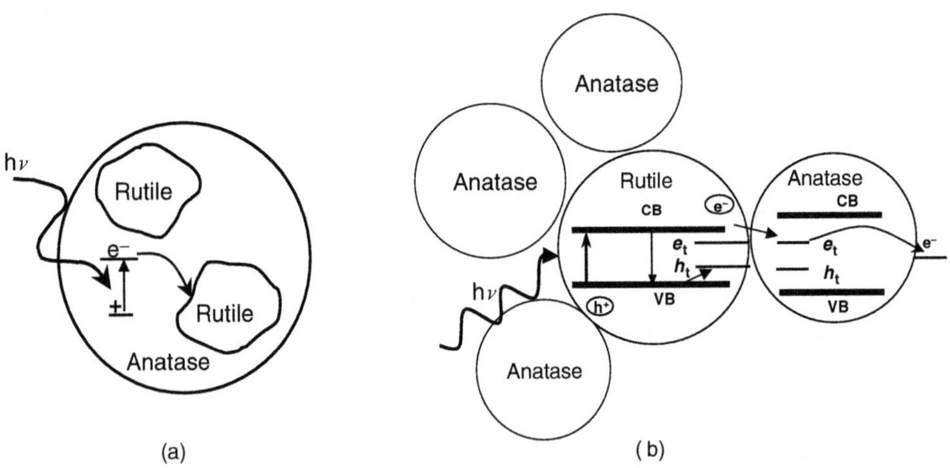

FIGURE 15.1 Models of P25: (a) conventional wisdom holds that rutile islands surround anatase particles, and rutile is an electron sink; (b) new picture involves a small rutile core surrounded by anatase crystallites, where electrons are shuttled from rutile to anatase.

blueshift varies with the size of the particles [73]. For example, 2-nm anatase particles exhibit a bandgap blueshift of 0.19 eV [74].

The other important effect is a dramatically slower relaxation of "hot electrons" and "hot holes," or charges produced by super-bandgap energy light. These effects were recently reviewed in depth by Nozik [75]. Briefly, charge separation triggered by photons with energy greater than the bandgap energy leaves the separated charges with excess energy. In a bulk semiconductor, the charges quickly relax by losing the excess energy. In a quantum dot, the relaxation can be prolonged by 1 to 3 orders of magnitude. A type of solar cell, discussed below (Section 15.4.1), relies on the transfer of an electron to TiO_2 from a photoexcited adsorbed dye molecule. Variants have replaced the dye with quantum dots of other semiconductors, for example, CdSe [76], InP [77], CdS [78], PbS [79], or Bi_2S_3 [80]. The bandgap of a quantum-sized sensitizer can be controlled by adjusting its particle size through sol–gel techniques [78]; this presents an alternative to adjusting dye properties through organic syntheses. Typically, the TiO_2 in such cells is of larger particle size; quantum-sized TiO_2 may have a CB edge that is too high to permit electron transfer from the sensitizer [73]. It is expected that hot electrons from the quantum dots will be very rapidly injected into TiO_2 due to their excess kinetic energy, which should result in higher efficiency, although this effect has not yet been demonstrated [75].

15.4 NOVEL APPLICATIONS

Once explored chiefly for catalytic properties, especially for environmental applications, TiO_2 has become the focus of a wide variety of semiconductor research. The photocatalytic properties of TiO_2 have been exploited not only for destroying organic compounds but also for killing microorganisms. Various other properties of the material have been utilized for applications as varied as photochromic devices, superhydrophilic and superhydrophobic surfaces, filtration, and corrosion protection. An especially intense area of study is photoelectrochemical cells for solar energy conversion, especially the so-called Grätzel cell, which relies on TiO_2 nanoparticles sensitized with a dye for the conversion of solar energy to electricity.

15.4.1 Solar Energy Conversion

The genesis of modern photocatalysis research was a 1972 report by Fujishima and Honda showing that illuminated TiO_2 electrodes could cleave water into H_2 and O_2. Solar production of H_2 from water has been called the Holy Grail of renewable energy, since the H_2 could be utilized as fuel for a fuel cell, producing electricity along with water vapor, resulting in a closed loop with solar energy as the only input and electrical energy as the only output. However, researchers were frustrated in their efforts to achieve efficient hydrogen generation using only TiO_2 and water, largely because the O_2 produced in the oxidation of water is subsequently reduced to $O_2^{\bullet -}$ and other species, effectively short-circuiting the overall process [3]. Work in this direction was largely abandoned in the 1980s as research became focused on photocatalytic destruction of pollutants.

In 1991, work toward a dye-sensitized TiO_2 solar cell culminated in a seminal paper by O'Regan and Grätzel [81], who reported a solar cell with an energy conversion efficiency of 7.1%. The Grätzel cell begins with a nanoparticulate TiO_2 film coated on a transparent conducting oxide (TCO) substrate. The TiO_2 is sensitized by a monolayer of a dye chemisorbed to the TiO_2 particles through carboxylate, phosphonate, or hydroxamate groups [82]. A liquid electrolyte provides electrical connectivity between the sensitized TiO_2 and the platinum counter-electrode. The electrolyte contains a redox couple that serves as a charge mediator. In the original paper, the dye was RuL_3, where L = 2,2′-bipyridyl-4,4′-carboxylic acid, and the redox couple was I_3^-/I^-.

The operation of the cell is illustrated in Figure 15.2. Absorption of visible light by the sensitizer (S) results in the excitation of an electron, which is subsequently injected into the CB of TiO_2. The ruthenium atom, initially in the +2 oxidation state, is thereby oxidized to Ru(III). Through a mechanism that is not entirely clear, the electron finds its way to the TCO collector and moves through the external circuit to the Pt electrode. There, it reduces I_3^- to I^-, which in turn reduces and thus regenerates the oxidized dye (S^+). The maximum voltage delivered by the cell is the difference between the Fermi potential of the TiO_2 (which is more negative under illumination than in the dark due to charge injection) and the potential of the redox couple.

Intense research activity has followed the 1991 paper and has been recently reviewed [82–85]. One area of progress is the development of new dyes [86–101]. For years, the best known dye was "N3", *cis*-$RuL_2(NCS)_2$ [87]. But in 2001 the "black dye", $RuL'(NCS)_3$ where L′ = 4,4′,4″-tricarboxy-2,2′:6′,2″-terpyridine, extended the visible absorption 100 nm to the red compared to N3 [89]. A dye-sensitized solar cell (DSSC) sensitized with the black dye set the record for DSSC efficiency at 10.4%.

The redox couple I_3^-/I^- leaves room for improvement: its redox potential is low, the oxidized form has some absorbance in the visible range, and conversion of one species to the other is a two-electron process. Numerous alternative redox mediators have been investigated including pseudohalogens [102], phenothiazine derivatives [103], and cobalt complexes [104, 105]. While these species generally have higher redox potentials, allowing for a greater open-circuit voltage (V_{OC}), their slow electron transfer kinetics generally lead to inferior photocurrent and efficiency under full sunlight compared to I_3^-/I^-.

The mechanism and kinetics of charge transfer between the semiconductor, dye, redox electrolyte, and collector have attracted great interest [106–126]. The kinetics determine the results of competition between the different electron transfer reactions that can occur within a DSSC. Desired charge transfers include charge injection (from the excited sensitizer, S*, to TiO_2), dye regeneration (from I^- to the oxidized dye, S^+), and reduction of I_3^- at the counterelectrode. Two types of recombination, from TiO_2 to either S^+ or I_3^-, are undesired. Fortunately, charge injection from the dye into TiO_2 is faster than the reverse process (recombination) by about five orders of magnitude [107]. It has been proposed that the former process may proceed directly between proximate states, whereas recombination rates are limited by diffusional encounters between the injected electron and the oxidized

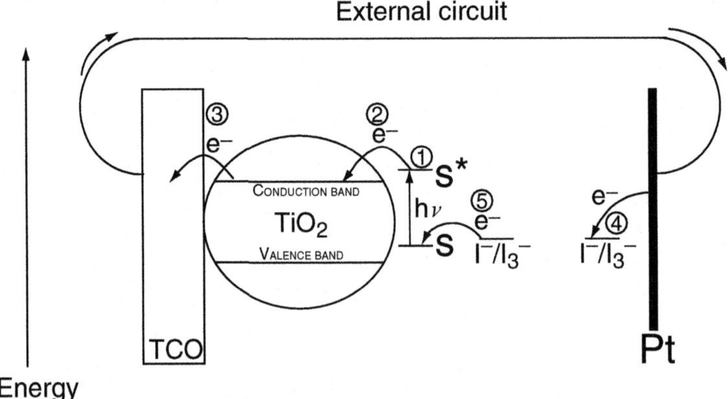

FIGURE 15.2 Schematic of the dye-sensitized solar cell, showing excitation of dye (1), charge injection (2), charge transfer to transparent conducting oxide (3) and flow through external circuit to Pt counter electrode, transfer of e^- to I_3^-(4), regeneration of dye by I_3^- (5).

dye [107]. This suggests that maintaining some distance between the chromophore and the semiconductor surface by an organic "spacer" may reduce the likelihood of diffusional contact and thus reduce recombination rates [114]. Charge injection rates are not adversely affected, as it has been shown than an electron can be injected across a distance of 24 Å with very fast (subpicosecond) kinetics [108].

Recombination from TiO_2 to S^+ is in competition with regeneration from I^- to S^+. While the recombination rate depends on the bias applied to the TiO_2 electrode, the regeneration rate does not [124]. In the absence of an applied bias, regeneration dominates as long as $[I^-] >$ 30 mM [124]. At high concentrations of I_3^-/I^-, electron transfer from TiO_2 to I_3^- is the dominant recombination process over back-transfer from TiO_2 to S^+ [121]. Understanding the kinetics of these processes allows configuration of the cell and electrolyte to optimize the solar energy conversion efficiency.

A practical DSSC must be sealed with sufficient durability so that the liquid electrolyte does not evaporate over the 20-year lifetime of the cell. The difficulty of this has prompted work toward a completely solid-state DSSC. The solid cell has a hole-conducting (p-type) material intervening between the dye-coated titania and the counterelectrode. Different polymers, such as OMeTAD [127] or poly(4-undecyl-2,2′-bithiophene) [128], and semiconductors such as CuI [128] and CuSCN [130] have been employed for this task. To date, all of the solid-state cells have significantly lower efficiencies than the liquid electrolyte cells; it remains for the market to sort out whether the advantages for manufacturing are worth the cost in efficiency. An intermediate solution involves cells with gel or molten salt electrolytes that are nonvolatile but not entirely solid-state [131–135].

While DSSCs are not as efficient as silicon-based solar cells, they have major cost advantages over the more mature technology. Crystalline silicon cells must be manufactured to high standards of purity, since defects act as recombination centers. In the DSSC, the semiconductor never supports simultaneously a hole and an electron; it merely serves to transport electrons from the dye molecule to the collector. Recombination can take other forms — e.g. back-transfer of an electron from TiO_2 to the dye, or reduction of I_3^- by the conduction-band electron — but it cannot occur within TiO_2 crystals, allowing less stringent materials processing. The DSSC has attracted considerable commercial interest. Most recently, STMicroelectronics, Europe's largest semiconductor maker, announced plans to develop Grätzel-type cells. Compared to silicon cells, their DSSCs are expected to be less efficient by a factor of 2, but less expensive by a factor of 20.

15.4.2 Disinfection

The use of photocatalysis for disinfection has attracted considerable interest since the initial discovery in 1985 that UV-illuminated slurries of TiO_2 could sterilize a solution of *Escherichia coli* [136]. A number of subsequent disinfection studies have used *E. coli* as the model microbe. Generally, it is found to be effectively degraded by UV-illuminated TiO_2, either in slurries or on immobilized films [136–150]. Under varying conditions, photocatalysis has been shown to be lethal to all of the microbes listed in Table 15.1. While the results are difficult to compare due to the different reaction variables, one study [151] found better than 3-log inactivation of *E. coli, Pseudomonas aeruginosa*, *Enterobacter cloacae*, and *Salmonella typhimurium* after 40 min irradiation of 0.1 g/L TiO_2 slurries with 5.5 mW/cm^2 near-UV light. An order-of-magnitude estimate of the quantum efficiency for photocatalytic inactivation of *P. aeruginosa* was calculated as between 10^{-11} and 10^{-9} cells killed per photon [152]. Some studies [143, 146] have found that application of a bias voltage to a TiO_2 electrode can improve antimicrobial performance, presumably by sweeping the conduction-band electron and thus inhibiting recombination, as in other photocatalytic applications. Doping the TiO_2 with Pt or Ir can improve germicidal activity [153], while Cu and Al have no effect or are deleterious [152].

TABLE 15.1
Microbes Shown to be Inactivated by Photocatalysis

Microbe	Refs
Salmonella choleraesus	[154]
Vibrio parahaemolyticus	[154]
Mutans streptococci	[155]
Pseudomonas aeruginosa	[152, 156]
Phage MS2	[157]
Poliovirus 1	[141]
Cancer cells	[158–162]

Photocatalysis has an advantage over other sterilization techniques in that illuminated titania can not only disinfect but also detoxify. For example, while UV illumination by itself is partially effective in killing *E. coli*, this merely results in the release of endotoxin, a pyrogenic constituent of the bacteria, from the dead cells. But illuminated TiO_2 can rapidly inactivate the bacteria while simultaneously destroying the endotoxin [142].

The mechanism of microbial inactivation by photocatalysis has been a subject of debate. It is, of course, a difficult question given the combined complexities of biological systems and the suite of radicals produced in photocatalysis. The issue can be subdivided into two components: the chemical species responsible for attacking the cells and the biological cause of death. Many researchers have assumed that the hydroxyl radical is the lethal chemical species, since it is in many cases thought to be the most important radical in the photocatalytic destruction of pollutants. Thus when Ireland et al. [138] observed an inhibition of photocatalytic disinfection by $S_2O_3^{2-}$, they attributed it to the oxidation of $S_2O_3^{2-}$ by $^{\bullet}OH$. The fact that photocatalytic sterilization is assisted by the addition of Fe(II) salts, which react with photocatalytically produced H_2O_2 to form additional $^{\bullet}OH$, supports this assumption [157].

The matter is complicated, however, by an investigation conducted by Kikuchi et al. [163], which concluded that hydrogen peroxide, not the hydroxyl radical, is the essential actor. Catalase, which decomposes H_2O_2, had a more inhibitory effect on disinfection rates than mannitol, which scavenges $^{\bullet}OH$. Moreover, their TiO_2 film could inactivate *E. coli* even when separated from the bacteria by a 50-μm-thick membrane. To kill at a distance requires that the active chemical species traverse the barrier. Given its high reactivity, $^{\bullet}OH$ was calculated to have a diffusion length of only a few micrometers, and would therefore be unlikely to reach the far side of the membrane, whereas H_2O_2 is far less labile. However, solutions of H_2O_2, even with concentrations up to three orders of magnitude higher than that produced by irradiated titania, do not exhibit the same bactericidal effect as photocatalysis. Moreover, even when the bacteria are exposed to both a peroxide solution and the black light UV irradiation, which is of insufficient energy to cleave H_2O_2 to $2^{\bullet}OH$, cell inactivation was not equivalent to the photocatalytic system. Therefore, the authors posit that disinfection relies on a reaction involving H_2O_2 and some other activated oxygen species, perhaps $O_2^{\bullet-}$, although their results leave some confusion as to the identity of this additional reactant.

The biological inactivation mechanism appears to involve structural damage to the cell more than specific toxicity. Linkous et al. [153] pointed out that photocatalytic destruction of D-(+)-glucose, which is ubiquitous in cellular structures, indicates a nonspecific germicidal mechanism. Observations have indicated a course of inactivation involving decomposition of the cell wall or cytoplasmic membrane, followed by leaking of cellular components such as K^+, RNA, and proteins [155, 164, 165]. TEM examinations of *P. aeruginosa* before and after photocatalysis showed numerous morphological defects including indented cell walls, abnormal cellular division, and bubble-like protuberances from the cell surface [156].

The ability of illuminated TiO_2 to inactivate microorganisms has led to many proposed antimicrobial applications. A silicone catheter has been coated with a TiO_2 film that renders it self-sterilizing and self-cleaning when exposed to light. Even under external illumination, the interior of the tube receives enough light to kill *E. coli* and bleach methylene blue [143]. TiO_2 has been incorporated into membranes for reverse osmosis, where it prevents biofouling when illuminated [166, 167]. A TiO_2-coated microfibrous mesh can filter *E. coli* from suspension in the dark. Occasional illumination of the filter not only sterilizes the *E. coli* but destroys the biofilm enough to significantly reduce the pressure drop across the filter [168]. TiO_2 has even been coated on concrete to prevent algae buildup with modest success [153]. A rare negative result is for a blacklit TiO_2 filter that did not exhibit good germicidal capability for the control of bioaerosols containing *E. coli*, *Bacillus subtilis* enspores, *Candida famata* var. *flareri* yeast, and *Penicillium citrinum* spores [169].

15.4.3 Sensors

Sensors capable of detecting and even quantifying gaseous compounds present a far more facile analytical method than capturing samples and analyzing them using conventional equipment. In addition, sensors allow for continuous monitoring. Numerous potential applications include environmental air quality monitoring, health and safety monitoring, and security. Various sensors based on TiO_2 have been devised for the detection of gaseous species. Interestingly, the different devices rely on entirely different properties of TiO_2.

The most sensitive TiO_2-based sensors yet designed rely on the change in resistivity of the semiconductor upon interaction of gases with the surface. Because the resistivity of pure TiO_2 is so high that the measurement of changes is difficult, the baseline resistivity has been reduced by doping with Nb [170]. Each lattice substitution of pentavalent Nb for tetravalent Ti effectively results in the addition of an electron to the CB, and hence increases the conductivity [171]. Doping TiO_2 films with 10 atom% Nb reduced the resistivity by about an order of magnitude [170]. Thick films (about 10 μm) of the pure TiO_2 maintained at 450°C showed high sensitivity toward CO, allowing the determination of CO concentrations as low as 0.5 ppm. The same sensor was almost insensitive to NO_2, indicating some selectivity in air quality monitoring, although other gases were not tested. Nb doping actually reduced the CO response but increased the NO_2 response.

In an opposite approach, Ruiz et al. [172] doped TiO_2 with a sufficiently high concentration (8.7 atom%) of trivalent Cr to render the TiO_2 p-type rather than n-type. This has the effect of reversing the resistivity response to CO: at 500°C, the Cr-doped TiO_2 experiences an increase in resistivity upon exposure to CO, unlike pure (n-type) TiO_2, which shows a resistivity decrease. Because the baseline resistivity of the Cr–TiO_2 is lower than that of the pure TiO_2 by two orders of magnitude, the change in resistivity is easier to measure. The Cr–TiO_2 also shows sensitivity to NO_2 roughly equal to that of Nb-doped TiO_2, although the sense of the resistivity change is opposite due to the p-type character of the Cr–TiO_2.

A unique sensor by Zhu et al. [173] takes advantage of chemiluminescence resulting from the thermocatalyzed degradation of certain gases. When maintained at temperatures between 400°C and 500°C, the TiO_2 surface causes decomposition of certain gases such that some products are formed in an electronically excited state. Relaxation to the ground state is accompanied by light emission, which can be detected and quantified. Gases result in different profiles of chemiluminescent intensity over time when probed at different wavelengths and reaction temperatures. The sensor has a limit of detection of 6.7 mg/L for acetone and 10.5 mg/L for ethanol.

The photocatalytic properties of TiO_2 have been exploited in a room-temperature sensor [174]. When organic gases come in contact with a UV-illuminated TiO_2-coated electrode, degradation ensues, causing a current in the electrode that depends on the applied voltage.

Sweeping the applied voltage from 0 to +5 V produced a current signature that was distinctive for different gases including ethanol, xylene, acetone, isopropanol, methylene chloride, and benzene. However, the gases were all measured at saturation, and no attempt was reported to detect less concentrated gases, so the sensor's utility in dilute environments is unknown.

TiO_2-based sensors have been developed for the biological sciences, including an aqueous glucose sensor [175] and a device for DNA detection in hybridization assays [176]. However, in both of these applications TiO_2 is used simply for its adsorptive properties, as oligonucleotides bind tightly to anatase.

15.4.4 Photochromic and Electrochromic Devices

By definition, photochromic or electrochromic devices change color in response to illumination or an applied electric potential, respectively. Applications of such devices include "smart" windows, that is, windows that darken upon exposure to light or application of a voltage. Windows with such properties have the potential to substantially reduce air-conditioning costs in commercial buildings [177]. Three types of smart window are shown schematically in Figure 15.3.

In contrast to smart windows that require an external power source to alter the transmittance of the glass, Bechinger et al. [177] devised a clever "self-powered" smart window, where the electrochromic current is generated from the incident sunlight. This window is essentially a Grätzel cell coupled to a film of WO_3 (see Figure 15.3a). Two parallel plates of conducting glass, one coated with TiO_2 and the other with WO_3, are arranged with the films facing each other, and between the films is a thin (~25 μm) layer of an electrolyte solution containing LiI. A wire connects the two plates. The TiO_2 surface is coated with a dye ($Ru^{II}L_2L'$, where L is 2,2′-bipyridine-4,4′-dicarboxylate and L′ is 4,4′-dimethyl-2,2′-bipyridine) that enters an excited state upon absorption of visible light and subsequently injects an electron into the CB of TiO_2. This electron travels to the conducting substrate, through the wire, and into the WO_3 film, causing a blue color. By carefully choosing the amount of adsorbed dye and the thickness of the WO_3 film, the authors constructed a window that is nearly transparent in its uncharged state, but achieves an average absorbance of about 1 in the spectral range 525–800 nm under illumination. Fully charging the WO_3 film requires about 100 s. If illumination is ceased while the two plates remain connected by a wire, the charge transfer is reversed, and the window decolorizes in about 125 s. However, if the circuit is opened before illumination is stopped, the blue color remains stable for more than 24 h.

Bechinger's smart window displays spatial resolution even though colorization occurs by the shuttling of electrons from one conducting glass plate through a wire to the other. That is,

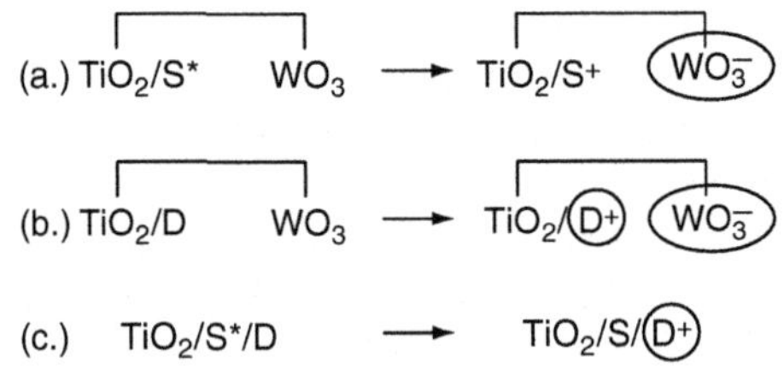

(a.) TiO_2/S^* WO_3 → TiO_2/S^+ WO_3^-

(b.) TiO_2/D WO_3 → TiO_2/D^+ WO_3^-

(c.) $TiO_2/S^*/D$ → $TiO_2/S/D^+$

FIGURE 15.3 Three types of smart windows, based on transfer of an electron from: (a) photoexcited sensitizer (S*) to WO_3 (photochromic, Ref. [176]); (b) donor (D) to WO_3 (electrochromic, Ref. [177]); (c) donor to photoexcited sensitizer (photoelectrochromic, Ref. [178]). Circled species are colored; overhead lines indicate electrical connectivity.

if one portion of the window is illuminated, only that part will darken, and the window will retain the illuminated pattern if the circuit is opened. This is the result of charge compensation involving the electrolyte, which contains the redox couple triiodide/iodide (I_3^-/I^-) and the ion Li^+. At all times, the net electrical charge must be zero in each of three zones: (1) the dye–TiO_2 system, (2) the WO_3 film, and (3) the intervening electrolyte solution. When the dye injects an electron into TiO_2, which relays it via the substrate across to the WO_3 film, an electron is transferred to the dye by I^-. The buildup of electrons on WO_3 is balanced by intercalation of Li^+ into the WO_3 film. The electrolyte remains uncharged as long as these two processes occur at equal rates. Spatial resolution, therefore, owes to the limited lateral ion mobility in the thin electrolyte layer, and to the intercalation of Li^+ ions, which effectively immobilizes them.

In a variant of the self-powered smart window, Bonhôte et al. [178] devised an electrochromic device, where an external voltage is used to colorize or decolorize the system (see Figure 15.3b). The setup is similar to that of the Bechinger device, but instead of coating with a dye, the TiO_2 is derivatized by a phosphonated triarylamine (PTA). Application of a voltage between the two plates causes electron transfer from the PTA to the WO_3, resulting in a blue color not only in the WO_3 film but also in the oxidized PTA. Because the reduced PTA does not absorb light, it can be coated on the TiO_2 electrode at full monolayer coverage without sacrificing transparency of the uncharged window. Since reduction of the dye by I^- would result in the loss of its blue coloration, the device instead relies on charge-compensation by electrolytic anions. The added absorbances of the PTA–TiO_2 plate and the reduced WO_3 would be expected to yield a greater total absorbance than the self-powered window. However, a drawback of the electrochromic window is that, when measured separately, the derivatized TiO_2 electrode reaches full coloration after only one-fifth the charge required to fully colorize the WO_3 plate. When the two plates are coupled and subjected to a potential difference of 1.5 V, the total charge transfer is limited by the lower capacity of the derivatized TiO_2 electrode, which limits the device's average visible absorbance to 0.85.

Photoelectrochromic systems, requiring both light and an applied voltage to become colored, have been described [179] (see Figure 15.3c). Here, both a sensitizer (S) and a donor (D; a triarylamine) are anchored to the TiO_2 surface, and in some configurations they are different parts of the same molecule. Photoexcitation of the sensitizer causes injection of a charge from S to the CB of TiO_2. S^+ is quickly reduced by D, which becomes a colored species D^+. If the applied voltage is more positive than the CB of TiO_2, the CB electron is swept, preventing recombination with D^+ and preserving its color.

A silver–TiO_2 composite showed a different kind of photochromic behavior [180]. When a TiO_2 film was illuminated with UV in the presence of Ag^+, reduction of the Ag^+ resulted in the formation of silver metal nanoparticles on the titania surface. This composite had a brownish-gray color, which was attributed to plasmon resonance of the Ag nanoparticles. Visible light illumination excites electrons in Ag, and in the presence of O_2, results in the oxidation of Ag back to Ag^+. Broadband visible illumination thus makes the composite colorless. The intriguing property of the material is that illumination with monochromatic light decreases the absorbance of the film in the same spectral region as the illumination. The effect is that the film takes on the color of the light shined on it. The authors attribute this phenomenon to a diversity of sizes and shapes for Ag nanoclusters, leading to different visible absorption wavelengths for different clusters. UV illumination restores the brownish-gray color.

15.4.5 Self-Cleaning and Superhydrophilic Surfaces

The development of transparent, photoactive coatings of TiO_2 on glass [181–182] has led to numerous potential consumer applications. Photocatalytic reactions under sunlight can render such glass self-cleaning for use in windows, windshields, etc. Furthermore, such films have been found to have the fortuitous property of "photoinduced superhydrophilicity" (PSH), whereby UV irradiation results in a dramatic increase in the hydrophilicity of the TiO_2 surface as evidenced by a decrease in the contact angle for water from anywhere between 30° and 70°, depending on the material, to 0°. This phenomenon imparts antifogging properties to TiO_2-coated glass and assists in the self-cleaning action for outdoor surfaces by allowing the sheeting of rainwater.

PSH was first observed as early as 1986 [184], and was initially attributed to the mineralization of organics on the surface [185]. However, more detailed investigations by Fujishima and coworkers [186–191] subsequently revealed a mechanism involving the photogeneration of defects at oxygen-bridging sites. Initially, they proposed that a surface Ti^{4+} is photoreduced to Ti^{3+} as a bridging O^{2-} is photooxidized, severing the bond between the bridging O and the reduced Ti atom [192]. Later, they showed that only the photoproduced holes were important [193]. In either case, one bond in Ti—O—Ti is severed, which is followed by dissociative adsorption of water at the insufficiently coordinated Ti atom, resulting in two neighboring surface-bound titanol groups (Ti—OH). This site is a small hydrophilic domain, and numerous such domains come to populate the TiO_2 surface, while the remainder of the film remains hydrophobic. The result is a checkered domain of hydrophilic and hydrophobic microdomains on the order of 100 nm, as revealed by friction force microscopy (FFM) [194]. Since a liquid drop is much larger than the scale of these domains, it can adhere to the domains to which it is attracted. That is, water drops can spread across and adhere to the hydrophilic microdomains, while a nonpolar liquid can bind to the hydrophobic microdomains. Thus, the irradiated TiO_2 film is not just hydrophilic but amphiphilic. This assists in its self-cleaning properties, since a water rinse can remove an oil drop wetted to the surface, and vice-versa [194]. If the film is left in the dark in air, the oxygen gradually returns to its original configuration, restoring the hydrophobic surface.

Transparency can be achieved by using thin films of titania nanoparticles that are sufficiently small (tens of nanometers) so as to not scatter visible light significantly. Sol–gel techniques have produced films that are transparent and mechanically stable [181, 182]. On the other hand, while films made from P25 show superior photocatalytic activity, they significantly scatter visible light and are easily abraded [182].

Photodegradation of stearic acid has emerged as a standard test for the self-cleaning activity of TiO_2-coated glass. As a fatty acid that is solid at room temperature, stearic acid is a good model for the oils that naturally deposit on outdoor surfaces. Furthermore, because stearic acid is quickly mineralized, it does not produce significant quantities of spectroscopically active byproducts, allowing for easy spectroscopic quantification of degradation kinetics [185]. Paz et al. [182] calculated that even though the measured efficiency of their sol–gel-produced TiO_2 film was only 0.03%, it was sufficient to mineralize a person's daily organic bioeffluents given 6 h sunlight on a 1 m^2 window.

A commercial self-cleaning glass produced by the British company Pilkington, Activ has a 15-nm-thick film comprised of nonspherical domes of TiO_2 about 30 nm in diameter [184]. The film has excellent mechanical stability and good consistency from lot to lot. It has been proposed [185] that Activ be used as the standard against which novel TiO_2-coated glass preparations are tested, as Degussa P25 has served as the standard for TiO_2 in slurry experiments.

Paradoxically, TiO_2 has also been incorporated into superhydrophobic surfaces [195]. Hydrophobic films may be prepared from boehmite (AlOOH) or silica. However, in outdoor applications, the accumulation of dirt on the surface reduces the hydrophobicity. If a small amount of TiO_2 is added, surface dirt can be photocatalytically destroyed while the TiO_2 itself has minimal impact on the surface hydrophobicity. At an optimal TiO_2 loading of 2%, the film maintains its hydrophobicity over time better than a film without TiO_2.

15.4.5 Corrosion Protection

Steel is often protected from corrosion by a coating of sacrificial zinc. While this shifts the potential of the substrate to the corrosion potential of zinc, which is more negative than that of steel, the zinc is consumed by corrosion and cannot offer indefinite protection. Ohko et al. [196] coated type 304 stainless steel with a thin film (1.2 μm) of TiO_2. In 1% NaCl, the TiO_2 coating actually increased the corrosion potential in the dark by about 100 mV, but under UV illumination, the corrosion potential was reduced by about 250 mV.

Since the corrosion protection offered by illuminated TiO_2 is due entirely to the reductive power of the CB electron, this system represents an ideal application for the TiO_2–WO_3 composite. Indeed, when immersed in aqueous 3% NaCl, a TiO_2–WO_3 film allows the corrosion potential to remain low for a few hours after illumination is ceased [196]. Interestingly, it makes little difference whether the TiO_2 and WO_3 are applied as a composite film or are coated in separate strips, as long as they are in electrical contact.

15.5 CONCLUSIONS

While this chapter is not comprehensive in scope, we have reviewed many novel and more commercial developments based on TiO_2 photocatalysis that are leading the way to more sustainable technologies by enhancing the security and quality of the ambient environment. We have highlighted recent efforts, both successful and unsuccessful, to make more active TiO_2-based materials, to improve supports, to probe fundamental behavior, and to create novel and efficient devices tailored to a wide range of environmental applications including promoting aesthetic environmental quality (widows, tiles), safety (sensors), and renewable energy production (solar cells). TiO_2 research has progressed on multiple tiers, whereby the study of fundamental processes promotes material development, allowing novel niche uses that promise to play a larger and larger role in engineering sustainable technologies.

REFERENCES

1. MR Hoffmann, ST Martin, W Choi, and DW Bahnemann. *Chem Rev* 95:69–96, 1995.
2. AL Linsebigler, G Lu, and JTJ Yates. *Chem Rev* 95:735–758, 1995.
3. A Mills and S Le Hunte. *J Photochem Photobiol A* 108:1–35, 1997.
4. U Stafford, KA Gray, and PV Kamat. *Heterogen Chem Rev* 3:77–104, 1996.
5. ME Zorn, DT Tompkins, WA Zeltner, and MA Anderson. *Environ Sci Technol* 34:5206–5210, 2000.
6. A Sirisuk, CG Hill, and MA Anderson. *Catal Today* 54:159–164, 1999.
7. TW Tibbitts, KE Cushman, X Fu, MA Anderson, and RJ Bula. *Instr Monit Control CELSS* 22:1443–1451, 1999.
8. XZ Fu, LA Clark, WA Zeltner, and MA Anderson. *J Photochem Photobiol A* 97:181–186, 1996.
9. MA Anderson, MJ Gieselmann, and Q Xu. *J Membrane Sci* 39:243–258, 1988.
10. T Murakata, R Yamamoto, Y Yoshida, M Hinohara, T Ogata, and S Sato. *J Chem Eng Jpn* 31:21–28, 1998.
11. S Komarneni, RK Rajha, and H Katsuki. *Mater Chem Phys* 61:50–54, 1999.
12. H Yin, Y Wada, T Kitamura, S Kambe, S Murasawa, H Mori, T Sakata, and S Yanagida. *J Mat Chem* 11:1694–1703, 2001.
13. Z Tang, J Zhang, Z Cheng, and Z Zhang. *Mater Chem Phys* 77:314–317, 2002.
14. RS Sonawane, SG Hegde, and MK Dongare. *Mater Chem Phys* 77:744–750, 2003.
15. DH Kim and MA Anderson. *Environ Sci Technol* 28:479–483, 1994.
16. ME Rincón, O Gómez-Daza, C Corripio, and A Orihuela. *Thin Solid Films* 389:91–98, 2001.
17. G Boschloo, H Linström, E Magnusson, A Holmberg, and A Hagfeldt. *J Photochem Photobiol A* 148:11–15, 2002.

18. H Lindström, A Holmberg, E Magnusson, SE Lindquist, L Malmqvist, and A Hagfeldt. *Nano Lett* 1:97–100, 2001.
19. H Lindström, A Holmberg, E Magnusson, L Malmqvist, and A Hagfeldt. *J Photochem Photobiol A* 145:107–112, 2001.
20. H Lindström, E Magnusson, A Holmberg, S Södergren, S-E Lindquist, and A Hagfeldt. *Sol Energy Mater Sol Cells* 73:91–101, 2002.
21. G Colón, MC Hidalgo, and JA Navío. *Catal Today* 76:91–101, 2002.
22. K Kobayakawa, C Sato, Y Sato, and A Fujishima. *J Photochem Photobiol A* 118:65–69, 1998.
23. Y Zhang, JC Crittenden, DW Hand, and DL Perram. *Environ Sci Technol* 28:435–442, 1994.
24. D Beydoun, R Amal, G Low, and S McEvoy. *J Nanoparticle Res* 1:439–458, 1999.
25. D Beydoun, R Amal, G Low, and S McEvoy. *J Phys Chem B* 104:4387–4396, 2000.
26. D Beydoun, R Amal, G Low, and S McEvoy. *J Mol Catal A* 180:193–200, 2002.
27. S Watson, D Beydoun, and R Amal. *J Photochem Photobiol A* 148:303–313, 2002.
28. SK Zheng, TM Wang, G Xiang, and C Wang. *Vacuum* 62:361–366, 2001.
29. T Takahashi, H Nakabayashi, N Yamada, and J Tanabe. *J Vac Sci Technol A* 24:1409–1413, 2003.
30. S-F Chen and C-W Wang. *J Vac Sci Technol B* 20:263–270, 2002.
31. MD Wiggins, MC Nelson, and CR Aita. *J Vac Sci Technol A* 14:772–776, 1996.
32. F Zhang, Z Zheng, X Ding, Y Mao, Y Chen, Z Zhou, S Yang, and X Liu. *J Vac Sci Technol A* 15:1824–1827, 1997.
33. XZ Ding, FM Zhang, HM Wang, LZ Chen, and XH Liu. *Thin Solid Films* 368:257–260, 2000.
34. N Martin, AME Santo, R Sanjinés, and F Lévy. *Surf Coat Tech* 138:77–83, 2001.
35. O Treichel and V Kirchhoff. *Surf Coat Tech* 123:268–272, 2000.
36. A Bendavid, PJ Martin, Å Jamting, and H Takikawa. *Thin Solid Films* 355–366:6–11, 1999.
37. Q Tang, K Kikuchi, S Ogura, and A MacLeod. *J Vac Sci Technol A* 17:3379–3384, 1999.
38. A De Giacomo, VA Shakhatov, GS Senesi, and S Orlando. *Spectrochim Acta B* 56:1459–1472, 2001.
39. D Guerin and SI Shah. *J Vac Sci Technol A* 15:712–715, 1997.
40. P Zeman and S Takabayashi. *J Vac Sci Technol A* 20:388–393, 2002.
41. R Asahi, T Morikawa, T Ohwaki, K Aoki, and Y Taga. *Science* 293:269–271, 2001.
42. T Ihara, M Miyoshi, Y Iriyama, O Matsumoto, and S Sugihara. *Appl Catal B* 42:403–409, 2003.
43. C Burda, Y Lou, X Chen, ACS Samia, J Stout, and JL Gole. *Nano Lett* 3:1049–1051, 2003.
44. H Irie, Y Watanabe, and K Hashimoto. *J Phys Chem B* 107:5483–5486, 2003.
45. MC Hidalgo, G Colón, and JA Navío. *J Photochem Photobiol A* 148:341–348, 2002.
46. M Lewandowski and DF Ollis. *J Adv Oxid Technol* 5:33–40, 2002.
47. M Lewandowski and DF Ollis. *J Catal* 217:38–46, 2003.
48. PV Kamat. *J Phys Chem B* 106:7729–7744, 2002.
49. N Chandrasekharan and PV Kamat. *J Phys Chem B* 104:10851–10857, 2000.
50. V Subramanian, E Wolf, and PV Kamat. *J Phys Chem B* 105:11439–11446, 2001.
51. V Subramanian, EE Wolf, and PV Kamat. *Langmuir* 19:469–474, 2003.
52. M Jakob, H Levanon, and PV Kamat. *Nano Lett* 3:353–358, 2003.
53. N Chandrasekharan and PV Kamat. *Nano Lett* 1:67–70, 2001.
54. PV Kamat, M Flumiani, and A Dawson. *Colloid Surf A* 202:269–279, 2002.
55. IM Arabatzis, T Stergiopoulos, D Andreeva, S Kitova, SG Neophytides, and P Falaras. *J Catal* 220:127–135, 2003.
56. A Dawson and PV Kamat. *J Phys Chem B* 105:960–966, 2001.
57. D Lahiri, V Subramanian, T Shibata, EE Wolf, BA Bunker, and PV Kamat. *J Appl Phys* 93:2575–2582, 2003.
58. PV Kamat. *Pure Appl Chem* 74:1693–1706, 2002.
59. C He, Y Xiong, J Chen, CH Zha, and XH Zhu. *J Photochem Photobiol A* 157:71–79, 2003.
60. T Tatsuma, S Saitoh, P Ngaotrakanwiwat, Y Ohko, and A Fujishima. *Langmuir* 18:7777–7779, 2002.
61. KB Eisenthal. *J Phys Chem* 100:12997–13006, 1996.
62. CT Williams and DA Beattie. *Surf Sci* 500:545–576, 2002.
63. Y Liu, JI Dadap, D Zimdars, and KB Eisenthal. *J Phys Chem B* 103:2480–2486, 1999.
64. JM Lantz, R Baba, and RM Corn. *J Phys Chem* 97:7392–7395, 1993.

65. JM Lantz and RM Corn. *J Phys Chem* 98:4899–4905, 1994.
66. JM Lantz and RM Corn. *J Phys Chem* 98:9387–9390, 1994.
67. RR Bacsa and J Kiwi. *Appl Catal B* 16:19–29, 1998.
68. T Ohno, K Tokieda, S Higashida, and M Matsumura. *Appl Catal A* 244:383–391, 2003.
69. RI Bickley, T Gonzalez-Carreno, JS Lees, L Palmisano, and RJD Tilley. *J Solid State Chem* 92:178–190, 1991.
70. DC Hurum, AG Agrios, KA Gray, T Rajh, and MC Thurnauer. *J Phys Chem B* 107:4545–4549, 2003.
71. S Leytner and JT Hupp. *Chem Phys Lett* 330:231–236, 2000.
72. DC Hurum, KA Gray, T Rajh, and MC Thurnauer. *J Phys Chem B*, 109:977–980, 2005.
73. PA Sant and PV Kamat. *Phys Chem Chem Phys* 4:198–203, 2002.
74. JJ Sene, WA Zeltner, and MA Anderson. *J Phys Chem B* 107:1597–1603, 2003.
75. AJ Nozik. *Ann Rev Phys Chem* 52:193–231, 2001.
76. JH Fang, JW Wu, XM Lu, YC Shen, and ZH Lu. *Chem Phys Lett* 270:145–151, 1997.
77. A Zaban, OI Micic, BA Gregg, and AJ Nozik. *Langmuir* 14:3153–3156, 1998.
78. LM Peter, DJ Riley, EJ Tull, and KGU Wijayantha. *Chem Commun* 1030–1031, 2002.
79. R Plass, S Pelet, J Krueger, M Grätzel, and U Bach. *J Phys Chem B* 106:7578–7580, 2002.
80. LM Peter, KGU Wijayantha, DJ Riley, and JP Waggett. *J Phys Chem B* 107:8378–8381, 2003.
81. B O'Regan, M Grätzel. *Nature* 353:737–741, 1991.
82. M Grätzel. *J Photochem Photobiol C* 4:145–153, 2003.
83. M Grätzel. *Curr Opin Coll Interface Sci* 4:314–321, 1999.
84. M Grätzel. *Pure Appl Chem* 73:459–467, 2001.
85. M Grätzel. *Nature* 414:338–344, 2001.
86. JJ Lagref, MK Nazeeruddin, and M Grätzel. *Synth Met* 138:333–339, 2003.
87. MK Nazeeruddin, A Kay, I Rodicio, R Humphry-Baker, E Müller, P Liska, N Vlachopoulos, and M Grätzel. *J Am Chem Soc* 115:6382–6390, 1993.
88. MK Nazeeruddin, R Humphry-Baker, M Grätzel, D Wohrle, G Schnurpfeil, G Schneider, A Hirth, and N Trombach. *J Porphyrins Phthalocyanines* 3:230–237, 1999.
89. MK Nazeeruddin, P Péchy, T Renouard, SM Zakeeruddin, R Humphry-Baker, P Comte, P Liska, L Cevey, E Costa, V Shklover, L Spiccia, GB Deacon, CA Bignozzi, and M Grätzel. *J Am Chem Soc* 123:1613–1624, 2001.
90. MK Nazeeruddin, R Splivallo, P Liska, P Comte, and M Grätzel. *Chem Commun* 1456–1457, 2003.
91. CR Rice, MD Ward, MK Nazeeruddin, and M Grätzel. *New J Chem* 24:651–652, 2000.
92. SM Zakeeruddin, MK Nazeeruddin, R Humphry-Baker, P Péchy, P Quagliotto, C Barolo, G Viscardi, and M Grätzel. *Langmuir* 18:952–954, 2002.
93. M Yanagida, A Islam, Y Tachibana, G Fujihashi, R Katoh, H Sugihara, and H Arakawa. *New J Chem* 26:963–965, 2002.
94. K Hara, K Sayama, Y Ohga, A Shinpo, S Suga, and H Arakawa. *Chem Commun* 569–570, 2001.
95. K Hara, H Sugihara, LP Singh, A Islam, R Katoh, M Yanagida, K Sayama, S Murata, and H Arakawa. *J Photochem Photobiol A* 145:117–122, 2001.
96. Q Dai and J Rabani. *J Photochem Photobiol A* 148:17–24, 2002.
97. S Ferrere and BA Gregg. *New J Chem* 26:1155–1160, 2002.
98. J He, G Benkö, F Korodi, T Polívka, R Lomoth, B Åkermark, L Sun, A Hagfeldt, and V Sundström. *J Am Chem Soc* 124:4922–4932, 2002.
99. J Ohlsson, H Wolpher, A Hagfeldt, and H Grennberg. *J Photochem Photobiol A* 148:41–48, 2002.
100. K Sayama, S Tsukagoshi, T Mori, K Hara, Y Ohga, A Shinpou, Y Abe, S Suga, and H Arakawa. *Sol Energy Mater Sol Cells* 80:47–71, 2003.
101. G Sauvé, ME Cass, SJ Doig, I Lauermann, K Pomykal, and NS Lewis. *J Phys Chem B* 104:3488–3491, 2000.
102. G Oskam, B Bergeron, GJ Meyer, and PC Searson. *J Phys Chem B* 105:6867–6873, 2001.
103. R Argazzi, CA Bignozzi, TA Heimer, FN Castellano, and GJ Meyer. *J Phys Chem B* 101:2591–2597, 1997.
104. H Nusbaumer, J-E Moser, SM Zakeeruddin, MK Nazeeruddin, and M Grätzel. *J Phys Chem B* 105:10461–10464, 2001.
105. H Nusbaumer, SM Zakeeruddin, JE Moser, and M Grätzel. *Chem Eur J* 9:3756–3763, 2003.

106. CA Kelly, F Farzad, DW Thompson, JM Stipkala, and GJ Meyer. *Langmuir* 15:7047–7054, 1999.
107. GM Hasselmann and GJ Meyer. *J Phys Chem B* 103:7671–7675, 1999.
108. P Piotrowiak, E Galoppini, Q Wei, GJ Meyer, and P Wiewiór. *J Am Chem Soc* 125:5278–5279, 2003.
109. F Farzad, DW Thompson, CA Kelly, and GJ Meyer. *J Am Chem Soc* 121:5577–5578, 1999.
110. CJ Kleverlaan, MT Indelli, CA Bignozzi, L Pavanin, F Scandola, GM Hasselman, and GJ Meyer. *J Am Chem Soc* 122:2840–2849, 2000.
111. P Qu and GJ Meyer. *Langmuir* 17:6720–6728, 2001.
112. E Galoppini, W Guo, P Qu, and GJ Meyer. *J Am Chem Soc* 123:4342–4343, 2001.
113. M Yang, DW Thompson, and GJ Meyer. *Inorg Chem* 41:1254–1262, 2002.
114. E Galoppini, W Guo, W Zhang, PG Hoertz, P Qu, and GJ Meyer. *J Am Chem Soc* 124:7801–7811, 2002.
115. AA Eppler, IN Ballard, and J Nelson. *Physica E* 14:197–202, 2002.
116. C Bauer, G Boschloo, E Mukhtar, and A Hagfeldt. *J Phys Chem B* 106:12693–12704, 2002.
117. N Beermann, G Boschloo, and A Hagfeldt. *J Photochem Photobiol A* 152:213–218, 2002.
118. G Boschloo, H Lindström, E Magnusson, A Holmberg, and A Hagfeldt. *J Photochem Photobiol A* 148:11–15, 2002.
119. A Hagfeldt, SE Lindquist, and M Grätzel. *Sol Energy Mat Sol Cells* 32:245–257, 1994.
120. SA Haque, Y Tachibana, RL Willis, JE Moser, M Grätzel, DR Klug, and JR Durrant. *J Phys Chem B* 104:538–547, 2000.
121. SY Huang, G Schlichthörl, AJ Nozik, M Grätzel, and AJ Frank. *J Phys Chem B* 101:2576–2582, 1997.
122. R Huber, JE Moser, M Grätzel, and J Wachtveitl. *Chem Phys* 285:39–45, 2002.
123. J Kruger, R Plass, M Grätzel, PJ Cameron, and LM Peter. *J Phys Chem B* 107:7536–7539, 2003.
124. I Montanari, J Nelson, and JR Durrant. *J Phys Chem B* 106:12203–12210, 2002.
125. S Pelet, M Grätzel, and JE Moser. *J Phys Chem B* 107:3215–3224, 2003.
126. Y Tachibana, MK Nazeeruddin, M Grätzel, DR Klug, and JR Durrant. *Chem Phys* 285:127–132, 2002.
127. U Bach, D Lupo, P Comte, JE Moser, F Weissörtel, J Salbeck, H Spreitzer, and M Grätzel. *Nature* 395:583–585, 1998.
128. S Spiekermann, G Smestad, J Kowalik, LM Tolbert, and M Grätzel. *Synth Met* 121:1603–1604, 2001.
129. VPS Perera and K Tennakone. *Sol Energy Mat Sol Cells* 79:249–255, 2003.
130. B O'Regan, DT Schwartz, SM Zakeeruddin, and M Grätzel. *Adv Mater* 12:1263-1267, 2000.
131. S Murai, S Mikoshiba, H Sumino, and S Hayase. *J Photochem Photobiol A* 148:33–39, 2002.
132. E Stathatos, R Lianos, SM Zakeeruddin, P Liska, and M Grätzel. *Chem Mater* 15:1825–1829, 2003.
133. P Wang, SM Zakeeruddin, I Exnar, and M Grätzel. *Chem Commun*:2972–2973, 2002.
134. P Wang, SM Zakeeruddin, JE Moser, MK Nazeeruddin, T Sekiguchi, and M Grätzel. *Nat Mater* 2:402–407, 2003.
135. P Wang, SM Zakeeruddin, P Comte, I Exnar, and M Grätzel. *J Am Chem Soc* 125:1166–1167, 2003.
136. T Matsunaga, R Tomoda, T Nakajima, and H Wake. *FEMS Microbiol Lett* 29:211–214, 1985.
137. T Matsunaga, R Tomoda, T Nakajima, N Nakamura, and T Komine. *Appl Environ Microbiol* 54:1330–1333, 1988.
138. JC Ireland, P Klostermann, EW Rice, and RM Clark. *Appl Environ Microbiol* 59:1668–1670, 1993.
139. C Wei, W-Y Lin, Z Zainal, NE Williams, K Zhu, AP Kruzic, RL Smith, and K Rajeshwar. *Environ Sci Technol* 28:934–938, 1994.
140. T Matsunaga and M Okochi. *Environ Sci Technol* 29:501–505, 1995.
141. RJ Watts, S Kong, MP Orr, GC Miller, and BE Henry. *Water Res* 29:95–100, 1995.
142. K Sunada, Y Kikuchi, K Hashimoto, and A Fujishima. *Environ Sci Technol* 32:726–728, 1998.
143. Y Ohko, Y Utsumi, C Niwa, T Tatsuma, K Kobayakawa, Y Satoh, Y Kubota, and A Fujishima. *J Biomed Mater Res* 58:97–101, 2001.
144. PSM Dunlop, JA Byrne, N Manga, and BR Eggins. *J Photochem Photobiol A* 148:355–363, 2002.

145. EJ Wolfrum, J Huang, DM Blake, PC Maness, Z Huang, J Fiest, and WA Jacoby. *Environ Sci Technol* 36:3412–3419, 2002.
146. JC Yu, HY Tang, JG Yu, HC Chan, LZ Zhang, YD Xie, H Wang, and SP Wong. *J Photochem Photobiol A* 153:211–219, 2002.
147. PA Christensen, TP Curtis, TA Egerton, SAM Kosa, and JR Tinlin. *Appl Catal B* 41:371–386, 2003.
148. MM Kondo, JFF Orlanda, MD Ferreira, and MT Grassi. *Quim Nova* 26:133–135, 2003.
149. T Sato, Y Koizumi, and M Taya. *Biochem Eng J* 14:149–152, 2003.
150. JC Yu, Y Xie, HY Tang, L Zhang, HC Chan, and J Zhao. *J Photochem Photobiol A* 156:235–241, 2003.
151. JA Ibáñez, MI Litter, and RA Pizarro. *J Photochem Photobiol A* 157:81–85, 2003.
152. P Amézaga-Madrid, GV Nevárez-Moorillón, E Orrantia-Borunda, and M Miki-Yoshida. *FEMS Microbiol Lett* 211:183–188, 2002.
153. CA Linkous, GJ Carter, DB Locuson, AJ Ouellette, DK Slattery, and LA Smitha. *Environ Sci Technol* 34:4754–4758, 2000.
154. B Kim, D Kim, D Cho, and S Cho. *Chemosphere* 52:277–281, 2003.
155. T Saito, T Iwase, J Horie, and T Morioka. *J Photochem Photobiol B* 14:369–379, 1992.
156. P Amézaga-Madrid, R Silveyra-Morales, L Córdoba-Fierro, GV Nevárez-Moorillón, M Miki-Yoshida, E Orrantia-Borunda, and FJ Solís. *J Photochem Photobiol B* 70:45–50, 2003.
157. JC Sjogren and RA Sierka. *Appl Environ Microbiol* 60:344–347, 1994.
158. DM Blake, PC Maness, Z Huang, EJ Wolfrum, J Huang, and WA Jacoby. *Sep Purif Meth* 28:1–50, 1999.
159. A Fujishima, K Hashimoto, Y Kubota, R Adachi, and T Kakeda, inventors; Tumor treatment apparatus. U.S. patent 5,855,595. 1995.
160. Y Kubota, T Shuin, C Kawasaki, M Hosaka, H Kitamura, R Cai, H Sakai, K Hashimoto, and A Fujishima. *Brit J Cancer* 70:1107–1111, 1994.
161. Y Kubota, M Hosaka, K Hashimoto, and A Fujishima. *Chem J Chin Univ* 16:56–62, 1995.
162. H Sakai, R Baba, K Hashimoto, Y Kubota, and A Fujishima. *Chem Lett* 185–186, 1995.
163. Y Kikuchi, K Sunada, T Iyoda, K Hashimoto, and A Fujishima. *J Photochem Photobiol A* 106:51–56, 1997.
164. K Sunada, T Watanabe, and K Hashimoto. *J Photochem Photobiol A* 156:227–233, 2003.
165. PC Maness, S Smolinski, DM Blake, Z Huang, EJ Wolfrum, and WA Jacoby. *Appl Environ Microbiol* 65:4094–4098, 1999.
166. SH Kim, SY Kwak, BH Sohn, and TH Park. *J Membr Sci* 211:157–165, 2003.
167. SY Kwak, SH Kim, and SS Kim. *Environ Sci Technol* 35:2388–2394, 2001.
168. JEO Lopez and WA Jacoby. *J Air Waste Manage* 52:1206–1213, 2002.
169. CY Lin and CS Li. *Aerosol Sci Tech* 37:162–170, 2003.
170. MC Carotta, M Ferroni, D Gnani, V Guidi, M Merli, G Martinelli, MC Casale, and M Notaro. *Sens Actuators B* 58:310–317, 1999.
171. M Ferroni, MC Carotta, V Guidi, G Martinelli, F Ronconi, O Richard, D Van Dyck, and J Van Lunduyt. *Sens Actuators B* 68:140–145, 2000.
172. A Ruiz, A Cornet, G Sakai, K Shimanoe, JR Morante, and N Yamazoe. *Chem Lett* 31:892–893, 2002.
173. YF Zhu, JJ Shi, ZY Zhang, C Zhang, and XR Zhang. *Anal Chem* 74:120–124, 2002.
174. LR Skubal, NK Meshkov, and MC Vogt. *J Photochem Photobiol A* 148:103–108, 2002.
175. S Cosnier, A Senillou, M Grätzel, P Comte, N Vlachopoulos, NJ Renault, and C Martelet. *J Electroanal Chem* 469:176–181, 1999.
176. KR Meier and M Grätzel. *Chem Phys Chem* 3:371–374, 2002.
177. C Bechinger, S Ferrere, A Zaban, J Sprague, and BA Gregg. *Nature* 383:608–610, 1996.
178. P Bonhôte, E Gogniat, M Grätzel, and PV Ashrit. *Thin Solid Films* 350:269–275, 1999.
179. P Bonhôte, J-E Moser, R Humphry-Baker, N Vlachopoulos, SM Zakeeruddin, L Walder, and M Grätzel. *J Am Chem Soc* 121:1324–1336, 1999.
180. Y Ohko, T Tatsuma, T Fujii, K Naoi, C Niwa, Y Kubota, and A Fujishima. *Nat Mater* 2:29–31, 2003.
181. T Negishi, K Iyoda, K Hashimoto, and A Fujishima. *Chem Lett* 1995:841–842, 1995.

182. Y Paz, Z Luo, L Rabenberg, and A Heller. *J Mater Res* 10:2842–2848, 1995.
183. Y Paz and A Heller. *J Mater Res* 12:2759–2766, 1997.
184. S Kume and T Nozu. Difficult to stain glass. Japan patent 63-100042. 1986.
185. A Mills, A Lepre, N Elliott, S Bhopal, IP Parkin, and SA O'Neill. *J Photochem Photobiol A* 160:213–224, 2003.
186. T Watanabe, A Nakajima, R Wang, M Minabe, S Koizumi, A Fujishima, and K Hashimoto. *Thin Solid Films* 351:260–263, 1999.
187. T Watanabe, S Fukayama, M Miyauchi, A Fujishima, and K Hashimoto. *J Sol–Gel Sci Technol* 19:71–76, 2000.
188. M Miyauchi, A Nakajima, A Fujishima, K Hashimoto, and T Watanabe. *Chem Mater* 12:3–5, 2000.
189. RD Sun, A Nakajima, A Fujishima, T Watanabe, and K Hashimoto. *J Phys Chem B* 105:1984–1990, 2001.
190. A Nakajima, S-i Koizumi, T Watanabe, and K Hashimoto. *J Photochem Photobiol A* 146:129–132, 2001.
191. N Sakai, A Fujishima, T Watanabe, and K Hashimoto. *J Phys Chem B* 107:1028–1035, 2003.
192. R Wang, N Sakai, A Fujishima, T Watanabe, and K Hashimoto. *J Phys Chem B* 103:2188–2194, 1999.
193. N Sakai, A Fujishima, T Watanabe, and K Hashimoto. *J Phys Chem B* 105:3023–3026, 2001.
194. R Wang, K Hashimoto, A Fujishima, M Chikuni, E Kojima, A Kitamura, M Shimohigoshi, and T Watanabe. *Adv Mater* 10:135–138, 1998.
195. A Nakajima, K Hashimoto, T Watanabe, K Takai, G Yamauchi, and A Fujishima. *Langmuir* 16:7044–7047, 2000.
196. Y Ohko, S Saitoh, T Tatsuma, and A Fujishima. *J Electrochem Soc* 148:B24–B28, 2001.
197. T Tatsuma, S Saitoh, Y Ohko, and A Fujishima. *Chem Mater* 13:2838–2842, 2001.

16 Nanoparticles in Environmental Remediation

Koodali T. Ranjit, Gavin Medine, Pethaiyan Jeevanandam, Igor N. Martyanov, and Kenneth J. Klabunde
Department of Chemistry, Kansas State University

CONTENTS

16.1 INTRODUCTION TO REACTIVE NANOPARTICLES

Nanoparticles deal with small crystallites having diameters in the range 1 to 10 nm and having ~125 to 70,000 atoms. Classic fields of chemistry or quantum chemistry deal with atoms (molecules) whose dimensions are generally less than 1 nm, while solid-state physics deals with particles having dimensions greater than 10 nm and having 100,000 to 6.02×10^{23} atoms. Thus, the regime of nanoparticles is unique in which neither the principles of quantum chemistry nor those of classical physics hold. This leads to a situation in which strong chemical bonding is present and the valence electrons are extensively delocalized. The delocalization varies with size, which, in turn, results in varying physical and chemical properties.

A host of properties depend on the size of particles in the nanoscale regime. These include magnetic, optical, and mechanical properties. Bandgaps can change in semiconductors, plasma resonance peaks in metals shift, melting points can change, coercive force in magnetic materials can be manipulated, mechanical properties such as hardness, ductility, and plasticity can change, and surfaces can become more reactive. Thus, considering the plethora of solid-state materials that have been synthesized, size-dependent properties opens an almost infinite number of new materials that can be made. It is desirable that the nanoparticles be prepared in a narrow size range. One of the challenges in the synthesis is that the nanoparticles are extremely reactive to water and air and this often leads one to resort to creative and skillful synthetic procedures to make pure nanoscale materials. Materials can be divided into metals, semiconductors, and insulators. In case of metals and semiconductors, changes in chemical and physical properties with size are well documented and several theoretical predictions have indeed been observed experimentally. However, in the case of insulators, size effects are not clearly understood and so far have been related to changes in surface properties.

Indeed, surface effects in nanoparticles play an important role in shaping new technologies for the present and future generations. The question that immediately comes into one's mind is whether the unique properties of insulator nanoparticles such as MgO, CaO, SrO, ZrO_2, etc. can be directly correlated with the particle size. Evidently from the numerous reports in literature, the answer is a resounding yes. In the following section, we will discuss the effects of nanosizing on the particle size and reactive surface sites in certain dielectrics such as MgO and CaO.

16.1.1 EFFECTS OF NANOSIZING ON SURFACE AREA AND REACTIVE SURFACE SITES

Nanoparticles have about 10^{19} interfaces/cm^3 and surface areas in the range from 400 to 1000 m^2/g. As the particle size decreases, the surface area increases and the reactivity is considerably enhanced. The increased reactivity is not simply due to the *enhanced surface areas alone*. The small crystallite sizes in nanocrystals are quite remarkable. For example, MgO prepared by the aerogel procedure (referred to as AP samples for aerogel-prepared sample) have crystallite size ~ 4 nm [1], AP-CaO ~ 7 nm [2], AP-TiO_2 ~ 10 nm [3], and AP-ZrO_2 ~ 8 nm [4]. Aerogel preparation of these nanoparticles involves a sol–gel approach where the methoxides are converted to hydroxide gels. This is followed by a supercritical drying and vacuum dehydration to yield very fine powders. The surface area of AP-MgO is ~500 m^2/g, whereas an MgO sample conventionally prepared (referred to as CP henceforth) has a surface area of ~200 m^2/g, while the surface area of a commercial (referred as CM) MgO sample is only around 30 m^2/g. The surface areas of AP-CaO, CP-CaO, and CM-CaO are ~150 m^2/g, ~100 m^2/g, and ~1 m^2/g, respectively. Similarly, AP-Al_2O_3 possesses surface area as high as 810 m^2/g, while a commercial Al_2O_3 prepared by high-temperature method has a surface area of only about 100 m^2/g [5]. These results clearly indicate that the aerogel-supercritical drying method followed by vacuum dehydration results in the formation of ultrafine particles with very high surface areas compared to commercial samples.

These nanocrystals exhibit higher surface chemical reactivities than CP samples. For example, in the adsorption of SO_2, AP-MgO adsorbed three times as much as CP-MgO/nm^2 [6, 7]. In the destructive adsorption of $CH_3(CH_3O)_2PO$, AP-MgO was found to adsorb $CH_3(CH_3O)_2PO$ about 50 times greater than CM-MgO [8, 9]. For the reaction, $2CaO + CCl_4 \rightarrow 2CaCl_2 + CO_2$, AP-CaO was found to react about 30 times higher than CM-CaO [1, 2]. A very high ratio of edge ion/surface ions, steps, kinks, and defects along with unusual morphology seem to be the factors responsible for the high reactivity of these nanocrystalline oxides. In the following section, we will discuss the results obtained from transmission electron microscopy (TEM) studies so that a better perspective of the morphology of the nanocrystalline samples can be gained.

16.1.2 Micrographs

Figure 16.1 shows the TEM of AP-MgO. The TEM picture shows porous weblike aggregates in the range of about 1400 nm. These are formed by the interaction of the 4 nm (average) polyhedral crystallites and their overall size distribution is narrow [10].

The AP-CaO and CuO [11] particles tend to coagulate and hence it was very difficult to obtain a very clear image. After numerous trial and errors, the best images were obtained after sonication in pentane solvent for 60 min. Figure 16.2 shows the TEM of AP-CaO. The crystallite sizes are ~7 nm and the TEM image reveals the weakly agglomerated porous particles as observed previously for AP-MgO.

Nanocrystalline AP-Al_2O_3 also consisted of weblike aggregates and the crystallite sizes were <2 nm as shown in Figure 16.3. High-resolution TEM (HRTEM) also confirmed that AP-Al_2O_3 consisted of small crystallites (<2 nm) and that the crystals were disordered [5].

16.2 MODIFIED AEROGEL PROCESS (MAP)

The aerogel-hypercritical drying approach for the preparation of metal oxides was first reported by Kistler in 1932 for the preparation of high surface area silica [12]. He observed that when supercritical pressure and temperature was used during the drying procedure, the pore structure was retained, thus resulting in the formation of high surface area silica. The major drawback of this procedure was the tedious preparation conditions. In order to hasten the process, Teichner and coworkers used organic solvents instead of water [13, 14]. This decreased the preparation time from weeks to a few hours. The metal oxides prepared are highly porous, having a large number of defect sites and very high surface areas. Aerogels find applications as insulators, insecticides, catalysts, detectors, etc. [15].

A further modification of the Kistler–Teichner aerogel method involved the addition of large amounts of aromatic hydrocarbons such as toluene to the alcohol–methoxide solution. This was done to further reduce the surface tension of the solvent mixture and help in the removal of the solvent during the alcogel to aerogel transformation [1, 16, 17]. This led to the development of high surface areas and smaller crystallite sizes for MgO, CaO, Al_2O_3, TiO_2, and ZrO_2.

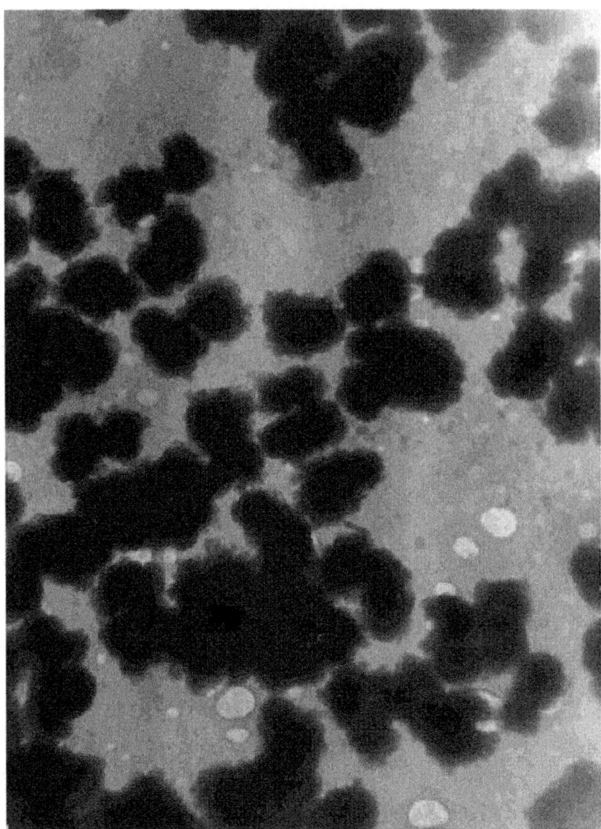

FIGURE 16.1 TEM image of AP-MgO secondary aggregate structures. (Reproduced from R Richards, W Li, S Decker, C Davidson, O Koper, V Zaikovski, A Volodin, T Rieker, and KJ Klabunde. *J Am Chem Soc* 122:4921–4925, 2000. Copyright 2000, American Chemical Society. With permission.)

16.2.1 Morphologies of AP-Nanoparticles

The MAP leads to the formation of free-flowing nanoparticles rather than monoliths. During the supercritical drying, the gel does indeed collapse, but very small stable nanocrystals are formed. It is well known that base catalysis leads to the formation of colloid-like gels, while acid catalysis leads to the formation of linear-chain polymer gels. Thus, the alkaline earth hydroxides, because of their basic character, lead to particulates rather than monoliths. We have seen the TEM images of MgO, CaO, and Al_2O_3 in the previous section. The MAP produces very small crystallites usually <10 nm with a very narrow size distribution. The microstructure of the nanocrystallites can be probed by small angle x-ray scattering (SAXS). The SAXS data are plotted as intensity versus q on log–log plots where $q = 4\pi/\lambda \sin(2\theta/2)$, where 2θ is the scattering angle. Scattering features at high q corresponds to small structures. For the AP-MgO sample, the SAXS data indicate a nonfractal morphology (q power law is -4) of the primary particles, which are about 100 Å. These primary particles aggregate to form a porous network of particles that are about 1 μm. The structure of AP-$Mg(OH)_2$, which is the precursor to AP-MgO, is also similar to that of AP-MgO. However the surface has fractal morphology (q power law is -3.3).

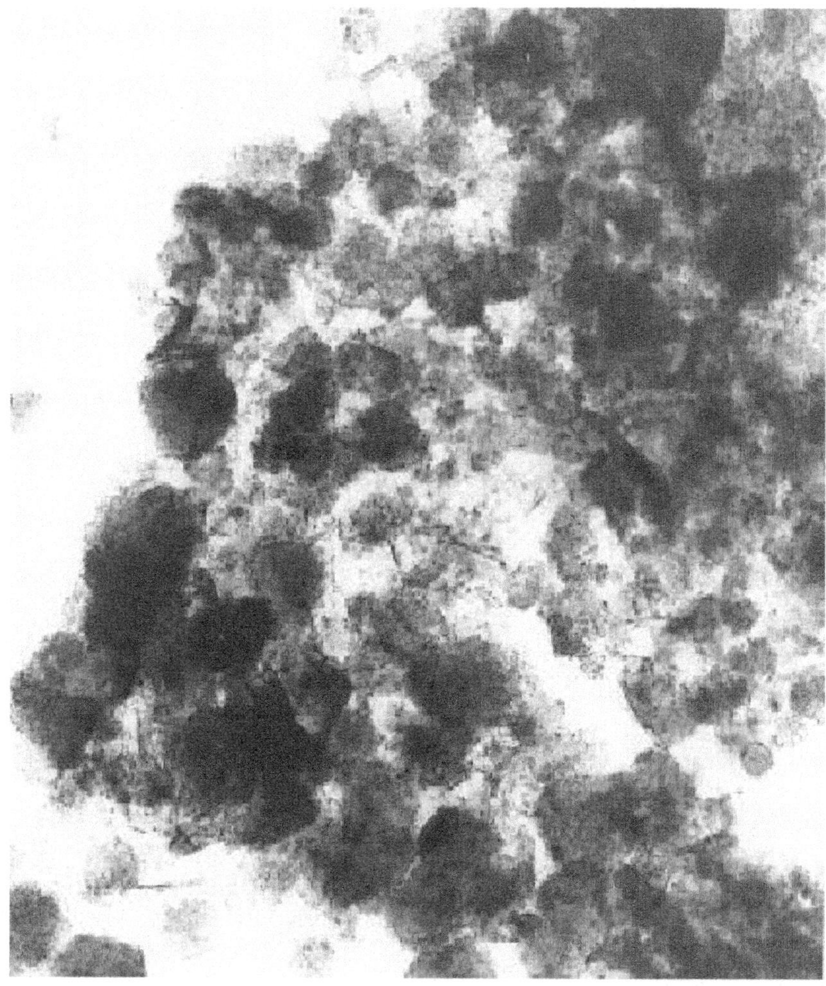

FIGURE 16.2 TEM image of AP-CaO (4.50 mm = 100 nm). (Reproduced from OB Koper, I Lagadic, A Volodin, and KJ Klabunde. *Chem Mater* 9:2468–2480, 1997. Copyright 1997, American Chemical Society. With permission.)

16.2.2 Intimately Mixed Bimetallic Oxides

Mixed metal oxides have been found to have several uses and there are many papers in the literature today devoted to these materials. For example, Chi and Chaung have studied NO and O_2 coadsorption on γ-Al_2O_3 supported Tb_4O_7, La_2O_3, BaO, and MgO [18]. Results show that BaO–γ-Al_2O_3 and MgO–γ-Al_2O_3 possessed a higher NO_x storage capability than Tb_4O_7–γ-Al_2O_3 and La_2O_3–γ-Al_2O_3. They also observed different types of adsorption depending on the particular mixed oxide. NO–O_2 was coadsorbed in the form of bridging bidentate, chelating bidentate, and monodentate nitrates on Tb_4O_7–γ-Al_2O_3, La_2O_3–γ-Al_2O_3, and BaO–γ-Al_2O_3 whereas MgO–γ-Al_2O_3 exhibited adsorption in the form of bridging bidentate and monodentate nitrates. Liotta et al. have studied the behavior of Pt supported on CeO_2–ZrO_2/Al_2O_3–BaO as a NO_x storage-reduction catalyst [19]. The catalyst was studied by reactivity tests and DRIFT experiments and compared with that of Pt–BaO on alumina. It was discovered that during calcination Ba^{2+} ions migrate over the surface of the catalyst, which shows a good NO_x storage-reduction behavior comparable with that of

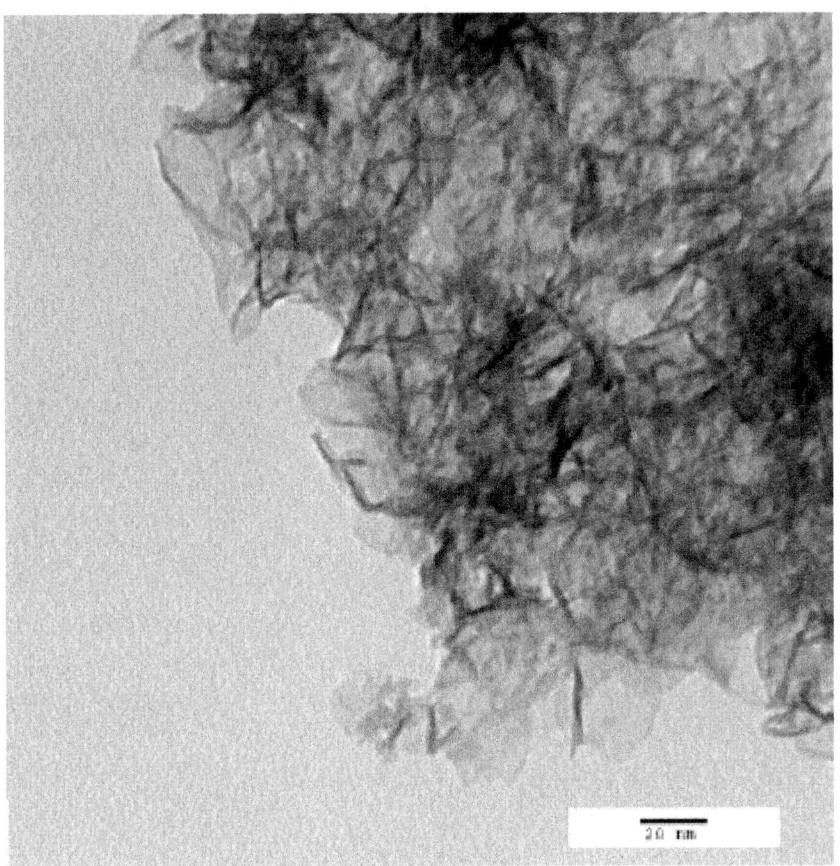

FIGURE 16.3 TEM image of AP-Al_2O_3. (Reproduced from CL Carnes, PN Kapoor, and KJ Klabunde. *Chem Mater* 14:2922–2929, 2002. Copyright 2002, American Chemical Society. With permission.)

Pt–BaO on alumina. The new catalyst was found to have a much better dispersion of Ba^{2+} ions throughout the sample. Ahmed and Attia reported a gel-prepared mixture of CaO, MgO, and SiO_2, which behaves as an excellent adsorbent for gases such as CO_2, SO_2, CO, NO, and H_2S [20].

An ideal way of preparing intimately intermingled mixed metal oxide nanoparticles is using a MAP. This procedure combines the sol–gel process with a supercritical removal step. The main modification is that a large portion of a spectator solvent that greatly enhances gelation rates is added, thereby increasing the chances that two hydrolyzing metal alkoxides will gel together. This results in the formation of more open-gel structures that eventually yield smaller, more reactive nanocrystals [21]. The co-gelation method can be simplified as follows:

$$Mg(OR)_2 + 2Al(OR)_3 \xrightarrow[\text{Organic solvent}]{H_2O} Mg(OH)_2/2Al(OH)_3 \longrightarrow MgAl_2O_4$$

The MAP is intended to prepare MgO–Al_2O_3 in molar ratios of 1:1 or any other desired ratios. The main point is to obtain intermingled mixtures that are essentially molecular in nature. Another advantage is to gain a highly reactive surface (MgO) along with a high surface area component (Al_2O_3). This technique was used to prepare a series of intimately intermingled mixed metal oxide nanoparticles. These mixed oxides were composed of alkaline earth oxide and aluminas. The surface area and reactivity of these materials decreased on going from magnesium to barium in these intimately intermingled mixed metal oxides. In the following paragraphs, we will discuss some of the salient features of the intimately mixed bimetallic sites.

16.2.2.1 XRD

It was found that only aerogel-prepared AP-$MgAl_2O_4$ and AP-$BaAl_2O_4$ showed any peaks in the XRD. In the XRDs for AP-$MgAl_2O_4$ and AP-$BaAl_2O_4$, it is possible to identify peaks corresponding to MgO and BaO, respectively. This gives a good indication that the samples are intimately mixed. As can be seen from Figure 16.4, the feature observed in the XRD for AP-$MgAl_2O_4$ is broad. Due to the broad features for AP-$MgAl_2O_4$, it is extremely difficult to determine the crystallite size of MgO in the intimately intermingled metal oxide. However, the crystallite size of BaO in AP-$BaAl_2O_4$ can be calculated and is found to be <2 nm, while the average crystallite size for a commercial sample of BaO is 30 nm. AP-$CaAl_2O_4$ and AP-$SrAl_2O_4$ are completely amorphous and no features were visible by XRD. When AP-Al_2O_3 is analyzed by XRD the spectrum is difficult to characterize as the sample is amorphous. The completely amorphous pattern obtained for AP-Al_2O_3 is attributed to particle size broadening.

16.2.2.2 Surface Area Analysis

When preparing samples of intimately intermingled mixed metal oxide samples one is able to retain high surface area and relatively large pore volumes when compared to that of the individual metal oxides. AP-$MgAl_2O_4$ and AP-$CaAl_2O_4$ have substantially higher surface areas and pore volumes than AP-$SrAl_2O_4$ and AP-$BaAl_2O_4$. This reduction in surface area and pore volume for AP-$SrAl_2O_4$ and AP-$BaAl_2O_4$ can be attributed to the large size and

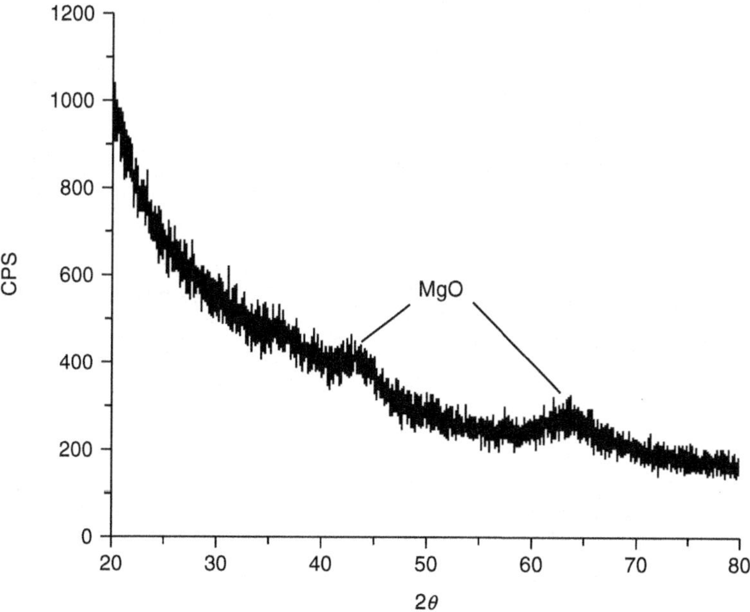

FIGURE 16.4 XRD of intimately mixed metal oxide, AP-$MgAl_2O_4$. (Reproduced from GM Medine, V Zaikovskii and KJ Klabunde. *J Mater Chem* 14:757–763, 2004. Copyright 2004, The Royal Society of Chemistry. With permission.)

atomic mass of Sr and Ba, the formation of larger agglomerates, and because SrO and BaO are not well dispersed throughout the Al_2O_3 framework. Table 16.1 gives the surface areas, pore diameter, and pore volume of intermingled metal oxides and metal oxides.

16.2.2.3 HRTEM

The extent of the intermingling of the two oxides is illustrated in Figure 16.5, which shows the HRTEM image of AP-$MgAl_2O_4$. Individual and aggregated Al_2O_3 boehmite planes mixed with AP-MgO nanocrystals can clearly be observed. Typical spacing between the aggregated boehmite planes is 15 Å. This gives a clear indication that the MgO and Al_2O_3 are intimately mixed throughout the entire material that result in some unique structures. The presence of MgO in the sample has led to an increase in the distance between the planes. Typically in AP-Al_2O_3 the spacing between planes is 6 Å. MgO is now sandwiched between the boehmite planes and the spacing has increased to 15 Å. The ability to disperse the basic nature of MgO throughout the high surface area framework of Al_2O_3 is clearly shown in these images and is an advantageous feature when dealing with surface adsorption behavior.

16.2.3 Relationship to Zeolites

In the previous sections, we have discussed the morphologies of the AP-nanoparticles and the newly synthesized homogeneously dispersed bimetallic oxides. In this section the morphology, pore structure, surface area, and engineered acid–base sites in the AP nanoparticles will be discussed.

Nanocrystalline metal oxides such as MgO, Al_2O_3, CaO, ZnO, and SrO, prepared by the MAP exhibit unprecedented adsorption properties compared to conventionally prepared metal oxides due to a combination of factors such as enhanced surface area and intrinsically higher surface reactivities [1–3]. These unique sorption properties are due to nanocrystal sizes and shapes, polar surfaces, and high surface areas. In addition, it was observed that on compaction of these oxides into pellets, the surface area and the surface reactivity are not degraded when moderate pressures are applied; hence these nanocrystalline oxides can be either used as loose powders or as pellets. An immediate question that comes into mind is how these materials are related to zeolites, which find extensive use as catalysts and adsorbents. Zeolites such as zeolite A find extensive use as adsorbents, ion-exchange resins, and detergents as "builders." Although zeolite Y finds widespread use as catalysts and particularly in oil refining as solid acid catalysts, their microporosity poses diffusional limitations on the reaction rate. Mesoporous molecular sieves such as MCM-41 offer promise but do not

TABLE 16.1
Surface Area Analysis of Intimately Intermingled Metal Oxides and Metal Oxides

Sample	Surface Area (m^2/g)	Pore Diameter (nm)	Pore Volume (cc/g)
AP-$MgAl_2O_4$	639	18	2.88
AP-$CaAl_2O_4$	517	28	3.56
AP-$SrAl_2O_4$	112	32	0.89
AP-$BaAl_2O_4$	135	31	1.05
AP-MgO	385	21	2.03
AP-Al_2O_3	637	20	3.17
AP-CaO	67	13	0.22
AP-SrO	16	13	0.05

Source: From GM Medine, V Zaikovskii and KJ Klabunde. *J Mater Chem* 19:757–763, 2004, The Royal Society of Chemistry. With permission.

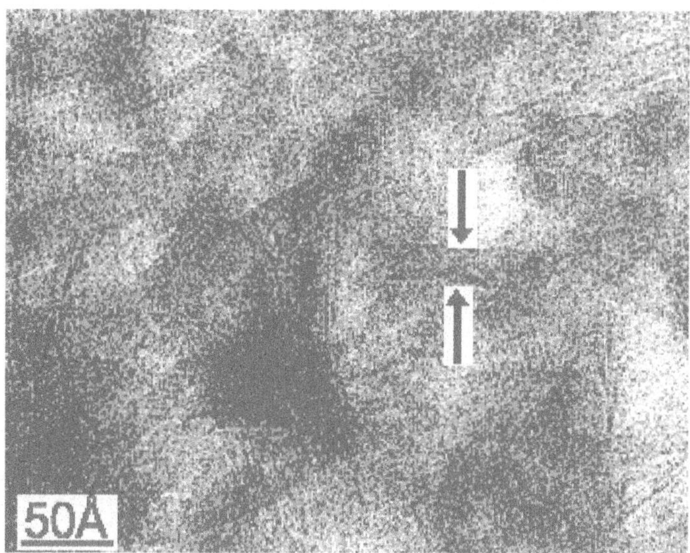

FIGURE 16.5 HRTEM of AP-$MgAl_2O_4$. (Al_2O_3 boehmite planes mixed with MgO nanocrystals). (Reproduced from GM Medine, V Zaikovskii and KJ Klabunde. *J Mater Chem* 14:757–763, 2004, Copyright 2004, The Royal Society of Chemistry. With permission.)

possess high acidity as some zeolites. However, these materials have poor hydrothermal stability, which can be attributed to the amorphous nature of the pore walls.

16.2.3.1 Morphologies

Nanocrystalline MgO, for example, prepared by the MAP, typically yields a fine white powder with ~4 nm average crystallite size. HRTEM imaging of a single MgO crystallite indicates the presence of high surface concentration of edge–corner sites and various exposed crystal planes (1 1 1, 0 0 1, 0 0 2, etc.) [10]. The TEM studies at low magnification show the presence of porous weblike aggregates in the range of 1400 nm and their size distribution is very narrow. Figure 16.6 shows the HRTEM of AP-MgO. The TEM picture reveals very interesting facets of the morphology of MgO. One can see numerous "cube-like" crystallites that aggregate into polyhedral structures with numerous edge–corner sites. A closer examination of the TEM picture also shows the embryonic formation of pores between the crystalline structures. SAXS data indicate a nonfractal surface morphology of the particles with about 100 Å diameter and these particles aggregate to porous particles of ~ 1 μm.

16.2.3.2 Pore Structure

The MAP nanomaterials exhibit adsorption–desorption isotherms typical of bottleneck pores (type E). An interesting aspect of these materials is their behavior toward compression to pellets. The AP-oxides, due to their polyhedral nanocrystal shapes and their tendency to form weblike aggregates, are resistant to collapse under pressure. These ultrafine powders on consolidation into pellet, retain their high surface areas and small crystallite size.

16.2.3.3 Surface Areas

As pointed out earlier, the specific surface areas of the AP oxides are very high compared to the conventionally prepared and commercially available metal oxides. In addition, the AP-oxides retain these high surface areas even upon compaction.

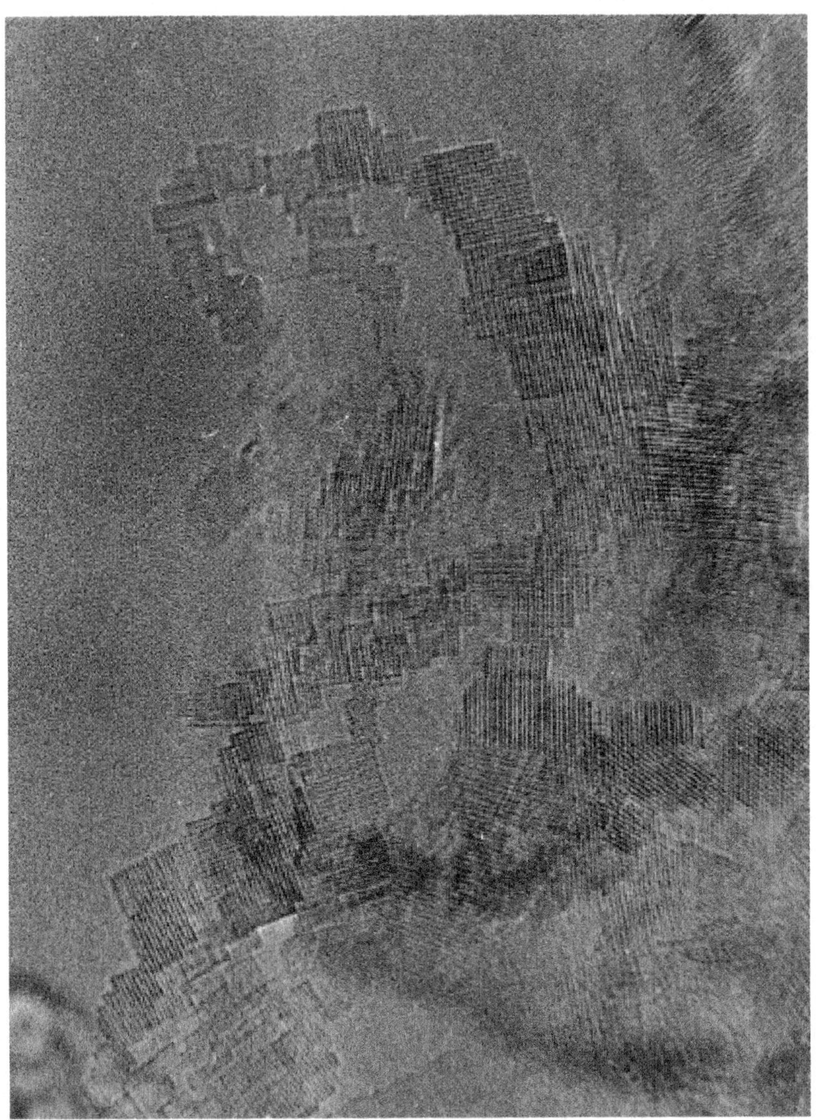

FIGURE 16.6 HRTEM of AP-MgO. (Reproduced from R Richards, W Li, S Decker, C Davidson, O Koper, V Zaikovski, A Volodin, T Rieker, and KJ Klabunde. *J Am Chem Soc* 122:4921–4925, 2000. Copyright 2000, American Chemical Society. With permission.)

16.2.3.4 Engineered Acid–Base Sites

Another interesting facet of the MAP is that one can engineer and tune acid–base sites in them. For example, intimately intermingled $MgO–Al_2O_3$ nanostructures exhibit enhanced capacity or activity over the pure forms of either AP-Al_2O_3 or AP-MgO. The enhanced reactivity is attributed to the Lewis base nature of the very small and very well dispersed MgO crystallites that are "housed" on and within the large surface area and pore volume of Al_2O_3. In addition, such intimately intermingled mixed oxides are highly thermally stable with minimum sintering after heating up to 973 K.

16.2.4 A New Family of Porous Inorganic Sorbents

We have thus far seen how the MAP leads to the formation of ultrafine powders that have very high surface area. HRTEM and SAXS data show that these nanocrystalline materials have a high concentration of surface edge–corner sites. For example, the percentage of edge ions in AP-MgO (4 nm crystallite) is ~20%, while a commercial MgO sample has only about 0.5% of edge ions. High surface area metal oxides are good sorbents, catalysts, and catalyst supports. Thus these nanocrystalline materials, MgO, CaO, Al_2O_3, ZnO, Al_2O_3–MgO, and other intimately mixed metal oxides constitute a new family of materials that are porous in nature and exhibit unparalleled destructive adsorption for polar organic compounds, acid gases, and chemical or biological warfare agents [22]. The chemical properties of these metal oxides can be adjusted by incorporating other oxides as monolayer films or as surface nanocrystals. Upon consolidation as pellets, their high surface area is maintained. Also, pore volumes and pore size openings can be controllably decreased with pelletization pressure increase. The pellets also retain the adsorbent affinities and capacities for a wide variety of organic molecules and acid gases.

16.3 DESTRUCTIVE ADSORPTION

16.3.1 High Temperatures

16.3.1.1 Chlorocarbons

The ever-increasing amount of chlorocarbons in our environment causes the need to find easy and convenient methods to destroy them without producing toxic by-products. The most widespread groups of chlorocarbons include polychlorinated biphenyls (PCBs) used as transformer fluids, heat transfer fluids, lubricants, plasticizers, and cleaning solvents such as CCl_4, $CHCl_3$, and C_2Cl_4. The toxicity and carcinogenic properties of the chlorocarbons have urged researchers worldwide to discover clean and effective means of destroying them. Reports on decomposition of chlorocarbons by catalytic oxidation to yield HCl and CO_2 as main products are available. Incineration methods are currently used and very high temperatures (1300 to 1500°C) are often required in order to avoid the formation of dioxins and furans. The desired reaction is the complete oxidation of chlorocarbons to water, CO_2, and HCl without the formation of any toxic by-products. In recent times, a noncatalytic approach, which depends on surface active agents that strip heteroatoms from the toxic material and allow only the release of nontoxic hydrocarbons or carbon oxides, has been suggested [22]. High surface area and high surface reactivity are desired properties of such destructive adsorbents and nanoparticles fall into this category. Nanoparticles will react to a greater extent than normal metal oxides because there are more molecules of metal oxide available for reaction with chemicals adsorbed on the surface of the particles. Also, nanocrystals exhibit intrinsically higher chemical reactivities owing to their unusual crystal shapes and lattice disorder [22]. A detailed discussion of destructive adsorption of chlorocarbons on nanocrystalline metal oxides is presented below.

Some chlorocarbons are destroyed during reactions with nanocrystalline metal oxides by becoming mineralized as environmentally benign metal chlorides and CO_2 gas [23–25]. The process can be written as $2MO + CCl_4 \rightarrow CO_2 + 2MCl_2$. Considering thermodynamics alone, the reaction of metal oxides with chlorocarbons are energetically favorable, e.g.,

$$2CaO_{(s)} + CCl_4 \longrightarrow 2CaCl_{2(s)} + CO_{2(g)},\ \Delta H^\circ = -573\,\text{kJ/mol}$$

$$2MgO_{(s)} + CCl_4 \longrightarrow 2MgCl_{2(s)} + CO_{2(g)},\ \Delta H^\circ = -334\,\text{kJ/mol}$$

However, since the reaction is a gas–solid reaction sensitive to surface effects, surface area and surface reactivity completely control the reaction pathways in spite of the exothermicity of the above reactions.

CaO and MgO destroy chlorocarbons such as CCl_4, $CHCl_3$, and C_2Cl_4 at temperatures around 400°C to 500°C in the absence of an oxidant, yielding mainly CO_2 and the corresponding metal chlorides. The capacities for destructive adsorption of CCl_4 on various metal oxides are shown in Table 16.2 [2]. The products of the destructive adsorption are nontoxic (CO_x and metal halides) and depend on the type of chlorocarbon decomposed as well as on the ratio between the metal oxide and the chlorocarbon.

As mentioned earlier, the reaction of CCl_4 with the metal oxides is exothermic but the metal oxide particles do not react completely. The mobility of oxygen and chlorine atoms in the bulk of the material becomes important and kinetic parameters involving the migration of these atoms are rate limiting. Generally, the order of reactivity of the oxide samples toward chlorocarbon destruction can be written as AP-CaO > CP-CaO ≫ CM-CaO. It was demonstrated that if MgO (CaO) particles were coated with transition metal oxides (e.g., Mn_2O_3, Fe_2O_3, CoO, or NiO), the reactivity could be enhanced substantially and a kinetic advantage might be gained [26]. Three of the transition metal oxides Fe_2O_3, V_2O_5, and NiO increases the reactivity by almost 1.5 times compared with that of the substrate alone (CaO). It is thermodynamically feasible that the iron oxide can first react with CCl_4 to form iron chloride, which in turn can then react with the substrate to regenerate the iron oxide and mineralize chlorine. The transition metal oxide enhances the O^{2-}/Cl^- exchange by continuously replenishing the oxide surface [27]. It can be noticed that the core shell nanoscale particles possess superior reactivity as shown in Table 16.2. With [Fe_2O_3]–CaO a complete stoichiometric conversion was achieved. It is worthwhile to realize that the nanocrystals are not just reactive at the surface alone; they allow the diffusion of CCl_4 molecules deep inside the particles.

Nanocrystalline MgO and CaO also allow the destruction of chlorinated benzenes (mono-, di-, and trichlorobenzenes) at lower temperatures (700°C to 900°C) than incineration [28].

TABLE 16.2
Efficiencies for Destructive Adsorption of CCl_4 with Metal Oxides at 425°C

Sample[a]	Surface Area (m^2/g)	Transition Metal Loading (wt%)	Performance Efficiency[b]
CM-MgO	10–30	0	0.12
CM-CaO	1.3	0	<0.01
CP-MgO	190	0	0.16
CP-CaO	100	0	0.23
AP-MgO	450	0	0.17
AP-CaO	140	0	0.31
[Fe_2O_3] CP-MgO	150	0.88	0.23
[Fe_2O_3] CP-CaO	90	1.6	0.44
[Fe_2O_3] AP-MgO	400	1.4	0.36
[Fe_2O_3] AP-CaO	130	1.6	0.51
[Mn_2O_3] AP-MgO	290	4.5	0.42
[NiO] AP-CaO	130	1.6	0.39
[CoO] AP-CaO	80	1.6	0.30

Source: From KJ Klabunde, JV Stark, O Koper, C Mohs, DG Park, S Decker, Y Jiang, I Lagadic, and D Zhang. *J Phys Chem* 100:12142–12153, 1996. With permission.

[a] CM, commercial; CP, conventionally prepared; and AP, aerogel prepared.

[b] Number of moles of CCl_4 destroyed according to the stoichiometric equation $2MO + CCl_4 \rightarrow 2MCl_2 + CO_2$; a maximum value of 0.5 is expected.

The presence of hydrogen as a carrier gas allows still lower temperatures to be used (e.g., 500°C). MgO was found to be more reactive than CaO and the latter induces the formation of more carbon. It was found that toxins as products (particularly the dibenzo-*p*-dioxin backbone with 0 to 3 chlorine substituents) are not produced with MgO but traces of toxins are produced on the surface of CaO. Also, chlorine-containing volatile products were never produced from the MgO reactor bed. Only easily combustible products such as benzene and CO along with CO_2 and H_2O are produced.

Mixed metal oxides such as $MgO–Al_2O_3$ also possess promising capabilities to destroy CCl_4 at high temperatures (Table 16.3) [5], and behave as stoichiometric reagents in the CCl_4 destructive adsorption, due to their unusual morphologies and crystal disorder.

16.3.1.2 Organophosphorus Compounds

Many organophosphorus compounds are toxic and some are members of group of chemical warfare (CW) agents. Usually charcoal filters are used to adsorb such toxic substances. Better alternatives such as thermal and catalytic methods have been sought for quite sometime. Development of effective catalysts would be better but heteroatom-containing pollutants are catalyst poisons. It is of great interest to develop and understand the surface chemistry of solid reagents that simultaneously adsorb, immobilize, and destroy such pollutants. Nanoparticles have been found to be a very good alternative for detoxifying organophosphorus compounds and one such example is discussed below.

Dimethyl methylphosphonate (DMMP) has been found to be a good mimic for toxic organophosphorus compounds and DMMP possesses some of the same types of bonds that exist in the more hazardous CW agents. It has been found that DMMP can be destructively adsorbed at 500°C by high surface area magnesium oxide, e.g., nanocrystalline MgO with a surface area of 390 m^2/g can destroy about 65 μl of DMMP [8]. The chemistry is limited to the surface of the particles and it was found that two surface moieties of MgO are required to decompose one molecule of DMMP. The decomposition process is a stoichiometric process rather than a catalytic one and high surface area of the metal oxide is crucial. DMMP decomposes on the surface of nanocrystalline MgO to give volatile products such as formic acid and methanol and the nonvolatile products formed are $[CH_3(CH_3O)P]_{ads}$ and $[CH_3(CH_3O)PO]_{ads}$, which are completely immobilized on the surface of MgO. The process requires two surface molecules of MgO. DMMP decomposes on the surface of MgO and the following products are obtained [27].

$$3[CH_3(OCH_3)_2P{=}O] \xrightarrow{6MgO_{(surface)}} [CH_3(OCH_3)PO]_{(a)} + 2[CH_3(OCH_3)P]_{(a)} + OCH_{3(a)} + 2HCOOH_{(g)} + 2H_{(a)}$$

TABLE 16.3
Reaction of CCl_4 with Oxide Samples at 500°C

Sample	Molar Ratio[a]
AP-Al_2O_3	1.44 mol of CCl_4:1 mol of Al_2O_3
CM-Al_2O_3	1 mol of CCl_4:16 mol of Al_2O_3
AP-(1/1) Al_2O_3–MgO	1.8 mol of CCl_4:1 mol of Al_2O_3–MgO
CM-MgO	1 mol of CCl_4:32 mol of MgO

[a]Theoretical molar ratio: 1 mol of CCl_4:2 mol of MgO, 1.5 mol of CCl_4:1 mol of Al_2O_3, 2 mol of CCl_4:1 mol of Al_2O_3–MgO.

16.3.2 Ambient Temperatures

It is well known that nanocrystalline materials exhibit a wide array of unusual properties and are considered as new materials that bridge molecular and condensed matter. One aspect that is particularly interesting is their enhanced surface chemical reactivity toward incoming adsorbates. It has been shown previously that nanocrystalline metal oxides possess surfaces that interact very strongly with polar organics. The method of adsorption is usually through dissociative chemisorption, and these are also examples of destructive adsorption.

16.3.2.1 Organophosphorus Compounds

Nanocrystalline metal oxides are extremely good destructive adsorbents and this type of adsorption is able to occur at ambient temperatures. Nanocrystalline metal oxides also have an advantage over typical adsorbents, such as activated carbon, which do not destructively adsorb the incoming species. Thus, these materials, with their high surface reactivities and polar nature, bring an extremely useful dimension to the adsorption of polar organics and CW agents. There are several organophosphorus compounds that are surrogates of CW agents and undergo destructive adsorption at ambient temperatures. Some common examples of these surrogates are diisopropyl phosphorofluoridate (DFP), DMMP, and paraoxan, as shown in Figure 16.7.

16.3.2.1.1 Paraoxon Adsorption

A comparison of the samples studied for the destructive adsorption of paraoxon is given in Table 16.4. It can be seen that both AP-MgO and AP-Al_2O_3 are very effective for paraoxon adsorption. AP-Al_2O_3 has nearly double the surface area of AP-MgO and performs marginally better than AP-MgO. Of the 16 μl paraoxon used in the test, AP-Al_2O_3 destructively

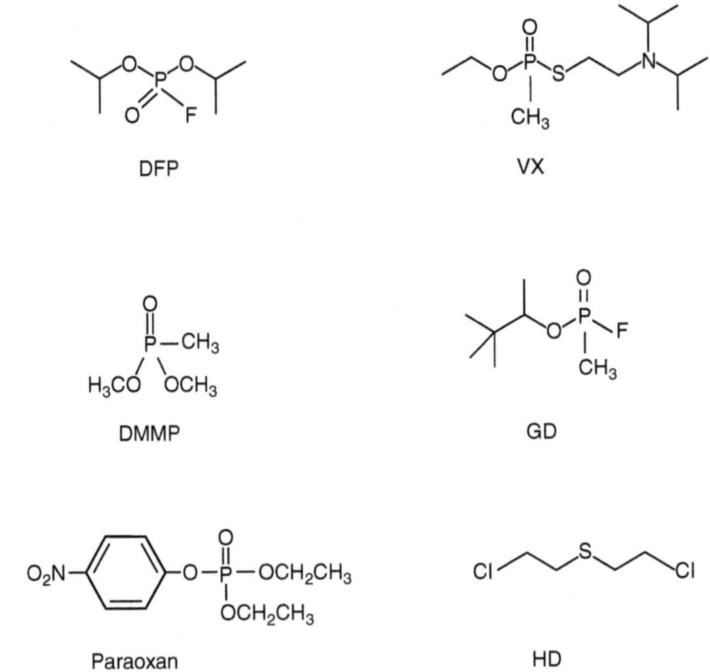

FIGURE 16.7 Structures of DFP, DMMP, paraoxan, VX, GD and HD.

adsorbed 16 μl and AP-MgO was able to adsorb 15 μl. AP-CaO and AP-SrO did not perform as well as AP-MgO and AP-Al_2O_3 due to their much lower surface areas.

The intimately intermingled metal oxides also exhibit varying performance toward the destructive adsorption of paraoxon. AP-$MgAl_2O_4$ performs the best out of these samples and was able to adsorb 16 μl in approximately 20 min, about the same as pure AP-Al_2O_3. However, AP-$MgAl_2O_4$ is much faster at adsorbing paraoxon than both AP-MgO and AP-Al_2O_3 individually. AP-$CaAl_2O_4$ also performs extremely well and adsorbs 16 μl. It behaves in a similar manner to that of AP-Al_2O_3. The presence of CaO in the intimately intermingled metal oxide does not seem to have such an important role as that of MgO in AP-$MgAl_2O_4$. It appears that the Lewis base nature of AP-MgO when incorporated into the very high surface area and pore volume of AP-Al_2O_3 is advantageous for the rapid destructive adsorption of paraoxon. However, it can also be seen that AP-$SrAl_2O_4$ and AP-$BaAl_2O_4$ do not perform well in the adsorption of paraoxon. These samples have lower surface area and also have smaller pore volumes. From TEM studies it was noted that these Sr and Ba analogs had larger agglomerates and were not "intermingled" very well.

By using paraoxon, which is less toxic than VX, it is possible to get a detailed explanation of the surface chemistry between MgO and paraoxon. The P—OAr bond is immediately cleaved to yield surface-bound *p*-nitrophenoxy anion (yellow) followed by the loss of the ethoxy groups. The paraoxon is destructively adsorbed and becomes dismantled and immobilized in large quantities. Calculations show that about one paraoxon molecule per nm^2 of MgO is destructively adsorbed. This is also the case for samples of Al_2O_3, $MgAl_2O_4$, and $CaAl_2O_4$ [5, 29].

16.3.2.1.2 DFP Adsorption

AP-MgO has also shown superior adsorbent capabilities when compared to CP-MgO and CM-MgO. AP-MgO is able to adsorb all the DFP, whereas the other MgO samples did not completely adsorb the DFP. IR and NMR results suggest that DFP is rapidly adsorbed on the AP-MgO through a bridge "POO" structure with the loss of the P—F bond, followed by a slower destructive adsorption of the P—O—C bonds [29].

16.3.2.2 Chemical Warfare Agents

Examples of CW agents include VX [*O*-ethyl-*S*-(2-diisopropylamino) ethylmethyl-phosphonothioate], GD (pinacolyl methylphosphono-fluoridate), and HD [bis(2-chloroethyl)sulfide] are shown in Figure 16.7.

TABLE 16.4
Summary of Samples Studied and Their Reactivity with Paraoxon

Sample	Surface Area (m^2/g)	Pore Volume (cc/g)	Amount of Paraoxon Adsorbed (μl)	Molar Ratio (Moles of Paraoxon to Moles of Material)
AP-MgO	385	2.03	15	1:36
AP-CaO	67	0.22	4	1:113
AP-SrO	16	0.05	2	1:104
AP-Al_2O_3	637	0.27	16	1:14
AP-$MgAl_2O_4$	639	2.88	16	1:10
AP-$CaAl_2O_4$	517	3.56	16	1:9
AP-$SrAl_2O_4$	112	0.89	6	1:19
AP-$BaAl_2O_4$	135	1.05	6	1:17

Source: From GM Medine, V Zaikovskii and KJ Klabunde. *J Mater Chem* 14:757–763, 2004, The Royal Society of Chemistry. With permission.

The presence of CW nerve agents in the world today and methods of safely decontaminating them are of great concern to the general public. These compounds are polar organic liquids under ambient conditions and are mostly P(V) organophosphorus esters. They tend to be very similar to insecticides that are known to irreversibly react with the enzyme acetylcholinesterase (AChE), which prevents the enzyme from controlling the central nervous system. Currently, the U.S. and Russia have declared CW agent stockpiles of 25,000 and 42,000 t, respectively [30].

There are many reactions that are employed to decontaminate CW nerve agents and their surrogates (compounds with similar structure but not as toxic). Only a few are feasible for practical neutralization because reactions must be simple and reactants must be stable, cheap, and relatively should have low molecular weight.

A majority of the work carried out on CW nerve agents has been done by the U.S. Army at the Aberdeen proving ground in Maryland. Several studies have been carried out by Yang and coworkers. Some of these methods for neutralizing CW agents will be mentioned here. Two common reactions that are employed to neutralize and detoxify CW agents under ambient conditions are nucleophilic and oxidation reactions [30–32].

Nucleophilic substitution reactions can be carried out with basic hydrogen peroxide solutions. It has been found that the HO_2^- ion is able to neutralize VX in basic peroxide solution at ambient temperature. Both sarin (GB) and soman (GD) also react with HO_2^- at a rate that is 20 to 30 times greater than that with OH^- [31]. The HO_2^- ion also reacts with mustard gas (HD) to produce two soluble sulfoxides. These studies have shown that HO_2^- ion is a more effective agent for neutralization than the nucleophile OH^-. One important note is that the reactor temperature must be kept low to prevent H_2O_2 decomposition.

The use of oxidation reactions is another approach that has been used to neutralize CW agents. Aqueous bleach (NaOCl or $Ca(OCl)_2$) and chlorine gas are examples of reagents that have been used in oxidation reactions. The reaction of OH^- with sarin and soman is very effective but this is not the case for VX and mustard gas. An alternative that has been studied is the oxidation of VX and mustard gas in aqueous hypochlorite (ClO^-) or hypochlorous acid (HClO). This method has also been proven very effective for G-agents (nerve agents), which were also detoxified with ease. The reaction of mustard gas and bleach occurs interfacially via a series of sulfur oxidation, substitution, elimination, and chlorination. In basic bleach VX has solubility problem, however, it becomes more soluble in neutral or acidic conditions. At lower pH values less active chlorine is consumed during reaction because the amino group is protonated and thus protected from oxidation [32]. For this type of neutralization the most useful reactants are simple nucleophiles such as OH^-, HO_2^-, and alkoxides; and stable oxidants such as bleach, H_2O_2, and persulfates [31].

A recent method that has been studied for the decontamination of CW agents is the use of inorganic oxides to destructively adsorb them. Dry powders have several advantages, some of which include that they are nontoxic, easy to handle and store, waterless, have low logistical burden, and dry products result so there is no liquid waste stream. These materials have also been used as adsorbents, catalysts, and catalyst supports. It has also been proposed that they are suitable for many decontamination applications. This ability to destructively adsorb materials makes metal oxide nanocrystals fundamentally different from normal adsorbents such as activated carbon.

16.3.2.2.1 Reactions of VX, GD, and HD with Nanosize MgO and CaO

Nanosize MgO and CaO can be used in room temperature reactions with CW agents such as VX, GD, and HD. Reaction with nerve agents VX and GD and metal oxides results in hydrolysis of the nerve agents. This is illustrated in Scheme 16.1 for the reaction of VX and GD with MgO. The reactions are analogous to their solution behavior with two differences. The first difference is that the corresponding nontoxic phosphonate products reside as surface-

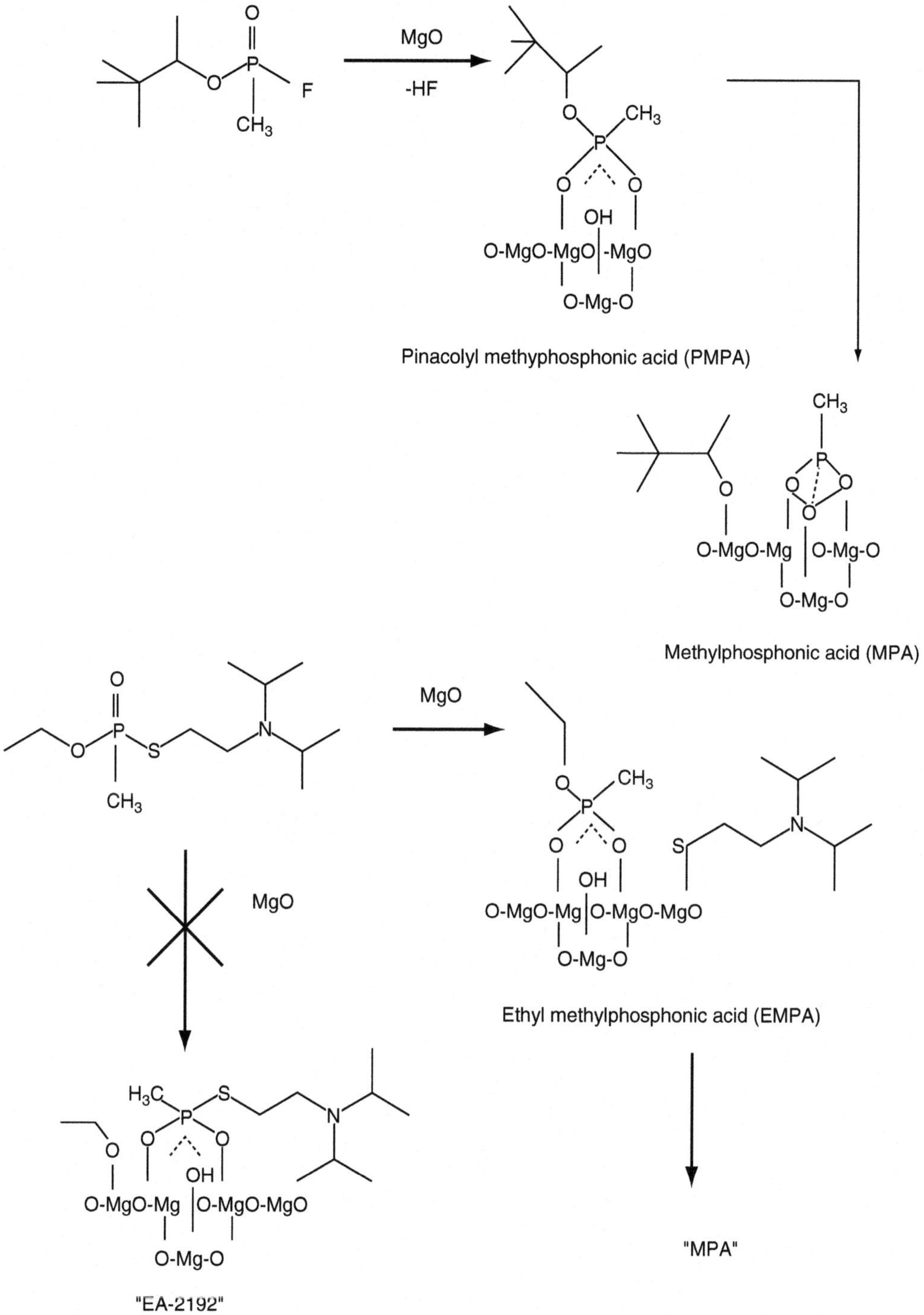

MgO

-HF

Pinacolyl methyphosphonic acid (PMPA)

Methylphosphonic acid (MPA)

MgO

MgO

Ethyl methylphosphonic acid (EMPA)

"EA-2192"

"MPA"

SCHEME 16.1 Reaction of VX and GD with AP-MgO. (Copyright GW Wagner, LR Procell, RJ O'Connor, S Munavalli, CL Carnes, PN Kapoor, and KJ Klabunde. *J Am Chem Soc* 123:1636–1644, 2001, American Chemical Society. With permission.)

bound complexes. The second noticeable difference is that toxic EA-2192, which is known to form under basic hydrolysis, is not observed on MgO or CaO. An example of the selectivity for VX hydrolysis is shown in the slow reaction of VX with an equimolar amount of water.

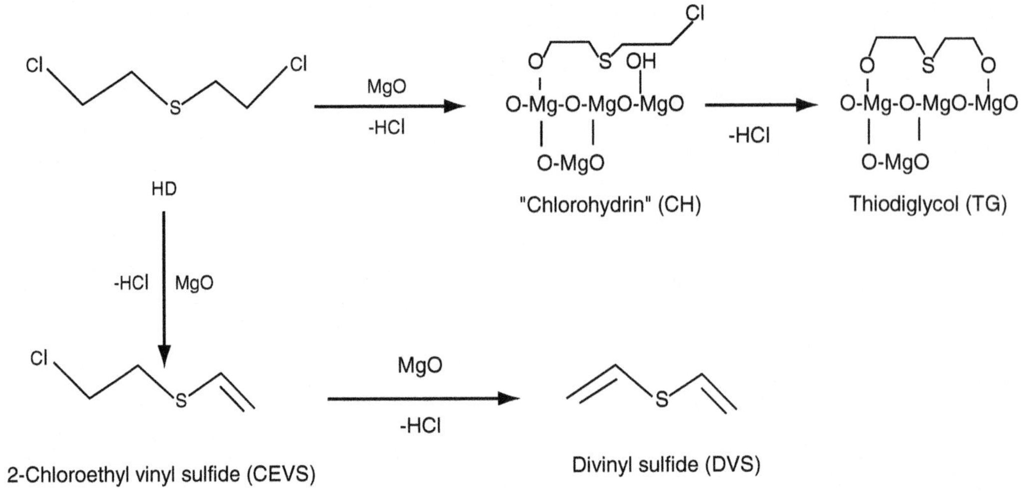

HD

MgO -HCl

"Chlorohydrin" (CH)

-HCl

Thiodiglycol (TG)

-HCl MgO

MgO -HCl

2-Chloroethyl vinyl sulfide (CEVS)

Divinyl sulfide (DVS)

SCHEME 16.2 Reaction of HD with AP-MgO. (Copyright Wagner 2001, American Chemical Society. With permission.)

Hydrolysis is also observed for HD, elimination of HCl is also a major reaction pathway as shown in Scheme 16.2. The product distribution on MgO is 50% thiodiglycol (TG) and 50% divinyl sulfide (DVS). Furthermore, HD on CaO exhibits autocatalytic dehydrohalogenation in which 80% of DVS forms along with TG and minor amounts of the sulfonium ion, CH-TG. Peaks observed by ^{13}C MAS NMR for TG on MgO and CaO suggest that the product resides as a surface-bound alkoxide [33–35].

16.4 BIOCIDAL ACTION OF NANOPARTICLE FORMULATIONS

Biological warfare agents include bacteria, fungi, viruses, and toxins. All these have significant differences in their behavior and hence different decontamination procedures have to be adopted to destroy them. Thus, a universal procedure to destroy them is not practically feasible. The most common and widely spread method of disinfecting bacteria, viruses, and certain fungi is to use a bleach solution or chloramine T solution. They quickly and completely destroy bacteria but have the disadvantage that they age quickly and hence their activity deteriorates significantly. Further, the bleach solutions are corrosive and cannot be used for cleaning sensitive surfaces and they also have a pungent smell that can be very irritating. Another method for decontamination is employing gases such as chlorine and chlorine dioxide. However, it too finds only limited use since the gases cannot be sprayed in open atmosphere and can damage sensitive surfaces. There are reports of various antimicrobial formulations and drugs in the form of nanoparticles that show promising results [36]. There has been a recent report in which TiO_2 nanoparticle embedded in a hybrid organic–inorganic membrane shows bactericidal activity under UV illumination [37].

An alternative approach to the above-mentioned procedures is the application of nanoparticles in the form of very fine powder. Solid materials have seldom been the choice as decontaminants since they have generally been observed to react slowly and incompletely with biological toxins. The advantages of using fine powders are that they can be applied on sensitive surfaces, are less corrosive, and after the decontamination procedure, they can be easily removed by vacuum cleaning. Nanoparticles increasingly find use for filtering and destroying biological species. For example, KES Science and Technology markets titanium dioxide filters that destroy airborne pathogens such as anthrax. Emergency filtration products produce 2-H Nano-enhanced Environmental Mask (Nanomask®), which contains reactive nanoparticles. Nanoparticles such as AP-MgO, CaO, and ZnO show some biocidal activity toward vegetative bacteria, but are not very effective toward spores [38]. However, nanoparticle metal oxides containing halogens have shown promise for the decontamination of several biological toxins including spores [39]. The sorption of halogens and the mode of action of the halogenated MgO nanoparticle will be discussed in the following paragraphs.

16.4.1 Sorption of Halogens

As discussed previously, nanoparticles possess very high surface areas and higher reactivity compared to their bulk counterparts. Thus, nanoparticles have high adsorption capacity per mass adsorbent. This allows for the adsorption of relatively high amounts of halogens. Halogens are excellent bactericides in general. However, in the free form their use is restricted because of their toxicity, high volatility, and corrosiveness. As discussed in a previous section, the presence of edge–corner sites and other defect sites such as vacancies allow the nanoparticles to possess high surface concentrations of reactive species. This presents an opportunity to prepare materials where reactive Lewis acid and Lewis acid–base sites can be augmented as biocidal materials by forming stable halogen adducts.

Thermogravimetic analysis (TGA) studies indicate that AP-MgO–halogen adducts are capable of adsorbing as much ~7 wt% free chlorine, 16 wt% free bromine, and as high as 20 wt% free iodine [8]. The amount of halogen released in the TGA experiments represents only the amount of adsorbed halogen on the MgO surface and does not indicate the presence of free halogen retained in the pores. In case of the AP-MgO–Cl_2 adduct, knowing the surface area of AP-MgO (~550 m^2/g) and the amount of Cl_2 released, the number of Cl atoms adsorbed is calculated to be in the range of 5 to 7 atoms/nm^2. In contrast a commercial CM-MgO sample adsorbs only ~1 wt% of chlorine, bromine, and iodine. Another important feature of the halogenated AP-MgO adduct is that the adducts are stable for weeks and have only a weak smell of halogens. The CM-MgO–halogen adducts release the adsorbates quite readily and have a strong smell of halogens. For example, the CM-MgO–I_2 adduct loses iodine crystals after about a week while the corresponding AP-MgO–I_2 adduct is stable for several months without any noticeable loss of iodine. Another important feature of the AP-MgO–halogen adducts is that they exhibit significant absorption in the UV–Vis region. The AP-MgO–Cl_2 adduct is off-white or light yellow in color but the AP-MgO–Br_2 adduct is bright yellow, while the AP-MgO–I_2 adduct is brown. In all experiments, the CM-MgO adducts lost their adsorbed halogen earlier compared to the AP-MgO adducts. These observations clearly suggest that the strength of adsorption is considerably larger in AP-MgO nanoparticles than that of the commercial CM-MgO material. FT-Raman spectra of the AP-MgO–halogen adducts show a downshift in the vibration frequency compared to the bare AP-MgO. This indicates that the halogen molecules adsorbed on the nanoparticle surface are in a more chemically active state than the corresponding gas-phase halogen molecules. In the following section, we will discuss the mode of bactericidal action.

16.4.2 Bactericidal Action

The bactericidal action of AP-MgO–halogen adducts are due to a combination of factors such as abrasiveness, basic character, electrostatic attraction, and oxidizing power (due to the presence of halogen).

16.4.2.1 Abrasiveness

Magnesium oxide prepared through an aerogel procedure yields square- and polyhedral-shaped nanoparticles with diameters varying slightly around 4 nm with considerable pore volume and having an extensive porous structure. Because of their high surface area and enhanced surface reactivity, the AP-MgO nanocrystals adsorb high amounts of active halogens. Their small crystallite size, allows many nanoparticles to cover bacterial cell walls and bring the active halogens in close proximity to the bacterial cell wall. Experiments have shown that AP-MgO–halogen adducts show excellent activity against *Escherichia coli* and *Bacillus megaterium* as well as very good activity against spores from *Bacillus subtilis*. Thus, AP-MgO is more abrasive than CM-MgO and this property leads to mechanical damage of the cell membranes.

16.4.2.2 Basicity

Several reports in the literature suggest that bactericides and sporicides show higher efficiency when paired with alkaline substances or when the spores are pretreated with such alkaline substances [40]. When AP-MgO—halogen nanoparticles are dissolved in water, an alkaline environment is created. Alkaline compounds dissolve the cell wall or the external part of the spore coat. The TEM results indicate that the cause of activity of the AP-MgO–halogen nanoparticles is in their disruptive action of the cell wall of the bacteria and spores.

16.4.2.3 Electrostatic Attraction

It is a well-established fact that the overall charge of the bacteria and the spore cells at biological pH values is negative, because of the excess number of carboxylic and other groups which on dissociation make the cell wall surface negative. Confocal laser microscopy studies show that the bacteria and the nanoparticles aggregates coagulate. TEM studies also confirm that the aggregates are composed of both nanoparticle aggregates and bacteria. The reason for the coagulation of the two is an electrostatic attraction between them. ξ-potential measurements show that AP-MgO and AP-MgO–halogen adducts are positively charged; AP-MgO (35.2 mV), AP-MgO–Cl_2 (33.0 mV), and AP-MgO–Br_2 (27.0 mV). Thus, electrostatic attraction is another reason for the activity of the AP-MgO nanoparticles, since this greatly facilitates the coming together of the nanoparticles and the bacteria or spores.

16.4.2.4 Oxidative Power

AP-MgO–halogen nanoparticles contain adsorbed halogen and also free halogen physically retained in the pore structure. Raman studies indicate that the halogen molecules adsorbed on AP-MgO nanoparticle are in a more "chemically active state" than the corresponding gas-phase halogen molecules. This "active chlorine" is capable of oxidizing bacteria much like chlorine or sodium hypochlorite.

The advantages of the powders are that they are stable for long periods of time, light and portable, and can hence be easily used in any environment and more importantly once they are dispersed for days in open air, they are converted to harmless common minerals. Furthermore, the AP-MgO–halogen adducts act rapidly, and ~20 min is enough for efficient decontamination.

16.4.3 Detoxification of Waterborne Toxins

As we have seen in previous chapters and in the preceding section in this chapter, nanoparticles exhibit unique ability to adsorb a wide range of polar organic molecules, toxic chemicals, a wide range of acids, CW agents, and their surrogates, and bacteria. Thus, the data generated indicate that nanoparticles offer promise in restoring air and water quality by offering more efficient remediation technologies. Another facet of these nanoparticles is that they can destroy even waterborne natural toxins. Dry contact of AP-MgO–halogen adducts with aflatoxins and contact with MS2 bacteriophage (surrogate of human enterovirus) in water also causes decontamination in minutes. All powder samples tested exhibited a dose-related response to AP-MgO–Cl_2 and AP-CaO–Cl_2. In addition, exposure time response of the bacteriophage to AP-MgO–Cl_2 and AP-CaO–Cl_2 was observed. All formulations inhibited bacteriophage MS2 plating at the 10 mg/ml concentration. In the 5 min exposure test AP-CaO–Cl_2 completely inhibited the MS2 bacteriophage's ability to form plaques at the formulation concentration of 1 mg/ml. Thus, metal oxide nanoparticles have disinfecting ability toward MS2 bacteriophage.

16.5 PHOTOCATALYSIS

In this section, we will discuss the application of nanomaterials in the field of photocatalysis. Although this is covered extensively in other chapters, a brief summary is described in the following paragraphs. The term "photocatalysis" has been introduced as early as 1930, but was largely the interest of a select group of researchers. It represented a branch of chemistry where catalytic reactions proceeded under the action of light. However following the pioneering work reported by Fujishima and Honda in the 1970s [41] and the subsequent energy crisis of the 1970s, several research groups have attempted to explore this largely neglected area of chemistry. Interest has also been stimulated in trying to mimic natural photosynthesis, a process that is responsible for sustaining all living organisms on Earth.

In view of the often contradictory and confusing terminology used by different scientists, IUPAC recently proposed the following definition of the term photocatalysis [42]. "Photocatalysis is a change in the rate of chemical reactions or their generation under the action of light in the presence of substances called *photocatalysts* that absorb light quanta and are involved in the chemical transformations of the reaction participants."

16.5.1 TiO_2

Several semiconductors possess bandgaps suitable to catalyze chemical reactions but titanium dioxide has become a "gold standard" semiconductor photocatalyst in the field of photocatalysis. This is because TiO_2 is cheap to produce and use and is chemically and biologically inert. However, the drawback of TiO_2 is that it does not absorb visible light since it has a bandgap of ~3.2 eV which means that it can only be excited by UV light. During the last 30 years since the first report of photosplitting of water by TiO_2 semiconductor over 25,000 articles have been published employing TiO_2 or modified TiO_2 in powder form as well as films as photocatalysts. Initial studies were focused on splitting of water but increasingly they are found to be effective in degrading water and air contaminants [43]. In recent years, applications to environmental remediation have been one of the most active areas in heterogeneous photocatalysis. TiO_2 has been successfully used for the remediation of a variety of organic compounds such as hydrocarbons, chlorinated hydrocarbons such as CCl_4, $CHCl_3$, C_2HCl_3, phenols, chlorinated phenols, surfactants, dyes, reductive deposition of heavy metals such as Pt^{4+}, Pd^{2+}, Au^{3+}, Rh^{3+}, Cr^{3+}, etc. from aqueous solutions to surfaces as well as destruction of biological materials such bacteria, viruses, and molds [44–50].

The basic principle operating in TiO_2 photocatalysis is as follows. Photoexcitation of TiO_2 with light energy greater than or equal to the bandgap ($h\nu \geq E_g$) leads to the formation of an electron–hole pair (e^-–h^+). The excited state conduction band electrons can recombine with the holes and dissipate the input energy (this process is promoted by amorphous defects and thus most amorphous TiO_2 materials show little if any photoactivity), get trapped in surface states or react with electron donors or electron acceptors adsorbed on the semiconductor surface.

The mechanism of TiO_2-photocatalyzed reactions has been the subject of extensive research. Although the detailed mechanism differs from one pollutant to another, it has been widely recognized that superoxide and specifically hydroxyl radicals $^{\bullet}OH$ act as active species for the remediation of the organic compounds. The radicals are formed by the following reactions (16.1) to (16.4).

$$O_2 + e^-_{cb} \longrightarrow O_2^{\bullet -} \tag{16.1}$$

$$H_2O + h^+ \longrightarrow {}^{\bullet}OH + H^+ \tag{16.2}$$

$${}^{\bullet}OH + {}^{\bullet}OH \longrightarrow H_2O_2 \tag{16.3}$$

$$H_2O_2 + O_2 \longrightarrow OH^- + {}^{\bullet}OH + O_2 \tag{16.4}$$

Insertion of $^{\bullet}OH$ radicals into C—H bond leads ultimately to the complete mineralization of the organic substrate. The reactive hydroxyl radicals can recombine and lead to H_2O_2; this however leads to incomplete mineralization.

TiO_2 exists in three different forms, anatase, rutile, and brookite, with anatase showing higher photocatalytic activity. Most of the previous study focused on the photodegradation of organics dissolved in aqueous solution and employed TiO_2 in the form of a powdered dispersion. The use of slurries of TiO_2 in commercial photoreactors is widespread since such dispersions are cheap, effective, and easy to replace. However, the product or reactants will have to be separated from the photocatalyst slurry and since this involves an additional step, attempts have been made to immobilize TiO_2 onto glass as a film. In most academic research such films are produced by sol–gel process, in which a Ti complex, such as titanium isopropoxide is hydrolyzed in a controlled manner with water. The hydrolyzed product is heated to produce TiO_2 thin films varying in thickness from 50 to 500 nm. However, most commercial devices that employ TiO_2 thin films produce them using chemical vapor deposition.

16.5.2 Visible Light Photocatalysts

An ideal photocatalyst would be one that is easy and cheap to produce and that can be activated by sunlight. Visible light accounts for about 43% of the incoming sunlight. However, no stable and efficient systems that utilize sunlight for practical use have been realized yet. Thus, the "holy grail" to efficiently utilize sunlight remains largely unfulfilled at this moment. Semiconductors such as CdS absorb in the visible region but are not stable under illumination. Unfortunately, TiO_2 absorbs light in the UV region and hence there has been considerable interest in developing semiconductor photocatalysts that can absorb visible light. Attempts have been made in this direction to extend the spectral response of TiO_2 into the visible region where less expensive sources of light can be used to activate TiO_2.

A common approach to solving this problem has been to dope TiO_2 with different transition metal ions in order to create a bathochromic shift of the bandgap energy. Doping of a transition metal ion into the TiO_2 bandgap creates d band states (donor or acceptor) between the conduction band and valence bandgap of TiO_2. For a random distribution of acceptor and donor dopants, a Gaussian density of states appears at the lower or upper part of the conduction and valence bands. The density of states is directly proportional to the dopant concentration and hence this gives one a convenient tool to fine-tune the absorption of TiO_2. The benefit of transition metal doping is the improved trapping of the electrons to inhibit electron–hole recombination. However, a high doping concentration may be detrimental since the photocatalytic activity decreases with growing level of dopant. The results obtained with transition metal ions-doped catalysts have not been entirely predictable. Transition metal dopants such as Cr^{3+} create sites, which increase electron–hole recombination and thus can reduce the activity [51]. For example, doping of Ti^{4+} ions with Co^{3+} ions had extended the spectral response of TiO_2 into the visible region but was found to be detrimental to the activity of TiO_2 for the liquid-phase degradation of $CHCl_3$; however, Fe^{3+}, Mo^{5+}, and V^{4+} doped TiO_2 were found to be active [52]. Recent reports indicate that Cr, V, and Pt containing TiO_2 are promising photocatalysts under visible light irradiation [53, 54]. TiO_2 doped with $PtCl_4$ during the sol–gel process was found to be visible light active for the degradation of 4-chlorophenol. Doped anatase-like binary oxides was found to be active in the gas-phase mineralization of toluene. Highly doped V and Cr containing anatase (dopant atom content ~20%) was found to be more active than only anatase [54].

Another approach to extend the visible light response of TiO_2 is via surface sensitization. Chemisorption or physisorption of dyes can extend light response to the visible region and at the same time increase the efficiency of the excitation process. Some common dyes that are used as sensitizers include rhodamine B, erythrosine B, thionine, and analogs of $Ru(bpy)_3^{2+}$.

Recently, a new strategy has been employed to modify TiO_2 to extend its spectral response to the visible region. Nitrogen-doped TiO_2 powders and thin films have been reported to show photocatalytic activity under visible light irradiation [55]. Sakatani et al. reported the decomposition of acetaldehyde to CO_2 under visible light irradiation in the presence of $TiO_{2-x}N_x$ powder, which is presumed to be TiO_2 with nitrogen substituted at oxygen sites [56]. The visible light sensitivity in $TiO_{2-x}N_x$ is presumed to be caused by narrowing of the bandgap by mixing N 2p and O 2p states, where the O 2p state forms the valence band of TiO_2.

Introduction of Ni into $InTaO_4$ was also found to extend the light response to the visible region [57]. The composite $In_{1-x}Ni_xTaO_4$ was found to be a visible active photocatalyst for splitting of water but so far there have been no reports on the use of this photocatalyst for environmental remediation.

In the following section, we will briefly discuss new nanoscale photocatalysts that are active in the visible and UV range. We will also discuss photocatalysis results pertaining to the degradation of 2-chloroethyl ethyl sulfide (2-CEES), which is a close simulant of bis (2-chloroethyl sulfide), the main component of mustard gas (HD) using TiO_2 photocatalysts [58]. In addition, the degradation of acetaldehyde (indoor pollutant, emitted from cooking hamburgers, building materials such as polyurethane foams, and consumer products such as adhesives, coatings, lubricants, inks, and nail polish) using transition metal-doped SiO_2, TiO_2–SiO_2, microporous ETS-10, and mesoporous AlMCM-41 will be discussed [59–62].

16.5.3 New Nanoscale Photocatalysts

We have developed transition metal incorporated titania–silica aerogels (M–TiO_2–SiO_2, M = transition metal is generally in the form of oxides) and M–SiO_2 that exhibit UV and visible light activity. The activity of Co, Cr, and Mn incorporated titania–silica aerogels is similar or higher under visible light (>420 nm) than commercial Degussa P25 utilizing UV light [62]. In the absence of silica, transition metal-doped titania did not exhibit any significant activity in the visible region. It has also been found that high surface area amorphous SiO_2 can serve as a support for transition metal ion doping for visible light photocatalysis [63]. Co and Cr oxides supported on SiO_2 were found to exhibit high activity for the gas-phase photooxidation of acetaldehyde to CO_2. In addition, transition metal ion-doped mesoporous M–Al–MCM-41 and microporous M-ETS-10 photocatalysts have been discovered. In the following sections, we will discuss some of the new photocatalysis results obtained in our laboratory.

16.5.4 New Photocatalysis Results

16.5.4.1 2-CEES

The ability of nanosized TiO_2 particles to mineralize organics under ambient conditions when illuminated with UV light has attracted much attention in the past two decades. The interest to this has been continually fed by numerous potential practical applications of TiO_2–UV system, such as self-cleaning, antifog windows, photoreactors for water and air purification, etc.

The possibility to use TiO_2-based photoactive substances for decontamination of surfaces exposed to CW agents has recently been recognized. Among all CWs, HD, a mustard gas with bis(2-chloroethyl) sulfide ($(ClCH_2CH_2)_2S$) as a major component, is found to be the most stable one.

The potential of water for nonphotolytic hydrolysis of HD has been assessed [63]. HD was found to undergo rapid and irreversible hydrolysis to thiodiglycol in a highly diluted solution. Hydrolysis of large amounts (~0.1 *M*) of HD, however, turned out to be slower, much more complicated and at some point a reversible process [63, 64]. Possible improvements [63] include the addition of organic liquid to water to increase the HD solubility, and usage of different nuclophiles (hydroxide, phenolate, thiosulfate, etc.) to suppress the side reactions and addition of agents able to bind forming HCl and products [33–35].

A number of water-free systems have been suggested for detoxification of HD through nucleophilic substitution or HCl elimination reactions [63, 65]. Such compositions include but are not limited to bifunctional polar solvents such as monoethanolamine ($HOCH_2CH_2NH_2$); a decontamination mixture of diethylenetriamine ($(H_2NCH_2CH_2)_2NH$), ethylene glycol monomethyl ether ($CH_3OCH_2CH_2OH$), and a small amount of sodium hydroxide (NaOH) with $CH_3OCH_2CH_2O^-$ as a key reactive component and a solution of KOH in methanol with CH_3O^- as a key reagent.

Another approach to HD detoxification is its oxidation [63, 66]. The list of compositions able to perform dark oxidation includes an aqueous bleach (NaOCl, $CaOCl_2$) or chlorine gas; water solution of oxone (mixture of $2KHSO_5$–$KHSO_4$–K_2SO_4); aqueous peroxydisulphate ($S_2O_8^{2-}$ active anion), or potassium permanganate. The disadvantages of the systems for HD detoxification are consumption of large amounts of oxidizing compounds and often formation of chlorosulfones as final products, which themselves have vesicant properties.

Prospective approaches to HD detoxification include incineration, supercritical water oxidation, steam gasification, plasma arc pyrolysis [66], biodegradation [67], electrochemical [68], and catalytic decomposition [69]. Recently, effective Au(III)Cl_2NO_3(thioether) homogeneous [70] and $Ag_5PV_2Mo_{10}O_{40}$ heterogeneous [71] catalysts for thioether oxidation have been discovered. In the presence of O_2 these catalysts were found to be able to convert 2-CEES to 2-chloroethyl ethyl sulfoxide with 100% selectivity.

When compared with other methods, the photocatalytic route for decontamination of HD has certain advantages. Indeed, the photocatalytic method is the only one that is heterogeneous and catalytic suggesting the potential of decomposing large amounts of pollutant under solar light in ambient conditions [72]. Nanosized TiO_2, Degussa P25, has been recently tested in our laboratories [58] for photocatalytic decomposition of gaseous 2-CEES — a mimic for the mustard gas. The reaction was carried out in air at 25°C. The injection of liquid 2-CEES in the photoreactor containing TiO_2 leads to redistribution of 2-CEES between the TiO_2 surface and surrounding gas phase with most part of 2-CEES residing on the TiO_2. The ability of 2-CEES to adhere strongly to the surfaces of many materials and its ability to infiltrate rubber septa impose significant restrictions on the type of materials that can be used for the photoreactor construction. Only, when a quartz cell sealed with a Teflon septum was employed, a reasonable dark stability of 2-CEES under ambient conditions was achieved.

In the UV–Vis spectrum, 2-CEES has a strong absorption band with a maximum around 210 nm and a tail extended to ca. 300 nm. To avoid direct photolysis of 2-CEES (that proceeds fairly fast under rigid UV light) a filter set transmitting UV-A light ($320\ nm < \lambda < 400\ nm$) was employed. As a result no photoreaction occurred without TiO_2 nanoparticles. The irradiation of the photoreactor containing nanosized TiO_2 and 2-CEES results in consumption of gaseous 2-CEES. Simultaneously, a number of primary intermediates form, all of which were separated, identified, and quantified with gas chromatograph mass spectrometer (GC-MS). Table 16.5 gives the list of primary intermediate products. It includes ethylene (CH_2CH_2), chloroethylene ($ClCHCH_2$), ethanol (CH_3CH_2OH), 2-chloroethanol ($ClCH_2CH_2OH$), acetaldehyde (CH_3CHO), chloroacetaldehyde ($ClCH_2CHO$), diethyl disulfide ($CH_3CH_2S_2CH_2CH_3$), 2-chloroethyl ethyl disulfide ($ClCH_2CH_2S_2CH_2CH_3$), and bis(2-chloroethyl) disulfide ($ClCH_2CH_2S_2CH_2CH_2Cl$).

The apparent quantum yield was determined taking into account only the gaseous fraction of compounds. For the purpose of apparent quantum yield determination, the reaction was carried out under monochromatic ($\lambda = 313\,nm$) light. Low volatility of some products that can be otherwise detected in the gas phase, when the reaction temperature is increased from 25°C to 80°C, or via extraction impedes determination of apparent quantum yields. In such cases, the apparent quantum yields are marked as below detection limit (BDL). The reaction of 2-CEES oxidation perhaps proceeds through abstraction of hydrogen atom by a photogenerated $OH^{\bullet}$ radical. The thioether α-carbon radical, $ClCH_2CH_2SC^{\bullet}HCH$, formed in this way is expected to be reversibly protonated giving ethylene:

$$ClCH_2CH_2SC^{\bullet}HCH_3 + H^+ \longrightarrow ClCH_2CH_2SC^{\bullet +}H_2CH_3 \longrightarrow ClCH_2CH_2S^{\bullet} + CH_2CH_2 + H^+$$

or after protonation being attacked by a hydroxide group to yield ethanol:

$$ClCH_2CH_2SC^{\bullet}HCH_3 + H^+ + OH^- \longrightarrow ClCH_2CH_2S^{\bullet} + CH_3CH_2OH$$

Oxidation of ethanol should yield acetaldehyde [73–75]:

$$CH_3CH_2OH \xrightarrow[TiO_2]{h\nu} CH_3C(O)H$$

Analogous reactions with the chlorine-containing branch of 2-CEES molecule should give chloroethylene, chloroethanol, and chloroacetaldehyde. Recombination of two types of thiol radicals gives three types of experimentally observed disulfides:

$$CH_3CH_2S^{\bullet} + CH_3CH_2S^{\bullet} \longrightarrow CH_3CH_2S_2CH_2CH_3$$

TABLE 16.5
List of Primary Intermediates Appearing during Photocatalytic Oxidation of 2-CEES over TiO_2 at 25°C

	Name of Compound	Structural Formula	Apparent Quantum Yield (%)
1.	2-Chloroethyl ethyl sulfide	$ClCH_2CH_2SCH_2CH_3$	−0.07685
2.	Ethylene	CH_3CH_3	0.02963
3.	Chloroethylene	$ClCH_2CH_3$	0.00574
4.	Ethanol	CH_3CH_2OH	0.0012
5.	2-Chloroethanol	$ClCH_2CH_2OH$	BDL
6.	Acetaldehyde	$CH_3C(O)H$	0.05556
7.	Chloroacetaldehyde	$ClCH_2C(O)H$	BDL
8.	Diethyl disulfide	$CH_3CH_2S_2CH_2CH_{3>}$	0.00426
9.	2-Chloroethyl ethyl disulfide	$ClCH_2CH_2S_2CH_2CH_3$	BDL
10.	*Bis*(2-chloroethyl) disulfide	$ClCH_2CH_2S_2CH_2CH_2Cl$	BDL

BDL, below detection limit.

$$ClCH_2CH_2S^\bullet + CH_3CH_2S^\bullet \longrightarrow ClCH_2CH_2S_2CH_2CH_3$$

$$ClCH_2CH_2S^\bullet + ClCH_2CH_2S^\bullet \longrightarrow ClCH_2CH_2S_2CH_2CH_2Cl$$

Among all intermediates formed, bis(2-chloroethyl) disulfide ($ClCH_2CH_2$ $S_2CH_2CH_2Cl$) is expected to be the most toxic one. Fortunately, this compound is also the least volatile and at 25°C can be detected only on TiO_2 surfaces.

Continuation of illumination results in further oxidation of the primary intermediates up to their complete conversion into CO_2, H_2O, HCl, SO_2, and SO_4^{2-}. Formation of nonvolatile SO_4^{2-} ions forming at the surface of TiO_2 may represent a significant challenge for the long-term stability of the photocatalyst. At the same time, the activity of TiO_2 is expected to be restored in outdoor applications of TiO_2 where sulfate ions can be removed through periodic raining.

16.5.4.2 Acetaldehyde Decomposition

Efforts are underway by several research groups to develop new photocatalysts that are active in the visible region. There are only a few reports concerning the decomposition of acetaldehyde under visible light irradiation. Nanosized MO_x–ZnO (M = Fe, W) composite powders were synthesized by spray pyrolysis and evaluated for the decomposition of acetaldehyde [76]. Co-doped nanometer-sized TiO_2 was found to decompose acetaldehyde under visible light [77]. As discussed in previous chapters, the aerogel process results in the formation of ultrafine powders that possess high surface area. We have seen that the metal oxides prepared by the aerogel route exhibit unparalleled activity as catalysts. Thus, it was of interest to see if TiO_2 and SiO_2 containing transition metal ions can exhibit photocatalytic activity. However, the potential of such aerogels as photocatalysts has largely been an unexplored area of research. Thus, it was our endeavor to see if the aerogel route can provide an alternate route to prepare nanostructured photocatalysts that are active in the visible region. With this objective the first experiments were carried out using M–TiO_2–SiO_2 as photocatalysts. Indeed as stated earlier, Mn, Co, and Cr containing titania–silica aerogels exhibited similar if not higher photoactivity for the gas-phase decomposition of acetaldehyde compared to commercial Degussa P25 TiO_2. Ni, Cu, Fe, and Mn containing M–TiO_2–SiO_2 and M–SiO_2 samples were however found to low activity compared to the Co, Cr, and V counterparts. Later, we found that M–TiO_2 did not exhibit any significant activity in the visible region, and thus, we also studied the decomposition of acetaldehyde employing M–SiO_2 photocatalysts. The aerogel catalysts were prepared by the MAP as discussed earlier. The M–TiO_2–SiO_2 samples were prepared by co-gelation of tetraethoxysilane and titanium isopropoxide and the corresponding transition metal precursors in methanol solvent. The XRD results indicate that the transition metal oxides are highly dispersed on the surface of SiO_2 and on TiO_2–SiO_2. The pure SiO_2 and TiO_2–SiO_2 samples showed negligible conversion of acetaldehyde to CO_2. In contrast, the transition metal ion containing M–SiO_2 and M–TiO_2–SiO_2 samples showed activity under visible light illumination. The apparent first-order rate constants are $k_{vis} = 0.013\,min^{-1}$ for Cr–SiO_2, $k_{vis} = 0.011\,min^{-1}$ for Co–SiO_2 while that of TiO_2–SiO_2 is $k_{vis} = 0.0028\,min^{-1}$. The SiO_2 and the Degussa P25 samples show only negligible activity in the visible region. The M–TiO_2–SiO_2 and M–SiO_2 (M = Co, Cr, V) samples show similar activity. These mesoporous SiO_2 and TiO_2–SiO_2 photocatalysts have high surface area (650 to 940 m^2/g) and large pore volumes (0.65 to 2.95 cc/g) and are thus able to finely disperse the transition metals. Thus, these aerogel materials are unique photocatalysts.

Since we had found that mesoporous M–SiO_2 and M–TiO_2–SiO_2 samples were active for the decomposition of acetaldehyde, we were interested to see if a microporous transition metal ion containing sample, having titanium and silicon can also function as an effective photocatalyst. For this purpose, we selected ETS-10 (Engelhard Titanosilicate) that has a large pore (12-ring) and a three-dimensional structure [78, 79]. The structure of ETS-10 consists of TiO_6 octahedra as in anatase and rutile forms of TiO_2. However, the bandgap of ETS-10 is 4.03 eV, which is higher than that of TiO_2. ETS-10, has been explored as a catalyst for the photodegradation of alcohols and in the shape-selective photocatalytic transformation of phenols in aqueous medium [80, 81]. However, there exists no report employing ETS-10 as a visible light photocatalyst.

ETS-10 has the formula $[(Na,K)_2TiSi_5O_{13}]$. The sodium and potassium ions occupy ion-exchange sites and do not constitute part of the framework ions. Thus, ETS-10 has a very high cation-exchange capacity, and hence a number of metal ions can be exchanged for the alkali ions (Na, K). Cr, Mn, Fe, Ni, Cu, and Ag ions were introduced in ETS-10 by performing ion exchange of ETS-10 with corresponding metal salt solutions. In addition, Co and Cr containing ETS-10 were prepared by adding the Co and Cr salt solutions along with the precursors necessary for the preparation of ETS-10. Scanning electron microscopy (SEM) and energy dispersive absorption x-ray fluorescence (EDAX) analyses confirm the presence of transition metal ion in these samples. Pure ETS-10 has absorption only in the UV region. UV–visible diffuse reflectance studies show a strong absorption in the visible region for the ion-exchanged M-ETS-10 (M = Co, Cr, Cu, Fe, Mn, and Ni) samples. Thus, it is expected that M-ETS-10 would be active as photocatalysts under visible light irradiation. As prepared samples of $(Na,K)_2(TiCr)Si_5O_{13}$ and $(Na,K)_2Ti(SiCo)_5O_{13}$ did not show any activity under visible light irradiation. However, when these samples were calcined at 500°C in static air for 1 h, and then employed as photocatalysts at room temperature, acetaldehyde was found to decompose to CO_2. The apparent first-order rate constants are $k_{vis} = 0.002\,min^{-1}$ for $(Na,K)_2(TiCr)Si_5O_{13}$, and $k_{vis} = 0.004\,min^{-1}$ for $(Na,K)_2Ti(SiCo)_5O_{13}$ while that of Degussa P25 TiO_2 is negligible under visible light irradiation. M-ETS-10 (M = Cr(III), Mn(II), Fe(III), Ni(II), and Cu(II)) were found to be active only under UV irradiation. Ag-ETS-10 and Ni-ETS-10 showed higher rate constants (0.038 and 0.028 min^{-1}) under UV irradiation. In general, the ETS-10 containing transition metal ions were also found to be better catalysts than Degussa P25 TiO_2 under UV irradiation. Also, the reaction rates for the transition metal incorporated ETS-10 samples are comparatively lower under visible light irradiation than under UV irradiation. Thus, the study involving ETS-10 as photocatalysts suggest that other Ti- and Si-containing microporous zeolite materials, such as ETS-4 are potential candidates as photocatalysts under both UV and visible light irradiation.

Transition metal ion containing mesoporous MCM-41 with well-ordered pores in the size range 2 to 10 nm has attracted the attention of several catalyst research groups. There are several reports of transition metals such as Ti, V, Cr, Mn, and Fe incorporated in mesoporous MCM-41 materials [82–84]. Cr–MCM-41 was found to be an efficient support for titania for the degradation of formic acid and 2,4,6-trichlorophenol [85]. Cr-incorporated MCM-41 is effective for the photoreduction of NO by visible light illumination [86]. However, there exists no report involving MCM-41 for photooxidation reactions using visible light. We show for the first time that transition metal incorporated Al–MCM-41 acts as an efficient photocatalyst for the degradation of acetaldehyde under visible light irradiation. A series of first row transition metal-doped M–Al–MCM-41 were synthesized and characterized and their photocatalytic activity evaluated. The Cr–Al–MCM-41 sample showed the highest activity among all transition metal ion incorporated MCM-41 photocatalysts. We attribute the activity of this catalyst to the presence of Cr^{3+}/Cr^{6+} redox couple. Such an explanation was offered for the photocatalytic degradation of 2,4,6-trichlorophenol by the composite Cr–TiO_2–MCM-41 catalyst. All the transition metal ion containing Al–MCM-41 catalysts in the study contain finely dispersed metal oxide on the Al–MCM-41 support. Thus, system with finely dispersed transition metal oxide moieties on Al–MCM-41 seems to be a promising candidate for the degradation of organic volatile compounds. Indeed Cr–Al–MCM-41 photocatalyst was found to efficiently degrade trichloroethylene in the gas phase [87].

16.6 SUMMARY

Rapid growth of population coupled with an increase in industrial output and depletion of natural resources has caused deterioration of our environment. Concerns regarding environmental hazards have prompted studies into benign methods for the treatment of toxic chemicals. This has resulted in an upsurge in the research activities in universities, institutes, government laboratories, and companies to find new technologies that restore air, soil, and water quality by offering efficient and cost-effective remediation and decontamination technologies. In this respect, nanotechnology has played a significant role in numerous environmental applications. In particular, reactive nanoparticles of metal or metal oxides can address a host of these issues. Nanoscale materials have been used for environmental remediation and "green" chemistry applications for the following reasons: (1) they possess high surface areas and have a large surface to bulk ratio compared to conventionally prepared materials, so that the nanomaterial is used efficiently; (2) they have unusual shape and high number of reactive edge, corner and defect sites such as steps and kinks that impart an intrinsically higher surface reactivity; (3) properties such as Lewis acid and Lewis base properties, oxidation and reduction potential can be tailored or "tuned" for a specific reaction; and (4) ionic metal oxides such as MgO, CaO, etc. can be pelletized while still maintaining the high surface areas of the ultrafine powders Thus, these nanomaterials represent a new family of porous inorganic sorbents that exhibit unparalleled activity toward the adsorption of polar organic molecules, ambient temperature detoxification of CW agents and their stimulants. In addition, we have also demonstrated that such porous metal oxide materials and their halogen adducts exhibit bactericidal, sporicidal, and viricidal properties and even are able to destroy waterborne toxins such as aflatoxins.

Nanoparticles with their high surface area, enhanced chemical activity, and rapid deployment offer a new perspective regarding detoxification and remediation. An added advantage of using these ionic metal oxides is that the products of the decontamination reaction are benign, mineral-like solids.

REFERENCES

1. OB Koper, I Lagadic, A Volodin, and KJ Klabunde. *Chem Mater* 9:2468–2480, 1997.
2. KJ Klabunde, JV Stark, O Koper, C Mohs, DG Park, S Decker, Y Jiang, I Lagadic, and D Zhang. *J Phys Chem* 100:12142–12153, 1996.
3. O Koper and KJ Klabunde. *Chem Mater* 5:500–505, 1993.
4. A Bedilo and KJ Klabunde. *Nanostruct Mater* 8:119–135, 1997.
5. CL Carnes, PN Kapoor, and KJ Klabunde. *Chem Mater* 14:2922–2929, 2002.
6. JV Stark, DG Park, I Lagadic, and KJ Klabunde. *Chem Mater* 8:1904–1912, 1996.
7. JV Stark and KJ Klabunde. *Chem Mater* 8:1913–1918, 1996.
8. YX Li and KJ Klabunde. *Langmuir* 7:1388–1393, 1991.
9. YX Li, O Koper, M Atteya, and KJ Klabunde. *Chem Mater* 4:323–330, 1992.
10. R Richards, W Li, S Decker, C Davidson, O Koper, V Zaikovski, A Volodin, T Rieker, and KJ Klabunde. *J Am Chem Soc* 122:4921–4925, 2000.
11. CL Carnes, J Stipp, KJ Klabunde, and J Bonevich. *Langmuir* 18:1352–1359, 2002.
12. SS Kistner. *J Phys Chem* 36: 52–64, 1932.
13. SJ Teichner. *Aerogels*. Berlin: Springer-Verlag, 1985, p. 225.
14. GA Nicolaon and SJ Teichner. *Bull Soc Chim Fr* 3107–3113, 1968.
15. HD Gesser and PC Goswami. *Chem Rev* 89:765–788, 1989.
16. S Utamapanya, KJ Klabunde, and JR Schlup. *Chem Mater* 3:175–181, 1991.
17. H Itoh, S Utamapanya, J Stark, KJ Klabunde, and JR Schlup. *Chem Mater* 5:71–77, 1993.
18. Y Chi and SSC Chaung. *J Phys Chem* 104:4673–4683, 2000.
19. LF Liotta, A Macaluso, GE Arena, M Livi, G Centi, and G Deganello. *Catal Today* 75:439–449, 2002.

20. MS Ahmed and YA Attia. *Appl Therm Eng* 18:787–797, 1998.
21. Y Diao, WP Walawender, CM Sorenson, KJ Klabunde, and T Ricker. *Chem Mater* 14:362–368, 2002.
22. E Lucas, S Decker, A Khaleel, A Seitz, S Fultz, A Ponce, W Li, C Carnes, and KJ Klabunde. *Chem Eur J* 7:2505–2510, 2001.
23. O Koper, I Lagadic, and KJ Klabunde. *Chem Mater* 9:838–848, 1997.
24. O Koper and KJ Klabunde. *Chem Mater* 9:2481–2485, 1997.
25. O Koper, YX Li, and KJ Klabunde. *Chem Mater* 5:500–505, 1993.
26. S Decker and KJ Klabunde. *J Am Chem Soc* 118:12465–12466, 1996.
27. SP Decker, JS Klabunde, A Khaleel, and KJ Klabunde. *Environ Sci Technol* 36:762–768, 2002.
28. YX Li, H Li, and KJ Klabunde. *Environ Sci Technol* 28:1248–1253, 1994.
29. S Rajagopalan, O Koper, S Decker, and KJ Klabunde. *Chem Eur J* 8:2602–2607, 2002.
30. YC Yang. *Acc Chem Res* 32:109–115, 1999.
31. YC Yang. *Chem Ind* 9:334–337, 1995.
32. YC Yang, JA Baker, and JR Ward. *Chem Rev* 92:1729–1743, 1992.
33. GW Wagner, PW Bartram, O Koper, and KJ Klabunde. *J Phys Chem B* 103:3225–3228, 1999.
34. GW Wagner, LR Procell, RJ O'Connor, S Munavalli, CL Carnes, PN Kapoor, and KJ Klabunde. *J Am Chem Soc* 123:1636–1644, 2001.
35. GW Wagner, O Koper, E Lucas, S Decker, and KJ Klabunde. *J Phys Chem B* 104:5118–5123, 2000.
36. F Forestier, P Gerrier, C Chaumard, AM Quero, P Couvreur, and C Labarre. *J Antimicrob Chemother* 30:173–179, 1992.
37. S Kwak, SH Kim, and SS Kim. *Environ Sci Technol* 35:2388–2394, 2001.
38. OB Koper, JS Klabunde, GL Marchin, KJ Klabunde, P Stoimenov, and L Bohra. *Curr Microbiol* 44:49–56, 2002.
39. PK Stoimenov, RL Klinger, GL Marchin, and KJ Klabunde. *Langmuir* 18:6679–6686, 2002.
40. JE Death and D Coates. *J Clin Pathol* 32:148–153, 1979.
41. A Fujishima and K Honda. *Nature* 238:37–38, 1972.
42. N Serpone and AV Emeline. *Int J Photoenergy* 4:91–131, 2002.
43. DM Blake. *Bibliography of Work on the Heterogeneous Photocatalytic Removal of Hazardous Compounds from Water and Air*. Golden, CO: NREL/TP 510-31319, 2001, pp. 1–272.
44. G Mills and MR Hoffmann. *Environ Sci Technol* 27:1681–1689, 1993.
45. DF Ollis and E Pelizzetti. *Environ Sci Technol* 25:1522–1529, 1991.
46. ST Martin, H Hermann, W Choi, and MR Hoffmann. *Trans Faraday Soc* 90:3315–3322, 1994.
47. S Hager and R Bauer. *Chemosphere* 38:1549–1559, 1999.
48. AG Rincon and C Pulgarin. *Appl Catal B: Environ* 44:263–284, 2003.
49. RJ Watts, S Kong, MP Orr, GC Miller, and BE Henry. *Water Res* 29:95–100, 1995.
50. A Fujishima. *Denki Kagaku Oyobi Kogyo Butsuri Kagaku* 64:1052–1055, 1996.
51. L Palmisano, V Augugliaro, A Sclafani, and M Schiavello. *J Phys Chem* 92:6710–6713, 1988.
52. W Choi, A Termin, and MR Hoffmann. *J Phys Chem* 98:13669–13679, 1994.
53. G Burgeth and H Kisch. *Coord Chem Rev* 230:41–47, 2002.
54. A Fuerte, MD Hernandez-Alonso, AJ Maira, A Martinez-Arias, M Fernandez-Garcia, JC Conesa, and J Soria. *Chem Commun* 2718–2719, 2001.
55. R Asahi, T Morikawa, T Ohwaki, K Aoki, and Y Taga. *Science* 293:269–271, 2001.
56. Y Sakatani and H Koike. Japan Patent, P2001-72419A, 2001.
57. Z Zou, J Ye, K Sayama, and H Arakawa. *Nature* 414:625–627, 2001.
58. IN Martyanov and KJ Klabunde. *Environ Sci Technol* 37:3448–3553, 2003.
59. J Wang, S Uma, and KJ Klabunde, *Micropor Mesopor Mater* 75:143–147, 2004.
60. J Wang, S Uma, and KJ Klabunde. *Appl Catal B: Environ* 48:151–154, 2004.
61. S Uma, S Rodrigues, IN Martyanov, and KJ Klabunde. *Micropor Mesopor Mater.* 67:181–187, 2004.
62. S Rodrigues, S. Uma, IN Martyanov, and KJ Klabunde. *J Photochem Photobiol A: Chem* 165:51–58, 2004.
63. GW Wagner and YC Yang. *Ind Eng Chem Res* 41:1925–1928, 2002.
64. YC Yang, LL Szafraniec, WT Beaudry, and JR Ward. *J Org Chem* 53:3293–3297, 1988.
65. GW Wagner, LR Procell, YC Yang, and CA Bunton. *Langmuir* 179:4809–4811, 2001.

66. U.S. Congress, Office of Technology Assessment, Disposal of Chemical Weapons: Alternative Technologies-Background Paper, OTA-BP-O-95, Washington, D.C.: U.S. Government Printing Office, 1992, pp. 1–35.
67. JJ Kilbame II and K Jackowski. *J Chem Technol Biotechnol* 65:370–374, 1996.
68. TC Franklin, R Nnodimele, and J Kerimo. *J Electrochem Soc* 140:2145–2150, 1993.
69. XL Zhou, ZJ Sun, and JM White. *J Vac Sci Technol A* 11:2110–2116, 1993.
70. E Boring, YV Geletii, and CL Hill. *J Am Chem Soc* 123:1625–1635, 2001.
71. JT Rhule, WA Neiwert, KI Hardcastel, BT Do, and CL Hill. *J Am Chem Soc* 1213:12101–12102, 2001.
72. DA Panayotov, DK Paul, and JT Yates Jr. *J Phys Chem B* 107:10571–10575, 2003.
73. DS Muggli, JT McCue, and JL Falconer. *J Catal* 173:470–483, 1998.
74. MR Nimlos, EJ Wolfrum, ML Brewer, JA Fennell, and G Bintner. *Environ Sci Technol* 30:3102–3110, 1996.
75. ML Sauer and DF Ollis. *J Catal* 158:570–582, 1996.
76. D Li and H Hajime. *J Photochem Photobiol A: Chem* 160:203–212, 2003.
77. M Iwasaki, M Hara, H Kawada, H Tada, and S Ito. *J Colloid Interf Sci* 224:202–204, 2002.
78. SM Kuznicki. U.S. Patent 4,853,202, 1989.
79. MW Anderson, O Terasaki, T Ohsuna, A Philippou, SP Mackay, A Ferreira, J Rocha, and S Lidin. *Nature* 367:347–351, 1994.
80. MA Fox, KE Doan, and MT Dulay. *Res Chem Intermed* 20:711–722, 1994.
81. P Calza, C Paze, E Pelizzetti, and A Zecchina. *Chem Commun* 2130–2131, 2001
82. S Biz and ML Occelli. *Catal Rev Sci Eng* 40:329–407, 1998.
83. A Corma, MT Navarro, and JP Pariente. *J Chem Soc Chem Commun* 147–148, 1994.
84. F Rey, G Sankar, T Maschmeyer, JM Thomas, and RG Bell. *Topic Catal* 3:121–134, 1996.
85. L Davydov, EP Reddy, P France, and PG Smirniotis. *J Catal* 203:157–167, 2001.
86. H Yamashita, K Yoshizawa, M Ariyuki, S Higashimoto, M Che, and M Anpo. *J Chem Soc Chem Commun* 435–436, 2001.
87. S Rodrigues, KT Ranjit, S Uma, IN Martyanov, and KJ Klabunde. *Catal*, in press.

17 Toward a Molecular Understanding of Environmental Catalysis: Studies of Metal Oxide Clusters and their Reactions

Elliot R. Bernstein and Yoshiyuki Matsuda
Department of Chemistry, Colorado State University

CONTENTS

17.1 INTRODUCTION

To say catalysis is ubiquitous in nature and for environmental systems is an understatement. Both homogeneous and heterogeneous catalysis are of natural fundamental importance to all three phases of our environment (gas, liquid, and solid) [1, 2]. Additionally, most biological reactions are of a catalytical nature and life on Earth is suggested to arise through a series of catalytic events [3–5]. The complete exploration of such environmental and biological processes will include an understanding and explication of the catalytic events that underlie these chemistries, but, at least at present, the natural processes are typically far too complicated and veiled by other events or systems to generate a detailed, fundamental picture of how the catalytical reactions proceed mechanistically. This is especially true for any heterogeneous process [1, 2].

Heterogeneous environmental catalysis is central to the chemistry of NO_x, O_3, hydrocarbons, organic carbon, and oxygen-centered radicals, and other species in the atmosphere and even in the condensed phase. Surfaces are thought to be the most important components of the fluid (gas or liquid)–solid interface, but their properties for reacting systems are not well characterized or understood [1, 2].

As is typical in complex systems, physical chemists have taken a reductionist approach to the explication of heterogeneous, environmental catalysis: high vacuum metal and metal–nonmetal surface studies [6–14], cluster studies [15–19], surface reaction studies with imaging [1, 2, 20–22], and high surface area studies under controlled conditions [23–27]. In order to present a manageable and readable review with practical, experimental, and theoretical procedures for the reader, the presented material will be limited to gas phase, neutral metal oxide clusters, and their reactions. In general, studies of neutral metal oxide clusters address issues of surface and bulk catalysis by metal oxide systems. Clusters of various sizes sweep through the electronic and structural properties of the condensed phase, as the cluster size changes from MO to M_xO_y ($x, y > 100$). Clusters do not often have the same structural arrangements as small pieces of solid or even surface, but catalytical events seem to occur not on perfect surfaces or for ordered solids. In fact, observed reactions on surfaces occur at either regular or irregular defect sites and even rearrange the surface as reactions occur [1, 2, 6–14].

This review of metal oxide catalytic behavior with regard to environmental problems and issues will focus on neutral clusters and their reactivity behavior. Many other groups have studied ionic systems [28–30], and in fact, most of the effort for elucidation of the properties and behavior of metal oxide clusters deals with ionic systems: a number of reasons can be suggested for this focus. First, positive and negative cluster ions can be made directly in an ablation source (see below) and used as generated, while neutral clusters, also made in higher concentration in the same source must be ionized to be detected by a mass spectrometer. Second, mass spectroscopy is the detection method of choice for metal oxide and other metal-containing clusters, because electronic states of these systems are very dense and radiationless decay of excited electronic states is much more rapid than radioactive decay. Thus, laser-induced fluorescence (LIF) spectroscopy cannot usually be applied to these species for the study of neutral clusters. Third, mass-selected species can be studied at present only for ionic systems, and ionic clusters have the apparent advantage of being selectable without fragmentation through the use of some sort of mass filter (e.g., time-of-flight, quadrupole, magnetic sector, etc.). Fourth, for the study of negative ion, photoelectron spectroscopy can sometimes be used on small clusters (M_xO_y, $x < 2$ or 3, $y < 6$ or 7) [31–36]. By this technique, information on the states of both the negative ion and the neutral clusters can, in principle, be generated. Fifth, one can argue that ionic clusters represent a model of some suggested defect sites in solids. Nonetheless, neutral clusters are at least as useful in terms of models for most bulk and surfaces sites of potential interest.

Neutral clusters, on the other hand, are more difficult to study because they cannot be size selected and the neutral cluster distribution generated by a growth mechanism must be determined through some ionization technique (e.g., multiphoton at infrared, visible, or ultraviolet energies, single photon with a vacuum ultraviolet (>10 eV) photon, or electron collision). No matter how the neutral clusters are ionized, one must be concerned about fragmentation of the clusters during the ionization process if the true neutral cluster distribution is to be identified. One needs this set of concentration if reactions and reactivity of the clusters are to be measured because the experiment requires the neutral cluster distribution both prior and subsequent to passing the clusters through a reaction cell to determine which clusters are catalytically or stoichiometrically reactive.

While the body of this review will mostly focus on work done in our laboratory, many others have contributed to the fundamental understanding of metal oxide systems (both neutral and ionic) and their chemistry. We briefly survey the results of these studies in the next few pages for both neural and ionic species to demonstrate what is known about metal oxide clusters in general.

Atomic clusters of refractory materials (metals, semiconductors, etc.) have been generated in the gas phase by ovens, sparks, rf discharges, lasers for over 40 years [37–41]. The technique that has survived and been the most productive over the years is laser ablation of a metal or solid systems into a high-pressure gas that expands into a vacuum system. This gas can contain reactant (N_2, O_2, CH_4, NO_x, SO_2, K) to generate the desired clusters [42–45]. Detection of the clusters is most general by time-of-flight mass spectroscopy (TOFMS), employing multiphoton or single-photon ionization of the neutral clusters in the expansion. Ionic clusters are also produced in this process and can be directly accessed by a number of mass spectroscopy and fluorescence (mostly for diatomics) techniques. A good deal of the history of this field nicely presented from the early 1980s to the late 1980s in Refs. [37–41].

As pointed out above, most of the studies of metal–nonmetal clusters are carried out on ionic systems. For metal oxides, the Castleman group has been very active in this area, and an overview of their efforts and results has recently appeared [46]. V, Ni, and Ti are the metal species in these cationic clusters and reactivity toward small hydrocarbons, CCl_4, HF, CH_2F_2, etc., is reviewed. An earlier comprehensive review of metal and metal–nonmetal cluster studies covers earlier efforts [47–49]. As pointed out, the specific cluster cation, for example $V_3O_7^+$, can be selected, reacted with, for example, CCl_4, and the products of this single-cluster species reaction explored.

The neutral metal oxides most heavily studied in the literature by other laboratories appear to be Al_xO_y, Zr_xO_y, Mg_xO_y, and Fe_xO_y. The studies are almost exclusively conducted through fragmenting multiphoton ionization (MPI) detection. Neutral iron oxide clusters were first studied by Riley and coworkers in the mid-1980s [47–49] using a flow tube reactor with O_2 into which Fe_n clusters are injected; the resulting Fe_xO_y clusters are ionized by 193 nm ArF excimer radiation. Interestingly, these techniques generate a very nearly identical ion distribution to that found by using laser ablation of Fe metal into a O_2–He gas mixture (ca. 1%) followed by supersonic expansion into a vacuum and subsequent 193 nm ionization and detection by a TOFMS. We will discuss this system further in the body of this review.

Neutral Al_xO_y clusters have been studied using a number of different ionization techniques [50–57]: ions directly from ablation source, and 245 nm, 193 nm, and 11 μm radiation for the neutral clusters. This latter technique is a new innovation for cluster studies using a free electron laser and getting some structural information from infrared spectra [58]. Each ionization method gives a somewhat different mass spectral intensity pattern. The ionization mechanism varies from multiphoton UV, direct ionization, to ionization by IR multiphoton (>30) absorption and thermionic emission from a hot cluster. The general form of these clusters is $AlO(Al_2O_3)_n$. The IR MPI spectrum looks similar to the ion distribution from the source and probably reflects the ion stability better than the UV MPI spectra. The free electron laser IR ionization method is resonant at the first photon and can give some limited (e.g., C—C versus M—C) bonding information and cluster structure. For example, AlO $(Al_2O_3)_n$ clusters are more structurally similar to solid γ-Al_2O_3 than α-Al_2O_3. This IR free electron laser is important because it has a high-output energy (100 mJ/pulse) and a high repetition rate; thus, it is able to sustain multiphoton absorption for even weak vibrational transitions and yield a reasonable signal-to-noise ratio for the collected spectra. $Nb_xO_y^+$ have also been studied by a variant of this technique to get their infrared spectra and some structural information [59].

Neutral $(MgO)_n$ clusters are also investigated by ionization techniques (UV and IR MPI) as well [60]. Again, the IR and UV ionizations produce different distributions indicative of

MPI approaches and not representative of the true neutral cluster intensity distribution. Nonetheless, the ion masses are consistent with a cubic MgO lattice structure.

Zr_xO_y neutral clusters have also been studied by the IR free-electron laser MPI approach [61] and, as will be seen, give very similar mass spectra to those generated by the UV MPI approach. Different mass spectra are generated with single-photon VUV ionization, and these reflect the true neutral distribution. Nonetheless, considering all the spectra and calculations taken together, a rather good understanding of the Zr_xO_y system can be achieved.

Before leaving this background presentation of earlier studies of neutral metal oxide species, a short excursion into metal carbide clusters is useful because it will show the variety and complexity of metal–nonmetal cluster systems, and it will show the usefulness of new techniques, however qualitative, that result in spectroscopic information on gas-phase clusters. Castleman and coworkers have discovered the very interesting system of M_xC_y clusters with particular stability for M_8C_{12} (M = Ti, V, Hf, Nb, Zr — called met-cars for metallo-carbohedranes) [62–69]. Both ions and neutrals appear to have a special stability. Highly symmetrical structures have been proposed for these "magic" clusters (similar to C_{60}), but the IR resonance MPI technique suggests C—C and M—M bond formation rather than M—C bond formation [70, 71]. Thus, the suggested most symmetrical T_d structure seems not to be the true symmetry of the cluster. On the other hand, the other clusters, such as $T_{14}C_{13}$, appear to have the cubic TiC rock salt structure. The M_8C_{12} structure is most consistently associated with a D_{2d}-type arrangement with few M—C bonds.

A research plan for the studies of catalytic processes involving metal oxides can be suggested as follows: (1) generate neutral clusters of the particular metal oxide of interest — these clusters will sweep through all possible regular and defect surface and bulk sites that may occur for solid-phase metal oxide catalysts; (2) identify, characterize, and quantify the neutral metal oxide cluster distribution present in the gas-phase collection of the clusters — accomplished by gentle near-threshold, single-photon ionization of all the clusters; (3) use calculations, or perhaps some spectroscopic (e.g., IR, microwave) techniques, to determine the electronic, vibrational, and structural properties of the observed neutral clusters; (4) identify any cluster fragmentation during the ionization process employing laser intensity and wavelength changes, linewidths of TOFMS features, different methods of ionization as appropriate, and covariance mapping [49, 72–77]; (5) pass entire neutral cluster beam distribution of M_xO_y species through a reaction cell filled with gas at different pressures (10^{-5} to 10^{-2} Torr) in order to determine how clusters behave as they collide with, react with, or facilitate reactions of (catalyze) a particular gas-phase species (e.g., NO_x, CO, SO_2, $C_n H_{2n+2}$, K); (6) observe the new distribution of clusters and products following step 5 and compare the unreacted or "empty cell" neutral cluster distribution to that obtained with the "filled cell" neutral cluster–product distribution [28, 78, 79]; (7) use calculational and spectroscopic techniques to learn as much as possible about the reactive or changed clusters in this process.

We will elucidate these steps as we present and discuss data on neutral metal oxide systems that have been studied in this manner in our laboratory and others. The focus in this review will be on Fe_xO_y, Cu_xO_y, V_xO_y, Ti_xO_y, and Zr_xO_y. We will limit our presentation to only neutral systems and deal specifically with these clusters, not because these neutral metal oxides are necessarily the most important clusters upon which to focus, but because we have studied these systems ourselves and know the most about them, and because the approach will force the presentation to remain within reasonable bounds.

The presentation will follow the guidelines set out for the goals of this volume. We will discuss at length our experimental techniques for generating the neutral cluster distribution for the various metal oxides discussed and will show how the neutral cluster distribution can be determined through single VUV photon ionization. We will also show how only neutral clusters can be selected for study. Then these determined cluster distributions are passed into and through the reaction cell and the single-photon detection process is again employed to look for products and changes in the neutral cluster distribution. Next, the calculational methods employed will be discussed to give some idea of their degree of success and difficulty, and care that must be exercised in their use and interpretation. Following the experimental and theoretical procedures, we will present data for the various clusters and discuss the extent of the reactivity results presently available on these neutral cluster systems. Note importantly that these efforts are an ongoing process that will more than likely (we are hopeful) be only part of the information available on these systems by the time this review is published (nearly a year after it is written). Finally, we will outline the conclusions and what remains to be accomplished with these studies.

17.2 EXPERIMENTAL PROCEDURES

17.2.1 Generation of Supersonic Expansion Beams of Neutral Metal Oxide Clusters

The various techniques for generating gas-phase clusters of nonvolatile species like metal oxides have been reviewed quite thoroughly in Refs. [42–45, 80–90]. Below are presented the specific methods and techniques we use to generate samples of gas-phase metal oxide cluster beams for catalysis studies.

A schematic diagram of the pulsed supersonic nozzle, laser ablation head attached to it, ion deflection plates, and reaction cell are all shown in Figure 17.1. The pulsed nozzle in this instance is an R. M. Jordan Co. nozzle, but this is not a unique feature of the system: other pulsed nozzles are also acceptable. The ablation head is a rather complex, three-dimensional structure that holds a rotating–translating drum about 3 cm in diameter, a small high gear ratio motor to rotate a nut that forces the threaded spindle on which the drum is fixed to rotate and translate, and has a 2 mm diameter through hole (~6 cm in length) in it that aligns with the orifice of the pulsed nozzle, the 1 mm hole through which the ablation laser beam passes, and a 1 m hole that exposes the drum surface to the gas flow from the nozzle in the 2-mm channel and the ablation laser beam. The entire assembly is designed to create a supersonic expansion from the nozzle through the 2 mm × 6 cm channel in the ablation head into the vacuum system in which the assembly is mounted. The ablation laser is a pulsed Nd–YAG laser with a doubling crystal to generate 1 to 5 mJ/pulse of 532 nm light at 10 Hz (~10 ns pulses). This light is focused

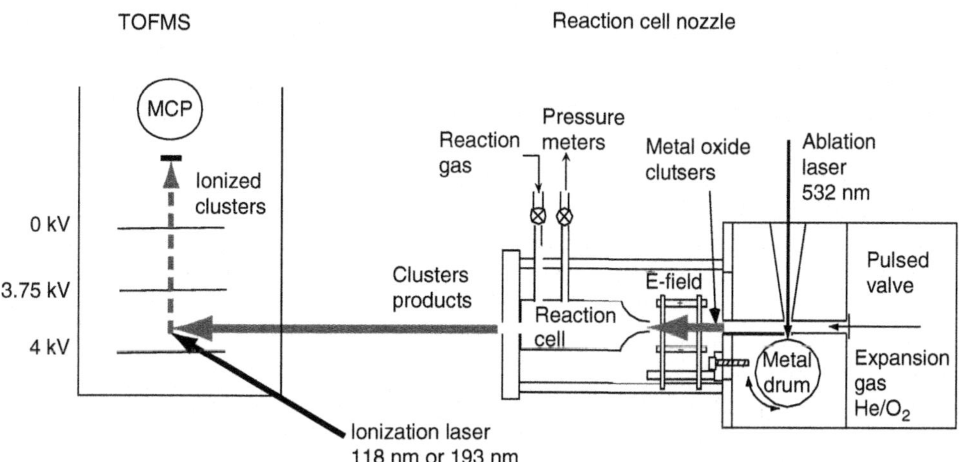

FIGURE 17.1 Schematic drawing of the laser ablation nozzle, electric field ion deflector, reaction cell, and TOFMS ionization region. This apparatus is enclosed in a vacuum chamber with two 6 in. diffusion pumps evacuating the chamber and flight tube.

to ca. 0.1 mm at the drum surface; after each ablation laser pulse, the drum is rotated to a new position such that the next pulse is directed to a mostly fresh part of the sample. The surface of the drum has a thread cut in it by this process that matches the thread on the threaded spindle upon which the drum is mounted. The motor drives the drum in one direction until the drum hits a reversing switch for the motor, and then drives the drum in the other direction until the drum hits the other reversing switch and the process begins again. The system can run for about 15 h without retracing any old groove cut by the laser. One metal sample can be used for many repeated tracings over old grooves cut in its surface.

The nozzle–ablation–expansion timing must be coordinated in order to obtain the best distribution of clusters and to emphasize particular components of the neutral cluster distribution. Timing is controlled by a laboratory clock — a time delay generator: first the nozzle is triggered, then the laser is triggered, and the ablation plasma plume is ejected from the drum surface and enters the expansion gas. The expansion gas is typically He but can be Ne, Ar, or different gas mixtures. The backing pressure of this expansion gas in the nozzle is ca. 100 psi, but the expansion into the vacuum system probably takes place from a pressure in the 2 mm × 6 cm channel of ca. 10 psi. The expansion gas typically contains 0.1% to 5% O_2 to synthesize the metal oxide clusters from the vaporized metal sample.

The nature of the sample itself depends on the particular system studied [44, 45, 72–77]. For iron, vanadium, zirconium, and titanium, we have found that the most convenient sample form is metal foil. This choice depends on the physical setup and the details of the nozzle structure. In these instances, we ablate metal into a few percent O_2 in the He expansion gas. Each metal has its own "best" O_2 concentration for a particular maximization of the distribution of neutral clusters. These exact concentrations depend on the chemical reactivity of the metal. For example, Fe is very reactive and only 1% O_2 is sufficient to generate a good cluster distribution. In fact, quite an acceptable concentration of Fe_xO_y clusters is generated from a fresh sample of Fe metal foil even with "no added" O_2 to the beam: apparent surface O_2 or Fe_mO_n on the surface of the metal foil is sufficient to generate a rich distribution of Fe_xO_y clusters. For Zr_xO_y clusters, ca. 5% O_2 is employed in the expansion gas. Up to 10% O_2 in the He expansion gas is still not sufficient to generate Cu_xO_y clusters as O_2 and Cu are apparently not that reactive. In this instance, we employ a Cu_2O or CuO sample for laser ablation and the plasma of various Cu and O species is reactive and well mixed enough to generate a good distribution of neutral Cu_xO_y species (to $x > 20$) such that studies of these clusters can be made. Pure CuO samples are difficult to employ in our setup because they are hard to stabilize on a cylindrical drum: our best samples here are made with either glue or sugar water–CuO mixtures that dry on the drum and generate cohesive and adhering films for laser ablation. In this way the ablated material in plasma contains both Cu and O species that are reactive with one another before they cool in the expansion. Even if O_2 is added to the expansion gas for a CuO sample, the neutral cluster distribution is not changed, supporting that the Cu plus O mechanism for Cu_xO_y formation within the plasma is probably a reasonable suggestion.

As the cluster beam exits the nozzle (see Figure 17.1), it expands and cools, perhaps forming larger clusters, into the vacuum system ($P \sim 10^{-7}$ Torr with nozzle off). Most workers in the field have assumed that the clusters reach a temperature of about 300 K or lower during this process, possibly because the nozzle body is ca. 300 K. We have measured the temperature of VO in a V_xO_y expansion following laser ablation of the metal [45]. The technique used to do this is spectroscopic, based on Franck–Condon factors and relative intensities for the (0,1) and (1,0) transition for VO. The vibrational temperature we obtain is $T_{vib} \sim 700 \pm 100$ K. This is a very rough estimate because intensities are hard to measure and Franck–Condon factors are approximate. Even so, this diatomic may be the hottest cluster in the beam because it has the smallest collision cross section and probably has a weak coupling between vibration–rotation–translation degrees of freedom during a collision due to low density of states and a high-energy vibrational mode.

The cluster beam, as it exits the ablation nozzle, first encounters an electric field to remove ions generated in the source from the beam. The field is applied by deflection plates placed above and below the beam with about 600 V between them. While these ions will not affect the mass spectrum of the neutral cluster beam because they would be deflected by the electric fields within the TOFMS ionization region, they can contribute to the background baseline signal for the mass spectral features, and they could react with gas in the reaction cell and distort the chemistry assumed due to neutral species.

The beam then passes through the reaction cell, the entrance to which is a skimmer, and the exit of which is a hole in the end plate. Both input and output holes are 2 mm, and the cell is ~6 cm in length including the skimmer. At ca. 1×10^{-3} Torr pressure in the cell, the mean free path for a ca. five-atom cluster is ~5 cm. These are quite qualitative estimates because structural parameters are basically unknown in general for the clusters. The clusters exit at the end of the cell and pass into the ionization region (ion source region) of a TOFMS of a Wiley–McLaren design [72–77]. Figure 17.1 shows this ionization region.

The beam of neutral metal oxide clusters is ionized by a laser pulse. This light can come from an excimer ArF laser at 193 nm, a 355 nm laser from the tripled output of a Nd–YAG laser (1064 nm), or a 118 nm laser from the ninth harmonic of a Nd–YAG fundamental output (355 nm /3 = 118 nm). This laser source is described in detail below and in Figure 17.2. Since most metal oxide clusters have ionization energies above 7 eV, using wavelengths 355 nm (3.49 eV) and 193 nm (6.42 eV) for ionization must generate ions by multiphoton absorption. Typical estimates of the number of absorbed photons for 193 nm ionization can run as high as nine to ten for iron oxide 72,73 and zirconium oxide [74] with other clusters thought to behave similarly. The 355-nm photoionization will also occur with at least three to four photon absorption during the ionization process. This process deposits lots of energy in the clusters and fragmentation is a typical outcome for ionization by multiphoton absorption. Typical laser pulse energies for these ionizations are a few mJ and each pulse contains 10^{15} to 10^{16} photons.

On the other hand, near threshold, single-photon ionization with a low energy/pulse laser will not cause neutral cluster fragmentation, and thereby loss of information about the species in the beam studied. This is exactly the situation for the 118 nm laser ionization. The method of generation for this light [73, 91–98] is to triplet the Nd–YAG fundamental output to 355 nm light using a KDP crystal and then triple this light to yield 118 nm light (e.g., 1064 nm $\xrightarrow[\text{KDP}]{}$ 355 nm $\xrightarrow[\text{Xe/Ar}]{}$ 118 nm). Starting with about 500 mJ/pulse at 1064 nm, ca. 25 mJ/pulse at 355 nm, one obtains ca. 1 μJ/pulse (~10^{11} to 10^{12} photons/pulse). If the laser is seeded, one gets a better, more stable in time, fundamental pulse and thus, a better signal using 118 nm radiation for the ionization. The scheme is depicted in Figure 17.2. Single-photon ionization yields a good representation of the neutral cluster distribution in the beam at the time during

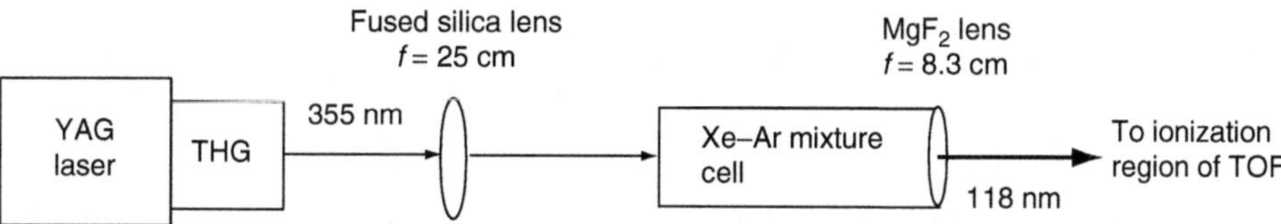

FIGURE 17.2 Schematic drawing of 118 nm laser generation system. A Xe–Ar mixture is the medium of 118 nm generation from 355 nm. Xe–Ar cell is attached to vacuum chamber. Generated 118 nm light is focused by the MgF_2 lens and directly enters the vacuum chamber. The 355 nm residual light is defocused by the MgF_2 lens.

which the beam is accessed. Another advantage of single-photon ionization with 10.5 eV photons is that product clusters, such as $V_3O_7\,gSO_2$, which might not be chemically bonded, could be accessed and detected in the TOFMS.

This ionization source has a number of advantages and additional features that are at first not obvious in the experimental design and interpretations of the data. First, one advantage is that the energy/pulse is low (1 μJ/pulse at 118 nm implies 6×10^{11} photons/pulse); this ensures that only one photon is absorbed by each cluster during the interaction time (ca. 10 ns) between the laser and the cluster beam. This low beam intensity also ensures that the signals are small, 5 to 50 mV/mass channel. Note that one detected ion/pulse generates ca. 5 mV into 50 Ω for a detector gain of 10^7. So the nature of the laser pulse for ionization has both pros and cons for the experiment: a factor of 10 to 100 more photons/pulse at 118 nm would be a plus for the detection process. Using more energy/pulse at 355 nm is not the appropriate approach because of breakdown of the Xe–Ar (1:10 at a total pressure of 100 to 200 Torr) mixture at the tightly focused beam waist in the cell. Second, due to the optical setup using two lenses, a quartz lens for 355 nm input and a MgF_2 lens for 118–355 nm output, one can adjust the focus of 118 nm such that a <50 μm focal width is achieved at this wavelength, while the 355 nm light is defocused to roughly 8 mm at the same point. This differential focusing behavior for the two lens systems is due mostly to the dispersion of the index of refraction of MgF_2 at these wavelengths. Third, the tight 118 nm focus, the 10 ns laser pulse width, and the single photon, nonfragmenting ionization process all conspire to generate a single ion peak width of ca. 2 ns for TOFMS features. Added together over a set of 500 to 2000 laser pulses, we observe the "envelope" of single pulse peaks to be 8 to 10 ns for a nonfragmenting ionization process. This observation is extremely helpful for determining which clusters are generated by 118 nm radiation versus residual 355 nm radiation, and which clusters appearing in the TOFMS are of fragmentation origin. We will discuss this point further when we present data for the various neutral metal oxide cluster systems accessed.

The TOFMS employed in the experiments detailed below is a linear one of the focusing Wiley–McLaren design. A reflectron TOFMS could possibly be employed to some advantage to be able to observe metastable decay that requires more than a few microseconds to occur. An ion of mass 100 amu would travel the 1.5 m distance to the detector in about 17 μs ($v = 8.7 \times 10^4$ m/s) and a reflectron would sort out the fragmented ions that would decay in the flight tube following acceleration to 4000 eV within the ionization region. Ions, once created by the light beam, exit the ionization region in ca. 1 μs.

A final word about the experimental apparatus. The cell has, as presently constituted, a 2 mm entrance and a 2 mm exit hole for the clusters: in order for this cell to be well evacuated (ca. 10^{-5} Torr with the chamber at ca. 10^{-7} Torr), a higher pumping speed is required to overcome even normal outgassing than is provided by these two apertures. The cell is thus connected back to the chamber pumping system through one of the input–output tubes attached to the cell to admit gas flow and sample pressure. The connection is made through a 1/4 in. tube connected through the gas handling–evacuation manifold that controls the pressure and fills in the cell.

As emphasized above, the important effort in this work with neutral metal oxide clusters as chemical reagents and catalysts is to determine the neutral cluster distribution before and after the clusters have interacted and reacted with various gas-phase species. The experimental method to ensure that such information is obtained is single photon, near-threshold ionization of the clusters such that they do not obtain enough excess energy over that needed for photoionization to fragment. Additionally, data analysis techniques can be employed to obtain the neutral cluster distribution even though fragmentation is occurring during the ionization process. One such approach is the line width data discussed above: parent ion signals are <10 ns full width at half maximum (FWHM), while daughter (fragment) ion signals are broad with FWHM approaching 100 ns depending on the time course of the multiphoton photodissociation. Another analysis technique employed to discover which peaks in a mass spectrum are due to parent and daughter clusters is called covariance mapping [44, 45]. This method of data analysis can be applied to mass spectroscopy features and is in principle quite simple and transparent. The covariance matrix is the off-diagonal component of the variance; that is, one asks for the correlation between the fluctuations $(x - \bar{x})$ in two spectral features (x,y). The covariance is defined as $C(x, y) = \frac{1}{n}\sum_{i=1}^{n}(x_i - \bar{x})(y_i - \bar{y})$ where x_i and y_i represent signal amplitude for peaks x and y for the ith laser shot, $\bar{x}, \bar{y}$ are averages of x_i and y_i for all laser shots in the experiment, and n is the total number of laser shots. The covariance can be -1 for two signals that anticorrelate (like daughter and parent ions), 0 for signals that do not correlate at all (like V_xO_y and Ti_xO_y) and $+1$ for signals that correlate (like V_xO_y and $V_{x+1}O_{y+2}$ for a $V_xO_y + VO_2 \longrightarrow V_{x+1}O_{y+2}$ growth reaction). The actual situation for a given data set and experiment is more complicated than this simple representation, however. Since the covariance depends on the fluctuations of the signals, details of the covariance matrix elements are somewhat dependent on the source of the fluctuation, which could be due to laser intensity fluctuations affecting the ionization or fragmentation probabilities, fluctuations in the growth process, and even fluctuations in the fragmentation behavior. Thus, careful study of the laser power, growth conditions, and other experimental parameters are required to model the correlations for the covariance matrix [44, 45]. Nonetheless, one can quite successfully learn much about the neutral cluster distribution and how to reduce cluster fragmentation during the ionization process by generating model and experimental covariance matrices as a function of experimental parameters. The process is discussed in detail and original references are given in Refs. [44, 45].

A caveat is appropriate to mention concerning the covariance analysis of mass spectral data and the identification of parent and fragment features in the mass spectrum of metal oxide clusters. Suppose the true neutral clusters in the supersonic expansion beam are all totally fragmented to smaller clusters by a MPI process (e.g., $M_mO_{2m} + nh\nu \longrightarrow M_nO_{2m-1}$, for all m). Given this situation, the fragmented species will be related to one another by a growth mechanism dominated correlation (positive correlation, $C > 0$) and one might assume that the M_mO_{2m-1} series were the true neutral cluster growth pattern, even though the real neutral clusters have the form M_mO_{2m}. One could only realize this latter fact if, at some low intensity for the ionization laser, a few of the $M_mO_{2m}^+$ species could be directly observed. If this series could be observed, one would then have the M_mO_{2m} and M_mO_{2m-1} anticorrelation indicative of a fragmentation-dominated mass spectrum.

Finally, one would like to know the structure of the observed clusters, how these structures change in a systematic fashion as cluster size changes, and whether or not structural patterns evolve in the cluster distribution as the clusters grow. Since, in general, spectroscopic techniques do not yield detailed structural information for metal oxide clusters, theoretical approaches must be enlisted to yield structures for at least special clusters of interest (e.g., clusters that are especially reactive or that facilitate special reactions). To this end one can employ "computer experiments" at the density functional theory (DFT) or Hartree–Fock (HF) level with appropriate basis functions and try to calculate bonding energies, structures, and relative stabilities for the various clusters of interest. Since both DFT and HF calculational schemes are highly approximate, and at best small basis sets can be used for clusters of more than ten atoms, one must be careful about taking such calculations too seriously. Nonetheless, they can be useful especially if certain precautions are followed: (1) many forms of the theory (DFT and HF) are used with widely different basis sets that all converge to the same ordered set of isomers; (2) calculated bonding energies and stabilities coincide for both calculations and experiment; and (3) the calculations and various algorithms are calibrated on one or two small clusters (e.g., MO, MO_2, etc.) for which actual spectroscopic and structural data are available. We have taken such an approach to the series Zr_mO_{2m} [44] and have generated some interesting data for this system for $m = 2$, K, 6. These results will be discussed below along with other calculations on this system.

17.3 RESULTS AND DISCUSSION

In presenting the results for the catalytic studies of neutral metal oxide clusters, we will break the presentation into two large sections: one dealing with cluster structure and the neutral clusters distribution, and the other dealing with cluster selective reactivity. Within each division, the individual metal oxides will be treated separately because even though there are some general principles that are followed by all the clusters (e.g., multiphoton fragmentation seems mostly to occur by loss of oxygen), each metal oxide has its own chemistry and its own stoichiometry with regard to the most stable and prevalent clusters found in the neutral distribution.

17.3.1 Iron Oxide

The neutral cluster distribution for iron oxide is most significantly dependent upon the concentration of oxygen in the expansion gas and the surface oxygen content of the iron metal sample employed in the ablation process [72, 73]. Iron is a very reactive metal with oxygen and 0.5% to 1.0% O_2 in the expansion gas is sufficient under our experimental conditions to generate saturated clusters of the form $Fe_mO_{m,m+1,2}$. These series are apparent even with 193 nm MPI for the detection process: Fe_mO_n clusters with up to 30 iron atoms can be identified and such a distribution is displayed in Figure 17.3. Even without oxygen in the expansion gas, enough surface oxygen is adsorbed or reacted with the Fe foil that a first pass of the ablation laser over a fresh Fe surface yields nearly the same distribution of Fe_mO_n clusters as with oxygen in the gas flow.

One observes (with 193 nm ionization) $Fe_mO_{m,m-1,2}$ clusters for $2 \leq m \leq 12$: fragmentation is still a concern here so we do not think of this yet as a determination of the neutral clusters distribution. A second and third pass over this same track on the Fe foil surface finally shows a significant difference in the cluster distribution under oxygen starvation conditions (Figure 17.4). One point is quite clear in this set of mass spectra, however; the

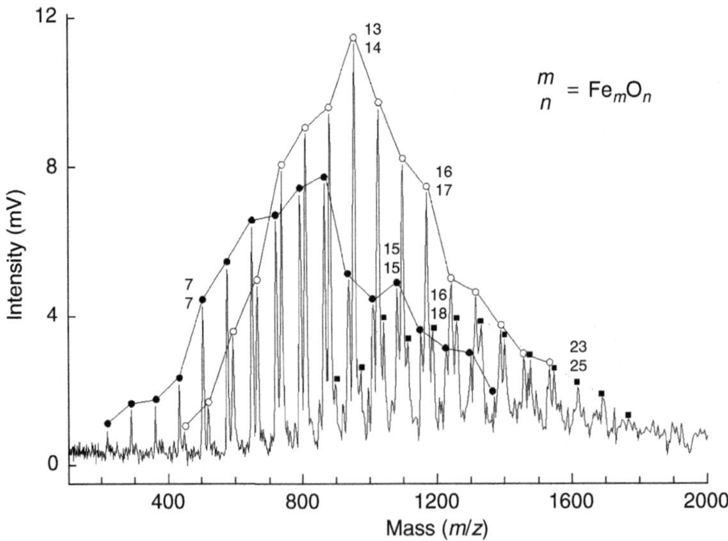

FIGURE 17.3 TOF mass spectrum of iron oxide clusters showing the series of clusters $Fe_mO_m^+$, $Fe_mO_{m+1}^+$, and $Fe_mO_{m-2}^+$ (0.75% O_2 in 20 psig He, ablation laser energy of 3.4 mJ/pulse, and ionization laser energy of 1.1 mJ/pulse). Closed circles, open circles, and closed squares show positions of $Fe_mO_m^+$, $Fe_mO_{m+1}^+$, and $Fe_mO_{m+2}^+$ clusters, respectively.

cluster fragmentation generated by MPI employing 193 nm radiation for cluster detection is not so severe as to mask completely the cluster growth conditions. One can observe the change in the mass pattern using 193 nm ionization in Figure 17.5, which shows how the detected distribution varies with laser power.

Very similar growth conditions give different mass spectra for different ionization laser wavelengths and powers. This situation is clear from Figure 17.6. At 355 nm ionization,

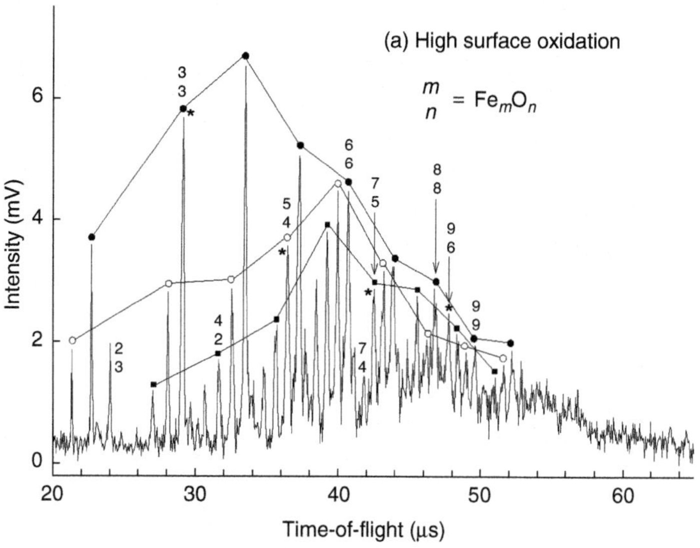

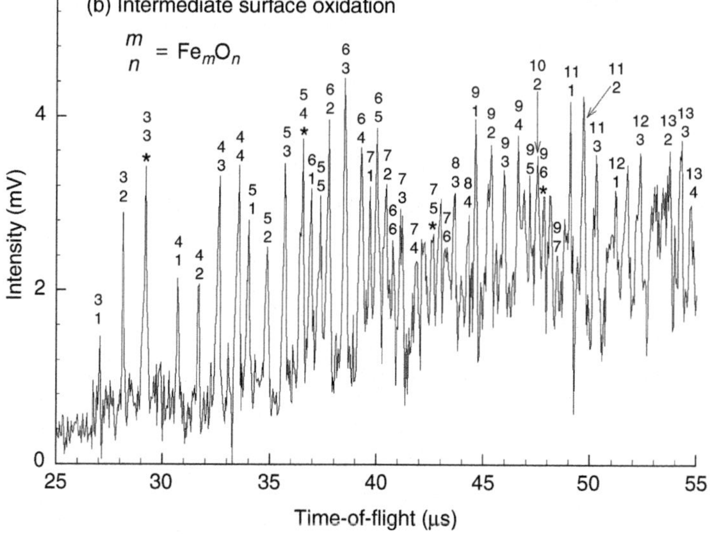

FIGURE 17.4 TOF mass spectra of iron oxide clusters obtained by ablation of three different surface oxidation conditions for an iron surface with no added oxygen in expansion gas (20 psig He and 5 Hz) (a) high oxidation — closed circles, open circles, and closed squares show positions of $Fe_mO_m^+$, $Fe_mO_{m-2}^+$, and $Fe_mO_{m-2}^+$ clusters, respectively; (b) intermediate oxidation — closed circles, open squares, and open circles show positions of Fe_mO_1, Fe_mO_2, and Fe_m clusters, respectively; and (c) low oxidation. Note that $Fe_{13}O_8$ and $Fe_{15}O_1$ (•) are degenerate in mass (c) and that (*) in (a) and (b) marks clusters of the form $Fe_mO_{(m+3)/2}^+$ for $m = 3, 5, 7, 9$.

continued

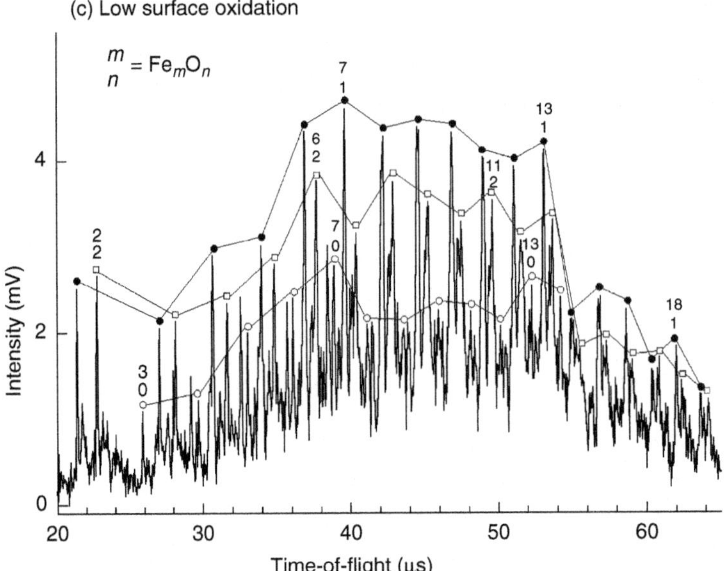

FIGURE 17.4 *continued*

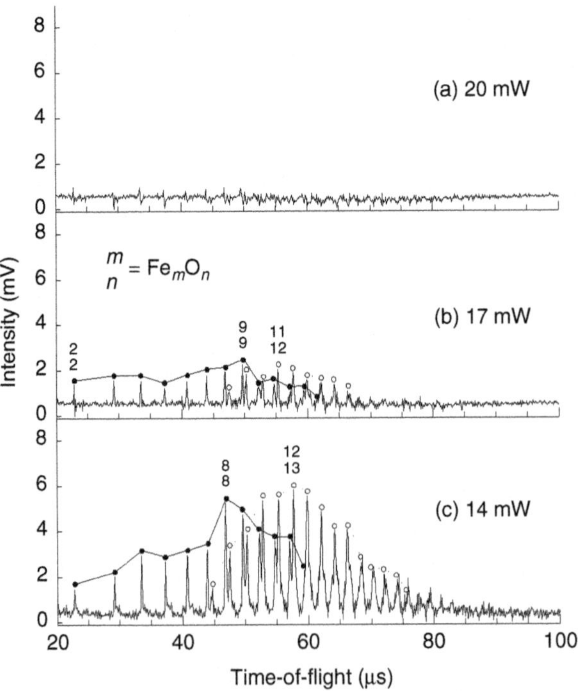

FIGURE 17.5 TOF mass spectra as a function of ionization laser power (0.75% O_2 in 110 psig He, and ablation laser power of 25 mW). Closed and open circles show positions of $Fe_mO_m^+$ and $Fe_mO_{m+1}^+$ clusters, respectively.

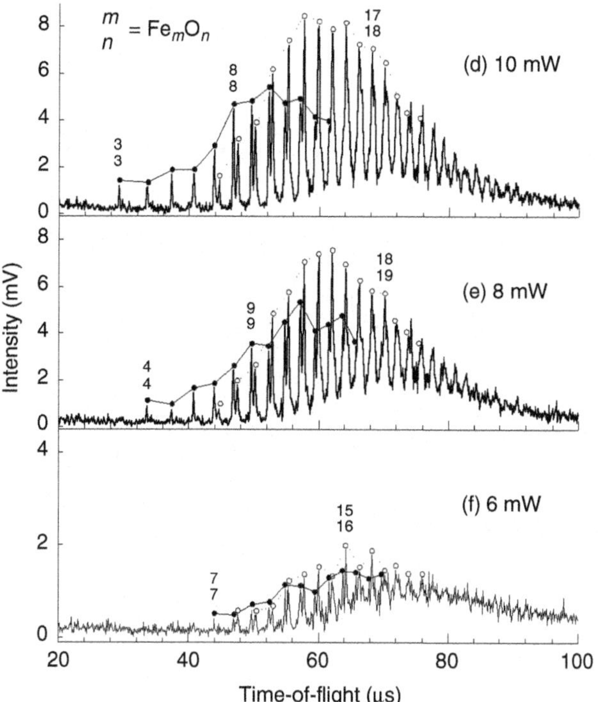

FIGURE 17.5 *continued*

multiphoton ionization or fragmentation dominates the mass spectrum and only small clusters with $Fe_mO^+_{m,m+1}$, $m \leq 13$, can be observed. Under the same experimental conditions, 193 nm ionization yields the distribution of ions $Fe_mO_{m\pm1,m\pm2}$ to $m \sim 20$, with the distribution

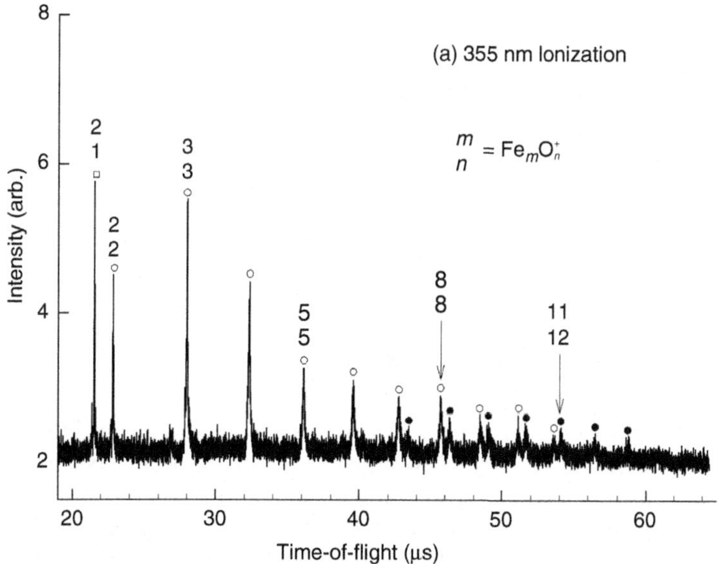

FIGURE 17.6 TOF mass spectra of iron oxide clusters ionized by three different ionization wavelengths: (a) 355 nm, (b) 193 nm, and (c) 118 nm (Stagnation pressure is 20 psig.) *continued*

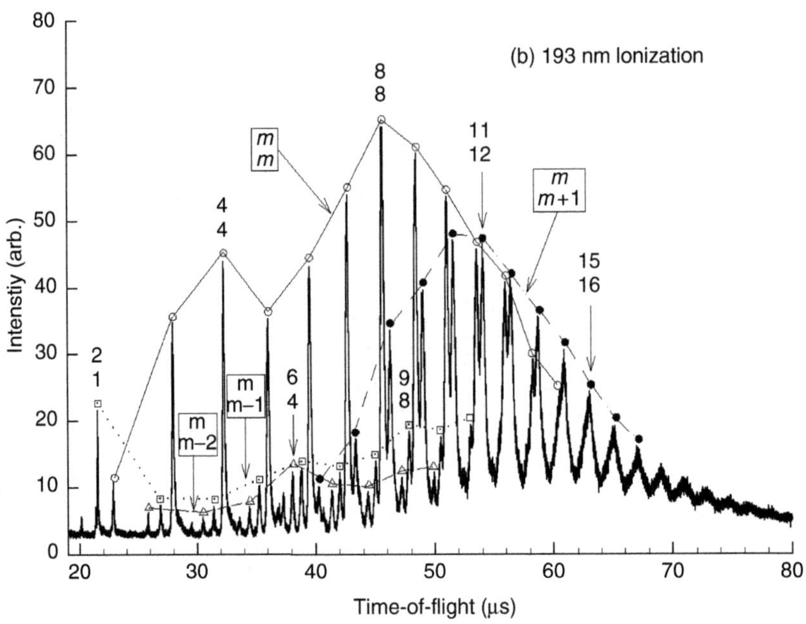

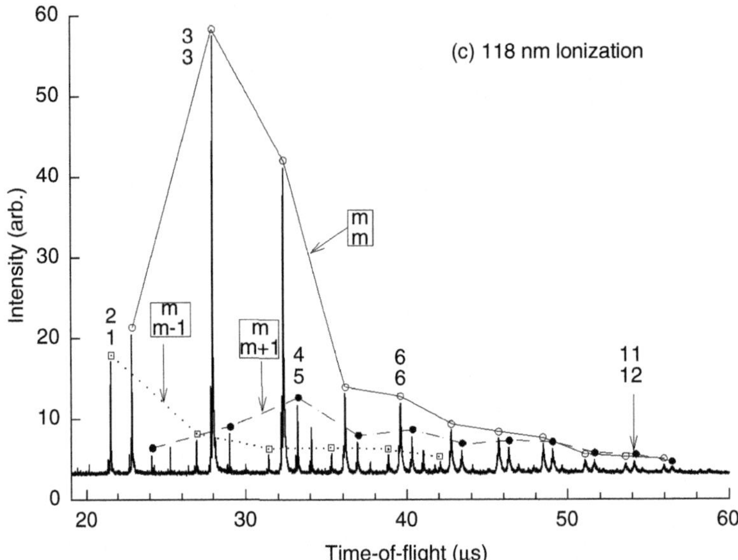

FIGURE 17.6 *continued*

maximum at Fe_mO_m, $m \sim 8$. For 118 nm single-photon ionization, one can obtain the true neutral cluster distribution of $Fe_mO_{m,m\pm1}$ to about $m \sim 12$ with $m = 3$ the most intense feature. A direct comparison between the 118 nm detected spectrum and the 193 nm detected spectrum is shown in Figure 17.7. Note two features of these spectra, obtained under growth and timing conditions conducive to generating a full set of medium-sized clusters. First, the 193 nm spectrum contains only broad features, while all the 118 nm ionized clusters have quite sharp peaks (see below for more detail). Second, the 193 nm ionized main features of this spectrum are $Fe_mO_{m,m-1,2,3}$ and Fe_mO_{m+1}, while those detected with 118 nm ionization are of the forms $Fe_mO_{m,m+1,2,3}$.

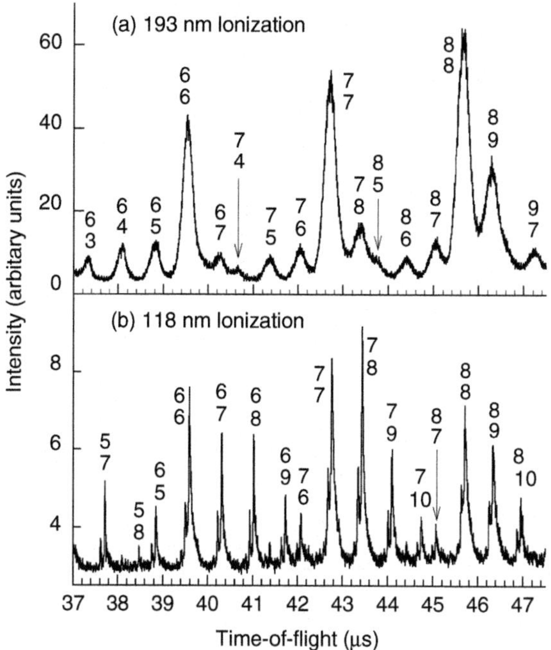

FIGURE 17.7 Expanded portions of the mass spectra of Figure 17.1b and Figure 17.3 for the region of $Fe_3O_4^+$ through $Fe_9O_7^+$ clusters. The clusters $Fe_mO_n^+$ are labeled as *mn*.

The proof that the 118 nm detected mass spectrum is representative of the true neutral cluster distribution with no fragmentation comes from the line width studies. Figure 17.8 shows the mass spectra of and clusters ionized with a combination of 355 and 118 nm radiation. One observes here that, as the 355 nm power is reduced, the spectral features sharpen, weaken, and in some instances, grow in relative intensity. These changes with 355 nm power are due to the relative reduction in the multiphoton 355 nm component of the ionization with respect to the single photon 118 nm component. In this instance, 355 nm derived fragmentation takes the forms $Fe_m\,O_n \xrightarrow[355\,nm]{} Fe_m\,O_{n-1,2}$ and $Fe_{m+1}O_n \xrightarrow[118\,nm\,+\,355\,nm]{} Fe_{m-1}O_n$. One can resolve the gaussian width σ ($\Delta t_{1/2} = \text{FWHM} = 2\sqrt{2\ln 2}\sigma = 2.355\sigma$) parameter by fitting these features with a gaussian function as shown in Figure 17.9. Features with FWHM $\leq$ 10 ns are generated without fragmentation.

Through a comparison of cluster ion distribution and TOFMS signal line widths observed for 355, 193, and 118 nm laser ionization, the neutral iron oxide cluster distribution is characterized for given conditions of laser ablation of the metal, oxygen concentration, and supersonic expansion pressures. We conclude from these studies the following: (1) for clusters with fewer than ten iron atoms, the most stable neutral clusters are of the form Fe_mO_m ($m \leq 10$); (2) for larger clusters, $m \geq 10$, oxygen-rich clusters begin to appear with significant intensity, $Fe_mO_{m+1,2}$; (3) clusters of the form $Fe_mO_{m-1,2}$ also can be observed but are generally less abundant than the others; (4) cluster features have gaussian widths as small as $\sigma \sim 4$ ns ($\Delta t_{1/2} \sim 10$ ns) for 118 nm ionization; and (5) line widths arise from laser spot size in the ionization region, laser pulsewidth, and fragmentation of neutral clusters. One can now compare these results with those following passage of the neutral clusters distribution through a reaction cell.

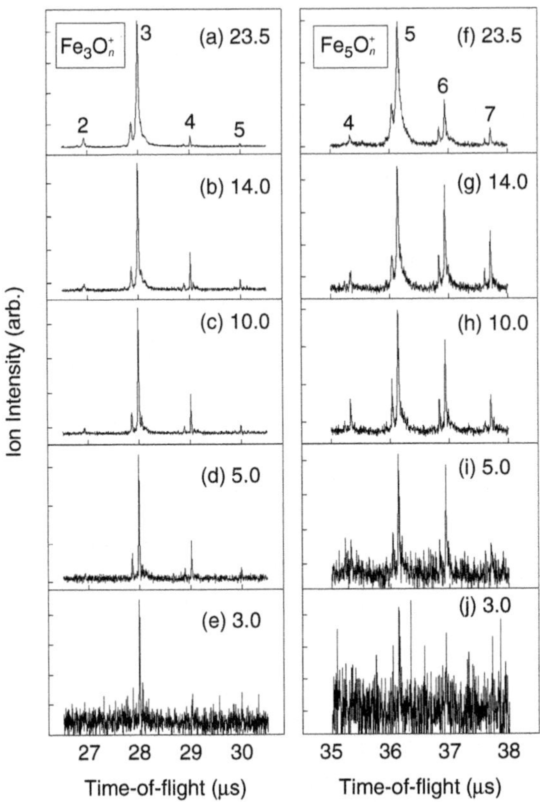

FIGURE 17.8 TOF mass spectra of $Fe_3O_n^+$ ($n = 2$ to 5) and $Fe_5O_n^+$ ($n = 4$ to 7) cluster ions (118 nm ionization) as a function of the pulse energy of 355 nm (at the range 3.0 to 23.5 mJ/pulse) for 118 nm generation. All y axes are scaled to the highest peaks, such as $Fe_3O_3^+$ and $Fe_5O_5^+$.

17.3.2 Copper Oxide

The situation described above for iron oxide is very different than that found for copper oxide [74]. With iron oxide, the metal is used as the ablated material to form the Fe_mO_n species in the gas phase, MPI (355 and 193 nm light) does cause fragmentation but primarily through loss of an oxygen atom or two, and a wide distribution ($2 \leq m \leq 30$, in our system) of clusters is observed. Iron is very reactive toward oxygen, and many clusters of the forms Fe_mO_m and $Fe_mO_{m\pm1,\pm2}$ have significant thermodynamic stability. Copper, on the other hand, is not very reactive and has a major change in stoichiometry if detected by MPI.

Copper oxide clusters generated by the most straightforward manner used for iron oxide and other metal oxides, gives only Cu_3O and Cu_4O_2 (very weak) clusters when ionized by either 193 nm or 118 nm radiation. In order to get a larger distribution of clusters Cu_mO_n ($m = 30$), one must ablate at CuO sample. This implies that the Cu metal plasma created by laser ablation is not as reactive with molecular oxygen as other metals are; as the plasma cools while it expands in the ablation–mixing process, Cu_m neutrals and ions do not react significantly with O_2. If CuO powder is placed on the drum with a glue (sugar water seems to work best) to hold it in place and to keep the film from cracking, laser ablation, in the same manner as with metal foil in general, generates a wide range of clusters to Cu_mO_n, $2 \leq m \leq 30$, with or without oxygen in the expansion gas. The essential feature here to generate a good distribution of thermodynamically controlled neutral clusters, and not only kinetically controlled small clusters, is that the copper and atomic oxygen components must be intimately mixed in the plasma and must react at a high temperature: apparently cooling is fast enough, and the molecular oxygen bond strong enough, that $Cu_n^{0,\pm1,K} + O_2 \longrightarrow Cu_m O_n$ is not a facile reaction under our ablation and flow conditions. Figure 17.10a presents a mass spectrum of the cluster ion distribution (Cu_mO_n, $m < 20$) created by 193 nm MPI of the Cu_mO_n clusters created by ablation of a CuO sample. The observed cluster stoichiometry is roughly $Cu_mO_{m/2}$. The same set of conditions for the generation of copper oxide clusters but simply changing the ionization laser (blocking the 193 nm beam, and unblocking the 118 nm beam), presents a much different picture of the neutral cluster distribution for Cu_mO_n. Figure 17.10b gives this mass spectrum. One should note three observations about this spectrum of Cu_mO_n clusters. First, the stoichiometry is now much more like Cu_mO_m, especially for the features that have a narrow line width ($\Delta t_{1/2} \sim 10$ ns for the smaller clusters growing to $\Delta t_{1/2} \sim 15$ ns for the larger ones, due to the fundamental nature of the TOFMS apparatus). Second, mass spectral features are mostly highly resolved and the broader ones with $\Delta t_{1/2} > 20$ ns are caused by residual 355 nm multiphoton, fragmenting ionization. Figure 17.11 and Figure 17.12 give an expanded view of some of these lineshapes. One can even observe that some clusters have one or two hydrogen atoms attached and that some of the widths of a few features may arise from fragmentation of these additional atoms. Hydrogen atoms attached to the Cu_mO_n clusters arise from the "glue" used to hold the sample on the drum: in this case, sugar and water. Third, one should note that the 118 nm mass spectrum ends at about Cu_6O_5, whereas the 193 nm spectrum continues to ~$Cu_{30}O_{15}$. This probably happens because the higher mass

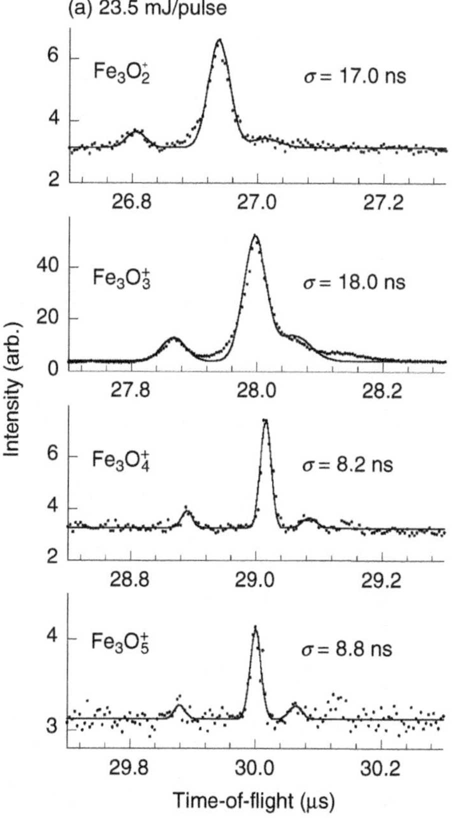

FIGURE 17.9 Gaussian widths, σ, of $Fe_3O_n^+$ cluster ion peaks at two different pulse energies of 355 nm light for the generation of 118 nm light; (a) 23.5 mJ/pulse and (b) 5 mJ/pulse. The fitted lineshape is in red.

continued

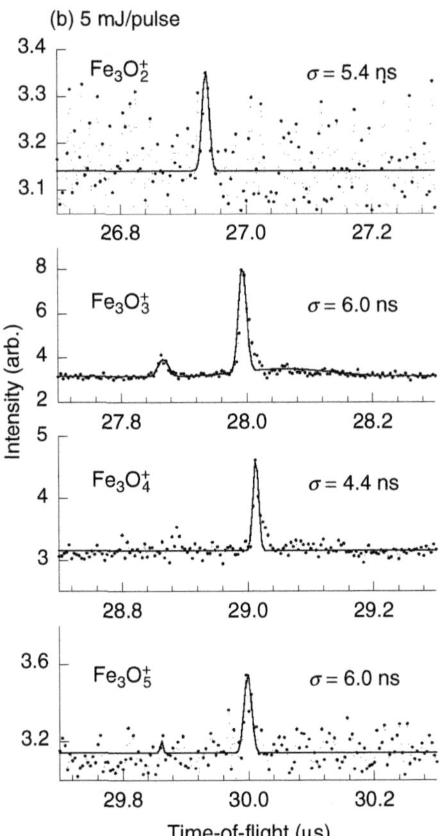

FIGURE 17.9 *continued*

clusters are in low concentration, the 118 nm light is very low energy/pulse (<1 fJ/pulse), and the larger clusters may have a low enough ionization energy such that they can be ionized by one photon of 6.4 eV energy (193 nm). From the high dispersion 118 nm spectrum of Cu_mO_n clusters, one can readily determine which features are due to 355 nm residual fragmentation and which are part of the neutral cluster distribution, even with one or two H atoms attached to the cluster. These latter species may be interesting with regard to chemical reactivity and even catalytic behavior.

To summarize the results for Cu_mO_n clusters, ionization with 118 nm radiation demonstrates that neutral cluster fragmentation does not occur by this single-photon process. We are thereby able to draw the following conclusions concerning the neutral cluster distribution created in the supersonic expansion.

1. Ablation of copper metal in the presence of molecular oxygen gives mostly Cu_m clusters, with only Cu_mO_n observed for $m < 4$ and $n \sim 1, 2$.
2. For ablation of CuO powder samples, the observed clusters in the distribution depend on the ionization laser wavelength. With 193 nm multiphoton, neutral cluster fragmenting ionization, the observed cluster ion distribution is roughly of the form $Cu_{2m}O_m^+$ for $4 \leq m \leq 20$ and at 355 nm multiphoton, neutral cluster fragmenting ionization, the observed cluster ion distribution is roughly of the form Cu_m and $Cu_mO_{1,2}$, for $m \leq 10$. For both of these laser wavelengths, extensive neutral cluster fragmentation occurs due to multiphoton absorption processes.

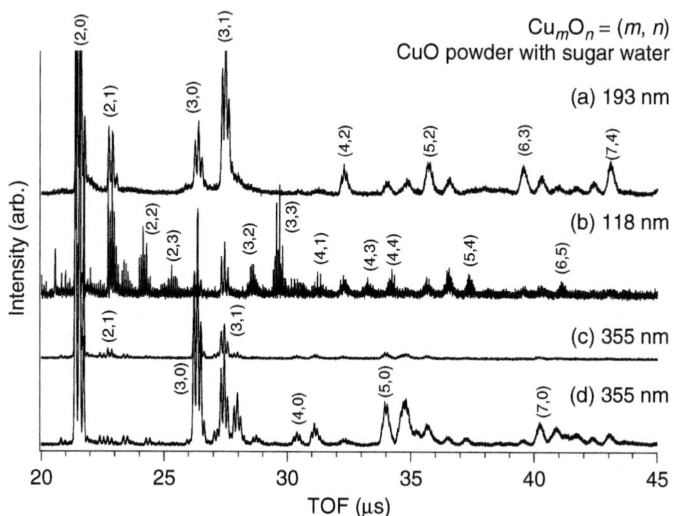

FIGURE 17.10 TOF mass spectra of CuO with 100 psig 10% O_2/90% He measured by (a) 193 (1 mJ/pulse), (b) 118 nm (generated by 355 nm of 24 mJ/pulse), and (c) 355 nm (24 mJ/pulse) ionization. These spectra are obtained under optimized conditions for 355 nm ionization with nozzle and ablation parameters unchanged. A CuO powder with sugar water sample is employed. The mass spectrum for 118 nm ionization also contains features generated by 355 nm ionization.

3. With 118 nm single-photon ionization for a copper oxide powder sample, the most abundant neutral clusters are of the form Cu_mO_m for $m \leq 4$ and Cu_mO_{m-1} for $m > 4$. Based on mass spectral feature line widths, the series of clusters uniquely observed with 118 nm ionization represent the true neutral cluster distribution for Cu_mO_n in the beam. These cluster series are then the thermodynamically stable ones for the neutral cluster distribution.

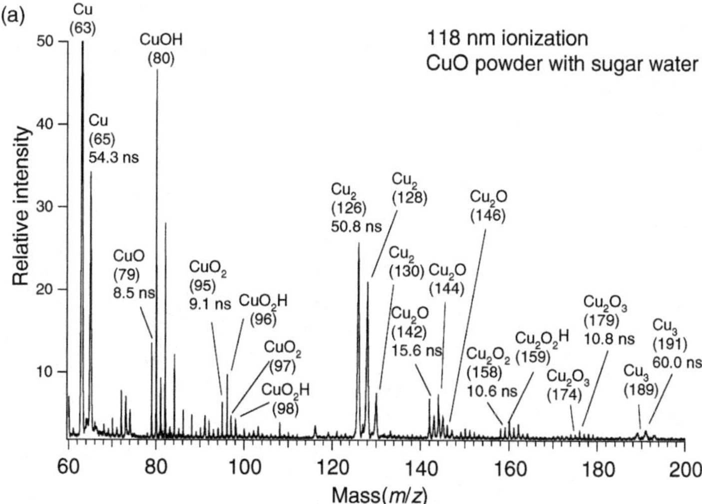

FIGURE 17.11 The 118 nm ionization TOF mass spectra with expanded dispersion and assignments of mass peaks. Cu_2^+, Cu_3O^+, $Cu_4O_2^+$, K, are generated by 355 nm and they are due to fragmentation. Numbers in parentheses are mass numbers of observed peaks. The ns indications give FWHM of mass peaks obtained by fitting a gaussian function to the features. *continued*

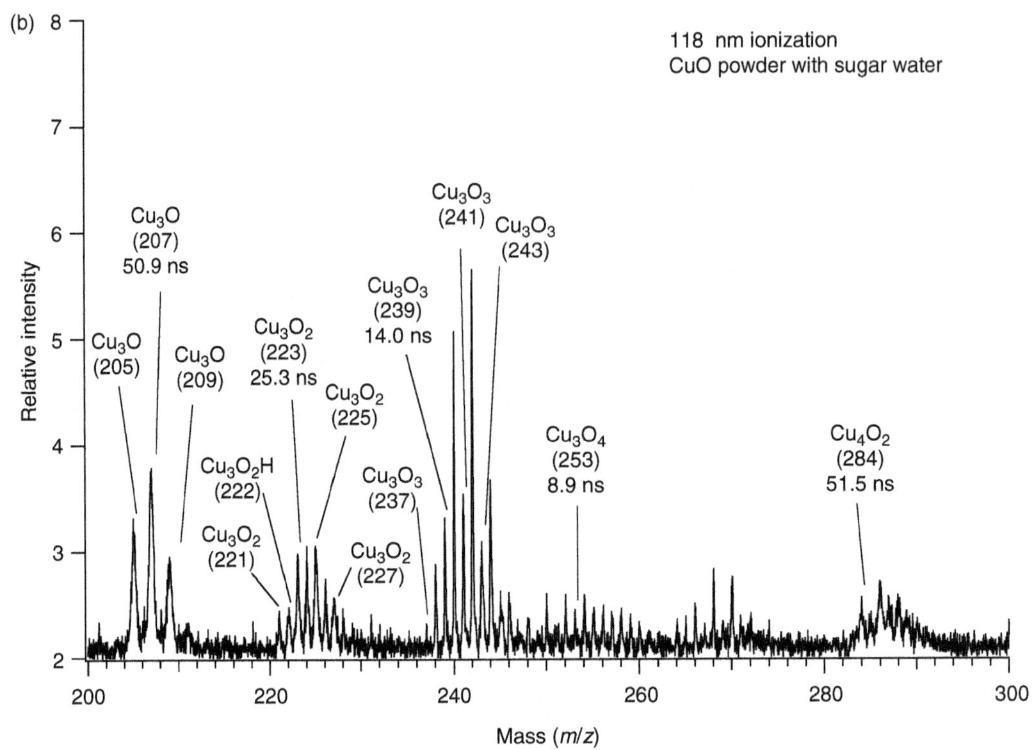

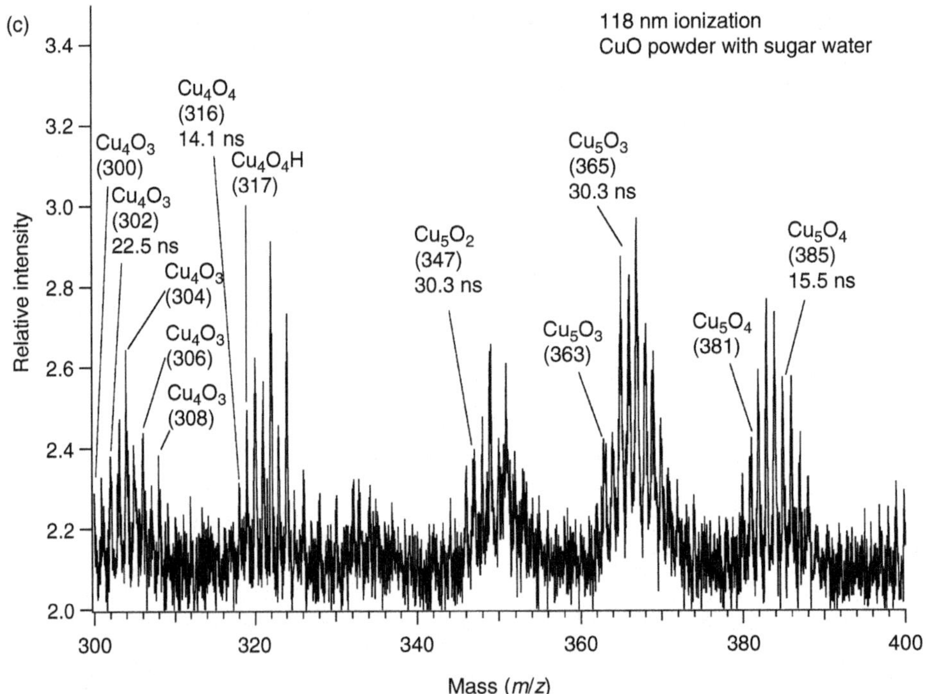

FIGURE 17.11 *continued*

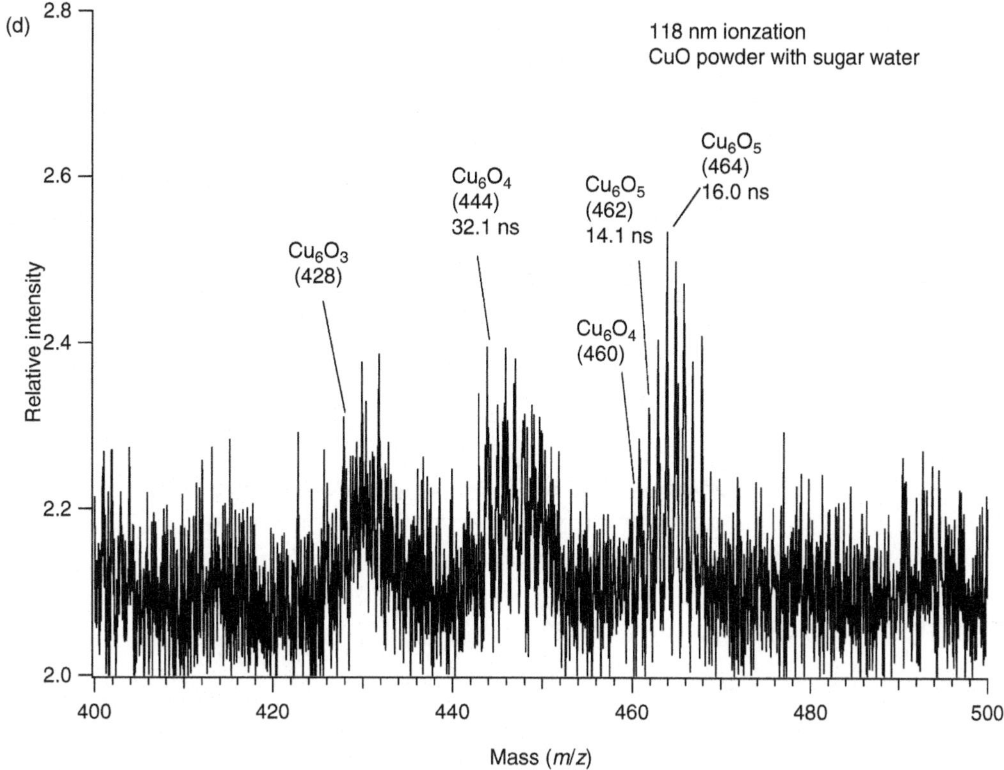

FIGURE 17.11 *continued*

4. The observed cluster fragmentation by 193 nm is therefore mainly of the form Cu_mO_m and $Cu_mO_{m-1} \longrightarrow Cu_mO_n$ with $n = m/2, m/2{+}1$.
5. The cluster growth mechanism for the copper oxide cluster system is suggested to be mCu + nO (ablated) $\longrightarrow Cu_mO_n$ or mCuO $\longrightarrow Cu_mO_{m,m-1,\ \ldots}$. In both instances, the reactions must occur with the copper and oxygen atoms generated in the plasma created by laser ablation. Oxygen atoms in the clusters come from the ablated sample and not the carrier gas.

17.3.3 Zirconium Oxide

Zirconium oxide cluster behavior and chemistry is much more like iron oxide than copper oxide [44, 75]: the metal can be used to generate a saturated distribution of neutral Zr_mO_n clusters; fragmentation of the neutral clusters through MPI (193 or 355 nm) results mostly in a loss of an oxygen atom; and small clusters Zr_mO_{2m}, $m \leq 5$, have a higher ionization energy than large ones and require more laser power at 355 and 193 nm to generate ions. These points are well illustrated in Figure 17.13 to Figure 17.15 for clusters as large as $Zr_{10}O_{20}$. Note, also, a general point that the pulsed cluster beams tend not to be homogeneous; that is, the neutral cluster distribution evolves as the relative timing between the ablation and ionization laser changes (Figure 17.15). One can even observe that the apparent most stable clusters evolve as this relative timing is varied. Figure 17.16 shows the beautiful and spectacular difference between fragmented and unfragmented features in the mass spectra of $Zr_2O_{3,4}$ and $Zr_5O_{9,10}$. One can almost estimate the fragmentation lifetime (1/e time) for Zr_2O_4 from the decay at the long-time tail of the Zr_2O_3 feature in Figure 17.16.

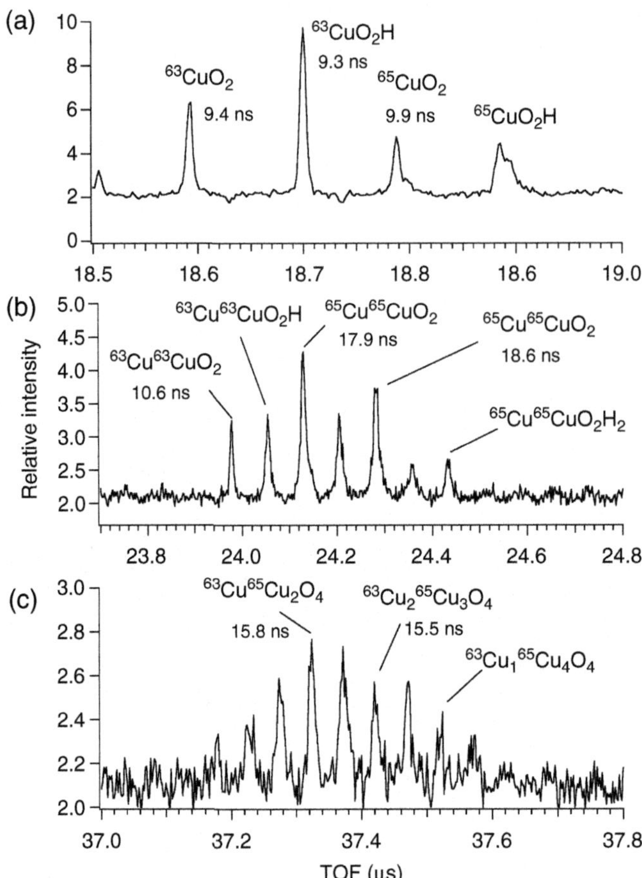

FIGURE 17.12 Expanded TOF mass spectra (118 nm ionization) in the (a) CuO_2, (b) Cu_2O_2, and (c) Cu_5O_4 mass region. The ns indications give FWHM of mass peaks obtained by fitting a gaussian function to the feature. Superscripts 63 and 65 indicate mass of isotopes of copper. Spectrum (a) shows the features are laser line width (in ns) limited and that the CuO_2H feature may contain a fragmentation component from CuO_2H_2. Spectrum (b) shows a $^{63}Cu_2O_2$ laser line width limited feature. Other features in this spectrum may have components from $Cu_2O_2H_{1,2}$ degenerate with $^{63}Cu^{65}CuO_2$ and $^{65}Cu_2O_2$ and may have line width contributions from $Cu_2O_2H_{2,3} \longrightarrow Cu_2O_2H_{2,1}^{+}$ fragmentation. Spectrum (c) shows the homogeneous line width broadening of the features due to TOFMS resolution for amu ~700.

The major result of this study is to show that high-energy, single-photon (10.5 eV, 118 nm), low-intensity (10^{11} photons/pulse) photoionization of transition metal oxide neutral clusters (zirconium oxide, Zr_mO_n) can be employed to determine the neutral cluster distribution in the gas phase. Even though one cannot, at present, select individual clusters for study, as can be done with ionic clusters, the neutral cluster relative populations could be obtained and how these clusters respond to reaction and scattering, and whether or not they evidence catalytic behavior can be studied.

The neutral cluster distribution for zirconium oxide clusters consists mainly of two series of clusters, Zr_mO_{2m} and Zr_mO_{2m+1}. MPI clusters with 193 and 355 nm laser radiation at moderate energy/pulse and focus (intensity, ca. 1×10^7 W/cm^2, ~1 mJ/pulse) generates fragmentation along with neutral cluster ionization. One can argue that the fragmentation is mostly, if not entirely, local in the sense that

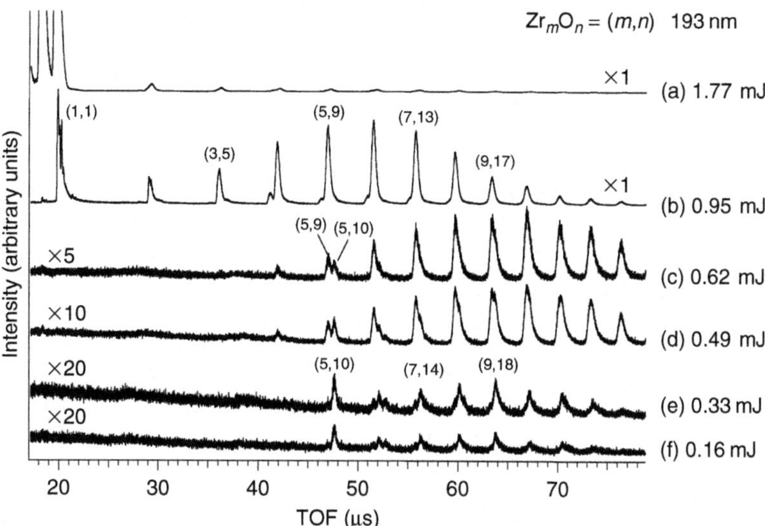

FIGURE 17.13 TOF mass spectra of zirconium oxide clusters measured by 193 nm laser ionization with different energies/pulse as indicated.

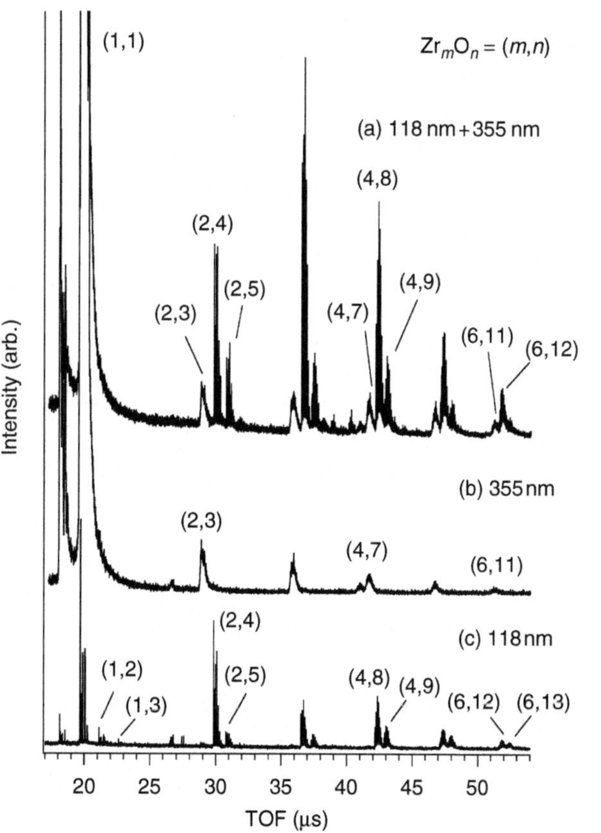

FIGURE 17.14 TOF mass spectra of zirconium oxide clusters that are ionized by (a) 118 nm and 355 nm (24 mJ/pulse), (b) 355 nm (24 mJ/pulse), and (c) 118 nm (17 mJ/pulse at 355 nm).

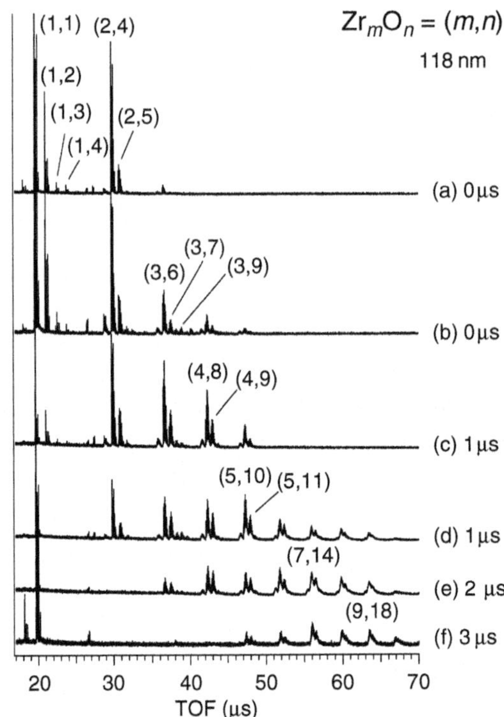

FIGURE 17.15 TOF mass spectra of zirconium oxide clusters detected by 118 nm. These spectra are measured by ionizing different regions of supersonic jet gas pulse of zirconium oxide clusters by displacing the laser horizontally a few mm and by delaying the ionization laser pulse with respect to the ablation process a few microseconds.

$$Zr_mO_{2m,\ 2m+1} \xrightarrow[\substack{193\,\text{nm}\\355\,\text{nm}}]{} Zr_mO_{2m-1} + 1 \text{ or } 2O$$

Mass spectral features generated by a tightly focused (≤50 μm), 118 nm ionization laser are very sharp and their line widths are governed by the duration of the laser pulse. Such line widths are the hallmark of nonfragmenting ionization of neutral clusters. The average full width for these features at half maximum is ca. 10 ns, but individual laser pulses can generate widths as small as a few nanoseconds or less due to the laser mode structure.

At low laser intensity for the MPI process, cluster ions with $m < 5$ are not observed due to higher cluster ionization energies and lower cluster density of states for the small clusters. These small clusters are also not generated by larger cluster fragmentation at these low laser intensities.

We have also performed extensive calculations for the structure of Zr_nO_{2m} for $m = 1$ to 6. Although the calculations are at a qualitative level, even for structure (DFT and HF), we have confidence in them because we have employed a number of difficult functionals and basis sets. All the calculations converge on the same lowest energy structure, predict the same higher energy isomer orders, give Zr_5O_{10} as the relatively most stable cluster of the group with suspect to loss of ZrO_2, and are calibrated for the known structure of ZrO_2. These calculations are described in Ref. [44]. The results of both calculations are similar, although the latter ones are not completed with so many basis sets and methods.

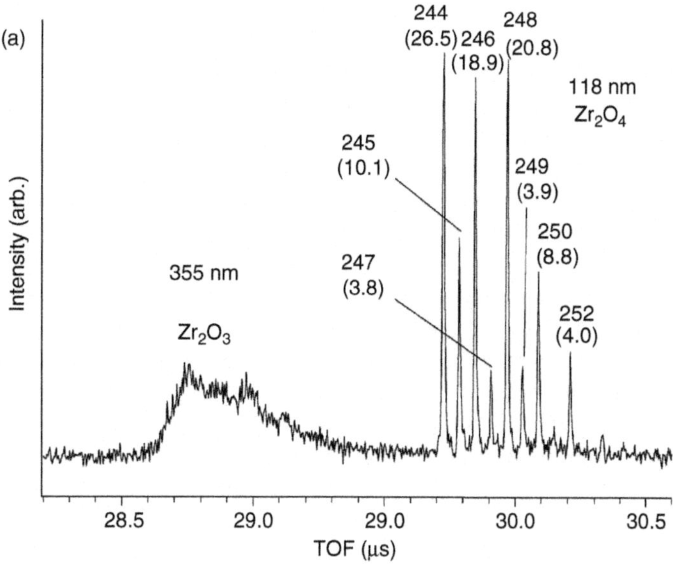

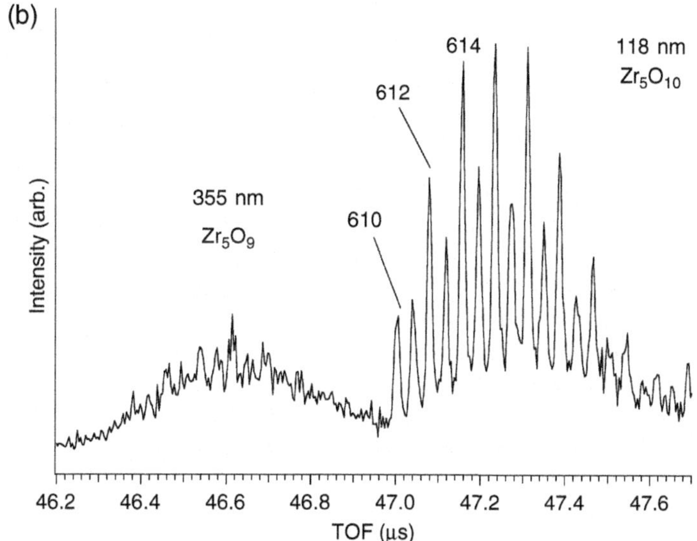

FIGURE 17.16 Expanded TOF mass spectra of zirconium oxide clusters for (a) $Zr_2O_3^+$ and $Zr_2O_4^+$, and (b) $Zr_5O_9^+$ and $Zr_5O_{10}^+$. These spectra are measured by 118 and 355 nm ionization simultaneously. Peaks for Zr_2O_4 are assigned to specific isotopic masses, as are some of those for Zr_5O_{10}. The numbers in parentheses are percentages of natural abundance Zr_2O_4 estimated by abundance of zirconium isotopies. Isotopic species with <1% abundance are ignored. The $Zr_5O_{10}^+$ species are indicated by a few mass numbers. The Zr_2O_4 and Zr_5O_{10} feature FWHM are both ca. 10 ns.

17.3.4 Vanadium Oxide

Vanadium oxide neutral clusters are identified in much the same manner that we have discussed for the three previous systems — Fe_mO_n, Cu_mO_n, and Zr_mO_n [45, 74]. Fragmentation of the neutral clusters is present in this system, as well, for 193 and 355 nm multiphoton processes. The distribution of clusters changes somewhat with increased oxygen content in the expansion gas over the range 0.5% to 4.0%. Laser timing also has an effect on the observed distribution. Again, cluster fragmentation is identifiable through line width studies with nonfragmented 118 nm spectra having a gaussian $\Delta t_{1/2} \sim 8$ to 10 ns. Figure 17.17 shows a timing sequence for 193 and 118 nm detected clusters for a 0.5% O_2 concentration on the expansion gas. Note the intense features for 193 nm radiation detection are V_2O^+, $V_2O_2^+$, $V_3O_3^+$, whereas for 118 nm nonfragmenting ionization, the intense features are V_2O_3, V_2O_4, V_2O_5, and V_4O_6. Figure 17.18 shows the same comparison for a 4% O_2 experiment. Here the abundant neutral clusters are VO_2, V_2O_5, V_3O_7, V_4O_9, V_5O_{12}, V_6O_{14}, and V_7O_{17}. The general trend for these neutral clusters is approximately $(V_2O_5)_x(V_2O_7)_y$. Clearly, the higher O content expansion pushes the clusters to be more oxygen rich. At 193 nm, detection of the 4% O_2 spectra show only $V_2O_3^+$, as the most intense species present.

An interesting note here is that, with 193 nm ionization, the main neutral clusters, such as V_2O_5, V_3O_7, V_4O_9, etc., do not show up in the mass spectrum. Thus, the fragment daughter clusters that do appear in the spectrum are all growth related as found by covariance mapping, because the daughter clusters arise from parent clusters that are no longer present in the spectrum and are growth related. If both parent and daughter clusters were present in the spectrum simultaneously, the daughter–daughter and parent–parent covariance would be expected to be positive, and the daughter–parent cluster covariance would be expected to be negative.

The TOFMS line widths for 193 nm ionized clusters are ca. 30 ns FWHM, and are ca. 10 ns for 118 nm ionized clusters. Figure 17.19 shows the 2000 shot averaged mass peak and single shot spectra for V_2O_3. These spectra are obtained by 118 nm ionization with both seeded and unseeded Nd–YAG lasers. The averaged mass peaks of gaussian shape are composed of clusters ionized at different times. Each mass peak generated by a single ion has a width of ca. 1.7 ns mainly due to the response time of the microchannel plate mass detector. The seeded and unseeded laser ionization pulses at 118 nm have different intensity versus time profiles. The seeded laser output is a smooth 8 to 10 ns FWHM gaussian profile, while that of the unseeded laser is typically a random collection of ~100 ps spikes (mode locked) that vary within a ~10 ns envelope from pulse to pulse. The ions generated by the seeded laser reflect the arrival time variations of the neutrals convolved with the smooth 8 to 10 ns laser pulse profile. The ions generated by the unseeded laser show this behavior in addition to the random intensity profile of the unseeded laser.

Again, the neutral cluster distribution is uncovered and this information and distribution can be employed to study chemistry in reaction cell experiments.

17.3.5 Titanium Oxide

Titanium oxide clusters show a chemistry that is similar to that of zirconium oxide clusters. [77] Iron oxide and copper oxide clusters have the general stoichiometry M_mO_m, vanadium oxide clusters have a stoichiometry of roughly V_mO_n, $n \sim 2m, 2m+1, 2, 3$, for saturated O_2 growth conditions, and Ti_mO_n and Zr_mO_n have typically a stoichiometry of $n = 2m$. The 118 nm ionization is again nonfragmenting, and the general formula for the entire series is Ti_mO_{2m} for m up to about 10. Some of the intense features are followed by a Ti_mO_{2m+1} peak at high O_2 content in the expansion. These two series of clusters are the thermodynamically most stable ones under saturated oxygen growth conditions for the neutral clusters. The main cluster growth mechanism under such conditions can be thought of as $TiO_2 + TiO_2 + L \longrightarrow Ti_m O_{2m} = (TiO_2)_m$. This behavior is shown in Figure 17.20. The mass spectral features obtained by 118 nm ionization are ca. 10 ns FWHM for a 2000 laser pulse averaged spectrum (see Figure 17.21). Individual laser pulses can produce line widths that are close to 2 ns FWHM depending on the laser focal spot size and mode structure for a given pulse, as presented in Figure 17.19 for vanadium oxide. Three competing effects occur to generate these spectra: at low

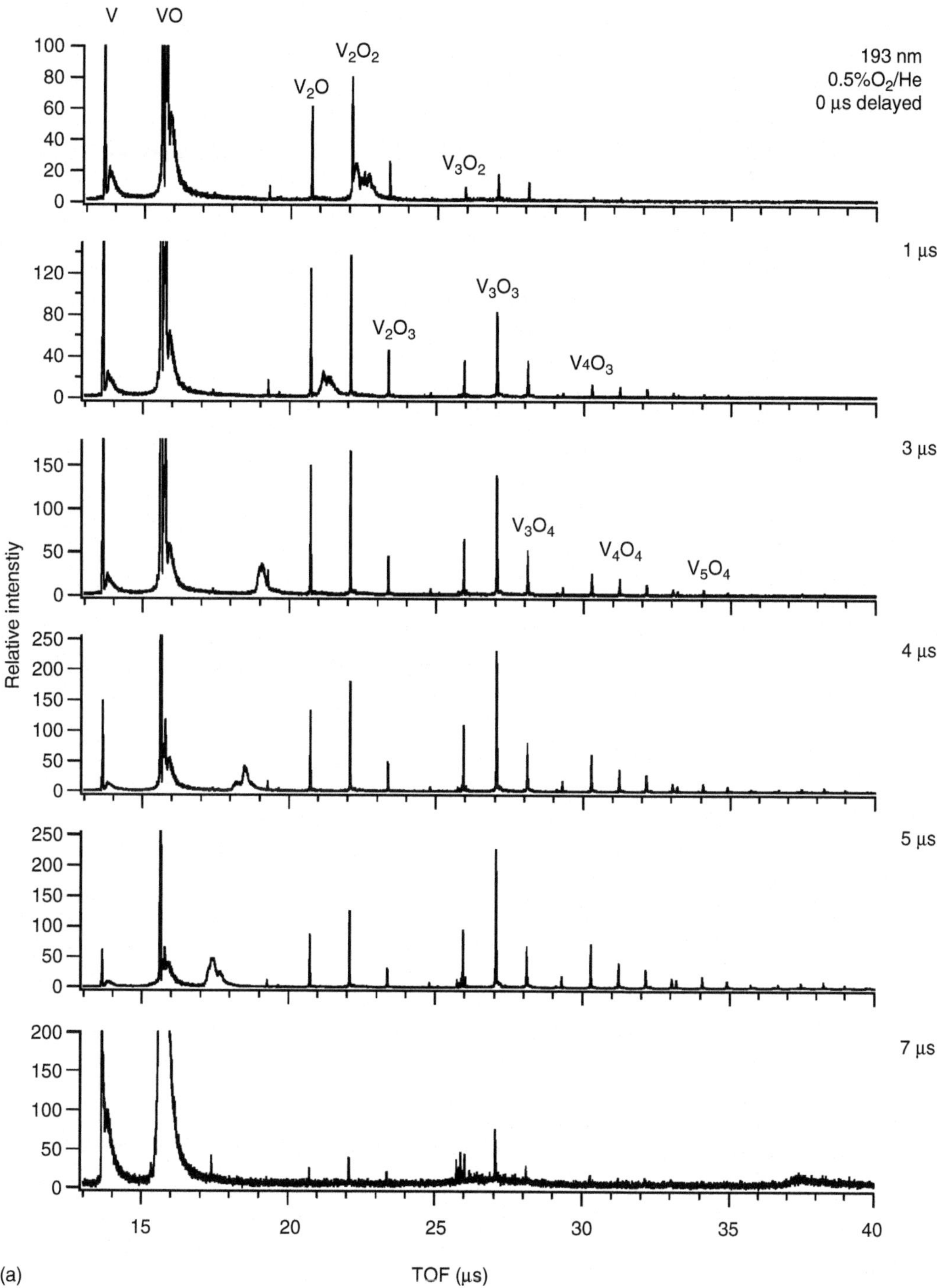

FIGURE 17.17 TOF mass spectra observed for different (a) 193 nm and (b) 118 nm ionization laser timings with respect to the nozzle opening and ablation laser firing times with 0.5% O_2/He reaction/expansion gas. Changes of the spectra with ionization laser timing imply that different regions of supersonic jet gas pulse of titanium oxide cluster are accessed and that the beam of clusters is not homogeneous. The delay time for the ionization laser relative to that for the top spectrum is given in μs.

continued

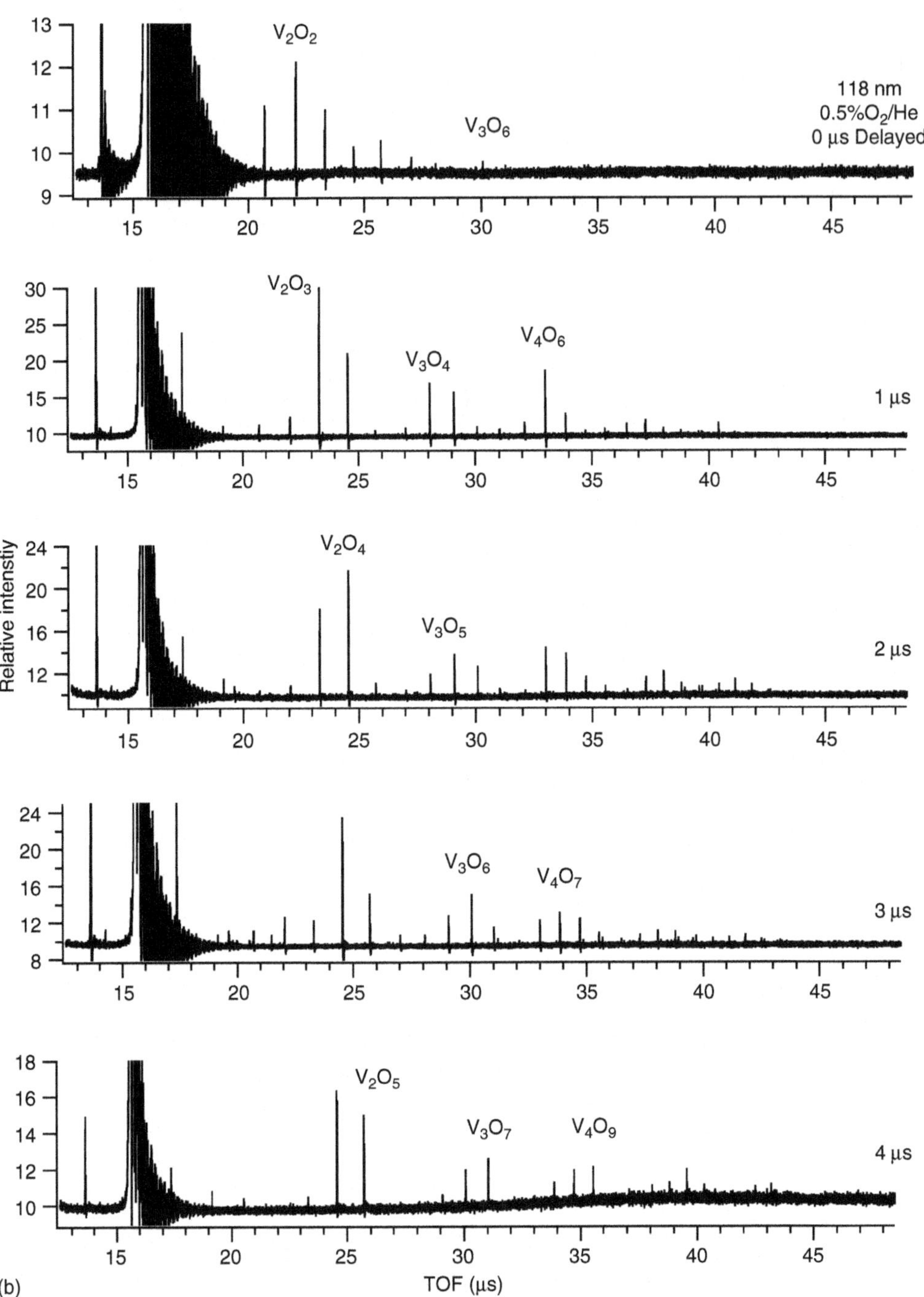

FIGURE 17.17 *continued*

oxygen concentration in the expansion gas, oxygen-poor clusters Ti_mO_{2m-1} are at a higher concentration because growth to the thermodynamically stable $Ti_mO_{2m,2m+1}$ series is kinetically controlled; for 193 nm radiation, fragmentation occurs for the oxygen-rich clusters, as $Ti_mO_{2m,2m+1} \xrightarrow{193\,\text{nm}} Ti_mO_{2m-1,2}$; kinetically controlled oxygen-poor clusters, Ti_mO_{2m-x}, $x > 2$,

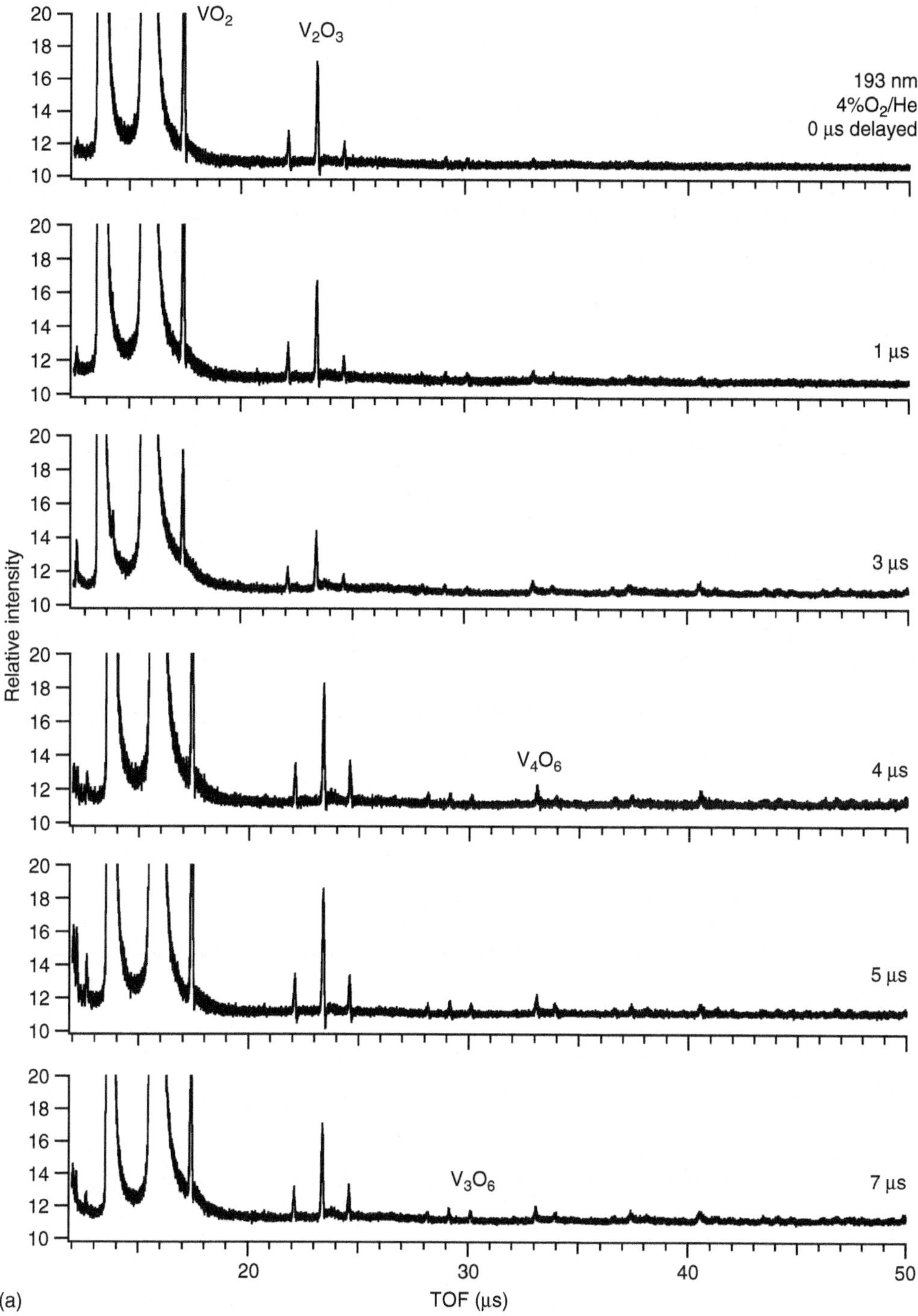

FIGURE 17.18 TOF mass spectra observed for different (a) 193 nm and (b) 118 nm ionization laser timings with respect to the nozzle opening and ablation laser firing times with 4% O_2/He reaction/expansion gas. The delay time for the ionization laser relative to that for the top spectrum is given in μs.

continued

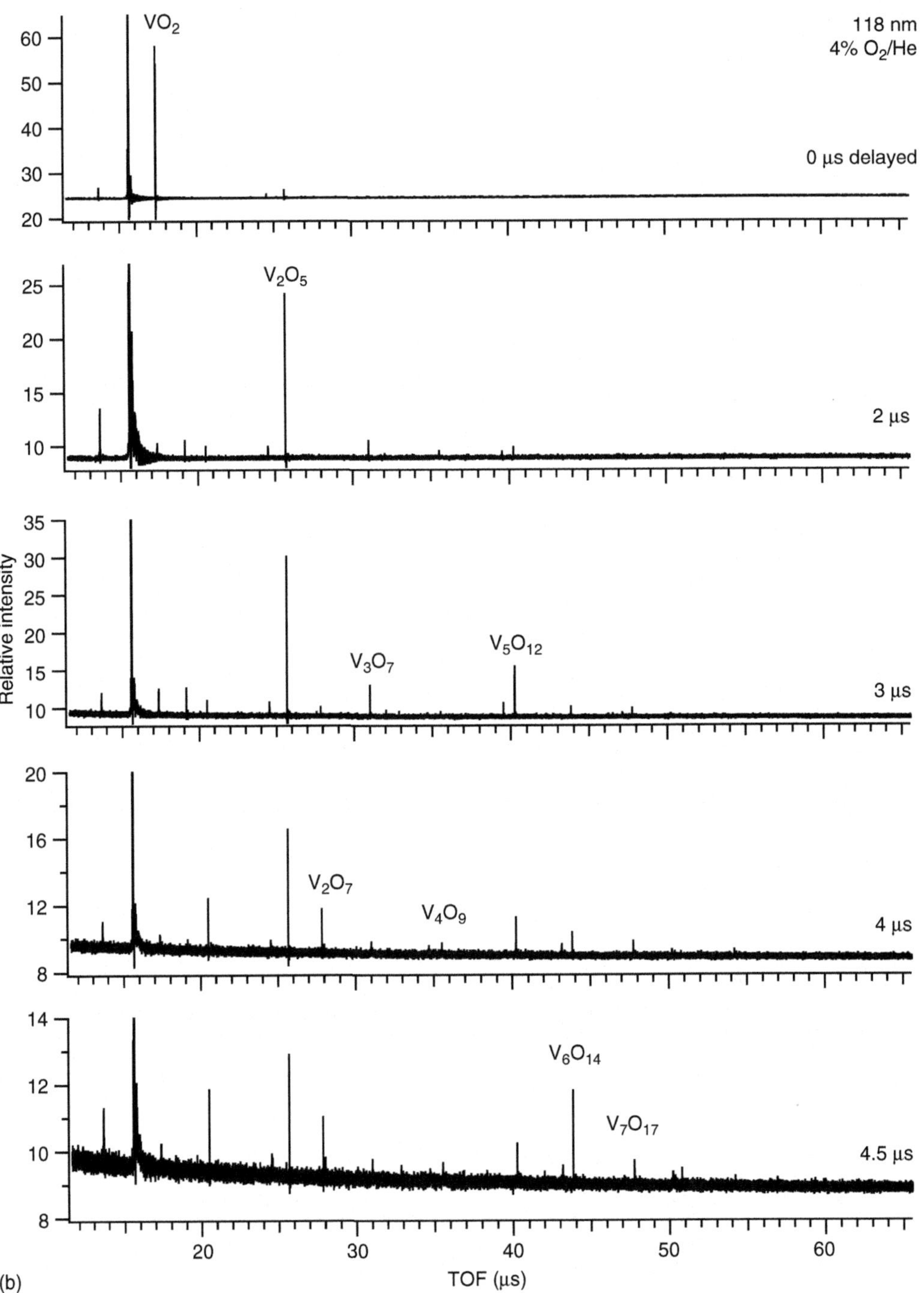

FIGURE 17.18 *continued*

are mainly ionized without fragmentation by 193 nm (multiphoton?) ionization. Thus, with increasing oxygen content in the expansion or reaction gas, the widths of observed mass peaks of clusters detected by 193 nm ionization become broad because of multiphoton fragmentation of oxygen-rich clusters, while the oxygen-poor cluster features remain sharp. Covariance mapping analysis can also help to recognize cluster growth patterns; however, the gentle

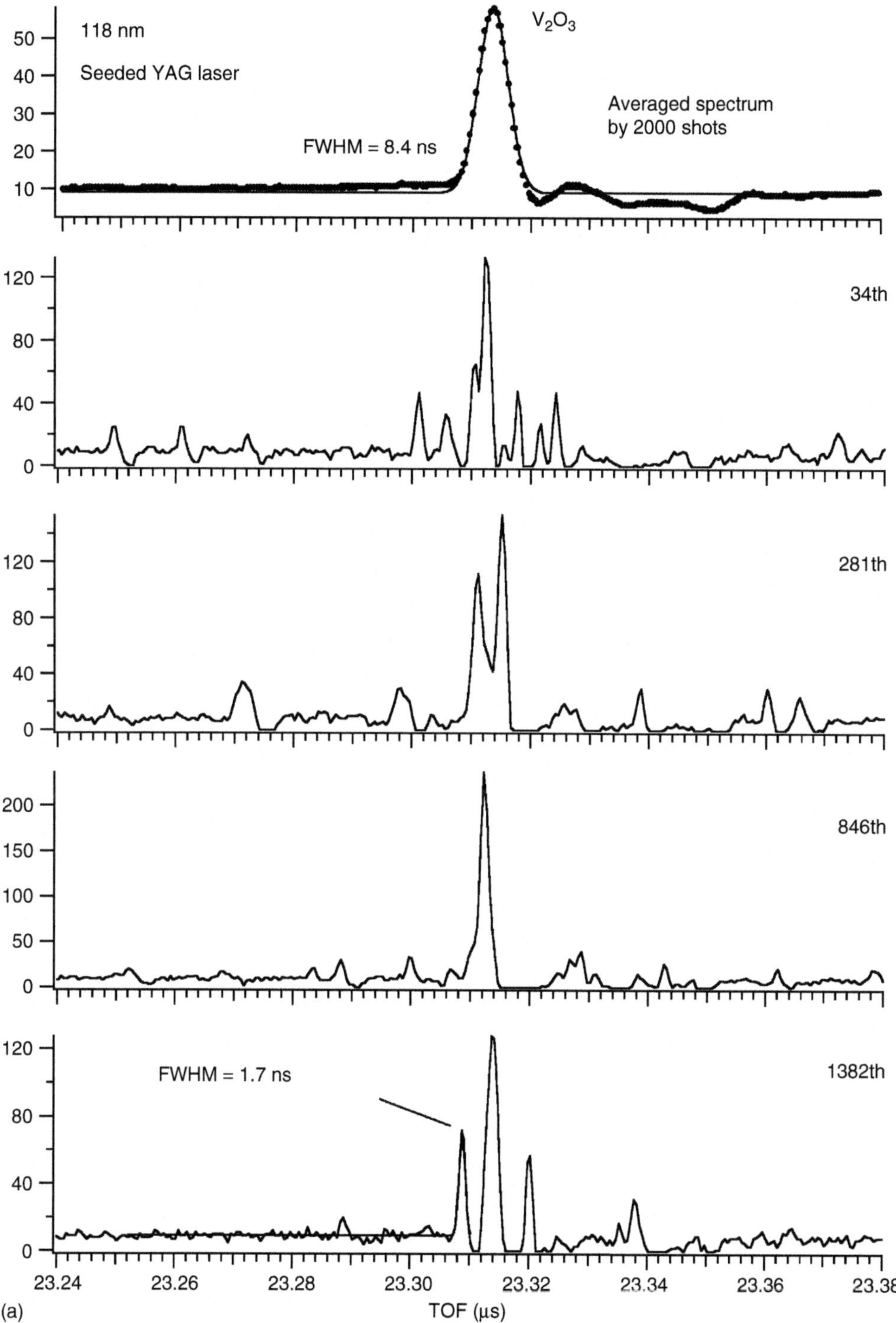

FIGURE 17.19 TOF mass spectra of V_2O_3 observed by 118 nm laser ionization, which is generated by 355 nm of (a) unseeded and (b) seeded YAG laser. All data are acquired at 2 G samples/s sampling rates. Each top spectrum is an averaged spectrum for 2000 laser pulses, and the four lower spectra are single pulse spectra. Averaged spectra are composed of single pulse spectra. *continued*

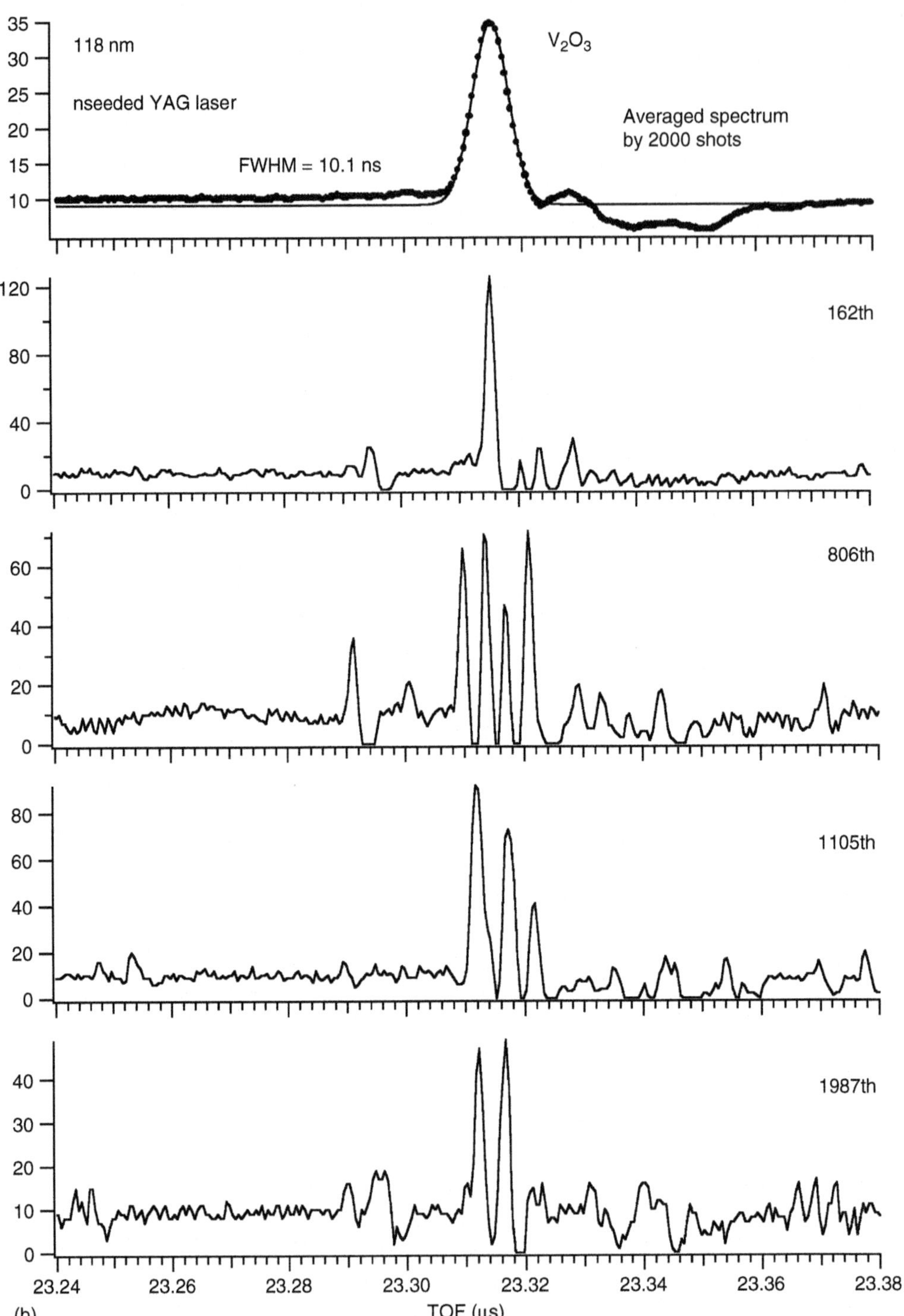

FIGURE 17.19 *continued*

118 nm, single-photon ionization is a more direct way to establish the neutral cluster distribution and to identify the thermodynamical stable neutral cluster species. These trends account for the spectra of Figure 17.20 and Figure 17.21.

We have demonstrated that 118 nm radiation can ionize metal oxide clusters, such as Zr_mO_n, Fe_mO_n, Cu_mO_n, V_mO_n, Ti_mO_n, gently without fragmentation. The 118 nm radiation will thus play a central role in the elucidation of the reactivity of these clusters through nonfragmenting ionization for detection. With these data and general trends, the Ti_mO_{2m} series and others perhaps, can be studied for both reactivity and photoreactivity in the reaction cell.

17.3.6 Reactivity of Metal Oxide Clusters

In this part of our chapter, we will present and discuss preliminary results for experiments in which the neutral cluster distribution of metal oxide species (Fe_mO_n and V_mO_n, so far) are passed through a reaction cell containing various coreactant gases. The general reacting systems of interest are ones for which the metal oxide is said to enhance the rate of a given conversion of reactants to products, that is, a reaction for which the metal oxide is a catalyst. The examples we will discuss are ones for which we have thus far initiated studies in our laboratory. First, we discuss our results for the reduction of NO to yield N_2, the oxidation of CO to yield CO_2, and the combined reaction of CO and NO to yield CO_2 and N_2. These reactions are observed as induced by the presence of Fe_mO_n clusters in the gas cell. Second, we will discuss the reactions induced by V_mO_n clusters for the conversion of SO_2 to SO_3 and for the CO–NO system.

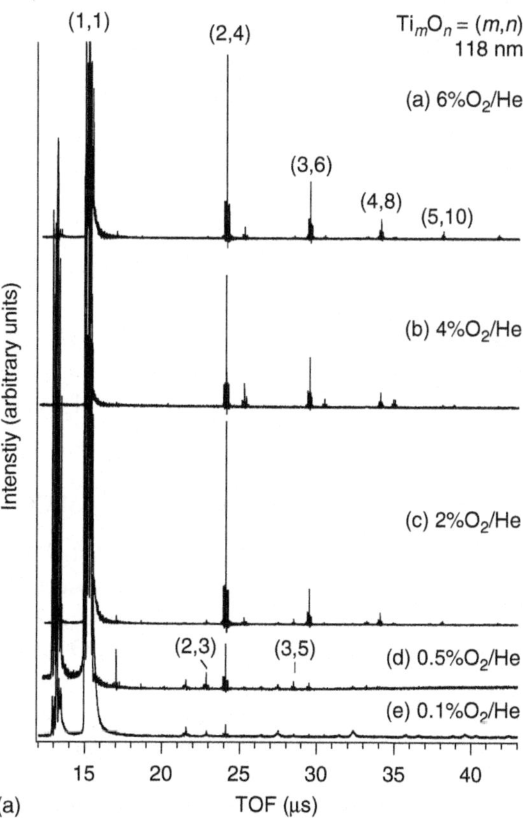

FIGURE 17.20 TOF mass spectra of titanium oxide clusters measured by (a) 118 nm and (b) 193 nm ionization. Each of five spectra is generated with a different composition of O_2–He reaction–expansion gas mixture: 6%, 4%, 2%, 0.5%, and 0.1% O_2 in He. Spectra (a) and (b) are observed under same experimental conditions, except for ionization laser wavelength. *continued*

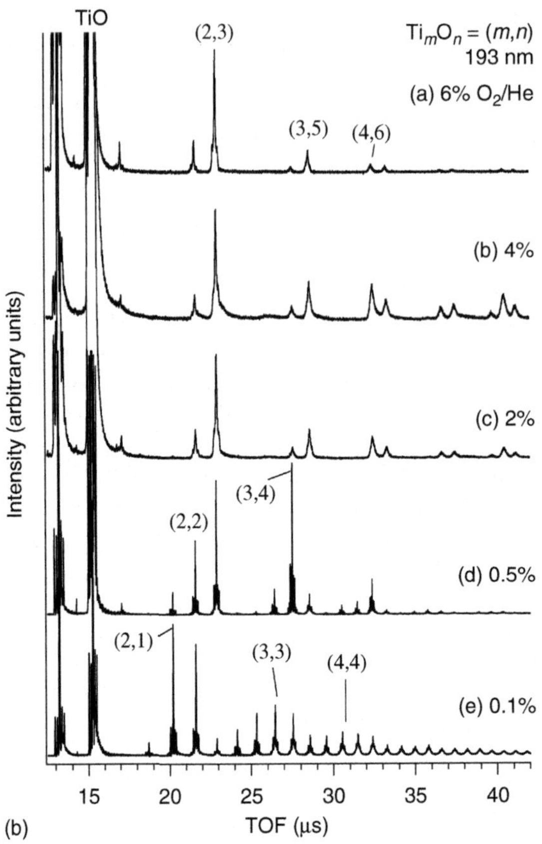

FIGURE 17.20 *continued*

The V_mO_n clusters are very rich in their range of oxidation and we have preliminary results for both oxygen-rich V_mO_n, $n \geq m$, and oxygen-poor V_mO_n, $n \leq m$ clusters.

Preliminary results are also available for cluster structure of some of the more interesting (reactive) small clusters and we will discuss these as well.

17.3.6.1 Iron Oxide Clusters — Catalysis for the Reactions of CO–NO to CO_2–N_2

Figure 17.22 show the change in the neutral cluster Fe_mO_n distribution (detected by 193 nm ionization) as it passes through a reaction cell with various pressures of CO. Note that for the TOFMS of Fe_mO_n the relative intensities of Fe_2O_1, Fe_2O_2, and Fe_3O_3 change significantly with respect to those for Fe_4O_4, Fe_5O_5, Fe_6O_6 clusters as the CO pressure increases. Note that all cluster TOFMS signals are reduced in intensity as the CO pressure increases, as can be expected for scattering, but only a few cluster signals behave differentially. Additionally, we would expect that larger cluster would be more affected by scattering than smaller ones, a trend that is the reverse of that presented in Figure 17.22a. Thus, the loss of signal intensity of $Fe_2O_{1,2}$ and Fe_3O_3 must represent some special behavior of these clusters with regard to CO. No such behavior can be observed for Ar used as a scattering gas in the reaction cell.

Figure 17.22b presents the low mass portion of the TOF mass spectra displayed in Figure 17.22a. These spectra show three important observations. First, with no CO in the cell we do see some small amount of CO_2 as a background signal in the chamber. This is not unexpected for outgassing of the system. Second, as the CO pressure increases in the cell, the amount of CO and CO_2 (broad features as marked) increases: both of these species come from fragmentation of the Fe_mO_n complexes, formed in the cell with CO, upon ionization. Third, both C and O features are also found in the mass spectra that increase with increasing CO pressure in the cell.

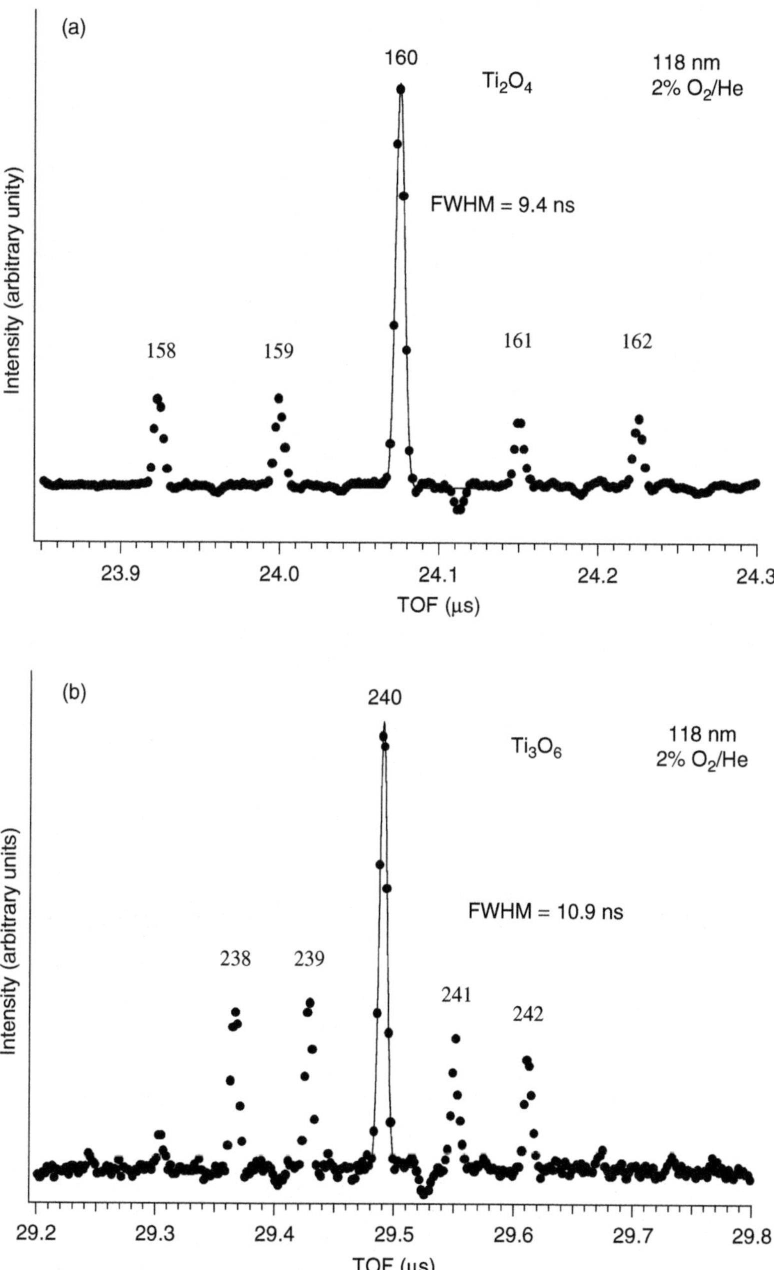

FIGURE 17.21 Expanded 118 nm ionization mass spectra for masses around the clusters (a) Ti_2O_4, (b) Ti_3O_6, and (c) Ti_4O_8 observed with 2% O_2–He expansion–reaction gas. Spectra (d) and (e) are obtained with 193 nm ionization for the gas mixtures indicated.

continued

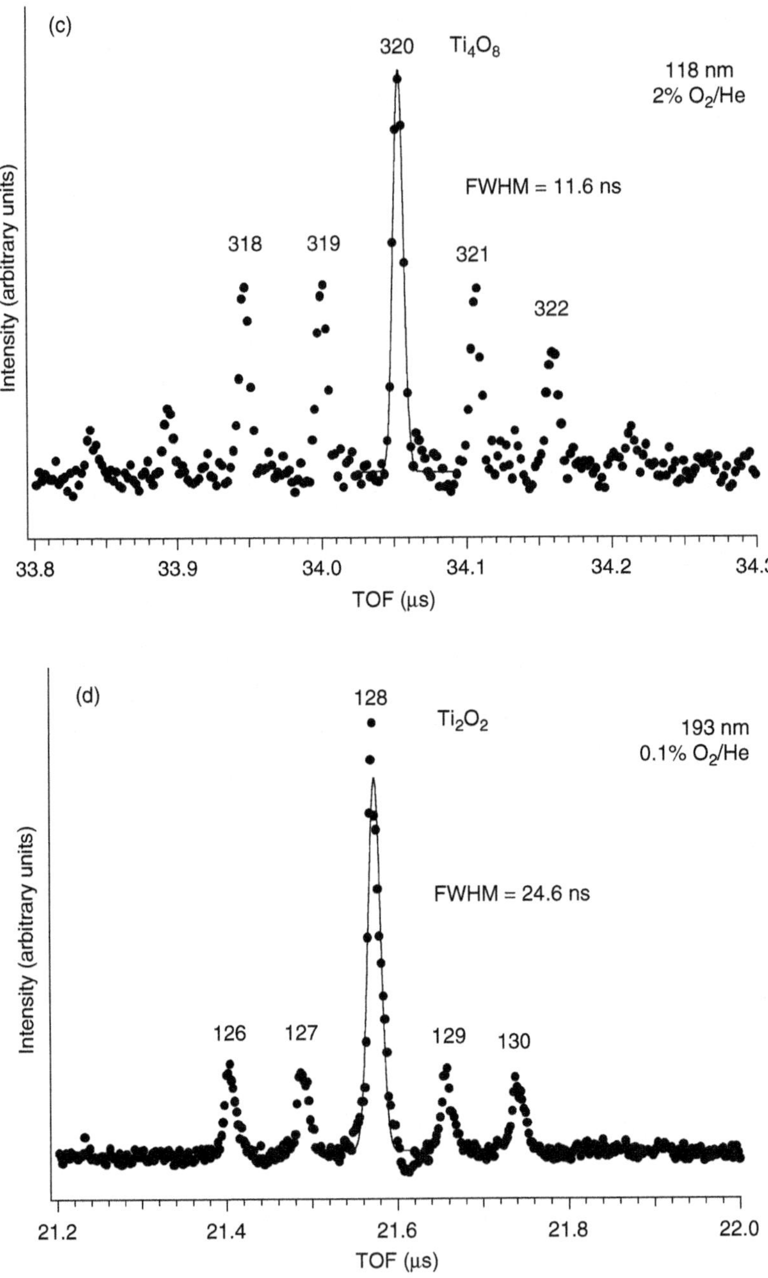

FIGURE 17.21 *continued*

One cannot compare absolute peak intensities between the various spectra; only relative intensities are meaningful within a given spectrum. Note that the "residual outgassing" sharp CO_2 feature and the fragmented CO_2 broad features are very different in relative intensities for the three traces presented in Figure 17.22b.

One possible (likely?) interpretation of these data is that the three Fe_mO_n clusters ($Fe_2O_{1,2}$ and Fe_3O_3) adsorb CO and their TOFMS intensity is diminished as CO_2 is released through

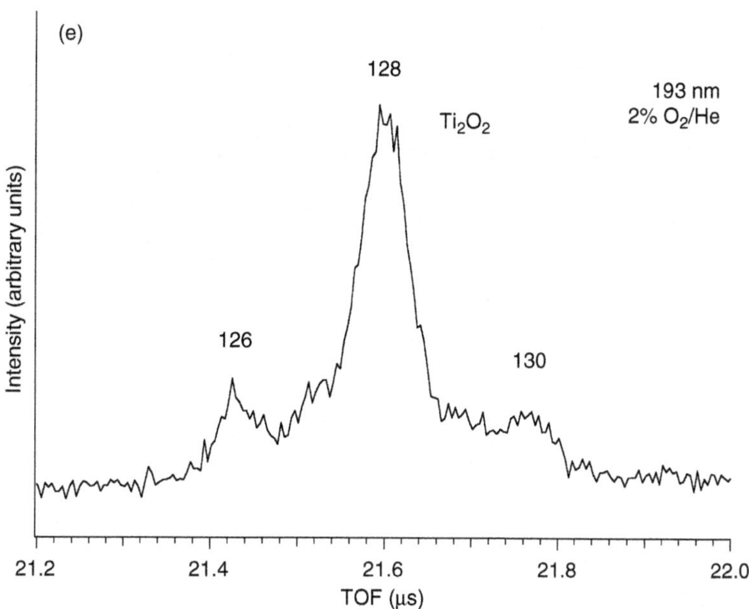

FIGURE 17.21 *continued*

the ionization process. The fragmented CO_2 derives from these clusters, which have formed complexes with CO to generate C, O, CO_2 perhaps upon ionization.

We observe similar behavior for iron oxide clusters with NO. Figure 17.23 show these data. Again the oxygen-deficient species, Fe_2O, Fe_2O_2, and Fe_3O_3, seem most affected by the presence of NO in the reaction cell. Compared to Fe_6O_6, Fe_7O_7, and Fe_8O_8 there clusters lose considerable intensity as the cell NO pressure increases from 7×10^{-4} Torr (background) to 29.8×10^{-4} Torr of NO. In the low mass region of these spectra, one finds the mass features N, NO, and N_2 fragmented from the clusters. This fragmentation is assigned by the large mass spectral feature line width. The residual "sharp" N_2 and NO fractures should have an extra 300 K width of ca. 7 ns.

We tentatively conclude from these results that the iron oxide clusters Fe_2O, Fe_2O_2, Fe_3O_3 are the active ones for the generation of N_2 from NO and CO_2 from CO. In the process these clusters lose or gain oxygen and possibly add to the baseline that underlies the lower cluster mass peak (e.g., $Fe_2O_1 \longrightarrow Fe_2$, $Fe_2O_2 \longrightarrow Fe_2O$, etc.). Signal-to-noise ratio is insufficient in these experiments, thus far, to detect the line shape changes suggested and anticipated. The above "mechanism" is not necessarily unique, and may not be correct, but at present it is consistent with the observations and does not appear to be unreasonable.

The next stage in these experiments is to generate the same spectra with 118 nm ionization. While this may not be essential because the 193 and 118 nm generated mass spectra for this ion mass region are similar for Fe_mO_n, it would still be useful and comforting to have these spectra.

In terms of interpreting these data, calculations would be very important. Small Fe_mO_n clusters have been calculated and their structure suggested [99–101]. We are presently in the process of calculating structures and energies for $Fe_2O_{1,2}$ and Fe_3O_3 with and without CO and NO in order to learn more about the chemistry of and on these species.

17.3.6.2 Vanadium Oxide Clusters — Catalysis for the Reaction SO_2–SO_3 and CO–CO_2

Figure 17.24 presents the data we thus far have for the reaction of SO_2 with V_mO_n clusters. We have studied this system for two laser ionization energies (193 and 118 nm). With 193 nm detection of a 0.5% O_2–He mixture of expansion gas, the clusters that appear active (have a differential intensity behavior as a function of SO_2 gas pressure in the reaction cell) are $V_4O_{4,5,6}$. We know these clusters (detected by 193 nm ionization) to be fragment-derived based on 118 nm ionization spectra and, as can be seen in Figure 17.24b V_4O_7, V_4O_8, V_4O_9 also appear to decrease in concentration (relative to the other clusters) as SO_2 gas pressure is increased in the cell. Thus, the V_4O_x $6 \leq x \leq 9$ series is certainly special with regard to its SO_2 collision behavior. At present one can suggest that the clusters V_4O_6, V_4O_7, V_4O_8, V_4O_9 are the potentially important species here. Again they are oxygen deficient with regard to the most stable V_2O_5, V_3O_7, V_5O_{12}, V_6O_{14}, and V_7O_{17} clusters. Additionally, Figure 17.24c shows that VO_2 is an important oxygen-deficient cluster as well for the interaction reaction of V_mO_n cluster with SO_2. We are not able to ionize, and thus detect, either SO_2 or SO_3 for this system with 118 nm light at ~10^{11} photons/pulse.

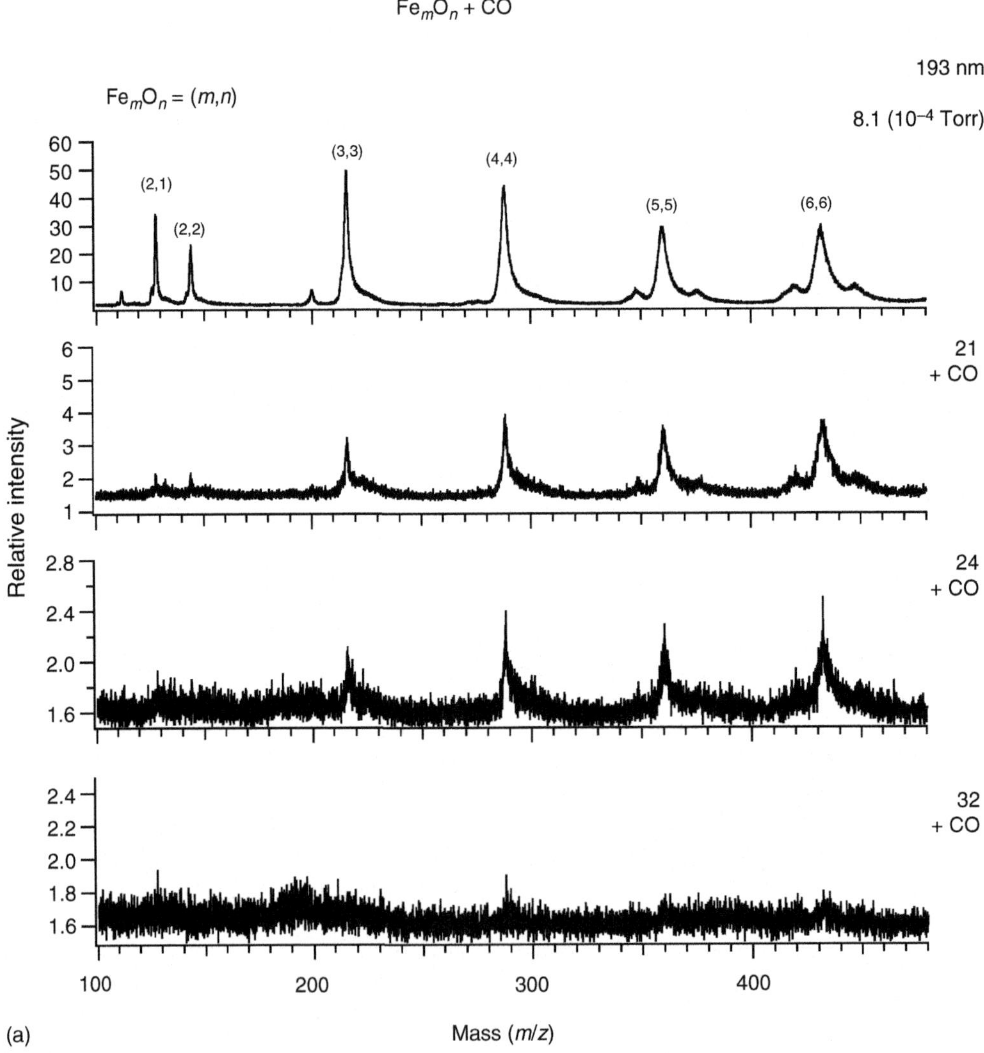

FIGURE 17.22 TOF mass spectra of iron oxide clusters in the region of (a) 100 to 480 (b) 10 to 50 *m/z* as function of CO pressure in reaction cell shown in Figure 17.1. Spectra are observed by 193 nm ionization with 0.75% O_2/He expansion gas. Pressure (10^{-4} Torr) in reaction cell is given on the right side of the traces. *continued*

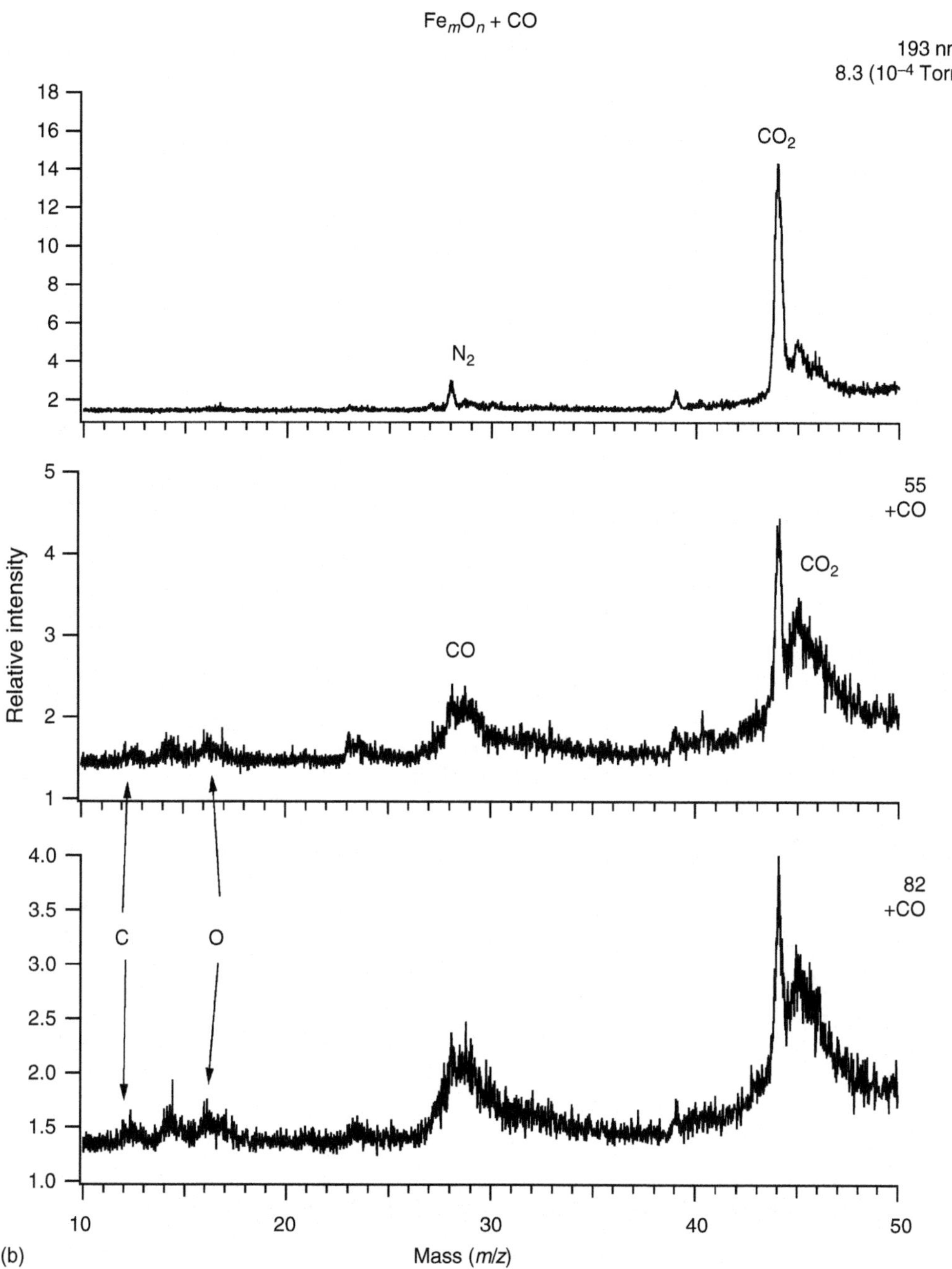

FIGURE 17.22 *continued*

We have observed CO reactions with specific oxygen-deficient V_mO_n clusters. Figure 17.25 displays these data. Note that $V_mO_n(CO)$ clusters are observed for $m = 8$ and $n = 2, 3, 4$. Again certain cluster stoichiometries are active while others are not, and oxygen deficiency seems to be an important factor in this activity for the clusters. In this instance, the reaction yields a complex between V_mO_n and CO which survives the ionization process.

We have undertaken calculation for the V_mO_n clusters to get structures in a systematic manner. Of particular interest here would be VO_2 (for SO_2) reactivity, V_4O_n ($n = 4$, K, 9 for

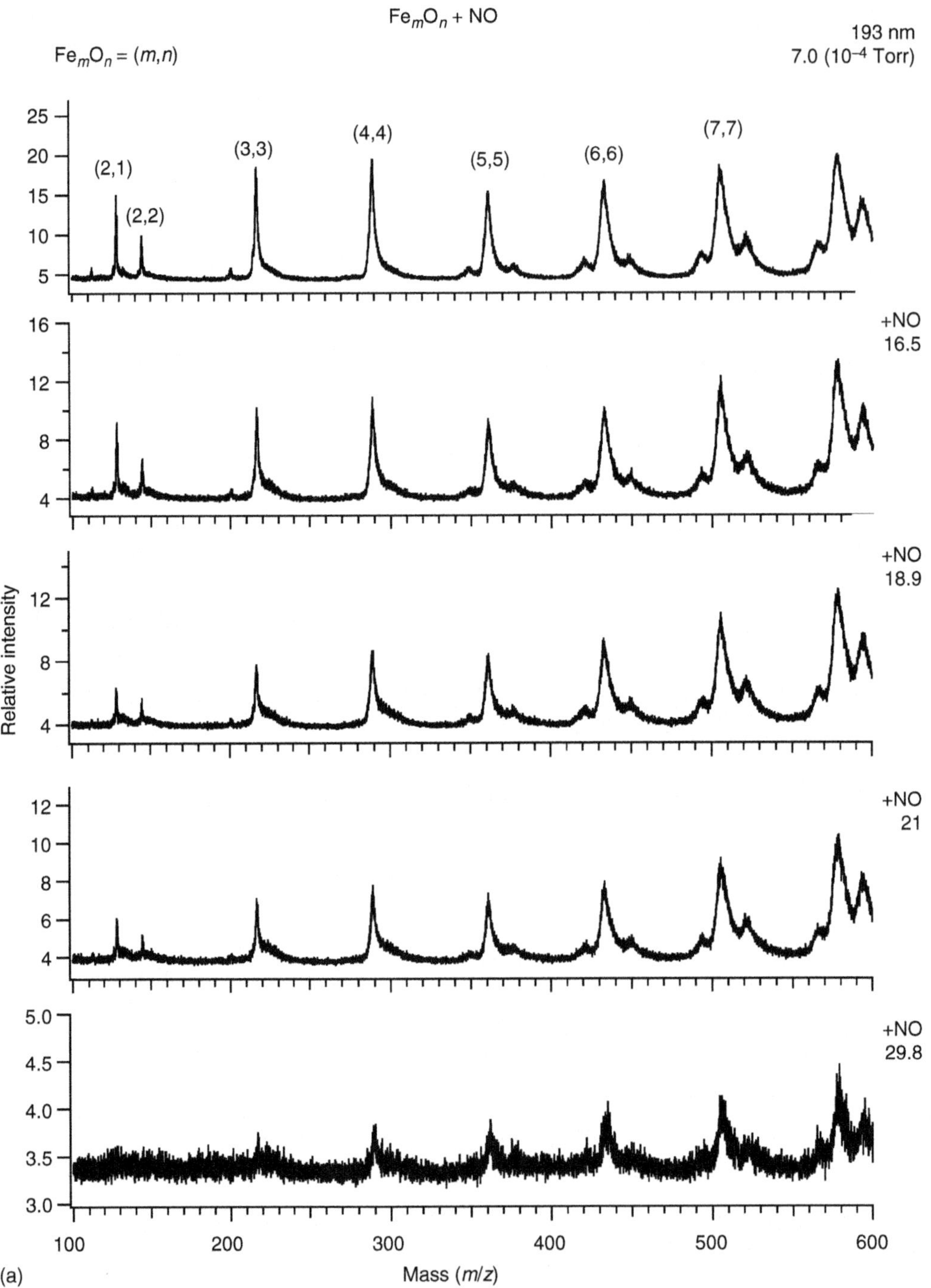

FIGURE 17.23 TOF mass spectra of iron oxide cluster in the region of (a) 100 to 600 (b) 0 to 100 *m/z* as function of NO pressure in reaction cell. Spectra are observed by 193 nm ionization with 0.75% O_2–He expansion gas. Pressure (10^{-4} Torr) in reaction cell is given on the right side of the traces.

continued

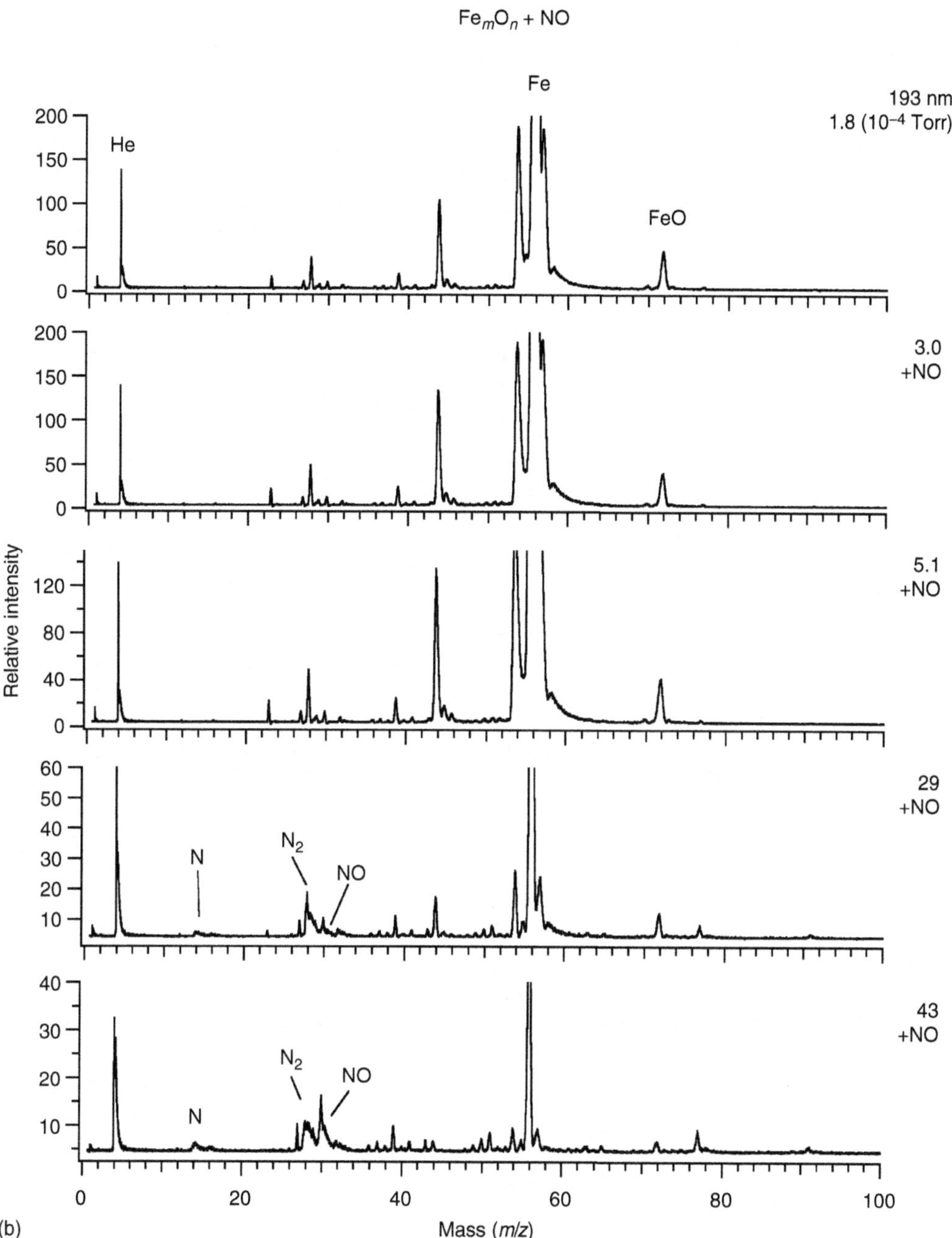

FIGURE 17.23 *continued*

SO_2 reactivity). At present we have structural data for a number of clusters as indicated in Figure 17.26.

17.3.7 Cluster Structure Calculations

As demonstrated in this chapter, single-photon ionization mass spectroscopy of neutral metal oxide clusters reveals both the distribution of neutral clusters and the stable stoichiometries.

From an experimental point of view, spectroscopy (IR, UV, visible, microwave, Raman, etc.) can only yield three constants describing cluster geometry, and thus experimental structures are only complete for small clusters. Theoretical calculations of cluster electronic

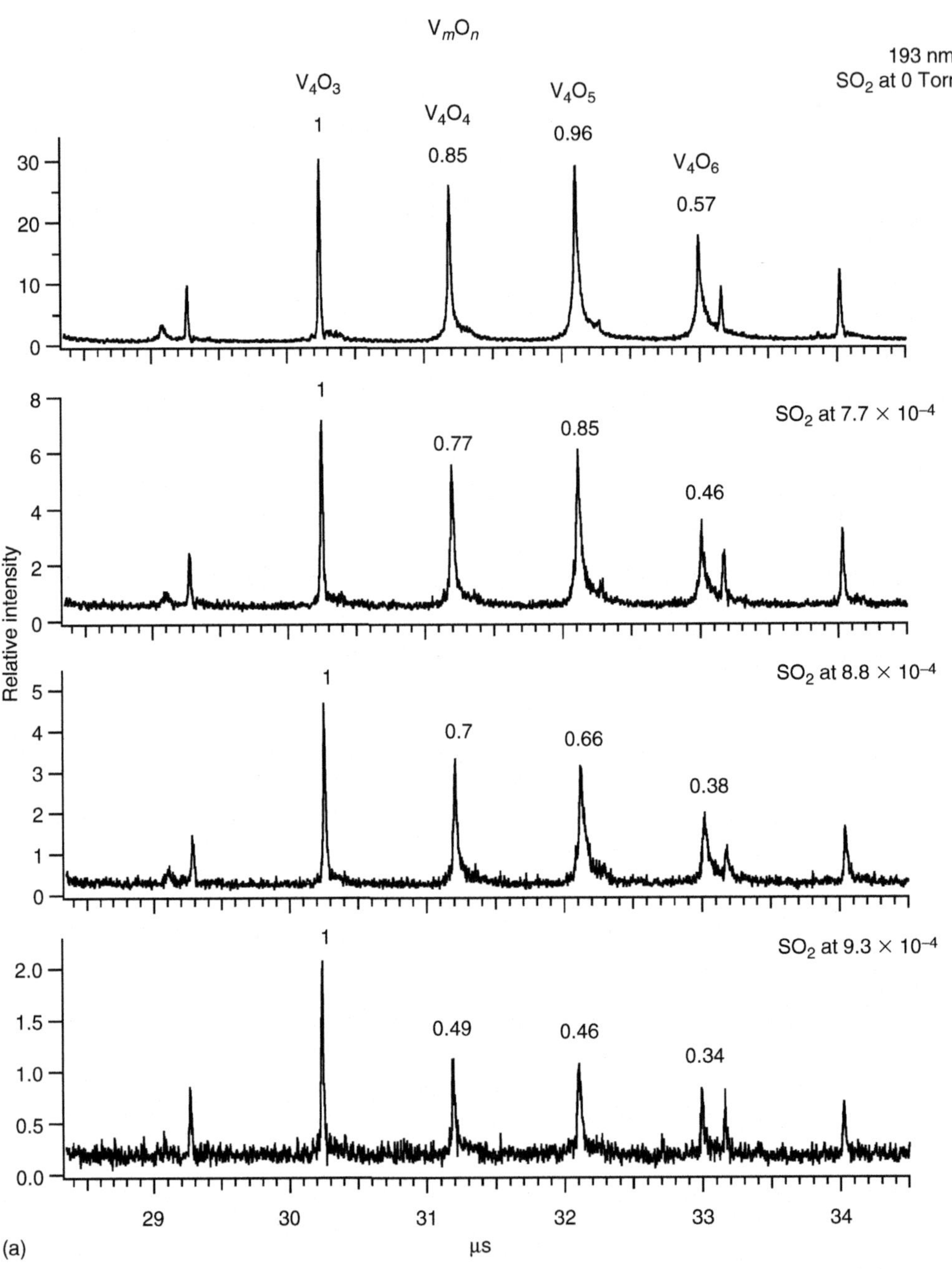

FIGURE 17.24 (a) The 193 nm ionization mass spectrum of vanadium oxide clusters for V_4O_n as function of SO_2 pressure in the reaction cell. The 118 nm ionization mass spectrum of vanadium oxide clusters as function of SO_2 pressure in reaction cell for (b) 0 to 400 and (c) 200 to 370 *m/z*. Spectra are measured with 0.5% O_2–He expansion gas. Pressure (10^{-4} Torr) in reaction cell is given on the right side of the traces.

continued

and geometric structures are an essential component of elucidating reactivity and catalytic properties. We have begun a calculational effort to generate such information with particular interest focused on those clusters identified to be active in cluster reaction studies. These calculations are carried out at the highest level consistent at present with the available computational power. Our previous studies for Zr_mO_n clusters are published in Ref. [10].

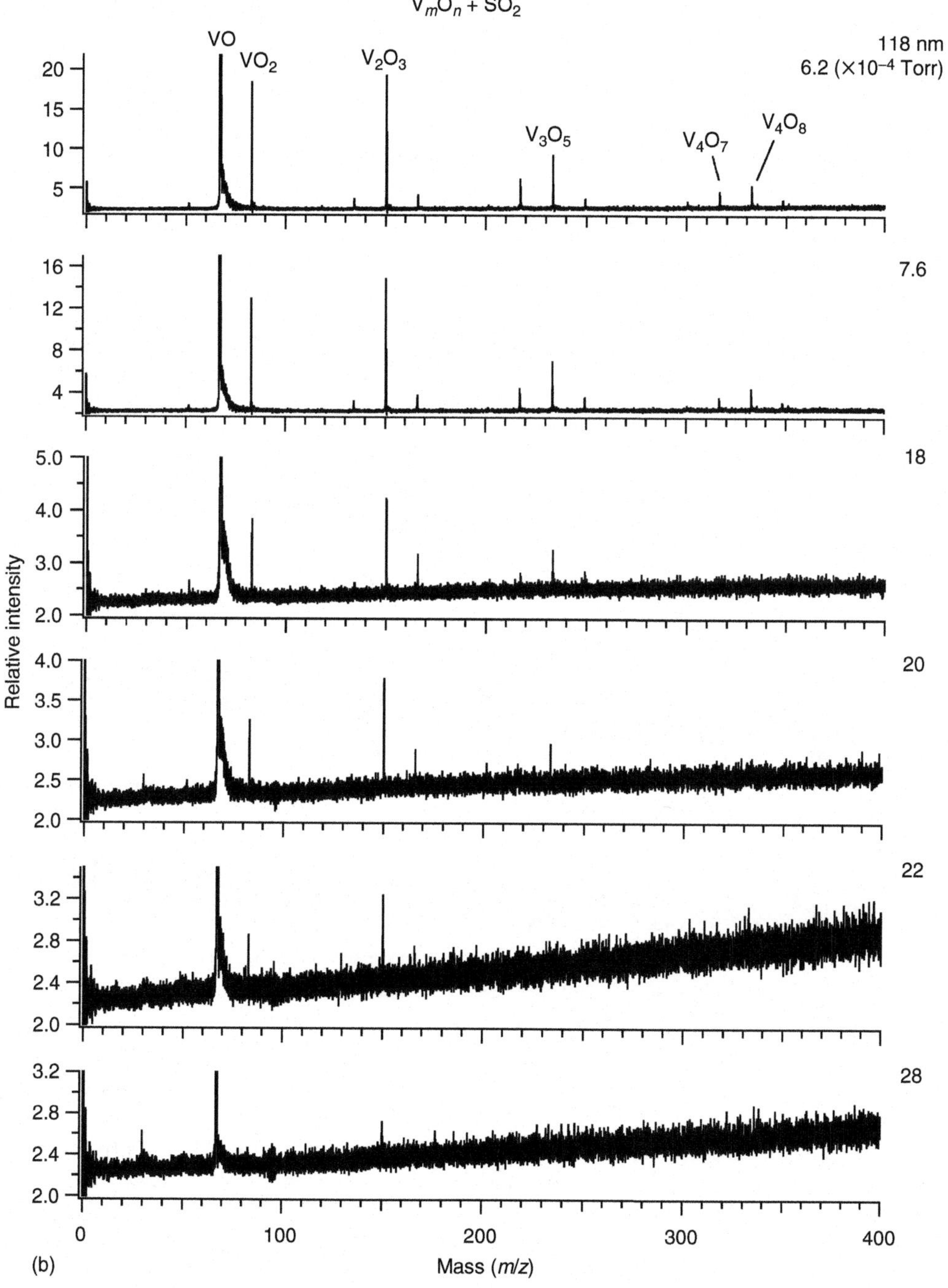

FIGURE 17.24 *continued*

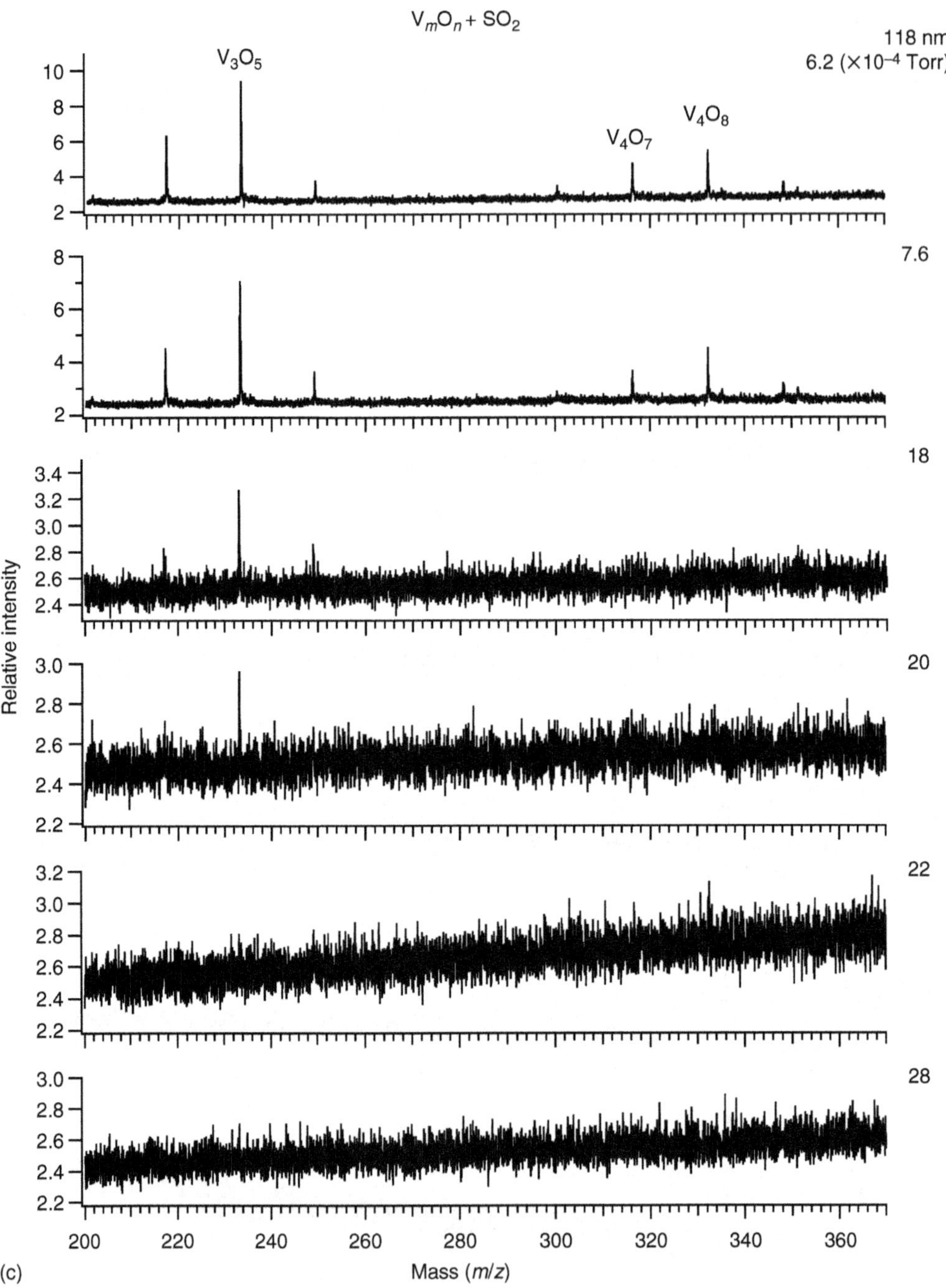

FIGURE 17.24 *continued*

Theoretical calculations have been reported for metal oxides at the HF and DFT levels for various basis sets: DFT levels BP86 and B3LYP with basis sets 6-31G*, DZP, TZVP, and D(T)ZVP. Previous DFT studies [99–113] for iron oxide [99–101], copper oxide [102], zirconium oxide [44, 61], titanium oxide [103–105], and vanadium oxide [106–109] are available. Our studies in this area are in progress and we focus on the clusters of most interest for reactivity studies: Fe_2O, Fe_2O_2, Fe_3O_3, and VO_2, V_4O_n, $4 \leq n \leq 9$. Figure 17.26 shows the

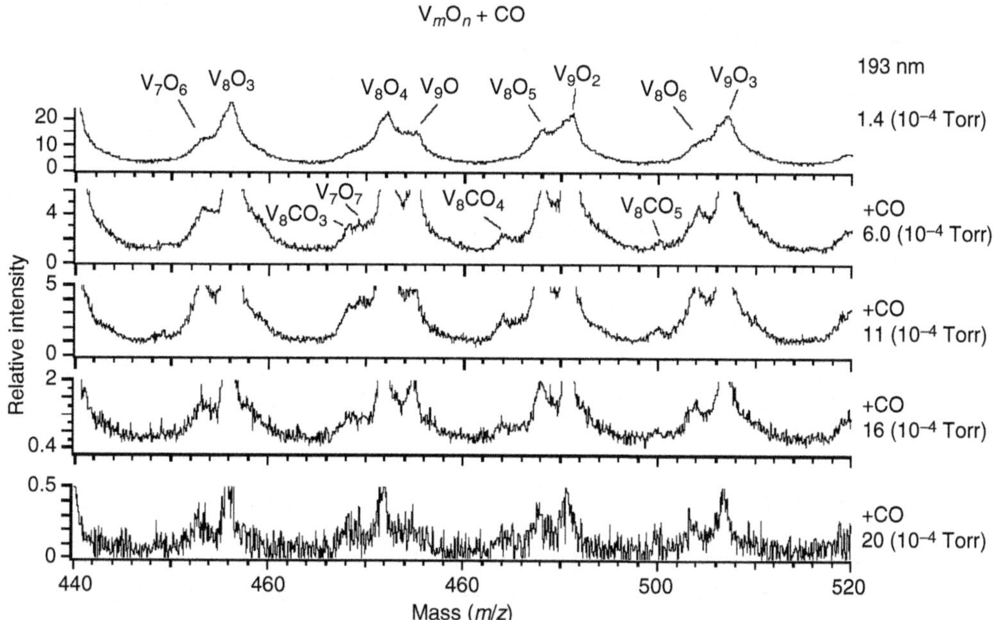

FIGURE 17.25 TOF mass spectra of vanadium oxide clusters in the region of 440 to 550 m/z as function of CO pressure in reaction cell. Spectra are observed by 193 nm with ~100% He expansion gas, which contains O_2 as residual gas in experimental system. Pressure (10^{-4} Torr) in reaction cell is given on the right side of the traces..

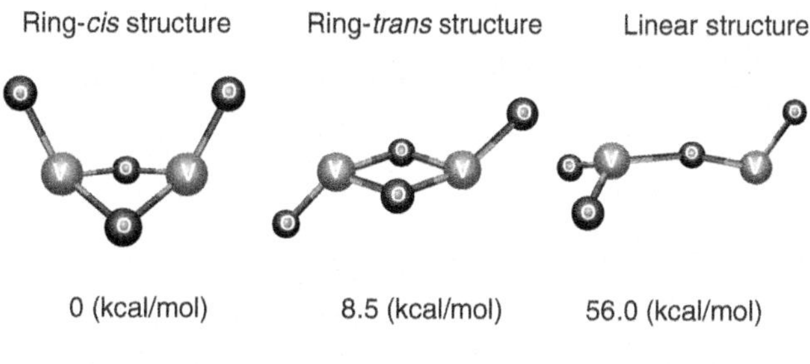

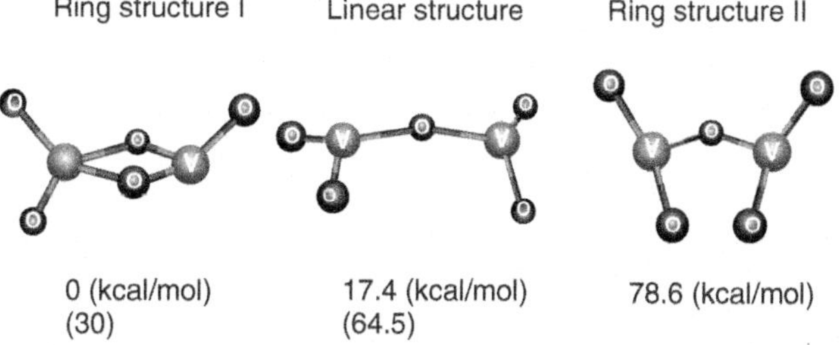

FIGURE 17.26 Calculated structures for (a) V_2O_4 and (b) V_2O_5 clusters at the DFT BPW91/TZVP level of theory. For V_2O_5 structures, the numbers in parentheses are energies for the lowest triplet state.

stable structures and their relative energies for V_2O_4 and V_2O_5 neutral clusters of singlet (and some triplet) electronic configurations at the BPW91/TZVP level calculated with Gaussian 98 [114] and visualized with Molekel 4.0 [115]. Most likely, structures that are more than 20 kcal/mol above the lowest energy structure at this calculational level are not contenders for the stable structures and are not populated in the beam. Both *cis* and *trans* ring structures are likely populated for V_2O_4 clusters. The V_2O_5 cluster may have two populated isomers, the four-membered ring structure and the higher energy linear one. Both of these cluster structures could contribute to the growth of larger clusters. We are in the process of generating cluster structures for the V_4O_n series and for the stable V_3O_7 cluster.

A note of caution is appropriate here. One does not know at this time what level of theory is required to generate reliable structures for vanadium oxide and iron oxide clusters. We are using the highest level we can at present and are comparing these new results with older ones referenced above to determine what results are stable with regard to "improved" levels of theory.

17.4 CONCLUSIONS

We summarize the work that is reported in this chapter in very general terms because we have given individual conclusions for each section as we have proceeded through the discussion. The two main points of these studies have been the following: (1) high energy, single-photon, low-fluence ionization of metal oxide clusters is the best way to detect the neutral cluster distribution of species using a time-of-flight mass spectrometer, as this approach assures that no fragmentation of the cluster will occur during the detection process; and (2) the clusters that appear to be reactive–reactive–catalytic for interacting with CO, NO, and SO_2 are small ($m \leq 8$) oxygen-deficient ones for Fe_mO_n and V_mO_n.

This latter result is important and interesting because it opens the opportunity for spectroscopic and theoretical studies of the most central clusters for reactivity and catalysis employing metal oxide systems. We are now working in these directions both experimentally and theoretically, in addition to searching for new reaction systems.

ACKNOWLEDGMENTS

Support for this work comes from the U.S. Department of Energy, Office of Energy, Basic Energy Science and Philip Morris USA. We wish to acknowledge Drs. G.J. Stueber, M. Foltin, and D.N. Shin for their expert experimental and theoretical efforts in these studies. Needless to say, the work could not have been done without them and has been originally published with their names. We also wish to acknowledge support from, and discussions with, Drs. B. Reddy, F. Rasouli, and M. Hajaligol.

REFERENCES

1. X-C Guo and RJ Madix. *Acc Chem Res* 36: 471–480 2003.
2. X-C Guo and RJ Madix. *J Phys Chem B* 107: 3105–3116, 2003.
3. JA Adams. *Chem Rev* 101: 2271–2290, 2001.
4. J Brunner, A Mokhir, and R Kraemer. *J Am Chem Soc* 125: 12410–12411, 2003.
5. A Fersht. *Structure and Mechanism in Protein Science — A Guide to Enzyme Catalysis and Protein Folding*. New York: Freeman, 1999, pp. 44–51.
6. VE Henrich and PA Cox. *The Surface Science of Metal Oxides*. New York: Cambridge University Press, 1994.
7. GA Somorjai. *Introduction to Surface Chemistry and Catalysis*. New York: Wiley Interscience, 1994.

8. RI Masel. *Principles of Adsorption and Reaction on Solid Surfaces*. New York: Wiley Interscience, 1996.
9. R Hoffmann. *Solids and Surfaces: A Chemist's View of Bonding in Extended Structures*. New York: VCH Publishers, 1988.
10. A Zangwill. *Physics at Surfaces*. New York: Cambridge University Press, 1988.
11. K Eller and H Schwarz. *Chem Rev* 91: 1121–1177, 1991.
12. MA Fox and MT Dulay. *Chem Rev* 93: 341–357, 1993.
13. B. Cornils and WA Hermann. *Applied Homogeneous Catalysis with Organometallic Compounds*, 2nd ed. Weinheim: Wiley-VCH, 2002.
14. RA Sheldon and H van Bekkum. *Fine Chemicals Through Heterogeneous Catalysis*. Weinheim: Wiley-VCH, 2001.
15. P Braunstein, LA Oro, and PR Raithby. *Metal Clusters in Chemistry*, Vol. 1–3. Weinheim: Wiley-VCH, 1999.
16. PA Hackett, SA Mitchell, DM Rayner, and B Simard. In *Metal–Ligand Interactions*, R Russo and DR Salahub, eds. Amsterdam: Kluwer Academic Publishers, 1996, pp. 289–311.
17. RE Smalley. *Advances Toward a Molecular Surface Science. CAMS Symposium on Molecular and Cluster Beam Science*, chapter 7, 1987, pp. 76–88.
18. RE Smalley. In *Metal–Metal Bonds and Clusters in Chemistry and Catalysis*. JP Fackler Jr, ed. College Station, TX: Texas A&M University, 1989, pp. 249–264.
19. JA Alonso. *Chem Rev* 100: 637–677, 2000.
20. AL Linsebigler, G Lu, and JT Yates Jr. *Chem Rev* 95: 735–758, 1995.
21. T Zubkov, GA Morgan Jr, JT Yates Jr, O Kuhlert, M Lisowski, R Schillinger, D Fick, and HJ Jansch. *Surf Sci* 526: 57–71, 2003.
22. G Ertl and H-J Freund. *Phys Today* Jan: 32–38, 1999.
23. G-F Xu, Y Carmel, T Olorunyolemi, IK Lloyd, and OC Wilson Jr. *J Mater Res* 18: 66–76, 2003.
24. G-F Xu, T Olorunyolemi, OC Wilson, IK Lloyd, and Y Carmel. *J Mater Res* 17: 2837–2845, 2002.
25. G-F Xu, IK Lloyd, Y Carmel, T Olorunyolemi, and OC Wilson Jr. *J Mater Res* 16: 2850–2858, 2001.
26. K McElroy, RW Simmonds, JE Hoffman, D-H Lee, J Orenstein, H Eisaki, S Uchida, and JC Davis. *Nature* 422: 592–596, 2003.
27. SH Pan, EW Hudson, J Ma, and JC Davis. *Appl Phys Lett* 73: 58–60, 1998.
28. KA Zemski, DR Justes, and AW Castleman Jr. *J Phys Chem B* 106; 6136–6148, 2002.
29. JB Griffin and PB Armentrout. *J Chem Phys* 106: 4448–4462, 1997.
30. JB Griffin and PB Armentrout. *J Chem Phys* 107: 5345–5355, 1997.
31. H Wu and L-S Wang. *J Chem Phys* 108: 5310–5318, 1998.
32. SR Desai, H Wu, CM Rohlfing, and L-S Wang. *J Chem Phys* 106: 1309–1317, 1997.
33. H-J Zhai and L-S Wang. *J Chem Phys* 117: 7882–7888, 2002.
34. DH Andrews, AJ Gianola, and WC Lineberger. *J Chem Phys* 117: 4074–4076, 2002.
35. A Pramann, K Koyasu, A Nakajima, and K Kaya. *J Phys Chem A* 106: 4891–4896, 2002.
36. A Pramann, K Koyasu, A Nakajima, and K Kaya. *J Phys Chem A* 106: 2483–2488, 2002.
37. ER Bernstein. In *Atomic and Molecular Clusters*. ER Bernstein, ed. New York: Elsevier, 1990, pp. 551–764.
38. J Fernandez, J Yao, JA Bray, and ER Bernstein. In *Structure and Dynamics of Excited States*, J Laane, ed. New York: Springer-Verlag, 1999, pp. 71–96.
39. ER Bernstein, ed. *Chemical Reactions in Clusters*. New York: Oxford, 1996, pp. 147–196.
40. SS Xantheas. *Recent Theoretical and Experimental Advances in Hydrogen Bonded Clusters NATO ASI Series C: Mathematical and Physical Sciences*, Vol. 561. Boston: Kluwer Academic Publishers, 2000.
41. JM Bowman and Z Bacic, eds. *Advances in Molecular Vibrations and Collision Dynamics*, Vol. 3, *Molecular Clusters Series*. Stanford: JAI Press, 1998.
42. MD Morse and RE Smalley. *Ber Bunsen Phys Chem Chem Phys* 88: 228–233, 1984.
43. SC Richsmeier, EK Parks, K Liu, LG Pobo, and SJ Riley. *J Chem Phys* 82: 3659–3665, 1985.
44. M Foltin, GJ Stueber, and ER Bernstein. *J Chem Phys* 114: 8971–8989, 2001.
45. M Foltin, GJ Stueber, and ER Bernstein. *J Chem Phys* 111: 9577–9586, 1999.
46. AW Castleman Jr and KH Bowen Jr. *J Phys Chem* 100: 12911–12944, 1996.

47. SJ Riley. In *Metal–Ligand Interactions: From Atoms, to Clusters, to Surfaces*, DR Salahub and N Russo, eds. The Netherlands: Kluwer Academic Publishers, 1992, pp. 17–36.
48. GC Nieman, EK Parks, SC Richtsmeier, K Liu, LG Pobo, and SJ Riley. *High Temp Sci* 22: 115–138, 1986.
49. SJ Riley, EK Parks, GC Nieman, LG Pobo, and S Wexler. *J Chem Phys* 80: 1360–1362, 1984.
50. DL Hildenbrand. *Chem Phys Lett* 20: 127–129, 1973.
51. C Yamada, E Cohen, and M Fujitake. *J Chem Phys* 92: 2146–2149, 1990.
52. JP Towle, AM James, OL Bourne, and B Simard. *J Mol Spectrosc* 163: 300–308, 1994.
53. H Ito and M Goto. *Chem Phys Lett* 227: 293–298, 1994.
54. K Chen, C Sung, J Chang, T Chung, and K Lee. *Chem Phys Lett* 240: 17–24, 1995.
55. N Sato, H Ito, and K Kuchitsu. *Chem Phys Lett* 240: 10–16, 1995.
56. B Bescós, G Moreley, and AG Ureña. *Chem Phys Lett* 244: 407–413, 1995.
57. DP Belyung and A Fontijin. *J Phys Chem* 99: 12225–12230, 1995.
58. G von Helden, D van Heijnsbergen, and G Meijer. *J Phys Chem A* 107: 1671–1688, 2003.
59. A Fielické, G Meijer, and G von Helden. *J Am Chem Soc* 125: 3659–3667, 2003.
60. D van Heijnsbergen, G von Helden, G Meijer, and MA Duncan. *J Chem Phys* 116: 2400–2406, 2002.
61. G von Helden, A Kirilyuk, D van Heijnsbergen, B Sartakov, MA Duncan, and G Meijer. *Chem Phys* 262: 31–39, 2000.
62. C Guo, KP Kerns, and AW Castleman Jr. *Science* 255: 1411–1413, 1992.
63. BC Guo, S Wei, J Purnell, S Buzza, and AW Castleman Jr. *Science* 256: 515–516, 1992.
64. H Sakurai and AW Castleman Jr. *J Phys Chem A* 101: 7695–7698, 1997.
65. SE Kooi and AW Castleman Jr. *J Chem Phys* 108: 8864–8869, 1997.
66. H Sakurai and AW Castleman Jr. *J Phys Chem A* 102: 10486–10492, 1998.
67. H Sakurai and AW Castleman Jr. *J Chem Phys* 111: 1462–1466, 1999.
68. JR Stairs, KM Davis, and AW Castleman Jr. *J Chem Phys* 117: 4371–4375, 2002.
69. H Sakurai, SE Kooi, and AW Castleman Jr. *J Cluster Sci* 10: 493–507, 1999.
70. D van Heijnsbergen, G von Helden, MA Duncan, AJA van Roij, and G Meijer. *Phys Rev Lett* 83: 4983–4986, 1999.
71. D van Heijnsbergen, A Fielicke, G Meijer, and G von Helden. *Phys Rev Lett* 89: 013401, 2002.
72. DN Shin, Y Matsuda, and ER Bernstein. On the iron oxide neutral cluster distribution in the gas phase. I. Detection through 193 nm multiphoton ionization. *J Chem Phys* 120: 4150, 2004.
73. DN Shin, Y Matsuda, and ER Bernstein. On the iron oxide neutral cluster distribution in the gas phase. II. Detection through 118 nm single photon ionization. *J Chem Phys* 120: 4157, 2004.
74. Y Matsuda, DN Shin, and ER Bernstein. On the neutral copper oxide cluster distribution in the gas phase: Detection through 355 nm and 193 nm multiphoton, and 118 nm single photon ionization. *J Chem Phys* 120: 4165, 2004.
75. Y Matsuda, DN Shin, and ER Bernstein. On the zirconium oxide neutral cluster distribution in the gas phase: Detection through 118 nm single photon, and 193 nm multiphoton ionization. *J Chem Phys* 120: 4142, 2004.
76. Y Matsuda and ER Bernstein. Identification, structure, and spectorseopy of nuteral vanadium oxide clusters. *J Phys Chem*, submitted.
77. Y Matsuda and ER Bernstein. On the titanium oxide neutral cluster distribution in the gas phase: Detection through 118 nm single photon, and 193 multiphoton ionization. *J Phys Chem*, 109, 314, 2005.
78. M Andersson and A Rosen. *J Chem Phys* 117: 7051–7054, 2002.
79. L Holmgren, M Andersson, and A Rosen. *Chem Phys Lett* 296: 167–172, 1998.
80. EA Rohlfing, DM Cox, and A Kaldor. *Chem Phys Lett* 99: 161–166, 1983.
81. DM Cox, DJ Trevor, RL Whetten, EA Rohlfing, and A Kaldor. *Phys Rev B* 32: 7290–7298, 1985.
82. EA Rohlfing, DM Cox, A Kaldor, and KH Johnson. *J Chem Phys* 81: 3846–6851, 1984.
83. L-S Wang, H Wu, and SR Desai. *Phys Rev Lett* 76: 4853–4856, 1996.
84. H Wu, SR Desai, and L-S Wang. *J Am Chem Soc* 118: 5296–5301, 1996.
85. L-S Wang, J Fan, and L Lou. *Surf Rev Lett* 3: 695–698, 1996.
86. M Sakurai, K Sumiyama, Q Sun, and Y Kawazoe. *J Phys Soc Japan* 68: 3497–3499, 1999.

87. M Sakurai, K Watanabe, K Sumiyama, and K Suzuki. *J Phys Soc Japan* 67: 2571–2573, 1998.
88. Q Wang, Q Sun, M Sakurai, JZ Yu, BL Gu, K Sumiyama, and Y Kawazoe. *Phys Rev B* 59: 12672–12677, 1999.
89. Q Sun, M Sakurai, Q Wang, JZ Yu, GH Wang, K Sumiyama, and Y Kawazoe. *Phys Rev B* 62: 8500–8507, 2000.
90. M Sakurai, K Watanabe, K Sumiyama, and K Suzuki. *J Chem Phys* 111: 235–238, 1999.
91. R Hilbig and R Wallenstein. *IEEE J Quantum Electron* QE-17: 1566–1573, 1981.
92. RH Page, RJ Larking, AH Kung, YR Shen, and YT Lee. *Rev Sci Instrum* 58: 1616–1620, 1987.
93. MP McCann, CH Chen, and MG Payne. *J Chem Phys* 89: 5429–5441, 1988.
94. PG Strupp, AL Alstrin, RV Smilgys, and SR Leone. *Appl Optics* 32: 842–846, 1993.
95. K Suto, Y Sato, CL Reed, V Skorokhodov, Y Matsumi, and M Kawasaki. *J Phys Chem A* 101: 1222–1226, 1997.
96. K Tonokura, T Murasaki, and M Koshi. *Chem Phys Lett* 319: 507–511, 2000.
97. RH Lipson, SS Dimov, P Wang, YJ Shi, DM Mao, XK Hu, and J Vanstone. *Instrum Sci Technol* 28: 85–118, 2000.
98. YJ Shi, S Consta, AK Das, B Maliik, D Lacey, and RH Lipson. *J Chem Phys* 116: 6990–6999, 2002.
99. GL Gutsev, CW Bauschlicher Jr, H-J Zhai, and L-S Wang. *J Chem Phys* 119: 11135–11145, 2003.
100. GV Chertihin, W Saffel, JT Yustein, L Andrews, M Neurock, A Ricca, and CW Bauschlicher Jr. *J Phys Chem* 100: 5261–5273, 1996.
101. Q Wang, Q Sun, M Sakurai, JZ Yu, BL Gu, K Sumiyama, and Y Kawazoe. *Phys Rev B* 59: 12672–12677, 1999.
102. C Massobro and Y Pouillon. *J Chem Phys* 119: 8305–8310, 2003.
103. T Albaret, F Finocchi, and C Noguera. *J Chem Phys* 113: 2238–2239, 2000.
104. M. Castro, S-R Liu, H-J Zhai, and L-S Wang. *J Chem Phys* 118: 2116–2123, 2003.
105. A Vijay, G Mills, and H Metiu. *J Chem Phys* 118: 6536–6551, 2003.
106. DR Justes, R Mitric, NA Moore, V Bonacic-Koutecky, and AW Castelman Jr. *J Am Chem Soc* 125: 6289–6299, 2003.
107. A Fielicke, R Mitiri, G Meijer, V Bonacic-Koutecky, and G von Helden. *J Am Chem Soc* 125: 15716–15717, 2003.
108. SF Vyboishchikov and J Sauer. *J Phys Chem A* 105: 8588–8598, 2001.
109. SF Vyboishchikov and J Sauer. *J Phys Chem A* 104: 10913–10922, 2000.
110. A Fielicke, G Meijer, and G von Helden. *J Am Chem Soc* 125: 3659–3667, 2003.
111. A Martinez, LE Sansores, R Salcedo, FJ Tenorio, and JV Ortiz. *J Phys Chem A* 106: 10630–10635, 2002.
112. M Calatayud, J Andres, and A Beltrán. *J Phys Chem* 105: 9760–9775, 2001.
113. M Bienati, V Bonacic-Koutecky, and P Fantucci. *J Phys Chem A* 104: 6983–6992, 2000.
114. MJ Frisch, GW Trucks, HB Schlegel, GE Scuseria, MA Robb, JR Cheeseman, VG Zakrzewski, JA Montogomeny, RE Stratmann, JC Burant, S Dapprich, JM Millam, AD Daniels, KN Kudin, MC Strain, O Farkas, J Tomasi, V Barone, M Cossi, R Cammi, B Mennucci, C Pomelli, C Adamo, S Clifford, J Ochterski, GA Petersson, PY Ayala, Q Cui, K Morokuma, DK Malick, AD Rabuck, K Raghavachari, JB Foresman, J Goslowski, JV Ortiz, BB Stefanov, G Liu, A Liashenko, P Piskorz, I Komariomi, R Gomperts, RL Martin, DJ Fox, T Keith, MA Al-Laham, CY Peng, A Nanayakkara, C Gonzalez, M. Challacombe, PMW Gill, BG Johnson, W chen, MW Wong, JL Andreo, M Head-Gordon, ES Reploge, and JA Pople. Gaussian 98 (Revision A) 6, Gaussian Inc., Pittsburgh, PA, 1998.
115. P Flukiger, HP Luthi, S Portmann, and J Weber. Molekel 4.0, Swiss Center for Scientific Computing, Manno, Switzerland, 2000.

18 Biocatalysis in Environmental Remediation–Bioremediation

Gene F. Parkin
Department of Civil and Environmental Engineering, University of Iowa

CONTENTS

18.1 INTRODUCTION

Environmental engineers have been using biological processes for the treatment of municipal wastewaters and sludges for over a century [1, 2]. Processes such as activated sludge and trickling filters are used to remove organic compounds from municipal wastewaters prior to discharge into receiving waters. Anaerobic digestion is used to stabilize municipal sludges, converting particulate organic matter into methane and carbon dioxide. Biological processes are increasingly used to reduce the nutrient content of wastewaters, primarily inorganic nitrogen and phosphorus, to prevent eutrophication (loosely defined as the overgrowth of algae and green plants), and dissolved oxygen depletion in receiving waters. In these processes, enzymes contained in bacteria catalyze a series of reactions that hopefully transform pollutants into less hazardous or environmentally benign products. In the case of biological phosphorus removal, the offending phosphorus is sequestered by bacteria and removed from the liquid stream by sedimentation prior to discharge. More recently, environmental engineers have applied biological processes to clean up ("remediate") contaminated waters, soils, and air. The term *bioremediation* is commonly used to describe this application of biological treatment.

The purpose of this chapter is to introduce the reader to how biocatalysis is used in environmental remediation. The primary focus will be bioremediation of contaminated groundwater aquifers. First, a description of the problems of environmental contamination that are most effectively treated using biocatalysis will be presented, followed by definitions of some general terms to provide a framework for subsequent discussion. Then, general requirements for effective bioremediation will be discussed. Then, specific examples of the use of biocatalysis in environmental remediation will be given with an emphasis on how fundamental knowledge of microbiology and biochemistry helps to develop effective remediation strategies.

18.1.1 The Problem

Until approximately 1985, the application of biological treatment was confined primarily to the treatment of municipal and industrial wastewaters and sludges, and perhaps landfill leachate. Although these wastes contained many different organic and inorganic compounds, they were (and still are) collected and represent a so-called "point source" of pollution. As such, treatment plants can be designed and built to treat these collected wastes to achieve specified removals. This is not to imply that such treatment is a trivial task. Nonetheless, the problems associated with treating contaminated groundwater aquifers are much more complex.

The difficulty in treating contaminated groundwater aquifers is that the contaminant is difficult to locate ("see"). The complexity of "underground, nonpoint-source" pollution is demonstrated in Figure 18.1. In Figure 18.1(a), a dense nonaqueous-phase liquid (DNAPL), for example tetrachloroethene (also called perchloroethene (PCE); $CCl_2{=}CCl_2$), is released into a homogeneous aquifer system. It moves down through the unsaturated (vadose) zone, leaving behind some residual DNAPL, and then enters the aquifer containing the groundwater. Some of the DNAPL dissolves in the water, some sorbs to the aquifer solids, some volatilizes into the soil gas, and some remains as free liquid. In Figure 18.1(b), which is much more typical, the DNAPL is released into a heterogeneous aquifer system with different types of soils having differing hydraulic conductivities including aquitards that essentially prevent the passage of liquids. Here, the DNAPL is much more dispersed over a much wider area and much more difficult to locate and treat. If the DNAPL is a mixture of compounds, which is many times the case, remediation becomes even more problematic. Bioremediation offers promise here because bacteria have the potential to, for lack of a better expression, go where the contamination is and to destroy (mineralize) the contaminants. They can remove target contaminants from the water which, due to equilibrium partitioning, causes release of contaminant sorbed to aquifer solids and soils, and enhanced dissolution of residual DNAPL.

Historically biological processes were used to treat readily degradable pollutants. For example, municipal wastewaters containing little industrial contribution are made up primarily of food organics — i.e., carbohydrates, proteins, and fats. Groundwater aquifers may be contaminated with difficult-to-degrade, hydrophobic, volatile, and potentially toxic compounds such as fuel hydrocarbons, chlorinated solvents, pesticides, heavy metals, and energetic compounds (e.g., explosives), to name a few. The presence of these anthropogenic and xenobiotic compounds in the complex subsurface environment described above makes treatment problematic.

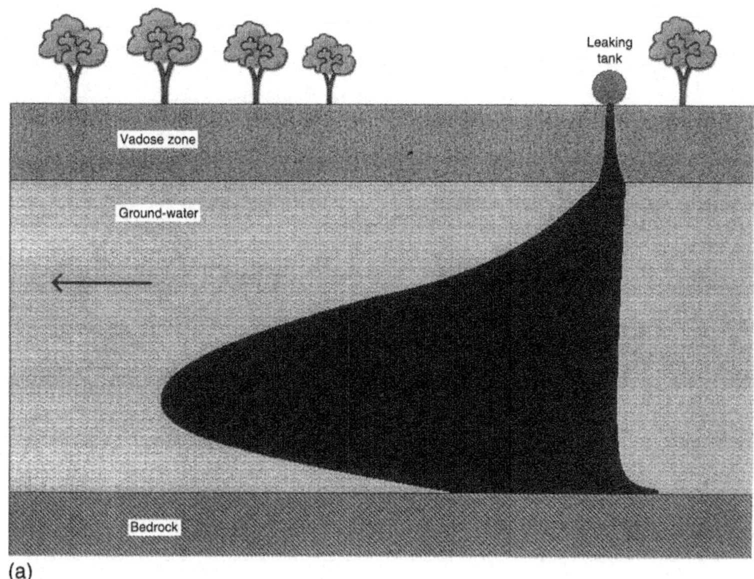

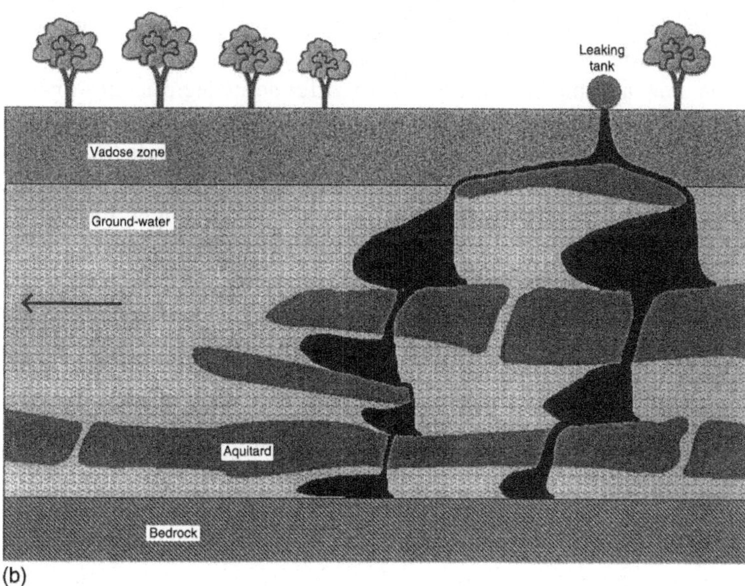

FIGURE 18.1 Contamination of a ground water equifer by a aense nonaqueous-phase liquid (DNAPL). (a) Homogenous aquifer system. (b) Heterogeneous aquifer system. (From BE Rittmann and PL McCarty. Environmental Biotechnology, New York: McGraw-Hill, 2001, redrawn with permission; original source: PL McCarty. "Scientific Limits to Remediation of Contaminated Soils and Groundwater." In Groundwater and Soil Contamination Remediation: Toward Compatible Science, Policy, and Public Perception, pp 38–52. Washington, D.C.: National Academy Press, 1990, with permission.)

18.1.2 Definition of Basic Terms and Scope

Table 18.1 contains brief definitions of important terms to be used in this chapter. The major focus will be on engineered *in situ* bioremediation and monitored natural attenuation (MNA),

TABLE 18.1
Brief Definitions of Important Terms

Bioaugmentation — addition of pure- or enriched-culture organisms to accelerate the rate of target contaminant biodegradation
Cometabolism — degradation of a contaminant in which the organism obtains no observable benefit, either because the concentration is too low to support growth or the organism cannot obtain energy from the transformation (also called co-oxidation or fortuitous degradation)
Engineered in situ bioremediation — enhancements (e.g., carbon source, electron donor, electron acceptor, nutrients, buffer) are added to the subsurface to stimulate the growth of indigenous bacteria
Intrinsic bioremediation — indigenous bacteria degrade target contaminants without intervention
Monitored natural attenuation (MNA) — comprehensive sampling (e.g., monitoring wells, etc.) is conducted to ensure that natural attenuation processes are reducing contaminant concentration and mass at a rate sufficient to protect public and environmental health
Natural attenuation — natural processes (physical, chemical, biological) that result in a decrease in contaminant concentration. These includes dilution, sorption, chemical oxidation and reduction, in addition to biodegradation
Phytoremediation — use of plants to remediate contaminated soil and groundwater
Primary substrate — compound serving as the electron donor or electron acceptor for growth

but the basic principles also apply to ex situ treatment. Phytoremediation will not be discussed here; the interested reader is referred to Refs. [3, 4].

Engineered *in situ* bioremediation and MNA offer some significant advantages in remediating subsurface contamination. They can be faster and cheaper than many alternatives and can result in the destruction or immobilization of toxic contaminants (as opposed to transfer to another phase — e.g., air stripping and carbon adsorption). They can attack hard-to-withdraw, hydrophobic compounds by affecting local equilibrium. Bioremediation causes minimum land disturbance and does not dewater the aquifer (i.e., conserves water). As no treatment alternative is a panacea, there are some significant disadvantages. For example, tight clay soils are difficult to bioremediate. It is a significant engineering challenge to deliver enhancements (i.e., electron donor, electron acceptor, nutrients, bacteria, etc.). There is potential for clogging the aquifer and changing the geochemistry of the aquifer (pH changes, mineral dissolution, etc.). Finally, there is the potential for forming more toxic metabolites (e.g., formation of vinyl chloride from reduction of trichloroethene). These advantages and limitations must be well understood if bioremediation is to be successful.

18.2 GENERAL REQUIREMENTS FOR EFFECTIVE BIOREMEDIATION

The general requirements for effective bioremediation will be briefly discussed below. For more detail, the reader is encouraged to consult Chapters 14 and 15 of Ref. [1] and Refs. [5–7]. There are two kinds of general requirements for effective bioremediation: (1) the contaminated site must be amenable to bioremediation and (2) requirements for bacterial growth must be met.

In order for bioremediation to be effective, the characteristics of the site must be amenable to microbial action. Characteristics of the "ideal" site for *in situ* bioremediation have been given as [1, 6]:

- Hydraulic conductivity is relatively homogeneous, isotropic, and large ($>10^{-3}$ cm/s)
- Residual concentrations of nonaqueous-phase-liquid (NAPL) contaminants should not be excessive (<10 g/kg to prevent aquifer permeability problems)
- Contamination is relatively shallow so that drilling and sampling costs are minimized.

Although such ideal conditions are rarely found, sites with consistent and high hydraulic conductivity, shallow contamination, and low residual NAPL offer the best chances for success.

The requirements and factors that affect bacterial growth can be summarized as:

- Carbon source
- Electron donor (energy source)
- Electron acceptor (e.g., O_2, NO_3^-, Fe(III), SO_4^{2-}, CO_2)
- Nutrients (N, P, perhaps Fe and S, trace metals)
- Proper pH (typically around neutral)
- Temperature (affects removal rates)
- Absence or control of toxic substances
- Adequate contact (between bacteria, nutrients, and contaminants)
- Adequate time
- Relevant bacteria (e.g., those that can degrade the target contaminant)
- Adequate moisture (important for vadose (unsaturated) zone remediation).

At most sites that are amenable to bioremediation, limitations are likely to be lack of adequate carbon source and electron donor, electron acceptor, nitrogen and phosphorus, or relevant bacteria. With MNA, a site assessment is conducted to determine whether these are in sufficient supply to ensure timely and effective biodegradation. With engineered remediation, one or more of these items will be added to the subsurface. Examples include, but are not limited to, electron donors such as acetate, glucose (sugar), lactate, vegetable oils, molasses, gaseous H_2, and proprietary compounds that release H_2 slowly; electron acceptors such as O_2 (typically added as air, gaseous O_2, hydrogen peroxide, or proprietary compounds that release O_2 slowly), nitrate, and sulfate; the "fertilizer" nutrients nitrogen and phosphorus; and perhaps buffers such as lime or sodium bicarbonate. If pure- or enrichment-culture bacteria are added to the subsurface, the process is termed *bioaugmentation*.

The following is a list of compounds and classes of compounds that have been successfully bioremediated at the pilot or field scale. It is not meant to be all-inclusive, but rather to give an indication of the variety of compounds amenable to bioremediation *if conditions are appropriate*:

- BTEX compounds (benzene, toluene, ethylbenzene, xylenes)
- Petroleum hydrocarbons (gasoline, diesel, oil, jet fuel)
- Polynuclear aromatic hydrocarbons (PAH, those with three rings or less)
- Creosote
- Alcohols, aldehydes, esters
- Chlorinated aliphatic hydrocarbons (CAH) and chlorinated benzenes
- Phenols and chlorinated phenols
- Polychlorinated biphenyls (PCBs)
- Nitroaromatics (e.g., nitrobenzene) and other explosives (e.g., RDX)
- Pesticides (e.g., alachlor, atrazine, etc.)
- Inorganics (e.g., nitrate, perchlorate, oxidized metals (Cr(VI) and U(VI)).

In the sections that follow, specific examples of bioremediation will be given.

18.3 BIOREMEDIATION OF FUEL HYDROCARBONS (BTEX)

As with other contaminants, fuel hydrocarbons find their way into the subsurface environment via accidental releases and improper disposal. Among the most common and problematic are the so-called BTEX compounds — benzene, toluene, ethylbenzene, and the three xylene isomers:

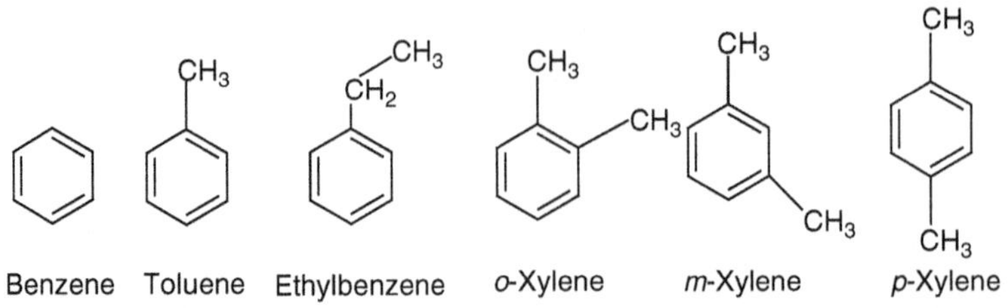

Benzene Toluene Ethylbenzene *o*-Xylene *m*-Xylene *p*-Xylene

These compounds are problematic water pollutants because they are the most soluble components of gasoline. They comprise approximately 20% by volume of typical gasoline [1]. These compounds pose human health concerns and because of this, drinking water standards, maximum contaminant levels (MCLs), have been set for each of them [8]. A primary source of BTEX is leaking underground storage tanks (UST). As of March 2002, over 420,000 confirmed releases from USTs have been documented [9]. The potential for bioremediating BTEX contamination is quite high because bacteria capable of degrading these compounds are fairly ubiquitous and their requirements are fairly well understood.

18.3.1 Basic Microbiology and Biochemistry of BTEX Degradation

A wide variety of soil bacteria, especially *Pseudomonas* strains, can degrade BTEX compounds under a variety of conditions. The general strategy is to hydroxylate or carboxylate the aromatic ring. In aerobic environments, the initial attack is usually by mono- and dioxygenase enzymes that are induced by the BTEX compounds. These enzymes use molecular O_2 as a cosubstrate, inserting oxygen and producing catechols that are easily amenable to ring cleavage (Figure 18.2). These pathways have been well known for quite some time [11]. Bacteria capable of aerobic degradation of BTEX compounds are ubiquitous in the environment [1].

BTEX compounds are also degradable under anaerobic conditions, although degradation rates are typically much slower [1, 12]. Anaerobic biodegradation of toluene, ethylbenzene, and the xylenes has been fairly well established [12, 13]. These organisms use strategies that end up in carboxylation of the aromatic ring prior to ring cleavage. One such pathway is shown in Figure 18.3 for toluene oxidation — the benzoyl-CoA produced is amenable to ring cleavage and further oxidation. For many years, benzene, the most toxic of the BTEX compounds, was thought to be resistant to anaerobic biodegradation. Recent research has shown that benzene can be degraded under nitrate-reducing [14–16], iron-reducing [14, 17], sulfate-reducing [14, 18, 19], and methanogenic conditions [14, 20].

18.3.2 General Requirements for BTEX Bioremediation

In concept, BTEX bioremediation is quite simple. Since BTEX-degrading bacteria are ubiquitous in the environment, if sufficient oxygen, nitrogen, and phosphorus can be supplied, rapid and complete remediation of all BTEX compounds is possible. Supplying adequate oxygen may be problematic due to its low solubility in water. And, even if adequate oxygen and nutrients are added, it may be difficult to achieve reductions in benzene concentration to its MCL of 5 μg/L. It may be that benzene concentrations are so low they will not support bacterial growth or enzyme induction [1]. In such cases, one suggested solution is to add a benign substrate such as benzoate to increase the population of BTEX-degrading bacteria [21]. Benzoate is an intermediate in the *tol* pathway (top pathway in Figure 18.2 — encoded in the *tol* plasmid), but does not itself induce the dioxygenase.

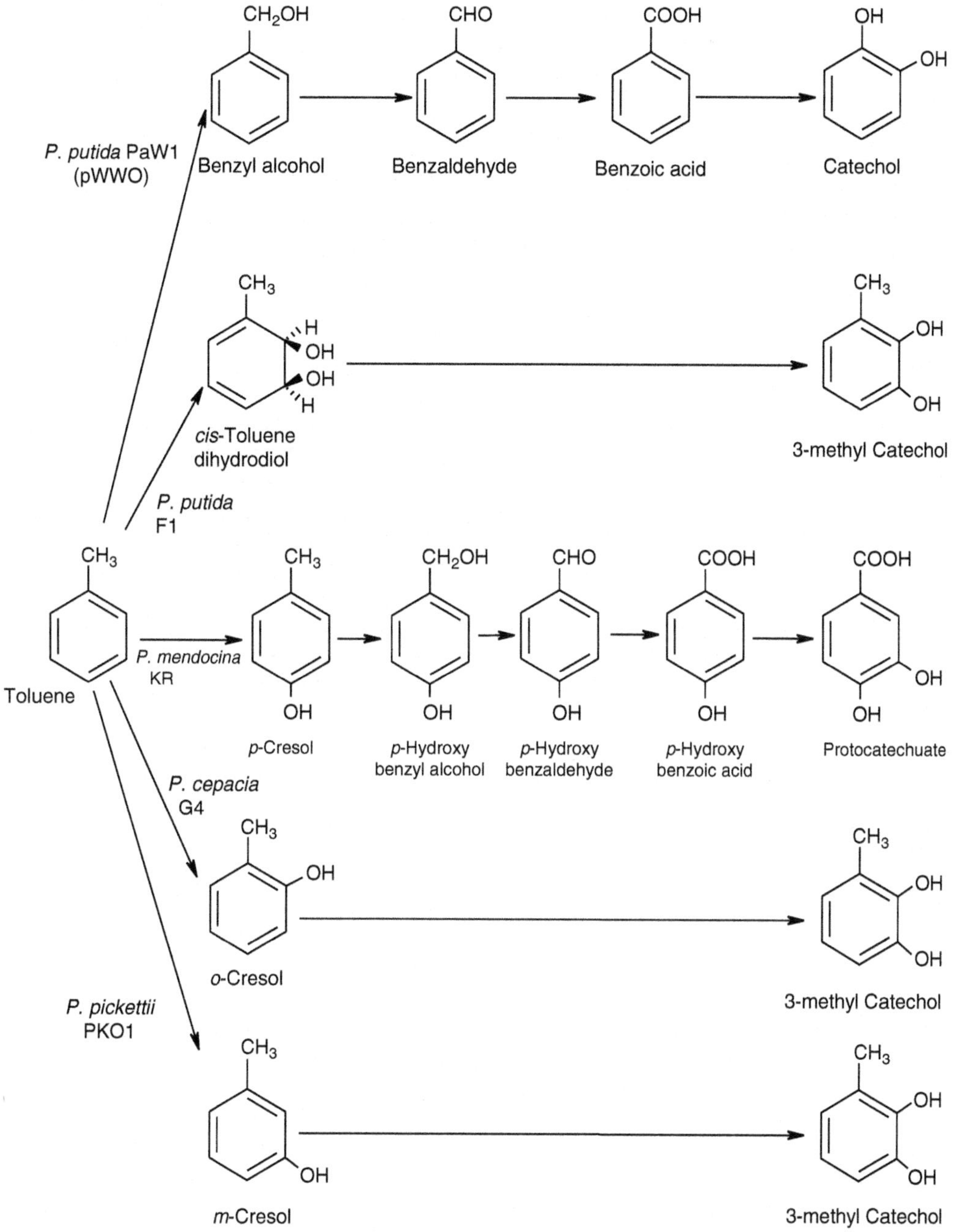

FIGURE 18.2 Aerobic toluene degradation pathways. (From BE Rittmann and PL McCarty. *Environmental Biotechnology*. New York: McGraw-Hill, 2001, redrawn with permission; original source Reference [10], with permission.)

Degradation rates for BTEX compounds are generally slower under anaerobic conditions. However, toluene degradation rates are quite high when nitrate is present as an electron acceptor [13]. A strategy advocated by some is addition of the more mobile, more soluble nitrate to serve as an electron acceptor for toluene degradation (and perhaps ethylbenzene and the xylenes) leaving oxygen to serve as the electron acceptor for benzene degradation [22, 23].

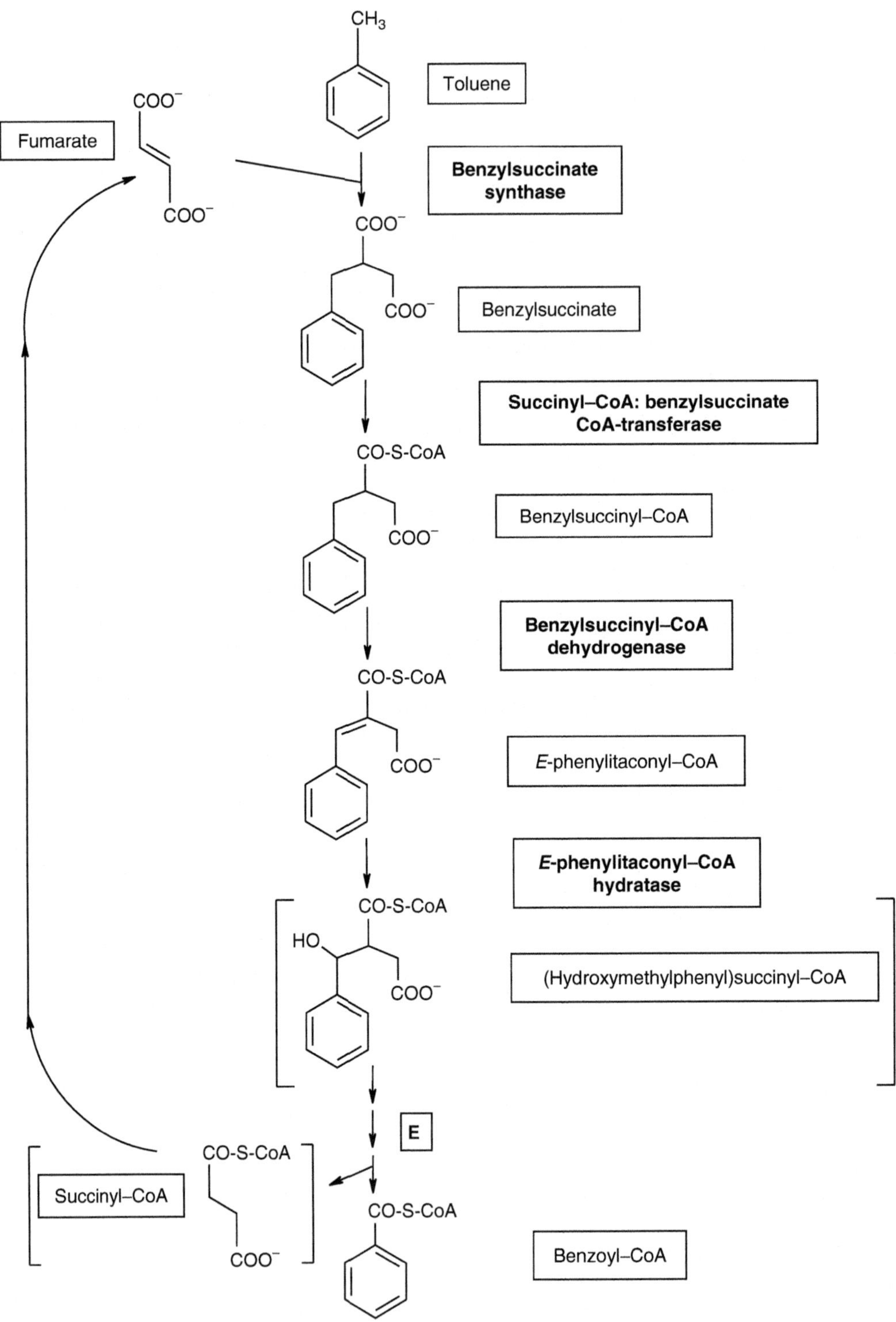

FIGURE 18.3 Anaerobic toluene degradation pathway. (*Source*: University of Minnesota Biocatalysis/Biodegradation Database (from Alfred Spormann and Eva Young), http://umbbd.ahc.umn.edu/.)

18.3.3 Examples of Successful BTEX Bioremediation

Bioremediation has been applied most successfully to petroleum hydrocarbon contamination [1, 6, 7]. The most common scenario is the addition of oxygen, nitrogen, and phosphorus to stimulate indigenous hydrocarbon-degrading bacteria. Perhaps the first reported success of bioremediation was by Raymond et al. [24] where a 100,000 gallon gasoline release was remediated by first recovering the free product and then adding ammonium sulfate (nitrogen source), phosphorus, and oxygen by air sparging to remove the remaining contamination. Successful bioremediation was confirmed by monitoring the disappearance of petroleum hydrocarbons with a concomitant increase in the concentration of bacteria. Similarly, Norris and Dowd [25] reported successful BTEX bioremediation of a site in California with the addition of hydrogen peroxide as an oxygen source, along with ammonium chloride and tri-polyphosphate. Bioremediation was confirmed by measuring the disappearance of ammonia along with the appearance of CO_2 in the soil gas and groundwater. A BTEX-contaminated site in Colorado was successfully bioremediated with addition of hydrogen peroxide, ammonium chloride, and phosphate salts [6, 26]. Bioremediation was confirmed using measurements of contaminant loss and O_2 consumption at the site in conjunction with laboratory experiments showing that appropriate BTEX-degrading bacteria were present at the site. Additional examples of successful BTEX remediation can be found in Refs. [6, 7, 27–31].

18.3.4 Challenges

Other than the above-mentioned challenges of supplying electron acceptors and nutrients, and meeting clean-up goals (e.g., MCLs), a current challenge is understanding the impact of added fuel oxygenates (primarily methyl *tert*-butyl ether [MTBE] and ethanol) on BTEX remediation (both engineered bioremediation and MNA). MTBE is volatile, very soluble in water, not strongly sorbed to aquifer solids, and relatively nonbiodegradable, although recent evidence suggests that it can be biodegraded under some conditions [8, 32, 33]. Due to its mobility and recalcitrance in groundwater systems, remediation of MTBE is in many cases more difficult than BTEX remediation. Because of suspected health concerns with MTBE, ethanol is likely to replace it as a fuel oxygenate [34]. However, ethanol presents its own set of problems with respect to remediation of BTEX contamination. For example, because ethanol is more degradable than BTEX, it will be preferentially degraded and consume the O_2 that is critical to effective benzene biodegradation [35, 36]. Such interactions need to be better understood if effective engineered bioremediation systems are to be designed or reliable predictions of the efficacy of MNA are to be made.

18.4 BIOREMEDIATION OF CHLORINATED ALIPHATIC HYDROCARBONS (CAH)

Chlorinated organic compounds are among the most problematic and common environmental contaminants. These include CAHs, chlorinated benzenes, chlorinated phenols, PCBs, and a variety of other pesticides (e.g., atrazine, alachlor, etc.) and industrial chemicals. All these chemicals can be at least partially biodegraded under the right conditions. The focus of this section will be the chlorinated ethenes, ethanes, and methanes (Table 18.2), which are the most commonly found chlorinated organics in contaminated groundwaters and at contaminated sites [37, 38]. In general, the highly chlorinated CAHs are more susceptible to anaerobic biodegradation (being highly oxidized, they are susceptible to reduction) while the less chlorinated CAHs tend to be susceptible to aerobic biodegradation (more reduced, more susceptible to oxidation) [1, 8].

18.4.1 Basic Microbiology and Biochemistry of CAH Degradation

CAHs can be degraded as *primary substrates* where the CAH serves as an electron donor or electron acceptor for growth. If the CAH serves as an electron acceptor for growth, the

TABLE 18.2
Biodegradation of Chlorinated Aliphatic Hydrocarbons

			Primary Substrate			Cometabolism	
CAH	**Acronym**	**Formula**	**Aerobic Donor**	**Anaerobic Donor**	**Anaerobic Acceptor**	**Aerobic**	**Anaerobic**
Ethenes							
Tetrachloroethene	PCE	$CCl_2{=}CCl_2$			Yes		Yes
Trichloroethene	TCE	$CHCl{=}CCl_2$			Yes	Yes	Yes
cis-1,2-Dichloroethene	*c*-DCE	$CHCl{=}CHCl$		Yes	Yes	Yes	Yes
trans-1,2-Dichloroethene	*t*-DCE	$CHCl{=}CHCl$		Yes		Yes	Yes
Chloroethene (vinyl chloride)	VC	$CH_2{=}CHCl$	Yes	Yes	Yes	Yes	Yes
Ethanes							
1,1,1-Trichloroethane	1,1,1-TCA	CH_3CCl_3			Yes	Yes	Yes
1,1,2-Trichloroethane	1,1,2-TCA	$CH_2ClCHCl_2$				Yes	Yes
1,1-Dichloroethane	1,1-DCA	CH_3CHCl_2				Yes	Yes
1,2-Dichloroethane	1,2-DCA	CH_2ClCH_2Cl	Yes	Yes		Yes	Yes
Chloroethane	CA	CH_3CH_2Cl	Yes			Yes	Yes
Methanes							
Carbon tetrachloride	CT	CCl_4					Yes
Chloroform	CF	$CHCl_3$				Yes	Yes
Dichloromethane	DCM	CH_2Cl_2	Yes	Yes		Yes	Yes
Chloromethane	CM	CH_3Cl	Yes			Yes	Yes

Yes — conclusive evidence for this pathway.
Blank — no evidence for this pathway.
Adapted from Ref. [1] and updated.

process is called *dehalorespiration*. These compounds can also be degraded by a process commonly called *cometabolism* in which the organism gets no observable benefit from the degradation. The terms *fortuitous* degradation and *co-oxidation* have also been used to describe this phenomenon. Cometabolic transformations can be oxidations or reductions and typically require a particular enzyme or series of enzymes to be induced, and that the organism be grown on other primary substrates. Examples are given below. The current status of knowledge regarding biodegradation of CAHs is given in Table 18.2.

18.4.1.1 Aerobic Cometabolism

The most common and well-understood cometabolic transformations are those catalyzed by oxygenase enzymes. For example, methanotrophic bacteria have methane monooxygenase (MMO) enzymes that catalyze the first step in the oxidation of their growth substrate, methane, by inserting an oxygen atom, eventually forming methanol [1]:

$$CH_4 + O_2 + NADH + H^+ \xrightarrow{MMO} CH_3OH + NAD^+ + H_2O$$

MMO exhibits relaxed substrate specificity and also catalyzes the first step in the cometabolic oxidation ("co-oxidation") of TCE, forming a reactive epoxide intermediate [1, 8]:

$$\text{(Cl)(H)C{=}C(Cl)(Cl)} + O_2 + NADH + H^+ \xrightarrow{MMO} \text{(Cl)(H)C{-}C(Cl)(Cl) (epoxide, O bridging)} + NAD^+ + H_2O$$

Toluene dioxygenase (TDO), an enzyme used by *Pseudomonas putida* F1 to oxidize toluene to *cis*-toluene dihydrodiol (second pathway in Figure 18.2), oxidizes TCE to a four-membered ring structure [39]:

$$\mathrm{ClHC{=}CCl_2} + O_2 \xrightarrow{TDO} \mathrm{ClHC(O{-}O)CCl_2}$$

Primary substrates that have been used to support these oxygenase-initiated, cometabolic oxidations of TCE include ammonia, ethane, ethene, propane, propene, butane, and phenol, among others [1].

These cometabolic oxidations of TCE require O_2 as a cosubstrate and reducing power (e.g., NADH). The organisms get no benefit from TCE cometabolism. In fact, the products of TCE cometabolism are in many cases toxic to the bacteria. In many cases, this toxicity is manifests itself in a *transformation capacity* — the organism appears to have a finite capacity for transforming TCE. This phenomenon has been shown for organisms using methane, toluene, propane, and phenol as growth substrates [40]. It has also been shown for cometabolic transformations of CF, 1,2-DCA, *c*-DCE, *t*-DCE, 1,1-DCE, and VC [41, 42]. In addition, the concept of transformation capacity has been demonstrated for anaerobic cometabolic transformation of PCE, CT, and 1,1,1-TCA by a mixed methanogenic enrichment fed acetate [43].

18.4.1.2 Dehalorespiration

Prior to the discovery of dehalorespiring bacteria in the early 1990s, remediation of groundwater aquifers contaminated with PCE and TCE was not promising. PCE is not degraded aerobically and the aerobic cometabolism of TCE described in the previous section is typically quite slow. In addition, it was discovered that anaerobic biodegradation of PCE and TCE could result in the production of the toxic (carcinogenic) metabolite, vinyl chloride (VC). *Reductive dechlorination* of PCE and TCE is a stepwise process in which chlorine atoms are sequentially removed and replaced with a hydrogen atom, a reduction requiring two electrons for each step:

$$\text{PCE} \longrightarrow \text{TCE:}\quad CCl_2{=}CCl_2 + H^+ + 2e^- \longrightarrow CHCl{=}CCl_2 + Cl^-$$

$$\text{TCE} \longrightarrow c\text{-DCE:}\quad CHCl{=}CCl_2 + H^+ + 2e^- \longrightarrow CHCl{=}CHCl + Cl^-$$

$$c\text{-DCE} \longrightarrow \text{VC:}\quad CHCl{=}CHCl + H^+ + 2e^- \longrightarrow CH_2{=}CHCl + Cl^-$$

$$\text{VC} \longrightarrow \text{ethene:}\ CH_2{=}CHCl + H^+ + 2e^- \longrightarrow CH_2{=}CH_2 + Cl^-$$

Thus, anaerobic biodegradation of PCE and TCE was thought to be undesirable due to potential build-up of VC.

In the early 1990s, field evidence suggested that PCE could be dechlorinated all the way to ethene [44]. There were reports in the literature of laboratory enrichment cultures that could completely dechlorinate PCE to ethene [45–47]. The first report of isolation of a bacterium that could grow using PCE or TCE as an electron acceptor was made in 1993 [48]. However, the isolated strain, *Dehalobacter restrictus*, reduced PCE and TCE to *c*-DCE and no further. Similarly, *Dehalospirillium multivorans* could not reduce beyond *c*-DCE [49]. *Dehalococcoides ethenogenes* strain 195 was the first bacterium that could dechlorinate PCE all the way to ethene [50]. This organism could grow using PCE, TCE, and *c*-DCE as electron acceptors. Reduction of VC to ethene was reported to be cometabolic, requiring the presence of another chloroethene as an electron acceptor for growth [51]. Recent reports have demonstrated that

Dehalococcoides-like bacteria can use *c*-DCE and VC, but not PCE or TCE, as electron acceptors for growth [52, 53]. However, as of this writing, there are no reports of a single bacterium that can use PCE, TCE, *c*-DCE, and VC as electron acceptors for growth. Nonetheless, it is clear that complete conversion of PCE to ethene is possible under the right conditions.

An exciting recent development is the discovery of an anaerobic bacterium, strain TCA1, which uses the chlorinated ethane, 1,1,1-TCA, as an electron acceptor for growth [54]. Strain TCA1 reductively dechlorinates 1,1,1-TCA to 1,1-DCA and CA, and requires H_2 as an electron donor. Phylogenic analysis indicated that strain TCA1 is closely related to a *D. restrictus* strain which grows on the chlorinated ethenes, PCE, and TCE. So, it appears that both chlorinated ethenes and ethanes can be used as electron acceptors for growth. Given these discoveries are relatively recent (within the past 10 years or so), it is likely than many additional halorespiring strains will be discovered in the near future as long as we keep looking.

18.4.2 General Requirements for CAH Bioremediation

As with most contaminants, successful CAH bioremediation requires an adequate supply of carbon, electron donor, electron acceptor, and the nutrients nitrogen and phosphorus. In some cases, buffers will be needed to maintain acceptable pH values. A review of Table 18.2 indicates that vinyl chloride, the problematic metabolite from the reductive dechlorination of PCE and TCE, is an electron donor for the growth of aerobic bacteria. If sufficient O_2 is present or supplied through exogenous addition, these organisms can complete the detoxification of PCE and TCE [55].

18.4.2.1 Aerobic Cometabolism

Accomplishing aerobic cometabolism in the field is quite complicated. The organisms of interest (those with the requisite oxygenase enzymes) must be supplied with adequate O_2 along carbon and energy for growth. Transformation of the CAH will likely be toxic to the bacteria — that is, the transformation capacity must be accounted for. In addition, the CAH, particularly TCE, may be a competitive inhibitor to the growth substrate [41]. Finally, there will be competition if other cometabolic substrates (i.e., other CAHs) are present. Many of these factors that demonstrate the complexity of aerobic cometabolism of CAHs can be seen by comparing the following two equations, the first describing "normal" bacterial growth and the second describing growth in the presence of cometabolism of a CAH [1]:

$$\mu = \left[\frac{Y\hat{q}_{\text{ed}}C_{\text{ed}}}{K_{\text{ed}} + C_{\text{ed}}}\right] - b \tag{18.1}$$

$$\mu = \left[\frac{1}{1 + \frac{C_{\text{ed}}}{K_{\text{ed}}} + \frac{C_{\text{CAH}}}{K_{\text{CAH}}}}\right]\left[\frac{Y\hat{q}_{\text{ed}}C_{\text{ed}}}{K_{\text{ed}}} - \frac{\hat{q}_{\text{CAH}}C_{\text{CAH}}}{T_{\text{c}}K_{\text{CAH}}}\right] - b \tag{18.2}$$

where μ is the specific growth rate of the bacteria (day^{-1}), C_{ed} and C_{CAH} are the concentrations of the electron donor and CAH, respectively (mg/L), K_{ed} and K_{CAH} the half-velocity coefficients for the electron donor and CAH, respectively (mg/L), Y is the bacterial yield (mg bacteria synthesized per mg electron donor), $\hat{q}_{\text{ed}}$ and $\hat{q}_{\text{CAH}}$ are the specific utilization rates for the electron donor and CAH, respectively (mg/mg bacteria per day), and b is the bacterial decay rate (day^{-1}). Please note that Equation (18.2) reduces to Equation (18.1) when $C_{\text{CAH}} = 0$.

Overcoming the toxicity, competitive inhibition, and the need for sufficient reducing power is a challenge. It has been suggested that the bacteria could be grown in a separate reactor in the absence of the CAH and then added to another reactor fed with the contaminated water [56]. This is not possible for *in situ* applications. One key to success is making sure enough electron donor is added to overcome competitive inhibition and toxicity associated with CAH transformation. Examples are discussed below.

18.4.2.2 Dehalorespiration

Since the CAH is meant to serve as the electron donor for growth, the major requirement for effective dehalorespiration is the effective delivery of sufficient electron donor. Of course, sufficient nitrogen and phosphorus must be present and the pH maintained near neutral. Molecular hydrogen (H_2) is generally the preferred, and in some cases obligate, electron donor for most chlorinated-ethene respiring bacteria, although organics such as acetate can work for some [1, 48–54, 57–59]. There will be competition between H_2-utilizing dehalorespiring bacteria and other organisms that use H_2 such as denitrifiers, dissimilatory iron reducers, sulfate reducers, and methanogens. It has been shown that dehalorespiring bacteria have a kinetic advantage over other H_2-utilizing bacteria at low H_2 concentrations [60, 61]. There is evidence that the dehalorespiring bacteria have complex nutritional needs that may be met in part by other organisms in the CAH-degrading consortium [50]. One study has shown that complete conversion of PCE to ethene was dependent on a healthy population of methanogenic bacteria [62].

The current paradigm for effective dehalorespiration is to add a more complex electron donor that will be slowly converted to H_2, providing the low levels of H_2 that should allow dehalorespiring bacteria to outcompete other bacteria. Soluble substrates that have been tried include, but are not limited to, benzoate [61, 63], lactate [64–66], pentanol [67], propionate [61, 63], and pyruvate [68]. More complex and solid substrates include, but are not limited to, oleate and tetrabutyl orthosilicate (TBOS) [63], biomass [63, 69], molasses [70, 71], vegetable oil [72], and proprietary compounds [73]. It is also possible to directly add molecular H_2 to the subsurface [74]. There is also some evidence that in some cases, acetate may be a more effective electron donor for conversion of PCE to ethene [58].

Although complete conversion of PCE to ethene has been observed at many field sites and with a variety of enrichment cultures in the laboratory, there are some sites and some cultures where dechlorination stops at *c*-DCE or VC [75–77]. Although there is still some debate in the literature as to why this occurs, it appears to involve one or more of the following factors: (1) lack of appropriate dehalorespiring strains, (2) competition for electron donor, (3) toxicity of PCE and TCE to *c*-DCE and VC degraders, and (4) kinetic limitations. These factors arise in part because it is felt that at least two populations of dehalorespiring bacteria are involved in the complete conversion of PCE to ethene — one that converts PCE and TCE to *c*-DCE and VC, and one that converts *c*-DCE and VC to ethene [75, 77]. Bioaugmentation with an enrichment that completely dechlorinates PCE has shown to be successful in some cases [76, 77]. However, there are indications that bioaugmentation near source zones where the concentrations of PCE and TCE may be quite high, will result in toxicity of the population that degrades *c*-DCE and VC [78]. In such cases, dechlorination beyond *c*-DCE may only occur downgradient (away) from these high-concentration zones. Recent work has suggested that remediation to low concentrations of *c*-DCE and VC may not be possible at some sites because of kinetic limitations [79, 80].

18.4.3 Examples of Successful CAH Bioremediation

As of this writing, very few sites contaminated with CAHs have been completely remediated. However, many clean-ups are underway and many field-scale studies have been completed. A few are described below.

18.4.3.1 Aerobic Cometabolism of TCE

The Moffett field site developed by Stanford University researchers has been used for a number of pilot studies on aerobic cometabolism. Methane was initially used as the primary substrate for aerobic cometabolism of TCE, *c*-DCE, *t*-DCE, and VC [81–85]. These studies found that while removals of the less-chlorinated CAHs (VC, *c*-DCE, and *t*-DCE) were quite good, removals of TCE were quite low (≈15%). Toluene and phenol were shown to give much higher removals of TCE (>90% for 250 μg/L of TCE) [86]. Based on these results, a larger, field-scale demonstration was conducted at site 19, Edwards Air Force Base, CA [87].

At Edwards Air Force Base, toluene, gaseous O_2, and hydrogen peroxide (H_2O_2) were added to groundwater that was recirculated between contaminated upper and lower aquifers using two treatment wells (Figure 18.4). Pure toluene was pulsed in every 30 min in order to control clogging. Hydrogen peroxide was added as a supplemental source of O_2 and, because it is toxic to bacteria in high concentrations, to prevent well clogging. During approximately 1 year of treatment, TCE was reduced from around 1000 μg/L to around 20 μg/L, a 98% reduction. Toluene was biodegraded to around 1 μg/L, far below its drinking water MCL of 1 mg/L. Once operation was stabilized, the major operational expense was associated with preventing clogging. This project demonstrated that aerobic cometabolism could be used to effectively treat TCE contamination.

18.4.3.2 Dehalorespiration of PCE Contamination

There are numerous examples where electron donors have been added to the subsurface and have resulted in decreases in PCE and TCE concentration, production of *c*-DCE and VC, and then ethene (e.g., Refs [70, 88–91]). The general response to electron donor addition can be summarized as the typical behavior shown in Figure 18.5. One of the first examples of a well-controlled field study was the addition of benzoate and sulfate to stimulate PCE conversion to ethene at a Victoria, Texas, sites [92]. Here, side-by-side plots were used to demonstrate that

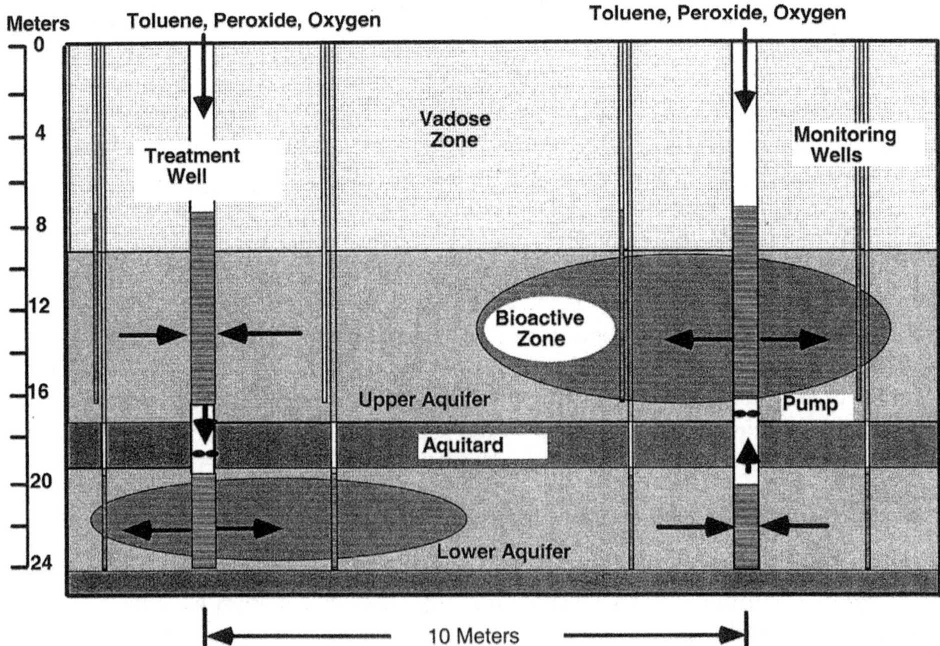

FIGURE 18.4 Schematic of cometabolic TCE treatment system at site 19, Edwards Air Force Base, CA. (Reprinted with permission from PL McCarty, MN Goltz, GD Hopkins, ME Dolan, JP Allan, BT Kawakami, and TJ Carrothers. *Environ. Sci. Technol.* 32: 88–100, 1998. Copyright 1998 American Chemical Society.)

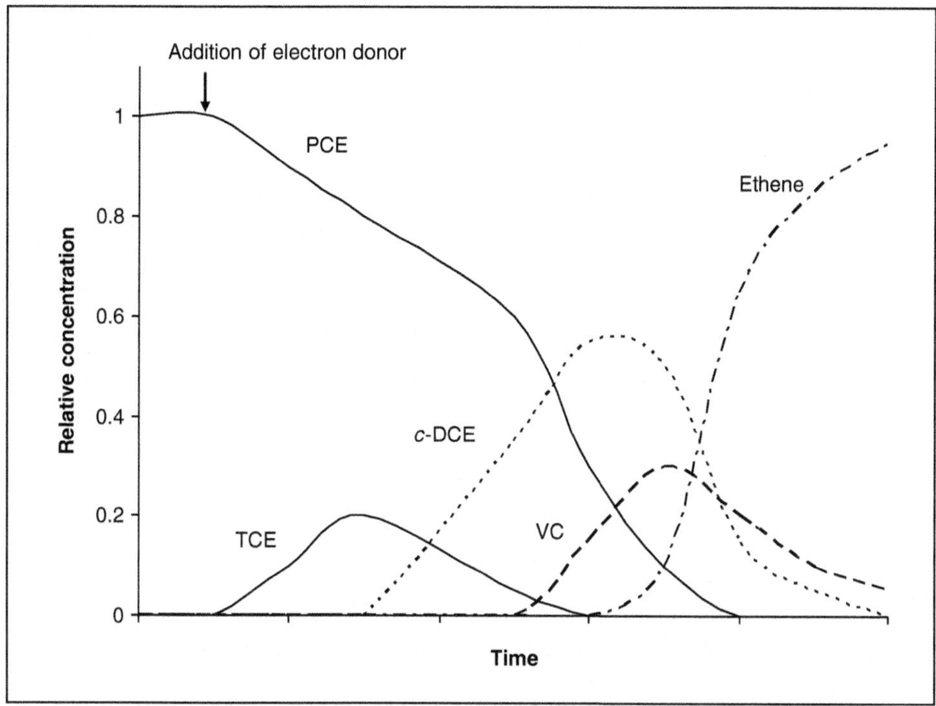

FIGURE 18.5 Typical response to electron donor addition at a PCE-contaminated site.

biodegradation was responsible for PCE removal. One plot received benzoate and sulfate while the other did not. No PCE removal was observed in the plot receiving no benzoate and sulfate while PCE conversion to ethene was observed when benzoate and sulfate were added. The success of this project led to a patent [93]. In addition, enrichment cultures and microcosms developed from site materials were found to contain *Dehalococcoides*-like organisms [52, 94].

A recent report reviewed the potential for enhanced anaerobic bioremediation of CAHs and raised some issues to be addressed [95]. A case study was presented describing remediation of an abandoned manufacturing facility in California by adding molasses as an electron donor to create a reducing environment that would support reduction of PCE to ethene and Cr(VI) to the less mobile and less toxic Cr(III). Anaerobic digester supernatant was added as a source of bacteria. Of the CAHs, TCE was present in the highest concentration, up to 17,000 μg/L, and Cr(VI) was found at concentrations in excess of 100,000 μg/L in some places. Based on pilot studies, a full-scale system was installed in 1997. After 1 year of operation, PCE, TCE, *c*-DCE, VC, and Cr(VI) concentrations were reduced to approximately 5 μg/L near the source zone [96]. Although addition of electron donors such as molasses have proven very effective in supporting the reduction of PCE and TCE to ethene, long-term effects on the biogeochemistry of the aquifer (e.g., organic carbon concentrations, pH, dissolution of metals, changes in microbial ecology, etc.) are largely unknown [95].

There are also examples of field-scale bioaugmentation with defined dechlorinating cultures. The example given above used anaerobic digester supernatant primarily to reduce the lag time associated with the start of significant dechlorination. The Pinellas culture, developed from a contaminated site in Florida, was used to completely dechlorinate TCE to ethene in field tests at Dover Air Force Base [76]. This enrichment culture was added because dechlorination seemed stalled at *c*-DCE. Similarly, 13 L of KB-1 culture was added to a contaminated site at Kelly Air Force Base in Texas to stimulate PCE dechlorination [77]. After approximately 200 days, PCE, TCE, and *c*-DCE concentrations were reduced to below

5 μg/L and ethene was produced. In addition, *Dehalococcoides*-like bacteria found in culture KB-1 were also detected in the field after bioaugmentation. Finally, a recent report indicates that bioaugmentation of DNAPL source zones holds promise [78].

18.4.4 Challenges

Many of the challenges in CAH bioremediation are associated with trying to stimulate complete dehalorespiration to ethene and to treat source zones (i.e., DNAPL). It is not clear why at many field sites dechlorination seems to stop at *c*-DCE. There is a need to better understand the requirements and capabilities of the various dehalorespiring populations. While enhanced anaerobic bioremediation offers promise in treating DNAPL (source zones), many questions are still unanswered [67, 68, 78, 95, 97, 98]. What is the effect of dehalorespiration on the dissolution of the DNAPL, partitioning of the reduction products into and out of the DNAPL, and mobility of the reduction products? What is the effect of toxicity on the different populations of dehalorespiring bacteria? How do we effectively deliver H_2 without significantly altering aquifer biogeochemistry? The current paradigm seems to be to add a complex organic carbon source (e.g., molasses) that is slowly fermented to release H_2. However, not all the electron equivalents in these complex donors are converted to H_2. Dissolved organic carbon concentrations over 200 mg/L are not uncommon, and short-chain volatile acids such as acetic acid and propionic acid can be produced and then converted to methane [66, 70]. The long-term consequences of these and other potential changes need to be better understood.

18.5 BIOREMEDIATION OF PERCHLORATE

Contamination of the environment by perchlorate (ClO_4^-), used as an oxidant in solid propellants and explosives, has become more widely known because of the improvement in analytical methods in 1997 that allow detection of perchlorate in μg/L concentrations [99]. Perchlorate has been detected in water samples collected from at least 20 states of the 44 states known to have perchlorate-utilizing industries, with groundwater concentrations as high as 3700 mg/L being reported [100–103]. Exposure to perchlorate has been linked to abnormal thyroid function and other human health problems [104–106]. As a result, the EPA is in the process of regulating perchlorate and a drinking water standard as low as 1 μg/L may be proposed based upon risk estimates [103]. The state of California, where widespread perchlorate contamination was first reported, has established an action level of 4 μg/L for drinking water [107]. As methods for effective remediation of perchlorate are sought, the most promising possibilities appear to involve the use of bacteria that can respire perchlorate, resulting in its degradation [108–111].

18.5.1 Basic Microbiology and Biochemistry of Perchlorate Degradation

The use of perchlorate as an electron acceptor for bacterial respiration is thermodynamically favorable [8]:

$$8H^+ + 8e^- + ClO_4^- \longrightarrow Cl^- + 4H_2O, \qquad \Delta G^{o'} = -93.48\,\text{kJ/eeq}$$

In fact, it is more favorable than oxygen [8]:

$$4H^+ + 4e^- + O_2 \longrightarrow 2H_2O, \qquad \Delta G^{o'} = -78.14\,\text{kJ/eeq}$$

The ability of bacteria to degrade perchlorate was reported as early as 1965 [112]. Bacterial degradation of chlorate, a perchlorate metabolite, was reported in the 1950s and 1960s

[112–115]. More recently, bacteria capable of complete reduction of perchlorate have been isolated and characterized [116–120]. Most of these perchlorate-degrading bacteria are within the genera *Dechloromonas* and *Dechlorosoma*; however, isolates capable of complete perchlorate reduction have been identified in the alpha, beta, and epsilon subclasses of Proteobacteria [118, 119, 121, 122].

Bacterial perchlorate degradation occurs via two identified enzymes (Figure 18.6): (1) a reductase that reduces perchlorate to chlorate and chlorate to chlorite, and (2) chlorite dismutase, an enzyme that disproportionates chlorite into chloride and oxygen. While both enzymes are required for perchlorate degradation, the chlorite dismutase appears unique to perchlorate- and chlorate-reducing bacteria [117, 119, 123–126]. More specifically, the DNA sequence of dismutase enzymes appears to be highly conserved among perchlorate and chlorate bacteria with differing phylogenies [124]. This dismutation step, which yields no energy for the cell, is hypothesized to have evolved for the purpose of transforming the highly reactive and toxic chlorite intermediate. While oxygen is produced by action of the dismutase, oxygen is not observed external to the cell in systems actively degrading perchlorate.

The perchlorate reductase enzyme of perchlorate-reducing bacteria appears to be distinct from the chlorate reductase of chlorate bacteria unable to reduce perchlorate. The chlorate reductase of *Pseudomonas* strain PK was more similar to selenate reductase of *T. selenatis* and dimethyl sulfide dehydrogenase from *R. sulfidophium* than to the perchlorate reductases of *Dechloromonas agitus*, *Dechloromonas aromatica*, and *Dechlorosoma suillum* [127]. It is note worthy that perchlorate degradation requires anaerobic conditions to proceed, and this appears due to the sensitivity of perchlorate reductase to oxygen [124, 125, 128]. Most perchlorate-reducing bacteria, however, are facultative. All identified perchlorate-reducing isolates display aero-tolerance, and most are capable of growing under aerobic conditions. Thus, perchlorate-degrading bacteria are more robust than their perchlorate-degrading activity. Additionally, while most perchlorate bacteria are capable of using nitrate as an alternate electron acceptor, some (e.g., *D. suillum*) also require conditions free from nitrate before perchlorate degradation will proceed [125].

18.5.2 General Requirements for Perchlorate Bioremediation

As interest in perchlorate bioremediation has increased, so have reports of isolation of perchlorate-degrading bacteria [102, 117–119]. It is known that these perchlorate degraders require an electron donor source to drive the respiration of perchlorate. Few studies have addressed the capabilities of these organisms to respire perchlorate under conditions likely to be found at perchlorate bioremediation sites. Optimal conditions are rarely observed *in situ*. Specifically, the electron donor needed to drive reduction of perchlorate is often in short supply in the subsurface. However, perchlorate-reducing bacteria have been shown to be fairly ubiquitous in the environment [119], and when sufficient electron donor is supplied, perchlorate reduction to chloride (Cl^-) and water can be quite rapid. Laboratory studies with both pure-strain and mixed-culture perchlorate-degrading bacteria have shown perchlorate degradation half-lives on the order of hours [102, 110 117, 119, 120, 129]. However,

$$ClO_4^- \xrightarrow[\text{Reductase}]{2H^+ + 2e^- \;\; H_2O} ClO_3^- \xrightarrow[\text{Reductase}]{2H^+ + 2e^- \;\; H_2O} ClO_2^- \xrightarrow[\text{Dismutase}]{} Cl^- + O_2$$

exogenous electron donor must often be added in significant quantities (at significant cost). Therefore, a challenge of perchlorate remediation is to provide sufficient effective electron donor, but not an excess, to promote perchlorate degradation.

Other requirements for effective perchlorate bioremediation are less well understood at this time. In general, an anaerobic environment with low redox potential is required for fast and complete perchlorate reduction [102, 122, 125]. However, slower and less complete perchlorate reduction can occur in more oxidizing environments [122]. Nitrate may compete with perchlorate as an electron acceptor depending on the species of bacteria present [125]. If such conditions exist, then enough electron donor must be present to satisfy competitive electron acceptor demand (i.e., O_2, NO_3^-).

18.5.3 Examples of Perchlorate Bioremediation

As of this writing, there is very little information about full-scale perchlorate bioremediation systems in operation. However, there are several field-scale pilot studies undertaken along with many laboratory-scale studies. Limited summaries of these activities are available [130, 131] as are some research reports [111, 132]. The major focus is on removal of perchlorate to the proposed drinking water action level of around 4 μg/L. At the present time, all systems tested are ex situ reactors of one type or another. Examples include fluidized-bed and packed-bed reactors, hollow-fiber membrane reactors (patented by Bruce Rittmann of Arizona State University), and anaerobic composting. All involve the addition of electron donors, ranging from acetate and corn syrup to H_2. It is clear that within the next few years bioremediation will be a major technology used for perchlorate remediation.

18.6 SUMMARY AND CHALLENGES

Bioremediation, the use of biological catalysts, has become an important tool in cleaning up environmental contamination. Such biocatalysis is also the predominant mechanism driving the use of MNA. Since 1990, much has been learned about the capabilities of a wide variety of bacteria that degrade environmental contaminants. Bacteria with new, heretofore unknown capabilities have been discovered (i.e., dehalorespiring bacteria). Bioremediation has been successfully used to remediate BTEX contamination, is currently widely used to remediate CAH contamination, and will soon be applied to perchlorate contamination. Keys to success include proper site characteristics and supplying adequate carbon and energy sources, electron acceptors, nitrogen and phosphorus, buffer, and when necessary and appropriate, bacteria with the desired catabolic capabilities. With MNA, it is a matter of determining if adequate supplies of these requirements are present.

Challenges to be faced are many, but the potential for success is great. A key task is to develop data sets that conclusively prove that biodegradation is in fact responsible for decreases in contaminant concentration and mass. Guidelines for doing this are available (e.g., Refs. [1–3]). Application of bioremediation to source zones will be a challenge. There is the ever-present engineering challenge of effectively and economically delivering enhancements to the contaminated area. There is a need to better understand the requirements of the indigenous organisms we wish to stimulate. The search for bacteria with "new" catabolic capabilities (e.g., dehalorespiring bacteria, perchlorate-reducing bacteria, MTBE-degrading bacteria, etc.) should be intensified. The discovery of chloroethene-respiring bacteria has changed our thinking about bioremediation of PCE and TCE dramatically. There is the challenge of using bioremediation to remediate sites contaminated with multiple chemicals (e.g,, mixtures of CAH, BTEX, metals, and perhaps other compounds). Finally, there is a need for molecular tools that will allow us to determine what organisms are present, what are they doing and with whom they are competing, and which catabolic genes are expressed or could be expressed.

ACKNOWLEDGMENTS

Dr. Joshua Shrout made significant contributions to the section on perchlorate bioremediation. Garrett Struckhoff helped with some of the graphics.

REFERENCES

1. BE Rittmann and PL McCarty. *Environmental Biotechnology*. New York: McGraw-Hill, 2001.
2. Metcalf & Eddy, Inc. *Wastewater Engineering: Treatment and Reuse*, 4th ed, revised by G Tchobanoglous, FL Burton, and HD Stensel. New York: McGraw-Hill, 2003.
3. SC McCutcheon and JL Schnoor. *Phytoremediation — Transformation and Control of Contaminants*. Hoboken, N.J.: Wiley-Interscience, 2004.
4. JL Schnoor, LA Licht, SC McCutcheon, NL Wolfe, and LH Carriera. *Environ. Sci. Technol.* 29: 318A–323A, 1995.
5. National Research Council. *Natural Attenuation for Groundwater Remediation*. Washington, D.C.: National Academy Press, 2000.
6. National Research Council. *In Situ Bioremediation: When Does It Work*? Washington, D.C.: National Academy Press, 1993.
7. GF Parkin. In: FB Rudolph and LV McIntire, eds. *Biotechnology: Science, Engineering, and Ethical Challenges for the 21st Century*. Washington, D.C.: John Henry Press, 1996, pp. 113–128.
8. CN Sawyer, PL McCarty, and GF Parkin. *Chemistry for Environmental Engineering and Science*. New York: McGraw-Hill, 2003.
9. U.S. EPA. http://www.epa.gov/swerust1/faqs/faq9a.htm (accessed June 17, 2004).
10. MD Mikesell, JJ Kukor, and RH Olsen. *Biodegradation* 4: 249–259, 1993.
11. DT Gibson and V Subramanian. In: DT Gibson, ed. *Microbial Degradation of Organic Compounds*. New York: Marcel Dekker, 1984, pp. 181–252.
12. EA Edwards, LE Wills, M Reinhard, and D Grbić-Galić. *Appl. Environ. Microbiol.* 58: 794–800, 1992.
13. J Heider, AM Spormann, HR Beller, and F Widdel. *FEMS Microbiol. Rev.* 22: 459–473, 1999.
14. DR Lovely. *Biodegradation* 11: 107–116, 2000.
15. JD Coates, R Chakraborty, JG Lack, SM O'Conner, KA Cole, KS Bender, and LA Achenbach. *Nature* 411: 1039–1043, 2001.
16. SM Burland and EA Edwards. *Appl. Environ. Microbiol.* 65: 529–533, 1999.
17. DR Lovely, JC Woodward, and FH Chapelle. *Appl. Environ. Microbiol.* 62: 288–291, 1996.
18. EA Edwards and D Grbić-Galić. *Appl. Environ. Microbiol.* 58: 2663–2666, 1992.
19. RT Anderson and DR Lovely. *Environ. Sci. Technol.* 34: 2261–2266, 2000.
20. D Grbić-Galić and TM Vogel. *Appl. Environ. Microbiol.* 53: 254–260, 1987.
21. PJJ Alvarez, LA Cronkhite, and CS Hunt. *Environ. Sci. Technol.* 32: 509–515, 1998.
22. ME Vermace, RF Christensen, GF Parkin, and PJJ Alvarez. *Water Res.* 30: 3139–3145, 1996.
23. PJJ Alvarez and TM Vogel. *Water Sci. Technol.* 31: 15–28, 1995.
24. RL Raymond, VW Jamison, and JO Hudson. *Am. Inst. Chem. Eng. Symp. Ser.* 73: 390–404, 1976.
25. RD Norris and KD Dowd. In: PE Flathman, DE Jerger, and JH Exner, eds. *Bioremediation: Field Experience*. Boca Raton: CRC Press, 1994, pp. 457–474.
26. C Nelson, RJ Hicks, and SD Andrews. In: PE Flathman, DE Jerger, and JH Exner, eds. *Bioremediation: Field Experience*. Boca Raton: CRC Press, 1994, pp. 429–456.
27. PE Flathman, DE Jerger, and JH Exner, eds. *Bioremediation: Field Experience*. Boca Raton: CRC Press, 1994.
28. RD Norris, RE Hinchee, R Brown, PL McCarty, L Semprini, JT Wilson, DH Kampbell, M Reinhard, EJ Bouwer, RC Borden, TM Vogel, JM Thomas, and CH Ward. *Handbook of Bioremediation*. Boca Raton: CRC Press, 1994.
29. RE Hinchee, JA Kittel, and HJ Reisinger. *Applied Bioremediation of Petroleum Hydrocarbons*. Columbus, OH: Battelle Press, 1995.
30. BC Alleman and A Leeson. *In Situ and On-Site Bioremediation: Vol. 1*. Columbus, OH: Battelle Press, 1997.

31. BC Alleman and A Leeson. *In Situ Bioremediation of Petroleum Hydrocarbons and Other Organic Compounds*. Columbus, OH: Battelle Press, 1999.
32. RA Deeb, HY Hu, JR Hanson, KM Scow, and L Alvarez-Cohen. *Environ. Sci. Technol.* 35: 312–317, 2001.
33. KT Finneran and DR Lovely. *Environ. Sci. Technol.* 35: 1785–1790, 2001.
34. SE Powers, D Rice, B Dooher, and PJJ Alvarez. *Environ. Sci. Technol.* 35: 24A–30A, 2001.
35. N Lovanh, CS Hunt, and PJJ Alvarez. *Water Res.* 36: 3739–3746, 2002.
36. MLB Da Silva and PJJ Alvarez. *J. Environ. Eng.* 128: 862–867, 2002.
37. National Research Council. *Alternatives for Ground Water Cleanup*. Washington, D.C.: National Academy Press, 1994.
38. PJ Squillace, MJ Moran, WW Lapham, CV Price, RM Clawges, and JS Zogorski. *Environ. Sci. Technol.* 33: 4176–4187, 1999.
39. S Li and LP Wackett. *Biochem. Biophys. Res. Commun.* 185: 443–451, 1992.
40. L. Semprini. *Curr. Opin. Biotechnol.* 8: 296–308, 1997.
41. H-L Chang and L Alvarez-Cohen. *Environ. Sci. Technol.* 29: 2357–2367, 1995.
42. JE Anderson and PL McCarty. *Appl. Environ. Microbiol.* 63: 687–693, 1997.
43. GF Parkin. *Water Environ. Res.* 71: 1158–1164, 1999.
44. PL McCarty and JT Wilson. In U.S. EPA R-92/126. *Bioremediation of Hazardous Wastes*. Cincinnati, OH: U.S. EPA Center for Environment Research Information, 1992, pp. 47–50.
45. TD DiStefano, JM Gossett, and SH. Zinder. *Appl. Environ. Microbiol.* 58: 3622–3629, 1992.
46. DL Freedman and JM Gossett. *Appl. Environ. Microbiol.* 55: 2144–2151, 1989.
47. WP de Bruin, MJJ Kotterman, MA Posthumus, G Schraa, and AJB Zehnder. *Appl. Environ. Microbiol.* 58: 1996–2000, 1992.
48. C Holliger, GSchraa, AJM Stams, and AJB Zehnder. *Appl. Environ. Microbiol.* 59: 2991–2997, 1993.
49. H Scholz-Muramatsu, A Neumann, M Messmer, E Moore, and G Diekert. *Arch. Microbiol.* 163: 48–56, 1995.
50. X Maymó-Gatell, Y-t Chien, JM Gossett, and SH Zinder. *Science* 276: 1568–1571, 1997.
51. X Maymó-Gatell, I Nijenhuis, and SH Zinder. *Environ. Sci. Technol.* 35: 516–521, 2001.
52. AM Cupples, AM Spormann, and PL McCarty. *Appl. Environ. Microbiol.* 69: 953–959, 2003.
53. J He, KM Ritalahti, MR Aiello, and FE Löffler. *Appl. Environ. Microbiol.* 69: 996–1003, 2003.
54. B Sun, BM Griffin, H L Ayala-del-Rio, S A Hashsham, and JM Tiedje. *Science* 298: 1023–1025, 2002.
55. NV Coleman, TE Mattes, JM Gossett, and JC Spain. *Appl. Environ. Microbiol.* 68: 6162–6171, 2002.
56. L Alvarez-Cohen and PL McCarty. *Environ. Sci. Technol.* 25: 1387–1392, 1991.
57. Y Sung, KM Ritalahti, RA Sanford, JW Urbance, SJ Flynn, JM Tiedje, and FE Löffler. *Appl. Environ. Microbiol.* 69: 2964–2974, 2003.
58. J He, Y Sung, ME Dollhopf, BZ Fathpure, JM Tiedje, and FE Löffler. *Environ. Sci. Technol.* 36: 3945–3952, 2002.
59. FE Löffler, Q Sun, J Li, and JM Tiedje. *Appl. Environ. Microbiol.* 66: 1369–1374, 2000.
60. DE Fennell and JM Gossett. *Environ. Sci. Technol.* 32: 2450–2460, 1998.
61. Y Yang and PL McCarty. *Environ. Sci. Technol.* 32: 3591–3597, 1998.
62. DT Adamson and GF Parkin. *Bioremed. J.* 5: 51–62, 2001.
63. Y Yang and PL McCarty. *Bioremed. J.* 4: 125–133, 2000.
64. CS Carr and JB Hughes. *Environ. Sci. Technol.* 32: 1817–1824, 1998.
65. DT Adamson and GF Parkin. *Environ. Sci. Technol.* 34: 1959–1965, 2000.
66. DP Leigh, CD Johnson, LA Bienkowski, and S Granade. In: GB Wickramanayake, AR Gavaskar, BC Alleman, and VS Magar, eds. *Bioremediation and Phytoremediation of Chlorinated and Recalcitrant Compounds*. Columbus, OH: Battelle Press, 2000, pp. 229–235.
67. Y Yang and PL McCarty. *Environ. Sci. Technol.* 34: 2979–2984, 2000.
68. N Cope and JB Hughes. *Environ. Sci. Technol.* 35: 2014–2021.
69. PE Haas, P Cork, and CE Aziz. In: GB Wickramanayake, AR Gavaskar, BC Alleman, and VS Magar, eds. *Bioremediation and Phytoremediation of Chlorinated and Recalcitrant Compounds*. Columbus, OH: Battelle Press, 2000, pp. 71–76.

70. CL Lutes, DS Liles, SS Suthersan, F Lenzo, M Hansen, FC Payne, JF Burdick, and D Vance. *Environ. Sci. Technol.* 37: 2618–2619, 2003.
71. MA Hansen, J Burdick, FC Lenzo, S Suthersan. In: GB Wickramanayake, AR Gavaskar, BC Alleman, and VS Magar, eds. *Bioremediation and Phytoremediation of Chlorinated and Recalcitrant Compounds*. Columbus, OH: Battelle Press, 2000, pp. 263–270.
72. KJ Boulicault, RE Hinchee, TH Wiedermeier, SW Hoxworth, TP Swingle, E Carver, and PE Haas. In: G.B Wickramanayake, AR Gavaskar, BC Alleman, and VS Magar, eds. *Bioremediation and Phytoremediation of Chlorinated and Recalcitrant Compounds*. Columbus, OH: Battelle Press, 2000, pp. 1–7.
73. SL Boyle, VB Dick, MN Ramsdell, and TM Caffoe. In: GB Wickramanayake, AR Gavaskar, BC Alleman, and VS Magar, eds. *Bioremediation and Phytoremediation of Chlorinated and Recalcitrant Compounds*. Columbus, OH: Battelle Press, 2000, pp. 255–262.
74. CJ Newell, PE Haas, JB Hughes, and T Khan. In: GB Wickramanayake, AR Gavaskar, BC Alleman, and VS Magar, eds. *Bioremediation and Phytoremediation of Chlorinated and Recalcitrant Compounds*. Columbus, OH: Battelle Press, 2000, pp. 31–37.
75. SJ Flynn, FE Löffler, and JM Tiedje. *Environ. Sci. Technol.* 34: 1056–1061, 2000.
76. DE Ellis, EJ Lutz, J M Odom, RJ Buchanan, Jr, CL Bartlett, MD Lee, MR Harkness, and KA Deweerd. *Environ. Sci. Technol.* 24: 2254–2260, 2000.
77. DW Major, ML McMaster, EE Cox, EA Edwards, SM Dworatzek, ER Hendrickson, MG Starr, JA Payne, and LW Buonamici. *Environ. Sci. Technol.* 36: 5106–5116, 2002.
78. DT Adamson, JM McDade, and JB Hughes. *Environ. Sci. Technol.* 37: 2525–2533, 2003.
79. ZC Haston and PL McCarty. *Environ. Sci. Technol.* 33: 223–226, 1999.
80. AM Cupples, AM Spormann, and PL McCarty. *Environ. Sci. Technol.* 38: 1102–1107, 2004.
81. PV Roberts, GD Hopkins, D M Mackay, and L Semprini. *Ground Water* 28: 591–604, 1990.
82. L Semprini, PV Roberts, GD Hopkins, and PL McCarty. *Ground Water* 28: 715–727, 1990.
83. L Semprini, GD Hopkins, PV Roberts, D Grbić-Galić, and PL McCarty. *Ground Water* 29: 239–250, 1991.
84. L Semprini and PL McCarty. *Ground Water* 29: 365–374, 1991.
85. L Semprini and PL McCarty. *Ground Water* 30: 37–44, 1992.
86. GD Hopkins and PL McCarty. *Environ. Sci. Technol.* 29: 1628–1637, 1995.
87. PL McCarty, MN Goltz, GD Hopkins, ME Dolan, JP Allan, BT Kawakami, and TJ Carrothers. *Environ. Sci. Technol.* 32: 88–100, 1998.
88. GB Wickramanayake, AR Gavaskar, BC Alleman, and VS Magar, eds. *Bioremediation and Phytoremediation of Chlorinated and Recalcitrant Compounds*. Columbus, OH: Battelle Press, 2000.
89. AR Gavaskar and ASC Chen, eds. *Remediation of Chlorinated and Recalcitrant Compounds — 2002*. Proceedings of the Third International Conf. on Remediation of Chlorinated and Recalcitrant Compounds, Monterey, CA. Columbus, OH: Battelle Press, 2002.
90. VS Magar, DE Fennell, JJ Morse, BC Alleman, and A Leeson, eds. *Anaerobic Degradation of Chlorinated Solvents*. Columbus, OH: Battelle Press, 2001.
91. Federal Remediation Technologies Roundable, http://www.frtr.gov/index.htm.
92. RE Beeman, JE Howell, SH Solazar, and EA Buttram. In: RE Hinchee, A Leeson, L Semprini, and SK Ong, eds. *Bioremediation of Chlorinated and Polycyclic Aromatic Hydrocarbons*. Boca Raton, FL: Lewis Publishers, p. 14, 1994.
93. RE Beeman. U.S. Patent 5 277 815, 1994.
94. ER Hendrickson, JA Payne, RM Young, MG Starr, MP Perry, S Fathnestock, DE Ellis, and RC Ebersole. *Appl. Environ. Microbiol.* 68: 485–495, 2002.
95. National Research Council. *Environmental Cleanup at Naval Facilities*. Washington, D.C.: National Academy Press, 2003.
96. U.S. EPA. FRTR cost and performance remediation case studies and related information. EPA-542-C-00-001. Washington, D.C.: Federal Remediation Technologies Roundtable, 2000.
97. CS Carr, S Garg, and JB Hughes. *Environ. Sci. Technol.* 34: 1088–1094, 2000.
98. HF Stroo, M Unger, CH Ward, MC Kavanaugh, C Vogel, A Leeson, JA Marqusee, and BP Smith. *Environ. Sci. Technol.* 37: 224A–230A, 2003.
99. ET Urbansky. *Bioremed. J.* 2: 81–95, 1998.
100. R Renner. *Environ. Sci. Technol.* 33: 110A–111A, 1999.

101. https://www.denix.osd.mil/denix/Public/Library/Water/Perchlorate/releases.html. U.S. Perchlorate Releases, accessed Jan. 5, 2004.
102. BE Logan. *Environ. Sci. Technol.* 35: 483A–487A, 2001.
103. U.S. EPA. http://www.epa.gov/safewater/ccl/perchlor/perchlo.html (accessed February 19, 2002).
104. RJ Brechner, GD Parkhurst, WO Humble, MB Brown, and WH Herman. *J. Occup. Environ. Med.* 42: 777–782, 2000.
105. JC Siglin, DR Mattie, DE Dodd, PK Hildebrandt, and WH Baker. *Toxicol. Sci.* 57: 61–74, 2000.
106. PN Smith, CW Theodorakis, TA Anderson, and RJ Kendall. *Ecotoxicology* 10: 305–313, 2001.
107. California Dept. of Health Services. http://www.dhs.ca.gov/ps/ddwem/chemicals/perchl/actionlevel.htm (accessed July 2, 2002).
108. DC Herman and WT Frankenberger. *J. Environ. Qual.* 28: 1018–1024, 1999.
109. BE Logan. *Bioremed. J.* 2: 69–79, 1998.
110. W Wallace, S Beshear, D Williams, S Hospadar, and M Owens. *J. Ind. Microbiol. Biotechnol.* 20: 126–131, 1998.
111. R Nerenberg, BE Rittmann, and I Najm. *J. Am. Water Works Assoc.* 94: 103–114, 2002.
112. E Hackenthal. *Biochem. Pharmacol.* 13: 195–206, 1965.
113. EH Bryan and GA Rohlich. *Sewage Ind. Wastes* 26: 1316–1324, 1954.
114. GN de Groot and AH Stouthamer. *Arch. Microbiol.* 66: 220–233, 1969.
115. F Pichinoty. *Arch. Microbiol.* 66: 315–320, 1969.
116. VI Romanenko, VN Koren'kov, and SI Kuznetsov. *Mikrobiologiia* 45: 204–209, 1976.
117. GB Rikken, AGM Kroon, and CG vanGinkel. *Appl. Microbiol. Biotech.* 45(3): 420–426, 1996.
118. W Wallace, T Ward, A Breen, and H Attaway. *J. Ind. Microbiol.* 16: 68–72, 1996.
119. JD Coates, U Michaelidou, RA Bruce, SM O'Connor, JN Crespi, and LA Achenbach. *Appl. Environ. Microbiol.* 65: 5234–5241, 1999.
120. BE Logan, H S Zhang, P Mulvaney, MG Milner, IM Head, and RF Unz. *Appl. Environ. Microbiol.* 67: 2499–2506, 2001.
121. LA Achenbach, U Michaelidou, RA Bruce, J Fryman, and JD Coates. *Int. J. Syst. Evol. Microbiol.* 51: 527–533, 2001.
122. JD Shrout. Characteristics and Electron Donor Requirements of Perchlorate Degradation by Mixed and Pure-Culture Bacteria. Ph.D. Dissertation. University of Iowa, Iowa City, IA, 2002.
123. K Stenklo, HD Thorell, H Bergius, R Aasa, and T Nilsson. *J. Biol. Inorganic Chem.* 6: 601–607, 2001.
124. KS Bender, SA O'Connor, R Chakraborty, JD Coates, and LA Achenbach. *Appl. Environ. Microbiol.* 68: 4820–4826, 2002.
125. SK Chaudhuri, SM O'Connor, RL Gustavson, LA Achenbach, and JD Coates. *Appl. Environ. Microbiol.* 68: 4425–4430, 2002.
126. A Wolterink, AB Jonker, SWM Kengen, and AJM Stams. *Int. J. Syst. Evol. Microbiol.* 52: 2183–2190, 2002.
127. KS Bender, R Chakraborty, SM Belchik, JD Coates, and LA Achenbach. The Genetics of (Per)chlorate Reduction. American Society for Microbiology 103rd General Meeting, Washington, D.C., May 18–22, 2003.
128. SW Kengen, GB Rikken, WR Hagen, CG van Ginkel, and AJ Stams. *J. Bacteriol.* 181: 6706–6711, 1999.
129. TL Gilbin, DC Herman, and WT Frankenberger. *J. Environ. Qual.* 29: 1057–1062, 2000.
130. U.S. EPA. http://clu-in.org/contaminantfocus/default.focus/sec/perchlorate/cat/Treatment_Technologies/ (accessed June 14, 2004).
131. California Environmental Protection Agency, Office of Pollution Prevention and Technology Development. Perchlorate Contamination Treatment Alternatives, Draft Report. 2004.
132. JP Miller and BE Logan. *Environ. Sci. Technol.* 34: 3018–3022, 2000.

19 Bioengineering for the *In Situ* Remediation of Metals

Jennifer L. Nyman, Sarah M. Williams, and Craig S. Criddle
Department of Civil and Environmental Engineering, Stanford University

CONTENTS

19.1 INTRODUCTION AND BACKGROUND

19.1.1 Microbial Bioremediation of Metals and Metalloids

The contamination of soil and groundwater by metals and metalloids* represents a significant threat to drinking water supplies and human health. Methods currently employed to manage such contamination — excavation, pump-and-treat, and immobilization using zero-valent iron as a reducing agent — are often expensive, have restricted ranges of application, and generate large volumes of hazardous waste for transport and disposal. As such, there is considerable interest in alternative treatment techniques for metal contamination. Use of bioremediation to immobilize metals *in situ* is attractive because microorganisms can transform metals into sparingly soluble forms, thus removing them from groundwater and alleviating the need for waste disposal. Microbial populations with the requisite capabilities are ubiquitous, so remediation efforts can usually focus on the addition of chemicals to stimulate the growth and activity of extant populations.

Though bioremediation has proven effective for the treatment of chlorinated solvents and hydrocarbons in groundwater, the immobilization of metals presents unique challenges. The fate of metals depends upon an extensive array of variables, including pH, types and concentration of ligands and co-contaminants, sorptive and diffusive processes, and microbial ecology and physiology. Manipulating these variables should result in the enhancement of the rate or extent of metal immobilization, but the situation is sufficiently complex to warrant careful experimentation. Bench-scale studies are a key element of initial characterization and planning for metal remediation. In scaling up for field-scale applications, chemical delivery and mixing issues become critically important, often necessitating pilot studies and groundwater flow modeling.

Although each metal-contaminated site has its own challenges, some general principles can be defined for treatment design, and we will elucidate many of those in this chapter. We first provide background on microbial metabolism and the types of species that mediate the immobilization of metals. We then address stoichiometry and review thermodynamic factors influencing metal transformation. A description of kinetic models follows, and we consider how sorption and desorption influence metal bioavailability and overall removal rates. We conclude with a discussion of chemical delivery and design considerations for the field.

19.1.2 Background

The physicochemical properties of metals are altered by bacteria via oxidation and reduction reactions, and these are initiated in the electron transfer reactions of microbial metabolism (illustrated in Figure 19.1). A substrate, typically an organic compound or hydrogen, is the source of electrons. This species is the *electron donor*, and its concentration is represented with the symbol S (for substrate). During growth, a microbial cell generates energy by oxidizing the electron donor and reducing a *primary electron acceptor*. In this chapter, we will represent the concentration of primary electron acceptor as A, and the minimum fraction of electrons transferred to it as f_e^o. A maximum fraction of electrons obtained from the donor used for the generation of cellular material is f_s^o, and f_e^o and f_s^o sum to one. In some cases, electrons can be transferred to other electron acceptors — *secondary electron acceptors*. The fraction transferred to such acceptors, f_o, decreases either f_s (the actual fraction of electrons used for cellular material), f_e (the actual fraction of electrons sent

*To avoid excessive repetition, future references to "metals" will implicitly include metalloids.

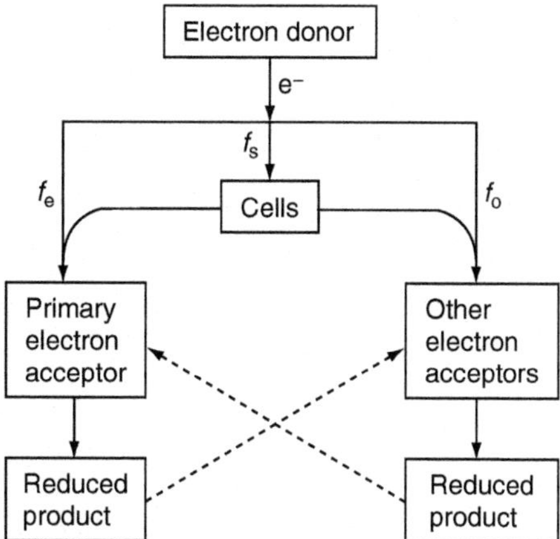

FIGURE 19.1 Biotic and abiotic pathways of electron flow in microbial systems. Solid lines indicate the flow of electrons by microbially mediated reactions, and dotted lines depict electron transfer by abiotic reactions. f_e is fraction of electrons sent to the primary electron acceptor, f_s is fraction of electrons used for cell synthesis, and f_o is fraction of electrons sent to other electron acceptors.

to the acceptor), or both, because an electron balance ($f_s + f_e + f_o = 1$) must be maintained. The reduced products of microbial reactions may also be capable of abiotically reducing oxidants in the system (Figure 19.1).

Bacterial species that reduce Fe(III) or other metals as primary or secondary electron acceptors are collectively referred to as dissimilatory metal-reducing bacteria (DMRB). As shown in Table 19.1, many metals are present as relatively insoluble forms in their lowest oxidation states, and thus microbial reduction represents a potential means of immobilizing metals in subsurface environments. Contaminants amenable to reduction by DMRB include uranium, cobalt, chromium, and technicium, and some DMRB are able to conserve energy for growth from the reduction of these metals [6–9]. The best-studied DMRB are members of the genera *Shewanella* and *Geobacter*.

Some sulfate-reducing bacteria (SRB) can also reduce metals as secondary or even primary electron acceptors. Primary electron acceptors for SRB include uranium, technicium, and chromium [8, 10], and certain SRB are able to reduce cobalt, selenium, and molybdenum as secondary electron acceptors [11–13]. SRB include the Gram-negative mesophilic bacteria (e.g., *Desulfobulbus*, *Delsulfovibrio*, and *Desulfobacter*) and Gram-positive spore formers (*Desulfotomaculum* and *Desulfosporosinus*). In addition to reductive immobilization, SRB can indirectly precipitate metals through the production of sulfide, and metal sulfides are often of limited solubility. Contaminants that can be removed as sulfides include cadmium, cobalt, copper, lead, nickel, and zinc.

For this chapter, we focus primarily on the reductive processes of anaerobic SRB and DMRB, as these reactions have been studied in the most detail and are of greatest potential importance to *in situ* metal bioremediation. We recognize that other promising approaches are available for microbial metal immobilization, including the coprecipitation of metals with microbially produced phosphate or carbonate [14, 15], and the basic principles elucidated herein still apply to such strategies. Though our focus is on the *in situ* bioremediation of metals, many of the same processes can be implemented ex situ.

TABLE 19.1
Metals and Metalloids that can be Reduced to Insoluble Forms and Their Regulatory Limits

Metal	Oxidation State	Important Soluble Species	Representative "Insoluble" Forms	EPA Maximum Contaminant Levels in U.S. Water (mg/l)	Common Sources
As[a]	–II		AsS	0.01	Erosion of natural deposits; runoff from orchards, runoff from glass and electronics production
	0		FeAsS, As		
	III	H_2AsO_3, $H_2AsO_3^-$, $HAsO_3^{2-}$, AsO_3^{3-} $HAsO_3^{2-}$, AsO_3^{3-}	As_2O_3		
	V	H_3AsO_4, $H_2AsO_4^-$, $HAsO_4^{2-}$, AsO_4^{3-} $HAsO_4^{2-}$, AsO_4^{3-}			
Cr	III		Cr_2O_3	0.1	Steel and pulp mills, erosion of natural deposits
	VI	H_2CrO_4, $HCrO_4^-$, CrO_4^{2-}			
Mo	IV		MoS_2, MoO_2	0.1	Natural deposits, mining wastes
	VI	H_2MoO_4, $HMoO_4^-$, MoO_4^{2-}	MoO_3		
Se	–II	H_2Se, HSe^-, Se^{2-}		0.05	Refineries, natural deposits, mines
	0		Se		
	IV	H_2SeO_3, $HSeO_3^-$, SeO_3^{2-}	SeO_2		
	VI	H_2SeO_4, $HSeO_4^-$, SeO_4^{2-}			
Tc	IV		$TcO_2 \cdot 2H_2O$	4 mrem/year for beta emitters, such as Tc-99	Radioactive waste produced by the decay of Zr-99, a product of U and Pu fission
	VII	TcO_4^-			
U	IV		UO_2, $USiO_4$	0.03	Mine tailings, atomic bomb fabrication sites, erosion of natural sources
	VI	UO_2^{2-}, $UO_2\,(CO_3)$, $UO_2\,(CO_3)_2^{2-}$, $UO_2\,(CO_3)_3^{4-}$			
V	II		VO, V_2O_4	No federal MCL, 0.05 (Minnesota limit)	Shale deposits
	V	VO_2^+, HVO_3, VO_3^-,H_3VO_4, $H_2\,VO_4^-$, HVO_4^{2-}, VO_4^{3-}	V_2O_5		

Adapted from S Simonton, M Dimsha, B Thomson, LL Barton, and G Cathey. Long-term stability of metals immobilized by microbial reduction. Conference on Hazardous Waste Research, Denver, CO, 2000, pp. 394–403.

[a]The normal valence states of arsenic are III and V, and As(III) is more mobile and toxic than As(V) [2]. Dissimilatory arsenate-reducing bacteria convert As(V) to As(III) (arsenite) [3–5], although other bacteria are known to reduce As(V) to As(–III) without the conservation of energy [2].

19.2 THERMODYNAMICS AND STOICHIOMETRY OF MICROBIAL GROWTH

19.2.1 The Thermodynamic Approach

The energy needed for cell growth and maintenance is obtained through either respiration or fermentation. During respiration, electrons are shuttled from the electron donor through one- and two-electron carriers in an electron transport chain to a terminal electron acceptor. The electron transfer is coupled to the creation of a proton gradient across the cell membrane, and this gradient can be tapped to produce the energy carrier adenosine triphosphate (ATP). This strategy for ATP production is termed oxidative phosphorylation. During fermentation, oxidation of the electron donor yields protons and partially oxidized organics that serve as terminal electron acceptors, and ATP is produced by substrate-level phosphorylation. ATP, made available through respiration or fermentation, is used for the synthesis and maintenance of cells.

Bacteria can use a variety of electron donors and acceptors for the generation of energy; some of the half-reactions for electron acceptors relevant to microbial metal remediation are given in Table 19.2. The amount of free energy made available by a given transformation can be calculated by coupling a half-reaction for oxidation of an electron donor with the half-reaction for reduction of an electron acceptor. Reactions are thermodynamically feasible if their computed free energy change is negative, in which case the redox potential (E_h) is positive.

TABLE 19.2
Redox Reactions for Microbial Electron Acceptors

Reaction	$\Delta G'^{a}$ (kJ/eq)	$E_h'^{a}$ (V)	Ref.
$MnO_2\ (s) + 4H^+ + 2e^- \longrightarrow Mn^{2+} + 2H_2O$	−53.01	0.549	[16]
$Mn_3O_4\ (s) + 8H^+ + 2e^- \longrightarrow 3Mn^{2+} + 4H_2O$	−58.14	0.603	[16]
$\gamma\text{-}MnOOH(s) + 3H^+ + e^- \longrightarrow Mn^{2+} + 2H_2O$	−53.38	0.555	[16]
$Fe^{3+} + e^- \longrightarrow Fe^{2+}$	−79.80	0.827	[16]
$FeOOH(s) + 3H^+ + e^- \longrightarrow Fe^{2+} + 2H_2O$	17.1	−0.177	[16]
$Fe(OH)_3(s) + 3H^+ + e^- \longrightarrow Fe^{2+} + 3H_2O$	1.14	−0.012	[16]
$Fe_2O_3\ (s) + 6H^+ + 2e^- \longrightarrow 2Fe^{2+} + 3H_2O$	14.82	−0.154	[16]
$Co(III) - EDTA^- + e^- \longrightarrow Co(II) - EDTA^{2-}$	−42.36	0.439	[17]
$HCrO_4^- + 7H^+ + 3e^- \longrightarrow Cr^{3+} + 4H_2O$	−42.07	0.436	[17]
$HCrO_4^- + 4H^+ + 3e^- \longrightarrow Cr(OH)_3(s) + H_2O$	−46.93	0.486	[16]
$UO_2\ (CO_3)_3^{4-} + 3H^+ + 2e^- \longrightarrow UO_2(s) + 3HCO_3^-$	−8.19[b]	0.085[b]	[18, 19]
$TcO_4^- + 4H^+ + 3e^- \longrightarrow TcO_2(s) + 2H_2O$	−10.48	0.109	[17]
$HAsO_4^{2-} + 4H^+ + 2e^- \longrightarrow HAsO_2 + 2H_2O$	−8.15	0.085	[17]
$SeO_3^{2-} + 6H^+ + 4e^- \longrightarrow Se(s) + 3H_2O$	−19.92	0.207	[17]
$SeO_4^{2-} + 2H^+ + 2e^- \longrightarrow SeO_3^{2-} + H_2O$	−49.35	0.512	[17]
$SO_4^{2-} + 10H^+ + 8e^- \longrightarrow H_2S + 4H_2O$	19.52	−0.202	[16]
$SO_4^{2-} + 8H^+ + 8e^- \longrightarrow S^{2-} + 4H_2O$	18.78	−0.195	[17]
$2NO_3^- + 12H^+ + 10e^- \longrightarrow N_2(g) + 6H_2O$	−70.19	0.727	[17]
$O_2(g) + 4H^+ + 4e^- \longrightarrow 2H_2O$	−77.69	0.805	[16]

[a]The following conditions were assumed for calculations: concentration of soluble, reduced species: 10^{-5} *M*; concentration of soluble, oxidized species: 10^{-4} *M*; concentration of bicarbonate: 30 m*M*; activity of solid phases: 1; concentration of H^+: 10^{-7} *M*; partial pressure of N_2: 0.78 atm; partial pressure of O_2: 0.21 atm; partial pressure of H_2: 10^{-4} atm.

[b]This value was calculated from data.

19.2.2 Stoichiometry of Microbial Reactions

Stoichiometry concerns the molar relationships among reactants and products in balanced reactions. Quantitative stoichiometric relationships enable estimates of biomass produced, electron donor required, reduced solid produced, and alkalinity or acidity needed for pH control.

An overall reaction to describe growth and the consumption of electron donors and acceptors combines three half-reactions: one for the reduction of the electron acceptor (R_a), one for the oxidation of the electron donor (R_d), and one for the synthesis of cells (R_c) [20]. For cells oxidizing acetate with soluble Fe(III)-citrate as an electron acceptor, the half-reactions are:

$$R_a\text{: Fe(III)}-\text{citrate} + e^- \longleftrightarrow \text{Fe(II)}-\text{citrate}^-$$

$$R_d\text{: } \tfrac{1}{8}\,CO_2 + \tfrac{1}{8}\,HCO_3^- + H^+ + e^- \longleftrightarrow \tfrac{1}{8}\,CH_3COO^- + \tfrac{3}{8}\,H_2O$$

$$R_c\text{: } \tfrac{1}{5}\,CO_2 + \tfrac{1}{20}\,HCO_3^- + \tfrac{1}{20}\,NH_4^+ + H^+ + e^- \longleftrightarrow \tfrac{1}{20}\,C_5H_7O_2N + \tfrac{9}{20}\,H_2O$$

The empirical formula for cells used in the synthesis equation, $C_5H_7O_2N$, is an approximation of the elemental content of a microbe and represents only its four major elements. It was developed for aerobic degradation by a mixed culture and is generally sufficient for practical purposes. Where possible, however, a representative formula should be determined empirically for analysis of a specific organism or community. The above equation assumes ammonium as the nitrogen source for growth. Nitrate could also be used, in which case the half-reaction for cell synthesis is:

$$R_c\text{: } \tfrac{5}{28}\,CO_2 + \tfrac{1}{28}\,NO_3^- + \tfrac{29}{28}\,H^+ + e^- \longleftrightarrow \tfrac{1}{28}\,C_5H_7O_2N + \tfrac{11}{28}\,H_2O$$

For each electron equivalent of electron donor utilized, f_e of the electrons is sent to the electron acceptor and f_s is used for synthesis, so the overall equation (R) is calculated as:

$$R = f_e R_a + f_s R_c - R_d$$

Assuming f_s is equal to 0.316 (see below for the calculation), the overall equation is:

$$f_e R_a\text{: } 0.684\text{Fe(III)}-\text{citrate} + 0.684e^- \longleftrightarrow 0.684\text{Fe(II)}-\text{citrate}^-$$

$$+f_s R_c\text{: } 0.063CO_2 + 0.016HCO_3^- + 0.016NH_4^+ + 0.316H^+ + 0.316e^- \longleftrightarrow 0.016C_5H_7O_2N + 0.142H_2O$$

$$-R_d\text{: } 0.125CH_3COO^- + 0.375H_2O \longleftrightarrow 0.125CO_2 + 0.125HCO_3^- + H^+ + e^-$$

$$R\text{: } 0.684\text{Fe(III)}-\text{citrate} + 0.125CH_3COO^- + 0.016NH_4^+ + 0.233H_2O \longleftrightarrow 0.684\text{Fe(II)}-\text{citrate}^- + 0.016C_5H_7O_2N + 0.062CO_2 + 0.109HCO_3^- + 0.684H^+$$

If more than one electron acceptor is used for energy generation, the relative proportion of electron equivalents transferred to each acceptor must be calculated. The proportion is represented in fractional form as e_{ai}. This fraction is a multiplier for each half-reaction, and all the half-reaction reduction equations added together give the half-reaction for the electron acceptor, R_a:

$$R_a = \sum_{i=1}^{n} e_{ai} R_{ai}$$

$$e_{ai} = \frac{\text{equiv}_{ai}}{\sum_{j=1}^{n} \text{equiv}_{aj}} \quad \text{and} \quad \sum_{i=1}^{n} e_{ai} = 1$$

Here, equiv_{ai} represents the electron equivalents sent to a given electron acceptor.

19.2.3 Microbial Energetics and Yield

Because the amount of energy available from the coupling of redox reactions varies considerably between different electron acceptors and donors, the amount of growth resulting from the oxidation of a given amount of electron donor varies. Microorganisms can direct a high percentage of the substrate electrons to cell growth (i.e., have a high f_s) when a large amount of energy is obtained by the coupled oxidation of an electron donor and reduction of a terminal electron acceptor. Observed yield, Y_{obs}, represents the quantity of biomass produced per unit substrate oxidized, per unit electron acceptor reduced, or per electron equivalent. When the substrate, acceptor, and other nutrients are not growth-limiting, observed yield approaches the theoretical yield — a value calculated by neglecting decay and maintenance considerations.

Rittmann and McCarty [20] present an approach for estimating theoretical yield from reaction energetics. First, the energy required to convert a carbon source to common organic intermediates, ΔG_p, is calculated. Pyruvate is used as a representative intermediate, and ΔG_p is the difference between the free energy of the pyruvate half-reaction, 35.09 kJ, and the free energy of the half-reaction for the carbon source:

$$\Delta G_c^{o'} = \text{free energy of the half-reaction for the carbon source}$$

$$\Delta G_p = 35.09\frac{\text{kJ}}{\text{e}^-\text{eq}} - \Delta G_c^{o'}$$

For heterotrophic bacteria, the carbon source is typically the electron donor, so $\Delta G_c^{o'}$ is the free energy of the half-reaction for the donor. When acetate is used as both the electron donor and carbon source:

$$\begin{aligned}\Delta G_p &= 35.09\frac{\text{kJ}}{\text{e}^-\text{eq}} - \Delta G_c^{o'} \\ &= 35.09\frac{\text{kJ}}{\text{e}^-\text{eq}} - 25.86\frac{\text{kJ}}{\text{e}^-\text{eq}} \\ &= 9.23\frac{\text{kJ}}{\text{e}^-\text{eq}}\end{aligned}$$

Pyruvate carbon must then be converted to cellular carbon, and this energy requirement, ΔG_{pc}, is 18.8 kJ/e$^-$ eq if ammonium is used as the nitrogen source and 13.4 kJ/e$^-$ eq if nitrate is the N source [20].

Energy is always lost in the transfer of electrons. The energy-transfer efficiency, ε, is used to account for this loss in the calculation of the overall energy required for cell synthesis, ΔG_s:

$$\Delta G_s = \frac{\Delta G_p}{\varepsilon^n} + \frac{\Delta G_{pc}}{\varepsilon}$$

McCarty [21] estimated a typical value of 0.6 for energy-transfer efficiency by comparing values of calculated yield versus observed yield for several aerobic and anaerobic microbial processes. When the comparison was limited to reactions relevant to the bioreduction of metals, we obtained a value of 0.44 for energy-transfer efficiency. The statistical deviation from this value is significant, indicating more data are needed describing the growth of metal-reducing bacteria, but it may be closer to the actual efficiency of metal reduction processes than 0.6. The published yield information used in this calculation is listed in Table 19.3.

The exponent n in the above equation accounts for the fact that energy is gained from some conversions of the carbon source to pyruvate, while from others it is lost. n is therefore 1 if ΔG_p is positive and -1 if ΔG_p is negative. For the example of acetate as a carbon source and ammonium as a nitrogen source:

$$\begin{aligned}\Delta G_s &= \frac{\Delta G_p}{\varepsilon^n} + \frac{\Delta G_{pc}}{\varepsilon}\\ &= \frac{9.23}{0.44^1} + \frac{18.8}{0.44}\\ &= 63.7\frac{\text{kJ}}{\text{e}^-\text{eq}}\end{aligned}$$

Theoretical yield neglects energy requirements for maintenance, and under these assumptions f_s is at its maximum, f_s°, and f_e is at its minimum, f_e°. The energy required for synthesis, $f_s^\circ\Delta G_s$, must be balanced by the energy released by the reduction of one or more electron acceptors. This quantity is the product of the energy available per mole electron acceptor reduced (ΔG_r), the energy-transfer efficiency, and the fraction of electrons sent to the electron acceptor. An equation for f_s° is obtained from this balance and the relationship between f_s° and f_e°:

$$\begin{aligned}f_s^\circ\Delta G_s &= -f_e^\circ\varepsilon\Delta G_r\\ f_s^\circ\Delta G_s &= -(1-f_s^\circ)\varepsilon\Delta G_r\\ f_s^\circ[\Delta G_s - \varepsilon\Delta G_r] &= -\varepsilon\Delta G_r\\ f_s^\circ &= \frac{-\varepsilon\Delta G_r}{-\varepsilon\Delta G_r + \Delta G_s}\end{aligned}$$

The energy available from a given electron donor and acceptor combination is the sum of the free energies of their half-reactions, so, for our example,

$$\Delta G_r = -25.86\,\frac{\text{kJ}}{\text{e}^-\text{eq}} + \left(-41.04\frac{\text{kJ}}{\text{e}^-\text{eq}}\right) = -66.9\frac{\text{kJ}}{\text{e}^-\text{eq}}$$

$$f_s^\circ = \frac{-\varepsilon\Delta G_r}{-\varepsilon\Delta G_r + \Delta G_s}$$

$$f_s^\circ = \frac{-0.44\left(-66.9\frac{\text{kJ}}{\text{e}^-\text{eq}}\right)}{-0.44\left(-66.9\frac{\text{kJ}}{\text{e}^-\text{eq}}\right) + 63.7\frac{\text{kJ}}{\text{e}^-\text{eq}}}$$

$$f_s^\circ = 0.316$$

$$f_e^\circ = 1 - f_s^\circ = 0.684$$

A balanced stoichiometric equation can then be written with these values of f_s° and f_e° (see example above).

Reactions useful for metal remediation are often electron-accepting, and for these processes considering yield with respect to the amount of electron acceptor reduced is useful. Y_A is yield in terms of biomass produced per electron equivalents sent to the electron acceptor. The theoretical yield can be calculated from f_s° with a unit conversion:

$$Y_A = \frac{f_s^\circ\,\frac{\text{e}^-\text{eq for biomass}}{\text{e}^-\text{eq from donor}}}{f_e^\circ\,\frac{\text{e}^-\text{eq to acceptor}}{\text{e}^-\text{eq to donor}}} \times \frac{5.65\text{ g VSS}}{\text{e}^-\text{eq for biomass}}$$

For our example,

$$Y_A = \frac{0.316 \frac{e^-\text{eq for biomass}}{e^-\text{eq from donor}}}{0.684 \frac{e^-\text{eq to acceptor}}{e^-\text{eq from donor}}} \times \frac{5.65\,\text{g VSS}}{e^-\text{eq for biomass}} = 2.61 \frac{\text{g VSS}}{e^-\text{eq}}$$

The above equations are all computed using free energy of formation values for reactants and products under standard conditions and pH 7. Such estimates should be used with caution. Standard conditions are never actually observed during a reaction, as reactant concentrations continuously decrease and product concentrations continuously increase until the system equilibrates.

Table 19.3 compares observed yield values to theoretical values for bacteria growing on a variety of electron acceptors and donors relevant to metal remediation. Generally, bacteria using electron acceptors that provide the least energy (such as sulfate) have the lowest theoretical and observed yield. Reported values for observed yield for a given electron donor–acceptor combination also vary between studies, which may be the result of variable experimental conditions such as concentration, temperature, and organism growth history that affect yield.

TABLE 19.3
Observed versus Theoretical Yield for Selected Anaerobic Processes, Based on Electron Equivalents Sent to the Electron Acceptor

Acceptor	Donor	Organism(s)	Observed Yield ($gVSS/e^-eq$)	Theoretical Yield[a] ($gVSS/e^-eq$)	Ref.
Fe(III)-citrate	Lactate[b]	*S. oneidensis* MR-1	0.55–12.6	4.39	[22]
	Lactate[b]	*S. oneidensis* MR-1	0.40	4.39	[23]
	Lactate	*S. putrefaciens* CN32	1.32	3.73	[24]
	Acetate	*G. metallireducens*	2.55[c]	2.61	[25]
	Glycerol	*Aeromonas hydrophila*	7.10[c]	4.70	[26]
Goethite	H_2	*S. alga* BrY	0.42	0.37[d]	[27]
	Lactate[b]		0.63	1.44	
	Lactate[b]	*S. oneidensis* MR-1	0.34	1.44	[23]
Fe(III) oxide	H_2	*S. alga* BrY	0.76[c]	1.11[d]	[28]
	Lactate[b]		1.48[c]	2.25	
	Acetate	*Desulfuromonas acetoxidans*	1.05[c]	0.97	[29]
	Acetate	*G. metallireducens*	1.01[c]	0.97	[25]
SO_4^{2-}	Lactate[b]	Mixed, w/ *D. desulfuricans*	0.26	1.31	[30]
	Ethanol[b]		0.07–0.11	0.92	
	Lactate[b]	*D. desulfuricans*	1.24	1.31	[31]
	Lactate[b]	*D. desulfuricans*	1.98	1.31	[32]
	Lactate[b]	*D. desulfuricans*	0.76–0.97	1.31	[33]
	Pyruvate[b]		1.40	3.20	
U(VI)	Acetate	*G. metallireducens*	2.17[c]	1.33	[7]
	H_2	*S. putrefaciens* CN32	0.83[c]	1.60[d]	[7]
Co(III)	Glycerol	*Aeromonas hydrophila*	2.63[c]	4.78	[26]

[a]Thermodynamic data were taken from Table 19.2 and Refs. [19, 34, 35].
[b]The electron donor was incompletely oxidized to acetate.
[c]Values were estimated from published growth curves.
[d]Value was calculated by assuming a hydrogen concentration of 0.5 n*M*.

19.2.4 Implications of Thermodynamics for Microbial Metal Remediation

19.2.4.1 Competitive Electron Acceptors and Potential Oxidants

Electron acceptors with half-reactions that have very negative $\Delta G^{o'}$ values (very positive $E_h^{o'}$ values) yield more energy in coupled whole reactions than those with less negative (or positive) $\Delta G^{o'}$ values (less positive $E_h^{o'}$ values). This translates into higher yields and higher specific growth rates. Consequently, microorganisms that use acceptors with highly positive $E_h^{o'}$ values can often outcompete and exclude microorganisms that use acceptors of lower potential. Based on thermodynamic considerations, the depletion of acceptors is expected to occur in the sequence of oxygen, nitrate, Fe(III), and sulfate. This pattern of acceptor use has been observed as spatial redox zonation in sediments [36]. If electron donor concentration is limiting, electron acceptors with high half-reaction reduction potentials can effectively compete for electrons from microbes and prevent the reduction of a metal with a lower potential. Such competitive electron acceptors must be removed or reduced to facilitate the reduction of the metal of interest.

Compounds with high redox potentials also have the potential to oxidize reduced metals abiotically. For example, manganese oxides oxidize U(IV) [37] and Cr(III) [38], increasing the mobility of these metals. These oxidants must be removed or reduced during bioremediation to prevent the reoxidation of reduced contaminants.

19.2.4.2 Speciation and Concentration

Speciation and concentration dramatically affect reaction thermodynamics. Different reduced and oxidized species of a metal have vastly different formation energies, which can change the free energy of a given process. Metal speciation is altered by pH, redox potential, ligand concentration, metal concentration, and total inorganic carbon concentration (see Section 19.4). The presence of 2.5 mM calcium, for example, shifts the dominant uranyl species in a bicarbonate buffer from $UO_2(CO_3)_3^{4-}$ to $Ca_2UO_2(CO_3)_3$ and decreases the calculated redox potential of the uranium reduction reaction by 0.119 V [39]. This change evidently led to inhibition of both the rate and extent of microbial uranium reduction [39]. Computer programs such as MINTEQA2 [40], PHREEQC [41], and HYDROGEOCHEM [42] can be used to obtain a preliminary assessment of the effects of speciation changes and the potential impacts of different remediation schemes.

For the general reaction $aA + bB \longrightarrow cC + dD$, the effects of concentration on reaction energetics can be computed from the relationship:

$$\Delta G = \Delta G^o + RT \ln \frac{C^c D^d}{A^a B^b}$$

As a reaction progresses, the driving force for the reaction decreases as reactants are removed and products accumulate. The treatment of metal contaminants to low levels may therefore require the continual removal of product or the optimization of operating conditions to maximize reaction rate.

The free energy available from microbial reactions can set the concentration of reaction intermediates in a system. Hydrogen is a key intermediate in biostimulated environments because it is both a product of initial anaerobic reactions and an electron donor for reactions in which terminal electron acceptors are reduced. Consequently, monitoring dissolved hydrogen provides useful information about redox conditions [43]. During metal remediation, the terminal electron-accepting process may be difficult to identify, since some anaerobic electron acceptors are surface-associated metal oxides. In such instances, a comparison of hydrogen concentration with values known to be characteristic for different terminal electron acceptors can aid in the identification. Nitrate reduction decreases hydrogen concentrations to <0.1 nM [44], Fe(III) reduction to 0.1 to 0.8 nM [44, 45], and sulfate reduction to 1.0 to 4.0 nM [46], while hydrogen concentrations of 5 to 25 nM are characteristic of methanogenesis [46]. In addition, maintaining hydrogen levels within the desired range provides a measure of process control, since the addition of excessive electrons can waste electron donor and may result in methane production, a potential explosion hazard.

19.3 KINETICS OF MICROBIAL GROWTH AND TRANSFORMATIONS

Once balanced reactions for microbial growth and reactant transformation are written, the rates of relevant reactions can be assessed with kinetic models. These models evaluate time-scales for different physicochemical and biological reactions and are therefore necessary in identifying potential rate-limiting processes. This knowledge can guide decision-making during remediation. Reaction rate equations can also be incorporated into larger transport models that describe the overall fate of metals for a given subsurface environment. Such models can be run repeatedly under a suite of operating scenarios, which can be evaluated and optimized to achieve the desired operational objectives, such as cost minimization [47].

Metal reduction rates are proportional to the concentration of the enzyme that catalyzes the reaction. Because enzyme concentration is typically proportional to biomass concentration, we begin with a discussion of microbial growth. Then, we describe kinetic parameters that can be used to predict rates of metal transformation when microbial growth rate is not limited and when it is constrained by the concentration of electron donor or electron acceptor.

19.3.1 MICROBIAL GROWTH AND DECAY

Bacteria grow by binary fission, in which a single cell splits into two. If we let X be cell concentration, the rate of increase in cell concentration depends upon the concentration of cells: $\mathrm{d}X/\mathrm{d}t = \mu X$. The coefficient μ is the specific growth rate, with units of inverse time (h^{-1} or day^{-1}). If μ is constant, $t = \ln(X_t/X_0)/\mu$, and a doubling time can be defined: $t_\mathrm{d} = \ln 2/\mu = 0.693/\mu$. For a microbial culture with constant μ, we can integrate from an initial biomass concentration X_0 to a final concentration X over time period t, to obtain $X/X_0 = \mathrm{e}^{\mu t}$, or $\ln X = \ln X_0 + \mu t$. A plot of $\ln X$ versus t gives a straight line with slope μ. If all substrates are abundant and there is no inhibition, cells will grow at a constant and maximum rate, so that $\mu = \mu_{\max}$. Table 19.4 gives some typical values of $\mu_{\max}$ and minimum doubling times for bacteria involved in metal reduction. A constant specific growth rate could also result if a limiting nutrient is provided at a constant rate.

When substrate levels become limiting, μ declines. Saturation kinetics are used to describe consumption of substrate and the resulting decrease in μ. The best-known example of saturation kinetics to describe growth is the Monod equation:

$$\mu = \mu_{\max}\left(\frac{S}{K+S}\right)$$

where S is the limiting substrate concentration, and K is the half-saturation coefficient — the substrate concentration where the specific growth rate is half its maximum. When $S = K$, $\mu = \mu_{\max}/2$, and when $S \gg K$, $\mu = \mu_{\max}$. Even though saturation kinetics expressions share the same form as the Michaelis–Menten expression for enzyme kinetics, the coefficients involved are purely empirical and not necessarily related to the activity of a specific enzyme. A plot of the Monod equation is shown in Figure 19.2a.

To obtain the specific rate of substrate utilization, q, we divide μ by the yield Y:

TABLE 19.4
Maximum Specific Growth Rates for DMRB and SRB

Organism Type	Electron Acceptor	Electron Donor	μ_{max} (day^{-1})	Minimum Doubling Time, t_d (day)	Ref.
DMRB	Iron	Lactate	7	0.1	[24]
SRB	Sulfate	H_2	0.29	2.4	[20]
		Acetate	0.5	1.4	[20]

$$q = \frac{\mu_{max}}{Y}\left(\frac{S}{K+S}\right) = q_{max}\left(\frac{S}{K+S}\right)$$

where q is the biomass-normalized rate of substrate use. The term μ_{max}/Y is a constant, and it is equal to the maximum specific rate of substrate utilization, q_{max}. Figure 19.2b shows a plot of the specific rate of substrate utilization versus substrate concentration. For $S \gg K$, $q = q_{max}$. For $S \ll K$, the ratio q_{max}/K is constant and equal to the pseudo second-order rate coefficient, k_1: $q = (q_{max}/K)S = k_1S$. Under these conditions, reaction rates decrease as the concentration of S decreases.

The above expressions describe cell growth and substrate utilization in the presence of electron donor. In its absence, decay becomes significant. The simplest kinetic description of decay assumes that in the absence of substrate, the specific growth rate μ becomes negative

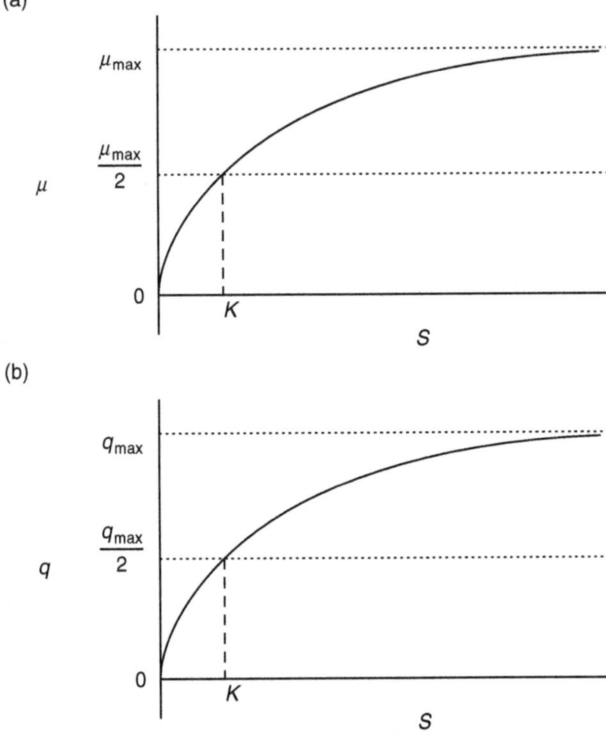

FIGURE 19.2 Kinetic curves for microbial growth and substrate utilization: (a) the Monod equation for microbial growth; (b) the specific rate of substrate utilization; and (c) the Herbert expression for microbial growth and decay.

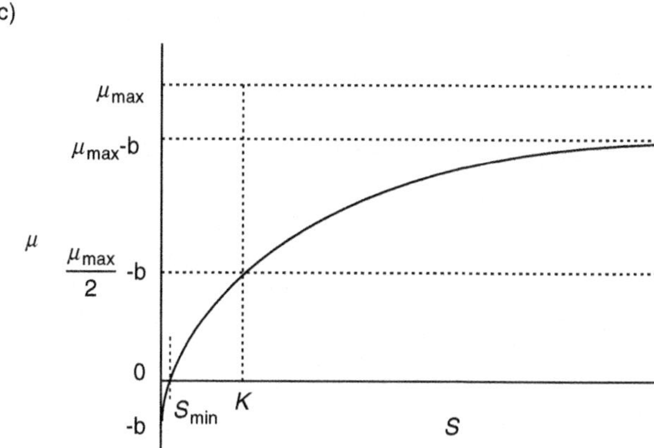

FIGURE 19.2 *continued*

and constant: $\mu = -b$, where b is the specific decay rate (units of biomass lost per unit of biomass per unit time). By adding this assumption to the Monod expression, we obtain the Herbert expression:

$$\mu = \mu_{\max}\left(\frac{S}{K+S}\right) - b$$

A plot of this expression is shown in Figure 19.2c. Of significance is the point where growth just balances decay. Here the specific growth rate is zero, and $S = S_{\min} = Kb/(\mu_{\max} - b)$. When the electron donor is present at concentrations above $S_{\min}$, net growth is possible, and net decay results at lower concentrations.

The above expressions assume that the sole rate-limiting substrate is the electron donor; however, this is often not the case. In situations where the primary electron acceptor is also growth-rate limiting, a second saturation kinetics term is included:

$$\mu = \mu_{\max}\left(\frac{S}{K+S}\right)\left(\frac{A}{K_A+A}\right) - b$$

When the electron donor is present in excess (i.e., $S \gg K$), we can set $\mu = 0$, and define a minimum concentration of electron acceptor needed for growth, $A_{\min} = bK_A/(\mu_{\max} - b)$. For concentrations above $A_{\min}$, the electron acceptor is a primary electron acceptor because it supports growth. For concentrations below $A_{\min}$, it is a secondary electron acceptor. Because half-saturation values for metals are typically large (Table 19.5), we can expect that the $A_{\min}$ values will also tend to be large. This suggests that metals will often behave as secondary electron acceptors. Calculations of energetics and stoichiometry should be based on the primary electron acceptor, as energy obtained from the reduction of secondary acceptors is typically insignificant. As an example, the contaminant U(VI) was found to be a secondary electron acceptor in acetate-amended sediment microcosms; Finneran et al. [57] observed that <1% of electrons were transferred to U(VI), and Fe(III) served as the primary electron acceptor for growth.

Either the electron donor or the biomass can serve as a source of electrons for reduction of electron acceptors. Accordingly, the specific rate of electron acceptor use q_A is given by:

$$q_A = \frac{q_{\max}\left(\frac{S}{K+S}\right)\left(\frac{A}{K_A+A}\right)}{Y_{S/A}} + \frac{b}{Y_{X/A}}\left(\frac{A}{K_A+A}\right)$$

TABLE 19.5
Kinetic Parameters for Selected Bacteria Involved in Metal Reduction

Acceptor	Donor	Growth Conditions	Organisms	$q_{max,A}$ (μmoles acceptor/mg VSS-h)	K_A (μM)	k_1 (l/mg VSS-day)	Ref.
U(VI)	Lactate	Nongrowth	*S. alga* BrY	3	132		[48]
	Lactate	Nongrowth	Mixed culture (SRB)	1556	248		[49]
	Lactate	Nongrowth	*D. desulfuricans*	2000	499		[49]
	H_2	Nongrowth	*S. putrefaciens* CN32	4	370		[50]
	Lactate	Nongrowth	*S. oneidensis* MR-1	8	54	0.563	[51]
	H_2	Nongrowth	*S. oneidensis* MR-1	14	297	0.569	[51]
TcO_4^-	Lactate	Nongrowth	*S. putrefaciens* CN32	846	53,200	0.006	[51]
	H_2	Nongrowth	*S. putrefaciens* CN32	98	1	0.043	[51]
Cr(VI)	Lactate	Nongrowth	*S. oneidensis* MR-1	5	74		[52]
	Lactate	Nongrowth	*S. putrefaciens* CN32	9	439	0.240	[51]
	Glucose	Nongrowth	*E. coli* ATCC 33456	0.035	166		[53]
	Glucose	Nongrowth	*Bacillus* sp.	0.008	104		[53]
	Molasses	Nongrowth	*E. cloacae* HO1	0.030	30		[54]
	Molasses	Nongrowth	Hanford enrichment	0.070	8		[54]
	Molasses	Nongrowth	Joseph enrichment	0.021	28		[54]
Co(III)	Lactate	Nongrowth	*S. oneidensis* MR-1	7	130	3.91	[51]
	Lactate	Nongrowth	*S. putrefaciens* CN32	2	46	1.6	[51]
Fe(III)-citrate	Lactate	Nongrowth	*S. oneidensis* MR-1	28,095	87,200	8	[51]
	H_2	Nongrowth	*S. oneidensis* MR-1	36,786	80,000	11	[51]
	Lactate	Nongrowth	*S. putrefaciens* CN32	156	700	4	[51]
	H_2	Nongrowth	*S. putrefaciens* CN32	35,952	80,000	11	[51]
		Growth	*S. putrefaciens* CN32	242	29,000		[24]
SO_4^{2-}	Lactate	Growth	*D. desulfuricans*	1	244		[55]
	Lactate	Growth	*D. sapovorans*	0.89	7		[56]
	Lactate	Growth	*D. salexigens*	3	77		[56]

where q_A is the specific rate of utilization of the electron acceptor, $Y_{S/A}$ is the ratio of the mass of electron donor consumed to the mass of electron acceptor used, and $Y_{X/A}$ is the ratio of the mass of biomass consumed during decay to the mass of electron acceptor used for biomass oxidation. By setting $q_{max,A} = q_{max}/Y_{S/A}$ and letting $b_A = b/Y_{X/A}$, we obtain:

$$q_A = q_{max,A}\left(\frac{S}{K+S}\right)\left(\frac{A}{K_A+A}\right) + b_A\left(\frac{A}{K_A+A}\right)$$

Under growth conditions, when $A \ll K_A$ and $S \gg K$,

$$q_A = \frac{q_{max,A}}{K_A}A + \frac{b}{K_A}A = k'A$$

where

$$k' = \frac{q_{max,A}}{K_A} + \frac{b}{K_A}$$

Under decay conditions, $S = 0$, and the value of k' decreases to b/K_A. Large half-saturation values for metal reduction mean that the concentration of the primary electron acceptor will often be much less than K_A in natural systems, and then the use of these first-order rate simplifications is appropriate. Moreover, as noted by Liu et al. [51], the use of a first-order expression can yield more precise coefficients than those obtained with a saturation kinetic expression that may be statistically over-parameterized.

The above relationships describe two periods of electron acceptor use: faster reduction during growth, followed by slower reduction during decay. Table 19.5 gives the values for $q_{max,A}$ and K_A for a variety of bacteria capable of metal reduction or precipitation of metals as sulfides. Note that these values were obtained in the presence of electron donor, but under variable growth conditions. Unfortunately, little is known about metal reduction parameters during periods of decay; research is needed in this area.

19.3.2 Cometabolism

Cometabolism is defined as the transformation of a nongrowth substrate in the presence of a growth substrate, or by resting cells in the absence of growth substrate [59]. Because the reduction of metals is not necessarily coupled to growth, metals can serve as nongrowth substrates. For example, SRB such as *Desulfovibrio desulfuricans* and *Desulfovibrio vulgaris* reduce uranium but do not appear to gain energy from it, and *Shewanella oneidensis* MR-1 reduces chromate cometabolically [52, 53]. A nongrowth substrate imposes an energy drain on the cell by diverting electrons from energy generation and cell synthesis and can be viewed as a form of toxicity.

Alvarez-Cohen and McCarty [60] described the effects of cometabolism on cell biomass using the expression $-\mathrm{d}C/\mathrm{d}X = T_c$, where C represents the concentration of cometabolized compound, and T_c is the transformation capacity, the mass of nongrowth substrate transformed per unit of resting cell biomass. Wang and Shen [53] accounted for the inhibitory effect of Cr(VI) reduction on biomass with the expression:

$$X = X_0 - \frac{C_0 - C}{T_c}$$

where X is the concentration of active cells, X_0 is the initial concentration of active cells, C_0 is the initial concentration of Cr(VI), and T_c is the Cr(VI) transformation capacity. T_c can also be incorporated into a kinetic expression for cell growth as a negative term [59]:

$$\mu = \mu_{max}\left(\frac{S}{K+S}\right)\left(\frac{A}{K_A + A}\right) - b - \frac{q_A}{T_c}$$

Table 19.6 gives Cr(VI) transformation capacities for several bacterial species.

Alternatively, researchers have accounted for the cometabolic inhibition of metal reduction by adding a noncompetitive inhibition term to the saturation kinetics expression [55, 57]:

$$q_C = q_{max,C}\frac{C}{(K_C + C)\left(1 + \frac{C}{K_i}\right)}$$

where K_i is the constant of noncompetitive inhibition for metal reduction.

19.3.3 Parameter Estimation

Several approaches may be used to derive kinetic parameters from laboratory data. The first involves measuring specific rates of metal reduction (q_A) at many different substrate concentrations and using nonlinear regression analysis to estimate kinetic parameters [48, 53].

TABLE 19.6
Cr(VI) Transformation Capacities for Selected Bacteria

Organism	T_c, Cr(VI) Reduction Capacity (mg Cr(VI)/mg VSS)	Ref.
S. oneidensis MR-1	0.47	[53]
S. putrefaciens CN32	0.68	[51]
S. alga BrY	0.31	[51]
E. coli ATCC 33456	0.03	[54]
Bacillus sp.	0.01	[54]

A second approach is to integrate the equation for specific rate of metal reduction (or some variation thereof, depending on growth conditions and inhibition effects) to obtain an equation in terms of A, $q_{max,A}$, K_A, and time. One can fit the integrated kinetic equation to progress curve data (metal concentration as a function of time), and parameters can be estimated by nonlinear regression [51, 54].

The parameters of saturation kinetics are highly correlated under many experimental conditions, however, which can result in large uncertainties in their estimation and preclude unique identification [61]. Sensitivity analysis [61] or a statistics-based approach [62] can be employed to determine experimental conditions that minimize parameter uncertainty and maximize identifiability. Alternatively, experimental data can be fit with a simplified expression. A study of metal reduction kinetics found a first-order approximation of a Monod model reduced parameter uncertainty [51]. In all cases, an engineer should seek the simplest model possible that accurately describes experimental data and verify the uniqueness of estimated parameters.

19.4 GEOCHEMICAL PROCESSES

Metals undergo numerous and significant geochemical reactions, including hydrolysis, complexation, redox transformations, precipitation–dissolution, and sorption–desorption. These reactions often control the mobility, transport, and reactivity of metals and are affected by factors such as pH, the presence of other chemical species, temperature, and relative concentrations. This section discusses the most important of these variables, pH, and then considers important geochemical reactions of metals in the context of equilibrium modeling.

19.4.1 Control of Solution pH: Alkalinity and Acidity

Solution pH is a critical control variable during metal remediation in subsurface environments. It affects speciation, biological activity, and the potential for chemical clogging. Before the addition of electron donor to initiate biostimulation, the pH of an aquifer must be adjusted to a range suitable for microbial activity. Although the limits on pH for metal-reducing bacteria are not yet well defined, microbial processes generally function best within the range of 6 to 9. After the initiation of biostimulation, pH and alkalinity should be monitored, and acidity and alkalinity added as needed to maintain the desired conditions. Sufficient alkalinity must be present to prevent low pH, and sufficient acidity must be present to prevent precipitation of calcium and magnesium at high pH and carbonate levels.

The bioremediation of uranium highlights challenges associated with pH control. Effective reduction requires bioavailable U(VI) and the prior or simultaneous reduction of competitive electron acceptors, the reduction of which influences pH, alkalinity, and acidity.

TABLE 19.7
Balanced Microbial Reactions for Ethanol Addition to an Aquifer with Mn(IV), Fe(III), and SO_4^{2-}, and Contaminated with NO_3^-

Electron Acceptor	Balanced Reaction[a]	Moles Alkalinity Produced per Mole Ethanol Consumed
NO_3^-	$C_2H_5OH + 2NO_3^- \longrightarrow 0.96N_2 + 0.086C_5H_7O_2N + 1.14HCO_3^- + 0.43CO_3^{2-} + 2.13H_2O$	2.0
Mn(IV)	$C_2H_5OH + 5.4MnO_2 + 0.043NO_3^- + 2.58H_2O \longrightarrow 5.4Mn^{2+} + 0.043C_5H_7O_2N + 1.79HCO_3^- + 9.06OH^-$	10.9
Fe(III)	$C_2H_5OH + 2.16Fe_5O_9H_3 + 0.04NO_3^- + 4.73H_2O \longrightarrow 10.8Fe^{2+} + 0.04C_5H_7O_2N + 1.79HCO_3^- + 19.9OH^-$	21.7
SO_4^{2-}	$C_2H_5OH + 1.35SO_4^{2-} + 0.043NO_3^- \longrightarrow 1.35HS^- + 1.52H_2O + 0.043C_5H_7O_2N + 1.35HCO_3^- + 0.44CO_2$	2.7

[a] f_s was assumed to be 0.2 for denitrification and 0.1 for Mn(IV), Fe(III), and SO_4^{2} reduction.

Moreover, the most promising U(VI) species for bioremediation is uranyl carbonate, but its formation is a complex function of pH, total carbonate concentration, and concentrations of other metals and ligands.

Issues associated with uranium remediation are illustrated by current efforts to bioremediate groundwater and soil at the former S-3 ponds of the Department of Energy (DOE) Field Research Center in Oak Ridge, Tennessee. Groundwater at this site contains high levels of U(VI) and has a low pH (~3.6) due to the disposal of nitric and sulfuric acid waste in unlined surface ponds. Alkalinity must be added initially to establish a pH suitable for biological activity. As shown by reaction stoichiometry (Table 19.7), however, the addition of ethanol to initiate U(VI) reduction will result in the formation of large amounts of alkalinity, making control of pH and alkalinity critical after biostimulation. Table 19.7 reveals that reductive processes generally result in a net production of alkalinity. Iron reduction, in particular, has a high yield of alkalinity, with nearly 22 moles of alkalinity produced per mole of ethanol consumed. This contrasts with aerobic processes in which acid production is the expected outcome, as occurs in environments with acid mine drainage problems.

Although addition of alkalinity is straightforward, the choice of the basic salt must be made with care due to the potential for adverse effects from the added cation, including soil deflocculation by sodium, the precipitation of calcium salts, and possible toxicity. Reduction of alkalinity can be accomplished by the addition of strong acids, but again, care must be taken in the choice of acid. Sulfuric and nitric acid are sources of competitive electron acceptors, and the use of phosphoric acid introduces an anion that can significantly alter uranyl sorption relationships.

Alkalinity is not necessarily tied to inorganic carbon levels, which can be manipulated independently — by carbonation to add CO_2 or by vacuum stripping to remove CO_2. The use of these techniques is especially valuable for process control during uranium remediation since soluble uranyl carbonate complexes are amenable to reduction.

19.4.2 Equilibrium Modeling

Geochemical equilibrium models are helpful in predicting metal behavior in biological systems, even though these models assume a system is homogeneous and at equilibrium. To use these models, the total concentrations of all system components are first specified. Components consist of the minimal set of chemical entities, such as minerals, molecules, and ions, necessary to describe a system. A list is compiled of possible interactions between components, including mass transfer reactions and biogeochemical reactions, along with relevant equilibrium constants and thermodynamic values. Finally, a distribution of components between species and phases is determined by minimizing the free energy of the system in an iterative process. A variety of computer codes are available to rapidly perform equilibrium calculations for a given thermodynamic database (see Section 19.2.4.2).

The value of thermodynamic models depends upon the accuracy of system characterization, the comprehensiveness of the reaction database, and the quality of thermodynamic and equilibrium values. Real waters contain a multitude of components that interact and compete with each other, so equilibrium calculations must consider the presence of all components and species. In addition, the geochemistry of a given wastewater, contaminated site, or experimental matrix should be thoroughly analyzed before using the model in a predictive fashion.

Speciation models based on thermodynamic data predict equilibrium conditions, but in real environments equilibrium is not achieved because of kinetic factors. Nevertheless, equilibrium assumptions appear adequate for the prediction of aqueous speciation and some redox reactions. Formation and dissolution of minerals is likely to be controlled by kinetics, however, as are gas transfer reactions. Such factors qualify predictions based on equilibrium assumptions.

The output of equilibrium modeling can be visualized in a variety of useful diagrams, such as speciation and prevalence charts (for example, Figure 19.3). These charts provide information on the conditions under which precipitation occurs, the speciation of a metal with pH, or the distribution of a metal between aqueous and solid phases. Modeling needs and remediation considerations for important geochemical processes are discussed in the following sections.

19.4.3 Precipitation and Solubility

The goal of metal remediation frequently is immobilization via the precipitation of an insoluble mineral, which can be directly or indirectly catalyzed by microbial activity. An important consideration for this approach is that the aqueous-phase solubility of the resultant mineral is below its regulatory limit (Table 19.1). The concentration of a metal contaminant in equilibrium with a given solid can be calculated from the solubility product of that solid. The reduced form of chromium, Cr(III), often precipitates as chromium(III) hydroxide, $Cr(OH)_3$. The solubility of chromium in equilibrium with $Cr(OH)_3$ at pH 7 can be determined from its solubility product:

$$Cr(OH)_3(s) = Cr^{3+} + 3OH^-$$

$$K_{sp} = 10^{-30.2}$$

$$K_{sp} = (Cr^{3+})(OH^-)^3$$

$$10^{-30.2} = (Cr^{3+})(10^{-7}M)^3$$

$$(Cr^{3+}) = 6.31 \times 10^{-10}M = 3.3 \times 10^{-5}\text{mg/l}$$

This concentration of chromium is well below the regulatory concentration limit for chromium, 0.1 mg/l (Table 19.1). The calculation assumes that soluble chromium is present as the free ion, Cr^{3+}, but the presence of various ligands can alter the aqueous speciation of a metal and increase its solubility. In addition, the contribution of other ions to ionic strength also affects solubility. Equilibrium modeling with consideration of mineral solubilities takes into account these factors, making solubility products important components of thermodynamic databases. Table 19.8 lists solubility products for many minerals relevant to environmental remediation.

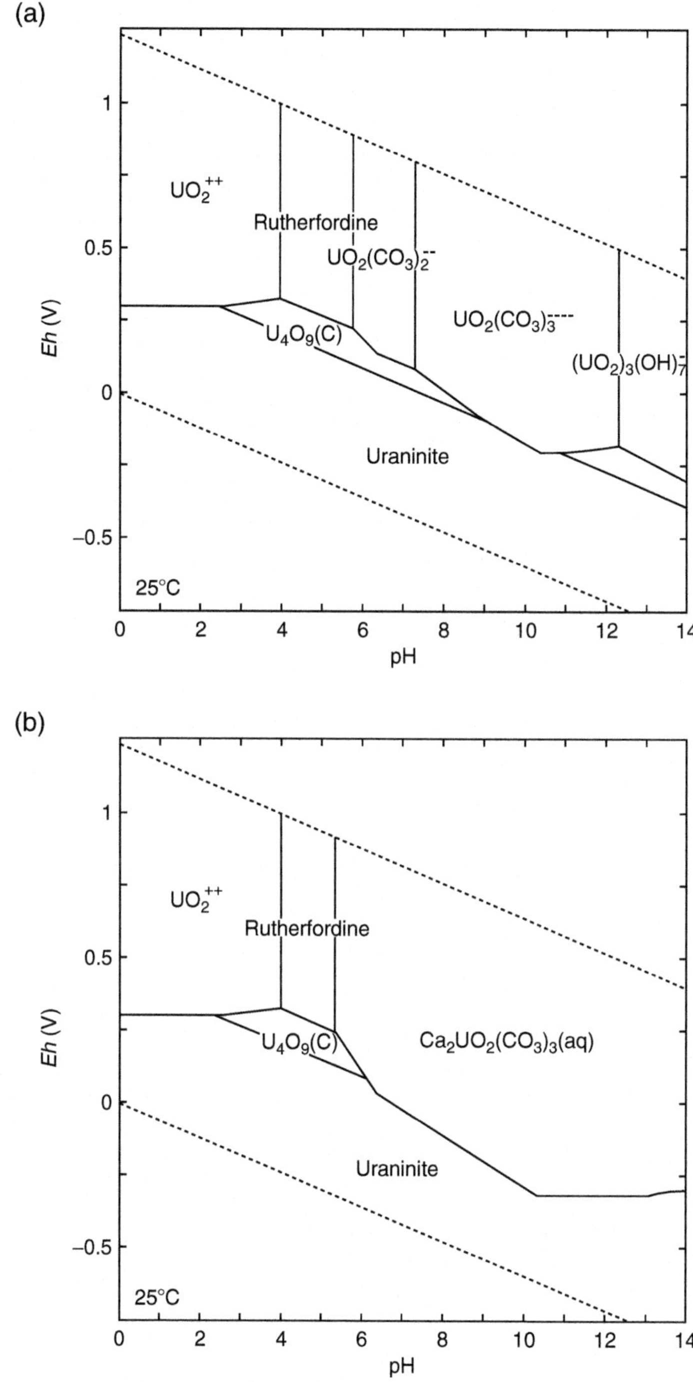

FIGURE 19.3 Uranium speciation in the presence and absence of calcium. The most prevalent species of uranium is shown for a range of pH and redox potential values. Speciation was calculated for 300 mg/l inorganic carbon and 40 mg/l uranium, and in the presence (a) or absence (b) of 1 g/l calcium.

TABLE 19.8
Mineral Solubility Products

Compound	Formula	pK_{sp}	K_{sp}	Ref.
Arsenic(III) sulfide	As_2S_3	21.68	2.1×10^{-22}	[63]
Cadmium sulfide	CdS	26.10	8.0×10^{-27}	[63]
Chromium(III) hydroxide	$Cr(OH)_3$	30.20	6.3×10^{-31}	[63]
Cobalt sulfide	CoS	20.40	4.0×10^{-21}	[63]
		24.70	2.0×10^{-25}	
Copper(I) sulfide	Cu_2S	47.60	2.5×10^{-48}	[63]
Copper(II) sulfide	CuS	35.20	6.3×10^{-36}	[63]
Ferrihydrite	$Fe(OH)_3$	39.5	3.16×10^{-40}	[64]
Goethite	FeOOH	40.7	2.00×10^{-41}	[64]
Hematite	Fe_2O_3	42.75	1.78×10^{-43}	[63]
Iron(II) sulfide	FeS	17.20	6.3×10^{-18}	[63]
Lead sulfide	PbS	27.10	8.0×10^{-28}	[63]
Manganese hydroxide	$Mn(OH)_2$	12.72	1.9×10^{-13}	[63]
Mercury(II) sulfide	HgS red	52.4	4×10^{-53}	[63]
	HgS black	51.8	1.6×10^{-52}	
Nickel				
α-Sulfide	NiS	18.5	3.2×10^{-19}	[63]
β-Sulfide	β-NiS	24.0	1.0×10^{-24}	
γ-Sulfide	NiS	25.70	2.0×10^{-26}	
Technicium	TcO_2	8	10^{-8}	[65]
Uraninite	UO_2	60.6	2.5×10^{-61}	[66]
Zinc sulfide				
Sphaelerite	ZnS	23.8	1.6×10^{-24}	[63]
Wurtzite	ZnS	21.6	2.5×10^{-22}	

19.4.4 Aqueous Speciation

Equilibrium modeling can provide information on how such factors as pH, redox potential, and concentration affect metal speciation, and speciation often determines substrate bioavailability and transformation rates. As discussed above, $Ca–UO_2–CO_3$ species have been shown to inhibit both the extent and rate of microbial U(VI) reduction, apparently because the reduction of $Ca_2UO_2(CO_3)_3$ is less energetically favorable than the reduction of uranyl carbonate species [39]. Figure 19.3 shows the speciation of uranium with pH and redox potential in the presence and absence of calcium. In the presence of millimolar levels of calcium, $Ca–UO_2–CO_3$ species dominate the aqueous speciation of uranyl. Also, uranyl complexation to organic ligands affects the rate of uranium reduction for various species differently. Ganesh et al. [66] found uranyl complexation to the monodentate ligand acetate decreased the rate of uranium reduction for *Shewanella alga* BrY, while complexation to multidentate ligands increased the rate. Conversely, *Desulfovibrio desulfuricans* reduced uranium more quickly from a complex with acetate than from complexes with multidentate ligands.

19.4.5 Abiotic or Surface-Catalyzed Reactions

The reaction of metals with the products of microbial processes can be a pathway of indirect bacterial transformation. As noted earlier, sulfide produced by SRB precipitates with many heavy metal contaminants, resulting in the formation of minerals of very low solubility (Table 19.8). This reaction can be utilized to immobilize metals *in situ*, or bioreactor systems can be employed to generate metal sulfides for disposal [68]. Also, Fe(II) produced by microbial iron reduction reduces Cr(VI) faster than direct enzymatic reduction [69]. Even though soluble Fe(II) is ineffective at reducing U(VI) in homogeneous solutions, U(VI) was reduced by Fe(II) sorbed to goethite [70]. The presence of an iron mineral surface catalyzes the reaction [71].

19.4.6 Sorption

19.4.6.1 Effect of Sorption on Bioremediation

Sorption of metals to solids and cells strongly impacts *in situ* remediation efforts. Iron and manganese oxides, common minerals in aquifers, have a high capacity for sorption of metals. This uptake can retard the transport of metals in groundwater, but the microbial reduction of iron and manganese oxides can release metal contaminants to the mobile phase [72]. Sorption to the surfaces of bacteria can also result in enhanced metal transport when bacteria detach from a solid surface and are mobilized in groundwater [73].

19.4.6.2 Factors Affecting Sorption

Variables that affect metal sorption include pH, metal concentration, water to solids ratio, concentration and type of ligands, concentration of competing cations, ionic strength, solid surface area, and surface reactive site density. A comprehensive treatment of these factors is beyond the scope of this chapter, and we refer the interested reader to reviews on this subject [74, 75]. The discussion in Section 19.5 suggests factors that can be manipulated to enhance metal bioavailability and rates of immobilization.

19.4.6.3 Modeling Sorption

Most models of metal sorption are empirical; parameters are typically fit to experimental data describing a decrease in aqueous metal concentration in the presence of a solid [76]. The simplest models use a distribution coefficient, K_d, to describe partitioning between solid and liquid phases. The incorporation of a single adsorption coefficient simplifies transport modeling, but its utility is limited to linear partitioning and to a single set of environmental conditions. K_d depends upon solid and solution chemistry and may be altered by modifying pH, ligand concentrations, or ionic strength.

Partitioning is also modeled using isotherms that relate aqueous concentration to sorbed concentration. Studies have fit uranyl sorption data to Langmuir [77, 78] and Freundlich isotherms [79, 80]. These isotherms do not reveal sorption mechanisms, however, so the results cannot be extrapolated to other systems.

The surface complexation model, or SCM, describes adsorption with a series of reactions similar to those for aqueous complex formation and includes additional terms to account for electrostatic interactions. Researchers generally agree that such an approach is a valid representation of adsorption mechanisms [81]. Many studies have used SCMs to fit macroscopic adsorption data for uranyl [77, 81, 83], and though they are superior to previous models in accurately depicting adsorption of uranyl, the fit may be the result of the large number of adjustable parameters in the SCM [84]. Thus SCMs should not be based on empirical fits to macroscopic adsorption data, but rather on independent measurements of variables such as cation-exchange capacity, surface site density, and surface acidity constants.

19.5 BIOAVAILABILITY AND OBSERVED REDUCTION RATES

The observed rate of metal reduction depends on (1) physicochemical phenomena that control metal bioavailability and (2) factors affecting the rate of the microbial reaction. Specifically, metal bioavailability depends upon pH, speciation, sorption, and location of the contaminant. Solids can be a major reservoir of metal contamination. Metal contaminants may be found naturally in solid form, they may have accumulated at sorptive sites of minerals, or they may have diffused into micro- and mesopores that are inaccessible to microorganisms (<0.2 μm). Under such conditions, desorption or diffusion can limit overall rates of reduction. Factors affecting the microbial reaction rate include concentration and types of bacteria present; concentration of electron donors, electron acceptors, and potential inhibitors; pH; and metal speciation.

Figure 19.4 shows a porous solid containing metals and microorganisms. Near the solid surface, equilibrium conditions prevail, and we can assume that an appropriate equilibrium model relates solution-phase concentration to solid-phase concentration. As noted above, the most flexible but complex model would be a verified SCM, while the simplest would be a linear partitioning model, such as $A_s = K_d A_{eq}$, where A_s is the concentration of oxidized metal on the solid phase, A_{eq} the concentration of metal in the aqueous phase after equilibration with the solid phase, and K_d is the distribution coefficient. Concentration within the bulk solution, A, may be larger or smaller than A_{eq}. When $A > A_{eq}$, aqueous metal species tend to sorb, and when $A < A_{eq}$, these species tend to desorb. Both tendencies can be represented using a single mass transfer expression: $J_A = k_w(A_{eq} - A)$, where J_A is the volumetric flux of metal (direction is positive moving away from the surface) and k_w is a mass transfer coefficient describing diffusion into the bulk solution.

Three mass transfer pathways are depicted in Figure 19.4: in pathway 1, contaminant mass is transported by advection with groundwater through the biologically active pore space; in pathway 2, the contaminant diffuses between the solid surface and the biologically active pore space; and in pathway 3, contaminant mass diffuses between meso- and micropores and the biologically active pore space. The concentration of the contaminant within the

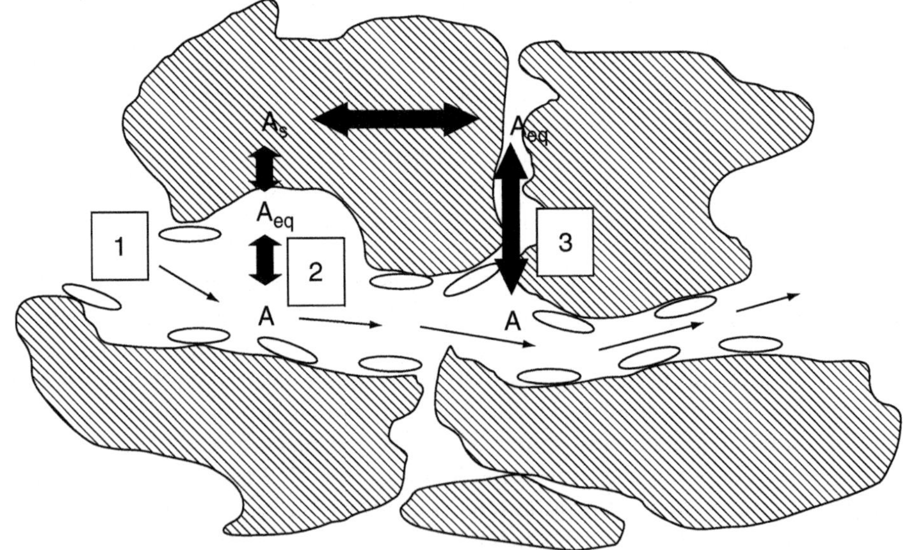

FIGURE 19.4 Mass transfer processes during microbial metal reduction in soil–water systems.

biologically active pore space is everywhere equal to A, so the specific rate of the microbial reaction is $q_A = k'A$ for a first-order reaction, and volumetric flux is $k'XA$.

Identifying which of the above processes control the reaction rate can guide actions to increase overall rates of reduction or to decrease the soluble concentrations of a contaminant below its regulatory limit. The approach of Ramaswami and Luthy [85] is useful in assessing controls on reaction rates. For each pathway in Figure 19.4, a dimensionless Damkohler number is defined that allows comparison of the relative rates of mass transfer and reaction. For pathway 1, the relevant Damkohler number is $D1 = k'Xl/v$, where v is the interstitial pore velocity and l is a characteristic length of the system. When a contaminant is carried into the biologically active pore space at a rate that just equals its removal rate, $vA/l = k'XA$ and $D1 = 1$. When $k'XA \ll vA/l$, then $D1 \ll 1$ and the time available for removal of contaminant is insufficient. A decrease in velocity (achieved by reducing the pumping rate, for instance) will increase overall removal rate. On the other hand, when $D1 \gg 1$, mass is delivered too slowly, and an increase in velocity will increase overall removal rate.

Both pathways 2 and 3 involve mass transfer coefficients, k_ws. These terms are a function of the diffusion coefficient for a given metal species and the distance over which the species diffuses. Because of the shorter diffusion path length, values of k_w for diffusion from the soil surface within the bioactive zone (pathway 2) will typically be larger than the values of k_w for diffusion through micro- and mesopores. In both cases, the relevant Damkohler number is $D2 = k'X/k_w$. When $D2 \gg 1$, mass transfer is much slower than the microbial reaction. Under such conditions, an increase in the rate of mass transfer will increase the overall removal rate. An increase in the reaction rate (say by addition of growth substrate to increase the concentration of microorganisms) will have little effect on the overall removal rate, but may significantly decrease solution-phase concentration of the target metal. In a given application, this could be important if escape of a contaminant during remediation must be minimized or if the effluent metal concentration must fall below a specific regulatory limit. Conversely, when $D2 \ll 1$, an increase in the microbial biomass concentration or the value of k' (say by addition of electron donor) will increase overall removal rate.

Figure 19.5 shows the flux relationships for diffusion (by either pathways 2 or 3) and microbial reduction of U(VI), where $k_w = 0.233$ day^{-1}, a value within the range of rates predicted for a site undergoing bioremediation at Oak Ridge, Tennessee. Here, U represents

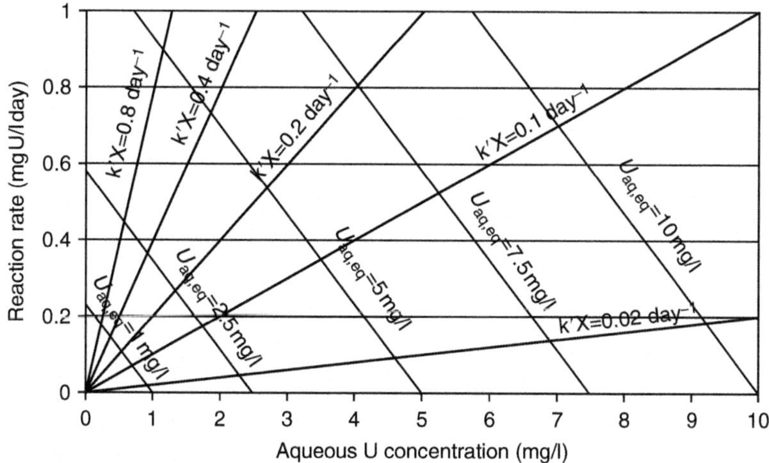

FIGURE 19.5 Diffusion and reaction flux lines for U(VI) transformation. The mass transfer coefficient, k_w, is 0.233 day^{-1}.

the soluble concentration of U(VI) in the bioavailable pore space. When $U = 0$, diffusive flux is maximum and equal to $k_w U_{eq}$, and as U increases, the driving force for diffusion decreases, dropping to zero when $U = U_{eq}$. The y-intercept term for these flux lines is $k_w U_{eq}$. Any change in solution chemistry that increases U_{eq} increases the y-intercept and thus increases the driving force for desorption. Figure 19.5 also shows the lines of reaction flux for microbial reduction at different values of $k'X$. Reaction flux is directly proportional to U, with a proportionality constant (slope) of $k'X$. Diffusive flux lines intercept reaction flux lines where the two fluxes balance one another: $k_w(U_{eq} - U) = k'XU$. At these points representing steady-state conditions, the corresponding overall rate of reaction can be read off the y-axis and the corresponding aqueous-phase U(VI) concentration can be read off the x-axis. Increasing values of $k'X$ correspond to increasing values of D2. Using Figure 19.5, we can define regions where diffusion is limiting (D2 > 10) and where reaction is limiting (D2 < 0.1).

For this example, microbial reaction rate is limiting over a wide range of uranium concentrations; diffusion only limits overall rate for a small concentration range at the low end. Thus, for this system, increasing X or k' will be most effective at increasing overall removal rates when U concentrations are high. Similarly, increasing pH or increasing total carbonate concentration to increase U bioavailability (by increasing the concentration of soluble uranyl carbonate species) will have greatest impact at low U concentrations.

A decrease in k_w occurs when contaminant mass is depleted on surfaces in the bioavailable zone. Under these conditions, the process that controls mass transfer is pathway 3, and the slope of the mass transfer flux lines can be expected to decrease.

19.6 CHEMICAL DELIVERY

Figure 19.6 shows the length and time scales important for bioremediation. Length scales vary over 12 orders of magnitude, from nanometer-scale interactions at soil surfaces to meter-scale interactions between wells. Of interest from the standpoint of bioavailability are the

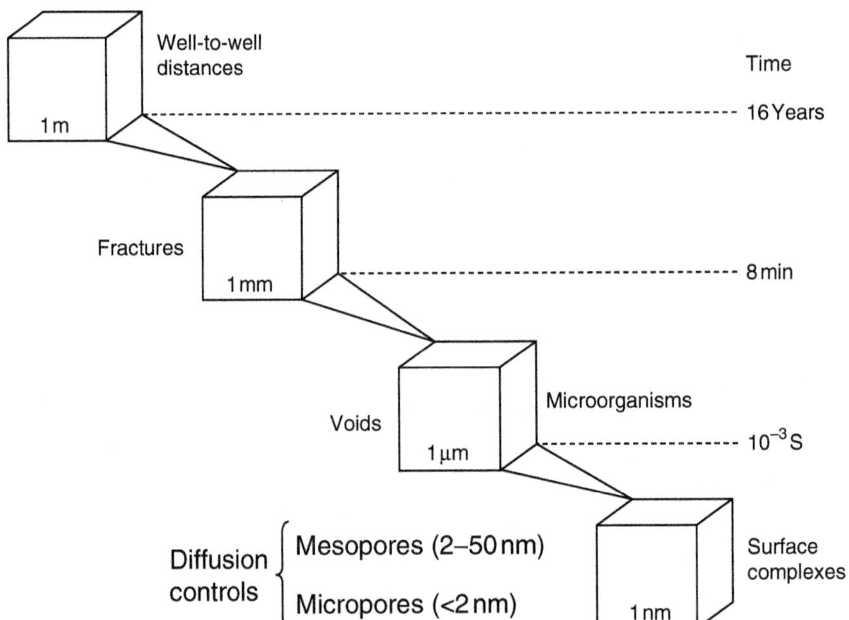

FIGURE 19.6 Time and length scales for diffusion. Time is approximately equal to $l^2/2D$, and D is assumed to be 10^{-9} m^2/s.

distances accessible by diffusion alone. At sites where contamination has been present for decades, contaminants can theoretically penetrate into regions that are meters from their original point of deposition. Remediation of such contamination may require a long-term removal strategy. In the short time periods normally planned for bioremediation, diffusion is only effective at delivering substrates and nutrients over a distance of millimeters, so other methods of chemical delivery must be considered.

A major challenge of chemical delivery within a porous medium is ensuring that the electron donor and primary electron acceptor are simultaneously available to microorganisms. This may be straightforward if either the donor or acceptor is already present in the aquifer in an insoluble form or sorbed to the solids. In such cases, addition of the soluble-limiting reactant by advection may be sufficient to stimulate growth. Fe(III), for example, is often naturally present as oxides on solid surfaces in the subsurface. In this case, advection may be an adequate means of delivering electron donor, assuming that Fe(III) is released into solution as water carrying the electron donor passes by. However, when the donor and acceptor are both soluble, mixing is best accomplished aboveground or in recirculation wells [47, 86]. Wells offer many advantages for chemical addition. They can be spaced and screened at different intervals and drilled to great depths, their pumping can be continuous or intermittent, and many different pumping configurations and recirculation schemes are possible, depending on local site conditions. Groundwater transport models provide useful insight into likely flow patterns between wells and fluid residence times, and are invaluable in the spacing of wells and selection of flow rates. Key parameters needed for these models are information about site geology, such as the boundaries on aquifers and the continuity of sand layers, and the spatial distribution of hydraulic conductivity. After wells are installed, residence time distributions can be measured with conservative tracers and the results used to calibrate and refine the transport model.

Chemicals are often easiest to add aboveground, and some contaminants are best removed aboveground. Examples of the latter include aluminum, which precipitates at pH values near 5, and nitrogen gas, the product of denitrification. If present at high concentrations, such species can easily cause clogging and alter flow paths, interfering with chemical delivery and hindering an assessment of process performance. Thus, a combination of aboveground elements (vessels for chemical addition and mixing, settling tanks, vacuum strippers, etc.) and belowground elements (wells and screens) is often desirable.

Figure 19.7 illustrates two possible options for well layout. Panel (a) shows an in-well mixing scheme in which two wells are screened at different levels and chemicals are added at different heights to stimulate remediation [86]. Such a scheme may include in-well purging to remove gases, but is best applied *in situ*ations where excessive chemical precipitate is not anticipated. Panel (b) describes a mixing scheme in which the mixing is carried out aboveground. Such a strategy provides considerable flexibility in design, allowing for continuous or intermittent recirculation and either continuous or intermittent addition of chemicals [47, 87].

19.7 SUMMARY

Biotransformation processes can be used to manipulate the solubility of metals, potentially resulting in their immobilization *in situ*, but the design and operation of such systems requires an understanding of a multitude of microbial, geochemical, and hydrological processes. The tools and concepts outlined in this chapter provide the fundamental framework for approaching the design and operation of such systems. The initial strategy is to develop balanced chemical reactions, model the essential geochemical processes, and derive kinetic expressions for the rates of microbial transformations and mass transfer. These efforts should be accompanied by bench- and pilot-scale studies. For field studies, hydrologic characterization is needed to assess the optimal strategy for chemical delivery. Transport models calibrated using

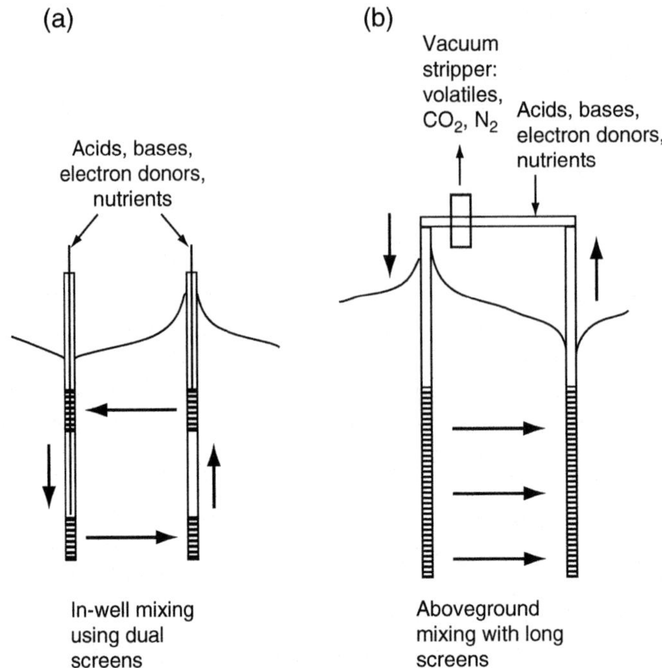

FIGURE 19.7 Examples of well configurations for chemical delivery: (a) an in-well recirculation scheme, and (b), a scheme for aboveground recirculation and mixing.

such data can be used to design effective chemical delivery schemes at a larger scale, and dimensionless numbers can be used to identify rate-limiting processes and to guide decision-making during operation.

REFERENCES

1. S Simonton, M Dimsha, B Thomson, LL Barton, and G Cathey. Long-term stability of metals immobilized by microbial reduction. Conference on Hazardous Waste Research, Denver, CO, 2000, pp. 394–403.
2. W Cullen and K Reimer. *Chem Rev* 89:713–764, 1989.
3. A Laverman, J Blum, J Chaefer, E Phillips, D Lovley, and R Oremland. *Appl Environ Microbiol* 61:3556–3561, 1995.
4. D Ahmann, A Roberts, L Krumholz, and F Morel. *Nature* 371:750, 1994.
5. D Newman, E Kennedy, J Coates, D Ahmann, D Ellis, D Lovley, and F Morel. *Arch Microbiol* 168:380–388, 1997.
6. F Caccavo, J Coates, R Rossello-Mora, W Ludwig, K Schleifer, D Lovley, and M McInerney. *Arch Microbiol* 165:370–376, 1996.
7. D Lovley, E Phillips, Y Gorby, and E Landa. *Nature* 350:413–416, 1991.
8. R Wildung, Y Gorby, K Krupka, N Hess, S Li, A Plymale, J McKinley, and J Fredrickson. *Appl Environ Microbiol* 66:2451–2460, 2000.
9. D Lovley. *Ann Rev Microbiol* 47:263–290, 1993.
10. B Tebo and A Obraztsova. *FEMS Microbiol Lett* 162:193–198, 1998.
11. T Blessing, B Wielinga, M Morra, and S Fendorf. *Environ Sci Technol* 35:1599–1603, 2001.
12. J Lloyd, A Mabbett, D Williams, and L Macaskie. *Hydrometallurgy* 59:327–337, 2001.
13. M Tucker, L Barton, and B Thomson. *J Environ Qual* 26:1146–1152, 1997.
14. L Macaskie, R Empson, A Cheetham, C Grey, and A Skarnulis. *Science* 257:782–784, 1992.
15. G Ferris. Calcite Precipitation and Trace Metal Partitioning in Groundwater and the Vadose Zone:

Remediation of Strontium-90 and Other Divalent Metals and Radionuclides in Arid Western Environments. Final Report to the U.S. Department of Energy, Project Number 70206, 2003.
16. R Bartlett and B James. *Adv Agron* 50:151–208, 1993.
17. AJ Bard, R Parsons, and J Jordan. *Standard Potentials in Aqueous Solution*. New York: Marcel Dekker, 1985.
18. I Grenthe, J Fuger, RJM Konings, RJ Lemire, AB Muller, C Nguyen-Trung, and H Wanner. *Chemical Thermodynamics of Uranium*. Amsterdam: North-Holland, 1995.
19. JA Dean, ed. *Lange's Handbook of Chemistry*. New York: McGraw-Hill, 1999.
20. B Rittmann and P McCarty. *Environmental Biotechnology: Principles and Applications*. New York: McGraw-Hill, 2001.
21. P McCarty. Energetics and bacterial growth. Fifth Rudolf Research Conference, Rutgers, The State University, New Brunswick, NJ, 1969.
22. C Myers and J Myers. *J Appl Bacteriol* 76:253–258, 1994.
23. J Kostka, D Dalton, H Skelton, S Dollhopf, and J Stucki. *Appl Environ Microbiol* 68:6256–6262, 2002.
24. C Liu, J Zachara, Y Gorby, J Szecsody, and C Brown. *Environ Sci Technol* 35:1385–1393, 2001.
25. D Lovley and E Phillips. *Appl Environ Microbiol* 54:1472–1480, 1988.
26. V Knight and R Blakemore. *Arch Microbiol* 169:239–248, 1998.
27. E Roden and J Zachara. *Environ Sci Technol* 30:1618–1628, 1996.
28. F Caccavo, R Blakemore, and D Lovley. *Appl Environ Microbiol* 58:3211–3216, 1992.
29. E Roden and D Lovley. *Appl Environ Microbiol* 59:734–742, 1993.
30. S Nagpal, S Chuichulcherm, A Livingston, and L Peeva. *Biotechnol Bioeng* 70:533–543, 2000.
31. S Okabe and W Characklis. *Biotechnol Bioeng* 39:1031–1042, 1992.
32. Y Konishi, N Yoshida, and S Asai. *Biotechnol Progr* 12:322–330, 1996.
33. M Cooney, E Roschi, I Marison, C Comninellis, and U von Stockar. *Enzyme Microb Technol* 18:358–365, 1996.
34. DR Lide, ed. *CRC Handbook of Chemistry and Physics*. Boca Raton, FL: CRC Press, 2002–2003.
35. RM Smith and AE Martell. *Critical Stability Constants*. New York: Plenum Press, 1982.
36. F Chapelle. *Ground-Water Microbiology and Geochemistry*. New York: John Wiley & Sons, 2001.
37. J Fredrickson, J Zachara, D Kennedy, C Liu, M Duff, D Hunter, and A Dohnalkova. *Geochim Cosmochim Acta* 66:3247–3262, 2002.
38. R Bartlett and B James. *J Environ Qual* 8:31–35, 1979.
39. S Brooks, J Fredrickson, S Carroll, D Kennedy, J Zachara, A Plymale, S Kelly, K Kemner, and S Fendorf. *Environ Sci Technol* 37:1850–1858, 2003.
40. JD Allison, DS Brown, and KJ Novo-Gradac. MINTEQA2/PROEDFA2: A Geochemical Assessment Model for Environmental Systems, U.S. Environmental Protection Agency, 1991.
41. DL Parkhurst and CAJ Appelo. User's guide to PHREEQC — A Computer Program for Speciation, Batch-Reaction, One-Dimensional Transport, and Inverse Geochemical Calculations: U.S. Geological Survey Water-Resources Investigations Report 99-4259, 1999.
42. G-T Yeh and V.S. Tripathi. HYDROGEOCHEM: A Coupled Model of HYDROlogic Transport and GEOCHEMical Equilibria in Reactive Multicomponent Systems, Oak Ridge National Laboratory, Oak Ridge, TN, 1990.
43. F Chapelle, P McMahon, N Dubrovsky, R Fujii, E Oaksford, and D Vroblesky (1995), *Water Resour Res*, *31*:359–371.
44. D Lovley and S Goodwin. *Geochim Cosmochim Acta* 52:2993–3003, 1988.
45. D Lovley, F Chapelle, and J Woodward. *Environ Sci Technol* 28:1205–1210, 1994.
46. D Vroblesky and F Chapelle (1994), *Water Resour Res*, *30*:1561–1570.
47. D Hyndman, M Dybas, L Forney, R Heine, T Mayotte, M Phanikumar, G Tatara, J Tiedje, T Voice, R Wallace, D Wiggert, X Zhao, and C Criddle. *Ground Water* 38:462–474, 2000.
48. M Truex, B Peyton, N Valentine, and Y Gorby. *Biotechnol Bioeng* 55:490–496, 1997.
49. J Spear, L Figueroa, and B Honeyman. *Environ Sci Technol* 33:2667–2675, 1999.
50. C Liu, J Zachara, J Fredrickson, D Kennedy, and A Dohnalkova. *Environ Sci Technol* 36:1452–1459, 2002.
51. C Liu, Y Gorby, J Zachara, J Fredrickson, and C Brown. *Biotechnol Bioeng* 80:637–649, 2002.
52. DE Lovely, EE Roden, EJP Phillips, and JC Woodward. Marine Geol 113:41–53, 1993.

53. SS Middleton, R Bencheikh Latmani, MR Mackey, MH Ellisman, BM Tebo, and CS Criddle. *Biotechnol Bioeng* 83:627–637, 2003.
54. Y Wang and H Shen. *Water Res* 31:727–732, 1997.
55. E Schmieman, D Yonge, M Rege, J Petersen, C Turick, D Johnstone, and W Apel. *J Environ Eng-ASCE* 124:449–455, 1998.
56. M Fukui and S Takii. *FEMS Microbiol Ecol* 13:241–247, 1994.
57. K Ingvorsen and B Jorgensen. *Arch Microbiol* 139:61–66, 1984.
58. K Finneran, R Anderson, K Nevin, and D Lovley. *Soil Sediment Contam* 11:339–357, 2002.
59. C Criddle. *Biotechnol Bioeng* 41:1048–1056, 1993.
60. L Alvarez-Cohen and P McCarty. *Environ Sci Technol* 25:1381–1387, 1991.
61. J Robinson and J Tiedje. *Appl Environ Microbiol* 45:1453–1458, 1983.
62. C Liu and J Zachara. *Environ Sci Technol* 35:133–141, 2001.
63. JA Dean, ed. *Lange's Handbook of Chemistry*. New York: McGraw-Hill, 2003.
64. RM Cornell and U Schwertmann. *The Iron Oxides*. Weinheim, German: VCH Publishers, 1996.
65. JA Rard, MH Rand, G Anderegg, and H Wanner. *Chemical Thermodynamics of Technicium*. Amsterdam: North-Holland, 1999.
66. D Langmuir. *Geochim Cosmochim Acta* 42:547–569, 1978.
67. R Ganesh, KG Robinson, GD Reed, and GS Sayler. *Appl Environ Microbiol* 63:4385–4391, 1997.
68. G Gadd. *Curr Opin Biotechnol* 11:271–279, 2000.
69. B Wielinga, M Mizuba, C Hansel, and S Fendorf. *Environ Sci Technol* 35:522–527, 2001.
70. J Fredrickson, J Zachara, D Kennedy, M Duff, Y Gorby, S Li, and K Krupka. *Geochim Cosmochim Acta* 64:3085–3098, 2000.
71. E Liger, L Charlet, and P Van Cappellen. *Geochim Cosmochim Acta* 63:2939–2955, 1999.
72. JS McLean, J-U Lee, and TJ Beveridge. In *Interactions between Soil Particles and Microorganisms*, Vol. 8, *IUPAC Series on Analytical and Physical Chemistry of Environmental Systems*, PM Huang, J-M Bollag, and N Senesi, eds. New York: John Wiley & Sons, 2002, pp. 227–261.
73. J McCarthy and J Zachara. *Environ Sci Technol* 23:496–502, 1989.
74. MF Hochella Jr and AF White, eds. *Mineral–Water Interface Geochemistry, Reviews in Mineralogy*, Vol. 23. Washington, D.C.: Mineralogical Society of America, 1990.
75. EA Jenne, ed. *Adsorption of Metals by Geomedia*. San Diego: Academic Press, 1998.
76. W Weber, P McGinley, and L Katz. *Water Res* 25:499–528, 1991.
77. U Gabriel, J Gaudet, L Spadini, and L Charlet. *Chem Geol* 151:107–128, 1998.
78. S Morrison, R Spangler, and V Tripathi. *J Contam Hydrol* 17:333–346, 1995.
79. E Voudrias and J Means. *Chemosphere* 26:1753–1765, 1993.
80. T Waite, J Davis, T Payne, G Waychunas, and N Xu. *Geochim Cosmochim Acta* 58:5465–5478, 1994.
81. EA Jenne. In *Adsorption of Metals by Geomedia*, EA Jenne, ed. San Diego: Academic Press, 1998, pp. 1–73.
82. M Kohler, G Curtis, D Kent, and J Davis, 1996. *Water Resour Res* 32:3539–3551.
83. G Turner, J Zachara, J McKinley, and S Smith. *Geochim Cosmochim Acta* 60:3399–3414, 1996.
84. TE Payne. In *Adsorption of Metals by Geomedia*, EA Jenne, ed. definitely tomorrow receive San Diego: Academic Press, 1998, pp. 75–97.
85. A. Ramaswami and R Luthy. In *Manual of Environmental Microbiology*, CJ Hurst, GR Knudsen, MJ McInerney, L Stetzenbach, and MV Walter, eds. Washington, D.C.: ASM Press, 1997, pp. 721–729.
86. P McCarty, M Goltz, G Hopkins, M Dolan, J Allan, B Kawakami, and T Carrothers. *Environ Sci Technol* 32:88–100, 1998.
87. MJ Dybas, DW Hyndman, R Heine, J Tiedje, K Linning, D Wiggert, T Voice, X Zhao, L Dybas, and CS Criddle. *Environ Sci Technol* 36:3635–3644, 2002.

Section III

Environmental Catalysis in Green Chemical Processing

20 Selective Oxidation

Rick B. Watson and Umit S. Ozkan
Department of Chemical Engineering, The Ohio State University

CONTENTS

20.1 INTRODUCTION TO SELECTIVE OXIDATION

20.1.1 Scope and Significance

Selective oxidation reactions are extensively used in the chemical industry to provide many useful intermediates including monomers, pharmaceuticals, fine chemicals, agrochemicals, fragrances, and flavorings. These are important reactions, providing a route to functionalize simple hydrocarbon molecules, to make them more useful. Examples of selective oxidation products include epoxides, aldehydes, ethers, esters, acids, and alcohols. More than 60% of the chemicals and intermediates synthesized via catalytic processes are products of oxidation and account for a net world-worth of US$20 to US$40 billion [1]. In particular, selective oxidation accounts for more than 20% of the total output of organic chemicals worldwide [2].

Selective oxidation reactions differ from other conversion processes because they aim to convert feed material to valuable products while minimizing the formation of by-products, such as carbon monoxide and carbon dioxide, which are the thermodynamically favored products. Carbon oxide formation could take place through further oxidation of the product as well as by complete oxidation of feed molecules or intermediate species. Selective oxidation reactions are carried out in the presence of a catalyst under controlled conditions to maximize the formation of the desired product. Often times, the selectivity of a reaction is far more important than its yield and processes may run at low conversion conditions to maximize selectivity. The selective oxidation reactions are carried out using both homogeneous and heterogeneous catalysts. Homogeneous catalysis may include oxidation with hydroperoxides and hydrogen peroxide and many other liquid-phase oxidation reactions. Heterogeneous catalytic oxidation processes are widely used to produce an array of chemicals and include aromatic oxidation to form acids and anhydrides; allylic oxidation to form aldehydes, nitriles, anhydrides and acids; epoxidation of olefins to oxides, methanol oxidation to formaldehyde; alkane oxidation to anhydrides and nitriles; and oxidative dehydrogenation (ODH) of alkanes to olefins.

20.1.2 Environmental and Economic Impact and Research Incentives

Vision 2020 [3] developed a 25-year vision for the chemical industry and outlined the challenges to be addressed in order to achieve this vision. The most significant areas for the application of catalyst technology in which improvements in homogeneous or heterogeneous catalyzed processes would help to achieve the goals of Vision 2020 are shown in Figure 20.1. Selective oxidation was chosen as the single most critical area in need for technological advancement. Furthermore, selective oxidation will also play a role in the second and third most crucial areas, namely alkane activation and the minimization of unwanted by-products. Development in these areas will be vital for the future of the chemical industry. In general, any environmentally friendly process should focus on lowering energy requirements via higher selectivity, more moderate temperature or pressure, and a reduced number of chemical unit operations.

While selective oxidation reactions account for a large portion of the chemical processes, there are many limitations and shortcomings about these processes. Many of these processes run at low selectivities and yields. Some processes pose serious environmental concerns and

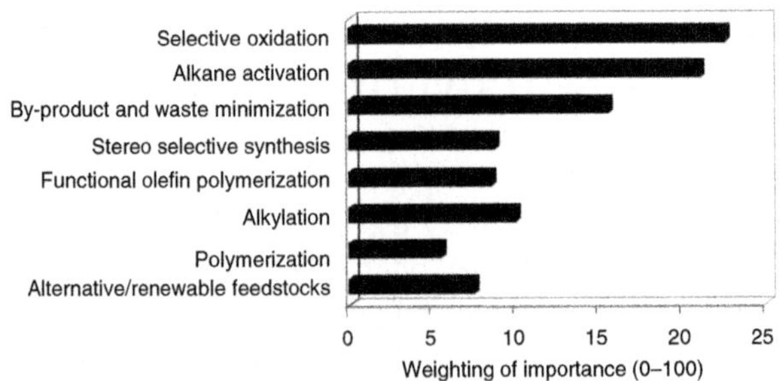

FIGURE 20.1 High impact areas for advancement in catalytic technology. (*Source*: Council for Chemical Research, U.S. D.O.E., and ACS. Catalyst Technology Roadmap Report: Build Upon Technology Vision 2020: The U.S. Chemical Industry. ACS Workshop, Washington, D.C., USA, 1997.)

challenges, such as carcinogenic feed materials or by-products. Use of certain oxidizing agents makes some processes economically unfeasible or environmentally unacceptable. For example, the most common methods in industrial oxidation for fine chemicals and pharmaceuticals use stoichiometric oxidizing agents, such as chromium (VI) compounds, hydrogen peroxide, potassium permanganate, chlorinated oxidizers, organic acids, and peroxides [4, 5]. While applicable to a wide range of reactions, the use of these materials suffer from major disadvantages, namely low efficiency, large amounts of toxic or hazardous waste, danger of explosive reagents, and high operating temperature and pressure. These major limitations strengthen the case for research into new processes, to ensure that chemical processes are better for the environment and also more economical for manufacturers. Current research in this area is focused on molecular oxygen as the oxidant, at low temperature and pressure with a heterogeneous catalyst, using environmentally friendlier feed materials. In general, research emphasis should be placed on the characterization of the different types of oxygen present on oxide surfaces and their role in alkane activation and subsequent oxidation. Identification of the factors controlling the selectivity in selective oxidation and ODH of alkanes and the selective oxidation of olefins and aromatics are also areas of fundamental research. Selective oxidation will be directed toward alkane activation through the study of factors influencing the controlled activation of C—H bonds and molecular oxygen on metals, metal oxides, and transition metal complexes.

The main drives and incentives for new and improved selective oxidation processes are listed below:

- Reduction and elimination of waste and solvent
- Increase selectivity to desired product
- Reduction and elimination of toxic and hazardous chemicals
- Use of renewable and alternative feedstocks.

Catalysis is a broad technical field and its great economic value is not as much the catalyst as a product but the reaction chemistry it provides. Similar targets could be set for unique processes and catalysts for selective oxidation, but the chemical industry is so large that to target just one catalyst or process would have little impact on overall industry energy usage or waste minimization. Instead, more general improvements within the field of selective oxidation could have deep economic, environmental, and energy usage impacts within the entire industry.

20.2 MECHANISTIC STEPS IN SELECTIVE OXIDATION REACTIONS

20.2.1 Homogeneous versus Heterogeneous Selective Oxidation

Interaction of a hydrocarbon molecule with oxygen at the surface of an oxide catalyst results in a network of parallel and consecutive elementary steps, in which hydrogen is abstracted, nucleophilic oxygen is inserted, electrophilic oxygen reacts with π-electrons, and C—C bonds are cleaved [6]. In the case of olefins, hydrogen abstraction followed by the nucleophilic addition of oxygen results in selective allylic oxidation, whereas interaction with the electrophilic oxygen leads to the cleavage of the C=C bond, to the formation of oxygenated alkane derivatives and to complete oxidation. For the oxidation to be selective, it must accelerate these steps that lead to the desired product and suppress side reactions.

A major subset of the selective oxidation reactions is the homogeneous, transition metal catalyzed transfer of oxygen from oxygen donors to organic substrates. Two types of mechanisms may prevail in these reactions. In the first, heterolytic activation and transfer of the activated oxygen, usually in the form of an alkylhydroperoxide [7], to a substrate by a metal center takes place. The substrate is usually ligated to the transition metal during the transfer. The second type of mechanism involves the formation of an oxo-metal intermediate, which then transfers the oxygen to the substrate. Examples of oxo-metal intermediates include Mn, Cr, Os, and Ru in their highest oxidation states [7]. This type of mechanisms involves a redox change at the metal center. Many transition metal cations fixed in organic or inorganic homogeneous matrixes are available for these types of reactions by the oxygen atom transfer from various kinds of oxidants. This scheme is presented in Figure 20.2 [8]. An example of this mechanism is the oxidation of ethylene to acetaldehyde by the Wacker process. A silica-supported $PdCl_2$—$CuCl_2$ molten salt catalyst is used. Bubbling ethylene and oxygen through an acetic acid solution of palladium and cupric chlorides yields acetaldehyde:

$$C_2H_4 + PdCl_2 + H_2O \longrightarrow C_2H_4O + 2HCl + Pd$$

$$Pd + 2CuCl_2 \longrightarrow PdCl_2 + 2CuCl$$

$$2CuCl + 1/2O_2 + 2HCl \longrightarrow 2CuCl_2 + H_2O$$

During the reaction, palladium forms a complex with ethylene, is reduced to Pd(0), and is then reoxidized by Cu(II) [9]. In a similar process, the oxidation of propylene to acetone is also accomplished with this catalyst. Furthermore, various noble metal substituted polyoxometalates with differing structures have been prepared and used in catalytic transformations with molecular oxygen, hydrogen peroxide, *t*-butylperoxide, and others [10].

Many important organic chemicals are also produced via heterogeneous selective oxidation catalysis. Generally, these processes are based on metallic oxides, either supported or unsupported by a refractory support. Heterogeneous oxidation catalysts mostly fall into two categories, depending on the method by which oxygen participates in the oxidation. One category represents a major class of selective oxidation catalysts, which are most often mixed oxides of transition metals, containing two or more metal cations. Here, the oxygen is readily transferred to and from the bulk structure of the catalyst. Examples include iron molybdates for methanol oxidation to formaldehyde, vanadium molybdates for acrolein oxidation to acrylic acid, and bismuth–molybdenum–cobalt–iron–potassium mixed oxides for propylene oxidation to acrolein. Another class of heterogeneous oxidation catalysts is supported metals

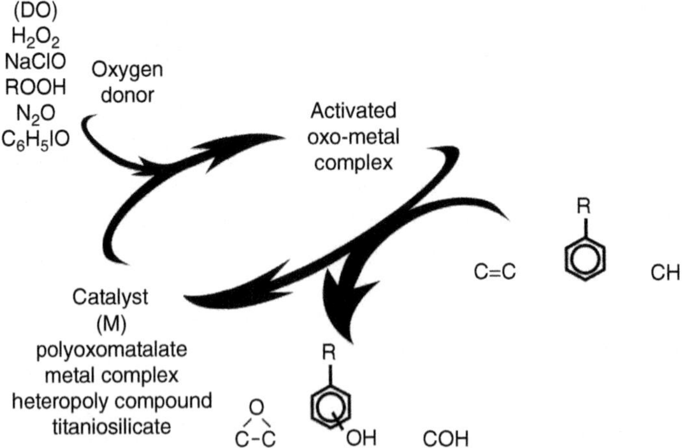

FIGURE 20.2 Oxygen atom transfer from oxygen donors. (*Source*: Y Moro-oka. *Catal. Today* 45: 3–12, 1998.)

on which oxygen can readily chemisorb. Examples of supported metals include supported silver catalysts for the ethylene-to-ethylene oxide process and supported palladium for vinyl acetate production. In addition to metals, these catalysts may also contain a variety of promoters to increase their activity, selectivity, structural strength, and lifetime.

20.2.2 Oxygen Species and Oxygen Insertion Mechanisms

In a heterogeneous selective oxidation, an oxygen-containing gas is introduced directly as a feed into the process, while in homogeneous reactions, oxygen is often incorporated into the reaction through a carrier, such as peroxides.

Oxides have been an important class of catalysts for many years, but the precise role of oxygen in these systems is often hard to characterize. The most commonly used oxidant is gaseous O_2, but it is generally believed that gaseous oxygen must be converted to a more active form or be incorporated into an oxide matrix before it will appear in oxidation products [11]. Several kinds of active oxygen species such as O^-, O^{2-}, and O_2^- have been detected on the surface of oxidation catalysts. The nature of these active oxygen species found on the catalyst surface depends strongly on the counter metal cation and structural conditions. Surface oxygen species can exist as mononuclear or molecular, neutral, or charged. It is often assumed that adsorbed molecular oxygen can accept electrons one by one until ultimately forming O^{2-} lattice ions [12]. The process is shown below:

$$O_{2ads} \longrightarrow O_{2ads}^- \longrightarrow O_{2ads}^{2-}/O_{ads}^- \longrightarrow O_{ads}^{2-} \longrightarrow O_{lattice}^{2-}$$

In particular, there has been great interest in the reactivity of the O^- ion for the role it may play in surface oxidation reactions [11]. However, oxide catalysts can be classified into two main groups according to the kind of oxygen species generated on the surface [13]. One group generates O^- and O_2^- species upon adsorption. These species are highly active and are believed to lead to total oxidation products. The second group of oxide catalysts does not adsorb oxygen readily and the lattice oxygen is the principal oxygen species used for product formation. In this second group, the adsorption of molecular oxygen is only possible if the catalyst exists in a partially reduced state, thereby replenishing the oxygen removed by the products.

The selective catalytic oxidation can be electrophilic in nature, in which the activation of oxygen to form one of the previously mentioned species is the first step. This is followed by nucleophilic oxidation, which proceeds through the activation of a hydrocarbon molecule. An example of a reaction in which oxygen activation is believed to occur first is the oxidation of olefins to form oxides. In the selective oxidation of ethylene, air or pure oxygen is fed into the reactor with ethylene to perform the following reaction:

$$H_2C{=}CH_2 + \tfrac{1}{2}O_2 \longrightarrow H_2C{-}CH_2 \text{ (with bridging O, ethylene oxide)}$$

The partial oxidation (POX) of ethylene competes with two complete oxidation reactions of the reactant ethylene and product ethylene oxide. The reaction occurs industrially on the surface of a silver catalyst. Initially, oxygen is activated and absorbed molecularly onto the surface of supported silver particles. It then reacts with ethylene forming ethylene oxide, leaving atomic chemisorbed oxygen. Atomic oxygen, left on the surface, is believed not to have the ability to generate ethylene oxide but to oxidize ethylene or ethylene oxide to CO and CO_2 [14]:

$$[Ag] + O_2 \longrightarrow [Ag].O_{2ads}$$

$$[Ag].O_{2ads} + H_2C{=}CH_2 \longrightarrow H_2C{-}CH_2 \text{ (epoxide, bridged by O)} + [Ag].O_{ads}$$

$$[Ag].O_{2ads} + 2\,CO \longrightarrow 2\,CO_2 + [Ag]$$

$$4\,[Ag].O_{ads} + H_2C{=}CH_2 \longrightarrow 2\,CO + 2\,H_2O + 4\,[Ag]$$

A schematic representation of the reaction between the adsorbed oxygen species and a gaseous ethylene is shown in Figure 20.3. Most authors agree that the epoxidation and complete combustion are coupled with each other and that the ratio of these rates depends on the interaction of the different species with the silver surface [15]. Three different oxygen species (molecular, atomic, and subsurface) can be found on the catalyst surface. Their role has not been clearly assigned, and although initially molecular oxygen was considered to be responsible for epoxidation and atomic oxygen for combustion, other studies have shown that atomic species, to some extent, may be responsible for both reactions. The selectivity and activity of Ag catalysts are primarily determined by the catalyst manufacturing process. Important parameters include silver particle size, impregnation method, support material, and existence of a cocatalyst or promoter.

Unlike epoxidation, in the oxidation of olefins to anhydrides and nitriles, activation of the hydrocarbon is believed to be the first step as well as the rate-determining step. Still, the elementary steps of a catalytic reaction over heterogeneous catalysts are most commonly the reaction between two adsorbed species or direct reaction of gaseous molecules with the adsorbed species. Well-understood mechanisms have been established where reactions consist of steps including adsorption of the reactant molecules, transport of the reactant molecules over the surface, a series of surface reactions, and desorption of the reaction products. In the Mars–van-Krevelen redox mechanism [16], oxygen originating from the catalyst bulk is introduced into the reactant molecule. Hence, the catalyst is reduced in this step. Subsequently, the bulk of the catalyst is replenished with oxygen from the gas phase, i.e., the catalyst is reoxidized. The rate-determining step in this mechanism is often, but not always, taken as the diffusion of oxygen in the catalyst bulk. The mechanism is represented schematically in Figure 20.4. In this mechanism, the reactant is adsorbed on a metal site to form a chemisorbed species and reacts with the lattice oxygen associated with metal A to produce a partially oxidized product. A lattice oxygen from a neighboring metal B site moves to replenish the lost oxygen while electrons move from metal A to metal B. Subsequently, molecular oxygen reacts with metal A to restore its original oxidation state.

Taking a simplified example of a hydrocarbon (HC) oxidation over a metal oxide catalyst, the reaction can be represented in terms of two elementary steps as follows:

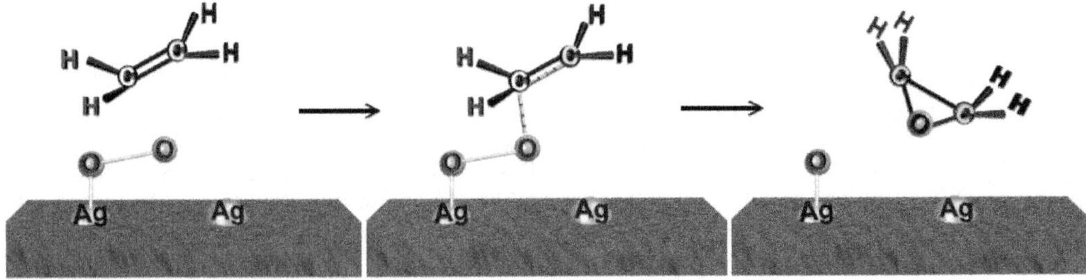

FIGURE 20.3 Reaction of ethylene with the molecularly adsorbed oxygen a silver surface, showing interaction of the oxygen with one of the carbons in ethylene. (*Source*: H Nakatsuji, ZM Hu, and H Nakai. *Int. J. Quantum Chem.* 65: 839–855, 1997.)

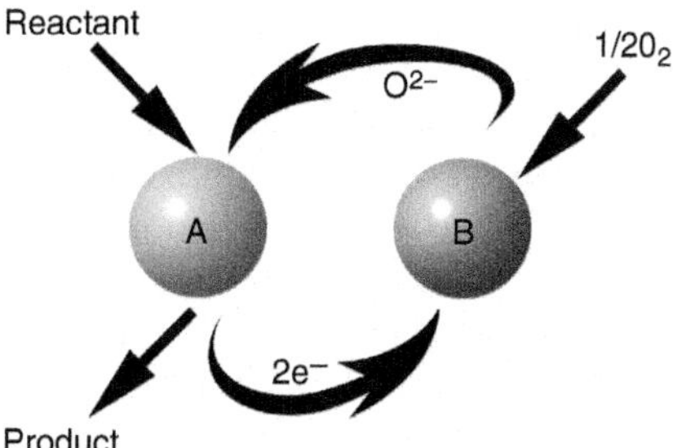

FIGURE 20.4 The oxidation–reduction mechanism of heterogeneous selective oxidation. (Mars–van-Krevelen.)

$$\mathrm{HC} + \mathrm{M\text{-}O} \xrightarrow{k1} \mathrm{HC\text{-}O} + \mathrm{M\text{-}}\square$$

$$2\mathrm{M\text{-}}\square + \mathrm{O_2} \xrightarrow{k2} 2\mathrm{M\text{-}O}$$

where M-$\square$ represents a reduced site.

The rates of surface reduction with hydrocarbon and reoxidation are then given by:

$$r_{\mathrm{red}} = (k1)(P_{\mathrm{HC}})q_{\mathrm{o}} \quad \text{and} \quad r_{\mathrm{ox}} = (k2)(P_{\mathrm{o_2}})q_{\mathrm{r}}^2$$

where θ_{o} and θ_{r} are the fraction of oxidized and reduced sites present on the catalyst, respectively. However, when the rate of catalyst reoxidation is much faster than reduction, the overall rate of hydrocarbon oxidation is equal to the rate of catalyst reduction, or:

$$-\frac{\mathrm{dHC}}{\mathrm{d}t} = (k1)(P_{\mathrm{HC}})$$

In this scenario, the rate of hydrocarbon oxidation is zero order in oxygen. Under certain conditions, however, the reverse can occur. That is, if the reoxidation is the rate-determining step, then the overall rate can be independent of hydrocarbon concentration and the rate is proportional to gas-phase oxygen concentration.

$$-\frac{\mathrm{dHC}}{\mathrm{d}t} = (k2)(P_{\mathrm{o_2}})$$

Both situations are commonly found in selective oxidation reactions and most strongly influenced by temperature.

20.2.3 Selectivity

During selective oxidation of hydrocarbons, nonselective mechanisms can exist in which oxygen, from the lattice or activated from the gas phase, can be inserted into the hydrocarbon, and several reaction steps advance ultimately to form carbon oxides. For selective catalysts, there should exist an optimal balance in the oxygen activity, acid–base characteristics, and lattice diffusivity. It is therefore essential for a catalyst to possess a certain degree of structural complexity [17]. Intermediate reducibility, weak Lewis acid centers, and oxygen mobility represent the essential requirements for selective oxidation. However, quantitative correlations between these properties and catalytic performance cannot easily be obtained and these characteristics are therefore usually expressed in the literature as a "good-mix" or "favorable balance" between acid–base characteristics and redox behavior. The "favorable" oxygen that can provide this requirement is the one which binds strongly enough to the surface to have attenuated oxidizing strength but weakly enough to oxidize the reactant molecule selectively. Over-supported transition metal oxides, the species of interest exist in the form of M=O, M—O—M, or M—O-support bonds, where M is the supported transition metal. The nature of the active oxygen will certainly depend upon transition metal loading, dispersion, support effects, and the addition of modifiers [18].

Extensive research has been performed regarding the selectivity in selective oxidation reactions. It has been observed that both product and feed hydrocarbons react to form carbon oxides, resulting in reduced selectivity. At low conversions, most of the carbon oxides are formed from oxidation of the feed molecule. At high conversion, though, most of the formation is due to oxidation of the product. This concept is perhaps best described in the case of the ODH of propane to form propylene. ODH selectivity is limited considering the high reactivity of propylene toward further oxidation. As discussed by Kung in a review article [19], the limitation arises from consecutive reactions, namely propane to propylene and propylene to carbon oxides. It was shown that, perhaps on all catalysts, the rate for the second reaction could be five to ten times higher than that of the first reaction. An examination of literature data for some of the most studied propane ODH catalysts is shown in Figure 20.5. Since the propylene formed can further decompose to carbon oxides, selectivity (defined as the ratio of propylene to total products) to propylene decreases. At very high propane conversions, selectivity to propylene drops to low levels.

20.3 SELECTIVE OXIDATION AND THE ENVIRONMENT

Selective oxidation reactions are extensively used in the chemical industry to provide many useful chemicals such as epoxides, aldehydes, esters, organic acids, and alcohols. Selective oxidation plays an essential role in improving a great number of processes, lowering the cost and energy requirements and minimizing the negative environmental impact. There are numerous opportunities to improve existing chemical processes, for environmental as well as economic reasons, some of which are currently researched. Below are some examples of the role selective oxidation catalysis can play in protecting the environment. While some of the examples represent processes already in place, others highlight the potential for future

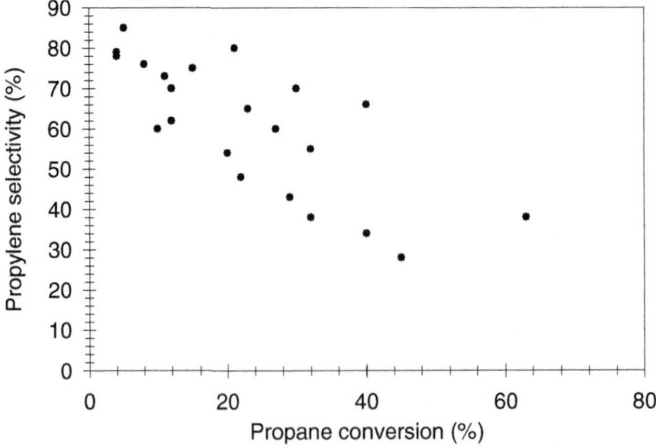

FIGURE 20.5 Decreasing selectivity for propylene with increasing propane conversion. (*Source*: HH Kung. *Adv. Catal.* 40: 1, 1994.)

processes that could lessen the negative impact on the environment, if they can be implemented in a cost-effective manner.

20.3.1 Production of Maleic Anhydride

Maleic anhydride was first commercially produced in the early 1930s by the vapor-phase oxidation of benzene [20]. The use of benzene as a feedstock for the production of maleic anhydride was dominant in the world market well into the 1980s and several processes have been used. Benzene, although easily oxidized to maleic anhydride with high selectivity, is an inherently inefficient feedstock since two excess carbon atoms are present in the raw material. Various C4 compounds have been evaluated as raw material substitutes for benzene in the production of maleic anhydride. The replacement of the benzene process with the cleaner reaction based on *n*-butane oxidation is an example of the way selective oxidation research benefits the environment.

Rapid increases in the price of benzene and the recognition of benzene as a carcinogen intensified the search for alternative process technology. Furthermore, the benzene process gave rise to large-molecule by-products, such as phthalic anhydride and benzoquinone. These factors led to the first commercial production of maleic anhydride from butane. By the early 1980s, the conversion of the U.S. maleic anhydride manufacturing capacity from benzene to butane feedstock was underway. One factor that inhibited the conversion of the installed benzene-based capacity was that early butane-based catalysts were not active and selective enough to allow the conversion of benzene-based plant without significant loss of capacity. Advances in catalyst technology, increased regulatory pressures, and continuing cost advantages of butane over benzene have led to a rapid conversion of benzene to butane-based plants. By the mid-1980s in the United States 100% of maleic anhydride production used butane as the feedstock.

One of the "overall" mechanisms of *n*-butane oxidation to maleic anhydride is shown in Figure 20.6. The formed maleic anhydride can also react with oxygen to form CO_2. As typically found in selective oxidation, as butane conversion increases, maleic anhydride selectivity decreases. Probing the reaction mechanism further, several surface intermediates in the reaction have been proposed in the literature [21, 22], such as crotonaldehyde, dihydrofuran, butadiene, furan, and crotonlactone.

From an environmental point of view, the butane process produces less gaseous products without benzene and benzene derivatives and less CO and CO_2; it uses less process water and, moreover, produces less liquid waste. From an economic point of view, maleic anhydride production cost is lower compared to that of the benzene process. A high-purity maleic anhydride grade can be achieved, as specifically required in some applications. The overall benefits of replacing the feedstock by butane include reduction in gaseous emissions, reduction in greenhouse gas emissions, reduction in aqueous effluents, reduction of process water, minimized solvent consumption, energy savings due to a less complex process with fewer unit operations, and reduction of involved environmental risks from the benzene cycle (production, storage, transport, and handling).

The increased importance of the butane-to-maleic anhydride conversion route has resulted in efforts made to understand and improve this process. The predominant area of

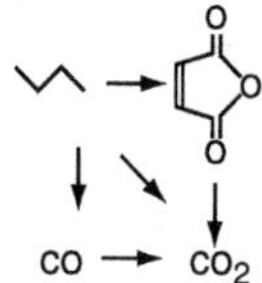

FIGURE 20.6 Overall maleic anhydride reaction mechanism from *n*-butane.

research concerns the catalyst because it is at the center of this process. The complexity of the butane-to-maleic anhydride reaction network that involves abstraction of eight hydrogen atoms, insertion of three oxygen atoms, and a ring closure necessitates a multifunctional catalyst that can promote all of these steps. The catalyst used in the production of maleic anhydride from butane is vanadium phosphorus oxide (VPO). This catalyst is the only commercially viable system that selectively produces maleic anhydride from butane. Several routes may be used to prepare the catalyst, but a frequent route cited in patent literature involves the reaction of vanadium (V) oxide and phosphoric acid to form vanadyl hydrogen phosphate. This material is then heated to eliminate water from the structure and irreversibly form vanadyl pyrophosphate, $(VO)_2P_2O_7$. Vanadyl pyrophosphate is believed to be the catalytically active phase required for the conversion of butane to maleic anhydride [23].

20.3.2 From Air to Oxygen

Conventionally, air is desired as the source of oxygen gas for a number of liquid- and gas-phase oxidation reactions. The air is usually delivered into the reactors via axial flow impellers or gas spargers in liquid-phase reactions, which break up the gas into tiny bubbles providing a high interfacial area for mass transfer between the gas (oxygen) and the liquid reactants in the reaction vessel [24].

Since air also contains the inert nitrogen gas, the reactor vent can contain significant amounts of solvent and reactants entrained in the gas as vapor. The recovery of such organic chemicals from the gas requires costly chemical treatment. In general, solvent loss is higher than it would be when using an oxygen-enriched gas. Additionally, the lower concentration of oxygen in air depresses the oxidation rate, necessitating higher reaction times and larger reactors. Oxidation with pure or a high-purity oxygen gas reduces the amount of the vent gases from the reactor and results in a much lower loss of solvent and reactants. Reaction time is reduced, and smaller-sized or a fewer number of reactors are needed. This gives a net saving in terms of capital expenditure and operating costs. But, equally important is the reduced impact on the environment that is achieved by substituting air with oxygen.

An example of selective oxidation improvements through the use of air is the production of formaldehyde by the selective oxidation of methanol. In the formaldehyde production process, methanol reacts with air in the presence of a catalyst to produce formaldehyde. Process yields are normally quite high, with over 95% to 98% of methanol ending up as formaldehyde [1]. Some secondary compounds are generated in the oxidation process due to catalyst inefficiencies. The emission by-products that require control are carbon monoxide, dimethyl ether, methanol, and unscrubbed formaldehyde. Both formaldehyde and methanol themselves are classified as hazardous air pollutants and require stringent emission reduction. Because of flammability reasons, the process was unable to operate under high concentrations of methanol (5–6%), largely limiting productivity and hindering separation processes. Even though the conversion and selectivity of the original process were very high and significant improvements did not seem possible, by simply adding and adjusting recycle ratios, the process is able to operate under a lower percentage of oxygen and a higher percentage of methanol, enabling higher productivity and greatly reducing the amount of emissions [25]. These types of improvements, economically and environmentally, have led to major modification of existing methanol oxidation processes.

20.3.3 Olefin Epoxidation

The classic manufacturing route for making epoxides (the chlorohydrin process), while no longer used for ethylene oxide formation, still accounts for 55% of global propylene oxide production. The overall selectivity to ethylene oxide by the chlorohydrin route was roughly 80% based on ethylene. However, chlorine gas is lost during the process into by-products.

With 1 kg of ethylene oxide formed, roughly 3 kg calcium chloride, 0.5 kg dichloroethane, and 0.1 kg of 2,2-dichlorodiethyl ether are produced [26]. The chlorohydrin process for ethylene oxide production is shown below:

$$CH_2{=}CH_2 + Cl_2 \xrightarrow{H_2O} \underset{OH}{CH_2}{-}\underset{Cl}{CH_2} + HCl \xrightarrow{Ca(OH)_2} \underbrace{CH_2{-}CH_2}_{O} + CaCl_2$$

The high chlorine gas requirement and significant by-products formation presented serious adverse effects for the environment while making the overall operation a costly process. Therefore, replacement of this process with an economic and environmentally more benign direct oxidation process became extremely important. Compared to the chlorohydrin process, direct oxidation eliminates the need for large volumes of chlorine. Also there are no chlorinated hydrocarbons or calcium chloride by-products, processing facilities can be made simpler, and operating costs are lower.

For propylene oxide, the chlorohydrin route involves reaction of propylene with dissolved chlorine (hydrochloric and hypochlorous acids) to form a chlorohydrin, followed by dechlorination using aqueous calcium hydroxide or sodium hydroxide. For each kg of propylene oxide manufactured, 40 kg of waste is produced in the form of an aqueous solution containing about 5% calcium chloride or sodium chloride; the chlorine contained in the waste is not recovered [24].

Alternatives to chlorohydrin process involve a two-step process, in which an organic reactant (isobutane or ethylbenzene) is first converted to a hydroperoxide, which is used to epoxidize the propylene. This process generates a coproduct (*t*-butyl alcohol or styrene). The economics of the process are then constrained by the need for a commercial demand for both products, which helps to explain why the chlorohydrin process is still widely used. The direct gas-phase oxidation of propylene by molecular oxygen has long been the most desirable as well as challenging goal of production of propylene oxide. However, despite considerable effort, no economically viable catalyst has yet been found. Most publications show either low selectivity or low conversion. Although the direct oxidation of ethylene by molecular oxygen to ethylene oxide has been commercialized with a silver catalyst, it is known that the analogous direct oxidation of propylene exhibits a low selectivity of less than 15% [27]. Thus, there is still a critical need in the selective oxidation for the development of an efficient direct route to propylene oxide through the use of heterogeneous catalysis.

A promising alternative to the previously mentioned methods is the use of a titanium silicalite catalyst (TS-1). TS-1 catalyzes the epoxidation of propylene with hydrogen peroxide at high conversion rate and selectivity under mild conditions in the liquid phase. The reaction occurs at low temperature using methanol as the solvent and aqueous hydrogen peroxide. Hydrogen peroxide is generally considered to be a "green" oxidant, as it is relatively nontoxic and breaks down in the environment to nontoxic by-products and thus, is highly desirable. However, the current method for production of hydrogen peroxide, which involves the sequential hydrogenation and oxidation of an alkyl anthraquinone, is capital-intensive, produces significant volumes of waste, and consumes sizeable quantities of energy during the purification and concentration of the product [28]. Therefore, there is considerable research in selective oxidation aimed toward the generation of hydrogen peroxide *in situ* by employing a mixture of O_2 and H_2. In these cases, the Ti-catalyst is modified with precious metals such as palladium and platinum, which catalyze the direct synthesis of hydrogen peroxide.

20.3.4 Photocatalytic Reactions

One of the major research goals of so-called green chemistry is to develop environmentally acceptable routes to important organic products. Methods by which this can be achieved are increasing product selectivity, aiming for 100% efficiency and replacing stoichiometric reagents with heterogeneous catalysts thus allowing easy separation of the catalyst from the product [29]. One of the possible solutions is photocatalytic reactions, which are applicable to a wide range of valuable industrial processes, including organic synthesis, photodestruction of toxic compounds, and purification of drinking water. Over the past two decades, there has been considerable work aimed at utilizing semiconductors as photocatalysts [30]. The feasibility of this technology on a commercial scale has also been demonstrated by the implementation of numerous small-scale applications, including treatment of air and water streams. The photocatalytic oxidation of many organic molecules, including saturated hydrocarbons, by optically excited semiconductor oxides is thermodynamically allowed in the presence of oxygen at room temperature. The oxide of choice is TiO_2 because it is inexpensive, nontoxic, and does not undergo photocorrosion. TiO_2 may be regarded from the point of view of semiconductor theory. From that perspective, absorption of a near UV photon can be regarded as exciting an electron from the valence band to the conduction band. A surface-trapped electron is readily captured by molecular O_2 and the hole left in the valence band is a strong oxidizing agent. It can even oxidize water to give the OH radical that attacks a very wide range of organic molecules. Thus, irradiation of TiO_2 can initiate the oxidation of most organic molecules by O_2.

For selective oxidations, photocatalytic oxidation selectivities different from those obtained by other oxidation means have been reported, showing the potential of the method for syntheses where the expected product is obtained with an acceptable yield. Gonzalez et al. [31] have reported the results for the photocatalyzed oxidation of toluene, ethylbenzene, cyclohexane, and methylcyclohexane to their corresponding oxygenates. The authors feel that further research must be performed to increase the efficiency and kinetics of the described reactions. In particular, the high oxidation potential of TiO_2 must be controlled to prevent over-oxidation of the primary oxygenates.

20.3.5 Partial Oxidation for Hydrogen Production

Hydrogen energy is likely to play an increasing role in meeting the future energy needs of our nation. Hydrogen provides a "cleaner" energy source since water is the only product from its combustion. Another important use of hydrogen is as a fuel for fuel cells, where the chemical energy of hydrogen can be converted to electricity far more efficiently than combustion. Market observers say that fuel-cell R&D budgets have climbed three orders of magnitude in the past several years and that annual spending in the industry has now reached the billion-dollar range [32]. However, for the hydrogen energy to fulfill its potential for protecting the environment, providing more efficient energy sources, and decreasing our nation's dependence on foreign oil, economical and efficient technologies for hydrogen production need to be developed.

The hydrogen production process takes place in several steps. First, a hydrocarbon feedstock (such as natural gas or a liquid fuel) is reformed at high temperature in the presence of a catalyst. Depending on the type of reformer, the feedstock reacts with steam or oxygen at high temperature to produce a synthetic gas or "syngas" composed of H_2, CO, CO_2, CH_4, and H_2O. The syngas is further processed to increase the hydrogen content (CO in the syngas is converted to hydrogen via the water gas shift reaction). Finally, hydrogen is separated out of the mixture at the desired purity, up to 99.999% for fuel cell applications. There are various types of reforming and chemical processes to produce hydrogen [33] including steam reforming, coal gasification, methanol dissociation, ammonia dissociation, and water electrolysis. Here, we will focus on POX, another commercially available method for deriving hydrogen from hydrocarbons. For example, methane (or some other hydrocarbon feedstock such as oil, diesel, or gas) is oxidized to produce carbon monoxide and hydrogen according to:

$$CH_4 + \frac{1}{2}O_2 \longrightarrow CO + 2H_2$$

The reaction is exothermic and no indirect heat exchanger is needed. Catalysts are not required because of the high temperature. However, the hydrogen yield per mole of input can be significantly enhanced by use of a precious metal catalyst [34]. A hydrogen plant based on POX includes a POX reactor, followed by a shift reactor and hydrogen purification equipment. Large-scale POX systems have been used commercially to produce hydrogen from hydrocarbons such as residual oil, for applications such as refineries. Large systems generally incorporate an oxygen plant, because operation with pure oxygen, rather than air, reduces the size and cost of the reactors.

20.3.6 Other Examples

20.3.6.1 Caprolactam Production

Caprolactam forms the basis for the production of Polyamide 6 production (nylon 6). Caprolactam is classically synthesized from cyclohexanol in three steps [35]. This process is illustrated in Figure 20.7. First, cyclohexanol is oxidized to cyclohexanone using chromium (VI). Cyclohexanone is converted into an oxime via reaction with hydroxylamine and sulfuric acid. The oxime is then converted into caprolactam through a Beckman rearrangement. However, this process yields undesirable ammonium sulfate in a greater amount than the desired caprolactam, which poses serious waste disposal challenges. Use of hydroxylamine, which is toxic, is another drawback for the existing process. As the profitability of caprolactam manufacturing strongly depends on the amount of ammonium sulfate, various processes to avoid this drawback have been investigated by many researchers.

Selective oxidation may offer a more environment-friendly alternative to this process. One possibility is ammoximation of cyclohexanone with ammonia and hydrogen peroxide over a heterogeneous catalyst [36]. This process eliminates the use of hydroxylamine and sulfuric acid as well as production of ammonium sulfate. An even more desirable alternative is the gas-phase ammoximation of cyclohexanone with ammonia and molecular oxygen, but the research literature reports low selectivities so far [37].

20.3.6.2 Reaction Engineering Solutions

Selective oxidation reactions combined with innovative reaction engineering approaches may also offer environment-friendly process alternatives. One nontraditional reactor design is simultaneous separation of the more reactive hydrocarbon products from the reaction medium

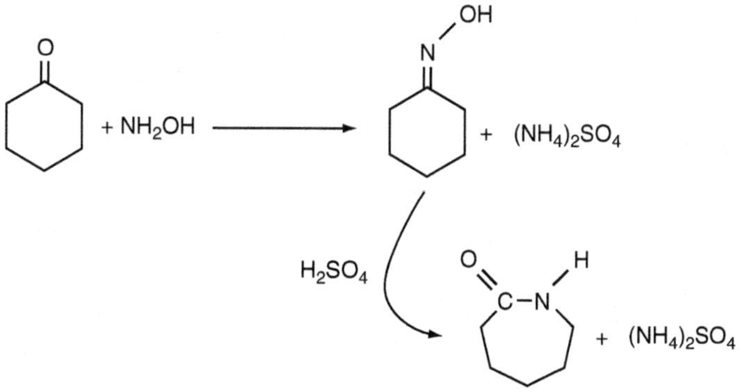

FIGURE 20.7 Commercial caprolactam production.

to minimize product over-oxidation. However, hydrocarbon separations, including alkanes from olefins and alcohols are not feasible at high temperatures. Another nontraditional reactor design improves POX yields by capitalizing on the oxygen concentration effect [38]. The extent of combustion may increase as the hydrocarbon to oxygen ratio decreases. Operation at a continuous low oxygen partial pressure by use of a distributed feed has distinct advantages. Reactors are designed to control the addition of oxygen along the reaction path. This limits the "runaway" oxidation reactions that generate high temperatures and unselective reactions observed with premixed feeds [39]. Careful adjustment of the oxygen addition profile allows the reaction to run in an oxygen-limiting mode, which results in very low waste product (CO_x) formation. Moderate temperatures improve catalytic oxidation selectivity, because high temperatures increase unselective gas-phase reactions faster than catalytic reactions.

Millisecond contact time reactors utilizing monolith or gauze catalysts present another example of how innovative reaction engineering approaches can make selective oxidation reactions friendlier to the environment. Millisecond contact time reactors have been previously used to catalyze conversion of methane to syngas [40, 41], ODH of ethane to ethylene [42, 43], and conversion of C_4 and higher alkanes to oxygenates [44, 45]. The reactions considered, including selective and POXs, are very fast and, therefore, limited by mass or diffusional limitations. Ethylene formation by catalytic ODH has been reported up to 70% selectivity at more than 80% ethane conversion over Pt-coated monoliths at 5 ms with no carbon formation [46, 47]. Product selectivity is highly dependent on reaction conditions. Low carbon to oxygen ratios lead to combustion products, while higher hydrocarbon to oxygen ratios can lead to POX products, dehydrogenation products, or oxygenates.

The millisecond contact time reactors are designed with extremely short contact times and product/feed heat exchange sections. Integrating the product/reactant heat exchange improves energy efficiency and eliminates the need for separate equipment to preheat the feeds or quench the products. Rapid preheating allows very little time for undesirable reactions to occur before reaching the reaction chamber. Rapid product cooling eliminates secondary cracking and polymerization reactions. Oxygen is similarly preheated, minimizing the volume that is hot at any given time and allowing the entire reactor to operate with relatively cool inlet gas temperatures. Understanding the millisecond contact time reactor requires coupling of reaction kinetics, mass transfer of reactants and products to and from the surface, transfer of heat from the surface to the gas phase and in the axial direction, and detailed fluid dynamics.

20.4 BASIC AND APPLIED RESEARCH DIRECTIONS

20.4.1 Alkanes as Alternative Feed Materials

Saturated hydrocarbons play a vital role in the world economy as the main constituent of crude oil. Over the past 15 years, many industrial and academic research efforts have been focused on the conversion of lower alkanes (C1 to C5) to petrochemical feedstocks. These are catalytic reactions that include methane oxidation to formaldehyde or methanol and methane oxidative coupling to C_2 hydrocarbons; ODH of ethane and propane to olefins and oxygenates (acetic acid, acrolein, acrylic acid, propylene oxide); oxidation of butane and pentane to maleic and phthalic anhydrides. Thus far, there are no industrially practical operations for such applications except for the production of maleic anhydride and acetic acid from butane. However, the incentive for such processes is high given the vast amounts of natural gas available containing significant quantities of ethane, propane, and butane in addition to methane.

In order to overcome poor reactivity, chemical industry relies on processes carried out at high temperature, noble metal catalysts, and liquid or solid strong acids [48]. Due to increased environmental legislation, research effort has been directed to replace conventional strong acids, such as HF or H_2SO_4, with more environment-friendly acids able to activate hydrocarbons.

The field of alkane activation and functionalization poses difficult challenges and is actively researched. The central problem is simply to develop ways to replace selected H atoms of alkanes by any of a variety of functional groups in an oxidative manner. Of the alkanes, methane is the most stable and most abundant. In one example of active research, conversion of methane to methanol is of significant practical interest in relation to energy production. Methane from natural gas is available in very large amounts at remote sites, but as a gas, methane cannot be transported economically. Oxidative conversion of the methane into a transportable liquid, such as methanol would make remote gas a viable energy source and lower the amount of deep-well injection or flaring of excess natural gas. A model reaction path for the selective oxidation of methane to methanol and formaldehyde is shown in Figure 20.8. This mechanism was suggested by Otsuka and Wang [49] to explain the observed behavior of methane oxidation over iron–molybdate catalysts. It was proposed that methane is activated by surface oxygen species derived from the dissociative adsorption of gaseous oxygen or from the bulk of the lattice. The possible reaction intermediates are adsorbed methyl species (CH_3) or methoxide species (CH_3O). In practice it is very difficult to protonate the adsorbed methoxide species to form methanol. This is explained in terms of the high reactivity of the methoxide species toward surface oxygen to form either formaldehyde or carbon oxides. For selective oxidation to methanol, a special structure of active sites possessing Brønsted acidity that can protonate the methoxide species without allowing successive oxidation to carbon oxides is required.

Another example in the direction of alkane activation for alternative feeds is the synthesis of ethylene and propylene by ODH. The present technology that is based on steam or thermal cracking involves heating hydrocarbon feedstock and steam at nearly atmospheric pressures in metal furnace tubes. The feed thermally decomposes to form ethylene, propylene, and other less valuable products. After reacting, the hot product gas mixture is cooled, compressed, refrigerated, and separated by distillation into ethylene and other products. Hydrogen and methane by-products are normally recycled to the furnaces as fuel. The catalytic ODH of propane and ethane is an attractive alternate route for the production of propylene–ethylene compared to the conventional cracking and dehydrogenation processes. This is because ODH is thermodynamically favored at lower temperatures and usually does not lead to the formation of coke and smaller hydrocarbons. Ethylene, in particular is a vital intermediate for the U.S. petrochemical industry. The conversion of ethane and propane (by-products of petroleum processing and present in natural gas) to olefins (ethylene, propylene) is in great demand. High operational costs and environmental issues associated with the nonoxidative route have made this conversion profitable only on a very large scale. With successful development of ODH, high yields of olefins will be possible through the conversion of much smaller volumes of alkanes. Compared with the conventional steam-cracking method of dehydrogenating

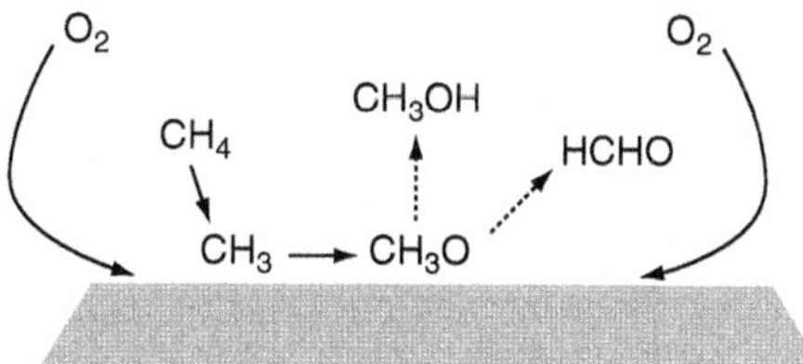

FIGURE 20.8 Possible reaction pathways in the selective oxidation of methane to methanol and formaldehyde.

alkanes to olefins and current catalytic dehydrogenation processes, ODH could reduce costs, lower greenhouse gas emissions, and save energy. Capital and operational efficiencies are gained by eliminating the need for providing external heat and for decoking shutdowns, lowering operating temperatures, lessening material demands, conducting fewer maintenance operations, and using a greater proportion of the alkanes in the olefin conversion process.

Fundamental research is needed to understand the key factors involved in alkane activation to selective oxidation. Such key factors have been summarized as activation of oxygen and of the alkane (i.e., role of adsorbed oxygen species and mode of alkane adsorption), reactivity of the reactant and products (i.e., the mechanism of C—H bond activation and stability of the formed products), mechanism of the transformation of the reactant (i.e., adsorption–desorption characteristics of intermediates, intermediate steps, and side reactions, the role of homogeneous reaction, the role of coadsorbates in facilitating the dissociative adsorption of saturated hydrocarbons) [1].

Through interaction with a catalyst surface, an alkane molecule can react to produce an alkyl radical or proceed through a beta-hydrogen elimination to form an alkene. Thereafter, the radical or alkene can either desorb from the surface or stay bound for further reaction. If the radical desorbs from the surface it can undergo homogeneous reaction in a desirable or undesirable way. If the radical stays on the surface it may go through an alkoxide or other intermediate to eventually be transformed either to an oxygenate, a lower hydrocarbon, or a carbon oxide. Which one of these is favorable depends on the alkane and the desired product. Thus, there are different catalysts capable of completing different functions.

The Lewis–Brønsted acidity of ODH catalysts is frequently characterized because of its inherent relation to the selectivity limits in selective oxidation of lower alkanes. Often a moderately basic catalytic site is required for the first hydrogen abstraction step. Consequently, the acidity will also play a role in the adsorption–desorption reaction behavior of the formed product. The consecutive reaction of the formed olefin is a matter of study in both ethane and propane reactions. These acidity features can be researched in various ways. By studying the adsorption–desorption behavior of probe bases (i.e., ammonia, pyridine) one may calculate acid strength or site density.

With the use of atomic-scale characterization and *in situ* methods, it is becoming easier to formulate mechanisms for catalytic reactions and deactivation processes. With the use of an active transition metal, such as molybdenum or vanadium, it is essential to study the redox behavior, interaction with a support, or actual bonding of intermediate during a reaction process. With the aid of surface sensitive spectroscopies and *in situ* techniques one can study the oxidation states before and after a reaction, identify intermediates on the surface of the catalyst, see what structural changes have taken place, and study redox behavior. Although alkane activation and oxidation has been studied extensively in the last several decades, many questions remain unanswered, making further research in this area a necessity. At the same time, some of the most desirable alkane transformation reactions continue to remain as technical challenges since they have not been developed into economically feasible industrial processes.

20.4.2 Oxygen Activation Strategies

Chemicals used as oxidizing agents have received much scrutiny because of their toxicity or environmental impact. The U.S. market for oxidizing and bleaching agents is estimated to be in excess of US$3 billion dollars and growing at a rate of 5% a year [50]. It is then relevant to discuss some advantages and disadvantages of commonly used oxidants for selective oxidation reactions and the drive to find environmentally and economically desirable alternatives.

In the manufacture of organic chemicals, the choice of oxidant is usually economically limited to molecular oxygen. However, since molecular oxygen could react with organic molecules even in the absence of a catalyst it does not, therefore, always exhibit high selectivity and inertness toward organic solvents. This makes the use of molecular oxygen not particularly attractive when high selectivity offsets economic factors. In the production of fine chemicals and pharmaceuticals, stoichiometric oxidation is used to produce aldehydes, ketones, and carboxylic acids. These processes use agents such as chromium compounds, in their various forms ($HCrO_4^-$, $Cr_2O_7^{2-}$, CrO_2Cl_2, $CrO_2(OAc)_2$), permanganate ions, and other metal salts to stoichiometrically oxidize hydrocarbons with high selectivity [51]. However, these processes are undesirable as they generate large volumes of waste salts, which have to be removed from the effluent streams. In fact, the production ratio between waste and fine chemicals may be very high, indicating the necessity to develop cleaner methods of oxidation [1]. Oxygen donors for oxygen transfer reactions provide somewhat easier handling for certain processes and include hydrogen peroxide, nitrous oxide, ozone, and a wide variety of oxo or peroxo metals or organic hydroperoxides. Soluble metal oxide clusters termed polyoxometalates are investigated and developed as homogeneous oxidation catalysts that have the ability to activate environmentally benign oxygen donors such as molecular oxygen, hydrogen peroxide, nitrous oxide, and ozone. Some of these oxidants may offer the possibility of inert by-products, such as H_2O in the case of hydrogen peroxide, while the ease of recycling or product separation may govern the use of other oxidants. The use of H_2O_2 in particular, in which water is typically the only by-product, would bring advantages over expensive organic peroxides and peracids or environmentally "unfriendly" chlorine-based oxidations [52].

Thus, with environmental as well as economic downsides to many industrial oxidants, there is a large amount of research devoted to the development of suitable oxidants based on solid catalysts. There is an increasing demand to eliminate organic solvents, as these are often noxious, decrease the reaction rates by dilution, or are involved in side reactions. Three-phase systems in which an aqueous-phase reacts with an organic phase over a solid catalyst are extensively studied. The development of titanium-substituted silicalite (TS-1) is an outstanding example of the research aimed at potential replacement of many industrial oxidations with the use of a solid catalyst [53]. TS-1 is a highly selective and active catalyst for the oxidation of olefins, alkanes, alcohols, and aromatic hydrocarbons using aqueous hydrogen peroxide as the oxidant. In particular, titanium silicalites may lead to new processes in alkene epoxidation, cyclohexanone ammoxidation, and phenol hydroxylation [1]. Titanosilicates combine the titanium activity in selective oxidation reactions with the shape selectivity of the zeolite framework. The catalytic activity of Ti framework substituted zeolites is limited by the fact that only those reactants that are small enough to reach the catalytic center can be converted. This may prevent some applications in the pharmaceutical industry where bulkier reagents are used; however, extensive research is underway. Other examples of research on novel selective and clean oxidation catalysts include transition metal incorporation into zeolite frameworks and amorphous redox oxides.

20.4.2.1 The Use of N_2O

Utilizing nitrous oxide (N_2O) in clean O-atom transfer reactions can reveal unique and selective oxidations of hydrocarbons on metals. This has been known for some time and has been exploited for dehydrogenation of lower alkanes. More recently, there has been a surge in activity studying zeolitic conversion of N_2O. Iron or cobalt exchanged ZSM-5 interacts with nitrous oxide in a number of ways. A unique form of adsorbed oxygen, known as alpha oxygen, can be produced on Fe-ZSM5 by N_2O decomposition [56]. Treatment in molecular O_2 leaves the catalyst unchanged. This unique form of oxygen is reported to have the ability to selectively oxidize methane to methoxide species at temperatures as low as 298 K [54, 55]. Several substrates may react with this form of oxygen, including alkanes, alkenes, arenas, and heterocyclic compounds. Thus, understanding the relationship between the structure of active centers in Fe- and Co-ZSM-5 and their catalytic performances is an active area of research. While it appears that N_2O is unique in its ability to form alpha oxygen, there has been considerable study of this phenomenon and much remains uncertain regarding the identity and other properties of alpha oxygen [56]. The question on the active iron state in Fe-zeolites responsible for alpha-oxygen formation appears to be of most interest. However, it has been proposed that N_2O reacts with binuclear iron centers [57, 58]. N_2O decomposition on the alpha sites is accompanied by the loading of alpha oxygen, which must change the properties of the active iron [59].

One area of interest using N_2O as an alternative oxidant is in the oxidative hydroxylation of aromatics. In this class, the direct conversion of benzene to phenol is cited as one of the most difficult problems in the field of organic synthesis [60]. Currently, the most widely used industrial route to prepare phenol is the cumene process, in which cumene hydroperoxide undergoes an acid-catalyzed cleavage to yield phenol and acetone [61]. A one-step process for the production of phenol, which does not depend on the market price of acetone, is highly preferred and actively researched. A large number of studies have been conducted on the ZSM-5 zeolite system as a catalyst for the one-step synthesis of phenol from benzene using nitrous oxide as the oxidant via gas-phase catalytic selective oxidation. Alpha oxygen has been identified as the active species for benzene hydroxylation by using kinetic isotope experiments [62]. The reaction exhibited high conversion of benzene (10 to 30%) and selectivity to phenol (>95%). However, there are opposing ideas regarding the mechanism of this reaction. While some agree that proton donors from the zeolite are responsible for the benzene hydroxylation, others relate the catalytic activity to the ability of extra framework metal to generate alpha-oxygen species.

20.4.2.2 Active Oxygen Species from Ozone

Active oxygen species can be generated from ozone at gas–solid interfaces. The use of ozone in the oxidation feed on oxide catalysts can result in a considerable decrease of the required reaction temperatures for the same catalyst efficiency [63]. The use of ozone is also well known for the neutralization of harmful components in industrial waste gases and has been found to increase combustion efficiencies [64]. Furthermore, it has been found that catalysts that have coked can be regenerated oxidatively at lower temperatures when treated in ozone [65]. At low temperatures, ozone has the ability to generate reactive peroxide species at the surface of solid catalysts that can be used for new low-temperature oxidation reactions. Despite this advantage, there is no widespread use of ozone in selective oxidation reactions. While it suffers from the disadvantage of requiring specialized equipment for its generation, it is likely to become increasingly important in the future as we research technologies that are cleaner, producing less salts. Oxidation experiments with secondary alcohols, 2-butanol, 2-hexanol, and cyclooctanol, showed only moderate product yields of 60–80% at total conversion [66]. Also, due to the inherent slow rate of alcohol oxidations with ozone, a successful industrial application seems unlikely. While carbohydrate oxidation with ozone has been reported to occur to some extent through both hemolytic and heterolytic pathways, neither appears to result in selective oxidation [67].

The development process for selective oxidation may reside in the development of suitable catalysts. Metalloporphyrins have been claimed to be a promising and effective oxidation catalysts used with ozone [68, 69]. Porphyrins are based on 16-atom rings containing four nitrogen atoms; they are of perfect size to bind nearly all metal ions of the periodic table. These metalloporphyrins have been the subject of many studies because of their potential application as selective catalysts. They are also model systems for oxidation of hydrocarbons in the liquid phase under mild conditions. Different oxygen donors such as iodosobenzene, hydrogen peroxide, organic hydroperoxides, hypochlorites, monopersulfates, and molecular oxygen with sacrificial coreductant or photochemical or electrochemical activation of the dioxygen were applied [70]. There is potential in this area of research for the use of ozone in the epoxidation of olefins and the hydroxylation of alkanes. One of the key areas of focus is preparation of highly stable metalloporphyrin catalysts in the presence of ozone.

20.4.2.3 *In Situ* Generation of H_2O_2

Hydrogen peroxide is commercially produced by the anthraquinone process. In the first stage, an alkylated anthraquinone (the carrier) is hydrogenated to its corresponding hydroanthraquinone, which in turn is oxidized with air to produce hydrogen peroxide and back again the starting anthraquinone (Figure 20.9). The direct synthesis of hydrogen peroxide is pursued via the catalytic hydrogenation of oxygen, carried out on Pd-based heterogeneous catalysts in aqueous solution and in the presence of strong mineral acids and halide ions. However, direct use of hydrogen peroxide has some serious drawbacks [71]. It is inherently corrosive and, especially in high concentrations, is prone to catalytic decomposition. Using dilute aqueous solutions can minimize these problems; however, the excess water often produces a biphasic mixture with the organic substrate and the oxidant residing in different phases. There are also problems associated with the use of molecular oxygen, for which there are few active and selective catalysts, and mono-oxygen donors, which are expensive oxidants leading to excess waste. A way around some of those obstacles is the *in situ* generation of active oxygen donor species, from molecular oxygen and a reducing reagent. One oxygen atom will be effectively used for the oxidation of the substrate (S), the other being discharged into a by-product, such as water in the case of hydrogen peroxide [72]:

$$\mathrm{S + O_2 + H_2 + Catalyst\ 1 \longrightarrow S + H_2O_2 + Catalyst\ 2 \longrightarrow SO + H_2O}$$

Therefore, the most attractive mono-oxygen donor for *in situ* production is hydrogen peroxide, on the grounds of following considerations: Methods by direct and indirect hydrogenation of oxygen are well established. Hydrogen is available in most industrial plants or can be produced by well-established processes. A robust chemistry of hydrogen peroxide, based on both homogeneous and heterogeneous catalysts, has been developed over the years. Finally, water is the only by-product of hydrogen peroxide.

The *in situ* epoxidation of propylene serves as an example where distinct advantages are present over the *ex situ* route. It avoids the shipment of large amounts of hydrogen peroxide when it is not available on site, is more flexible in terms of downscaling, and there is no dependence on external suppliers. The epoxide and water are produced by the reaction of an alkylated anthrahydroquinone with molecular oxygen and propylene. The corresponding anthraquinone is subsequently hydrogenated to close the cycle of reactions, as shown in Figure 20.10. Hydrogen peroxide, which is produced in the reaction medium, then migrates

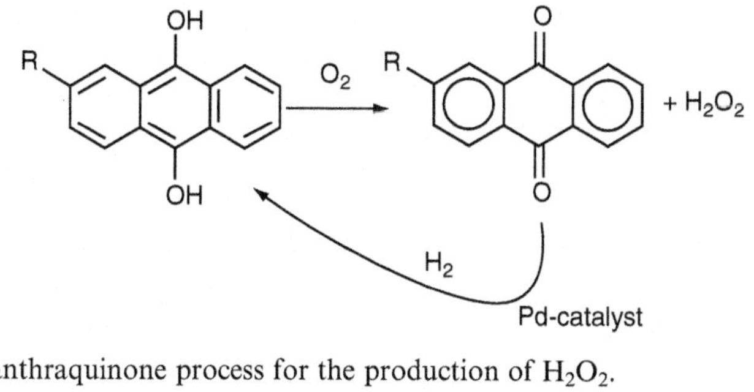

FIGURE 20.9 The anthraquinone process for the production of H_2O_2.

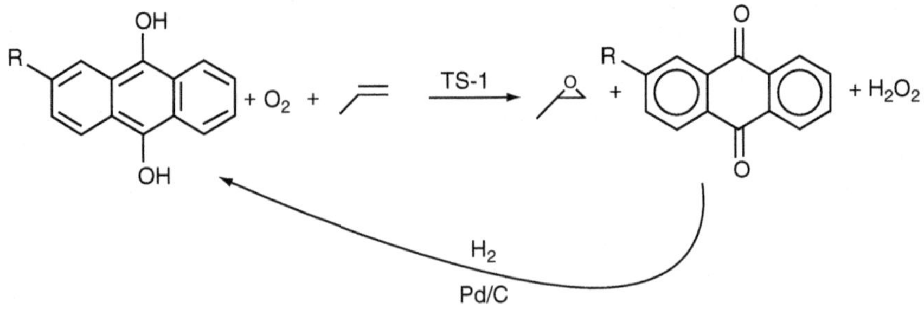

FIGURE 20.10 Epoxidation of propylene with *in situ*-generated H_2O_2.

in the channel system of TS-1 to form the active species at Ti-sites [73]. This reacts with propylene in the oxygen-transfer step. The organic carrier does not interfere in the catalytic process because its cross section is larger than the diameter of TS-1 pores. Propylene epoxidation over a TS-1 catalyst containing palladium and platinum metals with *in situ* formed hydrogen peroxide has been cited [74]. The reaction was carried out in a fixed bed reactor under high-pressure conditions. The continuous operation allowed the study of catalyst deactivation and changes in product distribution with time-on-stream. The initial propylene oxide selectivity was very high, 99% at 3.5% conversion, but the catalyst deactivated rapidly with time-on-stream and successively the formation of methyl formate became the prevalent reaction. When H_2O_2 is directly used as the oxidant, the yields of propylene oxide are significantly higher [75]. Therefore, it can be assumed that the *in situ* formation of H_2O_2 is the rate-determining step for the epoxidation of propylene with hydrogen and oxygen catalyzed by Pd–TS-1.

20.4.2.4 Singlet Oxygen

The inertness of molecular oxygen toward reaction with activated hydrocarbons is largely due to its triplet ground state with two unpaired electrons in two different molecular orbitals in parallel spins. Singlet oxygen $^1O_2$ is an excited form of molecular oxygen with two paired electrons in the same molecular orbital. Selective oxyfunctionalization of organic compounds by singlet molecular oxygen has received much attention since the 1960s, because of the unique reactivity of $^1O_2$ and its highly selective reaction with electron-rich organic substrates such as olefins [76]. Singlet O_2 can be generated from $^3O_2$ using a photosensitizer. Photosensitization refers to a light-activated process that requires the presence of a light-absorbing substance, the photosensitizer that initiates a chemical process in a nonabsorbing substrate. For example, an optically excited dye may transfer its energy to molecular oxygen at its ground energy level, leading to the formation of singlet oxygen. An alternative pathway is the catalyzed production of $^1O_2$ from alkaline H_2O_2. Several metal ions can act as homogeneous catalysts for this reaction, in particular molybdate and calcium [77]. Additionally, vanadium haloperoxidases have been reported to catalyze oxidation of hydrogen peroxide into singlet oxygen and water [78]. Industrially, singlet oxygenation is performed by dye-sensitized photooxidation based on molecular oxygen. Large-scale photooxidation entails hazardous processing conditions because of the combination of light, organic solvents, and dioxygen. As a result of the preceding disadvantages, industrial use of photooxidation is mainly limited to the manufacture of low-volume, high-value flavor and fragrance compounds [79].

It might be expected that surface oxygen would show the same reaction chemistry as singlet oxygen in homogeneous media. However, the possibility that singlet oxygen could be involved in reactions at oxide surfaces has not been considered seriously because the energy level is 22.64 kcal above that of the ground state of triplet oxygen [11]. Although it is difficult to transpose these results to practical examples of how the *in situ* generation of singlet oxygen could be used in more robust selective oxidation reactions, more research would be useful.

20.4.2.5 Charge Transfer O_2 Activation

There are reports in the literature about molecular oxygen activation at room temperature through an electron transfer from a hydrocarbon molecule to the oxygen to form a radical cation and O_2^- (superoxide) [80]. Charge transfer from hydrocarbon to oxygen can occur when a photon is absorbed by an oxygen–hydrocarbon collisional pair. High steady-state concentrations of collisional pairs of small alkenes, alkanes, or alkyl benzenes with O_2 are formed when the gases are loaded into dehydrated zeolites at ambient temperature. Alkali or alkaline earth exchanged, large-pore zeolites feature very high electrostatic fields in the vicinity of poorly shielded cations (zeolite type Y or L exchanged with Na^+, K^+, or Ba^{2+}). These electrostatic fields stabilize the charge-transfer state of collisional complexes, facilitating absorption of photons. The radical cations and superoxide ions formed through this process can lead to the formation of selective oxidation products. Aldehyde or ketone formation through a corresponding alkyl hydroperoxide intermediate is such an example. Zeolitic systems facilitate the process by providing strong adsorption sites for collision pair formation and high electrostatic fields for inducing photon absorption. Limited pore size and geometry may also be important in preventing side reactions, such as coupling.

Recent work [81] has found that higher alkanes like cyclohexane or isobutane can be oxidized selectively by O_2 to corresponding alkyl hydroperoxide or its dehydration product (ketone) in zeolite NaY or BaY under visible light without a photosensitizer. This effect has also been demonstrated with cyclohexane–O_2, toluene–O_2, and olefin–O_2 complexes in alkali and alkaline earth exchanged zeolite Y [82]. Photocatalyzed charge transfer in zeolite cages opens new areas of investigation for selective oxidation reactions at very low temperatures. However, research needs to be conducted to increase productivity which may be limited by the low desorption rates of oxygenated products.

20.4.2.6 Electrochemical O_2 Activation

Electrochemical activation of oxygen is the basis behind the operation of H_2/O_2 fuel cells. In these devices, the electrochemical oxidation of H_2 to e^- and H^+ occurs at the anode and the reduction of O_2 with e^- and H^+ to H_2O takes place at the cathode. The net reaction is the formation of water from H_2 and O_2.

$$\text{Anode: } H_2 \longrightarrow 2H^+ + 2e^-$$

$$\text{Cathode: } \frac{1}{2}O_2 + 2H^+ + 2e^- \longrightarrow H_2O$$

$$\text{Overall} = H_2 + \frac{1}{2}O_2 \longrightarrow H_2O$$

It is possible to use the same concept for oxygen activation in catalytic reactions. While the oxygen activation takes place at the cathode by transfer of electrons through the external circuit, hydrocarbons can also be introduced to the cathode to react selectively with activated oxygen species. The reaction rate can be controlled by monitoring the flow of electrons to the cathode and, in turn, the formation rate of active oxygen species.

Direct oxygenation of aromatics to phenols, alkanes to alcohols, and aldehydes through H_2/O_2 fuel cell reactions at room temperatures have been reported [83]. Selective POX of alkenes in the gas phase using alkene–O_2 cell systems has also been studied [84, 85]. When C_2H_4 was supplied to the gas–cell system, C_2H_4 was electrochemically oxidized with H_2O to acetaldehyde over the Pd-anode. The gas–cell system has several advantages compared with the current Wacker process in the aqueous HCl with $PdCl_2$ and $CuCl_2$ as catalysts [86].

A novel oxidation technology using solid oxide electrochemical cells and oxygen transport membranes that separates oxygen and performs POXs simultaneously in a single reactor has been reported [87]. The tubular reactor is constructed of the electrolyte material, yttria-stabilized zirconia, and proprietary cathode and anode materials. Oxygen is reduced at the cathode: $O_2 \longrightarrow O_2 \longrightarrow O^- \longrightarrow O^{2-}$, and the hydrocarbon never comes into contact with free oxygen. Studies of methane and methane–ethane mixtures were able to show good selectivity on syngas production and C_2 and higher coupling products by the use of appropriate electro-catalysts. While advanced cells integrating simultaneous oxygen separation and POX have been developed, work is needed to increase oxygen diffusional fluxes.

These new types of electrochemical cells and novel membrane catalyst shows promise for other catalytic reactions. However, drawbacks of high cost (membrane, electrolyte) and the development of better electro-catalysts limit commercial applications, although applications in the synthesis of fine chemicals may be possible.

20.4.3 Toward 100% Selective Processes

The ideal oxidation procedure must give yield with high selectivity without any by-products through a simple, safe operation using a clean, well-defined, and cheap oxidant. Catalysis is the key for making the chemistry of these reactions environmentally friendly, since one-step synthesis of complex transformations over solid catalysts reduces the process complexity, avoids solvents and decreases risks and energy use. Furthermore, 100% selectivity eliminates by-products and improves process economy. The development of the next-generation chemical industry requires an integrated effort in understanding how to control catalyst properties in order to make possible a selective multistep process in a single passage over a highly selective solid catalyst surface. This achievement requires an integrated effort between theory, modeling, surface characterization, and reactivity testing.

Development of novel catalysts that would enable 100% selective oxidation processes will have significant beneficial impact on the environment. Research investments to meet these critical needs are recommended in Ref. [3] and are shown below:

- New catalyst design through combined experimental and mechanistic studies and improved computational modeling of catalytic processes
- Development of techniques for high throughput synthesis and rapid testing of catalysts on diverse processes, and reduction of analytical cycle time by parallel operation and automation
- Better techniques for catalyst characterization under actual operating conditions, particularly at high temperature and pressure (>1 atm)
- New methods to synthesize stable, high productivity catalysts with control of active-site architecture.

20.5 CONCLUSIONS

Selective oxidation reactions are used extensively for production of many important chemicals such as epoxides, aldehydes, esters, organic acids, and alcohols. Oxidation catalysis has been playing an important role in safeguarding the environment through increased emphasis in green chemistry. Although the last several decades have brought significant improvements to existing processes and opened up new possibilities for developing novel selective oxidation processes, there is still need for catalytic and reaction engineering research to achieve higher selectivity, to minimize waste, to reduce or eliminate the use of toxic or hazardous chemicals, to use environmentally benign and renewable feedstocks while decreasing energy requirements and minimizing the cost at the same time. These are the challenges facing chemists and chemical engineers.

REFERENCES

1. G Centi, FC Cavani, and F Trifiro. *Selective Oxidation by Heterogeneous Catalysis.* New York: Kluwer Academic/Plenum Publishers, 2001.
2. WR Moser. *Catalysis by Organic Reactions.* New York: Marcel Dekker, 1981.
3. Council for Chemical Research, U.S. D.O.E., and ACS. Catalyst Technology Roadmap Report: Build Upon Technology Vision 2020: The U.S. Chemical Industry. ACS Workshop, Washington, D.C., USA, 1997.
4. JS Rafelt and JH Clark. *Catal. Today* 57: 33–44, 2000.
5. AB Crozon, M Besson, and P Gallezot. *New J. Chem.* 22(3): 269–273, 1998.
6. J Haber and W Turek. *J. Catal.* 190: 320–326, 2000.
7. CL Hill, AM Khenkin, MS Weeks, and Y Hou. In: *Catalytic Selective Oxidation*, ST Oyama and JW Hightower, eds. Washington D.C.: American Chemical Society, 1993.
8. Y Moro-oka. *Catal. Today* 45: 3–12, 1998.
9. CR Reilly and JJ Lerou. *Catal. Today* 41: 433–441, 1998.
10. AM Khenkin, LJW Shimon, and R Neumann. *Inorg. Chem.* 42: 3331–3339, 2003.
11. M Che and AJ Tench. *Adv. Catal.* 31, 77–133, 1982.
12. M Che and AJ Tench. *Adv. Catal.* 32, 1–148, 1983.
13. KP Peil, G Marcelin, JGG Goodwin, Jr. In: *Methane Conversion by Oxidative Processes*, EE Wolf, ed. New York: Van Nostrand Reinhold, 1992.
14. H Nakatsuji, ZM Hu, and H Nakai. *Int. J. Quantum Chem.* 65: 839–855, 1997.
15. D Lafarga, MA Al-Juaied, CM Bondy, and A Varma. *Ind. Eng. Chem. Res.* 39: 2148–2156, 2000.
16. CN Satterfield. *Heterogeneous Catalysis in Industrial Practice*, 2nd ed. New York: McGraw-Hill, 1991.
17. RB Watson and US Ozkan. *J. Phys. Chem. B* 106: 6930–6941, 2002.
18. RB Watson and US Ozkan. *J. Mol. Cat. A* 194: 115–135, 2003.
19. HH Kung. *Adv. Catal.* 40: 1, 1994.
20. A Bielanski and M Najbar. *Appl. Catal. A* 157: 223–261, 1997.
21. B Kubias, U Rodemerk, HW Zanthoff, and M Meisel. *Catal. Today* 32: 243–253, 1996.
22. Y Zhang-Lin, M Forissier, RP Sneeden, JC Vedrine, and JC Volta. *J. Catal.* 145: 256–266, 1994.
23. WH Cheng and W Wang. *Appl. Catal.* 156: 57–69, 1997.
24. A Bielanski and J Haber. *Oxygen in Catalysis.* New York: Marcel Dekker, 1990.
25. AE Comyns. *Dictionary of Named Processes in Chemical Technology.* New York: Oxford University Press, 1993.
26. DL Trent. *Encyclopedia of Chemical Technology*, 4th ed., Vol. 20. New York. Wiley, 1996.
27. A Kaddouri, C Mazzocchia, and E Tempesti. *Appl. Catal. A* 169: L3–L7, 1998.
28. EJ Beckman. *Green Chem.* 5: 332–336, 2003.
29. JS Rafelt and JH Clark. *Catal. Today* 57: 33–44, 2000.
30. A Mills and S Le Hunte. *J. Photochem. Photobiol. A* 108: 1–35, 1998.
31. MA Gonzalez, SG Howell, and SK Sikdar. *J. Catal.* 183: 159–162, 1999.
32. M Jacoby. *C$EN News* 81, No. 3: 32–36, 2003.
33. JR Rostrup-Nielsen. *Phys. Chem. Chem. Phys.* 3: 283–288, 2001.
34. RP O'Connor, EJ Klein, and LD Schmidt. *Catal. Lett.* 70: 99–107, 2000.
35. H Ichihashi and M Kitamura. *Catal. Today* 73: 23–28, 2002.
36. H Ichihashi and H Sato. *Appl. Catal. A* 221: 359–366, 2001.
37. DP Dreoni, D Pinelli, and F Trifiro. In: *New Developments in Selective Oxidation by Heterogeneous Catalysis*, Vol. 72, P Ruiz and B Delmon, eds. Amsterdam: Elsevier Science, 1992.
38. AL Tonkovich, JL Zilka, DM Jimenez, GL Roberts, and JL Cox. *Chem. Eng. Sci.* 51: 789–806, 1996.
39. FA Al-Sherehy, AM Adris, MA Soliman, and R Hughes. *Chem. Eng. Sci.* 53: 3965–3976, 1998.
40. LD Schmidt, EJ Klein, CA Leclerc, JJ Krummenacher, and KN West. *Chem. Eng. Sci.* 58: 1037–1041, 2003.
41. RP O'Conner, EJ Klein, and LD Schmidt. *Catal. Lett.* 70: 99–107, 2000.
42. BE Traxel and KL Hohn. *J. Catal.* 212: 46–55, 2002.
43. A Beretta, E Ranzi, and P Forzatti. *Chem. Eng. Sci.* 56: 779–787, 2001.
44. RP O'Connor, EJ Klein, D Henning, and LD Schmidt. *Appl. Catal. A* 238: 29–40, 2002.

45. JJ Krummenacher, KN West, and LD Schmidt. *J. Catal.* 2015: 332–343, 2003.
46. M Huff and LD Schmidt. *J. Catal.* 149: 127–141, 1994.
47. M Huff and LD Schmidt. *J. Phys. Chem.* 97: 11815–11820, 1993.
48. J Sommer, R Jost, and M Hachoumy. *Catal. Today* 38: 309–319, 1997.
49. K Otsuka and Y Wang. *Appl. Catal. A* 222: 145–161, 2001.
50. Business Communications Company, Inc., Oxidizing and Bleaching Agents, 1 January 2003. www.marketresearch.com
51. RA Sheldon and RA van Santen. *Catalytic Oxidation*. Singapore: World Scientific Publishing, 1995.
52. S. Ritter. *C&EN News*, Vol. 81, No. 19, 11, 2003.
53. G Langhendries, DE De Vos, BF Sels, I Vankelecom, PA Jacobs, and GV Baron. *Clean Products Process*. 1: 21–29, 1998.
54. VI Sobolev, KA Dubkov, OV Panna, and GI Panov. *Catal. Today* 24: 251–252, 1995.
55. PP Knops-Gerrits and WA Goodard III. *J. Molec. Catal. A* 166: 135–145, 2001.
56. C Sang and CRF Lund. *Catal. Lett.* 73: 73–77, 2001.
57. KA Dubkov, NS Ovanesyan, AA Shteinman, EV Starokon, and GI Panov. *J. Catal.* 207: 341–352, 2002.
58. GD Pirngruber. *J. Catal.* 219: 456–463, 2003.
59. NS Ovanesyan, KA Dubkov, AA Pyalling, and AA Shteinman. *J. Radioanal. Nucl. Chem.* 246: 149–152, 2000.
60. RA Sheldon and JK Kochi. *Metal Catalyzed Oxidations of Organic Compounds*. New York: Academic Press, 1981.
61. A Ribera, IWCE Arends, S de Vries, J Pérez-Ramírez, and RA Sheldon. *J. Catal.* 195: 287–297, 2000.
62. GI Panov, AK Uriarte, MA Rodkin, and VI Sobolev. *Appl. Catal.* 41: 365–385, 1998.
63. D Mehandjiev, K Cheskova, A Naydenov, and V Georgesku. *React. Kinet. Catal. Lett.* 76: 287–293, 2002.
64. A Gervasini, GC Vezzoli, and V Ragaini. *Catal. Today* 29: 449–455, 1996.
65. CL Peik, CA Querini, and JM Parera. *Appl. Catal. A* 165: 207–218, 1997.
66. JHF Stefan, EJM Mombarg, H van Bekkum, and RA Sheldon. *Synthesis* 6: 597–619, 1997.
67. SK Rakovsky, RA Sheldon, and van Rantwijk. *Oxid. Commun.* 19: 482–491, 1996.
68. K Perie, JM Barbe, P Cocolios, and R Guilard. *Bull. Soc. Chim. Fr.* 133: 697–702, 1996.
69. S Campestrini, A Robert, and B Meunier. *J. Org. Chem.* 56: 3725–3727, 1991.
70. J Haber, L Matachowski, K Pamin, and J Poltowicz. *J. Molec. Catal. A* 162: 105–109, 2000.
71. G Strukul and C Angew. *Int. Ed. Engl.* 37: 1199–1209, 1998.
72. MG Clerici and P Ingallina. *Catal. Today* 41: 351–364, 1998.
73. R Meiers and WF Holderich. *Catal. Lett.* 59: 161–163, 1999.
74. G Jenzer, T Mallat, M Maciejewski, F Eigenmann, and A Baiker. *Appl. Catal. A* 208: 125–133, 2001.
75. MG Clerici, G Belussi, and U Romano. *J. Catal.* 129: 159–165, 1993.
76. F van Laar, D DeVos, F Pierard, AK DeMesmaeker, L Fiermans, and PA Jacobs. *J. Catal.* 194: 139–150, 2001.
77. F van Laar, D Devos, D Vanoppen, B Sels, and PA Jacobs. *Chem. Commun.* 2: 267–268, 1998.
78. A Ligtenbarg, R Hage, and BL Feringa. *Coord. Chem. Rev.* 237: 89–101, 2003.
79. V Nardello, M Herve, PL Alsters, and JM Aubry. *Adv. Syth. Catal.* 344: 184–191, 2002.
80. F Blatter, H Sun, and H Frei. *Catal. Today* 41: 297–309, 1999.
81. H Sun, F Blatter, and H Frei. *Catal. Lett.* 44: 247–253, 1997.
82. H Sun, F Blatter, and H Frei. *J. Am. Chem. Soc.* 118: 6873–6879, 1996.
83. K Otsuka and I Yamanaka. *Catal. Today* 41: 311–325, 1998.
84. GA Stafford. *Electrochem. Acta* 32: 1137–1145, 1987.
85. K Otsuka, Y Shimizu, and I Yamanaka. *J. Electrochem. Soc.* 137: 2076–2086, 1990.
86. I Yamanaka, K Komabayashi, A Nishi, and K Otsuka. *Catal. Today* 71: 189–197, 2001.
87. TJ Mazanec. In: *The Activation of Dioxygen and Homogeneous Catalytic Oxidation*, DHR Barton, AE Martel, and DT Sawyer, eds. New York: Plenum Press, 1993.

21 Environmental Catalysis in Organic Synthesis

Jianliang Xiao
Department of Chemistry, University of Liverpool

CONTENTS

21.1 INTRODUCTION

Organic synthesis is a central theme in many disciplines of science ranging from chemistry through biology to materials. Over the last one and half centuries, organic synthesis has evolved to such a degree of sophistication that it allows molecules of almost any complexity to be constructed. However, traditional synthetic methodologies are dominated by stoichiometric reactions, multistep manipulation, and use of hazardous reagents, and as such, they are raw material and energy intensive and generate large amounts of waste. Some of the well-known examples in this regard include the use of stoichiometric metals, metal hydrides, and metal oxides, e.g., Zn, Na, $LiAlH_4$, $NaBH_4$, $K_2Cr_2O_7$, and $KMnO_4$, for reduction and oxidation, and the use of stoichiometric metal reagents, e.g., $AlCl_3$ and RMgX, for the formation of C—C bonds. A clear indication of the impact of organic synthesis on the environment is seen in Sheldon's *E* factor, which suggests that the pharmaceutical industry, a domain of synthetic organic chemistry, generates 25 to >100 kg waste for every kg of the product [1]. The key to dealing with these problems is to use catalysts.

Catalysis contributes to efficient and clean synthesis in a number of aspects. The most significant of them includes replacement of stoichiometric reagents, improving reaction rates, and selectivities, and delivering reactions and selectivity patterns that are impossible with traditional approaches [2]. In the past few decades, a great number of innovative catalytic processes that fall into these categories have been invented, attesting to the power of catalysis in responding to the rampant eco-economic issues facing synthetic organic chemistry. These reactions have been summarized in many excellent review articles, some of which can be found in the Refs. [3–5]. This chapter concentrates on catalytic reactions pertinent to organic synthesis with a view toward cleaner synthesis.

Whilst catalysis has been the most fundamental component of clean synthesis, cleaner catalysis is achievable when serious considerations are given to atom selectivity or efficiency, alternative solvents, and solid catalysts. In comparison with soluble catalysts, solid catalysts, including immobilized molecular metal catalysts, are easier to handle and in general less hazardous and more importantly, they allow for easier catalyst separation and recycle. Progress made in catalysis by solids and solid-supported metal complex catalysts can be found in Refs. [5–7]. In the sections to follow, selected examples of catalytic reactions performed in nonconventional media and under solvent-free conditions will be presented, aiming to illustrate the exciting possibilities for synthetic chemistry when environmental catalysis principles are implemented. The emphasis is placed on catalysis by soluble metal complexes; but where appropriate, examples of catalysis on solids will be drawn, and the issue of atom selectivity will be brought up. A personal account on the latter subject has recently been given by Trost [8]. Our discussion starts with a brief introduction to atom economy and alternative media. Selected reactions will be presented next, followed by conclusions.

21.2 ATOM ECONOMY AND ALTERNATIVE SOLVENTS

As with organic synthesis, catalysis has traditionally been associated with improving reaction rates and selectivities and discovering new reactions, with less regard to other issues that also have an impact on the environment. Thus, while near 100% chemo-, regio-, diastereo- and enantio-selectivity can be attained through catalysis, an important issue has too often been disregarded, that is atom selectivity or atom economy, which measures how much of what one puts into a reaction ends up in the product [8]. It may be quantified as the molecular weight of the desired product divided by the molecular weight sum of all products in the stoichiometric equation [1]. The examples shown in Scheme 21.1 is illustrative. The Heck reaction, while synthetically extremely useful for forming C—C bonds, normally produces equal molar salt waste relative to the product. An atom selectivity of 36% results if bromobenzene is olefinated by ethylene in the presence of NEt_3. Likewise, asymmetric epoxidation with iodosylbenzene generates one equivalent of iodobenzene for every equivalent of the product, giving rise to an

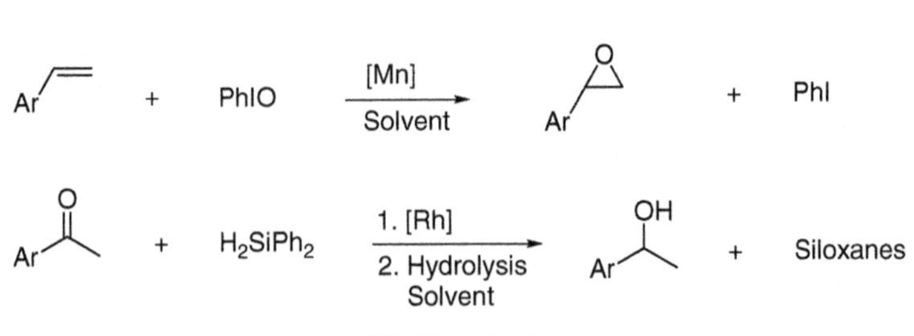

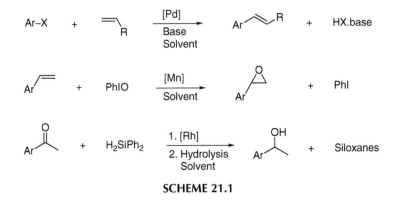

Ar–X + (=\R) —[Pd], Base, Solvent→ Ar/\=/R + HX.base

Ar/= + PhIO —[Mn], Solvent→ (Ar-epoxide) + PhI

Ar(C=O)CH₃ + H_2SiPh_2 —1. [Rh], 2. Hydrolysis, Solvent→ Ar(CHOH)CH₃ + Siloxanes

SCHEME 21.1

atom selectivity of 37% for the epoxidation of styrene. Hydrosilylation presents yet another example, in which unwanted siloxanes or polysiloxanes are produced.

Parallel to the issue of atom economy is the use of solvents. Hazardous and volatile organic solvents such as ethers and chlorinated hydrocarbons are widespread in catalysis, and as with hazardous reagents, these are under ever-increasing pressure to be replaced by environmentally benign alternatives. Underlining this trend, a recent study suggests that rigorous management of solvent use could make the greatest improvement in the greener production of pharmaceutical intermediates [9]. The best solvent is no solvent and if a solvent is needed, green or potentially green alternatives should be considered. Preferentially, such alternative solvents should also enable easy catalyst separation and recycle. The most significant of these alternatives includes water, supercritical CO_2 ($scCO_2$), ionic liquids, and fluorous carbons.

Among all the solvents, water is most environmentally friendly and least expensive. Water is protic and highly polar with $\varepsilon = 78$. It is thus ideally suited for reactions with dipolar transition states. The contribution of entropic effects due to hydrophobic interactions should not be overlooked, however. Apart from its total eco-compatibility, water is attractive for catalysis because its low miscibility with most organic compounds allows for facile catalyst separation and recycle through a biphasic catalysis mode [10–12].

CO_2 is nontoxic and nonflammable in addition to being inexpensive. $ScCO_2$ ($T_c = 31.1°C$, $P_c = 73.86$ bar, $d_c = 0.468$ g/mL) possesses excellent transport properties (low viscosity and high diffusivity) and is completely miscible with catalytically important gases such as H_2, O_2, and CO, a significant advantage for reactions with mass transfer problems. In contrast to water, $scCO_2$ is apolar; but it has a quadrupole moment and is capable of hydrogen bonding and donor–acceptor interactions with solute molecules [13]. For homogeneous catalytic synthesis, $scCO_2$ is in general only suitable for compounds of low polarity. An interesting aspect of $scCO_2$ is that, as with other supercritical fluids, its physical properties vary with pressure or temperature, or both. This can be harnessed to tune catalytic activity and selectivity and to separate a catalyst from product by simple precipitation [14–16].

Ionic liquids are salts of organic cations with a low melting point. Examples of those most often used in the literature are shown in Scheme 21.2 [17–21]. Many of these salts have liquid ranges over 300°C and dissolve a wide range of organic, inorganic, and organometallic compounds. However, they display low solubilities towards compounds of low polarity such as alkanes and ethers, and they are insoluble in $scCO_2$ [22]. Thus, biphasic catalysis can be easily performed with ionic liquids, and for monophasic reactions, $scCO_2$ or other eco-friendly solvents that do not dissolve the catalyst can be used to extract products. An attractive feature of ionic liquids is that their properties, such as polarity and hydrophilicity, are readily adjustable by varying the substitute R or the nature of the cations and anions. These liquids do not have effective vapor pressure and hence do not contribute to atmospheric emissions. Their biodegradability and toxicity are yet to be fully determined, however.

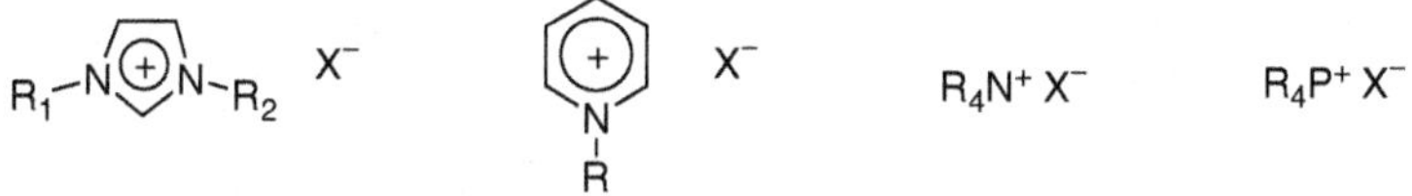

SCHEME 21.2

The term "fluorous solvents" generally refers to perfluorinated alkanes, dialkylethers, and trialkylamines. They show very low miscibility with common organic solvents; hence they can easily form biphasic catalytic systems [23–26]. Of particular importance to catalysis is the observation that a fluorous biphasic system may become one single phase when heated, thus allowing for homogeneous catalysis at high temperature and catalyst/product separation at ambient conditions. These solvents are exceedingly chemically inert, thermally stable, and nonflammable, and are thought to have no ozone-depletion potential. However, if not properly contained, their long atmospheric lifetime could make them potent greenhouse gases [27].

21.3 CLEAN CATALYSIS AND SYNTHESIS

21.3.1 HYDROGENATION

Hydrogenation is an atom-economic reaction and one of the most often used tools in synthetic chemistry. Hydrogenation is also one of the most studied reactions in the alternative solvents. Examples of both achiral and asymmetric hydrogenation reactions are numerous. Selected examples are presented below.

21.3.1.1 Achiral Hydrogenation

A number of simple and functionalized olefins have been hydrogenated in water, ionic liquids, $scCO_2$, and fluorous solvents, with most examples documented in water. Water has been shown to be a particularly good solvent for synthetically relevant functionalized olefins [28, 29]. An earlier example is the hydrogenation of various functionalized olefins using $RhCl_3$ in the presence of the exceedingly water-soluble phosphine, TPPTS [TPPTS = $P(m\text{-}C_6H_4SO_3Na)_3$; solubility in water: 1100 g/L] (Scheme 21.3) [30]. The hydrogenation reaction proceeded smoothly at room temperature and atmospheric pressure of H_2, with complete chemoselectivity towards the C=C bonds. In the case of dienes, the less hindered C=C bond was first saturated and the hydrogenation could be terminated at the monoene stage. While a soluble phosphine was used, the reaction was thought to be effected by rhodium colloids stabilized by oxidized TPPTS.

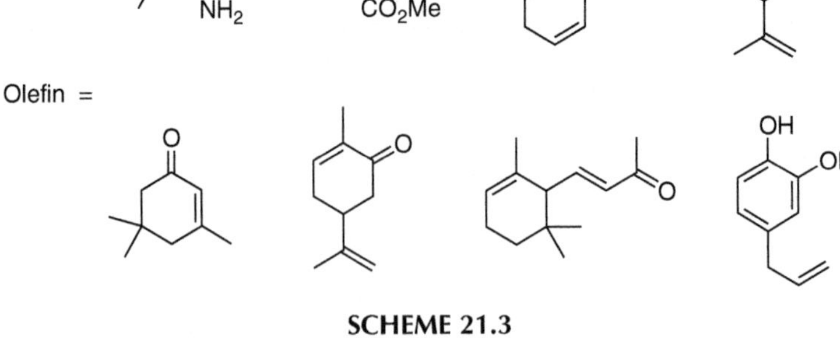

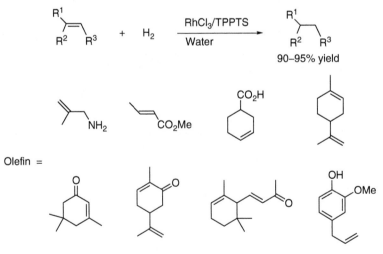

SCHEME 21.3

Chemoselective hydrogenation of the C=C bonds is also possible with other rhodium catalysts. For instance, Alper reported that the water-insoluble complex $RhHCl_2(PCy_3)_2$ was capable of hydrogenating a variety of α,β-unsaturated carbonyl compounds at the C=C bonds in a biphasic solvent mixture, in which water was indispensable [31]. The water-soluble tetranuclear complex $Rh_4(O_2CPr)Cl_2(MeCN)_4$ (Pr = *n*-propyl) catalyzed the selective hydrogenation of α,β-unsaturated alcohol, ketone, nitrile, carboxylic acid, and amide substrates at ambient conditions [32].

The presence of an aqueous phase can alter hydrogenation selectivities. For instance, in the hydrogenation of 3,8-nonadienoic acid by $RhClP[(p\text{-Tolyl})_3]_3$, addition of water to benzene shifted the regioselectivity from favoring the terminal C=C bond to the internal one [33].

Water is also a good solvent for the selective hydrogenation of C=O bonds in unsaturated carbonyl compounds. Ruthenium complexes are often the catalysts of choice [28, 29, 34]. In one of the earliest studies, Joo showed that keto-acids, e.g., pyruvic acid and phenylpyruvic acid, could be reduced to hydroxy-acids with Ru-TPPMS [TPPMS = $PPh_2(m\text{-}C_6H_4SO_3Na)$] catalysts at 60°C and 1 bar of H_2 with a turnover number (TON) of 1300 [35]. A practical application of this chemistry is the selective hydrogenation of α,β-unsaturated aldehydes for the production of allylic alcohols. As shown in Scheme 21.4, the hydrogenation of unsaturated aldehydes was effected by $RuCl_3$ in the presence of TPPTS in a biphasic water–toluene (1/1) mixture, with conversions over 93% and C=O selectivities over 97% [36]. The catalyst could be recycled and no leaching of the metal and ligand into the organic phase was detected. Interestingly, the chemoselectivity could be reversed to favor C=C bond saturation by using a Rh-TPPTS catalyst.

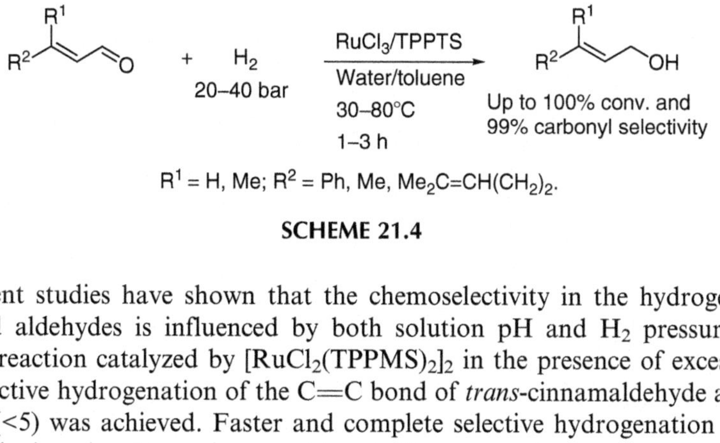

R^1 R^2 (α,β-unsaturated aldehyde) + H_2 (20–40 bar) $\xrightarrow[\text{Water/toluene, 30–80°C, 1–3 h}]{RuCl_3/TPPTS}$ R^1 R^2 (allylic alcohol, OH)

Up to 100% conv. and 99% carbonyl selectivity

R^1 = H, Me; R^2 = Ph, Me, $Me_2C{=}CH(CH_2)_2$.

SCHEME 21.4

More recent studies have shown that the chemoselectivity in the hydrogenation of α, β-unsaturated aldehydes is influenced by both solution pH and H_2 pressure [37]. Thus, although the reaction catalyzed by $[RuCl_2(TPPMS)_2]_2$ in the presence of excess of TPPMS was slow, selective hydrogenation of the C=C bond of *trans*-cinnamaldehyde at 1 bar of H_2 and low pH (<5) was achieved. Faster and complete selective hydrogenation of the C=O group resulted when the pH was increased to >7. However, selective saturation of the C=O bond could also be achieved at a higher H_2 pressure even though the aqueous phase pH was low. These observations can be accounted for by the interplay between two equilibrating species, $RuClH(TPPMS)_3$ and $RuH_2(TPPMS)_4$, their concentrations varying with pH and pressure. It has been shown that the former is selective toward C=C bonds whilst the latter toward the C=O groups.

Hydrogenation of ketones in water has been less studied. An example is the hydrogenation of acetophenones with rhodium and iridium complexes in the presence of a water-soluble 2, 2′-bipyridine (bpy) ligand. Acetophenone could be hydrogenated with a yield of >99% at room temperature and 40 bar of H_2; catalyst reuse was possible [38]. The water-soluble complex $[Cp^*Ir(H_2O)_3]^{2+}$ (Cp^* = $\eta^5\text{-}C_5Me_5$) was shown to be active for the hydrogenation of aldehydes and ketones in a pH range of −1 to 4 in water [39].

Transfer hydrogenation is also feasible in water. An early example is provided by Joo, who reported that benzaldehydes, cinnamaldehyde, crotoaldehyde, 1-citronellal, and citral could be reduced at the C=O bonds by HCOONa in the presence of $RuCl_2(TPPMS)_2$, with conversions typically over 98% [40]. Ogo and Watanabe have recently reported that ketones can be reduced by HCOONa using the water-soluble complex $[(\eta^6\text{-}C_6Me_6)Ru(bpy)(H_2O)][SO_4]$ as catalyst (Scheme 21.5) [41]. Both water-soluble and water-insoluble ketones were reduced at 70°C, with yields over 97% and turnover frequencies (TOFs) ranging from 21 to 153 h^{-1}. The active catalyst is believed to be a monohydride species. As in the case of hydrogenation, solution pH can play an important role in transfer hydrogenation as well [41, 42]. For the reactions in Scheme 21.5, the reaction rate showed a maximum at ca. pH = 4.

Aromatic nitro compounds could be cleanly reduced to the corresponding amines by CO in water. Thus, the nitro compounds shown in Scheme 21.6 were reduced with yields >99% in

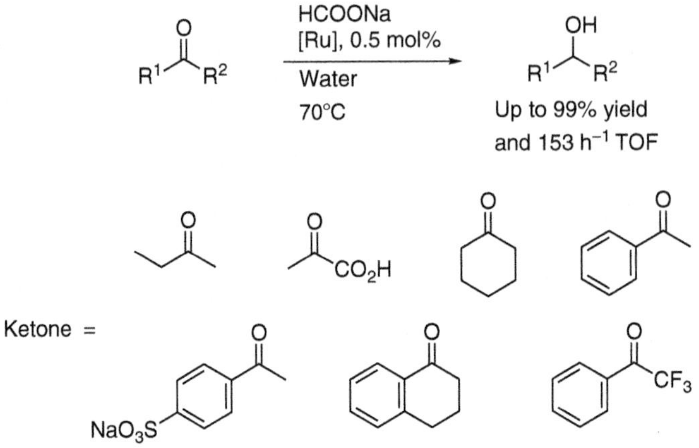

HCOONa
[Ru], 0.5 mol%
Water
70°C

Up to 99% yield
and 153 h^{-1} TOF

Ketone =

SCHEME 21.5

$ArNO_2$ + 3 CO + H_2O → $ArNH_2$ + 3 CO_2
20 bar
$Ru_3(CO)_{12}$, $HN(i\text{-}Pr)_2$, Water/diglyme, 150°C
>99% yield

$ArNO_2$ =

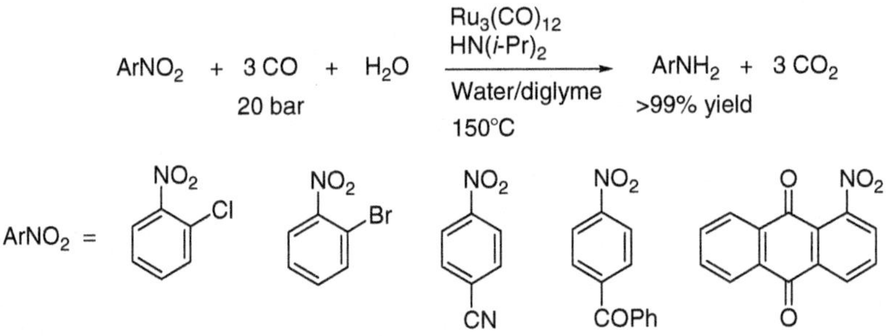

SCHEME 21.6

the presence of a catalytic amount of $Ru_3(CO)_{12}$ and excess $HN(i\text{-}Pr)_2$ at 20 bar of CO and 150°C in a mixture of water and diethyleneglycol dimethyl ether (diglyme). Whilst a stoichiometric quantity of CO_2 was formed at the end of the reduction, only a trace amount of H_2 was detected, suggesting that the reduction is not affected by H_2, which could be generated through the water–gas shift reaction [43].

In comparison with hydrogenation in water, much less has been reported on the achiral hydrogenation of functionalized olefins in the other alternative solvents. This partly reflects the longer history of the former. An example from ionic liquids is seen in the ruthenium-catalyzed stereoselective hydrogenation of sorbic acid in a biphasic system consisting of methyl *t*-butyl ether (MTBE) and [bmim][PF_6] (bmim = 1-butyl-3-methylimidazolium) (Scheme 21.7); *cis*-3-hexenoic acid could be obtained in 90% selectivity at 45% conversion at 10 bar of H_2 and 60°C [44]. More recently, Dupont reported that olefins, such as styrene, methyl methacrylate, and 4-vinylcyclohexene, could be readily hydrogenated in the same ionic liquid at 4 bar of H_2 and 75°C by iridium nanoparticles generated from $[IrCl(COD)]_2$ (COD = 1,5-cyclooctadiene) in the ionic liquid. TOF as high as 6000 h^{-1} was recorded and the catalyst could be separated via decantation and recycled more than seven times. In the case of 4-vinylcyclohexene, 4-ethylcyclohexene was obtained in 91% selectivity at 91% conversion [45].

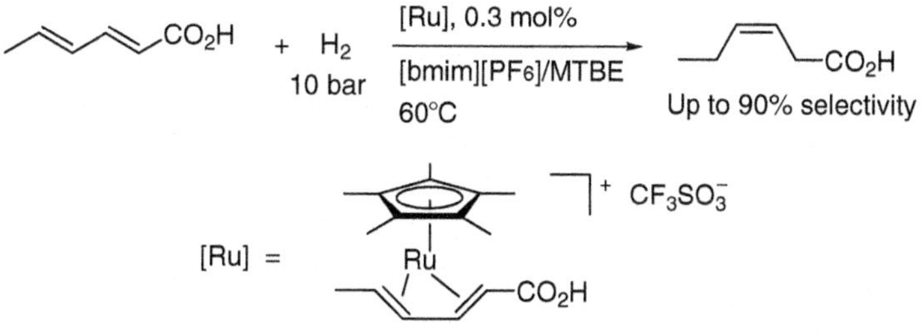

CO_2H + H_2 10 bar; [Ru], 0.3 mol%; [bmim][PF_6]/MTBE; 60°C; CO_2H; Up to 90% selectivity

[Ru] = Ru; CO_2H; + $CF_3SO_3^-$

SCHEME 21.7

A few examples of achiral hydrogenation in $scCO_2$ and fluorous solvents are known [46–52]. Unlike reactions in ionic liquids, catalysis by organometallic compounds in $scCO_2$ and fluorous media generally require the modification of ligands, which normally do not or only slightly dissolve in these solvents. The modification has been performed most often by introducing fluorous chains, and this extra step has to some degree deterred advancement of catalysis in these areas. Hydrogenation in $scCO_2$ can be effected by using heterogeneous catalysts without any modifications, however. A recent example is the hydrogenation of *trans*-cinnamaldehyde by a MCM48-supported Pt–Ru catalyst, which yielded exclusively cinnamyl alcohol at ca. 45% conversion in $scCO_2$ [53]. Earlier, Poliakoff reported that aldehydes, ketones, nitroarenes, and arenes could be continuously hydrogenated using aminopolysiloxane-supported Pd catalysts in $scCO_2$. For example, *m*-cresol was reduced to give four products and interestingly, the chemoselectivity could be easily tuned (Scheme 21.8). Thus, the selectivity to 3-methylcyclohexanol was maximized by using a temperature of ca. 250°C or a higher H_2/substarte ratio [54].

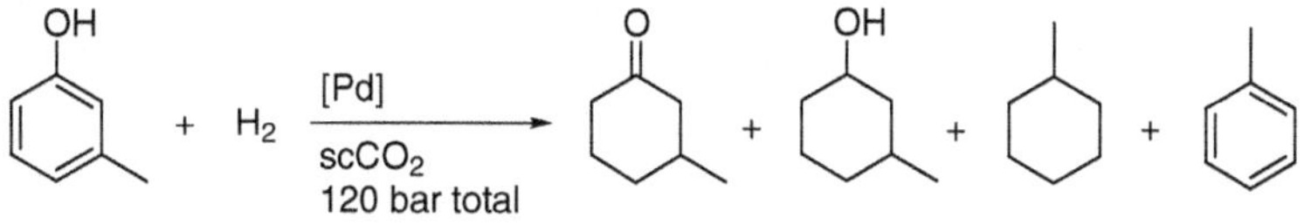

OH + H_2; [Pd]; $scCO_2$; 120 bar total; O + OH + +

SCHEME 21.8

Sinou and Gladysz reported that 2-cyclohexen-1-one could be readily hydrogenated to cyclohexanone by rhodium in combination with fluorous-soluble phosphines, e.g., $P(p\text{-}C_6H_4OCH_2CH_2C_7F_{15})_3$, in fluorous-organo biphasic media [55, 56]. With other functionalized olefins such as cinnamate and cinnamaldehyde, however, the hydrogenation was much slower [56].

21.3.1.2 Asymmetric Hydrogenation

Enantioselective hydrogenation in the alternative solvents is of particular interest, as they can provide a clean way for the synthesis of valuable fine chemicals and pharmaceutical intermediates and make easy the recycle of expensive chiral ligands and metals. In many cases, however, reduced enantioselectivities have been observed on going from organic solvents, such as methanol, to water, ionic liquids, and $scCO_2$. This is particularly so in the case of water in the early days [29, 57, 58].

The water-soluble, hydroxy-functionalized phospholane shown in Scheme 21.9 is one of the several exceptions [59]. This ligand has the backbone of Burk's well-established DuPhos ligands [60] and was first reported by Holz, who showed that in rhodium-catalyzed hydrogenation of 2-acetamidoacrylic acid and its methyl ester, the phospholane gave rise to enantiomeric excesses (ee) of up to 99.6% in pure water [61]. The work of Rajan-Babu shows that under the catalysis of [RhL(NBD)][SbF_6] (NBD = norbornadiene) (1 mol%), methyl 2-acetamidoacrylate can be reduced to methyl *N*-acetylalanate with 100% conversion

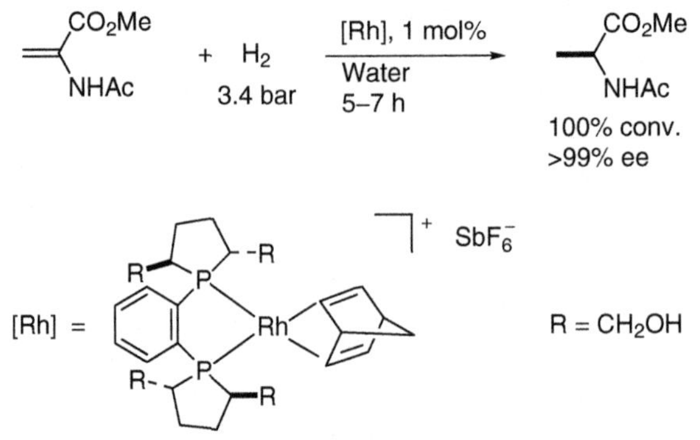

SCHEME 21.9

and >99% ee at ca. 3 bar of H_2 and room temperature in 5–7 h reaction time. Remarkably, the catalyst could be recycled four times without losing activity and enantioselectivity; the product was extracted with Et_2O. A similar ligand has been synthesized by Zhang and shown to give excellent ee's in the hydrogenation of itaconic acid in water in the presence of a small amount of MeOH (Scheme 21.10) [62].

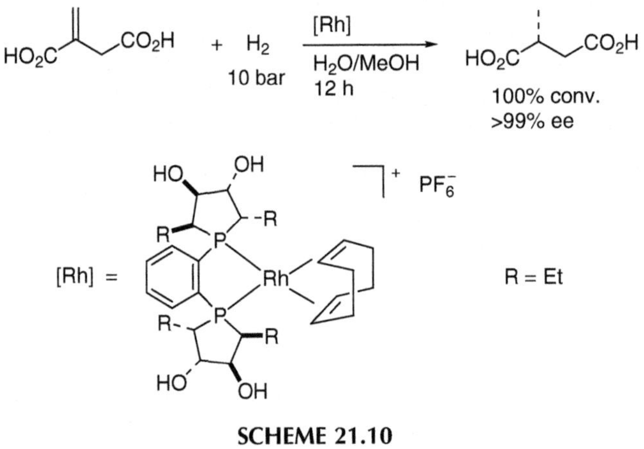

SCHEME 21.10

The low rates and enantioselectivities encountered in hydrogenation in water can be improved by the addition of amphiphiles in some instances, as shown by Oehme and Selke [29, 63]. An example is the asymmetric hydrogenation of phosphonates to give α-aminophosphonic acids by the water-insoluble Rh-BPPM catalyst (Scheme 21.11) [64]. Various phosphonates were readily reduced with ee's in the range of 96–99% in the presence of a surfactant, sodium dodecyl sulfate (SDS). The introduction of SDS resulted in both a higher enantioselectivity and a much improved reaction rate, due at least partly to increased catalyst and substrate solubility in water.

SDS has also been shown to influence asymmetric hydrogenation of dehydroamino acids by water-soluble catalysts. This is seen in the hydrogenation of methyl (*Z*)-α-acetamidocinnamate by a rhodium catalyst containing a trehalose-derived phosphinite ligand, which furnished a faster reaction and a higher ee value of >99% in the presence of SDS (10–200%) at 5 bar of H_2 and room temperature. A number of other enamides were also fully hydrogenated within 1 h with excellent ee's under such conditions [65]. While the precise role of amphiphiles remains speculative, it appears that the formation of micelles is important. A recent study has further shown that the micellar solution pH affects hydrogenation rates and enantioselectivities [66].

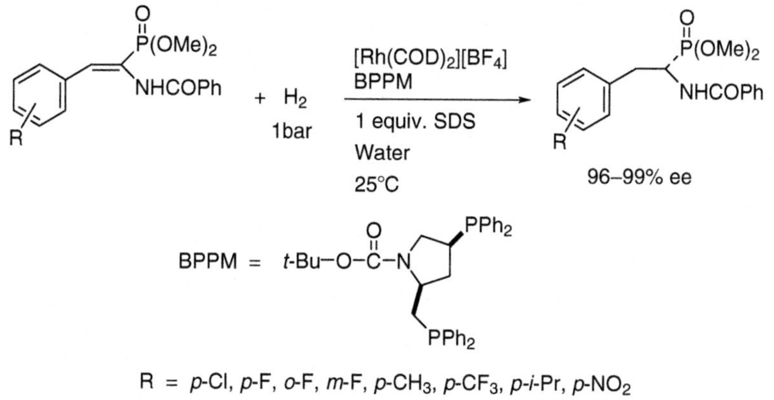

[Rh(COD)$_2$][BF$_4$]
BPPM
+ H_2 1bar
1 equiv. SDS
Water
25°C
96–99% ee

BPPM = *t*-Bu–O–C(=O)–N (pyrrolidine) PPh$_2$, PPh$_2$

R = *p*-Cl, *p*-F, *o*-F, *m*-F, *p*-CH_3, *p*-CF_3, *p*-*i*-Pr, *p*-NO_2

SCHEME 21.11

Enantioselective hydrogenation of prochiral ketones in water is less well studied [29, 58]. However, several examples of asymmetric transfer hydrogenation of ketones have recently appeared [67–69]. Scheme 21.12 shows that various aromatic ketones can be reduced by ruthenium in combination with a water-soluble vicinal diamine in either water or a H_2O/CH_2Cl_2 biphasic system, with ee's ranging from 80% to 98% [67]. A low conversion of 21% was obtained with tetralone, presumably due to its low solubility in water. Interestingly, the aqueous phase reduction of ω-bromoacetophenone yielded 2-bromo-1-phenylethanol in 94% ee without complications from potential formate displacement. The bromoalcohols could be converted to synthetically valuable styrene oxides with the same ee values.

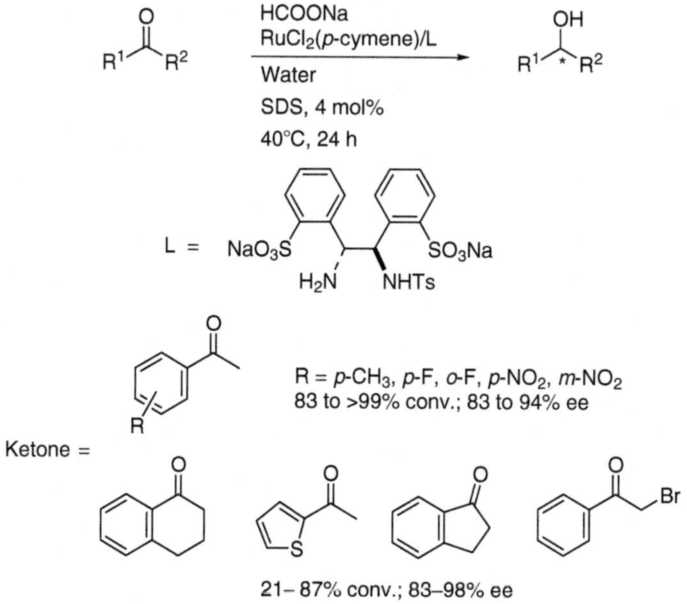

HCOONa
RuCl$_2$(*p*-cymene)/L
Water
SDS, 4 mol%
40°C, 24 h

L = NaO$_3$S ... SO$_3$Na, H$_2$N, NHTs

Ketone =

R = *p*-CH_3, *p*-F, *o*-F, *p*-NO_2, *m*-NO_2
83 to >99% conv.; 83 to 94% ee

21–87% conv.; 83–98% ee

SCHEME 21.12

Enantioselective hydrogenation in ionic liquids was first demonstrated by Chauvin and Olivier [70]. α-Acetamidocinnamic acid was hydrogenated to give (*S*)-phenylalanine in 64% ee by a rhodium catalyst containing an unmodified chiral diphosphine ligand in a mixture of [bmim][BF$_4$] and *i*-PrOH. The anti-inflammatory drug (*S*)-naproxen was obtained in 80% ee by the hydrogenation of 2-(6-methoxy-2-naphthyl)acrylic acid with a ruthenium catalyst, $[RuCl_2\{(S)\text{-BINAP}\}]_2NEt_3$, in the same solvent mixture. Phenylacrylic acid was similarly hydrogenated, with ee's up to 86%. Dupont reported that the product could be quantitatively separated, and the separated, catalyst-containing ionic liquid reused several times without significant loss of activity and enantioselectivity [71].

New examples of enamide hydrogenation have been reported [72, 73]. Scheme 21.13 shows the hydrogenation of methyl α-acetamidoacrylate and methyl α-acetamidocinnamate in a $[bmim][PF_6]$/*i*-PrOH biphasic mixture. The Rh-Duphos catalyst, without any modifications, dissolved in the ionic liquid phase, and catalyzed the hydrogenation at ambient temperature and 2 bar of H_2, affording the amino acids with 93–96% ee values. The product was easily separated by removing *i*-PrOH, and the catalyst-containing ionic liquid could be recycled five times. Although the conversion decreased from 83% to 58% during the recycle, no significant change in ee's was recorded. This stems partly from the enhanced stability of the rhodium catalyst in the ionic liquid in comparison with conventional solvents [73].

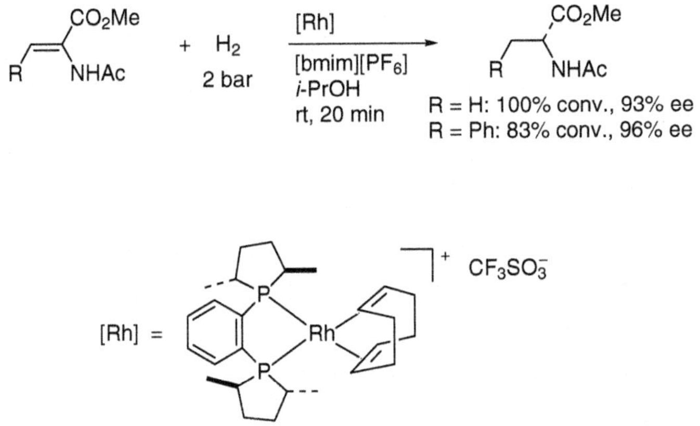

CO_2Me R NHAc + H_2 2 bar → [Rh] / $[bmim][PF_6]$ / *i*-PrOH / rt, 20 min → CO_2Me R NHAc

R = H: 100% conv., 93% ee
R = Ph: 83% conv., 96% ee

[Rh] = P, P, Rh, $CF_3SO_3^-$

SCHEME 21.13

Product separation in ionic liquids can sometime pose a problem, because of the strong solvent power of ionic liquids towards many organic compounds. This can be effectively addressed in a greener manner by using $scCO_2$ extraction, and has recently been demonstrated by Jessop in the enantioselective hydrogenation of unsaturated carboxylic acids [74]. This chemistry is discussed elsewhere in this book.

Asymmetric hydrogenation in $scCO_2$ was first demonstrated by two groups (Scheme 21.14) [75, 76]. Using a $scCO_2$-soluble Rh-DuPhos catalyst, Burk and Tumas reported that enamides could be hydrogenated to α-aminoacids with ee's up to 99.5% at 14 bar of H_2 and 40°C. The solubility of the catalyst is imparted by the $scCO_2$-philic, fluorinated BARF anion. Noyori, Ikariya, and coworkers used a partially hydrogenated BINAP ligand for the ruthenium-catalyzed hydrogenation of α,β-unsaturated carboxylic acids. The H_8-BINAP ligand is more soluble than BINAP in $scCO_2$, in which the latter hardly dissolves. Tiglic acid was hydrogenated with an ee value of 81%, which is comparable to that (82%) observed in the commonly used MeOH under similar conditions. The conversion and ee could be increased to 89% by adding fluorous alcohols to the reaction mixture. These studies illustrate some of the well-known challenges facing catalysis in $scCO_2$ and fluorous solvents, that is the low solubility of metal catalysts and polar substrates.

Using BARF as solubility promoter, Leitner and Pfaltz reported that the enantioselective hydrogenation of an imine by an cationic iridium catalyst proceeded to completion within

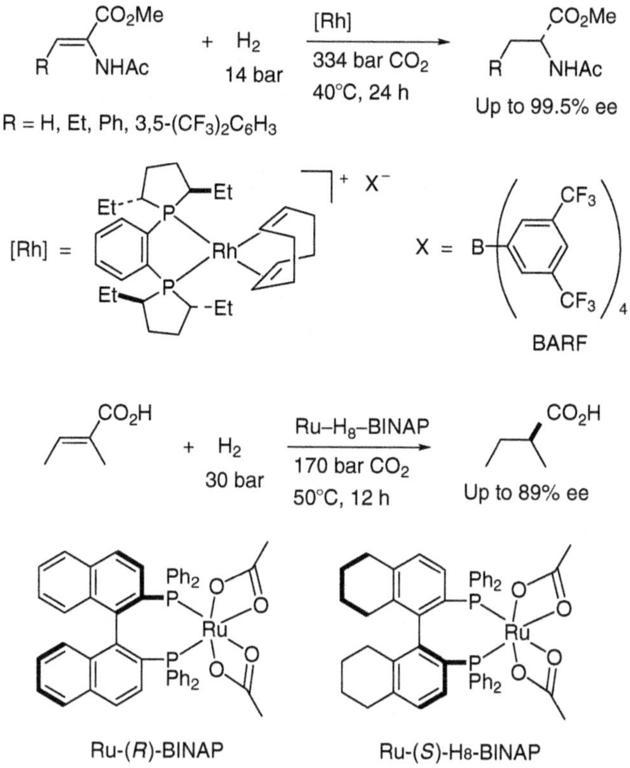

SCHEME 21.14

1 h at 30 bar of H_2 and 40°C in $scCO_2$, affording the amine in up to 81% ee with a catalyst loading of less than 0.1 mol% (Scheme 21.15) [77]. Whilst the observed enantioselectivity in $scCO_2$ is slightly lower than that obtained in CH_2Cl_2, the initial TOF (up to 2800 h^{-1}) with the former doubles that with the latter. In the end of the hydrogenation, the amine could be quantitatively extracted by $scCO_2$ purging, allowing for catalyst recycle several times without significant changes in catalyst activity and enantioselectivity. This study reveals that while attaching a fluorous ponytail onto the ligand may lead to enhanced catalyst solubility and hence a higher conversion, it does not necessarily result in a higher enantioselectivity.

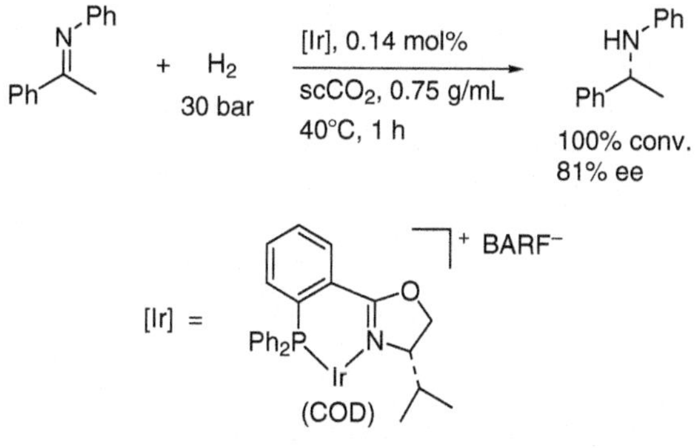

SCHEME 21.15

Asymmetric hydrogenation in $scCO_2$ can sometime be performed with unmodified catalysts in the presence of a cosolvent. Scheme 21.16 shows two examples [78]. MeOH was used

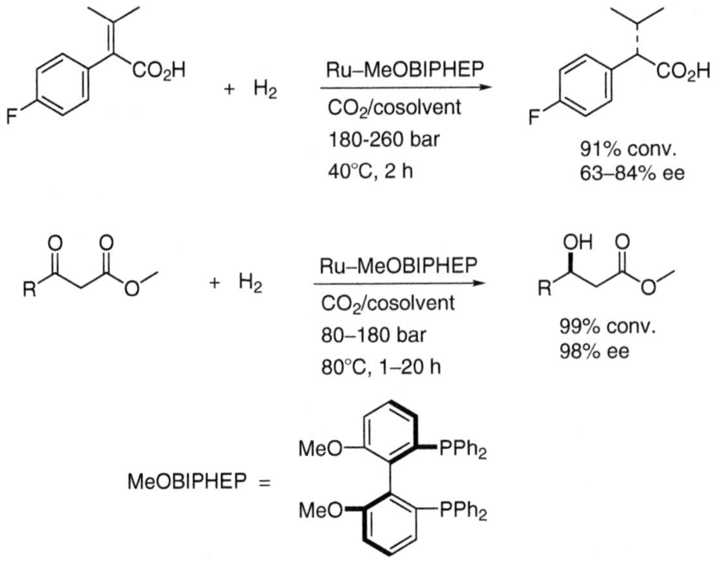

$$\text{(4-F-C}_6\text{H}_4\text{)C(=CMe}_2\text{)CO}_2\text{H} + H_2 \xrightarrow[\substack{CO_2/\text{cosolvent} \\ 180\text{-}260\ \text{bar} \\ 40^\circ\text{C, 2 h}}]{\text{Ru-MeOBIPHEP}} \text{(4-F-C}_6\text{H}_4\text{)CH(CHMe}_2\text{)CO}_2\text{H}$$

91% conv.
63–84% ee

$$\text{RC(O)CH}_2\text{CO}_2\text{Me} + H_2 \xrightarrow[\substack{CO_2/\text{cosolvent} \\ 80\text{–}180\ \text{bar} \\ 80^\circ\text{C, 1–20 h}}]{\text{Ru-MeOBIPHEP}} \text{RCH(OH)CH}_2\text{CO}_2\text{Me}$$

99% conv.
98% ee

MeOBIPHEP = (MeO, PPh_2 substituted biphenyl)

SCHEME 21.16

in both reactions. The carboxylic acid and alcohol were obtained in up to 84% and 98% ee, respectively. In the case of the ketoester hydrogenation in the absence of CO_2, the reaction was slow, accompanied with a much lower enantioselectivity. This is likely to be a result of improved H_2 mass transfer under the $scCO_2$ conditions.

21.3.2 Hydroformylation and Carbonylation

Hydroformylation of olefins by CO and H_2 results in the formation of a new C—C bond and introduces a carbonyl functional group into the original carbon chain. There are a number of catalysts, in particular those based on rhodium, which can be used to perform this reaction. A further appealing aspect for synthetic chemistry is that this is a totally atom-efficient reaction, with all the starting atoms incorporated into the product. The largest industrial application of this reaction is the hydroformylation of ethylene and propene, where both cobalt and rhodium are employed as catalyst [3]. For the more active, synthetically more useful rhodium catalysts, their separation and recycle can be a problem. In the case of propene, this has been most successfully addressed by Ruhrchemie/Rhone–Poulenc using an aqueous biphasic system and a water-soluble Rh-TPPTS catalyst [79]. For higher olefins, this simple approach is no longer viable, due to their low solubility in water. A great deal of effort has therefore been focused on the hydroformylation of higher olefins, using in particular nonconventional solvents as a means for catalyst immobilization and separation. This has been extensively reviewed elsewhere and will not be dealt with here [3, 80–83]. For the synthetically more relevant, functionalized olefins, their hydroformylation is less documented. This reflects to some degree the greater difficulties encountered in the reaction, namely slower rates, more side reactions, and less-controllable selectivities in comparison with that of simple olefins.

In aqueous-phase hydroformylation, some olefins have been shown to react with syngas exceedingly well, however. An example is the hydroformylation of acrylic esters to give bifunctional compounds. Mortreux and coworkers reported that when carried out in an aqueous biphasic system, the hydroformylation reaction shown in Scheme 21.17 proceeded with initial TOFs of up to 28 min^{-1}, which is 14-fold faster than in the commonly used toluene [84, 85]. The reaction was performed in a water–toluene mixture at 50°C and 50 bar of syngas, furnishing almost exclusively the branched aldehydes. It is believed that the low rates in

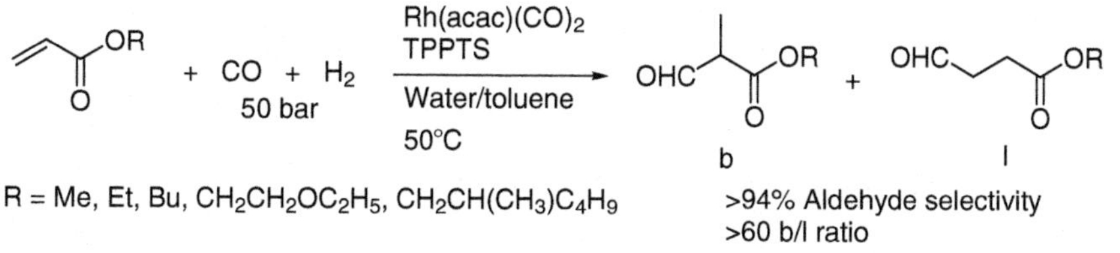

OR + CO + H_2 (50 bar) → [Rh(acac)$(CO)_2$, TPPTS, Water/toluene, 50°C] OHC...OR (b) + OHC...OR (l)

R = Me, Et, Bu, $CH_2CH_2OC_2H_5$, $CH_2CH(CH_3)C_4H_9$

>94% Aldehyde selectivity
>60 b/l ratio

SCHEME 21.17

common organic solvents result from the formation of thermodynamically stable five- or six-membered rings via the coordination of the acrylate carbonyl group to rhodium, with the reaction rate determined by the opening of the ring to give a coordinatively unsaturated intermediate. The ring-opening step could be facilitated by H_2O–carbonyl hydrogen bonding interactions, leading to faster rates in water.

Sheldon has reported that the hydrofromylation of *N*-allylacetamide can be efficiently carried out in water [86]. Using again the Rh-TPPTS catalyst, the reaction was shown to proceed with an outstanding TOF of 3891 h^{-1} at 70°C and 10 bar of syngas, although the regioselectivity, which could be improved with bidentate phosphines, was less satisfactory (Scheme 21.18). For comparison, TOFs ranging from 589 to 1342 h^{-1} were observed in THF, toluene, and MeOH using Rh-PPh_3 as catalyst under similar conditions. The catalyst could be separated from the water-soluble product by adopting an inverted aqueous biphasic system, in which a hydrophobic ligand (e.g., PPh_3) could be used and would reside in the organic phase. The higher rates in water probably result again from easier generation of the coordinatively unsaturated intermediate.

NHAc + H_2 + CO (10 bar) → [Rh(acac)$(CO)_2$, TPPTS, Water, 70°C, 45 min] OHC...NHAc (l) + CHO...NHAc (b)

98.9% conv.; 3891 h^{-1} TOF
1.3 l/b ratio

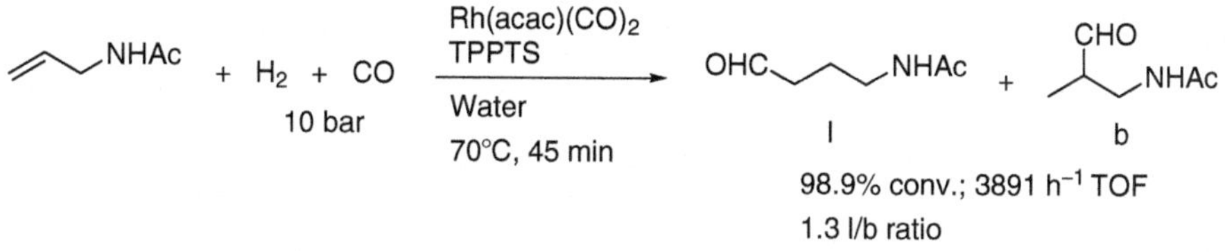

SCHEME 21.18

Functionalized olefins such as styrenes and vinyl ethers have also been successfully hydroformylated in aqueous media [87–89]. A recent example is the regioselective hydroformylation of 2-benzyloxystyrene to give a linear aldehyde, 2-chromanol, a structural moiety present in several interesting therapeutically active molecules [90]. The aldehyde could be obtained in up to 86% yield with a rhodium catalyst containing TPPTS or a water-soluble diphosphine at 100°C and 10 bar of syngas in a water–toluene solvent mixture. Aldehydes formed in hydroformylation reactions can be condensed with amines to form imines, which give amines upon hydrogenation. This has been demonstrated by Beller using water-soluble rhodium and iridium catalysts; the former catalyzes hydroformylation while the latter affects C=N bond hydrogenation [91].

By replacing H_2 with nucleophiles such as HOH, ROH, and HNR_2, olefins and aryl halides can be carbonylated to give carbonyl compounds such as carboxylic acids, esters, and amides. The most often used catalysts for carbonylation are those based on palladium. A earlier study has shown that palladium in combination with water-soluble phosphines is active and selective for the carbonylation of simple olefins to give carboxylic acids in water [92].

Pauson–Khand reaction is a variant of the common carbonylation reactions, where CO is reacted with an alkyne and alkene forming a cyclopentenone. A novel way for performing this reaction is by coupling aqueous-phase decarbonylation with aqueous micellar carbonylation, through which a variety of enynes have been converted into cyclopentenones, as shown in Scheme 21.19 [93]. Formaldehyde was used as the source of CO, and the decarbonylation was catalyzed by a water-soluble Rh-TPPTS species. The CO so generated was then incorporated into the Pauson–Khand cycle catalyzed by a Rh-DPPP species (DPPP = $Ph_2P(CH_2)_3PPh_2$). The catalysis is believed to occur in micelles formed by the surfactant SDS. The only by-product from this reaction is H_2.

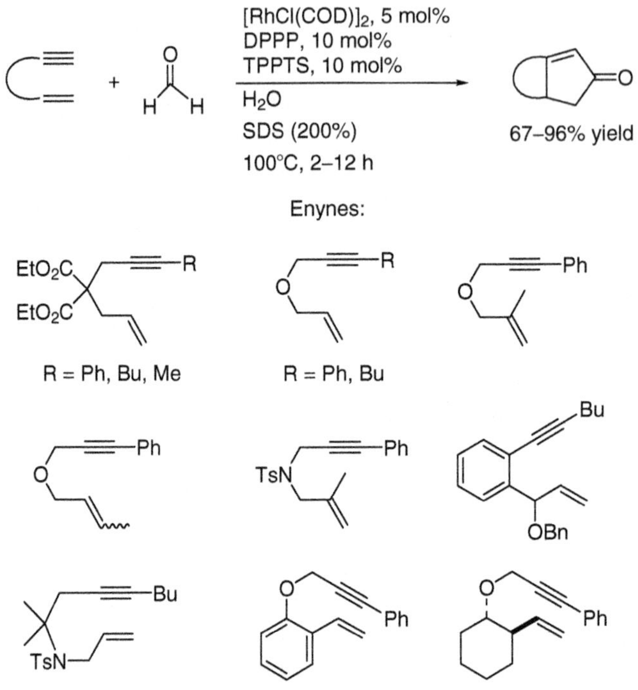

[RhCl(COD)]$_2$, 5 mol%
DPPP, 10 mol%
TPPTS, 10 mol%
H_2O
SDS (200%)
100°C, 2–12 h

67–96% yield

Enynes:

R = Ph, Bu, Me

R = Ph, Bu

SCHEME 21.19

Hydroformylation and carbonylation have also been demonstrated in ionic liquids, although most studies deal with nonfunctionalized substrates [70, 94, 95]. An example involving a functionalized olefin is seen in the hydroformylation of methyl 3-pentenoate in [bmim][PF_6] (Scheme 21.20) [96]. The ionic liquid-soluble catalyst was formed by combining Rh(acac)$(CO)_2$ with the ligand. The desired linear aldehyde, formed upon isomerization of the internal C=C bond to a terminal position, is a precursor of adipic acid. Of particular note is that the ionic liquid medium provides a more stable catalyst, allowing the catalyst to be recycled ten times furnishing a total TON of 6640, which is almost sevenfold greater than achievable in organic solvents (e.g., CH_2Cl_2). The product was removed by distillation (0.2 mbar/110°C).

Carbonylation of styrenes in the presence of an alcohol yields arylpropionic esters. The Pd-NMDPP-catalyzed alkoxycarbonylation of styrene derivatives in [bmim][BF_4]–cyclohexane afforded 2-arylpropionic esters in good yields and high regioselectivities (>99.5%) (Scheme 21.21) [97]. No enantioselectivity (<5%) was observed, however. The reaction proceeded under mild conditions in the presence of *p*-toluenesulfonic acid (TsOH), with the product separated by simple decantation.

More recently, the carbonylation of alcohols, alkynols, and aryl halides in ionic liquids have been reported [98–100]. Novel selectivities and better yields compared to conventional media were observed. For instance, the hydroesterification of *t*-butyl alcohol by palladium catalysis in the presence of ethanol and TsOH in an imidazolium ionic liquid yielded ethyl *t*-valerate, which does not form in organic solvents [98]. An interesting example of using ionic

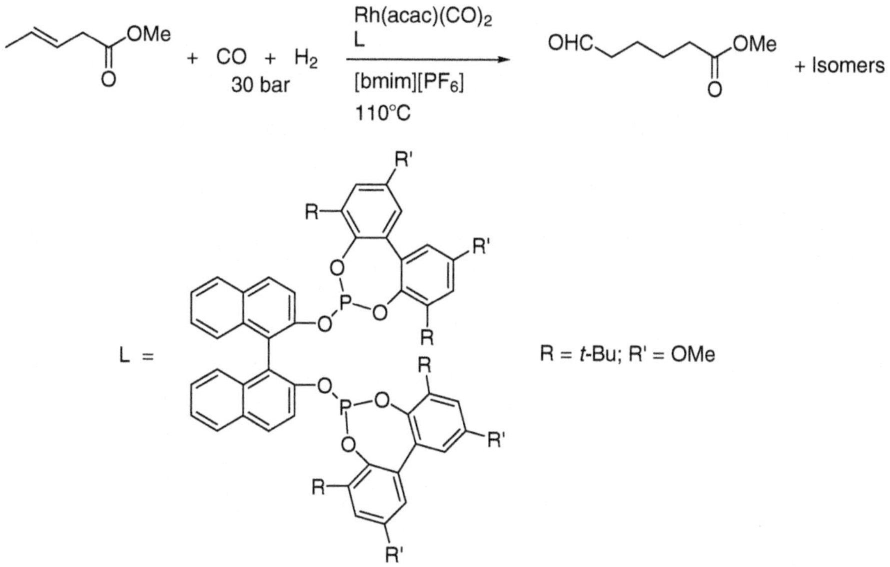

OMe + CO + H_2 30 bar; Rh(acac)$(CO)_2$, L, [bmim][PF_6], 110°C → OHC...OMe + Isomers

L = ; R = *t*-Bu; R' = OMe

SCHEME 21.20

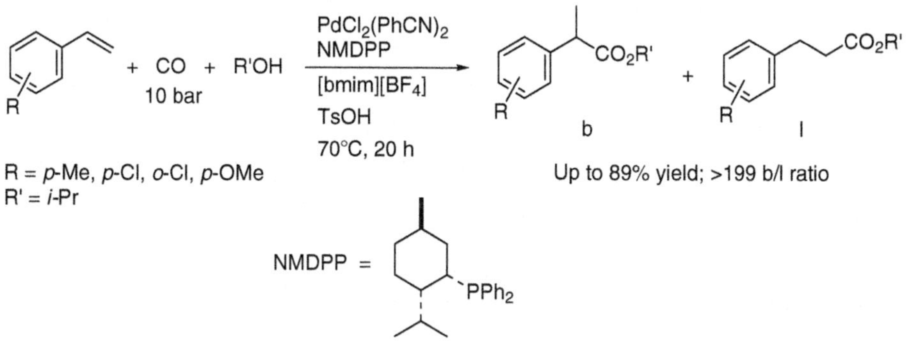

+ CO + R'OH 10 bar; $PdCl_2(PhCN)_2$, NMDPP, [bmim][BF_4], TsOH, 70°C, 20 h → CO_2R' (b) + CO_2R' (l)

R = *p*-Me, *p*-Cl, *o*-Cl, *p*-OMe
R' = *i*-Pr

Up to 89% yield; >199 b/l ratio

NMDPP = ; PPh_2

SCHEME 21.21

liquids as catalyst is found in the rapid oxidative carbonylation of aniline and related amines to give ureas. The catalyst was a selenite anion of an imidazolium ionic liquid; but the reaction was performed in MeOH not the ionic liquid [101].

Hydroformylation is among the first reactions to be investigated in $scCO_2$ and fluorous solvents. Most early examples were concerned with nonfunctionalized olefins such as 1-alkenes, which could be converted into aldehydes by rhodium catalysis, with rates surpassing those observed in conventional solvents. An example involving functionalized olefins is provided by the hydroformylation of acrylates, which can be effected very efficiently by $scCO_2$-soluble, molecular as well as polymeric rhodium catalysts in $scCO_2$ [102, 103]. Scheme 21.22 shows that acrylates can be hydroformylated with TOFs of over 2000 h^{-1} at 80°C and 30 bar of syngas using Rh(acac)$(CO)_2$ in combination with a fluorous polymeric phosphine. Excellent regioselectivities in favor of the branched aldehydes were recorded, with the linear aldehyde accounting for less than 1% of the product. Remarkably, the catalyst was totally chemoselective towards acrylates. This is made manifest in the inter- and intramolecular competition reactions shown in the scheme, where only the C=C bond adjacent to the carbonyl group reacted. The polymeric ligand was also shown to be effective for the hydroformylation of styrene and acrylates in fluorous media [104]. The high rates in $scCO_2$ are

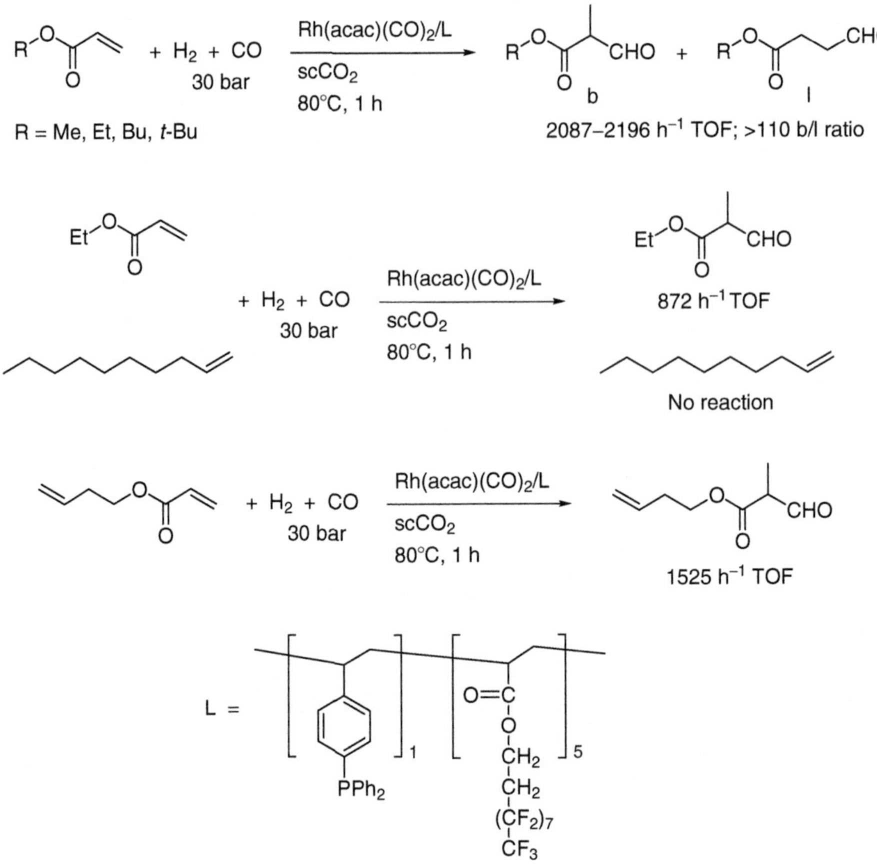

SCHEME 21.22

believed to be due to the ring-opening step, discussed earlier, being facilitated by CO_2–carbonyl interactions.

Asymmetric hydroformylation in $scCO_2$ was demonstrated by Leitner [105]. Scheme 21.23 shows that vinyl arenes and acetate could be hydroformylated by a rhodium catalyst containing a fluoroalkylated BINAPHOS ligand into the branched aldehydes with ee's ranging from 77% to 95% at 40–60°C and 40–60 bar of syngas. Lower enantioselectivities were observed with other substrates, however. The regioselectivities were typically over 90% in favor of the branched aldehydes, which are significantly higher than those achieved with unsubstituted BINAPHOS, due probably to a small electron-donating effect of the fluorous ponytail. To separate the product, the reaction mixture was converted into a two-phase system by reducing the CO_2 pressure, and the aldehydes were then extracted with dense CO_2, with the catalyst left behind for subsequent reactions.

Carbonylation in $scCO_2$ has been less studied. Ikariya demonstrated that 2-iodobenzyl alcohol could be carbonylated to give phthalide using a $scCO_2$-soluble palladium catalyst containing trialkyl or triaryl phosphite ligands. The reaction proceeded at 130°C and a low pressure of CO (1 bar), affording a TON of 4650 in 18 h (200 bar CO_2). A lower TON was obtained when the carbonylation was run in toluene [106].

Unlike most phosphine-based metal catalysts, metal carbonyl compounds often show good solubilities in $scCO_2$. This has been taken advantage of in the Pauson–Khand reaction. As shown in Scheme 21.24, $Co_2(CO)_8$ (2–5 mol%) catalyzed the smooth cyclization of enynes at 90–95°C and 30 bar of CO, affording cyclopentenones in yields of up to 91%.

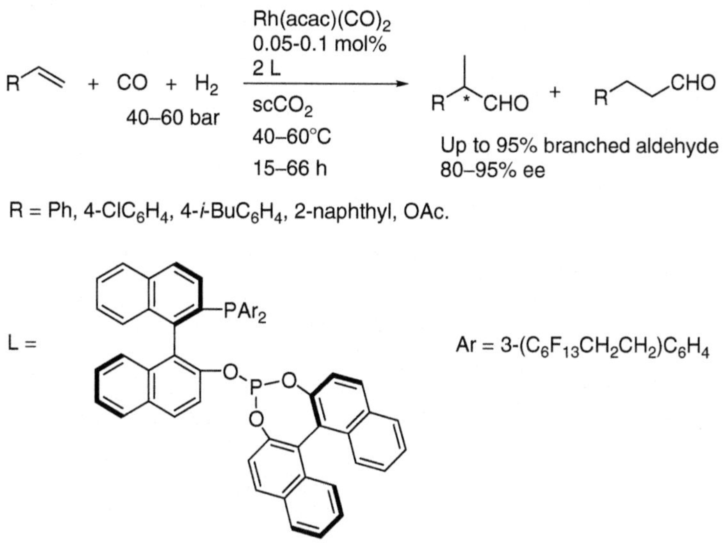

$Rh(acac)(CO)_2$ 0.05-0.1 mol% 2 L

R + CO + H_2 (40–60 bar) → $scCO_2$, 40–60°C, 15–66 h → R * CHO + R CHO

Up to 95% branched aldehyde
80–95% ee

R = Ph, 4-ClC_6H_4, 4-*i*-BuC_6H_4, 2-naphthyl, OAc.

L = PAr$_2$ … O–P–O … O

Ar = 3-($C_6F_{13}CH_2CH_2$)C_6H_4

SCHEME 21.23

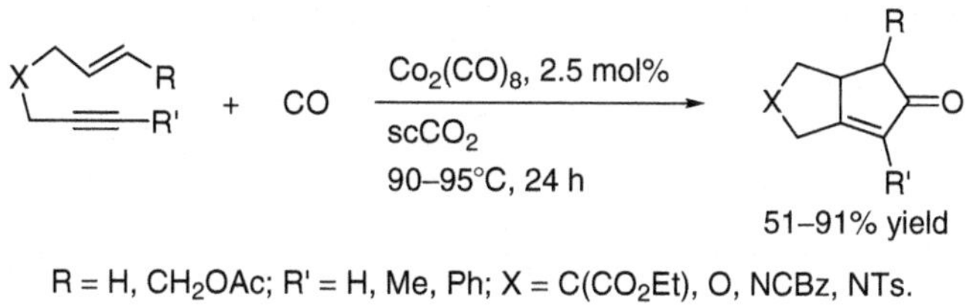

X R R' + CO → $Co_2(CO)_8$, 2.5 mol%, $scCO_2$, 90–95°C, 24 h → X R =O R'

51–91% yield

R = H, CH_2OAc; R' = H, Me, Ph; X = C(CO_2Et), O, NCBz, NTs.

SCHEME 21.24

Catalyst decomposition occurred at a lower CO pressure. Intermolecular cyclization was also shown to be feasible [107].

21.3.3 Catalytic C—C Coupling Reactions

C—C bond formation by catalysis is of central importance in synthetic chemistry. The reactions have traditionally been conducted in common organic solvents. In the past one decade or so, however, a great deal of effort has been invested in the alternative media, aiming at easy catalyst recovery and reducing the environmental impact of the reactions. There is an enormous amount of literature in this area; hence only selected examples could be discussed.

21.3.3.1 Heck Reactions

The Heck reaction is concerned with the coupling of an aryl or vinyl halides (or their derivatives) with an olefin (Scheme 21.1). Palladium in combination with a phosphine is usually used as catalyst for the reaction, which is usually conducted in polar organic solvents such as DMF. When a water-soluble phosphine (e.g., TPPTS) is used, it is possible to perform the Heck reaction in water [108]. A recent example using carboxylated phosphines as ligand is shown in Scheme 21.25 [109]. The palladium catalyst generated from $Pd(OAc)_2$ and *m*-TPPTC was effective for both inter- and intramolecular coupling in a homogeneous $MeCN–H_2O$ mixture, affording the desired products in high isolated yields. The organic solvent increases the solubility of substrates.

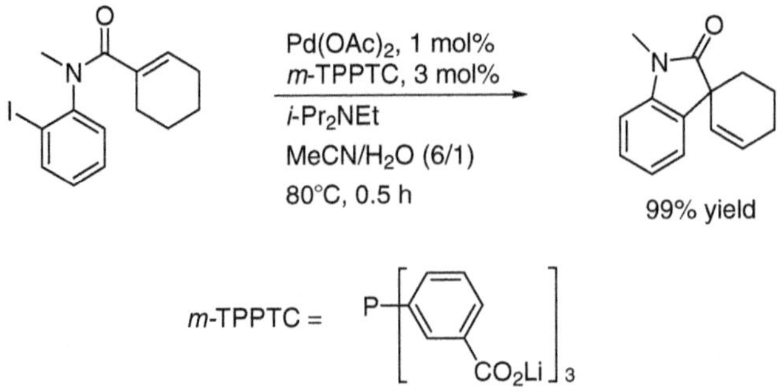

$Pd(OAc)_2$, 1 mol%
m-TPPTC, 3 mol%
i-Pr_2NEt
$MeCN/H_2O$ (6/1)
80°C, 0.5 h

99% yield

m-TPPTC = $P\left[C_6H_4CO_2Li\right]_3$

SCHEME 21.25

Heck reaction could also be effected in water by polymer-supported, amphiphilic phosphines. Uozumi reported that both electron-rich and electron-deficient olefins coupled with aryl iodides and bromides using the polystyrene–poly(ethylene glycol) resin-supported palladium catalyst as shown in Scheme 21.26 [110]. The products were obtained in good to excellent yields, although the yield (52%) was low with bromobenzene. The catalyst could be separated by simple filtration and was demonstrated to be recyclable, affording an average yield of 92% in five runs. Presumably, the poly(ethylene glycol) segment makes the catalyst hydrophilic and therefore more miscible in water, as in its absence the coupling was extremely slow.

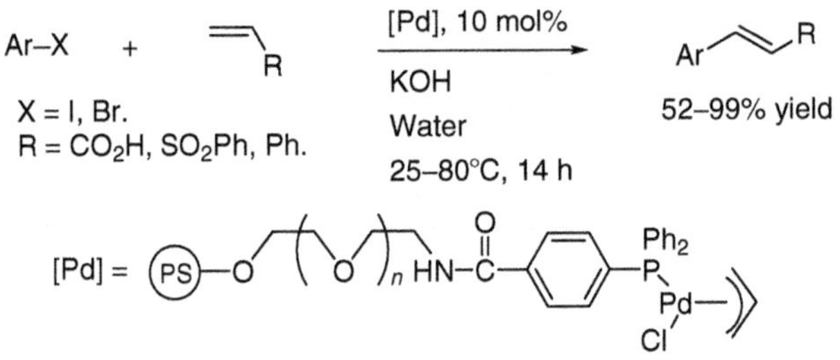

Ar–X + R

X = I, Br.
R = CO_2H, SO_2Ph, Ph.

[Pd], 10 mol%
KOH
Water
25–80°C, 14 h

Ar R

52–99% yield

[Pd] = (PS)–O–(CH_2CH_2O)$_n$–HN–C(=O)–C_6H_4–PPh_2–Pd(Cl)(allyl)

SCHEME 21.26

The introduction of water into an organic solvent can bring about significant alterations in catalyst activity or selectivity. For instance, in intermolecular Heck cyclization reactions that proceed via an *exo* or *endo* process, the presence of water has been demonstrated to reverse the usual *exo* preference, affording *endo* products instead [108]. More recently, Hallberg reported that aqueous DMF as solvent promoted smooth and highly regioselective Heck arylation of vinyl ethers by aryl halides including heteroaromatics (Scheme 21.27) [111]. Hydrolysis of the branched olefins led to ketones in good to excellent yields. In the absence of water, the Heck reaction was slow and yielded a mixture of α and β arylation products. With common solvents, the reaction is synthetically useful only when aryl triflates or halide scavengers, such as silver and thallium salts, are used instead of aryl halides [112]. This is believed to be due to the reaction proceeding via two pathways, with one ionic forming the

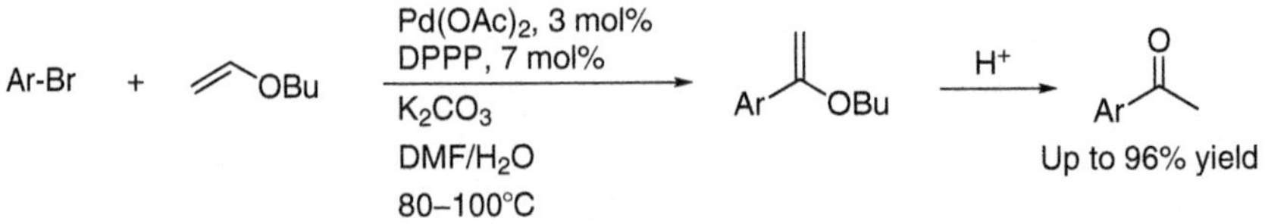

Ar-Br + OBu

$Pd(OAc)_2$, 3 mol%
DPPP, 7 mol%
K_2CO_3
DMF/H_2O
80–100°C

Ar OBu

H^+

Ar

Up to 96% yield

SCHEME 21.27

α product and one neutral forming the β product. The former is made favorable when a good leaving anion (e.g., triflate) is involved, or when the halide ion is abstracted from palladium. Most of Hallberg's examples display an α/β ratio of 99/1, indicating that the ionic pathway is promoted by a polar medium. A drawback of the aqueous reaction is the necessity for long reaction times, which could be made considerably shorter by microwave heating.

Heck reaction in ionic liquids was first reported by Kaufmann in 1996, who demonstrated the high yield coupling of bromobenzenes and an activated chlorobenzene with butyl acrylate in molten tetraalkyl ammonium and phosphonium bromide salts [113]. It should be mentioned, however, that the beneficial effect of quaternary ammonium salts as additives on the activity and stability of palladium catalysts in the Heck reaction has long been recognized [114]. A number of similar reactions have since been demonstrated in these and related imidazolium ionic liquids [19, 115]. A recent example is provided by the Pd nanoparticle-catalyzed stereospecific reaction of cinnamates with aryl halides to give trisubstituted cinnamic esters in tetrabutylammonium bromide (TBAB) (Scheme 21.28) [116]. The nanoparticles (2–6 nm in size) were generated from a Pd(II)–benzothiazole carbene complex or $Pd(OAc)_2$ in, and stabilized by, the bromide salt in the presence of tetrabutylammonium acetate. The excellent stereoselectivity may result from fast neutralization of Pd–H by the acetate ions; the hydride can trigger olefin isomerization. In the presence of sodium formate as reducing agent and sodium bicarbonate as base, the carbene complex was itself effective in coupling *trans*-cinnamates with both electron-rich and electron-poor aryl halides in TBAB at 130°C. The reaction was devoid of stereoselectivity, however [117]. The Heck coupling product could be separated from the ionic liquid by phase separation or distillation. In imidazolium ionic liquids, catalytically active N-heterocyclic carbene complexes of palladium could be *in situ* generated from the solvent and catalyst precursors [118]. Excellent results have also been obtained in coupling aryl chlorides with styrene using Pd(0) supported on a basic, Mg–Al layered double hydroxide in TBAB [119].

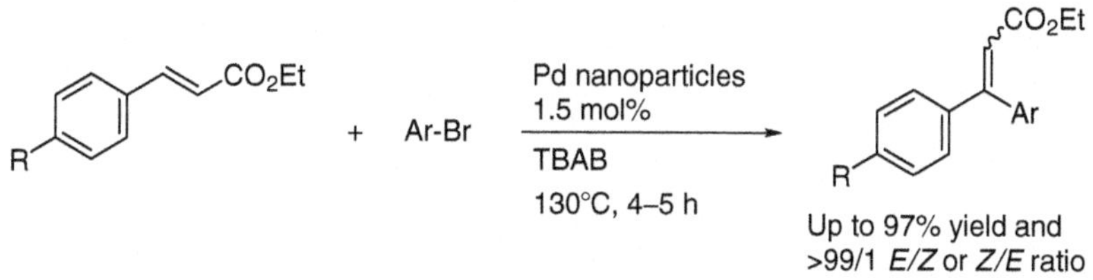

CO_2Et + Ar-Br → Pd nanoparticles 1.5 mol%, TBAB, 130°C, 4–5 h → CO_2Et, Ar

Up to 97% yield and >99/1 *E/Z* or *Z/E* ratio

R = H, OMe, Ac, NO_2; Ar = Ph, *p*-MeC_6H_4, *p*-$MeOC_6H_4$.

SCHEME 21.28

Using TBAB as solvent, Beller has shown that the novel monocarbene complex shown in Scheme 21.29 catalyzes efficient coupling of aryl chlorides with styrene and an acrylate at 140°C [120]. Up to 99% yields were obtained with electron-deficient arenes, but lower yields resulted with nonactivated analogs. In common dipolar aprotic solvents, only moderate conversions and selectivities could be obtained with activated chloroarenes. TBAB is also a good solvent for the coupling of aryl bromides and activated chlorides using other palladium catalysts (e.g., palladacycles and even $PdCl_2$) allowing for high TONs and ease of catalyst/product separation (via, e.g., distillation) [121]. However, highly productive Heck coupling of chloroarenes remains a challenge. The use of active palladium catalyst precursors coupled with a stabilizing high-temperature ionic liquid medium provides an additional option for addressing this problem.

As mentioned in the foregoing discussions, the Heck reaction can proceed via an ionic or a neutral pathway. The ionic pathway would be expected to be favored in an ionic medium. This chemistry was explored in the regioselective arylation of vinyl ethers in imidazolium ionic liquids. Indeed, butyl vinyl ether could be arylated by a wide variety of aryl halides in

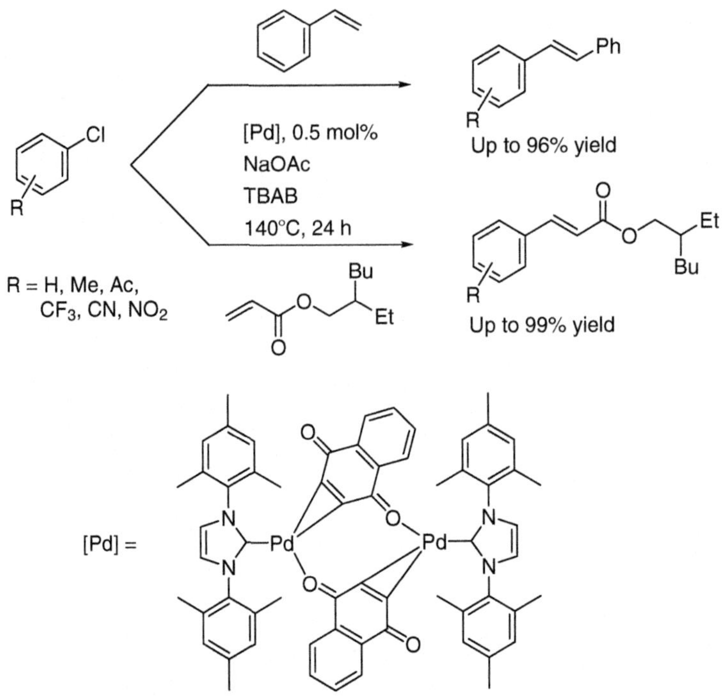

R = H, Me, Ac, CF_3, CN, NO_2

[Pd], 0.5 mol%
NaOAc
TBAB
140°C, 24 h

Up to 96% yield

Up to 99% yield

[Pd] =

SCHEME 21.29

[bmim][BF_4] in the presence of Pd$(OAc)_2$/DPPP, furnishing aryl methyl ketones after acidification in excellent regioselectivity and high isolated yields (Scheme 21.30) [122]. The molecular solvents, toluene, MeCN, DMF, and DMSO, were also examined, none affording an α/β ratio close to that observed in the ionic liquid. It was noted that while the decomposition of palladium complexes to various degrees into palladium black always is accompanied by arylation in the molecular solvents, palladium black was rarely observed in [bmim][BF_4]. Evidently, not only does [bmim][BF_4] promote the ionic pathway to preferentially give the α-arylated product, but also plays a role in stabilizing active palladium–phosphine species. Promotion of the ionic route by using aqueous DMF was later demonstrated by Hallberg (Scheme 21.27). A significant advantage of these two solvent systems is that aryl halides can be directly employed for the coupling reactions without recourse to halide scavengers.

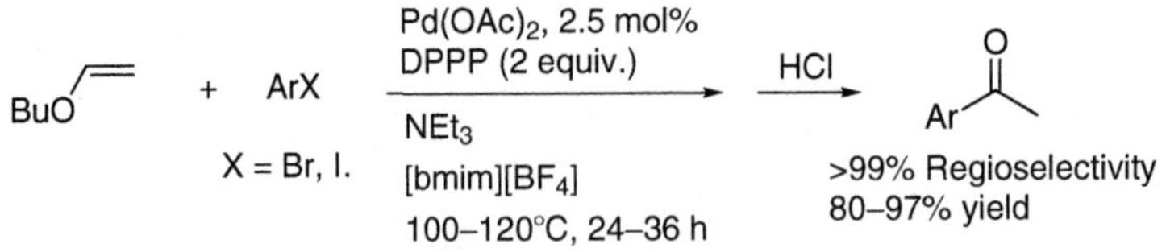

BuO + ArX, X = Br, I. → Pd$(OAc)_2$, 2.5 mol%; DPPP (2 equiv.); NEt_3; [bmim][BF_4]; 100–120°C, 24–36 h → HCl → Ar(C=O)Me

>99% Regioselectivity
80–97% yield

SCHEME 21.30

The Heck reactions in ionic liquids can be accelerated by high-temperature microwave heating. For instance, the regioselective coupling of 1-bromonaphthalene with butyl vinyl ether in [bmim][PF_6] catalyzed by Pd$(OAc)_2$/DPPP was complete in 2 h by microwave heating at 130°C, and the coupling with electron-deficient acrylates could be accomplished in 20 min at 220°C (see Scheme 21.30 for comparison) [123]. Ionic liquids appear to be ideally suited for microwave heating, as they interact very efficiently with microwaves through conduction and are rapidly heated without generating a significant pressure.

The Heck reaction has been less successful in $scCO_2$ and fluorous solvents. Only the active aryl iodides appear to produce some TONs (< 100) [124–129]. This is partly because the Heck reaction usually requires the use of dipolar solvents and partly because the reactions involving the less-reactive bromides and chlorides can only be effected with bulky and electron-rich ligands, such as $P(t\text{-}Bu)_3$ and N-heterocyclic carbenes, the fluorous version of which is not readily available. For instance, while easily prepared and active for the coupling of 4-bromobenzaldehyde with butyl acrylate in DMF, palladium complexes of fluoroalkylated N-heterocyclic carbenes were shown to display much reduced activity in $scCO_2$ [130], and although it can be directly used in $scCO_2$ without modification, $P(t\text{-}Bu)_3$ appears to be only active for aryl iodides in the coupling with butyl acrylate [126].

The use of $scCO_2$ as solvent can, in some instances, bring about beneficial effects. Thus, double bond isomerization, a serious competing side reaction in intramolecular Heck cyclizations in conventional solvents, could be suppressed to minimal levels in $scCO_2$ [131]. In contrast to that in conventional media, the coupling of aryl iodides with β-substituted α, β-enones in $scCO_2$ resulted in preferential hydroarylation instead of Heck vinylic substitution (Scheme 21.31) [132]. The hydroarylation products probably results from NEt_3 reduction of the Pd–alkyl intermediate generated from olefin insertion. The products were isolated in good to excellent yields after reactions at 100 bar CO_2 and 80°C, though the reaction time was long, at 60 h.

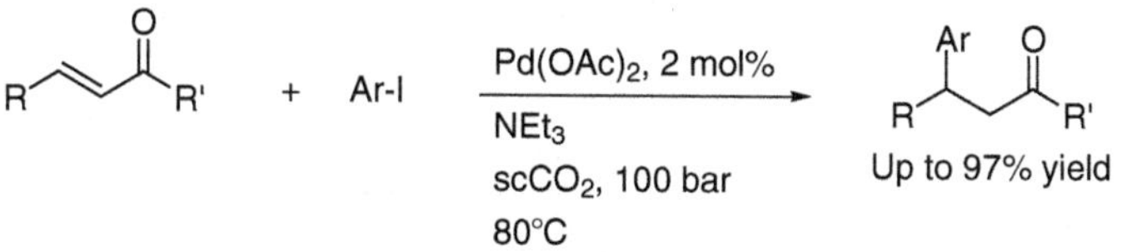

SCHEME 21.31

21.3.3.2 Suzuki Coupling

The Suzuki–Miyaura cross coupling is a most useful and efficient reaction for the synthesis of biaryl and heterobiaryl compounds using aryl halides and boronic acids. This is a highly selective reaction and, as with the Heck reaction, tolerates a variety of functional groups on either coupling partner. Boronic acids are generally nontoxic, thermally stable, and insensitive to air and moisture, and hence easier to handle than other commonly used cross-coupling reagents.

Amongst all the alternative solvents, water has proved to be the most effective for the Suzuki reactions. The coupling of arylbromides with boronic acids can be effected by palladium in water in the presence of the easily available, water-soluble TPPTS and related phosphines [133, 134]. For faster and more productive coupling or for reactions involving arylchlorides, however, bulky and electron-rich ligands are necessary. The water-soluble phosphines shown in Scheme 21.32 have been reported to form highly active palladium catalysts for both aryl bromides and chlorides. Thus, with the ammonium salt-based ligands **A** or **B**, electron-rich, electron-neutral, and electron-poor arylbromides could be coupled with various boronic acids in a mixture of H_2O–MeCN (1/1) at room temperature, affording the biaryls in yields usually over 90% within 1–2 h [135]. At 80°C, the catalyst loading could be lowered to a ppm level, giving rise to a TON of 734,000 after 4 h reaction time and an average TOF of 184,000 h^{-1}. In these reactions, the palladium catalysts partitioned predominately in the organic phase of the reaction mixture. Reactions carried out in water alone occurred more slowly than when a cosolvent was used. It was also shown that a 1:1 ratio of ligand/Pd gave a higher activity than a 2:1 ratio.

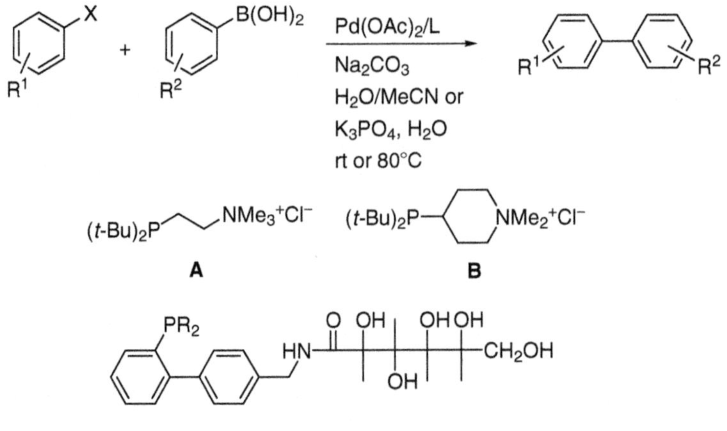

$X = Br$, R^1 = *p*-OMe, *p*-Me, *p*-Ac, *p*-OH, *p*-CO_2H; R^2 = H, *p*-OMe, *o*-Me
L = **A** or **B**, rt, 1-2 h, up to 96% yield.
X = Cl, R^1 = *p*-CN, *p*-CHO, *p*-Ac, *p*-OMe, *p*-CO_2Et, *p*-CO_2H, *p*-CO_2H-*m*-OH
p-CH=CHCO_2H, *m*-CO_2H-*p*-OH, *o*-CN, *o*-CO_2Et; R^2 = *p*-Me.
L = **C**, 80°C, 16 h, up to 98% yield.

SCHEME 21.32

The gluconamide phosphines **C** and **D** were active for aryl bromide coupling at ambient temperature and aryl chloride coupling at 80°C in neat water [136]. For the bromides, a PPh_3 analog was also effective; but **C** was required for more electron-rich substrates. Activated aryl chlorides were coupled to give the biaryls in high yields in 16 h with a $Pd(OAc)_2$/**C** (two equivalents of **C**) loading of 0.1 mol%; higher loading was necessary for more electron-rich chlorides. Chloropyridines, chloroquinolines, and aryltriflates were all coupled under these conditions.

The oxime palladacycle shown in Scheme 21.33 is effective for arylchloride coupling with phenylboronic acid in water [137]. Thus, at a catalyst loading of 0.01 mol% in refluxing water,

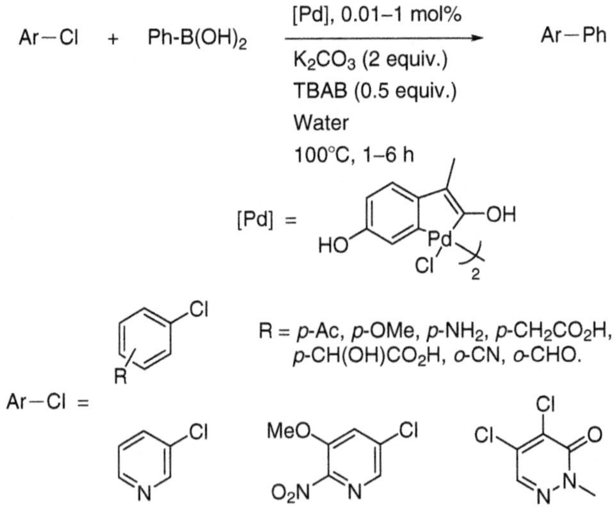

R = *p*-Ac, *p*-OMe, *p*-NH_2, *p*-CH_2CO_2H,
p-CH(OH)CO_2H, *o*-CN, *o*-CHO.

SCHEME 21.33

p-chloroacetophenone could be coupled with phenylboronic acid to give a TON of 7700 and TOF of 3850 h^{-1}. The reaction could also be conducted at room temperature, though at a much reduced rate. A variety of aryl and heteroaryl chlorides have been demonstrated to react with phenylboronic acid, giving the corresponding biaryls in yields ranging from 27% to 100%. Most of the reactions were complete within a few hours. For fast coupling, the presence of TBAB (0.5–1 equiv.) was crucial, presumably due to the formation of the more reactive $[PhB(OH)_3]^-[Bu_4N]^+$ and the increased solubility of substrates in water.

As with the Heck reactions, Suzuki coupling in water, with or without a ligand, can be greatly accelerated by microwave heating [138–141]. Leadbeater reported that in the presence of 0.4 mol% $Pd(OAc)_2$, 3 equivalents of Na_2CO_3, and 1 equivalent of TBAB, arylbromides including those that are sterically demanding (1 mmol) could be coupled with phenylboronic acid to give biaryls with yields of 70–96% by microwave heating (60 W) at 150°C for a few minutes [138]. A wide range of functional groups were tolerated in the reaction and not affected by the high temperature and aqueous conditions used. The reaction could be scaled up to, e.g., 3 mmol, without compromising yields. It was also possible to couple arylchlorides, although a higher temperature (175°C) was necessary and lower yields resulted. Recent studies by the same group has revealed that the reaction of bromides and iodides can be performed equally well using conventional heating (150°C), and the microwave heating could even be employed to effect Suzuki coupling of bromides and iodides without using any metal catalyst [140, 142].

Suzuki coupling has been demonstrated in ionic liquids; but the TONs achieved so far are lower than those observed with water, and electron-rich aryl chlorides are practically inactive. Welton reported that phenyl boronic acids could be coupled with aryl bromides in $[bmim][BF_4]$ under $Pd(PPh_3)_4$ catalysis at 110°C to give biaryls with TONs of up to 81 and TOFs of up to 465 h^{-1}. Under the same conditions, chlorobenzene failed to couple [143]. The reactions could be promoted with ultrasound, which allowed phenyl boronic acid to couple with various aryl halides, including chlorotoluene (58 TON), at a lower temperature of 30°C when using a palladium–bis(carbene) complex as catalyst [144].

Higher-yield coupling of aryl chlorides has been realized in the quaternary phosphonium salt tetradecyltrihexylphosphonium chloride containing a small amount of water and toluene [145]. Thus, phenyl boronic acid coupled with 4-chloroacetophenone in the presence of $Pd(0)—PPh_3$ and K_3PO_4, furnishing the biaryl in 84% isolated yield at 70°C in 30 h. The more electron-rich 4-methoxychlorobenzene showed a much slower reaction, however. Aryl bromides could be coupled to give yields of up to 99% at a lower temperature of 50°C and a short reaction time of 1–3 h. Iodides reacted with high yields even in the absence of PPh_3. As with the Heck reaction, product separation and catalyst recycle in the case of Suzuki coupling can be easily performed by phase separation or extraction with a solvent of low polarity.

Suzuki coupling of aryl iodides and bromides has also been demonstrated in fluorous media. Using as ligand a fluorous dialkyl sulfide, $[F_{17}C_8(CH_2)_n]_2S$ ($n = 2$ or 3), Gladysz reported that palladium-catalyzed coupling of aryl bromides could be performed in a mixture of $CF_3C_6F_{11}$, DMF, and H_2O (1/3/2) in the presence of K_3PO_4 at ambient temperature or 50°C (for more electron-rich bromides, e.g., *p*-bromoanisole) to give biaryls with TONs of up to 5000 in 24 h reaction time [146]. In recycling reactions via phase separation, however, TOFs diminished. For instance, in the first run of the coupling of bromobenzene with phenyl boronic acid, a conversion of 97% was reached in 0.5 h. In contrast, the third run called for ca. 20 h to give a conversion of 85%. The continuous decrease in catalyst activity results from the decomposition of recyclable, fluorous-soluble palladium complexes into catalytically active, but nonrecyclable, palladium nanoparticles; the former merely serves as a steady-state palladium source, until exhausted.

To avoid using the often expensive fluorous solvents, fluoroalkylated silica has been employed as a solid support to immobilize palladium complexes containing fluorous ligands

(Scheme 21.34). Bannwarth reported that fluorous palladium complexes could be immobilized on fluorous silica without covalent linking and were capable of catalyzing the coupling of a variety of aryl iodides and bromides with arylboronic acids in the presence of Na_2CO_3 in dimethoxyethane (DME) [147]. Conversions ranging from 48% to >98% were obtained at 80°C in 15 h at a catalyst loading of 0.1 mol%, with the lowest conversion associated with *p*-bromoanisole. The immobilized catalyst was prepared by simply mixing the fluorous silica with the palladium complex in diethyl ether and hexafluorobenzene followed by evaporation of the solvent. In the end of a reaction, the product could be separated by cooling the reaction mixture to 0°C and by removal of the liquid phase with a pipette, thus allowing for catalyst recycle. As with the fluorous sulfide catalyst, however, a significant decrease in yields was observed in subsequent runs, which may again stem form decomposition of recyclable palladium complexes. Recent work by Gladysz points to the importance of this route to the formation of palladium nanoparticles and the masking of activity losses by employing, for example, high catalyst loading and longer reaction times than required [148]. In addition to the immobilization approach, fluorous solvents could be eliminated by adopting the so-called thermomorphic fluorous catalysts, which dissolve and catalyze reactions in common solvents at elevated temperature and separate from the solvents upon cooling [148].

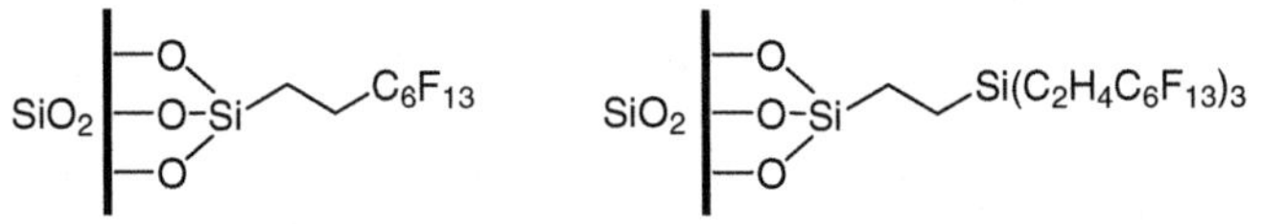

SiO_2 —O—Si—C_6F_{13} SiO_2 —O—Si—$Si(C_2H_4C_6F_{13})_3$

$PdCl_2[P(C_6H_4\text{-}R)_3]_2$

R = *p*-$C_2H_4C_8F_{17}$, *m*-$C_2H_4C_8F_{17}$, *o*-$OCH_2C_7F_{15}$.

SCHEME 21.34

The previously mentioned Pd–P(*t*-Bu)$_3$ catalyst (Section 21.3.3.1) has also been applied to the Suzuki coupling of aryl iodides and bromobenzene with boronic acids in $scCO_2$. However, the TONs were low, in particular for bromobenzene, which coupled with phenyl boronic acid to give biphenyl in ca. 50% yield at 5 mol% catalyst loading in 16 h at 100°C [126].

Suzuki coupling can be conducted under solvent-free conditions. A recent example is provided in the preparation of highly pure thiophene oligomers [149]. The reaction was performed by microwave heating of a mixture of thienyl boronic acids or esters with thienyl bromides on aluminum oxide, which acts as a dispersing reagent. Thus, α-conjugated quaterthiophene was obtained in 81% yield by coupling a dibromobithiophene with 2-thiopheneboronic acid on Al_2O_3 in the presence of KF and $PdCl_2$(DPPF) [DPPF = 1,1′-bis (diphenylphosphino)ferrocene] under microwave irradiation at 80°C for 4 min. Neumann reported that palladium nanoparticles (15–20 nm) supported on the polyoxometalate $[PW_{11}O_{39}]^{7-}$ in the presence of KF was active for the Suzuki coupling of phenylboronic acid with chloroarenes substituted with either electron-withdrawing or electron-donating moieties, furnishing the biaryls with yields of up to >99% at 130°C without solvent [150].

21.3.3.3 Stille Coupling

Alongside the Suzuki coupling, the reaction of organostannanes with aryl halides, namely the Stille coupling, is widely utilized for C—C bond formation as well. This reaction has been less investigated in the alternative solvents, however, partly due to the toxicity of organotin reagents and by-products. To address the issue of toxicity, a water-soluble organostannatrane has been developed, which permits removal and recycle of tin residues from a Stille reaction via aqueous extraction [151]. An example of aqueous-phase Stille coupling is provided by Neumann, who reported that the polyoxometalate-supported palladium nanoparticles (Section 21.3.3.2) were effective for the coupling of tetraphenyltin with 4-bromotoluene or 1-chloro-4-nitrobenzene, providing 4-methyl-1,1′-biphenyl or 4-nitro-1,1′-biphenyl in 92% yield at 110°C in 12 h [150].

Stille coupling reactions in other alternative solvents are known as well. For instance, α- and β-iodocyclohexenones and aryl iodides and bromides could be coupled with vinyl and phenyl-tributyltin by $PdCl_2(PhCN)_2$ or $Pd(PPh_3)_4$ in [bmim][BF_4] at 80°C with good to excellent yields of product [152]. As might be expected, the Pd(II) precatalyst worked for the iodides (in the presence of 10 mol% CuI in the case of iodoenones), while the Pd(0) precatalyst was preferred for the arylbromides. The product could be extracted with diethyl ether and catalyst recycled five times without significant loss of catalytic activity.

Bannwarth has reported that $PdCl_2$ in combination with fluoroalkylated PPh_3 is capable of catalyzing the Stille reactions of aryl bromides in $scCO_2$ and fluorous solvents [153, 154]. Thus, in a biphasic mixture of DMF and perfluoromethylcyclohexane, which turned into a single phase under the reaction conditions, the three palladium complexes shown in Scheme 21.35 catalyzed the coupling of aryl bromides with stannanes in the presence of 1 equivalent of LiCl, affording the coupled products in 40–98% yields at 80°C in a few hours, except with an *ortho*-substituted stannane, which required a prolonged time due to steric hindrance. The organic phase separated from the fluorous phase upon cooling the reaction mixture to room temperature, thus allowing for easy product/catalyst separation, and the palladium-containing fluorous phase could be reused several times with comparable yields in each run. Intriguingly, the three complexes displayed almost identical activities. In the closely related $scCO_2$, these precatalysts have also been shown to be effective for similar coupling reactions, and the nonfluorous $PdCl_2(PPh_3)_2$ was almost equally effective. Thus, as in the case of the Suzuki coupling with fluorous palladium complexes, the nature of the active palladium species in these Stille reactions remains to be delineated.

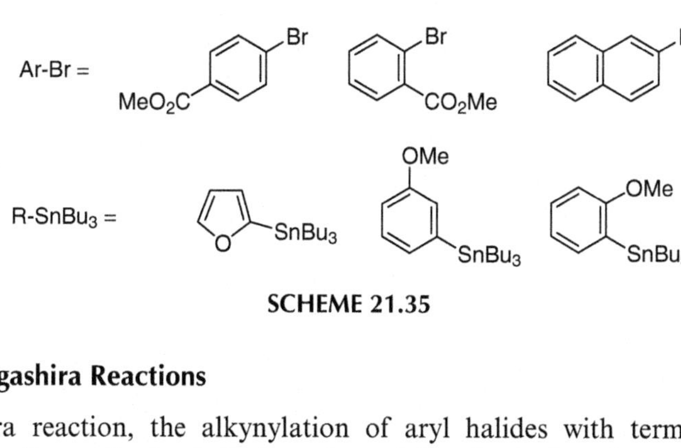

Ar–Br + R–$SnBu_3$ —[$Pd(PAr_3)_2Cl_2$, 1.5 mol%; LiCl (1 equiv.); $CF_3C_6F_{11}$/DMF; 80°C, 3 h]→ Ar-R

PAr_3 = P–[C_6H_4–R′]$_3$ R′ = *p*-C_8F_{17}, *p*-$C_2H_4C_8F_{17}$, *m*-C_8F_{17}.

Ar-Br = MeO$_2$C–C$_6$H$_4$–Br; Br–C$_6$H$_4$–CO$_2$Me; naphthyl–Br

R-$SnBu_3$ = furyl–$SnBu_3$; OMe–C$_6$H$_4$–$SnBu_3$; OMe–C$_6$H$_4$–$SnBu_3$

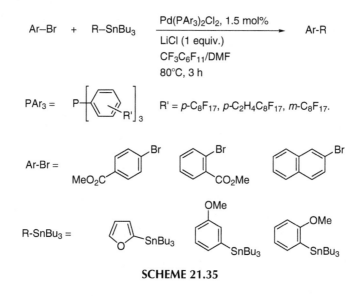

SCHEME 21.35

21.3.3.4 Sonogashira Reactions

The Sonogashira reaction, the alkynylation of aryl halides with terminal alkynes, is a powerful tool for the synthesis of alkyl-, aryl-, and diaryl-substituted acetylenes. An example of an aqueous-phase reaction is provided by the coupling of N-propargylamino acids with aryl and heteroaryl bromides in a DME–water mixture (Scheme 21.36). The reaction was conducted with 3 mol% Pd/C and 10 mol% CuI using K_2CO_3 as base, affording the arylated

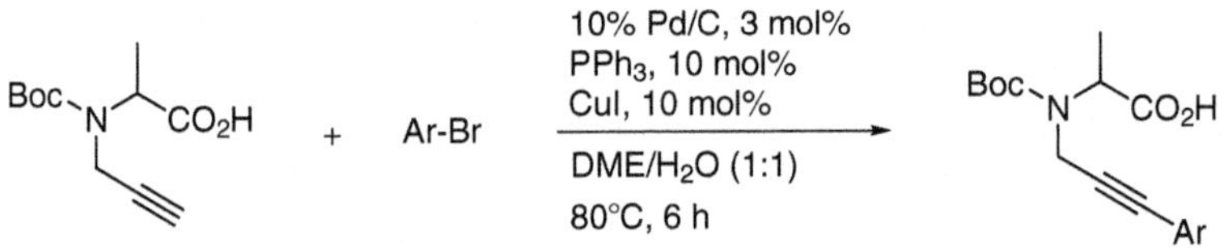

SCHEME 21.36

alkynes in yields of up to 96% at 80°C in 6h. Hydrogenation of the products led to N-alkylamino acids featuring aryl or heteroaryl substituents. This chemistry has been applied to the modification of peptide backbones [155]. Highly charged peptides ranging in length from 17 to 33 residues have also been coupled to a trialkyne nucleus by using a soluble Pd(0)–TPPTS catalyst in water. In addition to the reacting aryl iodide units, these synthetic peptides contain free amines, carboxylates, guanidines, hydroxyls, and thiolesters, demonstrating the utility of the aqueous catalytic system in dealing with multifunctional molecules [156].

The 2,2-dipyridylmethylamine-based palladium complexes in Scheme 21.37 have been used for the Sonogashira reaction of aryl iodides and bromides with terminal alkynes in both organic solvents and water. For example, 4-bromochlorobenzene coupled with phenylacetylene in the presence of the urea-containing precatalyst, 2 equivalents of pyrrolidine (as base), and 1 equivalent of TBAB, yielding 4-diphenylacetylene with 480 TON at 100°C in 4 h. Unlike that shown in Scheme 21.36, the reaction required no copper additives. The palladium complex was also shown to be active for the Heck and Suzuki reactions of aryl bromides and activated chlorides in water [157].

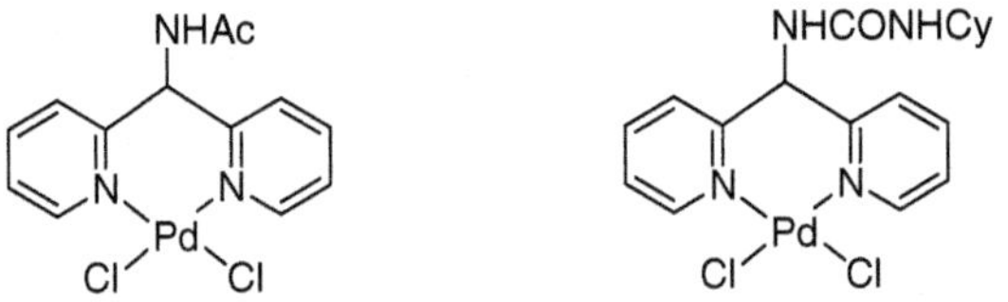

SCHEME 21.37

An example of Sonogashira reaction in ionic liquids is the coupling of aryl halides and terminal acetylenes in [bmim][PF_6] (Scheme 21.38) [158]. The reaction was catalyzed by the easily available $PdCl_2(PPh_3)_2$ with diisopropylamine as base, and proceeded efficiently, again without using a copper cocatalyst, to give the disubstituted acetylenes in yields of 85–97%. In the case of aliphatic acetylenes (e.g., 1-octyne), piperidine instead of diisopropylamine provided a faster reaction. The coupling products could be separated from the catalyst and solvent by extraction with hexane or ether, and the catalyst could be reused several times with only slight loss in activity. When carried out in conventional solvents, the same coupling reaction gave lower yields of the product.

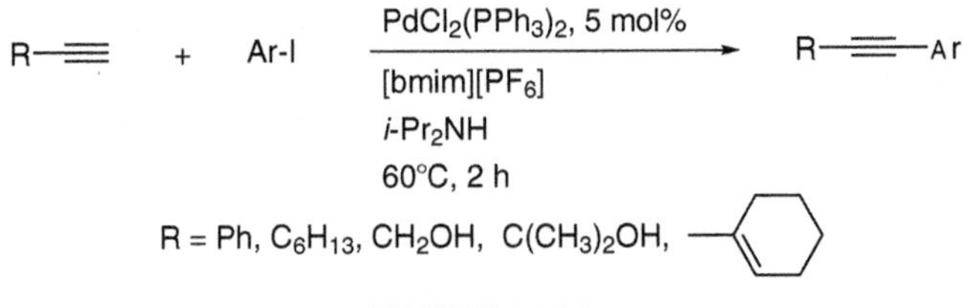

SCHEME 21.38

Fluorous palladium complexes similar to those shown in Scheme 21.35 are active for the Sonogashira reaction of arylbromides and acetylenes in an organo-fluorous solvent mixture in the presence of CuI with Bu_2NH as base [159]. Activated bromides tend to give higher yields (70% to >98%, 100°C, 4 h) and allow for easier catalyst recycle than electron-rich variants. Again, as in the case of the Stille reaction, the nature of the fluorous ponytails attached to the arylphosphines appears to have little effect on the catalytic activity.

21.3.3.5 Allylic Substitution

Palladium-catalyzed allylic substitution, often called the Tsuji–Trost reaction, is a well-established and widely used tool for the formation of C—C bonds. This reaction is usually carried out in an organic solvent such as THF but has been successfully demonstrated in some of the alternative solvents in recent years. Uozumi reported that in water, the P^N chiral ligand bound to an amphiphilic polystyrene–poly(ethylene glycol) resin (Scheme 21.26) catalyzed the asymmetric allylic substitution of both cyclic and acyclic substrates, with isolated product yields of up to 94% and enantioselectivity of up to 98% ee at 40°C in the presence of Li_2CO_3 (0.9 *M*) (Scheme 21.39) [160]. Of particular note is the high ee's achieved with the cyclic substrates (e.g., 98% ee in the substitution of cycloheptene ester by diethyl malonate), which often give relatively low ee's due to smaller size compared with acyclic substrates. The catalyst could be separated by filtration and reused with no loss in ee. Contrary to one might expect, the resin catalyst was less active in organic solvents. The higher activity in water may stem from the amphiphilic nature of the support; the allylic substrate and the anionic nucleophile could be brought to proximity more easily in water than in an organic solvent such as CH_2Cl_2 by the hydrophobic and hydrophilic segments of the resin.

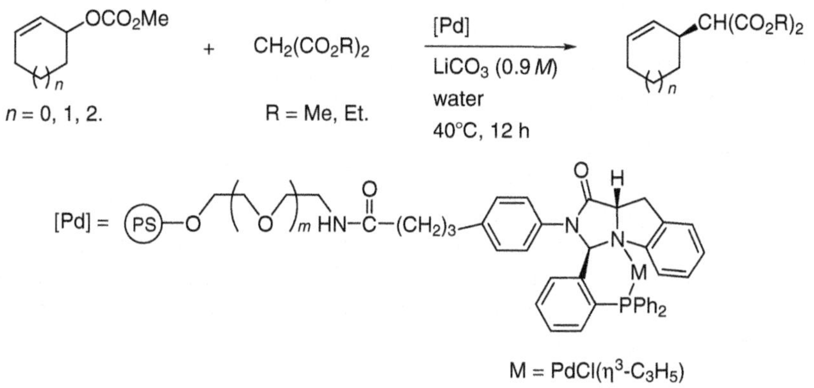

OCO$_2$Me + CH$_2$(CO$_2$R)$_2$ →[Pd], LiCO$_3$ (0.9 *M*), water, 40°C, 12 h → CH(CO$_2$R)$_2$

n = 0, 1, 2. R = Me, Et.

[Pd] = (PS)–O–(CH$_2$CH$_2$O)$_m$–HN–C(=O)–(CH$_2$)$_3$–C$_6$H$_4$–N ... M, PPh$_2$

M = PdCl(η^3-C$_3$H$_5$)

SCHEME 21.39

Earlier, Uemura and Ohe reported that 1,3-diphenyl-3-acetoxyprop-1-ene could be alkylated and aminated in good yields and ee's (up to 85%) by palladium catalysis using an amphiphilic, D-glucosamine-derived phosphinite–oxazoline ligand in a MeCN–H_2O biphasic medium. While soluble in MeCN, the catalyst could be transferred into the aqueous phase by introducing Et_2O in which it is insoluble [161]. Recent work by Sinou shows that asymmetric alkylation of the same substrate with dimethyl malonate can be readily performed in water by palladium catalysis using nonderivatized (*R*)-BINAP as ligand in the presence of a surfactant, with ee's of up to 91% and full conversion obtained in 1 h at room temperature (K_2CO_3 as base) [162]. An important observation is that both the nature and concentration of the surfactant impact on the reaction rates and enantioselectivity. The best results were obtained with a cationic surfactant at concentrations higher than its critical micelle concentration (cmc), while neutral, zwitterionic, and anionic variants gave poorer results or no reaction at all. The nucleophilic attack at the palladium–allyl intermediate by the anionic nucleophile appears to take place in micelles, which explains why a cationic surfactant was more effective.

The palladium-catalyzed allylic substitution with soft carbon nucleophiles can also be effectively performed in ionic liquids. Thus, in [bmim][BF_4], 1,3-diphenyl-3-acetoxyprop-1-ene was readily alkylated with active methylene compounds in the presence of 2 mol% $Pd(OAc)_2$ and 4 equivalents of $P(4\text{-}C_6H_4OMe)_3$ at room temperature with DBU as base (Scheme 21.40) [163, 164]. Amination was also demonstrated to be feasible. This study indicates that the rate of nucleophilic attack at the Pd(II)–allyl species in the ionic liquid is greatly enhanced by more electron-rich or σ-donating phosphines (e.g., $P(4\text{-}C_6H_4OMe)_3$ and PCy_3).

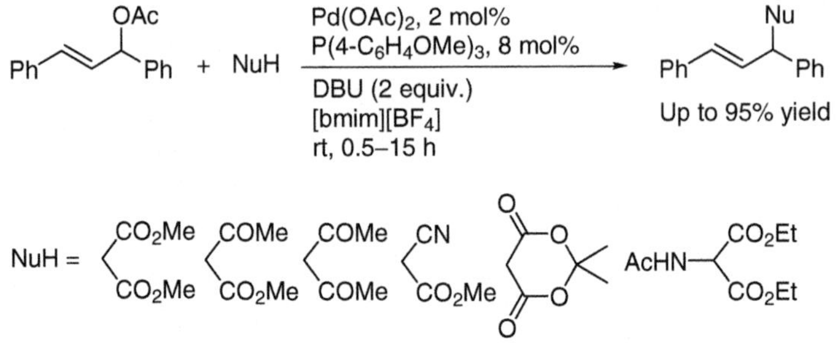

OAc, Ph, Ph + NuH; $Pd(OAc)_2$, 2 mol%; $P(4\text{-}C_6H_4OMe)_3$, 8 mol%; DBU (2 equiv.); [bmim][BF_4]; rt, 0.5–15 h; Nu, Ph, Ph; Up to 95% yield

NuH = CO_2Me/CO_2Me; COMe/CO_2Me; COMe/COMe; CN/CO_2Me; AcHN–CO_2Et/CO_2Et

SCHEME 21.40

With strong π-accepting phosphines (e.g., $P(OPh)_3$), little or no reaction took place. It was also shown that the reaction was faster in the ionic liquid than in the molecular solvent THF and a wider range of phosphine ligands could be effectively employed in the former than in the latter. The slower rates in THF can be attributed, at least partly, to the formation of tight ion pairs between the Pd(II)–allyl cation and anionic acetate and to reversible attack at the cation by the acetate [165, 166] both of which can be expected to diminish in the ionic environment provided by [bmim][BF_4].

Chiral fluorous bisoxazolines and 2-(diphenylphosphino)-2′-alkoxy-1,1′-binaphthyls have been prepared and examined for allylic substitution. Whilst high enantioselectivities could be realized in common solvents and benzotrifluoride, the palladium catalysts derived from these ligands lost activities in the fluorous media. However, catalyst separation can be made easy with such ligands, as fluorous extraction can be applied following a reaction in common solvents [167, 168].

21.3.3.6 Aldol and Michael Reactions

The aldol addition reaction allows for the construction of highly functionalized and stereochemically complex compounds and hence is widely practiced in synthetic chemistry. Lewis acid catalysis of this reaction has been successfully demonstrated in the alternative media, particularly in water [58, 169–171]. A recent example is the novel, boronic acid-catalyzed Mukaiyama aldol addition of silyl enol ethers to various aldehydes. Kobayashi reported that in the presence of a catalytic amount of Ph_2BOH and $PhCO_2H$, aromatic, α,β-unsaturated, and aliphatic aldehydes reacted in water with silyl enol ethers derived from aliphatic ketones and thioesters, furnishing β-hydroxy carbonyl compounds in yields ranging from 40% to 90% (Scheme 21.41) [169]. Diastereoselectivities (*syn/anti*) of up to 97/3 were obtained for *cis*-enolates, whereas the *trans* variants led to lower values. Presumably the reaction proceeds via an *in situ* Si–B exchange, leading to a boron enolate followed by reaction with an aldehyde. The Brønsted acid may promote the Si–B exchange by protonation of Ph_2BOH, which makes the hydroxy group easier to leave. While boron enolates can be directly used in stoichiometric aldol reactions, the new system allows boron to be used catalytically.

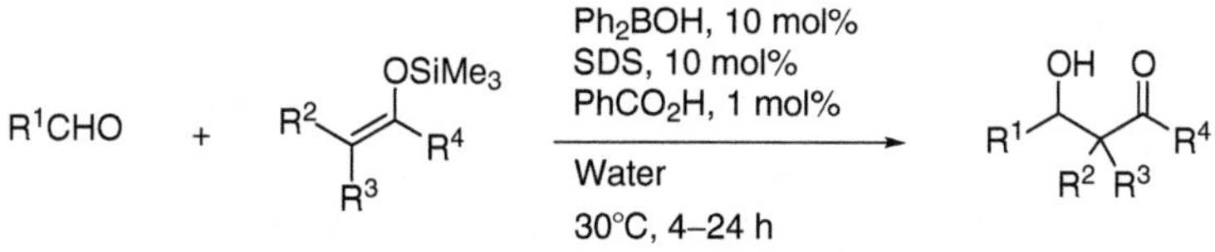

R^1 = Ph, 4-ClC_6H_4, 1-naphthyl, PhCH=CH, PhCH_2CH_2, cyclohexyl.
R^2 = H, Me.
R^3 = H, Me, $(CH_2)_4$.
R^4 = Et, *i*-Pr, Ph, S-*t*-Bu.

SCHEME 21.41

Allyl alcohols can be isomerized into enols that may undergo adol addition. This has recently been demonstrated by Li and coworkers, who reported that in the presence of $RuCl_2(PPh_3)_3$ in a solvent mixture of water and toluene, 3-buten-2-ol or α-vinylbenzyl alcohol reacted with aryl aldehydes, forming β-hydroxy carbonyl compounds in good yields with the *syn* product favored. Ruthenium-catalyzed isomerization of the allylic alcohol via a Ru–allyl intermediate is believed to be involved. The resulting ruthenium–enol species undergoes addition to the incoming aldehyde. In neat organic solvents, no reaction was observed. The catalytically generated enols also reacted with imines, giving rise to Mannich-type product [171].

It is also possible to run more traditional aldol condensation in water. A recent example is seen in the synthesis of chalcones and azachalcones by reacting acetophenone or 2-acetylpyridine with aromatic aldehydes using a catalytic amount of Na_2CO_3 in neat water. The reaction was heterogeneous and the resulting solid product (up to 98% isolated yield) could be readily separated by filtration [172].

Under the influence of a proper chiral Lewis acid, asymmetric aldol reaction can be viable in water [170]. Kobayashi recently reported that reactions similar to those shown in Scheme 21.41 could be run in aqueous medium using a rare earth metal triflate in combination with a chiral crown ether, furnishing the aldol adducts in high yields with good diastereo and enantioselectivities. An example is seen in Scheme 21.42 [173].

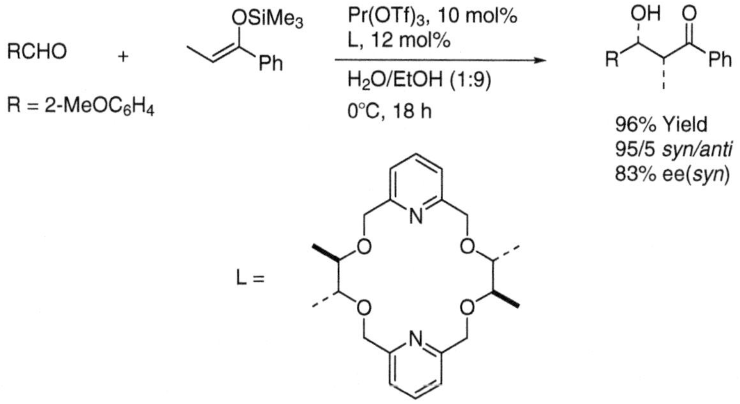

SCHEME 21.42

Small organic molecules have recently been intensively investigated in asymmetric direct aldol reactions, which require no activating groups and hence are completely atom-economic [174]. Some of these reactions have been demonstrated to be feasible in ionic liquids [175, 176]. For instance, in the presence of L-proline (30 mol%) in [bmim][PF_6], *o*- and *p*-substituted benzaldehydes reacted with unactivated ketones such as acetone, methyl ethyl ketone, and cyclopentanone, affording β-hydroxy ketones in up to 94% yields and 93% ee's in 24 h at room temperature. In most cases, the product could be separated by extraction and the remaining ionic liquid phase reused [176]. The ionic liquid appeared to have a bearing on the reaction and, in terms of both product yield and ee values, [bmim][PF_6] performed the best [175].

The Mukaiyama aldol reactions of silyl enol ethers can also be carried out in the apolar solvents fluorous carbons and $scCO_2$. When a normal metal Lewis acid is used as catalyst, its low solubility presents a problem, however. This can be alleviated by catalyst modification or introduction of surfactants. For example, a fluorous analog of scandium triflate, $Sc[N(SO_2C_8F_{17})_2]_3$ (<0.1 mol%), catalyzed the aldol reaction of benzaldehyde (0.1 *M*) with trimethylsilyl enol ether (0.2 *M*) in perfluoromethylcyclohexane in a nano flow reactor, furnishing the product in up to 97% yield in less than 1 min at 55°C [177]. In $scCO_2$, poly(ethylene glycol) derivatives and 1-dodecyloxy-4-perfluoroalkylbenzene have been demonstrated to accelerate the aldol reactions and improve product yields; they enhance the solubility of catalyst and substrates in $scCO_2$ by forming emulsions [178].

A related reaction is the conjugate addition of carbon nucleophiles to activated unsaturated systems, namely the Michael reactions. This reaction is traditionally mediated by strong bases in organic solvents. However, recent studies have demonstrated the possibility of effecting the reaction with Lewis acids in the alternative solvents [179–181]. An example is seen in the palladium–aquo complexes shown in Scheme 21.43, which catalyzed the enantioselective addition of β-ketoesters to methyl vinyl ketone in both water and ionic liquids [180]. Thus, in the presence of the catalyst in the ionic liquid [bmim][OTf], *t*-butyl-2-oxo-cyclopentanecarboxylate reacted with the Michael acceptor methyl vinyl ketone, affording the Michael adduct in excellent yield and good enantioselectivity. The catalyst was recyclable upon treatment with a fluorinating reagent and was recycled five times with no change in ee's.

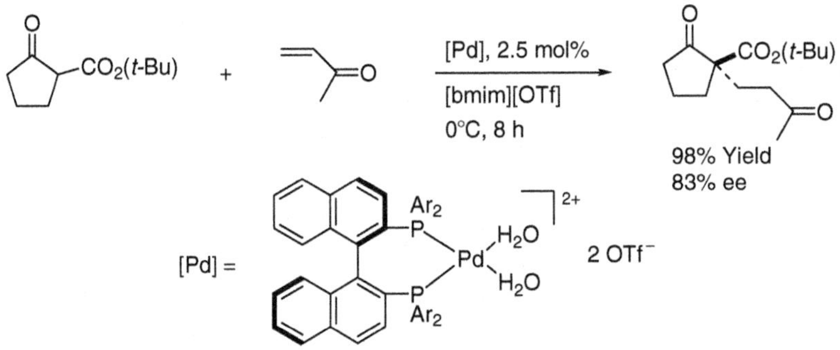

$CO_2(t\text{-Bu})$ + =O; [Pd], 2.5 mol%; [bmim][OTf]; 0°C, 8 h; $CO_2(t\text{-Bu})$; =O

98% Yield
83% ee

[Pd] = Ar_2 P, Pd, H_2O, H_2O, P Ar_2, 2+, 2 OTf^-

SCHEME 21.43

21.3.4 Diels–Alder Reactions

The atom-economic reaction of a diene and dienophile giving rise to a six-membered ring compound, that is the Diels–Alder (DA) reaction, is one of the most widely investigated reactions in the alternative media. This has particularly been the case with water since the pioneering work of Breslow, who showed that the DA reaction can be dramatically accelerated when run in water and its stereoselectivity can also be significantly improved therein [182, 183]. The rate acceleration results from two factors, the hydrophobic and hydrogen bonding interactions between water molecules and the two reactants or activated complex. Lewis acid catalysis can be exploited in conjunction with water to promote the reaction; this has been summarized in recent reviews [184, 185]. The first example of a DA reaction catalyzed by Lewis acids in water is from Engberts and coworkers. Their work shows that the DA reaction shown in Scheme 21.44 is catalyzed by metal cations, such as Cu^{2+}, Zn^{2+},

Co^{2+}, and Ni^{2+}, with Cu^{2+} being the most active, and when compared with uncatalyzed reactions in common solvents, the Lewis acid-catalyzed reaction can be considerably faster. For instance, the DA reaction (R = NO_2) in the presence of 0.02 *M* Cu^{2+} in water was 232,000 times faster than that in MeCN without the metal ion. Chelation of Cu^{2+} to the dienophile through its carbonyl oxygen and pyridyl nitrogen atom is critical to the catalysis.

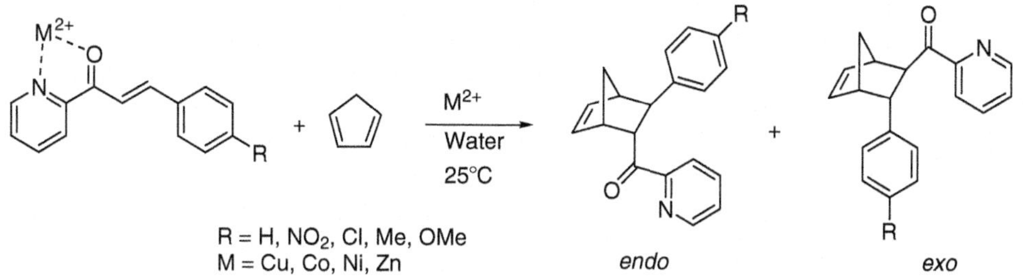

R = H, NO_2, Cl, Me, OMe
M = Cu, Co, Ni, Zn

SCHEME 21.44

When the chelation was impossible, no rate enhancement was observed. The *endo*/*exo* selectivity of the catalyzed reaction was not significantly affected by water, however [186]. In the presence of an α-amino acid which coordinates to the copper cation, the cycloaddition reaction could be made enantioselective. For example, the *endo* product (*endo*/*exo* = 93:7) was obtained in 74% ee in water when L-abrine was introduced [17.5 mol% L-abrine, 10 mol% $Cu(NO_3)_2$, 0°C, 48 h] [187].

There are a number of other Lewis acids, including lanthanides and $MeReO_3$, that are effective catalysts for the DA reaction in water. Organic hydrogen bond donors, e.g., substituted thioureas, could also be used to accelerate the reaction involving substrates containing Lewis basic site (e.g., α,β-unsaturated compounds), which act as hydrogen bond acceptor. As in the case of metal cation catalysis, the lowest unoccupied molecular orbitals (LUMO) of the dienophiles would be lowered in energy upon interaction with the hydrogen bond donor [188]. Even simple amines such as pyrrolidine have been shown to catalyze DA reactions in water. Presumably, the dienophile is activated through the formation of iminium ions [189].

In addition to the carbo-DA reactions, hetero-DA reactions in water have been documented as well [184]. A recent example of an aza-DA reaction is seen in Scheme 21.45 [190]. Kobayashi demonstrated that, although sensitive to water, Danishefsky's diene added to a range of imines in the presence of AgOTf in water, affording dihydro-4-pyridones in good to excellent yields. Interestingly, when run homogeneously in a mixture of THF and water, the reaction was less efficient, presumably due to faster hydrolysis of the diene. This reaction is also catalyzed by alkaline salts in water, such as NaOTf [191]. The imines could be *in situ* generated by reacting an aldehyde with a secondary amine, thus alleviating the stability problem of some imines and allowing the cycloaddition to be conducted in an one-pot fashion. For example, the relatively unstable imines derived from aliphatic aldehydes could be generated by mixing the aldehyde with an amine in water to feed the DA reaction initiated by slow addition of the dienophile. A carbo-DA reaction of the same diene catalyzed by $InCl_3$ is also known [184].

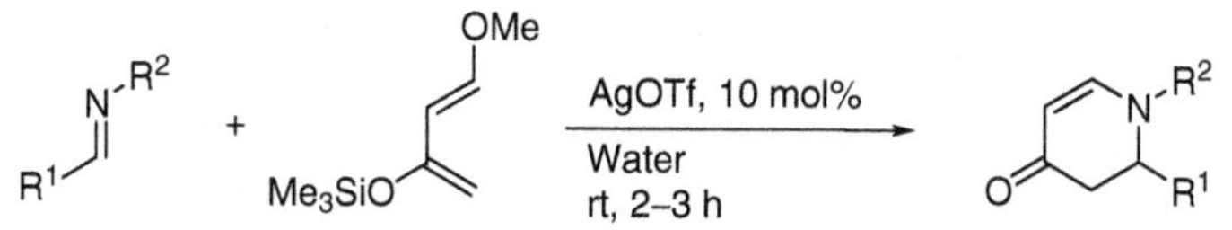

R^1 = Ph, *p*-$MeOC_6H_4$, *p*-BrC_6H_4, *p*-$NO_2C_6H_4$, PhCHCH-; R^2 = Ph.

SCHEME 21.45

It is well known that the DA reaction can be accelerated by the lithium salt, $LiClO_4$, in diethyl ether. Driven partly by the similarity of $LiClO_4$ to ionic liquids, a number of investigations into the DA reaction in imidazolium salts haven been undertaken [115, 192]. An impressive example is seen in the work of Song and coworkers. They showed that the addition of 2,3-dimethylbuta-1,3-diene to 1,4-naphthoquinone was complete within 2 h in the presence of only 0.2 mol% of $Sc(OTf)_3$ at room temperature in [bmim][PF_6], [bmim][SbF_6] or [bmim][OTf] (Scheme 21.46) [193]. In contrast, the same reaction in CD_2Cl_2 produced a much lower yield of 22% under otherwise identical conditions. In fact, the reaction in the ionic liquid became too fast to control when using a normal 10% catalyst loading, and it could be accelerated in CD_2Cl_2 by adding some [bmim][PF_6]. The reaction was also successful with other dienophiles such as methyl vinyl ketone and maleic anhydride and dienes such as 1, 3-cyclohexadiene and furan. The products are depicted in Scheme 21.46. The *endo*/*exo* selectivities observed in the ionic liquid were higher than those in CH_2Cl_2, and product/catalyst separation was easier with the ionic liquid reaction. Thus, upon completion of the reaction, the catalyst-containing ionic liquid phase was almost quantitatively recovered by simple extraction of product with Et_2O, and the recovered ionic liquid phase could be reused. In the case of 2,3-dimethylbuta-1,3-diene reacting with methyl vinyl ketone in [bmim][OTf], the catalyst and ionic liquid were recycled 11 times with no decrease in isolated product yield.

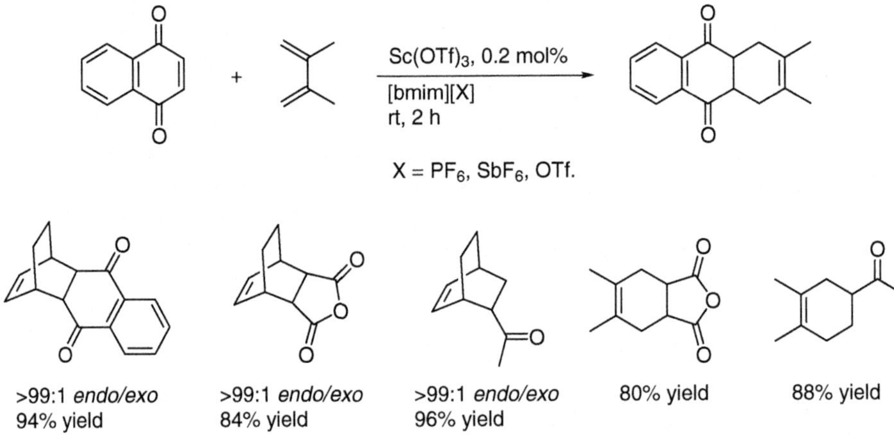

$Sc(OTf)_3$, 0.2 mol%
[bmim][X]
rt, 2 h

X = PF_6, SbF_6, OTf.

>99:1 *endo/exo* 94% yield | >99:1 *endo/exo* 84% yield | >99:1 *endo/exo* 96% yield | 80% yield | 88% yield

SCHEME 21.46

The DA reaction could be promoted in ionic liquids without using a metal Lewis acid. An example is the hetero-DA reaction shown in Scheme 21.47 [194]. The reaction proceeds via imines that are derived from the reaction of aldehydes with anilines; the imines act as heterodienes and undergo an imino-DA reaction with the dienophiles 2,3-dihydrofuran and 3,4-dihydropyran, leading to furano- and pyrano-quinolines. In [bmim][BF_4], the reactions went to completion at room temperature in a few hours, furnishing the quinolines in over 80% isolated yield, whereas in the absence of the ionic liquid or when carried out in a polar solvent such as DMF, no product was formed even after a prolonged reaction time. As with reactions in common solvents, the *endo* product was favored in the ionic liquid; but in the case of 2, 3-dihydrofuran, the *endo* isomers were exclusively obtained for a range of aldehydes and anilines. The ionic liquid could be reused without affecting the product yield. The promoting effect of the ionic liquids may stem from their hydrogen bonding capability. The C^2–H proton of the imidazolium cation is acidic and is known to be a good hydrogen bond donor [195]. Improved rates and selectivities have also been reported for other DA reactions run in ionic liquids in the recent literature [196, 197].

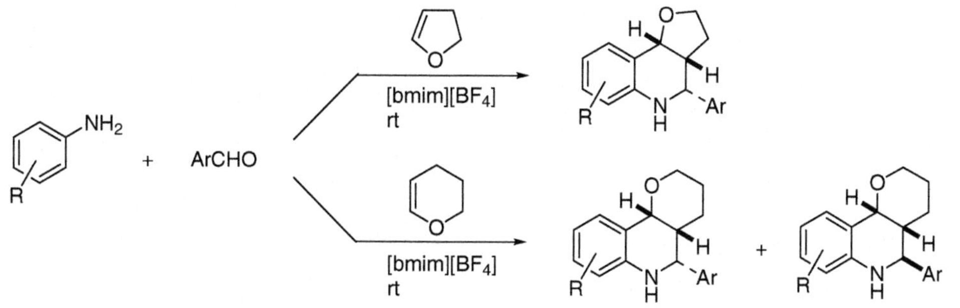

SCHEME 21.47

The Lewis acid-catalyzed DA reaction can also be effected in the apolar $scCO_2$ [198–200]. Rayner and Clifford reported that in the presence of $Sc(OTf)_3$ (6.5 mol%) in $scCO_2$, the reaction of cyclopentadiene and *n*-butyl acrylate went to completion in 15 h at 50°C. Most interestingly, the *endo/exo* selectivity (24:1) was considerably higher than that achievable in common solvents such as toluene and $CHCl_3$ (<11:1) and it was affected by the CO_2 density, reaching a maximum at ca. 1.03 g/mL [198]. The solubility of $Sc(OTf)_3$ in $scCO_2$ is limited, which can be enhanced by introducing fluorous chains. Using $Sc(OSO_2C_8F_{17})_3$ as catalyst, Kobayashi reported that dienes such as isoprene, 2,3-dimethylbuta-1,3-diene and 1,3-cyclohexadiene added to dienophiles such as methyl vinyl ketone and acrylates in $scCO_2$ (50°C, 150 atm, 24 h), giving rise to the cycloaddition product in high yields and selectivities. The same catalyst and solvent could also be applied to the aza-DA reaction of Danishefsky's diene [200].

The DA reactions have proved feasible under solvent-free conditions [201, 202]. For instance, Danishefsky's diene reacted with a variety of aldehydes under titanium catalysis to afford dihydropyranones with up to quantitative yield and 99.8% ee without using solvent (Scheme 21.48). A combinatorial screening identified the best titantium catalysts to be derived from the combination of $Ti(O\text{-}i\text{-}Pr)_4$ with either H_4-BINOL or a mixture of H_4-BINOL and H_8-BINOL. The catalyst loading could be reduced to as low as 0.005 mol%, although the reaction time tends to be long (24 to 144 h) [201].

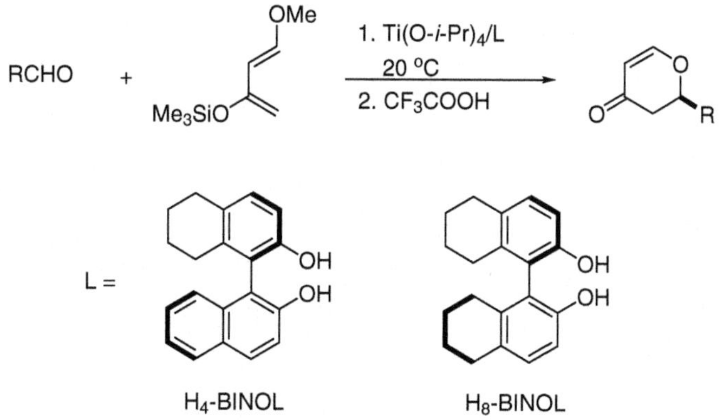

SCHEME 21.48

21.3.5 Friedel–Crafts Reactions

Friedel–Crafts (FC) reactions are catalyzed by a large number of Lewis acids. Whilst traditional Lewis acids such as $AlCl_3$ and $TiCl_4$ are deactivated and decomposed by water, some metal triflates are highly effective for the FC reactions in water [203]. The first example of such a reaction was reported by Kobayashi and coworkers, who showed that the FC-type reaction shown in Scheme 21.49, an atom-economic reaction reminiscent of the Michael conjugate addition, was complete in 1 h by using 2.5 mol% of a dodecyl sulfate (DS) salt of scandium, $Sc(DS)_3$ [204]. The Lewis acidic metal and the anionic surfactant are both indispensable for the reaction, because of the formation of catalyst-containing micelles that enhance the solubility of substrates in water. In line with this, $Sc(OTf)_3$ was much less effective. This chemistry was shown to be applicable to other substrates as well, such as substituted indoles, cyclopentenone, and β-nitrostyrene.

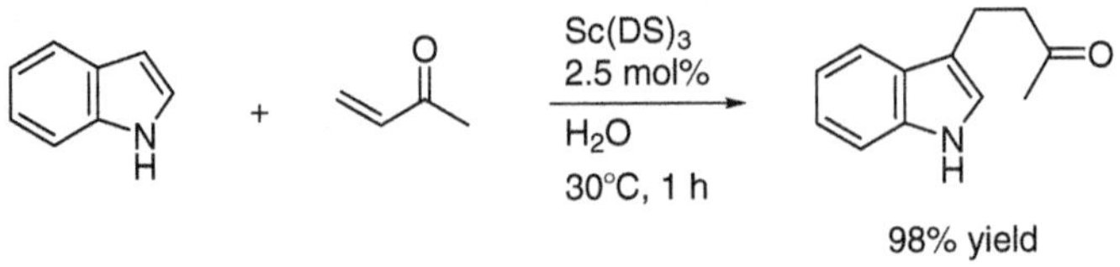

$Sc(DS)_3$ 2.5 mol%, H_2O, 30°C, 1 h; 98% yield

SCHEME 21.49

The use of metal Lewis acids may not be necessary for the FC reactions. Jørgensen recently reported that indoles or pyrroles could be brought to react efficiently with ethyl glyoxylate to give FC addition adducts in a $NaHCO_3$–H_2O or a buffered NaH_2PO_4–Na_2HPO_4 solution [205]. Although the mechanism of the catalysis is unclear, it was shown that the nature of salt and the solution pH affect the product. Thus, double addition product was exclusively formed when NaCl, $NaHSO_4$, or pure water was employed.

The FC reactions are well documented in ionic liquid. Early examples were mainly concerned with the strongly acidic imidazolium chloroaluminate ionic liquids, which were used as both solvent and highly active catalyst [17–19, 21]. They are difficult to handle, however, due to their air and moisture sensitivity. Some of the stable metal triflates have recently proved to work efficiently in ionic liquids that are easier to handle. For instance, $Cu(OTf)_2$ catalyzed the benzoylation of anisole by benzoyl chloride in [bmim][BF_4], affording methoxybenzophenone within 1 h in 81% isolated yield with an *para/ortho* product ratio of 96/4 (Scheme 21.50) [206]. The same reaction performed in the molecular solvents CH_3CN and CH_2ClCH_2Cl gave lower conversions and a reduced regioselectivity. The chemistry could be extended to the benzoylation and acetylation of anisole and derivatives by other acyl chlorides and their anhydrides. With the less-reactive toluene derivatives, conversions were low, although the benzoylation of mesitylene led to quantitative conversion to product after overnight reaction. The anhydrides generally gave rise to slower reactions, as a consequence of triflate being converted into the ionic liquid-insoluble $Cu(OAc)_2$. More recently, $Bi(OTf)_3$ has been reported to be effective, at a catalyst loading as low as 1%, for the acylation of less-active arenes such as benzene at a higher temperature of 150°C in imidazolium ionic liquids. Bi_2O_3 also catalyzed the reaction and allowed for catalyst recycle up to six times; $BiCl_3$ and BiOCl were less effective [207]. However, $BiCl_3$ has been reported to be an efficient catalyst for FC acylation under solvent-free conditions and can be recovered in the form of BiOCl after aqueous workup. BiOCl is water-insoluble and can be *in situ* converted into the active catalyst $BiCl_3$ by reaction with acyl chlorides [208].

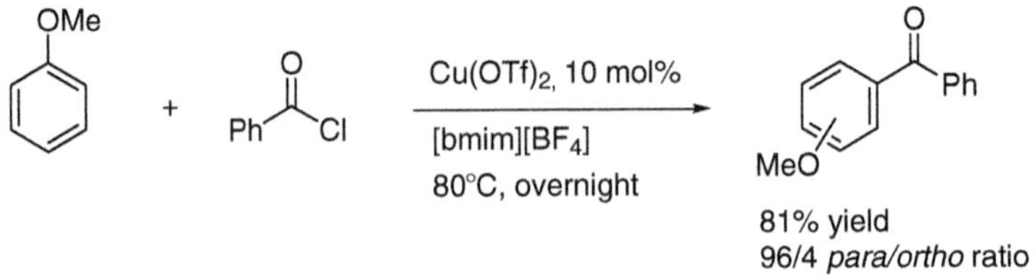

OMe + Ph(C=O)Cl → $Cu(OTf)_2$, 10 mol%; [bmim][BF_4]; 80°C, overnight → MeO–C$_6$H$_4$–C(=O)Ph

81% yield
96/4 *para/ortho* ratio

SCHEME 21.50

A good example of FC alkylation in ionic liquids was reported by Song and coworkers [209]. They showed that benzene, anisole, and phenol could be readily alkylated with acyclic and cyclic olefins under $Sc(OTf)_3$ catalysis in hydrophobic imidazolium ionic liquids under mild conditions. Scheme 21.51 illustrates two examples. These reactions were biphasic, with the catalyst forming a suspension in the ionic liquid phase. Product separation could thus be easily carried out by decantation and reuse of the catalyst proved feasible. Noteworthy is that the catalyst is sensitive to anions of the solvent, with those (e.g., BF_4^- and OTf^-) forming hydrophilic ionic liquids displaying no activity. Intriguingly, in these ionic liquids $Sc(OTf)_3$ is soluble. The results obtained in the hydrophobic ionic liquids contrast sharply with those observed in common organic solvents and water, in which neither alkylation nor olefin isomerization occurred.

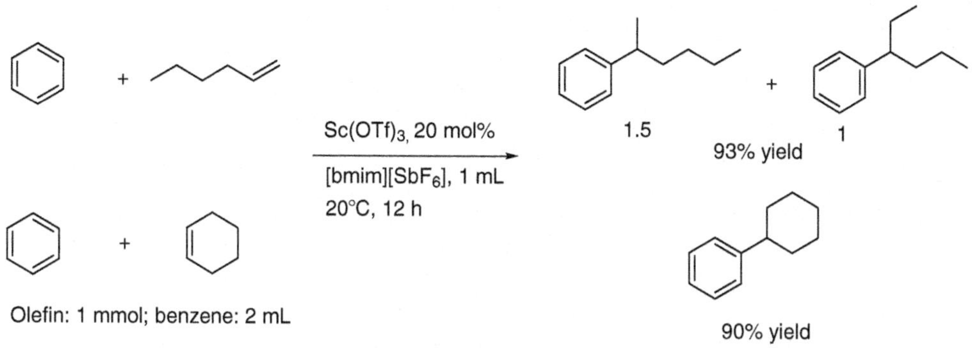

Sc(OTf)3, 20 mol%
[bmim][SbF6], 1 mL
20°C, 12 h
1.5
1
93% yield
Olefin: 1 mmol; benzene: 2 mL
90% yield

SCHEME 21.51

21.3.6 Olefin Metathesis

Olefin metathesis has become a powerful synthetic tool in the past several years as a result of the development of well-defined catalysts that deliver high catalytic activity and excellent functional group compatibility [210, 211]. Pioneering work by Grubbs and coworkers demonstrated that the water-soluble alkylidene complexes shown in Scheme 21.52 could promote ring-opening metathesis polymerization (ROMP) and ring-closing metathesis (RCM) reactions in protic solvents such as water and methanol [212]. These catalysts could initiate the

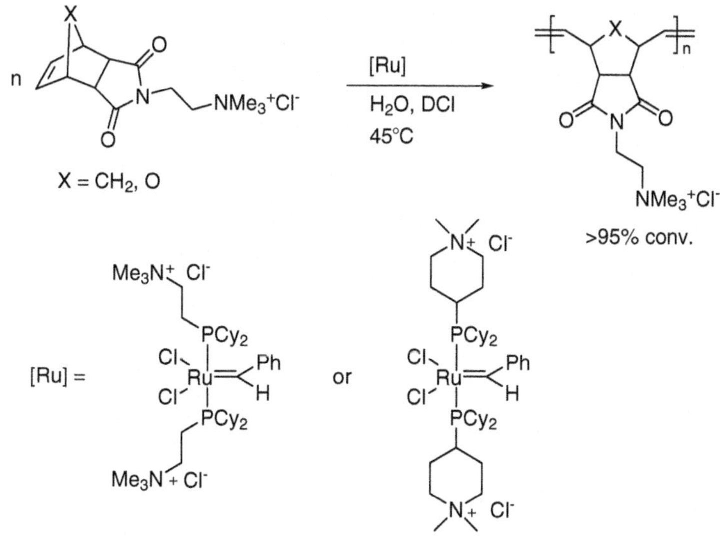

X
O
n
N
NMe3+Cl-
O
X = CH2, O
[Ru]
H2O, DCl
45°C
n
NMe3+Cl-
>95% conv.
[Ru] =
Me3N+ Cl-
PCy2
Cl
Ru
Ph
H
Cl
PCy2
Me3N+ Cl-
or
N+ Cl-
PCy2
Cl
Ru
Ph
H
Cl
PCy2
N+ Cl-

SCHEME 21.52

ROMP of functionalized norbornenes and 7-oxanorbornenes in water, and in the presence of a Brønsted acid (e.g., HCl), the polymerization was rapid and quantitative with a narrow polydispersity for the monomers shown in the scheme. Interestingly, introduction of additional monomer into the completed polymerization reactions resulted in further quantitative polymerization, suggesting that the polymerizations could be living.

RCM of acyclic olefins could be conducted in water with the conventional Grubbs's catalysts such as $RuCl_2(PCy_3)_2(CHPh)$ with or without a surfactant [213]. Although heterogeneous, the RCM reaction of a range of dienes in water was shown to give cyclic and heterocyclic olefins in good yields at short reaction times and room temperature. To enhance the stability of the metathesis catalysts in protic media, N-heterocyclic carbene-substituted ruthenium alkylidenes could be employed, some of which have been reported to be highly effective in RCM in methanol or its mixture with water in air [214].

The Grubbs's catalysts $RuCl_2(PCy_3)_2(CHPh)$ could also be used without modification for RCM in ionic liquids. However, recycle of the catalyst presents a problem if the product is to be removed by extraction, which can result in catalyst leaching [215]. A more attractive approach is to use ruthenium alkylidenes that bear ionic functionalities. A recent example is seen in Scheme 21.53 [216]. The ionic liquid-tagged alkylidene catalyst was reported to be highly active for a range of dienes in [bmim][PF_6], completing the RCM in less than 1 h at 60°C at a catalyst loading of 2.5 mol%. Catalyst recycle could be easily performed by extraction with toluene and in the case of *N,N*-diallyltosylamide, the catalyst-containing ionic liquid was shown to be capable of delivering high conversions even after the ninth run, indicating an enhanced catalyst life in the ionic liquid. Furthermore, once decomposed, the catalyst could be removed by treating the ionic liquid with black carbon; the ionic liquids could then be reloaded with catalyst and reused. An almost identical ionic alkylidene catalyst has also been reported and shown to be equally recyclable using imidazolium ionic liquids [217].

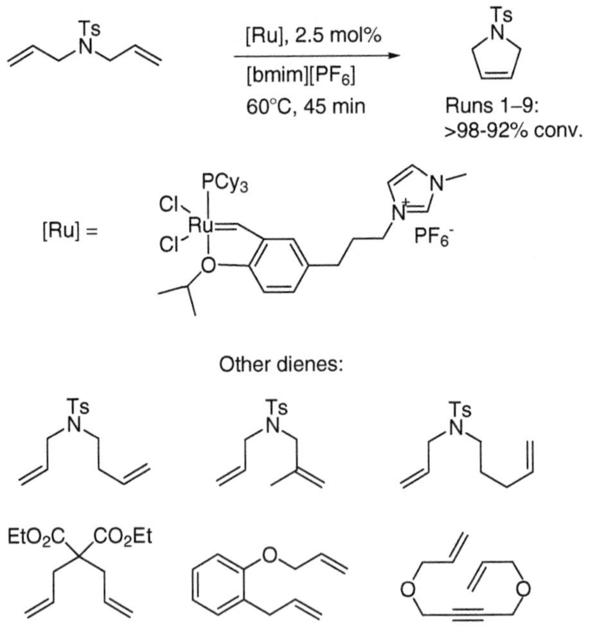

SCHEME 21.53

$ScCO_2$ is also an effective medium for RCM and ROMP. Fürstner and Leitner reported that the molybdenum and ruthenium alkylidene complexes shown in Scheme 21.54 catalyzed the RCM of a range of dienes with efficiency similar to that in chlorinated organic solvents

[218]. Two examples of heterocycle synthesis are provided in the scheme. For the seterically less-demanding diene, the original Grubb's catalyst (first on the left) was sufficiently active to furnish epilachnene in good yield, whereas for the more hindered substrate (leading to karahanaenone), the more active molybdenum and N-heterocycle-substituted ruthenium alkylidenes were desired. Of interest is the observation that the aminodiene was converted into a carbamic acid under the reaction conditions, thus obviating the need for N-protection during RCM. The fluoroalkylated molybdenum catalyst was soluble in $scCO_2$; but the

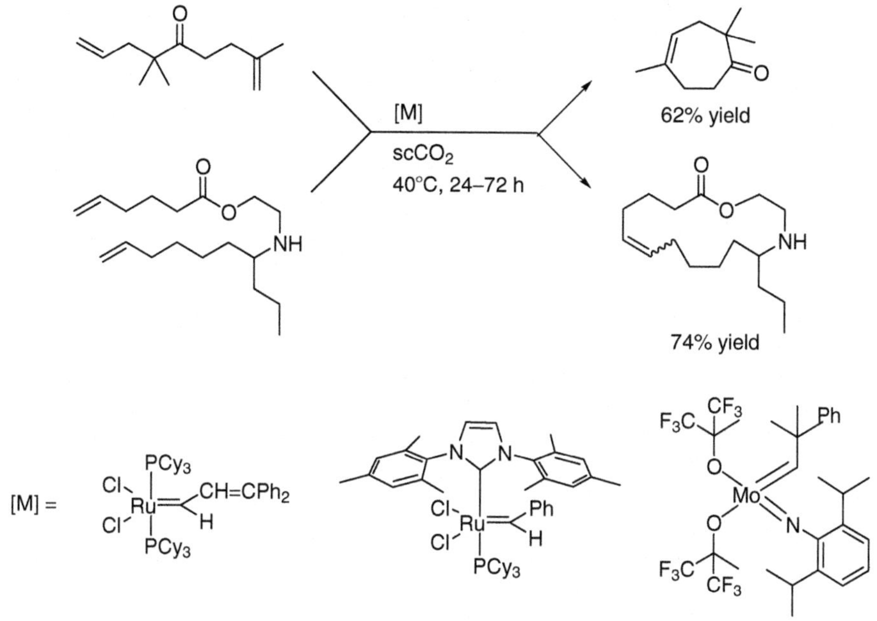

SCHEME 21.54

ruthenium complexes were not, suggesting catalysis with these complexes to be heterogeneous. The insolubility of the ruthenium alkylidens provides a simple means for catalyst/product separation, as the $scCO_2$-soluble product could be removed by CO_2 extraction. These catalysts were also shown to be highly active for ROMP of norbornene and cyclooctene in $scCO_2$. Interestingly, with the Grubb's catalysts, the metal species in the resulting polymers were comparted in a small highly colored portion that could be removed by cutting.

21.3.7 Olefin Epoxidation

Epoxides are one of the most versatile building blocks in organic synthesis due to the ease with which they can be converted into various bifunctional compounds. Epoxidation, leading to epoxides, is most often carried out in chlorinated solvents such as CH_2Cl_2, but has been demonstrated to be practical in the alternative media. For example, Burgess recently reported that simple manganese salts such as $MnSO_4$ (1 mol%) catalyzed the epoxidation of a variety of olefins in $NaHCO_3$-buffered water with H_2O_2 as oxidant and DMF or *t*-BuOH as cosolvent [219]. Aryl- and alkyl-substituted olefins including allylic alcohols gave high yields of up to 97%; but the more challenging terminal alkyl-substituted olefins did not react. The oxidizing species is believed to be the peroxymonocarbonate HCO_4^- formed between H_2O_2 and the bicarbonate ion. The less-reactive terminal aliphatic olefins could be readily epoxidized with aqueous 50% H_2O_2 by a Fe(II) complex containing a tetradentate pyridyl-amine ligand (within 5 min, in 60–90% isolated yields, with 3 mol% catalyst loading at 4°C). However, a cosolvent, CH_3CN, was again required [220].

Epoxidation with aqueous H_2O_2 without cosolvent is feasible and has previously been reported by Noyori, who showed that under the catalysis of Na_2WO_4 (2 mol%), terminal aliphatic alkenes such as 1-dodecene could be epoxidized in 99% yield with only 1.5 equivalents of 30% H_2O_2 at 90°C by means of rapid stirring [221]. Supported titanium and polyoxometalate catalysts have also been demonstrated to effect epoxidation of cyclic and acyclic olefins in aqueous H_2O_2 without additional organic solvents [222, 223]. The epoxidation chemistry of H_2O_2 has been summarized in a recent review [224].

While the Sharpless asymmetric epoxidation of allylic alcohols by titanium tartrates usually requires the use of molecular sieves to remove adventitious water, catalysts made of Mo(VI) and W(VI) salts allow the epoxidation reaction to be run in aqueous H_2O_2. Thus, in the presence of molybdic acid (10 mol%) in a buffered aqueous solution of sucrose (1 *M*), geraniol could be oxidized to 2,3-epoxygeraniol in 92% isolated yield at 2°C. The carbohydrate additive is amphiphilic and enhances solubility of the substrate in water. It could also orientate the substrate and hence induce faceselection during the epoxidation, as was observed in the case of geraniol (10% ee). A range of acyclic and cyclic allylic alcohols including terminal variants were epoxidized under these conditions in good to excellent isolated yields, albeit with relatively long reaction times (12–72 h) [225].

Epoxidation in ionic liquids has also been reported. For example, iron and manganese porphyrin complexes were immobilized in imidazolium ionic liquids and were shown to catalyze the epoxidation of various olefins by H_2O_2 or $PhI(OAc)_2$ [226, 227]. The manganese catalyst allowed for the epoxidation of terminal olefins such as 1-undecene with good yields [227]. Although these catalyst systems permit easy catalyst separation and recycle, the use of CH_2Cl_2 as cosolvent nullifies, to some degree, one of the incentives for choosing ionic liquids as solvent.

Asymmetric epoxidation with NaOCl catalyzed by a familiar Mn(III)–salen complex was earlier reported to give excellent ee's (up to 96%) in a mixture of [bmim][PF_6] and CH_2Cl_2 (1:4). The latter was necessary as the ionic liquid solidifies at the reaction temperature of 0°C. A faster reaction was observed in the presence of the ionic liquid, while the enantioselectivity was comparable with that obtained in chlorinated solvents. Recycle of the catalyst-containing ionic liquid was possible; but some decrease in catalyst activity was noticed [228].

The combination of ionic liquids and water provides an effective biphasic system for the epoxidation of α,β-unsaturated carbonyl compounds. As shown in Scheme 21.55, mesityl oxide could be readily epoxidized with H_2O_2 (5 equiv.) and in excellent selectivity to the epoxyketone under the influence of aqueous NaOH (20 mol%) in [bmim][PF_6] [229]. Benzylideneacetone and chalcone were equally suitable, but lower conversion was encountered with compounds such as methyl cinnamate. In comparison with the traditional CH_2Cl_2–H_2O system, the ionic liquid–water protocol enables much faster reaction and enhanced epoxide selectivity. The imidazolium cation presumably acts as a catalyst, transferring OH^- out of, and the active species OOH^- into, the ionic liquid phase where the epoxidation takes place. The relatively hydrophobic nature of [bmim][PF_6] renders the concentration of OH^- low, thus reducing the probability of ring opening of the product in the ionic liquid.

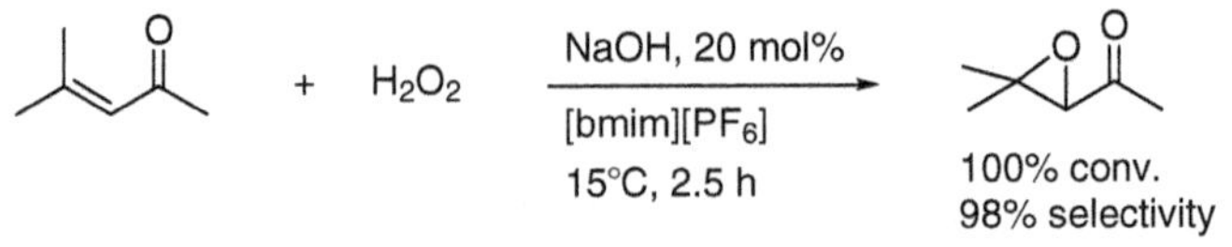

+ H_2O_2 NaOH, 20 mol% [bmim][PF_6] 15°C, 2.5 h

100% conv.
98% selectivity

SCHEME 21.55

The simple, yet efficient, manganese sulfate catalyst, which requires a cosolvent when used in aqueous phase [219], can be adopted for epoxidation in ionic liquids with no need for additional organic solvents. Chan recently reported that lipophilic olefins could be readily epoxidized in up to >99% conversion and >99% epoxide selectivity with 35% H_2O_2 (10 equiv.) by $MnSO_4$ in the presence of 1 equivalent of Me_4NHCO_3 in [bmim][BF_4] [230]. As is seen in Scheme 21.56, a range of olefins were oxidized. However, as with the aqueous-phase epoxidation, terminal aliphatic olefins failed to react. The use of Me_4NHCO_3 was critical; the ionic liquid-insoluble $NaHCO_3$ did not bring about the epoxidation. A nuclear magnetic resonance (NMR) study indicates that the epoxidation operates on a similar mechanism in both the ionic liquids and aqueous solutions. The epoxide product could be extracted with pentane and the remaining ionic liquid together with the manganese and bicarbonate salts were demonstrated to be reusable. Efficient epoxidation of similar olefins in [emim][BF_4] has previously been reported, with the urea–H_2O_2 adduct as oxidant and $MeReO_3$ as catalyst. Like the manganese sulfate catalyst, the rhenium-oxo species was also less effective towards more difficult olefins [231, 232].

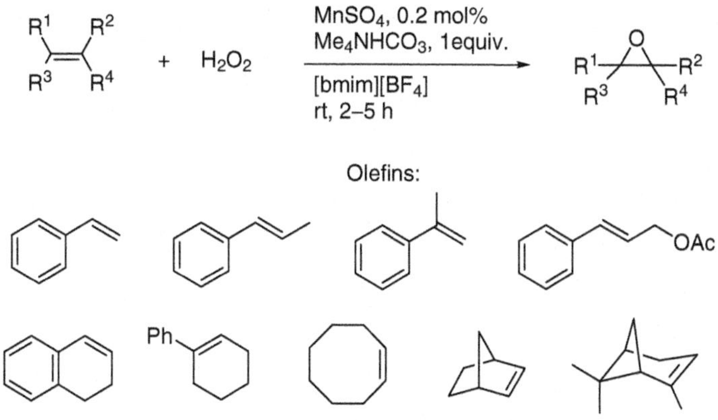

SCHEME 21.56

Epoxidation reactions in $scCO_2$ and fluorous media have also been reported. However, new examples of synthetic relevance are few. For earlier examples, the readers are referred to Refs. [16, 25, 26, 233, 234].

21.4 CONCLUSIONS

Catalysis is the cornerstone for "greening" synthetic organic chemistry. However, catalysis by itself is not sufficient to implement a sustainable synthetic process. Considerations must also be given to factors such as atom efficiency and the use of solvents when designing and developing a reaction. The need to be conscious of environmental impacts should be taken as a rewarding opportunity rather than an encumbrance in synthetic endeavors. Indeed, the many examples presented in this chapter unambiguously demonstrate that catalytic organic synthesis can benefit greatly from adopting the atom economy principle and employing alternative reaction media. These benefits include, not exhaustively, enhanced catalyst activity, selectivity and productivity, new selectivity patterns, reduced or eliminated waste by-products and solvent emissions, and ease of operation. The four classes of solvents that form the focal point of this chapter span the entire range of the solvent spectrum in terms of their chemical and physical properties, which can be further tuned to fulfill the specific demands of a synthetic task. Keeping these attributes in mind, many more environmentally benign synthetic reactions can be envisaged and undoubtedly, many more will materialize with these alternative media in the foreseeable future.

REFERENCES

1. RA Sheldon. *J. Mol. Catal. A: Chemical* 107: 75–83, 1996.
2. HU Blaser and M Studer. *Appl. Catal. A: General* 189: 191–240, 1999.
3. B Cornils and WA Herrmann, eds. *Applied Homogeneous Catalysis with Organometallic Compounds*. Weinheim: Wiley-VCH, 1996.
4. M Beller and C Bolm, eds. *Transition Metals for Organic Synthesis*. Weinheim: Wiley-VCH, 1998.
5. RA Sheldon and H van Bekkum, eds. *Fine Chemicals through Heterogeneous Catalysis*. Weinheim: Wiley-VCH, 2001.
6. DE de Vos, IFJ Vankelecom, and PA Jacobs, eds. *Chiral Catalyst Immobilization and Recycling*. Weinheim: Wiley-VCH, 2000.
7. JA Gladysz. *Chem. Rev.* 102: 3215–3216, 2002.
8. BM Trost. *Acc. Chem. Res.* 35: 695–705, 2002.
9. AD Curzons, DJC Constable, DN Mortimer, and VL Cunningham. *Green Chem.* 3: 1–6, 2001.
10. B Cornils and WA Herrmann, eds. *Aqueous-Phase Organometallic Catalysis. Concepts and Applications*. Weinheim: Wiley-VCH, 1998.
11. F Joo. *Aqueous Organometallic Catalysis*. Dordrecht: Kluwer, 2001.
12. CJ Li and TH Chan. *Organic Reactions in Aqueous Media*. New York: Wiley, 1997.
13. MA Blatchford, P Raveendran, and SL Wallen. *J. Am. Chem. Soc.* 124: 14818–14819, 2002.
14. W Leitner. *Acc. Chem. Res.* 35: 746–756, 2002.
15. PG Jessop, T Ikariya, and R Noyori. *Chem. Rev.* 99: 475–493, 1999.
16. T Ikariya and Y Kayaki. *Catal. Surv. Jpn.* 4: 39–50, 2000.
17. T Welton. *Chem. Rev.* 99: 2071–2083, 1999.
18. P Wasserscheid and W Keim. *Angew. Chem. Int. Ed.* 39: 3772–3789, 2000.
19. RA Sheldon. *Chem. Commun.* 2399–2407, 2001.
20. J Dupont, RF de Souza, and PAZ Suarez. *Chem. Rev.* 102: 3667–3692, 2002.
21. KR Seddon. *J. Chem. Technol. Biotechnol.* 68: 351–356, 1997.
22. LA Blanchard, D Hancu, EJ Beckman, and JF Brennecke. *Nature* 399: 28–29, 1999.
23. IT Horvath and J Rabai. *Science* 266: 72–75, 1994.
24. IT Horvath. *Acc. Chem. Res.* 31: 641–650, 1998.
25. JA Gladysz and DP Curran. *Tetrahedron* 58: 3823–3825, 2002.
26. RH Fish. *Chem. Eur. J.* 5: 1677–1680, 1999.
27. AR Ravishankara, S Solomon, AA Turnipseed, and RF Warren. *Science* 259: 194–199, 1993.
28. F Joo. *Acc. Chem. Res.* 35: 738–745, 2002.
29. T Dwars and G Oehme. *Adv. Synth. Catal.* 344: 239–260, 2002.
30. C Larpent, R Dabard, and H Patin. *Tetrahedron Lett.* 28: 2507–2510, 1987.
31. VV Grushin and H Alper. *Organometallics* 10: 831–833, 1991.
32. ZY Yang, M Ebihara, and T Kawamura. *J. Mol. Catal. A: Chemical* 158: 509–514, 2000.
33. T Okano, M Kaji, S Isotani, and J Kiji. *Tetrahedron Lett.* 33: 5547–5550, 1992.
34. P Smolenski, FP Pruchnik, Z Ciunik, and T Lis. *Inorg. Chem.* 42: 3318–3322, 2003.
35. F Joo and MT Beck. *React. Kinet. Catal. Lett.* 2: 257–263, 1975.
36. JM Grosselin, C Mercier, G Allmang, and F Grass. *Organometallics* 10: 2126–2133, 1991.
37. G Papp, J Elek, L Nadasdi, G Laurenczy, and F Joo. *Adv. Synth. Catal.* 345: 172–174, 2003.
38. V Penicaud, C Maillet, P Janvier, M Pipelier, and B Bujoli. *Eur. J. Org. Chem.* 1745–1728, 1999.
39. N Makihara, S Ogo, and Y Watanabe. *Organometallics* 20: 497–500, 2001.
40. A Benyei and F Joo. *J. Mol. Catal.* 58: 151–163, 1990.
41. S Ogo, T Abura, and Y Watanabe. *Organometallics* 21: 2964–2969, 2002.
42. S Ogo, N Makihara, and Y Watanabe. *Organometallics* 18: 5470–5474, 1999.
43. K Nomura. *J. Mol. Catal. A: Chemical* 130: 1–28, 1998.
44. S Steines, P Wasserscheid, and B Driessen-Holscher. *J. Fur Praktische Chemie-Chemiker-Zeitung* 342: 348–354, 2000.
45. J Dupont, GS Fonseca, AP Umpierre, PFP Fichtner, and SR Teixeira. *J. Am. Chem. Soc.* 124: 4228–4229, 2002.
46. I Kani, MA Omary, MA Rawashdeh-Omary, ZK Lopez-Castillo, R Flores, A Akgerman, and JP Fackler. *Tetrahedron* 58: 3923–3928, 2002.

47. ZK Lopez-Castillo, R Flores, I Kani, JP Fackler, and A Akgerman. *Ind. Eng. Chem. Res.* 41: 3075–3080, 2002.
48. B Richter, AL Spek, G van Koten, and BJ Deelman. *J. Am. Chem. Soc.* 122: 3945–3951, 2000.
49. DE Bergbreiter, JG Franchina, and BL Case. *Org. Lett.* 2: 393–395, 2000.
50. E de Wolf, AL Spek, BWM Kuipers, AP Philipse, JD Meeldijk, PHH Bomans, PM Frederik, BJ Deelman, and G van Koten. *Tetrahedron* 58: 3911–3922, 2002.
51. J van den Broeke, E de Wolf, BJ Deelman, and G van Koten. *Adv. Synth. Catal.* 345: 625–634, 2003.
52. SL Vinson and MR Gagne. *Chem. Commun.* 1130–1131, 2001.
53. M Chatterjee, Y Ikushima, and FY Zhao. *New. J. Chem.* 27: 510–513, 2002.
54. MG Hitzler, FR Smail, SK Ross, and M Poliakoff. *Org. Proc. Res. Develop.* 2: 137–146, 1998.
55. T Soos, BL Bennett, D Rutherford, LP Barthel-Rosa, and JA Gladysz. *Organometallics* 20: 3079–3086, 2001.
56. D Sinou, D Maillard, A Aghmiz, and AMM i-Bulto. *Adv. Synth. Catal.* 345: 603–611, 2003.
57. WA Herrmann and CW Kohlpaintner. *Angew. Chem. Int. Ed. Engl.* 32: 1524–1544, 1993.
58. D Sinou. *Adv. Synth. Catal.* 344: 221–237, 2002.
59. TV RajanBabu, YY Yan, and S Shin. *J. Am. Chem. Soc.* 123: 10207–10213, 2001.
60. MJ Burk. *Acc. Chem. Res.* 33: 363–372, 2000.
61. J Holz, D Heller, R Sturmer, and A Borner. *Tetrahedron Lett.* 40: 7059–7062, 1999.
62. W Li, Z Zhang, D Xiao, and X Zhang. *J. Org. Chem.* 65: 3489–3496, 2000.
63. G Oehme, E Paetzold, and R Selke. *J. Mol. Catal.* 71: L1–L5, 1992.
64. I Grassert, U Schmidt, S Ziegler, C Fischer, and G Oehme. *Tetrahedron: Asymmetry* 9: 4193–4202, 1998.
65. K Yonehara, K Ohe, and S Uemura. *J. Org. Chem.* 64: 9381–9385, 1999.
66. I Grassert, J Kovacs, H Fuhrmann, and G Oehme. *Adv. Synth. Catal.* 344: 312–318, 2002.
67. YP Ma, H Liu, L Chen, X Cui, J Zhu, and JE Deng. *Org. Lett.* 5: 2103–2106, 2003.
68. Y Himeda, N Onozawa-Komatsuzaki, H Sugihara, H Arakawa, and K Kasuga. *J. Mol. Catal. A: Chemical* 195: 95–100, 2003.
69. T Thorpe, J Blacker, SM Brown, C Bubert, J Crosby, S Fitzjohn, JP Muxworthy, and JMJ Williams. *Tetrahedron Lett.* 42: 4041–4043, 2001.
70. Y Chauvin, L Mussmann, and H Olivier. *Angew. Chem. Int. Ed. Engl.* 34: 2698–2700, 1995.
71. AL Monteiro, FK Zinn, RF de Souza, and J Dupont. *Tetrahedron: Asymmetry* 8: 177–179, 1997.
72. S Guernik, A Wolfson, M Herskowitz, N Greenspoon, and S Geresh. *Chem. Commun.* 2314–2315, 2001.
73. A Berger, RF de Souza, MR Delgado, and J Dupont. *Tetrahedron: Asymmetry* 12: 1825–1828, 2001.
74. RA Brown, P Pollet, E McKoon, CA Eckert, CL Liotta, and PG Jessop. *J. Am. Chem. Soc.* 123: 1254–1255, 2001.
75. JL Xiao, SCA Nefkens, PG Jessop, T Ikariya, and R Noyori. *Tetrahedron Lett.* 37: 2813–2816, 1996.
76. MJ Burk, SG Feng, MF Gross, and W Tumas. *J. Am. Chem. Soc.* 117: 8277–8278, 1995.
77. S Kainz, A Brinkmann, W Leitner, and A Pfaltz. *J. Am. Chem. Soc.* 121: 6421–6429, 1999.
78. S Wang and F Kienzle. *Ind. Eng. Chem. Res.* 39: 4487–4490, 2000.
79. EG Kuntz. *Chemtech* 17: 570–575, 1987.
80. F Ungvary. *Coord. Chem. Rev.* 228: 61–82, 2002.
81. F Ungvary. *Coord. Chem. Rev.* 241: 295–312, 2003.
82. F Ungvary. *Coord. Chem. Rev.* 218: 1–41, 2001.
83. F Ungvary. *Coord. Chem. Rev.* 213: 1–50, 2001.
84. G Fremy, E Monflier, JF Carpenter, Y Castanet, and A Mortreux. *Angew. Chem. Int. Ed. Engl.* 34: 1474–1476, 1995.
85. G Fremy, E Monflier, JF Carpentier, Y Castanet, and A Mortreux. *J. Mol. Catal. A: Chemical* 129: 35–40, 1998.
86. G Verspui, G Elbertse, G Papadogianakis, and RA Shedon. *J. Organomet. Chem.* 621: 337–343, 2001.
87. JH Chen and H Alper. *J. Am. Chem. Soc.* 119: 893–895, 1997.
88. AN Ajjou and H Alper. *J. Am. Chem. Soc.* 120: 1466–1468, 1998.

89. S Paganelli, M Zanchet, M Marchetti, and G Mangano. *J. Mol. Catal. A: Chemical* 157: 1–8, 2000.
90. C Botteghi, S Paganelli, F Moratti, M Marchetti, R Lazzaroni, R Settambolo, and O Piccolo. *J. Mol. Catal. A: Chemical* 200: 147–156, 2003.
91. B Zimmermann, J Herwig, and M Beller. *Angew. Chem. Int. Ed.* 38: 2372–2375, 1999.
92. MS Goedheijt, JNH Reek, PCJ Kamer, and P van Leeuwen. *Chem. Commun.* 2431–2432, 1998.
93. K Fuji, T Morimoto, K Tsutsumi, and K Kakiuchi. *Angew. Chem. Int. Ed.* 42: 2409–2411, 2003.
94. RPJ Bronger, SM Silva, PCJ Kamer, and P van Leeuwen. *Chem. Commun.* 3044–3045, 2002.
95. F Favre, H Olivier-Bourbigou, D Commereuc, and L Saussine. *Chem. Commun.* 1360–1361, 2001.
96. W Keim, D Vogt, H Waffenschmidt, and P Wasserscheid. *J. Catal.* 186: 481–484, 1999.
97. D Zim, RF de Souza, J Dupont, and AL Monteiro. *Tetrahedron Lett.* 39: 7071–7074, 1998.
98. K Qiao and YQ Deng. *New J. Chem.* 26: 667–670, 2002.
99. CS Consorti, G Ebeling, and J Dupont. *Tetrahedron Lett.* 43: 753–755, 2002.
100. V Calo, P Giannoccaro, A Nacci, and A Monopoli. *J. Organomet. Chem.* 645: 152–157, 2002.
101. HS Kim, YJ Kim, H Lee, KY Park, C Lee, and CS Chin. *Angew. Chem. Int. Ed.* 41: 4300–4303, 2002.
102. YL Hu, WP Chen, AMB Osuna, JA Iggo, and JL Xiao. *Chem. Commun.* 788–789, 2002.
103. YL Hu, WP Chen, AMB Osuna, AM Stuart, EG Hope, and JL Xiao. *Chem. Commun.* 725–726, 2001.
104. WP Chen, LJ Xu, and JL Xiao. *Chem. Commun.* 839–840, 2000.
105. G Francio, K Wittmann, and W Leitner. *J. Organomet. Chem.* 621: 130–142, 2001.
106. Y Kayaki, Y Noguchi, S Iwasa, T Ikariya, and R Noyori. *Chem. Commun.* 1235–1236, 1999.
107. N Jeong, SH Hwang, YW Lee, and JS Lim. *J. Am. Chem. Soc.* 119: 10549–10550, 1997.
108. JP Genet and M Savignac. *J. Organomet. Chem.* 576: 305–317, 1999.
109. R Amengual, E Genin, V Michelet, M Savignac, and JP Genet. *Adv. Synth. Catal.* 344: 393–398, 2002.
110. Y Uozumi and T Kimura. *Synlett* 2045–2048, 2002.
111. KSA Vallin, M Larhed, and A Hallberg. *J. Org. Chem.* 66: 4340–4343, 2001.
112. W Cabri and I Candiani. *Acc. Chem. Res.* 28: 2–7, 1995.
113. DE Kaufmann, M Nouroozian, and H Henze. *Synlett* 1091–1092, 1996.
114. T Jeffery. *J. Chem. Soc. Chem. Commun.* 1287–1289, 1984.
115. H Zhao and SV Malhotra. *Aldrichimica Acta* 35: 75–83, 2002.
116. V Calo, A Nacci, A Monopoli, S Laera, and N Cioffi. *J. Org. Chem.* 68: 2929–2933, 2003.
117. V Calo, A Nacci, A Monopoli, L Lopez, and A di Cosmo. *Tetrahedron* 57: 6071–6077, 2001.
118. LJ Xu, WP Chen, and JL Xiao. *Organometallics* 19: 1123–1127, 2000.
119. BM Choudary, S Madhi, NS Chowdari, ML Kantam, and B Sreedhar. *J. Am. Chem. Soc.* 124: 14127–14136, 2002.
120. K Selvakumar, A Zapf, and M Beller. *Org. Lett.* 4: 3031–3033, 2002.
121. VPW Bohm and WA Herrmann. *Chem. Eur. J.* 6: 1017–1025, 2000.
122. LJ Xu, WP Chen, J Ross, and JL Xiao. *Org. Lett.* 3: 295–297, 2001.
123. KSA Vallin, P Emilsson, M Larhed, and A Hallberg. *J. Org. Chem.* 67: 6243–6246, 2002.
124. DK Morita, DR Pesiri, SA David, WH Glaze, and W Tumas. *Chem. Commun.* 1397–1398, 1998.
125. MA Carroll and AB Holmes. *Chem. Commun.* 1395–1396, 1998.
126. TR Early, RS Gordon, MA Carroll, AB Holmes, RE Shute, and IF McConvey. *Chem. Commun.* 1966–1967, 2001.
127. S Fujita, K Yuzawa, BM Bhanage, Y Ikushima, and M Arai. *J. Mol. Catal. A: Chemical* 180: 35–42, 2002.
128. N Shezad, RS Oakes, AA Clifford, and CM Rayner. *Tetrahedron Lett.* 40: 2221–2224, 1999.
129. RS Gordon and AB Holmes. *Chem. Commun.* 640–641, 2002.
130. LJ Xu, WP Chen, JF Bickley, A Steiner, and JL Xiao. *J. Organomet. Chem.* 598: 409–416, 2000.
131. N Shezad, AA Clifford, and CM Rayner. *Tetrahedron Lett.* 42: 323–325, 2001.
132. S Cacchi, G Fabrizi, F Gasparrini, P Pace, and C Villani. *Synlett* 650–652, 2000.
133. C Dupuis, K Adiey, L Charruault, V Michelet, M Savignac, and JP Genet. *Tetrahedron Lett.* 42: 6523–6526, 2001.
134. S Parisot, R Kolodziuk, C Goux-Henry, A Iourtchenko, and D Sinou. *Tetrahedron Lett.* 43: 7397–7400, 2002.
135. KH Shaughnessy and RS Booth. *Org. Lett.* 3: 2757–2759, 2001.
136. M Nishimura, M Ueda, and N Miyaura. *Tetrahedron* 58: 5779–5787, 2002.

137. L Botella and C Najera. *Angew. Chem. Int. Ed.* 41: 179–181, 2002.
138. NE Leadbeater and M Marco. *Org. Lett.* 4: 2973–2976, 2002.
139. D Villemin, MJ Gomez-Escalonilla, and JF Saint-Clair. *Tetrahedron Lett.* 42: 635–637, 2001.
140. NE Leadbeater and M Marco. *J. Org. Chem.* 68: 888–892, 2003.
141. M Larhed and A Hallberg. *J. Org. Chem.* 61: 9582–9584, 1996.
142. NE Leadbeater and M Marco. *Angew. Chem. Int. Ed.* 42: 1407–1409, 2003. However, see: RK Arvela, NE Leadbeater, MS Sangi, VA Williams, P Granados, and RD Singer. *J. Org. Chem.* 70: 161–168, 2005.
143. CJ Mathews, PJ Smith, and T Welton. *Chem. Commun.* 1249–1250, 2000.
144. R Rajagopal, DV Jarikote, and KV Srinivasan. *Chem. Commun.* 616–617, 2002.
145. J McNulty, A Capretta, J Wilson, J Dyck, G Adjabeng, and A Robertson. *Chem. Commun.* 1986–1987, 2002.
146. C Rocaboy and JA Gladysz. *Tetrahedron* 58: 4007–4014, 2002.
147. CC Tzschucke, C Markert, H Glatz, and W Bannwarth. *Angew. Chem. Int. Ed.* 41: 4500–4503, 2002.
148. C Rocaboy and JA Gladysz. *New J. Chem.* 27: 39–49, 2003.
149. M Melucci, G Barbarella, and G Sotgiu. *J. Org. Chem.* 67: 8877–8884, 2002.
150. V Kogan, Z Aizenshtat, R Popovitz-Biro, and R Neumann. *Org. Lett.* 4: 3529–3532, 2002.
151. XJ Han, GA Hartmann, A Brazzale, and RD Gaston. *Tetrahedron Lett.* 42: 5837–5839, 2001.
152. ST Handy and XL Zhang. *Org. Lett.* 3: 233–236, 2001.
153. T Osswald, S Schneider, S Wang, and W Bannwarth. *Tetrahedron Lett.* 42: 2965–2967, 2001.
154. S Schneider and W Bannwarth. *Angew. Chem. Int. Ed.* 39: 4142–4145, 2000.
155. MP Lopez-Deber, L Castedo, and JR Granja. *Org. Lett.* 3: 2823–2826, 2001.
156. DT Bong and MR Ghadiri. *Org. Lett.* 3: 2509–2511, 2001.
157. C Najera, J Gil-Molto, S Karlstrom, and LR Falvello. *Org. Lett.* 5: 1451–1454, 2003.
158. T Fukuyama, M Shinmen, S Nishitani, M Sato, and I Ryu. *Org. Lett.* 4: 1691–1694, 2002.
159. C Markert and W Bannwarth. *Helvet. Chim. Acta* 85: 1877–1882, 2002.
160. Y Uozumi and K Shibatomi. *J. Am. Chem. Soc.* 123: 2919–2920, 2001.
161. T Hashizume, K Yonehara, K Ohe, and S Uemura. *J. Org. Chem.* 65: 5197–5201, 2000.
162. D Sinou, C Rabeyrin, and C Nguefack. *Adv. Synth. Catal.* 345: 357–363, 2003.
163. J Ross, WP Chen, LJ Xu, and JL Xiao. *Organometallics* 20: 138–142, 2001.
164. WP Chen, LJ Xu, C Chatterton, and JL Xiao. *Chem. Commun.* 1247–1248, 1999.
165. C Amatore, S Gamez, and A Jutand. *Chem. Eur. J.* 7: 1273–1280, 2001.
166. C Amatore, A Jutand, G Meyer, and L Mottier. *Chem. Eur. J.* 5: 466–473, 1999.
167. J Bayardon and D Sinou. *Tetrahedron Lett.* 44: 1449–1451, 2003.
168. J Bayardon, M Cavazzini, D Maillard, G Pozzi, S Quici, and D Sinou. *Tetrahedron-Asymmetry* 14: 2215–2224, 2003.
169. Y Mori, J Kobayashi, K Manabe, and S Kobayashi. *Tetrahedron* 58: 8263–8268, 2002.
170. K Manabe and S Kobayashi. *Chem. Eur. J.* 8: 4095–4101, 2002.
171. MW Wang, XF Yang, and CJ Li. *Eur. J. Org. Chem.* 998–1003, 2003.
172. Z Zhang, YW Dong, and GW Wang. *Chem. Lett.* 32: 966–967, 2003.
173. T Hamada, K Manabe, S Ishikawa, S Nagayama, M Shiro, and S Kobayashi. *J. Am. Chem. Soc.* 125: 2989–2996, 2003.
174. B Alcaide and P Almendros. *Angew. Chem. Int. Ed.* 42: 858–860, 2003.
175. TP Loh, LC Feng, HY Yang, and JY Yang. *Tetrahedron Lett.* 43: 8741–8743, 2002.
176. P Kotrusz, I Kmentova, B Gotov, S Toma, and E Solcaniova. *Chem. Commun.* 2510–2511, 2002.
177. K Mikami, M Yamanaka, MN Islam, K Kudo, N Seino, and M Shinoda. *Tetrahedron Lett.* 44: 7545–7548, 2003.
178. I Komoto and S Kobayashi. *Org. Lett.* 4: 1115–1118, 2002.
179. Y Mori, K Kakumoto, K Manabe, and S Kobayashi. *Tetrahedron Lett.* 41: 3107–3111, 2000.
180. Y Hamashima, H Takano, D Hotta, and M Sodeoka. *Org. Lett.* 5: 3225–3228, 2003.
181. MM Dell'Anna, V Gallo, P Mastrorilli, CF Nobile, G Romanazzi, and GP Suranna. *Chem. Commun.* 434–435, 2002.
182. R Breslow, U Maitra, and D Rideout. *Tetrahedron Lett.* 24: 1901–1904, 1983.
183. DC Rideout and R Breslow. *J. Am. Chem. Soc.* 102: 7816–7817, 1980.
184. F Fringuelli, O Piermatti, F Pizzo, and L Vaccaro. *Eur. J. Org. Chem.* 439–455, 2001.
185. S Otto and JBFN Engberts. *Pure Appl. Chem.* 72: 1365–1372, 2000.

186. S Otto, F Bertoncin, and JBFN Engberts. *J. Am. Chem. Soc.* 118: 7702–7707, 1996.
187. S Otto and JBFN Engberts. *J. Am. Chem. Soc.* 121: 6798–6806, 1999.
188. A Wittkopp and PR Schreiner. *Chem. Eur. J.* 9: 407–414, 2003.
189. DB Ramachary, NS Chowdari, and CF Barbas. *Tetrahedron Lett.* 43: 6743–6746, 2002.
190. C Loncaric, K Manabe, and S Kobayashi. *Adv. Synth. Catal.* 345: 475–477, 2003.
191. C Loncaric, K Manabe, and S Kobayashi. *Chem. Commun.* 574–575, 2003.
192. CC Tzschucke, C Markert, W Bannwarth, S Roller, A Hebel, and R Haag. *Angew. Chem. Int. Ed.* 41: 3964–4000, 2002.
193. CE Song, WH Shim, EJ Roh, SG Lee, and JH Choi. *Chem. Commun.* 1122–1123, 2001.
194. JS Yadav, BVS Reddy, JSS Reddy, and RS Rao. *Tetrahedron* 59: 1599–1604, 2003.
195. J Ross and JL Xiao. *Chem. Eur. J.* 9: 4900–4906, 2003.
196. I Meracz and T Oh. *Tetrahedron Lett.* 44: 6465–6468, 2003.
197. I Hemeon, C DeAmicis, H Jenkins, P Scammells, and RD Singer. *Synlett* 1815–1818, 2002.
198. RS Oakes, TJ Heppenstall, N Shezad, AA Clifford, and CM Rayner. *Chem. Commun.* 1459–1460, 1999.
199. SI Fukuzawa, K Metoki, Y Komuro, and T Funazukuri. *Synlett* 134–136, 2002.
200. J Matsuo, T Tsuchiya, K Odashima, and S Kobayashi. *Chem. Lett.* 178–179, 2000.
201. J Long, JY Hu, XQ Shen, BM Ji, and KL Ding. *J. Am. Chem. Soc.* 124: 10–11, 2002.
202. A DiazOrtiz, JR Carrillo, E DiezBarra, A delaHoz, MJ GomezEscalonilla, A Moreno, and F Langa. *Tetrahedron* 52: 9237–9248, 1996.
203. GKS Prakash, P Yan, B Torok, I Bucsi, M Tanaka, and GA Olah. *Catal. Lett.* 85: 1–6, 2003.
204. K Manabe, N Aoyama, and S Kobayashi. *Adv. Synth. Catal.* 343: 174–176, 2001.
205. W Zhuang and KA Jorgensen. *Chem. Commun.* 1336–1337, 2002.
206. J Ross and JL Xiao. *Green Chem.* 4: 129–133, 2002.
207. S Gmouth, HL Yang, and M Vaultier. *Org. Lett.* 5: 2219–2222, 2003.
208. S Repichet, C Le Roux, N Roques, and J Dubac. *Tetrahedron Lett.* 44: 2037–2040, 2003.
209. CE Song, WH Shim, EJ Roh, and JH Choi. *Chem. Commun.* 1695–1696, 2000.
210. TM Trnka and RH Grubbs. *Acc. Chem. Res.* 34: 18–29, 2001.
211. A Furstner. *Angew. Chem. Int. Ed.* 39: 3012–3043, 2000.
212. DM Lynn, B Mohr, RH Grubbs, LM Henling, and MW Day. *J. Am. Chem. Soc.* 122: 6601–6609, 2000.
213. KJ Davis and D Sinou. *J. Mol. Catal. A: Chemical* 177: 173–178, 2002.
214. SJ Connon, M Rivard, M Zaja, and S Blechert. *Adv. Synth. Catal.* 345: 572–575, 2003.
215. RC Buijsman, E van Vuuren, and JG Sterrenburg. *Org. Lett.* 3: 3785–3787, 2001.
216. N Audic, H Clavier, M Mauduit, and JC Guillemin. *J. Am. Chem. Soc.* 125: 9248–9249, 2003.
217. QW Yao and YL Zhang. *Angew. Chem. Int. Ed.* 42: 3395–3398, 2003.
218. A Furstner, L Ackermann, K Beck, H Hori, D Koch, K Langemann, M Liebl, C Six, and W Leitner. *J. Am. Chem. Soc.* 123: 9000–9006, 2001.
219. BS Lane, M Vogt, VJ DeRose, and K Burgess. *J. Am. Chem. Soc.* 124: 11946–11954, 2002.
220. MC White, AG Doyle, and EN Jacobsen. *J. Am. Chem. Soc.* 123: 7194–7195, 2001.
221. K Sato, M Aoki, M Ogawa, T Hashimoto, and R Noyori. *J. Org. Chem.* 61: 8310–8311, 1996.
222. T Sakamoto and CJ Pac. *Tetrahedron Lett.* 41: 10009–10012, 2000.
223. MD Skowronska-Ptasinska, MLW Vorstenbosch, RA van Santen, and HCL Abbenhuis. *Angew. Chem. Int. Ed.* 41: 637–639, 2002.
224. BS Lane and K Burgess. *Chem. Rev.* 103: 2457–2473, 2003.
225. C Denis, K Misbahi, A Kerbal, V Ferrieres, and D Plusquellec. *Chem. Commun.* 2460–2461, 2001.
226. KA Srinivas, A Kumar, and SMS Chauhan. *Chem. Commun.* 2456–2457, 2002.
227. Z Li and CG Xia. *Tetrahedron Lett.* 44: 2069–2071, 2003.
228. CE Song and EJ Roh. *Chem. Commun.* 837–838, 2000.
229. B Wang, YR Kang, LM Yang, and JS Suo. *J. Mol. Catal. A: Chemical* 203: 29–36, 2003.
230. KH Tong, KY Wong, and TH Chan. *Org. Lett.* 5: 3423–3425, 2003.
231. GS Owens and MM Abu-Omar. *Chem. Commun.* 1165–1166, 2000.
232. GS Owens, A Durazo, and MM Abu-Omar. *Chem. Eur. J.* 8: 3053–3059, 2002.
233. PG Jessop, T Ikariya, and R Noyori. *Chem. Rev.* 99: 475–493, 1999.
234. G Musie, M Wei, B Subramaniam, and DH Busch. *Coord. Chem. Rev.* 219–221: 789–820, 2001.

22 Catalytic Reactions of Industrial Importance in Aqueous Media

Nan Jiang
Department of Chemistry, Tulane University

Chao-Jun Li
Department of Chemistry, McGill University

CONTENTS

22.1 INTRODUCTION

The increasing environmental consciousness of academia and industry has led to the search for more efficient and environmentally benign products and processes [1], which, generally, can be guided by the 12 principles of green chemistry [2]. The 12 principles are (1) prevent waste before created, (2) be atom-economical, (3) use benign substrates, (4) make benign products, (5) use less or benign solvent, (6) use less energy, (7) use renewables, (8) avoid protecting groups, (9) use catalysts, (10) make things biodegradable, (11) analyze in real time, and (12) be safe.

Catalysis is sometimes referred to as a "foundational pillar" of green chemistry [3], due to reduced energy requirements and simplified separations from increased selectivity. The outstanding achievements in asymmetric catalysis have been recognized in the year 2001 by awarding Nobel Prize in Chemistry for this area, since asymmetric catalysis provides selective synthesis of the desired product without contamination by its undesired and often toxic enantiomer.

Water is not only the typical solvent in physiological processes, but also a favored solvent of green chemistry [4]. The use of water in organometallic catalysis is a fruitful development of the last two decades due to the synthesis and application of water-soluble and water-tolerant ligands. So far, the most outstanding achievement in aqueous organometallic catalysis is the Ruhrchemie–Rhône Poulenc process for the industrial hydroformylation of propene [5], accounting for over 600,000 tons of C_4-products every year.

22.2 HYDROFORMYLATION OF OLEFINS BY AQUEOUS BIPHASIC CATALYSIS

Hydroformylation is a major industrial process that produces aldehydes from olefins and syngas (CO_2 and H_2) (Equation (22.1)). The reaction was discovered in 1938 by Roelen [6], who detected the formation of aldehydes in the presence of a cobalt catalyst. In the 1950s, it was found that rhodium was more active than cobalt, and the use of rhodium–phosphine complexes proved to be even more effective. In general, 65 years of hydroformylation saw five quantum leaps of organometallic catalysis: (1) heterogeneous cobalt catalysis; (2) homogeneous cobalt catalysis; (3) rhodium catalysis; (4) ligand-modified rhodium or cobalt catalysis; and (5) aqueous biphasic catalysis with water-soluble rhodium catalysts [7]. As for all homogeneously catalyzed reactions, the problem of separating the homogeneously dissolved catalyst is inherently disadvantageous, and only homogeneous biphasic catalysis allows the chance to separate products and catalyst just by simple phase-separation techniques.

$$\mathrm{R{-}CH{=}CH_2} + CO + H_2 \xrightarrow{\text{“catalyst”}} \mathrm{R{-}CH_2CH_2{-}CHO} + \mathrm{R{-}CH(CH_3){-}CHO} \qquad (22.1)$$

22.2.1 Homogeneous Biphasic Catalysis

So far, homogeneous transition-metal catalysis has been applied in production of bulk chemicals through oligomerization, coordination polymerization and hydroformylation of alkenes, and many other processes. However, a wider use of homogeneous catalysis is hampered by the difficulties of catalyst recovery or recycling and a simultaneous isolation of a catalyst-free product. Catalyst–product separation can easily be achieved in heterogeneous catalysis, but many attempts of immobilizing catalysts on various solids have failed to produce catalysts that are stable, active, selective, and economical enough for large-scale industrial applications (although some are successful). A more successful approach is the two-phase process of homogenous catalysis whose development began with Shell's SHOP process for the production of higher olefins by the oligomerization of ethene [8]. In all two-phase processes the homogeneous catalyst is dissolving in a liquid phase immiscible from the phase of reaction reactants and products [9]. Consequently, the catalysis can still be regarded as a homogeneous one with molecularly dispersed catalyst and substrates. However, by easy and well-known phase-separation techniques (simple mechanical decantation), the catalyst can be recovered and recycled in one of the two liquid phases (Figure 22.1).

Biphase procedures are still under intensive investigations with the recent introduction of the *fluorous* [10a], supercritical fluids [10b], and ionic liquids [10c] as one phase of the biphase systems. Nevertheless, the use of water as one of the immiscible phases is more economical and environmentally benign, compared with organic solvents [11]. The wider use of

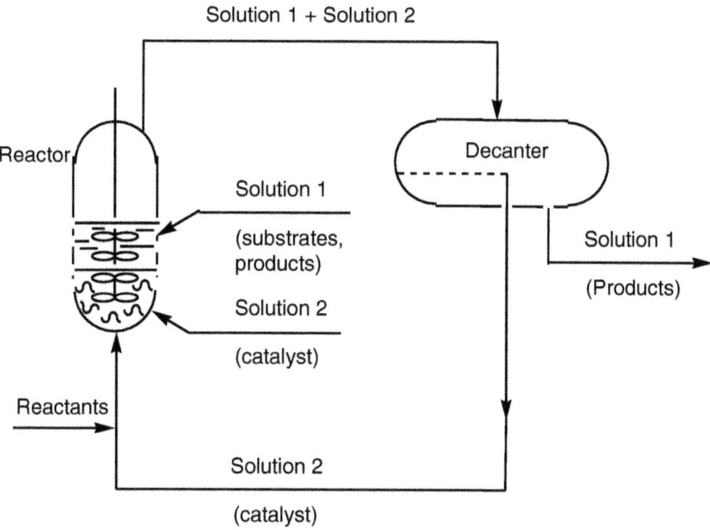

FIGURE 22.1 Principle of homogeneous biphasic catalysis.

water-soluble tertiary phosphine ligands played a crucial role on recent aqueous complex-catalyzed reactions. Of the several hundreds of water-soluble phosphine ligands investigated so far, sulfonated triphenylphosphine derivatives have attracted the most attention due to their relatively easy synthesis, good solubilities, and coordination properties in a wide range of pH value, in contrast to other water-soluble phosphines with amino, hydroxyl, or carboxylic functionalities. The sulfonation of triphenylphosphine with fuming sulfuric acid generates sulfonated derivatives with $—SO_3^-$ groups in *meta* positions, such as monosulfonated phosphine, $Ph_2P(C_6H_4$-*m*-$SO_3M)$ (*m*-TPPMS) [12a], bis-sulphonated phosphine $PhP(C_6H_4$-*m*-$SO_3M)_2$ (*m*-TPPDS) [12b], and tris-sulphonated phosphine $P(C_6H_4$-*m*-$SO_3M)_3$ (*m*-TPPTS) [12c], which are usually isolated as alkali metal salts ($M = Na^+$ or K^+) (Figure 22.2). *Ortho*- and *para*-sulfonated triphenylphosphines have also been used as water-soluble ligands in organometallic catalysis.

More often, rhodium is used as the central atom of complex catalysts for aqueous biphasic hydroformylation. For example, by using the water-soluble complex $RhH(CO)[Ph_2P(C_6H_4$-*m*-$SO_3Na)]_3$, RhH(CO)(*m*-TPPMS)$_3$, the hydroformylation can be carried out at >70°C [13]. However, with *m*-TPPMS as ligand, some leaching of rhodium into the organic phase was observed. In order to prevent the leaching of rhodium, the more water-soluble tris-sulfonated ligand, $P(C_6H_4$-*m*-$SO_3M)_3$ (*m*-TPPTS), was used [14]. A variety of terminal olefins were hydroformylated with this catalyst in high *n/iso* selectivity, affording the corresponding terminal aldehydes [15].

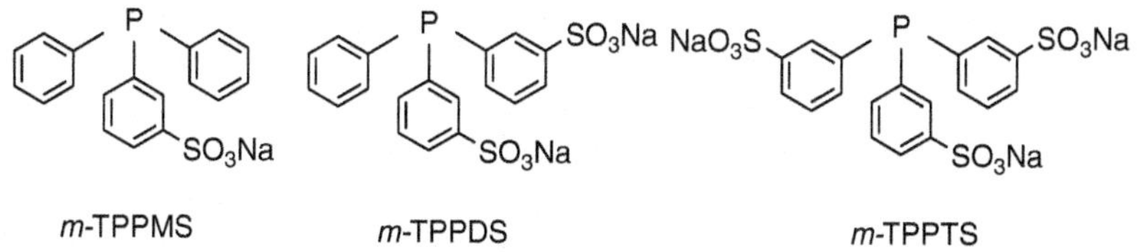

P SO$_3$Na *m*-TPPMS   P SO$_3$Na SO$_3$Na *m*-TPPDS   NaO$_3$S P SO$_3$Na SO$_3$Na *m*-TPPTS

FIGURE 22.2 Structures of water-soluble phosphine ligands.

22.2.2 Aqueous Biphasic Hydroformylation of Olefins

The state-of-the-art bulk production of *n*-butyraldehyde via aqueous biphasic hydroformylation of propylene catalyzed by RhH(CO)(*m*-TPPTS)$_3$ complex is known as the Ruhrchemie–Rhône Poulenc process (the RCH/RP process, Equation (22.2)) [16].

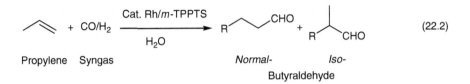

$$\text{Propylene} + \underset{\text{Syngas}}{CO/H_2} \xrightarrow[H_2O]{\text{Cat. Rh}/m\text{-TPPTS}} \underset{\textit{Normal-}}{R\text{–}CH_2CH_2\text{–}CHO} + \underset{\textit{Iso-}}{R\text{–}CH(CH_3)\text{–}CHO} \quad (22.2)$$

Butyraldehyde

Today, the RCH/RP process is successfully operated with the following characteristics: (1) high selectivity, affording almost exclusive linear aldehydes (*n/iso* > 96/4); (2) simple recycling of the homogeneous rhodium catalyst (immobilized by the water); and (3) excellent economics due to minimal losses of rhodium catalyst. Since the first unit went on stream in 1984, four plants at two locations produce over 600,000 tons of C$_4$-products per annum, which amounts for roughly 10% of the world capacity for the oxo synthesis (hydroformylation and hydrocarbonylation). The produced C$_4$-aldehydes are converted to either the corresponding C$_4$-alcohols or after aldolization and hydrogenation to 2-ethylhexanol.

Besides *m*-TPPTS, a number of other sulfonated phosphine ligands were also investigated and tested on a pilot plant scale. The most active and selective water-soluble catalyst for hydroformylation proves to be the system of rhodium and BINAS (BINAS = sulfonated NAPHOS, sulfonation degree between six and eight) [17], with high *n/iso* selectivity of 98/2 (*m*-TPPTS: >96/4) and around ten times higher reactivity than *m*-TPPTS. However, BINAS is far less stable than *m*-TPPTS, which is the major reason that only *m*-TPPTS has proven economically successful for industrial hydroformylation as far as handling, performance, and cost are concerned. Due to the sufficiently strong coordination of *m*-TPPTS to the rhodium metal center and high water-solubility of the complex, almost no rhodium leaching into the organic phase is observed ($<10^{-9}$ g Rh per kg *n*-butyraldehyde) [18].

In order to suppress side reactions of aldehydes, the RCH/RP process is usually operated at a pH-value between 5 and 6 for the best performance. A careful pH-control is necessary for better catalytic reactivity and selectivity. In summary, the major advantages of the Rh–*m*-TPPTS system are the simple, mild catalyst recovery and recycling, the catalyst-free products, and the high linear selectivity (>20).

With the present RCH/RP process, butene hydroformylation can also proceed with satisfying space time yields (STY). Compared with propylene, butene is less soluble in water, analogous to their higher homologs [19]. Thus, significantly lower reaction rate for butene than propylene is observed, and higher catalyst loadings are required to achieve satisfying STY for the manufacturing of valeric aldehydes. As a very selective catalyst, RhH(CO)(TPPTS)$_3$ does not catalyze 2-butene hydroformylation, and the 2-butene can easily be separated from the products in a stripping column. Consequently, a 12,000 tons unit was operated in 1995 to convert "raffinate-2" (a mixture of 1-butene and 2-butene) to C$_5$-aldehydes, which are mainly used to manufacture the corresponding alcohols and acids.

For higher olefins, the hydroformylation generally proceeds slowly and with low selectivity in the biphasic system, due to their poor solubility in water. Many modifications have been made to overcome the very low reaction rate. Chaudhari found that hydroformylation of 1-octene could reach turnover frequencies (TOF) of 180 to 800 h^{-1} (mol converted substrate/mol catalyst·h) with the system of Rh–TPPTS by using PPh$_3$ as promoter ligand [20], and these TOFs are comparable to the range desired for industrial processes (TOF > 200 h^{-1}). Monflier et al. reported a conversion of up to 100% and a high regioselectivity

(>95%) for the Rh-catalyzed hydroformylation of dec-1-ene in water, free of organic solvent, in the presence of partially methylated β-cyclodextrins (Equation (22.3)) [21].

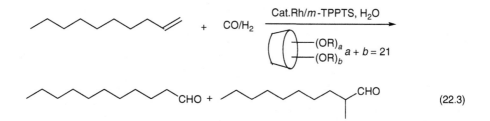

$$\text{dec-1-ene} + CO/H_2 \xrightarrow[\text{CD}-(OR)_a,\ (OR)_b,\ a+b=21]{\text{Cat.Rh}/m\text{-TPPTS},\ H_2O} \text{n-C}_{10}H_{21}CHO + \text{branched aldehyde (CHO)} \qquad (22.3)$$

On the other hand, Arhancet and coworkers developed the concept of supported aqueous phase catalysis (SAPC) [22]. A thin, aqueous film containing a water-soluble catalyst adheres to silica gel with high surface area, and the reaction occurs at the liquid–liquid interfaces. Through SAPC, the hydroformylation of very hydrophobic alkenes, such as octene or dicyclopentadiene, is possible with the water-soluble catalyst RhH(CO)(TPPTS)$_3$.

Hydroformylation of olefins containing a heteroatom functionality is also cumbersome since these substrates react slowly, require high catalyst loadings, and harsh reaction conditions, which leads to by-products, such as acetals, hemiacetals, imines, etc. [23]. Sheldon and coworkers reported that RhH(CO)(TPPTS)$_3$ catalyzed hydroformylation of *N*-allylacetamide in high yield, and the product could be used in a one-pot synthesis of *N*-acetyl-5-methoxytryptamine (melatonin) (Equation (22.4)) [24].

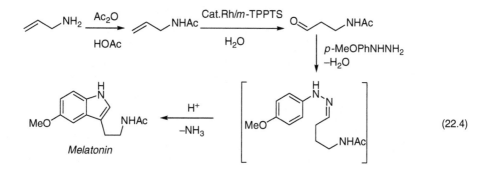

$$\text{allyl-}NH_2 \xrightarrow[HOAc]{Ac_2O} \text{allyl-NHAc} \xrightarrow[H_2O]{\text{Cat.Rh}/m\text{-TPPTS}} \text{OHC-(CH}_2)_3\text{-NHAc} \xrightarrow[-H_2O]{p\text{-MeOPhNHNH}_2} [\text{MeO-C}_6H_4\text{-NH-N=CH-(CH}_2)_3\text{-NHAc}] \xrightarrow[-NH_3]{H^+} \text{Melatonin} \qquad (22.4)$$

Besides rhodium, other transition metals, such as Pd, Ru, Co, or Pt, are also used occasionally. For example, Khan used the complex [Ru(EDTA)]$^-$ for the hydroformylation of hexene in water, and very high linearities were obtained for the C$_7$ aldehyde generated (98% to 100%) [25]. Methyl formate can also be used for hydroformylation instead of carbon monoxide and hydrogen [26].

Although unsuccessful, asymmetric hydroformylation in aqueous biphasic catalysis has also been investigated by several groups using styrene as substrate, and enantioselectivity of 34% ee was reported by Herrmann using chiral sulfonated NAPHOS ligand [27].

22.3 CATALYTIC HYDROGENATION IN AQUEOUS MEDIA

Catalytic hydrogenation is a very important process in organic synthesis [28], and the Wilkinson catalyst, RhCl(PPh$_3$)$_3$, has been the most widely employed since it was first introduced in 1960 [29]. So far, a large number of transition metals are more or less active in hydrogenation, but the most important are Rh, Ru, Ir, Pd, and recently Os [30]. Although catalytic hydrogenation involving aqueous media had been studied in the early 1960s in the presence of cobalt complex [31], the first successful hydrogenation in aqueous solution by a catalyst containing a water-soluble phosphine ligands was reported in 1975 [32].

22.3.1 Hydrogenation of C=C Bonds

Reaction of $RuCl_3 \cdot xH_2O$ and $RhCl_3 \cdot xH_2O$ with an excess of *m*-TPPMS generates $[RuCl_2(m\text{-}TPPMS)_2]_2$ and $RhCl(m\text{-}TPPMS)_3$ [33], which were both found to be very active in catalytic hydrogenation of olefins in aqueous media. For example, hydrogenation of crotonic and fumaric acids proceeded smoothly at elevated temperature (25°C to 60°C) with TOF of 100 to 700 h^{-1} (mol converted substrate/mol catalyst·h) [34], similar to those of olefin hydrogenations catalyzed by the related $RuCl_2(PPh_3)_3$ or $RhCl(PPh_3)_3$ complexes [35].

Larpent and coworkers used water-soluble complex $RhCl(m\text{-}TPPTS)_3$ for hydrogenation [36], and many olefins were hydrogenated with 100% conversion and complete selectivity on the C=C double bond. Various functionalities can survive the reaction conditions. Through ^{31}P NMR study, the active catalyst in a Rh–*m*-TPPTS system was found to involve the phosphine oxide, $OP(m\text{-}TPPTS)_3$, and this observation has been confirmed by further experiments in which no hydrogenation was observed without the presence of a sufficient amount of phosphine oxide [37]. Surprisingly, the hydrogenation of α,β-unsaturated aldehydes to saturated aldehydes proceeds without the involvement of *m*-TPPTS oxide [38].

An interesting observation is the selectivity switch for the rhodium-catalyzed hydrogenation of the diene when going from organic solvent to aqueous biphasic system (Equation (22.5)) [39]. This unusual selectivity could come from the coordination between the carboxylic group and the metal center.

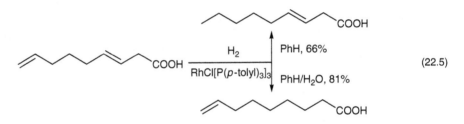

COOH

H_2 PhH, 66%

COOH (22.5)

$RhCl[P(p\text{-tolyl})_3]_3$ PhH/H_2O, 81%

COOH

As an interesting new alternative, Jacobson and coworkers reported a system consisting of water and supercritical CO_2. They observed a remarkable enhancement of the reaction rate with a number of special phase-transfer reagents (Equation (22.6)) [40].

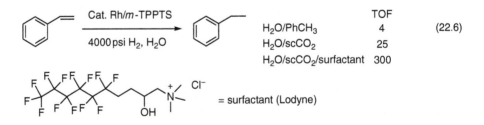

Cat. Rh/*m*-TPPTS

4000 psi H_2, H_2O (22.6)

	TOF
$H_2O/PhCH_3$	4
$H_2O/scCO_2$	25
$H_2O/scCO_2$/surfactant	300

F F F F F F F F F F F OH Cl^- = surfactant (Lodyne)

22.3.2 Hydrogenation of C=O and C=N Bonds

Various hydrogenations of C=O and C=N functionalities in aqueous media are known. Especially, ruthenium and rhodium are the active metals in these reactions. Most complexes are prepared *in situ* by the combination of the metal salt and a water-soluble phosphine ligand.

The combination of ruthenium and *m*-TPPTS in aqueous media shows good catalyst activity in the hydrogenation of aldehydes. Especially, the hydrogenation of unsaturated aldehydes has been well examined, and the combination of ruthenium and sulfonated phosphines seems to be successful to hydrogenate the carbonyl bond and not the olefinic double bond (Equation (22.7)) [41]. This is a rather unusual finding since there are only a few homogeneous catalysts that show good selectivities for the hydrogenation of an aldehyde in the presence of an olefinic bond. Basset and coworkers reported that enhanced reactivity was observed with the addition of NaI, which would assist the rapid formation of a metal–carbon bond [42].

$$\text{(CH}_3)_2\text{C=CHCHO} \xrightarrow[H_2]{[HRuCl(m\text{-TPPTS})_3]} \text{(CH}_3)_2\text{C=CHCH}_2\text{OH} \quad 99\% \tag{22.7}$$

Joó and coworkers reported that the pH value of the solution is decisive for the ruthenium-catalyzed selective hydrogenation of unsaturated aldehydes. Reaction of $[RuCl_2(m\text{-}TPPMS)_2]_2$ with the excess *m*-TPPMS under hydrogen atmosphere in aqueous media was studied in detail through both pH potentiometry and simultaneous ^1H and ^{31}P NMR investigations [43]. The experiments revealed the formation of three Ru(II) hydrides (Equation (22.8) to Equation (22.10)), governed by the concentration of proton. Thus, the distribution of ruthenium

$$[RuCl_2(m\text{-TPPMS})_2]_2 + 2H_2 \rightleftharpoons 2\,[HRuCl(m\text{-TPPMS})_2]_2 + 2H^+ + 2Cl^- \tag{22.8}$$

$$[RuCl_2(m\text{-TPPMS})_2]_2 + 2H_2 \xrightleftharpoons{2\,m\text{-TPPMS}} 2\,[HRuCl(m\text{-TPPMS})_3]_2 + 2H^+ + 2Cl^- \tag{22.9}$$

$$[HRuCl(m\text{-TPPMS})_3]_2 + 2H_2 \xrightleftharpoons{2\,m\text{-TPPMS}} 2\,[RuH_2(m\text{-TPPMS})_4]_2 + 2H^+ + 2Cl^- \tag{22.10}$$

in the three hydrides is governed by the pH-value of the aqueous media. With an excess of *m*-TPPMS, $[HRuCl(m\text{-}TPPMS)_2]_2$ is a minor species, whereas $HRuCl(m\text{-}TPPMS)_3$ and $RuH_2(m\text{-}TPPMS)_4$ are virtually exclusively generated in strongly acidic and in strongly basic solutions, respectively. As established earlier [34], the active catalyst for olefin hydrogenation is $HRuCl(m\text{-}TPPMS)_2$, generated from *m*-TPPMS dissociation of $HRuCl(m\text{-}TPPMS)_3$ in acidic conditions, and therefore, the rate of olefin hydrogenation decreases with increasing concentration of *m*-TPPMS. Conversely, $RuH_2(m\text{-}TPPMS)_4$ was found to be a selective catalyst for the hydrogenation of aldehydes. The obvious reason for this selectivity comes from the coordinative saturation of $RuH_2(m\text{-}TPPMS)_4$, which prevents the coordination between a olefinic bond and the metal center but allows the intermolecular nucleophilic hydride transfer to reduce aldehydes. Aldehyde hydrogenation catalyzed by the coordinatively saturated Ru species is also proved by the lack of inhibition by excess *m*-TPPMS; in fact, excess phosphine increases the reaction rate, probably due to its surfactant effect. Thus, selective hydrogenation of unsaturated aldehydes can be switched from C=O to C=C reduction by controlling pH-value of the aqueous media (Equation (22.11)).

$$\text{PhCH=CHCHO} \xrightarrow[m\text{-TPPMS},\ H_2]{[RuCl_2(m\text{-TPPMS})_2]_2} \begin{cases} \text{pH} = 3: \ \text{PhCH}_2\text{CH}_2\text{CHO} \\ \text{pH} = 9: \ \text{PhCH=CHCH}_2\text{OH} \end{cases} \tag{22.11}$$

Ketone reduction in aqueous media can be achieved mainly by rhodium, and to a lesser extent, by ruthenium and iridium in combination with water-soluble bipyridyl ligands. Lau and coworkers carried out ketone hydrogenation with up to 90% yield in a water–THF biphasic system [44]. Penicaud and coworkers obtained yields up to 99% of aryl methyl carbinols in rhodium- or iridium-catalyzed hydrogenation in water [45].

In contrast to the hydrogenation of C=O bonds in aqueous media there exist only few reports on imine hydrogenation. Beller and coworkers reported the iridium-catalyzed hydrogenation of terminal imines (intermediates from rhodium-catalyzed hydroformylation of terminal alkenes to the corresponding aldehydes and followed by conversion with ammonia to imines) to synthesize primary amines, and the overall selectivity of the primary amine was 91% with *m*-TPPTS as the ligand (Equation (22.12)) [46].

R⌒⟋ —(Rh/*m*-TPPTS; CO/H_2)→ R⌒⌒CHO —(Ir/*m*-TPPTS; R^1R^2NH, H_2)→ R⌒⌒⌒NR^1R^2 (22.12)

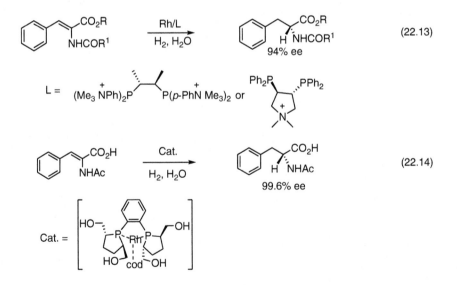

22.3.3 Asymmetric Hydrogenation

With the use of chiral phosphine ligands, hydrogenation of prochiral alkenes can afford optically active products. Similarly, asymmetric hydrogenation in aqueous media with water-soluble chiral catalysts has been intensively investigated with a variety of substrates. For example, by using a water-soluble chiral Rh catalyst, up to 94% ee (enantiometric excess) was obtained for hydrogenation of an α-acetamidoacrylic ester (Equation (22.13)) [47]. A real milestone in aqueous enantioselective hydrogenation was achieved with 99.6% ee by Holz and coworkers (Equation (22.14)) [48].

Ph–CH=C(CO_2R)($NHCOR^1$) —(Rh/L; H_2, H_2O)→ Ph–CH_2–CH(CO_2R)($NHCOR^1$), 94% ee (22.13)

L = $(Me_3\overset{+}{N}Ph)_2P$–CH($CH_3$)–CH($CH_3$)–$P(p\text{-}Ph\overset{+}{N}Me_3)_2$ or 3,4-bis(PPh_2)-N,N-dimethylpyrrolidinium (Ph_2P, PPh_2, $\overset{+}{N}$)

Ph–CH=C(CO_2H)(NHAc) —(Cat.; H_2, H_2O)→ Ph–CH_2–CH(CO_2H)(NHAc), 99.6% ee (22.14)

Cat. = [Rh(cod)(1,2-bis(phospholanyl)benzene bearing HO, OH hydroxymethyl groups)]

In a biphasic system of ionic liquid (butylmethylimidazolium hexafluorophosphate) and water, the high enantioselectivity of 92% ee was also achieved for the hydrogenation of α-methylcrotonic acid catalyzed by a ruthenium–BINAP complex (Equation (22.15)) [49].

High enantioselectivities have also been achieved in asymmetric hydrogenation of the prochiral imines and ketones. Especially, ruthenium–BINAP complexes are active hydrogenation catalysts for the ketones, and up to 94% ee was achieved for the hydrogenation of ethyl acetylacetate [50]. By using a water-soluble Rh catalyst the hydrogenation of a prochiral imine affords a product with up to 96% ee (Equation (22.16)) [51].

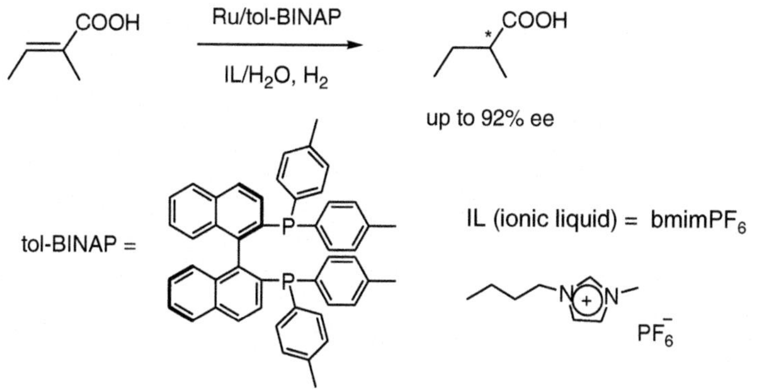

COOH — Ru/tol-BINAP, IL/H_2O, H_2 → COOH (22.15)

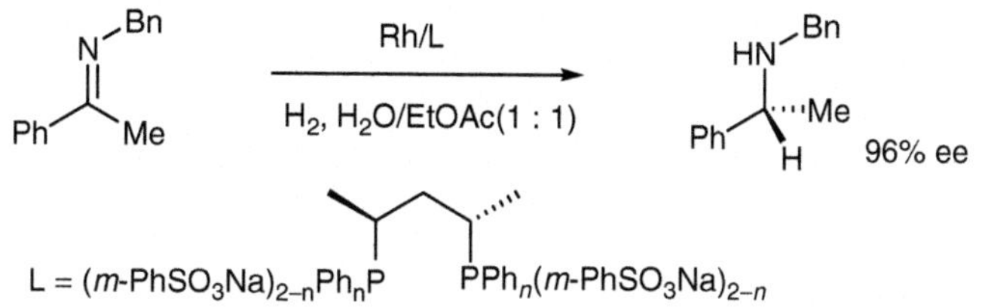

up to 92% ee

tol-BINAP =

IL (ionic liquid) = bmimPF_6

PF_6^-

Ph(Me)C=N–Bn — Rh/L, H_2, H_2O/EtOAc(1 : 1) → HN–Bn, Ph, Me, H (22.16)

96% ee

L = $(m\text{-PhSO}_3\text{Na})_{2-n}\text{Ph}_n\text{P}$ $\text{PPh}_n(m\text{-PhSO}_3\text{Na})_{2-n}$

However, the optical yield of the product for asymmetric hydrogenation in aqueous media is generally lower than those obtained in organic solvents, especially methanol [52]. The enantioselectivity often decreases with an increase in water content. Amrani and Sinou [53] explained the decrease of enantioselectivity with the increased water content by the use of Halpern's model [54] for asymmetric hydrogenation of (*Z*)-α-acetamidocinnamic acid (Figure 22.3). With two enantiomers generated from two different pathways, the enantiometric excess is determined by the transition-state energy difference. The lower enantiometric excess in water could come from the smaller energy difference between the two transition states.

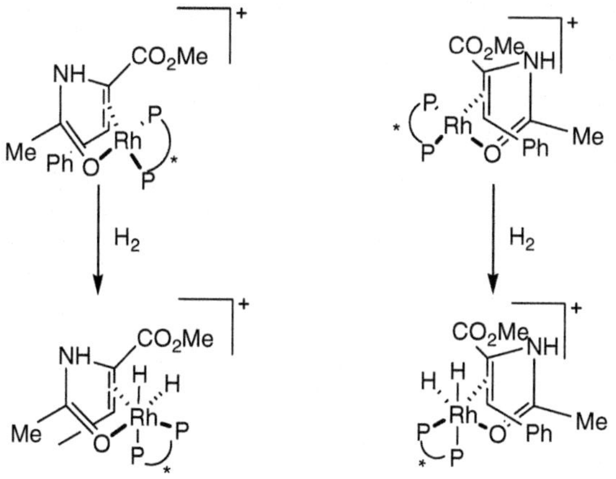

FIGURE 22.3 Halpern's model for asymmetric hydrogenation

Surprisingly, Toth and coworkers [55] found that the enantioselectivity in water was better for hydrogenating (*Z*)-α-acetamidocinnamic acid and (*Z*)-α-benzoylamidocinnamic acid and their methyl esters (95% ee) than under biphasic condition (67% ee) with DIOP, BDPP, and modified CHIRAPHOS as ligands.

In a biphasic system of ethyl acetate and D_2O, up to 75% deuterium is incorporated in the reduced product [56]. This result indicates that the role of water in catalytic hydrogenation is not only as solvent. Research on asymmetric hydrogenation in aqueous media is still intensively investigated.

22.3.4 Transfer Hydrogenation

Due to the high water solubility, 2-propanol and formates are frequently used as the hydrogen sources for the catalytic transfer hydrogenation in aqueous media. Various aldehydes were found to undergo reduction to the corresponding alcohols by hydrogen transfer from sodium formate catalyzed by ruthenium, rhodium, and iridium (with excess *m*-TPPMS) [57].

By using $[RuCl_2(m\text{-TPPMS})_2]_2$ together with an excess amount of *m*-TPPMS, TOF for transfer hydrogenation were generally over $100\,h^{-1}$. Most notably, hydrogen transfer hydrogenation of all unsaturated aldehydes afforded unsaturated alcohols with 100% selectivity. However, $RhCl(m\text{-TPPMS})_3$ efficiently catalyzed the formation of saturated aldehydes, showing negligible activity in hydrogenation of the aldehydes (Equation (22.17)).

Ph CH₂OH ← [RuCl₂(m-TPPMS)₂]₂ / HCOONa, H₂O, m-TPPMS — Ph CHO — RhCl(m-TPPMS)₃ / HCOONa, H₂O → Ph CHO (22.17)

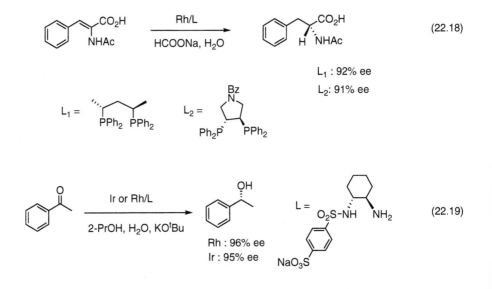

Silanes can also be used as hydrogen sources for the transfer hydrogenation of alkenes and alkynes in water. Tour and coworkers [58] reported that triethoxysilane reduced C=C unsaturated bonds to saturation with palladium acetate as catalyst. For the alkyne hydrogenation, the reactions can be stopped at the alkene stage with carefully controlled conditions.

The asymmetric transfer hydrogenation of olefinic substrates is also successful in aqueous media. Rocha Gonsalves and coworkers [59] reported that up to 92% ee for the transfer hydrogenation of (*Z*)-α-acetamido-cinnamic acid was obtained with the system of formic acid and sodium formate catalyzed by chiral rhodium complexes (Equation (22.18)). Thorpe and coworkers [60] reported the synthesis of enantiomerically enriched alcohols by transfer hydrogenation with 2-propanol catalyzed by rhodium or iridium combined with water-soluble chiral aminosulfonamide ligands, and up to 96% ee was reached for the hydrogenation of numerous methyl aryl ketones (Equation (22.19)).

CO₂H, NHAc — Rh/L / HCOONa, H₂O → CO₂H, H NHAc (22.18)

L₁ : 92% ee
L₂: 91% ee

L₁ = PPh₂ PPh₂ L₂ = Bz N, Ph₂P PPh₂

O — Ir or Rh/L / 2-PrOH, H₂O, KOᵗBu → OH (22.19)

Rh : 96% ee
Ir : 95% ee

L = O₂S-NH NH₂, NaO₃S

22.4 OXIDATION IN AQUEOUS MEDIA

Oxidation is a central process for converting petroleum-based materials to useful chemicals of a higher oxidation state. The use of aqueous media in oxidations is mainly due to the convenience of readily available aqueous reagent solution. In some cases, however, the use of aqueous media has some special effect in oxidations.

22.4.1 Oxidation of Olefins

22.4.1.1 Epoxidation

In the natural environment of water and air, biological oxygenations of organic compounds proceed with cytochrome P-450. One of the most important oxygenations by P-450 is alkene epoxidation. Recently, extensive studies have been carried out in synthesizing water-soluble matalloporphyrins as catalysts for epoxidation and other oxidations in aqueous media to mimic the properties of P-450. With the use of these water-soluble metalloporphyrins, alkene epoxidations can be carried out with a variety of oxidizing reagents, such as PhIO, NaClO, O_2, H_2O_2, ROOH, and $KHSO_5$, which has been reviewed in detail [61]. For example, carbamazepine was epoxidzed with $KHSO_5$ catalyzed by water-soluble iron and manganese porphyrins (**1** and **2**) in water (Equation (22.20)) [62].

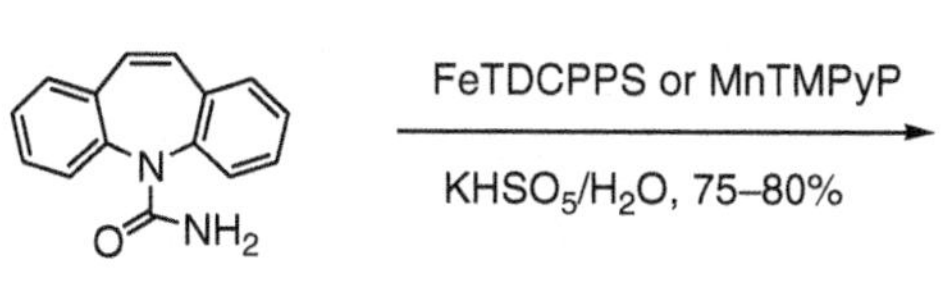

FeTDCPPS or MnTMPyP

$KHSO_5/H_2O$, 75–80%

(22.20)

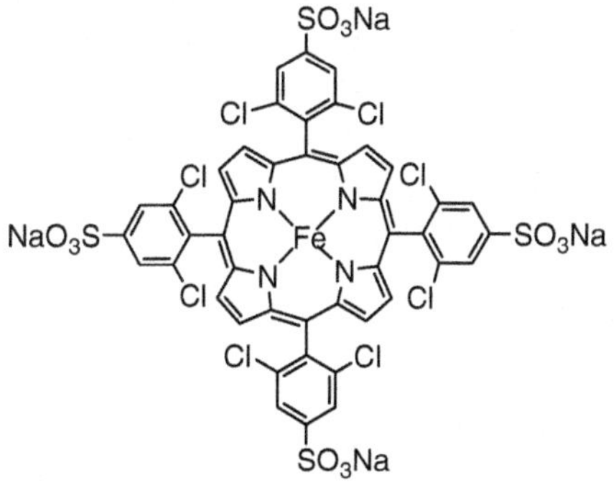

1 FeTDCPPS

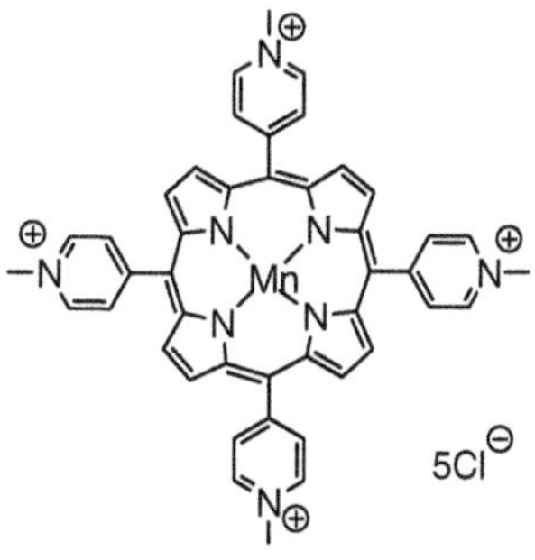

2 MnTMPyP

Alkene epoxidation can also be accomplished with a variety of organic peroxy acids or related reagents such as peroxy carboximidic acid, RC(NH)OOH, which is readily available through *in situ* oxidation of a nitrile with hydrogen peroxide. For example, epoxidation of alkenes with *m*-chloroperoxybenzoic acid in water at room temperature affords the epoxides in high yields [63]. With monoperoxyphthalic acid (MPPA) as oxidizing reagent and cetyltrimethylammonium hydroxide (CTAOH) as base to control the pH-value of the aqueous medium, highly regioselective epoxidation of allyl alcohols in the presence of other olefinic bond was reported (Equation (22.21)) [64].

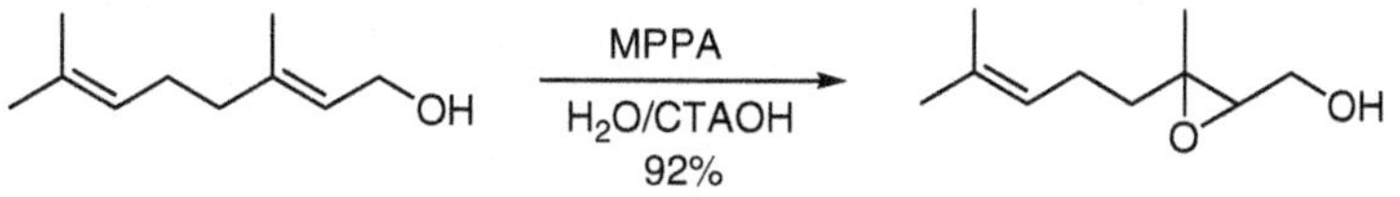

MPPA

H_2O/CTAOH

92%

(22.21)

Epoxidation of alkenes with a conjugated electron-withdrawing groups proceeds only very slowly or not at all with peroxy acids or alkyl peroxides. However, this epoxidation can be achieved with hydrogen peroxide under basic biphasic conditions, known as the *Weitz–Scheffer epoxidation* (Equation (22.22)) [65]. It has been applied successfully to many α,β-unsaturated aldehydes, ketones, nitriles, esters, sulfones, and other compounds. The reaction is first-order for both unsaturated ketones and hydrogen peroxide anion (HOO^-) through a Michael-type addition of the hydrogen peroxide anion to the conjugated system followed by ring closure of the intermediate enolate with expulsion of OH^-. The epoxidation of electron-deficient olefins can also be carried out by using a catalytic amount of sodium tungstate together with hydrogen peroxide [66].

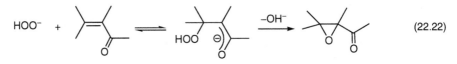

HOO⁻ + [α,β-unsaturated ketone] ⇌ [HOO-adduct enolate] —(−OH⁻)→ [epoxy ketone] (22.22)

22.4.1.2 Dihydroxylation

The *syn* dihydroxylation of alkenes can also be accomplished using aqueous media in a catalytic amount of osmium tetraoxide together with chlorate salts as the primary oxidants. This catalytic oxidation is usually performed in water–THF biphasic systems (Equation (22.23)) [67].

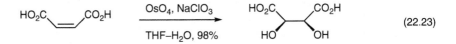

HO_2C–CH=CH–CO_2H —(OsO_4, $NaClO_3$; THF–H_2O, 98%)→ HO_2C–CH(OH)–CH(OH)–CO_2H (22.23)

Hydrogen peroxide can also be used as the primary oxidizing reagent for the *syn* dihydroxylation of alkenes with a catalytic amount of osmium tetroxide. Originally developed by Milas, the oxidation can be carried out with aqueous hydrogen peroxide combined with organic solvents such as acetone or diethyl ether [68]. Similarly, allyl alcohol can be quantitatively hydroxylated in water (Equation (22.24)) [69].

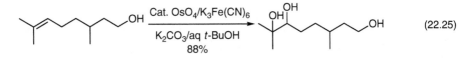

CH₂=CH–CH₂OH —(OsO_4, H_2O_2; H_2O, 100%)→ HO–CH₂–CH(OH)–CH₂–OH (22.24)

Minato and coworkers reported that $K_3Fe(CN)_6$ in the presence of K_2CO_3 in aqueous *tert*-butyl alcohol provides a powerful system for the osmium-catalyzed dihydroxylation of olefins (Equation (22.25)) [70]. This system overcomes the disadvantages of over-oxidation and low reactivity on hindered olefins compared with the previous processes.

[alkenol]–OH —(Cat. $OsO_4/K_3Fe(CN)_6$; K_2CO_3/aq *t*-BuOH, 88%)→ [diol]–OH (22.25)

The *anti* dihydroxylation of alkenes can proceed with hydrogen peroxide by a catalytic amount of some oxides, notably tungstic oxide (WO_3) or selenium dioxide (SeO_2) (Equation (22.26)) [69]. The WO_3 process is usually performed at elevated temperatures (50°C to 70°C) in aqueous hydrogen peroxide. Aqueous biphasic systems have also been used successfully to increase the solubilities of the alkenes. It has been proposed that peroxytungstic acid (H_2WO_5) is involved in the *syn* dihydroxylation of alkenes for the combination of the hydrogen peroxide–tungstic oxide.

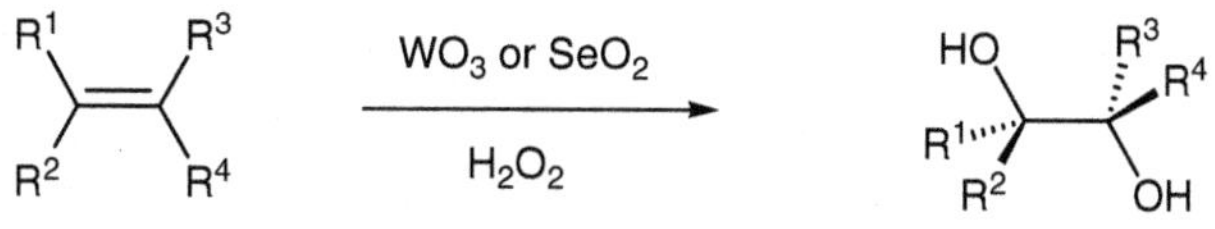

(22.26)

22.4.1.3 The Wacker Oxidation

The oxidation of ethylene to acetaldehyde by palladium chloride in water has been known since the 19th century [71]. However, the oxidation requires the use of a stoichiometric amount of $PdCl_2$, generating Pd(0) deposit at the same time. In the late 1950s, Smidt [72] of Wacher Chemie discovered that by using $CuCl_2$, Pd(0) can be reoxidized back to Pd(II) before it deposits out. The $CuCl_2$ itself is reduced to cuprous chloride (Cu^ICl), which is air-sensitive and readily reoxidized back to Cu(II) (Scheme 22.1). This oxidation, known as the Wacker oxidation, is now one of the most important transition-metal-catalyzed reactions in industry.

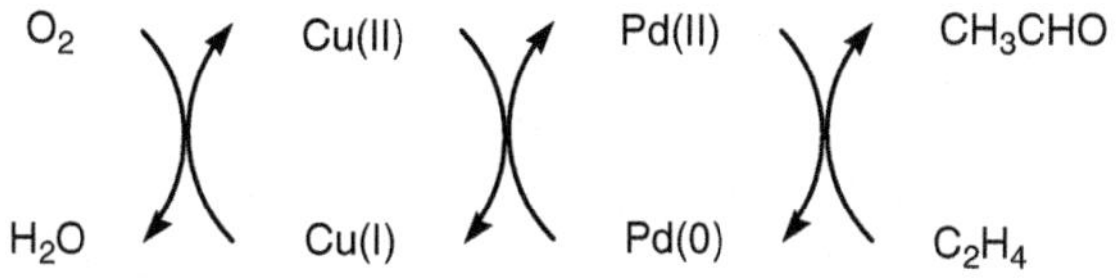

Scheme 22.1

The mechanism for the alkene oxidation was originally proposed to involve a *cis* transfer of OH to a metal-bound alkene [73]. However, stereochemical studies carried out by Bäckvall [74] and by Stille [75] proved the involvement of a *trans* addition (Scheme 22.2).

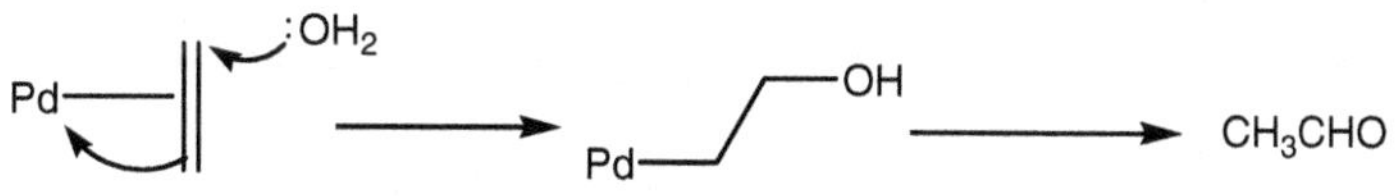

Scheme 22.2

For the oxidation of internal olefins, high regioselectivity can sometimes be observed. For example, oxidation of β,γ-unsaturated esters in aqueous dioxane or THF can be achieved selectively to generate γ-ketoesters (Equation (22.27)) [76]. In many cases, the Wacker oxidation can be used very efficiently in transforming olefins to ketones, and consequently, the olefins become masked ketones. An example of this application is the synthesis of (+)-19-nortestosterone carried out by Tsuji (Equation (22.28)) [77].

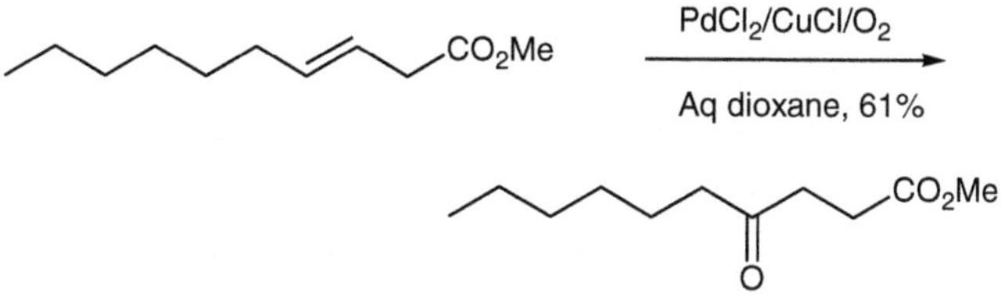

(22.27)

Although commercially successful, this process has some drawbacks. For example, high concentration of copper salts (ca. 1 *M*) and chlorides (ca. 2 *M*) are needed to achieve favorable redox potential of the Pd^{II}/Pd^0 and Cu^{II}/Cu^I couples and to solubilize Cu^I as $Cu^ICl_2^-$. Acid (HCl) is also required to prevent deposition of atomic palladium (Scheme 22.3). Thus, the reaction system is highly corrosive, and the formation of chlorinated by-products is often observed.

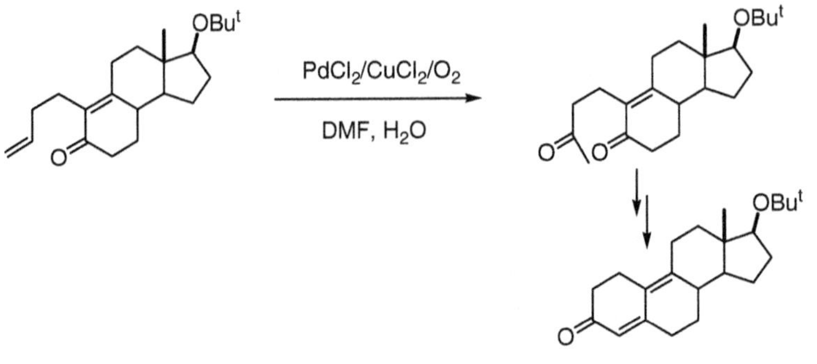

$$\text{(steroid with OBu}^t\text{, allyl side chain, ketone)} \xrightarrow[\text{DMF, } H_2O]{PdCl_2/CuCl_2/O_2} \text{(diketone)} \rightarrow \text{(enone, OBu}^t) \quad (22.28)$$

$$Pd^{II}Cl_4^{2-} + H_2O + R\text{CH=CH}_2 \longrightarrow Pd^0 + 2\,H^+ + 4\,Cl^- + RC(O)CH_3 \quad (1)$$

$$Pd^0 + Cu^{II}Cl^2 + 4\,Cl^- \longrightarrow Pd^{II}Cl_4^{2-} + 2\,Cu^ICl^- \quad (2)$$

$$2\,Cu^ICl_2^- + 2\,H^+ + 0.5\,O_2 \longrightarrow 2\,Cu^{II}Cl_2 + H_2O \quad (3)$$

$$R\text{CH=CH}_2 + 0.5\,O_2 \longrightarrow RC(O)CH_3 \quad (4)$$

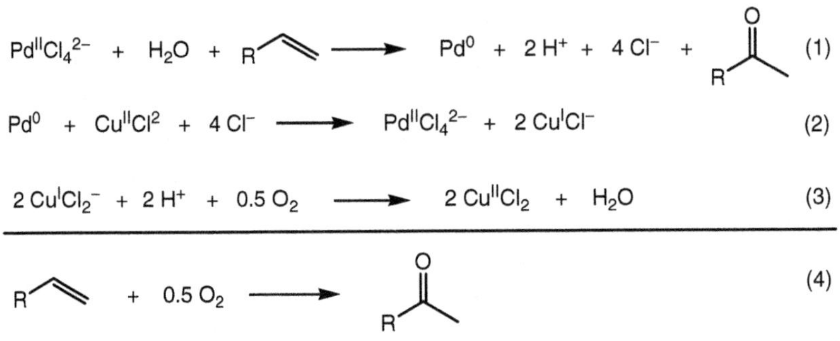

Scheme 22.3

Sheldon and coworkers [78] developed a copper- and chloride-free Wacker oxidation of terminal olefins to corresponding ketones in aqueous biphasic system at 100°C by using water-soluble ligand, bathophenanthroline disulfonate (PhenS), to stabilize Pd(0) (Equation (22.29)).

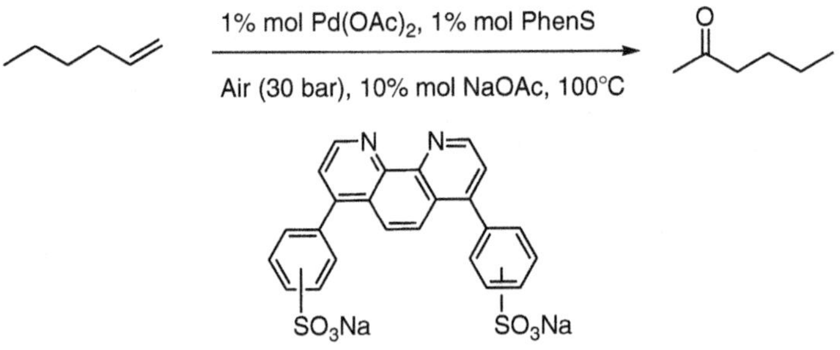

$$\text{1-hexene} \xrightarrow[\text{Air (30 bar), 10\% mol NaOAc, 100°C}]{\text{1\% mol Pd(OAc)}_2\text{, 1\% mol PhenS}} \text{2-hexanone} \quad (22.29)$$

PhenS

22.4.2 Oxidation of Alcohols

Oxidation of alcohols to carbonyl compounds is a fundamental process in synthetic organic chemistry [79]. Alcohol oxidation is typically performed with stoichiometric inorganic oxidants, notably chromium(IV) reagents [80], which are not only relatively expensive, but also generate heavy-metal waste and are usually performed in chlorinated solvents. Thus, using dioxygen (O_2) or hydrogen peroxide as oxidant and water as solvent provides the "cleanest" procedure for oxidation of alcohols, producing water as the only by-product [81].

By using the (PhenS)Pd^{II} complex in a biphasic system with no additional solvent and air as oxidation, a green catalytic oxidation of alcohol has been achieved, with secondary alcohols to be oxidized into ketones and primary alcohols into aldehydes or carboxylic acids (Equation (22.30)) [82]. Allylic alcohols can be oxidized selectively without competing (Wacker) oxidation of the double bond.

$$R^1R^2CH{-}OH + O_2 \xrightarrow[\text{Air (30 bar), 10\% mol NaOAc; } H_2O\text{, 100°C}]{\text{0.5\% mol Pd(OAc)}_2\text{/PhenS}} R^1R^2C{=}O + H_2O \qquad (22.30)$$

Surprisingly, when the alcohol group is further away from the olefinic double bond, the substrate undergoes selective oxidation in the double bond, even though alcohol is more easily oxidized than the olefinic double bond. A possible explanation is due to the olefinic double bond's coordination to the palladium(II) complex (Equation (22.31)).

$$CH_3CH(OH)(CH_2)_8CH{=}CH_2 + O_2 \xrightarrow[H_2O\text{, 100°C}]{\text{Pd/PhenS, air (30 bar)}} CH_3CH(OH)(CH_2)_8C(O)CH_3 \qquad (22.31)$$

Noyori and coworkers [83] found that aqueous 30% hydrogen peroxide can effectively oxidize alcohols to afford corresponding carbonyl compounds in high yield with a catalytic amount of sodium tungstate (Na_2WO_4) and methyl-trioctylammonium hydrogensulfate [$CH_3(n\text{-}C_8H_{17})_3N^+ HSO_4^-$] ($Q^+ HSO_4^-$). The oxidation is run at or below 90°C with only a slight excess of H_2O_2. The oxidation of secondary alcohols affords ketones, while the oxidation of primary alcohols gives carboxylic acids via an aldehyde intermediate (Equation (22.32) and Equation (22.33)). A remarkable advantage of this oxidation is its extremely high turnover number. For example, the oxidation of 2-octanol with a ratio of 200,000 (alcohol:H_2O_2:Na_2WO_4:$Q^+HSO_4^-$ = 200,000:300,000:1:100) gave a turnover number of 77,000 (40% yield), two orders of magnitude higher than previously reported H_2O_2 oxidations.

$$CH_3CH(OH)\text{-}n\text{-}C_6H_{13} \xrightarrow[\text{90°C, 4 h; 95\%}]{\text{1.1 eq. 30\% } H_2O_2\text{; 0.2\% mol } Na_2WO_4\text{; 0.2\% mol } Q_+HSO_4^-} CH_3C(O)\text{-}n\text{-}C_6H_{13} \qquad (22.32)$$

$$n\text{-}C_7H_{15}CH_2OH \xrightarrow[\text{90°C, 4 h; 80\%}]{\text{1.1 eq. 30\% } H_2O_2\text{; 2\% mol } Na_2WO_4\text{; 2\% mol } Q^+HSO_4^-} n\text{-}C_7H_{15}COOH \qquad (22.33)$$

22.5 CONCLUSION

As shown from this brief review, catalysis from aqueous media has great potentials, which can be summarized as: (1) economically, the low price of water as solvent and simplified product–catalyst separation in many processes (by using "water-immobilizing" catalysts); (2) environmentally, reducing the use of organic solvents, especially chlorinated solvents; and (3) unique properties in many reactions. It can be expected that many industrial processes based on catalysis in aqueous media will be developed in the future.

ACKNOWLEDGMENT

We are grateful to NSF and NSF-EPA joint program for a sustainable environment for support of our research. C.J.L. is a Canada Research Chair (Tier I) in Green Chemistry at McGill University.

REFERENCES

1. PT Anastas, LG Heine, and TC Williamson. Green chemical syntheses and processes: Introduction. In *Green Chemical Syntheses and Processes*, PT Anastas, LG Heine, and TC Williamson, ed. American Chemical Society: Washington, D.C., 2000.
2. PT Anastas and JC Warner. *Green Chemistry: Theory and Practice*. Oxford University Press: New York, 1998, 30pp.
3. PT Anastas, MM Kirchhoff, and TC Williamson. *Appl Catal A: Gen* 2001;221(1–2):3–13.
4. (a) P Tundo, PT Anastas, SSC Black, J Breen, T Collins, S Memoli, J Miyamoto, M Polyakoff, and W Tumas 4(b) CJ Li, TH Chan Organic Reactions in Aqueous media, Johnwiley: New York, 1997. *Pure Appl Chem* 2000;72:1207–1228.
5. B Cornils and WA Herrmann, eds. *Aqueous-Phase Organometallic Catalysis. Concepts and Applications*. Wiley-VCH: Weinheim, 1998; B Cornils and WA Herrmann, eds. *Applied Homogeneous Catalysis with Organometallic Compounds*. VCH Publishers: Weinheim, 1996.
6. D Roelen. DE 84584, 1938, Ruhrchemie.
7. J Falbe, ed. *New Syntheses with Carbon Monoxide*, chapter 1. Springer: Berlin, 1980; B Cornils. *J Mol Catal A* 1999;143:1.
8. B Cornils and WA Herrmann, eds. *Applied Homogeneous Catalysis with Organometallic Compounds*. VCH Publishers: Weinheim, 1996.
9. F Joó. *Aqueous Organometallic Catalysis*. Kluwer: Dordrecht, 2001; B Cornils and WA Herrmann, eds. *Aqueous-Phase Organometallic Catalysis. Concepts and Applications*. Wiley-VCH: Weinheim, 1998.
10. (a) IT Horváth. *Acc Chem Res* 1998;31:641–650; (b) PG Jessop, T Ikariya, and R Noyori. *Chem Rev* 1999;99:475–493; (c) R Sheldon. *Chem Commun* 2001:2399–2407.
11. F Joó and Z Tóth. *J Mol Catal* 1980; 8: 369–383; WA Herrmann and CW Kohlpaintner. *Angew Chem Int Ed Engl* 1993;32:1524–1544; *Angew Chem* 1993;105:1588–1609; F Joó and Á Kathó. *J Mol Catal A: Chem* 1997;116:3–26.
12. (a) F Joó, J Kovács, Á Kathó, ACS Bényei, T Decuir, and DJ Darensbourg. *Inorg Synth* 1998;32:1–8. (b) WA Herrmann and CW Kohlpaintner. *Inorg Synth* 1998;32:8–25.
13. AF Borowski, DJ Cole-Hamilton, and G Wilkinson. *Nouv J Chim* 1998;2:137.
14. E Kuntz. U.S. Patent 4,248,802, Phone-Poulenc Ind, 1981; *Chem Abstr* 1977;87:101944n.
15. J Jenick. Fr Patent 2,478,078 to Rhone-Poulenc Ind (03-12-1980); E Kuntz. Fr Patent 2,349,562 to Rhone-Poulenc Ind (04-29-1976).
16. B Cornils and J Falbe. In 4th International Symposium on Homogeneous Catalysis. Leningrad, September 1984, Proceedings, pp. 487; B Cornils and EG Kuntz. *J Organomet Chem* 1995;502:177; O Wachsen, K Himmler, and B Cornils. *Catal Today* 1998;1267:1.
17. WA Herrmann, H Bahrmann, and W Konkol. *J Mol Catal* 1992;73:191; WA Herrmann, CW Kohlpaintner, RB Manetsberger, H Bahrmann, and H Kottmann. *J Mol Catal* 1995;97:65; H Bahrmann, HW Back, and WA Hermann. *J Mol Catal* 1997;116:49.
18. EG Kuntz. CHEMTECH 17:570, 1987.
19. B Cornils. *Org Proc Res Dev* 1998;2:121.
20. RV Chaudhari, RM Bhanage, RM Deshpande, and H Delmas. *Nature* 1995;373:501–503.
21. E Monflier, G Fremy, Y Castanet, and A Mortreux. *Angew Chem Int Ed Engl* 1995;34:2269.
22. JP Arhancet, ME Davis, JS Merola, and BE Hanson. *Nature* 1998;339:454. J Haggin. *Chem Eng News* 1992;70(17):40. WA Herrmann. *Hoechst High Chem Magazine* 1992;13:14.
23. P Eilbracht, C Buss, C Hollmann, CL Kranemann, T Rische, R Roggenbuck, and A Schmidt. *Chem Rev* 1999;99:3329.
24. G Verspui, G Elbertse, FA Sheldon, MAPJ Hacking, and RA Sheldon. *Chem Commun* 2000:1363.
25. MMT Khna, SB Halligudi, and SHR Abdi. *J Mol Catal* 1988;48:313.
26. G Jenner. *Tetrahedron Lett* 1991;32:505.

27. FA Rampf, M Spiegler, and WA Herrmann. *J Organomet Chem* 1999;582:204–210.
28. RL Augustine. *Catalytic Hydrogenation*. Marcel Dekker: New York, 1965. PN Rylander. *Catalytic Hydrogenation in Organic Synthesis*. Academic Press: New York, 1979. PN Rylander. *Hydrogenation Methods*. Acedamic Press: Orlando, 1985.
29. JF Young, JA Osbourne, FH Jardine, and G Wilkinson. *J Chem Soc Chem Commun* 1965:131; MA Bennett and PA Longstaff. *Chem Ind (London)* 849, (1965); RS Coffey and JB Smith. U.K. Patent, 1,121,642, 1965.
30. F Lopez-Linares, MG Gonzales, and DE Paez. *J Mol Catal A Chem* 1999;145:61–68.
31. MS Spencer and DA Dowden. US Patent 3,009,969, 1961; J Kwiatek, IL Madok, and JK Syeler. *J Am Chem Soc* 1962;84:304.
32. F Joó and MTT Beck. *React Kinet Catal Lett* 1975;2:257.
33. F Joó, J Kovács, Á Kathó, AC Bényei, T Decuir, and DJ Darensbourg. *Inorg Synth* 1998;32:1–8.
34. Z Tóth, F Joó, and MT Beck. *Inorg Chim Acta* 1980; 42:153-161; F Joó, L Somsák, and MT Beck. *J Mol Catal* 1984; 24:71–75; F Joó, L Nádasdi, AC Bényei, and DJ Darensbourg. *J Organomet Chem* 1996;512:45–50.
35. PA Chaloner, MA Esteruelas, LA Oro, and F Joó. *Homogeneous Hydrogenation*. Kluwer: Dordrecht, 1994.
36. C Larpent, R Dabard, and H Patin. *Tetrahedron Lett* 1987;28:2507.
37. C Larpent and H Patin. *J Organomet Chem* 1987;335:C13.
38. JM Crosselin, C Mercier, G Allmang, and F Grass. *Organometallics* 1991;10:2126.
39. T Okano, M Kaji, S Isotani, and J Kiji. *Tetrahedron Lett* 1992;33:5547.
40. GB Jacobson, CT Lee, KP Johnson, and W Tumas. *J Am Chem Soc* 1999;121:11902.
41. JM Grosselin and C Mercier. *J Mol Catal* 1990;63:L25.
42. E Fache, F Senocq, C Santini, and JM Basset. *J Chem Soc Chem Commun* 1990:1776.
43. F Joó, J Kovács, AC Bényei, and Á Kathó. *Angew Chem, Int Ed Engl* 1998;37:969–970; F Joó, J Kovács, AC Bényei, and Á Kathó. *Catal Today* 1998;42:441–448.
44. CP Lau and L Cheng. *Inorg Chim Acta* 1992;195:133.
45. V Penicaud, C Maillet, P Jauvier, M Pipelier, and B Bujoli. *Eur J Org Chem* 1999:1745.
46. B Zimmermann, J Herweg, and M Beller. *Angew Chem, Int Ed Engl* 1999;38:2372.
47. I Toth, BE Hanson, and ME Davis. *Tetrahedron Asymmetry* 1990;1:913.
48. J Holz, D Heller, R Stürmer, and A Börner. *Tetrahedron Lett* 1999;40:7059.
49. RA Brown, P Pollet, E McKoon, CA Eckert, CL Liotta, and PG Jessop. *J Am Chem Soc* 2001;123:1254.
50. T Lamonille, C Saluzzo, R ter Halle, F le Guyader, and M Lemaire. *Tetrahedron Lett* 2001;42:663.
51. J Bakos, A Orosz, B Heil, M Laghmari, P Lhoste, and D Sinou. *J Chem Soc, Chem Commun* 1991:1684; C Lensink and JG de Vries. *Tetrahedron Asymmetry* 1992;3:235.
52. L Lecomte, D Sinou, J Bakos, I Toth, and B Heil. *J Organomet Chem* 1989;370:277.
53. Y Amrani and D Sinou. *J Mol Catal* 1986;36:319.
54. J Halpern. In *Asymmetric Synthesis*, Vol. 5, JD Morrison, ed. Academic Press: Orlando, 1985.
55. I Toth, BE Hanson, and ME Davis. *Catal Lett* 1990;5:183.
56. M Laghmari and D Sinou. *J Mol Catal* 1991;66:L15.
57. F Joó and A Bényei. *J Organomet Chem* 1989;363:C19–C21. A Bényei and F Joó. *J Mol Catal* 1990;58:151–163.
58. JM Tour and SL Pendalwar. *Tetrahedron Lett* 1990;31:4719. JM Tour, JP Cooper, and SL Pendalwar. *J Org Chem* 1990;55:3452.
59. AM d'A. Rocha Gonsalves, JC Bayon, MM Pereira, MES Serra, and JPR Pereira. *J Organomet Chem* 1998;553:199.
60. T Thorpe, J Blacker, SM Brown, C Bubert, J Crosby, S Fitzjohn, JP Muxworthy, and JMJ Williams. *Tetrahedron Lett* 2001;42:4041.
61. B Meunier. *Chem Rev* 1992;92:1411.
62. J Bernadou, AS Fabiano, A Robert, and B Meunier. *J Am Chem Soc* 1994;116:9375.
63. F Fringuelli, R Cermani, F Pizzo, and G Savelli. *Tetrahedron Lett* 1989;30:1427.
64. F Fringuelli, R Germani, F Pizzo, F Santinelli, and G Savelli. *J Org Chem* 1992;57:1198.
65. G Berti. In *Topics in Stereochemistry*, Vol. 7, NL Allinger and EL Eliel, eds. John Wiley: New York, 1967, 93pp.

66. KS Kirshenbaum and KB Sharpless. *J Org Chem* 1985;50:1979.
67. KA Hofmann. *Ber* 1912;45:3329; PA Grieco, Y Ohfune, Y Yokoyama, and W Owens. *J Am Chem Soc* 1979;101:4749.
68. NA Milas and S Sussman. *J Am Chem Soc* 1936;58:1302; R Daniels and JL Fischer. *J Org Chem* 1963;28:320.
69. M Mugdan and DP Young. *J Chem Soc* 1949:2988.
70. M Minato, K Yamamoto, and J Tsuji. *J Org Chem* 1990;55:766.
71. IC Phillips. *J Am Chem Soc* 1894;16:255.
72. J Smidt, W Jafner, R Jira, J Sedlmeier, R Sieber, R Ruttinger, and H Kojer. *Angew Chem* 1959;71:176; 1962;74:93.
73. PM Henry. *Adv Organomet Chem* 1975;13:363.
74. JE Bäckvall, B Åkermark, and SO Lijunggren. *J Am Chem Soc* 1979;101:2411.
75. JK Stille and P Divakaruni. *J Am Chem Soc* 1978;100:1303.
76. H Nagashima, K Sakai, and J Tsuji. *Chem Lett* 1982:859.
77. J Tsuji, I Shimizu, H Suzuki, and Y Naito. *J Am Chem Soc* 1979;101:5070.
78. GJ ten Brink, IWCE Arends, G Papadogianakis, and RA Sheldon. *Chem Commun* 1998:2359.
79. RC Larock. *Comprehensive Organic Transformations.* VCH Publishers: New York, 1989, pp. 604–834; M Hudlicky. *Oxidations in Organic Chemistry,* ACS Monogr Ser. 186. American Chemical Society: Washington, D.C., 1990, 114pp.; *Comprehensive Organic Functional Group Transformations,* Vols. 3 and 5, AR Katritzky, O Meth-Cohn, CW Rees, G Pattenden, and CJ Moody, eds. Elsevier Science: Oxford, 1995.
80. G Cainelli and G Cardillo. *Chromium Oxidants in Organic Chemistry.* Springer: Berlin, 1984; SV Ley and A Madin. *Comprehensive Organic Synthesis,* Vol. 7, BM Trost, I Fleming, and SV Ley, eds. Pergamon: Oxford, 1991, pp. 251–289.
81. K Sato, M Aoki, and R Noyori. *Science* 1998;281:1646.
82. GJ ten Brink, IWCE Arends, and RA Sheldon. *Science* 2000;287:1636.
83. K Sato, M Aoki, J Takagi, and R Noyori. *J Am Chem Soc* 1997;119:12386; K Sato, J Takagi, M Aoki, and R Noyori. *Tetrahedron Lett* 1998;39:7549; K Sato, M Aoki, J Takagi, K Zimmermann, and R Noyori. *Bull Chem Soc Jpn* 1999;72:2287.

23 Zeolite-Based Catalysis in Supercritical CO_2 for Green Chemical Processing

Yusuf G. Adewuyi
Department of Chemical Engineering, North Carolina A&T State University

CONTENTS

23.1 INTRODUCTION

The manufacture of fine and specialty chemicals in batch processes has commonly been associated with the production of large quantities of toxic wastes due to the widespread use of traditional reagents such as mineral acids, strong bases, stoichiometric oxidants, and toxic metal reagents. The practice also has other drawbacks including handling difficulties, inorganic contaminants of organic products, and poor reaction selectivity leading to unwanted isomers and side-products. The trend towards the integration of attractive economics with low environmental impact in chemical manufacture ("enviroeconomics") is expected to continue in the future as traditional environmentally unacceptable processes are replaced by cleaner alternatives. The widespread application of environmentally benign processes such as clean catalytic processes in benign solvents will continue to play a pivotal role in the drive towards environmentally responsible technologies with minimal waste generation.

The use of catalyst in principle is environmentally benign in that the ultimate goal is to maximize selectivity, minimize waste production and energy consumption. If these objectives are achieved in the absence of organic solvents and the catalysts are nontoxic, reusable or do not deactivate rapidly then the process would be benign and profitable. However, many catalytic processes use hazardous liquid catalysts such as mineral acids (HF, H_2SO_4, etc.) and organic solvents in order to achieve high selectivity [1, 2]. For example, acylation of aromatics via the Friedel–Crafts reactions uses Lewis acids (e.g., $AlCl_3$, BF_3) as catalysts and nitrobenzene as solvent or liquid HF in batch-type processes [1–4]. The use of these reagents and solvents make most Friedel–Crafts acylation processes inherently dirty and highly polluting. Typical Friedel–Crafts reaction sequence using Lewis acid catalyst ($AlCl_3$) is illustrated in Figure 23.1 for the acylation of anisole with acetic anhydride. The product forms a complex with catalyst, requiring a water wash, resulting in the formation of hydrochloric acid (HCl) and salt wastes. The use of the Lewis acid catalysts and solvents also results in by-product formation, thus increasing the cost of downstream processing. Moreover, an overall atom and mass balance suggests this method results in only about 30% conversion of raw materials. Thus, a simple calculation of atom efficiency identifies this method as a potential target for environmental improvement [5, 6]. Therefore, the development of new and greener technologies for commercial use is of high priority [7–9].

As alternatives to liquid mineral acids and Lewis acid catalysts, zeolites have been used as solid acid catalysts in heterogeneous reactions for the synthesis of organic compounds for over two decades, due to their high activity, stability, and selectivity [10–12]. The applications of zeolites as alternative acid catalysts for selective chemical synthesis are growing rapidly. Using the materials as solid acid catalysts contributes to the "greening" of process technology in three ways: first by replacing many hazardous acidic catalysts such as HF, HCl, and H_2SO_4. Second, by eliminating some intermediate steps in certain processes to reduce overall waste output and energy use (i.e., significant reduction in the by-products formed). Third, they are used in catalytic amounts, regenerable, reusable, and their disposal is not an environmental issue due to their eco-friendly nature (i.e., absence of pollution and corrosion effects). The use of solid acid catalysts in traditional organic processes is important to the chemical, petrochemical, and pharmaceutical industries such as aliphatic–aromatic alkylation, aromatic acylation, and nitration are under active research [10–15]. For example, heterogeneously catalyzed studies of 2-methoxynaphthalene (2-MNP) have been the subject of several studies because the product of the acylation in 6-position (6-acyl-2-methoxynaphthalene) is of particular interest for the production of the anti-inflammatory drug Naproxen [16]. The acylation of anisole with acetic anhydride using a zeolite catalyst has been commercialized and the acylation of isobutylbenzene [10, 16] over zeolite beta has been

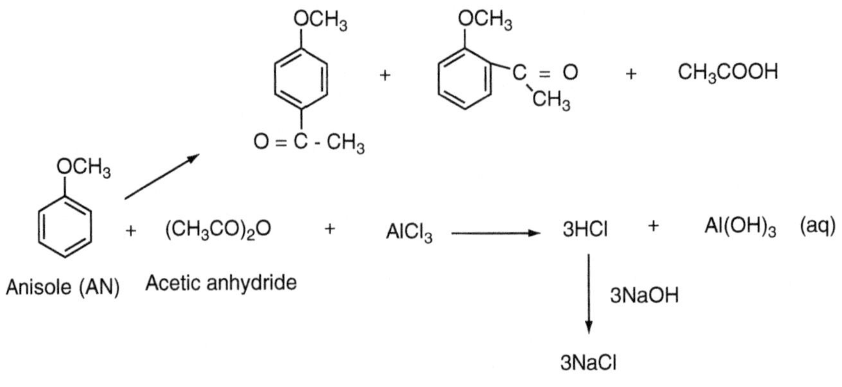

FIGURE 23.1 Friedel–Crafts acylation of anisole with Lewis acid showing starting materials and products.

reported by researchers at Rhône-Poulenc to be selective in the production of 4-isobutylacetophenone. However, there is a problem with maintaining the activity of the zeolite catalysts for longer period of time due to catalyst deactivation by deposition of heavy reaction products within the pores or the outer surface of the crystallites. For example, Bronsted activity of the zeolite often leads to oligomerization of the sensitive molecules (e.g., in alkylation reactions) and deactivation of the catalyst [11].

In order to obtain practical catalysts, it is essential to pay more attention to catalyst design (i.e., optimal balance of acidity, pore-size distribution, and hydrophilicity and hydrophobicity), deactivation mechanism, and possibilities to improve stability through technological aspects such as appropriate reactor choice and mode of operation. For heterogeneous catalytic reactions characterized by catalyst deactivation caused by coking or fouling, the catalysts can be reactivated by the adjustment of the pressure and temperature so that the reacting medium is in the supercritical (sc) state [17–20]. It is also known that in the presence of supercritical carbon dioxide ($scCO_2$) the catalyst deactivation could be thwarted and activity of the catalyst restored to a steady level due to the ability of supercritical CO_2 to strip out the coke, low-volatile compounds and deactivating products, prolonging and improving the activity of the catalyst [21–25]. There are several other reasons why supercritical fluids (SCFs) are potentially attractive for heterogeneous catalysis. In general, if properly used SCFs afford the following advantages in heterogeneous catalysis: (1) enhancement of the reaction rate, (2) enhanced mass and heat transfer; (3) increased catalyst lifetime and regeneration; (4) ease of separating the product from the SCF solvent by pressure reduction; and (5) environmental, health, and safety benefits [26–28]. This chapter focuses on the theoretical aspects of zeolite catalysis in supercritical CO_2, recent developments, and practical applications and implications.

23.2 THEORETICAL FUNDAMENTALS

23.2.1 ZEOLITES AND THEIR PROPERTIES IN HETEROGENEOUS CATALYSIS

Zeolites are crystalline microporous solids with high-internal-surface areas and internal cages and channels in one, two, or three dimensions, giving uniform pores with precisely defined molecular dimensions. Typical zeolite pore dimensions are 3 to 10 Å (synonymously called molecular sieve), although some so-called extra large pore zeolites have pore dimensions approaching 15 Å [29–33]. Also, new mesoporous materials with 25 to 100 Å pores are now available [13]. Zeolites are aluminosilicates with elementary building units made up of SiO_4 and AlO_4 tetrahedra. The structure of zeolite is made up of open, three-dimensional framework consisting of tetrahedral $(AlO_4)^{-5}$ and $(SiO_4)^{4-}$ units linked through shared oxygen [30–33]. Channels and cage openings defined by 8-, 10-, and 12-membered rings of oxygen atoms are possible, giving rise to small- (<0.5 nm pore diameter), medium- (ca. 0.55 nm), and large- (ca. 0.74 nm) pore zeolites [29]. Examples of zeolites commonly used in chemical synthesis include Beta (three-dimensional with intersecting elliptical channels, 0.76 × 0.64 nm), Faujasite (three-dimensional cage structure with 0.76 nm cage openings), mordenite (one-dimensional with elliptical channels, 0.65 × 0.70), and ZSM-5 (three-dimensional with intersecting channels, 0.53 × 0.56 nm) [29, 34].

Zeolite structures can be visualized by taking a neutral SiO_2 framework and isomorphously substituting AlO_2^- for SiO_2 (i.e., substitution of Al^{3+} for Si^{4+}). The zeolite structure can be represented by [30–32]:

$$\begin{array}{c} \mathrm{Si} \\ | \\ \mathrm{O} \quad \mathrm{O} \quad | \\ | \\ \mathrm{Al^- - O - Si - O - Al - O - Si - O - Al^-} \\ | \\ \mathrm{O} \end{array} \tag{23.1}$$

The resulting inorganic macromolecule exhibits a net negative charge on the framework aluminum, which are balanced by "nonframework" cations (e.g., Na^+, K^+, or NH_4^+) that reside in the interstices of the framework. The "nonframework" cations are relatively mobile and can in many cases be easily exchanged for other cations. The larger the substitution of Si^{4+} by lower valence aluminum cation (Al^{3+}) the higher the exchange capacity of the particular zeolite. The chemical composition of a zeolite can hence be represented by a formula of the type [33]:

$$A_{y/m}^{m+}\left[(\mathrm{SiO_2})_x \cdot (\mathrm{AlO_2^-})_y\right] \cdot z\mathrm{H_2O} \tag{23.2}$$

where A is a cation with the charge m, $(x + y)$ is the number of tetrahedron per crystalline unit cell, and x/y is the so-called framework silicon/aluminum ratio (Si/Al). Many aluminosilicate zeolites can be synthesized over a range of aluminum contents, for example zeolite ZSM-5 from Si/Al ~ 10 to ∞ [30].

Both Brønsted and Lewis acid sites occur in zeolites and are produced by the aluminum atoms. The chemical nature of the Brønsted acid sites is the hydroxyl group formed by a proton and framework oxygen in an AlO_4 tetrahedron (i.e., result of negative-charge compensation by protons). Upon severe heat treatment ($\geq$500°C), the Brønsted acid sites are degraded (dehydroxylation), water is split off with the formation of Lewis acid sites [30, 33]. The structures of the acid sites are shown in Figure 23.2. In general, the Brønsted sites are the external OH^- groups, while the Lewis sites are the exposed threefold coordinated Al^{3+} ions (such as any transition ion), substituting for the Si^{4+} ions in the tetrahedral sheets (i.e., coordinatively unsaturated cationic centers). The superacid sites are Brønsted sites having

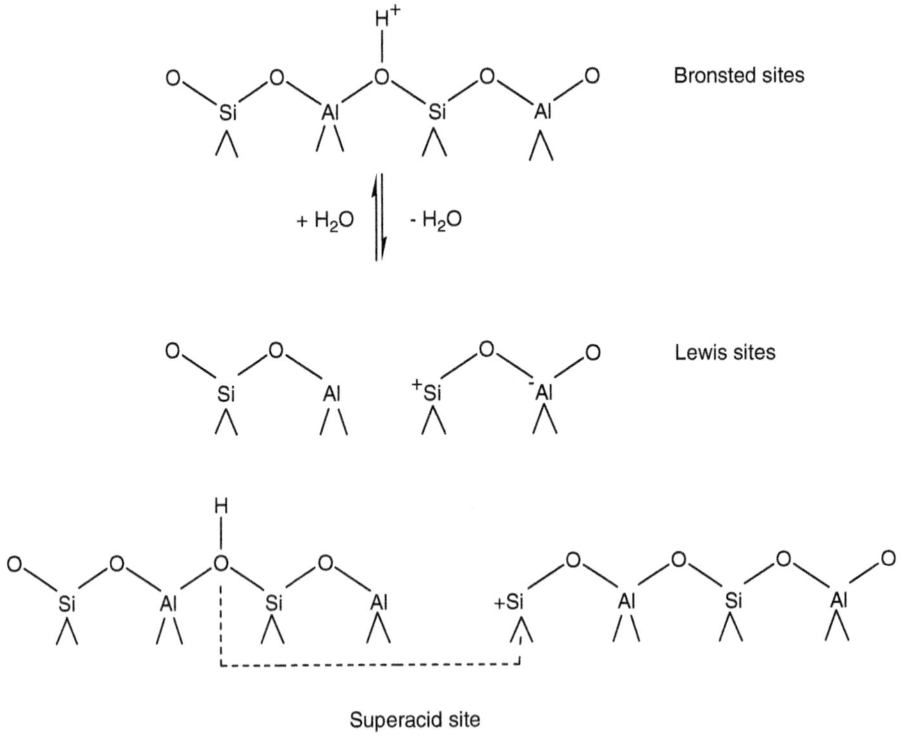

FIGURE 23.2 Structure of various acid sites in zeolites.

enhanced acidity due to inductive effects generated by neighboring Lewis sites [32]. Among the properties that are affected by the framework aluminum content are the density of the negative framework charges, the cation-exchange capacity, the density of Brønsted acid sites and their strength, the thermal stability, the hydrophilic or hydrophobic surface properties, and the unit cell dimensions. Zeolites have long been known as solid acid catalysts. It is generally accepted that the aluminum distribution in the zeolite framework is important for zeolite acidity. It has been shown that when the Si/Al ratio is increased, there is a transition in the surface selectivity from hydrophilic to hydrophobic. A more hydrophilic zeolite is probably favorable for the adsorption of the organic reactants into the zeolite [35]. The density of the Brønsted acid sites responsible for the catalytic activity in a zeolite is also related to the framework aluminum content and the strength of these sites increases with the extent of dealumination. Zeolites with high Si/Al ratio (i.e., limited Al substitution or reduced aluminum content) have lower acid site concentrations (i.e., catalytic site density) and hence lower total acidity and polarity but increased acid site strength. Increasing the zeolite aluminum (Al) content increases the number of active sites (kinetic effect) and simultaneously enhanced undesired competitive adsorption effects (thermodynamic effect) [36, 37].

Zeolites are considered to be ideal for environmentally benign processing because (i) they can perform shape-selective catalysis due to their well-defined pore systems to yield ultrahigh reaction selectivities, (ii) they have a high number of active sites giving high reaction rates, (iii) they can easily be separated, recovered, and regenerated, and (iv) their disposal is not an environmental issue since they are natural materials as well as synthetic. Zeolites are also known for shape-selective catalysis, which encompasses all effects where the selectivity of the heterogeneously catalyzed reaction depends unambiguously on the pore width architecture of the microporous catalyst [13, 29]. There are three main ways in which zeolite can affect shape selectivity based on the differences in molecular dimensions of reactants and products [13, 33]. In reactant shape selectivity the dimensions of the channels in the zeolite may preferentially prevent certain bulkier substrates (including isomers) from entering the interior of the zeolite and thus reacting. The less bulky molecules will react preferentially. In product shape selectivity certain bulkier products may be retained in the zeolite structure or have very slow diffusion rates out of the zeolite structure. In this case the less bulky product may be formed preferentially. Finally, in restricted transition state shape selectivity the transition state in a reaction between two substrates is sterically hindered so that the molecules are forced to interact in a particular conformation much like the action of an enzyme. Also, in this case the reaction that proceeds via a bulkier intermediate which cannot be accommodated in the zeolite pores is suppressed.

23.2.2 Supercritical Fluids

23.2.2.1 General Properties of Supercritical Fluid

SCFs are gases, the temperature and pressure of which are above their critical values T_c and P_c, respectively, as shown in the phase diagram for carbon dioxide (Figure 23.3). At these conditions, the fluid exists as a single phase, possessing favorable properties of both liquid and gas [38–40]. Typical properties of liquids, gases, and SCFs are shown in Table 23.1 [41]. It is obvious that diffusivity and viscosity of a SCF are more gas-like in the supercritical region, whereas density is comparable to liquids. In terms of reduced properties, SCFs are gases with reduced temperature, $T_r \sim 1.0$ to 1.1 and reduced pressure, $P_r \sim 1$ to 2. The properties of some SCFs most frequently used in chemical reactions are also shown in Table 23.2. The low viscosity and ease of diffusion in SCFs can reduce mass transfer limitations by eliminating gas–liquid transfer resistance and lowering external fluid film diffusion resistance. SCFs also have a surface tension close to zero, enabling them to penetrate the porous structure of the

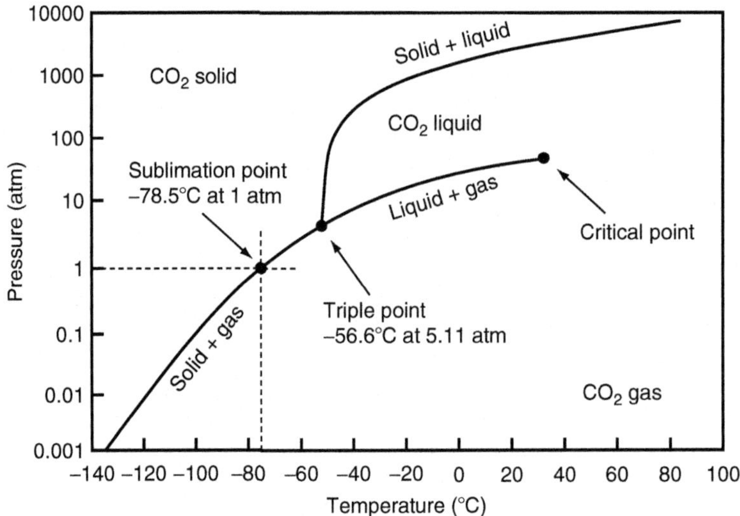

FIGURE 23.3 Phase diagram for carbon dioxide.

TABLE 23.1
Properties of Liquids, Gases, and SCFs Near the Critical Region [27]

Physical Quantity	Gas (ambient)	Supercritical Fluid (T_c, P_c)	Liquid
Density ρ ($kg\,m^{-3}$)	0.6–2	200–500	600–1600
Dynamic viscosity η (MPa s)	0.01–0.3	0.01–0.03	0.2–3
Kinematic viscosity ν ($10^6\,m^2\,s^{-1}$)	5–500	0.02–0.1	0.1–5
Diffusion coefficient D ($10^6\,m^2\,s^{-1}$)	10–40	0.07	0.0002–0.002

catalyst [27]. The high solvent power of SCFs compared with gases allows them to dissolve waxes and tars that might otherwise contaminate solid catalysts. This enhances reaction rates and increases the lifetime of catalysts. Moreover, their ability to dissolve both gases and solids allows substrate and gaseous reactants to mix at a more molecular level [25, 42]. A unique property of SCFs is the pressure-dependent density, which can be continuously adjusted from that of a vapor to that of a liquid. SCFs can also improve yield because of the increase of density of the reactants with pressure, which enhances the rate of intermolecular reaction. This tunability property and other characteristics of SCFs provide the opportunity to engineer the reaction environment (e.g., to maximize reaction yields and selectivity) by manipulating temperature and pressure [26, 27]. For example, a reaction which is diffusion controlled in the liquid phase can be enhanced by conducting it at supercritical conditions, due to higher diffusivity and elimination of gas–fluid and fluid–fluid interphases.

23.2.2.2 Pressure Effects and Kinetic Aspects

It has been demonstrated by many researchers that in supercritical reaction media pressure has a significant effect on the reaction rate and equilibrium constants in the region near the critical point [26–28]. The rate constant is a very strong function of pressure in the critical region and that the effect of pressure is reduced at higher temperatures. That is, elevated pressures could have either enhancing or inhibiting effect on rate and equilibrium constants. Pressure can also influence selectivity for at least some reactions. For example, in heterogeneous systems

TABLE 23.2
Critical Data (Temperature, Pressure, and Density) of Typical Solvents

Solvent	Temperature (°C)	Pressure (atm)	Density (g/ml)
Carbon dioxide (CO_2)	31.3	72.9	0.448
Ammonia (NH_3)	132.4	112.5	0.235
Water (H_2O)	374.2	218.3	0.315
Nitrous oxide (N_2O)	36.5	71.7	0.45
Xenon (Xe)	16.6	57.6	0.118
Krypton (Kr)	−63.8	54.3	0.091
Methane (CH_4)	−82.1	45.8	0.200
Ethane (C_2H_6)	32.3	48.1	0.203
Ethylene (C_2H_4)	9.21	49.7	0.218
Propane (C_3H_8)	96.7	41.9	0.217
Propylene (C_3H_6)	91.6	46.0	0.233
Pentane (C_5H_{12})	196.6	33.3	0.232
Methanol (CH_3OH)	240.5	78.9	0.272
Ethanol (CH_3CH_2OH)	243.0	63.0	0.276
Propanol ($CH_3CH_2CH_2OH$)	263.6	51.7	0.275
Isopropanol	235.3	47.0	0.273
Isobutanol	275.0	42.4	0.272
Cyclohexanol	356.0	38.0	0.273
Diethyl ether	193.6	36.3	0.267
Sulfur hexafluoride (SF_6)	45.5	37.7	0.735
Chlorotrifluoromethane	28.0	38.7	0.579
Trichlorofluoromethane	196.6	41.7	0.554
Monofluoromethane	44.6	58.0	0.300

higher-than-optimum pressure could introduce transport limitations that diminish catalyst effectiveness factor while lower-than-optimum pressure could limit the desorption of heavy hydrocarbons, adversely affecting catalyst activity and product selectivity [20].

On the basis of the transition-state theory, the rate constant for a bimolecular reaction where a chemical equilibrium is assumed between the reactants A and B and the transition state M (i.e., Equation (23.3)) is given by Equation (23.4):

$$A + B \Leftrightarrow M^{\pm} \longrightarrow \text{products} \tag{23.3}$$

$$k = \frac{k_B T}{h} K^* \chi = \chi \frac{k_B T}{h} \left(\exp\left(-\frac{\Delta G_o^*}{RT} \right) \right) \frac{\gamma_A \gamma_B}{\gamma^*} \tag{23.4}$$

where K^* is the equilibrium constant for the formation of the transition state, χ is the transmission coefficient expressing the probability of the transition state going forward to product as opposed to returning to reactants, and k_B and h are the Boltzmann and Plank constants, respectively, and ΔG_o^* is the Gibbs standard free energy of activation, e.g., the standard free energy changes between reactants and transition state. The transmission coefficient and the activity coefficient term are usually assumed to be unity for simplicity and because of lack of knowledge [26]. The volume of activation, Δv^*, a measure of the pressure dependence of the rate constant is given by partial differentiation of Equation (23.4) with respect to pressure and is given in Equation (23.5).

$$\left(\frac{\partial \ln k_x}{\partial P}\right)_{T,x} = \frac{1}{RT}\left(\frac{\partial \Delta G^*}{\partial P}\right)_{T,x} = -\frac{\Delta v^*}{RT} \tag{23.5}$$

where k_x is the molar fraction based rate constant, R is the universal gas constant, P is the pressure, T is the absolute temperature, and Δv^* is defined as the difference between the partial molar volume of the activated complex and sum of the partial molar volumes of the reactants (i.e., Equation (23.6)) [20, 26]

$$\Delta v^* = \nabla_M - \nabla_A - \nabla_B \tag{23.6}$$

Unlike activation energy for normal reaction (Equation (23.7)), the volume of activation, a measure of the so-called "kinetic pressure effect" can be either positive or negative:

$$\left(\frac{\partial \ln k}{\Delta T}\right) = \frac{E_a}{RT^2} \tag{23.7}$$

If the volume of activation is a positive quantity, the reaction will be hindered by pressure. However, if the volume of activation is negative quantity, then the reaction rate will be enhanced by pressure. The values reported in the open literature for activation and reaction volumes in *sc* reaction media are up to two orders of magnitude greater than those encountered in liquid-phase reactions, suggesting that pressure effects on rate and equilibrium constants can be significant in *supercritical*-phase reactions [27, 28]. The rather large values in *sc* media are ascribed to the fact that the partial molar volumes can assume large negative values near the critical point [20].

If the rate constant is related to a pressure-dependent measure of concentration (such as partial pressure), then the following equation is used:

$$\left(\frac{\partial \ln k_c}{\partial P}\right)\left(\frac{\partial \ln k_s}{\partial P}\right)_{T,c} = \frac{\Delta v^*}{RT} + \left(\frac{1-n}{P}\right)\left[1 - \left(\frac{\partial \ln \kappa_T}{\partial \ln P}\right)_{T,c}\right] \tag{23.8}$$

where k_c is the pressure-dependent rate constant, n is the molecularity of the reaction, and κ_T is the isothermal compressibility factor [20, 27]. The isothermal compressibility of any pure fluid at the critical point is given by:

$$\kappa_T = \frac{1}{\rho}\left(\frac{\partial \rho}{\partial P}\right)_T \tag{23.9}$$

where ρ is density, infinite and very large under conditions usually met in practical applications of SCFs. Both Equations (23.5) and (23.8) are based on the assumption that the transition-state transmission coefficient is not a function of pressure. Similarly, the effect of pressure on equilibrium constant is given by:

$$\left(\frac{\partial \ln k_x}{\partial P}\right)_{T,x} = \frac{\Delta v_r}{RT} \tag{23.10}$$

$$\left(\frac{\partial \ln k_c}{\partial P}\right) = \frac{\Delta v_r}{RT} + \kappa_T \sum \nu_i \tag{23.11}$$

where Δv_r is the reaction volume (defined as the difference between the partial molar volumes of reactants and products), k_x is the mole-fraction based equilibrium constant, k_c is the concentration-based equilibrium constant, and ν_i is the stoichiometric coefficient [20, 27].

Knowledge of the reaction volume Δv_r is crucial for predicting the effect of pressure on equilibrium constant. In high-pressure reactions in solution (liquid phase) the values of Δv_r are typically in the range -30 to $30\,cm^3\,mol^{-1}$. These values represent approximately equal contribution from structure (volume changes in the activated complex as a result from bond formation or breakage or other mechanistic features) and solvation and correspond to a rate acceleration or retardation of about a factor of 3.5 between atmospheric pressure and a pressure of about 100 MPa [27, 28]. The components in and composition of an SCF reaction mixture dictate the region of the pressure–temperature space in which a single phase exists, and pressure and temperature influence the reaction equilibrium. Furthermore a solvent's dielectric constant can influence the rates of reactions and density can be used as a lever to manipulate the dielectric constant and hence reaction kinetics. Therefore, knowledge of how both chemical composition and phase equilibrium vary with temperature and pressure is advantageous for the proper interpretation of experimental kinetics data and the rational design of processes for conducting reactions at supercritical conditions [27, 28].

23.3 SUPERCRITICAL FLUIDS IN HETEROGENEOUS CATALYSIS

As mentioned earlier there are several reasons why SCFs are potentially attractive for heterogeneous catalysis. The use of highly compressed SCF reaction phases for the reactivation of heterogeneous catalysts thus has considerable advantages over the usual methods employed in heterogeneous catalytic syntheses: (1) the reactivation conditions are relatively mild; (2) the reactivation or maintenance of the activity is performed by the fluid reaction phase itself; no other substances have to be added to the reaction system; (3) the reactivation can be easily carried out *in situ*; (4) the heterogeneous catalyst can be used over long working periods with constant catalytic activity; (5) at constant residual catalytic activity, more drastic reaction conditions can be used and higher space–time yields are obtainable [17]. Coke deposits cover active catalyst sites and plug the catalyst pore structure, rendering the catalyst useless. SCFs have been shown to be effective in controlling such catalyst deactivation by extracting coke compounds that are largely insoluble in the gas or liquid phase from the porous catalysts. When heterogeneous reactions are carried out in supercritical media, the solid acid catalyst deactivation slows down considerably [17–23]. Catalyst activity is stabilized if the rate at which coke precursors are formed is balanced by the rate at which the coke precursors are removed, thereby slowing down the accumulation of the coke precursors in the catalyst [24, 25].

One of the most important effects of using SCFs as reaction media is their ability to enhance the activity of the solid acid catalysts. It has been demonstrated that SCFs prevent solid acid catalysts (e.g., zeolite) from deactivating rapidly under certain process conditions (e.g., alkylation reaction conditions) by removing coke precursors from the zeolite, resulting in the extension of the catalyst lifetime [23, 24]. Figure 23.4 shows coke buildup and removal mechanisms in a single catalyst pore exposed to conventional gas or liquid-phase reaction medium and to a supercritical reaction medium [24]. For a conventional reaction media, coke accumulation via parallel reaction pathways results in progressively increasing diffusion limitations, eventually resulting in pore-mouth plugging. The coke precursor compounds like heavy hydrocarbons have low volatilities. In gas-phase reactions these coke precursor compounds (or heavy hydrocarbons) are not desorbed from the catalyst surface, leading to catalyst deactivation in time. In liquid-phase reactions these heavy hydrocarbons may be solubilized by the reaction medium, but then the pore diffusion limitations hinder the transport of the hydrocarbons out of the catalyst pore. The optimal reaction medium would be one that has liquid-like densities to solubilize (i.e., desorb) the coke precursors and gas-like transport properties to effectively transport the species out of the catalyst pores [24]. As discussed earlier, SCFs possess this optimum combination of fluid properties. The *in*

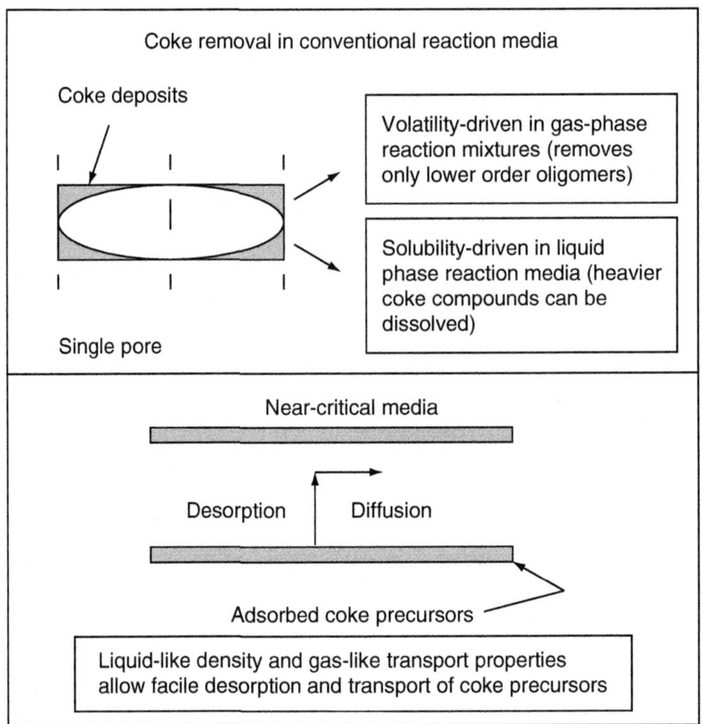

FIGURE 23.4 *In situ* mitigation of coke buildup with supercritical reaction media. (Adapted from B Subramaniam. *Appl Catal A: General* 212: 199–213, 2001. With permission.)

situ removal of the coke precursor by SCFs mitigates coke buildup and therefore alleviates pore diffusion limitations and elongates the life and activity of the catalyst.

The increase of pressure in connection with variation of physicochemical fluid properties results in two major effects: (1) increase of solvent power, in certain cases up to several orders of magnitude, in the supercritical region mainly at temperatures near T_c; (2) increase of mass transfer between catalyst and fluid phase when all other parameters of the system besides fluid properties remain constant. While pressurization at constant temperature of a gaseous reaction phase usually favors sorption processes in heterogeneous catalytic conversions and thus hinders desorption of low-volatile compounds, the increase of pressure in the supercritical region causes the opposite effect [17]. The higher the applied pressure is the faster the reactivation of the catalyst. However, it has been shown that while intercrystalline deposits on a precoke level can be removed by SCF-treatment, intracrystalline deposits cannot. Thus, the use of an SCF-reaction mixture offers the ability to study deactivation processes in micropores of catalysts with bimodal pore structure [26]. If precautions are made to suppress micropore deactivation, the lifetime of zeolite-type catalysts can be prolongated.

Temperature appears to be a double-edged parameter. Higher temperatures speed up the removal of coke precursors because of their increased solubility and faster diffusion in the catalyst pores. At the same time higher temperatures also normally enhance reaction rates including the coking rates and the formation of heavier coke compounds whose extraction may require much higher reaction densities. Hence, the maintenance or decay of activity in porous catalysts depends upon the relative rates of coke formation on the catalyst and of subsequent extraction of the coke compounds from the porous matrix. While the intrinsic rate of coke formation is primarily affected by concentration (i.e., density) and temperature, the extraction rate is affected by the solubility of the coke compounds in the reaction mixture and by the effective diffusivity of the extracted compounds through the porous matrix. The solubility and the effective diffusivity are in turn dictated by density and temperature [21, 43].

Therefore, it can be expected that optimum temperature exists at which the deactivation is significantly reduced but the reaction rate is still fairly high. A catalyst with high activity in its original state will produce very high conversion but will also deactivate very fast, resulting in a shorter average time of operation. In this sense, not only conversion temperature but also the activity and acidity of a catalyst may become optimization variables. Tiltscher et al. [17, 44] in the isomerization of 1-hexene on Al_2O_3 in sc medium observed that catalyst activity decreases, passes through a minimum, and then increases with an increase in temperature. In the case of zeolites with strongly acidic centers, the initial rate of conversion is higher but, at the same time, deactivation is faster. Also, a rapid initial deactivation will not only deactivate the strong acid centers but also other active sites by pore blocking. Therefore, it might be advantageous to block the most active sites (with highest activity) in the beginning by an irreversible deactivation, e.g., by exchanging hydrogen ions with metal ions [18].

23.3.1 Present Status of Research on Zeolite-Based Heterogeneous Catalytic Reactions in Supercritical Carbon Dioxide

A number of SCFs have been investigated as solvents/cosolvents for reaction media [38], but only a few of them have mild critical properties, reasonable cost, and would be considered green solvents. Carbon dioxide (CO_2) is often an ideal choice among other SCFs for chemical processing because it is nontoxic, unregulated, abundant and inexpensive, recyclable, non-flammable (actually a fire retardant) and has mild critical temperature and pressure (T_c = 31°C, P_c = 73.8 atm). Furthermore, the fact that reactions in $scCO_2$ produce very similar results to reactions in nonpolar solvents, $scCO_2$ has gained most attention as a solvent for application in heterogeneous catalysis. The use of supercritical CO_2 as a reaction medium includes the replacement of traditional organic solvents, which are hazardous, enhanced product selectivities that minimize waste and inherently safe reactor operation. A crucial advantage in using $scCO_2$ as a solvent for heterogeneous catalysis is the ease of phase separation of the organic liquid and the CO_2. Supercritical CO_2 has been investigated as a solvent in reactions like heterogeneous alkylation, amination, esterification, hydrogenation, and oxidation. These heterogeneous catalytic reactions at supercritical conditions are reported in recent comprehensive reviews [27, 28]. Unfortunately, strictly zeolite-based heterogeneous reactions in supercritical carbon dioxide reported to date in the open literature are mostly limited to alkylation reactions. In this section a review of zeolite and carbon dioxide based alkylation studies is provided and our current effort in developing zeolite-based Friedel–Crafts acylation reaction in supercritical carbon dioxide is summarized.

23.3.1.1 Alkylation and Acylation

Alkylation and acylation can be defined as the addition or insertion of an alkyl (R) or an acyl (RCO) group, respectively, into a molecule. These two types of reactions have so much in common that essentially the same catalysts can be applied, leading to related reaction intermediates. Alkylation and acylation of aromatics is of considerable industrial interest for making intermediates, which are used in the production of pharmaceuticals, insecticides, perfumes, etc. Some commonly used catalysts in alkylation and acylation reactions are metal halides ($AlCl_3$, BF_3, and $FeCl_3$), metal alkyls and alkoxides, acidic oxides and acidic sulfides [3, 4]. The most widely used alkylating agents are olefins and alkyl halides. Important olefins used are ethylene, propene, and dodecene. For example, alkylation of isobutene with C_3–C_5 olefins using sulfuric acid (H_2SO_4) or hydrofluoric acid (HF) as catalyst currently produces alkylates that are the highest-quality and cleanest-burning gasolines [45]. Despite numerous efforts aimed at developing solid-acid alkylation catalysts and solid-acid-based isobutene–olefin alkylation processes, such processes are yet to be commercialized due to one or more of the following drawbacks: rapid catalyst deactivation (often within hours or minutes) resulting from coke formation, unacceptable product quality (i.e., low alkylate fractions), and thermal degradation of catalyst during the regeneration step. However, it has recently been reported that zeolite-catalyzed alkylation processes for the bulk petrochemical ethylbenzene are operated commercially and are in an advanced stage of development for cumene [29].

The most common methods to introduce an acyl group into an organic compound are the Friedel–Crafts acylation reactions using Lewis acids (e.g., $AlCl_3$, BF_3) as catalysts and nitrobenzene as solvents or mineral liquid acids such as HF. The acylating agents most frequently used are acyl halides, carboxylic acids, and anhydride. In the traditional Friedel–Crafts acylation, an aromatic ketone is formed by the reaction of an aromatic compound with an acylating agent, such as an acyl halide, acid anhydride, an acid or an ester, in the presence of an acidic catalyst (e.g., $Alcl_3$ in liquid phase) [3, 4].

$$C_6H_6 + RCOX \xrightarrow{\text{catalyst}} C_6H_5COR + HX \tag{23.12}$$

where R is an alkyl or aryl group. These methods have numerous drawbacks including unpredictable regio-selectivity, atom inefficiency, use of large amount of hazardous solvents, and large catalyst requirements. Cleaner, heterogeneously catalyzed Friedel–Crafts reactions are now beginning to show promise, but few of these have been carried out in SCFs especially $scCO_2$.

23.3.1.2 Zeolite-Catalyzed Alkylation Reactions in Supercritical Carbon Dioxide

Some efforts aimed at developing solid acid-based alkylation processes in SCFs have been reported in the last decade. Hitzler et al. [39, 46] reported continuous Friedel–Crafts alkylation of mesitylene ($C_6H_3(CH_3)_3$) and anisole ($C_6H_5OCH_3$), propene, or propan-2-ol in supercritical propene or $scCO_2$ using a polysiloxane supported solid acid catalyst (DELOXAN, degussa AG) in a small fixed-bed reactor (10 ml) with stable catalyst activity for at least 15 h. The authors also observed an increase in catalyst life in the sc reaction medium, and the separation of product was achieved by simply expanding the fluid to atmospheric pressure. The work of this group demonstrated the feasibility of continuous and sustainable Friedel–Crafts alkylation in SCF solution. Sander et al. [47, 48] demonstrated that isomerization of undiluted supercritical *n*-butane over a commercial sulfated zirconia catalyst could be carried out between 443 and 533 K, pressures between 4.6 and 8.1 MPa and weight hourly space velocities between 17 and 170 h^{-1} (equivalent to residence times between 5,000 and 42,000 kg s m^{-3}). It was shown that the catalyst was stable under sc conditions but in a narrow temperature regime (<500 K). It was also shown that stable conversion levels obtained with pure *n*-butane under sc conditions did not exceed 20% but due to high throughputs, production capacity of isobutane was much higher under sc conditions than at atmospheric pressure. Also, the same catalyst suffered from rapid coke deactivation when the isomerization is carried out in the gas phase.

Fan et al. [49] carried out an investigation over Y-zeolite and sulfated zirconia under liquid, gas, and near-critical phases. Two types of supercritical-phase alkylation reactions were conducted on a zeolite catalyst with the sc fluids also acting as reactants, isopentane (T_c = 188°C, P_c = 3.3 MPa) with isobutene and isobutane (T_c = 135°C, P_c = 3.6 MPa) with isobutene. The liquid reaction exhibited initial high yield of alkylate (2,2,4-trimethylpentane) as high as 70%, but the activity declined to zero when the accumulated feed amount of olefin reached 15 mmol/cat g. Similar deactivation trend was also observed for the gas-phase reaction. However, it was shown that supercritical alkylation performed with excess isobutene (P_c = 36.5 bar; T_c = 408 K) above the critical temperature (T_c) and critical pressure (P_c) slowed down catalyst deactivation and extended alkylate activity compared to alkylation in liquid and gas phases. It was also shown that at reaction temperatures exceeding 408 K, undesirable side reactions such as oligomerization and cracking dominated resulting in unacceptable product quality. Shi et al. [50] carried out the alkylation of benzene by ethylene over a high surface area β-zeolite catalyst by monitoring the reaction at five different temperatures and pressures near the critical point of the binary system, one of which was in the sc region. It was shown that the maximum rate of reaction occurred close to the critical point, but the presence of a multiphase mixture was suggested to have resulted in the degradation of the catalyst.

Ginosar et al. [51] explored the alkylation of toluene with ethylene over microporous USY zeolite and mesoporous sulfated zirconia (S/ZrO_2) catalysts at liquid, near-critical liquid, and supercritical conditions using propane as the sc cosolvent. Both catalysts demonstrated high levels of alkylation activity under all the three reaction conditions explored. It was shown that while the S/ZrO_2 catalyst demonstrated alkylation activity almost exclusively, the USY catalyst showed both alkylation and cracking/disproportionation activities. Ginosar et al. [52] also explored the alkylation of isobutene with *trans*-2-butene over six solid acid catalysts (i.e., two zeolites; two sulfated metal oxides; two Nafion catalysts) in liquid, near-critical liquid, and supercritical regions through the addition of an inert cosolvent to the reaction mixture. For the conditions explored, it was shown that the zeolite catalyst had the best performance of the six catalysts. It was also shown that light hydrocarbons and trifluoromethane were better cosolvents whereas other nonhydrocarbon fluids (CO_2, SF_6, and CH_3F) were poor. Gao et al. [53] studied the effect of scCO_2 on catalyst deactivation using the alkylation of benzene with ethylene on a Y-zeolite. The study compared the reactions under three phases: in the liquid phase, at supercritical conditions of the reaction mixture, and under conditions where scCO_2 was used as a solvent. It was shown that both supercritical operation modes resulted in significantly lower catalyst deactivation and improved selectivity to ethylbenzene due to suppressed formation of by-products. The slower catalyst deactivation observed in supercritical reaction mixture and scCO_2 was attributed to the higher solubility and enhanced diffusivity of polynuclear aromatic compounds that deposited on the catalyst surface. These deposits act as precursors to coke formation, which reduce the activity of the catalyst. The higher selectivity of the reaction in scCO_2 was also ascribed to faster removal of the product, ethylbenzene, from the catalyst surface due to higher diffusivity. Glaser and Weitkamp [54] also employed scCO_2 as a reaction medium to significantly reduce the deactivation rates of a promoted Y-zeolite (LaNaY-73) and H-mordenite zeolite during the alkylation of naphthalene with alcohols.

Clark and Subramaniam [55] studied the isobutane/1-butane alkylation in gas, liquid, near critical with CO_2, and supercritical with CO_2 on zeolite catalysts. It was shown that, using a molar excess of a low T_c diluent such as carbon dioxide, 1-butene/isobutane alkylation could be performed at supercritical conditions at temperatures lower than the critical temperature of isobutane. The process resulted in a steady alkylate (trimethylpentanes and dimethylhexanes) production on both microporous zeolitic (USY) and mesoporous solid acid (sulfated zirconia) catalysts for experimental duration of nearly 2 days. It was suggested that the excess of CO_2 served the dual role of lowering the concentration of 1-butene and improving *in situ* extraction of coke-forming compounds, which resulted in lower oligomerization and coking rates. It was also shown that at the higher temperatures (>135°C, the T_c of isobutane) required for supercritical operation without CO_2, coking and cracking reactions were significant as inferred from wide product spectrum and extensive surface area/pore volume losses (up to 90%) in the spent catalysts. Santana and Akgerman [56] also investigated supercritical-phase alkylation on a USY zeolite using carbon dioxide as a diluent to lower the critical temperature of the reaction mixture in the ratio 27:9:1 for carbon dioxide/isobutene/1-butene. It was shown that the use of the diluent favored both alkylate selectivity and coke precursor removal. The authors also observed that the initial concentrations of C_9^+ in the liquid products were high, suggesting the supercritical reaction medium solubilized and removed high molecular weight species from the catalyst structure, preventing their accumulation on the catalyst surface.

23.3.1.3 Zeolite-Catalyzed Acylation Reactions in Supercritical Carbon Dioxide

Numerous studies have investigated the use of zeolite for heterogeneous acylation of aromatics. The mechanism for the acylation reactions using zeolites is similar to the one proposed by Olah [3, 4] for homogeneous catalysis, where an acylium ion or complex (i.e., the electrophile) is formed by the interaction of the surface sites of the zeolites with an acylating agent (e.g., acetic anhydride) as shown in Figure 23.5 for the acylation of anisole to *p*-methoxyacetophenone (*p*-MAP). However, the structure of the electrophile intermediate would depend upon the acylating agent and the nature of the zeolite acid sites. The anhydride would transform into the acylium cation, but the carboxylic acid would lead to the protonated and coordinated carboxylic acid as the electrophiles over the Brønsted and Lewis acid sites of the zeolites, respectively [57]. The external OH-groups on the zeolite are the Brønsted acid sites (i.e., proton donor or proton acidity). The acylation reaction with zeolites usually occur on the Brønsted acid sites according to a carbocation mechanism where in acylium species are the electrophiles that attack the aromatic ring, the reaction is an electrophilic aromatic substitution resulting in the formation of a corresponding ketone.

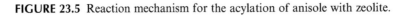

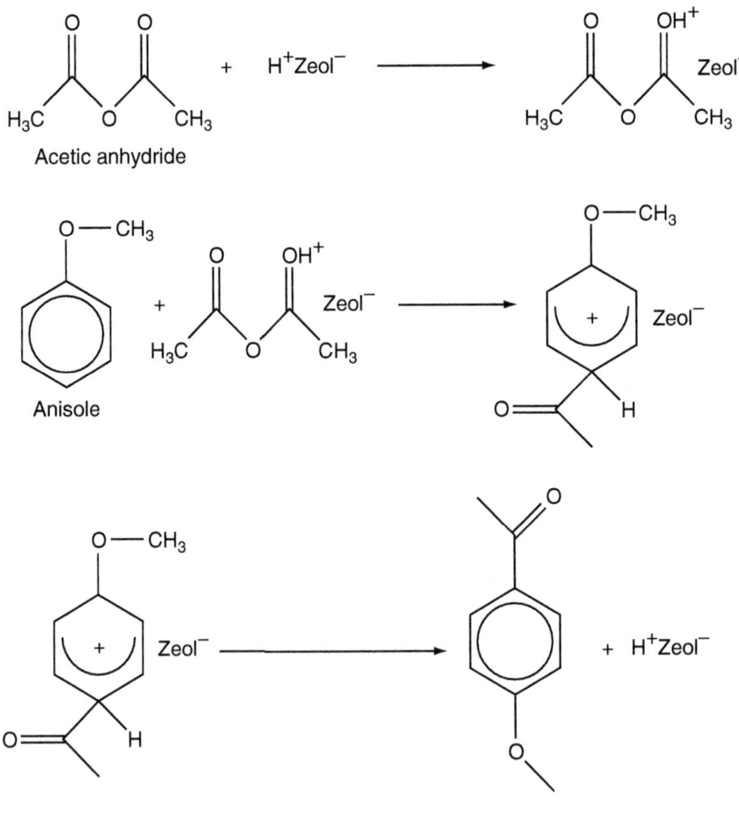

FIGURE 23.5 Reaction mechanism for the acylation of anisole with zeolite.

Das et al. [58] carried out the liquid phase batch acylation of 2-MNP in the temperature range 100–150°C using H-mordenite, H-beta, and H-Y zeolite as catalysts. Both acetyl chloride and acetic anhydride were used to study the effects of acylating agents using sulfolane (bp 285°C) as the solvent. It was observed that at 100°C, acetyl chloride showed much higher conversion than that observed with acetic anhydride. It was also shown that as the reaction temperature increased both acylating agents gave similar conversions and the maximum conversion of only 30% was attained after about 9 h time-on-stream (TOS) at 125°C. With further increase in temperature the conversion reduced due to reverse acylation. This decrease in conversion was more noticeable at higher reaction temperatures. Derouane et al. [36, 37] carried out the liquid-phase acylation of anisole by acetic anhydride (anisole to acetic anhydride molar ratio of 2:1) using beta catalyst (HBEA with Si/Al = 11) in a batch reactor at 90°C for up to 10 h under inert atmosphere (N_2) without and with addition of the product *p*-MAP. The authors observed that rate of production of 4-MAP was very high initially and decreased rapidly after about 50 min on stream when conversion had reached about 60%. It was shown from both the progressive decrease in initial rates and the lower overall conversions observed at long reaction times that catalyst deactivation was due to product inhibition, i.e., the competitive adsorption of the reactants and products in the zeolite intracrystalline volume. Rohan et al. [59] studied the liquid-phase acylation of anisole (35 mmol) with acetic anhydride in equimolar amount in a batch reactor at 60°C over a beta zeolite (HBEA) with Si/Al = 10. They also found *p*-MAP, the main reaction product, to be selectively and rapidly formed on the catalyst, deactivating the catalyst and inhibiting the acylation reaction and its own production due to its strong adsorption on the acid sites of the zeolite. Botella et al. [60] found that the acylation of toluene with acetic anhydride over beta zeolite (Si/Al = 12.5–93) at 150°C in a stainless steel stirred autoclave was limited by the poisoning of the active sites by adsorption of the product and pore blockage due to "coke" -type products, which adsorbed on the catalyst. The use of a nanocrystalline beta zeolite catalyst with nominal framework Si/Al ratio was used to minimize catalyst decay.

It is obvious from these studies that acid site accessibility is of paramount importance in the acylation reaction. Hence, the fast catalyst decay produced by coke deposition within the micropores of the zeolites limits the application of these materials in commercial acylation processes. However, limited work has been published in the area of acylation reactions in supercritical media. It has recently been shown that the acylation of anisole with acetic anhydride to mainly 4-methyl acetophenone as products could be accomplished more efficiently and benignly compared with conventional and other developmental methods using β-zeolite as a catalyst and supercritical CO_2 as a solvent [61–63]. Several experimental runs were conducted in a laboratory-scale, high-pressure 25 ml batch (SS 316) reactor using a zeolite catalyst with different Si/Al ratios (75, 150, and 300) at 30–90°C. The results of the study indicated that anisole conversion as high as 80% could be achieved (compared to 30% with some conventional methods) at 90°C and the conversion increased with temperature from 30°C to 90°C. The conversion also increased with Si/Al ratio but this effect was less significant at the high temperatures. The catalyst with the highest Si/Al ratio (HBEA 300) showed the highest initial activity. This was attributed to a higher acid strength of the Broøsted acid sites because of dealumination or to lower level of deactivation due to adsorption of acylation product or coke formation. Most significantly, the activity of the catalyst was maintained for a time period up to 30 h, compared with other investigators who reported loss of activity within 2 to 10 h using zeolite catalysts in organic solvents. The activity was also fully restored after calcination in air at 550°C for 6 h. The mechanistic pathways were proposed and the kinetics of reaction elucidated using the Langmuir–Hinshelwood–Hougen–Watson (LHHW) approach. It was found that surface reaction was the rate-controlling step with k_s values ranging from 0.0162 to 0.1189 (ml $gmol^{-1}$ s^{-1}) and varying linearly with catalyst loading at 1200 psi [63].

23.4 CONCLUDING REMARKS

This chapter has addressed an important problem related to green chemistry, green engineering, and sustainability in that traditional industrial processes utilizing inherently dirty and highly polluting catalysts, mineral acids, and organic solvents could be replaced by environmentally benign alternatives. The zeolite and carbon dioxide-based processes showed promise as an environmentally superior alternative to conventional processes based on liquid acids as reaction media. The zeolites are used in catalytic amounts, they are regenerable, reusable, and their disposal is not an environmental issue due to their eco-friendly nature (i.e., absence of pollution and corrosion effects). The drawbacks of carbon dioxide-based process include capital and operational cost associated with high-pressure vessels and equipment and the compression and recycle of carbon dioxide. However, the solid acid-based process in $scCO_2$ allows easy separation of solid catalysts from liquid products and offers the possibility of using continuous reactors rather than batch reactors with inherent advantages, which include pressure-tunable selectivities, minimization of waste formation, enhanced reaction, smaller and safer reactor (under pressure), reduced capital cost and easy scale-up. Also, replacing organic solvent with nonflammable CO_2 (with pressure-tunable heat capacity to avoid reactor runaway) in reactors is inherently safer.

The challenge is to demonstrate the feasibility of environmentally benign processes, which simultaneously display stability, high selectivity, improved product quality, and economic competitiveness. The recent construction of hydrogenation a pilot-scale plant by Thomas Swan Co. in the United Kingdom to generate 1000 t per year of product by Friedel–Crafts acylations and alkylations in SCFs and the construction of a plant by Dupont in North Carolina for producing fluoropolymers in supercritical carbon dioxide are strong indications of industries' willingness to embark on sc processes for green chemical processing [64–66]. It is hoped that this presentation has stimulated thinking beyond the cases presented and should spur future research and development of zeolite and carbon dioxide based green chemical and engineering-based processes.

ACKNOWLEDGMENT

This research was supported by grant from the STC Program of National Science Foundation under Agreement No. CHE-9876674.

REFERENCES

1. RA Sheldon. *Chem. Indust.* 7: 903–906, 1992.
2. JH Clark. *Green Chem.* 1: 1–11, 1999.
3. GA Olah. *Friedel–Crafts and Related Reactions*, Vol. III, Part 1. New York: Interscience, 1964, pp. 1–105.
4. PH Groggins and SB Detwiler. *Ind. Eng. Chem.* 43: 1970–1996, 1951.
5. DT Allen and DR Shonnard. *Green Engineering: Environmentally Conscious Design of Chemical Processes.* Upper Saddle River, New Jersey: Prentice-Hall, 2002, pp. 177–198.
6. MC Cann and ME Connelly. Real-World Cases in Green Chemistry. Washington, DC: American Chemical Society, 2000, pp. 1–24.
7. PT Anastas and JC Warner. *Green Chemistry: Theory and Practice.* New York: Oxford University Press, 1998, pp. 1–200.
8. AS Matlack. *Introduction to Green Chemistry.* New York: Marcel Dekker, 2001, pp. 1–300.
9. PT Anastas and TC Williamson, eds. *Green Chemistry: Frontiers in Benign Chemical Synthesis and Processes.* New York: Oxford University Press, 1998, pp. 1–50.
10. CB Dartt and ME Davis. *Ind. Eng. Chem. Res.* 33: 2887–2899, 1994.
11. RA Sheldon and RS Downing. *Appl. Catal. A General* 189: 163–183, 1999.

12. RA Sheldon. *Pure Appl. Chem.* 72: 1233–1246, 2000.
13. AJ Butterworth, SJ Tavener, and S.J. Barlow. In: *Chemistry of Waste Minimization*, JH Clark, ed. New York: Blackie Academics and Professional, 1995, pp. 522–543.
14. BM Choudary, M Sateesh, ML Kantam, KK Rao, KVR Prasad, KV Raghavan, and JARP Sarma. *Chem. Commun.* 25–26, 2000.
15. GD Yadav, PK Goel, and AV Joshi. *Green Chem.* 3: 92–99, 2001.
16. P Andy, J Garcia-Martinez, G Lee, H Gonzalez, CW Jones, and ME Davis. *J. Catal.* 192: 215–223, 2000.
17. H Tiltscher, H Wolf, and J Schelchshorn. *Angew. Chem. Int. Ed. Enbl.* 20: 892–894, 1981.
18. G Manos and H Hofman. *Chem. Eng. Technol.* 14: 73–78, 1991.
19. B Subramanian and MA McHugh. *Ind. Eng. Chem. Process. Des. Dev.* 25: 1–12, 1986.
20. B Subramaniam and DH Busch. In: *CO_2 Conversion and Utilization*, C Song, AF Galfney, and K Fujimoto, eds. Washington, D.C.: American Chemical Society, 2001, pp. 364–386.
21. S Saim and B Subramaniam. *J. Catal.* 131: 445–456, 1991.
22. S Baptist-Nguyen and B Subramaniam. *A.I.Ch.E. J.* 38: 1027–1037, 1992.
23. V Arunajatesan, KA Wilson, and B Subramaniam. *Ind. Eng. Chem. Res.* 42: 2639–2643, 2003.
24. B Subramaniam. *Appl. Catal. A: General* 212: 199–213, 2001.
25. B Subramaniam, CJ Lyon, and V Arunajatesan. *Appl. Catal. B: Environ.* 37: 279–292, 2002.
26. H Tiltscher and H Hofman. *Chem. Eng. Sci.* 42: 959–977, 1987.
27. A Baiker. *Chem. Rev.* 99: 453–473, 1999.
28. PE Savage, S Gopalan, TI Mizan, CJ Martino, and EE Brock. *A.I.Ch.E J.* 41: 1723–1778, 1995.
29. RS Downing, H van Bekkum, and RA Sheldon. *CATTECH* 95–109, 1997.
30. DEW Vaughan. *CEP* 25–31, 1988.
31. BK Marcus and WE Cormier. *CEP* 47–53, 1999.
32. J Dwyer. *Chem. Indust.* 258–269, April 2, 1984.
33. J Weitkamp. *Solid State Ion.* 131: 175–188, 2000.
34. JC Jansen, EJ Creygton, SL Njo, HV Koningsveld, and H van Bekkum. *Catal. Today* 38: 205–212, 1997.
35. K Gaare and D Akporiaye. *J. Mol. Catal. A: Chemical* 109: 177–187, 1996.
36. EG Derouane, CJ Dillon, D Bethell, and SB Derouane-Abd Hamid. *J. Catal.* 187: 209–218, 1999.
37. EG Derouane, G Crehan, CJ Dillon, D Bethell, H He, and SB Derouane-Abd Hamid. *J. Catal.* 194: 410–423, 2000.
38. GP Jessop and W Leitner. *Chemical Synthesis using Supercritical Fluids.* New York: Wiley-VCH, 1999, pp. 1–200.
39. M Poliakoff, NJ Meehan, and S Ross. *Chem. Indust.* 750–752, October 4, 1999.
40. R Noyori. *Chem. Rev.* 99: 353–354, 1999.
41. D Ambrose. In: *Handbook of Chemistry and Physics*, DR Lide, ed. Boca Raton: CRC Press, 1991, 2000pp.
42. JR Hyde, P Licence, D Carter, and M Poliakoff. *Appl. Catal. A: General* 222: 119–131, 2001.
43. S Saim and B Subramaniam. *J. Supercritical Fluids* 3: 214–221, 1990.
44. H. Tiltscher, H Wolf, and J Schelchshorn. *Ber. Bunsenges Phys. Chem.* 88: 897–900, 1984.
45. LF Albright. *Ind. Eng. Chem. Res.* 42: 4283–4289, 2003.
46. GM Hitzler, RF Small, KS Ross, and M Poliakoff. *Chem. Commun.* 359–360, 1998.
47. B Sander, M Thelen, and B Kraushaar-Czarnetzki. *Ind. Eng. Chem. Res.* 40: 2767–2772, 2001.
48. B Sander, M Thelen, and B Kraushaar-Czarnetzki. *Ind. Eng. Chem. Res.* 41: 4941–4948, 2002.
49. L Fan, I Nakamura, S Ishida, and K Fujimoto. *Ind. Eng. Chem. Res.* 36: 1458–1463, 1997.
50. YF Shi, Y Gao, WK Yuan, and YC Dai. *Chem. Eng. Sci.* 56: 1403–1410, 2001.
51. DM Ginosar, K Coates, and DN Thompson. *Ind. Eng. Chem. Res.* 41: 6537–6545, 2002.
52. DM Ginosar, DN Thompson, K Coates, and DJ Zalewski. *Ind. Eng. Chem. Res.* 41: 2864–2873, 2002.
53. Y. Gao, L-F Shi, Z-N Zhu, and W-K Yuan. *Proceedings of the Third International Symposium on High Pressure Chemical Engineering.*, Zurich, Switzerland, 1996, pp. 151–156
54. R Glaser and J Weitkamp. In: *Proceedings of the 12th International Zeolite Conference*, MMJ Treacy, ed. Warendale: Materials Research Society 2: 1447–1453, 1999.
55. MC Clark and B Subramaniam. *Ind. Eng. Chem. Res.* 37: 1243–1250, 1998.

56. GM Santana and A Akgerman. *Ind. Eng. Chem.* 40: 3879–3882, 2001.
57. Y Ma, QL Wang, W Jiang, and B Zuo. *Appl. Catal. A: General* 165: 199–206, 1997.
58. D Das and S Cheng. *Appl. Catal. A: General* 201: 159–168, 2000.
59. D Rohan, C Canaff, E Fromentin, and MJ Guisnet. *J. Catal.* 177: 296–305, 1998.
60. P Botella, A Corma, JM Lopez-Nieto, S Valencia, and R Jacquot. *J. Catal.* 195: 161–168, 2000.
61. YG Adewuyi and AS Gupta. Benign Catalytic Synthesis of Pharmaceutical Intermediates in Supercritical Carbon Dioxide. Green Chemistry Seventh Annual Proceedings, Washington, DC, 2003, pp. 146–148.
62. YG Adewuyi and J Mbah. Annual AICHE-Meeting, November, 2004, 204i.
63. GS Aakash. Development of Zeolite Based Friedel–Crafts Acylation Process in Supercritical Carbon Dioxide. MS thesis, North Carolina A&T State University, Greensboro, NC, 2002
64. EJ Beckman. *Environ. Sci. Technol.* 36: 347A–353A, 2002.
65. MG Hitzler, FR Smail, SK Ross, and M. Poliakoff. *Org. Proc. Res. Dev.* 2: 137–142, 1998.
66. P License and M Poliakoff. Chemical Reactions in Supercritical Carbon Dioxide: From Laboratory to Commercial Plant Green Chemistry Seventh Annual Proceedings, Washington, DC, 2003, pp. 1–5.

24 Green Biphasic Homogeneous Catalysis

Philip G. Jessop and David J. Heldebrant
Department of Chemistry, Queen's University

CONTENTS

24.1 INTRODUCTION

In homogeneously catalyzed processes, catalyst recovery and recycling are necessary in order to prevent heavy-metal contamination of products and the loss of expensive catalysts. Transition metals, especially Rh, Pd, and Pt, have become enormously expensive. Additionally, at least for asymmetric catalysis, the chiral organic ligand is even more valuable than the transition metal. Environmental considerations also dictate that the catalysts must be recovered; metal contamination of product or waste streams obviously must be avoided. Traditional methods for catalyst recovery include distillation of the product from the solution (only possible for thermally robust catalysts) and decomposition of the catalyst followed by filtration or extraction [1]. These methods add significantly to the complexity and expense of the process. The use of heterogeneous or heterogenized catalysts (e.g., polymer-supported complexes) is of course ideal because of the facility in which they can be separated from the product stream, but they often suffer from leaching and diminished tunability and selectivity. There is an industrial example of the use of a heterogenized catalyst, a Rh complex bound electrostatically to an ion exchange resin for methanol carbonylation, but severe leaching is observed [2, 3]. Thus there is a need in industry for an alternative method of homogeneous catalyst use and recovery. The need is met by biphasic catalysis.

Biphasic catalysis, in which two immiscible fluid phases partition the product and the catalyst from each other, can be performed in three general batch process schemes [4]:

(a) The products separate naturally, during catalysis or subsequent cooling
(b) An extraction solvent is used to extract the products after catalysis, or
(c) The extraction solvent is added before catalysis begins.

The last two of these methods are illustrated in Figure 24.1, in which supercritical CO_2 ($scCO_2$) is shown as the extraction solvent. Of course, there are more variations including continuous-flow versions, as will be described later. The choices currently available for the two immiscible solvents are listed in Table 24.1.

In many of these systems, the catalyst must be modified from its usual lipophilic form so that it will partition preferentially into the catalyst-bearing fluid rather than the product-bearing fluid. If the partition coefficient is not very high, the losses of catalyst or free ligand after each cycle will be unacceptably high. For example, for the organic–fluorous biphasic system, the catalyst must be highly fluorous, meaning that it partitions preferentially into fluorous rather than normal organic solvents. For systems in which water or another highly polar phase serves as the catalyst-bearing phase, it is preferable to use a catalyst which is a salt or which contains ligands that have charged functional groups (Scheme 24.1).

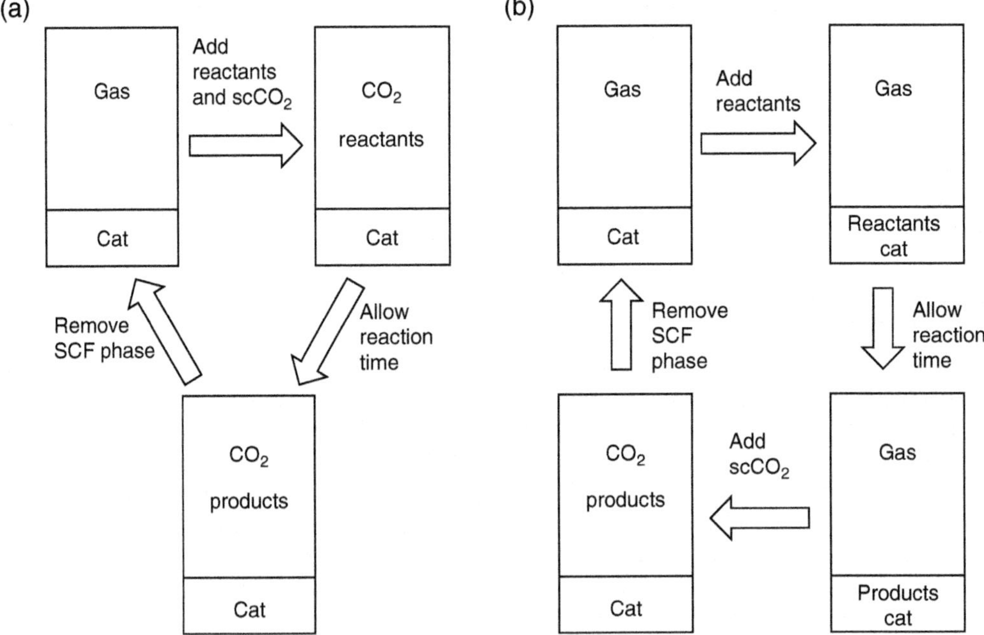

FIGURE 24.1 Bench-scale biphasic catalysis using $scCO_2$ as the product-bearing phase and a CO_2-insoluble liquid as the catalyst-bearing phase (represented by the small lower rectangle). The "gas" label represents air, inert gas, or residual CO_2. The $scCO_2$ can be added either (a) after the reaction is complete, or (b) before the reaction starts. In practice, the product does not partition exclusively into the $scCO_2$ phase, so that a flow of $scCO_2$ over some time is required to extract all the product. (From PG Jessop. *J Synth Org Chem* 61:484–488, 2003. With permission.)

TABLE 24.1
Binary Solvent Systems for Biphasic Catalytic Transformations

Product-Bearing Phase[a]	Catalyst-Bearing Phase[a]	Catalyst Modification	Ref.[b]
Organic	H_2O	Water soluble	[6]
Organic	Organic	None	[7]
Organic	Fluorous[c]	Fluorous[c]	[8, 9, 10]
Organic	Ionic liquid	None[d]	[11–14]
Organic	Liquid polymer	None[d]	[15]
H_2O	Ionic liquid	None	[16, 17]
H_2O	H_2O	Water soluble	None (see text)
$scCO_2$	H_2O	Water soluble	[18]
$scCO_2$	Ionic liquid	None[d]	[19–22]
$scCO_2$	Liquid polymer	None[d]	[23]

[a]Note that inverted systems are possible, in which the two solvents switch roles.
[b]Not intended to be a comprehensive list of papers.
[c]Having high fluorine content, such as perfluorinated alkanes or amines.
[d]Although no catalyst modification is strictly necessary, leaching of the catalyst, if it occurs, could be minimized by using modified catalysts.

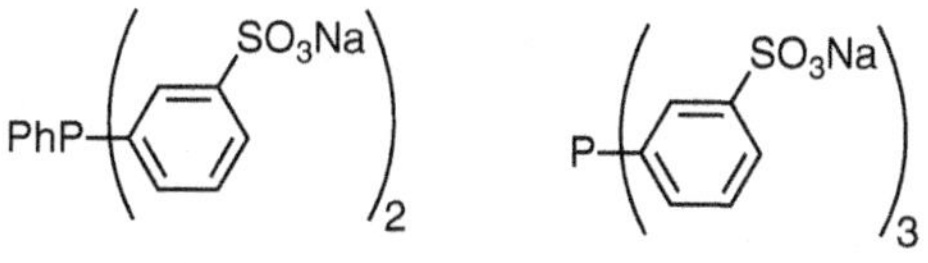

tppds

tppts

SCHEME 24.1 Two water-soluble phosphines commonly used for homogeneous catalysis in the aqueous phase.

Historically, early examples of biphasic catalysis included Kwiatek's 1962 report of homogeneous hydrogenations in H_2O–benzene [24], Shell's SHOP process (patents in 1969 to 1972) [4, 6], and Parshall's catalysis in ionic liquids (ILs) in 1972 [25]. Academic interest has flourished in the past decade, primarily because of the successes obtained industrially using this method.

Biphasic catalysis is in use industrially in several processes [4, 6, 26]. Ruhrchemie (now Celanese) operates a hydroformylation plant in Oberhausen, Germany using aqueous–organic biphasic catalysis; the aqueous phase contains the catalyst while the product aldehyde is itself the organic phase [4, 6, 26, 27]. The SHOP process uses a phosphine-modified nickel catalyst in 1,4-butanediol for the oligomerization of ethylene to α-olefins, which phase separate because they are immiscible with the solvent. In terms of capacity, this is the largest "single feed application of homogeneous catalysis" [4]. A variety of other processes use biphasic homogeneous catalysis including Kuraray's process for the preparation of octanol via butadiene hydrodimerization, which uses a water–hexane solvent pair [28, 29].

It is arguable that biphasic catalysis, even when it involves traditional organic solvents, should be considered an example of green chemistry (and green chemical engineering). The elimination of the many postreaction separation stages represents the elimination of significant energy consumption and waste material, and the use of catalysis avoids the use of more steps with stoichiometric reagents. The authors, therefore, hope that industry will continue to convert processes to the biphasic method.

Because the elimination of volatile organic solvents from chemical processes is a desirable goal of green chemistry, this review is concerned with those biphasic systems not containing a volatile organic solvent, which eliminates the first five of the solvent combinations in the table. Even remaining the combinations are not always likely to be green; the energy requirements, the toxicity and environmental impact of the solvents, and the efficiency of the separation are all factors that should be taken into consideration. One should also note that the use of the product itself instead of a second solvent is a green strategy and will also be discussed. Each of the VOC-free solvent combinations in Table 24.1 will be described in turn in the remainder of this chapter.

VOC-free solvent combinations are preferred because of the obvious economic and environmental advantages of reducing emissions, eliminating flammability, and in many cases reducing or eliminating toxicity. Of the solvents to be discussed, primarily H_2O, CO_2, poly(ethylene glycol), and ILs, the first three are nontoxic and preliminary results show that the toxicity is low for some of the ILs [30].

24.2 H_2O–SCF SYSTEMS

Aqueous-phase catalysis is now a major field on its own [6], having surmounted the early prejudice that organometallic chemistry could not be performed in water. It is particularly effective when the organic products, but not the reagents, are insoluble in water, for then an automatic phase separation is observed. However, for those systems in which the product does not automatically phase separate, one can use an added solvent to extract the product. The most obvious but not the only VOC-free choice of added solvent is $scCO_2$ [31, 32]. The use of $scCO_2$ ($T_c = 31.1°C$, $P_c = 73.8$ bar) [33, 34] as the product-bearing phase has not only the potential to decrease the environmental impact but also to improve the efficiency of the separation because it is particularly poor at dissolving water, salts, or hydrophilic organics. Other SCFs such as ethane, SF_6, or partially fluorinated methanes are also effective but not as desirable environmentally.

24.2.1 Phase Behavior

The solubility of water in $scCO_2$ is low but not zero at moderate temperatures. For example, at 50°C and 345 bar of CO_2, the solubility of water in the supercritical phase is 0.75 mol% or 0.31 mass% [35]. Thus, the H_2O–$scCO_2$ system is not perfect, in the sense that the CO_2 phase is contaminated by some of the lower solvent. However, the fact that neither solvent is toxic, environmentally damaging, or expensive more than makes up for this contamination for any application in which removal of water from the product is not necessary or is not difficult.

CO_2 also has significant solubility in water, which causes a significant drop in the pH of the aqueous phase. As a result, the pH of the aqueous solution drops to between 2.85 and 2.90 at 50°C and CO_2 pressures above 70 bar [36]. The pH drop can be somewhat ameliorated by the use of buffers [37] or reverse emulsions [38]. The pH inside the aqueous core of reverse micelles in $scCO_2$ continuous phase has been found to be 3.1 to 3.5 [38].

Partition coefficients, here defined as $[\text{solute}]_{CO_2}/[\text{solute}]_{aq}$, indicate the relative ease of extraction of the solute from water by $scCO_2$. The partitioning of solutes is dependent on the pH and ionic strength of the aqueous phase [39]. For example, C_6Cl_5OH, an acidic solute, partitions preferentially into $scCO_2$ but the value of the partition coefficient drops significantly with a rise in pH [39]. Hydrogen-bond donor molecules such as alcohols and primary amines have, as one would expect, smaller partition coefficients than nondonor

TABLE 24.2
Partition Coefficients of Selected Solutes at 40°C [41]

Solute	80 ± 4 bar	150 ± 4 bar	300 ± 3 bar
Benzyl alcohol	0.41	1.56	2.28
Benzoic acid	0.61	1.42	2.11
2-Hexanone	28.9	132.7	141
Cyclohexanone	5.6	45.9	62.9
Caffeine	—	0.13	0.22
Vanillin	—	1.41	2.66

molecules (3 for aniline, 47 for benzaldehyde at 50°C and 300 bar) [40], showing that extraction of products that are hydrogen donors is likely to be more problematic (see also Table 24.2).

The partition coefficients are also pressure and temperature sensitive, as one would expect [40, 41]. The data of Brudi et al. [41] are offered here as illustrative of the trends. Studying the solutes phenol, benzoic acid, benzyl alcohol, caffeine, and vanillin at temperatures of 40°C, 50°C, and 60°C, they showed that the partition coefficients increase rapidly until a pressure near 200 bar, after which they increase only slowly (Figure 24.2). At 150 bar, the partition coefficients decrease with an increase in temperature, but at 250 bar they increase with increasing temperature. This indicates that there is a crossover pressure. This phenomenon is unique to SCFs, and is the pressure at which a change in temperature will have no effect on the absolute solubility of the solute in the supercritical fluid [42]. At lower pressures, although a temperature rise elevates the vapor pressure of the solute, the dominating effect is the reduction in density of the CO_2 phase, which decreases the solubility of the substrate. At pressures above the crossover pressure, the temperature does not alter the density of the

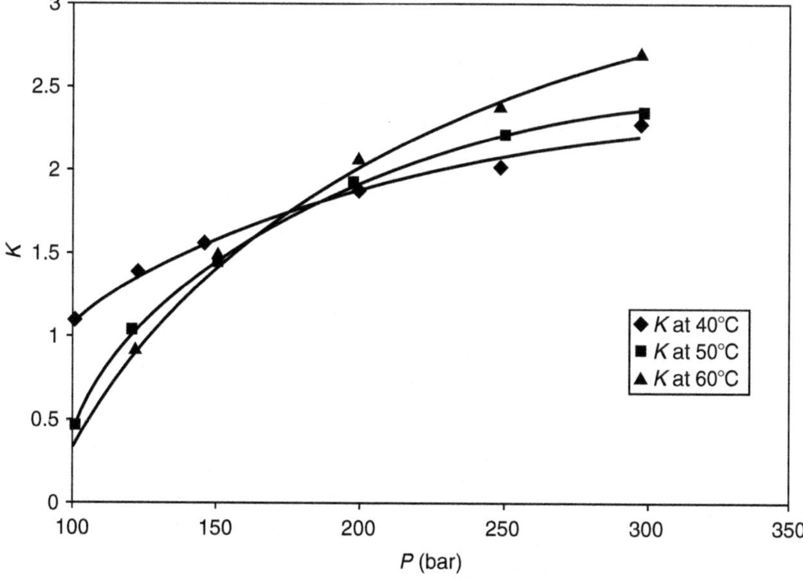

FIGURE 24.2 The $scCO_2$–water partition coefficient of benzyl alcohol as a function of the temperature and CO_2 pressure [41]. A crossover pressure is evident at approximately 180 bar.

CO_2 significantly, so that the dominating effect is the elevated vapor pressure of the solute, which increases the solubility of the solute in the CO_2 phase. The data of Brudi et al. show that there is a crossover pressure in partition coefficients as well.

Homogeneous catalyst precursors are typically insoluble in water, so they must be modified to render them soluble in the water and unextractable by $scCO_2$. The field of aqueous-phase homogeneous catalysis has been reviewed [6, 28, 43].

24.2.2 Applications

The CO_2–H_2O biphasic system has obvious advantages, including being environmentally and toxicologically benign, being inexpensive, and not having serious consequences from any cross-contamination of the two solvents. A major point of concern is the effect of the pH drop on reaction performance and catalyst stability [37]. The pH drop may not always be a problem; catalysts that either tolerate or perform better at low pH will not be adversely affected. The Kuraray process for the aqueous-phase hydrodimerization of water, with a Pd catalyst containing $Ph_2PC_6H_4$*m*SO_3Li ligands, uses CO_2 gas not as an extracting phase but as a reagent to generate carbonic acid with which to stabilize the phosphine as the phosphonium salt [26].

Bhanage et al. first described the use of water-soluble catalysts in an aqueous–$scCO_2$ biphasic system (Equation (24.1)) [18]. With the $RuCl_3$–TPPTS combination, they found that a H_2O–$scCO_2$ medium allowed for greater conversion and greater selectivity for the unsaturated alcohol (38% conversion, 99% selectivity) over 2 h than did H_2O–toluene (11 and 92). Switching to $RhCl_3$ and TPPTS, they obtained 100% selectivity for the saturated aldehyde.

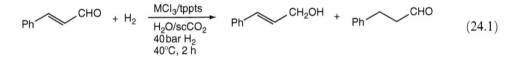

Ph⁀CHO + H_2 —[MCl_3/tppts; $H_2O/scCO_2$, 40 bar H_2, 40°C, 2 h]→ Ph⁀CH_2OH + Ph⁀CHO (24.1)

Colloidal catalysts have also been found to be effective in H_2O–SCF biphasic media. Bonilla et al. [37] used an aqueous suspension of Rh colloid as a catalyst for hydrogenation of arenes to cyclohexane derivatives in aqueous–scC_2H_6 biphasic medium (for example Equation (24.2), giving 99% conversion, 98% selectivity). The Rh colloid was generated *in situ* by the action of H_2 upon $[RhCl(cod)]_2$ (cod = 1,5-cyclooctadiene). Even substrates that are highly hydrophobic, such as *p*-xylene, were hydrogenated in good conversion. For these reactions, H_2O–$scCO_2$ is not appropriate because the catalyst is acid-sensitive. Catalyst reuse was not tested.

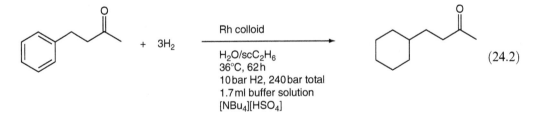

+ $3H_2$ —[Rh colloid; H_2O/scC_2H_6, 36°C, 62 h, 10 bar H2, 240 bar total, 1.7 ml buffer solution, $[NBu_4][HSO_4]$]→ (24.2)

Jacobsen et al. improved the concept by adding surfactants, thereby creating emulsions. This leads to a large increase in rate for styrene hydrogenation (Equation (24.3)) over that in the absence of surfactants. The rate increase was attributed to the greater interfacial surface area, which presumably increased the rate of mass transfer of substrate or hydrogen to the site of the catalysis (believed to be at the interface of the phases rather than in the aqueous core of the micelles) [44]. Even hydrophobic 1-eicosane was hydrogenated at a reasonable rate.

Breaking of the emulsion is as simple as lowering the pressure; thus postreaction separation of product and catalyst is not much more difficult than with traditional biphasic catalysis. Catalyst reuse was demonstrated for three cycles. With some engineering design, this kind of system could even be made into a continuous-flow process.

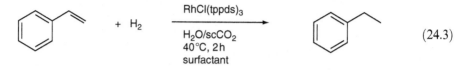

$+ H_2$ — $RhCl(tppds)_3$; $H_2O/scCO_2$, 40°C, 2 h, surfactant (24.3)

Emulsions can also be created with catalysts that are themselves surfactants. This has been tested for aerobic oxidation of aromatic hydrocarbons, an industrially important reaction that is normally rather hazardous owing to the combination of O_2, organic solvent (acetic acid), and heat. Zhu et al. [45] investigated the possibility of using an H_2O–$scCO_2$ biphasic system as a nonexplosive alternative. In an H_2O–$scCO_2$ emulsion, the oxidation of toluene was achieved with >99% selectivity and conversion, and a TOF (turnover frequency) of 22 h^{-1} (Equation (24.4)). No surfactant beyond the fluorinated catalyst was used. Catalyst recycling was not investigated.

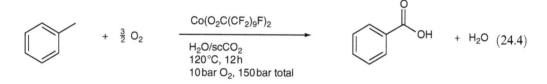

$+ \frac{3}{2} O_2$ — $Co(O_2C(CF_2)_9F)_2$; $H_2O/scCO_2$, 120°C, 12 h, 10 bar O_2, 150 bar total — OH $+ H_2O$ (24.4)

A CO_2–H_2O emulsion created without the use of surfactants was demonstrated by Timko et al. [46]. Those authors showed that 2 min of continuous ultrasound could create an emulsion that lasted for 30 min, and pulsed ultrasound could maintain the emulsion for as long as desired. The technique was demonstrated with an uncatalyzed test reaction.

Particularly water-soluble products, of course, will tend to partition into the aqueous phase along with the catalyst; for such products the method must be modified. Leitner [49] proposed inverting the method, using CO_2-soluble catalysts, so that the $scCO_2$ phase becomes the catalyst-bearing phase and the water becomes the product-bearing phase. Hâncu and Beckman [47, 48] used a CO_2-soluble catalyst, $PdCl_2(PAr_3)_2$ (Ar = $C_6H_4pCF_3$ or $C_6H_4pCH_2CH_2C_6F_{13}$), to promote the direct synthesis of H_2O_2 from air and H_2 in $scCO_2$. An aqueous phase was present during the reaction in order to sequester the product.

The inverted method was also used by Leitner [32, 49] for a hydroformylation reaction (Equation (24.5)), giving the desired aldehyde in 500 TON (turnover number) and complete conversion. The catalyst could be successfully recycled by draining out the aqueous phase, without depressurization of the CO_2, followed by addition of a fresh aqueous solution of substrate. Formation of Rh black was blamed on the low pH, which could be ameliorated by the addition of 0.1 *M* MOPS buffer (3-*N*-[morpholino]propane sulfonic acid) with the result that Rh black formation was prevented at the cost of a lower conversion.

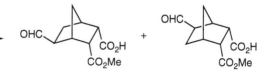

CO_2H, CO_2Me — $Rh(acac)(CO)_2$, $10P[C_6H_4pCH_2CH_2(CF_2)_6F]_3$; CO/H_2 20 bar, $H_2O/scCO_2$, 60°C, 20 h — OHC, CO_2H, CO_2Me + OHC, CO_2H, CO_2Me

(24.5)

24.3 IONIC LIQUID–SCF SYSTEMS

24.3.1 PHASE BEHAVIOR

ILs, salts that are liquid at or near room temperature, date from the 1914 discovery of $[NH_3Et][NO_3]$ by Walden [50]. Parshall [25] and others [51] in the 1960s and 1970s used ILs as media for homogeneous catalysis because the product would automatically separate as it formed. Parshall's IL was $[NEt_4][SnCl_3]$, which melts at 68°C. Chauvin et al. [52] used [bmim]Cl–$AlEtCl_2$ as an IL solvent for $NiCl_2(P^iPr_3)_2$-catalyzed dimerization of propene in 1990. ILs that are liquid at room temperature, metal-free, thermally stable, and water-insensitive were first published by Wilkes and Zaworotko in 1992 [53]. These contained anions such as BF_4^- and PF_6^-. The field has rapidly developed in the last several years, so that ILs of a wide variety of anions are now known and the number of future possibilities seems almost endless. A few commonly used ions are shown in Scheme 24.2. While most ILs are completely nonvolatile, dimethylammonium *N,N*-dimethylcarbamate (DIMCARB) and related liquids are exceptions, as they are distillable at moderate temperatures. Without exception (so far), the ILs are very polar solvents, at least as reported by solvatochromic probe molecules [54–56]. They are also extremely designable. A large number of combinations and structures of ions are possible.

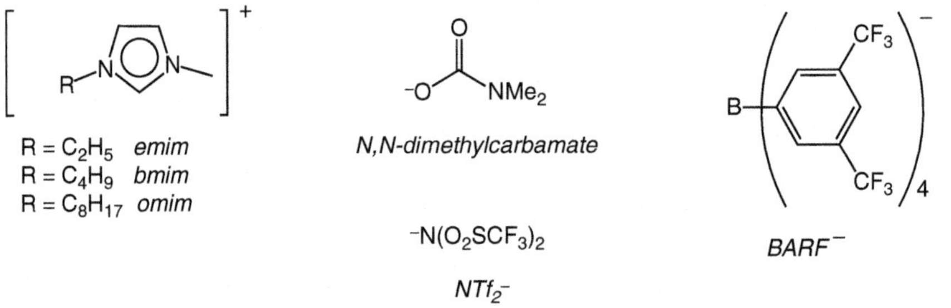

$R = C_2H_5$ *emim*
$R = C_4H_9$ *bmim*
$R = C_8H_{17}$ *omim*

N,N-dimethylcarbamate

$^-N(O_2SCF_3)_2$

NTf_2^-

$BARF^-$

SCHEME 24.2 Structures of selected cations and anions used in ionic liquids. Cations that need no illustration include tetraalkylammonium, tetraalkylphosphonium, and *N*-alkylpyridinium.

While ILs are portrayed in the literature as "green" reaction media as a result of their complete involatility, their qualifications as green solvents are somewhat questionable. Some ILs, particularly the imidazolium cations with longer alkyl chains, may have considerable toxicity [30]. As long as volatile organic solvents are used, as they typically are, to extract products from the ILs or to clean and purify the ILs, then the ILs are only as green as the extracting solvent. In some cases, the products are volatile enough to be distilled from the IL, but for applications in homogeneous catalysis this is unlikely to be practical for all but the most volatile products because of the resulting thermal damage to the catalyst. In the most desirable scenario, the product, as it is formed, naturally separates from the IL due to mutual immiscibility. However, in all other cases, what is most needed is a green method for removing product from the IL.

A green method was found when Brennecke and Beckman's groups discovered in 1999 that $scCO_2$ is completely incapable of dissolving $[bmim]PF_6$, a commonly used IL, even at enormous pressures [57]. This meant that it is possible to extract organic products from ILs using $scCO_2$, or to perform biphasic catalysis in IL–$scCO_2$ biphasic media, *with no contamination of the product-bearing fluid by the catalyst-bearing fluid.* This is impossible with any of the competing technologies. The research team continued with reports that evaluated how much CO_2 would be required to extract certain organic products from $[bmim]PF_6$ (Table 24.3) [58–60]. As expected, the organic compounds that have less polarity and thus dissolve CO_2 readily require less CO_2 to be extracted from the IL. The major drawback to

TABLE 24.3
Moles of CO_2 Required to Extract an Organic Solute from [bmim]PF_6 at 40°C, Compared to the Solubility of CO_2 in the Pure Liquid Solute [58–60]

Organic Solute	Mole Ratio CO_2: Solute for 95% Recovery	Pressure (bar) Required to Dissolve 70 mol% CO_2 in the Liquid Solute
Cyclohexane	1840	64.8
Hexane	2340	54.1
Benzene	2750	59.6
Anisole	4000	60.8
Chlorobenzene	4770	66.7
Methyl benzoate	5870	71.0
Benzaldehyde	6850	77.8
Acetophenone	20,300	75.4

extraction by CO_2 is of course the excessively large amount of CO_2 that is required to extract products of low volatility.

The inability of CO_2 to extract nonvolatile compounds from ILs is fortunate because most homogeneous catalysts are nonvolatile. Thus, in theory, unmodified catalysts can be dissolved in IL and will not be extracted with the products after the reaction. However, extraction of dissociated ligand could potentially lead to catalyst decomposition. Catalyst modification may therefore be advisable in IL–scCO_2 biphasic catalysis in order to greatly bias the partitioning of the catalyst and free ligand to the IL phase or to increase their solubility in the IL. Several strategies have been used in IL–VOC systems that are appropriate for adaptation to IL–scCO_2 systems:

(a) Using a cationic ligand, such as a phosphine attached to a cationic group [61]
(b) Using an anionic ligand such as TPPTS, with either an alkali metal counterion [11] or more preferably (for solubility reasons) an organic cation [21, 62], or
(c) Using a charged complex [63].

ILs are not always completely insoluble in scCO_2. Some ILs, such as DIMCARB, are volatile and therefore potentially extractable by CO_2. Although the tests have not been made, it seems likely that a highly fluorinated IL could have significant solubility in CO_2. Even nonvolatile ILs are known to be soluble in scCO_2/cosolvent mixtures. Wu et al. [64] showed that, if a polar cosolvent is present in scCO_2 at >10% mole fraction, the supercritical mixture is capable of dissolving significant quantities of [bmim]PF_6. This is unlikely to be a problem during extractions from ILs except for the most volatile and easily extracted solutes.

Carbon dioxide, although it does not dissolve ILs, does dissolve *into* ILs. Brennecke and coworkers [57, 59] found that the solubility of CO_2 appears to be greatest in ILs with fluorinated anions, especially the PF_6^- anion. The solubility of CO_2 in ILs increases dramatically with increasing pressure but not temperature. Most surprising is the lack of a large volumetric expansion of the ILs even at high pressures and CO_2 mole fractions as high as 75%. In contrast, nonionic organic solvents expand many fold when they are exposed to CO_2 at pressures over 50 bar [65–67]; these are then called "CO_2-expanded liquids." Kazarian et al. [68], using ATR-IR to study CO_2-expanded ILs, contradicted Blanchard and Brenneke's prediction that the CO_2 interacts more favorably with the PF_6^- anion than BF_4^-. It was discovered that the BF_4^- anion interacts more strongly with CO_2, through Lewis acid–base interactions. The difference between the solubilities of CO_2 in these two similar ILs is predicted to be due to a difference in the available free volume in the ILs.

The dissolution of CO_2 into ILs changes the physical properties of the liquid. Even at pressures as low as 60 bar of CO_2, it is visually obvious that the viscosity of ILs has dropped greatly. Although the effects have not all been measured, one would expect, based upon the behavior of other "expanded" liquids, that the dissolution of CO_2 in ILs would increase mass transfer rates, lower the viscosity [69, 70], increase the solubility of reagent gases, and decrease the strength of the solvent cage. Any of these effects could potentially alter the rate or selectivity of reactions performed in ILs in the presence of CO_2. Significant viscosity changes have been observed [69, 70]. Melting point depressions for ILs under CO_2 pressure were reported by Kazarian et al. [71]; [C_{16}mim]PF_6 generally melts at 75°C but under 70 bar of CO_2 (0.55 mole fraction) it is reduced to 50°C. The IL did not melt at 40°C even under pressures as high as 170 bar. Similar melting point depressions have been observed for nonionic organic compounds under CO_2 pressure [72, 73]. Freezing of the solvent upon removal of the CO_2 could have the side benefit of encapsulating catalysts within, thereby protecting them from oxidation until the next time they are used. Although one might have expected the solvent polarity to drop upon expansion of the IL, the polarity of the local environment around solvatochromic probe molecules does not significantly change [69, 70]. Overall, it is clear that there remains a great deal of work to be done before we can understand the nature and potential uses of "CO_2-expanded" ILs.

Although CO_2 dissolves well into ILs such as [bmim]PF_6 (the Henry's law constant is 53 bar at 25°C), the solubility of other gases into ILs is not nearly as high. Brennecke and coworkers [74] found that the Henry's law constant for dissolution of O_2 into [bmim]PF_6 at 25°C is approximately 8000 bar, while those of H_2 and CO are so high that the dissolution was completely undetectable. Such extremely low solubilities are potentially problematic for reactions involving those reagent gases. Fortunately, it appears that CO_2-expansion of the IL or the use of ILs with different anions are reasonable strategies for solving this problem.

24.3.2 Applications

Before the advantages of the IL–scCO_2 combination for biphasic catalysis were appreciated, there was an example of the use of the combination for catalysis without catalyst recovery. Noyori and coworkers [75, 76] used the IL DIMCARB as a reagent and second phase in the synthesis of *N,N*-dimethylformamide by $RuCl_2(PMe_3)_4$-catalyzed hydrogenation of scCO_2 (Equation (24.6), R = CH_3) The location of the catalyst during the reaction has never been positively identified, but the recent evidence that the hydrogenation proceeds by a cationic mechanism [77] suggests that the catalysis may have taken place in the IL rather than in the scCO_2 phase. Catalyst recovery and recycling was neither attempted nor contemplated because of the very high yields of the reaction (420,000 turnovers or mol product per mol catalyst).

$$R_2NH + H_2 + CO_2 \xrightarrow[100\ ^\circ C]{\text{Ru(II) catalyst}} [R_2NH_2][O_2CH] \xrightarrow{\Delta} R_2NC(O)H + H_2O \quad (24.6)$$

The advantages of the IL–scCO_2 combination for catalyst recovery became immediately obvious to several research groups after the publications of Brennecke and Beckman. The first to publish was Jessop's group [19], who described the asymmetric hydrogenation of α,β-unsaturated carboxylic acids in [bmim]PF_6 followed by product extraction with scCO_2 (Equation (24.7) and Equation (24.8)). The catalyst was highly aromatic and therefore far more soluble in the IL than in the supercritical fluid. Not only was the enantioselectivity high, but also the catalyst solution could be reused for a total of five cycles with no loss in selectivity and only a slow decline in conversion per cycle (99% in the first cycle, 97% in the last).

In a subsequent paper [78], the group described in more detail the effect of a number of parameters on the selectivity of the reaction. The inclusion of an alcohol or CO_2 dissolved in the IL caused an increase in the enantioselectivity for the hydrogenation of the atropic acids

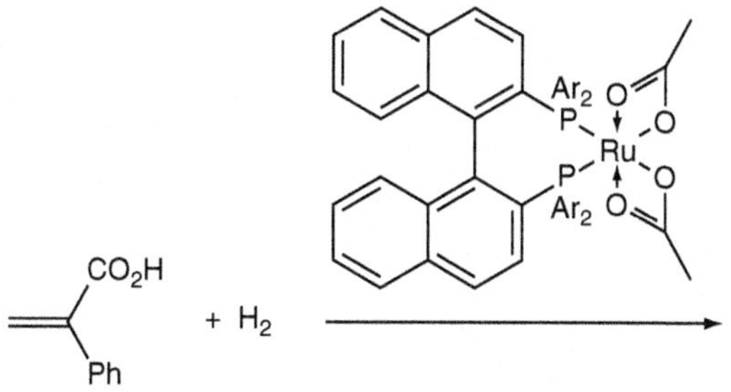

(24.7)

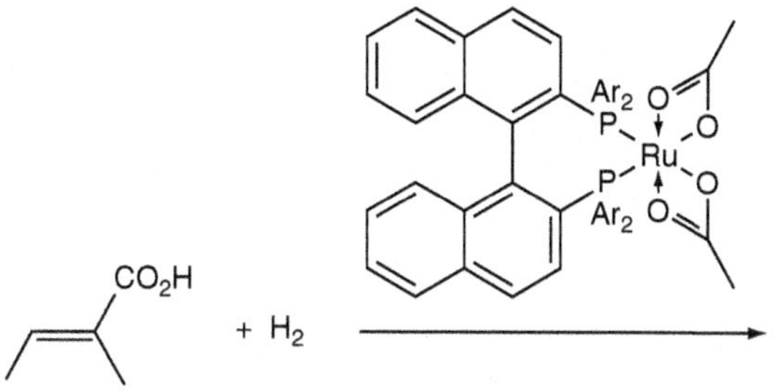

(24.8)

(Equation (24.7)) but a decrease in the enantioselectivity for the hydrogenation of tiglic acid (Equation (24.8)). These observations were explained with reference to the earlier findings by others for these reactions in liquid methanol; the kinetic and thermodynamic availability of large quantities of H_2 dissolved in the reaction phase strongly affects the enantioselectivity of these hydrogenations, positively in the case of the atropic acids and negatively in the case of tiglic acid. The dissolution of a cosolvent (either alcohol or CO_2) is believed to have increased the solubility or mass transfer of H_2 into the IL phase enough to significantly alter the enantioselectivity. Thus, for the atropic acid substrates, it is important that the $scCO_2$ be present during the reaction time, while for tiglic acid and like substrates it is better that the $scCO_2$ not be introduced until the reaction is complete.

Liu et al. [20] reported the use of [bmim]PF_6–$scCO_2$ biphasic medium for the hydrogenation of CO_2 in the presence of dipropylamine, to produce dipropylformamide (Equation (24.6), R = propyl, catalyst = $RuCl_2(dppe)_2$ where dppe = $Ph_2PCH_2CH_2PPh_2$) in high yield and selectivity. A total of four cycles were performed using one batch of the catalyst solution. Recovery of the dipropylformamide by extraction with $scCO_2$ was inefficient during the first two cycles but was nearly quantitative in subsequent cycles. Liu et al. also demonstrated the hydrogenation of 1-decene and cyclohexene with $RhCl(PPh_3)_3$ catalyst, achieving high conversion in each of four cycles.

During studies of the hydroformylation of 1-hexene using $[Rh_2(OAc)_4]$–$P(OPh)_3$ as the catalyst (Equation (24.9)), Cole–Hamilton's group observed chemical instability of the [bmim]PF_6 IL [21]. Hydrolysis of the PF_6^- anion generated $O_2PF_2^-$ and HF, which led to degeneration of the catalyst after two to three cycles. With charged ligands of the type [bmim][$Ph_2PC_6H_4SO_3$], however, the catalyst deactivation and leaching were not observed until after nine cycles.

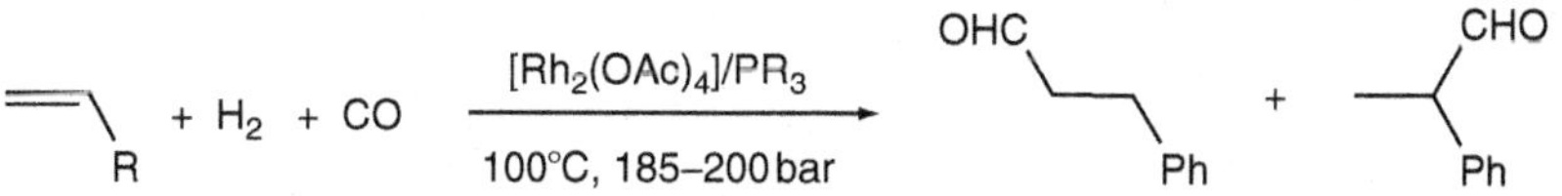

$$\text{R-CH=CH}_2 + H_2 + CO \xrightarrow[100°C,\ 185\text{–}200\,bar]{[Rh_2(OAc)_4]/PR_3} \text{OHC-CH}_2\text{CH}_2\text{Ph} + \text{CH}_3\text{CH(CHO)Ph} \quad (24.9)$$

From an engineering standpoint, continuous-flow processes are generally preferable and more efficient than batch processes, but are usually not possible for homogeneously catalyzed reactions. With catalyst immobilization in an IL and the use of $scCO_2$ as the mobile phase carrying substrate and product, continuous-flow homogeneous catalysis has been demonstrated [21, 22]. The Cole–Hamilton group first described this method for the hydroformylation of 1-octene (Equation (24.9) and Figure 24.3a). The ratio of linear (preferred) over branched product was a constant 3.1 for 33 h [21]. The researchers later reported that using the ligand [pmim][$PPh_2C_6H_4mSO_3$] and the IL [omim][NTf_2] allowed greater rates [2] and a new ligand (structure **1**) allowed a very impressive *l*:*b* ratio of 40 [79].

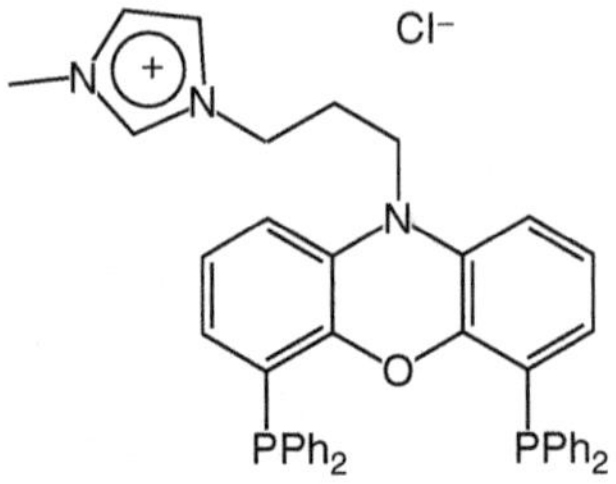

Cl⁻

PPh₂ PPh₂

Asymmetric hydrovinylation of styrene (Equation (24.10)) was found by Leitner's group to be particularly selective if performed in [emim][BARF] but not in ILs of other anions [22]. Complete conversion and 89% ee were achieved. The observed loss of catalytic activity upon recycling was blamed on the instability of the catalyst in the absence of substrate. Converting the process from batch mode to continuous flow (Figure 24.3b) prevented the catalyst deactivation. The catalyst is stable in the presence of substrate, and the use of a continuous-flow method meant that the substrate was never depleted [22]. Conversion was constant at 75% for 60 h.

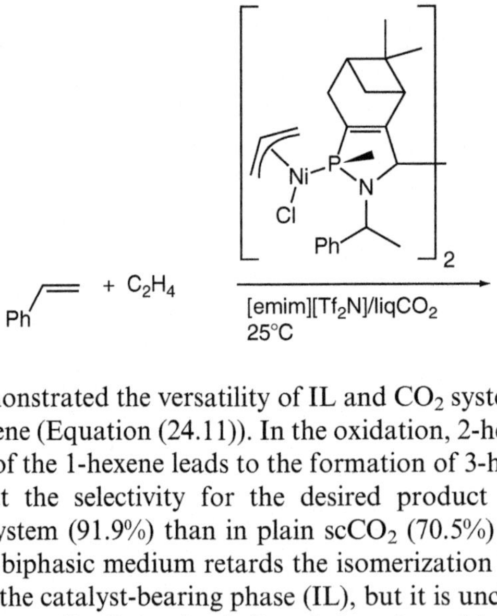

$$\text{PhCH=CH}_2 + C_2H_4 \xrightarrow[\text{[emim][Tf}_2\text{N]/liqCO}_2,\ 25^\circ\text{C}]{[\text{Ni catalyst}]_2} \text{PhCH(CH}_3\text{)CH=CH}_2 \tag{24.10}$$

Hou et al. [80] demonstrated the versatility of IL and CO_2 systems by performing the Wacker oxidation of 1-hexene (Equation (24.11)). In the oxidation, 2-hexanone is the desired product, but isomerization of the 1-hexene leads to the formation of 3-hexanone. In their studies, Hou et al. showed that the selectivity for the desired product was higher in their biphasic [bmim]PF_6–CO_2 system (91.9%) than in plain $scCO_2$ (70.5%) or just the IL (64.2%). It was suggested that the biphasic medium retards the isomerization by lowering the concentration of the substrate in the catalyst-bearing phase (IL), but it is unclear why this would not retard the oxidation rate equally. It is also believed that the oxidation benefits greatly from the increased mass transfer due to CO_2 expansion. The catalyst was recycled after the product was extracted with CO_2, but selectivity and conversion was lost after each cycle.

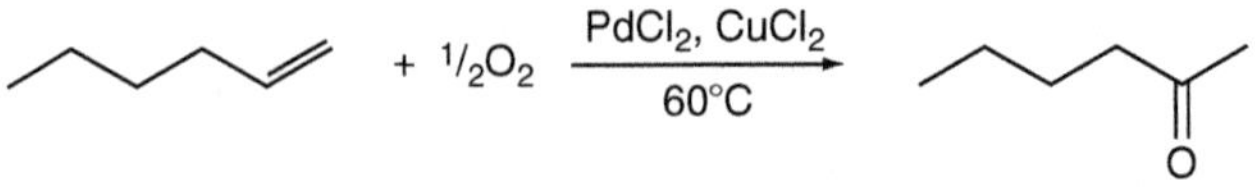

$$\text{1-hexene} + {}^1/_2O_2 \xrightarrow[60^\circ\text{C}]{PdCl_2,\ CuCl_2} \text{2-hexanone} \tag{24.11}$$

Homogeneously catalyzed dimerization of methyl acrylate has been tested as a batch process in IL–$scCO_2$ biphasic medium (Equation (24.12)), with the results nearly identical to the data in IL alone, suggesting that this reaction also could be performed in continuous-flow mode [81].

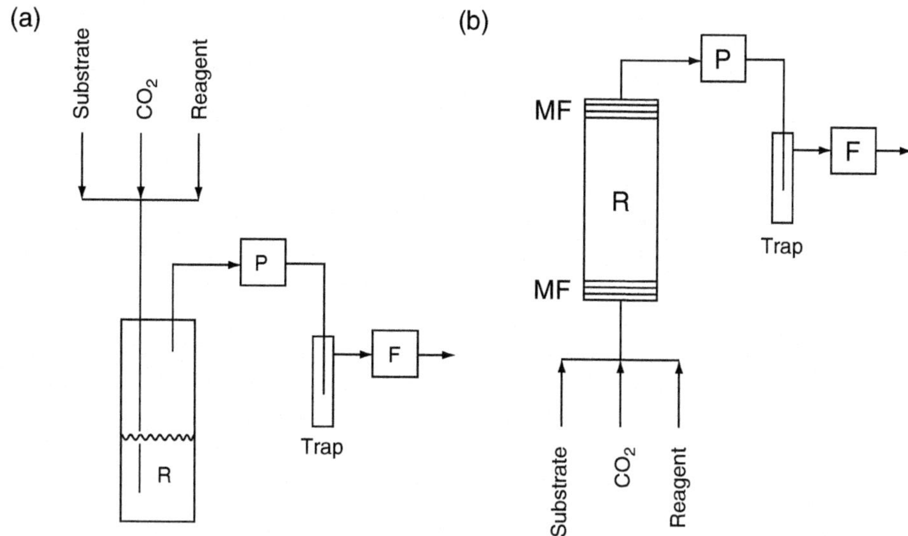

FIGURE 24.3 Flow charts for two methods for continuous-flow homogeneous catalysis using ionic liquids and $scCO_2$ [21, 22]. F = flowmeter, MF = metal filter, P = pressure reducer, and R = reactor. (From PG Jessop. *J Synth Org Chem* 61:484–488, 2003. With permission.)

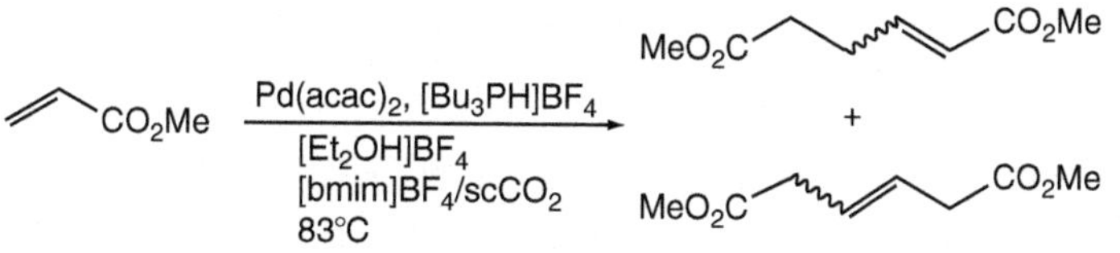

$$\text{CH}_2\text{=CHCO}_2\text{Me} \xrightarrow[\substack{[\text{Et}_2\text{OH}]\text{BF}_4 \\ [\text{bmim}]\text{BF}_4/\text{scCO}_2 \\ 83^\circ\text{C}}]{\text{Pd(acac)}_2,\ [\text{Bu}_3\text{PH}]\text{BF}_4} \text{MeO}_2\text{C}\sim\sim\text{CO}_2\text{Me} + \text{MeO}_2\text{C}\sim\sim\text{CO}_2\text{Me} \quad (24.12)$$

24.4 LIQUID POLYMER–SCF SYSTEMS

Liquid polymers can also serve as a lower liquid phase in combination with $scCO_2$ [23]. Polymers that could conceivably be used in this fashion must not be soluble in $scCO_2$ but must be able to dissolve reagents and catalysts and must be liquid at the desired operating temperature. Candidates appropriate for use at room temperature or above are shown in Scheme 24.3.

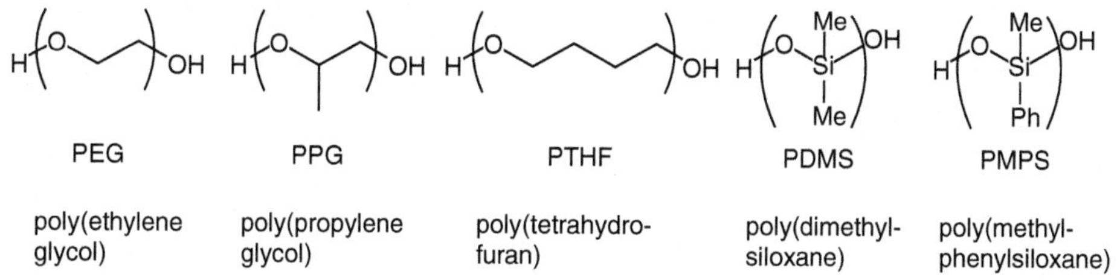

PEG	PPG	PTHF	PDMS	PMPS
poly(ethylene glycol)	poly(propylene glycol)	poly(tetrahydro-furan)	poly(dimethyl-siloxane)	poly(methyl-phenylsiloxane)

SCHEME 24.3 Liquid polymers.

Historically, a related method was called gas–liquid phase transfer catalysis (GL-PTC), in which thermally stable catalysts are dissolved in hot liquid polymers and gaseous reagents are passed over the liquid phase in a continuous-flow system. The catalyst is not readily leached in such a system, but the high temperature requirement limits the applications. Successful examples of the method have been reported by Tundo et al. [82].

The addition of $scCO_2$ to the gas phase enhances the volatility of many organic substrates so much that the method can be used at temperatures close to room temperature.

Advantages to the use of liquid polymers as the catalyst-bearing phase include their low toxicity and cost. The toxicity of poly(ethylene glycol) is known to be very low [83]; it is even approved for use in the U.S. as a food additive [84]. The cost of PEG in bulk is currently approximately U.S. $1/kg. PDMS is also used in food, but only at very low concentrations [85].

24.4.1 Phase Behavior

There are several types of liquid polymers that could potentially be considered here, but the focus will be on the polyglycols and the polysiloxanes. Poly(ethylene glycol) (PEG), otherwise known as poly(ethylene oxide), is liquid at room temperature unless the number average molar mass is over around 700 (Figure 24.4). The melts are highly viscous. Poly(dimethylsiloxane) (PDMS) is a liquid at any molar mass (pour points typically −60°C to −40°C) but the viscosity can be high [86]. PDMS is the most nonpolar of the liquid polymers in Scheme 24.3, having a dielectric constant of 2.8 at 20°C [86], only slightly more polar than toluene. Siloxanes, like poly(glycols), can be structurally varied to give them different physical and chemical properties. Aromatic siloxanes like PMPS and polar and coordinating siloxanes such as nitrilesiloxanes could also be used.

The solubility of CO_2 in these liquid polymers can be extremely high. Daneshvar et al. [88] have shown that the weight fraction of CO_2 in PEG-600 (i.e., PEG of molar mass 600) can be as high as 60% (Figure 24.5). The higher the molar mass of the polymer, the higher the mass% of CO_2 in the polymer phase. The weight fraction solubility of CO_2 in liquid PDMS (mw 308,000) [89] reaches 40% at 50°C, 23% at 80°C, and 20% at 100°C and 260 bar.

As mentioned previously, when pressurized CO_2 dissolves into a solvent, the solvent volume increases and its properties change; this is true for liquid polymers as well. The melting point of PEG drops somewhat upon expansion with CO_2 [90] and the viscosity drops enormously (89% reduction in viscosity for PEG-400 at 40°C, 251 bar) [91]. Liquid PDMS [89] expands by roughly 70% in volume when exposed to 200 bar of CO_2 at 50°C. The volumetric expansion of the polymer is pressure and temperature dependent. When the gas is

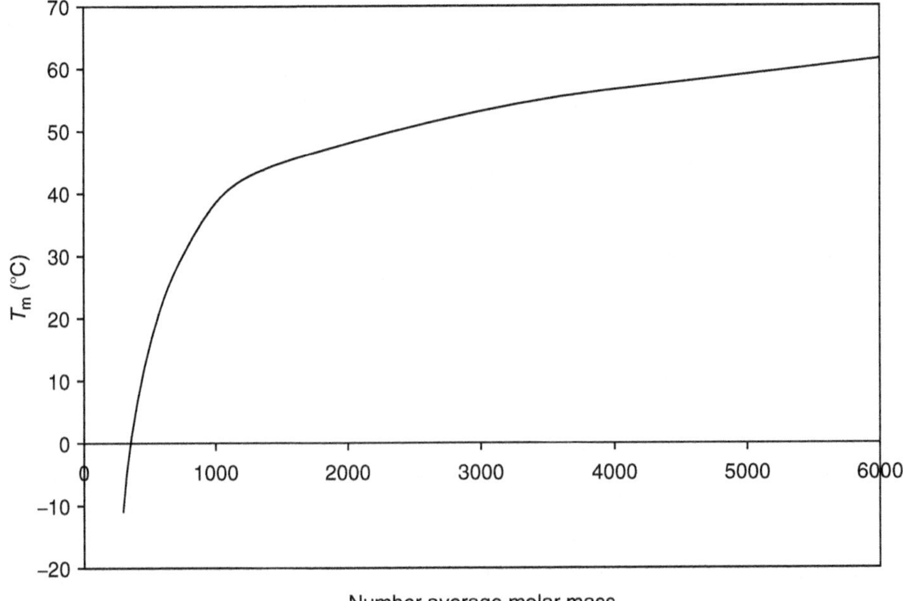

FIGURE 24.4 The melting point of PEG as a function of number average molar mass [87].

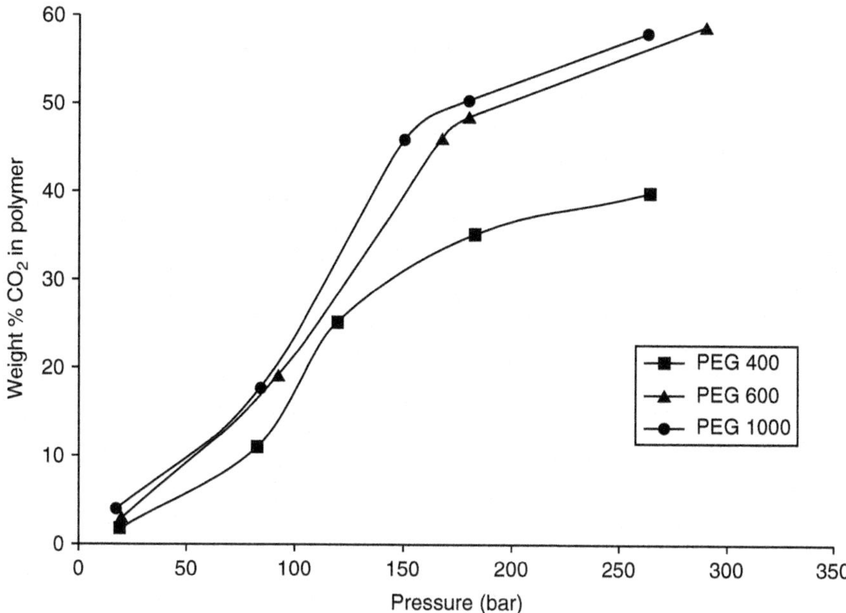

FIGURE 24.5 Weight % of CO_2 in PEG at 50°C [88].

subcritical, the expansion is rapid due to the solubility increase of the gas into the polymer. When the gas becomes supercritical, the mixture becomes less compressible and the rate of expansion decreases. The viscosity of PDMS-308,000 decreases by nearly an order of magnitude when exposed to 90 bar CO_2 at 50°C [92].

Poly(ethylene glycol) has extremely poor solubility in $scCO_2$ and the solubility drops rapidly with increasing molar mass (Figure 24.6). PEG-400 reaches a high of 2.1% by weight in $scCO_2$ at 270 bar and 50°C [88]. PEG-600 reaches only 0.3% under the same conditions. Garg et al. [89] found PDMS-308,000 to be insoluble in $scCO_2$, as demonstrated by the lack of polymer deposits on venting the gas phase. Unfortunately, low molar mass fractions of PDMS are quite soluble in $scCO_2$.

Eckert and coworkers [93, 94] studied the partitioning of some organic compounds such as naphthalene, acridine, and 2-napthol between CO_2 and cross-linked polymer phases including PDMS and poly(cyanopropylmethylsiloxane) (PCPMS). At low pressures (<75 bar), the partitioning of the three organic solutes favors the polymer phase by a factor of 10^2 to 10^3, but at 90 bar the partitioning drops to 1 to 5. This partitioning behavior was also observed when cosolvents such as methanol or 2-propanol (0.3 *M*) were added to the mixture.

Modification of catalysts may be necessary to enhance their solubility in PEG and decrease the extent of extraction of catalyst or ligand by $scCO_2$. Although an attempt to extract PPh_3 from a PEG solution of $RhCl(PPh_3)_3$ removed only 0.06 equivalents of PPh_3 per Rh within 4 h [23], this would be unacceptable in a continuous-flow system. Naughton and Drago used a water-soluble catalyst, RhH(CO)(tppts)$_3$, for hydroformylation in a supported PEG phase, but not with CO_2 as the extracting phase [95].

24.4.2 Applications

The first biphasic reaction using a liquid polymer paired with $scCO_2$ was done by Heldebrant and Jessop [23], who showed that CO_2-expanded PEG-900 can be used successfully as a medium for the hydrogenation of styrene with $RhCl(PPh_3)_3$, extraction of the ethylbenzene

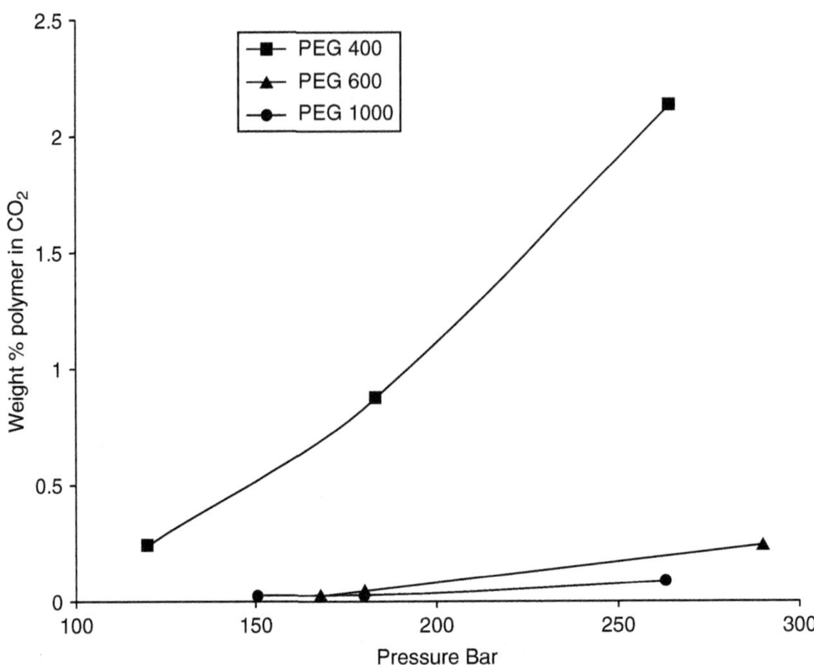

FIGURE 24.6 Weight % of PEG in $scCO_2$ at 50°C [88].

product with $scCO_2$, and catalyst recovery, allowing high catalytic activity through five successive cycles. Each cycle achieved greater than 99% conversion with minimal catalyst leaching (<1 ppm). Note that in order to have a PEG with a MW high enough to prevent significant extraction of the PEG by CO_2, the authors needed to use at least PEG-900 and preferably PEG-1500, both of which are solids at the reaction temperature (40°C). Fortunately, the dissolution of CO_2 into the PEG was found to cause the PEG to melt.

Whether or not PEG and the other liquid polymers will be found to be appropriate solvents for a variety of reactions or appropriate for continuous-flow systems is not yet known. PEG can form complexes with metal ions, possibly in a manner analogous to the crown ethers [87], and therefore may not act as a typical solvent for transition metal complexes. PEG alone cannot be expected to be suitable for every reaction, but a range of liquid polymers with a range of polarities, coordinating abilities, and other characteristics will be needed.

24.5 OTHER BIPHASIC VOC-FREE SYSTEMS AND TECHNIQUES

24.5.1 H_2O–IL Systems

Although many ILs are miscible in water in all proportions, there are many that are immiscible in water, including all alkylmethylimidazolium PF_6^- or NTf_2^- salts and those BF_4^- salts that have an alkyl chain longer than butyl [96]. It is possible, theoretically, to use H_2O–IL as a biphasic medium for catalysis. However, there is likely to be cross-contamination of both solvents by the other, and it is not clear whether partition coefficients for catalysts would be sufficiently large to prevent significant leaching.

A mixture of water and [omim]BF_4 is biphasic at room temperature but the phases merge upon heating. Dyson et al. [16] used this solvent pair to perform homogeneous hydrogenation of 2-butyne-1,4-diol (Equation (24.13)) at 80°C, at which temperature of the system is monophasic. Upon subsequent cooling and phase separation, the catalyst partitions preferentially to the IL phase because of its hydrophobic character, while the product is found in both phases. The aqueous phase was decanted and the catalyst-bearing IL phase recycled. After the first cycle, the IL phase was saturated with product diol and yields of product found in the aqueous phase were quite satisfactory. Some colloidal metal formation and metal contamination of the aqueous phase was observed after a few cycles.

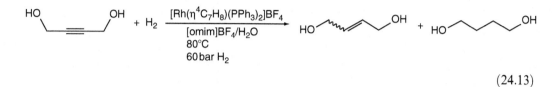

$$\text{HOCH}_2\text{C}\equiv\text{CCH}_2\text{OH} + \text{H}_2 \xrightarrow[\substack{[\text{omim}]\text{BF}_4/\text{H}_2\text{O} \\ 80°\text{C} \\ 60\,\text{bar H}_2}]{[\text{Rh}(\eta^4\text{C}_7\text{H}_8)(\text{PPh}_3)_2]\text{BF}_4} \text{HOCH}_2\text{CH}=\text{CHCH}_2\text{OH} + \text{HO(CH}_2)_4\text{OH} \quad (24.13)$$

Peng et al. [17] showed that the biphasic system H_2O–[omim]PF_6 was effective for the oxidation of benzene to phenol (Equation (24.14)), preventing overoxidation. In a monophasic system, the phenol is more readily oxidized than the benzene, leading to poor selectivity. In the biphasic system, with $Fe(O_3S(CH_2)_{11}CH_3)_3$ as catalyst, up to 60% selectivity could be obtained because the aqueous phase would extract the phenol from the catalyst-bearing IL phase. Ironically, however, the product was recovered from the aqueous phase by extraction with ether.

$$\text{C}_6\text{H}_6 + \tfrac{1}{2}\text{O}_2 \xrightarrow[\substack{\text{H}_2\text{O}\,/\,[\text{omim}]\text{PF}_6 \\ 50°\text{C, 6h}}]{\text{Fe(O}_3\text{S(CH}_2)_{12}\text{H)}_3} \text{C}_6\text{H}_5\text{OH} \quad (24.14)$$

24.5.2 H_2O–H_2O Systems

"Aqueous biphasic systems," in which both liquid phases are aqueous but the phases are mutually immiscible, can be created by the dissolution of mixtures of polymers or mixtures of salts and polymers into water [97]. This type of system, to our knowledge, has not yet been used for synthetic biphasic catalysis, but has been applied to wood delignification catalyzed by $MgSO_4$ or Li_2SO_4 [98].

24.5.3 Substrate or Product Immiscibility

Product immiscibility or insolubility in the reaction phase (whether aqueous, IL, SCF, or liquid polymer) naturally allows for facile product–catalyst separation without the use of a volatile organic solvent. The field is too large to attempt a comprehensive review of the topic here, but the following summary is offered.

The Ruhrchemie–Rhône–Poulenc hydroformylation process relies on the automatic separation of the water-immiscible product from the aqueous phase. In processes of this kind, this separation could be encouraged, if it does not occur naturally, by cooling the aqueous phase or possibly by the application of CO_2 gas.

There are many examples of biphasic catalysis in ILs in which the second phase is the liquid product [11, 12, 25, 52, 63, 99]. Decantation is all that is required for separation. For solid products, precipitation of the product can occur. Filtration can then be used [100]. If the product is volatile at the reaction temperature, then it can be recovered by concurrent distillation [101].

Commercially, the Difasol process developed by the Institut Français du Pétrole uses an IL as the solvent for dimerization of butenes to iso-octenes. The product iso-octenes naturally separate into an upper phase. Selectivity for dimers rather than trimers is improved because the phase separation of the dimers inhibits further oligomerization [102].

Substrate immiscibility is only useful if low conversions are expected and acceptable. Thus, the unreacted substrate serves as a solvent to extract the product from the catalyst-bearing phase. The Idemitsu process for the hydration of butene to 2-butanol is performed in a biphasic mixture of water and supercritical butene, the latter serving as both substrate and extracting solvent. The hydration is catalyzed in the aqueous phase by a solid heteropolyacid. The butene is separated from the product and then recycled. The biphasic design facilitates separation and prevents the buildup of polymeric residues in the reactor. The plant has been in production since 1985, producing 40,000 metric tons per annum of the eventual product 2-butanone [103–105].

24.5.4 Supported Aqueous-Phase Catalysis

Supported aqueous-phase catalysis (SAPC) differs from aqueous biphasic catalysis only in that the aqueous phase is a thin film on the surface of a solid support, which gives it greater surface area in contact with the reagent-bearing (continuous) phase [106, 107]. Again, the ligands of the catalyst are often modified to render them water-soluble and less likely to be extracted [106]. The continuous phase can be gaseous but is often a liquid solvent. Versions of SAPC that do not use volatile organic solvents include those that use the substrate or product as the continuous phase and those that use $scCO_2$ in that role. One of many examples of the former strategy is Davis' report of the hydroformylation of oleyl alcohol using $RhH(CO)(tppts)_3$ in a porous glass-supported aqueous phase. Leached rhodium was not observed in the product phase (<1 ppb Rh) [107].

The concept of SAPC has been tested with $scCO_2$ as the continuous phase. Bhanage et al. used the method for the hydrogenation of cinnamaldehyde (Equation (24.1)) and found $scCO_2$ to be superior to toluene as the nonpolar medium [18]. Their catalyst was prepared *in situ* by the combination of $RuCl_3$, TPPTS, water, and silica.

24.5.5 Supported Ionic Liquid Catalysis (SILC)

Water is not the only liquid that can be supported on silica or other solids. ILs, liquid polymers, or any liquid that is likely to bind to the surface of the support can be used. In some cases, it may be advisable to modify the surface of the support in order to enhance the binding of the liquid phase. Instead of using a solid such as silica to support the catalyst-bearing IL, it is also possible to use a polymeric filter membrane [108].

DeCastro et al. [109] used silica-supported [bmim]Cl–$AlCl_3$ IL as the catalyst for the Friedel–Crafts alkylation of aromatics with dodecene. No additional solvent was necessary for the reactions of liquid aromatics. No leaching and only slow deactivation was observed when the process was run in a continuous-flow arrangement.

Hydroformylation of 1-hexene using a Rh complex in a silica gel-supported $[bmim]PF_6$ layer was reported by Mehnert et al. [62] The catalyst precursor was $RhH(CO)(tppti)_3$, where tppti is $[bmim]_3[P(C_6H_4mSO_3)_3]$. The surface of the silica gel was modified as shown in structure **2** to enhance binding of the IL layer. The rate of hydroformylation was more than doubled by the use of SILC rather than conventional IL, possibly because of improved mass transfer as a result of the greater surface area.

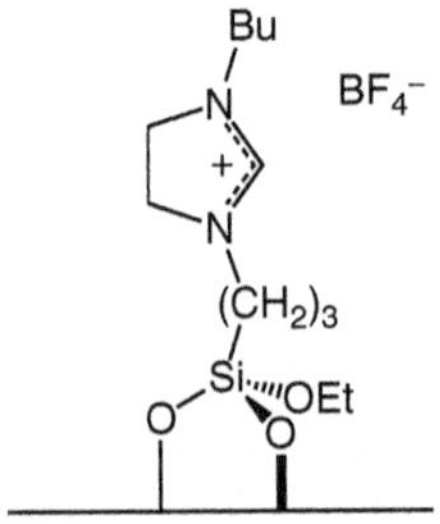

Bu

BF_4^-

$(CH_2)_3$

Si OEt

24.5.6 Supported Liquid Polymer Catalysis

In this method, a nonvolatile liquid polymer supported on a solid is used as the catalyst-bearing phase. The continuous phase can be water, substrate, or a volatile organic solvent. Naughton and Drago [95] have pioneered this technique and illustrated its use with hydroformylation of 1-hexene catalyzed by $RhH(CO)(tppts)_3$ in silica gel-supported PEG-600. Addition of a nonionic surfactant almost doubled the rate. The extent of leaching of catalyst was qualitatively determined by measuring the catalytic activity of the washes of the catalyst; there was no activity.

The use of supported liquid polymer phase catalysis with $scCO_2$ as the continuous phase should have positive effects on mass transfer and catalyst lifetimes.

24.6 CONCLUSIONS

Biphasic catalysis can be performed entirely without the use of volatile organic solvents. Processes of this type are already in use in industry, at least with water as the catalyst-bearing phase. Researchers have, in the past few years, invented many more ways in which catalyst recycling can be achieved with "green" solvents. Among these discoveries are viable examples of continuous-flow homogeneous catalysis. It seems unlikely that this rapid progress will stop in the near future, so we predict that there will be great versatility available in the method. This will make it possible for industry to move more toward homogeneous catalysis rather than heterogeneous catalysis, with significant gains in selectivity for some processes.

However, the biphasic solvent systems that have been reported do not yet constitute a sufficiently diverse group. Chemical reactions are often rather fastidious in the sense that they are sufficiently rapid and selective in one solvent and perform poorly, if at all, in most other solvents. If we as chemists hope or expect that industry will use green solvents in biphasic catalysis, or indeed in any chemical syntheses, we should be able to offer a better selection of green solvents than we have at present. In the near future, it would be best to have a nonvolatile and nontoxic solvent for every possible combination of solvent characteristics, including protic–nonprotic, aromatic–nonaromatic, coordinating–noncoordinating, polar–nonpolar, and viscous–nonviscous. It is clear that so far our set of potential catalyst-bearing solvents is not nearly this diverse. There is a good selection of very polar solvents (H_2O, IL, PEG) but not nonpolar solvents (only $scCO_2$ and perhaps some liquid polymers). Too many of these solvents (H_2O, IL, and perhaps liquid polymers) do not have the ability to dissolve nonpolar reagent gases such as H_2 to any great extent. Circumstantial evidence suggests that this latter problem, in addition to mass transfer limitations, may be ameliorated to some extent by the dissolution of CO_2 into the IL or liquid polymer, but this maneuver will not help the solubility of H_2 in water. There is greater diversity in the question of coordinating ability; both coordinating and noncoordinating solvents can be found within the class of ILs and liquid polymers. CO_2 can be used as a noncoordinating catalyst-bearing phase. Similarly, protic and nonprotic examples of both are available. Aromatic nonvolatile solvents are limited to the pyridinium and imidazolium ILs and perhaps poly(methylphenylsiloxane) alone among the liquid polymers.

Finally, in an analysis of the weaknesses of the methods described, one cannot help but mention the question of expense. The compression of CO_2 to supercritical pressures is expensive, as is the preparation and purification of ILs. Nevertheless, $scCO_2$ has been used economically in a variety of successful industrial processes and there is no doubt in our minds that the same future will be in store for several ILs. The major disadvantage of liquid polymers is not likely to be expense but rather long-term thermal stability.

With the laudable attention that academics have recently been paying to the field, one can expect many more significant advances including solutions to some of the concerns expressed above. Particularly encouraging are recent data on the leaching (or lack thereof) of catalysts from some of the catalyst-bearing phases. The future is bright for green homogeneous catalysis in biphasic systems.

ACKNOWLEDGMENTS

We acknowledge the kind assistance of colleagues in the field of biphasic catalysis, especially Dr. Charles Eckert and Dr. Charles Liotta (Georgia Institute of Technology) and Dr. Walter Leitner (RWTH Aachen). We acknowledge support from the Division of Chemical Sciences, Office of Basic Energy Sciences, Office of Science, U.S. Department of Energy (grant number DE-FG03-99ER14986), and Natural Sciences and Engineering Research Council, Canada. P.G.J., Canada Research Chair in Green Chemistry, acknowledges the support of the Canada Research Chair program.

REFERENCES

1. J Falbe and H Bahrmann. *J Chem Ed* 61:961–965, 1984.
2. DJ Cole-Hamilton. *Science* 299:1702–1706, 2003.
3. S. Hidecki. Japan Patent 235250, 1997, as cited in Ref. [2].
4. W Keim. *Green Chem* 5:105–111, 2003.
5. PG Jessop. *J Synth Org Chem* 61:484–488, 2003.
6. B Cornils and WA Herrmann, eds. *Aqueous-Phase Organometallic Catalysis.* Weinheim: Wiley-VCH, 1998.
7. A Durocher, W Keim, and P. Voncken. *Erdöl und Kohle, Erdgas, Petrochemie* 29.1:31, 1976, as cited in Ref. [4].
8. IT Horváth and J Rábai. *Science* 266:72–75, 1994.
9. B Cornils. *Angew Chem, Int Ed Engl* 36:2057–2059, 1997.
10. E de Wolf, G van Koten, and B-J Deelman. *Chem Soc Rev* 28:37–41, 1999.
11. Y Chauvin, L Mussmann, and H Olivier. *Angew Chem, Int Ed Engl* 34:2698–2700, 1995.
12. PAZ Suarez, JEL Dullius, S Einloft, RF Desouza, and J Dupont. *Polyhedron* 15:1217–1219, 1996.
13. T Welton. *Chem Rev* 99:2071–2083, 1999.
14. P Wasserscheid and W Keim. *Angew Chem, Int Ed* V39:3773–3789, 2000.
15. RG da Rosa, L Martinelli, LHM da Silva, and W Loh. *Chem Commun* 33–34, 2000.
16. PJ Dyson, DJ Ellis, and T Welton. *Can J Chem* 79:705–708, 2001.
17. J Peng, F Shi, Y Gu, and Y Deng. *Green Chem* 5:224–226, 2003.
18. BM Bhanage, Y Ikushima, M Shirai, and M Arai. *Chem Commun* 1277–1278, 1999.
19. RA Brown, P Pollet, E McKoon, CA Eckert, CL Liotta, and PG Jessop. *J Am Chem Soc* 123:1254–1255, 2001.
20. FC Liu, MB Abrams, RT Baker, and W Tumas. *Chem Commun* 433–434, 2001.
21. MF Sellin, PB Webb, and DJ Cole-Hamilton. *Chem Commun* 781–782, 2001.
22. A Bösmann, G Franciò, E Janssen, M Solinas, W Leitner, and P Wasserscheid. *Angew Chem, Int Ed* 40:2697–2699, 2001.
23. DJ Heldebrant and PG Jessop. *J Am Chem Soc* 125:5600–5601, 2003.
24. J Kwiatek, IL Mador, and JK Seyler. *J Am Chem Soc* 84:304–305, 1962.
25. GW Parshall. *J Am Chem Soc* 94:8716–8719, 1972.
26. B Cornils. *Org Proc Res Devel* 2:121–127, 1998.
27. E Kuntz. German Patent DE 2627354, 1976, Rhone–Poulenc.
28. B Cornils and E Wiebus. *CHEMTECH* 25:33–38, 1995.
29. N Yoshimura. In *Aqueous-Phase Organometallic Catalysis*, B Cornils and WA Herrmann, eds. Weinheim: Wiley-VCH, 1998, pp. 408–417.
30. B Jastorff, R Störmann, J Ranke, K Mölter, F Stock, B Oberheitmann, W Hoffmann, J Hoffmann, M Nüchter, B Ondruschka, and J Filser. *Green Chem* 5:136–142, 2003.
31. PG Jessop and W Leitner, eds. *Chemical Synthesis using Supercritical Fluids.* Weinheim: Wiley-VCH, 1999.
32. W Leitner. *Acc Chem Res* 35:746–756, 2002.

33. S Angus, B Armstrong, and KM de Reuck, eds. *International Thermodynamic Tables of the Fluid State: Carbon Dioxide*. Oxford: IUPAC, Pergamon Press, 1976.
34. R Span and W Wagner. *J Phys Chem Ref Data* 25:1509–1596, 1996.
35. K Jackson, LE Bowman, and JL Fulton. *Anal Chem* 67:2368–2372, 1995.
36. KL Toews, RM Shroll, CM Wai, and NG Smart. *Anal Chem* 67:4040–4043, 1995.
37. RJ Bonilla, BR James, and PG Jessop. *Chem Commun* 941–942, 2000.
38. ED Niemeyer and FV Bright. *J Phys Chem B* 102:1474–1478, 1998.
39. MSS Curren and RC Burk. *J Chem Eng Data* 45:746–750, 2000.
40. K-D Wagner, K Brudi, N Dahmen, and H Schmieder. *J Supercrit Fluids* 15:109–116, 1999.
41. K Brudi, N Dahmen, and H Schmieder. *J Supercrit Fluids* 9:146–151, 1996.
42. EH Chimowitz and KJ Pennisi. *AIChE J* 32:1665, 1986.
43. N Pinault and DW Bruce. *Coord Chem Rev* 241:1–25, 2003.
44. GB Jacobson, CT Lee, KP Johnston, and W Tumas. *J Am Chem Soc* 121:11902–11903, 1999.
45. J Zhu, A Robertson, and SC Tsang. *Chem Commun* 2044–2045, 2002.
46. MT Timko, JM Diffendal, JW Tester, KA Smith, WA Peters, RL Danheiser, and JI Steinfeld. *J Phys Chem A* 107:5503–5507, 2003.
47. D Hancu and EJ Beckman. *Green Chem* 3:80–86, 2001.
48. D Hancu, H Green, and EJ Beckman. *Ind Eng Chem Res* V41:4466–4474, 2002.
49. M McCarthy, H Stemmer, and W Leitner. *Green Chem* 4:501–504, 2002.
50. P Walden. *Bull Acad Imper Sci (St Petersburg)* 1800, 1914.
51. See references cited in Ref. [25].
52. Y Chauvin, B Gilbert, and I Guibard. *J Chem Soc Chem Commun* 1715–1716, 1990.
53. JS Wilkes and MJ Zaworotko. *J Chem Soc Chem Commun* 965–967, 1992.
54. SNVK Aki, JF Brennecke, and A Samanta. *Chem Commun* 413–414, 2001.
55. AJ Carmichael and KR Seddon. *J Phys Org Chem* V13:591–595, 2000.
56. JG Huddleston, GA Broker, HD Willauer, and RD Rogers. In *Ionic Liquid: Industrial Applications for Green Chemistry* (ACS Symposium 818), RD Rogers and KR Seddon, eds. Washington, D.C.: American Chemical Society, 2002, pp. 270.
57. LA Blanchard, D Hancu, EJ Beckman, and JF Brennecke. *Nature* 399:28–29, 1999.
58. LA Blanchard and JF Brennecke. *Ind Eng Chem Res* 40:287–292, 2001.
59. LA Blanchard, Z Gu, and JF Brennecke. *J Phys Chem B* 105:2437–2444, 2001.
60. LA Blanchard and JF Brennecke. *Ind Eng Chem Res* 40:2550, 2001.
61. CC Brasse, U Englert, A Salzer, H Waffenschmidt, and P Wasserscheid. *Organometallics* 19:3818–3823, 2000.
62. CP Mehnert, RA Cook, NC Dispenziere, and M Afeworki. *J Am Chem Soc* 124:12932–12933, 2002.
63. JEL Dullius, PAZ Suarez, S Einloft, RFd Souza, J Dupont, J Fischer, and AD Cian. *Organometallics* 17:815–819, 1998.
64. W Wu, J Zhang, B Han, J Chen, Z Liu, T Jiang, J He, and W Li. *Chem Commun* 1412–1413, 2003.
65. PS Gallagher, MP Coffey, VJ Krukonis, and N Klasutis. In *Supercritical Fluid Science and Technology*, Vol. 406, KP Johnston and JML Penninger, eds. Washington, D.C.: American Chemical Society, 1989, pp. 334.
66. CJ Chang and AD Randolph. *AIChE J* 36:939–942, 1990.
67. A Kordikowski, AP Schenk, RM Van Nielen, and CJ Peters. *J Supercrit Fluids* 8:205–216, 1995.
68. SG Kazarian, BJ Briscoe, and T Welton. *Chem Commun* 2047–2048, 2000.
69. SN Baker, GA Baker, MA Kane, and FV Bright. *J Phys Chem B* 105:9663–9668, 2001.
70. J Lu, CL Liotta, and CA Eckert. *J Phys Chem A* 107:3995–4000, 2003.
71. SG Kazarian, N Sakellarios, and CM Gordon. *Chem Commun* 1314–1315, 2002.
72. DM Lamb, TM Barbara, and J Jonas. *J Phys Chem* 90:4210–4215, 1986.
73. MA McHugh and TJ Yogan. *J Chem Eng Data* 29:112–115, 1984.
74. JL Anthony, EJ Maginn, and JF Brennecke. *J Phys Chem B* 106:7315–7320, 2002.
75. PG Jessop, Y Hsiao, T Ikariya, and R Noyori. *J Am Chem Soc* 116:8851–8852, 1994.
76. PG Jessop, Y Hsiao, T Ikariya, and R Noyori. *J Am Chem Soc* 118:344–355, 1996.
77. AD Getty, CC Tai, J Linehan, PG Jessop, MM Olmstead, and AL Rheingold. Submitted for publication.

78. PG Jessop, R Stanley, RA Brown, CA Eckert, CL Liotta, TT Ngo, and P Pollet. *Green Chem* 5:123–128, 2003.
79. DJ Cole-Hamilton. 226th ACS National Meeting. New York, NY, 2003.
80. ZS Hou, BX Han, L Gao, T Jiang, ZM Liu, YH Chang, XG Zhang, and J He. *New J Chem* 26:1246–1248, 2002.
81. D Ballivet-Tkatchenko, M Picquet, M Solinas, G Franciò, P Wasserscheid, and W Leitner. *Green Chem* 5:232–235, 2003.
82. P Tundo, G Moraglio, and F Trotta. *Ind Eng Chem Res* 28:881–890, 1989.
83. Toxicological Evaluation of Certain Food Additives. Food Additives Series 14, Geneva, World Health Organization, 1979.
84. Code of Federal Regulations, Title 21, Volume 3, CITE 21CFR172.820. Washington, FDA, 2001.
85. JL Friedman and CG Greenwald. *Kirk Othmer Encyclopedia of Chemical Technology*, Vol. 11. New York: John Wiley, 1994.
86. W Noll. *Chemistry and Technology of the Silicones*. Orlando: Academic Press, 1968.
87. FE Bailey and JV Koleske. *Poly(ethylene oxide)*. New York: Academic Press, 1976.
88. M Daneshvar, S Kim, and E Gulari. *J Phys Chem* 94:2124–2128, 1990.
89. A Garg, E Gulari, and CW Manke. *Macromolecules* 27:5643–5653, 1994.
90. DJ Heldebrant and PG Jessop. Unpublished material, 2002.
91. D Gourgouillon, H Avelino, J Fareleira, and MN da Ponte. *J Supercrit Fluids* 13:177–185, 1998.
92. LJ Gerhardt, CW Manke, and E Gulari. *J Polym Sci B: Polym Phys* 35:523–534, 1997.
93. NH Brantley, D Bush, SG Kazarian, and CA Eckert. *J Phys Chem B* 103:10007–10016, 1999.
94. SG Kazarian, MF Vincent, BL West, and CA Eckert. *J Supercrit Fluids* 13:107–112, 1998.
95. MJ Naughton and RS Drago. *J Catal* 155:383–389, 1995.
96. KR Seddon, A Stark, and MJ Torres. *Pure Appl Chem* 72:2275–2287, 2000.
97. HD Willauer, JG Huddleston, and RD Rogers. *Ind Eng Chem Res* 2591–2601, 2002.
98. Z Guo, JG Huddleston, RD Rogers, and GC April. *Ind Eng Chem Res* 42:248–253, 2003.
99. Y Chauvin, L Mussmann, and H Olivier-Bourbigou. *CHEMTECH* September:26–30, 1995.
100. XF Zhang, X Fan, H Niu, and J Wang. *Green Chem* 5:267–269, 2003.
101. W Keim, D Vogt, H Waffenschmidt, and P Wasserscheid. *J Catal* 186:481–484, 1999.
102. H Olivier. *J Mol Catal A — Chem* 146:285–289, 1999.
103. T Yamada and T Muto. *Sekiyu Gakkaishi* 34:201–209, 1991.
104. T Yamada, T Muto, and K Yamaguchi. AIChE 1987 Summer National Meeting, 1987.
105. PG Jessop and W Leitner. In PG Jessop and W Leitner, eds. *Chemical Synthesis using Supercritical Fluids*. Weinheim: Wiley-VCH, 1999, pp. 1–36.
106. JPea Arhancet, ME Davis, JS Merola, and BE Hanson. *Nature* 339:454–455, 1989.
107. ME Davis. *CHEMTECH* 22:498–502, 1992.
108. TH Cho, J Fuller, and RT Carlin. *High Temperature Material Processes* 2:543, 1998.
109. C DeCastro, E Sauvage, MH Valkenberg, and WF Holderich. *J Catal* V196:86–94, 2000.

25 Green Chemical Manufacturing with Biocatalysis

Jon D. Stewart
Department of Chemistry, University of Florida

CONTENTS

25.1 INTRODUCTION

While biocatalysis — the use of enzymes to carry out chemical conversions — has been employed for many years in various forms, its full potential has only recently begun to be appreciated as a means to make chemical synthesis more sustainable [1–5]. The use of enzyme-mediated steps in place of traditional synthetic methodology can address all of the goals of green chemistry.

A process described in 1934 for vitamin C (ascorbic acid) production provides one of the earliest examples of a biocatalytic step within a larger chemical synthesis. This method was used for many years as the basis for commercial ascorbic acid production. Glucose is a logical starting material for ascorbic acid **1** since both materials contain six carbons, all of which are oxygenated; the key differences between the two molecules lie in their oxidation states at C_1, C_5, and C_6. In principle, it should be a simple matter to convert glucose to ascorbic acid merely by adjusting the oxidation states of these three positions. Unfortunately, this task is complicated because multiple functional groups with similar chemical reactivities are present (Scheme 25.1).

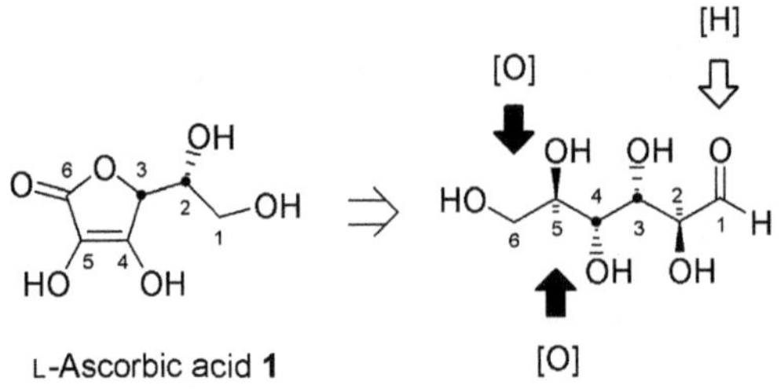

L-Ascorbic acid **1** D-Glucose **2**

SCHEME 25.1

The Reichstein and Grüssner solution to ascorbic acid production from glucose not only illustrates three different strategies for the three redox reactions, but also the types of selectivities associated with both chemical and biological methods. The first step involves reduction of D-glucose **2** to D-sorbitol **3** by catalytic hydrogenation (Scheme 25.2) [6]. This conversion can be carried out cleanly by a standard "chemical" method since glucose contains only a single functional group (the C_1 aldehyde) capable of reacting under these conditions. By contrast, carrying out the oxidations (at C_5 and C_6) is much more complex since other positions have nearly the same reactivities. Standard methodologies lack the necessary chemo- and regioselectivities and near-statistical mixtures of all possible oxidation products would be obtained unless steps were taken to ensure that only a single position could react. Enzymes that catalyze alcohol oxidations, on the other hand, often do so with exquisite positional and stereoselectivities. In this case, *Acetobacter suboxydans* cells, which express an enzyme that oxidizes only the C_5 hydroxyl of **3**, afford L-sorbose **4**, even in the presence of four chemically similar secondary hydroxyl moieties. Employing a biocatalytic strategy for this step allows it to proceed cleanly, with minimal waste generation and in a single step. Unfortunately, no biological catalyst was identified for the C_6 oxidation, and this step had to be carried out chemically using potassium permanganate. Because this oxidant lacks regio- and chemoselectivity, all other hydroxyl groups had to be protected first (**4** $\longrightarrow$ **5**), then the ketals had to be removed in a separate deprotection step (**6** $\longrightarrow$ **7**) following oxidation. This is a good example of how the use of an unselective "chemical" reagent can not only increase the length of the synthesis but also generate a significant waste stream. Final conversion of **7** to ascorbic acid was easily accomplished by heating in dilute aqueous acid.

Of the six steps in the Reichstein and Grüssner process, only four are essential. If all three of the required redox changes could be accomplished with high positional selectivity, the two protection and deprotection steps could be eliminated. Sonoyama et al. [7, 8] reduced this idea to practice in the mid-1970s by using two different bacterial strains (*Erwinia* sp. and *Corynebacterium* sp.) in sequence to carry out all three redox changes to produce **7** from glucose (Scheme 25.2). These biological catalysts obviated the need for hydroxyl protection and a heavy metal oxidant. Their use also cut the number of synthetic steps by half. Both of the strains employed by Sonoyama et al. were wild-type. The *Corynebacterium* 2,5-diketo-L-gulonic acid reductase was subsequently isolated and the corresponding gene was expressed in the same *Erwinia herbicola* strain that produced **8** from glucose, thus allowing a one-step synthesis of **7** from glucose [9].

The history of industrial ascorbic acid production illustrates several key features of biocatalysis applied to chemical synthesis. The major advantages over stoichiometric, chemical reagents are selectivity, mild reaction conditions, and minimized waste generation. Moreover, by augmenting classical strain isolation and improvement techniques with modern molecular biology, the impact of biocatalysis on chemical synthesis can be increased even further.

Biocatalysis is a well-developed field, and it is not possible to cover all aspects here. Instead, this chapter will focus on selected examples of enzymatic reductions and oxidations

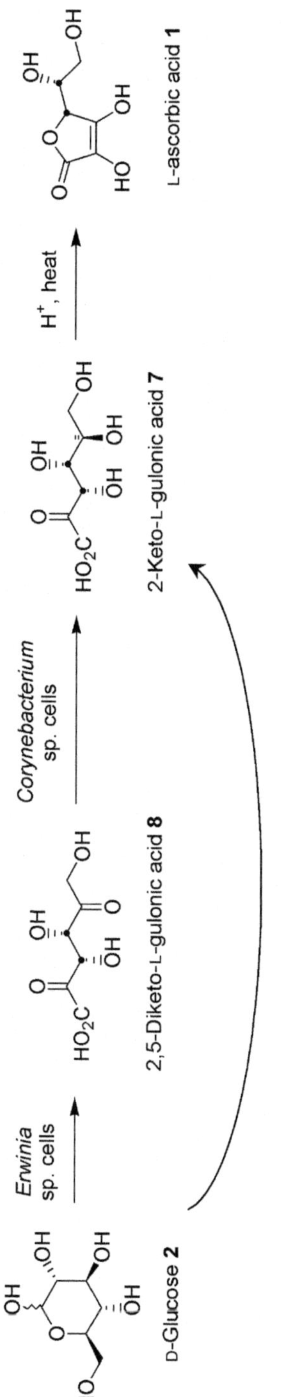

Reichstein process:
H_2
D-Glucose 2
D-Sorbitol 3
Acetobacter suboxydans cells
L-Sorbose 4
Acetone, H^+
Diacetone-L-sorbose 5
[O]
Diacetone-2-keto L-gulonic acid 6
H_3O^+
2-Keto-L-gulonic acid 7
H^+, heat
L-ascorbic acid 1
Sonoyama et al., 1975, 1976; Anderson et al., 1985:
Erwinia sp. cells
D-Glucose 2
2,5-Diketo-L-gulonic acid 8
Corynebacterium sp. cells
2-Keto-L-gulonic acid 7
H^+, heat
L-ascorbic acid 1
Engineered Erwinia herbicola cells

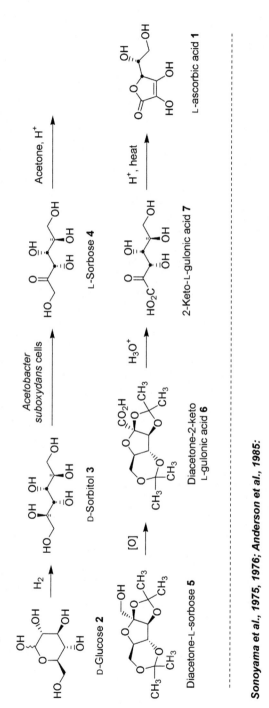

SCHEME 25.2

to illustrate the ways in which these catalysts can contribute to green chemical synthesis. These two classes of reactions offer many opportunities for improvements since traditional methods usually involve stoichiometric quantities of reactants, which are often heavy metals. Waste disposal is therefore a key issue. By contrast, by-products from enzyme-mediated redox reactions are generally benign materials such as water or carbon dioxide. The reactions are usually conducted in aqueous solutions (rather than organic solvents). It should be noted that this is not always an advantage, however, since many processes involving hydrophobic substrates and products result in a dilute product stream that requires treating large volumes of water.

25.1.1 Cofactors and Regeneration

As is true for all catalysts, enzymes mediating redox reactions must remain unchanged at the conclusion of each turnover. To alter the oxidation states of substrates, redox enzymes require stoichiometric quantities of coreactants referred to as cofactors in addition to the substrate of interest. The vast majority of redox enzymes applied to chemical synthesis require nicotinamide cofactors NAD($^+$/H) and NADP($^+$/H) (Scheme 25.3). These cofactors cycle between two-electron reduced and oxidized states and one equivalent of the appropriate form is consumed by each turnover cycle. In principle, stoichiometric quantities of nicotinamides could be used (Scheme 25.3, top); however, their high costs make this simple approach economically implausible except for exploratory reactions carried out on milligram scales. Instead, provision is always made for regenerating the required form of the cofactor by a second reaction that consumes an inexpensive, sacrificial cosubstrate (Scheme 25.3, bottom). Providing these nicotinamides is therefore an important consideration in process design [10]. In some cases, a single enzyme can be used for both the reaction of interest and cofactor regeneration. When restricted substrate specificity makes this simple approach impossible, a second enzyme tailored for economical cofactor regeneration can be added to the reaction mixture. Direct chemical reduction of oxidized nicotinamides has also been demonstrated recently [11]. The use of whole microbial cells that express the enzyme of interest in place of purified or semipurified enzymes opens another avenue to cofactor regeneration. Cellular metabolism of simple carbon sources such as glycerol, glucose, or sucrose yields reduced nicotinamides that can then be used for the desired redox reaction. Whole cells are thus self-contained bioreactors that can be grown on demand. The "best" strategy for cofactor regeneration depends on the reaction, process conditions, and catalyst stability.

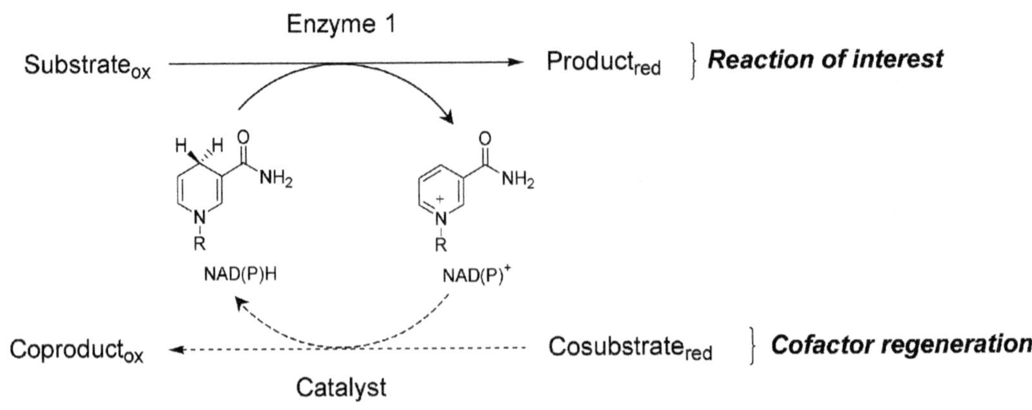

SCHEME 25.3

25.1.2 Process Economics and Practical Considerations

The goals of those engaged in biocatalytic process development are the same as those sought for all other synthetic routes: high volumetric productivity, high product titer, simple downstream processing, minimized waste and by-product formation, and rapid development time [2, 12, 13]. The relative importance of these goals varies with the current needs of the overall project and with the economic constraints imposed by the price that can be fetched by the final product. For example, process development in the pharmaceutical industry is dominated by time-to-market concerns; the very high values of the final products make even inefficient processes usable, particularly during clinical trial phases. Methods that use "off-the-shelf" catalysts that can be scaled up quickly are generally preferred over those requiring extended development times. Meeting the required delivery timelines with sufficient quantities of intermediates for clinical trials is of paramount importance. On the other hand, once the focus shifts to manufacturing, economic considerations in bioprocess design become much more important. This is also the case for biocatalysis applied to lower-value products.

Low-product titer often proves to be the Achilles heel of bioprocess economics and improving this situation is very often the top priority in process development. Biocatalytic reactions are normally carried out in aqueous solutions (or suspensions) on substrates that may have poor water solubilities. Low product concentrations require large reaction volumes that in turn demand undesirably large capital investment in process equipment. A product concentration of 0.10 *M* probably represents the minimal acceptable practical limit in the absence of special economic considerations. By contrast, traditional chemical processes are usually conducted in organic solvents, which routinely allow 1 to 10 *M* product concentrations to be achieved. Not being able to overcome this product titer problem is very often the key reason that "chemical" processes are ultimately chosen over biocatalytic routes, even when the latter show better performance in other categories (total catalyst turnover, stereoselectivity, etc.).

Several strategies to increase product titer in enzyme-catalyzed processes have been explored. Conducting bioconversions in organic solvents (rather than water) is the most straightforward approach. This approach has proven extremely successful for enzymes that catalyze acyl transfer, e.g., lipases, esterases, and proteases [14]. Unfortunately, the nicotinamide cofactors required by redox enzymes are only sparingly soluble in organic solvents, so this strategy is usually not applicable. A few biocatalysts are stable to relatively high concentrations of water-miscible solvents, and this can be used to increase the solubility of hydrophobic substrates and products. A more generally useful strategy is to use a two-phase reaction in which the organic milieu acts as a reservoir for the hydrophobic substrate and product while the biocatalytic reaction takes place in aqueous solution [15]. The solvent must be chosen carefully, however, since its partition constant (K_P value) for the substrate and product must be sufficiently high to keep their aqueous concentrations below levels that are toxic to the biocatalyst, but not so high that the substrate concentration in the aqueous phase is too low for efficient bioconversion. Moreover, when whole cells are employed as the biocatalyst, the organic solvent should not disrupt the cell membranes, unless cell permeabilization is desired. Several examples of bioprocesses that have successfully negotiated these constraints are described below. In some cases, the auxiliary organic phase is a water-immiscible solvent; in others, a solid, hydrophobic resin is employed. Regardless of the physical form of the second phase, it not only allows a much higher product titer to be reached during the reaction, but also facilitates product isolation after the conversion has been completed.

The physical form of the biocatalyst also has important ramifications for process economics. When purified or semipurified enzymes are employed, the catalyst cost usually represents a significant fraction of the total. It is therefore essential that it be reusable for multiple batches (or employed for extended periods in a continuous reactor). Because proteins are large compared to most substrates of interest, carrying out the bioconversion in a vessel equipped with a size-selective (ultrafiltration) membrane allows the product to be removed while the enzymes are retained. Nicotinamides, however, have relatively low molecular weights, so they are washed out with the product stream and must be replenished at the beginning of each reaction cycle. Alternatively, chemically modified forms of the cofactors with high molecular weights may be used. Covalent attachment of polyethylene glycol (PEG) has been used to increase the molecular weights of nicotinamides without affecting their acceptance by redox enzymes (vide infra).

Whole cells containing the enzyme of interest can also be used in place of purified enzymes. As noted above, intact whole cells contain their own supply of nicotinamide cofactors as well as metabolic pathways for their regeneration from simple carbon sources. They are thus self-contained biocatalysts that can be easily grown when needed. These advantages are tempered, however, by a lesser degree of control over reaction conditions since intact cells actively maintain specific internal conditions including pH, ion content, etc. On a per weight basis, substituting intact cells for purified enzymes adds significant biomass to the reaction mixture since the enzymes of interest represent only a subset of the total cell mass. Large quantities of biomass nearly always complicate product recovery and downstream processing. The proportion of active enzyme within whole cells can be increased by overexpressing the protein in an easily handled host organism by recombinant DNA techniques. This approach can also be used to overproduce an enzyme for cofactor regeneration in the same cells.

25.2 CASE STUDIES

The examples described below were chosen from the research literature spanning 1995 to 2003. The list is not meant to be all-inclusive. Instead, an attempt was made to highlight key examples that illustrate the general points described above and also provide picture of the current state-of-the-art in redox biocatalysis. All of the bioprocesses included below have been carried out on ≥2 L scales with explicit consideration of process scale-up and economic issues. While only a few were actually used for manufacturing, the lessons from each contribute to valuable information in evaluating the feasibility for future processes and also indicate the most productive directions for future research in making biocatalytic methods more competitive. When available, final product titers and volumetric productivities are listed for each.

25.2.1 REDUCTIONS

25.2.1.1 Cofactor Regeneration by Single Enzymes

Continuous enzymatic production of L-*tert*-leucine is currently practiced on a 10-ton scale (Scheme 25.4) [1]. The process uses purified leucine dehydrogenase to catalyze the desired reductive amination of trimethylpyruvic acid and *Candida boidinii* formate dehydrogenase to regenerate NADH. The reaction is carried out in the presence of an ultrafiltration membrane

Leucine dehydrogenase

9 + NH_3 → 10

PEG-NADH PEG-NAD^+

CO_2 ← HCO_2H

Candida boidinii formate dehydrogenase

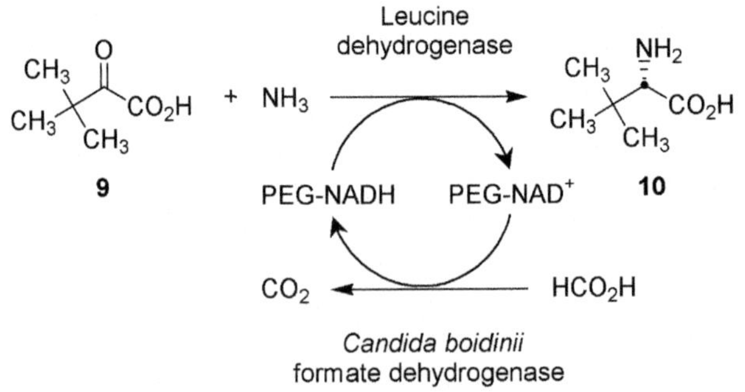

SCHEME 25.4

to retain the catalyst while allowing the product to be removed continuously. The cofactor is derivatized with PEG at N_6 to ensure >99.99% retention in the reaction vessel [16]. Computer modeling revealed that product inhibition was a serious limitation to the process, and that two continuous stirred tank reactors (CSTRs) in series could overcome this problem with negligible increase in biocatalyst costs. This example provides a very compelling argument for the feasibility of large-scale enzymatic redox conversions. Liese and coworkers have recently published a quantitative comparison between biological and chemical reductions using membrane reactors that highlights the strengths and weaknesses of both [17].

Workers at Pfizer recently used a similar strategy to produce a key α-hydroxy acid intermediate for a protease inhibitor undergoing clinical trials (Scheme 25.5) [18]. D-Lactate dehydrogenase was used to reduce α-keto acid **11** to the corresponding (*R*)-alcohol **12** in >99.9% ee. Reductases from both *Leuconostoc mesenteroides* and *Staphylococcus epidermidis* converted **11** to **12** with equal facility. A substrate concentration of 0.2 *M* was used in a membrane reactor that retained both enzymes to allow for a continuous process. Underivatized NAD^+ was used, which

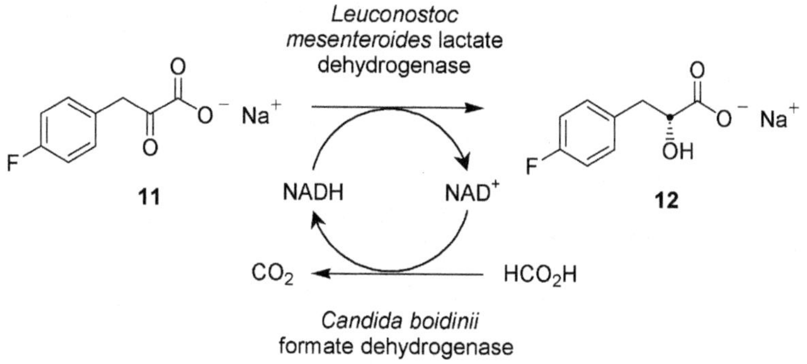

Leuconostoc mesenteroides lactate dehydrogenase

11 NADH NAD^+ **12**

CO_2 HCO_2H

Candida boidinii formate dehydrogenase

SCHEME 25.5

meant that fresh cofactor had to be added along with the α-keto acid substrate. Because the cofactor contribution was a negligible part of the total process cost, this was not a serious drawback. Using optimized conditions, the average conversion was >90% and the space–time yield was 23 g/L h in a 2.2 L reactor. A total of 14.5 kg of (*R*)-**12** was prepared by this method.

Daicel scientists investigating the production of (*R*)-1,3-butanediol by kinetic resolution of the racemate via enantioselective oxidation identified an alcohol dehydrogenase from *Candida parapsilosis* IFO 1396 that possessed very high stereoselectivity [19]. The gene encoding this enzyme was cloned and overexpressed in *Escherichia coli*, which allowed whole cells to be used for bioconversions. In one example, (*R*)-**14** was produced in 95% yield and 99% ee at a concentration of 36.6 g/L (Scheme 25.6). *The C. parapsilosis*

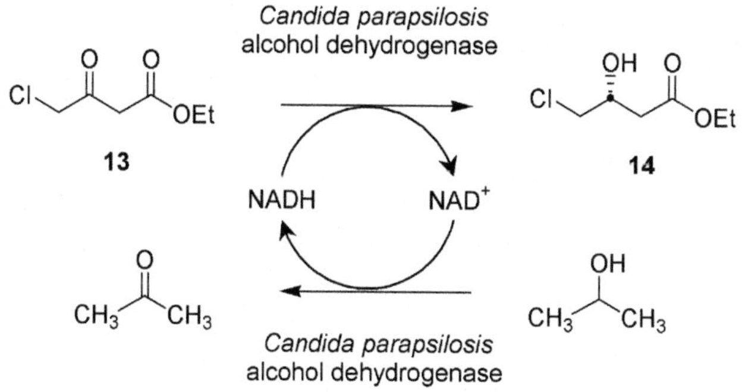

Candida parapsilosis alcohol dehydrogenase

13 **14**

NADH NAD^+

Candida parapsilosis alcohol dehydrogenase

SCHEME 25.6

dehydrogenase possesses broad substrate specificity that includes simple alcohols. This allowed isopropanol to be used as both a cosolvent to increase the solubilities of the starting material and product as well as a sacrificial cosubstrate for regenerating NADH.

Kroutil applied a conceptually similar approach to reduce a variety of methyl ketones **15a–j** with high stereoselectivities and product titers (Scheme 25.7) [20]. In this case, an organism (*Rhodococcus ruber* DSM 44541) was isolated after selection for tolerance to elevated levels of water-miscible organic solvents. A secondary alcohol dehydrogenase from this organism reduced a variety of methyl ketones efficiently when lyophilized whole *R. ruber* cells were suspended in phosphate buffer containing 22% (volume/volume) isopropanol [21]. As in the previous example, the organic cosolvent increased both the accessible substrate and product concentrations and also served as a hydride donor for cofactor regeneration. Substrate concentrations of 0.16 *M* were employed and the reductions proceeded with volumetric productivities ranging from 0.66 to 0.78 g/L h. In all cases, the % ee values were >99%.

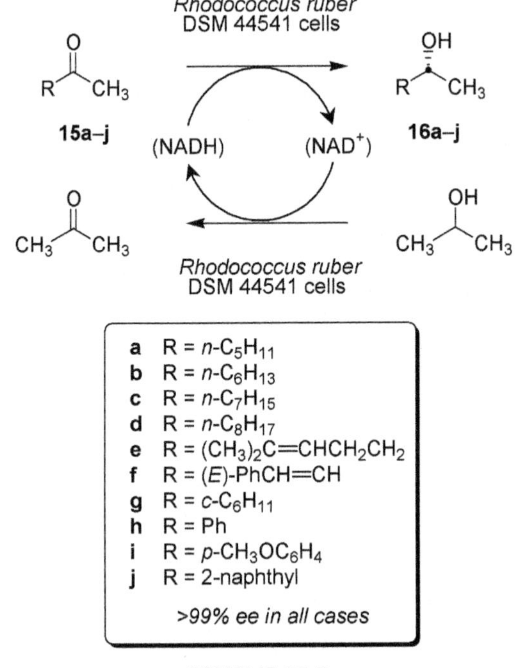

SCHEME 25.7

A two-phase system can also be used to increase the product titer dramatically. This strategy has been applied to the asymmetric reduction of β-keto ester **13** to yield both alcohol enantiomers, depending on the choice of reductase enzyme (Scheme 25.8). *E. coli* cells that overexpressed either the aldehyde reductase from *Sporobolomyces salmonicolor* [22, 23] or the carbonyl reductase from *Candida magnoliae* [24] were used in the bioconversions. Both of these dehydrogenases are specific for NADPH, and the cofactor was regenerated by glucose oxidation mediated by the $NADP^+$-linked *Bacillus megaterium* glucose dehydrogenase. This was either coexpressed along with the *S. salmonicolor* reductase or added separately. To carry out the reductions, aqueous suspensions of the appropriate engineered *E. coli* strain were mixed with an equal volume of *n*-butyl acetate, which allowed for a starting substrate concentration of ≥ 1 *M*. The organic solvent permeabilized the cell membranes, which necessitated adding exogenous $NADP^+$. The volumetric productivities were 16.7 and 5.9 g/L h for the strains overexpressing the *S. salmonicolor* [25] and *C. magnoliae* [24] reductases, respectively. These are very impressive space–time yields for biocatalytic reductions.

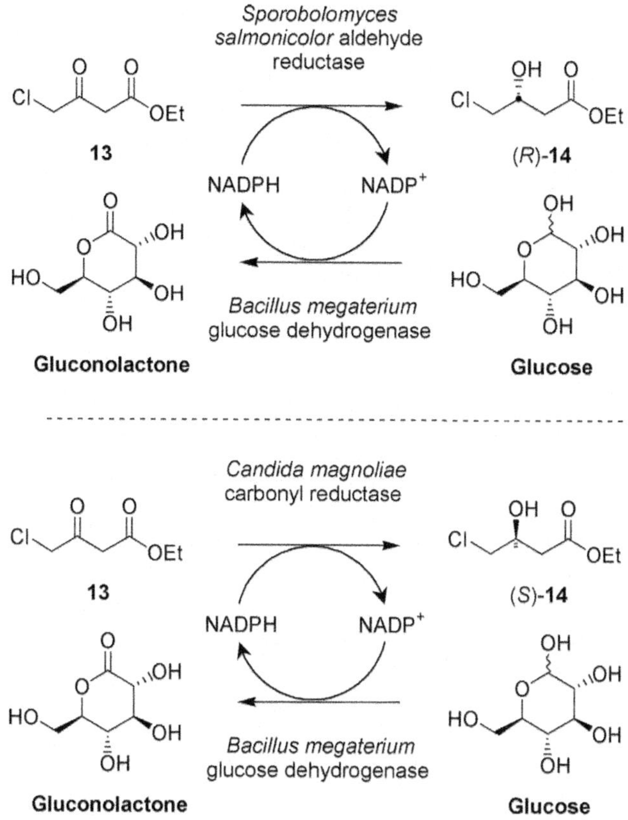

SCHEME 25.8

25.2.1.2 Cofactor Regeneration by Metabolic Pathways

In all of the abovementioned examples, the nicotinamide cofactor was regenerated by a single enzyme, e.g., formate dehydrogenase, glucose dehydrogenase or the same enzyme used to reduce the substrate of interest. The following examples utilize a different approach: the NAD(P)H required is produced by metabolism of simple carbon sources. While generally simpler than employing purified enzymes, this approach can result in somewhat lower volumetric productivities.

Liese and coworkers have described a continuous process for the production of (2*R*,5*R*)-hexanediol, an important building block for chiral ligands [26]. They used nongrowing cells of *Lactobacillus kefir* DSM 20587 to carry out a dual reduction of both ketone carbonyl moieties of **17** (Scheme 25.9). Both steps proceeded with >99% enantioselectivity. The reaction was carried out on a 2 L scale under anaerobic conditions with an external loop containing an ultrafiltration membrane with a 400,000 molecular weight cutoff (MWCO), which allowed the product to be removed continuously while retaining the cells. The continuous process achieved a space–time yield of 2.7 g/L h and represented a 30-fold improvement in the yield of product per gram of biocatalyst when compared to a batch process using the same cells.

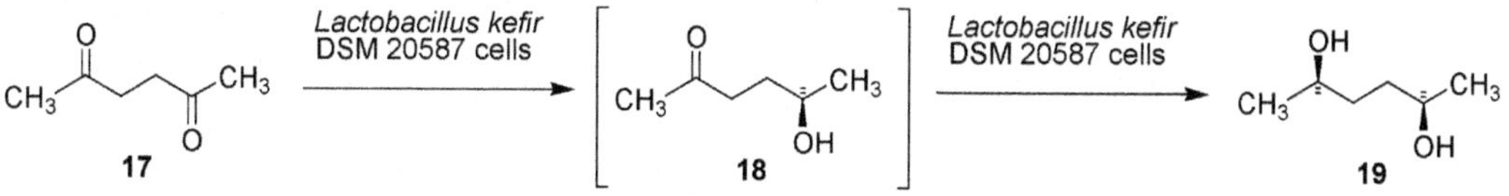

SCHEME 25.9

The asymmetric reduction of ketone **20** was a key step in the synthesis of a β3 adrenergic receptor agonist undergoing clinical evaluation by Merck (Scheme 25.10) [27]. Screening a variety of "chemical" strategies for the conversion of **20** to **21** revealed that none were suitable for large-scale use. By contrast, several whole cells were able to carry out this reduction and the yeast *Candida sorbophila* emerged as the most suitable (>99.5% ee and 75% yield). The reactions were carried out on kilogram scales at substrate concentrations of 5–50 g/L.

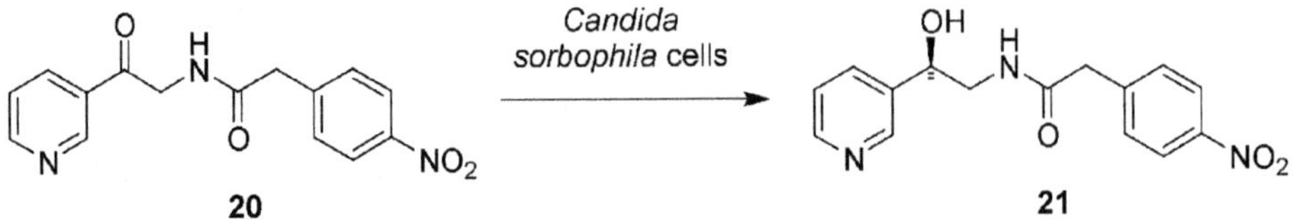

Candida sorbophila cells

20 **21**

SCHEME 25.10

The industrial production of trimegestone utilizes a baker's yeast-mediated reduction in the final step (Scheme 25.11) [28]. This is a difficult chemical problem since **22** contains three different ketone moieties and two olefins, all of which are susceptible to reduction. The D-ring side chain of triketone **22** was elaborated by chemical methods; unfortunately, standard chemical methods for reducing the desired ketone proceeded with insufficient selectivities. By contrast, whole baker's yeast cells reduced only the C_{21} carbonyl to the desired (*S*)-alcohol. Glycerol was used to regenerate the required nicotinamide cofactors. The sparing aqueous solubility of **22** necessitated a low substrate concentration (1 g/L). In addition, baker's yeast cells were relatively inefficient reduction catalysts, and large quantities of biomass were required (240 g/L). These factors limited the volumetric productivity of the conversion to 0.17 g/L h. Even with these disadvantages, the low cost of baker's yeast, combined with the high value of the final product, made this bioconversion economically feasible.

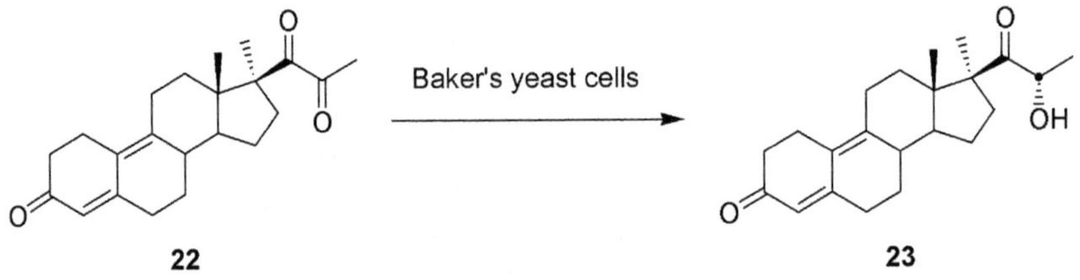

Baker's yeast cells

22 **23**

SCHEME 25.11

Lilly workers used a whole cell-mediated reduction as a key step in the synthesis of a psychoactive benzodiazepine [29]. A variety of organisms were screened for the ability to reduce methyl ketone **24** to the desired (*S*)-alcohol and *Zygosaccharomyces rouxii* emerged as the most promising biocatalyst (Scheme 25.12). Whole *Z. rouxii* cells could be used directly; unfortunately, the accessible substrate concentrations were limited by toxicity of **24** toward the cells, which restricted the volumetric productivity of the process to a value too low for practical utility. This problem was solved by adding a solid, nonpolar resin to the reaction mixture, which acted as a reservoir for both the hydrophobic substrate and product and limited the aqueous concentrations to levels compatible with cell survival. With this improvement, the volumetric productivity of the reaction was increased to 3.3 g/L h. This was one of the first examples of using a nonpolar resin directly in a bioconversion to act as a second phase. It was particularly successful in this case because the reduction proceeded under anaerobic conditions, which allowed the suspension of resin and cells to be stirred very gently. More vigorous agitation leads to cell lysis by mechanical abrasion.

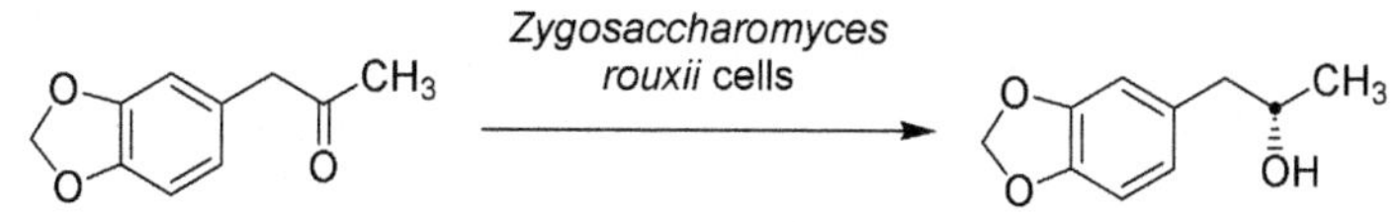

Zygosaccharomyces rouxii cells

24 25

SCHEME 25.12

Chemical methods for producing D-mannitol from D-fructose have several disadvantages including high feedstock costs, lack of specificity, and the need to isolate the product by chromatography. Biocatalytic alternatives have been explored; however, the volumetric productivities were too low to be practically useful (Scheme 25.13). A lactic acid bacterial strain (*L. mesenteroides* ATCC 9135) showed much higher volumetric productivity than those described earlier [30]. This strain was used under nongrowing conditions in nutrient-poor medium in a 100 L pilot plant reactor to achieve a final mannitol concentration of 91 g/L with a volumetric productivity of 19 g/L h [31]. The product stream was isolated by an external loop connected to an ultrafiltration membrane and analytically pure D-mannitol was isolated after crystallization with 98.6% purity. The cells were retained in the bioreactor, which was then recharged with fresh D-fructose for a semicontinuous process. When compared with the chemical process, which produces 1.6 kg of side products per kg of mannitol, the bioconversion afforded only 0.7 kg of side products per kg of mannitol. While this type of information

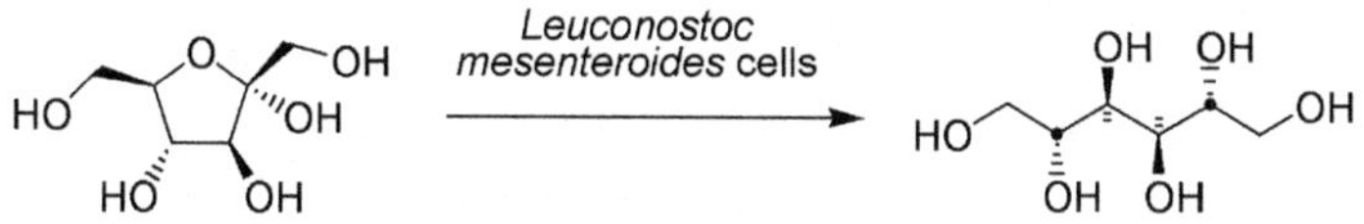

Leuconostoc mesenteroides cells

D-Fructose **26** D-Mannitol **27**

SCHEME 25.13

on waste streams is rarely reported in the literature, it speaks volumes about the ability of biochemical methods to achieve green chemistry aims.

25.2.2 OXIDATIONS

Biological oxidations can be broadly divided into oxygen-independent and oxygen-dependent conversions. The former category includes dehydrogenases operating the oxidation direction (at the expense of an oxidized nicotinamide cofactor). Biocatalysts in the latter category (oxygen-dependent reactions) can be further subdivided into monooxygenases, which transfer one atom of oxygen from O_2 into the product and release the other as water, and dioxygenases, which transfer both atoms of O_2 into the product. Both types of oxidation catalysts require two electrons that must be supplied by nicotinamide cofactors. Synthetically useful oxidation biocatalysts include dehydrogenases, cytochrome P-450s, Baeyer-Villigerases, and aromatic hydroxylases and dioxygenases. With the important exception of dehydrogenases, many enzyme-mediated oxidations involve several proteins, some of which are membrane bound. This makes it difficult to use the purified proteins for preparative reactions and processes involving whole cells are therefore more common. The examples cited below illustrate the variety of oxidations that can be carried out by biological methods on practical scales. Some of these reactions take advantage of the regioselectivity inherent in enzymatic oxidations while others involve the creation of chiral centers.

BASF has developed a process to convert (*R*)-2-phenoxypropionic acid **28** to its *para*-hydroxylated derivative **29**, which is a key intermediate in herbicide synthesis (Scheme 25.14) [32]. This would be a difficult transformation to achieve by traditional chemical methodology

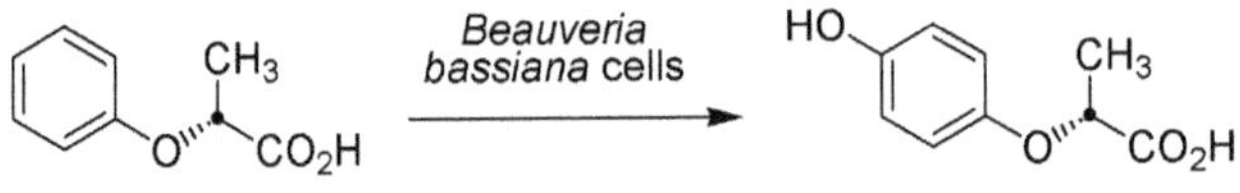

SCHEME 25.14

with its limited regioselectivity in aromatic ring hydroxylation. A collection of fungal and bacterial strains was screened for the ability to carry out the desired reaction, then the best candidate (*Beauveria bassiana* Lu 4068) was submitted to an iterative strain improvement procedure. Selection for high substrate tolerance (>100 g/L) was followed by screening for high productivity. This procedure resulted in a strain with a volumetric productivity of 0.29 g/L h in a small fermenter. Its performance was identical at a 100,000 L scale. The cell mass was the only waste product requiring disposal. This example nicely illustrates how enzymatic reactions can be useful for lower-value products.

2-Quinoxalinecarboxylic acid **32** is a key intermediate in pharmaceutical synthesis (Scheme 25.15). Pfizer workers had developed a "chemical" route to this compound in the 1970s; however, a key intermediate in this route was mutagenic and thermally unstable. By analogy to microbial degradation pathways for methylated aromatics, it was suspected that an enzymatic side-chain oxidation of **30** might yield the desired target. Screening a culture collection of bacterial and fungal strains yielded the fungus *Absidia repens* ATCC 14849. Process development studies revealed that the substrate was toxic to the organism at levels >1 g/L. While both the alcohol and aldehydes were anticipated intermediates in the overall conversion, only low levels of **31** were detected; the corresponding aldehyde did not accumulate. It is likely that the initial hydroxylation to **31** was catalyzed by a cytochrome P-450 and subsequent oxidations involved alcohol and aldehyde dehydrogenases. Because of the large number of proteins involved in the bioconversion, whole *A. repens* cells were employed and no attempt was made to isolate or characterize the enzymes.

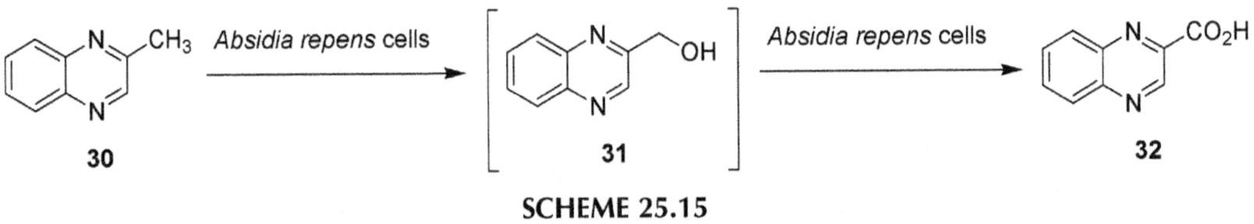

SCHEME 25.15

Despite the low-product titer, the project timeline required that the synthesis of **32** be scaled up to 14,000 L (three runs) with the substrate added in a fed-batch protocol to a final concentration of 1.5 g/L. These conversions yielded a total of 20.5 kg of **32** with a volumetric productivity of 9.4×10^{-3} g/L h. Product isolation required four extraction steps with methylene chloride; even with 70% recovery, 26,000 L of the solvent was used. This example dramatically underscores the point that time-to-market concerns dominate pharmaceutical bioprocess development and that this often means scaling up suboptimal processes to meet compound delivery dates. It also shows how low-product titer has a negative impact on almost every aspect of the process. This shortcoming was, of course, recognized by the authors, who also reported preliminary efforts to replace *A. repens* cells with a *Pseudomonas putida* strain grown on benzyl alcohol that gave tenfold higher concentrations of **32**. This strain, perhaps in conjunction with a hydrophobic resin to reduce the concentration of substrate and products to sublethal levels [29], could lead to a much improved process suitable for manufacturing.

Two related examples of aromatic heterocycle oxidations were described by Lonza workers (Scheme 25.16). A *P. putida* strain grown on xylenes converts 2,5-dimethylpyrazine **33** to the corresponding monocarboxylic acid **34** with no detectable over-oxidation [33, 34].

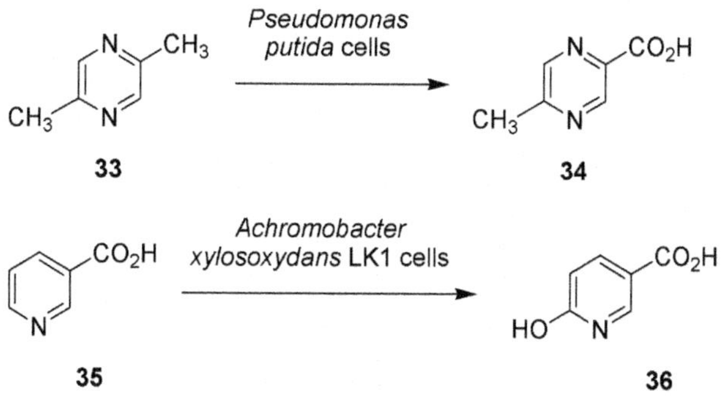

SCHEME 25.16

Standard chemical methodologies would be hard-pressed to match this exquisite selectivity. Acid **34** is a key intermediate in pharmaceuticals and its synthesis has been demonstrated on a 20 L scale with a volumetric productivity of 0.37 g/L h. The bioprocess is currently practiced by Lonza on a scale of 15,000 L.

A number of insecticides can be constructed from 6-hydroxynicotinic acid **36**. The direct oxidation of nicotinic acid **35** is an attractive route to this building block since the starting material is a vitamin that is produced in large quantities. Unfortunately, chemical strategies for the conversion of **35** to **36** yield a mixture of regioisomers that must be separated from the desired material. By contrast, *Achromobacter xylosoxydans* LK1 (DSM 2783) catabolizes nicotinic acid via a route involving an initial, regioselective 6-hydroxylation followed by oxidative decarboxylation. Interestingly, the enzyme catalyzing the second step is inhibited by high levels of **35** [35], which allows a fed-batch reaction to accumulate acid **36** in the culture medium so long as the concentration of starting material is maintained at >10 g/L. The volumetric productivity of this process is >8.3 g/L h and Lonza has produced more than 10 t of this key building block using this biocatalytic route.

Baeyer-Villiger monooxygenases are flavin-containing proteins that catalyze the insertion of an oxygen atom between a ketone carbonyl and the flanking carbon at the expense of molecular oxygen and a reduced nicotinamide. Chemical methods for carrying out Baeyer-Villiger oxidations generally require toxic or explosive peracids or related reagents [36], and with a few exceptions (see Ref. [37] and references therein), yield racemic products. By contrast, enzyme catalysts for this reaction use molecular oxygen, yield water as the only waste product and generally show very high stereoselectivities. The enzyme from *Acinetobacter* sp. NCIB 9871 has been characterized most thoroughly with respect to its substrate- and stereoselectivity [38, 39] and methods for applying the enzyme to laboratory-scale reactions have been well-developed (see Ref. [40] and references therein). What has been lacking until recently are strategies for carrying out enzymatic Baeyer-Villiger oxidations with high volumetric productivities. Two studies addressing this issue have recently appeared (Scheme 25.17). Woodley and coworkers used nongrowing cells of an *E. coli* strain that overexpresses *Acinetobacter* sp. NCIB 9871 cyclohexanone monooxygenase to convert racemic **37** to a mixture of regioisomeric lactones [41]. Glycerol was used as a carbon source and the mixture of lactone regioisomers was formed with a volumetric productivity of 0.76 g/L h and a final product concentration of 3.8 g/L on a 55 L scale. Walton and Stewart [42] used whole cells of a slightly different engineered *E. coli* strain expressing the same enzyme to convert 4-methylcyclohexanone **40** to the corresponding (*S*)-lactone in >98% ee. Cells were used under nongrowing conditions with both glucose and **40** added in fed-batch mode. The volumetric productivity of this process was 0.47 g/L h with a final lactone concentration of 11.3 g/L. These authors also showed that the longevity of the bioprocess was limited primarily by the limited stability of the intracellular Baeyer-Villiger monooxygenase during the bioconversion.

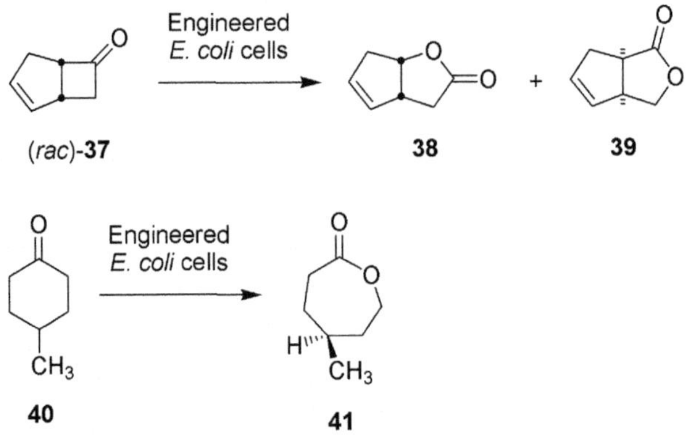

Engineered
E. coli cells

(*rac*)-**37** **38** **39**

Engineered
E. coli cells

CH_3 CH_3

40 **41**

SCHEME 25.17

Homochiral styrene oxide enantiomers are important pharmaceutical building blocks and Witholt has described an efficient biocatalytic route to the (*S*)-enantiomer **43** (Scheme 25.18) [43]. Substrate and product toxicity are severe problems with this reaction, which led to the choice of a two-phase bioprocess. The two-component styrene monooxygenase from *Pseudomonas* sp. VLB 120 was expressed in *E. coli* cells, which were grown in a minimal salts

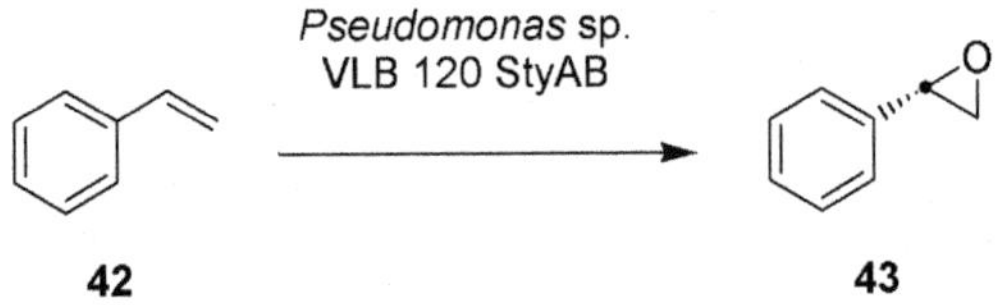

Pseudomonas sp.
VLB 120 StyAB

42 **43**

SCHEME 25.18

medium to stationary phase. An equal volume of a bis(2-ethylhexyl)phthalate solution containing 10 g/L *n*-octane (to induce styrene oxygenase production) and 20 g/L styrene was then added. The styrene oxide that formed partitioned strongly into the organic phase and the volumetric productivity of the process was 1.3 g/L h with a final (*S*)-styrene oxide concentration of 22.4 g/L in the organic phase. These values were significantly higher than could be achieved in the absence of the second phase and they compare very favorably with earlier biocatalytic routes to this compound based on kinetic resolutions of racemic **43** by epoxide hydrolases.

Glyphosate **48** is a highly successful preemergent weed killer that selectively targets an essential plant enzyme involved in aromatic amino acid biosynthesis. DuPont workers investigated a biocatalytic route to **48** from glycolic acid **44** that relied on spinach glycolate oxidase to catalyze the key step (formation of glyoxylic acid **45**; Scheme 25.19) [44]. Hydrogen peroxide is a by-product of this oxidation, and the inclusion of *Aspergillus niger* catalase, which catalyzed the disproportionation of hydrogen peroxide into O_2 and water, significantly improved the yield of the reaction, presumably by protecting the enzymes from toxic effects of H_2O_2. Moreover, by including aminomethylphosphonic acid **46** in the reaction mixture, the labile aldehyde **45** could be trapped *in situ* as Schiff's base **47**. Both of the required enzymes were coexpressed in the yeast *Hansenula polymorpha* GO1 and the permeabilized cells could be used for at least 30 reaction cycles. An excess of **44** (1.3 equivalents) was used relative to amine **46**, assuring that essentially all of the latter would react. After the cells had been removed by centrifugation, the crude biotransformation mixture was hydrogenated to afford **48**, which could be isolated from the reaction mixture by simple crystallization. This is a very ingenious process that takes advantage of *in situ* trapping to stabilize an intermediate in a synthetically productive fashion. To date, no reports detailing the scale-up of this process have appeared in the literature.

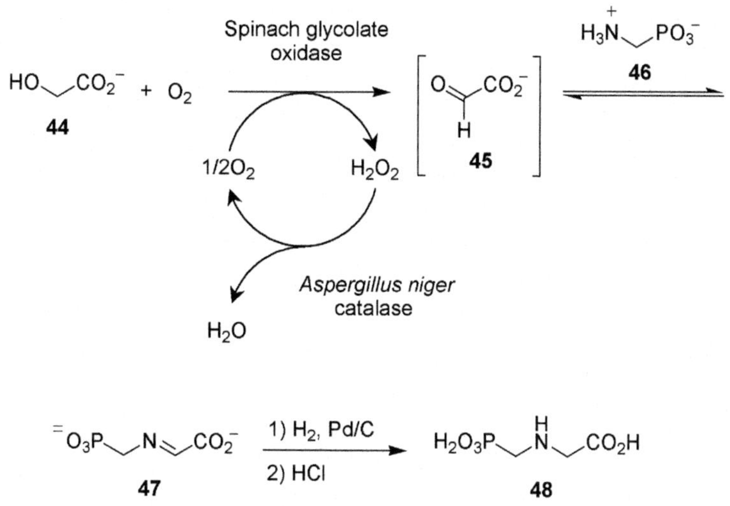

SCHEME 25.19

Landis and coworkers at Pharmacia used a conceptually related strategy to achieve a very short, efficient synthesis of *n*-butyldeoxynojirimycin **54**, a potential therapeutic agent for AIDS and related retroviruses (Scheme 25.20) [45]. Their strategy also echoes lessons learned from the ascorbic acid process, in which enzymes are used to adjust the oxidation states of polyols with very high selectivities. Compound **49** is the reductive amination product of *n*-butylamine and glucose. Cyclization to the final product can proceed by a second reductive

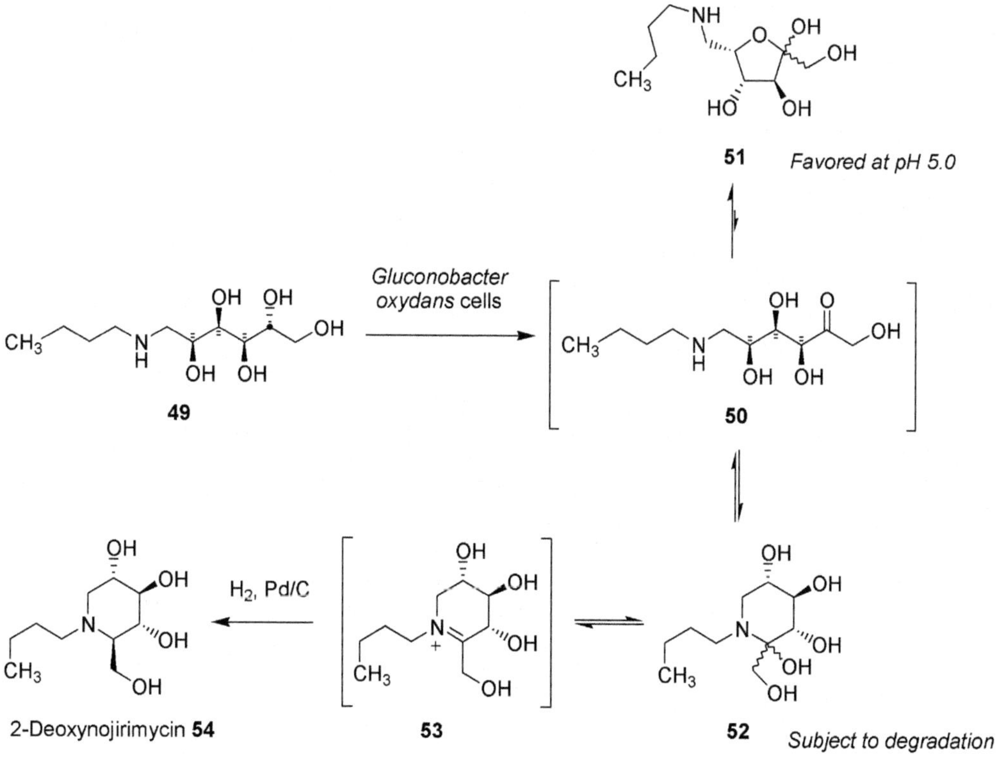

SCHEME 25.20

amination from ketone **50**. The similar reactivities of the hydroxyls of **49** makes chemical methods for selective oxidation at C_5 difficult; by contrast, *Gluconobacter oxydans* cells perform this conversion with high volumetric productivity (6.8 g/L h) on a 5500 L scale and a final product concentration of 190 g/L. When the bioconversion was carried out at pH 5.0, **51** was the major form of the oxidation product; at elevated pH values, **52** predominated, and this was irreversibly degraded during the bioconversion. The catalytic hydrogenation to form **54** was carried out directly on the bioconversion mixture (after cells had been removed). This is an exceptionally efficient process that highlights the power of biocatalysis, ingeniously combined with chemical methodology.

25.3 CONCLUSIONS

While all of the abovementioned examples focused on biocatalytic redox reactions, the key principles that emerge from the successful bioprocesses are applicable to other classes of chemical conversions. Maximizing both the volumetric productivity and final product titer are the two most important goals in development and these numbers often mean the difference between a process that moves forward versus one that is abandoned in favor of an alternative. High catalyst activity and stability to the reaction conditions play are key contributors to bioprocess efficiency and longevity. These properties can be maximized by genetic engineering or by screening organisms to find enzymes with the most suitable properties. Because the substrates or products are often show inhibition or toxicity toward the biocatalysts, slow substrate addition and *in situ* product extraction may be necessary to avoid these issues. Bioprocesses can be carried out with purified (or semipurified) enzymes or with whole cells that express the enzyme of interest. Both systems have advantages and disadvantages, and nicotinamide cofactor regeneration can be carried out effectively under either regime. Finally, rapid bioprocess development times are essential if these methodologies are to be competitive with alternatives. As more successful examples of large-scale bioprocesses are disclosed, the need to invent completely new solutions to problems encountered declines. It is hoped that the examples cited here can speed this process.

ACKNOWLEDGMENTS

Work in the authors laboratory in this area has been generously supported by grants from the NSF (CHE-0130315) and USDA (00-52104-9704).

REFERENCES

1. C Wandrey, A Liese, and D Kihumbu. *Org Proc Res Develop* 4:286–290, 2000.
2. A Schmid, JS Dordick, B Hauer, A Kiener, M Wubbolts, and B Witholt. *Nature* 409:258–268, 2001.
3. MJ Burk. *Adv Synth Catal* 345:647–648, 2003.
4. W-D Fessner. *Adv Synth Catal* 345:649–650, 2003.
5. HE Schoemaker, D Mink, and MG Wubbolts. *Science* 299:1694–1697, 2003.
6. T Reichstein and A Grüssner. *Helv Chim Acta* 17:311–328, 1934.
7. T Sonoyama, B Kageyama, and T Honjo. U.S. Patent No. 3,922,194, 1975.
8. T Sonoyama, H Tani, K Kageyama, K Kobayashi, T Honjo, and S Yagi. U.S. Patent No. 3,998,697, 1976.
9. S Anderson, CB Marks, R Lazarus, J Miller, K Stafford, J Seymour, D Light, W Rastetter, and D Estell. *Science* 230:144–149, 1985.
10. W Hummel. *Trends Biotechnol* 17:487–492, 1999.
11. F Hollmann, B Witholt, and A Schmid. *J Mol Catal B: Enzymatic* 19–20:167–176, 2002.
12. AJJ Straathof, S Panke, and A Schmid. *Curr Opin Biotechnol* 13:548–556, 2002.

13. GJ Lye, PA Dalby, and JM Woodley. *Org Proc Res Devel* 6:434–440, 2002.
14. AM Klibanov. *Nature* 409:241–246, 2001.
15. AJJ Straathof. *Biotechnol Prog* 19:755–762, 2003.
16. M-R Kula and C Wandrey. *Meth Enzymol* 136:9–21, 1987.
17. S Laue, L Greiner, J Wöltinger, and A Liese. *Adv Synth Catal* 343:711–720, 2001.
18. J Tao and K McGee. *Org Proc Res Devel* 6:520–524, 2002.
19. A Matsuyama, H Yamamoto, and Y Kobayashi. *Org Proc Res Devel* 6:558–561, 2002.
20. W Stampfer, B Kosjek, C Moitzi, W Kroutil, and K Faber. *Angew Chem Int Ed Engl* 41:1014–1017, 2002.
21. W Stampfer, B Kosjek, K Faber, and W Kroutil. *J Org Chem* 68:402–406, 2003.
22. K Kita, K Matsuzaki, T Hashimoto, H Yanase, N Kato, MC-M Chung, M Kataoka, and S Shimizu. *Appl Environ Microbiol* 62:2303–2310, 1996.
23. K Kita, T Fukura, K-I Nakase, K Okamoto, H Yanase, M Kataoka, and S Shimizu. *Appl Environ Microbiol* 65:5207–5211, 1999.
24. Y Yasohara, N Kizaki, J Hasegawa, M Wada, M Kataoka, and S Shimizu. *Biosci Biotechnol Biochem* 64:1430–1436, 2000.
25. M Kataoka, K Yamamoto, H Kawabata, M Wada, K Kita, H Yanase, and S Shimizu. *Appl Microbiol Biotechnol* 51:486–490, 1999.
26. J Haberland, W Hummel, T Daussmann, and A Liese. *Org Proc Res Devel* 6:458–462, 2002.
27. JYL Chung, G-J Ho, M Chartrain, C Roberge, D Zhao, J Leazer, R Farr, M Robbins, K Emerson, DJ Mathre, JM McNamara, DL Hughes, EJJ Grabowski, and PJ Reider. *Tetrahedron Lett* 40: 6739–6743, 1999.
28. V Crocq, C Masson, J Winter, C Richard, Q Lemaitre, J Lenay, M Vivat, J Buendia, and D Prat. *Org Process Res Devel* 1:2–13, 1997.
29. BA Anderson, MM Hansen, AR Harkness, CL Henry, JT Vicenzi, and MJ Zmijewski. *J Am Chem Soc* 117:12358–12359, 1995.
30. N von Weymarn, M Hujanen, and M Leisola. *Proc Biochem* 37:1207–1213, 2002.
31. FNW von Weymarn, KJ Kiviharju, ST Jääskeläinen, and MSA Leisola. *Biotechnol Prog* 19:815–821, 2003.
32. C Dingler, W Ladner, GA Krei, B Cooper, and B Hauer. *Pestic Sci* 46:33–35, 1996.
33. A Kiener. *Chemtech* 25:31–35, 1995.
34. NM Shaw, KT Robins, and A Kiener. *Adv Synth Catal* 345:425–435, 2003.
35. HG Kulla. *Chimia* 45:81–85, 1991.
36. GR Krow. *Org Reactions* 43:251–798, 1993.
37. C Bolm, O Beckmann, A Cosp, and C Palazzi. *Synlett* November 1461–1463, 2001.
38. JD Stewart. *Curr Org Chem* 2:211–232, 1998.
39. MD Mihovilovic, B Müller, and P Stanetty. *Eur J Org Chem* November 3711–3730, 2002.
40. MD Mihovilovic, G Chen, S Wang, B Kyte, F Rochon, MD Kayser, and JD Stewart. *J Org Chem* 66:733–738, 2001.
41. SD Doig, PJ Avenell, PA Bird, P Gallati, KS Lander, GJ Lye, R Wohlgemuth, and JM Woodley. *Biotechnol Prog* 18:1039–1046, 2002.
42. AZ Walton and JD Stewart. *Biotechnol Prog*, 20:403–411, 2004.
43. S Panke, MG Wubbolts, A Schmid, and B Witholt. *Biotechnol Bioeng* 69:91–100, 2000.
44. JE Gavagan, SK Fager, JE Seip, DS Clark, MS Payne, DL Anton, and R DiCosimo. *J Org Chem* 62:5419–5427, 1997.
45. BH Landis, JK McLaughlin, R Heeren, RW Grabner, and PT Wang. *Org Proc Res Devel* 6:547–552, 2002.

Index

C

H